APPLIED RELIABILITY ASSESSMENT IN ELECTRIC POWER SYSTEMS

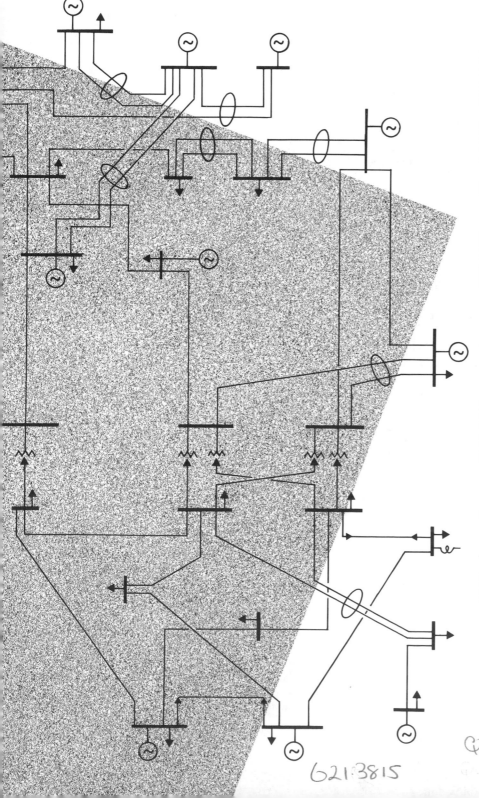

Edited by

ROY BILLINTON
Electrical Engineering Department
University of Saskatchewan, Canada

RONALD N. ALLAN
Electrical Energy and Power Systems
 Research Group
University of Manchester Institute of
 Science and Technology, United Kingdom

LUIGI SALVADERI
Planning Department
Ente Nazionale Per L'Energia Elettrica, Italy

A volume in the IEEE PRESS Selected
Reprint Series, prepared under the
sponsorship of the IEEE Power
Engineering Society.

**IEEE
PRESS**

The Institute of Electrical and
Electronics Engineers, Inc.,
New York

IEEE Order Number: PC0251-9

Library of Congress Cataloging-in-Publication Data

Applied reliability assessment in electric power systems / edited by
Roy Billinton, Ronald N. Allan, Luigi Salvaderi.
    p.   cm. — (IEEE Press selected reprint series)
Includes bibliographical references and index.
ISBN 0-87942-264-5
    1. Electric power systems—Reliability. I. Billinton, Roy.
II. Allan, Ronald N. (Ronald Norman) III. Salvaderi, Luigi.
TK1005.A739   1991
621.31—dc20   90-40180
                                          CIP

Contents

Preface and Acknowledgments vii

Introduction ix

Power System Reliability in Perspective, *R. Billinton and R. N. Allan* (*IEE Journal on Electronics and Power,* March 1984) 1

Part 1: General Capacity—Hierarchical Level I 7

Section 1.1 Installed Capacity Assessment 11

Generating Reserve Capability Determined by the Probability Method, *G. Calabrese* (*AIEE Transactions,* September 1947) 11

The Determination and Allocation of the Capacity Benefits from Interconnecting Two or More Generating Systems, *C. W. Watchorn* (*AIEE Transactions,* May 1950) 23

Determination of Reserve-Generating Capability, *H. Halperin, H. A. Adler* (*AIEE Transactions on Power Apparatus and Systems,* August 1958) 30

Mathematical Models for Use in the Simulation of Power Generation Outages: I—Fundamental Considerations, *C. J. Baldwin, D. P. Gaver, and C. H. Hoffman* (*AIEE Transactions on Power Apparatus and Systems,* December 1959) 45

Application of Probability Methods to Generating Capacity Problems, *Report of AIEE Subcommittee on Application of Probability Methods* (*AIEE Transactions on Power Apparatus and Systems,* February 1961) 53

Determination of Reserve Requirements of Two Interconnected Systems, *V. M. Cook, C. D. Galloway, M. J. Steinberg, and A. J. Wood* (*AIEE Transactions on Power Apparatus and Systems,* April 1963) 71

Effective Load Carrying Capability of Generating Units, *L. L. Garver* (*IEEE Transactions on Power Apparatus and Systems,* August 1966) 87

Frequency and Duration Methods for Power System Reliability Calculations: I—Generation System Model, *J. D. Hall, R. J. Ringlee, and A. J. Wood* (*IEEE Transactions on Power Apparatus and Systems,* September 1968) . 97

Frequency and Duration Methods for Power System Reliability Calculations: II—Demand Model and Capacity Reserve Model, *R. J. Ringlee and A. J. Wood* (*IEEE Transactions on Power Apparatus and Systems,* April 1969) 107

Generating Capacity Reliability Evaluation in Interconnected Systems Using Frequency and Duration Approach: Part I. Mathematical Analysis, *R. Billinton and C. Singh* (*IEEE Transactions on Power Apparatus and Systems,* July/August 1971) 121

A Four-State Model for Estimation of Outage Risk for Units in Peaking Service, *Report of the IEEE Task Group on Models for Peaking Service Units, Application of Probability Methods Subcommittee* (*IEEE Transactions on Power Apparatus and Systems,* March/April 1972) 130

A Frequency and Duration Method for Generating System Reliability Evaluation, *A. K. Ayoub and A. D. Patton* (*IEEE Transactions on Power Apparatus and Systems,* November/December 1976) 140

Reliability Evaluation in Energy Limited Generating Capacity Studies, *R. Billinton and P. G. Harrington* (*IEEE Transactions on Power Apparatus and Systems,* November/December 1978) 145

Multi-Area Generation Reliability Studies, *C. K. Pang, A. J. Wood, R. L. Watt, and J. A. Bruggeman* (*Proceedings of the IEEE Summer Power Meeting,* July 1978) 155

Application of Fourier Methods for the Evaluation of Capacity Outage Probabilities, *N. S. Rau and K. F. Schenk* (*Proceedings of the IEEE Winter Power Meeting,* January 1979) 164

The IEEE Reliability Test System—Extensions to and Evaluation of the Generating System, *R. N. Allan, R. Billinton, and N. M. K. Abdel-Gawad* (*IEEE Transactions on Power Systems,* November 1986) 172

A Monte Carlo Simulation Approach to the Reliability Modeling of Generating Systems Recognizing Operating Considerations, *A. D. Patton, J. H. Blackstone, and N. J. Balu* (*IEEE Transactions on Power Systems,* August 1988) 179

Criteria Used by Canadian Utilities in the Planning and Operation of Generating Capacity, *R. Billinton* (*IEEE Transactions on Power Systems,* November 1988) 186

Section 1.2 Operating Capacity Assessment 193

Application of Probability Methods to the Determination of Spinning Reserve Requirements for the Pennsylvania-New Jersey-Maryland Interconnection, *L. T. Anstine, R. E. Burke, J. E. Casey, R. Holgate, R. S. John, and H. G. Stewart (AIEE Transactions on Power Apparatus and Systems,* October 1963) 193

A Probability Method for Bulk Power System Security Assessment, I—Basic Concepts, *A. D. Patton (IEEE Transactions on Power Apparatus and Systems,* January/February 1972) 203

The Effect of Rapid Start and Hot Reserve Units in Spinning Reserve Studies, *R. Billinton and A. V. Jain (IEEE Transactions on Power Apparatus and Systems,* March/April 1972) 212

Interconnected System Spinning Reserve Requirements, *R. Billinton and A. V. Jain (IEEE Transactions on Power Apparatus and Systems,* March/April 1972) 217

Part 2: Composite Systems—Hierarchical Level II 227

Composite System Reliability Evaluation, *R. Billinton (IEEE Transactions on Power Apparatus and Systems,* April 1969) 229

Transmission Planning Using a Reliability Criterion, Part I: A Reliability Criterion, *R. Billinton and M. P. Bhavaraju (IEEE Transactions on Power Apparatus and Systems,* January 1970) 234

Quantitative Evaluation of Power System Reliability in Planning Studies, *P. L. Noferi and L. Paris (IEEE Transactions on Power Apparatus and Systems,* March/April 1972) 241

Monte Carlo Methods for Power System Reliability Evaluations in Transmission and Generation Planning, *P. L. Noferi, L. Paris, and L. Salvaderi (Proceedings of the 1975 Annual Reliability and Maintainability Symposium,* January 1975) 249

Reliability Indices for Use in Bulk Power Supply Adequacy Evaluation, *Report of the Working Group on Performance Records for Optimizing System Design, Power Systems Engineering Committee (IEEE Transactions on Power Apparatus and Systems,* July/August 1978) 260

Adequacy Indices for Composite Generation and Transmission System Reliability Evaluation, *R. Billinton, T. K. P. Medicherla, and M. S. Sachdev (Proceedings of the IEEE Winter Power Meeting,* February 1979) 267

IEEE Reliability Test System, *Report of the Reliability Test System Task Force of the Application of Probability Methods Subcommittee (IEEE Transactions on Power Apparatus and Systems,* November/December 1979) 276

Station Originated Multiple Outages in the Reliability Analysis of a Composite Generation and Transmission System, *R. Billinton and T. K. P. Medicherla (IEEE Transactions on Power Apparatus and Systems,* August 1981) . 284

Bulk Power System Reliability Assessment—Why and How? Part I: Why?, *J. Endrenyi, P. F. Albrecht, R. Billinton, G.E. Marks, N. D. Reppen, and L. Salvaderi (IEEE Transactions on Power Apparatus and Systems,* September 1982) 293

Bulk Power System Reliability Assessment—Why and How? Part II: How?, *J. Endrenyi, P. F. Albrecht, R. Billinton, G. E. Marks, N. D. Reppen, and L. Salvaderi (IEEE Transactions on Power Apparatus and Systems,* September 1982) 300

Terminal Effects and Protection System Failures in Composite System Reliability Evaluation, *R. N. Allan and A. N. Adraktas (IEEE Transactions on Power Apparatus and Systems,* December 1982) 311

Effect of Station Originated Outages in a Composite System Adequacy Evaluation of the IEEE Reliability Test System, *R. Billinton, P. K. Vohra, and S. Kumar (IEEE Transactions on Power Apparatus and Systems,* October 1985) 317

A Comparison between Two Fundamentally Different Approaches to Composite System Reliability Evaluation, *L. Salvaderi and R. Billinton (IEEE Transactions on Power Apparatus and Systems,* December 1985) 325

Requirements for Composite System Reliability Evaluation Models, *M. P. Bhavaraju, P. F. Albrecht, R. Billinton, N. D. Reppen, and R. J. Ringlee (IEEE Transactions on Power Systems,* February 1988) 332

Monte Carlo Approach in Planning Studies: An Application to IEEE RTS, *O. Bertoldi, L. Salvaderi, and S. Scalcino (IEEE Transactions on Power Systems,* August 1988) 341

Bulk System Reliability—Measurement and Indices, *C. C. Fong, R. Billinton, R. O. Gunderson, P. M. O'Neil, J. Raksany, A. W. Schneider, Jr., and B. Silverstein (IEEE Transactions on Power Apparatus and Systems,* August 1989) 350

Part 3 Transmission and Distribution System Reliability 357

Power System Reliability—I—Measures of Reliability and Methods of Calculation, *D. P. Gaver, F. E. Montmeat, A. D. Patton (IEEE Transactions on Power Apparatus and Systems,* July 1964) 359

Transmission System Reliability Evaluation Using Markov Processes, *R. Billinton and K. E. Bollinger (IEEE Transactions on Power Apparatus and Systems,* February 1968) 370

On Procedures for Reliability Evaluation of Transmission Systems, *R. J. Ringlee, S. D. Goode (IEEE Transactions on Power Apparatus and Systems,* April 1970) 379

Three-State Models in Power System Reliability Evaluations, *J. Endrenyi* (*IEEE Transactions on Power Apparatus and Systems,* July/August 1971) 389

A Computerized Approach to Substation and Switching Station Reliability Evaluation, *M. S. Grover and R. Billinton* (*IEEE Transactions on Power Apparatus and Systems,* September/October 1974) 397

Reliability Assessment of Transmission and Distribution Schemes, *R. Billinton and M. S. Grover* (*IEEE Transactions on Power Apparatus and Systems,* May/June 1975) 407

Reliability Evaluation of Electrical Systems with Switching Actions, *R. N. Allan, R. Billinton, and M. F. de Oliveira* (*Proceedings of the IEE,* April 1976) 416

Common Mode Forced Outages of Overhead Transmission Lines, *Report of the IEEE Task Force on Common Mode Outages of Bulk Power Supply Facilities of the Application of Probability Methods* (*IEEE Transactions on Power Apparatus and Systems,* May/June 1976) 422

Reliability Evaluation of the Auxiliary Electrical Systems of Power Stations, *R. N. Allan, R. Billinton, and M. F. de Oliveira* (*IEEE Transactions on Power Apparatus and Systems,* September/October 1977) 427

Modelling and Evaluating the Reliability of Distribution Systems, *R. N. Allan, E. N. Dialynas, I. R. Homer* (*IEEE Transactions on Power Apparatus and Systems,* November/December 1979) 436

MAPP Bulk Transmission Outage Data Collection and Analysis, *M. G. Lauby, K. T. Khu, R. W. Polesky, R. E. Vandello, J. H. Doudna, P. J. Lehman, D. D. Klempel* (*IEEE Transactions on Power Apparatus and Systems,* January 1984) 445

Proposed Terms for Reporting and Analyzing Outages of Electrical Transmission and Distribution Facilities, *D. W. Forrest, P. F. Albrecht, R. N. Allan, M. P. Bhavaraju, R. Billinton, G. L. Landgren, M. F. McCoy, N. D. Reppen* (*IEEE Transactions on Power Apparatus and Systems,* February 1985) 454

Part 4: Reliability Cost/Worth 467

Costs of Power Interruptions to Industry—Survey Results, *E. M. Mackay and L. H. Berk* (*Proceedings of the CIGRE International Conference on Large High Voltage Electric Systems,* August 30–September 7, 1978) 469

Comprehensive Bibliography on Electrical Service Interruption Costs, *R. Billinton, G. Wacker, E. Wojczynski* (*IEEE Transactions on Power Apparatus and Systems,* June 1983) 478

Interruption Cost Methodology and Results—A Canadian Residential Survey, *G. Wacker, E. Wojczynski, and R. Billinton* (*IEEE Transactions on Power Apparatus and Systems,* October 1983) 485

Comparison of Two Alternate Methods to Establish an Interrupted Energy Assessment Rate, *R. Billinton, J. Oteng-Adjei, and R. Ghajar* (*IEEE Transactions on Power Systems,* August 1987) 493

Appendix: IEEE Bibliographies 501

Bibliography on the Application of Probability Methods in Power System Reliability Evaluation, *R. Billinton* (*IEEE Transactions on Power Apparatus and Systems,* March/April 1972) 503

Bibliography on the Application of Probability Methods in Power System Reliability Evaluation, 1971–1977, *Report of the IEEE Subcommittee on the Application of Probability Methods, Power System Engineering Committee* (*IEEE Transactions on Power Apparatus and Systems,* November/December 1978) 515

Bibliography on the Application of Probability Methods in Power System Reliability Evaluation, 1977–1982, *R. N. Allan, R. Billinton, and S. H. Lee* (*IEEE Transactions on Power Apparatus and Systems,* February 1984) 523

Bibliography on the Application of Probability Methods in Power System Reliability Evaluation, 1982–1987, *R. N. Allan, R. Billinton, S. M. Shahidehpour, and C. Singh* (*IEEE Transactions on Power Systems,* November 1988) 531

Author Index 541

Subject Index 543

Editors' Biographies 549

Preface

THE reliability assessment of power systems has undergone continuous development and application since the early publications appeared in the 1930s. The initial techniques were very simplistic compared with present-day methods. These procedures, however, form the foundation of today's techniques and established various thought processes and philosophical concepts which are fundamental to the applications currently practiced.

The IEEE has published a series of bibliographies [1–4] that cite nearly 700 individual papers from worldwide sources, although most of them are from the IEEE and IEE. These bibliographies are not complete, as they only include papers published in English and exclude publications from some of the related conferences that have taken place in various parts of the world. Despite these limitations, they form a valuable source of reference in this important, and still developing, area of activity. For this reason, these bibliographies are reprinted in the Appendix in order to provide this extensive source of references in one volume.

Some of the bibliography papers consider rigorous mathematical approaches, others are empirical, pragmatic, and application-oriented, and a few are philosophical. The bibliographies, therefore, provide an excellent source from which to choose a wide range of classical papers, survey papers, and current papers. The bibliographies, as published papers, were also subjected to the peer review process conducted by the IEEE Power Engineering Society.

The basic aim of this book was to create a collection of reprinted papers from this extensive literature that highlight not only the present state of the art, but also provide an indication of the evolutionary process. Although many of the very early techniques have been either eclipsed or extensively revamped, they have a very distinctive place in the historical development and should not be forgotten. It was therefore considered essential to include a number of these early pioneering contributions. These considerations led to an attempt to create a fine balance between present-day practice and the developments that have led to this situation.

The overall objective was to present a reprint volume that gives a comprehensive review of power system reliability assessment and application that is useful to practicing engineers, academics, researchers, and students. This will provide both a valuable "instant library" and an overall picture of early and recent concepts in this important field.

The size limitations of this book dictated that less than ten percent of the papers cited in the bibliographies could be selected. This imposed considerable difficulties and there is no doubt that a number of excellent papers have been excluded in order to create the required balance of papers in the various sections. The papers included in each part of this book are, however, preceded by an overview of the relevant area, which includes the citation of several additional papers. The reader will find these additional references useful in supporting and enhancing the concepts and discussion contained in the selected papers.

The four main parts of this book are:

Part I—Generating Capacity: Hierachical Level I (22 papers)
Part II—Composite Systems: Hierachical Level II (16 papers)
Part III—Transmission and Distribution Systems (12 papers)
Part IV—Reliability Cost/Reliability Worth (4 papers)

In addition, the paper entitled "Power System Reliability in Perspective" [5] is included as an introductory paper to the book. This paper presents several philosophical aspects concerning power system reliability. It puts the reliability aspects in perspective, describes a hierarchical framework of analysis, and discusses how the economics of reliability should be compared.

Finally, it is worth noting that several important texts have been published that deal with various aspects of power system reliability evaluation and application. These texts [6]–[16] form a valuable set of additional material that consolidates the theoretical concepts and indicates how power system reliability evaluation techniques can be used.

ACKNOWLEDGMENTS

The material in this book has been selected from the four bibliographies on the application of probability methods in power system reliability evaluation prepared under the auspices of the IEEE Subcommittee on the Application of Probability Methods. We would like to thank the Subcommittee for its assistance in developing these bibliographies and for providing a focal point for power system reliability assessment. We would also like to thank the IEE and CIGRE for granting their permission to include several publications from these organizations in this collection of reprinted papers.

Our joint association over many years in the Subcommittee on the Application of Probability Methods and on the CIGRE Working Group 38.03 on Power System Reliability Analysis has provided the foundation for the mutual respect, appreciation, and understanding required in a project such as this. We have found the opportunity to work together, albeit by long distance and mostly by telecommunication links, to be a very interesting and rewarding experience. Agreement has been unanimous on the majority of the selected papers. The discussion on the selection of the remaining papers has been vigorous, stimulating, and educational. We are indebted to our respective organizations for their support in this project.

We would particularly like to thank Brenda Rowe and Tracy Garrett of the College of Engineering at the University of Saskatchewan for their skill and dedication in typing the basic material and the numerous lists which were created in the process of reducing the nearly seven hundred publications in the four bibliographies to the fifty-six papers that appear in this book. We wish to thank our respective wives, for their patience and their support in bringing this project to a satisfying and rewarding conclusion. Luigi Salvaderi would like to dedicate this book to the memory of his sister Eugenia and to her heritage of ideals and moral strength.

Roy Billinton
Ron Allan
Luigi Salvaderi

References

[1] R. Billinton, "Bibliography on the application of probability methods in power system reliability evaluation," *IEEE Trans. Power Apparatus Syst.*, vol. PAS-91, no. 2, pp. 649–660, Mar./Apr. 1972.

[2] IEEE Subcommittee on the Application of Probability Methods, "Bibliography on the application of probability methods in power system reliability evaluation 1971–1977," *IEEE Trans. Power Apparatus Syst.*, PAS-97, no. 6, pp. 2235–2242, Nov./Dec. 1978.

[3] R. N Allan, R. Billinton, and S. H. Lee, "Bibliography on the application of probability methods in power system reliability evaluation 1977–1982," *IEEE Trans. Power Apparatus Syst.*, vol. PAS-103, no. 2, pp. 275–282, Feb. 1984.

[4] R. N. Allan, R. Billinton, S. M. Shahidehpour, and C. Singh, "Bibliography on the application of probability methods in power system reliability evaluation 1982–1987," *IEEE Trans. Power Syst.*, vol. 3, no. 4, pp. 1555–1564, Nov. 1988.

[5] R. Billinton and R. N. Allan, "Power system reliability in perspective," *IEE J. Electron. Power*, vol. 30, pp. 231–236, Mar. 1984.

[6] R. Billinton, *Power System Reliability Evaluation*, NY: Gordon and Breach, 1970.

[7] R. Billinton, R. J. Ringlee, and A. J. Wood, *Power System Reliability Calculation*, Cambridge, MA: MIT Press, 1973.

[8] J. Endrenyi, *Reliability Modelling in Electric Power Systems*, Chichester, UK: John Wiley, 1978.

[9] H. Khatib, *Economics of Reliability in Electrical Power Systems*, Gloucester, UK: Technicopy, 1978.

[10] J. Vardi and B. Avi-Itzhak, *Electric Energy Generation: Economics, Reliability and Rates*, Cambridge, MA: MIT Press, 1981.

[11] R. Billinton and R. N. Allan, *Reliability Evaluation of Power Systems*, New York: Plenum Publishing, 1984.

[12] R. Billinton and R. N. Allan, *Reliability Assessment of Large Electric Power Systems*, Boston, MA: Kluwer Academic Publishers, 1988.

[13] CIGRE Working Group 38.03, "Power system reliability analysis—Application guide," Paris: CIGRE Publications, 1988.

[14] IEEE Tutorial Text No. 71 M 30 PWR, "Probability analysis of power system reliability."

[15] IEEE Tutorial Text No. 82 EHO 195-8-PWR, "Power system reliability evaluation."

[16] IEEE Tutorial Text No. 90EH0311-1-PWR, "Reliability assessment of composite generation and transmission systems."

Introduction

THE number of technical papers associated with all areas of science and technology is increasing rapidly year by year. Engineers, academics, and researchers already established in a particular field of activity can become overwhelmed by this continuous escalation of information. Those who are entering a field for the first time have even greater difficulty in knowing where to begin and what to read. This situation is certainly the case in the area of power system reliability assessment and its applications.

It is clearly of considerable benefit for both newcomers and established personnel to recognize the pioneering contributions, the important aspects associated with historical developments, and the present-day techniques and applications. Literature surveys using computerized resources and published bibliographies cannot entirely resolve this problem because they only provide complete—or frequently incomplete—listings of all known papers in the specified area. They do not generally give any perspective, real or otherwise, to the individual papers, nor do they indicate the contribution each paper makes. With these thoughts in mind, we set ourselves the task of compiling a volume of previously published papers that chronicle the developments in power system reliability assessment. The objective was to present the pioneering contributions together with the historical developments which form the basis of the present-day state of the art.

We realized before starting that it would be a formidable task and we would have considerable difficulty in restricting the papers to the number allowed. These difficulties were anticipated because of differences in international practices and viewpoints, in utility practices and applications, in data collection and processing, and in the perception of needs and requirements. We also recognized that many excellent papers are published in languages other than English or in sources from which it is either not easy or even impossible to retrieve material. This is particularly the case with material developed by utilities and university personnel who operate in countries having native languages other than English. It was important to try and minimize these difficulties so that the perceptions and techniques of such utilities and academics are included. These potential problems were discussed and resolved in a true spirit of international camaraderie. It was decided that only papers from the related IEEE bibliographies would be considered. The selection was then achieved from our individual experiences and national and international associations. The papers themselves have been divided into four parts. This itself caused some problems because some papers do not naturally fall easily into one of these areas. Our long-established working relationship and personal friendship ensured harmonious discussions and resolutions.

The four parts: generating capacity, composite systems, transmission and distribution, reliability cost/reliability worth, do not contain equal numbers of included papers. This reflects the total number of papers published in each of these areas, the length of time each has been actively pursued, and the qualities of the various papers as perceived by us as individuals.

We readily appreciate that a selection of this type can be a very personal choice and that, while many of the papers we have selected would also be chosen by others, there would be differences of opinion over the remaining ones. We trust however that those who do not agree with all of our selections will understand the personal element involved in such a choice and respect the quality of the papers we have selected.

Power-system reliability in perspective

This article discusses several philosophical aspects concerning power-system reliability. It puts the reliability aspects in perspective, describes a hierarchical framework of analysis and discusses how the economics of reliability should be compared

by R. Billinton and R.N. Allan

A power system serves one function only and that is to supply customers, both large and small, with electrical energy as economically as possible and with an acceptable degree of reliability and quality. Modern society, because of its pattern of social and working habits, has come to expect that the supply should be continuously available on demand. This is not physically possible due to random system failures which are generally outside the control of power-system engineers, although the probability of customers being disconnected can be reduced by increased investment during either the planning phase, operating phase or both. It is evident therefore that the economic and reliability constraints can conflict, and this can lead to difficult managerial decisions at both the planning and operating phases.

These problems have always been widely recognised, and it is not suggested that they have only recently come to the fore. Design, planning and operating criteria and techniques have been developed over many decades in an attempt to resolve and satisfy the dilemma between the economic and reliability constraints. The criteria and techniques first used in practical applications, however, were all deterministically based.

The essential weakness of deterministic criteria is that they do not respond to nor reflect the probabilistic or stochastic nature of system behaviour, of customer demands or of component failures.

The need for probabilistic evaluation of system behaviour has been recognised since at least the 1930s, and it may be questioned why such methods have not been widely used in the past. The main reasons were lack of data, limitations of computational resources, lack of realistic reliability techniques, aversion to the use of probabilistic techniques and a misunderstanding of the significance and mean of probabilistic criteria and risk indices. None of these reasons need be valid today as most utilities have reliability databases, computing facilities are greatly enhanced, reliability evaluation techniques are highly developed[1-5] and most engineers have a working understanding of probabilistic techniques. Consequently, there is now no need to artificially constrain the inherent probabilistic or stochastic nature of a power system into a deterministic one.

Need for power-system reliability evaluation

The economic, social and political climate in which the electric power supply industry now operates has changed considerably during the last 20-30 years.

During the period following the Second World War, and prior to the end of the 1950s, planning for the construction of generating plant and facilities was basically straightforward because it could be assumed that the load would at least double every 10 years (7-8% annual growth rate). Therefore past trends provided a relatively simple guide for the future. In addition, plant construction was relatively uncomplicated. The lead time for a coal-fired station was perhaps 3-5 years with only a relatively small period of that time associated with planning and environmental inquiries. Generating units were relatively small

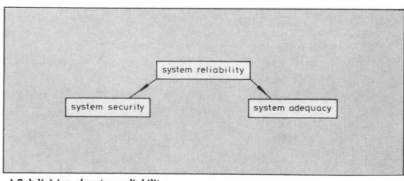

1 Subdivision of system reliability

Reprinted with permission from *IEE J. Electron. Power*, vol. 30, pp. 231–236, March 1984.

compared with total system capacity and costs were relatively stable. In the 1960s, unit sizes were increased and high-voltage transmission and interconnections between utilities were rapidly expanded to take advantage of the economies of scale. The timing of unit construction and the development of quantitative methods for determining the correct amount of spare capacity in both single and highly interconnected systems became more important. The problems were still manageable, however, because of the continued growth in consumer demand.

This situation changed abruptly in the mid-1970s. Inflation and the astronomical increase in oil prices created a rapid increase in consumer tariffs. This was a reversal of a long-standing trend. Their combined effects introduced considerable uncertainty in predicting future demand. Also conservation became a major issue which created a further reduction in forecast demand. Therefore construction plans had to be modified to recognise the new scenario.

In addition, it became evident in the late 1970s that nuclear power was not going to be the universal panacea for our electric energy needs. The nuclear programmes in the USA for instance have met considerable problems. Environmental considerations created by public concern have been added to construction difficulties and safety and reliability problems. Accidents such as the Three Mile Island incident have created a focal point for sometimes mindless but nevertheless vociferous public outcry. These factors mean that plant lead times have become more and more protracted and can now easily exceed 10 years. This requires extended estimation of demands

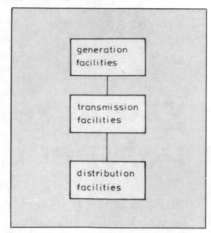

2 Basic functional zones

and costs. When this is coupled with persistent inflation and a highly uncertain growth in demand, the present situation arises. A key term in the minds of governments, power-system managers and the general public is 'excess capacity'; how much should there be and who should pay for it?

The immediate concerns therefore regarding the development and utilisation of reliability concepts are not in a growth scenario but are in a period of economic slowdown and restraint.

One very important outcome of the economic situation in which the electric power supply industry finds itself is that it is being scrutinised very closely by many different organisations and individuals. The industry is capital intensive; it plays a major role in the economic and social wellbeing of a nation and indeed on the quality of life. Governments, licensing bodies, consumer advocates, environmental and

conservation groups and even private citizens are expressing their concerns in ways which did not exist a decade ago. It is therefore with this background that our present reliability techniques and concepts are being developed, utilised and scrutinised. The industry has difficulty in communicating with external organisations on the subject of power system reliability, and this we believe is largely the industry's own fault. Indeed, difficulties also exist in communicating with each other within the power industry. This problem has been recognised fully by the IEE, the Power Division of which has helped the industry considerably by organising three international conferences on this very important topic.[6-8]

It is therefore very useful to give some thought to defining the problem zones in the general areas of overall power-system reliability evaluation and to discuss data needs, methodologies and techniques. It is necessary, however, to first create a common philosophy on a broad scale which can be articulated to those outside the industry.

Definition of power-system reliability

The function of an electric power system is to satisfy the system load requirement as economically as possible and with a reasonable assurance of continuity and quality. The ability of the system to provide an adequate supply of electrical energy is usually designated by the term reliability. The concept of power-system reliability, however, is extremely broad and covers all aspects of the ability of the system to satisfy the consumer requirements.

The term reliability has a very wide range of meanings and cannot be associated with a single specific definition such as that often used in the mission-oriented sense. It is therefore necessary to recognise its extreme generality and to use it to indicate, in a general rather than specific sense, the overall ability of the system to perform its function.

A simple but reasonable subdivision of the concern designated as system reliability is shown in Fig.1. This represents the two basic aspects of a power system: system adequacy and system security. These two terms can best be described as follows.

Adequacy relates to the existence of sufficient facilities within the system to satisfy the consumer load demand. These include the facilities necessary to generate sufficient energy and the associated transmission and distribution facilities required to transport the energy to the actual consumer load points. Adequacy is therefore associated with static conditions which do not include system disturbances.

Security relates to the ability of the system to respond to disturbances arising within that system. Security is therefore associated with the response of the system to whatever perturbations it is subject. These include the conditions

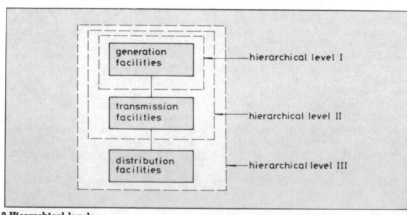

3 Hierarchical levels

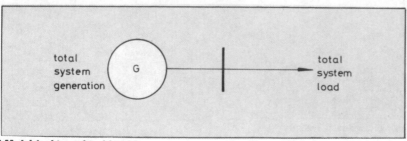

4 Model for hierarchical level I

associated with both local and widespread disturbances and the loss of major generation and transmission facilities.

It is important to realise that most of the probabilistic techniques presently available for power-system reliability evaluation are in the domain of adequacy assessment. Some work has been done on subsets of the security problem, such as quantifying spinning or operating capacity requirements[2] and transient stability assessment.[9]

Functional zones

The basic techniques for adequacy assessment can be categorised in terms of their application to segments of a complete power system. These segments are shown in Fig.2 and are defined as functional zones: generation, transmission and distribution. This division is the most appropriate as most utilities are either divided into these zones for purposes of organisation, planning, operation and/or analysis or are solely responsible for one of these functions. Adequacy studies can be, and are, conducted individually in these three functional zones.

Hierarchical levels

The functional zones shown in Fig.2 can be combined to give the hierarchical levels shown in Fig.3. These hierarchical levels can also be used in adequacy assessment. Hierarchical level I (HLI) is concerned only with the generation facilities. Hierarchical level II (HLII) includes both generation and transmission facilities and HLIII includes all three functional zones in an assessment of consumer load point adequacy. HLIII studies are not usually done directly due to the enormity of the problem in a practical system. Instead, the analysis is usually performed only in the distribution functional zone, in which the input points may or may not be considered fully reliable. Furthermore, functional zone studies are often done which do not include the functional zones above them. These are usually performed on a subset of the system to examine a particular configuration or topological change. These analyses are frequently undertaken in the sub-transmission and distribution system functional zones because these are less affected by the geographical location of the generating facilities.

Adequacy evaluation at hierarchical level I

At HLI the total system generation is examined to determine its adequacy to meet the total system load requirement. This is usually termed 'generating capacity reliability evaluation'. The system model at this level is shown in Fig.4.

In HLI studies, the reliability of the transmission and its ability to move the generated energy to the consumer load points is ignored. The only concern is in estimating the necessary generating capacity to satisfy the system demand and to have sufficient capacity to perform corrective and preventive maintenance on the generating facilities. The historical technique used for determining this capacity was the percentage reserve method. In this approach the required reserve is set as a fixed percentage of either the installed capacity or the predicted load. Other criteria, such as a reserve equal to one or more largest units, have also been used. These deterministic criteria have now been largely replaced by probabilistic methods which respond to and reflect the actual factors that influence the reliability of the system.

Criteria such as loss of load expectation (LOLE), loss of energy expectation (LOEE), and frequency and duration (F&D) — see panel (p. 4) — are now widely used by electric power utilities. These indices are generally calculated using direct analytical techniques although sometimes Monte Carlo simulation is used — see panel (p. 4).

The LOLE approach is by far the most popular and can be used for both single systems and interconnected systems. Expectation indices are most often used to express the adequacy of the generation configuration. These indices give a physical interpretation which cannot be provided by a value of probability. There is, however, considerable confusion within and without the power industry on the specific meaning of these expectation indices and the use that can be made of them. A single expected value is not a deterministic parameter: it is the mathematical expectation of a probability distribution (albeit unknown), i.e. it is therefore the long-run average value. These expectation indices provide valid adequacy indicators which reflect factors such as generating unit size, availability, maintenance requirements, load characteristics and uncertainty, and the potential assistance available from neighbouring systems.

The basic modelling approach for an HLI study is shown in Fig.5. The capacity model can take a number of forms. It is formed in the direct analytical methods by creating a capacity outage probability table. This table represents the capacity outage states of the generating system together with the probability of each state. The load model can either be the daily peak load variation curve (DPLVC), which only includes the peak loads of each day, or the load duration curve (LDC) which represents the hourly variation of the load. The risk indices are evaluated by convolution of the capacity

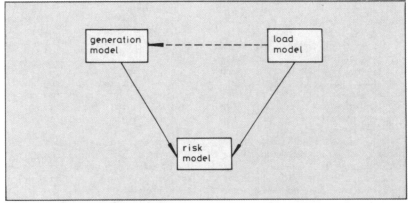

5 Conceptual tasks for HLI evaluation

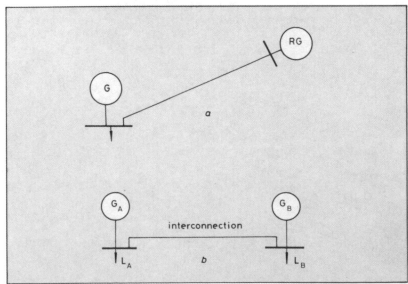

6 (a) Model of remote generation in HLI studies; (b) Model of interconnected systems in HLI studies

Typical reliability criteria

● **Loss of load expectation (LOLE)**

The loss of load expectation is the average number of days on which the daily peak load is expected to exceed the available generating capacity. It therefore indicates the expected number of days on which a load loss or deficiency will occur. It does not indicate the severity of the deficiency and neither does it indicate the frequency nor the duration of loss of load. Despite these shortcomings, it is the most widely used criterion in generation-planning studies.

● **Loss of energy expectation (LOEE)**

The loss of energy expectation is the expected energy that will not be supplied by the generating system due to those occasions when the load demanded exceeds available generating capacity. This is a more appealing index as it measures severity of deficiencies rather than just number of days. It is likely that it will

be used more widely in the future. The complementary value of energy not supplied, i.e. energy actually supplied, is sometimes divided by the total energy demanded. This gives a normalised index known as the energy index of reliability which can be used to compare the adequacy of systems that differ considerably in size.

● **Frequency and duration (F&D)**

The frequency-and-duration criterion is an extension of the LOLE index in that it also identifies the expected frequency of encountering a deficiency and the expected duration of the deficiences. It therefore contains additional physical characteristics which makes it sensitive to further parameters of the generating system, and so it provides more information to power-system planners. The criterion has not been used very widely in generating-system reliability analyses, although it is extensively used in network studies.

and load models. Generally the DPLVC is used to evaluate LOLE indices giving a risk expressed in number of days during the period of study when the load will exceed available capacity. If an LDC is used, the units will be the number of hours. If LOEE indices are required, the LDC must be used.

Limited considerations of transmission can be included in HLI studies. These include the modelling of remote generation facilities (Fig.6a) and interconnected systems (Fig. 6b). In the latter case, only the interconnections between adjacent systems are modelled; not the internal system connections.

In the case of remote generation, the capacity model of the remote source is modified by the reliability of the transmission link before being added to the system capacity model. In the case of interconnected systems the available assistance model of the assisting system is modified before being added to the capacity model of the system under study.

Adequacy evaluation at hierarchical level II

In HLII studies, the simple generation-load model shown in Fig.4 is extended to include bulk transmission. Adequacy analysis at this level is usually termed composite system or bulk transmission system evaluation. A small composite system is shown in Fig.7.

HLII studies are required to assess the adequacy of an existing system and the impact of various reinforcement schemes, at both the generation and transmission levels. In the case of the system shown in Fig.7, it may be necessary to evaluate the effects of such additions as lines 7 and 8. These effects can be assessed by evaluating two sets of indices: individual bus (load-point) indices and overall system indices. These indices are complementary, not alternatives. The system indices give an assessment of overall adequacy and the load-point indices monitor the effect on individual busbars and provide input

values to the next hierarchical level.

The HLII indices can be calculated for a single load level and expressed on a period basis, such as a year, or obtained using simplified load models for a specific period. Although these indices add realism by including bulk transmission, they are still adequacy indicators. They do not include the system dynamics or the ability of the system to respond to transient disturbances: they simply measure the ability of the system to adequately meet its requirements in a specified set of probabilistic states. There are many complications in this type of analysis such as overload effects, redispatch of generation and the consideration of independent, dependent, common-cause and station-associated outages. Many of these aspects have not yet been fully resolved and there is no universally accepted method of analysis. The important aspect therefore is for each utility to decide which parameters and effects are important in its own case and then to use a consistent method of analysis. Some of the aspects needing consideration are discussed below.

The single-line/busbar representation shown in Fig.7 is a conventional load-flow representation and does not indicate some of the complexities which should be included in the analysis. For instance, consider busbar 1 where there are three transmission lines and a generating facility. This could have a configuration as shown in Fig.8. The protection system of this ring substation is such that a single fault on some components will cause a multiple-outage event. It follows therefore that the total protection system should be included in the analysis.

It is also seen in Fig.7 that two transmission lines (1 and 6) leave busbar 1 and go to the same system load point (busbar 3). If these are on the same

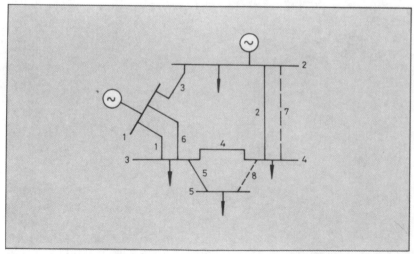

7 Composite system for HLII studies

tower structure or possibly the same right of way, there is the possibility of simultaneous outages due to a common-mode event. All these factors, plus others, can be included in an HLII evaluation. The primary indices are, however, expected values and are highly dependent on the modelling assumptions used in the computer simulation.

Adequacy evaluation at hierarchical level III

The overall problem of HLIII evaluation can become very complex in most systems because this level involves all three functional zones, starting at the generating points and terminating at the individual consumer load points.

For this reason, the distribution functional zone is usually analysed as a separate entity. The HLIII indices can be evaluated, however, by using the HLII load-point indices as the input values at the sources of the distribution functional zone being analysed. The objective of the HLIII study is to obtain suitable adequacy indices at the actual consumer load points. The primary indices are the expected frequency (or rate) of failure, the average duration of failure and the annual unavailability (or outage time) of the consumer load points. Additional indices, such as expected load disconnected or energy not supplied, can also be obtained.

The analytical methods for evaluating these indices are highly developed. The usual techniques are based on the minimal-cut-set method or failure-modes analysis in conjunction with sets of analytical equations which can account for all realistic failure and restoration processes.

Comments on hierarchical level assessment

The adequacy indices calculated at each hierarchical level are physically different. Those calculated at HLI are a measure of the ability of the generating system to satisfy the system load requirement. The indices calculated at the HLII level are a measure of the ability of the system to satisfy the individual load requirements at the major load points. They extend the HLI indices by including the ability to move the generated energy through the bulk transmission system. The individual customer adequacy is reflected in the HLIII indices. In most systems, the inadequacy of the individual load points is caused mainly by the distribution system. Therefore the HLII adequacy indices generally have a negligible effect on the individual load-point indices. Typical statistics show that HLII indices contribute less than 1% to customer unavailability. The HLI and HLII indices are very important, however, because failures in these parts of the system affect large sections of the system and therefore can have widespread and perhaps catastrophic consequences for both society and its environment. Failures in the distribution system, although much more frequent, have much more localised effects.

Analytical methods/ Monte Carlo simulation

Evaluation techniques can be classified either as analytical or as Monte Carlo simulation. Analytical techniques represent the system by a mathematical model and evaluate the reliability indices from this model using mathematical solutions. Monte Carlo simulation methods, however, estimate the reliability indices by simulating the actual process and random behaviour of the system. The method therefore treats the problem as a series of real experiments. There are merits and demerits in both methods. Generally Monte Carlo simulation requires a large amount of computing time and is not used extensively if alternative analytical methods are available. In theory, however, it can include any system effect or system process which may have to be approximated in analytical methods.

Reliability-cost/reliability-worth

Adequacy studies of a system are only part of the overall required assessment.

As discussed earlier, the economics, of alternative facilities play a major role in the decision-making process. The two aspects, adequacy and economics, can be consistently appraised by comparing adequacy cost (the investment cost needed to achieve a certain level of adequacy) with adequacy worth (the benefit derived by the utility, consumer and society).

This type of economic appraisal is a fundamental and important area of engineering application, and it is possible to perform this kind of evaluation at the three hierarchical levels discussed. A goal for the future should be to extend this adequacy comparison within the same hierarchical structure to include security, and therefore to arrive at reliability-cost and reliability-worth evaluation. The extension of quantitative reliability analysis to the evaluation of service worth is a deceptively simple inclusion which is fraught with potential misapplication. The basic concept of reliability-cost/reliability-worth evaluation is relatively simple and is summarised in Fig.9. This same idea can also be presented in the cost/reliability curves of Fig.10.

These curves show that the utility cost will generally increase as consumers are provided with higher reliability. On the other hand, the consumer costs associated with supply interruptions will decrease as the reliability increases. The total costs to society will therefore be the sum of these two individual costs. This total cost exhibits a minimum, and so an 'optimum' or target level of reliability is achieved.

This concept is quite valid. Two dif-

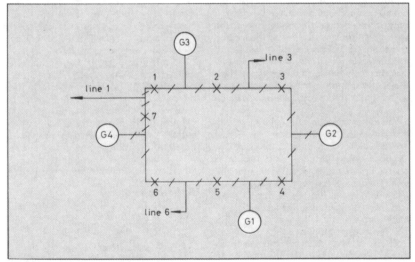

8 Typical substation configuration

cost to society of providing quality and continuity of electric supply } should be related to { worth or benefit to society of having quality and continuity

9 Reliability cost and reliability worth

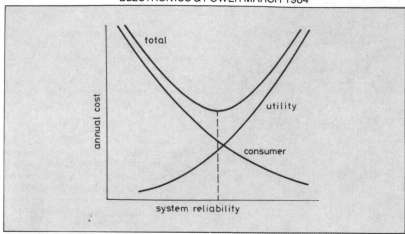

10 Reliability costs

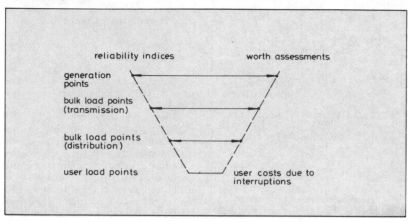

11 Disparity between indices and worth at different hierarchical levels

ficulties arise in its assessment. First the calculated indices are usually derived only from adequacy assessments at the various hierarchical levels. Secondly, there are great problems in assessing consumer perceptions of outage costs. The disparity between the calculated indices and the monetary costs associated with supply interruptions are shown in Fig.11.

The left-hand side of Fig.11 shows the calculated indices at the various hierarchical levels. The right-hand side indicates the interruption cost data obtained by user studies. It is seen that the relative disparity between the calculated indices at the three hierarchical levels and the data available for worth assessment decreases as the consumer load points are approached.

There is still considerable research needed on the subject of the cost of an interruption. This research should consider the direct and the indirect costs associated with the loss of supply, both on a local and widespread basis.

Reliability data

Any discussion of quantitative reliability evaluation invariably leads to a discussion of the data available and the data required to support such studies. Valid and useful data are expensive to collect, but it should be recognised in the long run that it will be even more ex-

pensive not to collect them. It is sometimes argued as to which comes first: reliability data or reliability methodology. Some utilities do not collect data because they have not fully determined a suitable reliability methodology. Conversely, they do not conduct reliability studies because they do not have any data. It should be remembered that data collection and reliability evaluation must evolve together and therefore the process is iterative. The point at which to stop on

either should be based on the economic use to be made of the tools, techniques and data. We firmly believe, however, that this point has not yet been reached in either sector.

Conclusions

This article has discussed and described various philosophical aspects concerning power-system reliability and, in particular, adequacy. The framework described in the article is one on which the discussions within the power industry and with external groups can be ideally based. The need for such a framework is already evident due to the growing number of people and organisations wishing to effect the planning decisions of power systems, and this trend will expand as the future progresses.

The framework described in this article is also suitable for assisting in the decision-making process itself. The main reasons for this conclusion are that:

- there should be some conformity between the reliability of various parts of the system, and a balance is required between generation, transmission and distribution. This does not mean that the reliability of each should be equal. Reasons for differing levels of reliability are justified, for example, because of the importance of a particular load, or because generation and transmission failures can cause widespread outages whereas distribution failures are very localised

- there should be some benefit gained from any improvement in reliability. The most useful concept for assessing this benefit is to equate the incremental or marginal investment cost to the incremental or marginal consumers' valuation of the improved reliability. The main difficulty with such a concept is the present uncertainty in the consumers' valuation. Until this problem is fully resolved, it is still beneficial for individual utilities to arrive at some consistent criterion by which they can assess the benefit of expansion and reinforcement schemes.

References

1 BILLINTON, R., and ALLAN, R.N.: 'Reliability evaluation of engineering systems, concepts and techniques' (Pitman Books, 1983)
2 BILLINTON, R., and ALLAN, R.N.: 'Reliability evaluation of power systems' (Pitman Books, 1984)
3 BILLINTON, R.: 'Bibliography on the application of probability methods in power system reliability evaluation', *IEEE Trans.*, 1972, **PAS-91**, pp.649-660
4 IEEE Committee Report: 'Bibliography on the application of probability methods in power system reliability evaluation, 1971-1977', *ibid.*, 1978, **PAS-97**, pp.2235-2242
5 ALLAN, R.N., BILLINTON, R., and LEE, S.H.: 'Bibliography on the application of probability methods in power system reliability evaluation, 1977-1982'. IEEE Winter Power Meeting, paper 83 WM 053-6
6 IEE conference on the economics of the reliability of supply, Oct. 1967
7 IEE conference on the reliability of power supply systems, Feb. 1977
8 IEE conference on the reliability of power supply systems, Sept. 1983
9 BILLINTON, R., and KURVEANTY, P.R.S.: 'A probabilistic index for transient stability', *IEEE Trans.*, 1980, **PAS-99**, pp.195-206

Dr. Billinton is with the University of Saskatchewan, Saskatoon, Canada, and Dr. Allan is with the University of Manchester Institute of Science & Technology, PO Box 88, Manchester M60 1QD, England. Robert Allan is an IEE Member

Part 1
Generating Capacity
Hierarchical Level I

INTRODUCTION

HIERARCHICAL level I (HLI) assessment refers to the generation system and its ability to satisfy the total system demand. Reliability evaluation at HLI is a quantitative assessment of this ability. This area of application can be divided into two classes: static capacity assessment and operating reserve assessment. Static capacity assessments are made at the planning stage in order to decide how much, and when, additional generating plants need to be installed. Operating reserve assessments are made at the operational planning stage or during actual operation in order to decide how much of the existing plant needs to be committed, the sharing of duties between thermal and hydro plant, and the effect on system operating cost.

STATIC CAPACITY RELIABILITY ASSESSMENT

This area of power system reliability assessment is the oldest and most extensively studied. Hence the greatest number of technical papers have been published in this area. This also reflects the number of papers reprinted in this book.

The first set of recognizable papers [A1]–[A5] envisaging the application of probability techniques appeared in the 1930s. These can be considered as pioneering papers but space restrictions prevent them from being included. However, it should not be forgotten that these papers preceded the use of reliability techniques in other applications such as military, which are often given the credit of pioneer status.

The first significant set of papers that added impetus to the application of probability theory to reliability assessments appeared in 1947. A recognized major contribution is that by Calabrese [1] although others made similar positive contributions [A6]–[A8] at the same time. The "Calabrese" method forms the basis of the loss of load approach which is still the most widely used probabilistic technique in the reliability evaluation of generating capacity. This basic method has been extended to include a loss of energy approach. Both of these methods concern evaluation of expectation using basic probability methods. These techniques are summarized in [2]. Another significant approach is the frequency and duration method. One of the most significant developments in this area was the application of recursive techniques published in a series of five papers by Ringlee, Wood, *et al.* [3], [4], [A9]–[A11]. Others have extended this technique, e.g. [5]. The frequency and duration method is still little used in the practical reliability assessment of generating capacity.

The techniques above are based on analytical approaches in which models of the system are analyzed using mathematical equations and direct numerical solutions. An alternative approach is simulation, generally known as Monte Carlo simulation. There has been a tendency for North America to use the former and for Europe and South America to use the latter. This distinction is now less clear but originally reflected the type of systems being studied and their requirements. Most of the papers included in this section are of the analytical type but several simulation references appear in Part 2. An early example of the simulation approach is found in [6]. A more recent paper is found in [7], in which it was proposed that load and generation should not be treated independently but in a related manner. It was found that simulation was appealing for this application. Other pioneers of simulation include ENEL (Italy) and EdF (France). Papers from these organizations, including [A12], [A13], and others appear in many CIGRE publications and various European conference material.

Initially, most applications related to single systems and the assessment of capacity requirements in such systems. This is typically illustrated in [8]. However, utilities recognized that there are significant economic benefits to be derived by sharing and allocating reserve capacity between neighboring systems. This encouraged the development of reliability analysis in interconnected systems. Examples are included in [9]–[11], which address the loss of load approach, and [12], which addresses the frequency and duration approach.

The majority of papers relating to analytical techniques have concerned systems without energy limitations, which is usually acceptable in thermal systems. However, energy limitations often occur in hydro systems and those using renewable energy sources [13], [A13], [A14]. Energy limitations are relatively easy to include in simulation techniques and several papers using this approach are included in Part 2. Several analytical papers address this energy limitation, including [13]. More recently others have appeared including the application of the approach to systems containing wind energy sources [A15].

Most analytical techniques are based on the Calabrese approach in which the generation model, represented by a capacity outage probability table, is constructed using a state enumeration method, e.g., the recursive technique [A16]. Various aspects of modeling have been the central theme of many papers, an example of which is given in [14]. This paper has formed the basis of several approximate modeling techniques including maintenance scheduling. However, alternative methods for evaluating the models do exist. One alternative is the cumulant method using Fourier transforms and Gram-Charlier expansion of a distribution. Its application to power system reliability was first proposed in [15] although several other applications have appeared since. This technique has been widely used for adequacy assessment, production costing, and maintenance scheduling. Care must be exercised in its use because considerable errors can occur, including apparent negative probabilities, under certain circumstances. Another alternative approach is based on the use of Fast Fourier Transforms [A17] which, although slower computationally than the cumulant method, does not exhibit the error problems.

Generating units form the basic building block in generation capacity reliability evaluation. The unit models are often simple ones representing base loaded units and are inapplicable and pessimistic for peaking units. This was recognized by the IEEE Application of Probability Methods Subcommittee which proposed the peaking unit model described in [16]. This, and derivatives of it, are now widely used.

Models and evaluation techniques are still being developed for generation systems. The IEEE Reliability Test System (RTS) [17] (see Part 2) was developed to test such techniques. This RTS has been extended in [18], which also includes important benchmark results against which other methods can be compared. This forms a valuable reference source.

The ultimate purpose of assessing the adequacy of generation systems is to help in deciding how much additional capacity to install and when. There is very little published material concerning the reliability criteria used by utilities in this decision-making process. One recent example that illustrates typical probabilistic criteria is given in [19]. This is restricted to Canadian utilities but serves as a very useful indication of present-day thinking.

OPERATING RESERVE RELIABILITY ASSESSMENT

There are very few papers dealing with the operating reserve problem compared with those on static generating capacity. The main reason is that very few utilities have felt the need for using probabilistic techniques but prefer to use long established deterministic techniques and criteria. This is likely to change with time, although not necessarily in the very near future.

There are two basic techniques that have been published: the PJM method and the security function approach. The basic PJM method was documented in [20]. This method evaluates the probability of the committed generation just satisfying or failing to satisfy the expected demand during the period of time that generation cannot be replaced, known as the lead time. An application of this method was given in [A18]. The security function approach is described in [21]. This evaluates the probability of breaches of security including inadequate spinning reserve using the concepts of conditional probability.

The PJM method has been extended in several respects. Reference [22] describes the inclusion of rapid start and hot reserve units. This paper made several significant developments which were modified and extended in [A19] and also recorded in [A16]. The PJM method was also extended by application to interconnected system evaluation in [23] and more recently in [A20].

Hierarchical level 1 assessment is an important area of power system reliability evaluation, and extensions, modifications, and new algorithms are being published continually. It is expected that this will continue as utilities attempt to minimize and optimize their reserve requirements while maintaining an acceptable level of system reliability.

References

Reprinted Papers

[1] G. Calabrese, "Generating reserve capability determined by the probability method," *AIEE Trans. Power Apparatus Syst.*, vol. 66, pp. 1439–1450, 1947.

[2] AIEE Subcommittee on Application of Probability Methods, "Application of probability methods to generating capacity problems," *AIEE Trans. Power Apparatus Syst.*, vol. 79, pp. 1165–1182, Feb. 1961.

[3] J. D. Hall, R. J. Ringlee, and A. J. Wood, "Frequency and duration methods for power system reliability calculations: Part I—generation system model," *IEEE Trans. Power Apparatus Syst.*, vol. PAS-87, no. 9, pp. 1787–1796, Sept. 1968.

[4] R. J. Ringlee and A. J. Wood, "Frequency and duration methods for power system reliability calculations: Part II—demand model and capacity reserve model," *IEEE Trans. Power Apparatus Syst.*, vol. PAS-88, no. 4, pp. 375–388, April 1969.

[5] A. K. Ayoub and A. D. Patton, "Frequency and duration method for generating system reliability evaluation," *IEEE Trans. Power Apparatus Syst.*, vol. PAS-95, no. 6, pp. 1929–1933, Nov./Dec. 1976.

[6] C. J. Baldwin, D. P. Gaver, and C.H. Hoffman, "Mathematical models for use in the simulation of power generation outages: I—fundamental considerations," *AIEE Trans. Power Apparatus Syst.*, pt. III, vol. 78, pp. 1251–1258, 1959.

[7] A. D. Patton, J. H. Blackstone, and N. J. Balu, "A Monte Carlo simulation approach to the reliability modeling of generation systems recognizing operation considerations," *IEEE Trans. Power Syst.*, vol. 3, no. 3, pp. 1174–1180, Aug. 1988.

[8] H. Halperin and H. A. Adler, "Determination of reserve-generating capability," *AIEE Trans. Power Apparatus Syst.*, vol. 77, pp. 530–544, Aug. 1958.

[9] C. W. Watchorn, "The determination and allocation of the capacity benefits resulting from interconnecting two or more generating systems," *AIEE Trans. Power Apparatus Syst.*, pt. II, vol. 69, pp. 1180–1186, 1950.

[10] V. M. Cook, C. D. Galloway, M. J. Steinberg, and A. J. Wood, "Determination of reserve requirements of two interconnected systems," *AIEE Trans. Power Apparatus Syst.*, pt. III, vol. 82, pp. 18–33, April 1963.

[11] C. K. Pang, A. J. Wood, R. L. Watt, and J. A. Bruggeman, "Multi-area generation reliability studies," IEEE Summer Power Mtg., Paper No. A 78 546-4.

[12] R. Billinton and C. Singh, "Generation capacity reliability evaluation in interconnected systems using a frequency and duration approach: Part I—mathematical analysis," *IEEE Trans. Power Apparatus Syst.*, vol. PAS-90, no. 4, pp. 1646–1654, July/Aug. 1971.

[13] R. Billinton and P. G. Harrington, "Reliability evaluation in energy limited generating capacity studies," *IEEE Trans. Power Apparatus Syst.*, vol. PAS-97, no. 6, pp. 2076–2085, Nov./Dec. 1978.

[14] L. L. Garver, "Effective load carrying capability of generating units," *IEEE Trans. Power Apparatus Syst.*, vol. PAS-85, no. 8, pp. 910–919, Aug. 1966.

[15] N. S. Rau and K. F. Schenk, "Application of Fourier methods for the evaluation of capacity outage probabilities," IEEE Winter Power Mtg., Paper No. A 79 103-3.

[16] IEEE Task Group on Model for Peaking Units of the Application of Probability Methods Subcommittee, "A four-state model for estimation of outage risk for units in peaking service," *IEEE Trans. Power Apparatus Syst.*, vol. PAS-91, no. 2, pp. 618–627, March/April 1972.

[17] IEEE Subcommittee on the Applications of Probability Methods, "IEEE reliability test system," *IEEE Trans. Power Apparatus Syst.,* vol. PAS-98, no. 6, pp. 2047–2054, 1979.

[18] R. N. Allan, R. Billinton, and N. M. K. Abdel-Gawad, "The IEEE reliability test system—extensions to and evaluation of the generating system," *IEEE Trans. Power Syst.,* vol. PWRS-1, no. 4, pp. 1–7, Nov. 1986.

[19] R. Billinton, "Criteria used by Canadian utilities in the planning and operation of generating capacity," *IEEE Trans. Power Syst.,* vol. 3, no. 4, pp. 1488–1493, Nov. 1988.

[20] L. T. Anstine, R. E. Burke, J. E. Casey, R. Holgate, R. S. John, and H. G. Stewart, "Application of probability methods to the determination of spinning reserve requirements for the Pennsylvania-New Jersey-Maryland interconnection" *IEEE Trans. Power Apparatus Syst.,* vol. PAS-68, pp. 726–735, Oct. 1963.

[21] A. D. Patton, "A probability method for bulk power system security assessment—I: basic concepts," *IEEE Trans. Power Apparatus Syst.,* vol. PAS-91, no. 1, pp. 54–61, Jan./Feb. 1972.

[22] R. Billinton and A. V. Jain, "The effect of rapid start and hot reserve units in spinning reserve studies," *IEEE Trans. Power Apparatus Syst.,* vol. PAS-91, no. 2, pp. 511–516, Mar./Apr. 1972.

[23] R. Billinton and A. V. Jain, "Interconnected system spinning reserve requirements," *IEEE Trans. Power Apparatus Syst.,* vol. PAS-91, no. 2, pp. 517–526, Mar./Apr. 1972.

Additional Papers

[A1] W. J. Lyman, "Fundamental consideration in preparing master system plan," *Electrical World,* vol. 101, no. 24, pp. 788–792, June 17, 1933.

[A2] S. A. Smith, Jr., "Spare capacity fixed by probabilities of outage," *Electrical World,* vol. 103, pp. 222–225, Feb. 10, 1934.

[A3] S. A. Smith, Jr., "Service reliability measured by probabilities of outage," *Electrical World,* vol. 103, pp. 371–374, March 10, 1934.

[A4] S. A. Smith, Jr., "Probability theory and spare equipment," Bulletin, Edison Electrical Institute (New York, N.Y.) March 1934.

[A5] P. E. Benner, "The use of theory of probability to determine spare capacity," *General Electric Review,* vol. 37, no. 7, pp. 345–348, July 1934.

[A6] W. J. Lyman, "Calculating probability of generating capacity outages," *AIEE Trans. Power Apparatus Syst.,* vol. 66, pp. 1471–1477, 1947.

[A7] H. P. Seelye, "Outage expectancy as a basis for generator reserve," *AIEE Trans. Power Apparatus Syst.,* vol. 66, pp. 1483–1488, 1947.

[A8] E. S. Loane and C. W. Watchorn, "Probability methods applied to generating capacity problems of a combined hydro and steam system," *AIEE Trans. Power Apparatus Syst.,* vol. 66, pp. 1645–1657, 1947.

[A9] C. D. Galloway, L. L. Garver, R. J. Ringlee, and A. J. Wood, "Frequency and duration methods for power system reliability calculations: Part III—generation system planning," *IEEE Trans. Power Apparatus Syst.,* vol. PAS-88, no. 8, pp. 1216–1223, Aug. 1969.

[A10] V. M. Cook, R. J. Ringlee, and A. J. Wood, "Frequency and duration methods for power system reliability calculations: Part IV—models for multiple boiler-turbines and for partial outage states," *IEEE Trans. Power Apparatus Syst.,* vol. PAS-88, no. 8, pp. 1224–1232, Aug. 1969.

[A11] R. J. Ringlee and A. J. Wood, "Frequency and duration methods for power system reliability calculations: Part V—models for delays in unit installations and two interconnected systems," *IEEE Trans. Power Apparatus Syst.,* vol. PAS-90, no. 1, pp. 79–88, Jan./Feb. 1971.

[A12] L. Salvaderi and L. Paris, "Pumped storage plant basic characteristics: Their effect on generating system reliability," *Proc. American Power Conf.,* vol. 35, pp. 403–418, 1974.

[A13] G. Manzoni, P. L. Noferi, and M. Voltorta, "Planning thermal and hydraulic power systems—relevant parameters and their relative influence," *CIGRE* Paper 32-16, 1972.

[A14] F. Insinga, A. Invernizzi, G. Manzoni, S. Panichelli, and L. Salvaderi, "Integration of direct probabilistic methods and Monte Carlo approach in generation planning," *Proc. 6th Power System Computational Conf.,* pp. 48–58, 1978.

[A15] R. N. Allan and P. Corredor-Avella, "Reliability and economic assessment of generating systems containing wind energy sources," *Proc. IEE,* vol. 132, part C, no. 1, pp. 8–13, Jan. 1985.

[A16] R. Billinton and R. N. Allan, *Reliability Evaluation of Power Systems,* New York: Plenum Press, 1984.

[A17] R. N. Allan, A. M. Leite da Silva, A. A. Abu-Nasser, and R. C. Burchett, "Discrete convolution in power system reliability," *IEEE Trans. Reliab.,* vol. R-30, pp. 452–456, Dec. 1981.

[A18] L. G. Leffler, G. A. Cucchi, R. J. Ringlee, N. D. Reppen, and R. J. Chambless, "Operating reserve and generation risk analyses for the PJM interconnection," *IEEE Trans. Power Apparatus Syst.,* vol. PAS-94, no. 2, pp. 396–407, Mar./Apr. 1975.

[A19] R. N. Allan and R. A. F. Nunes, "Modelling of standby generating units in short-term reliability evaluation," IEEE Paper A79 006-8.

[A20] R. Billinton and N. A. Chowdhury, "Operating reserve assessment in interconnected generating systems," *IEEE Trans. Power Syst.,* vol. 3, no. 4, pp. 1474–1487, Nov. 1988.

Generating Reserve Capacity Determined by the Probability Method

GIUSEPPE CALABRESE
MEMBER AIEE

A FUNDAMENTAL PROBLEM in system planning is the correct determination of reserve capacity. Too low a value means excessive interruption, while too high a value results in excessive costs. The greater the uncertainty regarding the actual reliability of any installation the greater the investment wasted.

In the typical case of system generating capacity reserve, the problem not only concerns the risk of outage but also the economic balance .between generator reserve and tie capacity in providing against local outage concentrations.

The complexity of the problem, in general, makes it difficult to find an answer to it by rules of thumb. The same complexity, on one side, and good engineering and sound economics, on the other, justify the use of methods of analysis permitting the systematic evaluations of all important factors involved. There are no exact methods available which permit the solution of reserve problems with the same exactness with which, say, circuit problems are solved by applying Ohm's law. However, a systematic attack of them can be made by a "judicious" application of the probability theory.

It is the purpose of this paper to describe this application, which resulted from a study, undertaken some ten years ago, for the purpose of rationalizing the reserve practice of the company with which the author is associated. More specifically, the application to the particular case of turbogenerator reserve will be referred to unless otherwise stated. Most of the conclusions, however, are general and may be extended to other cases.

The qualification "judicious" is essential because the results are subject to certain limitations which it is well to enumerate.

(a). It is assumed that forced outages of units are random events, independent from one another, governed by the same laws of chance governing the drawing of balls or the throwing of dice. A method will be given to check the general correctness of the assumed independence of forced outages of units.

(b). The probability theory aims only at foreseeing average performances. In the case under consideration, this means that the probability theory can predict only the average outage performance of a group of units during a long period of time.

(c). The basic information on which predictions for the future are made necessarily is obtained from previous experience of existing units. Some or all of these units may be of different design, or may have been operated under different operating conditions, or both, from those involved in the future.

With the approximation set by these limitations it will be shown that for a given system, or portion thereof, and for a given reserve value, probability calculations permit the prediction of

1. The fraction of time during which loss of load may be expected to occur during any future period.

2. The kilowatt-hour losses expected to result from forced outages.

The fraction of time as under (1) will be called the *loss of load duration* or the *loss of load probability*, and for convenience may be expressed in terms of "so many days upon which loss of load may be expected to occur during a given number of years," say 10 or 100. This number of days provides a first index for measuring and comparing service reliabilities.

Otherwise the energy losses as under (2) may be expressed in terms of so many kilowatt-hours per kilowatt of installed capacity, or of maximum load, per year or decade or century, thus obtaining two other indexes of service reliability.

It should be emphasized, however, that the calculated expected loss of load duration (whether expressed as a fraction of, or in number of days during a given period) and the calculated kilowatt-hour losses per kilowatt of installed capacity or of maximum load, refer to the long range prediction and cannot be expected to materialize year in and year out.

The choice of an appropriate level of service reliability, and thus of an appropriate reserve capacity, in any particular case, will be based on any one of these indexes or combinations thereof. Necessarily, this choice depends on personal judgment and is affected by local conditions.

Apart from the application of the probability theory to the determination of an appropriate absolute level of service reliability for a group of units connected to the same bus, there is another class of problem for the solution of which the probability theory seems especially well suited. This class of problems concerns the evaluation of the relative reliability of alternative plans. What is the reliability gained or lost by increasing or decreasing the reserve capacity of a group of units?—by changing the size and connections of ties?—by using a common header for a group of boilers? To the analysis of problems such as these, the probability theory is well adapted. It places, so to say, the alternative schemes in their proper relative positions from the standpoint of reliability.

Briefly, the method of reserve determination described in this paper proceeds from the assumption that forced outages of units are random events independent from one another and that

Paper **47-248**, recommended by the AIEE power generation committee and approved by the AIEE technical program committee for presentation at the AIEE Midwest general meeting, Chicago, Ill., November 3–7, 1947. Manuscript submitted March 25, 1947; made available for printing September 18, 1947.

GIUSEPPE CALABRESE is assistant engineer, system engineering department, Consolidated Edison Company of New York, Inc., New York, N. Y.

The author wishes to express his thanks to K. F. Bellows (M'38) and P. Doane (A'26) for their co-operation and contributions in the preparation of this paper.

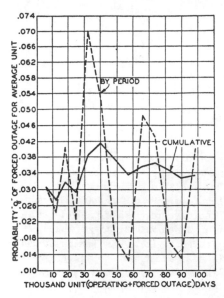

Figure 1. Graphical determination of the forced outage rate q

outage durations of units can be calculated by the well known binominal expansion. With this in mind, the paper includes the following major items:

1. Calculation of the average rate of forced outage of a group of units. This is the basic quantity of the method.

2. Probability calculations of various outage combinations. Binominal law.

3. Check of the independence of the individual outages. This check is necessary in order to establish the practical reasonableness of treating forced outages of units as random events obeying the laws of chance.

4. Probability of outages exceeding any given reserve. This is essential for the following three sections.

5. Effect of connections on the probability of outages.

6. Probability of loss of load.

7. Expected loss of kilowatt-hour output.

8. Effect of interconnections of stations on their reliability.

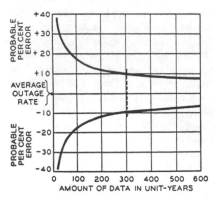

Figure 2. Probable error of average outage rate

1. CALCULATION OF THE AVERAGE RATE OF FORCED OUTAGES OF A GROUP OF UNITS

Consider the performance of a single unit during a long past period T. During this period the unit has been in operation the time S_1 and out of service, due to forced outages of the unit itself, and not for causes external to the unit, the time O_1. The average outage rate of the unit is

$$q_1 = \frac{O_1}{S_1 + O_1}$$

The sum $S_1 + O_1$ is the demand time.* If the unit has been in demand during the whole period, and there have been no scheduled outages, $S_1 + O_1$ is equal to T, otherwise $S_1 + O_1$ is smaller than T. The outage and service times O_1 and S_1 may be expressed in hours or days. Though the use of hours seems entirely proper, days are used in this paper unless otherwise stated. A fraction of one day is called one day.

Practical problems involve a group of n units more or less similar among themselves or which may be subdivided into a small number of subgroups containing similar units.

If the n units of the group are, or are assumed to be, similar in design and operating conditions the outage rate of the individual average unit of the group is calculated with the formula

$$q = \frac{O_1 + O_2 + \ldots + O_k + \ldots + O_n}{(S_1 + O_1) + (S_2 + O_2) + \ldots + (S_k + O_k) + \ldots + (S_n + O_n)}$$

$$= \frac{O}{S + O} \qquad (1)$$

where O_k is the total forced outage time and S_k the total service time of the Kth unit. O is the total forced outage time and S the total service time of all units. The service rate p of the average unit is obtained in a similar manner. Evidently $p + q = $ one.

With a given amount of the units past performance available a series of values of q are calculated by adding each year's forced outage and service times, respectively, to the cumulated forced outages and service times of the preceding years. If this leads to a stable value of q, that is, a value which changes little as $S + O$ increases, this value may be used to predict the expected future performance of the average unit. This is, this value of q may be taken as the average forced outage rate of all units, or simply the outage rate

* This definition of demand time as the sum of time in operation plus time out on forced outage will be adhered to in this paper.

that may be expected to obtain during a sufficiently long period in the future, assuming that the units will be operating under the same conditions as in the past and that their performance will not be affected by aging. In this sense q may be called the expected forced outage duration or, more simply, the *outage duration* of the average unit during any future demand period, expressed as a fraction of the period.

The stable value of q may also be looked upon as representing the probability that any unit of the group will be out due to forced outages, or more simply the *outage probability*, during any future demand period.

More accurately the stable value of q should be corrected for the effect of future aging. This effect, if already present, may be detected by formal statistical methods or, graphically, by plotting the series of values of q obtained as previously stated.

Needless to emphasize that the outage probability q is the ratio between time out on forced outage and demand time and yields no information whatever relative to the duration of each outage.

The service rate, or service probability, or service duration, p of the units is obtained in a similar manner.

The outage rate q is the basic quantity on which depends the whole prediction of the future performance of a group of units. It is, therefore, essential that the value of q be stable and corrected for trends, due to aging or other causes. It is also necessary that its derivation from past records be checked for the assumed homogeneity of the group of units. To

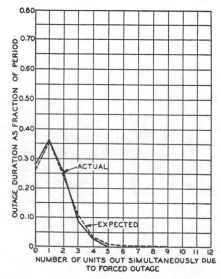

Figure 3. Actual versus expected experiences with simultaneous outages (44 units, 13 years, q = 0.03)

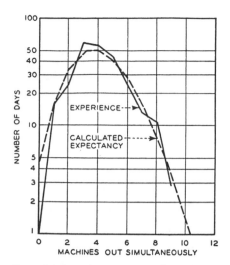

Figure 4. Actual versus expected experiences with simultaneous outages (309 units, one year q = 0.0164)

this end the group should be divided into subgroups according to some or all of the following characteristics: size, type, speed, age, demand factor, and pressure. A value of q should be obtained for each subgroup which should be tested for stability and trend as previously stated. How far this process of subgrouping should be carried depends on the amount of past information available. If there are no pronounced differences among the values and trends of the q's of the subgroups, one value of q may be used for the whole group, otherwise it may be necessary to consider two or more subgroups separately. Pioneering units as a rule form a class by themselves and need to be separated from the rest unless they are so diversified in size and location with respect to existing units that their effect may be neglected. Particular attention

should be paid to the largest units because their outages are the most significant from the standpoint of reserve.

Figure 1 illustrates the graphical method just described. The dotted curve gives the forced outage rate for each year. The solid curve gives the plot of the values of q obtained by adding the yearly forced outage days and the operating days of all units, respectively, to the corresponding cumulated values of the preceding years. Though the fluctuations of q decrease toward the end of the period it cannot be said definitely that a stable value has been attained. The data for this figure were obtained in the very early stage of this investigation, from the limited and, from the standpoint of this paper, incomplete records on outages of the Consolidated Edison Company. It was then impossible to segregate the real forced outages from outages which could have been postponed to week ends or even to the overhaul season. For this reason the forced outage rate $q = 0.03$ adopted at that time was high.

A further investigation of forced outages on several systems of boilers and steam turbogenerator units covering a period of six years resulted in the following outage rates:

Steam turbine generators

Below 1,000 pounds per square
   inch.......................1.0 per cent
1,000 pounds per square inch
   and over...................1.9 per cent

Boilers

Below 1,000 pounds per square
   inch.......................1.3 per cent
1,000 pounds per square inch
   and over...................2.1 per cent

The experience data available for the determination of q may be that of a relatively small number of units covering several years, or it may be that of a large number of units covering only a few years, say one or two. In the first case a plot such as shown in Figure 1 may be drawn to make a quick check of the stability of q. In the second case no such check can be made and recourse must be made to statistical methods in order to check the stability of q. This is illustrated in Figure 2 relative to 1-year experience with 303 units having a calculated averate outage of 1.98 per cent. Figure 2 gives the probable error of this average outage rate in function of the available data expressed in number of unit years.

If we have n units, having a forced outage rate q, what is the significance of q from the standpoint of future operation? Does it mean that during any future demand period D, in days, any given unit of

the group will be out Dq days, or does it mean something else? It means that during the period D, the n units will be out on forced outage nDq unit-days with a probability approaching unity, that is certainty, as nD approaches infinity. This means that as the number n of units in the group increases the length of the period D necessary to attain a stable value of q, decreases and vice versa. At one extreme when the number of units n is very large, a number of them equal to nq is out continuously day in and day out; as one unit is restored to service another one goes out on forced outage so that the total number of units out is always the same with a probability approaching unity as the number of units n approaches infinity. At the other extreme with only a few units in the group, the probability is that it will take a large number of years before a stable value of q is reached. In subsequent years the same average outage rate may be expected to obtain during periods of the same duration with a probability approaching unity as the length of the period approaches infinity.

From the standpoint of reserve capacity the average outage rate need not be determined very accurately, the question then arises: how accurate does it need be? The effect of variations of q on reserve requirements will be analyzed in the following section.

Before closing this section, it seems appropriate to make a final remark regarding the calculation of the rate of outage of a unit composed of several elements connected in a series relation so that the outage of one element causes the whole

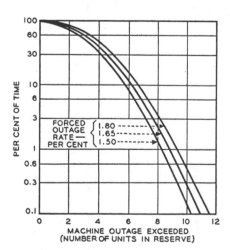

Figure 5. Effect of error in outage rate on calculated reserve (248 units)

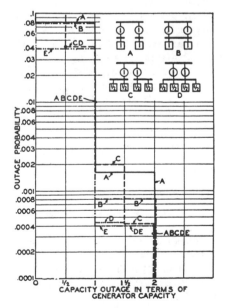

Figure 6. Effect of connections on reserve requirements

unit to be out. As an example, a turbo-generator outage may be caused by trouble in the boiler or boilers, or in the turbine or in the generator. In general a unit may be composed for several elements $a, b, c, d \ldots$ in series, having, respectively, the outage rates $q_a, q_b, q_c, q_d \ldots$ Each one of these outage rates is calculated considering only the outage caused by troubles in the particular element and not in the other elements. The corresponding service probabilities or durations are, respectively, $p_a = \text{one} - q_a$; $p_b = 1 - q_b$; $p_c = 1 - q_c$, and so forth. The quantity p_a is the probability or the expected total fraction of the time that the element a will be in service during any demand period. During the same fraction of time p_a it is expected that the element b will be in service the fraction p_b, so that the expected total fraction of time that both elements a and b will be in service during the whole demand periods is $p_a p_b$. By extending this reasoning to all the elements, the service probability p_1 of the units as a whole, that is the expected total fraction of time that all the elements will be in service, is $p_1 = p_a p_b p_c p_d \ldots$ The outage rate q_1 of the unit as a whole is thus $q_1 = 1 - p_1 = 1 - p_a p_b p_c p_d \ldots$

2. OUTAGE PROBABILITY OF COMBINATIONS OF UNITS CONNECTED IN PARALLEL—BINOMINAL LAW

From the outage rate of individual units, calculated as described in the preceding section, the outage probabilities or outage durations of the combinations of a group of units connected in parallel, such as a group of boilers connected to the same header, now can be calculated, assuming, as previously stated, that forced outages of individual units are independent from one another.

By applying the formal rules of the probability theory the outage probabilities of the combinations of n dissimilar units having individual outage rates q_1, $q_2, q_3 \ldots q_k \ldots q_n$ are obtained by developing the following product of binominal factors:

$$(p_1 + q_1)(p_2 + q_2)\ldots(p_k + q_k)\ldots(p_n + q_n) \quad (2)$$

where

$p_1 = 1 - q_1$, $p_2 = (1 - q_2)$, and so forth

The general term of the expansion of this product is of the form $p_j p_i \ldots q_e q_f \ldots$ and contains $m q$'s and $(n-m)$ p's. It represents the probability that the $e^{th}, f^{th} \ldots$ units (m in all) be out simultaneously due to forced outages, and the remaining $n-m$ in service, during any demand period. The quantity m takes, successively, all $(n+1)$ values: $0, 1, 2 \ldots n$ and for each

value of m there are $n!/[m!(n-m)!]$ terms in the expansion each corresponding to one combination of m units out of the total n.

If the n units of the group are, or are assumed to be, all similar, the same outage rate q is used for all of them. In this case the product of binominal factors (equation 2) degenerates into the following well known binominal formula:

$$(p+q)^n = p^n + np^{n-1}q + \frac{n!}{2!(n-2)!} p^{n-2}q^2 + \frac{n!}{m!(n-m)!} p^{n-m}q^m + \ldots q^n \quad (3)$$

where $p = 1 - q$.

The general term

$$\frac{n!}{m!(n-m)!} p^{n-m}q^m$$

of the expansion gives the probability, or expected duration, of having m units,

simultaneously, on forced outage and $n-m$ in service, during any demand period.

The probabilities, or durations, obtained from the two expansions (equation 2 or 3) are expressed as fractions; when multiplied by the length of any demand period in days they give the *expectancies* or the expected total number of days that m units will be out simultane-

ously due to forced outages, during that period.

In general, equation 2 should find applications in dealing with parallel feeders of different lengths, equation 3 in dealing with boilers, turbogenerators, and transformers.

3. CHECK OF THE INDEPENDENCE OF THE INDIVIDUAL FORCED OUTAGES

The assumed independence of outages of individual units now can be checked, with the aid of equation 2 or 3 by comparing past experience with calculated durations. The period covered by the records should be divided into subperiods by determining the number of days on which the same number of units were either operating or in a state of forced outage. For each period the expectancy of each combination of units should be calculated. The summations of the calculated expect-

ancies for all subperiods, respectively for $0, 1, 2, 3, 4 \ldots$ units out simultaneously, should be compared with past records and found in agreement. Figures 3 and 4 illustrate such comparison.

As previously stated, this independence of the individual forced outages is one of the basic assumptions of the method. Figure 3 pertains to the experience with a group of about 44 units over a period of

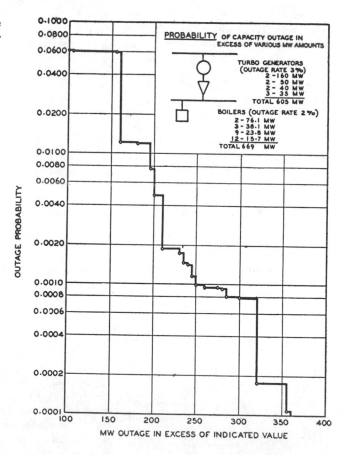

Figure 7. Example of outage probability curve

Table I. Probability in Millionths That Indicated Number of Units Would Be Out Simultaneously for Groups Having Given Number of Units When Outage Rate Is 0.02

Units Out	1	2	3	4	5	6	7	8	9	10	11	12	13	14	15	16	18
0	980000	960400	941192	922368	903921	885842	868125	850763	833748	817073	800731	784717	769022	753642	738569	723798	693115
1	20000	39200	57624	75295	92237	108471	124018	138900	153127	166749	179756	192175	204027	215526	226093	236142	255156
2		400	1176	2305	3764	5554	7593	9921	12501	15314	18543	21571	24985	28564	32299	36175	44297
3			8	32	77	151	259	405	595	833	1123	1467	1869	2332	2856	3445	4821
4					1	2	5	11	18	30	46	68	95	131	175	228	369
5										1	1	2	4	5	8	11	21
6																	1

Units Out	20	22	24	25	30	36	40	44	48	50	75	100	120	150	240	360	510
0	667608	641170	615780	603465	545484	483213	445700	411100	379185	364170	219763	132620	88558	48296	7839	694	33
1	272493	287875	301607	307890	333970	355013	363837	369151	371447	371602	336458	270652	216827	147845	38395	5099	349
2	52850	61687	70785	75402	98828	126791	144795	161974	178143	185801	253996	275414	263291	224785	93656	18680	1812
3	6469	8393	10594	11797	18825	29326	37429	46278	55746	60669	126154	182276	211549	226514	151602	45491	6262
4	561	814	1135	1324	2503	4938	7066	9681	12799	14548	46555	90208	126162	169735	183315	82859	16198
5	57	60	93	114	275	645	1058	1581	2298	2732	13428	35347	59754	101148	176581	120599	33454
6	2	3	6	8	23	68	124	210	336	418	3197	11421	23365	49886	141144	145380	57409
7					2	6	12	23	41	53	645	3130	7766	20944	96291	150062	84436
8							1	2	4	6	112	742	2238	7640	57234	135115	108565
9											17	155	569	2460	30109	107846	123532
10											2	29	129	708	14195	77254	126160
11											0	5	26	184	6057	50164	116976
12												1	5	43	2359	29775	99271
13													1	10	844	16267	77669
14														2	279	8228	56227
15															86	3873	37943
16															25	1704	23957
17															7	704	14207
18															2	274	7941
19																100	4196
20																35	2103
21																12	1001
22																4	454
23																1	197
24																	81
25																	33
26																	12
27																	4

13 years. It represents one of the first checks made of the method, at the very beginning of this investigation, on the basis of an outage rate $q=0.03$, obtained from the analysis of the outage records of the Consolidated Edison system, previously referred to. Figure 4 pertains to 1-year experience with approximately 300 units.

Both Figures 3 and 4 verify the assumed independence of the individual unit forced outages.

4. PROBABILITY OF OUTAGES IN EXCESS OF RESERVE

When the n units of a group are all of the same size C, the total capacity of any number m of them is mC. The outage probabilities of the various combinations of the units of the group can be calculated as described in section 2. The outage probability of a capacity larger than mC is obtained by adding successively the outage probability of the individual combinations from the one of maximum outage (all n units out) up to and including the $n!/[(m+1)!(n-m-1)!]$ ones of $(m+1)$ units out.

Tables I and II give the probabilities that, out of a group of n units, m be out on forced outage for values of q equal to 0.02 and 0.03, respectively, and for various values of m and n. Thus if $q=0.03$ the outage probability of having more than two units out in a group of eight units is 0.00135.

It will be understood that this does not mean that the actual number of days on which more than two units will be out simultaneously during say a 2-year period will be exactly $0.00135 \times 730 = 1$ day. It means that $0.00135 \times 365 = 1/2$ day per year is the *average* rate which outages of more than two units would be expected to approach as the length of the period is increased. Actually, of course, such a condition may exist 10 days in one year and 5 in the next and not at all in the next 28 years. From the probability standpoint the chances are 135 in 100,000 that on any future day the condition will exist.

Table II. Probability in Millionths That Indicated Number of Units Would Be Out Simultaneously for Groups Having Given Number of Units When Outage Rate Is 0.03

Units Out	1	2	3	4	5	6	8	10	13	15	20	30	40	50	100
0	970000	940900	912673	885293	858734	832972	783743	737424	673027	633251	543794	401007	295712	218065	47552
1	30000	58200	84681	109521	132794	154572	193916	228069	270599	293777	336358	372069	355830	337215	147069
2		900	2619	5080	8214	11952	20991	31742	50214	63601	98830	166855	220629	255518	225152
3			27	106	254	492	1299	2618	5694	8523	18339	48164	86432	126442	227473
4					4	12	50	142	440	791	2411	10055	24727	45949	170606
5						1	1	5	25	54	239	1617	5506	13074	101508
6									1	3	18	209	994	3033	49610
7											1	22	149	590	20604
8												2	19	98	7408
9														14	2542
10														2	639
11															167
12															38
13															8
14															3
15															

Table III. Array for Combining Outage Probabilities of Two Groups

q—0.03

Units	Mega-watts	Prob-ability	Units	Mega-watts	Prob-ability	Units	Mega-watts	Prob-ability	Units	Mega-watts	Prob-ability	Units	Mega-watts	Prob-ability
							Three 22-Megawatt Unit Groups Separately							
Three 15-Megawatt Unit Groups	0	0	0.912673	1	22	0.084681	2	44	0.002619	3	66	0.000027			
Separately							Combined Groups							
0	0	0.912673	0	0	0.832972	1	22	0.077286	2	44	0.002390	3	66	0.000025
1	15	0.084681	1	15	0.077286	2	37	0.007171	3	59	0.000222	4	81	0.000002
2	30	0.02619	2	30	0.002390	3	52	0.000222	4	74	0.000007	5	96	
3	45	0.000027	3	45	0.000025	4	67	0.000002	5	89		6	111	

We are now in a position to make a first check of the question previously raised regarding the accuracy of the average outage rate. The effect of variations of the average outage rate on the reserve required, for different values of outage durations, is illustrated in Figure 5 which gives the calculated reserve required by 248 units for different outage durations at three different outage rates, namely 1.80 per cent, 1.65 per cent, 1.50 per cent.

If the units of the group are similar but not all of the same size, it becomes necessary to calculate for each value of capacity rather than simply for number of units out. As an example, consider a group of three 15-megawatt and three 22-megawatt units. The outage probabilities for any number of 15-megawatt or 22-megawatt units may be calculated as previously shown. The probability of, say, one 15-megawatt and two 22-megawatt units out simultaneously, a total of 59 megawatts, is the product of the separate probabilities, namely, 0.08468 (one 15-megawatt unit out) by 0.002619 (two 22-megawatt units out). An array such as the one shown in Table III may be found useful. The process is completed by arranging the individual probabilities in order of capacity and cumulating the values as shown on Table IV.

From column 3 of Table IV the probabilities of outages exceeding any reserve value now may be determined for the case under consideration.

5. Effect of Connections on the Probability of Outages

Outage probabilites are affected by the manner in which the various elements of the equipment used for a given purpose are connected. Thus in the case of electric energy generation, outage probabilities are affected by the manner in which boilers, turbines, and generators are arranged. In calculating the reliability of the supply to a given load, the arrangement of the transmission and distribution equipment such as feeders and transformers must be considered in addition to the generating equipment. In the case of hydro plants the supply of water is in a series relation with the turbines. The procedure that may be followed in making calculations for such cases will be illustrated by a discussion of the case of steam electric generation mentioned before. In an arrangement commonly used all the boilers are connected to a common header which supplies all the turbines. This arrangement is more reliable than the one whereby each turbine has its own boiler and no common header is provided. That this is the case may be seen of course from the fact that if no common header is used, the loss of the boiler of one unit and the turbine or generator of another unit causes the loss of two units. If a common header is provided, under the same circumstances, only one unit would be lost. For the numerical evaluation of the increase in reliability due to the common header, the boilers are considered separately as a separate group and so are the turbogenerators. The outage rates of the individual boilers and turbogenerators are calculated as described in section 1. The outage probabilities for the various boiler and turbogenerator capacities are calculated as described in section 4. The outage probabilities of the station as a whole are calculated from these separate probabilities as shown in the following paragraphs.

For the sake of clarity three cases will be distinguished.

A. The total capacity of the boiler group equals the total capacity of the turbogenerator group. In this case a boiler outage of M_b kilowatts will cause a station outage of the same amount provided that any outage that may occur simultaneously in the turbogenerator group does not exceed M_b kilowatts. The corresponding station outage probability, or outage duration, is the product of the probability of losing M_b kilowatts in the boiler group by the probability of losing M_b kilowatts or less in the turbogenerator group. Vice versa, a turbogenerator outage of M_{tg} kilowatts will cause a station outage of M_{tg} kilowatts provided that any outage that may occur simultaneously in the boiler group does not exceed M_{tg} kilowatts. The corresponding station outage probability is given by the product of the probability of losing M_{tg} kilowatts in the turbogenerator group by the probability of losing *less* than M_{tg} kilowatts in the boiler group.

The outage probabilities, or outage durations of the whole station for all kilowatt values, are obtained by giving to M_b and M_{tg} all values from zero to the maximum in the boiler group and in the turbogenerator group, respectively. The qualification "less than" in the case of station outages M_{tg} caused by turbogenerator outages M_{tg} is essential in order to avoid adding twice the terms relative to $M_b = M_{tg}$.

B. The total boiler capacity exceeds the total turbogenerator capacity by M kilowatts. In this case the first statement in the foregoing must be modified in the sense that a boiler outage M_b will cause a station outage of $M_b - M$ kilowatts, provided that any outage that may occur

Table IV. Summary of Probability of Outages for Groups of Three 15-Megawatt and Three 22-Megawatt Units

(1) Megawatts	(2) Probability of Outage Exactly Equal to Indicated Megawatts	(3) Probability of Outage Exceeding Indicated Megawatts
0	0.832972	0.167028
15	0.077286	0.089742
22	0.077286	0.012456
30	0.002390	0.010066
37	0.007171	0.002895
44	0.002390	0.000504
45	0.000025	0.000480
52	0.000222	0.000258
59	0.000222	0.000036
66	0.000025	0.000012
67	0.000002	0.000009
74	0.000007	0.000002
81	0.000002	
89		
96		
111		

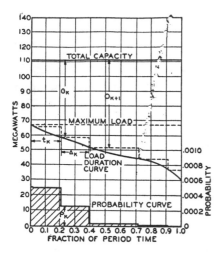

Figure 8. Calculation of the probability of loss of load

simultaneously in the turbogenerator group is equal to $(M_b - M)$ kilowatts or less. The product of the probability of losing M_b kilowatts in the boiler group by the probability of losing $(M_b - M)$ kilowatts or less in the turbogenerator group gives the probability of losing $(M_b - M)$ kilowatts in the station. The second statement as under A has to be modified in the sense that an outage of M_{tg} kilowatts in the turbogenerator group will cause a station outage of the same amount provided that any outage that may occur simultaneously in the boiler group is equal to $M + M_{tg}$ or less. Correspondingly the probability of losing M_{tg} kilowatts in the station is equal to the product of the probability of losing M_{tg} kilowatts in the turbogenerator group by the probability of *losing* less than $M + M_{tg}$ kilowatts in the boiler group. The qualification "less than" is necessary for the reason stated under A.

C. The total capacity of the turbogenerator group exceeds that of the boiler group by M_1 kilowatts. In this case a boiler outage of M_b kilowatts will cause a station outage of the same amount provided that any outage that may occur simultaneously in the turbogenerator group is equal to $M_1 + M_b$ or less. Correspondingly, the probability of losing M_b kilowatts in the station is equal to the product of the probability of losing M_b kilowatts in the boiler group by the probability of losing $(M_1 + M_b)$ kilowatts or less in the turbogenerator group. A turbogenerator outage of M_{tg} kilowatts will cause a station outage of $M_{tg} - M_1$ provided that any outage that may occur simultaneously in the boiler group is equal to or less than $M_{tg} - M_1$. The corresponding outage duration of $(M_{tg} - M_1)$ kilowatts in

the station is equal to the product of probability of losing M_{tg} kilowatts in the turbogenerator group by the probability of *losing less than* $(M_{tg} - M_1)$ kilowatts in the boiler group.

When $M_b - M$ or $M_{tg} - M_1$ are negative they should be replaced with zero. As an illustration, calculations were made for a group of two turbogenerator units with various boiler arrangements assuming, for the boilers and the turbogenerators, a forced outage rate $q = 0.02$. The arrangements considered were as follows:

A. Each unit is supplied by a single boiler of the same capacity as the turbogenerator. No common header is used for the boilers.

B. Two boilers are used as under A. The two boilers have a common header.

C. Each unit is supplied by two boilers. Each boiler has a capacity equal to one-half the capacity of the turbogenerator. No common header is used.

D. Four boilers are used as under C. The four boilers have a common header.

E. Same as under D assuming the boilers to be 100 per cent reliable.

The results of the calculations under case D are shown in Table V.

The outage probabilities under the various conditions A, B, C, D, and E are plotted in Figure 6. It is of interest to note that the curves for cases D and E are almost coincident, indicating that, under the proper circumstances when the number of boilers connected to a common header is large, the boiler outages may

be neglected altogether. Figure 7 gives the outage probabilities for the Hell Gate station of the Consolidated Edison Company as calculated in the early stages of this investigation. With reference to this figure, it is of interest to note the large variation in the reliability gains for given increases in reserve at the various points. For instance, increasing the reserve of the

Table V

Capacity Out	Outage Probabilities	
	Specified Amount	Specified Amount or Less
Boilers alone ($q = 0.02$)		
0	0.922368	0.922368
$1/2$	0.075295	0.997663
1	0.002305	0.999968
$1\,1/2$	0.000031	0.999999
2	0.000001	1.000000
Turbogenerators alone ($q = 0.02$)		
0	0.9604	0.9604
1	0.0392	0.9996
2	0.0004	1.000

Station (boilers and turbogenerators together)

Capacity Out	Factors of Outage Probabilities		Outage Probabilities	
	Boilers	Turbogenerators	Specified Capacity	More Than Specified Capacity
0	0.922368	0.9604	0.885942	0.114158
$1/2$	0.075295	0.9604	0.072313	0.041845
1	{0.002305 0.999968}	{0.9604 0.0392}	0.041413	0.000432
$1\,1/2$	0.000031	0.9996	0.000031	0.000401
2	{0.000001 1.000000}	{0.9996 0.0004}	0.000401	0.000000

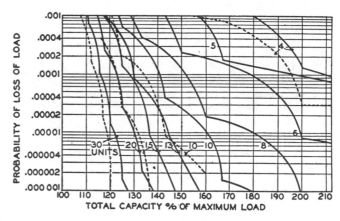

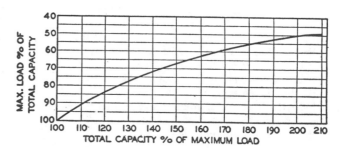

Figure 9. Probabilities of loss of load for groups of different number of units of the same size ($q = 0.03$ and $q = 0.02$), assuming the same load duration curve as in Figure 8

Solid lines = $q = 0.03$

Dashed lines = $q = 0.02$

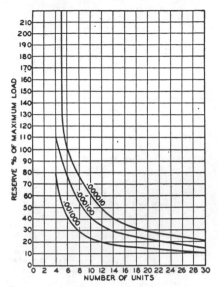

Figure 10. Effect of number of units on reserve requirements for three values of loss of load probability

From Figure 9

station from 200 megawatts to 210 megawatts decreases the outage probability from 0.0048 to 0.00185. However, further increasing the reserve from 210 to 230 megawatts decreases the outage probability from 0.00185 to 0.00175 only. In Figure 7 the probability curve is decidedly discontinuous. As the number of units increases this discontinuity decreases until, for very large systems, the probability curve is almost continuous.

6. PROBABILITY, OR EXPECTED DURATION OF LOSS OF LOAD

The ultimate purpose of providing reserve capacity is to guard against loss of load. The work thus far described has dealt with loss of capacity only. The load of a station, however, undergoes hourly, daily, and seasonal variations. Any capacity outages, therefore, may or may not result in loss of load depending on whether the remaining capacity suffices to carry the load while the outage exists. In order to take this fact into account a load duration curve is needed in addition to the outage probabilities previously calculated. One such a curve is shown in Figure 8, where the ordinates are the ranked values of daily maximum load and the abscissas are values of the number of days, in the entire period, on which the indicated load exists or is exceeded, expressed as a fraction of the entire period. If it is desired to make allowance for the possibility of carrying load at reduced voltage the load duration curve can be modified accordingly.

Each of the finite numbers of loss in

capacity values $O_1, O_2, O_3, \ldots O_k, O_{k+1} \ldots$, obtainable with the various unit combinations, may be plotted down from the total capacity lines as shown by the horizontal dashed lines in Figure 8. Let Δ_k be the time interval between the intercepts on the load duration curve of the two successive capacity values O_k, O_{k+1}. For loads within the interval Δ_k, loss of load occurs whenever the capacity outage exceeds O_k. Let P_k be the outage probability of outages in excess of O_k. The product $P_k\Delta_k$ is the probability of loss of load, during the whole period, contributed by capacity outages exceeding O_k. Otherwise $P_k\Delta_k$ represents the total expected fraction of the whole period during which loss of load will occur, due to capacity outages in excess of O_k. In this sense, $P_k\Delta_k$ may be called the *loss of load duration* due to capacity outages exceeding O_k. In the same manner the probabilities $P_1\Delta_1, P_2\Delta_2 \ldots$ of loss of load due, respectively, to capacity outages exceeding O_1, $O_2, O_3 \ldots$ may be calculated. The Δ_s' between capacity values equal to, or smaller than, the peak time reserve are zero and so are the corresponding $P\Delta_s'$ probabilities The sum

$$P_e = P_1\Delta_1 + P_2\Delta_2 + \ldots + P_k\Delta_k + \ldots$$

gives the total probability of loss of load, or the expected loss of load duration, due to all capacity outages, during the whole period. By plotting the outage probabilities $P_1, P_2, \ldots P_k \ldots$ a probability step curve is obtained as shown in the lower part of Figure 8. Evidently P_e is equal

to the area under this curve. The values of Figure 8 correspond to the group of units of Table IV assuming a peak load of 67 megawatts.

If the probability P_k varies continuously with Q_k, as in the case of a hydro plant, or, for all practical purposes, in the case of a steam plant with a large number of boilers and turbogenerators, the step curve P_k shown in the lower part of Figure 8 becomes a continuous curve and the expression of P_e is equal to

$$\int_0^1 P_k dt$$

Otherwise, P_e may be calculated noting that with a capacity outage O_k, loss of load occurs during the time t_k when the load exceeds the available remaining capacity. If p_k is the probability of losing exactly the capacity O_k, the product $p_k t_k$ gives the probability of loss of load contributed by the capacity O_k outage. By considering all capacity values $O_1, O_2, \ldots O_k, \ldots$ a second expression may be written for the total probability of loss of load as follows:

$$P_e = p_1 t_1 + p_2 t_2 + p_k t_k + \ldots$$

As before the t's corresponding to capacity outages equal to or smaller than the peak time reserve are all zero and contribute nothing to the total probability of loss of load.

The numerical calculations for the example of Figure 8 are given in Table VI. The first method is used in the upper part and the second in the lower one. The

Table VI.   Calculation of Probability of Loss of Load

Groups of Table IV With Reserve of 44 Megawatts

(1) Capacity Outage O_k	(2) t_k	(3) Interval of Time Δ_k	(4) Probability of Greater Outage, P_k	(5) Product of Column 3 × Column 4, $P_k\Delta_k$
44	0	0.018	0.000504	0.000009
45	0.018	0.192	0.000480	0.000092
52	0.210	0.185	0.000258	0.000048
59	0.395	0.318	0.000036	0.000011
66	0.713	0.003	0.000012	0.000000
67	0.716	0.196	0.000009	0.000002
74	0.912	0.088	0.000002	0.000000
81	1.000			
			Probability of loss of load	0.000162

(1) Capacity Outage O_k	(2) Number of Units Out	(3) t_k	(4) Probability of Specified Outage, p_k	(5) Product of Column 3 × Column 4, $p_k t_k$
44	2	0	0.002390	0.000000
45	3	0.018	0.000025	0.000000
52	3	0.210	0.000222	0.000047
59	3	0.395	0.000222	0.000088
66	3	0.713	0.000025	0.000018
67	4	0.716	0.000002	0.000001
74	4	0.912	0.000007	0.000006
81	4	1.000	0.000002	0.000002
			Probability of loss of load	0.000162

total probability of loss of load, with the assumed 44-megawatt reserve, is 0.00162. Calculations for other reserve values can be made in a similar manner.

In Figure 9, curves are given showing probabilities of loss of load as a function of total capacity in per cent of maximum load, and also as a function of maximum load in per cent of total capacity, for groups of different numbers of units of the same size, on the basis of a load duration curve of the same shape as that of Figure 8, and outage rates of $q = 0.03$ and $q = 0.02$, respectively.

From Figure 9, the curves of Figure 10 have been derived to illustrate the effect of the number of units, and thus of the unit size for a given load, on the reserve requirements, for three values of loss of load probability based on a forced outage rate $q = 0.03$.

The product of the probability of loss of load P_e by the length D of the period, in days, gives the expected total number of days D_e on which load will be lost during the period under consideration.

Changes in conditions due to changes in the total capacity (as may be caused by overhaul) or to wide seasonal variations of the load duration curve may be taken into consideration by subdividing the year into periods $D_1, D_2, \ldots$. The probability of loss of load of each period is calculated as just described and the values $P_1', P_2', \ldots$ obtained. The expected loss of load days of the periods are, respectively, $P_1'D_1, P_2'D_2, \ldots$ and their sum $D_y = (P_1'D_1 + P_2'D_2 + \ldots)$ gives the expected total number of days of loss of load for

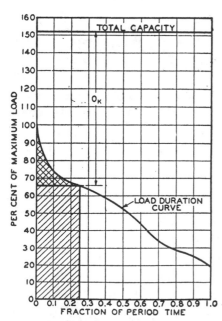

Figure 11. Calculation of expected loss of output for a capacity outage O_k

the whole year. This of course does not mean that under the assumed conditions load will be lost D_y days year in and year out. It means that D_y is the per year average number of days on which loss of load is expected to occur over a reasonably large number of years. The reasonableness of this number of years should be determined on the basis of the number of unit demand days necessary to obtain a stable value of the forced outage rate.

A last remark in this section concerns the possibility of using the probability of capacity outage at peak time to determine the reserve requirements of a group of units instead of the more elaborate method of loss of load just described. The probability of capacity outage at peak time depends only on the number of units and the rate of outage q. The probability of loss of load depends on these two quantities and the load duration curve as well. It is then evident that there is no fixed ratio between the two probabilities. From a practical standpoint, however, particularly when dealing with a large number of units, preliminary calculations can be made on the basis of the probability of capacity outages at peak time and a final check made using the more accurate method of loss of load.

Calculations may be simplified by assuming that the load duration curve is a straight line or portion of straight lines. Strictly speaking this assumption is not correct; it is justified, however, by the very nature of the problem.

7. PROBABILITY OF LOSS OF KILOWATT-HOUR OUTPUT

Some writers on service reliability have considered that kilowatt-hour output or sales lost should be taken as the measure of service reliability. If this basis is to be used an assumption must be made as to the effect of a shortage of capacity. At one extreme, as for a single network, the entire load must be dropped. At the other extreme load may be dropped or added in such a way as to match exactly the available capacity. Figure 11 shows a load duration curve on an hourly basis. Under the second assumption above, the double-hatched area is proportional to the kilowatt-hours which would be lost as a result of a capacity outage O_k were it to last throughout the period. The sum of the single and of the double-hatched area is proportional to the kilowatt-hour loss under the first assumption above caused by the same outage O_k under the same conditions. In either case the product of the probability, p_k, of losing exactly the capacity O_k, by the kilowatt-

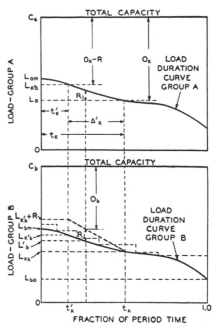

Figure 12. Calculation of the effect of interconnecting two groups of units on the loss of load probability of either

hours represented by the relative area, gives the expected kilowatt-hour loss due to the particular outage. The summation of the expected kilowatt-hour losses of all capacity outages $O_1, O_2 \ldots O_k \ldots$ gives the expected total loss of kilowatt-hour output. The effect of operation at reduced voltage easily may be taken into consideration.

In order to obtain indexes of reliability this total loss of kilowatt-hour output may be expressed as

(*a*). "Expected loss of kilowatt-hour output per kilowatt of installed capacity," by dividing it by the total installed capacity in kilowatts.

or as

(*b*). "Expected loss of kilowatt-hour output per kilowatt of maximum load," by dividing it by the maximum load in kilowatts.

8. EFFECT OF INTERCONNECTING GROUPS OF UNITS ON THEIR LOCAL RESERVE REQUIREMENTS

The preceding sections have dealt with the problem of determining the local reserve of a single group of units. In this section an analysis will be made of the effect of interconnecting a number of such single groups on their local reserve requirements.

The reserve requirements of two adjacent groups A and B can be fulfilled by either one of two alternative methods

(*a*). The two groups are isolated the one

from the other and each group is provided with sufficient reserve to give the desired level of service reliability.

(b). The two groups are interconnected, in which case the local reserve of each group can be decreased as each group can rely partially on the other group for its reserve; the more so the larger is the capacity of the interconnection, within certain limits.

Let C_a be the total capacity of group A, and C_b that of group B. In order to evaluate the effect of interconnecting the two groups with a tie of capacity R on the probability of loss of load of either group, say group A, it is necessary to have the load duration curves of the two groups as shown in Figure 12. It will be assumed that the capacity of either group will be available for the other group provided it is not needed locally.

If the two groups are not interconnected a capacity outage O_k, in A, contributes the amount $p_k t_k'$ (Figure 8) to the total probability of loss of load of group A, as described in section 6. When the two groups are interconnected with a tie of capacity R, the same outage O_k will cause a loss of load the probability Q_k of which is composed of three parts calculated as follows:

1. In the first place, irrespective of what happens to group B and to the tie, a loss of load will certainly occur in A during the time t_k' that the load of this group exceeds $C_a + R - O_k$. Evidently t_k' is the abscissa of the intersection on the load duration curve of the capacity $O_k - R$ measured down from C_a.

2. During the interval of time $\Delta_k' = t_k - t_k'$, group A will have to call upon group B for a transfer R_1, over the tie, varying from

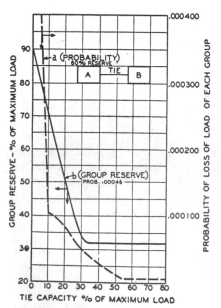

Figure 13. Illustration of the effect of tie capacity on the loss of load probability (curve a) and on the group reserve (curve b)

the maximum value R to zero. A loss of load will thus occur in group A, if simultaneously to the outage O_k in A, there occurs in B a capacity outage exceeding $O_b = C_b - (L'_b + R_1)$ irrespective of whether the tie is wholly or partially available or not. As seen from Figure 12, L_b' is the load of system B in the interval Δ_k'. The corresponding probability of loss of load in A is

$$p_k \int_{t_k'}^{t_k} P_b dt$$

where P_b is the probability of losing in group B a capacity exceeding O_b. With steam units, due to the discontinuous character of the capacities corresponding to the various unit combinations of the B group, this integral is actually equal to the sum of a few terms of the form $p_k P_b \Delta_t$ and thus easy to calculate.

3. During the interval Δ_k', even if the capacity outage in B is less than O_b, a loss of load occurs in A if, due to a tie outage, it should be impossible to effectuate the transfer R_1 from B to A. The corresponding probability of loss of load in group A is

$$p_k \int_{t_k'}^{t_k} (1 - P_b) p_{r1} dt$$

in which p_{r1} is the probability of a tie capacity outage exceeding $(R - R_1)$, R being to total tie capacity. As before due to the discontinuous character of both tie and B capacities, this integral breaks up into the sum of a few terms of the form $p_k (1 - P_b) p_{r1} \Delta_t$.

Thus the probability Q_k of loss of load is group A, with a capacity outage O_k in the same group, is

$$Q_k = p_k[t_k' + \int_{t_k'}^{t_k} \{P_b + (1 - P_b)p_{r1}\} dt]$$
$$= p_k[t_k' + \sum \{P_b + (1 - P_b)p_{r1}\} \Delta_t]$$

Evidently

$$\Sigma \Delta_t = \Delta_k'$$

and

$$(P_b + (1 - P_b)p_{r1}) < 1$$

for the different values of P_b and p_{r1} under Σ, so that

$$t_k' + \{P_b + (1 - P_b)p_{r1}\} \Delta t < t_k$$

Q_k takes the place of $p_k t_k$ of section 6. Its value decreases from the maximum $P_k t_k$ to a minimum, as the tie capacity R increases from zero to a maximum value R_{k0} which, for the specified outage O_k, can be calculated as follows.

From Figure 12 it is evident that as far as the specified outage O_k is concerned, the tie capacity R used should satisfy the inequality

$$R \leqslant C_b - L_{k'b} \tag{5}$$

where $L_{k'b}$ is the B load corresponding to t_k'. This is because B has no capacity available to fulfill any transfer requirement not satisfying the inequality, equation 5. The maximum tie capacity R_{k0} giving the minimum value of Q_k also must satisfy the inequality.

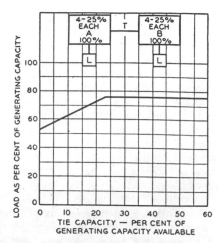

Figure 14. Effect of tie capacity on the load that can be carried by each one of two interconnected systems

From Figure 12 and the expression of Q_k it is seen that as R increases, t_k' and Q_k decrease. Evidently the minimum value that t_k' can attain is zero. Two cases are thus possible as R increases, starting from the zero value and t_k' decreases from the maximum value t_k:

(a). Before t_k' reaches the value zero, R becomes equal to $C_b - L_{k'b}$. In this case, the maximum tie capacity R_{k0} giving the minimum value of Q_k is

$$R_{k0} = C_b - L_{k'b} \tag{6a}$$

From this and Figure 12

$$O_k = C_a + C_b - (L_{k'a} + L_{k'b}) \tag{7a}$$

(b). The quantity t_k' becomes zero before R becomes equal to C_b and L_{kb}'. In this case the maximum value of tie capacity R_{k0} is that making $t_k' = $ zero.

From Figure 12

$$R_{k0} = L_{am} - L_a = O_k - (C_a - L_{am}) \tag{6b}$$

where $(C_a - L_{am})$ is the reserve at peak time.

The outages O_k, which give minimum probability of loss of load Q_k, for the tie capacity values making $t_k' = $ zero, thus may be obtained from equations 6b and 5. They are those satisfying the relation

$$O_k \leqslant C_b + C_a - (L_{am} + L_{bm}) \tag{7b}$$

where L_{am} is the peak load of the A group and L_{bm} is the corresponding load of the B group. From equations 7a and 7b it is evident, then, that for all outage O satisfying equation 7b the maximum tie capacity is given by equation 6b. For all larger outages O_k not satisfying equation 7b the maximum tie capacity is given by equation 6a. From equations 7a and 7b the value of t_k' for the maximum tie capacity R_{k0} may be obtained graphically by plotting a load duration curve sum of the two load duration curves of Figure 12 and measuring O_k down from the $(C_a + C_b)$ capacity line to this load duration curve. The abscissa of the intersection is the value of t_k' being sought.

As the value of the outage O_k increases, the maximum tie capacity R_{ko} increases also, being at the most equal to $C_b - L_{bo}$ where L_{bo} is the minimum load of group B. From a practical standpoint it will suffice to determine the tie capacity on the basis of the outages O_k giving the largest $p_k t_k$ products, as shown by inspection of calculations similar to those shown in the lower part of Table VI.

Calculations for the B group are carried in the same manner.

While in Figure 12 it has been assumed that the peaks of the two groups are coincident, this assumption is not necessary as any diversity between the two groups may be taken into consideration by drawing the B load duration curve in its proper relative position with respect to the A one.

By way of illustration Figure 13 shows two curves calculated for two identical groups of four units, serving identical loads, assuming all units of the same size. Curve *a* gives the variation of the probability of loss of load of either station in function of the tie capacity assuming a local reserve for each group equal to 80 per cent of the maximum load. In this case the values of R_{ko} for the various outages are

Outage	Tie Capacity	
$O_1 = 45...R_{10} = 0$		
$O_2 = 90...R_{20} = 10$	(From equations 7b and 6b)	
$O_3 = 135...R_{30} = 55$		
$O_4 = 180...R_{40} = 90$	(From equations 7a and 7b)	

Curve *b* gives the tie capacity necessary to obtain a probability of loss of load equal to 0.00046 with different local reserve capacities. Both curves *a* and *b* are based on an average outage rate $q = 0.03$ and a load duration curve of the same shape as that shown in Figure 8.

From curve *b* it will be noted that the reduction in the reserve requirements of each system is approximately equal to the capacity of the tie so that the reserve requirements of the two systems considered together are reduced by twice the tie capacity. Of course this is true up to the maximum useful value of the capacity (30 per cent of maximum load) but not beyond this value. In Figure 14 curve *b* of Figure 13 has been redrawn to show the load L that can be carried as a percentage of the generating capacity available.

Again it is noted that as the tie capacity increases from 0 to 24 per cent, the load that can be carried increases from 52 per cent to 76 per cent, that is, increases 24 per cent.

The method of analysis just discussed for the case of an interconnection between two groups of units may be extended to the case of a system of more than two groups. In general, it will be necessary to repeat the described calculations for each tie, or group of ties, and the two unit groups into which it divides the system. In order to simplify calculations it will be not only advisable but necessary to introduce some simplifications, the nature and extent of which must be judged by the case on hand.

Approximate Method of Calculating the Probability of Loss of Load

The method of determining the loss of load probability outlined in section 6 requires long calculations. The very nature of the problem, justifies, in most cases time-saving simplifications. For this purpose it is proposed that, whenever possible, the following assumptions be made:

1. The probability P_k of capacity outages in excess of O_k, when plotted on semilog paper, is replaced by either a straight line or portions of straight lines.

2. The load duration curve also is replaced by a straight line, or portions of straight lines.

Strictly speaking, neither one of these two assumptions is correct. However, it is believed that in most cases both assumptions can be made without greatly sacrificing the practical correctness of the results.

By the first assumption the probability P_k of outages in excess of a capacity O_k can be expressed by an exponential (or portions of exponentials) of the form

$$P_k = A \epsilon^{-aM}$$

where A and a are two constants and M has been used in place of O_k.

By both assumptions, the step probability curve shown at the bottom of Figure 8 is replaced by a continuous curve (or portions of continuous curves) which is related to the maximum load L_m, the system capacity C, the reserve at peak time $R_0 = C - L_m$, and the load factor l in per cent, by the relation

$$P_k = A \epsilon^{-aR_0} \epsilon^{-2a\left(1 - \frac{l}{100}\right)L_m t}$$

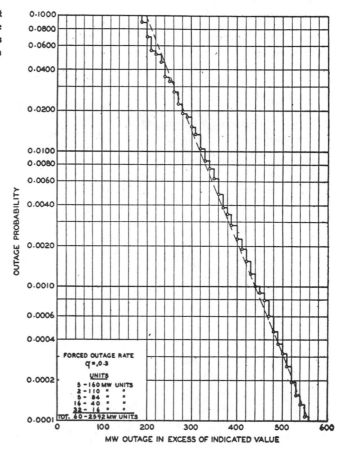

Figure 15. Exact and approximate probability curves for a large system

FORCED OUTAGE RATE
q = 0.3

UNITS
5 - 160 MW UNITS
2 - 110 "
6 - 84 "
16 - 40 "
32 - 16 "
TOT. 60 - 2592 MW UNITS

OUTAGE PROBABILITY (vertical axis)

MW OUTAGE IN EXCESS OF INDICATED VALUE (horizontal axis)

The probability P_e of loss of load becomes:

$$P_e = \int_0^{1.0} P_k dt = \frac{A\epsilon^{-aR_0}}{2a\left(1-\dfrac{l}{100}\right)L_m}\left\{1-\epsilon^{-2a\left(1-\frac{l}{100}\right)L_m}\right\}$$

which is easy to calculate.

The process is illustrated in Figure 15 relative to a large system with 60 units totaling 2,592 megawatts. In this figure the probability step curve, for all practical purposes and in the range of capacity outages shown, may be expressed by the equation

$$P_k = 5.3\epsilon^{-0.0192M}$$

By so doing the probability P_e of loss of load is given by

$$P_e = \frac{118.0\epsilon^{-0.0192R_0}}{\left(1-\dfrac{l}{100}\right)L_m}\left[1-\epsilon^{-0.0384\left(1-\frac{l}{100}\right)L_m}\right]$$

If the peak time reserve is 320 megawatts this becomes

$$P_e = \frac{0.0001315}{1-\dfrac{l}{100}}\left[1-\epsilon^{-86.7\left(1-\frac{l}{100}\right)}\right]$$

which for various values of l assumed the following values:

l, Per Cent	P_e
70	0.000435
90	0.001315
95	0.00265
100	0.01 (outage probability at peak time)

Conclusions

In concluding, it is believed that the method of probability should find a more extensive use than heretofore in the solution of reserve problems. In this respect it is hoped that the material presented in this paper may be of service to those engaged in the determination of system reserve.

As a large amount of past experience is required to make reliable predictions of the expected future performance of a particular group of units, it is hoped that the basic information collected under the auspices of the Edison Electric Institute soon will be made available.

The discussion of the preceding sections has been limited almost entirely to the case of generating unit reserve; it is evident, however, that the conclusions reached may be applied to a variety of other problems in the electrical field and in other fields.

References

1. CALCOLO DELLE PROBABILITÁ (book), Guido Castelnuovo. Nicola Zanichelli, editor, Bologna, Italy.

2. INTRODUCTION TO MATHEMATICAL PROBABILITY (book), T. V. Uspensky. McGraw-Hill Book Company, Inc., New York, N. Y.

3. PROBABILITY THEORY AND SPARE EQUIPMENT, S. A. Smith, Jr. *Bulletin*, Edison Electric Institute (New York, N. Y.), March 1934.

4. THE USE OF THE THEORY OF PROBABILITY TO DETERMINE SPARE CAPACITY, F. P. Benner. *General Electric Review* (Schenectady, N. Y.), July 1934.

5. THE PROBABLE NUMBER OF STEPS MADE BY AN ELEVATOR, Bassett Jones. *General Electric Review* (Schenectady, N. Y.), August 1923.

No Discussion

The Determination and Allocation of the Capacity Benefits Resulting from Interconnecting Two or More Generating Systems

C. W. WATCHORN
MEMBER AIEE

THE interconnection of two or more electric generating systems always reduces the amount of generating capacity required to be installed as compared with that which would be required without the interconnection. The amount of such reduction may be large or small, depending on the interrelation of the various factors and characteristics of the interconnecting systems and the desired level of service reliability. The installation of the transmission facilities that are necessary to bring about the interconnection may or may not be justified. This fact can be determined only by a proper economic comparison of the alternate installations.

Having once decided on interconnecting two or more generating systems, some plan or basis is necessary for the allocation of the resulting capacity benefit among the interconnecting systems.

This paper treats generally of the several considerations and factors involved in the determination of the capacity benefit resulting from interconnecting several generating systems as one factor in the determination of the justification of the interconnection, and also as a basis for determining the installed

Paper 50-188, recommended by the AIEE Joint Subcommittee on Application of Probability Methods to Power-System Problems and approved by the AIEE Technical Program Committee for presentation at the AIEE Summer and Pacific General Meeting, Pasadena, Calif., June 12–16, 1950. Manuscript submitted July 5, 1949; made available for printing May 1, 1950.

C. W. WATCHORN is with the Pennsylvania Water and Power Company, Baltimore, Md.

capacity obligations of the interconnecting systems. In addition, the article submits what is believed to be a logical and equitable basis for the determination of such obligations as the result of the allocation of such benefit among the interconnecting systems.

Determination of Capacity Benefits

The determination of the capacity benefits resulting from the interconnection of generating systems involves only the evaluation of the composite effect of all the diversities that may exist among them. These diversities, in addition to those resulting from any differences that may exist in the times of occurrence of daily peak loads, include those properly involved in "long-time load-forecasting errors," those resulting from the amount of variation in the magnitude of the daily peak loads, from maintenance requirements, seasonal reductions in the load-carrying capability of both the hydro and steam capacity and from forced outages of generating facilities, including the effect of the number, sizes, pressures, and arrangements of the turbines and boilers. The true composite effect of all these diversities can best and most logically be evaluated by means of probability methods, since such methods automatically give proper weight to each of these factors in accordance with the various conditions relating to the several systems. Various phases of the application of probability methods to this type of problem are dis-

cussed in several papers that have been presented during the past several years.[1-4]

Figure 1 illustrates certain of the general results obtained by an application of probability methods to the problem of determining the capacity benefit resulting from the interconnection of two systems. This illustration is developed with respect to Systems B and C treated in the Tables. The curve for the two systems combined, as well as those for the systems separately, are shown on the basis that there are no intra- nor inter-system transmission limitations. If two or more generating systems interconnect, assuming no transmission limitations, they operate, with some given amount of installed generating capacity, at the same level of service reliability regardless of what their separate ideas or desires may be. Thus, when two or more generating systems interconnect, the capacity benefits of the interconnection must be determined on the basis of a mutually agreeable level of service reliability.[7] In other words, regardless of any differences in the reliability that may have existed among the several systems for separate operation, some agreement must be reached as to the capacity requirements of the several systems predicated on a degree of reliability that will be the objective with interconnected operation.

It is believed that a reasonable level of service reliability, when the effect of probable load forecasting errors is included in the evaluation, is a probability of failure to carry the load of in the order of an average rate of one day in from eight to ten years. For the purpose of this illustration, a probability of failure to carry the load one-half the distance between probabilities of 0.001 and 0.0001 was used, which corresponds to the probability of failure to carry the load at the average rate of one day in 8.66 years. The computed capacity benefit resulting from interconnecting Systems B and C, or the reduction in the installed capacity requirements for this degree of reliability, is 170,000 kw with adequate interconnection capacity, which is the difference, along the dotted horizontal line, between the two solid line curves to the right of

Reprinted with permission from *AIEE Trans.*, vol. 69, pt. II, pp. 1180–1186, May 1950.

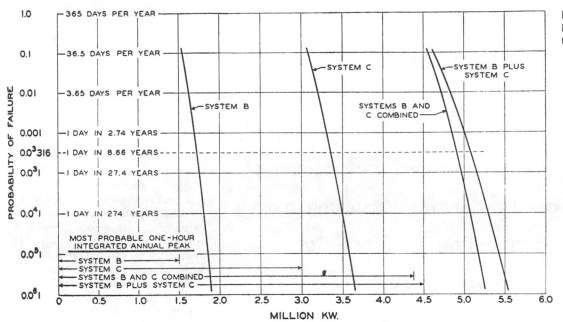

Figure 1. This is the basic or the indicated maximum capacity benefit, and results when the transmission capacity is equal to at least one-half the amount of such computed deferrable capacity. With interconnecting transmission capacity smaller than one-half the amount of such computed deferrable capacity, the indicated capacity benefit would be reduced to an amount equal to twice that of such smaller transmission capacity.

These results are modified by the fact that the units are of discrete sizes. Because of this fact it would be a remote coincidence if just exactly the desired level of service reliability were obtained. Although a slightly smaller level of service reliability than that selected as standard may be tolerated, what may otherwise appear to be a comparatively small capacity deficiency may result in a grossly inadequate degree of service reliability. This characteristic is indicated by the steepness of the curves in Figure 1. Consequently, as a practical matter the experience probably will be that a greater amount of installed capacity generally will be available than that indicated as being the minimum requirement. This experience would be the case regardless of whether or not the systems are interconnected. The result is that when two or more generating systems are interconnected, additional benefits over those computed in the manner illustrated by Figure 1 are obtainable as the result of staggering the installation of capacity among the interconnecting systems. The amounts of such additional capacity benefits are variable and depend on the differences between the indicated and actual

installed capacity requirements, the capacity of the individual units being installed, and the amount of the transmission capacity.

Figure 2 shows the average capacity benefits, over a period of years, resulting from interconnecting Systems B and C with various interconnecting transmission capacities and various capacities of the individual units being installed. The manner in which these results were determined are given in the Appendix.

It will be noted that the curves of Figure 2 are discontinuous for transmission capacity equal to half the capacity of the units being installed, half the capacity benefit computed in the manner illustrated by Figure 1, and transmission capacities equal to the sum of these two latter amounts. The reasons for these characteristics are seen from the Appendix and Figure 4 treated therein. These charac-

Figure 2. Interconnection capacity benefits versus transmission capacity (Systems B and C)

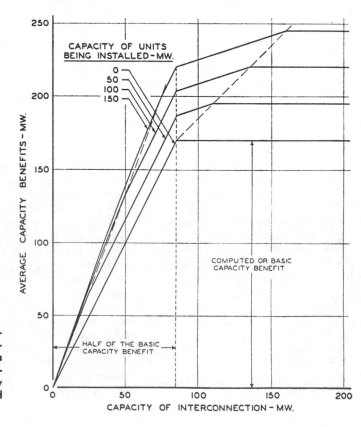

24

teristics mean that the incremental capacity benefit resulting from the interconnection decreases as the interconnecting transmission capacity increases. This relationship is shown in Figure 3.

Note the converging characteristic of the two curves to the right of Figure 1, as the service reliability decreases or as the capacity becomes tight. The reason for this characteristic is that interconnections do not create capacity. They rather only make possible greater effective use of the available generating capacity, which use decreases quite rapidly as the load, as related to some specified amount of installed generating capacity, increases and the amount of the installed generating capacity in excess of the load decreases. This is particularly true when the interconnecting facilities are substantial and are not the limitation. By contrast, there is no comparable decrease in the effectiveness with which the alternate capacity may be used as the capacity becomes tight. Thus, for a low level of service reliability, the capacity benefit of an interconnection is greatly reduced. Additional generating facilities are then obviously required to restore a reasonable level of reliability, but such additional generating capacity would be smaller with an interconnection than without. Thus, an interconnection is a substitute for installed generating capacity only where properly backed up. It is believed that a corollary to this characteristic is that the generating capacity always should be installed rather than the alternate transmission facilities, unless the economic comparison shows that the interconnection is clearly justified by an attractive margin.

Allocation of Capacity Benefits

When it is concluded that the interconnection is to be made, one of the questions to be decided is that of the allocation of the resulting capacity benefit

among the several participating systems. Several possible bases for allocating such benefits readily come to mind, for example:

1. The benefit should be divided equally among the participating systems.

2. The installed capacity requirements of each of the participating systems should be proportional to that of each for separate operation.

3. The reserve requirements of each should be proportional to that of each for separate operation.

4. The reserve requirements of each should be proportional to the installed capacity requirements of each for separate operation.

5. The reserve requirements of each should be proportional to the peak load of each for separate operation.

6. The benefit should be divided among the participating systems in proportion to the additional benefits obtained as the result of treating each participating system as a newcomer in the interconnected group, which method may be termed "the additional benefits method of allocation."

7. The benefit should be divided among the participating systems in proportion to the benefits for all combinations of two's among them, which may be termed "the mutual benefits method of allocation."

There may be still other bases.

When only two systems participate in an interconnection, the capacity benefit is a mutual benefit to which neither system by itself contributes any more nor less than the other, the benefit thus being

impossible of attainment without the other system, with neither the size, nor any other characteristic of one system by itself, contributing anything whatsoever to the ultimate capacity benefit. These facts lead to the conclusion that when only two systems are involved in an interconnection, the resulting capacity benefit should be divided equally between them. The application of suggested methods 1, 6, and 7 all give this desired result for a 2-system interconnection. It would only be under unusual circumstances that suggested methods 2, 3, 4, and 5 would give this same result. It would be a very unusual coincidence that the capacity benefit would be divided equally under these latter suggested methods unless the systems were exactly equal in size and otherwise similar in essential characteristics. When more than two systems are involved it would appear that a logical and equitable requisite would be that the benefit allocated to any one of the participating systems should not be reduced by the addition of any one or more of the other participants into the interconnection. Suggested method 1 generally will not meet this requirement because, if the result with the inclusion of certain of the participating systems in the interconnected group is that small, additional capacity benefits are obtained for all combinations with the other participating systems, while with the similar in-

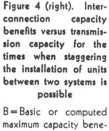

Figure 3 (left). Incremental interconnection capacity benefits versus interconnecting transmission capacity for the interconnection of two systems

Figure 4 (right). Interconnection capacity benefits versus transmission capacity for the times when staggering the installation of units between two systems is possible

B = Basic or computed maximum capacity benefit

C = Capacity of individual units being installed

T = Transmission capacity

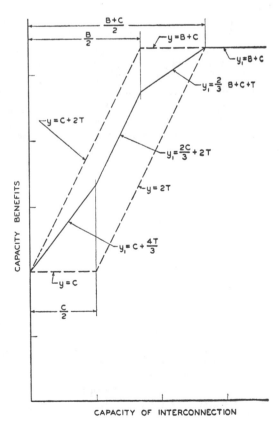

Table I. Basic Data for Three Interconnected Systems A, B and C

Systems and Combinations of Systems	Installed Capacity Requirements, Megawatts	Most Probable 1-Hour Integrated Peak Load, Megawatts	Sum of Separate Capacity Requirements, Megawatts	Capacity Benefits, Megawatts (Column 4 Minus Column 2)	Reserve Requirements, Megawatts (Column 2 Minus Column 3)
A	560	400			160
B	1,725	1,500			225
C	3,360	3,000			360
Total	5,645	4,900			745
A and B	2,220	1,890	2,285	65	330
A and C	3,845	3,385	3,920	75	460
B and C	4,915	4,400	5,085	170	515
A, B, and C	5,395	4,785	5,645	250	610

Table II. Allocation of Capacity Benefit to Systems A, B and C

Combinations of Systems by Two's	Benefits for Combinations of Two's			
	A	B	C	Total
A and B, megawatts	65	65		130
A and C, megawatts	75		75	150
B and C, megawatts		170	170	340
Total, megawatts	140	235	245	620
Allocation ratios, per cent (from Totals)	22.58	37.90	39.52	100
Allocated benefits, megawatts	56	95	99	250
Minimum allocable benefits, megawatts	37.5	85	85	

clusion of other participating systems the additional capacity benefits are large, and the total benefit were allocated equally among all, those systems for which the additional capacity benefits are large would be allocated a smaller amount of capacity benefit when those systems for which the additional capacity benefits are small are included than when such latter systems are not included. Such a result is inequitable to those systems for which the additional capacity benefits are large, and they would have a just basis for objecting to further interconnection that would increase their capacity requirements, even though the making of such interconnection would be justified from an over-all point of view. A somewhat opposite result would be obtained by the use of suggested method 6, "the additional benefits method of allocation," in that, if relatively larger additional benefits are obtained as the result of the inclusion of certain of the participating systems for all combinations with other systems, than with the inclusion of the other systems as additional participants such system would be allocated such a large portion of the total capacity benefit under this method that cases may well result for which other participating systems would be allocated a smaller amount of capacity benefit when such system is included in the interconnection than when it is not included.

These considerations then leave us with only suggested method 7, "the mutual benefits method of allocation," as apparently meeting the two requirements of equal division among the participants when two companies interconnect and of not reducing the allocated benefit to any of the participating parties with the addition of one or more of the other participating systems to the interconnection.

An allocation of the benefit resulting from the interconnection of three systems has been applied to the conditions set forth in Table I, and the results thereof have been tested by the two previously-

proposed basic requirements of a proper allocation. The method of making this allocation and the results obtained therefrom are shown in Table II. These calculations are shown with respect to the capacity benefits computed in the general manner illustrated by Figure 1, which we have termed "the basic capacity benefits," without any adjustment for any additional capacity benefits that may result from the staggering of capacity installations. It is believed that this situation is handled satisfactorily by the procedure suggested in the section called Discussion.

In Table I the allocation has been made on the basis that the portion of the total benefit (that is, the 250-megawatt saving shown as the last item of column 5) allocated to each should be proportional to the sum of the benefits for the various combinations of two's among all the participating systems.

Table II illustrates such calculations. Here the benefit allocated on the suggested "mutual benefits method of allocation" meets the stated requirement that the allocated benefits to any system not be reduced by the inclusion of an additional participating system (comparison of the results shown in the last two lines of the table).

This basis of allocation appears to be

equally satisfactory with the inclusion of a fourth participating system, System D, the basic data for which conditions are shown in Table III.

The allocation of the total capacity benefits of 350 megawatts resulting from the data of Table III for this group of four interconnected systems, in the same manner as for the group of three interconnected systems as shown on Table II, is shown in Table IV.

Table V shows data not previously shown for combinations of three's among the four systems considered in Tables III and IV, from which, together with prior data, as related to these various conditions, the allocations of capacity benefits for the various combinations of three's were made in the manner shown in Tables II and IV, with the results shown in Table VI.

Discussion

The tables have shown that "the mutual benefits method of allocation" has met the previously stated requirements for all the conditions studied. It is believed that this result would be ob-

Table III. Basic Data for Four Interconnected Systems A, B, C and D

Systems and Combinations of Systems	Installed Capacity Requirements, Megawatts	Most Probable 1-Hour Integrated Peak Load, Megawatts	Sum of Separate Capacity Requirements, Megawatts	Capacity Benefits, Megawatts (Column 4 Minus Column 2)	Reserve Requirements, Megawatts (Column 2 Minus Column 3)
A	560	400			160
B	1,725	1,500			225
C	3,360	3,000			360
D	950	800			150
Total	6,595	5,700			895
A and B	2,220	1,890	2,285	65	330
A and C	3,845	3,385	3,920	75	460
A and D	1,435	1,190	1,510	75	245
B and C	4,915	4,400	5,085	170	515
B and D	2,590	2,285	2,675	85	305
C and D	4,215	3,780	4,310	95	435
A, B, C, and D	6,245	5,565	6,595	350	680

Table IV. Allocation of Capacity Benefit to Systems A, B, C and D

Combinations of Systems by Two's	A	Benefits for Combinations of Two's B	C	D	Total
A and B, megawatts	65	65			130
A and C, megawatts	75		75		150
A and D, megawatts	75			75	150
B and C, megawatts		170	170		340
B and D, megawatts		85		85	170
C and D, megawatts			95	95	190
Total, megawatts	215	320	340	255	1130
Allocation Ratios, per cent (from Totals)	19.02	28.32	30.09	22.57	100
Allocated Benefits, megawatts	67	99	105	79	350
Minimum Allocable Benefits,* megawatts	56	98	102	70	

*From Table VI.

Table V. Capacity Benefits for Combinations of Three's of Systems A, B, C and D

Combinations of Systems	Installed Capacity Requirements, Megawatts	Most Probable 1-Hour Integrated Peak Load, Megawatts	Sum of Separate Capacity Requirements, Megawatts	Capacity Benefits, Megawatts (Column 4 Minus Column 2)	Reserve Requirements, Megawatts (Column 2 Minus Column 3)
A, B, and C	5,395	4,785	5,645	250	610
A, B, and D	3,085	2,675	3,235	150	410
A, C, and D	4,725	4,200	4,870	145	525
B, C, and D	5,765	5,180	6,035	270	585

Table VI. Allocated Capacity Benefits for Combinations of Three's of Systems A, B, C and D

Systems	A	B	C	D	Total
			Megawatts		
Allocated Benefits	56*	95	99		250
Minimum Allocable Benefits	37.5	85	85		
Allocated Benefits	47	50		53	150
Minimum Allocable Benefits	37.5	42.5		42.5	
Allocated Benefits	45		50	50	145
Minimum Allocable Benefits	37.5		47.5	47.5	
Allocated Benefits		98*	102*	70*	270
Minimum Allocable Benefits		85	85	47.5	

*Minimum allocable benefits for the combination of four for these systems (last line of Table IV).

tained for all conditions so long as the installed capacity requirements were determined on the basis of a consistent application of probability methods. ,Capacity requirements could be assumed or be determined by means other than by an application of probability methods, for which the mutual benefits method might reduce the allocated benefits to one or more of the participating systems with the inclusion of additional systems in the interconnection. If such should occur, it would be the result of inconsistencies of such stated capacity requirements among the various combinations of systems rather than of the method of allocation suggested herein.

It is possible that for certain conditions, particularly those for which there is marked seasonality in the load requirements of some of the participating systems and possibly also with considerable load diversity, that the allocated capacity benefits to one or more of the participating systems may be of such amounts that their resulting capacity requirements would be smaller than their respective most probable annual peak loads. It is not believed that there is anything illogical in such results. Greater reliability would be required of the facilities interconnecting such systems for such conditions than with the other participating systems for which this condition does not obtain.

The same considerations that apply to installed capacity apply also to the allocation of operating capacity benefits. At times the allocation of the operating capacity benefits might also result in the required operation of a smaller amount of capacity by some one or more systems than the amount of their expected loads. Under such circumstance more reliable interconnection facilities would again be required than otherwise.

In practice, the actual capacity installed usually would be greater than that actually required. This would be true of each of the participating systems for sep-

arate operation; but for the interconnection as a whole, if the full potential benefits of interconnecting are to be realized, a few of the participating systems and possibly only one would have excess installed capacity and the others would be deficient. This situation requires an adjustment so that the over-all savings may be equitably distributed to all the participating systems. The algebraic sum of these resulting capacity excesses and deficiencies would be equal to the interconnection capacity excess. This interconnection excess should be allocated to each of the participating systems, using the same allocation ratios used to allocate the capacity benefits as shown in Tables II and IV. The deficient capacity systems should be debited with the sum of their respective capacity deficiencies and allocated portions of the interconnection capacity excess, which totals should be credited to the excess capacity systems. Such adjustments could be made either by bookkeeping entries related to capacity or by periodic money settlements. If the latter is to be the practice, a further adjustment would be required to compensate the debit systems where there would be a loss in operating economies because of foregoing the earlier installation of additional economical generating capacity.

Situations may arise where one or more of the participating systems, for various reasons, may desire to install or operate more capacity than that determined as its requirements as a participant. In such case, it alone should be responsible for such additional capacity.

The successful and satisfactory operation of such interconnections require reasonable co-ordination and co-operation among the participating systems, particularly as related to maintenance schedules, which factor is possibly subject to wider control than any of the others that are involved in the determination of the installed capacity requirements. The question as to the loads to be used as a basis for the determination of the installed capacity requirements might present some difficulties. However, these difficulties would be greatly minimized by the use of a mutually agreeable statistical basis for developing the loads to be used for determining the installed capacity requirements. Such a basis would obviate any point of an after-the-fact adjustment for positive load forecasting errors. Such deviations from the forecast loads comprise one of the factors or contingencies for which reserve is provided and consequently their occurrence does not mean that insufficient capacity was provided any more than if the reduced margin between the available capacity and the load were the result of forced outages of generating facilities. All that is required is that the installed capacity be provided as related to reasonably stable information as related to load forecasts in so far as it is possible to do so, in very much the same general manner as for forced outages of generating facilities.[8] The result is that there would be no more basis or point for hindsight adjustments of the determinations of the capacity requirements after-the-fact with an interconnection than without, which latter is obviously impossible to attain.

There are special situations that may arise that will need to be considered in connection with the actual allocation of capacity benefits that it is not possible to consider in detail in a general paper. For example, the question may arise as to the disposition of the increase, if any, in the firm capacity[5] of the hydro plants that may be available on one or more of the interconnecting systems, that may result from the interconnection. It is believed that such capacity benefits are sufficiently different from the capacity benefits resulting from the interconnection of systems, upon which the installed capacity is all firm, to warrant different treatment to that suggested herein. The same considerations apply to peaking steam capacity. Likewise, after an interconnection is made it is very likely that the participating systems will develop the generating facilities with larger capacity units than if operated separately. This would be more particularly true for the smaller participating systems than the larger. Such a practice would be in line with attaining maximum economy for the interconnecting systems. On the other hand, this practice would tend to result in somewhat higher capacity obligations for the smaller systems than if they continued to install the smaller size units as for separate operation. These increased capacity obligations for the smaller systems, however, are at least largely offset by the greater economies, both as to annual charges per kilowatt of installed capacity and as to operating costs, resulting from the installation of such larger units, which, together with their share of the allocated capacity benefits and of the operating economies resulting from interchange purchases and sales, should, nevertheless, make interconnecting very attractive in many cases, even for the smaller systems.

Likewise, there is the problem as to the allocation among the systems participating in the interconnection of the costs of the interconnecting facilities. This problem is also beyond the scope of this paper, but it would appear that such allocation may well be on approximately the same basis as the allocation of the capacity benefits.

Conclusion

The capacity benefits that result from the interconnection of two or more electric generating systems can best and most logically be evaluated by means of probability methods, and such benefits are most equitably allocated among the systems participating in the interconnection by means of "the mutual benefits method of allocation," since it is based on the benefits mutually contributed by the several systems.

Appendix

After the capacity requirements are adjusted for the allocated computed or basic capacity benefit, two interconnected systems, for which the interconnecting transmission capacity is equal to or greater than half the capacity of the individual units being installed, would require one less unit to be installed 50 per cent of the time, as a matter of chance, as the result of staggering, and still result in the interconnected systems satisfying the combined capacity requirements. This situation results when, without the staggering of the capacity installations, the sum of the installed capacity would be in excess of the indicated combined installed capacity requirements by an amount larger than the capacity of the individual units being installed. The other 50 per cent of the time, the sum of the installed capacity without staggering would be in excess of the indicated combined installed capacity requirements by an amount smaller than the capacity of the individual units being installed, and it is not possible to stagger the capacity installations in such cases.

When staggering is possible, the best results or utilization of the available facilities are obtained when the system with the smaller capacity deficiency continues as the deficient system. In such case the capacity deficiency of the deficient capacity system would range from zero to the smaller of the capacity of the interconnecting transmission facilities and 50 per cent of the capacity of the individual units being installed. The total capacity benefit, for the 50 per cent of the time when staggering is possible, is then equal to the capacity of the individual units being installed and twice the excess of the tie capacity over the capacity deficiency of the deficient capacity system. We then may write

$$y = C + 2(T - x) \qquad (1)$$

where

y = the total capacity benefit

C = the capacity of the individual units being installed

T = the capacity of the interconnecting transmission facilities

x = the capacity deficiency of the deficient capacity system

The minimum value of y is C, at which value it remains constant for x equal to T, from T and x equal to zero up to T and x both equal to $C/2$. Since this is the maximum value x can have, y increases thereafter at a 2-to-1 ratio for increasing values of T up to T equal to $(B+C)/2$, where B is the basic capacity benefit with no transmission limitations, determined as discussed in the body of the paper, for which value of T, y is at its maximum value of $(B+C)$. The minimum value x can have is zero, at which value y is again equal to C for T also equal to zero; after which y increases again at a 2-to-1 ratio for increasing values of T, up to T equal to $B/2$, for which y is again at its maximum value of $(B+C)$. These values of y are the boundary values, and they describe a parallelogram as shown by the dash lines of Figure 4. All other values of y for values of x between zero and $C/2$ fall within this parallelogram.

The chances are equal that x will have any value between zero and $C/2$ for T equal to and between $C/2$ and $B/2$. The relative number of occurrences of y for some such value of T is then $(C/2 - x)$. We may then weight y with respect to this quantity, as follows

$$y_1 = \frac{\int_0^{C/2}(C/2 - x)\{C + 2(T - X)\}dx}{\int_0^{C/2}(C/2 - x)dx} \qquad (2)$$

from which we obtain

$$y_1 = \frac{2C}{3} + 2T \qquad (3)$$

This means that for values of T equal to and between $C/2$ and $B/2$, the weighted average value of y for the various values of T is two-thirds the distance between the minimum value of $2T$ and the maximum value of $(C + 2T)$. For T equal to $C/2$, y_1 is equal to $5C/3$. Then since for T equal to zero, y, and also y_1, are equal to C, we may write for values of T equal to and between zero and $C/2$, assuming a straight line

$$y_1 = C + \frac{4T}{3} \qquad (4)$$

This expression also may be obtained in the same general manner as equation 3, as follows, by similarity, for values of T equal to and between zero and $C/2$

$$y_1 = \frac{\int_0^T(T - x)\{C + 2(T - x)\}dx}{\int_0^T(T - x)dx} \qquad (5)$$

from which we have

$$y_1 = C + \frac{4T}{3} \qquad (6)$$

which is identical to equation 4.

Then since for T equal to $B/2$, we have from equation 3 that y_1 is equal to $(2C/3 + B)$, and for T equal to $(B+C)/2$, y and y_1 are both equal to $(B+C)$, we may write for values of T equal to and between $B/2$ and $(B+C)/2$, again assuming a straight line

$$y_1 = \frac{2}{3}(B+C+T) \qquad (7)$$

These values of y_1 of equations 3 and 4 or 6 and 7 are shown by the solid lines of Figure 4.

The weighted average total capacity benefits expressed by equations 3 and 4 or 6 and 7 are available 50 per cent of the time. The other 50 per cent of the time staggering is not possible and for the corresponding conditions for equations 3 and 4 or 6

$$y_1 = 2T \qquad (8)$$

and for the conditions for equation 7, the basic capacity benefit

$$y_1 = B \qquad (9)$$

For values of T equal to and smaller than $C/2$, the percentage of time the conditions of equations 4 or 6 exist, or for which staggering is possible, is

$$p' = \frac{T}{C} \qquad (10)$$

and for which the conditions of equation 8 exist, or for which staggering is not possible, is

$$p'' = 1 - \frac{T}{C} \qquad (11)$$

We then may write for the total over-all weighted average capacity benefit, for T equal to and between zero and $C/2$

$$y_2 = \frac{T}{C}\left(C + \frac{4T}{3}\right) + \left(1 - \frac{T}{C}\right)2T \qquad (12A)$$

$$= T\left(3 - \frac{2T}{3C}\right) \qquad (12)$$

for T equal to and between $C/2$ and $B/2$

$$y_2 = 0.5\left(\frac{2C}{3} + 2T\right) + 0.5(2T) \qquad (13A)$$

$$= \frac{C}{3} + 2T \qquad (13)$$

for T equal to and between $B/2$ and $(B+C)/2$

$$y_2 = 0.5\left(\frac{2}{3}\right)(B+C+T) + 0.5B \qquad (14A)$$

$$= \frac{5B}{6} + \frac{C}{3} + \frac{T}{3} \qquad (14)$$

and for T equal to and greater than $(B+C)/2$

$$y_2 = 0.5(B+C) + 0.5B \qquad (15A)$$

$$= B + \frac{C}{2} \qquad (15)$$

Equations 12, 13, 14, and 15 are shown plotted on Figure 2, as functions of T, for Systems B and C for which B of the equations is equal to 170,000 kw and for C of the equations equal to zero, 50,000 kw, 100,000 kw, and 150,000 kw.

It is interesting to note that equation 12 is an exponential and that equations 13, 14, and 15 are linear; that equations 12 and 13 are both equal to $4C/3$ for T equal to $C/2$; equations 13 and 14 are both equal to $(B+C)/3$ for T equal to $B/2$; and that equation 14 equals equation 15 for T equal to $(B+C)/2$.

References

1. DETERMINATION OF REQUIRED RESERVE GENERATING CAPACITY, W. J. Lyman. *Electrical World* (New York, N. Y.), volume 127, number 19, May 10, 1947, pages 92–95.

2. GENERATING RESERVE CAPACITY DETERMINED BY THE PROBABILITY METHOD, Guisepne Calabrese. *AIEE Transactions*, volume 66, 1947, pages 1439–50.

3. CALCULATING PROBABILITY OF GENERATING CAPACITY OUTAGES, W. J. Lyman. *AIEE Transactions*, volume 66, 1947, pages 1471–77.

4. OUTAGE EXPECTANCY AS A BASIS FOR GENERATOR RESERVE, Howard P. Seelye. *AIEE Transactions*, volume 66, 1947, pages 1483–88.

5. PROBABILITY METHODS APPLIED TO GENERATING CAPACITY PROBLEMS OF A COMBINED HYDRO AND STEAM SYSTEM, E. S. Loane, C. W. Watchorn. *AIEE Transactions*, volume 66, 1947, pages 1645–57.

6. PROBABILITY THEORY HELPS DETERMINE SYSTEM RESERVE GENERATING CAPACITY, G. Calabrese. *Power* (New York, N. Y.), volume 92, July 1948, pages 423–26.

7. PLANNING THE DEVELOPMENT OF A METROPOLITAN ELECTRIC SYSTEM, M. L. Waring. *AIEE Transactions*, volume 67, part II, 1948, pages 1467–73.

8. OUTAGE RATES OF STEAM TURBINES AND BOILERS AND OF HYDRO UNITS, AIEE Committee Report. *AIEE Transactions*, volume 68, part 1 1949, pages 450–57.

No Discussion

Determination of Reserve-Generating Capability

HERMAN HALPERIN
FELLOW AIEE

H. A. ADLER
MEMBER AIEE

RESERVE generating capability studies of the type described in this paper have been used for two main purposes: 1. to furnish management with evaluations of reserves corresponding to varying degrees of reliability for use in determining required additions to generating capacity, and 2. to provide a basis for comparison of alternative plans for system development, such as the effect of varying the size of units upon required reserve and the value of power supply over interconnections in place of increasing the owned generating capacity.

The following basic approach has been developed to make reserve studies for the conditions and requirements of a large system:

1. Determine by probability methods the expected frequency and duration of forced outages of generating capacity.

2. Schedule overhauling according to the "constant-risk" principle.

3. Evaluate the required reserve capacity on the basis of the estimated twelve monthly peak loads in a given year instead of the yearly maximum demand only.

4. Determine a monthly peaking factor to take into account the effect of variations in weekday daily peak loads within each month on the required reserve.

5. Evaluate help during emergencies from power supplied by other utilities, from emergency loading of generating-station equipment, and from the dropping of interruptible loads.

6. Establish criteria of reliability on the basis of frequency of emergencies which the

Paper 58-126, recommended by the AIEE Power Generation and System Engineering Committees and approved by the AIEE Technical Operations Department for presentation at the AIEE Winter General Meeting, New York, N. Y., February 2–7, 1958. Manuscript submitted October 21, 1957; made available for printing November 18, 1957.

HERMAN HALPERIN and H. A. ADLER are with the Commonwealth Edison Company, Chicago, Ill.

The authors appreciate the counsel of K. M. Smith, system planning engineer, and the assistance of J. P. Lynskey and E. E. Ciesielski in making the calculations.

system must be able to meet without voltage reduction or loss of load.

7. Make no specific allowance for possible deviations from forecast loads, with the understanding that calculated reserves are minimum recommended values, which may be increased by management, if desired, on the basis of over-all evaluation of all circumstances.

These principles and their application are discussed in detail in the main body of the paper. Mathematical methods, especially for calculating intervals, frequencies, and average durations of forced outages, are given in the Appendix.

Methods Used in Reserve Capability Studies

EXPECTATION OF FORCED OUTAGES

Expectations of forced outages are calculated by probability methods which give results in terms of frequencies, intervals, and average durations. While this method is more difficult (especially for a complex system consisting of units of different size and performance characteristics), it has been found to give more significant results than the usual probability calculations, which express the results only in terms of a ratio. A probability of 0.01 indicates an outage for 1% of the time, but it does not state, for example, whether the outage occurs once in 100 days with a duration of 1 day, or once in 500 days with a duration of 5 days. Such a difference in interpretation may be of considerable importance in the evaluation of the reliability of a system, as a large loss of capacity would create a more acute problem if it should last for 5 days than if it should last only 1 day.

The equations which apply to a uniform system are given in section I of the Appendix. A uniform system is one for which all parts are equal in size, outage rate, and average duration of outage. It is

usually not practicable to assume that a complex system is equivalent to a system with uniform components. Equations have been given for a system of completely diversified units of different sizes and outage rates.[1] However, even with the help of a digital computer, these equations for a completely diversified system would result in excessive calculations.

For practical calculations, it was found convenient and sufficiently accurate to subdivide the system into subgroups of units which could be considered reasonably uniform. The subgroups should be so selected that the deviations of individual sizes from the average within the group, as measured by the standard deviation, will give results within the desired accuracy. The standard deviation is a measure of the discrepancy in the magnitudes of multiple outages introduced by using multiples of average values for the group instead of the sum of individual values in their various possible combinations. A test should be made for the normality of the distribution as a basis for judgment of the probable error.

After the values for each subgroup have been calculated according to the methods given in section I of the Appendix, the results are meshed mathematically with each other to give the values for the composite system. The necessary steps are shown in section III of the Appendix. The result is an array of corresponding values of magnitude, frequency, or in-

Table I. Basic Values on Forced Outages

Items	Forced Outage Rate, Per Cent	Average Duration, Days	Average Interval, Days
Common header turbine generators	1.2	6	500
Boilers	0.5	2.5	500
Unit-type system: boiler and turbine generator	2.5	6	240

Note: Durations and intervals are in calendar days. Outage rates are based on unit days of forced outages divided by unit days of exposure, counting as full days all days (including weekends and holidays) on which the unit was out for any part of the heavy-load period of weekdays, but omitting weekends and holidays if the total outage was 3 days or less.

Reprinted with permission from *AIEE Trans. Power Apparatus Syst.*, vol. 77, pt. III, pp. 530–544, Aug. 1958.

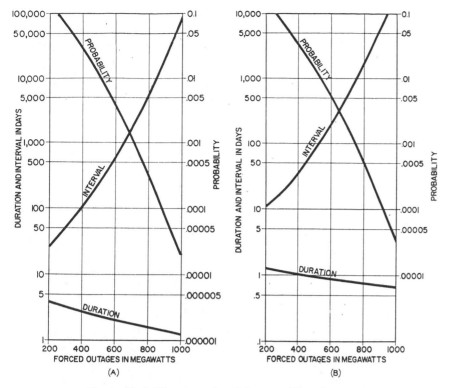

Fig. 1. Probability, interval, and duration of forced outages

A—For 2,645-mw unit-type system
B—For 5,303-mw system of common-header and unit-type installations

terval, and average duration of outages for the composite system. The meshed results are grouped and plotted. Procedures for these operations are shown in section IV of the Appendix. The methods of calculation are adaptable to the use of a digital computer.[2]

The outage rates used in recent studies for the various types of equipment were based both on national data[3] and on records for the Commonwealth Edison system. The values are shown in Table I. In the application of these outage rates, consideration must be given to the fact that most of the time the total system is not exposed to forced outages. Units out of service for overhauling or maintenance are, of course, not subject to failure. For this reason the outage rates must be corrected by an exposure factor to put the rates on the basis of the average capacity expected to be in operation.

Because unit-type installations within a few years will account for more than half of the system studied, the outage rate for such equipment is of special importance. Furthermore, the outage rate for unit-type installations requires careful consideration because the statistical experience of the performance of this equipment is relatively small and is practically unavailable for the extra-large sizes. The outage rate of 2.5% as used by the

authors has been the result of studies of all available information.

Figs. 1 and 2 show the results of calculations for the Commonwealth Edison system as it may be several years from now. At that time, the generating equipment will consist of 47 common-header turbine generators of 57-mw (megawatt) average size with 104 boilers ranging in size from 4 to 80 mw, and 14 unit-type installations with rated capacities of 53 to 305 mw.

Fig. 1 shows the results for the unit-type system of 2,645 mw and the total combined system of 5,303 mw, respectively. The three curves are the probability, average duration, and interval of forced outages. The interval is the duration divided by the probability. The frequency is the reciprocal of the interval and is omitted from the graphs. For the unit-type system a forced outage of a magnitude expected once in 30 days has an average duration of 3¾ days, and an outage expected once in 9,000 days (about 25 years) has a duration of 1¼ days. For the composite system, the average duration of a once-in-30-days outage is nearly 1.2 days; for a once-in-25-years outage, it is 0.7 day.

The difference in the results expressed in terms of intervals and probability is illustrated in the following comparison

derived from Fig. 1. For the unit-type system, a 220-mw outage has an expected interval of 30 days while the probability is 0.125, which corresponds in the usual interpretation to an outage of one in 8 days. An 845-mw outage has an interval of 9,000 days and a probability of 0.00014, which corresponds to an outage of once in about 7,000 days. Similarly for the total system, a 380-mw outage has an interval of 30 days and a probability of 0.039, which corresponds to an outage of once in 26 days. A 930-mw outage has an interval of 9,000 days and a probability of 0.000078, which corresponds to an outage of once in 13,000 days. The reason for the difference between the interval based on the probability ratio and the interval calculated by the method presented in this paper is that in the first method the duration of the outage is assumed to be exactly 1 day, while in the second method the duration of the outage is calculated.

In these calculations, the boilers and the turbine generators of the common-header system were each subdivided into two subgroups. The unit-type system was subdivided into three subgroups. As a consequence of these subdivisions, the probable deviation of the indicated magnitudes of outages was reduced to a relatively small value.

Fig. 2 shows the intervals for the combined boilers and turbine generators of the common-header system, for the unit-type system, and for the composite system. While the common-header and the unit-type systems are approximately equal in size, the expected outages, especially the less frequent ones, are much larger for the unit-type system. Furthermore, the effect of the common-header system on the magnitude of the forced outages of the

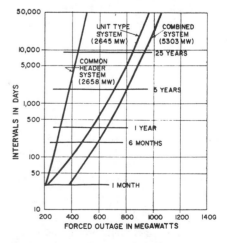

Fig. 2. Intervals of forced outages of common-header installations, unit-type installations, and combined 5,303-mw system

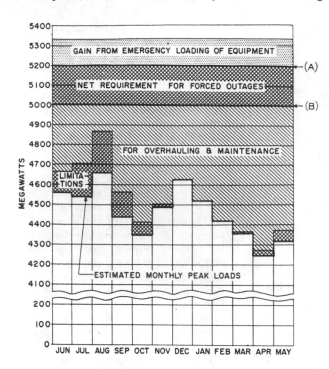

Fig. 3. Constant-risk principle in determining reserve generating capacity

A—Minimum required owned capacity on normal rating basis

B—Constant total of load + overhauling + limitations

Net requirement for forced outages is (magnitude of once-in-5-years' forced outage) minus (help from interconnections, emergency loading, dropping of interruptible load, and monthly peaking factor). "Reserve" on normal-rating basis is the sum of (requirements for overhauling) and (net requirement for forced outages)

combined system is relatively small. As the common-header system is replaced on account of obsolescence by unit-type installations, the magnitude of forced outages will increase, even assuming no increase in the size of the total system.

As indicated in Fig. 2, at least the following frequencies of occurrence of forced outage are given specific consideration in the reserve studies: once a month, once in 6 months, once a year, once in 5 years, and once in 25 years.

In establishing reserve requirements, the criterion for the adequacy of the system is the ability to meet all forced outages with frequencies up to and including once in 5 years without voltage reduction and concurrent loss of load. Like other similar proposed criteria of service reliability, this upper limit of once in 5 years is based essentially on judgment.

The forced-outage records of the Commonwealth Edison Company for a period of 16 years show three cases where the magnitude of forced outages equalled or exceeded the values expected once in 5 years on the basis of probability calculations. As three times in 16 years is equivalent to once in 5⅓ years, the agreement of actual experience with calculated expectations may be considered reasonable.

CONSTANT RISK PRINCIPLE

The constant-risk principle provides for arranging the overhauling schedule so that the sum of the monthly peak load, overhauling requirements, and limitations

(due to warmer condenser-circulating water) is constant throughout the year, as outlined previously.[4] This scheme has the effect of providing essentially the same amount of capacity to take care of forced outages on any monthly peak day throughout the year. In contrast, excess reserve would be available during most of the year if the reserve were planned with relation to only the maximum demand of the year.

Because ability to meet emergencies is essentially constant throughout the year with reference to the 12 monthly peaks, this scheme makes the most efficient use of reserve capacity. The result therefore is minimum required total capacity for a given degree of service reliability.

The development of the constant-risk principle has been a consequence of the change in the yearly load curve. Years ago, the December maximum demand greatly overshadowed the loads for the other months, and the summer valley was so deep that it could easily accommodate the necessary overhauling. The reserve problem was met therefore by the provision of adequate capacity for December to take care of the maximum forced outage for which protection was wanted. With increasing loads of nonlighting nature, especially large industrial and commercial loads and air-conditioning loads, the summer valley has disappeared and soon the summer peak is expected to exceed the winter peak of the same year. The two valleys in the annual load curve, in spring and in fall, are not enough to accommodate the required overhauling.

As a consequence, increased reserve capacity is required and careful planning of overhauling throughout the year has become a necessity.

It is recognized that in practice the ideal of the constant-risk principle cannot be fully achieved for the following reasons:

First, it is practically impossible to match the overhauling schedule with the load curve so as to achieve a perfect constant total of monthly peaks, overhauling, and limitations. This problem increases as large units with overhauling periods of long duration are added to the system. For this reason, the sum of the megawatt-month figures for the required overhauling and maintenance of the various pieces of equipment is increased by a stacking allowance. Studies have indicated that a stacking allowance of 15% applied to the calculated overhauling requirements is adequate for the next several years, and this percentage is used in calculating planned reserve requirements.

Second, the equipment exposed to forced outages is not constant throughout the year because the equipment in operation varies on account of overhauling and limitations. As a consequence, expected forced outages of a given frequency are not constant throughout the year, and therefore the risk is not strictly constant even if the margin in capacity provided for forced outages is constant. Calculations show that allowing for the variations in exposure results in about 5% in maximum deviation in magnitude of expected forced outages from the yearly average. It was felt that such a refinement was not justifiable.

The constant-risk principle is illustrated in Fig. 3.

The "net requirement for forced outages" is the owned capacity required throughout the year for the once-in-5-years forced outages after subtracting emergency help and the peaking factor allowance. As discussed later in the paper, this emergency help consists of assistance from power supply over interconnections with other utilities, from dropping of interruptible load, and from emergency loading of equipment. The help available from emergency loading is not exactly constant over the year, but varies by a small amount with variations through the year in the capacity available for operation after taking into account overhauling, limitations and forced outages. The effect of variations in available capacity upon emergency loading partly offsets the effect of these variations on forced outages. The more equipment there is in operation, the more there is which is subject to forced outages; but also the more there is which is available for emergency loading.

The minimum required owned capacity on the normal-rating basis is, as shown on the graph of Fig. 3 the sum of the "constant total" and of the net requirement for forced outages. The "reserve," in the usual sense of the term, for each month is the sum of the net reserve which is available for forced outages and the reserve for planned overhauling and maintenance.

It is possible that a situation may arise in which the sum of the August peak load and limitations, with no overhauling, may exceed the sum of monthly peak load, overhauling, and limitations for the other months. In such a case, the required capacity can be so evaluated that a somewhat smaller reserve in August is compensated by a somewhat larger but constant reserve in the 11 remaining months. In this way, the ability to meet forced outages expected up to once in 5 years, or any other standard, would still be maintained. It will be necessary to exercise caution in the application of this method so that the emergency reserve margin allowed for August does not become too small, with a resultant risk of deficient capacity for a few days in this month. In such a case, if the over-all protection is planned for once-in-5-years outages, the margin in August should be at least large enough to take care of outages with a frequency of once in 2 or 3 years.

MONTHLY PEAKING FACTOR

The overhauling schedule and the constant-risk principle are applied with reference to the peak loads of each of the 12 months. However, major forced outages are not likely to occur on the day of the monthly peak. Calculated average durations of multiple forced outages based on the equations in the Appendix, as well as actual service experience, indicate that the coincident maximum value of major multiple outages lasts only for durations of the order of a day. The probability of the coincidence of a major outage with a monthly peak day is relatively small as has been confirmed by actual service experience.

A study was made of the variations of daily peak loads within each month over several years, and the results were adjusted to the expected load conditions in future years under study. Fig. 4 shows the results of such a study. The abscissa shows the difference between monthly peak and daily peak, and the ordinate shows the probability that these load differences will be equal to or greater than the amount shown. The curves show that for the year under consideration in 90% of the cases, it may be expected that the load will be at least 65 mw below the monthly peak. As a matter of judg-

ment, it has been decided that protection for the 10% of cases where the load might be nearer to the peak is not justified. Therefore, a "peaking factor" of 65 mw is subtracted from the emergency requirements.

In principle, this method of reducing the reserve requirements is equivalent to the probability of loss-of-load method used by others,[5,6] but it is based on weekday maximum demands for each month instead of yearly load curves, and it has the advantage of great simplicity and flexibility in problems involving different conditions. If a different degree of protection is required, the necessary values may be readily determined from curves similar to those in Fig. 4.

In considering once-in-5-years forced outages, it may be expected that in 10% of the cases the load will be in excess of the available capacity, and in 90% of the cases it will be equal to or below the available capacity. In other words, in 50 years, 10 forced outages of a once-in-5-years magnitude may be expected, but only one of them will occur when the capacity is not adequate. Therefore, in terms of the "loss-of-load method" a capacity deficiency may be expected once in 50 years.

With a peaking factor based on 90% protection, the possible deficiency, if the emergency should fall on a peak day, is only of the order of $1^{1}/_{2}$% of the load. Such an occurrence would be rare. Therefore the deficiency could be readily taken care of by a small voltage reduction. If, however, a higher monthly peaking factor were allowed, not only the necessity of voltage reduction would arise more frequently, but also the magnitude of the reduction would be greater. For example, if planning were done so that the capacity were adequate for only half the days of the month, the deficiency, if the emergency occurred on a peak day of the month, would be so large as to require a system voltage reduction of 10% to 12%. Such operating conditions would be intolerable.

EMERGENCY HELP

As mentioned in connection with Fig. 3, there is available emergency help from three sources which can be evaluated and deducted from the reserve needed to take care of forced outages.

The first of these is the emergency excess capability available by loading of equipment in excess of normal capability. In the true sense, this is not a deduction, but is actually capacity available for usage for a limited time. Since generating requirements are determined in terms of normal capability for day-in-and-day-out operation, this excess appears as a compensating factor deductible from the total emergency requirements. The emergency excess capability is determined from actual operating experience, but consideration must be given to the fact that in an emergency in which this help is required, not all the plant capacity will be in operation and that some equipment may not be in condition to furnish more than normal capability.

The second factor deductible from the capacity requirements is an allowance for interruptible load. This is load of certain customers, which by contractual agreement may be reduced. In the evaluation of this deduction, not the arithmetical sum of the maximum demands of all these loads is considered deductible but only a portion of this load which, on a coincidental basis and in consideration of possible other conditions, can be reliably allowed for.

Finally, an allowance is made for the emergency help obtainable over interconnections with other utilities. The amount of such help is determined on a conservative evaluation of the available sources of supply, and full utilization of such power is counted on only for use in rare emergencies. For more frequent use, smaller portions of this help are considered to be mutually available. The main principle in determining the amount of power available over interconnections between two independent utilities (in contrast to utilities operating under

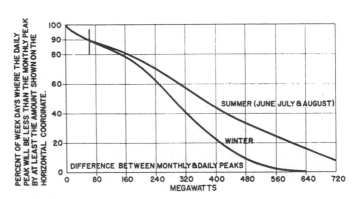

Fig. 4. Monthly peaking factor

pooling arrangements) is that of equal interchange; i.e., each of the two interconnected systems is expected to help the other in equal amounts. In the apportioning of savings in reserve from an interconnection, this principle of equal reciprocity is used as the main criterion. The corresponding equations for such an apportionment are given in section V of the Appendix.

In the determination of reserve requirements for an interconnected system, it is important that not only the reserve be adequate for the combined system, but also that each of the two systems be able to furnish the required help to the other without excessively reducing its own reserve. Theoretically, for a once-in-5-year demand for help from system Y, system X must be prepared only for its own coincident forced outage of average magnitude. Any outage of greater magnitude on system X would coincide with the demand from system Y less frequently than once in 5 years and, therefore, protection for such a contingency is not necessary.

The magnitude for an average outage is of the order of 2% of the system capacity. From a practical operating standpoint, however, it is generally desirable to protect for larger forced outage coincident with the demand from the other system. In reserve studies for the Commonwealth Edison system, one of the requirements is for the system to be able to furnish to any of the other interconnected utilities the same amount of help as is expected of them in an emergency and at the same time to have enough additional reserve for the loss of its largest unit. For the system represented in Fig. 2, the largest unit has a capacity of 305 mw, which is almost 6% of the system capacity. In the example covered by this pa-

per, there are three units of this size and one almost as large. While each of these is expected to be on forced outage once in 240 days with an average duration of 6 days, according to Table I, there are combinations of forced outages of other equipment also resulting in a loss of about 305 mw. On the average, a 305-mw loss of capacity may be expected to have an average duration of 1.2 days and an interval of 18 days.

This method of planning and checking calculations is proving satisfactory, but it may become impractical when the first of a new, very large sized unit is added to the system, such as 8% or 10% of the total system capacity. For such a situation, it may be considered sufficient to protect for a forced outage of 6% of the system capacity coincident with any emergency demand from the other system. No definite rule has been established for such protection. It is a matter of judgment and general operating conditions for the system under consideration, and the percentage will vary with systems.

Application of Methods

DETERMINATION OF RESERVE

The application of the methods discussed in a determination of the required reserve is illustrated in Table II. The values for forced outages are taken from Fig. 2. The values for overhauling and maintenance are as shown in Fig. 3 and include an allowance for special maintenance which may be necessary for new units and associated equipment of untried size or type.

Each of the columns in Table II for the various frequencies of occurrence is independent, and furnishes the basis for decisions by management for the degree of protection desired. In this case, pro-

Table III. Required Owned Generating Capacity*

For Peak Load of 4,660

Items	Average for 12 Months	August	December
Net owned reserve for forced outages†	195	195	195
Reserve for overhauling	480	135	370
Total required owned reserve	675	330	565
Peak load	4,460	4,660	4,630
Required available net capacity	5,135	4,990	5,195
Limitations	60	205	0
Required capacity in terms of normal winter capability	5,195	5,195	5,195

* All figures are in megawatts.
† From Table II for once-in-5-years' forced outage.

tection for the once-in-5-year forced outages requires only 10 mw more reserve than does protection for once-a-month outages. This is because for such frequent outages no help from emergency loading and little help from interconnections is contemplated.

If the system were planned as self-sufficient, i.e., without any help from interconnections, protection for the once-in-5-year forced outages would require 395 mw more reserve than protection for once-a-month outages. The increase in required reserve for protection for once-in-25-year forced outages compared with protection for once-in-5-year outages amounts to only 40 mw for the system with interconnections, but the increase would be 180 mw without them. The calculations resulting in these figures are not shown. They are based on item 4 of section V of the Appendix.

The requirements for owned capacity based on the once-in-5-year criterion are shown in Table III. They are constant throughout the year.

The various reserves, i.e., required emergency reserve for forced outages, total required owned reserve, and total reserve, are given in Table IV. In connection with the constant-risk principle, item A shows a constant owned emergency reserve of 330 mw available at the time of any monthly peak throughout the year to take care of forced outages. This amounts to about 7% of the load. On days other than the monthly peak-load days the reserve is higher. According to prevalent usage, the reserve is expressed as the difference between the established normal capability and maximum de-

Table II. Reserve Generating Requirements*

Peak Loads: Annual, 4,660; Average monthly, 4,460

Items	Once a Month	Once in 6 Months	Once a Year	Once in 5 Years	Once in 25 Years
A. Requirements for forced outages	380	590	655	805	935
B. Emergency help					
Interconnections	80	240	310	360	450
Emergency loading	0	90	135	135	135
Interruptible load	50	50	50	50	50
Total	130	380	495	545	635
C. Monthly peaking factor	65	65	65	65	65
D. Net requirements for forced outages: A−B−C	185	145	95	195	235
E. Required owned reserve: item D+overhauling					
Average for year: overhaul 480 mw	665	625	575	675	715
August: overhaul 135 mw	320	280	230	330	370
December: overhaul 370 mw	555	515	465	565	605

* All figures are in megawatts.

Table IV. Emergency Reserve, Owned Reserve, and Total Reserve*

Items	Average for 12 Months	August	Decembe
A. Required emergency reserve for forced outages			
Net owned reserve for forced outages	195	195	195
Available from emergency loading of generating capacity	135	135	135
Total emergency reserve	330	330	330
Total in per cent of peak load involved	7.4%	7.1%	7.1%
B. Required Owned Reserve			
Net owned emergency reserve plus overhauling	675	330	565
Available from emergency loading	135	135	135
Total required owned reserve	810	465	700
Total in per cent of load	18.2%	10.0%	15.1%
C. Total Reserve			
Owned	810	465	700
Interconnections: emergency help	360	360	360
Total	1170	825	1060
Total in per cent of load	26.2%	17.7%	22.9%

* All figures are in megawatts except as indicated.

mand. These values, which include the allowance for overhauling, are shown in the first line under item *B*. The owned reserve for the system studied actually includes also the margin available from emergency loading, and so the total owned reserves under item *B* become 10.0% for the August peak, 15.1% for the December peak, and 18.2% for the average monthly peak. With the help through interconnections taken into account, these figures finally rise to sizable amounts, i.e., 17.7%, 22.9%, and 26.2% respectively.

The reserve figures of special importance are the emergency reserve and the average owned reserve throughout the year. The emergency reserve is the capacity available to take care of forced outages. It is essentially constant throughout the year with reference to the monthly peaks as previously discussed. The average owned reserve is significant because it represents the over-all conditions under which the system operates and affords a more complete basis of comparison for a given system as well as between systems as the annual load curve changes. In addition to the average owned reserve, the reserve at the annual peak (or peaks) is, of course, of special interest.

The owned reserve consists of the margin available for forced outages and the margin provided for planned overhauling and maintenance. This latter margin provides a possible source of help in an emergency. Both records and theory show that multiple forced outages of large magnitude are very unlikely to start simultaneously. They almost always build up over a period of days as additional units fail, thus contributing to the magnitude of the total outage. Under such circumstances, it is frequently possible to return some equipment which

is out for planned overhauling or maintenance when the reserve in operation approaches a small value. For this reason the average margin for overhauling and maintenance and, therefore, the average owned reserves throughout the year are important criteria of the reliability of the system.

Studies of the type presented in this paper are made with reference to normal capabilities of equipment. Normal capabilities are understood to apply for winter conditions with 1-inch back-pressure for steam turbines and are expressed in terms of net output of stations after deduction for auxiliary power requirements. The normal net capability is considered to be the load which a unit is capable of carrying for at least 4 hours on each of 5 consecutive days. The limitations shown in Table III are reductions from this normal winter capability on account of warmer condenser-circulating water in the summer months. Emergency capabilities are based upon the full usage of

the maximum output once or twice a year for 1 hour in winter or 2 hours in summer; however, for planning purposes full usage is counted on only once a year. The frequency of emergency loading is limited because such operation subjects the turbines to extra stresses and increases the possibility of the fouling of boilers. The loads shown in Table III are estimated $1/_2$-hour integrated monthly peak values.

The results of such studies as shown in Tables II, III, and IV are considered the minimum values of reserve to be recommended to management. The conditions and results shown in each column of Table II apply specifically to a system adequate to take care of the forced outage of the frequency shown at the top of the column. If a system is planned to be adequate for once-in-5-years forced outages, the reserve will be large enough so that in the case of more frequent forced outages the full amounts of emergency help shown in the first three columns of Table II may not be required.

The figures in the three tables are based on the assumption that all available equipment may be utilized at the time of peak demands. In addition to these minimum requirements, provisions may be made for extra reserve to allow for special circumstances. An increase over the required reserve may result also from the selection of a unit of an attractive size for the expansion of the system which may produce some excess capability in any one year.

EFFECT OF INTERCONNECTIONS ON RESERVE

The effects of the interconnection of two systems on the generating reserve requirements are demonstrated in Table V for two assumed systems each consisting of 30 units of 100-mw capacity with

Table V. Effect of Interconnection on Generating Reserve Requirements*

Average Conditions for Year

Items	Before Interconnection		After Interconnection	
	Each System	Both Systems	Each System	Both Systems
A. Requirements				
Average monthly peak load	2,407	4,814	2,407	4,814
Overhaul and limitations	240	480	228	456
Forced outage: once in 5 years	448	896	438	607
Total	3,095	6,190	3,073	5,877
B. Emergency help				
Monthly peaking factor	36	72	36	72
Emergency loading	35	70	33	69
Interruptible load	24	48	24	48
Interchange	0	0	136	
Total	95	190	229	189
C. Required capacity: A less B	3,000	6,000	2,844	5,688

* All figures are in megawatts.

35

Table VI. Test of Ability to Reciprocate*

Items	Conditions for System X	
	Emergency on System X	Emergency on System Y
A. Owned capacity	2,844	2,844
B. Emergency help		
Monthly peaking factor	36	36
Emergency loading	33	33
Interruptible load	24	24
Interchanges	136	0
Total	229	93
C. Total available to meet requirements:		
A + B	3,073	2,937
D. Requirements		
Load	2,407	2,407
Overhaul and limitations	228	228
Forced outages	438	166†
Help to other system	0	136
Total	3,073	2,937

* All figures are in megawatts.
† The amount of 166 mw available for forced outages is the difference between the total available capacity as given in item C and the sum of the requirements for load, overhaul, and limitations, and help to the other system, as given in item D.

an outage rate of 2%. The interchange required in a once-in-5-years emergency is 136 mw. The calculated saving in owned capacity for each system is 156 mw. The saving in owned capacity for each system is 20 mw greater than the power received over the interconnection. This difference is the reserve each system would have to provide if it installed owned capacity to supply 136 mw throughout the year.

Table VI shows the conditions for a once-in-5-years emergency for each system after interconnection. The first column of numbers shows the situation for an emergency on system *X*, and has the same values as the third column of numbers in Table V. The second column of Table VI shows that the same system is able to supply 136 mw to the other system and at the same time take care of a forced outage on its own system of 166 mw. This is more than the loss of the largest unit and amounts to 5.8% of the owned capacity.

An analysis of two interconnected systems should include not only the determination of the required capacity of the tie line or lines, but also a study of the required reliability of the interconnections. By means of probability calculations, the amounts can be determined by which the owned reserve of each system must be increased to compensate for the outage rate of the tie line. The results give a basis for deciding on the value of design features such as making the tie lines sleet proof or lightning proof. Such probability calculations of the effect of reliability of ties upon reserve

requirements must be made not only for interconnections between systems, but also transmission lines within a system which may affect the reserve requirements. Such studies were made for the Commonwealth Edison system, which is very well tied together. All stations are connected to the system by at least three transmission lines. It was found that of all the transmission lines in the system there were only two where forced outages could cause possible limitation on the utilization of the output of generating capacity. However, the forced-outage rates for these lines are so low that the effect on reserve requirements according to probability calculations was negligible.

EFFECT OF SIZE OF UNITS ON RESERVE

A useful application of probability methods is the study of the effect upon the required reserve of size of units used in expanding a system. The results of such a study applied to a large composite system are shown in Fig. 5. The curves show in per cent the net usable capabilities for various sizes and number of units added to a 5,608-mw system consisting of units of 92-mw average size. The new installations were assumed to be of the unit type. The net usable capability is the increase in the average monthly peak load which the system is capable of supplying as a result of the addition of the new unit, after allowing for incremental requirements for forced outages, average overhauling, and limitations.

For purposes of this comparative study, the same outage rates and the same overhaul requirements were assumed for the various sizes of added units as for existing unit-type installations, even though the larger ones exceed any built to date. To isolate the effect of one variable, the size of the unit, and exclude variables which are independent of the unit size, no allowance was made for reduction in reserve requirements or for increase in net usable capabilities on account of help from interconnections, from dropping of interruptible load, and from peaking factor. Furthermore, no provision for the effect of emergency loading upon reserve was included in the calculations for Fig. 5 because the excess of emergency capability above normal, if any, depends on the design of the units, and it was not considered desirable to introduce such a further variable in this comparison.

The lowest curve shows that the increase in system load that can be supplied by the first unit added to the system becomes smaller in per cent of its own capability as the size of the added unit in-

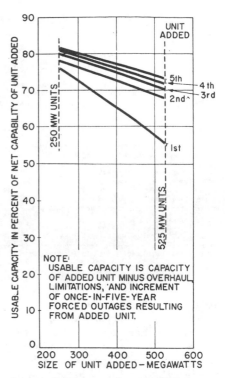

Fig. 5. Usable capability of additions to a system of 5,608 mw with average unit size of 92 mw

creases. The effect of size on the usable capability becomes smaller as more units of the new size are added to the system. For instance, the first 525-mw unit would increase the average usable system capability to supply load by only 55% of 525 mw, or by 289 mw, while the fifth would increase the usable system capability by 73% of 525 mw, or by 383 mw. If the new units have emergency capabilities above normal, these values of usable capability would increase because of the resulting reduction in the required reserve. Incidentally, a 525-mw unit would be 72% larger than the largest unit on the existing system. The first such unit would amount to 9.3% of the existing system, while the fifth would be 7.8% of the then existing system.

A similar study for the addition of units of the same size as the existing largest units, i.e., 305 mw, gives the following results: The first unit would increase the usable system capability by 72% of its normal capability, and the fifth unit by 80%. The system assumed in this study already contained 4 units of this size. It is of interest that the value of 72% of usable capability for an additional 305-mw unit is about the same as the 73% for the fifth 525-mw unit.

OTHER APPLICATIONS

Studies of the type described in this article have been applied also to the evaluation in terms of required reserve for certain types of loads with specific seasonal characteristics, such as air-conditioning or space-heating loads.

Summary

The following is a summary of the main problems to which the methods discussed have been applied and of points of special interest in connection with these problems:

1. Determination of the magnitude of required reserve in the planning of system additions to meet future load requirements.

The two main procedures are the use of probability methods for the determination of forced outages and the use of the constant-risk principle in scheduling overhauling to give the minimum required reserve for a desired degree of reliability. Factors considered include the following: monthly peaking factor, and help from emergency loading of equipment, from interconnections, and from dropping of interruptible load.

2. Evaluation of the savings in owned capability derived from an interconnection and determination of the required capability of the interconnection.

These studies presuppose reciprocity over the interconnections on an equal basis. The saving in owned capacity by a given utility is greater than the power interchange by an amount equal to the incremental reserve which would be required if the power were obtained from additions to owned generating equipment.

3. Study of the effect of size of units upon the required reserve in the planning of system additions.

The first new unit of a much larger size than existing ones adds to the usable capacity to supply load a much smaller fraction of its capability than that obtained if the added unit is equal in size to the largest existing one. This fraction increases as more units of the same new size are added.

Appendix. Mathematical Methods of Reserve Studies*

I. Frequency and Duration of Forced Outages of a Uniform System

Assume a uniform system of n units of equal size of M mw with a uniform outage rate of p. Then

$$p = t/T \qquad (1)$$

where t is the average duration of individual outages in days, and T is the average interval of individual outages in days. The frequency of individual events is

$$F = \frac{1}{T} = \frac{p}{t} \qquad (2)$$

*This Appendix was prepared by H. A. Adler.

The probability of r-fold simultaneous outages is expressed by the binomial formula

$$P_r = \frac{n!}{r!(n-r)!} \, p^r (1-p)^{n-r} \qquad (3)$$

The average duration in days of r-fold simultaneous outages is

$$t_r = \frac{Tp(1-p)}{r+p(n-2r)} = \frac{t(1-p)}{r+p(n-2r)} \qquad (4)$$

From equations 3 and 4 the average interval in days of r-fold simultaneous outages is derived by

$$T_r = \frac{t_r}{P_r} \qquad (5)$$

and the average frequency by

$$F_r = \frac{P_r}{t_r} \qquad (6)$$

The magnitude of the r-fold outage is

$$M_r = rM \qquad (7)$$

If the system is actually not uniform with regard to size of units, but consists of units with an average size M and a standard deviation σ, then the magnitude of an r-fold outage becomes rM with a deviation of $\pm \sigma \sqrt{r}$.

II. Derivation of Equation for Duration of r-fold Simultaneous Outages

The derivation of equation 4 is as follows: For a system of n machines with an individual outage rate of p, an average outage duration of t days, and an average interval between individual outages of T days, the average duration of r-fold outages is

$$t_r = \frac{t(1-p)}{r+p(n-2r)} \ \text{days}$$

Assuming first that the events can change position in time only in unit steps, then for each of the T possible positions of an event in the period T, there will be t time units occupied and $(T-t)$ unoccupied. There will be therefore for the T possible positions (Tt) occupied time units and $T(T-t)$ unoccupied time units. For each position of an event there are $t^{(r-1)}$ positions which the $(r-1)$ other simultaneous events can take, so that they overlap partly or completely with this event. Therefore, for the T possible positions of the event, there are $Ttt^{(r-1)}$ or Tt^r time units it can occupy together with the $(r-1)$ other events. That just r and not more than r events coincide, the $(n-r)$ other events, which are not involved in the r-fold simultaneous outage, must take positions not overlapping any part of the interval occupied by the r simultaneous events. In other words, the interval occupied by the r simultaneous events must be unoccupied by the $(n-r)$ other events. This is possible in $(T-t)^{n-r}$ ways. Finally, there are $n!/r!(n-r)!$ possible ways in which r events out of n events may be combined. Therefore, for all possible different positions of n events in the interval T, the total number of time units occupied in simultaneous r-fold occurrences is

$$T(n!/r!(n-r)!) \quad t^r (T-t)^{n-r}$$

Individual r-fold simultaneous events can have durations varying from the minimum overlap of one time unit to complete overlap of t time units. To obtain the average duration of overlap of r events, the total number of $T[n!/r!(n-r)!] \, t^r (T-t)^{n-r}$ time units which can be occupied must be divided by the number of possible r-fold individual events. An individual event is defined as a succession of occupied time units not affected by the start of a new event, nor the end of an event, nor the simultaneous end of one event and the start of another event, which, while not changing the number of simultaneous events, would change their identity. This latter requirement is important in calculations where events may involve, for example, outages of equipment of different capacities.

Of the t successive time units of an individual event there are $(t-1)$ units directly preceded by another unit. These may be called for simplicity "connected" units; and there is one unit not preceded by another which may be called the "leading" unit. Since the t successive units of an individual event should be counted only as one event, only the leading unit out of the t units should be counted and the connected units should not be counted. Similarly of the $(T-t)$ unoccupied units, only the leading unit should be counted in determining the number of unoccupied intervals and the $(T-t-1)$ connected units should not be counted.

In determining the number of r-fold simultaneous events the following observations apply: Only a leading unit can create a leading unit of a multiple event either by coincidence with other leading units or with connected units. The coincidence of connected units with connected units creates only connected units of multiple events. This applies both to intervals of occupied as well as to intervals of unoccupied units. The leading unit of an interval of unoccupied units starts a change in events as it is equivalent to the end of an interval of occupied units. The coincidence of connected unoccupied units with connected unoccupied units does not cause a change in events.

Now, for the

$$T \frac{n!}{r!(n-r)!} \, t^r (T-t)^{n-r}$$

units occupied in all possible r-fold simultaneous events in the interval T, there are $T(t-1)^r$ connected occupied units combined with $(T-t-1)^{n-r}$ connected unoccupied units in

$$\frac{n!}{r!(n-r)!}$$

possible combinations. Therefore, there are

$$T \frac{n!}{r!(n-r)!} \, (t-1)^r (T-t-1)^{n-r}$$

connected occupied units in all possible cases of r-fold simultaneous events in the interval T. As a consequence there are

$$T \frac{n!}{r!(n-r)!} \, t^r (T-t)^{n-r} -$$
$$T \frac{n!}{r!(n-r)!} \, (t-1)^r (T-t-1)^{n-r}$$

leading occupied units or different individual events in all possible cases of r-fold individual events in the interval T.

Table VII. Schematic Pattern of Meshing of Two Subgroups

System B	System A				
	A_0	A_1	A_2	A_3	A_4
B_0	A_0B_0	A_1B_0	A_2B_0	A_3B_0	A_4B_0
B_1	A_0B_1	A_1B_1	A_2B_1	A_3B_1	A_4B_1
B_2	A_0B_2	A_1B_2	A_2B_2	A_3B_2	A_4B_2
B_3	A_0B_3	A_1B_3	A_2B_3	A_3B_3	A_4B_3
B_4	A_0B_4	A_1B_4	A_2B_4	A_3B_4	A_4B_4

The average duration of an r-fold simultaneous event is the ratio of the number of all occupied units to the number of all individual events or

$$t_r = \frac{T\frac{n!}{r!(n-r)!}t^r(T-t)^{n-r}}{\left(T\frac{n!}{r!(n-r)!}t^r(T-t)^{n-r} - T\frac{n!}{r!(n-r)!}(t-1)^r(T-t-1)^{n-r}\right)}$$

or $t_r = \dfrac{t^r(T-t)^{n-r}}{t^r(T-t)^{n-r}-(t-1)^r(T-t-1)^{n-r}}$

The restriction that events can change position only by unit steps can be removed by expressing both t and t_r'as fractions of T and by letting T increase to infinity. By definition, $p=t/T$. Therefore substituting pT for t yields the following equation for the average duration of an r-fold event expressed as a fraction of T:

$$\frac{t_r}{T}=\frac{1}{T}\frac{(pT)^r(T-pT)^{n-r}}{[(pT)^r(T-pT)^{n-r}-(pT-1)^r(T-pT-1)^{n-r}]}$$

or, after dividing numerator and denominator by T^n,

$$\frac{t_r}{T}=\frac{1}{T}\frac{p^r(1-p)^{n-r}}{[p^r(1-p)^{n-r}-(p-1/T)^r(1-p-1/T)^{n-r}]}$$

Then, by expanding and omitting terms which will vanish if T becomes infinity

$$\lim_{T\to\infty}\frac{t_r}{T}=\frac{p(1-p)}{r+p(n-2r)}$$

or $t_r = \dfrac{T\,p(1-p)}{r+p(n-2r)}=\dfrac{t(1-p)}{r+p(n-2r)}$

III. Meshing of Two Subgroups of a Nonuniform System

Consider a system AB composed of two uniform subgroups A and B. For each of the subgroups the following values are determined for all values of r from 0 to r according to the equations in section I.

M_A and M_B, magnitude of r-fold outage

σ_A and σ_B, standard deviation for r-fold outage

P_A and P_B, probability of r-fold outage

t_A and t_B, average duration of r-fold outage

Then, the values for the composite system are calculated from the values for the subgroups for all possible combinations of r

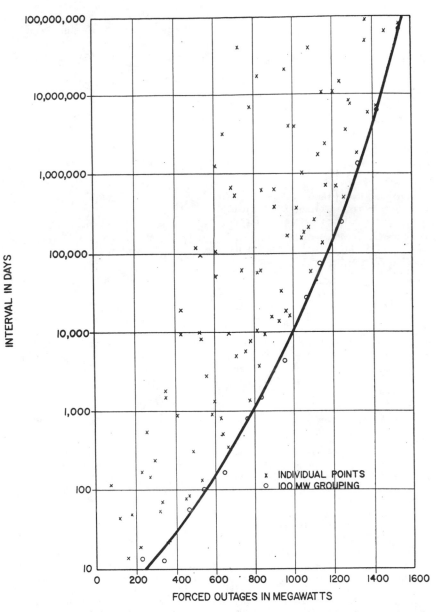

Fig. 6. Frequency of forced outages for a 5,608-mw system; individual and grouped results

according to the matrix pattern shown in Table VII.

The composite values are as follows:

$$M_{AB}=M_A+M_B \qquad (8)$$

$$\sigma_{AB}=\sqrt{\sigma_A{}^2+\sigma_B{}^2} \qquad (9)$$

$$P_{AB}=P_AP_B \qquad (10)$$

$$t_{AB}=\frac{t_At_B}{t_A+t_B} \qquad (11)$$

$$T_{AB}=t_{AB}/P_{AB} \qquad (12)$$

$$F_{AB}=1/T_{AB} \qquad (13)$$

The combination of a third subgroup C with a combined group AB is done in a similar way.

The derivation of equation 11 for t_{AB} is as follows:

Assume t_A and T_A are the duration and interval of an r-fold outage in system A and t_B and T_B are the duration and interval of an s-fold outage in system B. In general

T_A and T_B will be different, so that $T_A = fT_B$ or the event in system B will occur once in the interval T_B but f times in the interval T_A. According to equation 12 of reference 1, the combined duration t_{AB} is the x^2 term of the expression of

$$\frac{[t_Ax+(T_A-t_A)][ft_Bx+(T_A-ft_B)]}{[t_Ax+(T_A-t_A)][ft_Bx+(T_A-ft_B)]-[(t_A-1)x+(T_A-t_A-1)][(ft_B-f)x+(T_A-ft_B-f)]}$$

The x^2 term is

$$t_{AB}=\frac{t_Aft_B}{t_Aft_B-(t_A-1)(ft_B-f)}=\frac{t_At_B}{t_At_B-(t_A-1)(t_B-1)}=\frac{t_At_B}{t_A+t_B-1}$$

Then, removing the limitation of unit steps by making the size of the steps very small, or t_A and t_B very large compared with 1, yields

$$t_{AB}=\frac{t_At_B}{t_A+t_B}$$

IV. Grouping and Plotting of Results

A plot of the interval versus magnitude of forced outages for a complex composite system calculated according to the methods described in section III results in an array of points as shown by the crosses in Fig. 6. These results apply to a 5,608-mw system composed of eight subgroups of different sizes and outages rates. The apparent, irregular array of the points is due to the nature of the system. Certain combinations of outages are relatively more likely than other intermediate values. In evaluating these data, it must be recognized that the values are by nature discontinuous. Only certain combinations of magnitudes of outages are possible, and the frequency of intermediate values is zero. Nevertheless, to apply the results to reserve studies, it is necessary to express them by a continuous curve of interval versus magnitude of outages. This curve must represent the summation of the individual frequencies. This can be most readily accomplished by grouping the results in ranges of magnitudes, such as 50- or 100-mw steps and determining the total frequency for each group. In selecting the group width, it must be recognized that a large group width will result in smooth curves, but conservatively high values of expected outages. Grouping in narrow megawatt steps will result in more accurate magnitudes of outage but more irregular curves. A group width equal to the average size of all units of the combined system was found to give most accurate results. The grouped results in Fig. 6 are shown by circles. The grouping was done in 100-mw steps, which is close to the average size of the units of this system. The curve through the grouped points is slightly beyond the envelope curve of the individual points, as should be expected, because the curve represents the summation of the individual frequencies, and the frequencies of the envelope points have the greatest weight.

The parameters of a group are derived from the individual values of the group by the following equations:

$$F_g = \sum F \tag{14}$$

where F_g is the frequency of the group and F are the frequencies of individual terms in the group;

$$P_g = \sum P \tag{15}$$

where P_g is the probability of the group and P are the individual probabilities;

$$T_g = 1/F_g \tag{16}$$

where T_g is the interval for the group;

$$t_g = P_g T_g \tag{17}$$

where t_g is the average duration for the group;

$$M_g = \frac{\sum MP}{P_g} \tag{18}$$

where M_g is the average magnitude of outage for the group, M are the individual magnitudes, and P the corresponding probabilities, and

$$\sigma_g = \frac{\sum \sigma P}{P_g} \tag{19}$$

where σ_g is the standard deviation for the group, and σ is the standard deviation for the individual values.

V. Division of Gains from Interconnection of Two Systems

NOMENCLATURE

C = capacity
L = load
F = forced outages
OH = overhauling
J = limitations
E = excess of emergency load capability over normal load capability
I = interruptible load
T = help from previously existing ties
PF = peaking factor
X = interchange over new interconnection
$D = I + T + PF$

Furthermore, subscripts A and B apply to the two interconnected systems respectively, and subscript AB to the combined system.

Division of Required Capacity Between Two Systems Based on Equal Interchange

$$C_{AB} = C_A + C_B$$

$$C_A = L_A + (F_A + OH_A + J_A - E_A) - D_A - X$$

$$C_B = L_B + (F_B + OH_B + J_B - E_B) - D_B - X$$

Within relatively narrow limits, it is permissible to assume that $(F + OH + J - E)$ is proportional to C or equal to YC. Then

$$C_A = L_A + Y_A C_A - D_A - X$$

$$C_B = L_B + Y_B C_B - D_B - X$$

is obtained. Setting $Z = 1 - Y$ yields

$$C_A Z_A = L_A - D_A - X$$

$$C_B Z_B = L_B - D_B - X$$

which results in

$$C_A = \frac{(L_A - D_A) - (L_B - D_B) + C_{AB} Z_B}{Z_A + Z_B}$$

and

$$C_B = \frac{(L_B - D_B) - (L_A - D_A) + C_{AB} Z_A}{Z_A + Z_B}$$

Interchange over Tie

The value of X is obtained by substituting the value of C_A in the equation

$$C_A Z_A = L_A - D_A - X$$

or from equation

$$X = \frac{F_A + F_B - F_{AB}}{2}$$

which is derived from the following four equations:

$$C_A = L_A + F_A + OH_A + J_A - E_A - D_A - X$$

$$C_B = L_B + F_B + OH_B + J_B - E_B - D_B - X$$

$$C_{AB} = (L_A + L_B) + F_{AB} + (OH_A + OH_B) + (J_A + J_B) - (E_A + E_B) - (D_A + D_B)$$

$$C_{AB} = C_A + C_B$$

Gain from Interconnection

The reduction in capacity made possible by the interconnection is determined for each system respectively by

Required capacity before interconnection
$$C = L + CY - D$$

Required capacity after interconnection
$$C' = L + C'Y - D - X$$

Therefore

$$\text{Gain} = C - C' = \frac{X}{1 - Y} = \frac{X}{Z}$$

References

1. A NEW APPROACH TO PROBABILITY PROBLEMS IN ELECTRICAL ENGINEERING, H. A. Adler, K. W. Miller. *AIEE Transactions*, vol. 65, Oct. 1946, pp. 630–32.

2. THE USE OF A DIGITAL COMPUTER IN A GENERATOR RESERVE REQUIREMENT STUDY, Homer E. Brown. *Ibid.*, vol. 77, pt. I, Mar. 1958, pp. 82–85.

3. FORCED OUTAGE RATES OF HIGH-PRESSURE STEAM TURBINES AND BOILERS, AIEE Joint Subcommittee Report. *Ibid.*, vol. 76, pt. III, June 1957, pp. 338–43.

4. *Discussion* by Herman Halperin of AN INVESTIGATION OF THE ECONOMIC SIZE OF STEAM-ELECTRIC GENERATING UNITS, L. K. Kirchmayer, A. G. Mellor, J. F. O'Mara, J. R. Stevenson. *Ibid.*, vol. 74, pt. III, Aug. 1955, pp. 609–10.

5. DETERMINATION OF RESERVE CAPACITY BY THE PROBABILITY METHOD, G. Calabrese. *Ibid.*, vol. 69, pt. II, 1950, pp. 1681–89.

6. ELEMENTS OF SYSTEM CAPACITY REQUIREMENTS, C. W. Watchorn. *Ibid.*, vol. 70, pt. II, 1951, pp. 1163–85.

Discussion

C. D. Galloway and L. K. Kirchmayer (General Electric Company, Schenectady, N. Y.): The paper is an excellent demonstration of how frequency and duration of capacity outages may be adapted as a criterion of reliability. Unquestionably, this index of system reliability has long merited the careful analysis made by Mr. Halperin and Mr. Adler.

As the aim of any probability technique is to predict required system reserve, it would be interesting to know if the authors have data showing how the reserve requirement called for by their method compares with that obtained by conventional techniques which do not recognize the frequency and duration of capacity outages. Such a comparison would point up the significance of the method.

One of the assumptions of the method is the selection of a single value of t (duration)

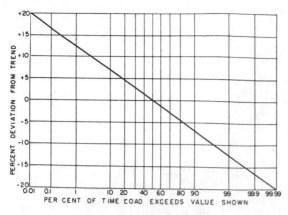

Fig. 7. Load-forecasting error

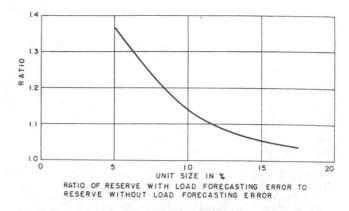

Fig. 8. Effect of load-forecasting error upon reserve required

and T (interval) to describe the behavior of a machine group. These values are not discrete numbers but rather were drawn from sets of data constituting probability distributions in themselves. Have the authors considered this point? Although the assumption is commonly made in probability calculations, it is not rigorously correct.

Concerning the mechanics of the method, we agree that to give consideration to each machine rating results in extremely involved arithmetic, and consequently the establishment of equivalent groups is necessary. We have not, however, attempted to determine the magnitude of the error that is introduced by this. Can the authors tell us how much error results in the reserve requirement because of this approximation?

The authors' method makes no specific allowance for possible deviations from forecast loads with the understanding that calculated reserves are minimum recommended values which may be increased by management. It should be pointed out that when conventional loss of load probability methods are used, the consideration of load forecasting errors increases the reserve more for small units than for larger. Consider, for example, the effect of including a load-forecasting error such as that shown on probability paper in Fig. 7. In Fig. 8 we have plotted the ratio of reserve with this load-forecasting error considered to reserve with the load-forecasting error neglected, as a function of unit size. It will be noted that for extremely large units the load forecasting error consideration has a negligible effect upon the reserve requirement.

An important consideration on all probability studies is the choice of the magnitude of the index to which the system is designed. It would appear that selection of a particular magnitude of index requires consideration of both the cost to provide facilities to obtain the given reliability, and the worth to the customer of such reliability. What are the authors' comments on choosing a particular value of index reliability?

It is realized that certain of these comments apply not only to the method outlined by the authors but also to any of the probability techniques in current usage, and it is realized that all such techniques require some approximations. The authors' method offers much promise and merits careful study of its assumptions and its field of application.

E. S. Loane (General Public Utilities Corporation, Reading, Pa.): This excellent paper covers a number of aspects of the reserve problem. The following discussion concerns the development of procedures for determination of durations and frequencies of multiple-unit forced outages.

It is pointed out that durations and frequencies of outages have more significance in the evaluation of reliability than do the simple probabilities of forced outage. As a general proposition, this statement may be open to question, and an answer for any particular situation will require exploration of other available measures of generating system reliability. Apart from its possibly more significant results, however, the complex computation of durations and intervals does appear to have particular applicability to several conditions encountered in the Commonwealth Edison system. In Table II, for example, the emergency loading of units and help from interconnected companies are shown as being dependent on the outage intervals.

A similar situation would exist in a combined hydro and steam system, where essentially run-of-river hydro is backed up by a limited storage of water. Under certain conditions of adverse river flow, the value of the storage and of the hydro capacity as emergency reserve will depend upon the durations and intervals of the emergencies in which the hydro might be called upon for maximum capability.

Some years ago, Mr. Watchorn and Mr. Loane were interested in such a situation[1] and, to avoid handling the mathematics, resorted to an experiment. In effect, a model of a generating system was set up and operated, with International Business Machines cards and a tabulator. Each unit was assigned its proper capacity, had the proper probability of outage occurrence with these occurrences being governed entirely by chance, and could have an outage of any duration, with the durations distributed according to the then available data. A 100-year synthetic record of forced outages for this generating system was produced within a relatively short time, but subsequent manual analysis of these data was somewhat tedious. The experimental results checked well against everything that could be computed concerning multiple unit outages, including a computation by Mr. Adler of the average durations for coincident outage of various numbers of units.[2]

With improvements in computing procedures, what was done slowly by hand could probably now be done by machine in minutes. Consequently, it is suggested, for consideration of those better able to judge the time and costs involved, that a computation of multiple forced outage durations and frequencies might well be done in the form of a probability experiment, in which the need for approximations and simplifying assumptions could be avoided.

One of the authors' assumptions is that the turbine outages have an average duration of 6 days, which is very close to the average indicated by the published data. These same data, however, show a very much skewed distribution of durations. The median and the minimum outage have the same value, one day, and about 1% of the outages are over 100 days. If the authors have evaluated the possible effect of this assumption on their results, it would be of interest to know their opinion of its importance.

The authors are not concerned in their computations with any outage of less than one day's duration, but the combination of several outages may result in average durations involving whole days plus fractional days. It is not apparent, however, how an average duration can be less than a whole day. Although it is undoubtedly a necessity of their computing method, the authors may want to comment further on the statement that the average duration of a once-in-25-years outage is 0.7 day, or on the durations of less than one day shown in Fig. 1(B).

In covering completely the various applications of probability computations and reserve determinations in the Commonwealth Edison system, the authors have adequately explained why they made certain assumptions or presented results in a certain way. Concerning such specific applications, they are best able to evaluate their methods and assumptions. On the other hand, for general application, it would be interesting to know if they would agree as to the desirability of treating both load-forecasting errors and emergency assistance from interconnections on a probability basis. In my opinion, these elements, which are subject to probability analysis, are too important to be considered generally in any other way.

REFERENCES

1. PROBABILITY METHODS APPLIED TO GENERATING CAPACITY PROBLEMS OF A COMBINED HYDRO

AND STEAM SYSTEM, E. S. Loane, C. W. Watchorn. *AIEE Transactions*, vol. 66, 1947, pp. 1645–54.

2. *Discussion* by H. A. Adler of reference 1, p. 1657.

M. J. Steinberg and **V. M. Cook** (Consolidated Edison Company of New York, Inc., New York, N. Y.): We extend our congratulations to the authors for the constructive value of their stimulating paper. The method for a practical application of probability mathematics for the determination of installed reserve capacity requirements utilizes the probable frequency of forced capacity outages as described by Adler and Miller in an earlier paper (see reference 1 of the paper). The fundamental concepts are expressed by equations 3, 4, and 5, which correlate the outage rate P, the frequency of occurrence F, and the interval between outage occurrence T. Because this is one of few available papers which utilizes the concept of frequency of occurrence it merits serious consideration by those interested in the application of probability mathematics to power system problems.

The methods developed by such earlier contributors as Calabrese, Lyman, Seelye, and Watchorn make use of the probable loss of load as the criterion for establishing objective reserve capacity requirements. The objections by the authors to the methods now in use is well taken, and conforms with our own views. We have felt for some time that the frequency of outage occurrence might be a more rational approach to the problem. General acceptance of the authors' method would depend upon the degree of variation of results and magnitude of the computations involved. We plan to apply both methods to our system for the purpose of evaluating them in this respect, at the earliest opportunity.

The application of probability mathematics involves extensive calculations even with the aid of electronic computers. To keep the calculations within practical limits, the authors have subdivided their system into unit groupings according to size, and substituted for each grouping the same number of units of equal size. Standard deviations are calculated to measure the degree of discrepancy introduced, and it is even suggested that a test of the normality of distribution be made to serve as a basis for judging the probable error. We wonder whether the reduction in the extent of the calculations, resulting from the simplifying assumption of equal unit sizes, has not been offset by the computations required to establish values of the standard deviation and to correlate the computed results. Would the authors supply a numerical illustration of their application of the standard deviation? If refinement of results is desired, we recommend that the calculations be based upon actual distribution of unit sizes.

In connection with the authors' philosophy on the constant-risk principle, we agree that if reserve capacity requirements are established in relation to only the maximum annual peak demand, the amount will be in excess of what can be justified by the short exposure time of one hour or less. The authors' method has an advantage over this, but it disappears when compared with the established methods which establish reserve capacity requirements in relation to all of the daily peak demands over an exposure period of one year. In the latter case the margin of reserve capacity above the sum of the daily peak loads and scheduled outages for overhaul will also be at a relatively constant value in a manner duplicating Fig. 3.

With reference to the monthly peaking factor, which the authors use to reduce the installed reserve capacity requirement, we note that this is justified on the grounds of what amounts to a limited exposure time, for which it is reasoned that a capacity deficiency may be expected only once in 50 years. Since daily peaks are not usually expressed as average values in excess of one hour's duration, the exposure time for a 20-weekday month would amount to a total of only 20 hours. This reasoning is valid when the daily peak demand is large in comparison with the demands at other times of the day. This would apply to the winter season. In the summer season, the variation of demand is not as pronounced, so the exposure time to a probable capacity deficiency is more than for one hour per day. This follows from the fact that off-peak loads on a particular day may exceed the on-peak loads of other days. Thus the curves of Fig. 4, which reflect only the daily on-peak demands, do not properly represent the exposure time as contemplated by the authors.

K. L. Hicks (Sargent & Lundy, Chicago, Ill.): We heartily agree with the authors' views on the constant-risk principle, particularly in those regions where summer peak loads and winter peak loads are in the same order of magnitude.

In applying this principle to the calculation of reserve requirements for a medium-sized system, we assumed that the maintenance outages would be scheduled in such a manner as to provide the same reserve for forced outages for each month. We found that the reserve required for scheduled outages for systems with both a summer and a winter peak is great enough to require some units to be scheduled out during the two annual peaks. It is possible therefore to express the reserve required for scheduled outages and the reserve required for forced outages as two separate entities, as the authors have done.

The paper presents a new concept of service reliability which is based on probability of outage of generation rather than the usual concept of probability of loss of load. Correlation between these two concepts depends on the characteristics of the system's daily peak loads and is by no means an easy matter. The authors' approach appears to be an oversimplification.

They state that since the once-in-5-years generation outage can cause a loss of load only during the 10% of time when it is not considered economical to provide protection, the resulting loss-of-load probability is 10% of the probable loss in generation. However, this seems to neglect the probability that more than the specific amount of outage for which protection is provided will occur, and that some of these outages will cause a loss of load during the 90% protected time as well as during the 10% "unprotected time"; also some additional loss of load will occur during the 10% un-protected zone for outages less than the once-in-5-years outage.

This correlation may be made by the more or less conventional calculation of loss of load. We agree that this calculation should be made on a daily peak load basis rather than an hourly basis and that it is much more convenient when applying the constant-risk principle to use a monthly period with a proper allowance for scheduled outages each month than to use an annual peak load curve corrected for scheduled outages.

The curves in the paper do not give enough information about the probability of specific outages from which a curve of more than specific outages can be constructed. The addition of such a curve would be helpful.

D. M. Sauter, **C. J. Baldwin**, and **K. M. Dale** (Westinghouse Electric Corporation, East Pittsburgh, Pa.): A number of papers have set out to determine the probability of loss of load expressed as a ratio or probable fraction of unit time during which there will be loss of load for a given set of conditions. The authors, however, consider that this ratio is not a sufficient evaluation of the reliability of a system. They align themselves with those who consider that the probable duration and probable interval between outages of a given magnitude is a better criterion of system reliability.

To compare the results obtained by these two alternative methods, we used the model system used recently in a paper by Arnoff and Chambers,[1] and for which the probable loss of load was given as 0.072 day per year, or approximately one day in 14 years.

Particulars of this model system are detailed in Tables VIII and IX. The average duration of an individual outage was assumed to be 0.06 year and the average interval between individual outages 3 years, giving the national average forced outage probability rate of 2% for the unit type of

Table VIII. Model Generating System*

No. of Units	Unit Size, Mw	Totals, Mw
1	250	250
3	150	450
2	100	200
4	75	300
9	50	450
3	25	75
22		1,725

* Assumed forced outage probability rate 2%.

Table IX. Expected Daily Peak Customer Demand for Typical Month

Estimated No. of Weekdays	Peak Customer Demand, Mw
2	1,300
2	1,325
5	1,350
6	1,375
3	1,400
2	1,425
1	1,450

Scheduled maintenance 100 mw; no emergency loading; no interruptible load; interconnections 200 mw.

Table X. Summation of Outage Frequencies

Magnitude of Outage, Mw	Probability of an Outage Exactly Equal to Column 1 Values	Probable Duration of Outage of Column 1 Values, Years	Frequency of an Outage Exactly Equal to Column 1 Values, Per Year	Frequency of Outages Equal to and Greater Than Column 1 Values, Per Year
(1)	(2)	(3)	(4)	(5)
0	0.641173	0.133634	4.797978	
25	0.039255	0.041999	0.934665	8.829087
50	0.118567	0.043048	2.754298	7.894422
75	0.059555	0.039826	1.495380	5.140124
100	0.038543	0.034220	1.126330	3.644744
125	0.011870	0.021956	0.540627	2.518414
150	0.046763	0.037136	1.259236	1.977787
175	0.005755	0.022794	0.252479	1.718551
200	0.008832	0.024159	0.365578	1.466072
225	0.004195	0.022575	0.185825	1.100494
250	0.015671	0.036197	0.432936	0.914669
275	0.001614	0.019386	0.083256	0.481733
300	0.003928	0.022068	0.177995	0.398477
325	0.001476	0.021905	0.067382	0.220482
350	0.001004	0.020445	0.049107	0.153100
375	0.000350	0.015921	0.021984	0.103993
400	0.001622	0.022462	0.045499	0.082009
425	0.000169	0.015074	0.011211	0.036510
450	0.000205	0.017000	0.012059	0.025299
475	0.000089	0.016270	0.005470	0.013240
500	0.000054	0.015574	0.003467	0.007778
525	0.000017	0.012628	0.001346	0.004311
550	0.000026	0.013538	0.001921	0.002965
575	0.000005	0.012254	0.000408	0.001044
600	0.000005	0.013095	0.000382	0.000626
625	0.000002	0.012309	0.000162	0.000244
650	0.000001	0.012220	0.000082	0.000082

system. With the procedure outlined by the authors, the probability, duration, and frequency of all possible outages, taken in 25-mw steps, were determined (see Table X, columns 2, 3 and 4). These values were then grouped together, as suggested by the authors, into groups having widths of 50 mw, 100 mw, and 150 mw, and the resultant intervals are plotted in Fig. 9, curves *A*, *B*, and *C* respectively. As explained by the authors, curve *A* is irregular in shape because of the relatively narrow group width used. Also plotted are the individual probabilities from which these curves were derived.

Clearly, the size chosen for the group width has considerable bearing on the value

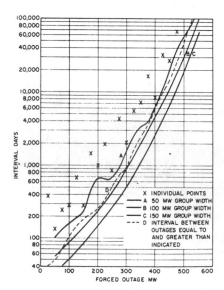

Fig. 9. Probable intervals between outages of various magnitudes

of the probable interval between outages of any given magnitude. What criterion have the authors used in deciding that a group width approximately equal to the average unit size gives the most accurate results? Since outages of magnitudes greater than the outage above which loss of load occurs will also contribute to loss of load, we believe that a more useful result would be the determination of the average interval between outages of a given magnitude and more. This may readily be evaluated by summating the frequencies of all outages equal to and greater than each particular outage considered (see Table X, column 5). The intervals between various outages are then the reciprocals of the various sums of the frequencies and are plotted in Fig. 9, curve *D*, which lies between curves *A* and *B*.

With a monthly peaking factor of 25 mw, the reserve generation available to meet forced outages on the model system amounts to 400 mw, sufficient to meet outages having intervals of:

19.2 years according to curve *A* (50-mw group width)

11.5 years according to curve *B* (100-mw group width)

7.4 years according to curve *C* (150-mw group width)

12.4 years according to curve *D*

The average unit size in the model system is 78 mw, and had we plotted a curve having a group width of this magnitude it would have approached our curve *D* much more closely. However, in our opinion, curve *D* is the more correct on theoretical grounds and is, moreover, easier to calculate from the individual probability and frequency values. We would appreciate the author's comments on this point.

The monthly peaking factor defined by the authors is a perfectly legitimate reduction of the monthly peak load and is comparable to the weighted probability method

used by others.[2,3] The 90% protection proposed by the authors is, however, on the conservative side compared with the weighted probability method, which gives in effect a 75% protection factor for the system considered. This latter figure might be more or less for other systems, depending on the relative values of the outage probabilities for outages contributing to loss of load (tending to be higher than 75% for systems having a smaller number of units than the one considered).

We agree that the probable duration of an outage is a most useful indication of the system reliability, but would not the probable duration of loss of load be an even better indication? Could authors suggest a means of determining such a duration? We should like to congratulate the authors on their straightforward way in which they evaluate the division of gains from the interconnection of two systems.

REFERENCES

1. OPERATIONS RESEARCH DETERMINATION OF GENERATION RESERVES, E. Leonard Arnoff, John C. Chambers. *AIEE Transactions*, vol. 76, pt. III, June 1957, pp. 316–28.

2. GENERATING RESERVE CAPACITY DETERMINED BY THE PROBABILITY METHOD, Giuseppe Calabrese. *Ibid.*, vol. 66, 1947, pp. 1439–50.

3. ELEMENTS OF SYSTEM CAPACITY REQUIREMENTS, C. W. Watchorn. *Ibid.*, vol. 70, pt. II, 1951, pp. 1163-85.

Andrew L. Miller (The Cleveland Electric Illuminating Company, Cleveland, Ohio): The authors are to be commended for undertaking one of the most complete and detailed programs of generation reserve calculations presented to date.

Two questions arise when these ideas, which were developed for a large metropolitan system, are applied to smaller systems, perhaps one third as large in capacity and numbers of units. For illustration, if each of 92 generators in a large system is down for 24 consecutive days of scheduled maintenance every 3 years, then there is an average of two units down every calendar day. Under these conditions, there is a good possibility of "playing checkers" with the maintenance schedule in such a way as to approximate either a flat rate of maintenance outage, or a program which will closely fit into the peaks and valleys of the predicted weekly load curve.

For a smaller company, with perhaps one third this number of units, such manipulation may become impossible. Where the largest unit covers 10% to 12% of the system capacity, half of the outage and loss-of-load risks of the whole year may occur in the month when the unit is out for maintenance. Do the authors have a suggestion on how to treat this situation?

The second question may be simply a restatement of the first. In most calculations of system reserves, it has been customary to relate the risks of successive values in forced outage to the duration curve of the load to be served. The combination of these two factors is necessary for the calculation of risks of loss of load.

Does Fig. 8 of the paper imply that the sum of load duration curve, plus system limitations, plus maintenance schedule, did equal a constant value in the sample calcu-

Table XI. Numerical Basis for Standard Deviation and Test for Normality: System A

Unit No.	Mw Size, X_n	$X_n{}^2$	$X_n - Avg$	$(X_n - Avg)^2$	$(X_n - Avg)^3$	$(X_n - Avg)^4$
1	160	25,600	−19.43	377.52	−7,335.31	142,525.05
2	160	25,600	−19.43	377.52	−7,335.31	142,525.05
3	160	25,600	−19.43	377.52	−7,335.31	142,525.05
4	160	25,600	−19.43	377.52	−7,335.31	142,525.05
5	185	34,225	5.57	31.02	172.81	962.54
6	214	45,796	34.57	1,195.08	41,314.08	1,428,227.92
7	217	47,089	37.57	1,411.50	53,030.24	1,992,346.08
	1,256	229,510		65,175.89		3,991,636.74

Table XII. Meshing of Systems A and B into System AB: Meshing Process of Standard Deviation*

System B Values	System A Values				
	σ_{A0} 0	σ_{A1} 24.3	σ_{A2} 34.35	σ_{A3} 42.1	σ_{A4} 48.6
$\sigma_{B0} = 0$	0	24.3	34.35	42.1	48.6
$\sigma_{B1} = 28.1$	28.1	37.15	44.38	50.62	56.14
$\sigma_{B2} = 39.7$	39.7	46.55	52.5	57.87	62.75
$\sigma_{B3} = 48.7$	48.7	54.43	59.6	64.37	68.80
$\sigma_{B4} = 21.1$	21.1	32.18	40.31	47.09	52.98
$\sigma_{B5} = 35.0$	35.0	42.61	49.04	54.75	59.89
$\sigma_{B6} = 45.0$	45.0	51.14	56.61	61.62	66.23

* $\sigma_{AB} = \sqrt{\sigma_A{}^2 + \sigma_B{}^2}$

lations? If so, a "peaking factor" of 65 mw (1.5% variation from the monthly average load) seems inadequate to adjust for the variations in load duration for such power systems as may have 20% load variation through the days of a single month. What changes in procedure would be recommended in cases where monthly load variations are so great as to make this consideration necessary?

Herman Halperin and H. A. Adler: It is gratifying to note the increased interest in the determination of reserve requirements by mathematical methods, as shown in the discussions of this paper and by industry in general. When we made our first use of probability methods for generating reserve planning in 1943, general interest in the subject was limited.

In connection with Mr. Casazza's discussion, a comparison of the reserve as a function of the largest unit size, the agreement is surprising in view of the fact that size and composition of the systems, forced outages rates, overhauling schedules, and many other factors, which have an important influence upon the reserve requirements, have not been considered in this comparison. As pointed out in the Galloway-Kirchmayer discussion, an exact comparison of the results obtained by various methods would be desirable, and this comparison is being investigated by the AIEE Subcommittee on the Application of Probability Methods.

Mr. Loane, and the Galloway-Kirchmayer discussion state correctly that the average duration t of forced outage is not a discrete value, but is itself subject to wide variations. It should be pointed out that this comment applies equally to the value p of outage rate. This was recognized in our early efforts, and attempts were made to evaluate the "spread in time" as well as

the "spread in magnitude" of forced outages. This turned out to be involved and has not been pursued further. We agree that this problem is worthy of further attention.

In answer to the request contained in the Steinberg-Cook discussion, the procedure for evaluating the probable deviation in magnitude of outages caused by grouping is as follows, based on the numerical values shown in Table XI.

Average capacity = 179.43

Standard deviation $(\sigma) = \sqrt{\dfrac{(X_n)^2}{n} - (Avg)^2}$

$= \sqrt{\dfrac{229,510}{7} - (179.43)^2} = 24.3$

$\mu_3 = \dfrac{\sum (X_n - Avg)^3}{n} = \dfrac{65,175.89}{7} = 9,310.84$

$\mu_4 = \dfrac{\sum (X_n - Avg)^4}{n} = \dfrac{3,991,636.74}{7}$
$= 570,233.82$

Skewness $= \dfrac{\mu_3}{\sigma^3} = \dfrac{9,310.84}{(24.3)^3} = 0.649$

Kurtosis $= \dfrac{\mu_4}{\sigma^4} = \dfrac{570,233.82}{(24.3)^4} = 1.625$

A skewness of zero and a kurtosis of three would indicate a normal distribution. The positive value of 0.649 for the skewness indicates that the distribution is asymmetric with a preponderance of values smaller than the average, and the kurtosis of less than three indicates a flat-topped curve or lack of extreme deviations. Therefore, the use of the standard deviation as an indication of the positive error is conservative.

The following shows how the standard deviations for double, triple, etc., outages

are determined for a subgroup, for system A, where $\sigma = 24.3$:

$\sigma_0 = 0$

$\sigma_1 = 24.3$

$\sigma_2 = \dfrac{2\sigma_1}{\sqrt{2}} = \sqrt{2}\ \sigma_1 = 34.35$

$\sigma_3 = \dfrac{3\sigma_1}{\sqrt{3}} = \sqrt{3}\ \sigma_1 = 42.1$

$\sigma_4 = \dfrac{4\sigma_1}{\sqrt{4}} = \sqrt{4}\ \sigma_1 = 48.6$

$\sigma_5 = \dfrac{5\sigma_1}{\sqrt{5}} = \sqrt{5}\ \sigma_1 = 54.25$

$\sigma_6 = \dfrac{6\sigma_1}{\sqrt{6}} = \sqrt{6}\ \sigma_1 = 59.5$

$\sigma_7 = \dfrac{7\sigma_1}{\sqrt{7}} = \sqrt{7}\ \sigma_1 = 64.4$

Finally, Table XII shows how the standard deviations for two subgroups are combined. In answer to Mr. Galloway and Mr. Kirchmayer, the standard deviation for a once-in-5-years outage of 847 mw is 41 mw and the probable deviation 28 mw. This probable error introduced by grouping is considered quite acceptable, especially since the value, as pointed out, is on the conservative side regarding positive deviations. Calculations of a completely diversified system would be very laborious.

Regarding Mr. Hicks' comments, it was not the intent to oversimplify the problem by the statement that a once-in-five-years outage would cause a once-in-50-years loss of load. It is recognized that other occurrences can also result in loss of load and that the figures are not directly comparable with the loss-of-load method. For the example given in the paper, evaluation of the forced outages of various frequencies on the basis of variation of daily peak loads with relation to monthly peaks results in an indicated inadequacy of capacity of once in 13 years.

Regarding the questions on grouping raised in the Sauter-Baldwin-Dale discussion, it should first be stated that the scattering of points shown in Fig. 9 is due to the fact that the results are inherently discontinuous and only certain combinations possible. These discontinuities would also result with the usual probability methods. In Fig. 9 there are no points in the interval between 2,000 and 4,000 days. Nevertheless, the question of the magnitude of outages for 3,000 days is significant. It might appear that expectancy of outages of X mw or more may be the solution, but actually the problem of discontinuity remains. The logical and practical solution is found in grouping of the points.

In selecting the width of the group, the following considerations are important; incidentally, they apply to the grouping of any statistical data:

1. Frequencies of magnitudes with insignificant differences should be added together.

2. Gaps with zero frequencies should be eliminated.

3. The grouping should remove irregularities.

4. Since wide grouping results in greater frequencies for a given magnitude of outage, the group width should be the smallest compatible with requirements 2 and 3.

A group width equal to the average size of the components was found to approximate these requirements best.

The effect of grouping width is not as great as it may appear when the magnitude of outage is determined for a given interval instead of intervals for a given outage. For an interval of 10,000 days in the example of the discussion, group widths of 50 and 100 mw result in outages of 425 and 450 mw respectively, less than 6% difference.

Regarding the values of X mw or more in the last column of the table and curve D in the Sauter-Baldwin-Dale discussion, it is true that the probability of X mw or more is the cumulative sum of all events equal to or greater than X, but the frequency of X mw or more is not the sum of the individual frequencies. As a simplified example, consider the partly overlapping outages of two machines. A period of a single outage is followed by a period of a double outage, followed by another single outage period. The frequencies are then two single outages and one double outage, but only one outage of one or more machines.

Lack of space prevents a detailed discussion, but the following will serve as an answer to the Sauter-Baldwin-Dale discussion and to Mr. Hicks: The frequency of one or more outages is equal to the frequency of zero outages. The frequency of two or more outages is the difference between the frequency of exactly one outage and the frequency of one or more outages. Similarly, the frequency of three or more outages is equal to the frequency of exactly two outages minus the frequency of two or more outages, etc.

As Mr. Loane points out, an experimental computer study, made by him and Mr. Watchorn giving a 100-year synthetic record of forced outages, produced results in good agreement with the computations. It would be interesting to make such studies with the improvements in computing procedures as suggested by Mr. Loane.

Several comments refer to the constant total and the peaking factor. With regard to Mr. Miller's question, the constant total is not the sum of the load duration curve, overhaul, and limitations, but is the sum of monthly peak loads, overhaul, and limitations. The peaking factor allows not for the total monthly variation of daily peaks, but only for the upper 10%. The Sauter-Baldwin-Dale discussion points out that this is conservative. Mr. Miller is correct in stating that a constant total is easier to achieve for a large system, especially if it contains a large number of small units, than for a small system with few units. However, in any case an attempt can be made to approach the constant total, and it is the purpose of the stacking factor to make allowance for the deviation from the ideal constant total. For a small system the stacking factor may be quite large.

Mr. Steinberg and Mr. Cook suggest that the margin above the sum of the daily peak loads and scheduled outages could be made constant. This would be an ideal situation. However, it appears impossible to achieve, because overhauling requires outages of the order of weeks and is not flexible enough to compensate for the relatively large variations of daily peak loads within a month. Their comments on the monthly peaking factor imply that it should be based on portions of a day rather than full days. This is equivalent to basing the loss of load method on a hourly load duration curve.

However, from a practical viewpoint, the frequency of inability to carry the load appears more important than the per-cent time of loss of load, especially since a dropped load may not be readily restored during a day even if the system load does drop to the value of the available capacity. This again emphasizes the importance of determining frequencies rather than probabilities as used in the loss-of-load method.

Mr. Loane and the Galloway-Kirchmayer discussion raise questions regarding error in load forecasting. Regarding Fig. 8, the smaller effect of error in load forecast for larger units is apparently caused by the relatively larger per-cent reserve required by larger units. No allowance is made for error in load forecast in the studies of the authors, because of the policy that management decides on the desirable margin, if any, over the recommended reserve, after giving consideration to the specific circumstances and contributing factors in each case rather than making a constant allowance. The reserve recommended to the management is a minimum figure. We agree with Mr. Loane that emergency assistance from interconnections should be based on probability calculations, and we have used this method.

As pointed out in the Galloway-Kirchmayer discussion, the selection of an index of reliability must consider a balance between cost of providing the desired reliability and its worth, but it must include such considerations as customers' good will, permissible frequency of voltage reductions, and other emergency measures. It is therefore not simply an economic problem. The loss of revenue due to loss of load is a small factor in these considerations. It is for this reason that we prefer planning for outages of a given frequency to determining the loss of load.

Mathematical Models for Use in the Simulation of Power Generation Outages

I—Fundamental Considerations

C. J. BALDWIN
MEMBER AIEE

D. P. GAVER
NONMEMBER AIEE

C. H. HOFFMAN
MEMBER AIEE

CONSIDERABLE attention has been given in recent years to the use of probability mathematics in the determination of power system generating reserve requirements. This mathematical approach[1-16] has for the first time permitted quantitative analysis of the question formerly answered by rule of thumb, "How much reserve generating capacity should be installed to maintain adequate service?" The use of probability mathematics to help answer such a question represents a significant step forward in applying the tools of science to the art of system planning. Besides the reserve problem, there are a number of system planning areas still unexplored where mathematical methods can be used to advantage. These may add to the consistency and logic of planning and point the way to planning automation.

What amount of installed reserve required is only one of a number of questions facing system planners today. Other questions have been outlined in an introductory paper[17] and may be summed up in the single query, "What is the optimum plan to follow in expanding system generation and transmission to meet increasing loads?" The answer to this can be found by economic evaluation of different plans; this concept has been presented by numerous authors, and detailed calculations have been made using various approximations and simplications. Now a co-operative investigation is under way to develop new methods of system planning and economic evaluation of alternate plans.[17] The technique under study is that of operational gaming, or system simulation.

Operational Gaming

Expansion patterns designed to investigate various planning policies may be evaluated by comparing present worths of all future revenue requirements[18] for the pattern. To do this the date of each step of the expansion is needed. Ideally, the dates should be determined by a method that recognizes the operating rules of the system, the management policies, and the uncontrollable random events of the future. One method including all of these is operational gaming, or system simulation. It provides solutions to problems with so many variables that they cannot be described by specific mathematical formulas or by single probability equations.

Operational gaming employs a combination of a system analog, Monte Carlo techniques, and simulated human decisions. The system analog is a mathematical and logical model of the system. With it can be represented sequences of system events that might occur in the future. Monte Carlo methods are applied to the model to simulate random occurrences. In the planning problem these might be unit forced outages, deviations of daily load estimates, random variations in daily peak load, and others. Built into the model is the logic of system operation; this is the human element. The model, then, simulates the random events that occur and the human decisions made;

Paper 59-851, recommended by the AIEE Power Generation and System Engineering Committees and approved by the AIEE Technical Operations Department for presentation at the AIEE Summer and Pacific General Meeting and Air Transportation Conference, Seattle, Wash., June 21–26, 1959. Manuscript submitted March 23, 1959; made available for printing April 24, 1959.

C. J. BALDWIN and D. P. GAVER are with the Westinghouse Electric Corporation, East Pittsburgh and Pittsburgh, Pa., respectively; and C. H. HOFFMAN is with the Public Service Electric and Gas Company, Newark, N. J.

therefore the system is operated and planned by model in a manner that closely approaches reality.

With the model actual system events are simulated day after day; if a digital computer is used, this simulation is accomplished at relatively high speed. Evaluation of the risk of losing load determines the dates of each generator addition, thereby permitting the final economic evaluation of the particular pattern of expansion. Finally, pattern comparisons evaluate planning policy changes.

SYSTEM MODEL

The power system model needed for simulation must properly represent both elements of capacity and elements of load. Furthermore, it must be capable of developing on a daily basis both "actual" and forecast system loads, and it must provide for a daily disposition of the generating capacity. Having load and capacity, it must be capable of evaluating the service standard criteria; and it must take corrective action should the criteria not be met. Details of the over-all system model and its use in simulation will be discussed in a future paper.

This paper is concerned only with that mathematical part of the model used to represent one of the random events, power generation outages. Of special interest are new extensions to old ideas that must be made to construct an outage model suitable for use in simulation. This paper will summarize outage models used in the past, present the model needed for simulation, and discuss its significance. A companion paper[19] will show how the outage model has been constructed for an actual power system. These papers form the foundation and justification for techniques used in the gaming process to represent outages. Future papers will present other parts of the system model and,

finally, the planning results obtained by gaming analysis.

Models Used in the Past

The most widely known model for forced outages was first presented by Calabrese.[1] Basically, it is a model to predict only average performance of a group of units during a long period of time. With it and a long-time load-duration curve, one can calculate the fraction of time during which loss of load might be expected to occur during any long future period.

The Calabrese model makes use of the parameter q, the average outage rate of a unit. In words, q is the fraction of a long time during which a unit is on outage, or "down," i.e.,

$$q = \frac{O}{S+O} \qquad (1)$$

where O is the time a unit is on forced outage during a long period t, and S is the time it is in service during the same period. Outages for economy reasons are excluded from both O and S. The q for each member of a group of homogeneous units is estimated by adding the individual outage times to get O and the individual in-service times to get S. If a long enough period t is taken, q may be thought of as representing the probability that any unit of a group will be on forced outage during a future period when it is needed. National surveys have been made to determine the probability q.[21-22]

With q defined for a group of units as just described, it has been used for predicting the average performance of a group over a long period. To do this, coincident outages must be considered.

Suppose there are n units with an average outage rate q. Then the binomial formula gives the probabilities of outages of various combinations:

$$(p+q)^n = p^n + \binom{n}{1}p^{n-1}q + \binom{n}{2}p^{n-2}q^2 +$$
$$\dots + \binom{n}{m}p^{n-m}q^m + \dots + q^n \quad (2)$$

where

$$\binom{n}{m} = \frac{n!}{m!(n-m)!}$$
$$p = 1 - q$$

and m is the number of units simultaneously on forced outage. Once the probabilities of coincident outages are known, they can be added successively and related to capacity to give the probabilities of outages of capacity C_k or more.

Capacity outages are significant from a reserve viewpoint only if they cause loss of load. The Calabrese model recognizes this, in a way, by relating the probabilities to a load-duration curve and computing a probability of loss of load. Fig. 1 illustrates how an outage C_k with probability of equal or greater outage P_k can be scaled on a load-duration curve to find the interval of time Δ_k that load loss might occur. Summing the $P_k\Delta_k$'s for each interval gives the well-known probability of loss of load. This is usually interpreted as the total days of load loss in a long period of time and is called a risk of shortage. One day in 10 years (0.000274) is a typical accepted risk computed in this way.

The Calabrese model has been widely used; it provided the first quantitative analysis of the effect of forced outages on reserve requirements. It has inspired additional work, simplifications of com-

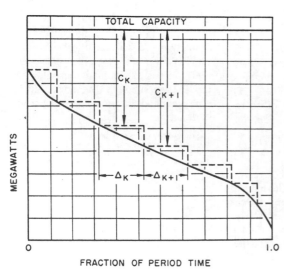

Fig. 1 (left). Use of the load-duration curve to calculate the probability of loss of load

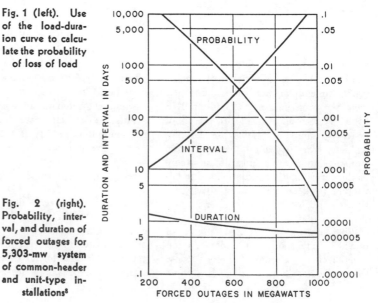

Fig. 2 (right). Probability, interval, and duration of forced outages for 5,303-mw system of common-header and unit-type installations[8]

putation, and various approximations. Arnoff and Chambers developed one approximation in which risk is analyzed monthly instead of annually.[12] However, all work based on it is restricted to the basic limitations of the model, several of which are cited in the original paper.[1] The limitations include the following:

1. The probability q yields no information about the duration of individual outages. Thus an outage rate of 0.02 represents both a 2-day outage in a 100-day period and a 10-day outage in a 500-day period.

2. Use of a load-duration curve in evaluating loss of load arranges the days in such a way that, should duration information be available, it could not be correlated directly with loads on successive days for analyzing the risk of load loss.

3. The model assumes in effect that the average thing happens every day and, for this reason, is good ·only in evaluating very long periods of time. It does not describe properly risks during shorter periods within the long period. Since the model evaluates only average performance, the dispersions of actual experience about average are not considered.

4. The model does not recognize the variability in exposure of individual units in a group.

Each of these limitations will be discussed in turn.

Outage Duration

The outage rate q approximates the fraction of a long time t during which a unit is on forced outage or down. It gives no information about outage duration. Halperin and Adler[8] recognized this deficiency in the Calabrese model and correctly observed that a long outage might have more serious consequences than a short outage repesented by the same outage rate q. They developed a new model to include the effect of duration.

Halperin and Adler have shown that that the average duration in days of r simultaneous outages in a uniform group is

$$t_r = \frac{t(1-q)}{r+q(n-2r)} \qquad (3)$$

where t is the average duration of individual outages and n is the number of units in the system. Their outage model uses this average duration, computed for the nonuniform group of units, with the group probability based on outage rate q and equation 2, to compute an average interval of forced outages in the form shown in Fig. 2. Here the interval is the duration divided by the probability.

With outages represented as in Fig. 2, Halperin and Adler establish the system forced-outage reserve requirements for

different intervals, e.g., 805 mw (megawatts) for an interval of once in 5 years. While their model does include duration and, as such, represents a significant contribution, it has the other limitations of the Calabrese model. This, in part, is because it bases its extensions on q and its restriction to average performance over a very long period.

Load-Duration Curve

The Calabrese model and others evaluate risk using a load-duration curve in which peaks are represented in descending order rather than chronologically. The consequences of specific outage durations cannot be evaluated with load-duration curves because capacity and load cannot be related for specific time intervals. Miller has recognized this and has taken some steps to correct it.[9]

Halperin and Adler recognized this deficiency, also, with their introduction of a monthly peaking factor in the evaluation of reserve requirements. In essence, this factor is a megawatt credit, reducing the reserve to account for the fact that the duration of a large multiple forced outage is unlikely to coincide exactly with a succession of high peak load days. Their analysis of load data showed that for 90% of the days in a month on their system the peak load might be expected to be 65 mw below the monthly peak. After noting that, on the average, large multiple forced outages lasted only about one day, they decided not to protect for 10% of the days on the average and reduced their reserve by 65 mw. This means that if their reserve is sized for once-in-5-year intervals of forced outage, only 10% would result in load loss. This is a risk of shortage of one day in 50 years.

While the Halperin-Alder model recognizes the difficulty of correlating outage duration with load and allows for it with a correction factor, it does not permit a direct correlation. It is wholly dependent on averages over long periods of time.

Dispersions of Actual Performance About Average

All models described in the literature assume in effect that the average thing happens every day. This is basically because they all use the average outage rate q which restricts them to consideration of long time periods. To understand this the meaning of the number q requires some discussion.

Equation 1 may be reworded as

$$q = \frac{\text{sum of down-period lengths}}{\text{sum of down-period lengths} + \text{sum of up-period lengths}} \qquad (4)$$

where down-period lengths do not include periods a unit is down for economy reasons. If numerator and denominator are divided by the number of down periods (or up periods, which tend to be equal as time approaches infinity),

$$q = \frac{\text{average down-period length}}{\text{average down-period length} + \text{average up-period length}} \qquad (5)$$

To understand the meaning of q, note first that q approximates the fraction of a long time t during which a unit is down, omitting economy outages. This approximation gets better, basic conditions remaining the same, as t increases. It gets better because deviations of the fraction (total time out to total time of observation) from the true fraction, which is based upon an infinite observation time, are reduced roughly inversely to the square root of the total observation time t. However, this does not mean that the total down time during a long observation time t tends to stabilize if deviations are measured in terms of days and not in per cent. Suppose many histories of unit availability are observed for t days. Then, each such history will have a different total number of days spent down. A more precise result can be obtained by using the methods of mathematical statistics. If t is reasonably long (equal or greater than about 25 times the sum of an average up period plus an average down-period duration), then the total number of days spent down during a history of length t will be approximately normally distributed with a mean qt and standard deviation proportional to $\sqrt{t}$. This means that if observation time t is increased, average total down time is increased, as are fluctuations of total down time.

The increase in average total outage time and its fluctuations are clearly in conformity with intuition, since longer histories permit more diverse happenings. If histories are obtained for several units, n in number, then the total number of unit days on outage for n simultaneous histories is normally distributed with mean qnt and standard deviation proportional to $\sqrt{nt}$. Total unit days spent down fluctuate more and more as exposure, measured by nt, increases.

Models used in the past have made no allowance for the fluctuations of forced outages, in terms of magnitude, frequency, or duration, about the average performance. The amount of possible fluctuation and its significance have not been investigated. Using just the average rate q in the usual manner permits no confidence limits to be placed on

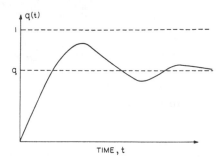

Fig. 3. Outage probability as a function of time

answers derived therefrom, because nothing is known about deviations from average performance or about the effects of errors in estimation of outage rates.

The situation can be likened to a study involving depreciation accounting in which all carrying charges are based only on average lives with no regard for the different dispersions. Just as equipment mortality may be described by an average life and characteristic dispersion, so forced outage performance can be described. However, existing models of forced outages do not permit such an evaluation. A new model is needed to include this effect.

Results of significance to installed reserve policy have been obtained by interpreting q as a probability. Again the meaning of this probability deserves some discussion. Suppose a unit is found in or is put into an in-service (up) condition at an initial time, and a history is allowed to develop. For at least a very short period after it is first observed, the unit will continue in an up state, not counting start-up failures which effectively prolong down times. The probability that it remains up decreases as time elapses. Then comes an outage, and a down period. Knowing that the unit is initially up, then, means that for at least a short period it will very likely continue up, and the probability of finding it down at the end of the period is small. If many histories are observed, all beginning with the unit initially up, one can count the number of those for which the unit is down at any particular time $t=t_a$ and obtain an estimate of the probability that the unit will be down at t_a. Letting the particular time t_a increase, the resulting probability varies in time, eventually approaching q as a limit as t_a approaches infinity. Fig. 3 shows a possible shape of the dependence of outage probability on time.

The number q is thus the probability that, if the unit is examined at the end of a long period, it will be found on outage.

This long-run probability is independent of the initial state of the unit; q is also equal to the probability that, if the unit is examined at a random (uniformly distributed) time during a long period, it will be found on outage.

In summary, then, q is the probability that a unit will *be* down at some remote moment either fixed or random, after an initial instant. It is not the probability that a unit will *go* down tomorrow or next week if it is known to be up today or this week. It does not provide information, given certain initial conditions, about availability for only moderately distant futures. Hence it cannot be used directly to establish policies for a short time in the future.

VARIABILITY IN UNIT EXPOSURE

A number of the models used for forced outages do not take into account the variability over a year or a month in the exposure of the unit to outage. There is no exposure to outage durings periods when the unit is already down for economy or maintenance reasons. By using the outage rate q, which is the average outage duration over the average demand time duration, they do recognize that failure is of importance only, and can occur only, when the unit is needed and run. However, error is introduced when probabilities P_k (see Fig. 1) computed from equation 2 are used with a load-duration curve to calculate risk of shortage.

The probabilities P_k are usually computed for the entire group of units. Maintenance is considered in a gross way, often on the constant-risk principle, by adjusting the load-duration curve upwards for maintenance "load" in the off-peak periods. This procedure fails to recognize the fact that during one period, Δ_k, certain units are on maintenance and therefore not exposed; hence they should not be included in equation 2. Likewise, during a later period, Δ_{k-1}, another set of units is unavailable, and a different binomial expansion with different P_k's results. Actually, a different expansion of equation 2 is needed for each combination of units available and running. This was recognized by Miller[9] who assigned maintenance outages at least to specific months and recomputed the P_k's for each unit mix.

The error introduced by neglecting changes in P_k with change in unit mix is compounded by the fact that the Δ_k's on the load-duration curve do not represent continuous time periods at all. The adjustment of the load-duration curve for maintenance is quite an approximation in itself.

A final fact that has been overlooked concerning unit exposure is that exposure is a function of the daily load forecast. The units that are run daily for the peak, and therefore exposed to outage, are only those which are both available (not on maintenance outage) and needed. The needed units are determined by the system operator's forecast of the daily peak. Since forced outages are a function of exposure, and exposure is a function of the daily load forecast, then any study of the effect of forced outages perhaps should include as a parameter the daily load forecast. At least its relevance should be evaluated. No existing outage models permit consideration of this aspect.

In summary, a number of forced-outage models have been presented in the literature. All of them are based on certain assumptions and simplifications. It is quite possible that satisfactory results can be obtained using these relatively simple models. However, in developing an over-all system model for the study of planning policy, a more exact forced-outage model was formulated to permit evaluating the improvements possible with more accuracy and to help develop more valid simplifications.

A New Model for Outages

The preceding section discussed forced-outage models used in the past and some of their assumptions and limitations. This section describes a new way of setting up the basic probability model for forced outages. It relates the availability of a unit at an arbitrary time to the distribution functions of down times and up times of the unit. These distributions can be estimated from data usually kept by electric utilities. Discussion of such estimation and statistical examples are comtained in Part II of this paper.[19] With the use of these distribution functions, outage histories for single units and groups of units can be simulated on a digital computer and may be used to evaluate the consequences of decision. In some cases analytical calculations can be made that eliminate the need for much of this computation and increase of precision of the results. Some of these calculations are described in the Appendixes; they include an account of the transient or time behavior of a unit's outage probability. Other analytical extensions are possible.

DESCRIPTION OF THE MODEL

For the purpose of describing outages, the performance of a single generation unit may be described in terms of a sequence of periods of time: up periods that alternate with down periods. During an up period the unit is available to meet the load, while during a down period the unit is undergoing repair and is unavailable. Units may also be idle for economy reasons, but these economy outages need not be included in the forced-outage model. Economy outages are considered in detail in the gaming process. An initial simplification will be that a single unit has a capacity that is either entirely available or entirely unavailable. This latter assumption will be relaxed to allow a unit to be "half out."[19] Thus an available capacity history of the unit can be graphically expressed as a function of the time t that has elapsed since some initial instant; see Fig. 4.

A simple example similar to Fig. 4 might be a unit that is available and put into operation on January 1, from which date time is measured. It is up through January 18, or 18 days, goes down on January 19 for 3 days, is up again on January 22, staying up until February 16, etc. Such a graph is said to represent an "outage history." In more technical language, an outage history is a random function of time. The collection of all possible such histories is called a stochastic (chance, or random) process.

In practice such an outage history is very unlikely to be repeated exactly. Even if the same unit is put into service on January 1 of the following year, its first outage will more than likely not occur on January 19, and the subsequent periods will not repeat themselves. This means that if $t_d(i)$, $(i=1,2,3,\ldots)$ denotes the time of the ith outage, and $t_u(i)$, $(i=1,2,3,\ldots)$ denotes the time of the ith return to service (outage termination), these times will exhibit considerable variability on different occasions, presenting a different detailed history and bearing out the adage that history never—precisely—repeats itself. In order to describe the variation observed, i.e., to assign probabilities to various possible histories, the following basic model may be used:

1. For a single unit the lengths of successive up periods are statistically independent random quantities (technically, random variables) all coming from the same statistical population.

2. For a single unit the lengths of successive down periods are independent random quantities all coming from the same statistical population. The up-period lengths and down-period lengths also are assumed independent of each other.

The concepts introduced are discussed in reference 23. A statistical investigation of this model is presented in Part II.[19]

The preceding paragraphs describe a

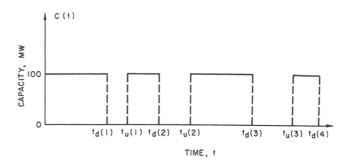

Fig. 4. Single-unit available capacity as a function of time (single-unit outage history)

new kind of model for forced outages. To specify the model fully, it is necessary to set down explicitly the probabilities to be associated with the populations of 1 and 2, above. These are

$$A(x) = Pr(T_u \leq x) \qquad (6)$$

$$B(x) = Pr(T_d \leq x) \qquad (7)$$

where T_u is the duration of a random up period and T_d is the duration of a random down period. $A(x)$ is the distribution function of the up periods; i.e., if x is any positive number, ordinarily a number of days, $A(x)$ is the probability that an up period, chosen at random, has a duration less than or equal to x. $B(x)$ is the distribution function of the down periods. The functions $A(x)$ and $B(x)$ are monotone nondecreasing functions of x, with $A(0)=B(0)=0$, and $A(\infty)=1$. Reference 24 provides further explanation. Examples of these functions have been estimated from actual data of the Public Service Electric and Gas Company system.[19]

THE MODEL AS A BASIS FOR SIMULATION

Given estimates of the distribution functions of up-period duration and down-period duration, the new model described provides a basis for simulating future outage histories. Constructed as it is, it permits inclusion of the many variables, particularly human decisions, that influence outages.

An outage history for each unit in a utility system can be created as follows: Supposing that the unit is up initially, draw an up-period duration from $A(x)$ by well-known techniques used to obtain random numbers with a prescribed distribution. Call the number obtained $T_u'(1)$. Then $T_u'(1)$ is the first realized value of the random variable T_u representing the length of an up period. This means that at time $t_d(1) = T_u'(1)$ the unit goes down. Next draw a down-period length from $B(x)$; call its value $T_d'(1)$. Then, for this history, $t_u(1) = T_u'(1)+T_d'(1)$; $t_d(2) = T_u'(1)+T_d'(1)+T_u'(2)$, etc. Continue this procedure to generate the successive times $t_u(i)$

and $t_d(i)$; these times specify a function of the form shown in Fig. 4. Of course, if the numbers $T_u'(i)$ and $T_d'(i)$ were drawn again, they would be different; and the resulting function describing available capacity would change. A number of such histories can be created for each unit in the system. If these histories are adjusted for economy outages, scheduled outages, and periodic overhauls and compared to simulated load histories, they then provide a basis for estimating the frequency of load loss, the average size and distribution of load loss, and other quantities of operational interest. A detailed simulation program including the forced-outage model described will be discussed in a further paper.

THE MODEL SIMPLIFIED: REASONING ADOPTED

Well-known probability methods[25] permit simplification of the basic model by adopting the following reasoning, often justified in practice:[19]

1. If a unit is up at time t, the probability that it goes down in the next short time interval of length dt is $a\,dt$, where a is independent of the elapsed length of the current up period; in general, a could depend upon the total elapsed time since the beginning of the history and is called the failure rate. If a is constant, it is called the failure rate parameter. Thus $a\,dt$ is the probability that an operating unit will go down at time t. This event must be distinguished from a unit *being down* at t, the probability of which is q in the long run.

2. If a unit is down at time t, the probability that it comes up in the next short time interval of length dt is $b\,dt$, b being independent of the time that has elapsed during the current down period; like a, b could depend upon the total elapsed time. The parameter b may be called the repair rate. If several units are creating histories at the same time, they may each be characterized by different values of a and b. It is convenient to assume at first that each of the units is perfectly independent of the other in the sense that, for

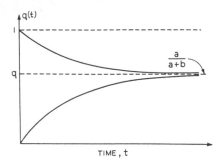

Fig. 5. Outage probability for exponential distributions

example, an outage of one unit does not alter the probability that another will either go down if up, or up if down. However, using the model described it would be possible to consider interaction between units caused by maintenance saturation.[26]

The discussions in paragraphs 1 and 2 mean that if a and b are constants, independent of time, the distribution of up-period duration and down-period duration are exponential:

$$A(x) = 1 - \epsilon^{-ax} \qquad (8)$$

$$B(x) = 1 - \epsilon^{-bx} \qquad (9)$$

The relationships between paragraphs 1 and 2 and equations 8 and 9 are given in Appendix I.

PROPERTIES

Some special properties of the exponential distribution are of interest:

1. The exponential distribution is memoryless. This is simply another way of saying that the failure rate does not depend upon the time the unit has been in service, where time refers either to the total time the unit has been in use or the total time since the last failure was repaired. Specifically, if the duration of an up period T_u is known to have reached x days, then the probability that it lasts for at least another y days is ϵ^{-ay}, i.e., symbolically:

$$Pr[T_u > (x+y) \mid T_u \geq x] = \epsilon^{-ay} \qquad (10)$$

2. The average (mean, expected value) of an exponentially distributed random variable is equal to the (rate parameter)$^{-1}$. If $E(T_u)$ is the average (expected value) of an up-period duration, then

$$E(T_u) = \int_0^\infty t\epsilon^{-at}a\,dt = \frac{1}{a} \qquad (11)$$

Thus the failure rate is the inverse of the mean time to failure, or the mean duration of an up period.

3. The standard deviation (root-mean-square error) of an exponentially distributed random variable is equal to the (rate parameter)$^{-1}$

$$\sigma(T_u) = \left\{ E(T_u^2) - [E(T_u)]^2 \right\}^{1/2}$$

$$= \sqrt{\frac{2}{a^2} - \frac{1}{a^2}} = \frac{1}{a} \qquad (12)$$

Properties 2 and 3 provide a convenient rough check of the supposition that a set of data comes from an exponential population.[19]

Paragraph 1 in the section "Reasoning," with a constant, is based on the chance failure hypothesis. This terminology is suggested by the fact that the probability of unit failure in a given time interval is independent of the length of time that it has operated prior to the beginning of the interval. Therefore it is also presumably independent of the accumulated wear the unit has received. Some kind of randomly occurring catastrophe "causes" failure. An alternative hypothesis, not adopted in this model, is that of dominant wear-out failure: sheer length of service finally causes failure, as when a flashlight battery wears out. In practice some failure mechanisms are not as simple as the foregoing two terms would imply. Failures, for example, may very likely be caused either by wear-out or catastrophe, whichever occurs first. Failures may be contagious in the sense that if one part gives way, several others may be weakened and will themselves fail before they normally would. An understanding of failure reasons and mechanisms cannot be obtained by simple examination of the distribution function $A(x)$; a variety of basically different causes could produce a distribution function of the same form. Only a combination of engineering study and sound statistical analysis will lead to a good understanding of likely failure mechanisms, to improvements in reliability, and incidentally, to an explanation of up-period and down-period distributions. In the simplified outage model $A(x)$ and $B(x)$ are assumed to be exponential. Power system data show that this simplification is often quite reasonable,[19] and the mathematical convenience introduced is considerable.

If the distributions $A(x)$ and $B(x)$ are exponential, a simple derivation (see Appendix II) will give the probability $q(t)$ that a single unit is down at elapsed time t after an initial instant:

$$q(t) = q(0)\epsilon^{-(a+b)t} + \frac{a}{a+b}(1 - \epsilon^{-(a+b)t}) \qquad (13)$$

where a is the failure rate, b the repair rate, and $q(0)$ the probability that the unit is down at the initial instant; if the unit is known to be up at $t=0$, then

$q(0) = 0$. Time-varying failure and repair rates, perhaps occurring during the shakedown period of the unit, can be included as shown in Appendix II. Group availability is handled by methods given in Appendix III.

Equation 13 supplies quantitative information about unit availability in both the short run and the long run. In the long run

$$\lim_{t \to \infty} q(t) = \frac{a}{a+b} = \frac{E(T_d)}{E(T_d)+E(T_u)} \qquad (14)$$

the last expression coming from equation 11. Thus the long-run probability agrees with equation 5 and the formula of Calabrese. Equation 14 for q holds true even when up-period and down-period distributions are not exponential. This may be shown by methods of renewal theory.[27] In the short run

$$\lim_{t \to 0} q(t) = q(0) \qquad (15)$$

Thus for a short time after an initial instant, the unit is very likely to be in the same state that it was initially.

From equation 13 $q(t)$ approaches the long-run probability q exponentially and monotonically, i.e., without oscillation. An oscillating approach, as shown in Fig. 3, is possible for different models but not when both up-period and down-period distributions are exponential. Thus if the unit is up initially, $q(0) = 0$, the probability that it is down at time t will appear as the lower curve in Fig. 5. Likewise, if $q(0) = 1$, the unit is down initially, and the probability that it is down at time t is represented by the upper curve.

Equation 13 shows that the long-run probability is determined by the ratio of failure rate to repair rate, a/b, the rate of approach to the long-run value is determined by the sum of these two rates, $a+b$. Thus two units with equal long-run probabilities may be decidedly different from the point of view of short-run availability. Since a great deal of maintenance planning is conducted on a short-run basis, a week to a month ahead, information of the type furnished by equation 13 easily can be used to sharpen the decisions that are made by giving a better assessment of the short-run risks involved.

Thus the model illustrated in Fig. 4 can be simplified by assuming exponential distribution functions which are substantiated by utility data. However the basic model is the same, one which allows the construction of outage histories of individual units or groups, with full recognition of scheduled maintenance, annual overhauls, and other human decisions affecting the outage performance of the system. Any time unit can be selected for the outage history so that even daily comparisons of capacity and load are possible.

Conclusions

1. Probability methods have been used to describe forced outages of electric generation equipment for more than 10 years. The usefulness of this approach to the description of outage behavior lies in the possibility of establishing maintenance policies, rationalizing reserve capacity requirements, and evaluating long-term capital expansion plans. In general, it has the potentiality of sharpening utility decisions by bettering the understanding of a major area of uncertainty.

2. In general, forced-outage models used in the past have been concerned with long-time average performance. Outage durations have been included in a few models but have not been related in detail to specific load periods. Variations in unit exposure because of maintenance have usually been overlooked. Finally, the dispersion of actual performance about average has not been investigated nor evaluated as to significance.

3. While existing models may be adequate to describe forced outages, there is no way to evaluate their adequacy; there is no standard of comparison. A new detailed model is needed that includes more of the relevant factors.

4. A new model for forced outages has been developed for use in a general system simulation program. It provides the basis for the development of outage histories that include both random events and the consequences of human decisions affecting outages. The model is versatile enough to allow daily comparisons of available capacity and peak load should they be desired.

Nomenclature

a = failure rate, per day
a_n = nonmature unit failure rate, per day
a_m = mature unit failure rate, per day
$A(x)$ = probability distribution function of up-time durations
b = repair rate, per day
$B(x)$ = probability distribution function of down-time durations
C_k = capacity of kth unit outage, mw
E = expected value (average) of argument
i, k, m, n, N = numerics (integers)
O = time a unit is on forced outage during a long period, days
$p = 1-q$
Pr = probability of
P_k = probability of an outage equal to or greater than C_k

q = probability that a unit will be down at some remote moment
r = number of simultaneous outages
$R(t)$ = function describing time dependence of failure and repair rates
S = time a unit is in service during a long period, days
t = time, days
$t_d(i)$ = time of ith outage, days
$t_u(i)$ = time of ith return to service, days
t_r = average duration of r simultaneous outages, days
T_d = down-period duration, days
T_u = up-period duration, days
T_d' = realized value of the random variable T_d, days
T_u' = realized value of the random variable T_u, days
x, y = possible values of random durations, days
z = variable of integration
α = nonmature unit rate of approach to maturity, per day
Δ_k = load-duration curve time interval, per unit
$\sigma(T_u)$ = standard deviation of up-period duration, days

Appendix I. Derivation of the Exponential Distribution

Equation 8 follows from the reasoning under paragraph 1 in the section "Reasoning Adopted" by observing that a unit put into service at time t is still up at time $t+x+dx$ if: 1. it survives to time $t+x$ without going down, and 2. it survives for a further time of length dx. Since $1-A(x)$ is the probability that the unit survives for a period of length x, then, because of the independence of these two events

$$1-A(x+dx) = [1-A(x)](1-adx) \qquad (16)$$

Multiplying out, collecting terms, dividing by dx, and taking the limit as $dx \to 0$, yields

$$\frac{dA}{dx} = [1-A(x)]a \qquad (17)$$

This can be solved to give equation 8; equation 9 is obtained in the same way.

Appendix II. Derivation of q(t)

The term $q(t)$ means the probability that a unit is down at elapsed time t after an initial instant. For the unit to be down at time $t+dt$ it can either 1. have been down at time t and remained down during dt, or 2. have been up at t and gone down during dt. There are other possibilities but, as dt shrinks to zero, their probabilities become negligible.

The probability of event 1 in the foregoing paragraph is just $q(t)(1-bdt)$; and that of event 2, $[1-q(t)]adt$. Adding the probabilities of these mutually exclusive events yields $q(t+dt) = q(t)(1-bdt) + [1-q(t)]adt$. Multiplying out, collecting terms, dividing by dt, and taking the limit as $dt \to 0$, a differential equation is obtained:

$$\frac{dq}{dt} = -(a+b)q(t)+a \qquad (18)$$

The most straightforward approach to

the solution is by way of the integrating factor:

$$\frac{dq}{dt}+(a+b)q(t) = \frac{d}{dt}[q(t)\epsilon^{(a+b)t}]\epsilon^{-(a+b)t}$$

Thus

$$\frac{d}{dt}[q(t)\epsilon^{(a+b)t}] = a\epsilon^{(a+b)t}$$

and, upon integrating both sides, equation 13 results.

Time Dependence of Failure and Repair Rates

If failure and repair rates actually depend upon the time since the beginning of the history, as they may in paragraphs 1 and 2 in the section "Reasoning Adopted," equation 13 is replaced by

$$q(t) = q(0)\epsilon^{-R(t)} + \epsilon^{-R(t)}\int_0^t a(z)\epsilon^{R(z)}dz \qquad (19)$$

where

$$R(t) = \int_0^t [a(x)+b(x)]dx \qquad (20)$$

Dependence of a and b upon t, where t is the time since the particular unit was first put into service, is a reasonable description of the "shakedown period" for a new unit. During such a period, failure rate would first tend to be high and repair rate low, these values gradually falling and rising respectively to mature values. Explicitly, for example, a dependence such as the following might be appropriate:

$$a(t) = (a_n-a_m)\epsilon^{-\alpha t}+a_m \quad a_n>a_m \qquad (21)$$

where a_n is the new unit failure rate, a_m is the mature unit failure rate, and α is the rate of approach to maturity. A similar expression may be written for $b(t)$. Time is measured from the moment of installation of the unit.

Appendix III. Group Availability

Using equation 13 derived for $q(t)$ and the assumed independence of units, one may write immediately the probability that, from a group of N units, $k \leq N$ are down at a time t after an initial instant. Suppose that all units have identical failure and repair rates and that initially they are all up, then this probability is given by the binomial expression

$$\binom{N}{k}[q(t)]^k[1-q(t)]^{N-k} \qquad (22)$$

where

$$\binom{N}{k} = \frac{N!}{k!(n-k)!}$$

and

$$q(t) = \frac{a}{a+b}(1-\epsilon^{-(a+b)t})$$

Equation 22 can be used to furnish the probability distribution of total capacity available from a group at time t, by either straightforward binomial methods or approximate methods.

References

1. GENERATING RESERVE CAPACITY DETERMINED BY THE PROBABILITY METHOD, G. Calabrese. *AIEE Transactions*, vol. 66, 1947, pp. 1439–50.

2. PROBABILITY METHODS APPLIED TO GENERATING CAPACITY PROBLEMS OF A COMBINED HYDRO AND STEAM SYSTEM, E. S. Loane, C. W. Watchorn. *Ibid.*, pp. 1645–57.

3. CALCULATING PROBABILITY OF GENERATING CAPACITY OUTAGES, W. J. Lyman. *Ibid.*, pp. 1471–77.

4. OUTAGE EXPECTANCY AS A BASIS FOR GENERATOR RESERVE, H. P. Seelye. *Ibid.*, pp. 1483–88.

5. A CONVENIENT METHOD FOR DETERMINING GENERATOR RESERVE, H. P. Seelye. *Ibid.*, vol. 68, pt. II, 1949, pp. 1317–20.

6. DETERMINATION OF RESERVE CAPACITY BY THE PROBABILITY METHOD, G. Calabrese. *Ibid.*, vol. 69, pt. II, 1950, pp. 1681–89.

7. ELEMENTS OF SYSTEM CAPACITY REQUIREMENTS, C. W. Watchorn. *Ibid.*, vol. 70, pt. II, 1951, pp. 1163–85.

8. DETERMINATION OF RESERVE-GENERATING CAPABILITY, H. Halperin, H. A. Adler. *Ibid.*, pt. III (*Power Apparatus and Systems*), vol. 77, Aug. 1958, pp. 530–44.

9. DETAILS OF OUTAGE PROBABILITY CALCULATIONS, A. L. Miller. *Ibid.*, pp. 551–57.

10. PROBABILITY CALCULATIONS FOR SYSTEM GENERATION RESERVES, C. Kist, G. J. Thomas. *Ibid.*, pp. 515–20.

11. A SIMPLIFIED BASIS FOR APPLYING PROBABILITY METHODS TO THE DETERMINATION OF INSTALLED GENERATING CAPACITY REQUIREMENTS, C. W. Watchorn. *Ibid.*, vol. 76, Oct. 1957, pp. 829–32.

12. OPERATIONS RESEARCH DETERMINATION OF GENERATION RESERVES, E. L. Arnoff, J. C. Chambers. *Ibid.*, June 1957, pp. 316–28.

13. GENERATOR UNIT SIZE STUDY FOR THE DAYTON POWER AND LIGHT COMPANY, W. J. Pitcher, L. K. Kirchmayer, A. G. Mellor, H. O. Simmons. *Ibid.*, vol. 77, Aug. 1958, pp. 558–63.

14. DETERMINATION OF RESERVE AND INTERCONNECTION REQUIREMENTS, H. D. Limmer. *Ibid.*, pp. 544–50.

15. FACTORS INFLUENCING THE TREND TOWARDS LARGER TURBINE-GENERATOR UNIT SIZES, J. K. Dillard, C. J. Baldwin. *Bulletin Scientifique AIM* (Association des ingénieurs electriciens, Sortis de l'institut electrotechnique Montefiore), Liege, Belgium, no. 9, Sept. 1958, pp. 899–928, 945–48.

16. DETERMINATION OF GENERATOR STAND-BY RESERVE REQUIREMENTS, H. T. Strandrud. *AIEE Transactions*, vol. 70, pt. I, 1951, pp. 179–88.

17. AN INTRODUCTION TO THE STUDY OF SYSTEM PLANNING BY OPERATIONAL GAMING MODELS, J. K. Dillard, H. K. Sels. *Ibid.*, see pp. 1284–90 of this issue.

18. THE CRITERION OF ECONOMIC CHOICE, P. H. Jeynes, L. Van Nimwegen. *Ibid.*, pt. III (*Power Apparatus and Systems*), vol. 77, Aug. 1958, pp. 606–32.

19. MATHEMATICAL MODELS FOR USE IN SIMULATION OF POWER GENERATION OUTAGES—II.

POWER SYSTEM FORCED OUTAGE DISTRIBUTIONS, C. J. Baldwin, J. E. Billings, D. P. Gaver, C. H. Hoffman. *Ibid.*, see pp. 1258–72 of this issue.

20. OUTAGE RATES OF STEAM TURBINES AND BOILERS AND OF HYDRO UNITS, AIEE Committee Report. *Ibid.*, vol. 68, pt. I, 1949, pp. 450–57.

21. FORCED OUTAGE RATES OF HIGH-PRESSURE STEAM TURBINES AND BOILERS, AIEE Committee Report. *Ibid.*, pt. III-B (*Power Apparatus and Systems*), vol. 73, Dec. 1954, pp. 1438–42.

22. FORCED OUTAGE RATES OF HIGH-PRESSURE STEAM TURBINES AND BOILERS, AIEE Joint Subcommittee Report. *Ibid.*, vol. 76, June 1957, pp. 338–43.

23. MATHEMATICAL METHODS OF STATISTICS (book), H. Cramer. Princeton University Press, Princeton, N. J., 1946, p. 144.

24. *Ibid.*, p. 57.

25. AN INTRODUCTION TO PROBABILITY THEORY AND ITS APPLICATIONS (book), W. Feller. John Wiley & Sons, Inc., New York, N. Y., 1950, p. 363.

26. *Ibid.*, p. 416 ff.

27. SOME OPERATIONAL CONSEQUENCES OF SYSTEM UNRELIABILITY, D. P. Gaver. *Proceedings*, Working Conference on the Theory of Reliability, New York University, New York, N. Y., 1957, pp. 41–45.

Application of Probability Methods to Generating Capacity Problems

AIEE COMMITTEE REPORT

PROGRESS has been made in the development and application of probability methods to generating capacity problems to the point where some review of the subject appears worth while. Although the subject is now at least 25 years old, it is still doubtful that it has reached maturity; and it is certain that only recently has its significance and worth been appreciated by large segments of the power industry.

It is difficult, of course, to say with assurance when many useful ideas were first suggested, for such suggestions were not always recorded or preserved in published form. This may be true of the general application of probability to capacity problems, and it is certainly true of some of the details of that application. Furthermore, in giving credit for early use of probability methods, it has been necessary to limit this review to domestic publications, although some early dated foreign references do appear in the references. Lyman, of Duquesne Light Company, suggested that probability be applied to generating capacity and other reserve problems as early as 1933.[1] About the same time Smith, of Public Service Electric and Gas Company, made similar suggestions in a 1933 McGraw Prize paper, submitted to the Edison Electric Institute (EEI). At least one other discussion appeared during this early period;[5] but these several papers seemed to be ignored for a long time.

The first Institute paper on the subject was delivered at Toronto in 1941 by Forbes and Bellows of Consolidated Edison Company of New York, Inc. (Con Ed). Unfortunately, this was considered a conference paper and was never published. Actually, the paper and its discussions set forth the basis for many of the procedures and assumptions which we are still using and improving today.

During the early 40's, outage data and probability applications were discussed in various meetings of EEI committees, and interest in the subject was slowly developed.

A group of papers, presented at Chicago in 1947 and published in the AIEE *Transactions*,[6-9] was the first visible evidence of an active Institute interest in the subject. The Subcommittee on Application of Probability Methods was organized soon thereafter, with G. O. Calabrese as chairman. Its first meeting was held in New York on June 11, 1948.

Progress had indeed been slow. It would have been even slower, had not the basic data on forced outages been collected by other agencies; first by Con Ed and later by several committees of the EEI. Early analyses of these data were made by Doane of Con Ed. Such information was not generally available; and one of the first tasks undertaken by the Subcommittee on Application of Probability Methods was the collection and publication of the available forced-outage data. The first such publication appeared in 1949.[10] It was followed by two additional reports on outage experience in 1954[11] and in 1957.[12]

The first and simplest criterion of reliability was the computed probability that the outage of generating capacity would exceed the reserve available at the time of peak load. It was soon realized, however, that although this was a useful step in analysis of reserve problems, it could not anwer all the questions that were being asked. What were the effects on reliability of daily and seasonal load shapes, of maintenance outages, of interconnections, of uncertainties in load forecasts, and of river flow and storage where hydroelectric capacity was involved? All these factors had been suggested and discussed by Calabrese, Watchorn, and others prior to 1947. There were differences of opinion, naturally, as to which of these factors should be recognized, how the recognition should be made, and how reliability should be measured. It is this last question with which we are primarily concerned here.

The 1947 group of papers presented several different criteria of reliability. Seelye[8] determined by computation the interval in years between outages of various magnitudes. Lyman[7] presented a short-cut method of computing probability of outage of any given magnitude. Several authors (Calabrese,[6] Loane, and Watchorn[9]) combined probability of outage with peak durations to determine the probability (per cent of time, or days per year) that the load might exceed the available capacity. Calabrese extended his computations to the determination of the kilowatt-hours (kw-hr) of load that might be interrupted; and then, to provide an index, he related the probable loss of

Paper 60-1185, recommended by the AIEE System Engineering Committee and approved by the AIEE Technical Operations · Department for presentation at the AIEE Fall General Meeting, Chicago, Ill., October 9–14, 1960. Manuscript submitted November 2, 1959; made available for printing August 2, 1960.

The personnel of the Working Group, AIEE Subcommittee on Application of Probability Methods of the System Engineering Committee are: L. K. KIRCHMAYER, *chairman;* H. A. Adler, A. K. Falk, C. D. Galloway, K. L. Hicks, C. Kist, H. D. Limmer, E. S. Laone, C. W. Watchorn, and H. S. Worcester.

Reprinted with permission from *AIEE Trans. Power Apparatus Syst.*, vol. 79, pt. III, pp. 1165–1182, Feb. 1961.

energy to some dimension representative of the system size. Loane and Watchorn were also interested in durations and intervals of multiple outages, in connection with use of hydro storage, but derived their data by experiment rather than by computation.

The several criteria and methods of computation suggested in 1947 were subsequently developed by the same and other authors. The availability of machine computation removed the burden of laborious and time-consuming work from probability computations and contributed greatly to their more widespread use, and more and more applications to actual system problems were reported. More papers were written, and eventually a group of papers presented in 1958 and published together in *Power Apparatus and Systems*, August 1958,[13-21] brought again into focus the same several criteria of reliability that that had been proposed in 1947. Halperin and Adler's papers[15,22] had some similarity to that of Seelye, in that it was the interval between events, rather than the probability of these events, that was of prime interest in measurement of reliability. Papers by Kirchmayer and others,[18-20] Limmer,[16] and Miller[17] each used the probability that the load would exceed the available capacity as the measure of system reliability. Kist and Thomas[13] used a kw-hr measure of reliability, somewhat similar to that suggested earlier by Calabrese.

With reliability measured in at least three different ways, and with more measures available, if needed, it appeared desirable that some effort be made to compare the results obtained from these different methods. Not only are there several different criteria of reliability; but each author in the presentation of his method has heretofore used different symbols, different systems of generating units, sometimes different forced-outage data, and frequently, different assumptions as to what factors (e.g., maintenance schedule and effect of load-forecast deviations) should be considered and how these should be handled in computation. It appeared, therefore, that if the different criteria were to be compared, the comparison should be made by application of each to the same problem. Furthermore, such factors as maintenance, uncertainties in load forecasts, and the effects of interconnection should be removed as differences by having similar treatment, so far as is possible, in each application. The Subcommittee, believing that a comparison of the several criteria was worth while, appointed a Working Group to carry out and report upon such a project.

The Several Measures of Reliability

This paper presents the results of three separate measures of reliability applied to the same problem. The three bases of measure are identified as follows and are described in greater detail in the indicated appendixes:

Method 1: loss of load probability, Appendix I.
Method 2: loss of energy probability, Appendix II.
Method 3: interval between outages, Appendix III.

The results of method 1 are stated in terms of the probable number of days per year that the system load may be expected to exceed the available generating capacity. This is a probability exactly similar in its significance to the statement that the probability of throwing 11 on any throw of two dice is 2/36, or 0.056, or 5.6%. Probability is expressed as a fraction or decimal in the range of 0 to 1. Days per year is also a fraction, for 1 day per year represents a probability of 1/365 or 0.274%. In fact, the results of method 1 computations have been presented by some authors in terms of "per cent of time," the percentage representing the probability of the condition in question being found to exist at any particular time. Because an acceptable probability of load in excess of available capacity will usually be less than 1 day per year, other authors have presented their results in terms of "1 day per indicated number of years." For example, the probability of load exceeding capacity on 0.2 day per year may also be stated as: 1. 1 day in 5 years, 2 days in 10 years, etc. 2. 0.055% of time, or 3. a probability of 0.00055. Regardless of the form in which the result is presented, its significance as a probability of existence of a specified condition, and nothing more, must be recognized.

Method 1 has probably been more widely applied than either of the other methods, and various papers have described its application or contributed to its development, first for manual[6,9,23-25] and later for machine computation.[16,19,20,26] The method sometimes referred to as the "Calabrese method" can no longer be identified with any one or two authors.

The results of method 2 are stated in terms of the probable ratio of load energy that must be curtailed (because of load in excess of available generating capacity) to the total amount of load energy that should be supplied to serve the full system load requirements. For any given load-duration curve, this ratio is independent of the duration of the period for which the computation is made, usually a month or a

year. Because the probability of each loss of load (which is itself small within the range of acceptable reliability) is multiplied by the ratio of the peak energy that is affected to the total energy under the load curve (which ratio is also usually small), the final answers of method 2 for acceptable system conditions will be a percentage, which is numerically much smaller than comparable results for method 1. This energy index of reliability can also be expressed as 1 minus this very small percentage; and in this form the index measures the probable ratio of load energy that will be supplied by available generating capacity to the total that should be supplied.

Method 2 was suggested first by Calabrese[6] and was subsequently developed and applied by Kist and Thomas.[13]

Methods 1 and 2 use the same computations of probability of available capacity; but this probability is combined with the load data in different ways. Each occurrence of daily peak load in excess of available capacity is given equal weight in method 1, regardless of the amount of the deficiency in energy or capacity. In method 2, each occurrence of available capacity less than the load is weighted by the percentage of load energy that must be interrupted. It is obvious that there could be an additional method, in effect located between 1 and 2, in which the occurrences of available capacity less than the load would be weighted by the amount that the load must be interrupted.

Method 3 measures the reliability of system generation in terms of the interval between outages and the corresponding average outage duration. These criteria, interval and duration, are related to the outage of capacity equal to or greater than some given amount, which may be more or less than the system reserve (when defined as installed capacity less forecast peak load). Interval and duration are measures of probability in exactly the same terms as are the days per year criterion of method 1. Average duration of outage divided by the average interval between outages is the probability of finding the specified outage conditions of method 3 in existence at any particular time. For example, the probability of an outage with an average duration of 2 days occurring at average intervals of 4 years (the measures used in method 3) corresponds to the probability that the same outage would occur at the rate of 0.5 days per year (using the same terms as for the measure of method 1). The results of methods 3 and 1 cannot be equated by such simple division when these methods are applied to a practical capacity prob-

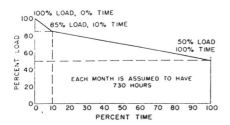

Fig. 1.   Monthly load-duration curve

lem, for method 1 takes into consideration load characteristics that are not recognized in method 3.

The statement of both duration and interval for a given outage condition adds a certain physical significance to the results of method 3, which may at times be lost in other methods. The other methods, however, have a much greater flexibility, particularly in recognition of load conditions, than does method 3.

Method 3 was first discussed by Seelye[8,27] and later by Strandrud.[28] A somewhat similar measure of reliability was suggested by Adler and Miller[22] and applied by Halperin and Adler,[15] based apparently on quite different computations.

The comparison between methods 1, 2, and 3 will direct attention to: 1. the similarities and differences in the basic computation methods; and 2. the response of the various criteria to typical changes in system conditions. Some idea of the reasons for the different values that are obtained is desirable, but the equivalence of numerical results is, of itself, meaningless. Certain advantages and disadvantages of each method will be evident, but no attempt will be made to recommend one method as preferable to another, for each can serve as a useful tool, provided its significance and its limitations are well understood. The choice between these several measures of reliability has been, and will remain, a matter of opinion.

No one of these methods, or any other method, can mechanically tell, without the guidance of human judgment, "How much generating capacity should be installed?" or "What is an acceptable level of reliability?" However, each can consistently indicate the optimum time for installing generating capacity that is needed to provide for any level of reliability that is acceptable to management. An acceptable level can be fixed arbitrarily, or decision may be supported in a more logical fashion. In any case, there comes a point where judgment must be exercised to differentiate between what is and what is not adequate reliability. Each of the several criteria of reliability can be

used to do this job, and none can do much more.

Use of an index of reliability also permits consistent measurement of the effect on system capacity requirements of:

1. Larger unit size.[15, 18, 23, 25, 29-31]

2. Less (or possibly more) reliable generation, as related to characteristics of particular units.[23, 25, 26, 28, 29]

3. Variability of available hydro capacity.[9, 21, 32]

4. Maintenance requirements and scheduling.[6, 15, 25, 31, 33]

5. Change in load characteristics.[25]

6. Interconnections.[6, 14-16, 31, 33, 34]

7. Uncertainties of load forecasts.[16, 17, 25]

It cannot be expected that each of these measures of reliability will produce exactly the same answer as the others when applied to a particular problem. The answers should at least be consistent and differences explainable in terms of the known characteristics of the several methods. No one answer is "right" and no answer is "wrong," and any of them should lead to system planning that is better than can be expected where decisions are made arbitrarily or by rule of thumb.

Nomenclature

The symbols common to the various methods are defined as:

p = average forced-outage rate

$$= \frac{\text{days on forced outage}}{\text{days on forced outage} + \text{days operated}}$$

q = average availability rate

$= 1 - p$

$$t = \frac{\text{days on forced outage}}{\text{number of forced outages}}$$

$$T = \frac{\text{days on forced outage} + \text{days operated}}{\text{number of forced outages}}$$

(Note that $p = t/T$.)

F = average frequency of outages per-unit (pu) time

$= 1/T$

n = total number of machines on system

r = number of machines simultaneously on forced outage

P_r = probability of existence of exactly r units on forced outage

O_k = megawatts (mw) on forced outage for kth entry in probability table

P_k = probability of existence of exactly O_k mw on forced outage

The additional symbols for method 3 are given in Appendix III.

Statement of Problem

The following initial hypothetical system was conceived to provide a common basis of comparison of the various methods.

1. Total system capacity = 4,000 mw.

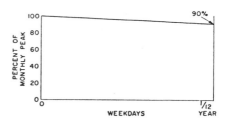

Fig. 2.   Variation of weekday hourly peak load

2. All boiler turbine generators are assumed to be of the unit type.

3. The configuration of the system is given as:

Number	Size, Mw	Total, Mw
1	400	400
2	300	600
2	250	500
3	200	600
3	150	450
5	100	500
6	75	450
10	50	500

4. Average forced-outage rate for each unit = p = 0.02.

5. Average duration of outage = t = 6 days.

6. Planned maintenance outage = 1 month/year/unit.

7. Planned maintenance outage schedule:

		Total
January	1–50, 1–150	200
February	1–50, 1–75, 1–200	325
March	1–50, 1–75, 1–300	425
April	1–75, 1–100, 1–300	475
May	1–50, 1–100, 1–250	400
June	1–50, 1–75, 1–200	325
July	1–50, 1–75, 1–200	325
August	1–50, 1–150	200
September	1–50, 1–100, 1–250	400
October	1–75, 1–400	475
November	1–50, 1–100, 1–150	300
December	1–50, 1–100	150

8. The maximum monthly peak loads in per cent of annual hourly peak are:

January	97.5%
February	95%
March	92.5%
April	90%
May	92.5%
June	95%
July	95%
August	97.5%
September	92.5%
October	90%
November	95%
December	100%

9. The monthly load duration curve is given in Fig. 1.

10. The variation of weekday hourly peak load within a given month is given in Fig. 2.

11. Interruptible load: none.

In order to determine the relative answers given by the various methods for different system configurations, a new system was studied which was assumed to have:

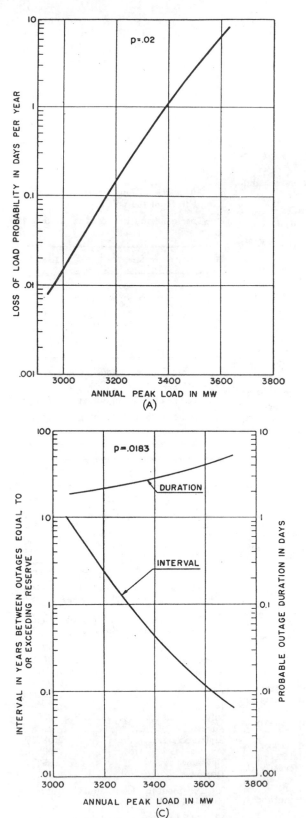

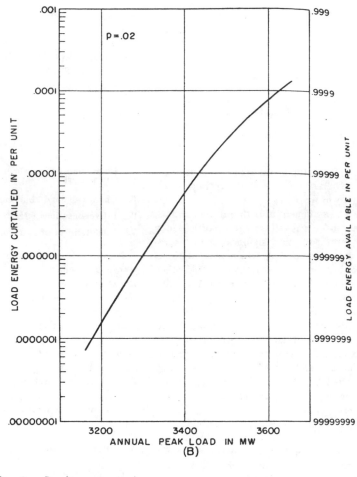

Fig. 3. Results, system of 4,000-mw capacity

A—Method 1
B—Method 2
C—Method 3

12. An installed capacity of 8,000 mw.

13. Twice the number of units of that of the various sizes given under item 3.

14. Maintenance scheduled as in item 7 with twice the number and amount of capacity out on planned maintenance for each period.

15. Load characteristics as previously described.

Results of Methods

The results obtained by the various methods for the 32-unit 4,000-mw system

expressed in terms of the respective yearly indexes are given in Figs. 3(A)–(C) for methods 1, 2, and 3, respectively. For purposes of comparison the values of the various indexes are given in Table I for selected values of system reserve. For purposes of this paper, reserve is expressed as a per cent of capacity.

Results similar to those described but based on a 64-unit 8,000-mw system are given in Figs. 4(A)–(C). A tabulation of results for selected values of system reserve is given in Table II.

If the 8,000-mw system is regarded as two 4,000-mw systems joined by a very large and perfectly reliable tie, the reduction in system reserve requirements made possible by this interconnection is shown in Table III for methods 1, 2, and 3. These results were based on an assumed annual peak load of 3,200 mw for each of the two 4,000-mw-capacity systems, and on a determination of the peaks that could be served by the 8,000-mw system at the same level of reliability as existed in the separate systems. Each method of computation leads to a slightly different result for the larger system, and the benefits accruing to each separate system are half the total benefits shown in Table III.

Fig. 5 shows corresponding benefits to

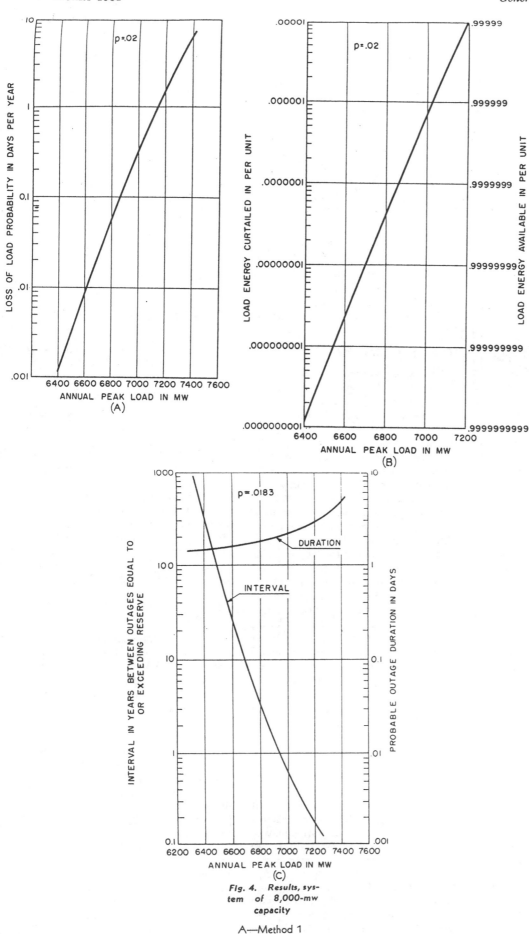

Fig. 4. Results, system of 8,000-mw capacity

A—Method 1
B—Method 2
C—Method 3

Table II. Comparison of Results for 8,000-Mw 64-Unit System

Reserve in Per Cent of Installed Capacity	Annual Peak Load Corresponding to Reserve Indicated, Mw	Method 1, Loss of Load Probability, From Fig. 4(A), Days/Year	Method 2, Loss of Energy Probability, From Fig. 4 (B), Pu	Method 3, Outage Frequency and Duration, From Fig. 4.(C)		
				Interval, Years	Frequency, 1/Years	Duration, Days
10	7,200	1.56	0.0000110000	0.17	5.9	2.85
15	6,800	0.057	0.0000000410	3.20	0.31	1.79
20	6,400	0.0012	0.0000000001	290	0.0035	1.48

Table I. Comparison of Results for 4,000-Mw 32-Unit System

Reserve in Per Cent of Installed Capacity	Annual Peak Load Corresponding to Reserve Indicated, Mw	Method 1, Loss of Load Probability, From Fig. 3(A), Days/Year	Method 2, Loss of Energy Probability, From Fig. 3(B), Pu	Method 3, Outage Frequency and Duration, From Fig. 3(C)		
				Interval, Years	Frequency, 1/Years	Duration, Days
10	3,600	6.50	0.00008000	0.12	8.3	4.10
15	3,400	1.08	0.00000550	0.45	2.2	2.85
20	3,200	0.149	0.00000015	2.4	0.42	2.15

Table III. Reduction in System Reserve Resulting from Interconnection

Method	Value of Index	4,000-Mw System			8,000-Mw System			Reserve Required for Two Independent 4,000-Mw Systems Minus Reserve Required for Single 8,000-Mw System, Mw
		Annual Peak Load that Corresponds to Index Shown, Mw	System Reserve of Installed Capacity Mw	%	Annual Peak Load that Corresponds to Index Shown, Mw	System Reserve of Installed Capacity Mw	%	
1	0.149 days/year	3,200	800	20	6,908	1,092	13.7	508
2	0.00000015 pu	3,200	800	20	6,890	1,110	13.9	490
3	2.4 years/outage	3,200	800	20	6,830	1,170	14.6	430

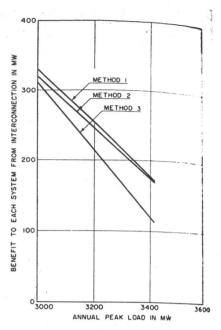

Fig. 5. Interconnection benefit as determined by methods 1, 2, and 3

each system resulting from the interconnection for loads ranging from 3,000 mw to 3,400 mw. This plot shows that all three methods indicate a decreasing benefit from interconnection as the reliability of the systems being joined decreases.

In order to compare the manner in which the indexes vary with load, consider the 32-unit 4,000-mw system and let the load of 3,600 mw (10% reserve) define the base value of each index. The ratio of each index to its respective base value is plotted in Fig. 6(A). The value of 10% was arbitrarily chosen and no significance should be attached to the selection of the base values for comparison of several indexes. It is not intended that 10% be considered as acceptable reserve; on the contrary, for the conditions of the assumed problem, there is little doubt that 10% reserve provides an inadequate level of reliability.

Fig. 6(B) indicates the results obtained when the procedure outlined in the preceding paragraph was applied to the 64-unit 8,000-mw system. Here the load of 7,200 mw (10% reserve) was chosen to define the base value of each index.

In Fig. 6(A) it can be seen that the method 1 index that corresponds to a load of 3,600 mw (10% reserve) is 6.50 days/year. In Fig. 6(C) the load chosen to define the base value of each index was 7,400 mw (7.5% reserve) so that the method 1 index was again 6.50 days/year. Using this load to define the base value of each index, Fig. 6(C) was constructed in a manner similar to that described for Figs. 6(A) and (B).

General Observations

In earlier papers on probability methods and applications, consideration has been given to various adjustments of load or capacity that are either necessary or desirable, although few, if any, of these are inherently part of one method or another. To keep the comparison of the several methods as simple as possible, only such

of these adjustments have been made in the Appendixes as were necessary; but all of them are here discussed without attempt at numerical expression of their effects.

The adjustments can be catalogued as shown in the sections that follow.

EXPOSURE OF UNITS

The exposure of units to forced outage is affected by the maintenance schedule and by system practice with respect to spinning reserve. The daily shutdown of units when not needed for economical operation is not the cause for a similar adjustment, however, because of the basis (daily rather than hourly) on which the forced-outage data have been compiled.

Only maintenance has been made the basis of an adjustment. Methods 1 and 2 can handle this in several different ways. When manual computations were necessary, maintenance was treated as an addition to the load, all units were considered to be operated or on forced outage, and an adjustment of the forced-outage rate may or may not have been made (if each unit were on scheduled outage for 10% of the time, the forced-outage rate could be multiplied by 90%). When programed for machine computation, recognition can be actually given to the different number and size of units available at different periods of the year. This was done here in the application of methods 1 and 2 on a monthly basis, using the specified schedule of maintenance outages. For method 3, maintenance was added to the load, and an adjustment made in the outage rate as already indicated.

Ordinary practice does not require that all available capacity be operated, whether needed or not. Spinning reserve may be about half, more or less, of installed reserve. This means that some of the older and less efficient units will seldom be called upon for operation, and consequently are less exposed to forced outage than other units. This fact is usually neglected in all three methods. Usually

all units are assumed to be operated, unless on forced (or maintenance) outage, with the result that a slightly larger installed capacity requirement is indicated. The effect of this practice on major forced outages and on capacity requirement has been generally treated as being negligible.

Seelye made an adjustment for the exposure of units, as affected by operating requirements, but this was predicted on the use of forced-outage data based on hours rather than on days of exposure and outage. No similar adjustment was needed in method 3.

VARIATIONS IN UNIT CAPACITIES AND LOADS

Seasonal variations in capacity, if small, can be treated as adjustments of load under any method. If these variations are substantial, they can be handled, along with maintenance, on a monthly basis of computation.

Emergency or overload ratings of equipment should be the basis of the ratings used in the probability computations. However, if the computations are based on normal ratings, the overload capacity for the units remaining in service, as a distinct component, can be used in meeting the indicated capacity or reserve requirement.

Reductions in loads from shedding of interruptible loads, or from voltage or frequency reduction, if permitted by company practice, can also be recognized to the extent desired. The amount of relief obtained from these sources may be stated as a separate factor in the comparison of load and capacity requirements. However, the more general practice seems to be to consider such measures for what they really are, i.e., loss of load.

In the treatment of emergency ratings and load reductions, one method has no advantage over the others. Method 1 results can be interpreted directly to show the probability that any given amount of such emergency help will be sufficient to meet probable capacity deficiencies. Method 3 can do the same in terms of interval and duration.

LOAD CHARACTERISTICS

The characteristics of load specified in the common problem are recognized in different ways by the several methods. These characteristics are the shape of the load-duration curve and the variation in daily peaks, both within any month and from month to month.

Method 1 makes no use of the load duration, for its concern is with the probability that available capacity will be less than the daily peak. Method 1 uses only load data with respect to variation in daily peaks.

Method 2 gives full recognition to all these characteristics in its measurement of the probability of interruption of supply to some part of the load energy.

Method 3 also uses these peak load data, but in less direct fashion. The mechanics of method 3 are such that it cannot be readily applied in a way that directly

recognizes variations in load, for the results are based on the outage of capacity equal to or greater than some specified amount. In the problem considered here, the variation in monthly peaks is very nearly cancelled out by the maintenance schedule, and there is essentially equal risk in each month. The variation of peaks within the month is recognized only in the application of a peaking factor, selected on the basis of judgment. The value of this peaking factor should vary, depending on the nature of the variations in daily peak loads.

Method 3 as applied by Halperin and Adler was based on the monthly peak loads and the constant-risk principle which provides for the sum of the monthly peaks, overhauling, and seasonal limitations being constant throughout the year. This results in a uniform margin of capacity available for emergencies. The variations of daily peaks from the monthly peak is recognized by a peaking factor which reduces the required reserve by an amount corresponding to the range of the upper 10% of the daily peak loads in a month. This in effect reduces the protection to 90% of the days, considered adequate by Halperin and Adler in view of the rarity of coincidence of large outages with heavy load days.

UNCERTAINTIES OF LOAD FORECASTS

Uncertainties of load forecasts (sometimes erroneously called load forecasting

errors) have a major effect on capacity requirements. The importance of a forecast made three or four years in advance cannot be minimized or eliminated by leaving the forecasting, in effect, to management. Even their forecasts are known not to be certainties, and the actual loads may be more or less within a fairly wide range. How wide the range is, and whether or not distribution of probable results within the range is symmetrical, will depend on the nature of the forecast (i.e., deliberatively or actually high or low, or based on projection of a long-time trend). In any event, probabilities can be assigned to the "official" estimate of load and to higher and lower values.

Methods 1 and 2 can readily recognize the uncertainties of load forecasts by computing their respective results at various load levels and then combining these on a basis which is weighted by the probability of reaching these various load levels in the forecast year. The result will be that, for any given level of reliability, a greater generating capacity will be required to serve a future uncertain peak load than would be required to serve that same nominal peak, were it considered as a certainty. Method 2 will show a somewhat larger increase in capacity requirements on this account than will method 1.

As far as is known, method 3 has not been applied in a situation where uncertainties of forecast loads were to be

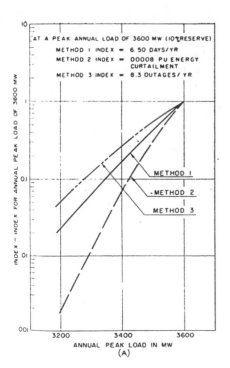
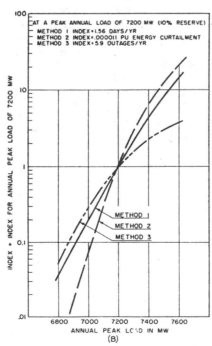
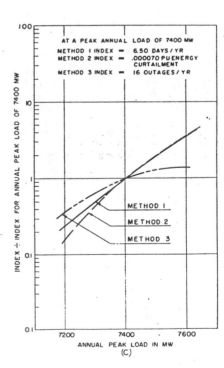

Fig. 6.   Rate of change of reliability index with load for methods 1, 2, and 3

A—4,000-mw system
B, C—8,000-mw system

recognized. Because the results of this method are inherently related to outage of a given amount of capacity, they cannot for any given capacity installation be readily related to varying load levels. This limitation is apparent in the treatment of both seasonal load characteristics and load forecasts.

INTERCONNECTIONS

The value of an interconnection, in effecting capacity savings, can be measured best by application of the same probability procedures to the separate systems and to the interconnected group. Any one of the three methods is available for this purpose, but each will give somewhat different results.

The results of an interconnection study will have to be interpreted in the light of transmission limitations and, perhaps, of contractual limitations as well. These are, of course, independent of the method used; but the results of a particular method may depend on the nature of these limitations.

When a probability computation is made for the separate systems and again for the group, it is for the purpose of determining the increase in load-carrying capability for the group, as compared with the sum of the individual systems, at the same over-all level of reliability. Here it is necessary to determine how the group will operate: will it be 1. the "one-company" concept in which each component company shares the difficulties of the others, or 2. will any company help another only so far as it can without sharing in the difficulties of the others?

Generally speaking, it would appear that the one-company concept would produce the better results, for what might cause interruption of load in a single company may be only a comparatively minor voltage or frequency reduction for the entire group. Nevertheless, method 1, because it gives no consideration to the severity of the capacity deficiency, would show a greater value for interconnection in the absence of the one-company concept. On the other hand, method 2, which measures severity, would properly favor the one-company idea. Since data are readily available in method 1 for measuring severity directly in terms of capacity, this situation would appear to require departure from the methods that have heretofore been recognized as method 1.

SPINNING RESERVE

The treatment of spinning reserve is not an adjustment in the sense of the discussed conditions, but it is an added element of information that can be developed in probability studies.

Installed reserve is related to the probable existence of coincident forced outages, while spinning reserve is related to their probable occurrence within some short period of time. The basic data on forced outage provide the necessary outage rates, for the probability of occurrence of an individual outage is equal to the probability of its existence divided by its average duration.

So far as spinning reserve is concerned, methods 1 and 2 use the same computation of probability of multiple forced-outage occurrences, but combination with the load may be different. The literature shows no application of method 2 to the spinning reserve problem, but there is no reason why a consistent policy should not be guided by a measure which recognizes the amount of energy that might be interrupted by a shortage of spinning reserve. Method 1 simply computes the probability of outage occurrences plus probable day-to-day load forecasting uncertainties greater than various amounts of spinning reserve which might be provided in excess of the expected peak on any given day.

It should be noted particularly in the selection of a spinning reserve requirement that consideration must be given to the experienced excess or deficiency of actual loads as compared with the estimates, for such variation of load in a large system may be of far greater importance than the forced-outage probabilities. Experience as to loads is accumulated very rapidly and will show definite dispersion patterns, which may be different for some days of the week and some seasons of the year. The combination of load variations with outage probablities presents no problems in application of either method 1 or 2.

The application of method 3 to this situation, so far as is known, has not been attempted.

Conclusions

The measurement of system reliability should be an essential part of the planning for the time and character of generating unit additions, of the determination of the benefits of interconnection, and of the development of optimum spinning reserve practices. At least three distinct measures of generating system reliability are available; and the appropriate application of any one of these measures would be a major advance (measured either in dollars or improved reliability) over rule-of-thumb methods of timing generator additions or establishing reserve policies. Consistent use of any of the three methods would provide essentially the same schedule for capacity additions with respect to both installation dates and unit size.

No numerical calculations can alone establish what is a satisfactory level of generating system reliability, or determine a proper value of the index of system reliability. The selection of a satisfactory level of reliability or a corresponding index, appropriate to any method of measurement, requires at some point the exercise of informed judgment. Once that selection has been made, however, the use of probability methods, with due recognition of their nature, can maintain the desired system reliability with minimum costs under widely changing conditions.

The several measures of reliability have here been compared in their application to a single problem. The numerical values of the three indexes of reliability show differences which are to be expected from their definitions. In spite of these differences in value, the indexes generally react in similar fashion to changed system conditions, so that comparable results would be obtained by consistent use of any of the methods as an aid to system planning.

The application of probability methods to generating capacity problems has reached a stage where it should be accepted as a normal tool of the system planner; but that does not imply that all problems have been solved. Comparison of the basic method of measuring reliability and discussion of refinements in computation and of the adjustments common to each method reveals a number of limitations and possibilities of improvement that require study beyond the scope of this report. In addition, still other methods of probability application await investigation and development.

Appendix I. Loss-of-Load Probability

To obtain this criterion,[6,9,23–25] consideration is given to the reliability of the generation sources, the characteristics of the load which is to be served, and the uncertainty of load forecast.

Calculation of the Probability of Capacity Outages

In order to show how this index of reliability is obtained, consider first the behavior of the power supply which serves the system. What is desired is some quantitative means of expressing the degree of unreliability inherent in this supply due to the fact that each of the

Table IV. Unit Data

Unit	Capacity Out, Mw	Probability of Outage Existence
1	{ 0	0.98
	{ 100	0.02
2	{ 0	0.97
	{ 100	0.03

Table VI. Final Results

Capacity Out (O_k), Mw	Probability of Outage Existence (P_k)
0	(0.98)(0.97) = 0.9506
100	(0.02)(0.97) + (0.98)(0.03) = 0.0488
200	(0.02)(0.03) = 0.0006
	1.0000

Table V. Method of Calculation

	Unit 2	
	0.97 .... 0 mw	0.03 .... 100 mw
Unit 1	**Units 1 and 2**	
0 mw 0.98	(0.98)(0.97) .... 0 mw	(0.98)(0.03) .... 100 mw
100 mw 0.02	(0.02)(0.97) .... 100 mw	(0.02)(0.03) .... 200 mw

generators which make up the supply is itself subject to failure.

The use of probability methods makes it possible to compute the over-all source reliability from the known forced-outage rate of each of the generators which comprise this system supply. To illustrate how this computation proceeds, consider the following example where a 100-mw unit will be added to an existing 100-mw unit to form a system of 200-mw capacity. What is desired is a measure of the reliability of the 200-mw system supply based on the known outage characteristics of the two component units.

Unit 1: forced-outage rate (p) = 0.02
availability rate (q) = 0.98
capacity, mw = 100

Unit 2: forced-outage rate (p) = 0.03
availability rate (q) = 0.97
capacity, mw = 100

These statements defining units 1 and 2 can be rewritten in the form shown in Table IV.

The method of combining the data shown in Table IV to obtain information describing the 200-mw system supply is shown in Table V.

From this table note that the probability of 0 mw being out on forced outage is equal to the product of the probability that unit 1 is available (0.98) and the probability that unit 2 is also available (0.97). The probability that unit 1 is available and unit 2 is out, which would result in a total outage of 100 mw, is equal to the product of the probability of unit 1 being available (0.98) and the probability that unit 2 is out (0.03). Similarly, the proba-

bility that unit 1 is out and unit 2 is available, which also corresponds to a capacity outage of 100 mw, is equal to 0.02×0.97. The probability that both units are out on forced outage is equal to 0.02×0.03 which corresponds to a total forced outage of 200 mw. Note that the capacity outage corresponding to the product of any row and column probabilities is equal to the sum of the corresponding capacity outages. The final results are presented in Table VI. As a check, note that the sum of the probabilities is unity, thus indicating that all possible happenings have been accounted for.

The procedure outlined can obviously be extended, so that by means of adding one generator at a time to an existing system whose characteristics are known it is possible to obtain, for any number of generators, a listing showing every possible outage that could occur, and the corresponding probability of its existence. This is the information which was sought, a quantitative expression of the reliability of the power supply for the system. This procedure of adding units one at a time is well suited to computer solution as it has been programed to minimize card handling and no sorting is necessary.[19]

Calculation of Loss-of-Load Probability

Forced outages are of concern only when they are of such a value that insufficient capacity remains to meet the system load.

Consequently, the next step to be taken is to combine the previously described probability of existence of forced outages with the probabalistic nature of the system load. Consider Fig. 7 which indicates a typical daily peak load variation curve over a 1-year period.

The contribution to the system loss-of-load probability made by any particular capacity outage is the probability of existence of the outage in question multiplied by the number of days per year that loss of load would occur if such a capacity outage were to exist continuously throughout the year. An examination of Fig. 7 will show that capacity outages which are equal to or less than the system reserve value R will produce no loss of load. Capacity outages of magnitude greater than R can produce outages varying in length from 0+ days to 260 days depending on the amount by which the outage in question exceeds the system reserve. To express these thoughts mathematically:

Contribution to system loss of load probability made by an outage of magnitude $O_k = P_k t_k$ in days/year
System loss of load probability = $\Sigma P_k t_k$ in days/year

The summation covers all possible capacity outages that can occur.

The loss-of-load probability is a chance figure similar to the chance figure of 1/36 which corresponds to obtaining snake eyes (2) when throwing two dice. The practice of the industry has been to express this chance figure in units of days per year.

As noted earlier, the daily peak load variation curve shown by Fig. 7 represents a time interval of 1 year. If no maintenance were done, this representation would be accurate as the number of units available to meet the load would be constant from month to month barring the occurrence of forced outages. Consequently, one table of outage (O_k) versus probability (P_k) would apply throughout any given year.

In practice, however, the number of units available to serve the load is not constant, but varies from one month to another due to the withdrawal of units for maintenance. The maintenance pattern followed by the illustrative problem is typical of what may be expected in practice. To account for this effect without approximation, it is necessary to construct a

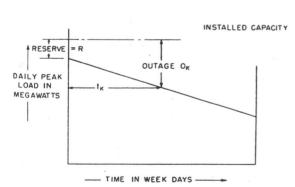

Fig. 7. Daily peak load variation versus time

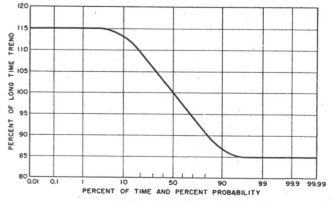

Fig. 8. Uncertainty of load forecast

Table VII.　Method of Calculation to Treat Uncertainties of Load Forecast

% Deviation from Forecast Load (1)	Load Increased or Decreased by Amount of Deviation (2)	Exact Probability of Deviation (3)	Loss of Load Probability, Days/Year (4)	Exact Probability of Deviation × Days/Year Outage if It Should Occur (5)
15	3,640	0.050	9.1	0.455000
14	3,608	0.03	6.8	0.204000
13	3,576	0.03	5.2	0.156000
12	3,545	0.03	4.1	0.123000
11	3,513	0.03	3.1	0.093000
10	3,482	0.03	2.42	0.072600
9	3,450	0.03	1.8	0.054000
8	3,418	0.03	1.3	0.039000
7	3,387	0.03	0.97	0.029100
6	3,355	0.03	0.73	0.021900
5	3,323	0.03	0.54	0.016200
4	3,292	0.03	0.39	0.011700
3	3,260	0.03	0.28	0.008400
2	3,228	0.035	0.195	0.006825
1	3,197	0.035	0.14	0.004900
0	3,165	0.040	0.10	0.004000
− 1	3,134	0.035	0.071	0.002485
− 2	3,102	0.035	0.05	0.001750
− 3	3,070	0.03	0.034	0.001020
− 4	3,038	0.03	0.023	0.000690
− 5	3,007	0.03	0.016	0.000480
− 6	2,975	0.03	0.011	0.000330
− 7	2,943	0.03	0.0075	0.000225
− 8	2,912	0.03	0.005	0.000150
− 9	2,880	0.03	0.0033	0.000099
−10	2,849	0.03	0.0022	0.000066
−11	2,817	0.03	0.0014	0.000042
−12	2,785	0.03	0.00095	0.000029
−13	2,754	0.03	0.00065	0.000020
−14	2,722	0.03	0.00041	0.000012
−15	2,690	0.050	0.00028	0.000014
				1.307037

different table of O_k versus P_k for each month of the year and to use with this table the appropriate monthly daily peak load variation curve to obtain $\Sigma P_k t_k$. A yearly value for $\Sigma P_k t_k$ can then be obtained by summing the 12 monthly $\Sigma P_k t_k$ quantities thus determined.

By varying the magnitude of peak load for a given assumed total capacity, it is possible to obtain a curve such as shown in Fig. 3(A).

Uncertainties of Load Forecast

If the annual system peak load is specified, Fig. 3(A) can be used to find the corresponding loss-of-load probability in days per year. Future loads, however, cannot be predicted exactly. For purposes of illustrating the method of computation, it is assumed that the uncertainty of load forecasts are as shown in Fig. 8.

What is desired then is a plot similar to that shown by Fig. 3(A) except that adjustment for the load forecasting deviation shall be included. To obtain this end, use the following procedure.

1. The curve shown in Fig. 3(A) is extended to cover a wider range of probability values than would normally be required.

2. From the data presented in Fig. 8, a distribution function of probable deviations is drawn up (column 3 of Table VII) and the loss-of-load probability in days per year is found by referring to Fig. 3(A). These values are shown as column 4 of Table VII. The entries found in columns 3 and 4 are now multiplied to determine the expected number of days per year out that is contributed by each possible variation. Summing all such expectancies gives the number of days per year of outage

expected for the particular value of system peak load chosen, in this case 3,165 mw.

3. The procedure outlined is repeated for several assumed values of system peak load and these points then become the basis for a curve similar to that of Fig. 3(A) but in which the effect of load forecasting error is included.

Appendix II. Loss of Energy Probability

The purpose of this method is to provide a means for evaluating the effect outages have on the customers of a system. The sum of the evaluation of each possible outage may be used to determine a figure of merit associated with the system under study. The effect of outages on customers, in the sample problem, is confined to kw-hrs curtailed.

The procedure[6,13] followed in this method has many steps similar to the method of Appendix I. First, a table of probability of existence of outage (O_k) is prepared for the capacity not currently undergoing maintenance for the time period under consideration. For purposes of the problem of this paper this time period is taken to be 1 month. By reference to the monthly load-duration curves (see Fig. 1) the expected pu energy curtailments resulting from overlaps of equipment outages and loads may be determined by the relation:

E = expected pu energy curtailment

$$= \frac{\Sigma P_k A_k}{X}$$

where

X = total kw-hrs under load-duration curve

A_k = area defined in Fig. 9

The results may also be expressed in terms of an energy reliability index by the relation:

Energy reliability index
$$= [1 - \text{Expected pu energy curtailment}]$$
$$= [1 - E]$$

In order to obtain an equivalent annual figure, the expected pu energy curtailments for each period are weighted according to the number of kw-hrs in each period.

The effect of load forecasting uncertainty may be treated in a manner similar to that described in Appendix I.

Appendix III. Frequency and Duration of Outage

In this method, the average expected frequency and the average expected duration of all possible forced-capacity outages are calculated for a given power system. In the region of greatest interest, however, the calculated frequency is extremely low so that it is preferable to define the frequency in terms of its period, the interval between forced outages. On the basis of experience and judgment, a reasonable time interval in the range of 5 to 20 years is preselected as the criterion of system protection. Using this method, calculations are then made to determine the amount of installed reserve necessary to protect the system against forced outages equal to or exceeding the installed reserve not more than once per selected time interval.

An explanation of the many variations accompanying this method is beyond the scope of this Appendix. Instead, the method will be developed in its simplest form in order to show the basic ideas essential to the technique. For a more complete development of this method and its variations, reference may be made to papers by Seelye,[8,27] Strandrud,[28] or Halperin and Adler.[15] The following description is based on Seelye's method.

Correction of Outage Rate for Machine Exposure

Outage rates are derived from the ratio of the product of machines and time on forced outage divided by the products of machines and time on forced outage plus time actually running. In other words, outage rates are predicated on machine exposure. Since the results of probability calculations are desired on a full-time basis including scheduled maintenance time, it is necessary to modify either the number of machines or the outage rate. For example, a machine with an outage rate of 0.02, running continuously, would be expected to be on forced outage 7.3 days per year. If, however, the machine were run only one half of the time, it would be expected to be on forced outage only 3.65 days per year. It would then be equivalent to one-half machine with an outage rate of 0.02 or a full machine with an effective outage rate of 0.01. Since every machine in the sample problem runs only 11 months per year, the outage rate applicable to all machines is $11/12 \times 0.02 = 0.0183$.

Each machine is exposed to the possibility of forced outage 11/12 or 91.7% of the time.

Calculation of Capacity-Outage Probability

Development of the final results must be preceded by the preparation of a table similar to Table VIII. Construction of this table will begin by listing every possible capacity outage that could occur. Next, for each capacity outage entry listed, the various ways in which this event could occur are shown, and the probability of existence of each of the capacity outage combinations is computed in a manner similar to the procedures described in Appendixes I and II.

To illustrate how the probability of a capacity outage event may be calculated, consider the condition of a capacity outage of 175 mw caused by the outage of two units rate 50 mw and one unit rated 75 mw. The probability of the stipulated outage occurring is obviously the product of the following three probabilities:

A = probability of finding two units on outage in a group of 10 units. Each unit in this group is rated at 50 mw and has a 0.0183 outage rate. $A = 0.001304$.

B = probability of finding one unit on outage in a group of 6 units. Each unit in this group is rated at 75 mw and has an 0.0183 outage rate. $B = 0.100279$.

C = probability of finding zero units on outage in the remaining 16 units of the system. The capacity ratings are immaterial since no units are on forced outage. The outage rate is again equal to 0.0183. $C = 0.743746$.

The probability of losing 175 mw in exactly the manner specified is denoted as P_r, where r represents the total number of units on forced outage for this condition.

$$P_r = ABC = 0.000972414$$
$$r = 2 + 1 + 0 = 3$$

It can be seen that in addition to the condition just described, it is also possible to obtain a capacity outage of 175 mw by losing one unit rated 75 mw and one unit rated 100 mw. By following the same procedure as has been outlined, it can be found that for such a condition the value of P_r is 0.00578644 and the associated value of r is 2.

As there are no other possible ways to lose 175 mw, the sum of the two foregoing values of P_r represent the probability of finding exactly 175 mw on outage regardless of how the event came about. Thus, the summation of P_r for a particular value of outage O_k produces P_k, the familiar exact outage existence probability that was discussed in Appendixes I and II.

Calculation of Frequency of Capacity Outages

The next step will require the calculation of D_r for each of the items for which a value of P_r is listed. By definition, D_r is a measure of the frequency of an r-fold outage:

$$D_r = P_r \left[\frac{r}{t} - \frac{(n-r)}{(1-p)T} \right]$$

The derivation and physical significance of D_r is contained in references 8, 27, and 28. For each capacity outage entry O_k, a value of D_r is computed for every possible way in which this capacity outage could be brought about. These values are then summed to produce one value, D_k, which can be associated with the outage magnitude O_k. To illustrate this procedure, refer to Table VIII and consider the entry for which $k=7$, $O_k = 175$ mw. It can be seen that this outage can arise in two fashions, so two values of D_r are computed:

D_r for 175 mw out (two 50-mw units, one 75-mw unit) = 0.158687
D_r for 175 mw out (one 75-mw unit, one 100-mw unit) = 0.553021
Then D_k for $k=7$, $O_k = 175$ mw = 0.711708

Frequency of Capacity Outages Which Equal or Exceed Specified Amount

At this point, it is desired to introduce the quantity F_{k+} and to define and describe this quantity as follows:

F_{k+} = frequency of occurrence of outages of exactly O_k mw or more
$$F_{(k+1)+} = F_{k+} - D_k$$

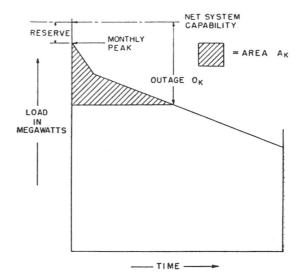

Fig. 9. Monthly load-duration curve

Table VIII. Illustration of Method 3

Index, k	Capacity Outage, Mw, O_k	50 (10)	75 (6)	100 (5)	150 (3)	200 (3)	250 (2)	300 (2)	400 (1)	Units Out, r	Units In, n-r	P_r	P_k	P_{k+}	D_r	D_k	F_{k+}	t_{k+}	T_{k+}
1	0									0	32	0.553242	0.553242	1.000000	−21.932825	−21.932825	0	∞	∞
2	50	1								1	31	0.103298	0.103298	0.446758	2.88775	2.88775	21.932825	0.02037	0.04556
3	75		1							1	31	0.061979	0.061979	0.343460	1.73265	1.73265	19.045078	0.01803	0.05251
4	100	2								2	30	0.008680	0.060329	0.281481	0.829533	2.273406	17.31243	0.01626	0.05776
				1						1	31	0.051649			1.443873				
5	125	1	1							2	30	0.011573	0.011573	0.221153	1.106043	1.106043	15.039024	0.01470	0.06649
6	150	3								3	29	0.000432	0.043959	0.209580	0.070527	2.137065	13.932981	0.01504	0.07177
			2							2	30	0.002893			0.276511				
		1		1						2	30	0.009644			0.921703				
					1					1	31	0.030990			0.866324				
7	175	2	1							3	29	0.0009724	0.006759	0.165621	0.158687	0.711708	11.797916	0.01404	0.08476
			1	1						2	30	0.0057864			0.553021				
8	200	4								4	28	0.0000141	0.040069	0.158862	0.003259	1.828001	11.086207	0.01433	0.09020
		1	2							3	29	0.0005402			0.088816				
		1			1					2	30	0.0057864			0.553022				
		2		1						3	29	0.0008103			0.132239				
				2						2	30	0.0019288			0.184341				
						1				1	31	0.0309895			0.866324				

$n = 32$, $p = 0.0183$, $r = 5$, $t = 0.01507$, $T = 0.8219$.

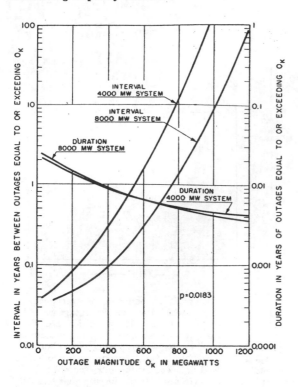

Fig. 10. Results of method 3 for 4,000- and 8,000-mw systems

By definition, the value of F_{k+} that applies for the condition of zero capacity outage ($k=0$, $O_k=0$), is zero. This fact permits the evaluation of F_{k+} for all other values of k as D_k is known and the recurrence relationship shown above applies. This is illustrated by the following material extracted from Table VIII.

k	F_k	D_{k+}	$F_{(k+1)+}$
0	−0	−21.93	21.93
1	21.93	2.89	19.04
2	19.04	1.73	17.31
3	17.31	2.27	15.04
4	15.04	1.11	13.93

It is desirable to know not only the frequency of capacity outages but also the expected average duration of a capacity outage, should it occur. P_{k+}, the cumulative value of P_k, expresses the amount of time in unit time that O_k mw or more can be expected to be on outage. Knowing F_{k+} and P_{k+}, a value can be obtained for t_{k+} as follows:

t_{k+} = average time duration of an outage of O_k mw or more

$$t_{k+} = P_{k+}/F_{k+}$$

Interval Between Outages

As interval is the reciprocal of frequency, the interval between occurrences of outages of O_k mw or more may be found as follows:

$$T_{k+} = 1/F_{k+}$$

These values are shown in Table VIII in the last column.

The results of the foregoing calculations may be plotted by making the interval between outages of amount O_k or greater (T_{k+}) and the duration of outages of amount O_k or greater (t_{k+}) the ordinates, and the outage magnitude (O_k) the abscissa. This plot is shown in Fig. 10.

Frequency and Duration of Outages as Function of Peak Annual Load

The result that is most desired, however, is a plot similar to that just described, but in which the abscissa is the annual peak load rather than outage magnitude. For the 32-unit system having an installed capacity of 4,000 mw, this can be done in the following manner:

1. From Fig. 10, choose a value of T_{k+} and record the associated values of O_k and t_{k+}.

2. If maintenance were not considered, and if for a given year the system load was at all times equal to the peak load, then it is apparent that the annual peak load that could be met with the value of T_{k+} found in item 1 would be $4{,}000-O_k$.

3. From the data given in the sample problem, it can be seen that if maintenance is considered as an increase in load, then the monthly peak load plus capacity on maintenance is approximately constant. As the annual peak load occurs, in the month of December, and 150 mw are out for maintenance that month, it can be said that the peak annual load consistent with the value of T_{k+} specified in item 1 is:

Peak annual load$_k = 4{,}000-O_k-150$

If the procedure described in steps 1, 2, and 3 is followed for various values of T_{k+}, data will be obtained to permit plotting the curves shown in Figs. 3(C) and 4(C). This procedure, in effect, displaces the curves by an amount equal to the capacity on maintenance at time of peak load along the abscissa.

References

1. FUNDAMENTAL CONSIDERATIONS IN PREPARING MASTER SYSTEM PLAN, W. J. Lyman. *Electrical World*, New York, N. Y., vol. 101, no. 24, June 17, 1933, pp. 788–92.

2. SPARE CAPACITY FIXED BY PROBABILITIES OF OUTAGE, S. A. Smith, Jr. *Ibid.*, vol. 103, Feb. 10, 1934, pp. 222–25.

3. PROBABILITY THEORY AND SPARE EQUIPMENT, S. A. Smith, Jr. *Bulletin*, Edison Electric Institute, New York, N. Y., vol. 2, no. 3, Mar. 1934, pp. 85–89.

4. SERVICE RELIABILITY MEASURED BY PROBABILITIES OF OUTAGE, S. A. Smith, Jr. *Electrical World*, vol. 103, Mar. 10, 1934, pp. 371–74.

5. THE USE OF THE THEORY OF PROBABILITY TO DETERMINE SPARE CAPACITY, P. E. Benner. *General Electric Review*, Schenectady, N. Y., vol. 37, no. 7, July 1934, pp. 345–48.

6. GENERATING RESERVE CAPACITY DETERMINED BY THE PROBABILITY METHOD, G. Calabrese. *AIEE Transactions*, vol. 66, 1947, pp. 1439–50.

7. CALCULATING PROBABILITY OF GENERATING CAPACITY OUTAGES, W. J. Lyman. *Ibid.*, pp. 1471–77.

8. OUTAGE EXPECTANCY AS A BASIS FOR GENERATOR RESERVE, Howard P. Seelye. *Ibid.*, pp. 1483–88.

9. PROBABILITY METHODS APPLIED TO GENERATING CAPACITY PROBLEMS OF A COMBINED HYDRO AND STEAM SYSTEM, E. S. Loane, C. W. Watchorn. *Ibid.*, pp. 1645–54.

10. OUTAGE RATES OF STEAM TURBINES AND BOILERS AND OF HYDRO UNITS, AIEE Committee Report. *Ibid.*, vol. 68, pt. I, 1949, pp. 450–57.

11. FORCED OUTAGE RATES OF HIGH-PRESSURE STEAM TURBINES AND BOILERS, AIEE Committee Report. *Ibid.*, pt. III-B (*Power Apparatus and Systems*), vol. 73, Dec. 1954, pp. 1438–42; also, *Combustion*, New York, N. Y., Oct. 1954, pp. 57–61.

12. FORCED OUTAGE RATES OF HIGH-PRESSURE STEAM TURBINES AND BOILERS, AIEE Joint Subcommittee Report. *Ibid.*, vol. 76, June 1957, pp. 338–43; also, *Combustion*, vol. 28, Feb. 1957, pp. 43–46.

13. PROBABILITY CALCULATIONS FOR SYSTEM GENERATION RESERVES, Carl Kist, G. J. Thomas. *Ibid.*, vol. 77, Aug. 1958, pp. 515–20.

14. USE OF PROBABILITY METHODS IN THE ECONOMIC JUSTIFICATION OF INTERCONNECTING FACILITIES BETWEEN POWER SYSTEMS IN SOUTH TEXAS, A. Pat Jones, A. C. Mierow. *Ibid.*, pp. 520–30.

15. DETERMINATION OF RESERVE-GENERATING CAPABILITY, Herman Halperin, H. A. Adler. *Ibid.*, pp. 530–44.

16. DETERMINATION OF RESERVE AND INTERCONNECTION REQUIREMENTS, H. D. Limmer. *Ibid.*, pp. 544–50.

17. DETAILS OF OUTAGE PROBABILITY OF CALCULATIONS, Andrew L. Miller. *Ibid.*, pp. 551–57.

18. GENERATOR UNIT SIZE STUDY FOR THE DAYTON POWER AND LIGHT COMPANY, W. J. Pitcher, L. K. Kirchmayer, A. G. Mellor, H. O. Simmons, Jr. *Ibid.*, pp. 558–63.

19. DIGITAL COMPUTER AIDS ECONOMIC-PROBABILISTIC STUDY OF GENERATION SYSTEMS—I, M. K. Brennan, C. D. Galloway, L. K. Kirchmayer. *Ibid.*, pp. 564–71.

20. DIGITAL COMPUTER AIDS ECONOMIC-PROBABILISTIC STUDY OF GENERATION SYSTEMS—II, C. D. Galloway, L. K. Kirchmayer. *Ibid.*, pp. 571–77.

21. TREATMENT OF HYDRO CAPABILITY DURATION CURVES IN PROBABILITY CALCULATIONS, K. L. Hicks. *Ibid.*, pp. 577–80.

22. A NEW APPROACH TO PROBABILITY PROBLEMS IN ELECTRICAL ENGINEERING, H. A. Adler, K. W. Miller. *Ibid.*, vol. 65, Oct. 1946, pp. 630–32, 1118.

23. THE RELATION OF THERMAL PLANT DESIGN TO RESERVE CAPACITY REQUIREMENTS, M. J. Steinberg. *Electrical Engineering*, vol. 69, Jan. 1950, pp. 64–67; also, *Electrical World*, vol. 133, Jan. 2, 1950, pp. 58–60; also, *Power Generation*, Barrington, Ill., vol. 54, Feb. 1950, pp. 76–79.

24. DETERMINATION OF RESERVE CAPACITY BY THE PROBABILITY METHOD, G. Calabrese. *AIEE Transactions*, vol. 69, pt. II, 1950, pp. 1681–89.

25. ELEMENTS OF SYSTEM CAPACITY REQUIREMENTS, C. W. Watchorn. *Ibid.*, vol. 70, pt. II, 1951, pp. 1163–85.

26. NEW ANALYTICAL TOOLS PERMIT INTEGRATED SYSTEM PLANNING, L. K. Kirchmayer, A. G. Mellor. *Electrical World*, June 1, 1959, pp. 52–54, 59, 60.

27. A CONVENIENT METHOD FOR DETERMINING GENERATOR RESERVE, Howard P. Seelye. *AIEE Transactions*, vol. 68, pt. II, 1949, pp. 1317–20.

28. DETERMINATION OF GENERATOR STAND-BY RESERVE REQUIREMENTS, H. T. Strandrud. *Ibid.*, vol. 70, pt. I, 1951, pp. 179–88.

29. AN INVESTIGATION OF THE ECONOMIC SIZE OF STEAM-ELECTRIC GENERATING UNITS, L. K. Kirchmayer, A. G. Mellor, J. F. O'Mara, J. R. Stevenson. *Ibid.*, pt. III (*Power Apparatus and Systems*), vol. 74, Aug. 1955, pp. 600–14; also, *Combustion*, Feb. 1955, pp. 57–64.

30. EVALUATION OF UNIT CAPACITY ADDITIONS, M. J. Steinberg, V. M. Cook. *Ibid.*, vol. 75, Apr. 1956, pp. 169–79.

31. THE EFFECT OF INTERCONNECTIONS ON ECONOMIC GENERATION EXPANSION PATTERNS, L. K. Kirchmayer, A. G. Mellor, H. O. Simmons, Jr. *Ibid.*, June 1957, pp. 203–14.

32. FIRM OR DEPENDABLE HYDRO CAPACITY, E. S. Loane, C. W. Watchorn. Pennsylvania Electric Association, Harrisburg, Pa., Oct. 1947.

33. DETERMINATION OF RESERVE CAPACITY BY THE PROBABILITY METHOD EFFECT OF INTERCONNECTIONS, G. Calabrese. *AIEE Transactions*, vol. 70, pt. I, 1951, pp. 1018–20; also *Electrical Engineering*, vol. 72, 1953, p. 100 for correction.

34. THE DETERMINATION AND ALLOCATION OF THE CAPACITY BENEFITS RESULTING FROM INTERCONNECTING TWO OR MORE GENERATING SYSTEMS, C. W. Watchorn. *Ibid.*, vol. 69, pt. II, 1950, pp. 1180–86.

35. PROBABILITY AND ITS ENGINEERING USES (book), T. C. Fry. D. Van Nostrand Company, Inc., Princeton, N. J., 1928.

36. A METHOD FOR THE ECONOMIC ESTIMATION OF THE OPERATING STANDBY RESERVE IN ELECTRICAL SYSTEMS (in Russian), R. A. Ferman. *Elektrichestvo*, Moscow, USSR, no. 20, 1932.

37. INDICES AND METHODS OF CALCULATING RELIABILITY IN POWER ECONOMY, B. M. Yakub. *Ibid.*, no. 18, 1934.

38. CONSIDERATIONS INVOLVED IN MAKING SYSTEM INVESTMENTS FOR IMPROVED RELIABILITY, S. M. Dean. *Bulletin*, Edison Electric Institute, vol. 6, no. 7, Nov. 1938, pp. 491–94.

39. RATIO BETWEEN INSTALLED CAPACITY AND MAXIMUM ALLOWABLE LOAD OF A PLANT AND OF A GROUP OF INTERCONNECTED PLANTS, G. J. Th. Bakker, J. C. Van Staveren. *Paper no. 331*, CIGRE, Paris, France, 1939.

40. DETERMINATION OF DEGREE OF RELIABILITY OF A HYPOTHETICAL STEAM ELECTRIC GENERATING SYSTEM, E. S. Loane, C. W. Watchorn. Pennsylvania Electric Association, Harrisburg, Pa., Nov. 1945.

41. SYNTHETIC DETERMINATION OF FREQUENCY AND DURATION OF MULTIPLE FORCED OUTAGES OF ELECTRIC GENERATING UNITS, E. S. Loane, C. W. Watchorn. Pennsylvania Electric Association, Harrisburg, Pa., Nov. 1945.

42. RESERVE REQUIREMENTS FOR INTERCONNECTED SYSTEMS, H. W. Phillips. *Electrical World*, May 11, 1946, p. 78; also, *Bulletin*, Edison Electric Institute, vol. 14, Apr. 1946, pp. 117–20; also, *Mechanical World*, Manchester, England, vol. 120, Sept. 27, 1946, pp. 351–53.

43. DETERMINATION OF REQUIRED RESERVE GENERATING CAPACITY, W. J. Lyman. *Electrical World*, vol. 127, no. 19, May 10, 1947, pp. 92–95.

44. SERVICE CONTINUITY STANDARDS AID ECONOMICAL DISTRIBUTION PLANNING, W. J. Lyman. *Electric Light and Power*, Chicago, Ill., vol. 25, no. 5, May 1947, pp. 66–70, 112–13; also, *Bulletin*, Edison Electric Institute, Sept. 1947, pp. 325–39.

45. PROBABILITY THEORY HELPS DETERMINE SYSTEM RESERVE OPERATING CAPACITY, G. Calabrese. *Power*, New York, N. Y., vol. 92, July 1948, pp. 423–26.

46. TABLES OF BINOMIAL PROBABILITY DISTRIBUTION TO SIX DECIMAL PLACES, AIEE Committee Report. *AIEE Transactions*, pt. III (*Power Apparatus and Systems*), vol. 71, Aug. 1952, pp. 597–620.

47. A CLASSIFICATION OF OUTAGES OF FACILITIES IN ELECTRIC POWER SYSTEMS, Arnold Rich. *Combustion*, vol. 24, Mar. 1953, pp. 43–47.

48. DETERMINATION OF FORCED OUTAGES OF CAPACITY OF POWER PLANT INSTALLATIONS, C. K. V. Rao. *Power Engineer*, Simla, India, vol. 4, Jan. 1954, pp. 15–22; Apr., pp. 70–79.

49. POWER AND ENERGY PRODUCTION, C. W. Watchorn. *AIEE Transactions*, pt. III-B (*Power Apparatus and Systems*), vol. 73, Aug. 1954, pp. 901–08.

50. CALCUL DE LA PUISSANCE DE RESERVE DANS LES COMPLEXES ELECTRIQUES PAR LA METHODE DES PROBABILITES, G. Calabrese. *Bulletin Scientifique A.I.M.*, Association des Ingenieurs Electriciens, Institut Electrotechnique Montefiore, Liege, Belgium, no. 11, Nov. 1955.

51. OPERATIONS RESEARCH DETERMINATION OF GENERATION RESERVES, E. L. Arnoff, J. C. Chambers. *AIEE Transactions*, pt. III (*Power Apparatus and Systems*), vol. 76, June 1957, pp. 316–28.

52. A SIMPLIFIED BASIS FOR APPLYING PROBABILITY METHODS TO THE DETERMINATION OF INSTALLED GENERATING CAPACITY REQUIREMENTS, C. W. Watchorn. *Ibid.*, Oct., pp. 829–32.

53. QUICK CALCULATIONS OF FORCED-OUTAGE PROBABILITIES OF GENERATING PLANT, N. T. van der Walt. *Journal*, South African Institution of Mechanical Engineers, Johannesburg, Union of S. Africa, vol. 7, 1957, pp. 75–100.

54. LE CALCUL DES PROBABILITES APPLIQUE AUX RESERVES DES GRANDS RESEAUX ELECTRIQUES, A. Calvaer. *Bulletin Scientifique A.I.M.*, Association des Ingenieurs Electriciens, Institut Electrotechnique Montefiore, no. 9, Sept. 1957.

55. PROBABILITY IN PLANNING, K. L. Hicks. *Electric Light and Power*, Chicago, Ill., July 15, 1958, pp. 43–46, 59.

56. THE USE OF A DIGITAL COMPUTER IN A GENERATOR RESERVE REQUIREMENT STUDY, Homer E. Brown. *AIEE Transactions*, pt. I (*Communication and Electronics*), vol. 77, Mar. 1958, pp. 82–85.

57. EVALUATING CHANCE IN PLANNING, C. J. Baldwin, C. H. Hoffman. *Electric Light and Power*, Aug. 15, 1959, pp. 55–57, 91.

58. AN INTRODUCTION TO THE STUDY OF SYSTEM PLANNING BY OPERATIONAL GAMING MODELS, J. K. Dillard, H. K. Sels. *AIEE Transactions*, pt. III (*Power Apparatus and Systems*), vol. 78, Dec. 1959, pp. 1284–90.

59. MATHEMATICAL MODELS FOR USE IN THE SIMULATION OF POWER GENERATION OUTAGES; I—FUNDAMENTAL CONSIDERATIONS, C. J. Baldwin, D. P. Gaver, C. H. Hoffman. *Ibid.*, pp. 1251–58.

60. MATHEMATICAL MODELS FOR USE IN THE SIMULATION OF POWER GENERATION OUTAGES; II—POWER SYSTEM FORCED-OUTAGE DISTRIBUTIONS, C. J. Baldwin, J. E. Billings, D. P. Gaver, C. H. Hoffman. *Ibid.*, pp. 1258–72.

61. STRATEGY FOR EXPANSION OF UTILITY GENERATION, D. N. Reps, J. A. Rose. *Ibid.* (Feb. 1960 section), pp. 1710–20.

62. THE APPLICATION OF PLANNING CRITERIA TO THE DETERMINATION OF GENERATOR SERVICE DATES BY OPERATIONAL GAMING, C. A. De Salvo, C. H. Hoffman, R. G. Hooke. *Ibid.*, pp. 1752–59.

63. MATHEMATICAL MODELS FOR USE IN THE SIMULATION OF POWER GENERATION OUTAGES—III. MODELS FOR A LARGE INTERCONNECTION, C. J. Baldwin, D. P. Gaver, C. H. Hoffman, J. A. Rose. *Ibid.*, pp. 1645–50.

Discussion

H. A. Adler (Commonwealth Edison Company, Chicago, Ill.): In order to avoid misunderstandings and in view of questions which have been raised, it appears worthwhile to point out the differences between method 3, as described in the paper, which is based on procedures described by Seelye and Strandrud (references 8, 27, and 28) and the method described by Halperin and Adler (references 22 and 15). The two methods are similar in that they determine frequencies or expectations of multiple outages rather than their probabilities, but they are derived from different premises and arrive at different formulas.

Basically, the Seelye method determines the expected frequency of "r-or-more" fold outages and from it derives their average duration after determining their probability. The Halperin-Adler method is based upon the determination of the average duration of r-fold outages and derives from this the frequency and intervals of r-fold outages. From these durations and frequencies, the durations and frequencies of r-or-more outages can be derived. Strandrud has shown that the

Halperin-Adler formulas are equivalent to the Seelye formulas.

Practically, there is little difference between the frequencies or intervals for r-fold and r-or-more fold outages, at least for the larger combinations which are of interest in reserve planning. However, the average duration of r-or-more fold outages should be greater than the average duration of r-fold outages as is apparent from theoretical considerations. This is illustrated in Table IX which compares the results of method 3 in Table I of the paper with the corresponding results of the Halperin-Adler method.

The question has been raised whether the loss-of-load method can be used in connection with method 3. This can be readily done and the loss-of-load method has been used in a number of cases in connection with method 3 as shown, for example, in the closing discussion of the Halperin-Adler paper (reference 15). The answers are, however, in terms of expected frequency instead of probability of loss of load. Loss of load in terms of frequency is considered more significant than in terms of probability.

One reason in favor of the use of method 3 without determination of the correspond-

Table IX

Reserve in Per Cent of Installed Capacity	Method 3, Outage Frequency and Duration					
	Interval, Years		Frequency, 1/Years		Duration, Days	
	Seelye	Halperin-Adler	Seelye	Halperin-Adler	Seelye	Halperin-Adler
10	0.12	0.12	8.3	8.3	4 10	3.0
15	0.45	0.42	2.2	2.4	2.85	2.3
20	2.4	2.3	0.42	0.43	2.15	1.8

ing loss of load is the desire to separate the system characteristics from the load characteristics. The expectation of an outage of a given magnitude is only a function of the system composition and, therefore, is especially valuable in comparing reserve requirements for alternate schemes of expansion, interconnections, size of units, etc. The expectation of loss of load is a function of both the system composition and load characteristics and separation of the effects of the two variables is not readily possible. It was preferable, therefore, to protect for forced outages of a given frequency and reduce the resulting required margin by an amount proportional to the load and dependent on the shape of the load curve. This separation of the effects on the reserve of system composition and load characteristics is especially helpful not only where different system compositions are studied but also where changes in load characteristics are considered, such as those caused by the development of certain types of loads.

L. T. Anstine (Baltimore Gas and Electric Company, Baltimore, Md.): While the paper deals primarily with the application of probability methods to the study of installed reserve requirements, it also refers briefly to their application to the study of spinning reserve requirements. It is this latter application that I wish to discuss.

In studying spinning reserve requirements it should be assumed that there is sufficient installed capacity so that as capacity is forced out of service other capacity can normally be started to take its place. With this assumption it is obvious that the longer the start-up time of marginal equipment the greater will be the chance of a second outage occurring before the first has been replaced; however, the part of the duration of a single forced outage which extends beyond the start-up time of the marginal equipment can be ignored in determining the probability of simultaneous outages. The effective duration of any outage is, then, the time required to start up additional equipment.

Using the outage occurrence rate determined as proposed in the paper (i.e., by dividing the probability of existence of a forced outage by its average duration in days) to compute the probability of simultaneous outages assumes 1 day as the effective duration of a single outage. Obviously this exaggerates the probability of occurrence of simultaneous outages.

The probability that a particular unit will be effectively on forced outage during any hour (forced out and not yet replaced) can be determined as the rate of occurrence of forced outages for a single unit (outages per hour of operation) multiplied by the start-up time, in hours, of the marginal equipment. The true duration of the individual forced outages is not a factor so long as the duration exceeds the start-up time, as it does in the majority of the cases of forced outage. The hour should be the unit of time used in compiling basic forced-outage data for use in connection with studies of spinning reserve requirements. This fact has been recognized and outage data are now being collected on both an hourly and daily basis.

C. J. Baldwin, C. A. DeSalvo (Westinghouse Electric Corporation, East Pittsburgh, Pa.) and **C. H. Hoffman** (Public Service Electric and Gas Company, Newark, N. J.): The Working Group which produced this Committee Report should be complimented for the excellent job it performed. The various methods of making probability calculations are explained in clear, concise terms, and the use of a common example to illustrate the techniques helps to establish firmly the various methods in the reader's mind.

The three methods described in the Committee Report differ in some details, but essentially all three methods are based on probability mathematics. Recently a fourth method of making probability calculations has been proposed. This method is the result of a joint study undertaken by Public Service Electric and Gas Company and Westinghouse Electric Corporation. The fourth method also uses probability mathematics, but in a slightly different form than the first three methods. Basically, the earlier methods analytically combined probability distributions representing loads and capacities and other factors to obtain the probability of losing load. The new method used the simulation process.

Simulation is particularly valuable in solving problems where many of the variables are random or depend on probability. In general, if there are many probability variables, the simulation approach tends to give a result more quickly than the analytical methods. On the other hand, if only a few probability variables are involved, or if one is willing to restrict the probability variables to a small number, the analytical methods produce an answer more quickly.

The simulation approach has been used to develop a computer program. The program is the most extensive ever written with regard to the generating capacity problem. Some of the variables it will handle include:

1. Long-range load forecast errors.
2. Short-range forecast errors.
3. Dispersion of daily peaks in a month.
4. Forced outages both from a going out and being out viewpoint. (These can be made different for different units and can be changed with time.)
5. Scheduled outages.
6. Maintenance outages.
7. Economy outages.
8. Availability of water and water limitations.
9. Effects of operation with interconnected neighboring utilities.

These variables and others were included so that it is possible to determine quantitatively how important the variables are, both singly and in combination. In less extensive studies it is necessary to decide, with little evidence at hand, which variables are important and which are not. Based on the quantitative experience gained from the simulation method, we are developing additional new approaches and expect to continue work in this area.

Hilton U. Brown III (Puget Sound Power and Light Company, Bellevue, Wash.), **Alfred R. Caprez** (Light Division, The Department of Public Utilities, Tacoma, Wash.) and **Lawrence A. Dean** (Bonneville Power Administration, Portland, Oreg.): A service of great value to all persons attempting to apply probability methods to utility problems has been performed by the

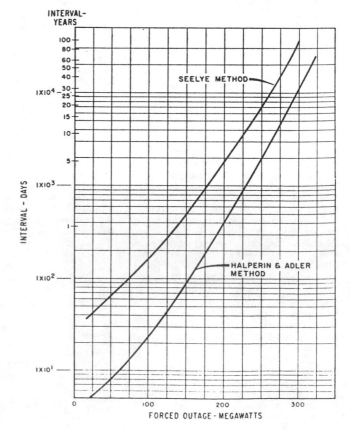

Fig. 11. Comparison of Halperin-Adler and Seelye methods for a system of 315 units totaling 3,734 mw

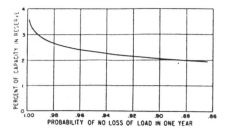

Fig. 12. Loss-of-load probability as function of generation reserve

Working Group in publishing this bibliography and in presenting the comparative analysis of three of the more widely used methods for arriving at generating capacity reserve requirements.

In the Northwest Power Pool we have had occasion to compare two of the three methods. In 1947, a study of reserve requirements in the area was made using the Seelye method. For comparative purposes, the study was recently repeated using the Halperin and Adler method. The system studied was made up of 315 hydroelectric units totaling 3,734 mw. The results are presented in Fig. 11. Considering the differences in the techniques used, the agreement between the two methods appears quite good.

In investigating the duration of outages for the Northwest Power Pool, it was found that they are distributed approximately as a "log-normal" function. Hence, the average duration does not very satisfactorily represent the "typical" duration. For this reason, although the computations for interval and duration are of considerable interest, primary emphasis in our studies has been placed upon the probability of outages calculations, and the interval curves are used only as rough indicators of the expected interval and duration of events.

The Northwest Power Pool has in service 443 hydro units. To add these units one by one to the probability matrix, as contemplated in Appendix I, would have taxed the capability of the available computer. Considerable experiment in this area confirms the opinion that partitioning the generators on the system into a number of subsystems in which the units are all of about the same size and have about the same characteristics does not noticeably reduce accuracy, particularly if extra care is taken to maintain the definition of the largest units. We would suggest that the sizes of the units comprising a subsystem should be distributed symmetrically about the average size for the subsystem, and that this criterion is more important than the criterion of "normality" that has been suggested elsewhere.

The probability computed by any of the usual methods is the probability of exactly a certain amount of capacity being on forced outage at a given time. We feel that a more useful quantity is the probability of a given amount or more capacity being on forced outage at a given time. This latter probability is easily computed from the former by summation.

Finally, we would suggest still another way in which the results of reserve studies can be presented. The probability of dropping load due to forced outages of generating equipment can be computed for each day of a typical year. One minus this probability is the probability of no loss of load on the given day, and the product of all such probabilities for a year is the probability of no loss of load for a year. A curve showing reserve as a per cent of total capacity and probability of no loss of load as the axis can then be constructed; see Fig. 12.

G. O. Calabrese (Argonne National Laboratory, Lemont, Ill.): In the section Several Measures of Reliability, it is stated: "The statement of both duration and interval for a given outage condition adds a certain physical significance to the results of method 3, which may at times be lost in other methods."

This may be construed to mean that methods 1 and 2 do not permit obtaining the duration expectancy. This, of course, is not correct. The manner in which the expected duration of loss of load may be computed was given under Article 32 of a 112-page monograph by the discusser, entitled "System Reserve Capacity by the Probability Method." This monograph (unpublished) was distributed in 1952 or 1953 to members of the AIEE Power Generation Committee, System Engineering Committee, and Subcommittee on Application of Probability Methods.

Since this question of frequency has often been raised in recent times and since the monograph was not published, the article relative to the frequency of outages and of loss of load is reproduced in the following section in its entirety for the benefit of those who are interested in the problem.

A first edition of my 1947 paper (reference 6 of the paper) was actually prepared some 8 or 9 years earlier, and a synopsis was forwarded to the Institute at its request. At the last minute, however, the paper was withdrawn for reasons outside the control of the author and so was not presented at the earlier time.

32. FREQUENCY OF OUTAGES AND OF LOSS OF LOAD*

In the preceding pages methods have been given for the calculation of capacity outage or loss-of-load probabilities. From these probabilities the expectancies can be calculated during any period. Of course, these expectancies represent average conditions expected to materialize over long periods. At this point let us raise another question, how often can loss of load be expected to occur? In other words, what is the expected frequency of loss of load expressed in terms of one loss-of-load occurrence every so many years. To the probability of loss of load, discussed in the preceding pages, corresponds an expectancy of loss of load of 1 day every so many years. This expectancy, however, represents an average value which may obtain as a result of a large number of short loss-of-load occurrences or through a small number of long ones. The loss-of-load expectancy yields no information in this respect excepting that, on the average, loss of load will occur 1 day every so many years, or D days per year, in general, D being a fraction of 1 day.

In order to answer the above question relative to the expected frequency of occurrence of loss of load, it is necessary to introduce the concept of *continuance probabilities* for simultaneous outages of units. Briefly, these are the probabilities that outages involving simultaneously a given number of units, say k, should last 1, 2, or 3, etc., days once they have occurred. From past records, continuance probabilities for outages of k units are calculated as follows: Let N_k be the total number of k-unit outages that have occurred in a past period. Of these N_k outages n_{k1}, n_{k2}, n_{k3}, ... have lasted respectively 1, 2, 3, ... days. From these recorded values, the continuance probabilities π_{k1}, π_{k2}, π_{k3}, ... are now calculated.

$$\pi_{k1} = \frac{n_{k1}}{N_k} = \text{continuance probability that a}$$
k-unit outage shall last 1 day once it has occurred

$$\pi_{k2} = \frac{n_{k2}}{N_k} = \text{continuance probability that a}$$
k-unit outage shall last 2 days once it has occurred

and so on.

Let $d_k = n_{k1} + 2n_{k2} + 3n_{k3} + \ldots$ be the total number of k-unit outage days involved. It is easily seen that:

$$n_{k1} = \frac{\pi_{k1} d_k}{\pi_{k1} + 2\pi_{k2} + 3\pi_{k3} + \ldots}$$

$$n_{k2} = \frac{\pi_{k2} d_k}{\pi_{k1} + 2\pi_{k2} + 3\pi_{k3} + \ldots} \qquad (32\text{-}1)$$

That is, once the continuance probabilities for k-unit outages are obtained from past records, the number of k-unit outages lasting respectively 1, 2, 3, ... days, that can be expected to occur during a reasonably long period, can be calculated if the expected total number of k-unit outage days is known.

The continuance probabilities π_{k1}, π_{k2} ... should be calculated for as many values of k as necessary and should be checked for stability, following the same general method outlined in article 2.

In order to determine the expected frequency of loss of load from the expected number of D days of loss of load, it is first necessary to determine the number of units k of the outages causing loss of load, for instance as shown in column 2 of the lower part of Table 7-1.* In this particular case (Table 7-1), it is seen that loss of load is caused almost entirely by 3-unit outages. In the more general case, loss of load will be caused by outages of 1, 2, 3 ... units. In such a case, a complete analysis would require the fractionalization of D into portions D_1, D_2, D_3, ... D_k ... according to the number of units involved. For each class of outages of k units, the number of outages n_{k1}, n_{k2}, ... lasting 1, 2 ... days, respectively, that may be expected during a sufficiently long period of years T, may now be calculated by replacing d_k with TD_k in equation 32-1. The sufficiency of the length of T for each value of k should be determined by the number of k-unit outage days necessary to obtain stable values of the continuance probabilities π_{k1}, π_{k2} ...

---

* From the monograph.

Table 32-1. Distribution of Certain Turbine-Generator and Boiler Actual Outages by Length*

Outage Duration in Days	Actual Turbine Forced Outages		Actual Boiler Emergency Outages†	
	Less than 1,000 Psig	1,000 Psig or More	Less than 1,000 Psig	1,000 Psig or More
1	66.5	40.0	52.3	37.0
2	17.2	31.0	22.6	26.6
3	4.3	6.9	8.1	13.5
4	2.2	4.7	3.8	7.9
5	1.4	3.3	3.2	5.1
6–10	3.5	6.3	6.3	7.0
11–15	1.2	1.9	1.4	1.6
16–25	1.4	1.9	1.1	0.9
26–50	1.1	2.7	0.9	0.4
51–100	0.6	0.5	0.3	
101 up	0.6	0.8		
	100.0	100.0	100.0	100.0
Total no. of outages	2,123	364	1,803	752.5
Average duration of 1 outage, days	3.7	5.0	3.0	3.0

* Abstracted from Table III of reference 10 of the paper.
† Full plus one half of partial.

Values of continuance probabilities for boilers and turbines, disregarding the number of units out simultaneously, may be obtained from Figs. 5, 6, 7, and 8 of Outage Rates of Steam Turbines and Boilers and of Hydro Units (reference 10 of the paper).

In the particular case of Table 7-1, the calculated loss-of-load probability of 0.00162 corresponds to 1 day outage every 6,200 days. Since loss of load is caused almost exclusively by 3-unit outages, in order to obtain the expected frequency of loss of load it would be necessary to have a continuance probability for 3-unit outages. Lacking this information, for the sake of discussion let it be assumed that the average duration of a unit outage is 4 days. *If it is assumed* that this figure is applicable to 3-unit outages, the expected frequency of loss of load would be once every 24,800 days. Actual distribution of outages are shown in Table 32-1.

Frank C. Poage (Ebasco Services, Inc., New York, N. Y.): Two questions are often put to us as consultants. The exact phrasing may differ in each instance but the questions are about as follows: (1) "What savings in reserve generating capacity can be expected with the proposed (X) interconnection?" and (2) "Which of the alternative proposed installations (or plans) A, B, or C can do the most toward maintaining or increasing the reliability of service from the system and what are the relative values in this respect?"

If the X interconnection of question 1 is one of the alternatives A, B, or C of question 2, then the two questions become one, but with much broader scope and complexity.

Usually, thinking concerning system reserve requirements is based upon experience-derived rule-of-thumb criteria such as:

Y per cent of load, or Z times the capacity of largest unit. Such criteria are not uniformly used or accepted, and even if uniformly applied would not result in equal protection to service due to differences in composition of the capacity resources. For the examples studied by the Working Group and reported in the paper the interconnection of two identical systems reduces by one half the percentage by which the largest unit is represented in the total capacity. This certainly does not mean that the combined system reserves can be reduced to half of those which should be carried by the individual systems operating alone, since the number of units and consequently the total number of opportunities for outage are doubled by the interconnection. These points are very well illustrated by the values reported for all three methods in Tables I, II, and III.

In approaching questions of future system capacity needs we feel that as a step beyond these rule-of-thumb criteria allowance for reserve is best treated as insurance using the mathematics of the insurance business, namely, statistics and probability. This insurance is carried to protect a company and its customers against the consequences of a capacity shortage. These consequences must be considered as affecting public relations as well as the financial aspects of the company's business. The insurance analogy can be carried still further in the sense that system capacity reserves are a paid-up endowment which can be returned to the

company if the system load continues to grow. The cost or premium for this insurance is the annual carrying charges against the capacity which exceeds that required for load-carrying purposes. This premium is diminished by the net operating economies achieved through use of the reserve facilities provided. Just as in the purchase of other forms of insurance it is a matter for management to determine at what point the cost of carrying these reserves exceeds the values risked by not providing them. In other words, at what level is the insurance worth the net premium to be paid.

In preparing our answers to the two questions posed, we are now using method 1, loss-of-load probability, to make the necessary estimates of reliability, thus enabling us to place logical numbers and values opposite alternative situations. To handle the calculations which, although relatively simple, are so voluminous and repetitious that manual computations have not often been attempted, we have developed a set of digital computer programs similar in some respects to those employed by the authors of references 14 and 56.

Our programs are written in one of the languages of the Bendix Aviation Corporation G-15 computer. By employing "double precision" techniques the computations are carried out to 12 significant decimal digits. The final results are printed out in 2-column tables with figures corresponding to columns 2 and 3 of Table I in the paper. As many values of annual peak load may be used as are deemed necessary.

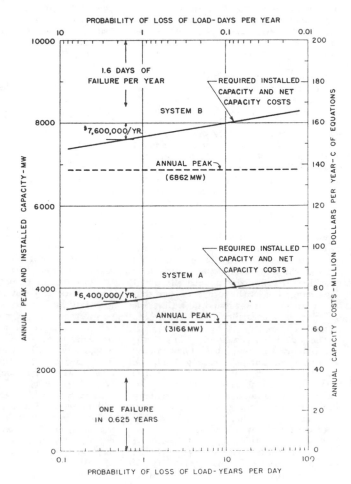

Fig. 13. Systems A and B, derived from Figs. 3(A) and 4(A) of paper

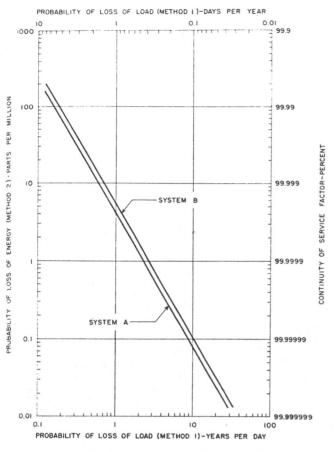

PROBABILITY OF LOSS OF LOAD (METHOD 1)-DAYS PER YEAR

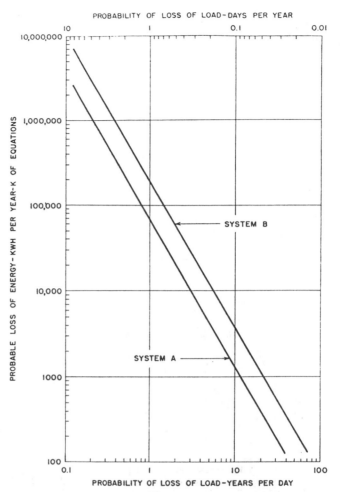

PROBABILITY OF LOSS OF LOAD-DAYS PER YEAR

Fig. 14 (above). Standard of service reliability of method 2 as function of method 1, from Figs. 3(A) and (B) and 4(A) and (B) of paper

Fig. 15 (right). Annual load factors of 60%, from Fig. 14

The results include the effects of the following factors:

1. Daily peak loads and their distribution by the calendar for the system and time period under study: week, month, season, or year.

2. Scheduled maintenance, also by the calendar for the same period.

3. Periodic or seasonal variations in capacity due to weather, stream flow, or other conditions during the same calendar period.

4. Any combination of number and size of generating units and other sources of capacity.

5. Relative reliability of individual sources.

From our experience to date we are confident that this implementation of probability methods by the fast and accurate arithmetic of the digital computer has now placed within reach of system planners powerful and useful new tools. In learning to use these tools we heartily recommend study of the conclusions, cautions, and observations of the Working Group as embodied in this truly significant paper.

C. W. Watchorn (Pennsylvania Power and Light Company, Allentown, Pa.): In this paper there are shown, for the first time, the results from application to a single problem of several different measurements of the

reliability of electric power generating systems. Comparison of these results is, in itself, a valuable contribution to the literature on application of probability methods. These results, in addition, provide the basis for a study of a possible economic criterion or index of reliability, as has been suggested in an earlier paper.[1]

An economic criterion or index is one in which the additional reliability, resulting from added generating capacity, is valued in dollars, so that comparison can be made between the incremental worth of added reliability and the incremental cost of obtaining it.

Fig. 13 of this discussion has been derived from Figs. 3(A) and 4(A) of the paper. Annual peak loads for each of systems A and B were determined so that the installed capacities of 4,000 and 8,000 mw for these systems, respectively, would result in a probability of loss of load of 1 day in 10 years. The installed capacity was then assumed to be varied and the peak load held constant. It was assumed that the same percentage reserves would produce the same degrees of service reliability as are shown in the paper, where capacity is held constant and the load is assumed to be varied. The net cost of installed capacity was assumed to be $20 per kw per year, so that dollars as well as mw can be directly related to probability of loss of load.

Fig. 13 indicates that tremendous annual savings, amounting to many millions of

dollars per year, are available from reasonable reduction in the standards of service reliability. Such savings, for example, amount to $6,400,000 and $7,600,000 per year for systems A and B, respectively, for a standard of service reliability equal to 1.6 days of failure per year, or 1 day of failure in 0.625 year, as compared with the more generally accepted standard of service reliability of 0.1 day of failure per year or 1 day of failure in 10 years. The possibility of even a portion of such savings justifies a further look at the generally accepted standards. Some indication of the significance of these standards can be obtained from consideration of the cost of reducing the energy losses.

Fig. 14, which was obtained from Figs. 3(A) and (B) and 4(A) and (B) of the paper, shows the standard of service reliability of method 2 as a function of that of method 1, that is, probability of loss of load is related to the probability of loss of energy. Fig. 15 was obtained from Fig. 14 on the basis that systems A and B both have annual load factors of 60%.

The difference in probable annual energy loss for a probability of loss of load of 1 day in 0.625 year as compared with 1 day in 10 years is 152,000 and 416,000 kw-hr per year for systems A and B, respectively. Thus, the average costs of providing this greater reliability compared with the smaller, are $42 and $18 per kw-hr for systems A and B, respectively. It is very doubtful if it would

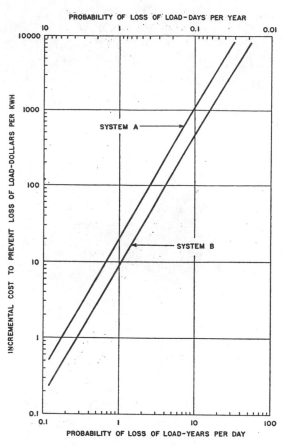

PROBABILITY OF LOSS OF LOAD-DAYS PER YEAR

INCREMENTAL COST TO PREVENT LOSS OF LOAD-DOLLARS PER KWH

SYSTEM A

SYSTEM B

PROBABILITY OF LOSS OF LOAD-YEARS PER DAY

Fig. 16. Incremental costs plotted as functions of service reliability corresponding to values of K in Fig. 15

be possible to justify such costs economically, for it is unlikely that the value of the probable energy loss is as high as $42 or even $18 per kw-hr. The limit to which we would be economically justified in installing capacity is that for which the probable economic loss, i.e., the cost of the loss of supply to the whole service area and not the revenue lost to the utility alone, is equal to the net cost of providing capacity to prevent such economic loss.

A basis for determination of such a limit can be derived from Figs. 13 and 15. The total net capacity costs per year of Fig. 13, designated as C, can be expressed as a function of the probable energy loss per year of Fig. 15, designated as K. Similar results could also have been obtained directly from Figs. 3(B) and 4(B) of the paper. In either case they are most useful shown in the form of equations as follows:

$$C = 89,940,000 - 3,140,000 \log K$$

for system A, and

$$C = 173,900,000 - 3,900,000 \log K$$

for system B, where log is to the base 10.

Obtained from differentiating these equations with respect to K:

$$\frac{dC}{dK} = -\frac{1,364,000}{K}$$

for system A, and

$$\frac{dC}{dK} = -\frac{1,694,000}{K}$$

for system B.

These values of incremental costs, plotted as functions of the service reliability corresponding to the value of K that are exhibited in Fig. 15 are shown in Fig. 16.

If the economic loss to the service area from loss of load were between $3 and $10 per kw-hr, it would be justified to install

capacity up to the point where the net capacity costs of preventing such economic loss were also $3 to $10 per kw-hr. Fig. 16 shows that such installation would result in probabilities of loss of load of between 1 and 3 days per year. Fig. 14 shows that this level of reliability corresponds to continuity of service factors in the order of 99.999%. This 99.999% is well in excess of the service reliability normally experienced for the other components of electric power systems, which is more generally in the order of 99.99%, and even smaller in some cases. As shown by Fig. 14, a continuity of service factor of only 99.99% corresponds to a reliability for generating capacity equivalent to a loss-of-load probability of about 5 to 6 times per year.

These results indicate that tremendous economic advantages are possible as the consequence of further rationalization of the basis for timing generating capacity additions. Because a move in this direction requires another break with tradition and the exercise of considerable judgment and imagination, it will take considerable time and additional investigation. However, an important step in the right direction could be made now by reducing the objective level of service reliability by just a small margin. For example, a loss-of-load probability of 1 day in 2 to 3 years results in incremental costs of additional reliability, as shown by Fig. 16, of about $30 and $130 per kw-hr for systems A and B, respectively, as compared with $500 to $1,100 per kw-hr, respectively, for a loss-of-load probability of 1 day in 10 years.

REFERENCE

1. POWER AND ENERGY PRODUCTION, C. W. Watchorn. *AIEE Transactions*, pt. III-B (*Power Apparatus and Systems*), vol. 73, Aug. 1954, pp. 901–08.

L. K. Kirchmayer: As Chairman of the Probability Applications Working Group, I wish to thank the discussers for their comments. These discussions are indeed helpful in supplementing and clarifying various aspects of this report.

We are pleased to note the complimentary nature of the discussions as well as the wide-spread interest in applying the methods discussed to system planning problems. We anticipate increasing usage of these methods by the industry. It is to be noted that all of the methods described in the report and the discussions may be applied with various degrees of refinement as desired for the particular analysis.

Determination of Reserve Requirements of Two Interconnected Systems

V. M. COOK
SENIOR MEMBER IEEE

C. D. GALLOWAY
MEMBER IEEE

M. J. STEINBERG
FELLOW IEEE

A. J. WOOD
SENIOR MEMBER IEEE

Summary: The interconnection of two power systems offers the opportunity to achieve an appreciable gain in the reliability of the two generation systems and planned economies as a result of capacity sharing. Use of the probabilistic methods described permits evaluation of these reliability and reserve benefits. The methods described extend the single system probability techniques to two interconnected systems. A digital computer program has been developed to implement these methods.

THE BENEFITS of interconnecting two power systems or areas to form an integrated power pool derive from the

following: (1) the interchange of energy; and (2) the gain in the reliability of the generating systems when an interconnection is constructed. The interconnection offers an opportunity for the two areas to share capacity reserves by taking advantage of the load diversities (daily and seasonal) in the two systems, the diversity of forced outages, and the opportunity for integrating planned maintenance outages on a pool basis. Interconnections also may afford the opportunity to take advantage of the economics of larger unit sizes. The

proper use of probabilistic techniques permits evaluation of these reliability and reserve benefits on a rational basis. This paper presents probability methods for calculating the generation system reliability levels for two interconnected power systems and describes a digital computer implementation of these methods.

Application of probability methods in power system planning to date has been primarily in the area of planning generating capacity requirements for single integrated power systems.[1-6] Such application, no matter what particular technique is used, assigns a probability to the generating capacity available, describes the load demands in some manner, and provides a numerical measure of the probability of failing to supply the demanded power or energy. By defining a standard "risk level" (i.e., a standard or maximum probability of failure) and allowing system load demands to grow as a function of time, these probability methods have been utilized to calculate the time when new generating capacity will be required.

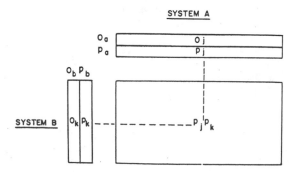

Fig. 1. Two-area interconnection array

O_a, O_b = capacity outages in systems A and B respectively
P_a, P_b = forced outage existence probabilities of exactly O_a or O_b

Paper 62-1344, recommended by the AIEE Power System Engineering Committee and approved by the AIEE Technical Operations Department for presentation at the AIEE Fall General Meeting, Chicago, Ill., October 7–12, 1962. Manuscript submitted May 31, 1962; made available for printing August 1, 1962.

V. M. Cook and M. J. Steinberg are with the Consolidated Edison Company of New York, Inc., New York, N. Y.; C. D. Galloway and A. J. Wood are with the General Electric Company, Schenectady, N. Y.

The authors wish to acknowledge the contributions of Mrs. M. K. Stuvland of the General Electric Company, Schenectady, N. Y., to the development of the digital computer program described in the text.

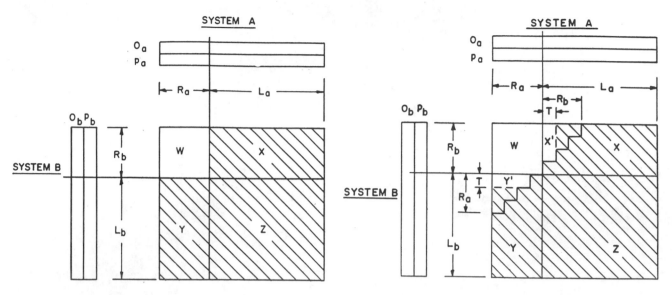

Fig. 2. Two-array with no tie and system loads of L_a and L_b mw

Fig. 3. Two-area array with an infinite tie

Since the probabilities associated with the generating system are a function of the particular combination of machines installed, these methods may be used as a tool to study different generation expansion patterns to establish an economic course of action for a particular system.

The evaluation of loss of load probabilities for interconnected systems is a more complex problem. The introduction of a tie line between two previously unintegrated systems to form a power pool generally has the effect of reducing the loss of load probabilities in each of the systems and reducing the total installed generating reserve capacity required to maintain a given reliability level.[7,8] This paper presents a rigorous method for calculating the probability of loss of load for two areas interconnected with a tie line of finite capacity and discusses how the reserve and reliability benefits of a tie line may be evaluated.

In order to assess the value of interconnections and to determine the installed reserve capacity reduction caused by pooling, a digital computer program has been developed and is discussed. This program is used to study the expansion of two interconnected generation systems and calculates the timing of generation additions for the two areas based on maintaining either the over-all system loss-of-load probability (i.e., days per year) below a preselected standard or by maintaining both system risk levels below the standards selected. The program may be used with other programs (viz., investment and energy production cost calculations) to evaluate the economics of the various power pool designs.

Mathematical Theory

CALCULATION PROCEDURE FOR LOSS OF LOAD PROBABILITY FOR A SINGLE SYSTEM

With only a single integrated system the calculation of the probability of loss of load is quite straightforward.[1,2] The well-known forced outage existence probability technique requires a representation of the system capacity configuration or arrangement and the expected loads. The capacity "model" consists of an ordered list of capacity outages (O_a) in mw (megawatts) and P_a, the corresponding probability of existence of outages equal to or greater than O_a. The construction of this model from generating unit forced outage existence rate data is a simple arithmetic operation which may be accomplished quite rapidly, even for very large systems, by the use of digital computer programs.[9] The load "model," say for a period of a year, consists of an ordered list of the daily peak hourly integrated loads (L_a) in mw. For any given day the difference between available capacity C_a and the system load is the available reserve R_a. For day k the loss-of-load probability P_k corresponds to the probability of existence of outages exceeding the reserve R_a. For this technique the measure of the generation system reliability is defined as the cumulative sum of the loss-of-load probabilities for a given period. The yearly loss of load probability is

$$LOLP = \sum_{k=1}^{\text{year}} P_k \qquad (1)$$

where $LOLP$ is the loss-of-load probability in days per period, in this case, 1 year.

Equation 1 should be considered as the definition of the measure of generation system reliability used in this outage existence probability technique.

INTERCONNECTION OF TWO SYSTEMS

When two systems are interconnected, the calculation of the daily loss-of-load probabilities becomes more complex. Besides accounting for the generation and load within each area, it may be possible for one system to assist the second in case the second area has insufficient reserves. In accounting for this latter situation the question of whether or not the interconnection capacity is large enough to transmit the required power between the areas must be answered.

Inherent in the outage existence probability calculation is the assumption that forced outages are independent random events. For any given time period three loss-of-load probabilities will exist, one for each of the two separate areas and one for the interconnected system as a whole.[10] The significance of the pool loss-of-load probability for a given day may be illustrated by the following. If P_A is the loss of load probability in System A, P_B is the loss-of-load probability in System B, and P_{AB} is the probability of the simultaneous loss of load in Systems A and B, then for the given day the loss of load probability in the total pool is

$$P_S = P_A + P_B - P_{AB} \qquad (2)$$

That is, the pool probability is the sum of the probabilities in each of the two areas, but since both of these probabilities contain the overlapping simultaneous probability P_{AB}, it must be taken out of the sum.

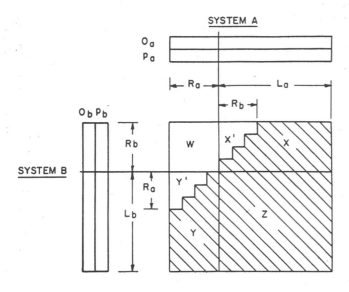

Fig. 4. Two-area array with a tie line of capacity "T" interconnecting systems A and B

Two Systems With No Tie Line

In order to illustrate the calculation of the loss-of-load probabilities for two systems, A and B, the following terms are defined:

O_a, O_b = capacity outages in System A or B, mw

P_a, P_b = probability of the existence in A or B of forced outages equal to or greater than O_a or O_b

p_a, p_b = probability of existence in A or B of forced outages of exactly O_a or O_b

C_a, C_b = capacity installed in A or B, mw

L_a, L_b = daily peak load in A or B, mw

R_a, R_b = available reserve capacity in A or B, mw

T = interconnection capacity between A and B, mw

The calculation of the loss-of-load probability is perhaps best illustrated by the probability array shown in Fig. 1. The probability listings for each of the systems contain all possible outage conditions in an ordered fashion along with the corresponding exact probability of existence. The large matrix-like probability array in the center contains an enumeration of all possible forced outage conditions which could occur simultaneously in both areas. That is, all possible events are represented in this array with the probability of simultaneous outages of O_j in System A and O_k in System B being given by $p_j p_k$. This array is the "sample space" for all forced outages in the two systems and may be used to develop the probability of loss of load for a given configuration of loads and capacities.

With no interconnection between the two systems and peak loads of L_a and L_b the array appears as shown in Fig. 2. For System A the loss-of-load probability corresponds to the shaded areas X and Z on the matrix. That is, for all outages greater than the reserve R_a the system loses load for any and all conditions in System B. Similarly, the loss-of-load probability in System B is the sum of the probabilities represented by the shaded areas Y and Z. That is,

$$LOLP_a = X + Z \tag{3}$$

and

$$LOLP_b = Y + Z \tag{4}$$

The numerical value of the sum of all of the probabilities of all possible events represented in the array is 1 (= a certainty). That is,

$$1 = W + X + Y + Z \tag{5}$$

Now, considering that the pool loss-of-load probability is given by 1—(the sum of the probabilities of all possible events where load is *not* lost), the pooled system loss-of-load probability is

$$LOLP_s = 1 - W \tag{6}$$

Or in terms of equations 3, 4, and 5,

$$\begin{aligned} LOLP_s &= 1 - (1 - X - Y - Z) \\ &= X + Y + Z \\ &= LOLP_a + LOLP_b - Z \end{aligned} \tag{7}$$

The probability represented by Z is the sum of the probabilities of all events corresponding to the simultaneous loss of load in both of the areas. This definition of the system outage probability is equivalent to that described previously.

Infinite Interconnection

If now it is assumed that the two systems or areas are connected to a common bus (i.e., an "infinite" interconnection), then there is no loss of load in the system until the sum of the capacity outages in both of the two areas is greater than the sum of the reserves in both of the areas.

This alters the probability array to that shown in Fig. 3.

In Fig. 3 the individual area loss-of-load probabilities correspond to the sums of the probabilities shown shaded. That is,

$$LOLP_a = X + Z \tag{8}$$

and

$$LOLP_b = Y + Z \tag{9}$$

The pool loss-of-load probability is then

$$\begin{aligned} LOLP_s &= 1 - (W + X' + Y') \\ &= LOLP_a + LOLP_b - Z \end{aligned} \tag{10}$$

The stepped boundary on the array of Fig. 3 between the shaded and unshaded areas is the boundary between the events corresponding to pool outages and those events which are not pool outages. In terms of reserve and outage conditions this boundary satisfies the relationship $O_a + O_b = R_a + R_b$. It is important to note that even with an interconnection of infinite capacity, three distinct loss-of-load probabilities exist, one for each system and one for the pool as a whole, and that these probabilities are generally different values.

Finite Interconnection

In practical situations the interconnection capacity between two systems will not be infinite and this fact must be considered in the analysis of the loss-of-load probabilities. With a finite tie of zero forced outage rate, two questions must be asked whenever assistance is needed by one system. For example, if a shortage of capacity exists in System A,

1. Is there reserve available in Area B of sufficient magnitude to relieve the shortage in Area A?

If the answer to 1 is "yes," then

2. Is the tie line capacity sufficiently large to carry this reserve into Area A?

The probability array corresponding to the previous figures must now be modified to that shown in Fig. 4. The basic difference between this situation and that shown in Fig. 3 for the infinite tie is that the areas X' and Y' (corresponding to the situations when one area is able to help the other in relieving a capacity shortage) have been reduced because of the tie line limitation. The loss-of-load probabilities remain as previously defined, but the calculation procedure must recognize the limitations imposed by the finite tie capability.

Fig. 4 also indicates a definition of an "infinite" tie line capability. Since the tie line can transmit only as much reserve

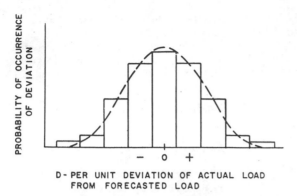

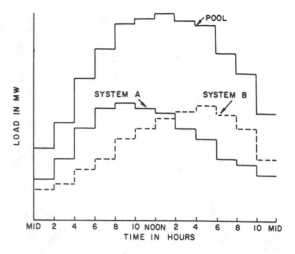

Fig. 5 (left). Probability of the deviation of actual peak loads from forecasted peak loads

Fig. 6 (right). Daily load cycles for two interconnected systems

capability as is available, an "infinite" tie is any tie equal to or larger than the maximum available reserves in either of the two systems.

The introduction of a forced outage rate for the tie line capacity increases the dimensionality of the sample space but does not materially complicate the theoretical or computational problem. For example, assume in Fig. 4 that the interconnection capability of T mw between Systems A and B consists of a single circuit with an outage existence rate of P_T. The loss-of-load probabilities for the individual systems are then

$$LOLP_a = (X+Z) + P_T X' = (X + P_T X') + Z \quad (11)$$

and

$$LOLP_b = (Y+Z) + P_T Y' = (Y + P_T Y') + Z \quad (12)$$

The pool loss-of-load probability is

$$LOLP_s = 1 - [W + (1-P_T)X' + (1-P_T)Y']$$
$$= (X + P_T X') + (Y + P_T Y') + Z$$
$$= LOLP_a + LOLP_b - Z \quad (13)$$

This technique may be extended to consider more than one circuit interconnecting the two areas.

Appendix I illustrates the loss-of-load probability calculation for one day for two sample systems interconnected by a finite-sized tie line.

Loss-of-Load Probability Calculation

The development in the previous section was based on a single set of loads, and capacity configuration. This analysis will now be extended to consider time-varying loads and capacity configurations, as the risk levels which are of value in planning capacity additions are obtained by considering long periods of time.

This calculation of expected risk levels requires the use of load and capacity models, the nature of which will be discussed now.

LOAD MODELS

The loss-of-load probability calculation for a single area utilizes a load model which simulates the integrated 1-hour daily peak loads for the system.[2] The load models for the two systems' probability calculation are similar but should reflect all expected degrees of load diversity between the two areas.

The loss-of-load probability calculation may be performed with two load models which are either in the form of yearly peak load variation curves or else models constructed on a chronological basis. However, if yearly peak load variation curves are utilized, seasonal and daily diversities are apt to be neglected. The use of load models constructed for weekly or monthly intervals offers the opportunity to reflect expected seasonal load diversities between the systems. These models may then be in the form of interval peak load occurrence curves or peak loads on a chronological basis.

These peak load variation curves for interconnected systems may be constructed from historical load data. In some cases it may be desirable to incorporate a consideration of the deviation of actual from forecasted loads in the calculation of the loss-of-load probabilities. The probability method presented here may be directly adapted to consider deviations from forecasted load in the same manner that these deviations are considered in the application of the outage existence method to single areas.

As an illustration of how this technique may be modified to incorporate these deviations, consider the probability distribution shown in Fig. 5. This curve is the probability density function for the deviation of actual load from forecasted load. The per-unit deviation D is defined as

$$D = \frac{\text{Actual load} - \text{Forecasted load}}{\text{Forecasted load}}$$

For example, assume that this same form of distribution applies to both of the two interconnected systems and that the deviations from forecasted loads on both systems will rise and fall simultaneously. This curve may then be represented by N discrete steps so that N distinct loads will be considered for each system for each day. The loss-of-load probability for the day is the summation of the product of the loss-of-load probability for each of the N loads times the probability of the occurrence of that load.

The actual application of the load forecasting deviation function is somewhat hindered by the lack of statistically significant and consistent data on load forecasts. The loss-of-load probability method, however, can be applied in situations where this deviation is to be considered.

Another factor which may have to be considered in the calculation of the probability of loss of load for interconnected systems is the effect of daily load diversity between the two systems. Fig. 6 is an example of system daily load cycles for two interconnected areas and the cycle for the total pooled systems. In this particular example, System A has a load cycle which peaks during the morning while System B has its peak load early in the evening. The total pool load, or system peak, occurs at neither of these times. Therefore, the maximum risks of loss of load for A, B, and the pooled system may occur at different hours of the day.

The method of loss-of-load probability calculations described previously lends itself quite conveniently to the inclusion of daily diversity. In this case the load model for each day must include three

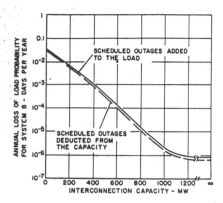

Fig. 7. Interconnection reliability benefits to one system for fixed annual load patterns

sets of loads corresponding to the three conditions listed here:

1. Time of System A peak.
2. Time of System B peak.
3. Time of combined system peak.

In order to account for the effect of daily load diversity, the probability calculations of equations 11 through 13 are repeated each day for these three different hourly periods, resulting in three different loss-of-load probabilities for each system and the pool. As a conservative measure of the risks involved, the daily loss-of-load probability for each element of the pooled system (viz., areas A and B and the total system) is taken as the maximum of the three probabilities calculated for that element for that day. Appendix II gives a numerical example of the effect of daily load diversity on the risk levels of two interconnected systems with the daily loss-of-load probabilities defined as previously.

CAPACITY MODELS

The basic form of the capacity model is the well-known capacity forced outage table such as is shown in Table I. These data may be generated from the capability and forced outage existence rate data for each individual element of the generation system by well-documented methods.[1,2,9] Theoretically, these models should be revised to reflect any change in the capacity configuration resulting from scheduled maintenance outages or changes in unit capabilities. This requirement has the effect of greatly increasing the number of combinations which must be calculated.

In the practical application of the outage existence technique it has been found that a very satisfactory approximation is obtained if scheduled capacity outages for maintenance are added to the load rather than subtracted from the capacity model. This has the effect of reducing the required number of capacity models for any given time period while

maintaining the correct value of available reserve margin. Fig. 7 illustrates the type of variation encountered. These data were calculated using the method described in this paper for two interconnected systems with a total installed capacity of approximately 16,000 mw. The data in Fig. 7 show the annual loss-of-load probability first calculated with the theoretically correct capacity models and then recalculated with scheduled outages added to the load. Adding maintenance capacity outages to the load results in an annual probability value which is not materially higher than that resulting from the theoretically correct method. For tie capacities up to 1,000 mw, the discrepancy between the tie capabilities indicated by the two techniques at a given probability level is less than 25 mw. It has been found that as long as the scheduled outages are a reasonable percentage of the installed capacity, the two techniques will yield nearly identical results.

The use of probabilistic methods for planning generation system expansions requires that the capacity model incorporate a means for scheduling future maintenance outages. Various possible techniques may be utilized for scheduling outages and allocating these maintenance outages to the intervals in the year. The method of calculation for the loss-of-load probabilities described previously does not depend upon the technique used for scheduling maintenance. With interconnected systems a choice must be made as to whether the calculation should be postulated on the total pool capacity and reserves or on the basis of the individual system data. For example, different schedules of maintenance outages will result if these outages are assigned to the intervals throughout a year on the basis of leveling over the year individual system reserves or if they are assigned on the basis of leveling the total pool reserves.

In brief, the capacity models used to calculate the loss-of-load probabilities for interconnected areas consist of ordered capacity outage-probability tables. These tables theoretically should be modified to reflect expected maintenance outages. Practically, the difference between adding the capacity outages to the respective daily loads and modifying the outage probability tables generally has a negligible effect on the annual loss-of-load probabilities.

Digital Computer Programs

In order to implement and facilitate the calculations described previously, two

digital computer programs have been developed. The first of these is a program which may be used to compute the initial capacity models required in order to study the generation expansion of interconnected systems. The methods and data requirements of this program are similar to those which are described in reference 9.

The second program is used to execute the expansion of two generation systems interconnected by a tie line of finite capability. The program calculates the timing of generation additions for the two areas based on maintaining either the over-all pool loss-of-load probability below a preselected standard or else by maintaining each system's loss-of-load probability below a standard. The expansion program may be used to study expansions directly with interconnection capacity determined or may be used to study the effect of different tie sizes in order to establish the optimum tie capacity for a given period of time.

In the program, the yearly maintenance scheduling is done automatically either on the basis of area loads and capacities or on the basis of system-wide data. Basically, the capacity model used for each area in the expansion program consists of an outage probability table plus sufficient data to permit automatic scheduling of maintenance. Maintenance outages and capacity variations are usually treated by load modifications but may be deducted from the capacity model in those instances where it is deemed necessary. The load models are ordered lists of the expected 1-hour integrated peak loads for each weekday of the year arranged and segregated according to maintenance intervals. These load data are used by the program in per unit and are converted internally to mw through the use of yearly peaks or by the use of a starting value and load growth rates.

The input to this program consists basically of a description of the loads expected, the starting generation systems, and tie capacity and a list of machines to be added whenever the standard risk level is exceeded. The annual input data also may include units independently scheduled to be retired or to be added during the year. The output of the expansion program is a running record of the system conditions for each year. The initial capacity models are read as input into the expansion program and kept up to date as units are added or removed from the systems. Maintenance schedules are printed out for each year.

At the end of each year of the expansion the program will test the pool or each

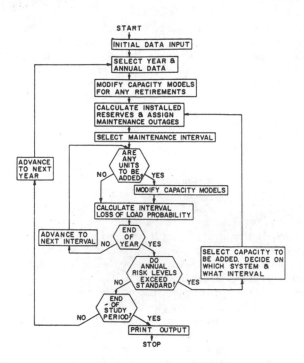

Fig. 8. Flow chart for interconnected systems expansion program

day" or "days per period," and may also, upon the user's option, be printed out for each day of each of the maintenance periods. The results of the automatic maintenance scheduling and probability calculations are summarized annually. Data are given which show by interval for each year the peak loads, capacities and reserves. These annual data also indicate what new capacity has been added, where it was added, and when during the study period it was added.

ILLUSTRATIVE RESULTS

The major output of the expansion program includes the following:

1. Up-dated capacity models for each permanent change in installed capacity.

2. A year by year summary of the scheduled maintenance outages.

3. A year by year summary of the loss of load probabilities in each system and in the pool.

Other data may be obtained at the user's option. A sample of the yearly output summary sheet for the pool or total system is shown in Fig. 9. These data are summarized on a monthly basis and give the loss-of-load probabilities, peak loads, installed capacities, scheduled outages, and installed and available reserves for each month. In addition the yearly risk level is given and a message is printed to show that in this example, the standard risk level had been exceeded, a 200-mw unit would be added to area B in maintenance interval 6, and the year would be recalculated.

Fig. 7 illustrates for a different power pool the effect of varying the interconnection capacity on the annual loss-of-load probability of one of the systems. The reliability benefits of an interconnection to this system are clearly indicated by the probability method. Fig. 10 illustrates for this same system the reserve benefits of various tie line capacities. These data were obtained by maintaining the system reliability fixed at one day in 10 years and adjusting the tie capacity and reserve margins. As might be expected, the relatively small tie capabilities reduce the system reserve requirements by the same amounts as the tie capacity. However, as the interconnection capability approaches or exceeds the reserve margin in the other system, the incremental benefits of increased tie capacity become negligible. The methods described in this paper provide a means whereby interconnection reliability and reserve capacity benefits may be evaluated for both member systems of the pool as well as from the pool standpoint.

---

individual system loss-of-load probability, and if the standard risk level is exceeded, it will add a unit to the area with the worst risk level. This addition is made in a preselected maintenance period and the year recalculated. This process continues until an acceptable annual loss-of-load probability for that year is achieved. As the calculation proceeds, yearly data are stored so that if it is desired to revise the expansion from any arbitrary point in time, the expansion may be restarted from any given year rather than from the beginning.

EXPANSION PROGRAM LOGIC

Fig. 8 shows an outline of the logical arrangement of the interconnected system expansion program. This logic is illustrated for a version which assigns maintenance outages on the basis of leveling available interval reserves over a year and calculates the loss-of-load probabilities with these capacity variations added to the loads. Automatic maintenance scheduling is accomplished through the use of maintenance cycle data for each machine. These data must specify for each machine the number of consecutive maintenance intervals that the machine is to be taken out of service for each year of the cycle. For a given machine this cycle may be repeated until the end of the study period, or it may be repeated as many times as desired and then automatically converted to a different cycle.

Provision has been made in the program to accommodate generating units which do not have a fixed maximum capability throughout the year, such as

hydroelectric machines. These capacity variations are specified by maintenance interval for each machine with a variable maximum capability. The program will then utilize these data in calculating interval reserves and automatically scheduling maintenance outages.

The expansion program proceeds on a chronological basis calculating the loss-of-load probabilities for each day of the maintenance interval. At the end of the year the yearly risk levels (i.e., annual loss-of-load probabilities) are compared with the standards and if either the pool or individual system standards are exceeded, a unit is automatically selected to be added to the system in difficulty. The unit selection process may use a preselected expansion pattern for each system or constant percentage unit size expansions may be run. When capacity is to be added (i.e., the calculated risk levels have exceeded the standard) the program generates a message informing the user what is being done and then automatically recycles to the start of the year just calculated, reassigns the maintenance taking into account the change in installed capacity in the year, and recalculates the loss-of-load probabilities. The program will not advance to the next year until the reliability levels are satisfactory.

The program output may be obtained in several modes which are selected by input indices. Initially in the program, the input data are printed in order to facilitate the checking of system data and to provide a positive method of identifying the case run. The loss of load probabilities may be printed in units of "years per

YEAR 1962 RUN 1 EXPANSION NUMBER 1

GENERAL ELECTRIC CO. EUAEO PROGRAM E-EP1 SYSTEM

PERIOD	LOSS OF LOAD PROBABILITY DAYS/PERIOD	LOSS OF LOAD PROBABILITY DAYS/YEAR	PERIOD PEAK LOAD (MW)	INSTALLED CAPACITY (MW)	MAINT. OUTAGE (MW)	INSTALLED RESERVE (MW)	INSTALLED RESERVE PRCT.OF PD.LOAD	AVAILABLE RESERVE (MW)	AVAILABLE RESERVE PRCT.OF PD.LOAD
1	0.36324662E-02	0.41938473E-01	1702.4	2300.0	125.0	597.6	35.10	472.6	27.76
2	0.12251959E-02	0.15559988E-01	1639.7	2300.0	150.0	660.3	40.27	510.3	31.12
3	0.96314339E-02	0.11649448E-00	1574.1	2305.0	325.0	730.9	46.43	405.9	25.78
4	0.80907762E-02	0.10816090E-00	1552.0	2305.0	355.0	753.0	48.52	398.0	25.64
5	0.92081035E-02	0.97452429E-01	1664.5	2300.0	155.0	635.5	38.18	480.5	28.87
6	0.91158127E-02	0.11025792E-00	1798.0	2295.0	105.0	497.0	27.64	392.0	21.80
7	0.42833070E-02	0.49452727E-01	1823.3	2295.0	55.0	471.7	25.87	416.7	22.86
8	0.86370602E-02	0.10446730E-00	1845.1	2290.0	55.0	444.9	24.11	389.9	21.13
9	0.44406003E-02	0.56395625E-01	1858.2	2290.5	4.5	432.3	23.27	427.8	23.03
10	0.12875249E-02	0.18168408E-01	1774.1	2300.0	0.	525.9	29.64	525.9	29.64
11	0.35566006E-02	0.36135063E-01	1757.0	2300.0	25.0	543.0	30.91	518.0	29.48
12	0.89067946E-02	0.10772980E-00	1819.7	2300.0	25.0	480.3	26.39	455.3	25.02
WORST PERIOD 3	0.96314339E-02	0.11649448E-00	1574.1	2305.0	325.0	730.9	46.43	405.9	25.78
YEARLY SUMMARY	0.	0.72015675E-01	1858.2						

MINIMUM INSTALLED RESERVE
PERIOD 9 432.3 MW
23.27 PRCT. OF ANNUAL PEAK LOAD

MINIMUM ACTUAL RESERVE
PERIOD 8 389.9 MW
20.98 PRCT. OF ANNUAL PEAK LOAD

TIE SIZE 80 MW

Fig. 9. Output sheet showing annual summary for total system

Table I. Probability of Loss of Installed Capacity (Installed Capacity 600 Mw)

Forced Outage, Mw	Probability*		Capacity Available, Mw
	Exact, p_k	Cumulative, P_k	
0	297,553	1,000,000	600
10	364,351	702,447	590
20	219,354	338,096	580
30	86,548	118,742	570
40	25,169	32,194	560
50	5,753	7,025	550
60	1,077	1,272	540
70	169	195	530
80	23	26	520
90	3	3	510

* Probabilities are given in millionths.

Table II

	Installed Capacity, Mw	Independent Peak Loads, Mw	Installed Reserve, Mw
System A isolated	5,400	4,906	494
System B isolated	3,600	3,225	375
Total	9,000	8,131	869

The load data for this day show that the reserve situation at the time of the pool peak load is as shown in Table III.

Table III

	System A, Mw	System B, Mw	Total, Mw
Installed capacity	5,400	3,600	9,000
Load	4,756	3,165	7,921
Installed reserve	644	435	1,079
Tie capacity	90	90	
Reserve with tie	734	525	1,079

Table IV

System A:
Without interconnection....0.000835
Reduction due to interconnection.............0.000638

With interconnection................0.000197
System B:
Without interconnection....0.001020
Reduction due to interconnection.............0.000913

With interconnection................0.000107

Pool:
Before correction for overlap..........................0.000304
Correction for overlap...............−0

Total outage probability
with tie available 100%
of time..........................0.000304

Conclusions

Besides the economies obtainable from the interchange of energy, the interconnection of two power systems to form a power pool offers the participating systems the opportunity to capitalize on the savings afforded by sharing capacity

reserves resulting from the seasonal and daily load diversities, the sharing of reserves to maintain or improve generating system reliability levels, and the possible benefits which may occur from integrated capacity maintenance planning. This integration will permit and encourage economies in energy production and the expansion of the systems utilizing larger units sizes. By utilizing probabilistic methods and suitable computer implementation, the capacity reserve, load diversity and maintenance scheduling benefits may be studied. These techniques will permit the evaluation of the reliability benefits of a tie and the expansion of interconnected systems to study unit size economics.

The paper presents a straightforward method of analyzing the loss-of-load probabilities of two systems interconnected by a tie-line of finite capability. This method is theoretically sound and may be adapted to include the effects of tie line forced outage rates, load forecasting deviations, daily diversity considerations.

In order to implement the study of the expansion of interconnected generation systems a digital computer program utilizing this technique has been developed. This program may be efficiently employed to time capacity additions and to establish the reliability and reserve benefits which may be obtained by the interconnection of two power systems to form a pool. The probabilistic method described permits the segregation of these data for each interconnected system as well as for the pool as a whole.

Appendix I. Example of Probability of Loss-of-Load Calculations for Two Systems Having a Finite Interconnection

The two systems, A and B, are assumed to be interconnected with a single tie line

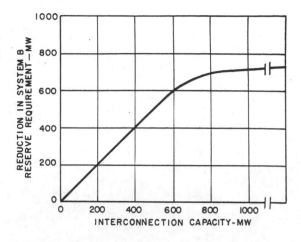

Fig. 10. Interconnection reserve benefits to one system in 1 year of expansion. Annual pool loss-of-load probability maintained at 1 day in 10 years

of 90 mw capacity. Either system may supply aid to the other system up to the capacity of the tie line, provided the sending system has capacity available to the receiving system. System A has 70 generating units of different capacity ratings and System B has 60 units of different sizes. The total installed capacity of System A is 5,400 mw and that of System B is 3,600 mw.

Fig. 11 shows the probability array for combining the forced outage probabilities of the two systems. The outage probability data for System A area are shown in the first three columns on the left side of the illustration and the System B data are given on the three top horizontal lines. The numbers in the matrix-like array are the exact probabilities (expressed in millionths) of the simultaneous existence of the outages in Systems A and B. Calculations are illustrated for one day only with the data for this day given in Table II. The reserve values of 644, 435, 1,079, 734, and 525 mw determine the position in Fig. 11 of the marked horizontal, vertical, and stepped lines in accordance with the theory as presented in the text.

Assuming that the forced outage existence rate of the tie line is zero, the loss-of-load probabilities for the two systems and the pool at the time of the pool peak may be calculated as indicated in Table IV.

Appendix II. Summary of the Results of Calculations Including Daily Peak Load Diversity

The effect of daily peak load diversity of the calculated annual loss-of-load probabilities is illustrated by using the systems of Appendix I. Using 252 weekdays, loads are arranged in descending order of magnitude. For illustrative purposes a variation has been assumed such that the minimum peak in System A is 80% of the annual System A peak, 90% in area B, and 84% for the pool. Appendix I indicates the load, capacity and reserve situation assumed to occur at the time of the annual pool peak. The installed capacities of 5,400 and 3,600 mw for Systems A and B respectively, as well as the interconnection capability of 90 mw are assumed to remain constant throughout the year. At the time of occurrence of the System A,

435 525

	System "A" 70 Units MW OUT	OUTAGE PROBABILITY	CUMULATIVE PROBABILITY	System "B" 60 Units 0 MW	30 MW	60 MW	90 MW	120 MW	150 MW	180 MW	210 MW	240 MW	270 MW	300 MW	330 MW	360 MW	390 MW	420 MW	450 MW	480 MW	510 MW	540 MW	570 MW
				193 540	107 411	135 728	171 737	104 199	93 707	73 185	43 793	31 387	19 645	11 135	6 775	3 715	1 966	1 057	527	260	126	59	27
				1 000 000	806 460	599 049	563 321	391 584	287 385	193 678	120 493	76 700	45 313	25 668	14 533	7 758	4 043	2 077	1 020	493	233	107	48
	0	158 136	1 000 000	30 606	16 986	21 464	27 158	16 478	14 818	11 573	6 925	4 964	3 107	1 761	1 071	587	311	167	83	41	20	9	4
	30	87 763	841 864	16 986	9 427	11 912	15 072	9 145	8 224	6 423	3 844	2 755	1 724	977	595	326	172	93	46	23	11	5	2
	60	110 899	754 101	21 463	11 912	15 052	19 045	11 555	10 392	8 116	4 857	3 481	2 179	1 235	751	412	218	117	58	29	14	7	3
	90	140 323	643 202	27 158	15 072	19 046	24 099	14 622	13 149	10 270	6 145	4 404	2 757	1 563	951	521	276	148	74	36	18	8	4
	120	85 137	502 879	16 477	9 144	11 556	14 621	8 871	7 978	6 231	3 728	2 672	1 672	948	577	316	167	90	45	22	11	5	2
	150	76 566	417 742	14 819	8 224	10 392	13 149	7 978	7 175	5 603	3 353	2 403	1 504	852	519	285	151	81	40	20	10	5	2
	180	92 069	341 176	17 819	9 889	12 495	15 811	9 593	8 627	6 738	4 032	2 890	1 809	1 025	624	342	181	97	49	24	12	5	3
	210	53 694	249 107	10 392	5 767	[illegible]	9 221	5 595	5 031	3 930	2 351	1 685	1 055	598	364	199	106	57	28	14	7	3	2
	240	48 278	195 413	9 344	5 186	6 551	8 291	5 031	4 524	3 533	2 114	1 515	948	538	327	179	95	51	26	12	6	3	1
	270	44 688	14[illegible] 135	8 649	4 800	6 065	7 674	4 656	4 188	3 270	1 957	1 403	878	497	303	166	88	47	24	12	6	3	1
	300	26 473	102 447	5 123	2 844	3 593	4 547	2 758	2 481	1 937	1 159	831	520	295	179	98	52	28	14	7	3	2	1
	330	21 161	75 974	4 096	2 273	2 872	3 634	2 205	1 983	1 549	927	664	416	235	143	79	42	22	11	5	3	1	1
	360	18 204	54 813	3 523	1 955	2 471	3 126	1 897	1 706	1 332	797	571	358	203	123	68	36	19	10	5	2	1	1
	390	10 553	36 609	2 043	1 134	1 432	1 812	1 100	989	772	462	331	207	117	72	39	21	11	6	3	1	1	
	420	8 176	26 056	1 582	878	1 110	1 404	852	766	598	358	257	161	91	55	31	16	9	4	2	1	1	
	450	6 337	17 880	1 227	681	860	1 088	660	594	464	278	199	124	70	43	24	12	7	3	2	1		
	480	3 664	11 543	709	394	497	629	382	343	268	161	115	72	41	25	14	7	4	2	1			
	510	2 668	7 879	517	286	362	458	278	250	195	117	84	53	30	18	10	5	3	1	1			
	540	1 950	5 211	377	209	265	335	203	183	143	86	61	38	22	13	7	4	2	1	1			
	570	1 110	3 261	215	119	151	191	116	104	81	49	35	22	12	7	4	2	1	1				
	600	780	2 151	151	84	106	134	81	73	57	34	24	15	9	5	3	2	1	1				
644	630	536	1 371	104	58	73	92	56	50	39	23	17	10	6	4	2	1	1					
	660	302	835	58	32	41	52	32	28	22	13	10	6	3	2	1	1	1					
	690	204	533	40	22	28	35	21	19	15	9	6	4	2	2	1							
734	720	134	329	26	14	18	23	14	12	10	6	4	3	2	1	1							
	750	74	195	14	8	10	13	8	7	6	3	2	1	1	1								
	780	49	121	9	5	7	8	5	5	4	2	2	1	1									
	810	30	72	6	3	4	5	3	3	2	1	1	1	1									
	840	17	42	3	2	2	3	2	2	1	1	1											
	870	10	25	2	1	1	2	1	1	1	1												
	900	7	15	1	1	1	1	1	1	1													
	930	3	8	(1)			(1)			(1)													
	960	2	5		(1)				(1)														
	990	1	3				(1)																
	1020	1	2				(1)																
1079	1050	1	1				(1)																
	1080																						
	[illegible]																						

Fig. 11. Probability of forced outage in excess of reserve for two interconnected systems at time of combined two systems peak load

annual peak load of 4,906 mw, the load in B is 2,985 mw. When B has its independent annual peak of 3,225 mw, the System A load is 4,636 mw.

Three calculations are made for each day.

1. At the time of System A peak.
2. At the time of System B peak.
3. At the time of the pool peak.

This results in 756 calculations similar to the one pool peak computation illustrated in Fig. 11. The annual results are given in Table V. The first three rows of this table show the annual loss of-load-probabilities calculated and summed for the daily peak load conditions as indicated (i.e., times of A peak, B peak, and the pool peak). The last row gives the probabilities summed on the basis of taking the daily loss-of-load probability for each system and the pool as the maximum value which exists during each day, and these values are taken as the measure of service reliability. In this example, the annual risk levels in the individual systems correspond to those values which were found by considering only the time of the day when the individual system peaks occurred. The pool annual risk level, however, corresponds to a combination of daily loss-of-load probabilities which occur at various times of the day.

References

1. GENERATING RESERVE CAPACITY DETERMINED BY THE PROBABILITY METHOD, Giuseppe Calabrese. *AIEE Transactions*, vol. 66, 1947, pp. 1439–50.

2. APPLICATION OF PROBABILITY METHODS TO GENERATING CAPACITY PROBLEMS, AIEE Subcommittee on Application of Probability Methods. *Ibid.*, pt. III (*Power Apparatus and Systems*), vol. 79, 1960 (Feb. 1961 section), pp. 1165–77.

3. THE APPLICATION OF PLANNING CRITERIA TO THE DETERMINATION OF GENERATOR SERVICE DATES BY OPERATIONAL GAMING, C. A. De Salvo, C. H. Hoffman, R. G. Hooke. *Ibid.*, vol. 78, Dec. 1959, pp. 1752–59.

4. AN INVESTIGATION OF THE ECONOMIC SIZE OF STEAM-ELECTRIC GENERATING UNITS, L. K. Kirchmayer, A. G. Mellor, J. F. O'Mara, J. R. Stevenson. *Ibid.*, vol. 74, 1956, pp. 600–14.

5. ELEMENTS OF SYSTEM CAPACITY REQUIREMENTS, C. W. Watchorn. *Ibid.*, pt. II (*Applications and Industry*), vol. 70, 1951, pp. 1163–80.

6. EVALUATION OF UNIT CAPACITY ADDITIONS, M. J. Steinberg, V. M. Cook. *Ibid.*, pt. III (*Power Apparatus and Systems*), vol. 75, 1956, pp. 169–79.

7. THE EFFECT OF INTERCONNECTIONS ON ECONOMIC GENERATION EXPANSION PATTERNS,

Table V. Annual Loss-of-Load Probabilities

Condition	Annual Loss-of-Load Probabilities, (Days Per Year)		
	System A	System B	Pool
Time of System A daily peak loads	0.036437	0.000012	0.036449
Time of System B daily peak loads	0.000367	0.015749	0.016116
Time of pool daily peak loads	0.003086	0.003250	0.006336
Daily selection of data to yield maximum probability	0.036437	0.015749	0.037917

L. K. Kirchmayer, A. G. Mellor, H. O. Simmons Jr. *Ibid.*, vol. 76, 1957, pp. 203–14.

8. ECONOMIC CHOICE OF GENERATOR UNIT SIZE, L. K. Kirchmayer, A. G. Mellor. *Transactions*, American Society of Mechanical Engineers, New York, N. Y., vol. 80, 1958. pp. 1015–23.

9. DIGITAL COMPUTER AIDS ECONOMIC PROBABILISTIC STUDY OF GENERATION SYSTEMS—I, M. K. Brennan, C. D. Galloway, L. K. Kirchmayer. *AIEE Transactions*, pt. III (*Power Apparatus and Systems*), vol. 77, 1958, pp. 564–71.

10. AN INTRODUCTION TO PROBABILITY THEORY AND ITS APPLICATIONS (book), William Feller. John Wiley & Sons, Inc., New York, N. Y., vol. I, second edition, 1957, pp. 23 and 89.

◆

Discussion

W. D. Masters and **R. C. Craft** (The Cleveland Electric Illuminating Company, Cleveland, Ohio): Increased emphasis is currently placed on achieving the benefits of interconnected operation of power systems. With more widespread acceptance of the application of probability mathematics to generation planning, the method proposed by the authors for determining reserve requirements has great potential value. We have applied the theories presented in an approximate manner. We would like to comment on several points which proved to be of concern to us.

UNIT SIZE AND REQUIRED RESERVE

One of the potential benefits of interconnected operation is the reduction in required reserve while maintaining a specified reliability. Interconnections also may afford the opportunity to take advantage of the economies of larger unit sizes. Studies of reserve requirements by probability methods demonstrate that the reserve required to maintain a specified risk level increases with unit size. Increased unit size, in essence, sacrifices some part of the reduction in required reserves made possible through interconnection. The program discussed in this paper permits a detailed examination of these effects.

MAINTENANCE ADDED TO LOAD

The authors mentioned that scheduled capacity outages for maintenance are added to the load rather than subtracted from the capacity model. The latter method is more accurate, but it requires considerably more computer running time. They note that, as long as the scheduled outages are a reasonable percentage of the installed capacity, the two techniques yield very nearly identical results. As individual unit sizes approach 12% of total installed capacity, the difference in results is of more concern. By applying both techniques to several different system sizes, it is possible to develop a table which relates an effective maintenance value for each large unit to total installed capacity. This effective maintenance value, which is less than the unit size, when added to load, again yields a loss-of-load probability nearly identical to that obtained by subtracting the unit from the capacity model.

TIE CAPACITY

The authors have not attempted to clarify the definitions of "tie capacity" and

"tie capability." Does "tie capability" refer to the potential reduction in required reserves? The value of "tie capacity," used as input to the program (labeled "interconnection capacity" in Fig. 10), would properly represent the maximum continuous power which can be transmitted from one system to the other, as distinguished from the thermal capacity of the interconnecting facilities. The exact value of tie capacity at any time can be governed by thermal and voltage limitations internal as well as external to the systems under study. Tie capacity, when thought of as the transferability of power, could quite conceivably vary with load level. It could also take on different values for the transfer of power in opposite directions, especially when more than one location of interconnecting circuits between the two systems results in circulating power.

The program described in this paper gives the user the freedom to select a value of tie capacity for each year. A change in tie capacity, as used by the program, implies a change in the configuration of interconnecting facilities. We suggest that the effect of changes in tie capacity for daily variations in load level and direction of power transfer can be as significant as the effects of tie line forced outage rate, load forecasting deviation, and daily load diversity. Digital load flows could be used to develop a table relating tie capacity to direction of flow and a weighted combination of the two system loads.

MEASURES OF RISK

One of the more important observations made possible by the theory presented here is that, generally, Systems A and B, and the combined pool have three different loss-of-load probabilities. This effect is contingent upon the assumption that, if System B is short on operable capacity, System A will assist System B to the extent of the tie capacity limitation or up to the point at which System A loses load, whichever is smaller. If each of two interconnected systems is operating at a risk level of 1 day per 10 years loss of load in year X, the probability that the loss-of-load event will occur in both systems on the same day is very small. The pool risk level is very nearly equal to the sum of risks for Systems A and B, or approximately 2 days per 10 years. At the opposite extreme, if each system had a risk level of 365 days per year the overlap would be 100%, and the pool risk would also be 365 days per year. This represents the only case in which all three risks can be identical.

With the idea of pool risk differing from the individual system risks a re-examination is in order of the whole concept of loss-of-load probability as a planning criterion. It is evident that the popular frequency related measure (days per year) does not incorporate adequately all of the phenomenon that contribute to over-all service quality. It seems entirely possible that two companies sharing the same over-all service quality objectives could achieve their goal by applying different standards of loss-load probability in their generation planning. Average customer demand, for example, just one factor which requires greater recognition of magnitude of shortage in a risk

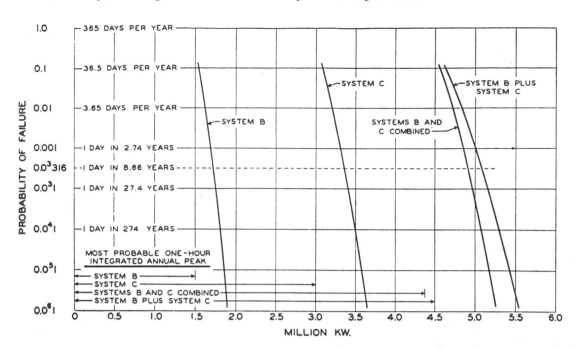

Fig. 12. Probability of failure to carry the load versus installed capacity

measure. Perhaps, as a compromise, the frequency-based risk measure (days per year) could be made proportional to system load or capacity, or to the number of customers. It will be desirable to develop, eventually, other risk measures which will incorporate more completely the concepts of duration and magnitude of outage as well as frequency.

FUTURE APPLICATIONS

The procedures outlined by the authors seem directly applicable to the planning of internal transmission. Using probability calculations in the manner described will permit the construction of transmission and generation location alternates while complying with the desired risk levels for various parts of the system. This application, however, is highly contingent upon resolving the difficulties, mentioned in the preceding paragraph, with regard to the establishment of risk standards.

With some judgment the two-area program can be applied to multi-area problems. For example, to evaluate the reserve benefits to System A for a pool consisting of Systems A, B, and C, either of the following two methods could be employed:

Method A: (1). Solve for reserve benefits to System A in a pool of A and B. (2). Solve for reserve benefits to System A in a pool of A and C. (3). Add the results of steps 1 and 2.

Method B: Solve for reserve benefits to System A in a pool of System A and a system equal to the sum of the generation and loads of B and C.

Method A will indicate a greater reduction in required reserve than method B. However, method B is a more realistic evaluation, and should be used even though an infinite interconnection must be assumed between systems B and C. An arbitrary reduction in the effective tie capacity between system A and the combination of B and C might also be employed to represent

the effect of a tie capacity limitation between B and C.

SUMMARY

There are many benefits to be derived from the interconnection of power systems. This paper focuses attention upon the reduction in required reserve. The installation of an interconnection between two previously independent systems has the effect of reducing loss-of-load probability in each of the systems, or reducing the total installed reserve capacity required to maintain a given reliability level. The choice must be made between one effect or the other; both cannot be achieved at the same time.

In the planning of interconnected systems, digital programs as described in this paper can provide answers to questions such as the following:

1. Can any generator installation dates in the present expansion be deferred by virtue of a proposed interconnection, without exceeding the standard risk level?
2. In what year should an interconnection be built?
3. What size should an interconnection be initially?
4. Can any additional benefits be derived from the construction of additional interconnecting facilities at a later date?
5. When and where should these additional facilities be built?

In summary, the theory and applications presented can be utilized to optimize, on an economic basis, the expansion of interconnected systems with respect to the installation of interconnecting facilities as well as generating capacity.

H. A. Adler (Commonwealth Edison Company, Chicago, Ill.): Evaluation of the gains from interconnections by means of probability studies is becoming increasingly important. This paper is an important contribution since it presents a computer pro-

gram which makes it practical to apply the loss-of-load method to a two-system interconnection. The loss-of-load method applied to two interconnected systems was described by G. Calabrese as early as 1947 and again in 1951. His method appears in principle to be the same as that described in the present paper but the use of computer techniques makes an otherwise complicated procedure more practical.

It appears that the method shows only the existence of excess or deficient capacity. The magnitude of these values must be determined by iteration. In general, it is important to know the magnitude of excess or deficiencies in each month because this information can serve as the basis of arrangements of fixed purchases between neighboring companies.

It appears that this method has been developed only for two-system interconnections. Studies requiring the evaluation of the value of interconnection between more than two systems are not uncommon. The loss-of-load method seems to increase rapidly in complexity as studies are extended beyond two-system interconnections.

One of the more important problems in such interconnection studies is the question of the allocation of the gains to the various participants in the interconnection. The problem is of special importance when a relatively small company joins an existing large interconnection. There are a number of related problems in connection with allocation of gains, such as when several systems of different size or with different reserve requirements interconnect, or how to apportion incremental gains from additional interconnections. The value of an interconnection may be greatly affected for a given company, depending on how much it is required to carry on owned reserve and how much it can depend on its neighbors. The author's ideas on this problem would be of interest.

Carl Kist and **H. S. Worcester** (Los Angeles Department of Water and Power, Los

Angeles, Calif.): The authors have taken an important step forward in the advancement of the solution of power system generation reserve capacity requirements by the application of probability methods.

This solution enables full use to be made of the generator reliability characteristics on both sides of a tie point as well as taking into account the tie line capacity limitations. It is a major step forward from the concept of calculating reserve requirement capacity for a single system which considers the far side of a tie point merely as an emergency generator or an intermittent load.

The method described can be applied within a utility having limited tie line capacity between parts of its system as well as between two utilities or two connected power pools having limited tie line capacity. In addition to such situations, cases exist where more than one limited capacity tie line becomes involved between or among interconnected utilities or pools. A particularly difficult problem of this kind would occur where two or more systems or pools are connected in series. It is believed that an actual solution to such a problem could be obtained by extension of the principles outlined in the paper. The authors' comments on this would be interesting.

The authors' method of handling generating capacity out for overhaul is very useful in reducing the amount of computations required. Recognition by the authors of discrepancies introduced by not subtracting the amount of reserve allocated to the capacity out for overhaul from the total reserve requirements, is noted. The text indicates that this discrepancy would be of the order of 25 mw for tie capacity of 1,000 mw. That is a relatively small amount, but 25 mw of additional capacity on a system represents a sizable investment. It may be desirable to make an approximation of the subtraction from the reserve required resulting from capacity out for overhaul based upon experience gained from probability calculations, and correct the reserve requirement accordingly.

Mr. Worcester and I will not pass up this opportunity to insert a commercial of our own. The criterion for reliability in the method selected by the authors is days out in a period of time. As you know, we like the method of kilowatt-hours of load interrupted because we think it provides more realistic results.

F. C. Poage, W. E. Slemmer, and **V. A. Thiemann** (Ebasco Services, Inc., New York, N. Y.): This paper concerning the effect of tie line capacity on the reserve requirements of interconnected areas is a very useful and interesting contribution to the subject. The authors are to be congratulated on an excellent paper presenting a straightforward method of analyzing the loss-of-load probabilities of two systems interconnected by a tie-line of finite capability.

The use of probability methods is the only way of computing the reserve required for a given degree of risk of loss of load. The programs described in this paper offer a very practical procedure for computing the loss-of-load probability. Means are afforded for taking into account diversities, deviations from forecasted loads, scheduled maintenance outages, changes in unit capabilities,

the expansion program for the generating system, and the effects of tie line forced outage rates.

The methods presented in this paper appear to be sufficiently flexible to permit incorporation in system planning programs to determine the conomics of various schemes of generating capacity expansion, and the optimum tie line capability.

We have approached this problem from the point of view of determining the value of an interconnection in terms of the increase in load-carrying capability that it produces. Our program develops the amount of load that can be carried with a given set of facilities operating as two or more independent systems and alternatively operated as an interconnected system either with a finite transmission tie or as a completely integrated system. The criterion used is to provide the same service reliability in either case.

Like the program discussed in the paper our program provides for evaluating diversities in load requirements. Scheduled maintenance is treated as a part of load requirements as are changes in unit capabilities. The procedure also evaluates deviations from forecast loads if probability data relating actual loads to forecasts are available.

While determination of the probability of loss of load is the basis for both programs, we believe that an answer which states the value of an interconnection in terms of resulting comparable load-carrying capabilities of the systems in megawatts is more readily usable than an abstract number representing service reliability.

C. W. Watchorn (Pennsylvania Power & Light Company, Allentown, Pa.): The paper has many interesting and important features, among which are (1) the an-

nouncement of the availability of a computer program that can handle large numbers of units in a rigorous manner, (2) the demonstration that it is entirely practical under certain circumstances to include the capacity of a unit that is out for maintenance in both the load and the available capacity as a computational technique, and (3) the pointing out of the lack of statistically significant and consistent data on load forecasting deviations.

It is noted that the program is such that the effect of maintenance may also be accounted for rigorously if necessary or desired. It would be interesting to learn if a criterion has been devised to decide or indicate whether the calculations involving maintenance of units should be made on the approximate or rigorous basis.

Neither Fig. 9 nor the text relating to it make it clear as to whether the decision to add more capacity is based on the probability of loss of load obtained as that for the worst period multiplied by approximately 12, as set out immediately below the listing of the results for each of the 12 periods, or whether the decision is based on the probability of loss of load obtained as the sum of those for the 12 periods, and as set out in the next line as the yearly summary. This should be made clear, since one of the considerations that gave impetus to the application of probability methods to power system problems during its early infancy, about a quarter of a century ago, was the realization that two systems, substantially the same in all respects, including the sizes of units and the same annual peak loads, could require different amounts of installed capacity if one had a shallower summer valley than the other. This realization gave rise to the desire to provide a means for measuring any resulting differences there may be in the

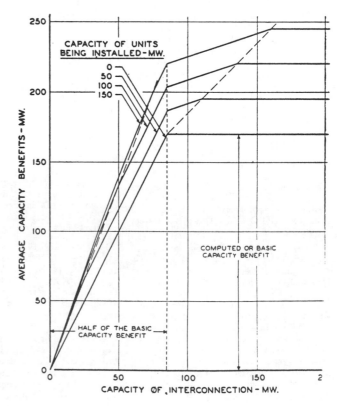

Fig. 13. Interconnection capacity benefits versus transmission capacity (Systems B and C)

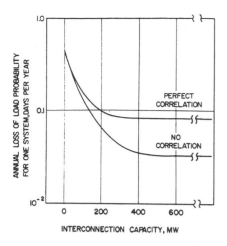

Fig. 14. Interconnection benefits to one of two identical systems for two correlating assumptions

installed capacity requirements. The first of these two possible criteria, related to Fig. 9, would be of no help in this regard, since it considers only the most critical period in the year without regard to the differences that may exist during the rest of the year. Such practice could greatly overstate the loss-of-load probability as compared with the results of the correct general method of evaluation, as shown by the line of Fig. 9 setting out the yearly summary and as obtained in accordance with equation 1 of the paper and the description therein of its application. However, if the latter is the intention of the paper, the purpose or significance of multiplying the period loss-of-load probability by approximately 12 is not at all clear.

I believe that a capacity year, which would be the 12 periods beginning with that for which additional capacity would be provided, that is, the sixth in the instant case, would be more significant than a calendar year as used in the paper since it relates to the same total amount of capacity for the whole period for which the loss-of-load probability is desired to be determined.

Although, as pointed out in the paper, there is no statistically significant and consistent data on load forecasting deviations, it is nevertheless necessary to incorporate provision for the effect of the variability of load in any proper determination of installed generating capacity requirements. This problem can be handled satisfactorily on a statical basis without regard to load forecasts as such or in the more generally understood sense. A method for doing just this is suggested in part in reference 1.

Incomplete studies as to the economic size of units have indicated that the economic size is somewhere between a constant percentage size and a constant size. This means that the economic size of unit is a decreasing percentage size with time. It would be of interest, and well worth-while, to determine, if possible, the most economical pattern of unit sizes as contrasted with merely the most economical among preselected expansion patterns of which none may be the truly most economical.

It is understood that the method of evaluating the capacity benefits of an interconnection, which is described in the

paper as being a more complex problem compared with the straight-forward calculations required for a single system, is limited to the interconnection of but two systems. I presented a method about 12 years ago, reported in reference 26 of reference 2 of the paper, by means of which a straight-forward determination of the benefits from interconnecting any number of systems may be evaluated, including a rigorous determination of the effect of load diversity rather than only approximating such effect, as it appears in the paper.

This method is simply illustrated for two systems by Fig. 12, which is Fig. 1 of the reference paper, where the probability of loss of load is shown as a function of installed generating capacity for (1) each of Systems B and C separately, (2) their sum, and (3) combined Systems B and C. The difference between the sum of the installed generating capacity requirements for Systems B and C and that for combined Systems B and C, i.e., the difference between items (2) and (3), is the capacity benefit from interconnecting Systems B and C, including a correct evaluation of the effect of load diversity. Such an evaluation is for no transmission limitations and zero forced outage rate for the transmission lines. This latter assumption is justified in most cases, since, generally, the forced outage rates of transmission lines are so small as to have very little practical significance. This evaluation was called the basic capacity benefit in the reference paper and is shown by the lower horizontal line in Fig. 13. The effect of tie line limitations as shown in Fig. 13 is confirmed by Fig. 10 of the subject paper. Since this latter figure is for but one system, the portion for which the transmission limitations are effective has a 1:1 slope, whereas, as should be expected, since there are two systems and the transmission line

forced outage rate is zero, Fig. 13 has a 2:1 slope for the corresponding portion of the curve of capacity benefits.

Actually, in most cases, there should be a smooth transition between the sloped and horizontal lines of Fig. 13 and as shown by Fig. 10 of the subject paper. This was omitted from Fig. 13 as being of negligible practical significance.

The effect of probable forced outages of transmission lines can be accounted for quite simply by treating each system separately and each line to it as a generator with the desired forced outage rate assigned to each of them. A full discussion of this problem is omitted here since it would require more space than would be appropriate for a discussion.

The actual reductions in installed capacity requirements correspond, of course, to the discrete sizes of the units that would otherwise be installed, save for the interconnection. An investigation that was prompted by the preparation of this discussion has indicated that from year to year such savings may be smaller than the basic capacity benefit with no transmission line limitations by possibly as much as the capacity of one unit of the size being installed and larger by possibly as much as the capacity of two such units. However, presently, I see no reason for revising the average effect with various size units being installed, as shown by Fig. 13, assuming that the basic capacity benefit is independent of unit size.

Nevertheless, such possibility should provide a further fruitful field of future investigations, particularly as related to the sound determination of minimum economic transmission line capacity requirements.

REFERENCE

1. LOAD GROWTH CHARACTERISTICS AS RELATED

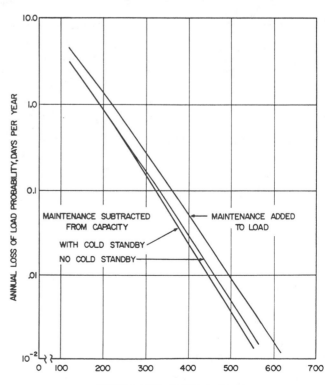

Fig. 15. Reserve requirements on a single system for various maintenance assumptions

TO GENERATING CAPACITY ADDITIONS. C. W. Watchorn. *AIEE Transactions*, pt. III (*Power Apparatus and Systems*), vol. 82, Apr. 1963, pp. 110–16.

A. K. Falk (The Detroit Edison Company, Detroit, Mich.): At the present time there exist approximately a hundred different papers on the application of probability techniques to capacity reserve determinations. Upon investigation by a special working group, the AIEE found that the entire host of various methods could be consolidated into but three basic classifications. Of the three, perhaps the loss-of-load classification is the most popular among those who prescribe to probability methods. The authors have prepared a straightforward and clear explanation of this method, particularly directed to evaluation of interconnection capacity benefit, which should be most helpful to those embarking upon pooling studies.

Some people do not subscribe to probability techniques. They say, from their experience, that they will protect against loss of their two largest units, for example, and they may or may not maintain additional reserve margin for regulation. These same people, however, often encounter a difficult problem, requiring arbitrary decision, of determining the capacity benefit of a proposed interconnection. More often than not, their decision in this difficult area is correct in terms of protection against loss of whole units. Dealing in whole units, however, their prescribed level of protection may be somewhat high or low, since the more correct level as established from probability techniques, may include a fractional unit either by derating or by combination of small unit sizes. With tomorrow's forthcoming large unit size this deviation could be costly either in terms of depressed security or in terms of added investment. This paper should promote greater acceptance among those who presently prefer experience, which incidentally is synonymous with probability.

The authors have explained two items of great practical importance: (1) the future capacity expansion program; (2) introduction of interconnection capability and reliability.

The future capacity expansion pattern of any company or pool must include the effects on reserve requirements of larger unit sizes and changing forced outage rates. Such expansion patterns must also provide for normal scheduled maintenance. The program described by the authors accomplishes these objectives.

The inherent reliability and capability of an interconnection including supporting system components are important considerations. From the results of this paper it appears that capability is the dominant factor and that interconnection reliability if within normal limits, is of secondary importance.

Students of probability have often disagreed as to whether or not it is preferable to add maintenance outages to load or to deduct them from capacity. In Fig. 7 the authors have demonstrated that the difference between these options is negligible. In Fig. 8 the authors show that the capacity benefits of interconnection are proportional to interconnection capability until the interconnection becomes ' infinite,' as defined by the authors. After this condition

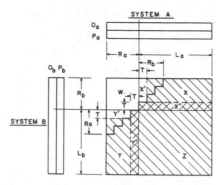

Fig. 16.   Two-area array with a tie line of capacity "T" interconnecting Systems A and B

is reached the benefit is 700 mw. If this value is in the order of 5 to 7% it would agree with recent studies made by the writer for systems of similar size as those described by the authors. Agreement or disagreement would, of course, be quite dependent upon assumed outage rates in each case. Because recent outage data are not available for probability studies, I made the following assumptions concerning forced outage rates:

For hydraulic units: 0.005
For header-type boiler-turbine units: 0.016
For unit-type boiler-turbine units: 0.024
For new units: 0.040 for first 2 years

It would be helpful to know how the results compare in per cent, and to know the assumed outage rates utilized by the authors.

E. S. Bailey, Jr. (Baltimore Gas and Electric Company, Baltimore, Md.): The authors have made another important contribution in the field of using probabilistic methods for determining generation reserve requirements. This paper logically presents the extension of the complex application of probability methods into a two-area interconnected system. Based upon either overall system or individual area loss-of-load risk levels, the proposed method will develop the timing for a listing of proposed generating units. By assuming various finite tie sizes, the developed digital computer program can also pinpoint a proper "two-way" interconnection rating for maximum installed generating capacity benefits on a pooled basis. This is particularly valuable when the interconnected areas are nearly equal in size and experience similar fuel cost conditions. It should be remembered that the determination of an absolute optimum tie size would also include a consideration of energy interchange and operating conditions.

The mathematical theory presented in the paper and explanatory Venn diagrams substantiate the existence of three distinct loss-of-load probabilities even with infinite tie capacity between the two systems. These would be one for each system and one for the combined pool. For finite tie capacities connecting systems of quite different sizes, a question arises as to the effect on the individual area risk levels and generation timing if unequal opposite direction tie sizes are introduced. This could be of importance to either one of the two individual areas.

In using the described digital computer program to establish generation timing on the two systems, it is not clear whether or not provision is made to incorporate each system's loss-of-load standard at different values as the criteria. Such a feature would allow the respective systems to incorporate their own established security policies.

Given a finite interconnection and a shortage of capacity in System A, the program in developing the generation schedules seeks the answer to two questions. First, is there reserve in System B to help out area A and if the answer is yes, is the tie capacity sufficiently large to carry this reserve into area A? It should be remembered that in reality System B might help beyond this point by possibly instituting a voltage reduction, thus shedding load, and providing additional aid to System A up to the tie line rating.

It is pointed out in the paper that satisfactory results can be obtained if scheduled capacity outages for maintenance are added to the load rather than subtracted from the capacity model. This is certainly true when working with very large systems. However, when the size of added units becomes an appreciable percentage of the system load, as it can in some smaller systems, caution should be exercised when applying this principle. The maintenance of the new unit as added to the load in conjunction with its forced outage rate can severely penalize the security representation of one area. Although additional computer time would be required, it would be more correct to revise the capacity model for scheduled maintenance outages. If this is not done for situations involving smaller systems, oversize tie requirements between the two areas may result.

H. G. Houser, J. A. Lechner (Westinghouse Electric Corporation, East Pittsburgh, Pa.): The authors are to be commended on their application of time-proved probability methods to the solution of a problem that is becoming increasingly important. They have properly recognized most of the basic problems that arise when one attempts to expand two systems tied through a finite interconnection. Their straightforward presentation leads us to but a few questions.

A logical initial assumption for two systems in geographical proximity is that their loads may be statistically dependent. That is, the factors that cause load variations, weather, primarily, would be expected to cause the loads on the two systems to fluctuate together. In their consideration of the deviation of actual loads from forecasted loads the authors have recognized this. In assuming that these deviations rise and fall on both systems simultaneously, however, they have assumed perfect correlation (complete dependence). This is convenient since, for N discrete steps in the probability distribution, only N calculations need be made for each system each day; whereas, consideration of real correlation, less than perfect, would require N^2 calculations for N discrete steps.

Moreover, assuming perfect correlation tends to overstate the loss-of-load probabilities, while an assumption of no correlation (independence) leads to understatement. An illustration of the differences in results obtained with these two extreme assumptions is provided in Fig. 14. Two identical, theoretical systems of 5,600 mw installed capacity each were assumed interconnected by the various values shown. Fig. 14 shows the results for either system for the two correlation assumptions, and appreciable differences can be seen. We are currently studying the interconnection of two actual systems whose monthly load correlation coefficients vary from 0.17 to 1.0 with a median of 0.49. (A correlation coefficient of zero indicates no correlation; 1.0 indicates perfect correlation.) For these two systems an assumption either of perfect correlation or of no correlation would be poor. Moreover, other methods make use of the calculated correlation. We should like to know then if the authors have considered including the effect of actual load correlation in their computations. If so, how is the load model constructed to account for it?

It seems that a practical minimum maintenance interval for the authors' probability method must be approximately 1 month, so that there are at least 20 points in the load variation curve. Indeed, a maintenance interval of 1 week, with only five points in the load curve, would lead to questionable accuracy. Have the authors encountered a practical minimum in their development?

The authors have shown that maintenance outages may usually be added to the load rather than subtracted from the capacity model with little loss of accuracy. However, our work has indicated that this is not always true One should not attempt to apply the results of Fig. 7 of the paper universally, particularly to smaller systems. Fig. 15 was drawn from results for an actual system of 1,600 mw installed capacity. It was obtained by first calculating the annual loss-of-load probabilities when maintenance outages were added to the load. Next, the probabilities were recalculated with maintenance outages subtracted from the installed capacity, but without allowing for cold standby. Finally, the probabilities were calculated again with maintenance outages subtracted from the installed capacity, and with units that could be shut down for cold standby made unavailable for forced outage. The same maintenance plan was used for all three curves. Reserve was varied by increasing loads according to the growth pattern predicted by the utility's system planners. For an annual probability of 0.1 day per year (1 day in 10 years), which is a widely accepted minimum service level, the method of adding maintenance to the load overstates the reserve by 35 mw, more than 10% of the indicated reserve required.

We must take exception to the authors' statements concerning an infinite interconnection. They say that for two systems connected to a common bus three distinct and, in general, different values of probability exist for the two systems and the pool. Now, it seems logical to expect that calculations for two such systems treated as one single system should yield the same probability as calculations considering the two as separate, but infinitely tied. Indeed, as tie between two systems is increased, the probabilities for each system and the pool should approach a common value; that common value is the probability obtained for the single combination of the two.

The approach adopted by the authors, however, will not lead to these results. The reason it will not is that it assumes that each system will share its surplus generation with the other, but that neither will ever share a loss with the other in order to reduce the magnitude of loss on the other system, even within the capacity limit of the interconnection. Thus, if the tie between two systems were 1,500 mw and System A had a reserve of 500 mw, while System B had a deficit of 1,200 mw, the authors would say that A will supply 500 mw to B but that B must sustain the full 700 mw net loss of load itself. A more realistic approach to true co-ordinated planning demands that A not only send 500 mw to B, but that it shed load and share the net loss with B, thus reducing the magnitude of loss to B.

This is, of course, a matter of policy that must be decided before pooling arrangements are made. These ideas can be incorporated in the authors' method with only slight alteration. Referring to Fig. 16, which is an alteration of Fig. 4 of the paper, and using the notation of the paper, the loss-of-load probabilities are

$$LOLP_a = (X+Z) + P_T X' + (1-P_T) Y''$$

and

$$LOLP_b = (Y+Z) + P_T Y' + (1-P_T) X''$$

where X includes X'' and Y includes Y''.

The authors' description of the computer program indicates that it has the ability either to determine new generator installation dates for given values of tie or to determine tie size for a fixed complement of units. Can it simulate system expansion and determine tie sizes simultaneously?

V. M. Cook, C. D. Galloway, M. J. Steinberg, and A. J. Wood: We appreciate the many excellent comments and additions to the paper offered by the various discussers. One of the unique advantages of the method described in the paper is the determination of separate loss-of-load probabilities for each system and the pool formed by the interconnected systems. These separate loss-of-load probabilities may be viewed in a nonmathematical way through the following "thought experiment."

THREE LOSS-OF-LOAD PROBABILITIES

Let us postulate an observer who can "see" an entire power pool for a period of 100 years. It is his task to record the number of days during the 100-year period when either of the two companies fails to meet its load. This information is conveyed to our observer by the flashing of a light in either of the two areas upon the occurrence of a loss of load. The observer then tabulates for 100 years the number of days on which load is lost in Company A, Company B, and anywhere in the pool. At the end of the 100-year period, he has recorded 10 days for Company A and 12 days for Company B. He has recorded 20 days for the pooled combination instead of 22, because during the passage of the 100 years both companies failed to meet their load simultaneously on two particular days.

Now, let these data be interpreted as representing the average loss-of-load probabilities of each of the two companies and the pooled combination. In units of days per year, the loss-of-load probability for Company A is

$$P_A = 0.10 \text{ days per year}$$

For B it is

$$P_B = 0.12 \text{ day per year}$$

and for the pool it is

$$P_S = 0.20 \text{ day per year}$$

Thus, with no assumption about the interconnection size (or for that matter, the existence of an interconnection), our observer has obtained three separate and distinct measures of the reliability of supply for the separate companies and the pooled combination.

Several of the discussers have raised the question of calculating loss-of-load probabilities for more than two interconnected companies. The theoretical complications for a completely rigorous analysis increase quite rapidly when more than two companies are considered. As indicated in the comprehensive discussion by Messrs. Craft and Masters, an approximate analysis may be made by grouping the companies to form various two-party pools. For a rigorous treatment, in order to calculate the reserve assistance which one participant may receive from the other members of the pool, one must account for a large number of tie line flows, and then restrict the flows whenever line capabilities are exceeded. Thus, besides the inherent constraints due to tie line capability limits, some means must be included to estimate the intercompany power flows under a large number of different conditions.

A fair amount of work has been done in this area and, for special cases of interconnection configurations, the computational problems are not insurmountable. For the general case, it would appear that additional research effort will be required before a completely rigorous and economically practical solution is obtained.

Capacity outages for maintenance or seasonal capacity variations may be conveniently handled *either* by adding them to the load *or* by modifying the outage probability tables when using the computer program. The difference in computation time is not significant when using the *7090* programs. As long as unit sizes are small, relative to the system size, the difference in loss-of-load probability is small.

Our present program permits the recognition of different "tie capabilities" depending upon the direction of flow. The definition of the interconnection capability does require some careful study. If, as Messrs. Masters and Craft suggest, the tie capability is to be a function of the load levels in the system, then it would be well to develop (using load flow studies) maximum 1-hour power transfer possible. The probability method presented does not depend upon this definition while loss-of-load probabilities calculated will depend on it.

The authors recognize the many contributions of Mr. Calabrese as indicated by reference 7 of the paper. The work of this paper extends Calabrese's technique to yield loss-of-load probabilities for both interconnected systems and the pool as a whole.

In reply to Mr. Adler, we should like to defer detailed discussion of the allocation of reserve benefits and gains from interconnections since it is somewhat outside the scope of the intent and objectives of the paper. Several approaches are possible using the technique and program described in the paper but detailed discussion of this interesting area would probably occupy as much time and space as the paper itself.

Messrs. Kist and Worcester have suggested the use of an energy-lost criterion. The technique and program could easily be modified to incorporate this reliability measure.

We appreciate the complimentary remarks of Messrs. Poage, Slemmer, and Thiemann and particularly their suggestion concerning seasonal capacity variations. This problem is currently handled in the manner suggested by the computer program.

Mr. Watchorn raises several questions pertaining to the application of the method as illustrated in Fig. 9. This illustration of the type of output obtainable from the computer program shows the loss-of-load probabilities expressed in several fashions. Both maintenance period risk levels and the yearly risk level are shown on this output sheet. Contrary to the impression which may be given by this illustration, the criterion which is to be used in the program to add capacity is almost completely at the discretion of the user. The program has in it a decision module which examines the risk levels calculated for a given period of time. Capacity additions may be made or the tie line capability increased on the basis of pool annual risk, individual system annual risk, or maintenance period values. The choice may be made by the user. The predominant use of this technique to date has based system capacity additions on annual risk levels. The use of a calendar year beginning in January is not mandatory either when using the program. There is no conceivable reason why expansions cannot be run using "capacity years" beginning in any month.

We disagree with Mr. Watchorn's statement that the suggested technique for including load forecasting deviations is approximate. *If* these probability distributions can be defined in a meaningful fashion, then this technique will permit the rigorous incorporation in the loss-of-load probability.

The calculations illustrated by Mr. Watchorn are of interest and are similar to the kind of results shown in Fig. 10 of the paper. In general, the capacity reserve benefits (i.e., the reduction in reserve requirements to maintain a given reliability level) caused by an interconnection of two systems, are *not* equal for the two interconnected companies. The technique discussed in the paper permits the calculation of these reserve benefits for each of the interconnected companies.

Mr. Watchorn suggests that a tie line may be adequately represented by calling it a generator with an appropriate forced outage rate. We have found this to be true as long as the tie capacity is much smaller than the reserve available in the other system.

Mr. Falk has raised a question concerning the tie benefits illustrated by the example of Fig. 10. This example was calculated using outage rates of 0.01 for each boiler and each turbine generator, and 0.007 for large hydroelectric units. The 700-mw figure mentioned by Mr. Falk is in the order of 10% of the company capacity.

We should like to caution, however, against the generalization of these specific data. The tie benefit to a company, measured as the allowable reduction in reserve capability to maintain a given reliability level, is a function of many parameters. It will vary from year to year and is more apt to be dependent upon the reserve capability available from the interconnected system than upon the particular spectrum of unit sizes and forced outage existence rates used.

Mr. Bailey has pointed out several interesting considerations particularly with reference to the computer program. As stated previously, the expansions calculated with this program may be made with different reliability standards for each of the companies and with tie capabilities which depend upon the direction of flow.

Messrs. Lechner and Houser have presented worthwhile data showing the effect of the correlation of the load occurrences between interconnected areas. We should like to point out that there are really two distinct problem areas:
1. Correlation of daily peak loads.
2. The treatment of the probabilistic nature of forecasted loads.
In the first case, the correlation of daily peak loads, historical data may, of course, be utilized to develop correlation coefficients for given time intervals. In the second case, the deviations of actual loads from forecasted probability distributions will depend upon the methods used to forecast the loads. As yet, very little seems to have been done in the study of forecast uncertainty.

We have tended to use monthly or four-week intervals for maintenance outage calculations. There is no reason why the probability method could not be used with shorter intervals. As a practical matter it would hardly seem economical to use less than weekly intervals.

We wish to take exception, however, to the statement in the discussion by Messrs. Houser and Lechner that "it seems logical to expect that calculations for two such systems treated as one single system should yield the same probability as calculations considering the two as separate but infinitely tied." This is a matter strictly of the definition of the loss-of-load probability and of the basic assumption made with regard to the assistance one system will give the other in time of emergency. When the assumption is made that one system will help the other only to the limit of its available reserves, then as shown in the paper, three distinct and calculable probabilities exist no matter what the tie capacity. Even with the assumption made in the discussion (i.e., equal load loss sharing without limit)

this is true unless the tie has a zero forced outage rate.

Consider the Venn diagram shown by Messrs. Houser and Lechner. Under their assumption that load losses are shared up to the limit of the tie capability, the loss-of-load probabilities for A, B, and the pooled system become the three distinct values,

$$LOLP_a = X + Z + P_T X' + (1 - P_T) Y''$$

$$LOLP_b = Y + Z + P_T Y' + (1 - P_T) X''$$

and

$$LOLP_s = X + Y + Z + P_T (X' + Y')$$

Now, if the tie becomes "infinite" in the sense that it is larger than either of the individual company reserves,

$$X'' = X$$

$$Y'' = Y$$

and

$$LOLP_a = X + Z + P_T X' + (1 - P_T) Y$$

$$LOLP_b = Y + Z + P_T Y' + (1 - P_T) X$$

$$LOLP_s = X + Y + Z + P_T (X' + Y')$$

It is only in the case where P_T is zero and the tie is "infinite" that the three loss-of-load probabilities are equal.

It would seem logical to us that the reliability of supply may be, and is, different for different parts of even the *same* system. For instance, power supply systems are commonly designed with elements (i.e., generators) which have outage probabilities of 0.02, but because of redundancy and reserve capability, the over-all system outage probability is 0.004. Therefore, it seems just as logical that two interconnected systems may have different reliabilities and that these may in turn be different than the reliability of pooled combination.

If a contractual obligation actually exists between two systems to share load losses up to a given limit (either arbitrarily assigned or caused by tie limits), then the calculation of the loss-of-load probability may easily be modified to accommodate this assumption. The net results of this condition, as opposed to the assumption of not sharing load losses, is to increase the generation reserves required to maintain a given reliability level if the two systems are each planned to hold a given reliability level.

We hope in the future to describe the computer programs in somewhat more detail. These particular programs will expand two systems, holding pool or individual system reliability levels as desired, and add capacity and/or change tie sizes. This particular technique facilitates the simultaneous calculation of loss-of-load probabilities with different tie capabilities and, therefore, permits the relatively easy expansion of systems with tie sizes being determined automatically.

Again, we wish to thank the many discussers for adding to the paper by their comments.

Effective Load Carrying Capability of Generating Units

L. L. GARVER, MEMBER, IEEE

Abstract—The theory of loss-of-load probability mathematics
has been generalized so that the effective load carrying capability of
a new generating unit may be estimated using only graphical aids.
A parameter m is introduced to characterize the loss-of-load probability as a function of reserve megawatts.

Once m is known or estimated, the effective load carrying capability of a new generating unit may be related to its rating and its
forced outage rate. Alternate unit additions may be compared on the
basis of their effective capabilities. Comparable expansion patterns
may be developed on the basis of equal load carrying capabilities.

Numerical examples are used to illustrate the application of the
effective capability concept to the evaluation of changes in the
rating of a new unit and to the strategical design of expansion plans.

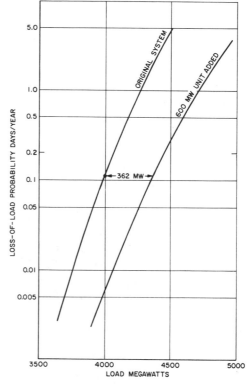

Fig. 1. Annual risk before and after adding a 600-MW unit with
five percent forced outage rate.

INTRODUCTION

THE THEORY of loss-of-load probability mathematics is generalized in this paper resulting in a
graphical method for estimating the effective load carrying capability of a new generating unit. The concept of
effective load carrying capability is best illustrated graphically as in Fig. 1. It is the distance in load megawatts between the annual risk functions before and after a unit
addition. The measurement of effective load carrying
capability is made at some designated level of reliability,
often the level calculated for the system in a previous
year. The effective capability of a new unit is, therefore
the load increase that the system may carry with the
designated reliability.

The graphical method for estimating effective capability
presented in this paper will aid in the preliminary investigation of generation expansion plans. Illustrations of
preliminary planning are presented along with examples
of parametric investigations to estimate the effects of a
change in the size or forced outage rate of one unit. The
estimating method provides insights into how much of a
unit's capability is needed to maintain system reliability.

As shown in Fig. 1, the system reliability will be measured in terms of the annual loss-of-load probability.
Other measures of reliability could be used to determine
the effective capability of a new unit [1]–[9]. The estimating procedure should also give comparable results for
these methods [8]–[11].

It is best to begin with a review of the method for
establishing the effective load carrying capability of a
unit from the results of a series of loss-of-load probability

Paper 31 TP 66-51 recommended and approved by the Power
System Engineering Committee of the IEEE Power Group for
presentation at the IEEE Winter Power Meeting, New York, N. Y.,
January 30–February 4, 1966. Manuscript submitted October 25,
1965; made available for printing December 2, 1965.

The author is with the General Electric Company, Schenectady,
N. Y.

calculations. The remainder of the paper will present the
new method of estimation and illustrate its use.

EFFECTIVE CAPABILITY

Other authors have presented figures similar to Fig. 1
and have also referred to load carrying capability or its
counterpart, increased reserve requirements [6], [12], [13].
The steps followed to obtain these results were similar to
those given below.

Steps to Determine Effective Capability

1) Determine the annual risk for the year *before* the unit
is to be added. This requires a loss-of-load probability
calculation based on data describing: a) the capability of
each generating unit and its forced outage rate, b) the
daily hourly-integrated peak loads, c) maintenance requirements for each unit, and d) other special features
such as seasonal deratings, energy interchange contracts.

2) Vary the annual peak load and each daily peak in
percent of the annual peak. Calculate the annual risk for
a range of loads such as ±20 percent. The graph of the
annual risk as a function of the annual peak will produce
a curve similar to the original system curve in Fig. 1.

3) Add the new unit into the loss-of-load probability

calculation data, keeping all other data fixed. Again vary the annual peak load with the daily peaks as a percentage and calculate the annual risks for a range of values, perhaps a zero to 40 percent increase over the previous midpoint load. The result when plotted, will form a second curve such as shown in Fig. 1.

4) The megawatt distance between these curves at the risk level determined in Step 1 is the amount of load growth the system can accept and still retain the same reliability the next year as in the starting year.

Illustration

Suppose that a certain system with a 4000-MW annual peak load and a reserve of 600 MW, or 15 percent, has an acceptable level of reliability. The loss-of-load probability calculation for this system resulted in an annual risk of 0.111 days per year. By varying the load up and down in 200-MW steps the *original system* curve in Fig. 1 was determined.

Then without changing any other data, a 600-MW unit with a forced outage rate (f.o.r.) of five percent was added to the generating system. Beginning with the 4000-MW load point the load was increased in 200-MW steps to obtain the *unit added* curve of Fig. 1. At a risk level of 0.111 days per year, i.e., the risk for the original system with a 4000-MW load; the system is able to carry a load of 4362 MW. The effective capability of a 600HMW five percent f.o.r., unit on this system is, therefore, 362 MW—the increase in load carrying capability of the system.

Further additions of 600-MW units to this system will result in larger effective capabilities for each successive unit. Figure 2 illustrates the results of adding five 600-MW five percent f.o.r. units. Table I summarizes the system load carrying capability, effective capability of each unit, the megawatts of reserve required, and the percent reserve. Kirchmayer and Mellor [12] illustrated the effect of adding a new unit in a figure very similar to Fig. 1. Steinberg and Smith [6] have noted the increases in system reserve requirements due to larger unit additions similar to those shown in Table I. Other examples of the increases in load carrying capability with repetitive additions have been published by Baldwin [13].

In this paper, the intention is to concentrate on the effect of the next unit addtion. The purpose of illustrating the effects of adding five new units is to stress the point that the initial effect, while important, is not the whole story. A complete evaluation of the investment and pro-

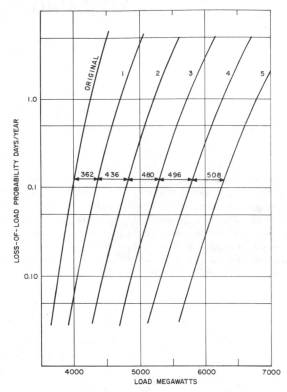

Fig. 2. Annual risk functions adding one to five 600-MW units with five percent forced outage rates.

duction cost economics over a period involving the addition of many units is necessary before the study of the next unit can truly be called complete [14], [15].

Effect of Risk Level

The selected risk level has only a minor effect on the load carrying capability. For example, in Fig. 1 if the risk to maintain was 0.2 days per year, nearly twice the previous risk, then the effective capability would be 385 MW, an increase of 23 MW or five percent of the unit rating. If the risk to maintain was 0.05, one-half the original risk, the effective capability would be 340 MW, a decrease of 22 MW. Thus, for estimating purposes, there is no need to be greatly concerned about selecting a precise risk level. If the system is felt to have insufficient reserve, part of the new unit's effective capability may be allocated to improving the deficiency. Similarly, if the system seems to be over built, then part of the load growth may be carried by the present system.

TABLE I

EFFECTS OF ADDING 600-MW UNITS WITH FIVE PERCENT FORCED OUTAGE RATES

No. of Units Added	Installed Capability, MW	Load for Risk 0.111	Effective Capability, MW	E.C. as percent of 600 MW	System Reserve, MW	Reserve in percent of Load
	4600	4000			600	15.0
1	5200	4362	362	60.4	838	19.2
2	5800	4798	436	72.6	1002	21.2
3	6400	5278	480	80.0	1122	21.3
4	7000	5774	496	82.7	1226	21.2
5	7600	6280	508	84.7	1320	21.0

ESTIMATING EFFECTIVE CAPABILITY

The material presented thus far is a review of work already presented by others. Even the term *load carrying capability* is not new. What *is* new is the technique for estimating the effect of a unit addition without making the entire set of probability calculations that are necessary for Fig. 1.

Procedure

The estimating procedure begins after completing the first two steps to determine effective capability, i.e., the determination of the original system risk function.

1) Graph the annual risk as a function of installed annual reserve using semi-log paper. Figure 3 presents the data of the original system function in Fig. 1 plotted vs. reserve.

2) Approximate the annual risk function by a straight line at the designated risk level.

3) Characterize the slope of this straight line by m, the megawatts of load increase necessary to give an annual risk e times larger than the designated risk, where e is the base of the system of natural logarithms, 2.718...

4) Calculate the ratio of the new unit's capability, c, to the characteristic m, giving the parameter c/m.

5) Refer to the generalized graph in Fig. 4 which relates the effective capability c^* to the parameter c/m, to the forced outage rate of the new unit r and to the characteristic m. A multiplication of the result from Fig. 4 by m completes the estimate of effective capability.

The derivation of the functions plotted in Fig. 4 and also in Figs. 5 and 6, is contained in Appendix I.

Examples of Procedures

The effective load carrying capability of a new unit may be estimated once the characteristic m has been determined for the system. The value of m is related to the straight line approximation of the annual risk function plotted on semi-log paper as shown in Fig. 3. The straight line is fit through the risk level to be maintained at a value e times above it, 0.111 and 0.302. The value of m determined graphically is 118 MW. A method for calculating m is given in Appendix II.

To use Fig. 4 in estimating the effective capability c^* of a 600-MW unit, first calculate the c/m ratio.

$$c/m = 600/118 = 5.09.$$

Entering Fig. 4 with this value and the assumed forced outage rate of five percent allows the effective capability to m ratio of 2.89 to be read off. The multiplication by m completes the estimate.

$$c^* = 2.89 (118) = 341 \text{ MW.}$$

This estimate is within six percent of the value determined in Fig. 1—362 MW.

Estimates of effective capability can be made which are close to the correct value by fitting the approximating straight line, which determines the value of m, through a point farther up on the risk function than e times designated

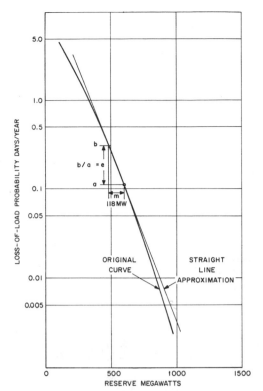

Fig. 3. Approximation of annual risk function by linear exponential function.

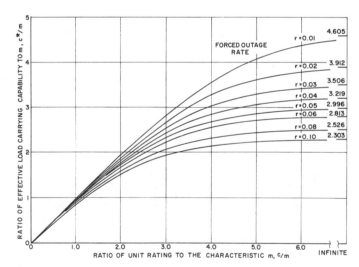

Fig. 4. Ratio of effective capability to m for various forced outage rates.

risk. The risk caused by a five percent load increase has proven more successful as the second point in determining m and produced results with five percent of the actual effective capability.

Alternate Procedure

As an alternate for Step 5 in the estimating procedure when the c/m ratio is 3.0 or less, Fig. 5 may be used to determine the reserve increase as a percent of the capacity of the unit. The effective capability is then the capacity less the reserve increase. Figures 4 and 5 present the same information in two different forms.

For the example of the 600-MW unit with a five percent forced outage rate added to a system with $m = 118$ MW,

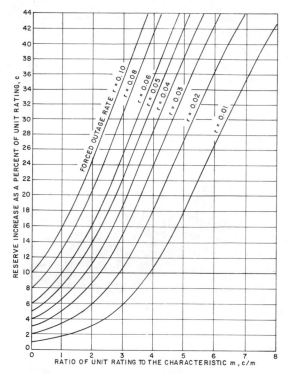

Fig. 5. Increase in reserve as percent of unit rating for various forced outage rates.

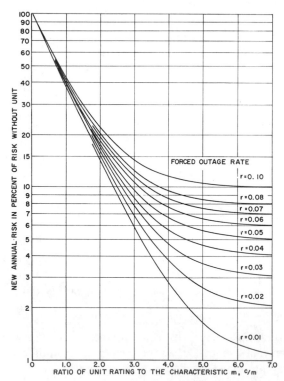

Fig. 6. Annual risk after unit addition as percentage of risk without addition.

Fig. 5 relates a 43.2 percent reserve increase to the c/m ratio of 5.09. The effective capability is thus 56.8 percent of the capacity, again, 341 MW.

Change in Reliability

In these examples, it was assumed that the annual risk was fixed and the load changed when the new unit was added. If we assume that the load remains fixed after a unit is added then the risk will be decreased. The amount of this decrease may be estimated with the aid of Fig. 6. This figure expresses the new loss-of-load probability as a percentage of the previous value when the only change to the system is the addition of a unit of a given c/m ratio and forced outage rate.

For example, if the 600-MW unit were added to the original system in Fig. 1 and the load did not increase, i.e., remained at 4000 MW, then from Fig. 6 a risk which is 5.6 percent of the previous value ($c/m = 5.09$ and $r = .05$) would be estimated. This compares favorably with the 5.1 percent determined from the digital computer calculation of the annual risk.

APPLICATIONS

The estimating procedure provides a tool to supplement the use of the digital computer programs which are normally used in calculating system reliability measures for generation expansion studies. Two types of preliminary investigations related to expansion planning will be illustrated by means of numerical examples.

1) Estimate the effects of a change in one unit's capacity or forced outage rate.

2) Prepare preliminary generation expansion plans which are estimated to maintain the present degree of reliability.

Effects of a Change in Assumed Unit Size or Outage Rate

The change in the assumed characteristics of a unit may be viewed as having either of two possible effects: 1) a change in system reliability for a given load level, or 2) a change in load carrying capability for a given risk level. Both of these alternate effects may be estimated using the information in Figs. 4 through 6.

Assume that for a certain system the calculated annual risk in 1972 is 0.053 days per year; the first step is to investigate how sensitive this risk is to the forced outage rate of the 1971 unit and second what the risk would be if the unit were reduced in size.

Suppose in the calculation that the 1971 unit was a 600-MW five percent f.o.r. unit. The 1972 system characteristic m may be found by a straight line approximation to the 1972 curve, as in Fig. 3. Assuming $m = 155$, the c/m is $600/155 = 3.87$. The following information is then available from Figs. 4 through 6:

effective capability = $2.67\,m$ = 2.67 (155) = 413 MW (constant risk)

percent capacity for reserve = 31.2 percent of 600 MW = 187 MW

new LOLP in percent of old = 7.0 percent (constant load).

If the forced outage rate were one percent higher, six percent instead of five percent, then these same figures would yield this information:

effective capability = $2.52\,m$ = 2.52 (155) = 392 MW

percent capacity for reserve = 34.6 percent of 600 MW = 208 MW new LOLP in percent of old = 8.0 percent.

Thus the one percent increase in the forced outage rate would reduce the effective capability by 413 − 392 = 21 MW or raise the annual risk to

$$0.053\,(0.08/0.07) = 0.053\,(1.14) = 0.0605\text{ days per year.}$$

If the capacity were reduced by 100 MW, 500 MW instead of 600 MW, and r = 5 percent, then the pertinent information would be, (c/m = 500/155 = 3.23):

effective capability = 2.43 m = 2.43 (155) = 376 MW
percent capacity for reserve = 24.7 percent of 500 MW = 124 MW
new LOLP in percent of old = 8.7 percent.

Thus the 100-MW decrease in capacity would reduce the effective capability 413 − 376 = 37 MW or raise the annual risk to

$$0.053\,(0.087/0.07) = 0.053\,(1.24) = 0.066\text{ days per year.}$$

The general data contained in Figs. 4 and 5 allow a system planner to develop curves comparing the effects of unit size and forced outage rate on the load carrying capability of the next addition to a particular system. The curves in Figs. 7 and 8 illustrate two possibilities for comparison. In Fig. 7 the effective load carrying capability is shown as a function of unit size. In Fig. 8 the effects of the forced outage rates are shown for both a 400-MW and an 800-MW unit. Both curves are for a system with a characteristic m of 125 MW.

Generation Expansion Planning

Generation expansion planning involves the development of alternate patterns of unit sizes and installation dates. Each alternate expansion is designed to meet a criteria of reliability [8], [9]. The estimated effective load carrying capabilities of future units offer a new guide for the preliminary design of alternate expansions.

The system planner may match the effective capability of the unit additions with the forecasted load growth to design alternatives based on constant reliability. The implementation of this concept will be illustrated by designing preliminary expansion plans to meet these three design strategies: 1) add one unit a year matching the effective capability to the load growth, 2) add one unit every two years, 3) add three units of the same size whose combined effective capabilities match four years of load growth.

The implementation of a strategy to add a unit to match the load growth for each year is illustrated by the data in Table II. A system growing at seven percent a year and an initial load of 4000 MW has been assumed. The initial system characteristic m should be determined from a loss-of-load probability calculation as in Fig. 3 or Appendix II. However, assume that such a calculation has not been made and m must be estimated.

An approximate method for estimating the value of m is given in Appendix III. Suppose that the generating system is composed of 4600 MW of capability: 3000 MW in

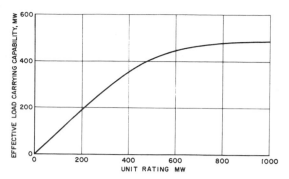

Fig. 7. Effective capability of new unit with two percent forced outage rate added to system, m = 125 MW.

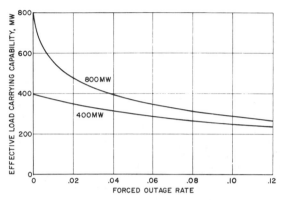

Fig. 8. Effects of outage rate on effective capability, for system with m = 125 MW.

units less than 299 MW in rating, 1200 MW in units between 300 MW and 399 MW, and a 400-MW unit. The forced outage rates assumed for existing and future units are shown in Table III. The estimate of m from (22) of Appendix III is

$$m \approx 3000\,(0.02) + 1200\,(0.03) + 400\,(0.04) = 112\text{ MW.}$$

In Table II each increment of load growth is first converted into the required load carrying capability to m ratio, c^*/m. For example the load growth for the first year is forecasted to be 280 MW and the required c^*/m is thus 2 80/112 = 2.50. The rating of the unit was found by looking up the c/m ratio in Fig. 4 corresponding to a c^*/m of 2.50 and an f.o.r. of three percent. The result was a c/m of 2.90 which when multiplied by the current value of m, 112 MW, gave a unit size of 325 MW.

Each addition of the system will change the value of m, (note Fig. 2). The amount of change was approximated using (22), which for the addition of one unit is just the rating times its forced outage rate. The change in m for the first year was estimated to be (325) (0.03) = 9.75.

In the fifth year, a three percent f.o.r. unit would have to be rated at 421.5 MW. Since this is in the four percent f.o.r. range, as shown in Table III, the c/m ratio was determined from the four percent line in Fig. 4. Similarly, in the eighth year the units become so large that the five percent line must be used. Table IV presents the reserve and percent reserve for the expansion of Table II.

TABLE II
SELECTING UNITS TO MATCH THE LOAD GROWTH

Year	Load, MW	Load Growth, MW	Needed, $c*/m$	Forced Outage Rate	c/m from Fig. 4	Unit Rating, MW	Change in m	New m, MW
0	4000							112
1	4280	280	2.50	0.03	2.90	325	9.75	121.75
2	4580	300	2.47	0.03	2.85	347	10.41	132.16
3	4900	320	2.42	0.03	2.80	370	11.10	143.26
4	5243	343	2.41	0.03	2.78	390	11.70	154.96
5	5610	367	2.37	0.03	2.72	422	(must change f.o.r.)	
				0.04	2.90	450	17.98	172.94
6	6003	393	2.27	0.04	2.72	470	18.80	191.74
7	6423	420	2.19	0.04	2.60	498	19.92	211.66
8	6873	450	2.13	0.04	2.51	532	(must change f.o.r.)	
				0.05	2.62	555	27.75	239.41
9	7354	481	2.01	0.05	2.43	582	29.10	268.51
10	7869	515	1.92	0.05	2.28	625		

TABLE III
FORCED OUTAGE RATES ASSUMED FOR EXAMPLES

Range of Unit Ratings, MW	Forced Outage Rates
0–299	0.02
300–399	0.03
400–499	0.04
500–700	0.05

TABLE IV
EXPANSION PLAN TO MATCH CAPACITY TO LOAD GROWTH

Year	Load, MW	Capacity Added, MW	Total Capacity, MW	Installed Reserve, MW	Reserve in Percent of Load
0	4000		4600	600	15.00
1	4280	325	4925	645	15.07
2	4580	347	5272	692	15.11
3	4900	370	5642	742	15.14
4	5243	390	6032	789	15.05
5	5610	450	6482	872	15.54
6	6003	470	6952	949	15.81
7	6423	498	7450	1027	15.99
8	6873	555	8005	1132	16.47
9	7354	582	8587	1233	16.77
10	7869	625	9212	1343	17.07

TABLE V
ATTEMPT TO MATCH THREE UNITS TO FOUR YEARS OF LOAD GROWTH

Unit size, MW	650
Forced outage rates, percent	5
Change in m, MW	32.5

Unit Number	Value of m	c/m	$c*/m$ (Fig. 4)	Effective Capability
1	112	5.8	2.9	329
2	144.5	4.5	2.8	405
3	177	3.7	2.6	460

Total effective capability, MW	1194
Four year load growth, MW	1213
Mismatch, MW	19

Suppose the strategy were changed to a new unit every *two* years. The load growth for two years is $280 + 300 = 580$ MW. The effective capability to m ratio, $c*/m$, of the new unit must be $580/112 = 5.18$. However, it is evident from Fig. 4, that no single unit with a forced outage rate of even one percent could carry that load growth with present system reliability. Thus, two units must be installed in the first two years to maintain system reliability.

A strategy of adding three identical units in four years will require a cut and try procedure. Assume a unit size and calculate the load carrying capabilities for all three. If the total load carrying capability is below the 4-year load growth, increase the unit size and try again. In Table V, the calculations are shown for the addition of three 650-MW units at five percent forced outage rate. The effective capability of the three fell 19-MW short of the 4-year load growth, 1213 MW. Another trial at 675-MW may prove successful.

The examples illustrate the manner in which the concepts of effective capability and the general curves of Figs. 4 through 6 can aid in generation expansion planning.

CONCLUSION

A procedure for estimating the effective load carrying capability of a generating unit has been presented. Its use makes possible preliminary investigations to supplement the detailed calculation of system reliability associated with a generation expansion study. The estimating procedure uses a graphical relationship between effective capability and the characteristics of the unit with a system parameter m. The parameter m has been introduced as a single number to characterize the annual risk function of the system. Though the value of m should be determined from a calculation of the loss-of-load probability function, a method for approximating its value is also presented.

The estimating procedure has been applied to two types of investigations associated with generation expansion planning—changing one unit's characteristics in a plan already studied in detail, and preparing expansion plans to meet certain design strategies while maintaining constant reliability. Numerical examples illustrated these two uses of the estimating procedure.

NOMENCLATURE

A_x annual risk or loss-of-load probability in days per year for a reserve of x MW

A_x' new annual risk for reserve x with a new unit added to the system

a_i ratio of day i peak load to annual peak load

B a constant in the approximation of A_x by an exponential function

c rating of a new generating unit, megawatts

c^* effective load carrying capability of a unit with rating c

e base of the system of natural logarithms, 2.718...

$\ln(\ )$ natural logarithmic function

m the MW of load increase that will give annual risk increase e times greater than before—called the system characteristic m

n total number of generating units in a system

k number of daily hourly–integrated peak loads considered in an annual risk calculation

P_{x_i} probability of having x_i MW or greater capacity on forced outage

r forced outage rate of a generating unit

x annual installed reserve, megawatts

x_i installed reserve for day i, megawatts

y increase in annual installed reserve in megawatts necessary after a unit is added to maintain the same reliability as before the addition

APPENDIX I
ESTIMATING THE EFFECTS OF A UNIT ADDITION

Annual Risk as a Function of Reserve

The first step in the development of the estimating procedure for the effective load carrying capability of a new unit is to change the independent variable of the annual risk from the load to the reserve. Reserve is a function of both the installed capacity and the annual peak load. The annual risk is also a function of both of these variables.

In Fig. 3 the original system curve of Fig. 1 is plotted vs. system reserve instead of system load.

Risk Function After a Unit Addition

The second step in the development is to express the annual risk after a unit addition in terms of the annual risk function before the addition. Recall from the fundamentals of the loss-of-load probability calculation that the annual risk is the sum of the daily risks. Each daily risk is the cumulative probability of having a total amount of capacity on forced outage greater than the reserve for that day

$$\text{annual risk} = \sum_{i=1}^{k} \text{daily risk}_i$$

$$\text{daily risk}_i = P_{x_i}$$

where P_{x_i} is the cumulative probability of having x_i MW or greater on forced outage. The expression for the annual

risk becomes

$$A_x = \sum_{i=1}^{k} P_{x_i} \tag{1}$$

where x is the annual installed reserve and x_i is the available reserve on day i.

When a new unit with rating c and forced outage rate r is added to the system the cumulative probability of outage x is

$$P_x' = (1-r) P_x + rP_{x-c}. \tag{2}$$

The new cumulative outage probability is the sum of the two components corresponding to the two possible conditions for the new unit, i.e., in service or on forced outage. The first component assumes that the new unit is in service, a probability of $(1-r)$, and the capacity outage of x MW or greater if it is to occur will be in the old system. The second component assumes that the new unit is on forced outage, a probability of r, and thus an outage of only $x-c$ MW or greater in the old system will cause a total outage of x or greater.

Expression (2) may be substituted into (1) for each daily reserve x_i giving the following expression:

$$A_x' = \sum_{i=1}^{k} [(1-r) P_{x_i} + rP_{x_i-c}]$$
$$= (1-r) \ (\sum_{i=1}^{k} P_{x_i}) + r(\sum_{i=1}^{k} P_{x_i-c}). \tag{3}$$

The first term in (3) is the old annual risk for reserve x multiplied by the innage rate for the new unit. The second term is the annual risk for reserve x but with each day's reserve decreased by the rating of the new unit c and multiplied by the forced outage rate of the unit.

Approximation 1: The second term in (3) will be replaced by the annual risk for a load increase of c MW multiplied by r. This amounts to replacing

$$x_i - c$$

with the expression

$$x_i - a_i c.$$

Thus each day's reserve is decreased by a percentage of c instead of the full amount, causing a smaller value for the second term in expression (3). The result of the first approximation is the following expression for the new annual risk in terms of the old values:

$$A_x' = (1-r) A_x + rA_{x-c}. \tag{4}$$

To gauge the error introduced by this approximation, we note that in the example of Fig. 1 the annual risk for a load increase of 200 MW was 0.604 days per year while the annual risk for a capacity decrease of 200 MW was 0.631 days per year.

Straight Line Approximation

The third step in the development of the estimate of load carrying capability requires that the annual risk as a function of reserve be approximated by a straight line on semi-log paper.

Approximation 2: Assume that the annual risk expressed in terms of the installed reserve has the following form:

$$A_x = Be^{-x/m} \qquad (5)$$

where

B = a constant that need not be evaluated

m = the system characteristic and has the dimension of megawatts

e = the base of the natural system of logarithms, 2.718…

x = the installed reserve on the system, megawatts.

This assumption is shown graphically in Fig. 3. To determine the constant m we require only two points on the straight line. Assume that the two points are

$$A_{600} = 0.111 \text{ days per year}$$

$$A_{400} = 0.604 \text{ days per year.}$$

Dividing the larger risk by the smaller yields

$$\frac{A_{400}}{A_{600}} = \frac{Be^{-400/m}}{Be^{-600/m}} = e^{-(400-600)/m} = e^{200/m}.$$

Also,

$$\frac{A_{400}}{A_{600}} = \frac{0.604}{0.111} = 5.44.$$

Equating the two expressions and taking the natural logarithm of both sides gives

$$200/m = \ln(5.44).$$

Solving for m gives

$$m = \frac{200}{\ln(5.44)} = \frac{200}{1.69} = 118 \text{ MW.}$$

The general expression for m in terms of any two given points y MW apart, A_x and A_{x-y}, is

$$m = \frac{y}{\ln(A_{x-y}/A_x)}. \qquad (6)$$

To gauge the effect of this approximation, refer to Fig. 3. The approximation gives higher values of annual risk than those actually calculated. The effects of Approximations 1 and 2 tend to offset one another since 1 gives values too low while 2 gives values too high. Experience in selecting the straight line approximation will enable a planner to estimate the effective capabilities quite close to those determined from a computer calculation of annual risks.

Derivation of Generalized Expressions

The fourth and final step in the development is to derive expressions for the new annual risk, the reserve increase in percent of the new capacity and the load carrying capability based on the previous annual risk function.

The annual risk after the addition of a unit and no change in load will be the value at a reserve of $x+c$ MW. Substituting $x+c$ for x in (4) yields

$$A'_{x+c} = (1-r) A_{x+c} + rA_x. \qquad (7)$$

Expressing A'_{x+c} in terms of A_x may be accomplished by using (5).

$$A_{x+c} = Be^{-(x+c)/m} = (Be^{-x/m}) e^{-c/m} = A_x(e^{-c/m}). \qquad (8)$$

Substituting (8) into (7) gives the desired expression for the new annual risk:

$$A_{x+c}' = [(1-r) e^{-c/m} + r] A_x. \qquad (9)$$

The general curves in Fig. 6 were calculated by solving (10) for the new annual risk in percent of the old value and evaluating the expression for a range of c/m and r.

$$\frac{100 A_{x+c}'}{A_x} = 100 [(1-r) e^{-c/m} + r]. \qquad (10)$$

To derive the expression for the reserve increase as a percentage of the new capacity, let y be the reserve increase necessary to maintain the same annual risk.

$$A_{x+y}' = A_x. \qquad (11)$$

The expression for the new annual risk in terms of the old function is found by substituting $x+y$ into (4):

$$A_{x+y}' = (1-r) A_{x+y} + rA_{x+y-c}. \qquad (12)$$

Express each term in this equation as a product involving A_x by using the assumption in (5).

$$A_{x+y} = Be^{-(x+y)/m} = e^{-(y/m)} A_x \qquad (13)$$

$$A_{x+y-c} = Be^{-(x+y-c)/m} = e^{-(y-c)/m}A_x. \qquad (14)$$

Substituting (13) and (14) into (12) and collecting terms yields

$$A_{x+y}' = [(1-r)e^{-y/m} + re^{-(y-c)/m}]A_x. \qquad (15)$$

Substituting into (11) and dividing through by A_x gives

$$[(1-r) + re^{c/m}]e^{-y/m} = 1. \qquad (16)$$

Take the natural log of both sides and recall that the log of a product is the sum of the logs.

$$\ln[(1-r) + re^{c/m}] + (-y/m) = \ln(1) = 0.$$

Solving for y:

$$y = m \ln[(1-r) + re^{c/m}] \qquad (17)$$

and expressing the reserve increase y as a percentage of the new capacity c

$$100 y/c = 100 (m/c) \ln[(1-r) + re^{c/m}]. \qquad (18)$$

Expression (18) has been evaluated for a range of values for c/m and r with the results plotted in Fig. 5.

The load carrying capability of the new unit is the difference between its capacity c and the required reserve of (17).

$$c^* = c - y = c - m \ln[(1-r) + re^{c/m}]. \qquad (19)$$

To generalize this expression it was divided by the system characteristic m.

$$c^*/m = (c/m) - \ln[(1-r) + re^{c/m}]. \qquad (20)$$

Expression (20) has been evaluated for a range of values of c/m and r and the results are shown in Fig. 4.

APPENDIX II
CALCULATING THE VALUE OF m

The value of m may be determined analytically instead of graphically as illustrated in Fig. 3. The value of m for the line through any two points on the annual risk curve may be calculated by assuming that the two known points are 0.397 days per year for a reserve of 450 MW and 0.111 days per year for a reserve of 600 MW. Thus, the value of m is

$$m = (600\text{-}450)/\ln(0.397/0.111) = 150/\ln(3.57)$$
$$= 150/1.272 = 118 \text{ MW}.$$

In general,

$$m = (b - a)/\ln(A_a/A_b) \qquad (21)$$

where

A_a = the annual risk for a reserve of a MW
A_b = the annual risk for a reserve of b MW
$\ln$ = the natural logarithmic function.

APPENDIX III
ESTIMATING THE VALUE OF m

In order to design a generation expansion alternative before the initial calculation of the loss-of-load probability, some estimate of the characteristic m is necessary. Also the value of m changes after each unit addition and these changes should also be approximated.

The value of m is determined by the change in the annual risk due to a change in the forecasted loads. The annual peak load change which gives an annual risk e times greater than before is the value of m for the system.

The annual risk is a function of every individual unit on the system: its rating and forced outage rate and the annual load shape. Any simple estimate of how this function will change with a change in the load forecast can only be a very rough estimate and should be checked with a computer calculation.

It appears from experience that a rough gauge of the value of m is the sum of each unit's rating times its forced outage rate.

$$\text{estimate of } m = \sum_{i=1}^{n} c_i r_i \qquad (22)$$

where

c_i = the megawatt capacity of the ith unit
r_i = the forced outage rate of the ith unit
n = the number of units presently in the system.

This expression, though only a rough approximation, does illustrate the behavior of the characteristic m. A unit with no forced outage rate does not affect the slope of the annual risk characteristic. The larger the unit or the larger its forced outage rate, the greater its effect on the slope m.

Thus, the first large unit on a system while not having a large percentage of load carrying capability, will have a great effect on the characteristic m and prepare the system to make better use of the second and third units.

ACKNOWLEDGMENT

The author wishes to acknowledge the assistance of Misses G. A. Cary and R. B. Roginska for their work in computing the data for this paper.

REFERENCES

[1] G. Calabrese, "Determination of reserve capacity by the probability method," *Trans. AIEE*, pt. II, vol. 69, pp. 1681–1989, 1950.
[2] C. D. Galloway and L. K. Kirchmayer, "Digital computer aids economic-probabilistic study of generation systems," *AIEE Trans. (Power Apparatus and Systems)*, vol. 77, pp. 564–577, August 1958.
[3] C. Kist and G. J. Thomas, "Probability calculations for system generation reserves," *AIEE Trans. (Power Apparatus and Systems)*, vol. 77, pp. 515–520, August 1958.
[4] H. D. Limmer, "Determination of reserve and interconnection requirements," *AIEE Trans. (Power Apparatus and Systems)*, vol. 77, pp. 544–550, August 1958.
[5] A. L. Miller, "Details of outage probability calculations," *AIEE Trans. (Power Apparatus and Systems)*, vol. 77, pp. 551–557, August 1958.
[6] M. J. Steinberg and V. M. Cook, "Evaluations of unit capacity additions," *AIEE Trans. (Power Apparatus and Systems)*, vol. 75, pp. 169–179, April 1956.
[7] C. W. Watchorn, "Elements of system capacity requirements," *Trans. AIEE*, vol. 70, pp. 1163–1185, 1951.
[8] AIEE Working Group Report, "Application of probability methods to generating capacity problems," *AIEE Trans. (Power Apparatus and Systems)*, vol. 79, pp. 1165–1182, 1960.
[9] FPC Special Technical Subcommittee, "Determination of reserve requirements for interconnected systems," Federal Power Commission National Power Survey Advisory Committee, Rept. 5, February 1963.
[10] H. Halperin and H. A. Adler, "Determination of reserve-generating capability," *AIEE Trans. (Power Apparatus and Systems)*, vol. 77, pp. 530–544, August 1958.
[11] E. S. Vassell and N. Tibberts, "An approach to the analysis of generation-capacity reserve requirements," *IEEE Trans. on Power Apparatus and Systems*, PAS-84, pp. 64–79, January 1965.
[12] L. K. Kirchmayer and A. G. Mellor, "Economic choice of generator unit size," *ASME Trans.*, vol. 80, pp. 1015–1026, July 1958.
[13] C. J. Baldwin "Planning for power pools and EHV ties," *Proc. of the American Power Conference*, vol. 25, pp. 823–833, 1963.
[14] C. D. Galloway and L. L. Garver, "Computer design of single-area generation expansions," *IEEE Trans. on Power Apparatus and Systems*, vol. 83, pp. 305–311, April 1964.
[15] E. S. Bailey, Jr., C. D. Galloway, E. S. Hawkins, A. J. Wood, "Generation planning programs for interconnected systems," *IEEE Trans. on Power Apparatus and Systems*, special supplement, pp. 761–788, 1963.

Discussion

J. H. Ashby (Technical Services Inc., Dallas, Texas): This paper is a very timely and valuable contribution to the methods of analyzing power system performance by the use of probability mathematics. The author has given the system planner a most useful tool by which he may extrapolate the results of a given probability study, or perhaps, what is more important, acquire a better perspective of his system's projected performance under the influence of many variables.

Manuscript received February 3, 1966.

The companies with which the author and I are associated have had several recent occasions to make good use of the methods presented in this paper. In 1964 and 1965 eight loss-of-load probability studies were performed, using the General Electric Program, to examine the Dallas Power and Light Company, Texas Electric Service Company, and Texas Power and Light Company. This examination concerned the companies system service reliability in the past and under projected conditions of variable forced outage rates, as well as constant and increasing generating unit size. After these studies were made, certain changes in forecast demands and the size of two future generating units called for a restudy of the affected years. In each case, the methods of this paper permitted an easy determination of change in the system's risk index from the available computer output of previous studies. When the system plan later became firm, another computer study was made and confirmed the validity of the approximating methods.

In all of these studies, two load levels were prescribed for each year—one for the normal forecast, and the other for 106 percent of this to account for the effect of an extremely hot summer. The computer output then gave the two points of data to permit plotting curves for each year similar to those of Fig. 3 of this paper. The reserve required each year to maintain the selected risk level of 0.1 day per year could then be read off and used as a measure of the relative reliability of the planned expansion pattern. In this process, it was noted that as unit size increased, loss-of-load probability became less sensitive to change in reserve, i.e., the slope of the system characteristic became less steep. This bears out the increasing value of m shown in Table II of the paper. It has been found in subsequent calculations using the more precise statement of (6) that the value of m can vary from 150 MW to nearly 300 MW as unit size increases, depending on the timing of such units and their forced outage rates.

C. W. Watchorn (Pennsylvania Power and Light Company, Allentown, Pa.): The paper presents a very interesting and valuable basis for extending the results of a computer study of installed generating capacity requirements to future conditions other than the specific ones for which the study was made without necessitating further computer studies. When a report is being studied such latitudes are always valuable because of the many questions that often arise about situations having conditions different from those considered in the initial study, for which it is highly desirable to obtain

Manuscript received February 17, 1966.

quick and, at least, reasonably accurate answers. The method described in the paper provides a means for doing just this with respect to capacity to be added in the future.

A simplified method for doing the same thing, plus a great deal more, i.e., providing a simple basis for making the basic study itself, for studying the effect of the removal, and for changing the forced outage rates of already installed as well as future units, with a simpler computer program than the one upon which this paper is based, was presented about ten years ago.[1] This method has been found to be very valuable not only for making investigations along the lines described in the subject paper but also for making the basic primary studies of installed capacity requirements.

The two principal draw backs to this method are: 1) it requires so few computer computations that the results can hardly be said to have been determined by a computer, and 2) no theoretical basis has been presented for the validity of the method although the results have been found to almost exactly check those determined by computer calculations for the same problem to well within the degree of practical accuracy requirements. The derivation of the method was entirely empirical, but it is no less believed to be the result of some rapidly converging mathematical process.

[1] C. W. Watchorn, "A simplified basis for applying probability methods to the determination of installed generating capacity requirements," *AIEE Trans.* (*Power Apparatus and Systems*), vol. 76, pp. 829–832, October 1957.

L. L. Garver: I appreciate learning from Mr. Ashby that the results of this paper have been successfully applied to their generation expansion planning problem. Mr. Watchorn puts his finger on the real need for an estimating procedure—data changes. Input data assumptions are forever being adjusted, and it is desirable to quickly extrapolate to new situations from the results of previous investigations.

The pioneering work of Mr. Watchorn is in evidence here with the reference to his 10-year-old publication on a method of estimating system reliability. However, there does not appear to be any simularity between our two methods.

Finally, I want to thank J. H. Ashby and C. W. Watchorn for their interest in this paper as evidenced by their discussions.

Manuscript received March 4, 1966.

Frequency and Duration Methods for Power System Reliability Calculations: I—Generation System Model

J. D. HALL, R. J. RINGLEE, SENIOR MEMBER, IEEE, AND A. J. WOOD, SENIOR MEMBER, IEEE

Abstract—As a goal, the evaluation and computation of electric power system reliability requires that a consistent technique be used for all portions—generation, transmission, and distribution. At present, a number of different methods are used for the generation system, while the frequency and duration of outages seems to be developing as a standard measure for the analysis of the distribution system. This paper and a subsequent one will present a reliability calculation method for the generation system that incorporates the frequency and duration of unit outages and includes consideration of the loads. This method leads to calculated generation reliability measures which are the availability, frequency of occurence, and mean duration of reserve states. These are cumulative states in that they specify system reserve conditions of a given magnitude or less. This paper is concerned with the procedure for calculating the availability, frequency, and outage duration for a number of generating units connected in parallel to form a single system. Numerical data are used to illustrate the technique and make comparisons with other methods.

INTRODUCTION

Background

ANALYSIS of power system reliability is a problem area which has received considerable attention over the years. Most of this effort has been expended in the reliability analyses and reserve requirement planning of generation system [1]–[3]. A renewed interest in this particular problem naturally arises when efforts are made to integrate this work with the more recent efforts that have been devoted to the evaluation of the reliability of the transmission and distribution systems [4]–[8]. Immediately, one is confronted with the divergence of reliability measures commonly used in the various portions of the system and with their apparent lack of compatibility.

In the study of generation reserve requirements, two reliability techniques are found in widespread use. The loss-of-load method is basically a calculation of the probability of the failure to be able to serve the expected peak loads over a specified time period. This technique includes the nature of the expected load and usually characterizes each individual generation unit by a maximum capability and a long-run probability of being in service, i.e., its availability. The other technique widely applied is the frequency and duration technique presented by Halperin and Adler [1]. This method utilizes more data about each

Paper 68 TP 56–PWR, recommended and approved by the Power System Engineering Committee of the IEEE Power Group for presentation at the IEEE Winter Power Meeting, New York, N. Y., January 28–February 2, 1968. Manuscript submitted September 8, 1967; made available for printing November 13, 1967.

The authors are with the General Electric Company, Schenectady, N. Y.

generation unit in that the average time durations of available and not-available, or repair, periods are used as well as the unit availability. The method allows the computation of the long-term probability of the generation system suffering an outage state of exactly a given amount and the expected frequency with which this state will reoccur. The second reference discusses these two methods in somewhat greater detail and gives a comparison of the numerical results obtainable by using each.

System Reliability

Increasing attention is being given to the development of reliability measures that are common to the entire system. At the distribution end of the system, reliability assessments and statistical performance records are apt to be kept on the basis of the number of outages, their frequency, their duration, and the number of customers affected. Methods of predicting these measures as an aid to more reliable transmission and distribution design have been developed [4]–[8]. Therefore, it would seem highly desirable to reexamine the methods used for the generation system to see if a technique compatible with these methods may be evolved.

This paper sets up a model of the generation system which may be used for generation reserve studies and which may be used to compute availability, frequency, and duration for both exact and cumulative capacity outage states. It thus offers the opportunity for the eventual integration of the power generating system model into an overall power system reliability model.

Generation System Model

In modeling the generation system, the units are assumed to be connected in parallel. Each unit is defined by a given maximum capability and by a long-run behavior pattern with regard to the occurrence of the available-repair cycles through which it passes. Each unit, in turn, may be merged into the generation system to permit the development of a capacity model characterized by the existence of various amounts of capacity available (or, conversely, on outage), the expected availability of exactly this capacity, and the expected recurrence, or cycle, time of this state.

The technique to be presented differs from previous techniques since, in this model, each unit may be described by its own capability and durations of available and repair periods, and further, the model with exact capability states is readily transformed into one characterized by occurrences of a given amount of capacity-available or more. These are the same as the cumulative outage states in the loss-of-load probability model. As shown later, this transformation may be accomplished without any restrictive, unnecessary, or simplifying assumptions. Thus,

Reprinted from *IEEE Trans. Power Apparatus Syst.*, vol. PAS-87, no. 9, pp. 1787–1796, Sept. 1968.

the model developed may be used to implement both of the commonly used generation system reliability computation techniques, it will generate information concerning the frequency and duration of cumulative outage states, and it provides frequency and duration measures of the generation system compatible with the transmission and distribution system reliability measures.

The basic goals of the remainder of this paper are to: 1) present the logical development of a reliability model of a power generation system; 2) illustrate how recursive relationships may be derived and used to develop numerical data; and 3) compare the results of this method with other techniques by means of examples.

FREQUENCY AND DURATION BY RECURSIVE TECHNIQUES

There are certain concepts that will be of value as this method is developed. The ability of a generator to provide power is equal to its instantaneous capacity; this is a value changing in time and dependent upon the state of the auxiliary equipment associated with the generator and the environment about the plant, such as the condenser cooling water temperature. The capacity may be at full machine rating for certain periods of time, changing suddenly to a partial rating due to the loss of certain auxiliary equipment, or it may be at zero value when the unit is taken out of service. The transitions from one capacity state to another are assumed to take place instantaneously and to occur at any time. The amount of time that the capacity remains at a certain value is the time in residence for the given state. The availability of a given state is then the mean-time-in-residence divided by the mean cycle of time for this particular state to occur or reoccur.

Fig. 1(a) shows a sketch of this phenomenon. Here it must be recognized that we are presenting the occurrence of only one state. The space between the shaded blocks represents transition into other states which may be one or many. It should be emphasized at the outset that one need not represent a generator as simply a binary machine with the capability of full output or none. More practically, one might consider representation of a partial capacity output state as well. The theory to be presented is applicable to the binary, ternary, or multiple state system. The examples are selected for binary machines.

Stochastic Processes, Markov Processes, and Transitions

A single repairable device which is either available (up) or in repair (i.e., down) may be characterized by its expected or mean behavior. It is assumed that repair and failure rates are constant. It is further assumed that the mean time to failure m and mean time to repair r are finite. The assumption of constant failure and repair rates puts the machine state description into the more restricted class of Markov processes. With finite r and m we say that both up and down states are "accessible," and that over a long interval of time, the availability, or fraction of time the machine will be in an up state, is a number greater than zero and less than one.

The mean cycle depicted on Fig. 1(a) defines the following terms.

$T = 1/f$, cycle time (days)
f = frequency (cycle per unit time)
$m = 1/\lambda$, mean up time (days)
$r = 1/\mu$, mean repair time (days).

In addition, these parameters may be used to define mean rates of occurrence and long-term availabilities as follows.

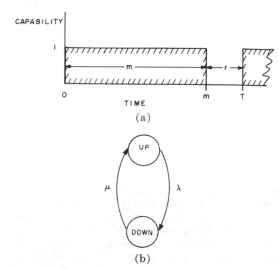

Fig. 1. (a) Average history of unit capability. (b) Two-state transition diagram for a repairable machine.

λ = failure rate (failures per unit time)
μ = repair rate (repairs per unit time)
$A = m/(m + r) = m/T$, availability (steady state)
$\bar{A} = 1 - A = r/T$, unavailability (steady state).

Note that the term availability is used to mean the steady-state or long-time average availability. The availabilities, transition rates, and mean cycle time are related by

$$\lambda = 1/AT \tag{1}$$

$$\mu = 1/\bar{A}T \tag{2}$$

$$f = A\lambda = \bar{A}\mu. \tag{3}$$

Fig. 1(b) indicates the state transition diagram drawn for the two-state device. The arrows indicate entries or exits from the states, and the quantities λ and μ designate the rates of departure and entry. It should be reasonably self-evident that frequency with which a state is encountered in the long run is as follows.

$$f_{(up)} = A\lambda$$
= (steady-state probability of being in state)
$\times$ (rate of departure) (4)

or

$$f_{(up)} = \bar{A}\mu$$
= (steady-state probability of not being in the state) $\times$ (rate of entry). (5)

Example 1: A single, repairable generator unit:

$$\text{Capacity} = 20 \text{ MW}$$

$$A = 0.98$$

$$r = 2.040816 \text{ days}.$$

Therefore

$$\mu = 1/r = 0.4900$$

$$\lambda = \bar{A}/rA = \mu\bar{A}/A = 0.0100$$

so that

$$T = 1/\bar{A}\mu = 102.0408 \text{ days}$$

is the mean cycle time for encountering either the up or down states.

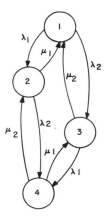

Fig. 2. Four-state transition diagram for repairable two machines in parallel.

Two Machines in Parallel

Equations (4) and (5) are perfectly general even if there is more than one mode of entering or leaving a state. The case of two repairable machines in parallel may be used to illustrate this. The number of possible states is $2^2 = 4$. The definitions of the four possible states are given below. Fig. 2 shows the transition diagram for these states.

State Number	Machine 1	Machine 2	Rate of Departure
1	up	up	$\lambda_1 + \lambda_2$
2	down	up	$\lambda_2 + \mu_1$
3	up	down	$\lambda_1 + \mu_2$
4	down	down	$\mu_1 + \mu_2$

For this model the run-repair process for each machine is independent of the processes for the other machines. The last column indicates the rates of departure from each of the states. Note that the mean time in residence in a state is equal to the reciprocal of the rate of departure. As an example, the cycle time between encountering state 2, on the average, is

$$T_2 = 1/[A_{\text{state } 2} (\lambda_2 + \mu_1)].$$

Example 2: Two generators in parallel.

Unit	Capacity (MW)	Availability	r (days)	μ	λ
1	20	0.9800	2.040816	0.49	0.01
2	30	0.9800	2.040816	0.49	0.01

Referring to the state transition diagram of Fig. 2, the availabilities and mean times between encountering the states are as follows.

State Number	Capacity Available (MW)	A (per unit)	Rate of Departure (per day)	Cycle Time (days)
1	50	0.9604	$\lambda_1 + \lambda_2 = 0.02$	52.0616
2	30	0.0196	$\mu_1 + \lambda_2 = 0.50$	102.0408
3	20	0.0196	$\lambda_1 + \mu_2 = 0.50$	102.0408
4	0	0.0004	$\mu_1 + \mu_2 = 0.98$	2551.02

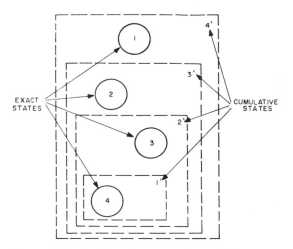

Fig. 3. Relationship of exact and cumulative state descriptions for the two-machine example.

Cumulative-Event Cycle Time

In the loss-of-load probability method, it is convenient to redefine the events considered so that an outage "event" is the occurrence of an outage of a given magnitude or greater. The same reasoning holds true in the frequency-duration method. It is interesting to know that the mean time between encountering an outage of exactly 30 MW is 102.0408 days, but it would be of more value to know the frequency of encountering an outage of 30 MW or more. That is, how often (frequency) will the outage change from a value less than 30 MW to an outage of 30 MW or more?[1]

To accomplish this it is merely necessary to redefine the state so that each state is now the occurrence of a given capacity outage or larger. The previous, two-parallel machine transition diagram may be used to illustrate this transformation procedure and the steps necessary to obtain the frequency of encountering the newly defined states. Fig. 3 shows the new states superimposed over the old. The new, or cumulative, states are denoted by primes. Notice that the new states are numbered differently than the old: state 1' = state 4, state 2' = states 3 and 4, and so on.

The frequency of encountering state 1' is the same as that of encountering state 4:

$$f_{1'} = A_4(\mu_1 + \mu_2) = f_4.$$

The frequency of encountering the new state 2' is equal to the sum of the frequencies with which transfers take place from old state 3 to old state 1, $A_3 \mu_2$, and from old state 4 to old state 2, $A_4 \mu_2$. Clearly, this result is less than the sum of the frequencies of encountering state 3 or 4 by the sum of the encounters of state 4 from 3, $A_3 \lambda_1$, and of state 3 from 4, $A_4 \mu_1$. Note that transfers between states 3 and 4 represent failure and repair of machine 1. The frequency of transfer either way is given by the product of the unavailability of machine 2 and the frequency of encounter of machine 1 "up" state. Here is a method of recursively determining the frequency of encountering cumulative states. To illustrate, the frequency of encounter of cumulative

[1] A Markov chain composed of independent two-state elements, such as the up–down process assumed for each machine, is irreducible. That is, all states are accessible; no matter what the initial state, given enough time, the system would assume all states. This will enable the transformation of the previously defined states into cumulative capacity states.

TABLE I

TWO MACHINE EXAMPLE

State Number	Exact Capacity States Capacity (MW)	Availability	Departure Rates λ_{up}	λ_{down}	Cumulative Capacity States State Number	Capacity (MW)	Availability	Cycle Time (days)
1	50	0.9604	0	0.02	4	50	1.0000	
2	30	0.0196	0.490	0.01	3	30	0.0396	52.0616
3	20	0.0196	0.490	0.01	2	20	0.0200	102.0408
4	0	0.0004	0.980	0	1	0	0.0004	2551.02

state 2′, $f_{2'}$ is given by the sum of the frequency of encounter of state 1′, $f_{1'}$, plus the frequency of encounter of old state 3 from old state 1, $A_3 \mu_2$, less the frequency of encounter of old state 4 from old state 3, $A_3 \lambda_1$:

$$f_{2'} = f_{1'} - A_3\lambda_1 + A_3\mu_2.$$

The following transition rates are defined so that the example may be generalized. Let

$\lambda_{+k} = \lambda_{up}$ = rate of transition out of a given capacity state k to one in which more capacity is available

and

$\lambda_{-k} = \lambda_{down}$ = rate of transition out of a given capacity state k to one in which less capacity is available.

The frequency of encountering a state with a given capacity or less is given by the recursive relationship below. In this relationship exact (i.e., unprimed) state k is being added to cumulative state $n - 1$ (i.e., primed) to obtain the new cumulative capacity state n.

$$f_n = f_{n-1} - A_k\lambda_{-k} + A_k\lambda_{+k}. \tag{6}$$

In (6),

A_k = availability of the exact capacity state k and the primes have been discarded and replaced by the subscripts, n and $n - 1$.

Example 3: Two generators in parallel.

This example continues Example 2 to calculate the availability and cycle times of the cumulative capacity states. For the sake of completeness, the availability of a cumulative capacity state n may also be found from the following well-known relationship

$$A_n = A_{n-1} + A_k \tag{7}$$

where again, the exact capacity state k is being appended to the cumulative state $n - 1$ to arrive at n.

The necessary data have been generated in the first two examples. The transition diagram for the two machines in parallel plus these data permit the calculation of λ_{up}, λ_{down}, A_n, and f_n for the four cumulative capacity states. Table I gives these results except that cycle time is given in place of the frequency.

Identical Capacity States

In the construction of the exact capacity-state availability and frequency tables for larger systems, identical capacity states may be generated by different combinations of units. In the sense of the transition diagrams, there is no direct linkage between

these states. That is, the only way that a system may transit within a given instant of time from one exact capacity state to another state with the same capacity available, is to have one machine repaired and another fail within the same instant. The probability of this occurring is of second order. That is, it is so unlikely that it is ignorable relative to the occurrence of a single event. Therefore, the two capacity states may be merged as states separated in time.

Since transfer cannot occur directly from one state to the other, their availabilities and frequencies of encountering will add directly. Let j and i designate two states which have exactly the same capacity available and k designate the merged state. Then the capacity, availability, and cycle frequency of the merged state are as follows.

Capacity:

$$C_k = C_i = C_j. \tag{8}$$

Availability:

$$A_k = A_i + A_j. \tag{9}$$

Frequency:

$$f_k = f_i + f_j. \tag{10}$$

Therefore, the total rates of departure to greater and lesser capacity states may be found from

$$A_k\lambda_{up,k} = A_i\lambda_{up,i} + A_j\lambda_{up,j} \tag{11}$$

and

$$A_k\lambda_{down,k} = A_i\lambda_{down,i} + A_j\lambda_{down,j}. \tag{12}$$

These relationships complete the set of those that are required to permit construction of nonredundant, exact capacity-availability tables. With the previous developments, cumulative capacity data may then be generated recursively. This model has been implemented in several digital computer programs, and the next section will illustrate some numerical results for larger systems.

NUMERICAL RESULTS

Since the practical applications of probabilistic models of generation systems tend to make more use of the data for the cumulative capacity-available (or outage) states, the numerical results will emphasize these data. The first data are presented to illustrate the type of information that may be obtained for a collection of dissimilar units without the necessity of making any simplifying approximations. The second example verifies the recursive method developed here by comparison with mean times to failure and repair obtained for a group of identical machines. These later results were obtained by application of formulas developed as an outcome of an investigation of general system reliability [9]. Finally, the last example is a 22-machine

TABLE II
FIVE-MACHINE EXAMPLE

Capacity Outage (MW)	Exact Outage States		Cumulative Outage States	
	Availability	Cycle Time (days)	Availability	Cycle Time (days)
0	0.88105	30.659	1.00000	—
20	0.01798	107.57	0.11895	30.659
30	0.01798	107.57	0.10097	41.166
40	0.02259	194.35	0.08299	62.627
50	0.02762	153.90	0.06040	81.514
60	0.02771	154.84	0.03278	133.81
70	0.10171×10^{-2}	1392.0	0.50726×10^{-2}	366.87
80	0.11122×10^{-2}	1275.3	0.40554×10^{-2}	487.56
90	0.12642×10^{-2}	1441.4	0.29432×10^{-2}	761.24
100	0.71004×10^{-3}	3290.0	0.16790×10^{-2}	1470.9
110	0.86836×10^{-3}	2671.5	0.96897×10^{-3}	2424.4
120	0.28518×10^{-4}	3.9145×10^{4}	0.10061×10^{-3}	1.2339×10^{4}
130	0.31458×10^{-4}	3.5546×10^{4}	0.72090×10^{-4}	1.7728×10^{4}
140	0.17490×10^{-4}	6.3452×10^{4}	0.40632×10^{-4}	3.4188×10^{4}
150	0.21900×10^{-4}	7.4362×10^{4}	0.23142×10^{-4}	7.1375×10^{4}

case. The data for this system are used to illustrate the results obtained by the use of the exact formulation and compares them with methods that have been previously suggested.

Example 4: A five-machine system.

The system for this example is composed of five machines with the following characteristics.

Capacity (MW)	Mean Repair Time (days)	Availability (per unit)
20	2.040816	0.980
30	2.040816	0.980
40	5.1082	0.975
50	5.1543	0.970
60	5.1543	0.970
Total 200 MW		

The results of using the relationship developed above are given in Table II. These data include the cycle times and the availabilities for both the exact and cumulative outage states.

Fig. 4 illustrates the cycle time and frequency of recurrence of the cumulative outage states. These data along with the more well-known information about the existence probability (i.e., the availability) of the cumulative outage states provide a comprehensive reliability picture of the generation system. Note that the frequency of occurrence of a zero outage, or more, is zero, meaning that the system is always in this state. The data of Fig. 4 are shown as smooth curves for the sake of clarity. Actually, they should be stepped to represent the changes that take place as specific units change their status.

Example 5: A system of 20 identical units.

Einhorn, in a general study of the reliability of a system of binary state, repairable elements, each with independent, exponentially distributed times in service and times on repair, has developed relations for the reliability of systems of paralleled units [9]. Specifically, the relationships for the availability, mean time in service (MTTF), and mean time to repair (MTTR) of *r* units out of a total of *n* may be utilized to verify the recursive relationships. Einhorn's analysis assumes constant failure and repair rates; he develops the expected (or mean) values of various system reliability measures. Hence, his results form the basis for checking the relationships derived here.

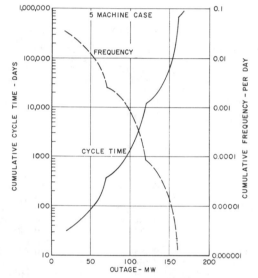

Fig. 4. Cumulative state data for five-machine system. Cycle time is mean time to recurrence of an outage of a given magnitude or more. Frequency is reciprocal of recurrence time.

The example assumes 20 identical machines each of which has a mean repair time of five days and an unavailability (i.e., forced outage rate) of 0.02. The computations were made in two ways. Table III gives the numerical results obtained using Einhorn's equations [9]. The sum of the mean time to failure plus the mean time to repair is the mean cycle time. For example, for the 16-machine cumulative state, the MTTF is 2226.48 days and the MTTR is 1.34 days, resulting in a mean cycle time of 2227.82 days.

This same division of the total cycle time between mean times to repair and failure have been computed using the methods of this paper by making appropriate use of (4) and (5) along with the availabilities and rates of transfer up and down out of the cumulative states. The total cycle times and availabilities of the cumulative capacity states were computed with the recursive equations and found to check exactly the results obtained with Einhorn's method.

As may be seen from the data of Table III, the MTTF is very much larger than the MTTR for all the states except those where only a few units are out of service. Thus, in all practical cases, the total cycle time effectively measures the MTTF.

TABLE III

20 IDENTICAL MACHINES

Number in Service	Availability	Cumulative State Results Cycle Time (days)	Mean Time to Failure—MTTF (days)	Mean Time to Repair—MTTR (days)
20	1.0	—		
19	0.33239	18.349	12.250	6.099
18	0.05990	47.321	44.487	2.834
17	0.70687×10^{-2}	257.64	255.82	1.821
16	0.59968×10^{-3}	2227.82	2226.48	1.336
15	0.38591×10^{-4}	2.7291×10^4	2.7290×10^4	1.053
14	0.19484×10^{-5}	4.4575×10^5	4.4575×10^5	0.868
13	0.78908×10^{-7}	9.3607×10^6	9.3607×10^6	0.739
12	0.26010×10^{-8}	2.4698×10^8	2.4698×10^8	0.642
11	0.70434×10^{-10}	8.0680×10^9	8.0680×10^9	0.568

TABLE IV

RESULTS FOR 22-MACHINE SYSTEM

Outage (MW)	Availability of Exact Outage	Frequency of Exact Outage Halperin and Adler [1] (per year)	Recursive (per year)	Frequency of Outage or More Sauter et al. [10] (per year)	Recursive (per year)
0	0.641171	4.797978	4.79788	—	—
25	0.039255	0.934665	0.934651	8.829087	4.79788
50	0.118567	2.754298	2.83611	7.894422	4.42402
75	0.059556	1.495380	1.53590	5.140124	3.28128
100	0.039136	1.126330	1.14588	3.644744	2.59117
125	0.011871	0.540627	0.487157	2.518414	1.99545
150	0.046756	1.259236	1.25372	1.977787	1.66536
175	0.005757	0.252479	0.251611	1.718551	1.07773
200	0.008414	0.365578	0.354191	1.466072	0.90359
225	0.004161	0.185825	0.184019	1.100494	0.66320
250	0.015668	0.432936	0.432576	0.914669	0.53508
275	0.001597	0.083256	0.078599	0.481733	0.32384
300	0.003702	0.177995	0.158219	0.398477	0.26637
325	0.001478	0.067382	0.067440	0.220482	0.15811
350	0.001020	0.049107	0.049960	0.153100	0.11044
375	0.000349	0.021984	0.020688	0.103993	0.07398
400	0.001015	0.045499	0.045120	0.082009	0.05776
425	0.000137	0.011211	0.008512	0.036510	0.02626
450	0.000188	0.012059	0.011169	0.025299	0.01948
475	0.000090	0.005470	0.005544	0.013240	0.01072
500	0.000055	0.003467	0.003543	0.007778	0.00632
525	0.000017	0.001346	0.001315	0.004311	0.00347
550	0.000027	0.001921	0.001718	0.002965	0.00237
575	0.000006	0.000408	0.000437	0.001044	0.00099
600	0.000005	0.000382	0.000359	0.000626	0.00062
625	0.000002	0.000162	0.000178	0.000244	0.00031
650	0.000001	0.000082	0.000102	0.000082	0.00016

Example 6: A 22-unit model system.

A system that has been used in previous publications to illustrate various techniques is the system of 22 units suggested by Arnoff and Chambers [10], [11]. The relevant data for the units in this system are listed below.

Number of Identical Units	Unit Size (MW)	Mean Repair Time r (years)	Mean Cycle Time T (years)
1	250	0.06	3.0
3	150	0.06	3.0
2	100	0.06	3.0
4	75	0.06	3.0
9	50	0.06	3.0
3	25	0.06	3.0
Total 22	1725 MW		

Table IV gives the results of applying different techniques to calculating the frequencies of the exact and cumulative states.

The Halperin-Adler method and the method of this paper were used to compute the frequency of occurrence of the exact outage states. These are the data in the third and fourth columns of Table IV. The deviations between the two methods rarely go beyond 20 percent, indicating the applicability of the technique of Halperin and Adler [1] for obtaining frequency and duration data for the exact outage states.

The calculation of the cumulative frequency has been done in three ways. The fourth and fifth columns of Table IV compare the results obtained by Sauter et al. [10] with those obtained by the recursive technique. Sauter et al. [10] suggested a technique which effectively ignored the transfer of the system within a given cumulative outage state. That is, they suggested in effect that (6), for the recursive development of the frequency of the cumulative state, be approximated by

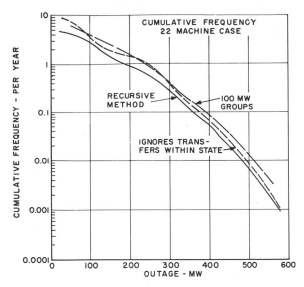

Fig. 5. Cumulative frequency curves for the 22-unit system. The exact data are compared with two approximations.

$$f_n = f_{n-1} + f_k$$

where f_k is the frequency of the exact state being added to the $(n-1)$th cumulative state to get the nth cumulative state. Halperin and Adler, on the other hand, suggested a machine grouping technique for obtaining a smooth curve estimating cumulative outage frequency as a function of the outage [1]. Their method was used with a 100-MW group size for this system.

The results of applying all three methods are shown in Fig. 5. Three curves are shown: 1) the curve obtained using the recursive formulation; 2) the "100-MW grouping" obtained by using the method suggested by Halperin and Adler [1]; and 3) the curve obtained using the technique of Sauter *et al.* [10]. These results indicate the inherent error in using either approximation to the cumulative frequency relationship.

Comments

The last three examples illustrate the results obtainable by modeling the generation system by a frequency and duration method; they substantiate the validity of the equations developed in the previous section, and they point out the effect of approximations on the calculations for the cumulative capacity states. The present modeling method develops the capacity outage—availability data as well as the cycle times. It may also be used to generate information concerning the mean duration of both exact and cumulative outage states. However, as would be expected and as shown by the results of Example 5, the mean repair times are very much less than the mean up times (i.e., MTTF). Perhaps the most significant result of all of these examples is the demonstration that frequency and duration of outages may be generated recursively in a quite straightforward fashion.

CONCLUSION

It has been the purpose here to present and illustrate a probabilistic model of the power generation system that may be used to calculate the availabilities, occurrence frequencies, and cycle durations for both exact capacity outage states and cumulative outage states. This model assumes constant failure and repair rates but does not require further assumptions to generate the frequency and mean duration of cumulative outage

states. Further, these methods yield a recursive technique that is easily implemented on a digital computer.

In a subsequent paper this model will be combined with a model of the load in order to provide generation margin risk indices. Finally, this frequency-duration model offers an approach to combined generation-transmission system risk assessment.

REFERENCES

[1] H. Halperin and H. A. Adler, "Determination of reserve generating capability," *AIEE Trans. (Power Apparatus and Systems)*, vol. 77, pp. 530–544, August 1958.
[2] AIEE Subcommittee Report, "Application of probability methods to generating capacity problems," *AIEE Trans. (Power Apparatus and Systems)*, vol. 79, pp. 1165–1182, 1960 (February 1961 sect.).
[3] V. M. Cook, C. D. Galloway, M. J. Steinberg, and A. J. Wood, "Determination of reserve requirements of two interconnected systems," *IEEE Trans. Power Apparatus and Systems*, vol. 82, pp. 18–33, April 1963.
[4] C. F. DeSieno and L. L. Stine, "A probability method for determining the reliability of electric power systems," *IEEE Trans. Power Apparatus and Systems*, vol. 83, pp. 174–181, February 1964.
[5] D. P. Gaver, F. E. Montmeat, and A. D. Patton, "Power system reliability: I—Measures of reliability and methods of calculation," *IEEE Trans. Power Apparatus and Systems*, vol. 83, pp. 727–737, July 1964.
[6] F. E. Montmeat, A. D. Patton, J. Zemkoski, and D. J. Cummings, "Power system reliability: II—Applications and a computer program," *IEEE Trans. Power Apparatus and Systems*, vol. PAS-84, pp. 636–643, July 1965.
[7] S. A. Mallard and V. C. Thomas, "A method for calculating transmission system reliability," *IEEE Trans. Power Apparatus and Systems*, vol. PAS-87, pp. 824–834, March 1968.
[8] R. J. Ringlee, "Steps in measuring service continuity," *Distribution* (General Electric Co., Schenectady, N. Y.), vol. 28, pp. 14–18, Fourth Quarter, 1966.
[9] S. J. Einhorn, "Reliability prediction for repairable redundant systems," *Proc. IEEE*, vol. 51, pp. 312–317, February 1963.
[10] D. M. Sauter, C. J. Baldwin, and K. M. Dale, discussion of Halperin and Adler [1], pp. 541–542.
[11] E. L. Arnoff and J. C. Chambers, "Operations research determination of generator reserves," *AIEE Trans. (Power Apparatus and Systems)*, vol. 76, pp. 316–328, June 1957.

Discussion

A. D. Patton (Electric Power Institute, Texas A&M University, College Station, Tex.): I wish to compliment the authors on their excellent paper. The methods presented seem to be a significant improvement over those published previously. Presumably the authors will, in due time, incorporate a load model of some kind to yield "loss-of-load" figures of merit. I hope that expected loss-of-load frequencies and durations will be among the figures of merit calculated, as I feel they are superior to loss-of-load probability as measures of system reliability. The authors have concentrated their discussion on expected failure rates and cycle times. It seems worthwhile, therefore, to state that, using the authors' assumptions, the expected duration of a period spent in a state is given as

$$r = \frac{1}{\Sigma \mu_i + \Sigma \lambda_j}$$
$$\text{units down} \quad \text{units up}$$

where μ_i is the repair rate of ith unit, and λ_j is the failure rate of jth unit.

It should be noted that the authors assume that departure from a state is possible due to repair of a unit that is down or failure of a unit that is up. Since a given state *may* result in system shutdown due to insufficient capacity, the authors' assumption essentially is that units can fail whether they are running or not. This may or may not be a good assumption. The numerical significance of the assumption depends on the relative magnitudes of μ_i and λ_j and the number of generators in the system.

Manuscript received February 14, 1968.

TABLE V

Machines Forced Out	In Service	$p_n{}^m$	$\dfrac{n}{t}$	$\dfrac{m-n}{T-t}$	$\dfrac{n}{t} + \dfrac{m-n}{T-t}$	$F_n = p_n{}^m\left[\dfrac{n}{t} + \dfrac{m-n}{T-}\right]$
0	20	667 607 96	0.000	0.081 632 653	0.081 632 653	0.054 498 609
1	19	272 493 04	0.200	0.077 551 020	0.277 551 020	0.075 630 721
2	18	52 830 28	0.400	0.073 469 388	0.473 469 388	0.025 013 520
3	17	6 469 06	0.600	0.069 387 755	0.669 387 755	0.004 330 310
4	16	561 08	0.800	0.065 306 122	0.865 306 122	0.000 485 506
5	15	36 65	1.000	0.061 224 490	1.061 224 490	0.000 038 894
6	14	1 86	1.200	0.057 142 857	1.257 142 857	0.000 002 338
7	13	7	1.400	0.053 061 224	1.453 061 224	0.000 000 102
						0.160 000 000

$p_{n+}{}^m$	$\dfrac{n}{t} - \dfrac{m-n}{T-t}$	$p_n{}^m\left[\dfrac{n}{t} - \dfrac{m-n}{T-t}\right]$	F_{n+}	T_{n+}	t_{n+}
1 000 000 00	$-0.081\ 632\ 653$	$-0.054\ 498\ 609$	0.000 000 000	—	—
332 392 04	0.122 448 980	0.033 366 495	0.054 498 609	18.349 092	6.099 092
59 899 00	0.326 530 612	0.017 250 704	0.021 132 114	47.321 342	2.834 501
7 068 72	0.530 612 245	0.003 432 562	0.003 881 410	257.638 332	1.821 173
599 66	0.734 693 878	0.000 412 222	0.000 448 848	2 227.925 712	1.335 998
38 58	0.938 775 510	0.000 034 406	0.000 036 626	$2.730\ 3 \times 10^4$	1.053 350
1 93	1.142 857 143	0.000 002 126	0.000 002 220	$4.504\ 5 \times 10^5$	0.869 369
7	1.346 938 776	0.000 000 094	0.000 000 094	$10.638\ 3 \times 10^6$	0.744 681

Note: $F_{n+} = \Sigma p_n{}^m\left[\dfrac{n}{t} - \dfrac{m-n}{T-t}\right]$.

Veazey M. Cook (Consolidated Edison Company of New York, Inc., New York, N. Y.): Only a few papers have been published that give methods of calculating the average frequency and duration of forced outage states that occur in a given generation system of units. This paper is very interesting and very welcome at this time. It is concerned only with units connected in parallel. The study of units connected in series and various combinations of parallel-series systems will have to be made to develop the reliability calculations for an entire generation, transmission, and distribution system.

There is a need for probability methods on both the exact capacity states and the cumulative capacity states. That this is possible by the methods suggested in the paper is illustrated by the two-machine example. A three-machine example would have been better because it would have illustrated more details of the calculations involved. Only those having some experience with change-of-state probability theory will be able to grasp the full meaning of the paper's suggested procedure.

In the two-machine example, the forced outage existence rates for the four states are obtained from the binomial expansion $(0.98 + 0.02)^2 = 0.9604 + 0.0196 + 0.0196 + 0.0004$, so that $A_1 = 0.9604$, $A_2 = 0.0196$, $A_3 = 0.0196$, and $A_4 = 0.0004$. Multiply each of these states by their respective rates of departure, 0.02, 0.50, 0.50, and 0.98, and get the frequency rate of occurrence of the four states.

$$F_1 = 0.02 \times 0.9604 = 0.019208 \text{ times per day}$$
$$F_2 = 0.50 \times 0.0196 = 0.009800 \text{ times per day}$$
$$F_3 = 0.50 \times 0.0196 = 0.009800 \text{ times per day}$$
$$F_4 = 0.98 \times 0.0004 = \underline{0.000392} \text{ times per day.}$$
$$0.039200$$

The sum of these frequencies, 0.039 200, should be equal to $(2 \times \text{number of machines})/T$ as a check on the arithmetic. And $(2 \times \text{number of machines})/T = 4/102.0408$ or 0.0392. The cycle times are

$$T_1 = 1/F_1 = 1/0.019208 = 52.06164 \text{ days}$$
$$T_2 = 1/F_2 = 1/0.009800 = 102.04082 \text{ days}$$
$$T_3 = 1/F_3 = 1/0.009800 = 102.04082 \text{ days}$$
$$T_4 = 1/F_4 = 1/0.000392 = 2551.02041 \text{ days.}$$

In Fig. 1, the frequency of encountering state 1 is $\lambda_1 A_1 + \lambda_2 A_1 = 2m/T^2$. This is the same as $\mu_1 A_2 + \mu_2 A_3 = 2m/T^2$. The exit and entering frequencies to state 1 are the same. This principle holds for the other states.

The example of 20 identical machines has been checked by a process different from that of the paper. The calculations involve the following definitions of symbols:

Manuscript received February 15, 1968.

m	number of machines installed
n	number of machines out of service due to forced outage
t	average duration of an individual forced outage
T	average interval of time between starts of forced outages of an individual machine
$p_n{}^m$	general term of the binomial expansion $(q + p)^m$, which is equal to $C_n{}^m p^n q^{m-n}$ where $C_n{}^m = m!/(m-n)!n!$
$p_{n+}{}^m$	summation of the binomial terms $p_n{}^m + p_{n+1}{}^m + \cdots + p_m{}^m$
F_n	occurrence frequency with exactly n machines out
F_{n+}	occurrence frequency with n or more machines out
T_n	average interval of time between starts of forced outages with exactly n machines out, $T_n = 1/F_n$
T_{n+}	average interval of time between starts of forced outages with n or more machines out, $T_{n+} = 1/F_{n+}$
t_n	average duration of periods with exactly n machines out, $t_n = p_n{}^m/F_n$
t_{n+}	average duration of periods with n or more machines out, $t_{n+} = p_{n+}{}^m/F_{n+}$.

The calculations are shown in Table V.

The values of T_{n+} and t_{n+} agree closely with the corresponding values in Table III of the paper. These might be exactly the same if sufficient decimals are carried in the calculations.

The assumption that failures should be considered successive, so that no two failures occur at the same instant of time, is quite important in the derivation of equations. An example is the case of two boilers connected to a single turbine-generator. If one boiler goes out, the other boiler or the turbine-generator can go out. But with two boilers out, the turbine-generator cannot go out. Also, if the turbine-generator goes out first, then neither one of the boilers can go out. The method of the paper should be extended to cover restricted transitions.

Herman Halperin and **Hans A. Adler** (Consultants, Menlo Park, Calif., and La Grange Park, Ill., respectively): We consider that the general proposals of the authors to use probability studies of standard or similar nature for all the parts of an electric utility system constitute an approach in the right direction. As indicated in their reference to our 1958 paper [1], we have preferred the use of frequency and duration of outages as the standard measures for bulk-

Manuscript received February 16, 1968.

power facilities, and we favor it for the other portions of electric systems—all of which is in accord with the authors' ideas.

We have found that the method of using frequency and duration, rather than simple probability methods, is readily adaptable to loss-of-load calculations for generating systems.

In our early probability studies of availability of generating system capacity, we included the effect of outages of important transmission lines. Our later studies regarding minemouth stations included the effects of possible outages of various links in the long transmission of power, such as lines, breakers, and transformers.

In today's systems, there are frequently complex series-parallel arrangements of transmission links, thus making the availability problem more difficult. For example, the outage of one of several parallel lines may or may not affect the system capacity available to serve load. We applied probability methods to parts of the system such as ac network areas, but found that calculations for a large bulk-power system require more complicated procedures.

Our studies over some 23 years lead to agreement with the authors on the importance of making allowance for partial outages and deratings. Adler has developed methods for including partial outages in calculations [12].

Table IV compares the results based on the method proposed by the authors with the results based on the Halperin-Adler method. As stated in the paper, the frequencies of exact outages agree very closely.

Regarding the determination of frequencies of outages of *n* or more, we refer to the closure of our 1958 paper [1], which contained a reply to the discussion by Sauter *et al.* [10]. We showed there a procedure for deriving frequencies of outages of *n* or more from outages of exactly *n*. As stated there, the frequency of outages of 25 MW or more is equal to the frequency of zero outages or 4.798 per year, which agrees with the authors' result. The figures in the second to last column of Table IV do not represent the results of our procedure. Neither does the 100-MW group line of Fig. 5 represent cumulative frequencies.

Our practice of grouping sizes does not have the purpose of determining cumulative frequencies. Rather, our purpose is to reduce the burden of the voluminous calculations for a large system and to produce greater uniformity of results. This is explained in detail elsewhere [13].

References

[12] H. A. Adler, "Principles and methods of calculation of probability of loss of generating capacity," *IEEE Trans. Power Apparatus and Systems*, vol. PAS-86, pp. 1467–1479, December 1967.
[13] E. E. Ciesielski and J. P. Lynskey, "Application of method: probability of loss of generating capacity," presented at IEEE Winter Power Meeting, New York, N. Y., January 28–February 2, 1968.

J. D. Hall, R. J. Ringlee, and **A. J. Wood:** We are pleased to respond to the thoughtful and helpful questions proposed by the discussers. It would appear that we are all agreed that average frequency and average duration of events are very useful measures from which to construct indices for measuring power system reliability. We feel it important to stress that the use of average frequency and duration does not imply any particular distribution of times-to-failure or times-to-restore. All that is implied is that the probability distribution for times-to-failure and times-to-restore possess mean values. The generality of this point is well discussed in Parzen [14]. It is interesting to note that the repair model used by Halperin and Adler (a fixed duration restoral time) and the restoral model assumed in our paper (an exponential distribution of repair times) lead to precisely the same results, provided that the average restoral time for the exponential distribution is equal to the constant restoral time of the fixed-duration model. For machine sets composed of similar units, the results reported in this paper and the results given previously by Halperin and Adler are identical for the "exact" outage states.

Another point that needs reemphasis is that the models usually employed for evaluating installed generation margins treat the states

Manuscript received March 18, 1968.

of each machine as statistically independent of all other machine states. This model has been used for a long time: more than thirty-five years to be precise [15], [16].

Mr. Patton's question pertains to the basis of the model employed. The model is based upon independent machine behavior and binary machine states, "running" or "failed." We presume that Mr. Patton is referring to the fact that the model contains no reserve shutdown states nor does it recognize dependent or coupled effects of massive generation or transmission outages when he suggests that the "... author's assumption essentially is that units can fail whether they are running or not." This inference is not correctly drawn within the framework of the model cited, as illustrated in Fig. 1(b) of the paper. There is no reason why such a model as Mr. Patton proposes cannot be developed or why analytical techniques cannot be prepared, particularly if the approach is taken through a system-state description [17], [18].

We are most grateful for the valuable contributions offered by Mr. Cook, and per his request, a three-machine example is illustrated in Table VI. We agree with Mr. Cook that the methods proposed should be extended to cover the restricted transition cases.

We wish to thank Messrs. Halperin and Adler for the perspective offered in their discussion. There is a semantic point, however, for which further discussion is required. Frequencies of outages of *n* or more can be taken to mean either *n* machines or more or, as in their discussion, *n* megawatts or more. As we understand the specific work referred to by Halperin and Adler, the procedure that they have presented yields, in general, the frequency of occurrence of outages of *n* or more machines and not for outages of *n* or more megawatts. It is interesting to note that the method of the paper and the method of Halperin and Adler as proposed in their discussion, agree for the prediction of the outages of *n* or more machines of the same capacity.

Halperin and Adler [1] state in their closure: "The frequency of one or more outages is equal to the frequency of zero outages. The frequency of two or more outages is the difference between the frequency of exactly one outage and the frequency of one or more outages. Similarly, the frequency of three or more outages is equal to the frequency of exactly two outages minus the frequency of two or more outages, etc." Table VII illustrates the development of these statements into a recursive algebraic formula for the generation of cumulative frequencies. Also shown are the relationships for the methods of the current paper. In this table, *n* denotes a cumulative state, *k* is the exact state, and the frequency for exactly *k* units failed is $A_k(\lambda_{+k} + \lambda_{-k})$.

From this listing, generalized cumulative frequency relationships may be written for state $(n - j)$ which has *j* machines failed. These are as follows:

Halperin-Adler:

$$f'_{(n-j)} = A_{(j-1)}[\lambda_{+(j-1)} + \lambda_{-(j-1)}] - f'_{(n-j+1)}$$
$$= A_{(j-1)}\lambda_{-(j-1)} + [A_{(j-1)}\lambda_{+(j-1)} - f'_{(n-j+1)}]. \quad (13)$$

Hall-Ringlee-Wood:

$$f_{(n-j)} = f_{(n-j+1)} + A_{(j-1)}[\lambda_{-(j-1)} - \lambda_{+(j-1)}]$$
$$= A_{(j-1)}\lambda_{-(j-1)} + [f_{(n-j+1)} - A_{(j-1)}\lambda_{+(j-1)}]. \quad (14)$$

Equations (13) and (14) are equal under special conditions. As pointed out before, they are always equal for the zero and one-machine outage cases. The only other condition is if $f'_{(n-j+1)} = A_{(j-1)}\lambda_{+(j-1)} = f_{(n-j+1)}$, or $f'_{(n-j)} = A_j\lambda_{+j} = f_{(n-j)}$. This states that the cumulative frequency of departure from state $(n - j)$ is equal to the exact frequency of leaving state *j* in an upward direction (i.e., fewer machines out).

Now, when the system states are identified by the number of machines on outage, the system transitions are always between states that differ only by a single unit outage, since only one unit transition may take place in a small unit of time. Therefore, in this case, between any two states $(n - j)$ and $(n - j + 1)$, $A_j\lambda_{+j} = A_{(j-1)}\lambda_{-(j-1)}$ and the cumulative frequency of transition from state $(n - j + 1)$ downward is equal to the frequency of transition from $(n - j + 1)$ to $(n - j)$. That is, the system is restricted from skipping any states, so that when the number of machines is the attribute used for state classification,

$$f'_{(n-j)} = A_j\lambda_{+j}.$$

TABLE VI
THREE MACHINES IN PARALLEL

Exact State Number	Machine Status			Rate of Departure from State		Cumulative State Number
	Number 1	Number 2	Number 3	Up	Down	
1	Up	Up	Up	0	$\lambda_1 + \lambda_2 + \lambda_3$	8'
2	Down	Up	Up	μ_1	$\lambda_2 + \lambda_3$	7'
3	Up	Down	Up	μ_2	$\lambda_1 + \lambda_3$	6'
4	Up	Up	Down	μ_3	$\lambda_1 + \lambda_2$	5'
5	Down	Down	Up	$\mu_1 + \mu_2$	λ_3	4'
6	Down	Up	Down	$\mu_1 + \mu_3$	λ_2	3'
7	Up	Down	Down	$\mu_2 + \mu_3$	λ_1	2'
8	Down	Down	Down	$\mu_1 + \mu_2 + \mu_3$	0	1'

$$f_{1'} = f_8 = A_8 (\mu_1 + \mu_2 + \mu_3).$$
$$f_{2'} = A_8 (\mu_2 + \mu_3) + A_7 (\mu_2 + \mu_3)$$
$$= A_8 (\mu_1 + \mu_2 + \mu_3) - A_7\lambda_1 + A_7 (\mu_2 + \mu_3)$$
$$= f_{1'} - A_7\lambda_1 + A_7 (\mu_2 + \mu_3).$$

$$f_{3'} = A_8 (\mu_3) + A_7 (\mu_2 + \mu_3) + A_6 (\mu_1 + \mu_3)$$
$$= A_8 (\mu_1 + \mu_2 + \mu_3) - A_7\lambda_1 + A_7 (\mu_2 + \mu_3) - A_6\lambda_2 + A_6 (\mu_1 + \mu_3)$$
$$= f_{2'} - A_6\lambda_2 + A_6 (\mu_1\mu_3).$$

TABLE VII

State Number	Number of Failed Machines	Exact Frequency	Cumulative Frequency	
			Halperin and Adler	Hall, Ringlee, and Wood
n	0	$A_0\lambda_{-0}$	$f'_0 = 0$	$f_0 = 0$
$n-1$	1	$A_1(\lambda_{-1} + \lambda_{+1})$	$f'_{n-1} = A_0\lambda_{-0}$	$f_{n-1} = f_n + A_k(\lambda_{-k} - \lambda_{+k}) = A_0\lambda_{-0}$
$n-2$	2	$A_2(\lambda_{-2} + \lambda_{+2})$	$f'_{n-2} = A_1(\lambda_{+1} + \lambda_{-1}) - f'_{n-1}$	$f_{n-2} = f_{n-1} + A_1(\lambda_{-1} - \lambda_{+1})$
$n-3$	3	$A_3(\lambda_{-3} + \lambda_{+3})$	$f'_{n-3} = A_2(\lambda_{+2} + \lambda_{-2}) - f'_{n-2}$	$f_{n-3} = f_{n-2} + A_2(\lambda_{-2} - \lambda_{+2})$

TABLE VIII
COMPARISON OF METHODS FOR CALCULATING THE FREQUENCY OF CUMULATIVE EVENT OCCURRENCES

State Number n	Outage Conditions		State Availability A_j	Departure Rates		Cumulative Frequency	
	Number of Machines j	Capacity		Up λ_+	Down λ_+	Recursive f_n	Halperin and Adler [1] f_n'
Two Identical Units: $C_1 = C_2 = 25$ MW, $\lambda_1 = \lambda_2 = 0.01$ per day, $\mu_1 = \mu_2 = 0.49$ per day.							
2	0	0	0.9604	0	0.02	0	0
1	1	25	0.0392	0.49	0.01	0.019208	0.019208
0	2	50	0.0004	0.98	0	0.000392	0.000392
Two Dissimilar Machines: $C_1 = 20$ MW, $C_2 = 30$ MW, $\lambda_1 = \lambda_2 = 0.01$ per day, $\mu_1 = \mu_2 = 0.49$ per day.							
3	0	0	0.9604	0	0.02	0	0
2	1	20	0.0196	0.49	0.01	0.019208	0.019208
1	1	30	0.0196	0.49	0.01	0.009800	N.A.
0	2	50	0.0004	0.98	0	0.00392	N.A.

N.A. means not available.

This still does not mean that the Halperin-Adler method will yield results identical to those in the present paper, since the system states proposed in this paper are classified by the magnitude of the *capacity* outages, not the number of units. These two state classifications are obviously equal only for the case of identical machines. A simple illustration on Table VIII shows, as expected, identical results for the two methods for two identical machines. However, when the two-machine capacities are different and the system states are classified by the *magnitude* of outage, the cumulative frequency data applicable for these states are not directly obtainable from the Halperin-Adler method.

It is our feeling that both calculating techniques are aimed at developing basically the same generation capacity reliability indices. However, we would caution against trying to extend too far the assumption that capacity outage states and states classed by the number of machine outages are equivalent.

As to computational efficiency, it has been our experience that probability data such as these are more efficiently and more ac- curately calculated on digital computers by recursive relationships (where each state's data is built on the preceding states) rather than by binomial theorem expansions and the subsequent merging of tables of data.

REFERENCES

[14] E. Parzen, *Stochastic Processes*. San Francisco, Calif.: Holden Day, 1962, pp. 170–182.
[15] W. J. Lyman, "Fundamental consideration in preparing master system plan," *Elec. World*, pp. 788–792, June 17, 1933.
[16] S. A. Smith, "Spare capacity fixed by probabilities of outage," *Elec. World*, pp. 222–225, February 10, 1934.
[17] R. Billinton and K. E. Bollinger, "Transmission system reliability evaluation using Markov processes," *IEEE Trans. Power Apparatus and Systems*, vol. PAS-87, pp. 538–547, February 1968.
[18] B. A. Kozlov, "Determination of reliability indices of systems with repair," *Engrg. Cybernetics*, pp. 87–94, July–August 1966.

Frequency and Duration Methods for Power System Reliability Calculations: II—Demand Model and Capacity Reserve Model

ROBERT J. RINGLEE, SENIOR MEMBER, IEEE, AND ALLEN J. WOOD, SENIOR MEMBER, IEEE

Abstract—This paper is a continuation of the work started in [1] and is aimed at incorporating a model of the power system load with the generation system model developed previously. Combination of this load and the generation model permits computation of the availability, frequency of occurrence, and mean duration of generation reserve, or margin states. The results of this work are illustrated by continuation of a simple numerical example begun in Part I.

The most widely applied of the previously developed techniques for assessing generation system reliability, the loss-of-load and loss-of-capacity methods, assume fixed outage or load duration intervals. The present model, on the other hand, uses an exponential distribution of durations. The reserve margin states developed using the exponential distributions contain data giving both the availability of each margin state and the expected frequency of recurrence. Previous methods yield only the availability of the reserve margin states, or else availability and frequency data for generating capacity states, not considering the load. The method presented and illustrated may be extended to consider the calculation of operating reliability or the inclusion of the effects of a simple transmission system.

Paper 68 TP 621–PWR, recommended and approved by the Power System Engineering Committee of the IEEE Power Group for presentation at the IEEE Summer Power Meeting, Chicago, Ill., June 23–28, 1968. Manuscript submitted February 12, 1968; made available for printing April 2, 1968.

The authors are with the General Electric Company, Schenectady, N. Y.

INTRODUCTION

THIS PAPER is a direct continuation of Part I in which a reliability model of a generating system was developed [1]. The generation system model is based on a Markov chain analysis and assumes statistically independent, stationary, exponential distributions of available and repair times for each machine. The model results of importance are the data describing the availability of cumulative capacity states and the expected frequency of occurrence of these conditions. These results are obtainable by use of an efficient recursive formulation which removes the necessity for grouping different machines into clusters of hypothetical, identical units. The Markov chain model of Part I is based on assumed exponential distributions of "available" and "repair" times for each unit as contrasted with the fixed time type of assumption implied in the usual application of the loss-of-capacity method.

Reprinted from *IEEE Trans. Power Apparatus Syst.*, vol. PAS-88, no. 4, pp. 375–388, Apr. 1969.

System Reliability

The results of this description of the generation capacity model are useful in themselves as measures of the reliability of the generation system, but it would seem that a more adequate measure would be one which incorporates a consideration of the expected load pattern. It appears to the authors that the reliability model used for planning generation systems should incorporate calculations of both the availability (i.e., probability of existence) and the occurrence frequency of cumulative capacity reserve margin states. The system model should result in predictions describing the conditions of generating capacity in excess of the load by a given amount or more. This, of course, requires that the previously developed capacity model be combined with (i.e., "convolved") a suitable load model.

In this paper a load model is suggested and data describing the load states, state availabilities, and frequencies of occurrence of the various states are given. The model permits the consideration of the fact that daily peak loads usually do not persist for a 24-hour period. The load model is based on a Markov chain.

The load model may be merged with the generating capacity model to permit the calculation of system reserve and reliability measures. This is done assuming that the load occurrence and capacity reliability processes are statistically independent. The recursive procedure suggested has proven to be very efficient computationally. These developments are illustrated by combining the two-machine, 50-MW system of Part I with the load model describing the loads for a 20-day interval.

The technique has been extended to consider relatively simple transmission networks and this is illustrated also. Another successful extension has been made in assessing operating reserve reliability. Consideration of this problem area must await a subsequent part as must a specific consideration of how this frequency and duration model might be used in planning generation systems.

The specific goals of the remainder of this part of this sequence are to present the results of the analysis of the Markov chain model for the peak load sequence of a power system, combine this load model with the previously developed generating capacity model to yield a model of the generating margin, and illustrate these developments with numerical examples.

LOAD MODEL

It is highly desirable to develop a technique which incorporates consideration of the expected loads in the frequency and duration technique, as is done in the use of the loss-of-load probability method. For purposes of planning generation system expansions it is generally felt that the loads may be adequately represented by daily peak load conditions.

The sequence of daily peak loads will be assumed to be a stationary, random process [2]. A Markov chain model is offered for consideration as a model of the daily peak load sequence. This model seems to be suitable for the problems associated with long range generation planning, it is compatible with the load model used in the loss-of-load method, and is amenable to analytic treatment. Further, it provides a means for extending the usual loss-of-capacity technique to incorporate a consideration of load distributions. The model represents the daily load cycle as a sequence of peak loads L_i, each of a mean duration of e days interspersed with periods averaging $(1 - e)$ days of a fixed, light load L_0. As illustrated in Fig. 1, the sequence of peak loads is random.

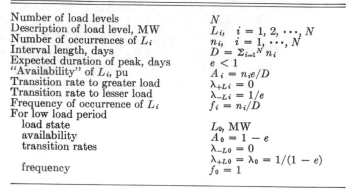

Fig. 1. Sequence of loads for basic load model.

TABLE I
SUMMARY OF LOAD MODEL

Number of load levels	N
Description of load level, MW	$L_i, \quad i = 1, 2, \cdots, N$
Number of occurrences of L_i	$n_i, \quad i = 1, \cdots, N$
Interval length, days	$D = \Sigma_{i=1}^{N} n_i$
Expected duration of peak, days	$e < 1$
"Availability" of L_i, pu	$A_i = n_i e / D$
Transition rate to greater load	$\lambda_{+Li} = 0$
Transition rate to lesser load	$\lambda_{-Li} = 1/e$
Frequency of occurrence of L_i	$f_i = n_i / D$
For low load period	
load state	L_0, MW
availability	$A_0 = 1 - e$
transition rates	$\lambda_{-L0} = 0$
	$\lambda_{+L0} = \lambda_0 = 1/(1 - e)$
frequency	$f_0 = 1$

Load Model for Planning

The load model is based on the following:

1) Daily loads in a period will be represented by a set of N load levels or load states.

2) The sequence of daily peak loads is a random sequence of the N load states.

3) The load model is statistically stationary.

4) The distribution of residence times in a given load state is exponential.

5) At each transfer to a new load state the probability of transfer to a particular state is directly proportional to the long term, average (i.e., quiescent or steady-state) probability of existence of the new state (i.e., its "availability") A_i.

6) Load state transitions occur independently of generation state transitions.

7) The mean duration of peak loads is the fraction e days, with each succession of peak loads separated by a low load period of $(1 - e)$ days mean duration.

Appendix I discusses some of the mathematical aspects of the load model when it is considered as a Markov chain. Table I gives a summary of the results of the mathematical analysis of the basic model.

These seven assumptions allow the construction of a stochastic load model which may be merged with the capacity availability model to produce an overall system description in terms of reserve margin states, their probability of existence (i.e., availability), and the expected time between recurrences.

The same general procedure may be followed in developing the load model as that used in analyzing the various capacity states. We need relationships establishing the availability of each load level, the cycle time or frequency of each exact load level, the rates of departure from each state, and the availability and frequency of encountering a given load level or less. An alternative load model which assumes direct transfers between peak load states is developed in Appendix II.

TABLE II
LOAD MODEL DATA

State Number	Load L_i (MW)	Occurrence n_i (days)	Availability A_i ($en_i/365$)	Cycle Time T_i (days)	Departure Rates per Day	
					λ_{up}	λ_{down}
1	40	2	0.00273973	182.5	0	2
2	25	5	0.00684932	73.0	0	2
3	20	8	0.01095890	45.625	0	2
4	15	5	0.00684932	73.0	0	2
0	0	20	0.02739727	18.25	2	0

$e = 1/2$ day.

Annual Quantities

To represent a year for either the generation or load, some means must be found to recognize the nonstationary effects of scheduled maintenance and seasonal load changes. A reasonable approach which has been frequently used in the past is to divide the year into "maintenance" intervals. These intervals, say 4 weeks duration or less, are short enough so that within the interval the lists of machines on scheduled maintenance may be reasonably assumed not to change. This period is also short enough so that the load model may be represented by a stationary stochastic process.

To relate the interval representation to a year, the values of A_i must be multiplied by the fraction of the year represented by the interval, $D/365$. The mean duration of residence of a given load level will not change, nor will there be a change in the transition rates to higher or lower loads from a given load level.

Each peak load lasts e fraction of a day on the average and is always followed by a low load period. The availability of the particular peak load level L_i in the interval D, days long is given by

$$A_i = en_i/D \qquad (1)$$

where n_i is the expected number of occurrences of L_i in the interval of D days. Therefore if one relates the interval representation to a yearly base of 365 days, the availability of the load state L_i becomes $en_i/365$.

Regardless of interval or yearly representation, the mean duration at load level L_i is e days. Since the model precludes transfer from one peak load state to another, the probability, given the load is in state L_i, that the next transfer is to a low load state is 1, i.e., a certainty. Accordingly, the transition rate to a lower load state is $1/e$ and to a higher load state, zero. The load model of Appendix II will permit the use of hourly load data.

Load Model Results

To illustrate the procedure and form of the basic load model, assume that during a 20-day period, peak loads of 40, 25, 20, and 15 MW are expected to occur 2, 5, 8, and 5 days, respectively. Let the duration of each peak load e be 0.5 fraction of a day. Results are given in Table II.

Data for the cumulative load states could be developed in a manner similar to that for the capacity model. Instead, the state availability and frequency of recurrence of various reserve, or margin, states are developed next.

RESERVE MARGIN STATES

Reserve, or margin, is the difference between the available capacity and the load. The capacity data from Part I and the load statistics developed above may be combined to develop data on the availability and frequency of recurrence of various degrees of reserve condition [1]. A margin state m_k is the combination of the load state L_i and the capacity state C_j.

$$m_k = C_j - L_i. \qquad (2)$$

In order to calculate data on a cumulative margin basis it is necessary to calculate the rates of departure from m_k to larger and smaller margin states. Let subscripts $+$ and $-$ designate up and down transitions, and the subscripts m, c, and L designate margin, capacity, and load states. Then

$$\lambda_{+m} = \lambda_{+c} + \lambda_{-L}$$

and

$$\lambda_{-m} = \lambda_{-c} + \lambda_{+L}.$$

That is, given the independent load and capacity states, the rate of transfer from a given margin state to one with larger margin is equal to the rate of transfer upward in capacity plus the rate of transfer down in load.

The data from the previous example may be used to illustrate the construction of a margin-availability table. The capacity data refers to Example 2 of Part I in which the two-machine system represented consists of 20- and 30-MW units.

Margin-Availability Tables

Table III illustrates the construction of margin-availability tables for the "exact" margin states which occur in the two-machine, four load-level example. The data in the table include margin in MW, availability, and transfer rates. The states shown include all combinations of load and capacity; as a result there are entries in the table which represent identical margin values but which are the result of differing combinations of loads and capacities. The data are for a 365-day year and assume the exposure factor e is 0.5. Entries for the zero, or low load, level are not shown. In the generation of these exact states the availability of the margin state is

$$A_k = A_j A_i \qquad (3)$$

and the departure rates are given by

$$\lambda_{\pm k} = \lambda_{\pm j} + \lambda_{\mp i}. \qquad (4)$$

TABLE III
Margin States

	Generation Data				Load Data			
					i = 1	2	3	4
					L_i = 40	25	20	15
					A_i = 0.00273973	0.00684932	0.0109589	0.00684932
					λ_+ = 0	0	0	0
j	C_j	A_j	λ_+	λ_-	λ_- = 2	2	2	2
1	50	0.9604	0	0.02	m = 10	25	30	35
					A = 0.002631	0.006578	0.010525	0.006578
					λ_+ = 2	2	2	2
					λ_- = 0.02	0.02	0.02	0.02
2	30	0.0196	0.49	0.01	m = −10	5	10	15
					A = 0.000054	0.000134	0.000215	0.000134
					λ_+ 2.49	2.49	2.49	2.49
					λ_- 0.01	0.01	0.01	0.01
3	20	0.0196	0.49	0.01	m = −20	−5	0	5
					A = 0.000054	0.000134	0.000215	0.000134
					λ_+ 2.49	2.49	2.49	2.49
					λ_- 0.01	0.01	0.01	0.01
4	0	0.0004	0.98	0	m = −40	−25	−20	−15
					A 0.000001	0.000003	0.000004	0.000003
					λ_+ 2.98	2.98	2.98	2.98
					λ_- 0	0	0	0

$e = 1/2$ day.

These may be combined to yield the occurrence frequency of these exact states by means of a modification of [1, eq. (3)],

$$f_k = A_k(\lambda_{+k} + \lambda_{-k}). \tag{5}$$

The margin and capacity states may be arrayed in tabular form prior to elimination of identical margin entries. The identical margin states may be combined as follows. For a given state m_k made up of N identical margin states such that $m_k = m_1 = m_2 = \cdots = m_N$,

$$A_k = \sum_{l=1}^{N} A_l \tag{6}$$

$$f_k = \sum_{l=1}^{N} f_l \tag{7}$$

and

$$\lambda_k = \sum_{l=1}^{N} A_l \lambda_l / A_k \tag{8}$$

where the last relationship applies to either the up- or the downward departure rates. In the development of (7) and (8) use was made of the observation that transfers cannot take place between two identical margin states without an intervening state of differing margin.

Cumulative Margin States

As with the capacity data, it is highly desirable to have availabilities and frequencies for cumulative margin states. With distinct, ordered tables of data for the exact margin states available, the cumulative data may be developed exactly as was done in Part I for the cumulative capacity data. That is, [1, eqs. (6) and (7)] may be used directly with the understanding that they are to be applied to the reserve margin data.

It has been found that more computationally efficient techniques result if the data describing cumulative generation states are combined with the information for the exact load states. These algorithms, or recipes, are developed below. In order to distinguish between exact and cumulative state descriptions, the following definitions are made.

m exact margin of m, MW
M margin of M, MW or less
L exact load of L, MW
C exact capacity of C, MW
G cumulative capacity of G, MW or less.

For a given cumulative margin state the availability is

$$A_M = \sum_{m \leq M} A_m. \tag{9}$$

For any exact margin state, $m = C - L$, the availability is

$$A_m = \sum_{C,L} A_C A_L \tag{10}$$

where the sum is over all events where the relation between C, L, and m is true. The cumulative margin state availability for a margin of M, or less, MW is

$$A_M = \sum_{m \leq M} \sum_{C,L} A_L A_C. \tag{11}$$

Since load and capacity states are independent, (11) may be rearranged by summing over all load levels. That is,

$$A_M = \sum_{L} A_L \sum_{C \leq L+M} A_C. \tag{12}$$

This last sum is the availability of the cumulative capacity state G, defined by the set of events

$$\{G\} = \{C \leq L + M\} \tag{13}$$

thus,

$$A_M = \sum_{L} A_L A_G. \tag{14}$$

For example, consider the four-state load and two-machine sample problem developed previously. Let $M = -10$ MW be the cumulative margin state under consideration. The availability of that cumulative margin state is found from the availabilities of the following load and capacity states:

Margin of M: -10 MW or less
Load: 40   25   20   15   MW exactly
Capacity: 30   0   0   0   MW or less

An efficient recursive relationship may also be developed for the frequency of occurrence of the cumulative margin states in terms of the exact load state conditions and the cumulative capacity data. The general relationship for the frequency of occurrence of cumulative states is given in [1, Eq. (6)]. By a minor modification of this relation, and in terms of exact margin states, the frequency of occurrence of a margin of M, or less, MW is

$$f_M = \sum_{m \leq M} A_m (\lambda_{+m} - \lambda_{-m}). \qquad (15)$$

The definitions of the transition rates and the independence of the load and capacity events permit writing this as

$$f_M = \sum_{m \leq M} \sum_{L,C} A_L A_C (\lambda_{+C} - \lambda_{-C} + \lambda_{-L} - \lambda_{+L}) \qquad (16)$$

where the exact load, capacity, and margin states are again related by $m = C - L$.

The same technique of summing over each exact load state may be used. That is,

$$f_M = \sum_L A_L \sum_{C \leq L+M} [A_C(\lambda_{+C} - \lambda_{-C}) + A_C(\lambda_{-L} - \lambda_{+L})]. \qquad (17)$$

If use is made of the definition of the cumulative capacity state G, this relationship may be simplified since

$$f_G = \sum_{C \leq L+M} A_C (\lambda_{+C} - \lambda_{-C}) \qquad (18)$$

is the frequency of the cumulative capacity state G. Therefore, the frequency of occurrence of the cumulative margin state is

$$f_M = \sum_L A_L [f_G + A_G(\lambda_{-L} - \lambda_{+L})]. \qquad (19)$$

These relationships have been found to be much more efficient computationally than combination of the exact margin states.

Equations (19) and (14) give frequency and availability, or probability of existence of the cumulative margin states. That is, the availability number gives the probability of finding a specified, cumulative margin state at any randomly selected time. This may be converted to the usual annual loss-of-load probability index by multiplying by the number of days in a year and dividing by the exposure factor e.

As an example, the frequency of occurrence of the cumulative margin state of -10 MW may be computed for the example using the load data given previously and the capacity data from [1, Table I]. That is,

$$f_{-10} = A_{40}(f_{30} + A_{30}\lambda_{-L}) + (A_{25} + A_{20} + A_{15})(f_0 + A_0\lambda_{-L})$$

$$= 0.00273973(1/52.0616 + 0.0792)$$

$$+ 0.02465754(1/2551.2 + 0.0008)$$

$$= 0.000299 \text{ per day}$$

which is the reciprocal of the cycle time given in Table IV.

TABLE IV
CUMULATIVE MARGIN DATA

Margin (MW)	$e = 1/2$ day Availability of Cumulative State (pu)	Cycle Time of Cumulative Events (days)
35	0.0273973	18.25
30	0.0208192	23.94
25	0.0102942	47.78
15	0.0037162	126.5
10	0.0035819	132.0
5	0.00073589	546.3
0	0.00046740	858.6
-5	0.00025260	1 582.4
-10	0.00011836	3 344.4
-15	0.000064658	6 030.3
-20	0.000061918	6 342.5
-25	0.000003836	87 488
-40	0.000001095	306 208

Rationalizations and Justifications

At this point, one might have reservations about the adequacy of the Markov model for representation of the sequence of daily peak loads. The times of residence in a given peak load state L_i form an exponentially distributed random sequence with mean value equal to e. Similarly, the times of residence in the low load state L_0 form an exponentially distributed random sequence with mean value equal to $(1 - e)$. Thus, only in the mean sense is the Markov model representative of the sequence of daily loads. But the calculations for frequency of occurrence and for probability of existence of a given state are, by definition, mean values. Therefore, we are asking no more from the Markov model than it can provide; nor, conversely does the model restrict the development of mean values.

There are many stochastic models which would give the same probability and mean duration in a given state as the Markov model, but the Markov model happens to be an especially simple one to deal with. A pertinent question is whether or not these other types of models would lead to the same durations of the margin states as the Markov model. A check was made by computing the expected duration of margin states assuming that the peak load state durations are exactly e rather than being an exponentially distributed random variable with a mean residence time of e. The result was exactly the same as for the Markov model. The mean duration of the margin state is $r_g e/(r_g + e)$, where r_g is the mean duration of the generation state.

The authors submit, therefore, that the Markov model proposed for representing daily peak loads will generate the proper expected duration, frequency, and probability of existence of given states. The margin model follows directly from the linear combination of the Markov load and generation models. Consistent frequency and probability indexes follow from this margin model.

EXTENSIONS

This development has concentrated on the step-by-step evolvement of a reliability model for use in generation system planning. This model permits the computation of margin state availabilities and cycle times and permits the use of an exposure factor. Time and space limitations preclude the consideration of extensive numerical relationships, appropriate index levels, or the application of the technique to the operating reliability area. This last application has been done for an actual system using a load model based on Appendix II, and it is the authors' intention to present a summary of the problems and results in this area along with a consideration of some of the numerical factors involved in the planning problem in a subsequent part of this series.

Annual Quantities from Interval Data

The data of Table IV applies for a single 20-day interval. The results are expressed in equivalent annual values since the load state availabilities were expressed on an annual basis. If there are data for the several consecutive intervals whose lengths sum up to the 365-day year, these interval data may be combined quite readily into annual values by utilizing the assumptions of statistical independence.

For any two intervals in a year, the peak load levels which occur in one are independent of those in the other. By ignoring the possibility of nonstationary, causal effects introduced by assigning units for maintenance outages, the same statistical independence may be asserted for the generating capacity states which occur in these two intervals. Thus the total availability and frequency of occurrence over the combined two-interval period of a specific margin state are respectively, the sums of the availabilities and frequencies which occur in each of the separate intervals. This line of reasoning may be extended over all of the intervals making up the entire year. Equations (6)–(8) may be used to create annual values of the availability, frequency, and state transition rate if the kth state is taken as the annual value, the lth state taken as denoting the value of the particular data describing this kth state in the lth interval, and N is the number of intervals considered to make up the year.

Transmission Line Effects

To indicate the type of analysis required to incorporate simple transmission line effects into what is basically a generation reserve reliability model, two simple examples will be shown. In both these examples the following is assumed.

1) The transmission line is a two-state device characterized by having a capability T (MW) in one state and zero in the other, i.e., the "available" and "in repair" states.

2) The distributions of time in each state are exponentially distributed.

3) The transition rate from available to repair state is λ_t and from repair to available state is μ_t.

4) The behavior of the transmission line is statistically independent of the generating system.

These are similar to those used in other studies of transmission effects [3]–[6].

The first example is the system in Fig. 2. This is a simple, single-generator, single-line example. The generator is also taken as a two-state element with state capabilities of C_g and zero and transition rates of λ_g and μ_g. What is of interest is the capacity available at the load bus, the availability of the various states, and frequencies of their occurrence.

The state transition diagrams for the independent elements and the two merged generation–transmission states as seen at the load bus are shown in Fig. 3. Since the generator and the line are in series and statistically independent, the state C, which represents having the capability C at the bus, is dependent upon having both the generator and line available.

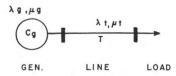

Fig. 2. Simple generation–transmission system.

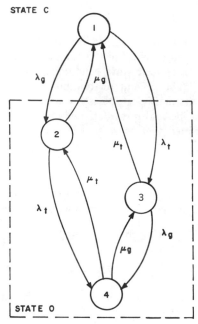

Fig. 3. State transition diagram for the generation–transmission system. 1—both up; 2—generator down, line up; 3—generator up, line down; 4—both down.

The availability of C is

$$A_C = A_g A_t. \tag{20}$$

The capability is

$$C = \min (C_g, T). \tag{21}$$

Since in this merged state the generation and transmission states are independent, and the elements are in series, the transition rate to a lower capacity state is the sum of the rates downward in capacity for generator and line. Therefore

$$\lambda_{-C} = \lambda_g + \lambda_t$$

and

$$\lambda_{+C} = 0.$$

For state zero which is the merger of states 2–4 in Fig. 3, the frequency of departure upward in capacity is the sum of the frequencies of departure upward in capacity for states 2 and 3. Therefore

$$A_0\lambda_{+0} = A_2\mu_g + A_3\mu_t \tag{22}$$

or

$$\lambda_{+0} = (\bar{A}_g A_t \mu_g + A_g \bar{A}_t \mu_t)/(1 - A_g A_t).$$

Second Transmission Example

For a second transmission example consider that the two paralleled generators, the 20- and 30-MW units, of the previous sections are connected to a load bus by a single, 40-MW transmission line. Again, the problem is to define the capacity available states at the load bus. The example is carried out by means of the data shown in Table V.

TABLE V
GENERATION–TRANSMISSION EXAMPLE

	State Number	Capacity Available (MW)	Transition Rates		Availability (pu)
			λ_+	λ_-	
			(per day)	(per day)	
Generation only	1	50	0	0.02	0.9604
	2	30	0.49	0.01	0.0196
	3	20	0.49	0.01	0.0196
	4	10	0.98	0	0.0004
Transmission only	5	40	0	0.022	0.999
	6	0	1.998	0	0.001
Transmission available plus 1–4	7	40	0	0.022	0.959440
	8	30	0.49	0.012	0.019580
	9	20	0.49	0.012	0.019580
	10	0	0.98	0	0.000400
Transmission down plus 1–4	11	0	1.998	0	0.0009604
	12	0	1.998	0	0.0000196
	13	0	1.998	0	0.0000196
	14	0	0	0	0.0000004
Merger of 10–14	15	0	1.70678	0	0.0013996

The first four states refer to the generation system only and the next two to the transmission line. The line is assumed to have an availability of 0.999 and a mean up time of 500 days. States 7–10 are the combinations of the four generating states with state 5 while states 11–14 consist of the merging of the line out condition, state 6, with states 1–4. The last eight states are then all of the eight possible conditions.

The zero capability states may be merged into a state designated as number 15. The availability of state 15 is

$$A_{15} = \sum_{i=10}^{14} A_i$$

and transition rates are

$$\lambda_{-15} = 0$$

and

$$\lambda_{+15} = \left(\sum_{i=10}^{15} \lambda_{+i} A_i \right) / A_{15}.$$

The data in Table V for states 7, 8, 9, and 15 form the complete description of the capacity available states that exist at the load bus.

The Markov chain model has been employed to consider simple transmission networks. The example above is worked out in an elementary fashion for the exact generating capacity states available at a load bus. This can be extended by the methods given previously to consider cumulative capacity states, to combine these results with a load model, and to refine the computational technique to be more efficient.

What is not done, however, is to consider a more realistic network with multiple source and load buses and with transmission loops. The methods indicated here permit a consideration of transmission limits in a radial type of network or perhaps in two interconnected systems. Further development is required for numerical assessment of the reliability of supply at many individual load buses in a system with several sources of supply and a transmission network.

CONCLUSIONS

The purpose of this paper has been to present and describe a probabilistic model of power system loads and generation reserve margins which may be used to calculate the availabilities, occurrence frequencies, and cycle durations for specified load and margin states, and for the various cumulative event states. Demonstrations have been offered of an elementary stochastic load model derived from a Markov chain. This model permits the straightforward development of methods in which the convolution or merging of the independent states of generation, load, and transmission lines may be carried out to give an algorithm for determining the load-carrying capability or reserve margin for a generation system. The load model presented in the body of the paper would seem to be particularly appropriate for generation system planning.

It is the authors' belief that these methods offer an opportunity to use a single technique to obtain the results given by both the loss-of-load probability technique and the loss-of-capacity method. The load model given may be combined with a capacity model to yield information concerning both the availability (or, if you prefer, probability of existence) and the frequency of recurrence of the reserve margin states.

In a subsequent part of this sequence, it is planned to present methods and models for studying operating reserve margins recognizing deviation of load from forecast and unit commitment procedures. It is also intended that reliability indexes and numerical relationships for system planning will be examined.

APPENDIX I
LOAD MODEL I

In order to represent the transfers of the system load from peak to lower values, the load model of Fig. 1 will be developed as a Markov chain. Since peak periods are normally less than a day, the mean duration of this daily peak period is taken as $e < 1$ day. The N load levels are ordered from highest to lowest. That is,

$$L_i > L_{i+1}, \quad i = 1, 2, \cdots, N - 1,$$

113

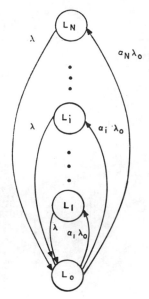

Fig. 4. State transition diagram for basic load model.

TABLE VI
INTERVAL LOAD MODEL DATA

State Number	Load L_i	Occurrences n_i (days)	Probabilities α_i	P_i (pu)	Transfer Rates λ_+	λ_- (per day)
1	40	2	0.10	0.025	0	4
2	25	5	0.25	0.0625	0	4
3	20	8	0.40	0.10	0	4
4	15	5	0.25	0.0625	0	4
0	L_0	20	1	0.75	4/3	0

$e = 1/4$ day, $\quad \lambda = 4$ per day, $\quad \lambda_0 = 4/3$ per day.

TABLE VII
VARIATION OF CYCLE TIME WITH e

Cumulative Margin (MW)	Cumulative Cycle Time e (days) 1/8	1/4	1/2	1*
30	24.0	24.0	23.9	23.9
20	132.4	130.4	126.5	119.3
10	137.6	135.7	132.0	125.3
0	1 007.8	952.6	858.6	717.1
−10	3 963.8	3 733.3	3 344.5	2 767.8
−20	7 559.0	7 104.8	6 342.5	5 222
−30	406 459	366 466	306 208	230 429

* Limiting values as e approaches unity.

and between each pair of peak load periods the system returns to a period of low load L_0. The mean duration in state L_0 is $(1 - e)$ days.

Each load state may occur on the average n_i times during an interval which is D days long. The loads occur in a random sequence. When the system is in state L_0, the probability that the next peak will be L_i is α_i, where

$$\alpha_i = n_i / D, \quad i = 1, 2, \cdots, N.$$

Recalling the assumptions concerning the load models, when the system is in state L_0, the probability that it will transfer to state L_i is proportional to this value of α_i. When it is in L_i the probability of transfer to L_0 is unity. With these considerations the transition rates downward from L_i is

$$\lambda = 1/e$$

and upward from L_0 to load state L_i is,

$$\alpha_0 \lambda_0 = \alpha_i / (1 - e), \quad i = 1, 2, \cdots, N.$$

With these rates, the state transition diagram is as shown in Fig. 4.

A Markov chain analysis may be carried out [7]. Let P_i denote the time-dependent probability that the system is in the state L_i for $i = 0, 1, \cdots, N$. Then the matrix differential equation of the system is

$$
\begin{bmatrix} \dot{P_0} \\ \dot{P_1} \\ \vdots \\ \dot{P_i} \\ \vdots \\ \dot{P_N} \end{bmatrix}
=
\begin{bmatrix}
-\lambda_0 & \lambda & \lambda & \cdots & & & \lambda \\
\alpha_1 \lambda_0 & -\lambda & 0 & & & & 0 \\
\vdots & & & \ddots & & & \vdots \\
\alpha_1 \lambda_0 & 0 & 0 & \cdots & -\lambda & \cdots & 0 \\
\vdots & & & & & & \vdots \\
\alpha_N \lambda_0 & 0 & 0 & & & & -\lambda
\end{bmatrix}
\begin{bmatrix} P_0 \\ P_1 \\ \vdots \\ P_i \\ \vdots \\ P_N \end{bmatrix}.
$$

(23)

In the steady state, the $\dot{P}$ vector is zero and

$$\sum_{i=0}^{N} P_i = 1$$

must hold. The solution gives

$$P_0 = \lambda / (\lambda + \lambda_0) = 1 - e$$

and

$$P_i = \alpha_i \lambda_0 / (\lambda + \lambda_0) = n_i e / D.$$

The remainder of the items in Table I follow directly.

Examples

As a first example consider the 20-day interval discussed previously. Let $e = 1/4$, day. The data for this load model on an interval basis are shown in Table VI. This set of loads is for the 20-day interval. On an annual basis the loads occur 20/365th of the time and, therefore, the steady-state probabilities must be multiplied by 20/365.

The purpose of the next example is to illustrate the effect of changing the exposure e on the cycle of recurrence of cumulative margin states. For comparative purposes, we have chosen to extend the results shown in Table IV for cumulative margin states with $e = 1/8$ up to the limit as e approaches unity.

The cycle times for recurrence of cumulative margin states are shown in Table VII. The availability of the margin state is directly proportional to e and, therefore, the values were not included in the table.

Appendix II

Alternative Load Model

In the basic load model presented in the body of the paper, the system must pass through a low load state between each peak period. This is true even as the exposure factor approaches unity. A fundamentally different model will result if the system is permitted to go from one peak load to the next directly. Fig. 5 indicates this type of load sequence.

In considering this model the 7th assumption listed in the paper is replaced by: "the mean duration of peak loads is normally 1 day." The model will be developed by considering one in which there are N distinct load states all equally likely. In this model all states communicate directly and using the conventions of Appendix I, the differential equations are

$$
\begin{bmatrix} \dot{P}_1 \\ \dot{P}_2 \\ \cdot \\ \cdot \\ \dot{P}_N \end{bmatrix} = \begin{bmatrix} -\lambda & \lambda/(N-1) & \cdots & \lambda/(N-1) \\ \lambda/(N-1) & -\lambda & & \lambda/(N-1) \\ \cdot & & \cdot & \\ \cdot & & & \cdot \\ \lambda/(N-1) & & & -\lambda \end{bmatrix}
$$

$$
\times \begin{bmatrix} P_1 \\ P_2 \\ \cdot \\ \cdot \\ \cdot \\ P_N \end{bmatrix}. \quad (24)
$$

where $\lambda = 1$ per day. The solution with $\dot{P} = 0$ gives

$$P_i = 1/N, \quad i = 1, 2, \cdots, N.$$

It is desirable to allow for the situation where several of these load states are merged into an equivalent state. The states are assumed to be ordered from highest to lowest. It is desired to merge states K to J, where $K > J$, into a combined single state L. This can be done by developing the availability and transition rates upward and downward out of the new state. The availability of the new state is

$$A_L = \sum_{i=J}^{K} A_i = (K + 1 - J)/N. \quad (25)$$

The transition rate of a state i to all states less than J (to greater load) is

$$\lambda_{+i} = (J - 1)/(N - 1)$$

and to all states of lesser load than K is

$$\lambda_{-i} = (N - K)/(N - 1).$$

The frequency of encountering higher (lower) states than the new merged state is the sum over each state i of the frequency of transition to the corresponding higher (lower) states. Therefore,

$$\lambda_{\pm L} = \frac{1}{A_L} \sum_{i=J}^{K} A_i \lambda_{\pm i}$$

or

$$\lambda_{+L} = (J - 1)/(N - 1)$$

and

$$\lambda_{-L} = (N - K)/(N - 1).$$

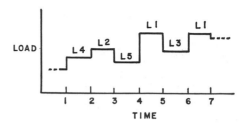

Fig. 5. Sequence of loads for alternative load model.

These may also be written as

$$\lambda_{+L} = \frac{(J - 1)/N}{1 - 1/N}$$

$$= \sum_{i=1}^{J-1} \frac{A_i}{1 - 1/N} \quad (26)$$

and

$$\lambda_{-L} = \frac{(N - K)/N}{1 - 1/N}$$

$$= \sum_{J+1}^{N} \frac{A_i}{1 - 1/N}. \quad (27)$$

The total departure rate from merged state L is

$$\lambda_L = \frac{1 - A_i}{1 - 1/N} \quad (28)$$

and the frequency of occurrence is

$$f_L = \frac{A_L(1 - A_L)}{1 - 1/N} \quad (29)$$

As may be seen, the assumption of peak load states which communicate directly does result in a fundamentally different load model. The procedure shown may be extended to merge further load states.

For this model the expected length of time of residence in a state such as L in an interval of D days is

$$n_L = A_L D. \quad (30)$$

A practical difficulty in developing data for this model results when (30) is used as the definition of the availability. The previous development from N equally likely states to the merged states assumes that N is a known integer. This will not be the case when actual load samples are used in conjunction with (29) to define the availabilities. It is necessary to use the observations to determine the basic number of load states which may or may not be equal to D.

For illustration, assume that a set of load availabilities A_i have been determined by using the duration of load L_i over an interval D days in length. Referring back to (25), there must exist some number I which makes all of the values (A_i/I) integers. This is the "greatest common divisor" of the set $\{A_i\}$, and its reciprocal is the basic number of equally likely load states.

The 20-day load interval forms an example. The availabilities are 0.1, 0.25, 0.4, and 0.25 for the loads of 40, 25, 20, and 15 MW, respectively. The greatest common divisor of the availability set is 0.05 and the basic number of load states is 20.

REFERENCES

[1] J. D. Hall, R. J. Ringlee, and A. J. Wood, "Frequency and duration methods for power system reliability calculations: I—generation system model," *IEEE Trans. Power Apparatus and Systems*, vol. PAS-87, pp. 1787–1796, September 1968.

[2] A. Papoulis, *Probability, Random Variables and Stochastic Processes.* New York: McGraw-Hill, 1965.

[3] S. A. Mallard and V. C. Thomas, "A method for calculating transmission system reliability," *IEEE Trans. Power Apparatus and Systems*, vol. PAS-87, pp. 824–834, March 1968.

[4] R. Billinton and K. E. Bollinger, "Transmission system reliability evaluation using Markov processes," *IEEE Trans. Power Apparatus and Systems*, vol. PAS-87, pp. 538–547, February 1968.

[5] D. P. Gaver, F. E. Montmeat, and A. D. Patton, "Power system reliability: I—measures of reliability and methods of calculation," *IEEE Trans. Power Apparatus and Systems*, vol. 83, pp. 727–737, July 1964.

[6] R. Billinton, "Composite system reliability evaluation," this issue, pp. 276–281.

[7] J. G. Kemeny and J. L. Snell, *Finite Markov Chains.* Princeton, N. J.: Van Nostrand, 1959.

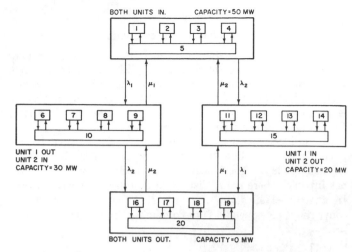

Fig. 6. State space diagram of capacity and load models.

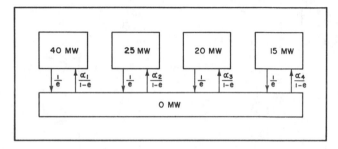

Fig. 7. Load states in each capacity state shown in Fig. 6.

Discussion

R. Billinton and **M. P. Bhavaraju** (University of Saskatchewan, Saskatoon, Sask., Canada): The authors are to be congratulated on a second excellent paper on the application of frequency and duration methods in power system reliability evaluation. In this paper the load model and the generation model are represented by two separate Markov processes. The load states and the capacity reliability states are statistically independent and the probability of the simultaneous occurrence of a load state and a capacity state is given by the product of the two independent probabilities. The load and capacity states can also be shown in a single-state space diagram. This is shown in Figs. 6 and 7 for the system shown in the paper. The steady-state probabilities for this system were calculated, and as expected are identical to those shown in Table III. In the combined system state space diagram, states 5, 10, 15, and 20 correspond to the 0-MW load condition at the various capacity levels. Table III does not include a 0 load level column, and in fact this point is noted in the paper itself. We would like to ask the authors why the 0 load level is not included as it adds additional margin states as shown below.

Capacity (MW)	Margin (MW)	Availability on Annual Basis
50	+50	0.026312
30	+30	0.000537
20	+20	0.000537
0	+0	0.000011

These values will not influence the negative margin states which of course are the most important; however, they will affect the positive margin states. The cumulative availabilities and cycle times of zero and higher margin states will be modified due to the 0-MW load level. The availability of the cumulative highest margin state of +50 MW will be $(1.0 \times 20/365)$ rather than the cumulative availability of $(0.5 \times 20/365) = 0.0273973$ shown for the +35-MW margin state. As previously noted, the 0-MW load states' availabilities could be easily added to Table III. There may be some difficulty, however, in adding the corresponding λ_+ values. We would appreciate the authors' comments on this point. The cumulative cycle times shown in Table VII would also be modified slightly. We would like to ask if there would be more variation in the cumulative cycle times for the positive margin states as the exposure e changes if the 0-MW margin were included.

It should be possible to use a load model containing selected discrete levels described by expected durations in hours determined from a chronological load variation curve in a similar manner to that shown in the paper. This may be more meaningful than assuming only two load levels, a peak and a low level, in any one day. The state space diagram shown in Fig. 7 would then be modified by added transitions within the different load levels rather than by the given transitions between the zero load level and other possible levels.

The extension to composite system reliability evaluation is shown in the paper for a simple series system. Under these conditions, the transmission line can be represented by an overall annual failure rate consisting of the appropriate storm and normal weather components. If parallel transmission facilities are involved, then the state space diagram is complicated considerably by the required storm and normal weather states. These states must be included if the sole criterion is continuity at the load point. The storm and normal aspect in a parallel or networked configuration becomes less vital if the conditional probability of the line loading exceeding its capability is incorporated in the analysis [6]. We would like to ask the authors if they have considered the inclusion of storm and normal weather states or conditional failure probabilities in their studies.

In conclusion, we would again like to compliment the authors on an excellent paper and on their contribution to quantitative reliability evaluation.

Manuscript received July 12, 1968.

Veazey M. Cook (Consolidated Edison Company, New York, N. Y.): The most interesting feature of this paper is the development of exact margin states and cumulative margin states. The data for the exact margins are shown in Table III where the generation data from [1] is combined with the load model data shown in Table II. It is to be noted that Table III is on a 365-day basis and with only one

Manuscript received June 25, 1968.

month of the year shown. Also, the low load periods for the 20 days are not shown. For Table III to be complete, the summation of the A_i's should equal unity. Then the sum of the calculated frequencies of occurrence for all exact margin states would be equal to the sum of the calculated frequencies of occurrence for all margin states based on the low load periods. A table of exact margin state frequencies might well have been included in the paper. The frequency of occurrence for a -10-MW exact margin is equal to $2.5 \times 0.000\ 054$, or $0.000\ 135$ times per day.

The frequency of occurrence for a margin of -10 MW, or less, is calculated as follows using (15):

for capacity = 30 MW:
    frequency = $2.48 \times 0.000\ 054 = 0.000\ 133\ 92$
for capacity = 20 MW:
    frequency = $2.48 \times 0.000\ 054 = 0.000\ 133\ 92$
for capacity = 0 MW:
    frequency = $2.98 \times 0.000\ 001 = 0.000\ 002\ 98$
    frequency = $2.98 \times 0.000\ 003 = 0.000\ 008\ 94$
    frequency = $2.98 \times 0.000\ 004 = 0.000\ 011\ 92$
    frequency = $2.98 \times 0.000\ 003 = 0.000\ 008\ 94$

              Total = $0.000\ 300\ 62.$

It appears that the method of the paper may be extended to cover all hourly loads for each day in the year. This could be done with a computer program. However, in most cases, the reserve margin at the time of daily peak loads is all that is required.

Just below (25) in Appendix II the derivations of λ_{+i} and λ_{-i} are shown which make use of $(N-1)$ as a probability factor. This assumes that one of the possibilities out of N is occupied so that all possible transitions are to the remaining $(N-1)$ load states. Would it be possible to develop the equations with the assumption that no particular load state is occupied initially?

It is important to note that the solutions to the two transmission examples depend a great deal on the assumption of statistical independence for the components. Otherwise the solutions would be quite different from those shown.

The paper is something new in the application of probability methods to generation reserve problems and contains abundant material for careful study.

Manuscript received July 2, 1968.

C. W. Watchorn (R.D. 3, Bethlehem, Pa.): The authors discuss a very interesting approach to the application of probability methods to power system problems. They propose renewal of the proposition that mathematical representations of the characteristic of both the occurrence of load and of forced outages of generating capacity would provide a better basis for calculating probable overall system performance than previously developed methods that do not require such representations. It is rather doubtful if such would be the case from a practical point of view, unless in the most unusual situations, for which such mathematical representations were both approximately valid at the same time. Added to these difficulties are the uncertainties of future loads and of generating unit forced outage characteristics which are both common problems to all methods. It is thus seen that it is impossible to obtain precise answers in such calculations. Nevertheless, since probability methods do provide us with powerful tools to obtain valuable answers to power system problems that can be obtained in no other way, it behooves us, as engineers, as with so many other engineering problems, to do the best we can under the circumstances.

The next consideration, before going any further, is to raise the question as to the uses that are to be made of the results of the calculations. Normally, they should be used by engineers as a basis for judgment to make recommendations to management for the most over all economical future development of the system. Then, before the engineer is in a position to make valid recommendations he should have an appreciation of the influence on the calculations of the uncertainties of future loads, of generating unit forced outage characteristics, and of various assumptions that may be made in order to facilitate the various methods of calculations.

One of the most difficult factors to determine for use in probability studies of power system problems is the generating unit forced outage characteristics. This is largely the result of the large number of unit-years of operating experience required for the various types of units involved. It has been found that analyses of the sensitivity of the results over a reasonable range of generating unit forced outage characteristics provide a reasonable basis for judgment in regard to the uncertainties with respect to this factor.

The appropriate manner to incorporate the uncertainties of future loads in the calculations is somewhat more involved than is usually set forth in previous papers on this general subject, if at all. The problem is to forecast future loads and then the distribution of the deviations of the probable actual future loads from such forecasts. The usual procedure for doing this, as has been suggested in some previous papers, has been to plot the deviations on probability paper. Such plots, almost invariably, indicate a straight line as being a good fit in the central portion, from which it has usually been concluded that the normal frequency distribution of the deviations is indicated. Actually, it is rather another beautiful illustration of the power of the central limit theory to resolve mixtures of all types of distributions so that they approach very closely the normal frequency distribution about the central or mean value.

Probable actual loads as of any given future time fall within a range depending upon the various conditions that can exist at such time. Likewise, estimates of probable future loads also fall within a range. Such forecasts may be within the range of probable actual loads or above or below it. If the forecasts are above the range of probable actual loads, the deviations are all negative, if below, the deviations are all positive, i.e., a certainty the actual loads will be below and above the forecasts, respectively, and if within the range some of the deviations will be positive and some negative, i.e., with varying degrees of probability of the actual loads being above and below the forecasts, depending on the level of the forecasts. Such characteristics are quite different from those for the normal frequency distribution. The distribution of the probable deviations with respect to any given forecast will depend on the length of time into the future it is made.

Short range forecasts, such as made a day or so in advance for the determination of operating capacity requirements, should normally be quite close to the actually experienced loads with the range of deviations with respect to a given level of forecast increasing as the time into the future for which the forecast is made increases. Also, particularly for forecasts three to four years and longer into the future, the distribution and levels of probable actual loads are independent of the levels of the forecasts.

These characteristics mean that the use of a normal frequency distribution based on the standard deviation obtained from the differences between various levels of forecasts and actual experienced loads can result in substantial overstatements of operating or installed capacity requirements, as the case may be, because of both the very much larger spread of the deviations thus obtained as compared with those for any given levels of forecasts and also the effect of the upper tail that is inherent in the use of the normal frequency distribution. Such overstatements are particularly significant with respect to the higher levels of the forecasts as related to the range of probable future loads.

These analyses mean that a lot more work than has been done heretofore is required before we will be in a position to properly incorporate the factor of load forecasting deviations into probability calculations, and, that, at least until such time as this can be done, the use of sophisticated calculation methods as compared with the basic or primitive methods that were developed some thirty odd years ago is academic unless they contribute to calculation advantages.

Such earlier methods of calculation assumed no cold reserve with respect to studies of installed capacity requirements the same as it is understood for the proposed method. This assumption results in indicating a slightly larger installed capacity requirement than if this factor were taken into account, but well within the limits of practical accuracy of the calculations on any basis.

Also, earlier methods, primarily as a computing convenience because of the nonavailability of automatic computers and not because of being an inherent requirement of those methods, assumed the load, with respect to studies of installed capacity requirements, to be increased from time to time by an amount equal to the sum of the seasonal capacity reductions and of the capacity of the units down for maintenance as of such times. Such a procedure is not so important now as in the early days of making probability calculations, because of the availability of computers and, if desired, the calculations can be made on the basis of the actual capacity available for operation and subject to forced outage. However, this early practice also resulted in slight increases in the indicated installed capacity requirements which likewise are well within the limits of practical requirements. In addition, this general procedure is particularly amenable to making calculations involving hydro capacity for which there may be limited amounts of hydro energy available and also with respect to measurable hydro capacity changes during the day with changes in head. Does the authors' proposed method provide a direct basis for handling these latter hydro problems?

Further, it is not necessary with such earlier methods to group "different machines into clusters of hypothetical identical units," as stated in the paper, since a method has been available for a long time to adjust the table of the probability of availability of capacity very accurately, computation-wise, to any reasonable discrete capacity steps desired [8].

Also, it is not only possible to compute the probabilities of loss of capacity, of the loss of load, and of energy by means of the older methods, but it is also possible to compute the probabilities of loss of both load and energy with respect to various levels of both generating capacity and load and, in addition, with respect to various magnitudes of load lost [9], i.e., of negative margins in the sense shown by Table IV.

Lastly, I would be very negligent if I let this opportunity pass without again suggesting that the economic criterion is by far the most meaningful aid to judgment, as has been discussed in earlier papers [9], [10]. Such an approach would be particularly advantageous since it would not only permit the independent evaluation of the most economical design of the generating, transmission, and distribution systems, but also of the separate elements of each, thus greatly simplifying the necessary calculations.

REFERENCES

[8] C. W. Watchorn, "Elements of system capacity requirements," *AIEE Trans.*, vol. 70, pp. 1163–1185, 1951.
[9] ——, "A review of some basic characteristics of probability methods as related to power system problems," *IEEE Trans. Power Apparatus and Systems*, vol. 83, pp. 737–743, July 1964, and pp. 868–873, August 1964.
[10] ——, "Power and energy production," *AIEE Trans. (Power Apparatus and Systems)*, vol. 73, pp. 901–908, August 1954.

H. A. Adler (Consulting Engineer, La Grange Park, Ill.): This paper is an interesting sequel to the earlier paper by the authors, as it adds procedures for determining loss of load to the procedures for calculating loss of capacity.

In evaluating the advantages of either method, consideration must be given to the nature of the problem. The loss of capacity method is of special advantage where studies are made of different schemes of system expansion, such as increasing the system capacity by one large unit as compared with two or more small units. In studies of this type the nature of the load duration curve is a common factor and it is, therefore, not necessary to involve this variable in the calculations. For studies of this type the loss of capacity method gives directly the net system capacity available for a given criterion of reliability. This results in a great simplification and flexibility of the studies.

On the other hand, there are problems where information on expectation of loss of load is desirable, especially where unusually steep or flat load duration curves for some seasons may affect the assessment of the reliability of the system. The usual loss of load calculations express the results by the sum of the probabilities of all combinations of capacity outages and load levels which result in a deficiency of capacity. In order to determine the extent of the deficiency or available excess capacity, the calculations must be repeated, changing the capacity in steps. For example, Table IV shows the magnitude and probabilities of deficiencies, but the expected probability of loss of load is the sum of all probabilities of possible deficiencies. In order to determine the required system capacity for a desired degree of reliability, the calculations must be repeated for various system capacities. The ideas of the authors on this subject would be of interest.

As the authors point out, determination of loss of load is more significant in terms of frequency of occurrence than in terms of probability of existence. Therefore, our loss of load studies were made in terms of frequency.

It would be of interest to learn whether the authors propose a model for generating load duration curves for use in loss of load calculations. Our studies were based on forecast monthly peak loads and on decrement curves of daily peak loads for each month based on studies of daily peak loads over a period of several years. We found that the slope of the daily peak load–decrement curves varied considerably for different months and may be quite steep for temperature-sensitive months. A model of expressing daily peak loads as a function of predicted monthly peaks simplifies the loss of load calculations considerably. The parameters of such a function should be derived from statistical experience. For example, by replacing the daily peak load–decrement curve by an equivalent straight line we could express the frequency of loss of load by a relatively simple function. We then converted the results of a series of calculations into curves which permit direct conversion of loss of capacity results into loss of load frequencies.

The two alternative models providing transition of daily peak loads from one level to the other, directly or through a low load period, are of interest. We used the first method which in effect assumes that the peak load lasts a day, as pointed out by the authors. The reasoning was that the occurrence of an outage, which in combination with the daily peak would result in a deficiency, will likely require emergency measures. Therefore such a day should be counted as a day of loss of load, even if the outage should not coincide with the peak.

R. J. Ringlee and **A. J. Wood:** Several discussers have raised the point that the technique, as presented, omits consideration of the condition of the low or zero load state so that the availabilities of all of the margin states used do not sum to unity. The authors feel that their inclusion would add little to the results while increasing the number of computations required. This question is rather similar to that of whether or not to include loads for weekend periods.

The inclusion of the low load state in the calculation adds no difficulty since the relations in the paper contain all of the necessary factors. Its inclusion influences primarily the positive margin states as pointed out by Mr. Billinton and Mr. Bhavaraju. The effect on the availability of the margin states may be shown by considering (12). If the low load level is considered to be L_0 MW, the contribution of this low load state to a given reserve margin state of M, MW, is as follows:

$$A_{M'} = A_0 \sum_{C \leq M+L_0} A_C$$

$$A_{M'} = (1 - e)A_{G=M+L_0} \tag{31}$$

where G is now the cumulative capacity state defined by

$$\{G\} = \{C \leq M + L_0\}.$$

A similar relationship for the contribution to the frequency of a cumulative margin state yields

$$f_{M'} = (1 - e)f_G - A_G \tag{32}$$

from (19) and the above definition of the cumulative capacity state G. A simple numerical example (Table VIII) will illustrate the effects of omitting or including this low load state. The data for this example assume $L_0 = 0$ MW.

The results for the availabilities and cycle times for five different cumulative margin states with the zero load state first excluded and then included are tabulated in Table IX. These data illustrate the numerical effects of including the low load state. We are still of the opinion that this low load state can be omitted since the primary concern is with outages of capacity which overlap peak load periods.

Manuscript received July 8, 1968.

Manuscript received August 8, 1968.

TABLE VIII
DATA FOR 3-MACHINE EXAMPLE

Capacity Model	Unit Capacity (MW)	Availability (pu)	Failure Rate (per day)
	20	0.98	0.01
	30	0.98	0.01
	30	0.98	0.01

Load Model	Load (MW)	Occurrence (days)
$e = 1/2$ day	65	15
	55	25
	35	305
	15	20
	0 (low load)	365

TABLE IX
RESULTS FOR 3-MACHINE EXAMPLE

Cumulative Margins (MW)	0 Load Omitted		0 Load Included	
	Availability (pu)	Cycle Time (days)	Availability (pu)	Cycle Time (days)
50	0.474214	1.0535	0.494014	1.0879
25	0.079397	5.8609	0.075974	5.8679
0	0.003059	127.572	0.003063	127.606
−25	0.000068	4 889.9	0.000068	4 889.9
−50	0.0000004	657 421	0.0000004	657 421.

Some of the discussers mentioned the possible use of a load model which includes a number of different load levels in a single day and which permits transfer directly between these states. Appendixes I and II present an outline of the Markov process analysis of one possible variety of this type of model. In answer to Mr. Cook, we have not been able to conceive of a load model which does not start with the system initially in one of the allowable load states. The model discussed would seem more suitable for use in a study of the reliability of generation operating practices and unit commitment policies for meeting specified reliability levels.

In reply to Mr. Billinton and Mr. Bhavaraju, the authors have not specifically considered storm and normal weather conditions in the generation–transmission example given. This question deserves more attention than can be devoted to it here.

Mr. Adler and Mr. Watchorn present a number of welcome observations on the use of these methods in particular, and probability techniques used for generation planning in general. It is the authors' feeling that the frequency and duration method being presented offers a number of conceptual and practical advantages in generation planning. For the first time, a single technique is available which yields both the availability of capacity and reserve margin states and at the same time gives expected frequencies of occurrence of these same capacity and reserve margin states. Previously, two separate and, unfortunately, somewhat unrelated techniques were required to supply these same data. One now may inquire if a loss-of-load probability of "one day in 10 years" means the same as a 10-year mean cycle time of the cumulative zero-MW reserve margin state.

We concur with the observations of Mr. Adler and Mr. Watchorn to the effect that sensitivity analyses are important, load forecasting problems must be analyzed further, and common sense must be used in applying any mathematical technique.

Mr. Watchorn has offered several comments regarding representation of load forecasting deviations in probability calculations. He is well aware that the "basic or primitive" methods cited by him and the extensions of them developed in this paper permit direct recognition of the probabilistic models of the load forecasting deviation process. We concur with his warning concerning uncritical use of continuous, unbounded probability distributions. We also agree with him that load forecasting and forecast deviation estimation are fruitful areas for continuing effort.

We must respectfully disagree with his comment that the "use of sophisticated calculation methods . . . is academic unless they contribute to calculation advantages." Here we assume he is classifying our method among the "sophisticated calculation methods."

Calculation improvement is certainly one reasonable measure for gauging the value of a new procedure; it is not the only one, however. An equally if not more significant measure would be the ability of a method to unify existing methods and indexes. The frequency and duration method represents an attempt to put a number of previously developed measures on a common theoretical foundation to yield compatible results, to suggest certain mathematical formulations that greatly facilitate numerical calculations, and to advance a suggested model of the power system load which is suitable for generation planning. It is the authors' opinion that none of the previously available methods represented the load in a fashion that permitted the computation of the expected frequency of occurrence of various reserve margin levels.

It is planned that a later part of this sequence will discuss the use of the frequency and duration method in generation planning and illustrate the calculated numerical values of the frequencies of various reserve margins which are found to correspond to commonly used system loss-of-load probability standard risks. This forthcoming work will also discuss the results of statistical analyses of actual system load cycles.

Mr. Adler points out the necessity for repeating calculations with various capacity levels when planning expansions. This is generally true and is routine with a well-designed computer program. If the annual availability versus reserve margin function has been computed for the basic system, a single computation will yield the new annual availability function after a single, new machine is added, assuming that nothing else is altered. Let

C	capacity of new machine, MW
$A(M)$	annual availability of the basic system as a function of the margin M
M	reserve margin
a	availability of the new machine.

Thus, the new availability function becomes

$$A'(M) = aA(M - C) + (1 - a)A(M). \qquad (33)$$

This relationship may be used to consider the effect of adding different machine capacities on the annual availability (i.e., the loss-of-load probability). A similar relationship may be worked out for the frequency calculations. These are approximate since they assume no change in maintenance outages for the old machines and that the new machine is not maintained during the year.

The technique for constructing the load model suggested by Mr. Adler is similar to that used by the authors and their associates. That is, historical load data are examined statistically to establish periodic (i.e., monthly or seasonal), per unit peak load variation curves, and monthly or seasonal peaks in per unit of the annual peak.

The data requirements are then primarily for the annual peak load forecasts. It might be observed that the load model of the paper does not require that the occurrence N of a load level of L_i MW be an integer value. This might be useful in specifying the peak load variation curve from historical data since, for instance, a model for a 30-day period might include 35 load levels. This would permit using more data to define the highest load levels in the peak variation curve. The load model also requires the specification of the mean duration e of the peak periods. This value is a matter of judgment, to be arrived at after a suitable study of load cycle data. Values of from 1/4 to over 1/2 of a day would seem to be appropriate for various different systems.

Concerning Mr. Watchorn's question about representing hydro units and plants (energy limits and head effects), we have not included these provisions in the analysis. However, we would like to note that the frequency and duration method will handle the same problems that may be treated by loss-of-load probability.

We agree with Mr. Watchorn that the "economic criterion" would be "by far the most meaningful aid to judgment." Even separate economic criteria for bulk power supply and for distribution systems would be of much value for system planning.

Again we wish to thank the various discussers for their contributions. It is gratifying to observe the continuing interest in this area.

GENERATING CAPACITY RELIABILITY EVALUATION IN INTERCONNECTED SYSTEMS
USING A FREQUENCY AND DURATION APPROACH
PART I. MATHEMATICAL ANALYSIS

Roy Billinton   Chanan Singh
Power System Research Group
University of Saskatchewan
Saskatoon, Canada

INTRODUCTION

Equations for computing the frequency and duration of various capacity outages were first given by Halperin and Adler[1]. The application to the generation area was subsequently extended by the publication of a recent series of papers[2-6] which presented a far more general approach to the problem. The general method is extremely powerful and can be applied to a wide range of problems[7-9]. This paper utilizes the basic concepts previously presented[2,3] and extends them to develop a frequency and duration approach to interconnected system reliability evaluation. Equations are developed for one system connected to another system and are subsequently extended to the case of a system connected to two or more systems. The application of the theoretical technique is illustrated in Part II of this paper entitled "System Applications".

An important aspect of reliability evaluation of interconnected systems is the type of agreement which exists between the systems. The assumption made in the approach presented in the paper is that one system will help the other or others as much as possible without curtailing its own load. The basic principles involved, however, are quite general and can easily be modified to cover other types of agreement. The usual assumption that the load and generation models in each system are stochastically independent is also made. The basic theory required in the recursive approach[2,3] is covered in considerable detail in the available literature[10] and is therefore not elaborated on in this paper. The extension of this approach to the interconnected system problem proceeds as follows.

System A Connected to System B

The capacity and the load in each system are assumed to exist at a discrete number of levels[3] and therefore the margin state, which is the capacity available less the load on the system, will also exist at a discrete number of levels in each system. In Figure 1, the arrays M_a and M_b contain the unaffected margin states of Systems A and B respectively, arranged in the order of decreasing reserve. The term unaffected implies that the interconnection is not effective. The effective margin states in System A, i.e., when the tie line is in operation, are given by the elements of the matrix M,

$$m_{ij} = m_{ai} + h_{ij} \qquad (1)$$

where h_{ij} is either the help available to System A from System B or it is the help required by System B from System A. In the latter case h_{ij} has a negative sign. If no help can be rendered by one system to the other $h_{ij} = 0$. The maximum value of h_{ij} is limited by the tie line capability.

Paper 71 TP 116-PWR, recommended and approved by the Power System Engineering Committee of the IEEE Power Engineering Society for presentation at the IEEE Winter Power Meeting, New York, N. Y., January 31-February 5, 1971. Manuscript submitted September 14, 1970; made available for printing November 30, 1970.

The effective margin states in System A, while the tie line is on forced outage are given by the elements of matrix M′,

$$m_{ij} = m_{ai} \qquad (2)$$

Equations 1 and 2 define the boundaries in M and M′ respectively of any effective cumulative margin state. In this presentation, m with proper subscript represents an exact margin and M with the same subscript denotes the corresponding cumulative margin, e.g., M_{ij} is a margin equal to or less than m_{ij}.

The principles involved in the determination of the availability and frequency of a particular state, say M_{34} will be examined and then condensed into generalized equations. It is assumed that in the matrix M, the thick line bcfhi demarcates the effective reserve equal to or less than m_{34}. The boundary for this margin state in M′ may be theoretically anywhere – in line with hi as $j_1 k_1$, above hi as $j_2 k_2$ or below hi as $j_3 k_3$. In practice the boundary in M′ can be:

1. Above hi only if there are no zero or negative margins in System B. Such a condition is improbable in actual practice.

2. Below hi only if m_{34} is a positive margin.

3. The boundary in M′ will be always in line with hi for a negative m_{34}.

Availability:

The effective margin states are separated in time and the contribution to the availability of M_{34} by the rows corresponding to m_{a3} for the condition $(m_{34} \geq m_{a3})$ will be given as

$$= (A_{a(3)} - A_{a(3+1)})(A_{ab} \sum_{j=4}^{NB} (A_{b(j)} - A_{b(j+1)}))$$

$$+ (A_{a(3)} - A_{a(3+1)})(\overline{A}_{ab} \sum_{j=1}^{NB} (A_{b(j)} - A_{b(j+1)}))$$

$$= (A_{a(3)} - A_{a(3+1)})(A_{ab} \cdot A_{b(4)} + \overline{A}_{ab})$$

If m_{34} were less than m_{a3}, the row in M′ corresponding to m_{a3} will not contribute to the availability of M_{34} since those states being equal to m_{a3} would not be included in M_{34}. Contribution under this condition would be given as

$$= (A_{a(3)} - A_{a(3+1)})(A_{ab} \cdot A_{b(4)})$$

Generalizing from the above

$$A_N = \sum_{L,K} (A_{a(\ell)} - A_{a(\ell+1)})(A_{ab} \cdot A_{b(k)} + \beta \overline{A}_{ab}) \qquad (3)$$

where $A_{a(\ell)}, A_{b(k)}$ = The availabilities of $M_{a\ell}$ and M_{bk} respectively

A_{ab} = The availability of the tie line between A and B

Reprinted from *IEEE Trans. Power Apparatus Syst.*, vol. PAS-90, no. 4, pp. 1646–1654, July/Aug. 1971.

A_N = The availability of an effective margin state $\leq N$.

L, K = The indices defining the corner states of the boundary of N. For example, the indices for $N = m_{34}$ are $(NA,1) - (6,1), (5,2), (4,3)$ and $(3,4)$.

and β = 1 for $N \geq m_{aL}$
0 for $N < m_{aL}$

Frequency:

Let

$f_{a(\ell)}, f_{b(k)}$ = The frequencies of encountering $M_{a\ell}$ and M_{bk} respectively.

λ_{ab}, μ_{ab} = The mean failure and the mean repair rates of the tie line.

f_N = The frequency of encountering an effective margin $\leq N$.

System A can transit from one effective margin state to another in any of the following ways.

1. The capacity or load transitions in System A itself

When the interconnection is in the up state, given that System A is in the effective state m_{ij}, it can transit to m_{kj}, $k \neq i$ and when the interconnection is in the down state, given that the system is in the effective state m'_{ij}, it can transit to m'_{kj}, $k \neq i$.

2. The capacity or load transitions in System B.

When the interconnection is in the up state, given that System A is in the effective state m_{ij}, it can transit to m_{ik}, $k \neq j$ and when the interconnection is in the down state, given that System A is in the effective state m'_{ij}, it can transit to m'_{ik}, $k \neq j$.

3. The failure or the repair of the tie line

Transition due to the failure of the interconnection will be from m_{ij} to m'_{ij} and transition due to the repair of the interconnection will be from m'_{ij} to m_{ij}.

The frequency of encountering an effective cumulative margin state in System A equals the "expected transitions per unit time" across the boundary defining that state plus the "expected transitions per unit time" associated with the deserting states. The deserting states are defined as the states which leave the domain of the cumulative margin state as a result of the failure or the repair of the tie line. As an example, if $m_{34} \geq m_{a3}$, the states m'_{31}, m'_{32} and m'_{33} desert M_{34} as a result of the repair of the tie line. Similarly, if $m_{34} < m_{a3}$, the states $m_{34}, m_{35}, \ldots, m_{3NB}$ desert M_{34} as a result of the failure of the tie line.

In determining the contributions to f_N due to the three different modes of system transition, it is helpful to keep in mind the following equations,

$$f_{a(\ell)} = \sum_{i=\ell}^{NA} E_{i\,(<\ell)}^a$$

$$f_{b(k)} = \sum_{j=k}^{NB} E_{j(<k)}^b$$

$$f_{a(\ell)} - f_{a(\ell+1)} = \sum_{i=1}^{\ell-1} E_{\ell i}^a - \sum_{i=\ell+1}^{NA} E_{\ell i}^a$$

and $f_{b(k)} - f_{b(k+1)} = \sum_{j=1}^{k-1} E_{kj}^b - \sum_{j=k+1}^{NB} E_{kj}^b$

where NA, NB = The total number of discrete levels in the capacity reserve models of Systems A and B respectively.

and E_{ij}^x = The expected number of "transitions per unit time" from an exact state "i" to an exact

state "j" in the system "x".

= (Steady state probability of being in state i) (The rate of departure from state i to state j).

As in the case of the availability, the contribution of the rows corresponding to m_{a5} to the frequency of M_{34} will be determined first and then out of this the generalized formulation will be evolved.

The total "E" (expected transitions per unit time) out of the boundary b_j in M and the corresponding boundary in M' is

$$= \sum_{j=1}^{NB} A_{ab} \cdot A'_{b(j)} \sum_{i=6}^{NA} E_{i(<6)}^a + \sum_{j=1}^{NB} \bar{A}_{ab} \cdot A'_{b(j)} \sum_{i=6}^{NA} E_{i(<6)}^a$$

The primed designation has been used to denote the availabilities of the exact states.

The total E out of the boundary abcde in M and the corresponding boundary in M' is

$$= \sum_{j=1}^{NB} A_{ab} \cdot A'_{b(j)} \sum_{i=6}^{NB} E_{i(<6)}^a + \sum_{j=1}^{NB} \bar{A}_{ab} \cdot A'_{b(j)} \sum_{i=6}^{NA} E_{i(<6)}^a$$

$$+ \sum_{j=2}^{NB} A_{ab} \cdot A'_{b(j)} \sum_{i=1}^{4} E_{5i}^a + \beta \sum_{j=1}^{NB} \bar{A}_{ab} \cdot A'_{b(j)} \sum_{i=1}^{4} E_{5i}^a$$

$$- \sum_{j=2}^{NB} A_{ab} \cdot A'_{b(j)} \sum_{i=6}^{NA} E_{5i}^a - \beta \sum_{j=1}^{NB} \bar{A}_{ab} \cdot A'_{b(j)} \sum_{i=6}^{NA} E_{5i}^a$$

$$+ A_{ab} \cdot A'_{a(5)} \sum_{j=2}^{NB} E_{j(<2)}^b$$

Here $\beta = 1$, if $m_{34} \geq m_{a5}$ since then the corresponding row in M' will be included in the boundary,

$= 0$, if $m_{34} < m_{a5}$ since then the corresponding row in M' will not be included in the boundary.

Rearranging the expression for total E out of abcde

$$= \sum_{j=1}^{NB} A_{ab} \cdot A'_{b(j)} \sum_{i=6}^{NA} E_{i(<6)}^a + \sum_{j=1}^{NB} \bar{A}_{ab} \cdot A'_{b(j)} \sum_{i=6}^{NA} E_{i(<6)}^a$$

$$+ \sum_{j=2}^{NB} A_{ab} \cdot A'_{b(j)} (\sum_{i=1}^{4} E_{5i}^a - \sum_{i=6}^{NA} E_{5i}^a)$$

$$+ \beta \sum_{j=1}^{NB} \bar{A}_{ab} \cdot A'_{b(j)} (\sum_{i=1}^{4} E_{5i}^a - \sum_{i=6}^{NA} E_{5i}^a)$$

$$+ A_{ab} \cdot A'_{a(5)} \sum_{j=2}^{NB} E_{j(<2)}^b$$

$$= f_{a(6)} + (f_{a(5)} - f_{a(5+1)})(A_{ab} \cdot A_{b(2)} + \beta \cdot \bar{A}_{ab})$$

$$+ (A_{a(5)} - A_{a(5+1)}) \cdot A_{ab} \cdot f_{b(2)} \tag{4}$$

Let $E_{mm'}^{ij}$ = The expected transitions per unit time from m_{ij} to m'_{ij}.

The contribution due to deserting states

$$= \sum_{j=1} E_{mm'}^{5j}, \text{ if } m_{34} \geq m_{a5}$$

$$= \sum_{j=2}^{NB} E_{mm'}^{5j}, \text{ if } m_{34} < m_{a5}$$

$$= \beta [(1 - A_{b(2)})(A_{a(5)} - A_{a(5+1)}) \bar{A}_{ab} \cdot \mu_{ab}]$$

$$+ \gamma [A_{b(2)}(A_{a(5)} - A_{a(5+1)}) A_{ab} \cdot \lambda_{ab}] \tag{5}$$

122

where γ = 1, for $m_{34} < m_{a5}$

$\quad\quad\quad$ = 0, for $m_{34} \geq m_{a5}$

Adding up 4 and 5, substituting ℓ and k for 5 and 2 respectively, and rearranging the general expression for f_N becomes

$$f_N = \sum_{L,K} [(f_{a(\ell)} - f_{a(\ell+1)})(A_{ab} \cdot A_{b(k)} + \beta \overline{A}_{ab})$$

$$+ (A_{a(\ell)} - A_{a(\ell+1)})(f_{b(k)} + \beta(1 - A_{b(k)})\lambda_{ab}$$

$$+ \gamma \cdot A_{b(k)} \cdot \lambda_{ab})A_{ab}] \quad\quad\quad (6)$$

where β = 1 for $N \geq m_{aL}$

$\quad\quad\quad$ = 0 for $N < m_{aL}$

and $\quad \gamma$ = 1 for $N < m_{aL}$

$\quad\quad\quad$ = 0 for $N \geq m_{aL}$

System A Connected to Two or More Than Two Systems

The reliability evaluation of System A connected to Systems B and C is illustrated first and later it is shown that this can be extended to the case where System A is connected to more than two systems. The techniques for determining the availability and the frequency of the failure state, i.e., the first negative cumulative margin in System A is outlined. The techniques are quite general and may be applied to any negative cumulative margin.

Technique 1

The state transition diagram of System A is shown in Figure 2 where "i", "j" and "n" are the indices for the margin states in Systems A, B and C respectively, arranged in decreasing order of magnitude.

Let $\lambda_{ab}, \lambda_{ac}$ = The mean failure rates of the tie lines AB and AC respectively

and μ_{ab}, μ_{ac} = The mean repair rates of the tie lines AB and AC respectively

Reference frames 1 – 4 are the three dimensional arrays representing the effective margin states in System A under the following conditions:

The R.F. #	Tie Line AB	Tie Line AC
1	UP	UP
2	DN	UP
3	UP	DN
4	DN	DN

The boundary of any effective cumulative margin state can be defined using the equation

$$m_{ijn} = m_{ai} + b \cdot h_{ij} + c \cdot h_{cn}$$

where $\quad m_{ijn}$ $\quad$ = The effective margin state in System A with m_{ai}, m_{bj} and m_{cn} as the unaffected margin states in the Systems A, B and C respectively

$\quad\quad h_{cn}$ = The help available from System C at its nth margin state

$\quad\quad\quad$ = $\min(m_{cn}, T_{ac})$ = 0 if m_{cn} is negative

$\quad\quad T_{ac}$ = The capability of the tie line AC

$\quad\quad b$ = 1 for the tie line A – B up

$\quad\quad\quad$ 0 for the tie line A – B down

and c $\quad$ = 1 for the tie line A – C up

$\quad\quad\quad$ 0 for the tie line A – C down

The Availability:

Let A_{xn} $\quad$ = The availability of the failure state, given the reference frame x and margin m_{cn} in System C

$\quad\quad\quad$ = $\displaystyle\sum_{Lx, Kx} (A_{a(\ell)} - A_{a(\ell+1)})A_{b(k)}$

$\quad$ Lx, Kx $\quad$ = The indices defining in the reference frame x, the corner states of the boundary of the failure state

$A'_{c(n)}$ = The availability of m_{cn}

and A_{ac} = The availability of the tie line AC

The expression for the availability is a straight forward summation:

$$A_- = \sum_{n=1}^{NC} A_{1n} \cdot A'_{c(n)} \cdot A_{ab} \cdot A_{ac}$$

$$+ \sum_{n=1}^{NC} A_{2n} \cdot A'_{c(n)} \cdot \overline{A}_{ab} \cdot A_{ac}$$

$$+ \sum_{n=1}^{NC} A_{3n} \cdot A'_{c(n)} \cdot A_{ab} \cdot \overline{A}_{ac}$$

$$+ \sum_{n=1}^{NC} A_{4n} \cdot A'_{c(n)} \cdot \overline{A}_{ab} \cdot \overline{A}_{ac}$$

$$= \sum_{n=1}^{NC} A_{1n} \cdot A'_{c(n)} \cdot A_{ab} \cdot A_{ac} + \sum_{n=1}^{NC} A_{a(U)} \cdot A'_{c(n)} \cdot \overline{A}_{ab} \cdot A_{ac}$$

$$+ A_{31} \cdot A_{ab} \cdot \overline{A}_{ac} + A_{a(FM)} \cdot \overline{A}_{ab} \cdot \overline{A}_{ac} \quad\quad (7)$$

where $\quad$ NC $\quad$ = The total number of discrete levels in the margin state matrix of System C

$\quad\quad A_{a(U)}$ = The cumulative availability of the margin state in M_a corresponding to the uppermost corner of the boundary of the failure state in the reference frame 1

and $\quad A_a(FM)$ = The cumulative availability of the first negative margin in M_a

The frequency:

System A can transit from one effective margin state to another in any of the following modes:

1. The capacity or the load transitions in System A
2. The capacity or the load transitions in System B
3. The capacity or the load transitions in System C
4. The failure or the repair of the tie line AB
5. The failure or the repair of the tie line AC

The contributions to the frequency of failure by the modes listed above are evaluated as follows.

(a) The contribution due to modes "1" and "2"

$$f_{c10} = \sum_{n=1}^{NC} f_{1n} \cdot A'_{c(n)} \cdot A_{ab} \cdot A_{ac}$$

where f_{cx0} = The contribution to the frequency of encountering the failure state in System A by the capacity or load transitions in System A or System B in reference frame x

and $\quad f_{xn}$ = The frequency of encountering the failure state in System A by modes 1 and 2, given the reference frame x and the margin m_{cn}.

123

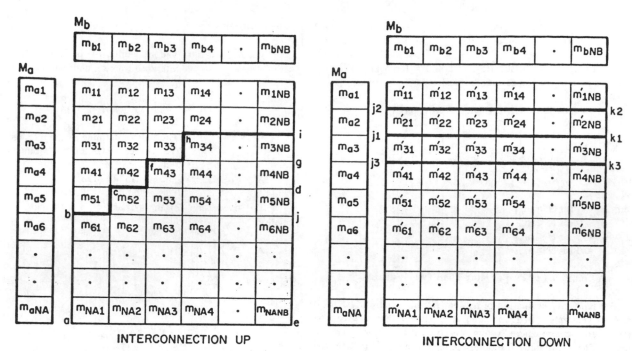

Fig. 1. Effective Margin States Matrices for System A Connected to System B

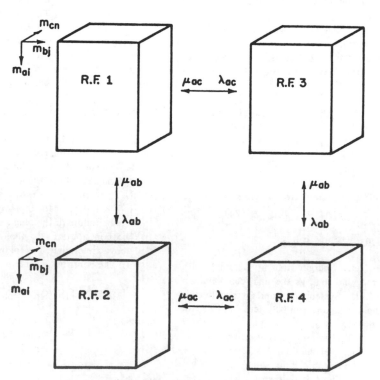

Fig. 2. The State Transition Diagram of System A Connected to Systems B and C where B and C are Not Connected Directly

$$= \sum_{Lx,Kx} [(f_{a(\ell)}-f_{a(\ell+1)})A_{b(k)}+(A_{a(\ell)}-A_{a(\ell+1)})f_{b(k)}]$$
$$\text{at } n$$

$$f_{c20} = \sum_{n=1}^{NC} f_{2n} \cdot A'_{c(n)} \cdot \overline{A}_{ab} \cdot A_{ac}$$

where $f_{2n} = f_{a(U)}$ = The frequency of the cumulative margin state in M_a corresponding to the uppermost corner of the failure state in the reference frame 1.

$$f_{c30} = \sum_{n=1}^{NC} f_{3n} \cdot A'_{c(n)} \cdot A_{ab} \cdot \overline{A}_{ac}$$

$$= f_{3n} \cdot A_{ab} \cdot \overline{A}_{ac}$$

The value of f_{3n} is the same at all values of n as help from System C is rendered ineffective due to the failure of the tie line AC.

$$f_{c40} = \sum_{n=1}^{NC} f_{4n} \cdot A'_{c(n)} \cdot A_{ab} \cdot \overline{A}_{ac}$$

where $f_{4n} = f_{a(FM)}^{CU}$ = The frequency of encountering the first cumulative negative margin in System A as no help is available from System B or System C

(b) The contribution due to mode 3

$$f_{c11} = \sum_{n=1}^{NC} A_{1n}(f_{c(n)}-f_{c(n+1)})A_{ab} \cdot A_{ac}$$

where f_{cx1} = The contribution to the frequency of encountering the failure state by mode 3 in reference frame x

$$f_{c21} = \sum_{n=1}^{NC} A_{2n}(f_{c(n)}-f_{c(n+1)})\overline{A}_{ab} \cdot A_{ac}$$

where $A_{2n} = A_{a(U)}$

$$f_{c31} = \sum_{n=1}^{NC} A_{3n}(f_{c(n)}-f_{c(n+1)})A_{ab} \cdot \overline{A}_{ac}$$

$$= A_{3n} \cdot A_{ab} \cdot \overline{A}_{ac} \sum_{n=1}^{NC} (f_{c(n)}-f_{c(n+1)}) = 0$$

$$f_{c41} = \sum_{n=1}^{NC} A_{4n}(f_{c(n)}-f_{c(n+1)})\overline{A}_{ab} \cdot \overline{A}_{ac}$$

$$= A_{4n} \cdot \overline{A}_{ab} \cdot \overline{A}_{ac} \sum_{n=1}^{NC} (f_{c(n)}-f_{c(n+1)}) = 0$$

(c) The contribution due to modes 4 and 5

Denoting the contribution due to transitions between the reference frames x and y by f_{cx-y}

$$f_{c1-2} = \sum_{n=1}^{NC} A'_{c(n)}(A_{2n}-A_{1n}) A_{ac} \cdot A_{ab} \cdot \lambda_{ab}$$

$$f_{c3-4} = \sum_{n=1}^{NC} A'_{c(n)} (A_{4n}-A_{3n}) \overline{A}_{ac} \cdot A_{ab} \cdot \lambda_{ab}$$

$$f_{c1-3} = \sum_{n=1}^{NC} A'_{c(n)}(A_{3n}-A_{1n}) A_{ab} \cdot A_{ac} \cdot \lambda_{ac}$$

$$f_{c2-4} = \sum_{n=1}^{NC} A'_{c(n)}(A_{4n}-A_{2n}) \overline{A}_{ab} \cdot A_{ac} \cdot \lambda_{ac}$$

The frequency of encountering the failure state in System A is given by

$$f_- = \sum_{i=1}^{4} (f_{ci0}+f_{ci1}) + f_{c1-2} + f_{c1-3} + f_{c2-4} + f_{c3-4} \qquad (8)$$

Technique 2

This is a very convenient technique for evaluating the availability and the frequency of the effective capacity reserve margins in System A connected to two or more than two systems. The effective capacity reserve model of System A connected to System B is developed first and this is then modified by successive reactions with the other systems using equations 3 and 6. The evaluation of the availability and the frequency of the failure state by this technique is as follows.

The effective margin state matrices for System A connected to Systems B and C are shown in Figure 3. The column vector M_{ab} represents the effective negative margin states in System A connected to System B. The boundary of the failure state in matrices MM and MM' can be determined using the equations

$$mm_{ij} = m_{abi} + hc_{ij}$$
and $$mm'_{ij} = m_{abi}$$

where hc_{ij} = The help available from System C at (i,j)
= min (m_{cj},T_{ac}) = 0 if m_{cj} is negative

The equations for the availability and the frequency of the failure state are given below:

$$A_- = \sum_{P,N} (A_{ab(p)}-A_{ab(p+1)}) (A_{ac} \cdot A_{c(n)}+\overline{A}_{ac}) \qquad (9)$$

and $$f_- = \sum_{P,N} [(f_{ab(p)}-f_{ab(p+1)}) (A_{ac} \cdot A_{c(n)}+\overline{A}_{ac})$$
$$+ (A_{ab(p)}-A_{ab(p+1)})(f_{c(n)}+(1-A_{c(n)}) \lambda_{ac}) A_{ac}] \qquad (10)$$

where $A_{ab(p)},f_{ab(p)}$ = The availability and the frequency of M_{abp}
$A_{c(n)},f_{c(n)}$ = The availability and the frequency of M_{cn}

and P,N define the indices p,n of the corners of the boundary of the failure state in matrix MM.

By suitable manipulation it can be shown that equations 7, 9 and 8, 10 are equivalent.

System A Connected to Other Interconnected Systems

When the system whose reliability is being evaluated is connected to several interconnected systems, the problem requires careful analysis, which includes the establishment of priorities for emergency help. Some simplifying assumptions may also be required. To illustrate this situation, the configuration shown below has been selected.

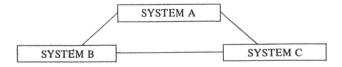

The following priority constraints have been assumed.
Constraints:
1. When both System A and System B need emergency help

125

from System C, System B will get preference. Similarly when both System A and System C need help from System B, System C will get preference.

2. When the tie line AB is on forced outage, emergency help by System B will be supplied to System A via BC-CA. Similarly when tie line AC is on forced outage, emergency help by System C will be supplied to System A via CB-BA.

3. When the tie line AB is in, emergency help from System B will only be supplied via BA and similarly when AC is in, emergency help from System C will only be supplied via CA.

The reliability evaluation of System A is made in the following steps.

Step 1.

The availabilities and frequencies of the effective positive margins in System B, from the point of view of emergency help to System A, are determined. The effective margin state matrices for System B are shown in Figure 4. The boundaries of positive margins in matricies M and M′ are defined using equations 1 and 2 respectively. The same equations are used to define the boundaries in matrices MX and MX′ except that $h_{ij} = 0$ when no help is needed by System B, i.e., when m_{bi} is zero or more.

The equation for the frequency of a cumulative effective margin N, in System B is given as

$$f_N = \overline{A}_{ac} \sum_{L,K} f_Y + A_{ac} \sum_{LX,KX} f_Y + \sum_{LX,KX} (A_{b(\ell)}$$

$$-A_{b(\ell+1)}) (A_{c(k)} - A_{c(KK)}) A_{bc} \cdot A_{ac} \cdot \lambda_{ac}$$

where L,K = The indices defining the corners of the boundary of the effective margin N in the matrix M

LX,KX = The indices defining the corners of the boundary of the effective margin N in the matrix MX

KK = The index defining the corner of the boundary of the effective margin N in the matrix M corresponding to LX

$A_{ac}, \overline{A}_{ac}$ = The availability and the unavailability of the tie line AC

λ_{ac} = The mean failure rate of the tie line AC

$$f_Y = (f_{b(\ell)} - f_{b(\ell+1)}) (A_{bc} \cdot A_{c(k)} + \overline{A}_{bc}) + (A_{b(\ell)} - A_{b(\ell+1)})$$

$$(f_{c(k)} + \beta(1 - A_{c(k)}) \lambda_{bc} + \gamma \cdot A_{c(k)} \cdot \lambda_{bc}) A_{bc}$$

This is equation 6 as applied to System B and System C.

Similarly the expression for the availability of a cumulative effective margin N, in System B is given as

$$A_N = \overline{A}_{ac} \sum_{L,K} A_Y + A_{ac} \sum_{LX,KX} A_Y$$

where $A_Y = (A_{b(\ell)} - A_{b(\ell+1)}) (A_{bc} \cdot A_{c(k)} + \beta \cdot \overline{A}_{bc})$

This is equation 3 as applied to Systems B and C.

Step 2.

The availabilities and frequencies of the effective positive margins in System C from the point of view of emergency help to System A are evaluated in exactly the same manner.

The effective positive margins of System B are combined with the negative margins of System A to obtain the availabilities and frequencies of the effective negative margins of System A. The effective positive margins of System C are then combined with the effective negative margins of System A to obtain the availability and frequency of the failure state in System A.

CONCLUSIONS

This paper has illustrated the theoretical development of the equations used to evaluate the reliability indices for interconnected systems. The basic concepts used for two interconnected systems have been extended by use of some simplifying assumptions to the case of more than two interconnected systems. The required assumptions are minimal in the case of a system which is connected to other systems that are not interconnected with each other. The general concepts, however, can be applied in any multi-system interconnection configuration after the transmission and interconnection constraints are established. The equations presented in the paper have been used in the development of digital computer programs for interconnected system reliability evaluation. Part II of this paper entitled "System Applications" illustrates the numerical application of these concepts.

ACKNOWLEDGEMENT

The authors would like to express their appreciation to the National Research Council of Canada for providing the necessary financial support for this work.

REFERENCES

1. Halperin, H. and Adler, H. A., "Determination of Reserve Generating Capability", AIEE Transactions (Power Apparatus and Systems), Vol. 77, August 1958, pp. 551-57.

2. Hall, J. D., Ringlee, R. J., and Wood, A. J., "Frequency and Duration Methods for Power System Reliability Calculations — Part I — Generation System Model", IEEE Transactions, PAS. 87, No. 9, September 1968, pp. 1787-96.

3. Ringlee, R. J. and Wood, A. J., "Frequency and Duration Methods for Power System Reliability Calculations — Part II — Demand Model and Capacity Reserve Model", IEEE Transactions, PAS. 88, No. 4, April 1969, pp. 375-88.

4. Galloway, C. D., Garver, L. L., Ringlee, R. J. and Wood, A. J., "Frequency and Duration Methods for Power System Reliability Calculations — Part III — Generation System Planning", IEEE Transactions, PAS. 88, No. 8, August 1969, pp. 1216-23.

5. Cook, V. M., Ringlee, R. J. and Wood, A. J., "Frequency and Duration Methods for Power System Reliability Calculations — Part IV — Models for Multiple Boiler Turbines and for Partial Outage States", IEEE Transactions, PAS. 88, No. 8, August 1969, pp. 1224-1232.

6. Ringlee, R. J. and Wood, A. J., "Frequency and Duration Methods for Power System Reliability Calculations — Part V — Models for Delays in Unit Installations and Two Interconnected Systems", IEEE Transactions, Paper No. 70 TP 158-PWR.

7. Billinton, R. and Prasad, Vipin, "Quantitative Reliability Analysis of HVDC Transmission Systems — Part I — Spare Valve Assessment in Mercury Arc Configurations", IEEE Transactions, Paper No. 70 TP 502-PWR.

8. Billinton R. and Prasad, Vipin, "Quantative Reliability Analysis of HVDC Transmission Systems — Part II — Composite System Analysis", IEEE Transactions, Paper No. 70 TP 503-PWR.

9. Ringlee, R. J. and Goode, S. D., "On Procedures for Reliability Evaluation of Transmission Systems", IEEE Trans., PAS. 89, No. 4, April 1970, pp. 527-536.

10. Billinton, R., Power System Reliability Evaluation (book), Gordon and Breach Science Publishers Ltd., New York, N. Y., 1970.

APPENDIX

An Alternative Approach

As an alternative to the margin states approach described in the body of this paper, the effective generation system model for System A can be developed by combining the generation system model of System B with that of System A, under the constraints of the load model of System B. The load model of System A can then be combined with its new generation system model to generate the effective capacity reserve model for System A. For two interconnected systems, this approach is as convenient as the margin states approach. It does, however, become cumbersome in the case of multi-interconnected system configurations.

The development of the effective generation system model is analogous to the development of the effective margin state model. Referring to Figure 5, with the interconnection in the up state

$$\bar{c}_{ij} = \bar{c}_{ai} - h_{ij} \qquad (11)$$

where $\bar{c}_{ij}$ = The effective outage of capacity in System A with $\bar{c}_{bj}$ as the capacity outage in System B and $\bar{c}_{ai}$ as the unaffected capacity outage in System A

h_{ij} = The help available to System A from System B or the negative of help required by System B

= $\min(R_b - \bar{c}_{bj}, T_{ab})$

or = $-\min(R_a - \bar{c}_{ai}, \bar{c}_{bj} - R_b, T_{ab})$ in the latter case

or = 0, if no help can be rendered by one system to the other

R_a, R_b = The operating reserves in Systems A and B respectively

= The installed capacity — The load on the system

T_{ab} = The capability of the interconnection between System A and System B.

With the interconnection in the down state

$$\bar{c}'_{ij} = \bar{c}_{ai} \qquad (12)$$

Equations 11 and 12 define the boundaries of the cumulative capacity outages in the matrices $\bar{C}$ and $\bar{C}'$. Whereas $\bar{c}$ with proper subscript denotes an exact outage of capacity, the corresponding cumulative outage of capacity is represented by $\bar{C}$ with the same subscript, e.g. $\bar{C}_{ai}$ means a capacity outage equal to or greater than $\bar{c}_{ai}$.

Equations 3 and 6 apply to the availability and the frequency of the effective cumulative capacity outage states with some modification of the qualifying statements. These are restated below:

$$A_0 = \sum_{L,K} (A_{a(\ell)} - A_{a(\ell+1)})(A_{ab} \cdot A_{b(k)} + \beta \cdot \bar{A}_{ab}) \qquad (13)$$

and $f_0 = \sum_{L,K} [(f_{a(\ell)} - f_{a(\ell+1)})(A_{ab} \cdot A_{b(k)} + \beta \cdot \bar{A}_{ab})$

$+ (A_{a(\ell)} - A_{a(\ell+1)})(f_{b(k)} + \beta(1 - A_{b(k)})\lambda_{ab}$

$+ \gamma \cdot A_{b(k)} \cdot \lambda_{ab}) A_{ab}] \qquad (14)$

where A_0, f_0 = The availability and the frequency of a capacity outage equal to or greater than "0"

L, K = The indices defining the boundary of this cumulative capacity outage

$A_{a(\ell)}, f_{a(\ell)}$ = The availability and the frequency of $\bar{C}_{a\ell}$

$A_{b(k)}, f_{b(k)}$ = The availability and the frequency of $\bar{C}_{bk}$

$$\beta = 1 \text{ for } 0 \le \bar{c}_{ai}$$
$$= 0 \text{ for } 0 > \bar{c}_{ai}$$

and $$\gamma = 1 \text{ for } 0 > \bar{c}_{ai}$$
$$= 0 \text{ for } 0 \le \bar{c}_{ai}$$

If this effective generation system model of System A is required to generate the availability and frequency of only the negative margins in System A, only those capacity outages in excess of the operating reserve in System A will be required. For this "0" is always less than or equal to $\bar{c}_{ai}$ and equations 13 and 14 simplify to

$$A_0 = \sum_{L,K} (A_{a(\ell)} - A_{a(\ell+1)})(A_{ab} \cdot A_{b(k)} + \bar{A}_{ab}) \qquad (15)$$

and $f_0 = \sum_{L,K} [(f_{a(\ell)} - f_{a(\ell+1)})(A_{ab} \cdot A_{b(k)} + \bar{A}_{ab})$

$+ (A_{a(\ell)} - A_{a(\ell+1)})(f_{b(k)} + \lambda_{ab}(1 - A_{b(k)}))A_{ab}] \qquad (16)$

Only a limited portion of the effective generation system model is required to determine the availability and the frequency of any particular cumulative negative margin state. The steps involved in determining the availability and the frequency of the failure state are outlined below.

Let

L_{ap} be the load levels in System A, $p = 1,2,3...na$, $L_{a1} > L_{a2} > ... > L_{ana}$

L_{bq} be the load levels in System B, $q = 1,2,3...nb$, $L_{b1} > L_{b2} > ... > L_{bnb}$

v be a small positive value (say 0.0000001)

IC_a, IC_b be the installed capacities on systems A and B respectively

and

L_{ao}, L_{bo} be the low load levels in systems A and B respectively

Step #1

$$R_{bo} = IC_b - L_{bo}$$
$$R_a = IC_a - L_{a1}$$

With these values of the operating reserves, the availability and the frequency of the effective capacity outages equal to or greater than $[(IC_a - L_{ap} + v), p = 1,2,3...na]$ are calculated.

Step #2

$$R_{bq} = IC_b - L_{bq}, q = 1,2,3...nb$$

For each of these operating reserves in System B, the availability and frequency of the effective capacity outages equal to or greater than $[(IC_a - L_{ap} + v), p = 1,2,3...na]$ are computed. The availabilities can be calculated directly using equation 15 but to compute the frequencies a term, to take care of the load transitions in System B must be added to equation 16. The complete expression for frequency is given by

$$f_{\bar{c}p(q)} = \sum_{L,K} [(f_{a(\ell)} - f_{a(\ell+1)})(A_{ab} \cdot A_{b(k)} + \bar{A}_{ab})$

$+ (A_{a(\ell)} - A_{a(\ell+1)})(f_{b(k)} + \lambda_{ab}(1 - A_{b(k)}))A_{ab}]$

$+ (A_{\bar{c}p(o)} - A_{\bar{c}p(q)})/e$

where $A_{\bar{c}p(o)}$ = The availability of the effective capacity outage equal to or greater than $(IC_a - L_{ap} + v)$ given the low load state in System B.

127

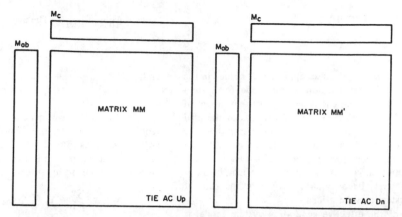

Fig. 3. The Effective Margin States Matrices of System A Connected to Systems B and C

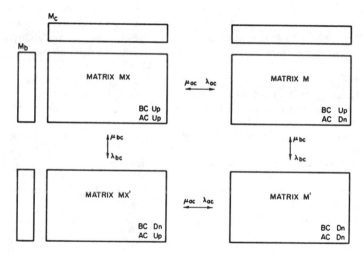

Fig. 4. The Effective Margin States Matrices for System B

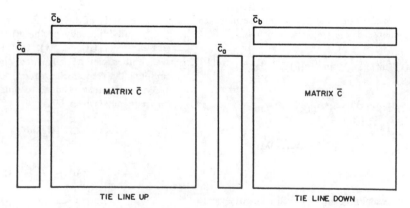

Fig. 5. The Effective Generation System Model of System A Connected to System B

$A_{\bar{c}p(q)}$ = The availability of the effective capacity outage equal to or greater than $(IC_a-L_{ap}+v)$ given that the load in System B is equal to L_{bq}.

$A_{L_{bq}}$ = The availability of the load level L_{bq}.

$f_{\bar{c}p(q)}$ = The frequency of encountering an effective capacity outage equal to or greater than $(IC_a-L_{ap}+v)$ given only two load levels in system B, i.e., L_{bq} and L_{bo}.

and e = The exposure factor in System B

Then

$$f_{(N\leq-v)q} = \sum_p A_{Lap} [f_{\bar{c}p(q)} + A_{\bar{c}p(q)} (\lambda_{-L}-\lambda_{+L})]$$

where $f_{(N\leq-v)q}$ = The frequency of encountering an effective margin equal to or less than "-v" given the load level in System B as L_{bq}

A_{Lap} = The availability of the load level L_{ap}

and $\lambda_{+L}, \lambda_{-L}$ = The transitions to higher and lower load levels respectively in System A

Finally the frequency of encountering an effective margin equal to or less than "-v" is given by

$$f_{(N\leq-v)} = \sum_{q=1}^{nb} A_{Lbq} \cdot f_{(N\leq-v)q}$$

and availability

$$A_{(N\leq-v)} = \sum_{q=1}^{nb} A_{Lbq} \cdot A_{(N\leq-v)q}$$

when

$$A_{(N\leq-v)q} = \sum_{p=1}^{na} A_{Lap} \cdot A_{\bar{c}p(q)}$$

The results obtained by this approach are identical with those obtained using the margin states approach.

A FOUR-STATE MODEL

FOR ESTIMATION OF

OUTAGE RISK FOR UNITS IN PEAKING SERVICE

Report of the IEEE Task Group on Models for Peaking Service Units,

Application of Probability Methods Subcommittee

Abstract—The conditional probability of a unit not being available given a demand occasion is developed using a four-state Markov model. The authoring Task Group recommends that the developed conditional probability be used in place of Forced Outage Rate for capacity planning studies, especially when the application is for units in peaking or cycling service. The Forced Outage Rate parameter has been recognized to be unsuitable as a measure of outage risk when unit annual service hours are low. This paper describes other models which

Paper 71 TP 90-PWR, recommended and approved by the Power System Engineering Society for presentation at the IEEE Winter Power Meeting, New York, N.Y., January 31-February 5, 1971. Manuscript submitted September 21, 1970; made available for printing November 17, 1970.

Task Group Members: A. B. Calsetta, Long Island Lighting Company, Hicksville, New York, Chairman; P. F. Albrecht, General Electric Company, Schenectady, New York; V. M. Cook, Consultant, Scarsdale, New York; R. J. Ringlee, Power Technologies, Inc., Schenectady, New York; J. P. Whooley, Public Service Electric and Gas Company, Newark, New Jersey; all Members IEEE.

have been used or offered in addressing the problem. After reviewing available models and known associated work, the Task Group developed the recommended four-state model.

INTRODUCTION

Outage statistics on generating units in base load, cycling, and peaking operation have been collected for a number of years and reported by the Edison Electric Institute.[1] Outage information from base load operation has been applied in generation capacity planning studies for a number of years for base load units. A parameter commonly used to assess the risk of unit forced unavailability for service is the Forced Outage Rate, FOR. This parameter is the ratio of the forced outage hours to the sum of forced outage hours plus service hours. For base load operation, this parameter provides a good estimate of the risk of a unit not being available at any time during a span between successive periods of scheduled maintenance. As such, it is a good

Reprinted from *IEEE Trans. Power Apparatus Syst.,* vol. PAS-91, no. 2, pp. 618–627, Mar./Apr. 1972.

measure to use in generation capacity planning studies involving base load type applications.

Difficulties have been encountered in applying the parameter FOR to statistics derived from peaking operation and cycling operation. In these instances, periods of service are frequently interrupted by periods of economy shutdown such that relatively small numbers of service hours are accumulated in the course of a year. The frequent startup and shutdown subjects the unit to additional starting stresses compared to units in base load operation. This additional starting stress has been recognized and reported as failure-to-start risk for gas turbines and diesel units, for example.[2] In addition, FOR makes use of two sets of statistics, service hours and forced outage hours, collected over different bases. Forced outage hours are counted from the time the unit enters the forced outage state and accumulated until the unit is ready for service or has been placed back in service; by contrast, service hours count only the time that the unit was synchronized to the bus. Thus, it is possible to greatly affect the Forced Outage Rate by the duty cycle required of the generating unit. Further, forced outage rate does not measure the probability that a peaking unit is ready to serve load for a duty cycle composed of successive periods of demand and reserve shutdown. Intuitively, a better measure of the risk of not being available to serve during a demand period would be the ratio of the forced outage hours during the demand period divided by the sum of the forced outage hours during demand period plus the service hours during demand period. To this end, a four-state model is offered to account for both the effects of the demand or duty cycle upon the unit and the failure and repair processes for units operating in cycles of in-service and reserve shutdown.

PREVIOUS MODELS

Two-State Model (Classical Model)

If demand is continuous, then the Run-Fail-Repair process can be represented by a two-state model, with steady state probabilities

$$P_{up} = \frac{m_s}{m_s + r} \qquad P_{down} = \frac{r}{m_s + r} \qquad (1)$$

where

P_{up} = probability of up state (in service)

P_{down} = probability of down state (forced outage)

m_s = mean service time per forced outage

r = mean forced outage time per forced outage

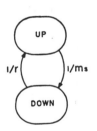

If the operation of one or several units are observed for some period of time, then estimates of these quantities can be formed as

$$\hat{m}_s = \frac{SH}{N} \qquad \hat{r} = \frac{FOH}{N} \qquad (2)$$

$$\hat{P}_{up} = \frac{\hat{m}_s}{\hat{m}_s + \hat{r}} = \frac{SH}{SH + FOH} \qquad (3)$$

$$\hat{P}_{down} = \frac{\hat{r}}{\hat{m}_s + \hat{r}} = \frac{FOH}{SH + FOH} \qquad (4)$$

where

FOH = forced outage hours during period

SH = service hours during period

N = number of forced outage occurrences

It can be seen that the number of forced outages, N, cancels out in the estimate of P_{up} and P_{down}. Thus, the state probabilities can be estimated directly from the corresponding observed state residence times during the same period.

The sample statistic FOH/(SH + FOH) is commonly called the "Forced Outage Rate." Equation (4) therefore states that Forced Outage Rate is an estimate of the probability of outage at any randomly selected instant of time.

It is worth noting that the *statistic*, Forced Outage Rate, can be calculated from operating data on any type of unit in any kind of service, and as a statistic it is well defined. However, whether this statistic, FOR, is a good estimate of the probability of forced outage will depend on the type of operation involved.

Eight-State Markov Model

The fact that FOR does not adequately estimate outage risk for units in peaking, cycling, standby, or other non-constant demand conditions has been recognized for several years. In 1968, R. J. Ringlee[3] developed a Markov Chain model employing eight states to "account for the demand and reserve periods and the unit service and reserve periods as well as the starting and running failure states."

Four pairs of states described equivalent unit conditions during periods of either (1) unit demanded for load or (2) unit not required for load. The four unit conditions included were:

Unit ready for operation
Unit synchronized
Unit failed to start
Unit failed during running

The model required separate estimates of repair time after a failure to start and after a failure during running. This was considered to be a potential problem area since no such repair statistics were known to be available. This model was submitted to the I.E.E.E. Application of Probability Methods Subcommittee.

The following conclusions were drawn from the eight-state model:
a) "Forced Outage Rate" as presently defined is not a suitable measure of the probability of a unit being on forced outage during a demand period.
b) Suitable "probability of forced outage" expressions may be constructed of terms derived frm observable outage data and from forecasted "duty."
c) Outage data reports for peaking and standby units ought to be made in terms of basic data such as mean-time-to-failure, mean-time-to-repair, and probability of unsuccessful start.

Modified Two-State Model

In 1969, EEI published[1], for the first time, specific data for gas turbines and other types of generation designed for short operating cycles. This report incorporated some of the recommendations regarding publication of basic data in addition to FOR. The 1969 report included mean-time-to-failure and number of forced outage occurrences. However, the report did not give sufficient data to permit application of the eight-state model.

The EEI report again focused attention on the need for methods for estimating outage risk from available data. One approach to the problem was described by P.F. Albrecht et al[2]. In much the same way that maintenance time had always been excluded in forced outage risk calculation for the two-state model, they formulated a model in which reserve shutdown time was also excluded. Basically the two-state model was applied, but with both m_s and r measured in *demand* hours. This required that forced outage hours, in *calendar* time, be reduced to include only forced outage hours during demand time. The resulting expression for risk of forced outage was

$$\hat{P}_{mod} = \frac{FOH(D/24)}{SH + FOH(D/24)} \tag{5}$$

This model was based on the assumption of a 24-hour demand cycle, i.e., D hours demand every 24 hours. The model is easily generalized to cover any other demand cycle as

$$\hat{P}_{mod} = \frac{FOH(K)}{SH + FOH(K)} \tag{6}$$

where $K = D/(T + D)$ and T is the reserve shutdown time between demand periods.

Modified FOR Model

Using an approach similar to that for the Modified Two-State Model, A. M. Adamson[4] suggested that FOR be defined in terms of service *days* rather than *hours* when applied to units in peaking service. The equation given was

$$FOR = \frac{FOD}{FOD + SD} \tag{7}$$

where

FOD = forced outage days

SD = service days

No assumption regarding a demand cycle is made with this model.

FOUR-STATE MODEL DEVELOPMENT

The EEI Equipment Availability Task Force, under Equipment Availability Research Project RP76, was interested in improved definitions and procedures. Goal No. 2 of the research project was "to seek general agreement on definitions and calculation procedures" and a specific area of study (No. 10) was[5] "Analysis Techniques For Different Types of Units (nuclear, peaking, etc.)."

In July, 1970, the IEEE Application of Probability Methods Subcommittee appointed a Task Group on Models for Peaking Service Units to study the problem and, as possible, to develop a more appropriate treatment of peaking service units. It was not clear at the outset whether a model could be developed which would be compatible with existing capacity planning study methods or whether a new study approach would be necessary. Model development was preferred in order to minimize changing established methods of study and data collection, and to avoid methods which might be prohibitive regarding computer time and memory requirements.

In selecting a model to be used for peaking service units it was desirable that the model should:

1. Have application using currently available data.
2. Incorporate starting failure experience.
3. Provide results which could be used directly with established methods for performing reliability calculations.
4. Be relatively easy to use.

An analysis of known work in progress on treatment of peaking service units was made by the Task Group. This analysis resulted in selection of the four-state model as described in the following. This model utilizes available operating statistics to produce a parameter to replace FOR.

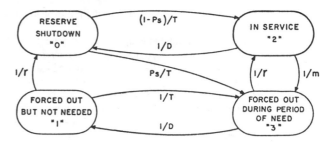

Figure 1.

Four-State Model

States of the model are:

"0" Reserve shutdown
"1" Forced out but not needed for load
"2" In service when needed for load
"3" Forced out during a period of need.

The symbology used in Figure 1 and throughout this paper is:

T = average reserve shutdown time between periods of need, exclusive of periods for maintenance or other planned unavailability (hrs.).

D = average in-service time per occasion of demand (hrs.).

m = average in-service time between occasions of forced outage, excluding forced outages as a result of failure to start (hrs.).

r = average repair time per forced outage occurrence (hrs.).

P_s = probability of a starting failure resulting in inability to serve load during all or part of a demand period. Repeated attempts to start during one demand period should not be interpreted as more than one failure to start.

With the exception of P_s, the above quantities can be calculated from regularly reported data.[1] This will be demonstrated in the section of this paper on Applications. The P_s parameter must at present be estimated since industry data is currently collected only on total starts. However, companies with peaking service units often keep records on the proportion of forced outages resulting as a consequence of failure to start.

As a Markov process[6] the equation for the model is

$$
\begin{vmatrix} \dot{P}_0(t) \\ \dot{P}_1(t) \\ \dot{P}_2(t) \\ \dot{P}_3(t) \end{vmatrix}
=
\begin{vmatrix}
-\frac{1}{T} & \frac{1}{r} & \frac{1}{D} & 0 \\
0 & -\left(\frac{1}{r}+\frac{1}{T}\right) & 0 & \frac{1}{D} \\
\frac{(1-P_s)}{T} & 0 & -\left(\frac{1}{m}+\frac{1}{D}\right) & \frac{1}{r} \\
\frac{P_s}{T} & \frac{1}{T} & \frac{1}{m} & -\left(\frac{1}{r}+\frac{1}{D}\right)
\end{vmatrix}
\begin{vmatrix} P_0(t) \\ P_1(t) \\ P_2(t) \\ P_3(t) \end{vmatrix}
\tag{8}
$$

or

$$\underline{\dot{P}}(t) = \bigwedge \underline{P}(t)$$

For the steady-state solution, $\underline{\dot{P}} = 0$ and we use the symbology

$$P_n = \lim_{t \to \infty} P_n(t)$$

to denote the particular solution for long-term probability of each state. Noting that

$$P_0 + P_1 + P_2 + P_3 = 1$$

the solution of $\Lambda P = 0$ yields

$$P_0 = T/r \, (1/r + 1/T + i/m + 1/D) / \Delta \qquad (9)$$

$$P_1 = (1/m + P_s/D) / \Delta \qquad (10)$$

$$P_2 = (1/r)[D(1/r + 1/T) + 1 - P_s] / \Delta \qquad (11)$$

$$P_3 = D(1/r + 1/T)(1/m + P_s/D) / \Delta \qquad (12)$$

where

$$\begin{aligned}
\Delta = \ & T/r(1/r + 1/T + 1/m + 1/D) \\
& + (1/m + P_s/D)[1 + D(1/r + 1/T)] \\
& + (1/r)[D(1/r + 1/T) + 1 - P_s]
\end{aligned} \qquad (13)$$

We see that in terms of this model Forced Outage Rate estimates the ratio:

$$(P_1 + P_3)/(P_1 + P_2 + P_3) \qquad (14)$$

However, the probability given in expression (14) does not measure the risk that a peaking unit is unavailable to serve load given a demand occasion. The four-state model does permit determination of this measure as given in equation (15):

P = probability that the unit will not be available given a demand occasion.

$$= P_3/(P_2 + P_3) \qquad (15)$$

The authoring Task Group submits that an estimate of the conditional probability, P, should be used in place of Forced Outage Rate, for studies of system reliability. Note that the qualification "for peaking service units" is not made in the foregoing statement. Although the work of the Task Group was initially directed towards developing a model for use with peaking service units, it can be seen from the model that expression (14) tends to expression (15) as service hours become larger, i.e., as $P_1 \rightarrow 0$.

A discussion of the philosophy in applying the probability, P, for system studies follows in the section on Applications. An approximate equivalent method for calculating P is also given.

APPLICATIONS

Two objects to be explored in this discussion are:
a) What should be the measure used for peaking type operation for outage risk measures in capacity planning studies? and
b) How should the interpretation of the risk indices used in capacity planning studies be changed, if at all, to recognize peaking type operation?

Concerning techniques for use of industry statistics to develop risk measures for outages for peaking operation, let us first consider the way in which such outage information has been used. Consider the risk of overlapping independent outages of several generators: for base load type operation it has been the practice to estimate risk of independent outage of generating units by the Forced Outage Rate; this number has been treated freely as a probability number. The risk

*The model as originally proposed by G. Calabrese[7] considered an array of hourly loads

of overlapping outage of two generators, then, is measured by the product of the two outage probabilities. This measure applies at any randomly selected time within the span of time between scheduled maintenance periods for the two generating units. Coupled with the average duration of forced outage periods, the probability of capacity outage, and the frequency of occurrence of this capacity outage have been used in the loss of load probability method[7], the loss of capacity method[8], and in the frequency and duration method[9]. All three methods assume the unit is at all times either available or on forced outage. None recognizes peaking operation in which the unit is shut down during light load periods and restarted when the demand requires.

The loss of load probability method, for example, develops a risk index which measures the expected time in days in which the generation available will be insufficient to meet the demand.* One full day is counted for each event in which capacity was deficient; an array of daily peak loads is used in the model. Since peaking operation does not demand continuous duty of the generating unit and since duty cycle does affect the risk of the unit not being available, it seems appropriate to suggest that attention be focused on the demand period of the day and risk measurements assessed during this period. For example, for the loss of load probability method, it seems appropriate to consider the demand period and to count again "one day" if a capacity deficiency occurs during that demand period. This is not to say that a unit could not be regarded as being potentially available during a period of low demand but rather to suggest that the method used to measure the risk of the unit not being available be sensitive to recognize the duty cycle. With this slight change in interpretation the loss of load probability index will represent the same numerical value but be interpreted as applicable to the demand period. In this case, the proper measure to use for the risk of a unit in peaking operation being unavailable during the demand period is the conditional probability that the unit is not available given that a demand period has occurred. This is the value, P, given in equation (16) which results from substitution of P_2 and P_3 from equations (11) and (12) into (15).

$$P = \frac{\left(\frac{1}{r} + \frac{1}{T}\right)\left(\frac{D}{m} + P_s\right)}{\frac{1}{r}\left[D\left(\frac{1}{r} + \frac{1}{T}\right) + 1\right] + \frac{D}{m}\left(\frac{1}{r} + \frac{1}{T}\right) + \frac{P_s}{T}} \qquad (16)$$

While P may be calculated from the model as in equation (16), a more convenient estimate may be obtained as follows:
Over a long period of time, P.H., the estimate of P_2 would be obtained by the ratio of service hours, S.H., to available hours + forced outage hours AH + FOH

$$\hat{P}_2 = SH/(AH + FOH) \qquad (17)$$

Note we are measuring events given that the unit is not on maintenance outage or planned outage.
The probability of being in the forced outage state $(P_1 + P_3)$ would be estimated by the ratio of Forced Outage Hours, FOH, to available hours plus forced outage hours:

$$(\hat{P}_1 + \hat{P}_3) = FOH/(AH + FOH) \qquad (18)$$

The risk of being unavailable when needed is given by equation (15). In order to use equations (17) and (18) in estimating P, let us define a factor f which expresses the ratio $P_3/(P_1 + P_3)$, then substituting f into equation (15) yields

$$P = f(P_1 + P_3)/[P_2 + f(P_1 + P_3)] \qquad (19)$$

An estimate of P may now be given in terms of SH and FOH using equations (17) and (18).

133

$$\hat{P} = f(FOH)/[SH + f(FOH)] \qquad (20)$$

The factor f has the interesting property that it is a weighting factor on forced outage hours to reflect the cumulative forced outage hours occurring during periods of demand. Depending upon duty cycle and outage duration, the factor f reflects that weighting to be applied to total forced outage hours to obtain the forced outage hours during the period of demand.

There is no easy way to obtain the number of forced outage hours occurring during the demand period from operating records but f may be expressed in terms of the Markov model.

$$f = (1/r + 1/T)/(1/D + 1/r + 1/T) \qquad (21)$$

Thus, a means is available for estimating the weighting by which forced outage hours should be reduced to reflect the effects of duty cycle and of duration of outage for peaking type operation.

An example utilizing the information on Figure 1 in the 1969 report of the EEI Prime Mover's Committee[1] is illustrated for the gas turbine installation. Referring to Figure 1 of the reference the service hours, SH, for this unit are 640.73 on an annual basis. The available hours for this unit are 6,403.54 on an annual basis. The total number of starts on the unit is assumed to be the number of economy outages, 34.2, plus the number of forced outages, 3.87, or 38.07. Let us assume zero starting failures, i.e., zero for P_s. The model parameters and results for this case are estimated as follows:

$$\hat{D} = \frac{(\text{Service Hours})}{(1 - P_s)(\text{Total Starts})} = \frac{640.73}{(1)(38.07)} = 16.8 \text{ hrs.}$$

$$\hat{D} + \hat{T} = \frac{(\text{Available Hours})}{(\text{Total Starts})} = \frac{6403.54}{38.07} = 168 \text{ hrs.}$$

$$\hat{r} = \frac{(\text{Forced Outage Hours})}{(\text{Number of Forced Outages})} = \frac{205.03}{3.87} = 53 \text{ hrs.}$$

$$\hat{m} = \frac{(\text{Service Hours})}{(\text{Number of Forced Outages}) - (P_s)(\text{Total Starts})}$$

$$= \frac{640.73}{3.87 - 0} = 166 \text{ hrs.}$$

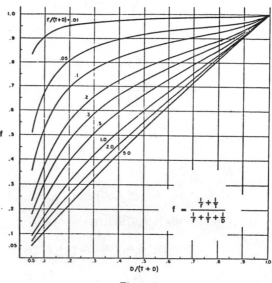

$$f = \frac{\frac{1}{r} + \frac{1}{T}}{\frac{1}{r} + \frac{1}{T} + \frac{1}{D}}$$

Figure 2.

A duty cycle of 16.8 hours every 168 hours and a mean repair time of 53 hours leads to a duty cycle $D/(T + D)$ equal to .10 and a ratio of repair to cycle time of .316. Referring to Figure 2 one reads a value of f approximately 0.3, thus, approximately 61.5 hours is the corrected forced outage hours during demand period estimated from the product of f times the forced outage hours. For the data and duty cycle such as are to be found in the report of EEI, Figure 1[1], one may determine a Forced Outage Rate of 24.24% and a probability of not being available when called of 8.765%.

The estimates of P and FOR for other values of P_s are given in Table I. Some concern may be raised about means for estimating the effects of change in duty cycle on the risk of the unit not being available when needed. For example, the EEI data, Figure 1, for gas turbines would indicate that a unit was called upon to serve about once a week on the average. Under this duty cycle, there is a little under 9% risk that the unit will not be available to serve at any randomly selected time during the demand period. Suppose one wished to know the risk of being unavailable to serve on a daily basis for a span of time of, say, 12 hours. Two options appear potentially available. One would be to search outage records for units exhibiting duty cycles corresponding to the type of application envisioned, the other would be to attempt to utilize the outage statistics and to modify them through the use of estimates of the parameters for the four-state

ALL NUMBERS IN PER CENT

P_s	P_0	P_1	P_2	P_3	FOR	P	f	$\hat{P}$
0.0	87.94	2.046	9.138	0.878	24.243	8.765	30.019	8.765
0.5	87.88	2.058	9.174	0.887	24.303	8.819	30.125	8.793
1.0	87.82	2.069	9.210	0.896	24.354	8.869	30.231	8.821
1.5	87.77	2.079	9.248	0.905	24.396	8.916	30.338	8.849
2.0	87.71	2.089	9.286	0.914	24.428	8.960	30.446	8.878
2.5	87.66	2.096	9.324	0.922	24.452	8.999	30.554	8.907
3.0	87.60	2.103	9.364	0.930	24.466	9.035	30.663	8.936
3.5	87.55	2.109	9.404	0.938	24.472	9.067	30.773	8.965
4.0	87.50	2.115	9.445	0.945	24.467	9.094	30.884	8.994
4.5	87.44	2.119	9.487	0.952	24.453	9.118	30.996	9.024
5.0	87.39	2.122	9.530	0.958	24.430	9.138	31.108	9.054
6.0	87.29	2.126	9.617	0.970	24.354	9.164	31.335	9.114
7.0	87.19	2.126	9.709	0.980	24.237	9.172	31.566	9.175
8.0	87.09	2.121	9.804	0.989	24.080	9.162	31.800	9.236
9.0	86.99	2.111	9.903	0.995	23.879	9.132	32.037	9.299
10.0	86.90	2.097	10.007	1.000	23.634	9.082	32.278	9.362

MODEL RESULTS FOR GAS TURBINE UNITS USING EEI DATA[1]

Table I

model in Figure 1 of this paper. In this instance, estimates of the mean time to failure, m, and the starting probability, P_s, would be necessary as would the mean repair time, r. The reader is cautioned not to utilize Figure 2 for different duty cycles than the collection of outage statistics actually show. Figure 2 is developed to assist in analyzing given sets of statistics.

As such, f from Figure 2 may be used in equation (20) only for the case of the duty cycle for which the outage statistics apply. If the reader wishes to apply a given set of statistics to a different duty cycle he is advised to estimate the parameters r, m, P_s, T and D and to substitute them into equation (16). The parameters r, m and P_s may well be affected by duty cycle. Extrapolation of outage statistics collected under conditions of one duty cycle should recognize changes in m and P_s due to changes in relative severity of stresses in starting and service. Changes in r may occur due to a change in emphasis in returning the unit to service as the duty cycle changes.

Discussion of the application of the four-state model has been focused upon peaking operation. The same model may also be applied to cycling operation and base load operation in which the periods of demand and reserve shutdown differ markedly from the duty cycle and duration of demand period for peaking operation. The distinction between cycling and base load operation is not sharp; base load operation may be regarded as the limit as the demand period becomes arbitrarily long. Note that by equation (21) the value of f tends to unity as the value of the demand period becomes arbitrarily large compared to the reserve shutdown period or the repair period. In this case, the conditional probability given by equation (16) tends to the ratio of r/(r+m) which is the analytic equivalent of the Forced Outage Rate under the condition of long demand period; as noted in equation (20), the estimate for the conditional probability as f approaches 1 becomes identical with the Forced Outage Rate.

Let us turn now to consideration of the application of the model to the loss of capacity and the frequency and duration models. The frequency of failure events shown in the four state model of Figure 1 is given by the product of the probability P_3 and the sum of the transition rates 1/r plus 1/D as illustrated in equation (22).
Frequency of failure events =

$$P_3 (1/r + 1/D) \qquad (22)$$

We claim that this value is exactly equal to the product of the probability of being unavailable when called, P, times the probability of a demand period D/(T + D) times the transition rate out of the failure state, 1/r + 1/D. This is illustrated in equation (23).
Frequency =

$$P\left(\frac{D}{T+D}\right)\left(\frac{1}{r}+\frac{1}{D}\right) \qquad (23)$$

The proof of this equivalence is rather short and is illustrated in equations (24), (25), and (26).
From Figure 1:

$$P_2 + P_3 = D/(T+D) \qquad (24)$$

$$\therefore \ P\left(\frac{D}{T+D}\right) = \left(\frac{P_3}{P_2+P_3}\right)\left(\frac{D}{T+D}\right) = P_3 \qquad (25)$$

$$\therefore \ P\left(\frac{D}{T+D}\right)\left(\frac{1}{r}+\frac{1}{D}\right) \equiv P_3\left(\frac{1}{r}+\frac{1}{D}\right) \qquad (26)$$

Accordingly, the equivalent failure probability and repair time for the peaking model for the frequency and duration and loss of capacity models are P and r.

CONCLUSIONS

The four-state model proposed addresses itself to a solution of an industry problem of representation of outage risk for units in peaking service. It is recommended that the probability of a unit not being available given a demand occasion, P, be used in place of the Forced Outage Rate in the methods commonly used for capacity planning studies. It is noted that the conditional probability developed has wider application to units in cycling and base load service.

REFERENCES

[1] EEI Prime Mover Committee "Report on Equipment Availability for the Nine-Year Period 1960-1968", EEI Publication No. 69-33, September 1969.
[2] P. F. Albrecht, W. D. Marsh, F. H. Kindl, "Gas Turbines Require Different Outage Criteria", Electrical World, April 27, 1970, pp. 38-40.
[3] R. J. Ringlee, "Proposed Availability Models for Peaking and Standby Generation", June 14, 1968, (unpublished).
[4] A. M. Adamson, "Gas Turbine and Diesel Forced Outage Rates and Their Application to Reliability Calculations", IEEE-ASME Joint Power Generation Conference, September 27-30, 1970.
[5] R. I. Smith, letter to Members of the EEI Equipment Availability Task Force, September 28, 1966.
[6] J. G. Kenny, J. L. Snell, *Finite Markov Chains* (book), D. VanNostrand Company, Inc., Princeton, New Jersey, 1960.
[7] G. Calabrese, "Determination of Reserve Capacity by the Probability Method", AIEE Transactions, Vol. 69, Part II, 1950.
[8] H. Halperin, H. Adler, "Determination of Reserve Generating Capability", AIEE Transactions, Vol. 77, 1958, pp. 530-544.
[9] R. J. Ringlee, A. J. Wood, "Frequency and Duration Methods for Power System Reliability Calculations, Part II, Demand Model and Capacity Reserve Model", IEEE Transactions, Vol. PAS-88, 1969, pp. 375-388.

Discussion

A. M. Adamson (Long Island Lighting Company, Hicksville, N.Y.): The subcommittee should be commended for their development of a model which I believe accurately portrays generating unit outage risk. My discussion deals with the application of this model and the performance data used.

There has been unquestionably a need for better methods for estimating the probability of peaking unit unavailability, particularly, since these units are increasingly forming a significant portion of system capability. I have seen gas turbine outage risks, for example, used for planning studies ranging from 1 to 25%. I published a paper on this subject in 1970 (Reference 4) where I suggested a method and some typical outage risks.

The subcommittee indicates in its paper that the four-state model was developed based on currently available data. However, the example in the paper, based on data from the EEI Prime Mover's Committee (Reference 1 in the paper), requires that the two quantities be assumed: the total number of starts per year and the starting reliability, P_s, of the unit.

The authors assumed that the total number of starts is equal to the number of forced and economy outages, which are published quantities. This appears to be a valid assumption, until we saw that the resultant average in-service time D is 16.8 hours. Intuitively, this seems to me too much for a gas turbine peaking unit, particularly in light of Long Island Lighting Company's experience of 3.5 hours. If this lower value of D were used instead of 16.8 hours, the resulting probability, P, would decrease to a risk more in line with the one I calculated in my paper.

Table I in the paper shows values of P based on starting reliabilities ranging from 90 to 100%, partly because this quantity is also unavailable.

If the four-state model is to be used for determining outage risk for planning studies, it would seem prudent that the EEI collect additional data required to more accurately determine the number of starts and starting reliability of peaking service units.

Because of the high outage risk of units when they are new, many power systems use separate outage rates in planning studies for representing immature and mature service states. Data for this is unavailable from EEI, but my studies using Long Island Lighting data

Manuscript received February 18, 1971.

shows that outage risks for the first three years of operation of gas turbines is about double that of a mature unit or about 10% for immature and 5% for mature units.

S. B. Bram and **R. C. Chan** (Consolidated Edison Company of New York, Inc., New York, N.Y. 10003): The members of the IEEE Task Group on Models for Peaking Service Units are to be commended for their excellent exposition on existing models for use in estimating the outage risk of peaking units and for the development of their own four-state model for this purpose. With regard to the model which has been developed and its application to practical problems involving peaking units, we offer the following comments for consideration.

The four-state model is a plausible and realistic representation for peaking units in a generation system. The Markov equations for the model, more elegantly called Chapman-Kolmogorov equations, are correctly written as a set of simultaneous linear equations. The equations in (8) contain no product of the state probabilities, for example, a product such as $P_1(t)P_2(t)$. Clearly if the system happens to be in state "1", it cannot also be in state "2" simultaneously. In stochastic processes, the definitions of system states and system state probabilities preclude the forming of products of state probabilities, when these state probabilities pertain to the same specific system. It is the linearity of equation (8) that permits its steady-state solution by matrix method.

In describing the four-state model as a Markov process, the Task Group must assume exponentially distributed times of residence in each state. Although justified, such assumption is not necessary. The steady-state results of the Markove process are identical to those of the frequency and duration methods, which specify no restriction on the times of residence in the states.

The Task Group did not take advantage of the fundamental properties of the model to relate its transition rates to statistics. We would like to explain how this might be accomplished, or rather how operating records may be kept in order to yield conveniently the transition rates necessary to define the model.

Consider the state-space diagram in Figure 1. The four-state model is specified completely by the five parameters, T, D, m, r and P_s. It should be noted that there is no transition from state "0" to state "1", since it is reasonably assumed that the unit cannot fail when it is idle. For convenience in terms of frequency and duration methods, let $t = 1/T$, $d = 1/D$, $\lambda = 1/m$ and $\mu = 1/r$. These are transition rates, and have a physical dimension of 1/time. The frequency of occurrence of state "2" is

$$P_2 (d + \lambda) = P_2 d + P_2 \lambda$$

which is a sum of two frequencies: $P_2 d$ is the frequency of transition from "2" to "0" and $P_2 \lambda$ is the frequency of transition from "2" to "3". Within a long period of time, clearly the ratio $P_2 d : P_2 \lambda = d : \lambda = m : D$ is the ratio of the number of transitions from "2" to "0" to the number of transitions from "2" to "3". The duration of state "2" is given as follows:

$$\text{duration of "2"} = \frac{1}{d + \lambda} = \frac{m D}{m + D}$$

Numerical data on the duration of "2" and the ratio of the transitions "2" to "0" to the transitions "2" to "3" enable us to solve for λ and d, or m and D. In practice, records of state "2" are kept for a long period of time. The cumulative times of residence in "2" are recorded. Within this period, if N_0^2 is the number of transitions from "2" to "0" and N_3^2 is the number of transitions from "2" to "3", then the total number of transitions out of "2" is $N_0^2 + N_3^2$, which is also the number of occurrences of state "2". Thus

$$\text{duration of "2"} = \frac{\sum \text{times of residence}}{N_0^2 + N_3^2}$$

$$= \frac{m D}{m + D} = \frac{1}{d + \lambda}$$

and

$$\frac{N_0^2}{N_3^2} = \frac{d}{\lambda} = \frac{m}{D}$$

We can now readily solve for m and D, two of the five parameters that define the model. In a similar manner, within a long period of time, knowledge of the total time of residence in state "0" and of the numbers of transitions from "0" to "2" and to "3" respectively, enables us to calculate P_s and T. If N_2^0 and N_3^0 are the numbers of transitions from "0" to "2" and to "3" respectively, then

Manuscript received February 19, 1971.

$$\frac{\text{total time in "0"}}{N_2^0 + N_3^0} = \frac{1}{(1 - P_s) t + P_s t}$$

$$= \frac{1}{t} = T$$

and

$$\frac{N_2^0}{N_3^0} = \frac{(1 - P_s)}{P_s}$$

P_s and T can now be found very simply. To find r, we consider the two forced outage states, "1" and "2". The frequency of forced outages, regardless of whether the unit is needed (as in "3") or not (as in "1"), and outage duration are given by:

$$\mu P_1 + \mu P_3 = \mu (P_1 + P_3)$$

$$= \text{frequency of forced outage}$$

$$\frac{1}{\mu} = r = \text{duration of outage}$$

$$= \frac{\text{total forced outage time}}{\text{number of outages}}$$

Existing statistics undoubtedly contains the total forced outage time and the number of such forced outages.

Thus without reference to any information or data on state "3", we have already calculated the five parameters, T, D, m, r and P_s, which define the four-state system. We could then use equations (14), (15) or (16) to get the desired probabilities. If we so desire, additional statistical information on state "3" will serve as an independent check on the calculations.

Having related the use of statistics to the model parameters through frequency and duration methods, which are exact and theoretical correct, we can provide an illustration by using the numerical example given in the paper. Using the methods and discussion above, our calculations proceed as follows. Total service hours in "2" is 640.73 hours. The number of economy outages, i.e. transitions from "2" to "0", is 34.2. The number of forced outages, i.e. transitions from "2" to "3", is 3.87. Total forced outage time is 205.03 hours. Our results are:

$$\frac{640.73}{34.2 + 3.87} = \frac{\hat{m} \hat{D}}{\hat{m} + \hat{D}} \quad \text{and} \quad (A)$$

$$\frac{34.2}{3.87} = \frac{\hat{m}}{\hat{D}}$$

$$\hat{D} = \frac{640.73}{34.2} = 18.8 \text{ hours}$$

$$\hat{m} = \frac{640.73}{3.87} = 166 \text{ hours}$$

$$\hat{r} = \frac{205.03}{3.87} = 53 \text{ hours}$$

Our value for $\hat{r}$ agrees perfectly with the $\hat{r}$ calculated by the Task Group. Our $\hat{m}$ is numerically the same as the $\hat{m}$ in their calculation, but not for the same reason. The values we have for $\hat{m}$ and $\hat{D}$ are results of solving the two equations in (A), and are independent of the assumption $P_s = 0$. Our $\hat{D}$ is different from 16.8 hours by about 2 hours, a considerable amount of time. Shutting down a 100 MW machine by two hours too soon could mean 200 MW-HR of needed energy not supplied to the system. We have found that the values for T, $\hat{m}$ and $\hat{D}$ calculated by the Task Group are not compatible with frequency and duration considerations, on which the four-state model is based.

Even with $P_s = 0$, the number of starts on the units, i.e. the number of transitions out of state "0", cannot be estimated correctly as $34.2 + 3.87 = 38.07$. Because there is a transition path from "3" back into "2", the number of transitions out of "0" is not necessarily the same as the total number of transitions out of state "2", or even approximately so. Simple path tracing on a state-space diagram will verify this assertion.

Without the correct number of starts on the units, we do not believe that an accurate calculation of T can be made. We need the number of starts on the unit, and the relative numbers of entry into "2" and "3" to complete the picture for the four-state model. This data, if not already being recorded, should be included in future statistics.

REFERENCES

[a] R. C. Chan and S. B. Bram, "An Examination of Probability Methods in the Calculation of Power System Performance," Presented at the 1970 IEEE Summer Power Meeting, Paper No. 70 CP 512 - PWR

[b] J. D. Hall, R. J. Ringlee and A. J. Wood, "Frequency and Duration Methods for Power System Reliability Calculations, Part I: Generation System Model," IEEE Trans. Power Apparatus and Systems, Vol. PAS-87, pp 1787-1796, September 1968.

Murty P. Bhavaraju (Public Service Electric and Gas Company, Newark, N.J.): We compliment the task group for developing an extremely interesting model to evaluate forced outage rate applicable to the period of demand. The PJM companies use an approach similar to that described in Reference 4 of the paper using days to calculate outage rates.

We have attempted to use the proposed method for calculating outage rates for our gas turbines. We had some difficulty due to the number of economy outages not being recorded in the data. It appears desirable to keep a separate record of number of starts on these units.

We had some 140-MW peaking units which had partial outages. We assumed that the unit had a demand during those periods of partial outages. Therefore, we added the equivalent partial forced outage hours (EPFOH) to the numerator of Equation (20) in the paper to calculate Equivalent Forced Outage Rate.

$$P = \frac{f(FOH) + EPFOH}{SH + f(FOH)}$$

We would appreciate the authors' comments on this extension of their equation.

Manuscript received February 19, 1971.

M. W. Mason, Jr. (General Electric Company, Schenectady, N.Y. 12305): This is an important paper for it serves to bring into focus some of the prevalent although sometimes conflicting points of view in assessing outage risks in general and peaking units in particular.

A major step forward could be achieved if agreement could be reached on the definition and method of measurement of forced unavailability. If this turned out to be defined as "P = the probability of a unit not being available given a demand occasion", as given in the papers conclusion then the task lessens to optimizing the measurement aspects. It is important to defer the development of the math models and measurement schemes until agreement on the precise definition of the dependent variable is arrived at.

The Markovian model seems to have omitted an important transition function and the corresponding occupancy state, i.e. the case where a unit placed in a "Reserve Shutdown" state transcends to the "Forced Unavailability State". This often occurs when defective conditions arise which must be corrected before the unit is permitted back into service. It seems possible that units "in service" could revert to the "Forced out, but not needed State" in certain circumstances. An additional transition function for this perhaps should be considered. A similar function apparently is needed to permit transition from the "Forced out during period of need" to the "Reserve shutdown State". The transition function from the former state to the "in service state" should include the parameter P_S. It may also be worthwhile to consider replacing the transition functions specified with empirically determined ratios of the actual frequencies of transitions from a given state to all other allowed transition states.

Manuscript received February 22, 1971.

L. L. Garver, and **W. D. Marsh** (General Electric Company, Schenectady, N.Y. 12305): The task force has prepared a solid mathematical basis for overcoming the difficulties involved in using statistics from past performance to estimate forced outage probabilities for peaking service equipment. We would like to point out the close parallel between the results of this four-state development and the modified two-state model of reference 2.

The four-state model and the modified two-state model use the same basic equation for calculating the probability of forced outage during a period of need, compare equation (6) and (20). Each equation includes a factor which represents the fraction of forced outage hours during the period of demand, i.e, the lost demand time. The adjustment factor K in (6) and the f in (20) are closely related as may be seen by the following:

$$K = D/(T + D) \tag{6a}$$

$$f = \frac{D(1 + T/r)}{T + D(1 + T/r)} \tag{21a}$$

Equation (21a) was obtained by rearranging terms in (21).

Note the similarity of equation (6a) and (21a). The difference arises because the two-state model does not account for the fact that every forced outage starts coincident with or during a demand period. This results in underestimating the adjustment factor. The error is small if the repair time is large relative to the average shutdown time (r>>T). Conversely, if the repair time is very small, the f factor approaches unity. Figure 2 shows the relation between f and the parameters T, D, and r. A diagonal line approximated by the "50" line of Figure 2 represents the K = D/(T + D) assumption used in the two-state model.

In reference 2 an example was shown using the values D = 4 hours and T + D = 24 hours. A repair time of r = 53 hours was calculated from EEI data. For these parameters, the four-state model gives f = 0.21 as indicated on Figure 2, whereas the two-state model uses K = 0.167. The resulting outage probabilities, using EEI data of FOH = 205 and SH = 641 hours, are:

Four-state model	P = 6.3%
Modified two-state model	P = 5.05%

The approximate nature of the two-state model was recognized when (6) was being developed, and a correction factor was determined to account for the fact that every outage starts during a demand period. However, in the interest of simplicity, it was decided to omit the correction factor from reference 2. This gives a lower value of outage risk. But it was also assumed that all service days occur consecutively, and this gives a higher value of outage risk. Hence the two approximations tend to balance each other.

Manuscript received February 22, 1971.

A. V. Jain and **Roy Billinton** (University of Saskatchewan, Saskatoon, Canada): The authors are to be congratulated on a very interesting and timely publication. We completely agree with the authors that the conventional definition of Forced Outage Rate does not provide a valid statistic when applied to the peaking equipment. The basic definition was developed for use with base load type equipment and as such does give a reasonable estimate of the probability of finding the unit on outage at some time in the future. It does not, however, give a good estimate in the case of a peaking unit due to the short shut down time. The desirability of using a simple formula such as that of Forced Outage Rate expressed in terms of hours or days should not cloud the fact that the result is not valid if it does not provide a good estimate of the probability of finding the unit on outage in the future.

Our studies in the Canadian Electrical Association have indicated gas turbine forced outage rates in excess of 20% using the conventional definition and hourly values. The approach suggested by the authors should prove extremely valuable in estimating the probability that the unit will be unavailable given a demand occasion. One difficulty which may arise, however, in our particular case, is the problem of obtaining the additional data requirements from the existing reporting schemes. The authors have done an excellent job of developing a valid model while minimizing the amount of additional data required. In some reporting schemes this data may be immediately available while others may require some modification. It is hoped that the benefits associated with a valid peaking unit representation will be appreciated and the necessary modification made.

Manuscript received February 23, 1971.

A basically similar model to that used in the paper has been developed at the University of Saskatchewan for use in spinning reserve studies. In a reliability study of operational system capacity requirements, peaking units are considered as standby capacity elements capable of picking up load within a relatively short period of time. The relevant statistics in a spinning reserve study are the probabilities of the unit starting or failing to start, and staying on the line or failing during the demand period. The model used in this case is shown in the following figure.

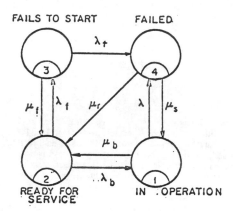

Fig. 1 State Space Diagram of a Rapid Start Unit for Spinning Reserve Purposes.

The time dependent probability values associated with the model states can be found by simple matrix multiplication[1]. The "forced out but not needed" state (#1) in Figure 1 is not required in a spinning reserve study. The starting failure rate λ_f in the above figure corresponds to the transfer rate P_s/T in Figure 1 of the paper. Similar models have been postulated for conventional thermal equipment operating in a hot reserve model. It is hoped that the results of these studies will be published in the near future.

We would again like to congratulate the authors on providing an excellent paper on a very practical problem.

REFERENCE

[1] Billinton, R., Jain, A.V., "Unit Derating Levels in Spinning Reserve Studies", IEEE Paper No. 71-TP-120-PWR, IEEE Winter Power Meeting, New York, N.Y., January 31 - February 5, 1971.

IEEE Task Group: The Task Group wishes to express appreciation for the contributions provided by discussers of the paper. In particular, the Task Group feels the following comments warrant expression in response to the specific questions raised by the discussers.

Messrs. Bram and Chan have provided a careful parallel analysis of the problem of obtaining estimates of model parameters from operating statistics. The Task Group concurs with their analysis but wishes to offer the observation that the transitions to the reserve shutdown state are not usually kept in outage records. A review of the reporting format for the EEI Prime Movers Task Force reveals that the following information are collected on the yearly summary of operating data:

1. Annual operating hours, SH
2. Total number of starts inititaed, N_{si}
3. Total number of successful starts, N_{ss}
4. Number of forced outages, N_f

Information relating to the total scheduled outage hours, SOH, and forced outage hours, FOH, are available as well as the number of the scheduled outages, N_{so}.

Proposed means by which the parameters necessary for the four-state model may be determined from the data of the type recorded by EEI are illustrated below.

First, the available hours, AH, are equal to the period hours, PH, less the forced outage hours, FOH, and the scheduled outage hours, SOH.

$$AH = PH - FOH - SOH$$

Manuscript received April 5, 1971.

The average duration of forced outage, r, is equal to the forced outage hours (accumulated duration in states '1' and '3') divided by the total number of forced outages.

$$\hat{r} = FOH/N_f$$

The duration in state '2', D_2, is terminated by transfer to reserve shutdown or to state '3'. The transfers into state '2' include the number of successful starts, N_{ss}, plus the number of repairs from state '3' to state '2', N_{32}. Thus

$$\hat{D}_2 = SH/(N_{ss} + N_{32})$$

We submit that $N_{32} \ll N_{ss}$ and may be neglected. Hence,

$$\hat{D}_2 \approx SH/N_{ss}$$

Referring to Figure 1, the duration in state '2', D_2, is related to m and D as,

$$\frac{1}{D_2} = \frac{1}{m} + \frac{1}{D}$$

Now, m may be estimated by

$$\hat{m} = SH/(\text{number of failures from state '2'})$$

$$= SH/(N_f - (N_{si} - N_{ss}))$$

Hence,

$$\hat{D} = SH/(N_{si} - N_f)$$

The probability of unsuccessful start is estimated from

$$\hat{P}_s = 1 - (N_{ss}/N_{si})$$

and the average reserve shutdown period, T, is estimated by

$$\hat{T} = (AH - SH)/N_{si}$$

These estimates are predicated on the following assumptions:

1. That a forced outage record is made for every event where the unit is not available to serve due either to a starting failure or a running failure or a failure discovered during normal inspection.

2. That one starting initiation is counted (and only one) until the cause of the starting failure is corrected. That is to say that repeated attempts at pushing a start button are not to be counted as individual starting failures.

On the basis of informal studies of outage reporting practices, it appears that there is inconsistency in the reporting of starting failures. There is an evident need for sharpening the definitions used in reporting startups, failures to start, and forced outages for peaking service equipment. The Task Group submits that the reporting means are already in existence within the EEI format and encourages efforts to improve uniformity of reporting of outage information. One of the objects of this paper was to illustrate the significance of and the kinds of information required for the analysis of peaking service applications.

Dr. Bhavaraju's principal point relates to the method of handling partial outages in the risk calcualtion. This is a problem common to all types of units. One way to handle partial forced outages would be to simply add more states. Six states would be required to model one level of forced outage. This would be analogous to the three-state (UP-DOWN-DERATED) model which has been proposed for continuous operation.

There is a question to be resolved regarding start and transfer immediately into a partial capacity state as against start into a full capacity state and then transfer into a partial capacity state. The Task Group is not aware of records which would indicate the number of starts with transfer direct to the partial capacity state. Accordingly, the extension of the model is deferred.

Another approach would be to simply modify the four-state model along lines suggested by Dr. Bhavaraju. We note that, as the partial forced outage durations are reported in clock hours, a weighting factor similar to f needs to be considered for partial forced outage hours. These approaches merit further study if partial forced outages are a significant contributor to down time. The EEI data indicates that partial outages have not been significant in the past. For example, in the 1969 report, we find for gas turbines FOR = 24.24% versus EFOR = 24.28%. The first figure counts full outages only, while the second counts all outages. In the future, partial outages may become more significant as multiple unit or combined cycle units become more common.

The Task Group wishes to express its appreciation for the comments made by Mr. Adamson both with regard to his question concerning the average up-period, D, contrasting the experience of Long Island Lighting Company with the statistics reported for the EEI survey and his comments on unit maturity. Of the many reasons for the wide dispersion between reported average operating hours the most likely candidates for consideration are one, methods of reporting information to EEI and two, widely diverse applications. There seems little question that performance of peaking service units should also reflect maturity effect.

The Task Group wishes to thank Mr. Mason for his comments. Mr. Mason is correct in his observation that the model proposed does not include a transition from reserve shutdown to forced unavailable state. Within the proposed model such events of discovery of conditions leading to forced unavailability as might occur during inspection and overhaul would be lumped within the total occurrence of forced outages. This tends to compensate for the omission of a transition from state '0' to state '1'. It is observed that such a transfer event occurs very infrequently.

The Task Group wishes to express appreciation of the discussion offered by Messrs. Garver and Marsh pointing out the similarity of results obtained with the two-state and four-state models. One of the objectives of the four-state model was to reflect the sensitivity of peaking service to risks of failure to start.

Messrs. Jain and Billinton have offered a contribution in a highly significant area not covered by the paper. This concerns the application of starting and running models for operating and spinning reserve studies. The Task Group encourages the publication of the models as indicated by the discussers.

A FREQUENCY AND DURATION METHOD FOR GENERATING SYSTEM RELIABILITY EVALUATION

A. K. Ayoub   A. D. Patton
Electric Power Institute
Texas A&M University
College Station, Texas

ABSTRACT

The paper presents a method for computing exactly the frequency and duration of load loss events as measures of generating system reliability. The method utilizes an exact state capacity model together with a cumulative state load model which requires no idealization. Efficient computer algorithms for building the exact state capacity model and computing the system reliability indices are presented.

INTRODUCTION

Methods for computing the frequency and duration of load loss events as measures of generating system reliability have received much attention in recent years.[1-6] All methods consist basically of two parts: a capacity model and a load model. The general approach is to compute the required parameters of the capacity model from the parameters of individual generating units; then combine the capacity model with the load model to compute the desired reliability indices. The parameters of the load model are determined in one way or another from the load cycle of the system being studied.

Two basic types of load models are possible: exact state and cumulative state. The most widely known load models are exact-state models.[2,7] These models idealize the actual system load cycle by approximating it by a sequence of discrete load levels. Probably the most common load model[2] represents the load cycle by an alternating sequence of peak and off-peak loads in which the durations of peak and off-peak loads are the same every day. The discrete load levels of the exact-state model are treated as a renewal process with parameters consisting of state probabilities, frequencies, and transition rates to states of higher and lower load. The exact-state load model can be combined with either an exact or cumulative-state capacity model[2] to yield the frequency and duration of load loss events (or, more generally, of margin states). The more efficient procedure of the two uses the cumulative-state capacity model.

The alternative type of load model is termed a cumulative-state load model. Here load is characterized as being greater than or less than a particular value and therefore discrete, specific, load levels need not be identified. This permits load to be treated as a continuous variable without any idealization whatever. The cumulative-state load model to be discussed in detail in this paper was originally suggested for use in transmission and distribution systems,[8] but is equally applicable to generation systems. The cumulative-state load model presented is completely defined by two characteristics which are directly and easily obtained from the load cycle being studied. The first of these characteristics is the familiar load-duration curve and the second is a characteristic referred to in this paper as the load-frequency curve. The cumulative-state load model can be combined with an exact-state capacity model in a manner which will be described to yield the frequency and duration of load loss events or of cumulative margin states in general.

Efficient algorithms for building an exact-state capacity model by adding generators to the model one at a time are implied, but not explicitly described in the literature. Hence, it is believed timely to derive and present the expressions required for efficient capacity model building and modification. Reference [3] describes a very efficient method for building and modifying cumulative capacity models.

EXACT-STATE CAPACITY MODEL

The exact-state generating capacity model for use in frequency and duration methods is defined by the following basic parameters for each of the possible capacity outage states: probability and effective departure rates to higher and lower capacity outage states. Parameters which can be readily calculated from these basic parameters are capacity outage state frequency and duration.

The capacity model is most efficiently constructed by adding one generator at a time to the model. In the expressions which follow, the common two-state representation of a generator is assumed. That is, each generator is presumed to be either fully available or on total forced outage. The extension of the expressions to handle generators with more than two capacity states when forced deratings are to be modeled is trivial.

Probability of Exact Capacity Outage State

The expression for the probability of the state "exactly X MW of capacity on forced outage" is well known and is

$$P(X) = P'(X)(1-R) + P'(X-C)R \qquad (1)$$

where:

$P(X)$ = probability of capacity outage of X MW after unit is added,

$P'(X)$ = probability of capacity outage of X MW before unit is added,

R = forced outage rate (FOR) of unit being added,

λ = average forced outage occurrence rate[7] of unit being added,

μ = average forced outage restoral rate[7] of unit being added,

C = capacity of unit being added.

In the above expression $P'(X-C)$ is zero if X is less than C since a state of negative capacity outage is obviously impossible. The recursive expression of equa-

Paper F 75 421-8, recommended and approved by the IEEE Power System Engineering Committee of the IEEE Power Engineering Society for presentation at the IEEE PES Summer Meeting, San Francisco, Calif., July 20-25, 1975. Manuscript submitted February 3, 1975; made available for printing April 28, 1975.

Reprinted from *IEEE Trans. Power Apparatus Syst.*, vol. PAS-95, no. 6, pp. 1929–1933, Nov./Dec. 1976.

140

tion (1) is initialized by setting $P(0) = 1-R_1$, $P(C_1) = R_1$, and all other state probabilities equal to zero where the first unit added to the capacity model has capacity C_1 and forced outage rate R_1. Note particularly that equation (1) takes into account the two mutually exclusive ways that a capacity outage of X MW may arise after a unit is added:

    (1)   system in capacity outage state X before unit added and the added unit up, and

    (2)   system in capacity outage state X-C before unit added and the added unit down.

Solving equation (1) for $P'(X)$ gives an expression for the probability of capacity outage state X after a unit is removed from the capacity model as might be required for the modeling of scheduled maintenance.

Effective Departure Rate From Exact Capacity Outage State

Let $\rho_+(X)$ be the effective departure rate from an exact capacity outage state X to states having less capacity out (i.e. to higher available capacity states). Similarly let $\rho_-(X)$ be the effective departure rate from exact capacity outage state X to states having more capacity out. The departure rates $\rho_+(X)$ and $\rho_-(X)$ may be computed by adding one generating unit at a time in a manner similar to that used in calculating $P(X)$. Consistent with the symbolism of equation (1) let $\rho_+'(X)$ and $\rho_-'(X)$ be the departure rates from capacity outage state X before a new unit is added. Now, recall the two mutually exclusive states which form state X after a unit is added. It is clear that the departure rates from these two mutually exclusive states are:

    (1)   system in capacity outage state X before unit added and added unit up

$$\rho_+(X) = \rho_+'(X)$$

$$\rho_-(X) = \rho_-'(X) + \lambda$$

    (2)   system in capacity outage state X-C before unit added and added unit down

$$\rho_+(X) = \rho_+'(X-C) + \mu$$

$$\rho_-(X) = \rho_-'(X-C)$$

Now, the departure rates from capacity outage state X after the unit is added are just the weighted averages of the departure rates from the mutually exclusive constituent states. Hence

$$\rho_+(X) = \frac{P'(X)(1-R)\rho_+'(X)+P'(X-C)R[\rho_+'(X-C)+\mu]}{P(X)} \quad (2)$$

and

$$\rho_-(X) = \frac{P'(X)(1-R)[\rho_-'(X)+\lambda]+P'(X-C)R\,\rho_-'(X-C)}{P(X)} \quad (3)$$

In the above expressions $P'(X-C)$, $\rho_+'(X-C)$, and $\rho_-'(X-C)$ are zero if X is less than C since states having negative capacity cannot exist. The recursive expressions of equations (2) and (3) are initialized by setting $\rho_-(0) = \lambda_1$, $\rho_+(C_1) = \mu_1$, and ρ_+ and ρ_- equal to zero for all other capacity outage states where the first unit added has capacity C_1 and parameters λ_1 and μ_1.

Solving equations (2) and (3) for $\rho_+'(X)$ and $\rho_-'(X)$ gives expressions useful when removing a unit from the capacity model to reflect scheduled maintenance.

Frequency and Duration of Exact Capacity Outage State

Once the quantities $P(X)$, $\rho_+(X)$, and $\rho_-(X)$ have

been found using equations (1), (2), and (3), the frequency $f(X)$ and duration $D(X)$ of the exact capacity outage state X are easily found:

$$f(X) = P(X)[\rho_+(X) + \rho_-(X)] \quad (4)$$

$$D(X) = 1/[\rho_+(X) + \rho_-(X)] \quad (5)$$

LOAD MODEL

The load model to be presented here can be used to represent any given load cycle without the necessity of idealizations or approximations of any kind. The load model presented can be characterized as a cumulative state model as opposed to load models which are exact state models.[2,6] It is this distinction which avoids the necessity for load cycle idealization. The load model presented is designed for use with an exact state generating capacity model as will be discussed in the next section.

The load model can be viewed as a two-state process as illustrated in Figure 1. Here one state is the event (load $\geq$ L) and the other state is the event (load < L) where L is a specified load level. The parameters of the load model can be simply determined for any given load cycle. The basic parameters needed are: (1) the probability of the state (load $\geq$ L) as a function of L, and (2) the frequency of the state (load > L) as a function of L. The characteristic giving the probability (or proportion of time) that load is greater than or equal to a specified load level is just the well-known load duration curve which requires no explanation. The load-frequency characteristic, the frequency of the state (load $\geq$ L) as a function of L, is obtained from the given load cycle by counting the number of transitions from the state (load < L) to the state (load $\geq$ L) for a range of values of L. The number of transitions counted for a particular value of L is the frequency of the state (load $\geq$ L), and also of the state (load > L), expressed on the time base of the load cycle. Converting frequency to some other time units such as days or hours requires that the number of transitions be divided by the number of days or hours in the load cycle.

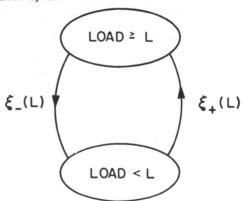

Fig. 1. Load model

The load-frequency characteristic for a load cycle will have the general shape illustrated in Figure 2. Frequency will be zero for L less than the minimum or more than the maximum value of the load cycle since no transitions between load model states occur for these conditions.

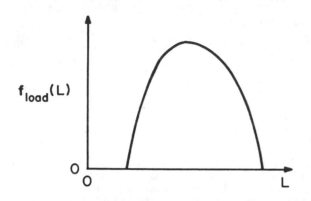

Fig. 2. Load-frequency characteristic

The load model state transition rates $\xi_+(L)$ and $\xi_-(L)$ can be easily found from the load-duration and load-frequency characteristics, if required. That is

$$\xi_-(L) = f_{load}(L)/P_{load}(L) \qquad (6)$$

and

$$\xi_+(L) = f_{load}(L)/[1 - P_{load}(L)] \qquad (7)$$

where: $f_{load}(L)$ = frequency of the state (load $\geq$ L) or of the state (load < L),
$P_{load}(L)$ = probability of the state (load $\geq$ L).

Expressing the load-duration and load-frequency characteristics in terms of load in per unit of load cycle peak allows use of the characteristics for any load cycle having the same shape. Thus, the load characteristics can be determined from observed load cycles and used in planning studies for future years provided the basic shape of the load cycle remains the same.

MEASURES OF RELIABILITY

The exact state capacity model and the cumulative state load model which have been described can be combined to yield such measures of reliability as probability of load loss, frequency of load loss, and average duration of load loss. More generally, the models can be combined to yield the probability, frequency, and duration of cumulative margin states. Margin is defined to be available capacity minus load and a cumulative margin state contains all states with margins less than or equal to the specified margin. Using this nomenclature, a negative margin constitutes a load loss or capacity deficiency equal in magnitude to the margin and a cumulative negative margin state corresponds to load loss greater than or equal to the specified margin.

Assuming that capacity and load states are independent as is reasonable, the probability of a cumulative margin state M, P_M, is given by

$$P_M = \sum_{all\ X} P(X)P_{load}(C-X-M) \qquad (8)$$

Here C is the installed capacity of the system less any capacity or scheduled outage, P(X) is the probability of a capacity outage of magnitude X and $P_{load}(C-X-M)$ is the probability that the load is greater than or equal to C-X-M. Note that M is a negative number if the specified margin is negative. Note also that when M is zero, equation (8) gives the loss-of-load probability in the traditional sense.

Two equivalent expressions for the frequency of a

cumulative margin state M are given below. Both of these expressions are derived in the Appendix. The first expression is

$$f_M = \sum_{all\ X} P(X)\{[\rho_+(X)-\rho_-(X)]P_{load}(C-X-M)+f_{load}(C-X-M)\} \qquad (9)$$

Recall that $\rho_+(X)$ and $\rho_-(X)$ are the effective departure rates from exact capacity outage state X and $f_{load}(C-X-M)$ is the frequency of the state (load $\geq$ C-X-M). Note in equation (9) that $\rho_+(X)$, $\rho_-(X)$, and $f_{load}(C-X-M)$ must be expressed in the same time units. The second expression for f_M is

$$f_M = \sum_{all\ X}\sum_y f_y - 2\sum_{all\ X} P(X)P_{load}(C-X-M)\rho_-(X) \qquad (10)$$

where: f_y = frequency of a state defined by the two events: capacity outage = X, and load $\geq$ C-X-M (i.e. the frequency of margins $\leq$ M due to the capacity outage state X).
$= P(X)\{[\rho_+(X)+\rho_-(X)]P_{load}(C-X-M)+f_{load}(C-X-M)\}$

Equation (10) is offered primarily to show the difference between the present expression for cumulative margin frequency and that offered in a previous publication[4]. The previous method, called the EPI method, in effect approximated f_M by the terms contained in the first summation in equation (10). The error associated with this approximation will be examined later for a sample system. In any case the approximation is no longer warranted as it is now as easy to calculate f_M exactly as approximately. It should be noted here that the expressions for f_M given in equations (9) or (10) are theoretically exact only if the capacity outage model is not truncated. Practically, values of f_M computed using truncated capacity outage models can be expected to be very accurate if the probabilities of the discarded capacity outage states are small.

The mean duration of a cumulative margin state M, D_M, is readily computed once P_M and f_M are known:

$$D_M = P_M/f_M. \qquad (11)$$

NUMERICAL EXAMPLE

Table I shows the generating unit parameters of a hypothetical system. A study period of 28 days was arbitrarily chosen made up of the daily load cycles shown in Figure 3 for a weekday, a Saturday, and a Sunday.

Table I
Generating Unit Parameters

Unit Capacity (MW)	λ (per day)	μ (per day)	FOR
20	0.00822	1.645	0.00497
10	0.00685	1.370	0.00497
10	0.00685	1.370	0.00497
5	0.00548	1.920	0.00285

The generating capacity model as computed using the methods of the paper is shown in Table II. The load-duration and load-frequency characteristics of the load cycle are given in Table III. Load frequency is expressed on a per day basis to agree with the time units used for generators λ's and μ's.

The frequency of load loss (frequency of a cumulative margin M = 0) for the example system as obtained using the method of the paper is 0.3857 per 28 days. The loss-of-load probability is 0.001565 and the mean load loss duration is 2.726 hours. By way of comparison the load loss frequency obtained using the approximate EPI method is 0.3874 per 28 days.

142

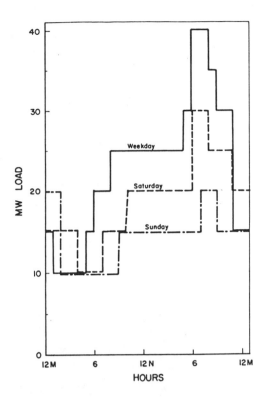

Fig. 3. Load cycle of example

Table II
Generating Capacity Model

Exact Capacity Outage, X (MW)	P(X)	$\rho_+(X)$ (per day)	$\rho_-(X)$ (per day)
0	0.98234254	0.0	0.0274
5	0.00280650	1.920	0.0219
10	0.00982310	1.370	0.0206
15	0.00002806	3.290	0.0151
20	0.00493622	1.650	0.0192
25	0.00001410	3.550	0.0137
30	0.00004912	3.015	0.0123
35	0.00000014	4.935	0.0069
40	0.00000012	4.385	0.0055
45	0.00000000	6.305	0.0

Table III
Load Model

Load Level, L (MW)	$P_{load}(L)$	$f_{load}(L)$ (per day)
>40	0.0	0.0
40 to 35+	0.0595	0.0298
35 to 30+	0.0893	0.0298
30 to 25+	0.1905	0.0357
25 to 20+	0.4762	0.0357
20 to 15+	0.6131	0.0416
15 to 10+	0.8214	0.0416
≤10	1.0	0.0

CONCLUSIONS

The paper has presented a method for calculating the frequency and duration of load-loss events (or cumulative margin states in general) as measures of generating system reliability utilizing a cumulative-state load model and an exact-state capacity model. Because the cumulative-state load model requires no

idealization of the load cycle being studied, the method permits the frequency and duration indices to be calculated exactly. All previous methods are, in one way or another, approximate.

It is shown that the cumulative-state load model is completely described by two characteristics: the load-duration durve and the load-frequency curve, which are directly and easily obtained from the load cycle being studied. Algorithms for the efficient construction and modification of the exact-state capacity model are also given. Thus, the paper provides a fairly self-contained reference for those interested in writing a program for computing frequency and duration reliability indices.

APPENDIX

Define a state y to be the intersection of the two events: available capacity = C-X, and load ≥ C-X-M. Here C is the installed capacity of the system less any capacity on scheduled outage, X is a capacity outage, and M is a specified margin. State y can be interpreted to be the set of margins ≤ M due to capacity outage X. It follows readily, that the probability of the available capacity state (C-X) is P(X) and the departure rates to higher and lower available capacity states are $\rho_+(X)$ and $\rho_-(X)$, respectively. Presuming capacity and load states to be independent it follows that the probability of state y, P_y, is

$$P_y = P(X) \, P_{load}(C-X-M) \qquad (12)$$

Departures from state y can occur by departures from the capacity state or from the load state. Let ρ_{+y} denote the departure rate to states of larger margin and ρ_{-y} denote the departure rate to states of smaller margin. Then

$$\rho_{+y} = \rho_+(X) + \xi_-(C-X-M) \qquad (13)$$

$$\rho_{-y} = \rho_-(X) \qquad (14)$$

Now consider a cumulative margin state M containing all states having margins ≤ M. The frequency of state M can be written as

$$f_M = \sum_{\text{all } y} P_y(\rho_{+y} - \rho_{-y}). \qquad (15)$$

Substituting the expressions of (12), (13), and (14) into (15) yields

$$f_M = \sum_{\text{all } X} P(X)P_{load}(C-X-M)[\rho_+(X) + \xi_-(C-X-M) - \rho_-(X)] \qquad (16)$$

Rearranging (16) and noting that

$$P_{load}(C-X-M) \, \xi_-(C-X-M) = f_{load}(C-X-M) \qquad (17)$$

yields

$$f_M = \sum_{\text{all } X} P(X)\{[\rho_+(X) - \rho_-(X)]P_{load}(C-X-M) + f_{load}(C-X-M)\} \qquad (18)$$

which is the desired expression of equation (9).

Equation (15) can also be written as

$$f_M = \sum_{\text{all } y} P_y(\rho_{+y} + \rho_{-y}) - 2 \sum_{\text{all } y} P_y \rho_{-y} \qquad (19)$$

Substituting the expressions of (12), (13), and (14) into (19) yields equation (10).

REFERENCES

(1) J. D. Hall, R. J. Ringlee, and A. J. Wood, "Frequency and Duration Methods for Power System Reliability Calculations, Part I - Generation System Model," _IEEE Transactions on Power Apparatus and Systems_, vol. 87, 1968, pp. 1787-1796.

(2) R. J. Ringlee and A. J. Wood, "Frequency and Duration Methods for Power System Reliability Calculations, Part II - Demand Model and Capacity Reserve Model," _Ibid_, vol. 88, 1969, pp. 378-388.

(3) R. J. Ringlee and A. J. Wood, "Frequency and Duration Methods for Power System Reliability Calculations, Part V - Models for Delays in Unit Installations and Two Interconnected Systems", _Ibid_, vol. PAS 71, Jan/Feb 1971, pp. 79-88.

(4) A. K. Ayoub, J. D. Guy, and A. D. Patton, "Evaluation and Comparison of Some Methods for Calculating Generation System Reliability," _Ibid_, vol. 89, 1970, pp. 521-527.

(5) Roy Billinton, _Power System Reliability Evaluation_, New York: Gordon and Breach, 1970.

(6) R. Billinton, R. J. Ringlee, and A. J. Wood, _Power-System Reliability Calculations_, Cambridge, Mass.: The MIT Press, 1973.

(7) R. Billinton and C. Singh, "System Load Representation in Generating Capacity Reliability Studies, Part I - Model Formulation and Analysis," _IEEE Transactions on Power Apparatus and Systems_, vol. 91, 1972, pp. 2125-2132.

(8) W. R. Christiaanse, "Reliability Calculations Including the Effects of Overloads and Maintenance _Ibid_, vol. PAS 71, July/Aug 1971, pp. 1624-1677.

(9) IEEE Committee Report, "Suggested Definitions Associated with the Status of Generating Station Equipment and Useful in the Application of Probability Methods for System Planning and Operation," IEEE publication C 72 599-9, 1972.

RELIABILITY EVALUATION IN ENERGY LIMITED GENERATING CAPACITY STUDIES

Roy Billinton   P.G. Harrington
Power System Research Group
University of Saskatchewan
Saskatoon, Canada

Abstract - Most generating capacity reliability studies assume that there are no inherent energy limitations and therefore concentrate on considering the effect of unit forced outages and uncertain load requirements. It appears, however, that the era of abundant energy is disappearing and that limitations must be included in conventional studies. This paper illustrates an extension of a well known technique for generating capacity evaluation which includes limited energy systems. The approach is illustrated by a simple numerical system example.

INTRODUCTION

The most popular technique at the present time for assessing the adequacy of an existing or proposed generating capacity configuration is the Loss of Load Probability or Expectation Method[1]. In its basic form this technique has been in existence for over thirty years. Another technique which has been in existence for almost as long is the Loss of Energy Approach[1]. In this technique, the capacity outage probability model is convolved with the period load duration curve to calculate the expected energy not supplied due to unit forced outages. In its basic form the generating units are not considered to be energy limited other than by their capacity and availability. In recent years attention has been given to production cost models for generating unit systems which permit the calculation of expected unit energy outputs and therefore the associated fuel costs[2,3]. These methods vary somewhat in detail but basically utilize the same concepts as in the fundamental Loss of Energy Method of generating capacity reliability evaluation.

This paper illustrates by a simple numerical example how the expected energy output of each unit in a system can be evaluated and then extends these concepts to include several basic types of energy limitations. Not all units with energy limitations can be considered in the same manner, as different operating procedures are used for different types of limitation. A general approach is considered in regard to the term "energy limitation". Five years ago this would have been applied only to hydraulic generating facilities. At the present time, fuel restrictions to gas turbines, oil and coal fired thermal units and potential uranium shortages to nuclear units can be added to the list of potential energy limiters. Three basic types of energy limitation have been considered in this study. They are,

(a) Units which have large amounts of storage available to them such that they can operate for more than a

few days on the stored energy i.e. hydro units with large reservoirs.

(b) Units which have fairly restricted storage facilities but with sufficient storage so that the available energy can be stored long enough to be used during the daily peak demands.

(c) Units which have no storage at all or not enough storage to allow the energy to be held until the next daily peak period i.e. run-of-the-river hydro facilities, limited storage gas fired units.

As previously noted, methods have been published for the calculation of expected unit energy output and expected energy not supplied when energy limitations are not considered[2,3]. These methods use in some form the period load duration curve and order the units in terms of their operating costs. "Peak shaving" methods are used to model energy-limitations. This applies in certain cases but not all limitation types can be modelled in this way. The actual operating procedure may vary according to the basic type of limitation and therefore the modelling technique must be capable of responding to these changes.

The basic method used for calculating the expected energy output of the units in the system is that proposed by Bram, Jaeger and Chan in their discussion of Reference 3. Using this approach, unit forced outage rates including derating levels and associated maintenance schedules can be incorporated in the analysis. Units can also first be partially loaded to one level in the priority order and then loaded to a higher level later in the priority order allowing for consideration of units whose energy production costs are not constant with respect to output capacity. Most utilities have digital computer programs for development of unit capacity outage probability tables. The bulk of these use a recursive algorithm for table development[1]. This is the principal element in expected energy output evaluation. Each time a unit is added to the capacity-probability table, the table is convolved with the period load duration curve to determine the expected energy not supplied. The expected energy supplied by the last unit to be added is equal to the expected energy not supplied before the unit was added minus the expected energy not supplied after the addition of the unit to the capacity outage probability table.

The effect of partial unit loading can be considered by an example. Consider a 25 MW unit which is first loaded to 20 MW and then, due to a higher cost of energy at the 25 MW output level, two other units are loaded before the 25 MW unit is loaded to full capacity. In calculating the expected energy output of the unit at each capacity level, the 25 MW unit would first be added to the capacity-probability table as a 20 MW unit. This table is convolved with the load model and the expected energy not supplied determined. The expected energy supplied by the unit at the 20 MW output level is then the expected energy not supplied before its addition to the system capacity-probability table minus the expected energy not supplied after its addition. The other two units can be added to the table one at a time and their expected energy outputs calculated in the same manner. The capacity-

F 78 004-4. A paper recommended and approved by the IEEE Power System Enginering Committee of the IEEE Power Engineering Society for presentation at the IEEE PES Winter Meeting, New York, NY, January 29-February 3, 1978. Manuscript submitted August 16, 1977; made available for printing November 1, 1977.

Reprinted from *IEEE Trans. Power Apparatus Syst.*, vol. PAS-97, no. 6, pp. 2076–2085, Nov./Dec. 1978.

probability table of the system can then be formed without the 20 MW unit (the 25 MW unit loaded to 20 MW) and the 25 MW unit, loaded to its full capacity is then added. The capacity-probability table thus formed would then be convolved with the load duration curve and a value of expected energy not supplied determined. This value is subtracted from the previous value, calculated after the addition of the other two units, to arrive at the expected energy output of the unit at the 25 MW level. Maintenance can be included by splitting the total period under consideration into periods of constant maintenance and calculating the expected unit energy outputs for each of these periods.

The basic simplicity of this approach can be best illustrated by a simple numerical example. It is then extended to include the various types of energy limitations previously noted.

System Example

A system contains three units with no inherent energy restrictions. The unit descriptions and their loading order are as follows. Unit #1 is a unit with three capacity states. The capacity levels and associated probabilities are shown in Table 1.

Table 1. Unit 1 Model

Capacity	Probability
0.0	0.05
15.0	0.30
25.0	0.65

This unit, due to differences in energy cost at different output capacities, appears in two places in the priority list. It is loaded to 20 MW at level #1 in the priority order and loaded to full capacity, 25 MW, at priority level #3. Unit #2 is a 30 MW unit with a forced outage rate, F.O.R., of 0.03 and appears second in the priority list. Unit #3 appears fourth in the priority list, has a capacity of 20 MW and its F.O.R. is 0.04.

The load duration curve for a period of 100 hours is shown in Figure 1. The energy required by the load is represented by the area under the load duration curve.

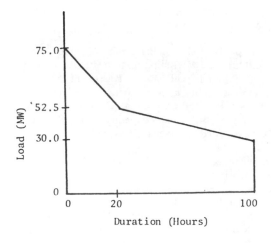

Figure 1. System Load Model

The actual load duration curve may not be of a simple straight line for shown in Figure 1. It can be represented in its segments. A two segment representation is used in this example to demonstrate the technique used. The model is sufficiently simple that only a hand calculator is required to repeat the procedure.

The formula used to determine the expected energy not supplied after each unit has been added to the capacity probability table of the system is:

$$EENS = \sum_{i=1}^{N} E_i \times P_i \tag{1}$$

where N = total number of capacity states in the current system capacity-probability table

E_i = area under load duration curve above a load equal to the capacity of the ith capacity state

P_i = probability of the ith capacity state

The capacity-probability table before any units have been added contains one level (a capacity of 0.0 with a probability of 1.0). The expected energy not supplied before any units have been considered is therefore equal to the expected energy of the load for the 100 hour period under consideration represented by the area under the load duration curve (above a load of 0.0 MW).

$$EENS_0 = 4575.0 \text{ MW-hr} \times 1.0 = 4575 \text{ MW-hr}$$

The expected energy output of the first level of the priority list is obtained by adding the capacity model of unit #1 loaded to 20 MW to the capacity-probability table to produce Table 2.

Table 2. System Capacity Model at Priority Level 1

Capacity	Probability
0.0	0.05
15.0	0.30
20.0	0.65

The expected energy not supplied is then determined using equation (1).

$$EENS_1 = 4575.0 \times 0.05 + 3075.0 \times 0.30 + 2575.0 \times 0.65$$

$$= 2825.0 \text{ MW-hr}$$

4575.0 MW-hr, 3075 MW-hr and 2474.0 MW-hr are the areas under the load duration curve of Figure 1 above loads of 0.0 MW, 15.0 MW and 20.0 MW respectively. The expected energy produced by unit #1 when loaded to 20 MW is therefore 4575.0 – 2825.0 MW-hr.

The next unit in the priority list, unit #2 is then added to the system capacity model to produce Table 3.

Table 3. System Capacity Model at Priority Level 2

Capacity	Probability
0.0	0.0015
15.0	0.0090
20.0	0.0195
30.0	0.0485
45.0	0.2910
50.0	0.6305

146

The expected energy not supplied at this priority level is determined using equation (1).

$$EENS_2 = 4575.0 \times 0.0015 + 3075.0 \times 0.0090 + 2575.0$$
$$\times 0.0195 + 1575.0 \times 0.0485 + 475.0 \times 0.2910$$
$$+ 286.111 \times 0.6305$$

$$EENS_2 = 479.756 \text{ MW-hr}$$

The expected energy supplied by unit #2 is therefore $EENS_1 - EENS_2 = 2825.0 - 479.756 = 2345.244$ MW-hr.

Next the loading on unit #1 is increased from 20 MW to its full capacity of 25 MW. The capacity probability table representing this situation is made up of unit #2 and unit #1 loaded to 25 MW and is shown in Table 4.

Table 4. System Capacity Model at Priority Level #3

Capacity	Probability
0.0	0.0015
15.0	0.0090
25.0	0.0195
30.0	0.0485
45.0	0.2910
55.0	0.6305

The expected energy not supplied at priority level 3 is:

$$EENS_3 = 4575.0 \times 0.0015 + 3075.0 \times 0.0090 + 2075.0$$
$$\times 0.0195 + 1575.0 \times 0.0485 + 475.0 \times 0.2910$$
$$+ 177.778 \times 0.6305$$

$$EENS_3 = 401.7014$$

The expected energy supplied by unit #1 when loaded at 25 MW is:

$$EENS_2 - EENS_3 = 479.756 - 401.7014 = 78.0546$$

The capacity model of unit #3 is then combined with the capacity model of Table 4 to determine the final system capacity model shown in Table 5. The expected energy not supplied is then:

$$EENS_4 = 4575.0 \times 0.00006 + 3075.0 \times 0.00036 + 2575.0$$
$$\times 0.00144 + 2075.0 \times 0.00078 + 1575.0$$
$$\times 0.00194 + 1119.444 \times 0.00864 + 475.0$$
$$\times 0.03036 + 286.111 \times 0.04656 + 177.778$$
$$\times 0.02522 + 44.444 \times 0.27936$$

$$EENS_4 = 64.0774$$

The expected energy output of unit #3 is then $EENS_3 - EENS_4 = 401.7014 - 64.0774 = 337.624$ MW-hr.

The expected load energy not supplied by the above system is then 64.08 MW-hr and the Energy Index of Reliability, EIR as described in Reference 1 is given by:

$$EIR = 1 - \frac{64.08}{4575.0}$$

$$EIR = 0.985993$$

Table 5. System Capacity Model at Priority Level #4

Capacity	Probability
0.0	0.00006
15.0	0.00036
20.0	0.00144
25.0	0.00078
30.0	0.00194
35.0	0.00864
45.0	0.03036
50.0	0.04656
55.0	0.02522
65.0	0.27936
75.0	0.60528

Production cost modelling is not the primary concern of this paper. The expected energy output of a non-restricted unit is however a relatively simply computation which has been over complicated in some of the publications available on this subject.

Energy-Limited Units With No Storage

The energy from units whose storage facilities do not have the capacity to save the energy until the next peak load period must be used as it becomes available. An example of this type of energy limitation is a run-of-the-river hydro installation. In this case, the rate of water flow determines the capacity of the unit and therefore the river flow rate probability distribution must be correlated to the unit capacity distribution. The unit can be represented as a multi-state unit with its capacity states corresponding to the flow rates. This would also apply to variable flow availabilities of natural gas if the storage facilities are inadequate. This multi-state distribution of unit capacity can then be added to the system capacity-probability table and the expected energy output determined in the usual manner.

Consider that the system used in the previous example has a 10.0 MW unit with a capacity distribution, due to its flow rate probability distribution, as described in Table 6.

Table 6. Capacity Model of the Unit With Energy Limitations and No Storage Facilities

Capacity	Probability
0.0	0.040
2.5	0.192
5.0	0.480
10.0	0.288

This unit can be placed in the priority list in an appropriate place according to its energy cost. If the per unit cost of energy for the unit described in Table 6 is less than the other units in the system then it would appear first in the priority list and thus be loaded first. If the same example system is used with this new unit appearing first in the priority order, the expected energy output of each unit in the system using the load duration curve of Figure 1 would be as shown in Table 7.

The expected load energy not supplied by the above system is then 35.53 MW-hr and the EIR is 0.992236.

147

Table 7. Expected Unit Energy Outputs With One Energy-Limited Unit

Priority Level	Unit	Expected Energy Not Supplied (MW-hr)	Expected Energy Output (MW-hr)
-	-------	4575.00	-------
1	Energy Limited Unit	3999.00	576.00
2	Unit #1 (loaded to 20 MW)	2249.00	1750.00
3	Unit #2 (30 MW)	306.96	1942.04
4	Unit #1 (loaded to 25 MW)	250.37	56.59
5	Unit #3 (20 MW)	35.52	214.85

Energy Limited Units With Limited Storage

The energy from those units with storage facilities which permit energy to be stored for a couple of days can be used during the peak load periods and thus reduce the requirement from more expensive peaking-units. An example of this type of energy limitation is a hydro plant with a small storage reservoir which can store incoming water for a day or two or a fossil fired plant with limited fuel supply. A "peak-shaving" technique can be used in which the capacity and energy probability distributions of the unit are used to modify the load duration curve. The modified curve is then used with the remaining units of the system to determine the expected unit energy outputs and the expected energy unsupplied.

The "peak-shaving" technique first modifies the original load duration curve using a conditional probability approach considering only the capacity probability distribution of the unit. This "capacity-modified" load duration curve is the equivalent load curve for the rest of the units in the system if the unit used to modify it was first in the priority list. That is, the expected energy output of the "peak-shaving" unit if it had no energy limitations and appeared first in the priority list is equal to the reduction in expected load energy caused by the "capacity-modification" of the original load duration curve. In addition, the expected energy outputs of the rest of the units of the system when this modified curve is used would be the same as if the original load duration curve were used with the modifying unit loaded first in the priority order. After the capacity probability distribution of the "peak-shaving" unit has been used to "capacity-modify" the original load duration curve, the conditional probability approach is again used when the energy probability distribution of the unit is considered. The final modified curve is then used in the normal manner with the rest of the units of the system to determine their expected energy outputs.

In the case of a hydro station consisting of several units and with a limited-storage reservoir, the capacity probability distribution of the units are combined. The combined capacity probability table, together with the energy probability distribution of the reservoir, is then used to "peak-shave" the load duration curve.

Consider the original example three unit system with no energy limitations to have two additional hydro units with the type of energy limitations just described. The first such unit (unit #4) has a capacity probability distribution as shown in Table 8.

Table 8. Capacity Model of Unit #4

Capacity	Probability
0.0	0.03
10.0	0.25
15.0	0.72

A hydro unit would not normally exhibit a derated state but often two or more hydro units share the same reservoir giving rise to a capacity-probability table for the plant with more than two capacity states. In order to demonstrate how the peak-shaving technique is used with units or plants with more than two capacity states, while keeping the example simple enough to be calculated by hand, a three state unit is used. The energy limitations of unit #4 are given in Table 9, where values of energy stored are tabulated with their respective probabilities of being equalled or exceeded.

Table 9. Energy Distribution for Unit #4

Energy MW-hr	Cumulative Probability
200.0	1.00
350.0	0.70
500.0	0.20

This model assumes that there are only three possible values of stored energy for the period being considered, each with a certain probability of occurrence. More energy states can be incorporated for a more exact model if more accurate data are available on the energy distribution. The second unit with a similar type of energy limitation (unit #5), is a 10 MW hydro unit with an F.O.R. of 0.04 and an energy distribution as shown in Table 10.

Table 10. Energy Distribution for Unit #5

Energy MW-hr	Cumulative Probability
70.0	1.00
150.0	0.60

As previously noted, the peak-shaving technique involves two stages of load duration curve modification. In the first stage, the load duration curve is modified by the capacity distribution of the unit assuming no energy limitations. In this example unit #4 is used to peak-shave first. If unit #4 had a capacity of 15 MW with zero probability of any other capacity the load duration curve would simply be reduced by 15 MW at every point (assuming no energy limitations). If it was a 10 MW unit with zero probability of any other capacity the load duration curve would be reduced by 10 MW and if it was a 0.0 MW unit the load duration curve would remain unchanged. These three situations are illustrated in Figure 2. The duration of each value of load on the "capacity-modified" load duration curve is then the duration given that unit #4 has a capacity of 15 MW times the probability of that capacity state (0.71), plus the duration given a capacity of 10 MW times its probability plus the duration given a 0.0 MW capacity times the probability of that capacity state (0.03). This calculation is given by equation (2).

148

Load	(a) 0.03 Duration on original load duration curve	(b) 0.25 Duration on original load duration curve reduced by 10.0 MW	(c) 0.72 Duration on original load duration curve reduced by 15.0 MW	(d) Duration on "capacity-modified" load duration curve
75.0	0.0	0.0	0.0	0.0
65.0	(8.889)	0.0	(0.0)	0.2667
60.0	(13.333)	(4.444)	0.0	1.5111
52.5	20.0	(11.111)	(6.667)	8.1778
42.5	(55.556)	20.0	(15.556)	17.8667
37.5	(73.333)	(37.778)	20.0	26.0444
30.0	100.0	(64.444)	(46.667)	52.7111
20.0	(100.0)	100.0	(82.222)	87.20
15.0	(100.0)	(100.0)	100.0	100.0
0.0	100.0	100.0	100.0	100.0

Table 11. Co-ordinates of Points of Curves of Figure 2

Table 12. Coordinate Points of the Load Duration Curve After Peak-Shaving with Unit #4.

Duration (hours)	Load (MW)
0.0	75.0
0.267	65.0
1.511	60.0
8.778	52.5
9.409	51.229
13.942	51.229
25.127	44.819
40.972	44.819
48.018	42.500
53.619	40.734
61.835	40.734
100.0	30.0
100.0	0.0

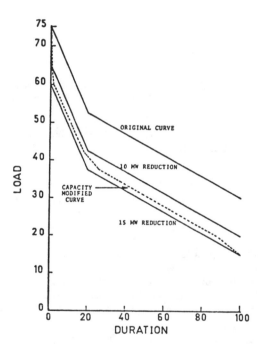

Figure 2. Original and Capacity Modified Load Duration Curves

$$D(L) = \sum_{i=1}^{N} d_i(L) \times P_i \qquad (2)$$

where: $D(L)$ = duration of load L on the "capacity-modified" curve

N = number of capacity states of the peak-shaving unit

C_i = output capacity of ith capacity state of the peak-shaving unit

P_i = probability of capacity C_i

$d_i(L)$ = duration of load L on the original load dura-tion curve when reduced by C_i MW.

It can be seen that $D(L)$ need only be calculated for values of load L, where an intersection in one of the load duration curves used in the calculation occurs. Table 11 gives the co-ordinates of the points on the curves of Figure 2. The numbers in brackets indicate values located between end points of line segments. As an example of the calculation consider a load level of 37.5 MW. The duration corre-sponding to 37.5 MW on the "capacity-modified" curve is given by:

$$D(37.5 \text{ MW}) = 73.33 \times 0.03 + 37.78 \times 0.25 + 20.0 \times 0.72$$

$$D(37.5 \text{ MW}) = 26.044 \text{ hours}$$

The second stage of the peak-shaving technique considers the energy distribution of the unit. The difference between energy in the "capacity-modified" load curve and that of the original load duration curve is equal to the expected energy output of the peak shaving unit if it had no energy limitations and was first in the priority list. As the word peak-shaving implies, the energy available to the unit is used during peak load periods. If more energy is available more of the peak load will be reduced. The energy modification starts at the top of the load duration curve. At each value of load, the area between the original load duration curve and the "capacity-modified" curve above that load value is calculated. This area represents the energy output of the peak-shaving unit if it is operated whenever the system load equals or exceeds the value of load being considered. If the unit has that much or more energy available to it, the value of duration at the load point being considered corresponds to the value of duration at the point on the "capacity-modified curve. If not, then the value for duration will correspond to the duration on the original load duration curve. The probabilities associated with each of these two events are taken from the cumulative energy distribution for the unit (which in this case is given by Table 4) and used to weight the two values of load dura-tion. The final peak-shaved load duration curve is then de-termined by:

$$D(L) = d_c(L) \times P[E(L)] + d_o(L) \times [1 - P[E(L)]] \qquad (13)$$

where: $D(L)$ = value of duration of final peak-shaved curve corresponding to a load of L MW.

$d_c(L)$ = value of duration on "capacity-modified" curve corresponding to a load of L MW.

$d_o(L)$ = value of duration on original load duration curve corresponding to a load of L MW.

$E(L)$ = expected energy output of unit if loaded where-ever system load equalled or exceeded L MW, i.e. the area between the "capacity-modified" and original load duration curves above L MW.

P [E(L)] = probability of energy equalling or exceeding E(L) MW.

Consider the 42.5 MW point. The area between to original and "capacity-modified" load duration curves above a load of 42.5 MW is 430.44 MW-hr. From the energy distribution of unit #4, Table 7, the probability of unit #4 having 430.44 MW-hr of energy or more is 0.20. Using equation (3):

D(42.5) = 17.866667 x 0.20 + 55.555556 x 0.80

D(42.5) = 48.017778 hours

Figure 3 shows the load duration curve after peak-shaving using unit #4. The horizontal segments of the curve are due to the discrete nature of the energy distribution. The coordinates of the load duration curve after peak-shaving with unit #4 are given in Table 12. The same technique can be used to peak-shave using unit #5. The only difference being that the load duration curve after being peak-shaved by unit #4 would be considered the original curve. The final load duration curve is given in Table 13. Loading the non-energy limited units in the same manner as before using the modified curve of Table 13 results in the expected energy outputs shown in Table 14. The expected load energy not supplied is 15.728 MW-hr and the EIR of the system for the 100.0 hour period considered is 0.996562.

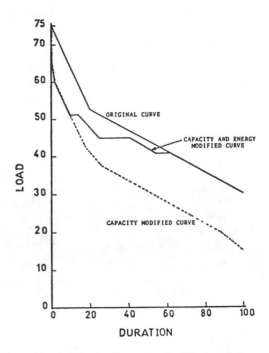

Figure 3. Original, Capacity Modified and Energy Modified Load Duration Curves

Table 13. Load Duration Curve Points After Peak-Shaving Using Units #4 and #5.

Duration (Hours)	Load (MW)
0.0	75.0
0.011	65.0
0.188	60.0
0.494	55.0
1.180	52.5
1.533	51.229
1.715	51.229
2.094	50.000
2.528	49.529
8.280	49.529
14.177	44.819
20.895	44.819
21.064	44.725
41.257	44.725
48.018	42.5
53.619	40.734
61.835	40.734
100.0	30.0
100.0	0.0

Energy Limited Unit Partially Base Loaded and Partially Used for Peak-Shaving

Consider a unit (unit #6) with capacity and energy distributions as shown in Table 15 (a) and (b) respectively was base loaded to 10 MW with the remaining capacity and energy used to peak-shave. The peak-shaving technique can be adpated to model this situation.

The first step is to "capacity-modify" the original load duration curve (Figure 1), assuming the unit to be a 10 MW unit with an FOR of 0.3. This "capacity-modified" curve is given in Table 16. The resulting decrease in expected load energy (area under original curve minus area under "capacity-

Table 15(a). Capacity Distribution of Unit #6

Capacity-Probability Table

Capacity	Individual Probability
0.0	0.3
15.0	0.7

Table 15(b). Energy Distribution of Unit #6

Energy Probability Table

Energy (MW-hr)	Cumulative Probability
750.00	1.00
850.00	0.70
1000.00	0.30

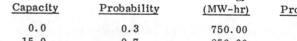

Table 14. Expected Unit Energy Outputs with Two Peak-Shaving Units

Unit	Expected Energy Not Supplies After Unit Added (MW-hr)	Expected Energy Output of Unit (MW-hr)
----	4575.00	
Unit #4 (Peak-Shaving)	4240.00	335.00
Unit #5 (Peak-Shaving)	4122.00	118.00
Unit #1 (Loaded to 20 MW)	2372.00	1750.00
Unit #2 (30 MW)	148.29988	2223.70
Unit #1 (Loaded to 25 MW)	134.66691	13.63
Unit #3 (20 MW)	15.72759	118.94

Table 16. Load Duration Curve "Capacity-Modified" by 10 MW.

Duration (Hours)	Load (MW)
0.0	75.0
2.667	65.0
13.778	52.5
30.667	42.5
75.111	30.0
100.000	20.0
100.000	0.0

modified" curve) would be the expected energy supplied by the base loaded part of unit #6. The energy distribution would then be reduced by this value, 700 MW-hr in this case, to give the energy distribution available for peak-shaving. The next step is to "capacity-modify" the original load duration curve assuming unit #6 to be a 15.0 MW unit with an FOR of 0.3 This "capacity-modified" curve is shown in Table 17. The peak-shaving technique is then employed assuming the data of Table 16 to be the original load duration curve and that of Table 17 to be the "capacity-modified" curve. The energy distribution used in the final step of the peak-shaving method is given by Table 15(b) but with each state energy level reduced by 700-MW-hr.

When the non-energy limited units (units #1, #2, #3) are then loaded in the usual manner using the load duration curve of Table 18 the expected unit energy outputs of Table 19 result. The expected load energy not supplied by the system in this case is 29.561 MW-hr and the EIR is 0.993539.

guided by a rule curve which dictates how much energy is to be using during a certain period. Any energy allotted for a certain period which is not used to reduce energy lost during thermal forced outages is used to reduce the output of more expensive units. Each period therefore has a different energy limitation. To determine the expected energy not supplied and the EIR, the peak shaving technique can be used but in a different manner. The capacity-probability table is built up for the non-energy limited units of the system and used to "capacity-modify" the load duration curve. This gives the equivalent load for the rest of the units in the system. Each unit with the above described energy limitations can then be used to "peak-shave" this equivalent load duration curve. The area under the load duration curve after all the energy limited units have been dispatched in this way represents the expected load energy not supplied by this system. The reduction in expected load energy after each energy limited unit has been used to "peak-shave" the equivalent load curve is the expected energy output of that unit when it is

Table 17. Load Duration Curve "Capacity-Modified by Unit #6 (15 MW)

Duration (Hours)	Load (MW)
0.0	75.0
4.0	60.0
10.667	52.5
36.0	37.5
62.667	30.0
100.0	15.0
100.0	0.0

Table 19. Expected Unit Energy Outputs With a Unit Partially Base Loaded and Partially Used to Peak Shave

Unit	Expected Energy Not Suppled After Unit Added (MW-hr)	Expected Energy Output of Unit (MW-hr)
----	4575.00	-- -----
Unit #6	3710.00	865.00
Unit #1 (Loaded at 20 MW)	1960.00	1750.00
Unit #2 (30 MW)	236.765	1723.235
Unit #1 (Loaded at 25 MW)	192.673	44.092
Unit #3 (20 MW)	28.187	164.486

Table 18. Final Load Duration Curve After Being Modified by Unit #6.

Duration (Hours)	Load (MW)
0.0	75.0
1.333	70.0
2.667	65.0
4.0	60.0
8.444	55.0
10.667	52.5
14.889	50.0
20.921	46.429
21.854	46.429
24.2667	45.0
28.489	42.5
34.111	40.0
39.733	37.5
48.622	35.0
53.702	33.571
58.679	33.571
71.378	30.0
83.822	25.0
92.489	21.518
96.222	21.518
100.0	20.0
100.0	0.0

Table 20. Equivalent Load After Dispatching Unit #8.

Duration (Hours)	Load (MW)
0.0	75.0
0.0002	60.0
0.0010	55.0
0.0023	52.5
0.0038	50.0
0.0077	45.0
0.0154	40.0
0.0272	37.5
0.0400	35.0
0.0541	32.5
0.0710	30.0
0.1110	27.5
0.1526	25.0
0.2386	22.5
0.3303	20.0
0.4387	17.5
0.5711	15.0
0.7494	12.5
0.9314	10.0
1.3682	7.5
1.8748	5.0
2.4185	2.5
2.8402	0.9430
6.3132	0.9430
6.8618	0.0

Energy Limited Units With Substantial Storage Facilities

A storage facility may be large enough to store energy inflows for several weeks, i.e. a hydro installation with a large reservoir. The operation of such an installation can be

151

loaded at the bottom of the priority list, i.e. only loaded when necessary to meet the system load. The remaining energy of each unit can be used to reduce the energy output of more expensive units. As an example consider the 3 non-energy limited units of the previous examples in addition to two energy limited units of the type described above. The first such unit, unit #8, has a capacity of 15 MW and a forced outage rate 0.30. It is energy limited to 33.23 MW-hr, i.e. its energy probability distribution contains one value, 33.23, with a probability of 1.0. The second unit, unit #9 is a 10 MW unit with a forced outage rate of 0.10. It is energy limited to 12.34 MW-hr. The capacity-probability table made up of the non-energy-limited units of the system is given in Table 5. When this table is used to "capacity-modify" the original load duration curve the result is Table A1 in the Appendix which is the equivalent load once the non-energy limited units have been dispatched. Unit #8 is then used to "peak-shave" this equivalent load. The difference between expected energy in the curve before unit #8 is used to "peak-shave" and that after is the expected energy output of unit #8 when loaded after the non-energy-limited units which in this case is 33.23 MW-hr, since it is energy limited to that value. Had the expected energy output been less than 33.23 MW-hr the extra energy could be used to partially off-load the more expensive units in the system. After unit #8 has been used to "peak-shave" the equivalent load duration curve, the resulting curve, given in Table 20, is then "peak-shaved" by unit #9. The difference in expected energy not supplied before and after this "peak-shaving" is the expected energy supplied by unit #9. The final value of expected energy not supplied, 18.51 MW-hr, is the expected load energy not supplied by the system. The EIR, therefore, is 0.995954.

Conclusions

This paper has illustrated the development of a technique for examining the reliability associated with a generation configuration using an energy based index. The approach is based upon the Expected Loss of Energy Approach and extends the technique to include the consideration of energy limitations associated with generation facilities. The most obvious form of energy limitations are those associated with hydraulic facilities and therefore the types of energy limitation considered have been formulated with this in mind. Those are (a) units with no energy storage facilities such as run-of-the-river installations (b) units with facilities capable of storing energy until a peak load period (c) units whose energy supply makes them suitable for part base load and part peak shaving and (d) units with substantial storage facilities in which specified quantities of energy are available within certain periods.

Energy limitations are not restricted to hydraulic generating units. This situation can easily arise with gas and oil-fired units and could also be appreciated to possibly exist for coal and nuclear stations. This paper has attempted to emphasis the reliability implications rather than the production cost evaluation aspects which are also performed during the calculation. The approach is basically simple as it retains the physical appreciation associated with the capacity model in the form of a capacity-probability table and the load model in the form of the load duration curve.

Acknowledgement

This work was supported by the National Research Council of Canada.

References

1. R. Billinton, "Power System Reliability Evaluation" (book) Gordon and Breach Science Publishers Inc., New York, 1970.

2. R. R. Booth, "Power System Simulation Based On Probability. Analysis", IEEE Trans. Vol. PAS-91, 1972, pp. 62 – 69.

3. M. A. Sager, R.J. Ringlee, A.J. Wood, "A New Generation Production Cost Program To Recognize Forced Outages", IEEE Trans. PAS-91, 1972. pp. 2114 – 2124.

Appendix

Table A1. Equivalent Load for Unit #8 When Loaded After Non-Energy Limited Units

Duration (Hours)	Load (MW)
0.00	75.0
0.00080	60.0
0.00267	55.0
0.00680	52.5
0.01130	50.0
0.02387	45.0
0.04502	40.0
0.07480	37.5
0.13916	32.5
0.18093	30.0
0.28964	27.5
0.40356	25.0
0.62093	22.5
0.85124	20.0
1.13760	17.5
1.48156	15.0
2.16307	10.0
3.11182	7.5
4.26298	5.0
5.40720	2.5
6.86182	0.0

Discussion

R. N. Allan (UMIST, Manchester, England): May I first compliment the authors on presenting a very instructive paper, the tutorial nature of which permits readers to readily adapt the methodology to their own particular problem. I would like to discuss two specific points, which the authors may like to consider.

The first concerns derated states. In the first example (no energy restrictions), the 15 MW capacity state in Table 1 implies a derated state assuming that the output of the unit is a maximum given that a failure or failures have occurred in a component or components of the unit system. If it is cost-ineffective to operate the unit at full capacity before other units are loaded, it is equally likely that it will be cost-ineffective to fully load the derated state before the other units are loaded. In these cases, not only should the full capacity state be truncated (20 MW in this example) but the derated state, although less than 20 MW, should also be truncated. The model shown in Table 2 would then be modified and the value of EENS$_1$ would increase. This does not alter the philosophy only the implications of the example being used.

The second point concerns energy-limited units with no storage. The analysis assumes that there is total independence between energy limitations and the system load. Although this may be acceptable for run-of-the-river installations it could be less sound for primary energy sources that have multi-applications. For example, the demand on natural gas supplies for direct energy usage (heating, etc.) is likely to increase at the same time as the demand for electrical energy increases. This may impose additional restrictions on the availability of gas for generation which is not accounted for if total independence is assumed.

Both of the above points can be resolved but only at the expense of increased data collection or analysis. For instance, all the states of the capacity model associated with the first point could be truncated as appropriate. This would however necessitate increased data collection and/or interpretation. Also the time period associated with energy-limited units with no storage could be divided into much smaller intervals so that the capacity model (flow rate distribution) would be much more appropriate to the correlated system load level. This would however increase the amount of analysis required. Perhaps the authors would like to comment on these points.

Manuscript received February 14, 1978.

Michael F. McCoy (Bonneville Power Administration, Portland, OR): Our compliments to the authors on a clear, understandable, and timely paper.

As the authors are well aware, large hydro complexes do not operate in parallel. Storage projects in headwater areas have inflows which would lead to a calculable energy distribution, but downstream plants are controllable and do not lend themselves to an easily calculated energy capacity when river optimization is taken into account. Do the authors intend that the controlled reservoirs be treated as independent random variables? If so, would they comment on the implications of so doing.

The procedures being developed for evaluating the quantitative reliability of bulk transmission systems must also account for energy availability. This is especially true for systems with huge amounts of hydro. If possible, would the authors comment briefly on the interaction of an energy limited generation model and a contingency evaluation type of bulk power reliability program.

Manuscript received February 14, 1978.

M. A. Sager and **A. J. Wood** (Power Technologies, Inc., Schenectady, NY): The authors have presented an interesting discussion of the treatment of energy limited generating units in the development of a generation reliability index based upon the expected failure to supply load energy. The extension of these ideas to the development of a failure probability related to peak load power demands (i.e., the conventional loss-of-load-probability index), rather than reliability indices based on energy, raises some interesting conceptual problems as does the consideration of energy storage generation such as pumped storage hydro.

Do the authors have any suggestions and/or experience in the treatment of limited energy supply generation in the calculation of the loss-of-load-probability index? It has been our experience that these units are not usually treated any differently than any other generating units in the system if and when it is assumed that the capacity is available for emergency supply.

The treatment of pumped storage plants in capacity related and energy related reliability studies is also necessary and possible. The probability of having a given amount of energy available for generation with this type of unit is, of course, dependent upon the probability that it was possible to store that amount of energy. We have found that this requires the use of either a frequency and duration reliability computation or else a probabilistic production cost model. Have the authors extended their analysis to this type of unit?

We should also like to observe that the various algorithms we have used over the years for the treatment of limited energy storage units are all variations of the basic probability table construction algorithm. "Peak shaving" is based upon the convolution of a limited energy unit with a load duration curve considered as a probability distribution that has been multiplied by the interval duration. For each step in the process of convolving a separate unit state, the state's energy and capacity constraints must be observed, leading to the peculiarly shaped curves of the authors' Figure 3 or Figure 3 of the authors' Reference 3.

There are some questions which we should like to suggest for possible further study. How should hydraulically coupled hydro-electric plants be treated in both generation reliability studies and probabilistic production cost studies? Is there any fundamental and philosophical difference in the analysis of the effect of pumped storage-like plants on (1) generation reliability and, (2) probable production costs? (If reliability were the only consideration, a pumped storage plant might be operated in a very defensive mode by keeping the maximum possible amount of energy stored for emergency use at all times. On the other hand, if economic operating principles are followed, the pumped storage-like plant may be in a state with practically zero available stored energy after a period of peak demands.) Should energy constraints be considered to be upper bounds or equality constraints? (The answer clearly appears to be an equality constraint for a hydro plant with a large storage capability, but seems to be less clear-cut as the type of plant is changed to one in which there is a true fuel cost, fuel consumption is optional but limited, and storage is non-existent. Finally is there any difference in an energy related reliability index, such as the expected unsupplied energy computed without regard to the economic dispatch of the generation system, as discussed by the authors and a similar index computed using a probabilistic production cost approach?

The authors may be inadvertently giving an erroneous impression concerning previous publications in this area. The original paper discussing probabilistic production cost computations by Jamoulle and his associates contained a treatment of limited energy units [1].

We have found that much of the published material relating to the probabilistic production cost technique becomes clearer if the interval load duration curve is mentally transformed into a probability distribution of needing a given amount of capacity or more by dividing the duration data by the length of the interval. Subsequent convolutions then follow more organized and formal mathematical treatments.

In the preprint that we reviewed it appears that equation (2) has a minor error.

REFERENCE

1. H. Balerioux, E. Jamoulle, and Fr. L. de Guertechin, "Establishment of a Mathematical Model Simulating Operation of Thermal Electricity Generating Units Combined with Pumped Storage Plants," revue E (Edition SRBE) S.A. EBES, Brussels, Belgium, Vol. 5, No. 7, pp. 1–24.

Manuscript received February 14, 1978.

R. Billinton and **P. G. Harrington**: The authors would like to thank all the discussers for their comments. In regard to the point raised by Dr. Allan in connection with the derating state, the loading should be treated as dictated by cost-effectiveness. As noted, this does not alter the basic philosophy of the approach. The analysis does assume that there is independence between energy limitations and the system load. This is an interesting point which could be included by using a conditional load model associated with each expected energy level and combining the results. As noted, by Dr. Allan, these points can be incorporated if sufficient data is available. The reduction of the time period into smaller intervals would aid in correlating the capacity model with system load level. It would increase the analysis requirement and could also pose difficulties in including the potential unit outage overlaps from one period to the next.

Dr. McCoy has raised the question of treating the controlled reservoirs as independent random variables. In general, the utilization of independence in reliability evaluation gives an optimistic appraisal of system adequacy. Dependence should be considered by developing a multi-unit hydraulic system model which includes the coupling and dependent aspects. This multi-level equivalent unit would then be incorporated in the analysis using the method proposed in the paper. At the present time it appears that composite system reliability evaluation using a direct analytical approach does not include energy limited capacity considerations. This could possibly be included using a conditional probability approach but could be computationally unmanageable. It may however be possible to include these constraints using a Monte Carlo or simulation technique.

In regard to the points raised by Messrs. Sager and Wood we would like to reply as follows. Our experience has also been that in most studies, limited energy units are treated the same way as non-limited energy units in the calculation of a loss-of-load probability index. This obviously gives an optimistic answer since the possibility that the energy limited units could not supply the required energy is not considered. A technique which might merit further study would be to use the load duration curve modification techniques described in the paper to find the equivalent load remaining after the energy limited units have been considered. Then perform a conventional loss-of-load probability study with the remaining units and the equivalent load modal. The loss of load expectation index in this case would be in terms of hours rather than the unit of days obtained by using the daily peak load variation curve.

153

The inclusion of pumped storage units unlimited energy system evaluation is an important consideration. This has however not been extensively examined in our present studies. We believe however that it should be possible to consider this by relating the energy available from pumped storage units to the excess energy available for pumping from base load units. The actual situation would however depend almost entirely on the system operating philosophy which is not adequately treated by our method or by an expected frequency and duration approach. A change in the rules by which a unit operates will change the actual system reliability. If an energy limited unit is run in a "defensive" mode, i.e. one in which the reservoir is always kept basically full until an emergency arises then the system reliability is enhanced but the overall production cost is higher than that experienced by running the units over peak to offset more expensive generation. Techniques developed to determine reliability should recognize the system operating philosophy, but it should be appreciated that as energy limitations arise, the economic dispatch philosophy of the system could change drastically to accommodate the limitations. Daily operating economic evaluation is done in the light of the conditions in which the system finds itself, which includes existing energy limitations. The authors did not intend to give the impression that energy limited units had never been previously considered. We do believe, however, that their consideration has been somewhat over complicated in earlier publications.

In conclusion we would again like to thank the discussors for their interest in our work and for their valuable comments.

Manuscript received June 15, 1978.

MULTI-AREA GENERATION RELIABILITY STUDIES

C.K. Pang, Member, IEEE
A.J. Wood, Fellow, IEEE
Power Technologies, Inc.
Schenectady, NY

R.L. Watt, Sr. Member, IEEE
SWEPCO
Shreveport, LA

J.A. Bruggeman, Member, IEEE
Central & South West Services
Dallas, TX

Abstract – This paper presents the results of a comprehensive multi-area generation reliability study of various possible interconnections of several power systems in the Southwestern portion of the U.S. Ten utility systems were represented in the study in distinct areas. System equivalent models were developed to represent these areas in five and six area models. A multi-area reliability program which incorporates the limitations of the transmission system on reserve and load-loss sharing was used to compute the loss-of-load probability (LOLP) of each area for various, different interconnections configuration and capacities. This reliability computation recognizes the limitations on transfer capabilities between the areas even when the network contains loops and the individual lines have non-zero forced outage rates.

Summaries of the models and computational methods used are given in the paper. Overall system data and interconnecting configurations are also provided. In general the calculated results show improvements in the various area reliability levels with increasing interconnection strength. Various studies were performed in order to test the models and system equivalents used. The paper presents a summary of the results of these tests and points out the usefulness and relative accuracy of various simplifications which may be used to increase the computational efficiency of the method.

The authors believe that this is the first multi-area generation reliability study where more than three areas are interconnected without restrictions on the network configuration.

INTRODUCTION

The computation of generation system reliability indices, particularly the failure to supply the expected load, has been done for single areas on a regular basis for a number of years. Extensions to two and three interconnected systems have been made, but in general, the analytical computation of loss of load probability (LOLP) values for a larger number of systems has been restricted to configurations without loops in the transmission network. The studies discussed in this paper use a computational method which computes the LOLP for all the areas simultaneously and recognizes explicitly the transfer capability limitations between areas in a multi-area system, regardless if the topology of the network configuration.[1,2] This method does not require many of the approximations needed when single-area or two-area models are used to study larger interconnected systems. To the knowledge of the authors, this is the first published report of an actual comprehensive multi-area reliability study in which limitations on the inter-area transfer capabilities are observed simultaneously with no restrictions on the network configuration.

The main objective of the reported study was to evaluate quantitatively the effect of various levels of transfer capabilities of the interconnection on the reliability of the individual generation-load areas. Loss of load probability is the reliability index computed throughout the study and values of LOLP with and without load loss sharing were computed simultaneously. An operating policy of sharing load losses implies for this study that all areas in the system will share their generation up to the limits of the transfer capabilities and, if necessary, an area will shed its own load to reduce load losses in other areas. On the other hand no load loss sharing implies that an area will share its generation only up to the limit of the transfer capabilities or its own available reserve, whichever is smaller. This corresponds to the policy of islanding an area short of capacity whenever the shortage exceeds the capabilities of the interconnected systems. There are an infinite set of possible operating policies that could conceivably be modeled. The authors have found that these two sets of LOLP values, bracketing the possible policies, are useful in interpreting the results.

A fairly large region of the electric power system in the United States is represented in this study. The period studied is from 1980 to 1986 during which time the total projected annual peak load of the region went from 54.7 GW in 1980 to 80.4 GW in 1986. Ten separate areas are represented in the study with each area corresponding to either a single power system or a natural grouping of systems located in a single geographical area.

MODELS AND METHODS

This section gives a summary of the models and computational method used in the studies reported herein. Certain essential definitions are also provided. The term "area" is used to denote a generation and load connected to a single bus, while the "system" is the entire collection of areas and the interconnecting transmission network.

Definitions

In considering more than a single area or bus it is necessary to define specifically what is meant by a failure. The following three definitions are used.

1. System Failure and Success – When the loads in at least one area are not satisfied, a failure state is said to have occurred in the system. A success state or condition is one in which loads are satisfied in all of the areas.

A 78 546-4. A paper recommended and approved by the IEEE Power System Engineering Committee of the IEEE Power Engineering Society for presentation at the IEEE PES Summer Meeting, Los Angeles, CA, July 16-21, 1978. Manuscript submitted January 31, 1978; made available for printing April 25, 1978.

Reprinted from *Proc. IEEE Summer Power Mtg.*, Paper A78, 546-4, July 1978.

2. Area Failure or Success - An area success state means that all the load in that area is satisfied while some load in the area is curtailed when an area failure occurs. Thus a system success state implies that all area loads are supplied, while a system failure means failure to supply loads in one or more of the areas within the system.

3. Load Loss Sharing - The area failure state is also dependent on the type of load loss sharing policy. When a failure occurs, the system can frequently be divided into a surplus region in which all area loads are satisfied, and a deficient region in which some area loads are not satisfied. The division, or "cut," occurs because the power transfer capability between the two regions is being fully utilized. All areas in the surplus region are in success states and all self-deficient areas in the deficient region are in failure states. A self-sufficient area in the deficient region is in a success state for a no load loss sharing policy since this policy states that an area will not shed its own load to help a deficient area. On the other hand, this self-sufficient area is in a failure state for the complete load loss sharing operating policy since this policy states that it is required to share its generation even to the point of curtailing some of its own load so as to reduce the load loss in a deficient area. It can be seen that for a system with "infinite" ties (i.e., tie capabilities greater than any that might be required to transfer generation reserves and share load losses) and under the complete load loss sharing policy, the system LOLP and LOLP values for all the areas will be identical.

System Models

A two-state model was used for all generating units in this study. The required data for each generating unit are its maximum capacity and its forced outage rate plus information on installation, maintenance and retirement. These unit data are used to construct the cumulative probability distribution table for each area by a well-known, numerical convolution algorithm. Generally the number of entries in the cumulative probability table is apt to be large for a large system with many units. This number is reduced to the neighborhood of some specified value by sampling at equal capacity levels in order to reduce the computational effort. The size of the table is strictly constrained by the amount of core storage and computing speed of the available computer. The resultant smaller table is then used to obtain the available capacity-exact probability table for the area concerned which is used in the reliability computation.

All transmission limitations internal to any area are ignored. This, in essence, is the definition for an "area." The areas are connected by the transmission lines and the system is represented by a linear flow network. The power flow pattern in such a network model is constrained only by the network topology and the capacity of the links or lines. Thus, Kirchoff's current law is observed and the voltage law ignored. The capacity of each link corresponds to the transfer capability between the connecting areas. In this particular study, the transfer capability of interconnections between areas is equal to the sum of the surge impedance loading capabilities of the transmission lines. Forced outage rates may be assumed for the individual tie lines if desired.

Seven years, 1980 through 1986 were studied with each year divided into three seasons. A finer subdivision was unnecessary because this particular region is strongly summer peaking. Load diversity among the areas within a season is ignored and area peak load distributions for individual seasons are used. Load diversity can be treated by dividing the year into more seasons so that the peak loads of the areas would fall into different seasons. The power system analyzed in this study has summer peak loads in all the areas so that ignoring diversity is a reasonable assumption in this particular case.

In order to keep the computing times within practical limits, it is necessary to limit the number of entries or load levels in the peak load distribution. The reduction in the number of load levels is done by sampling the load distribution at some specified interval and combining consecutive load levels which are close together (i.e., for each area, the difference of the loads in the two levels does not exceed some specified value).

The effects of maintenance of generating units upon reliability should be considered. Generating units are unavailable for meeting load demands when they are on maintenance, hence calculated system reliability indices will be worse than the case in which maintenance is ignored. The multi-area program allows the modification of the load or capacity model to simulate maintenance effects.

In this study, maintenance was accounted for by adding the maintenance capacity out of service to the load, permitting the use of the same available capacity-probability tables for these periods during which the installed capacity does not change. A test case in which maintenance was represented by modifying the available capacity-probability tables yielded practically the same computed annual LOLP values as those computed when maintenance capacity was added to the load. In this case, this excellent agreement is due to the fact that practically all of the LOLP contribution comes from the peak summer season during which little or no maintenance is carried out.

LOLP Computation

Only a brief description of the computational method used will be given here since a full description has been reported in References 1 and 2. Essentially, a systematic and efficient classification process identifies all the system success and failure states. All possible system states or contingencies are evaluated.

The maximal flow-minimal cut labeling algorithm (sometimes referred to as the Ford and Fulkerson algorithm) is used to determine the network flow pattern such that all load, or as much load as possible is satisfied.[3] When some load is not satisfied, the minimal capacity cut across the network is identified and this cut contains the necessary information regarding the cause of the failure and the associated relevant probability values.

System Description

The main objective of the study was to evaluate the reliability benefits of varying degrees of interconnection between two major systems in the Southwestern part of the United States. For purposes of this paper, one major system is designated Northeast (NE) and the other Southwest (SW). Each of the two major systems is divided into five areas where one area constitutes either a single utility company or a group of companies. Areas 1 through 5 are in the NE system while the SW system consists of areas 6 through 10. Area 10 is split up into 10A and 10B for the third interconnecting configuration studied. Area 4 is rather remote from other NE areas and was therefore included for only some sensitivity studies.

Four basic interconnection configurations between the two systems are examined.

The first, Mode 1, is shown on Figure 1 where the two major systems NE and SW are not interconnected. In the Mode 2 configurations shown on Figure 2, Areas 6 and 9 are intertied to the NE system leaving Areas 7, 8 and 10 to form a reduced SW system. The third configu-ration (Mode 3) shown on Figure 3 is similar to Mode 2 except that Area 10 has been split into two sub-areas, 10a and 10b and 10b is connected to the NE system by ties to Areas 6 and 9. Finally, Figure 4 shows the complete interconnection of both the SW and NE systems. The numerical data shown on the various interconnecting lines are the transfer capabilities in MW of forecast for the Year 1980.

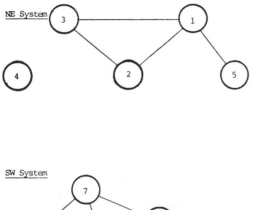

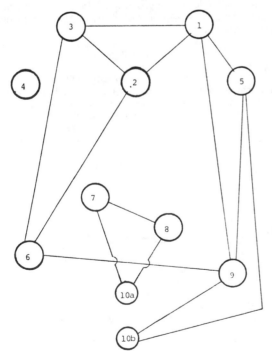

Figure 1. The Mode 1 Interconnection Configuration - Area 4 Remains Isolated and Areas 10a and 10b Are Treated as a Single Area (10).

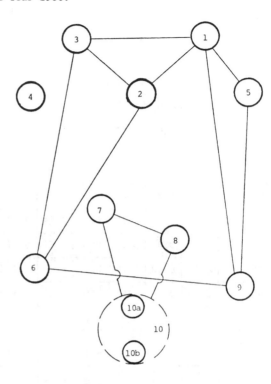

Figure 2. The Mode 2 Configuration - Area 10 Includes Both 10a and 10b.

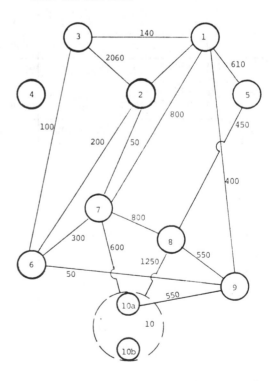

Figure 3. In the Mode 3 Interconnection Configuration Area 10 is Split Into Two Sub-areas.

Figure 4. The Mode 4 Interconnection Configuration - Area 10 Includes 10a and 10b. MW Transfer Capabilities Expected in 1980 Are Indicated.

157

The approximate peak load forecasts used for the years studied (i.e., 1980 through 1986) are given in Table I. Capacity forecasts and projected new transmission lines were obtained from published data. Identical capacity forecasts are used in all the four modes of interconnection except the capacities of Areas 1 2, 6 and 9. Generation expansion plans for these four areas for each of the four modes are shown in Table II where in some instances portions of larger units are indicated. The total projected capacities are approximately equal for the four modes studied.

TABLE I

Annual Peak Loads (GW)

(Forecast in 1976)

Areas	Year						
	1980	1981	1982	1983	1984	1985	1986
1	2.99	3.26	3.56	3.84	4.14	4.47	4.83
2	3.57	3.80	4.11	4.36	4.68	5.07	5.34
3	6.11	6.68	7.28	7.95	8.70	9.47	10.30
4	2.65	2.81	2.96	3.12	3.29	3.45	3.62
5	5.36	5.62	5.90	6.20	6.51	6.84	7.18
Subtotal for NE	20.67	22.17	23.81	25.47	27.32	29.29	31.36
6	0.90	0.94	1.00	1.05	1.11	1.17	1.23
7	15.68	16.82	18.06	19.37	20.78	22.28	23.89
8	9.90	10.50	10.95	11.60	12.15	12.60	13.05
9	2.87	2.96	3.08	3.20	3.33	3.46	3.64
10	9.33	10.10	10.85	11.67	12.49	13.39	14.38
Subtotal for SW	34.00	36.27	38.52	41.06	43.61	46.20	49.00
Total	54.68	58.44	62.33	66.53	70.93	75.49	80.36

When Area 4 is excluded from consideration, the planned generation capacity reserves for Mode 4 for each year are as follows:

Year	Reserves (% of Peak Load)		
	NE	SW	Total
1980	16.4	24.5	21.7
1981	9.9	28.4	21.9
1982	15.8	31.7	26.1
1983	15.2	26.6	22.6
1984	16.2	21.9	19.9
1985	14.3	20.2	18.1
1986	18.2	19.2	18.9

The planned reserves for the other configurations are not significantly different.

TABLE II

Generation Plan for Areas 1, 2, 6 and 9

for Different Modes of Interconnection

Year	Unit Data			Mode			
	MW	Type*	Area	1	2	3	4
1980	550	C	9	X	X	X	X
	450	C	2	X	X	X	X
	530	C	1	X	X	X	X
1981	320	N	9	X			X
1982	530	C	1	X	X	X	X
	320	N	9	X			X
	550	N	9			X	
	640	C	9		X		
1983	700	N	2	X	X	X	X
1984	640	L	1	X	X	X	X
1985	700	N	2	X	X	X	X
1986	640	L	1	X	X	X	X
	420	L	1		X	X	X
	550	C	9	X			

* Fuel Type: C = Coal, N = Nuclear, L = Lignite

Actual outage rates for existing generating units are used wherever available. Outage rates for new units and those existing units for which outage rates are not available were taken to be equal to industry average data as reported in 1975. Transmission line outage data were estimated from a variety of studies on line outages.

RESULTS

The results of a number of different studies are discussed below. LOLP values with and without load loss sharing were computed but detailed discussions of results are restricted to those data computed assuming no load loss sharing. Nevertheless, reliability indices computed with both assumptions have been very useful in interpreting the adequacy of generating capacities between areas. This study provided the opportunity to investigate the applicability of the multi-area reliability approach to a number of different situations involving all, or some of the 10 areas studied.

Effects of the Total Number of System States

As discussed in Reference 1, multi-area generation reliability assessment involves the consideration of a very large number of system states. For a specific study system, the total number of system states is given approximately by the number of load levels times the product of the number of capacity steps in the capacity-probability table for each area. The algorithm used for computing success and failure state probabilities is very efficient but even so, in this particular study it was necessary to reduce the number of possible states by restricting the number of possible discrete load levels and the size of the individual area available capacity-probability tables. A series of annual reliability calculations for the 1983 NE system (excluding area 4) were made with various load models and probability table lengths in order to provide a quantitative measure of the effect of the maximum number of system states on computed LOLP values and the required computation effort.

LOLP values for eight different, 4 area cases were computed. The system configuration is that shown on Figure 1 for the NE system. Each case represents a combination of a different number of steps in the area capacity-probability tables and a different number of load levels so that a wide range of the total number of possible system states was achieved.

Figure 5 shows plots of the relative computer time, the system LOLP, and LOLP of Area 1 versus the total number of system states. Variations of LOLP for the other three areas show essentially the same form as that for Area 1. Computing time increases rapidly with the total number of possible system states while both the system and individual area LOLP values remain fairly constant.

As the models were altered to increase the number of states the same relative reliability levels for all the four areas were maintained. The simpler models (smaller number of system states) yielded conservative LOLP values with respect to the more detailed models. Thus meaningful and conservative results can be obtained from simpler models at considerable savings in computational effort.

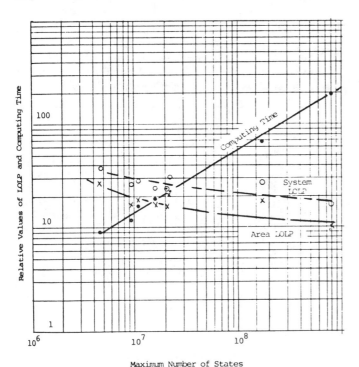

Figure 5. Results of 4-Area, NE System Test Case Showing Effect of Reduced Number of System States on LOLP Values and Required Computing Time

System Equivalents

The computational effort for the multi-area analysis increases exponentially with the number of areas in the system. Figure 6 shows the relative computer effort versus the number of areas modeled for various configurations for this particular system. The required computing effort appeared to be excessive for

the nine area systems corresponding to the Mode 4 interconnection, when all of the various cases were taken into account. The effect on LOLP values of using system equivalents for several of the areas was studied.

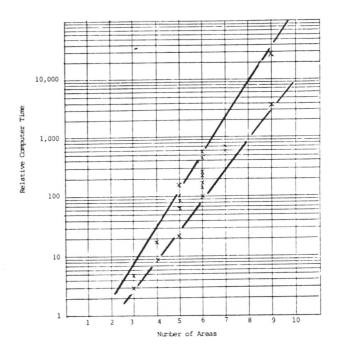

Figure 6. Relative Computational Effort for Multi-Area Reliability Calculation

The Mode 4 interconnection and generating system data for 1980 were used, but only the annual peak load was examined. The LOLP reliability computation was performed with the 1980 nine-area system shown on Figure 4. The transmission circuit capabilities between areas are shown on this diagram. A second computation was then carried out with the five-area system (Figure 7) in which all the five areas in system SW had been grouped together as a single area. That is, the total load and generation of the SW system was assumed connected to a common bus. This provided the LOLP values for the areas in system NE as well as a system LOLP value for the SW equivalent. Next the NE system (excluding Area 4) was treated as a single area

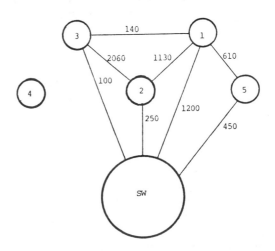

Figure 7. Equivalent System Model Used to Calculate Area LOLP Values for the NE System Areas

159

and a third computation was done with the resulting six-area system as given in Figure 8. LOLP values were computed for the five SW areas and the four area, NE system equivalent.

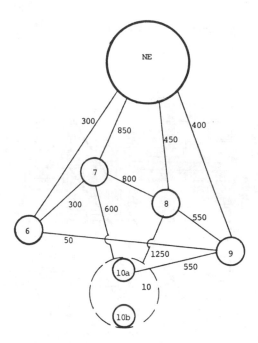

Figure 8. Equivalent System Model Used to Compute the LOLP Values for the SW System Areas. Area 10 Includes Both 10a and 10b.

Individual area and system LOLP values for the three computations are given on Table III. The area LOLP values computed by using the system equivalents were increased or remained approximately the same as those computed for the nine-area system. The largest differences occurred for Areas 9 and 10 which had the lowest LOLP values so that these discrepancies are of little or no significance in the overall reliability evaluation. The computing time for the 9-area case is almost 40 times the sum of the computing times for the two cases using the equivalent systems. It may be concluded that meaningful LOLP values may be obtained at significant savings of computing effort by performing two computations using equivalent systems rather than one computation on a system with a larger number of areas.

Results from Different Modes of Interconnection

Computations were performed for each of the four modes of interconnection shown on Figures 1 through 4 for each year of the study period. The LOLP values were obtained for each of the 9 areas and were generally much higher for the latter part of the study period (1984-1986) compared to those for the earlier period (1980-1982). These variations in the LOLP values were expected and all due to the overall, planned lower percentage capacity reserves and the installation of larger units with significantly higher forced outage rates in the later part of the study period.

Mode 4 provided the most reliable operating condition, that is the lowest LOLP values, for all the areas in all years of the study period. In certain instances, the calculated LOLP values for Mode 4 were

TABLE III

Area LOLP's (Per Unit) for the 1980 Peak Loads
Mode 4 Interconnection: No Load Loss Sharing

Case / Area or System	Per Unit LOLP Values		
	9 Areas	4 Areas & SW Equivalent	5 Areas & NE Equivalent
NE System			
Area 1	0.410×10^{-5}	0.680×10^{-5}	
Area 2	0.375×10^{-5}	0.440×10^{-5}	0.918×10^{-5}
Area 3	0.124×10^{-4}	0.152×10^{-4}	
Area 5	0.154×10^{-2}	0.155×10^{-2}	
SW System			
Area 6	0.125×10^{-5}		0.109×10^{-5}
Area 7	0.246×10^{-5}		0.570×10^{-5}
Area 8	0.357×10^{-5}	0.183×10^{-5}	0.457×10^{-5}
Area 9	0.463×10^{-7}		0.847×10^{-6}
Area 10	0.460×10^{-8}		0.256×10^{-7}
Total System	0.156×10^{-2}	0.156×10^{-2}	0.172×10^{-4}
Relative Computer Running Time	175.2	1.00	3.52

about three orders of magnitude lower than the corresponding values for the worst modes. These results were the most significant ones for this particular study.

The LOLP data computed for each area each year studied includes its LOLP as an isolated area as well as two annual LOLP indices (i.e., with and without load loss sharing) for each of the four modes of interconnection. In addition, system LOLP indices were computed for the NE, SW and total interconnected systems where appropriate.

Figure 9 illustrates the type of individual area results obtained and shows several of the yearly LOLP values calculated for Area 8. Data are shown for Area 8 considered as an isolated system without any effects of ties to other systems and for two of the interconnection configurations, Modes 1 and 4. Data for Modes 2 and 3 generally follow similar patterns but show generally higher calculated LOLP values for these modes since in these configurations area 8 is interconnected with fewer systems.

Two different annual LOLP values are shown for each of the two interconnected modes. The solid curves show the risks calculated assuming no sharing of load losses while the dashed curves show the results assuming an operating policy of sharing load losses.

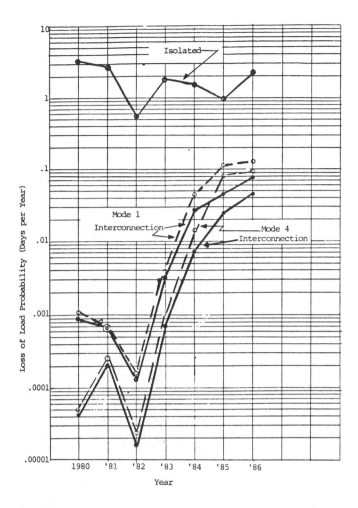

Figure 9. LOLP Values for Area 8

These data are higher as expected. The difference for a given interconnection is relatively insignificant, however, until the last three years of the study when several area reserve levels are low and the tie capabilities have increased to the point where there are significant possible load losses that may be shared.

In all areas it was generally found that the calculated LOLP values decreased as the strength and extent of the interconnection increased. The rare exceptions occur when a major system with very low reserves (and a high LOLP) is initially interconnected with other, more reliable areas and the LOLP values are computed assuming load loss sharing.

Delay of Units

One of the more frequent types of sensitivity studies that keep recurring concern the reliability effects of possible delays in unit installation. Several patterns of possible unit delays were studied in order to assess the quantitative effects on area reliabilities.

The area LOLP data calculated illustrate the impacts on adjacent area reliability levels when specific units are delayed. For example, in a Mode 4 study the delay of a single 750 MW unit in Area 7 affected primarily the reliabilities in Area 7 and in the most closely tied Areas 1, 6, 8, 9, and 10. The LOLP values in 2 and 3 were almost unaffected because of the relative remoteness of these two areas from the point of view of sharing generation capacity.

Another observation is that the delay of one unit in the SW in area 7 had little impact on the LOLP values of the areas in the NE system while the reverse was not true when a unit in area 1 was delayed. This can be explained by the fact that the generating capacity reserves in the SW system were several times the total transfer capability between the two systems. On the other hand, the generating reserves in the NE system were only about 70 percent larger than the tie capacity so that delaying a unit in area 1 in the NE system effects the entire interconnected system.

Inclusion of Another Area

This study was undertaken to examine the sensitivity of the area LOLP values with the inclusion of an additional area. Practical constraints on the allowable computing time did not permit the inclusion of all areas which are connected in some way to the study system since this could include the entire Eastern group of interconnected utility systems. Figure 10 illustrates the 1985 system interconnected in

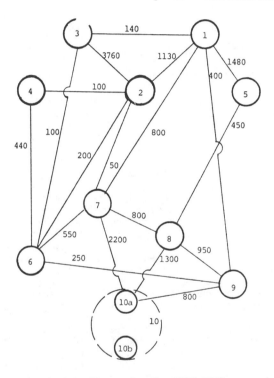

Figure 10. Mode 4 Configuration for 1985 With Area 4 Included in the NE System. Area 10 Includes Both 10a and 10b.

the Mode 4 configuration with inter-area tie capabilities shown.

The area LOLP values for 1985 computed with and without the inclusion of area 4 are shown on Table IV. It may be seen from these data that all area LOLP values were reduced almost proportionally when Area 4 was added to the system model. The relative reliability levels between the other study areas was still preserved.

The reliability of individual areas will tend to improve when more areas are added to the study system as long as the new areas are not exceptionally low in generating capacity reserves and it is feasible to transfer these reserves to other systems. Obviously, the addition of a new area consisting only of a new load will increase the LOLP index levels. However, as

TABLE IV

Effect of Including Area 4 on 1985 Mode 4 LOLP's

LOLP Values (Days Per Year)

Area	Without Load Loss Sharing		Including Load Loss Sharing	
	Without Area 4	With Area 4	Without Area 4	With Area 4
1	.088	.069	.107	.083
2	.053	.040	.124	.096
3	.102	.078	.124	.096
4	NA	.122	NA	.150
5	.059	.048	.110	.087
6	.084	.071	.084	.071
7	.071	.053	.084	.062
8	.024	.018	.082	.061
9	.061	.045	.082	.061
10	.011	.009	.082	.061

NA: Not Available

more areas with generating reserves are added, computational effort increases and the absolute reliability (LOLP) levels will decrease. Engineering judgment and experimentation will indicate the extent of the interconnection required to study a particular problem.

Inter-Area Transfer Tie Requirements

The reliability of a sub-system comprising of only Areas 1, 2, 6 and 9 interconnected in the form of a bridge network with a diagonal circuit from Area 1 to Area 6 was studied in order to estimate the transfer capability requirements such that complete sharing of reserves between them can be achieved. A series of reliability computations was performed for different sets of transfer capabilities which were gradually increased. The need for increased tie capacity was based on the computed probabilities of any particular tie line being limiting. The availability of these probability data from the multi-area calculation provides an efficient means of studying the need for added tie capacities. The probability data showing the probability that any specific tie is limiting are computed considering all possible outage events and not just for single or double contingencies or selected outages.

Figure 11 shows the plots of the LOLP values of Areas 6 and 9 calculated with no load loss sharing for the year 1980 versus the transfer capability into each respective area. As expected these LOLP values decreased with an increase in transfer capability and eventually saturate at some value. Area 6 experienced a more rapid rate of reliability improvement than Area 9, attributable to the fact that Area 9 is about three times as large as Area 6.

The LOLP data obtained in the various studies of the Mode 4 interconnection using the equivalent systems (shown on Figures 7 and 8) provide an approximate indication of the limiting intersystem capability needed to share all available generation reserves between the NE and SW systems. Each LOLP computation utilizing one or the other of these equivalents provides a computation of the LOLP of the equivalent considered as an isolated system and with the equivalent system intertied to the other areas. Both of the equivalent system LOLP data (see Table III), computed isolated and with load

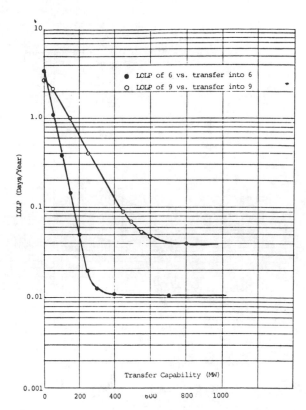

Figure 11. Results of 4 Area Test to Investigate Transfer Capabilities For Areas 1, 2, 6 and 9. Data are Shown For No Load Loss Sharing.

loss sharing, may be plotted versus the total transfer capability connected to each of the systems and extrapolated to estimate the inter-systems transfer capacity needed so that both equivalent systems will have the same calculated LOLP indices. A log-linear extrapolation appears to be most appropriate for this purpose.

This procedure is similar to the well-known techniques used in two area reliability studies.[4],[5] Its use in this multi-area reliability study provided a useful guide to the total, intersystem transfer capabilities which would be effective for sharing generation reserves. In practically all instances these capabilities were close to the levels established from other design criteria.

Reduction of Sub-System LOLP in the Interconnection

The multi-area method facilitates the study of reliability benefits due to interconnection. The 4-area inter-area transfer capability study discussed above provides another set of LOLP data points to illustrate the improvement in reliability for areas 1, 2, 6 and 9 by means of further interconnection.

1980 LOLP (Days per Year)

Area	Isolated	4 Area Interconnection 1,2,6, & 9	Mode 1 Interconnection	Mode 4 Interconnection
1	4.85	0.06	0.03	0.00004
2	0.47	0.03	0.002	0.00002
6	3.38	0.01	0.002	0.00002
9	2.70	0.04	0.0004	0.00002

The computed area LOLP values for these four systems decrease by several orders of magnitude as the strength and extent of the interconnection are increased. This may be illustrated by the 1980, area LOLP data tabulated below for four different interconnections. These data were computed with no load loss sharing.

As a further indication of the dependence of sub-areas upon their ties for reliability of supply, area 7 was divided into four independent sub-systems for purposes of computing LOLP values. Data of a similar nature to that above was computed for 1980 and are as follows:

1980 LOLP (Days per Year)

Area	Sub-Areas Isolated	Area 7	Mode 1	Mode 4
7A	0.13	0.10		
7B	30.8	0.33	0.005	0.0003
7C	17.6	0.28		
7D	20.8	0.29		

Note that the data for Modes 1 and 4 are computed assuming all of the sub-systems are connected to an "infinite bus." For generation reliability studies this single LOLP value applicable to all four sub-systems implies no internal tie restrictions and load loss sharing.

These data as well as the preceding graphically illustrate the reliability improvement potentially available from extensive interconnections. There are, of course, practical limitations to the degree to which remotely located generation reserves may be shared. These limitations involve voltage and VAR support problems, operating and control difficulties and questions of system security.

CONCLUSIONS

The results of this study showed that for the given forecasts and system configurations, area LOLP values are expected to increase during the latter part of the study period due primarily to a decline in the expected generation reserves. The installation of large units having significantly higher outage rates than those for the existing units is also a contributing factor. In the region studied, the complete interconnection of all areas in a Mode 4 configuration (see Figure 4) was the most reliable of all the different configurations. Even though the SW system was much more reliable than the NE system when each was considered independently, interconnecting them reduced the LOLP values of the individual areas in the SW system as well as benefiting those in the NE system.

The interconnection of all the areas will greatly improve their ability to achieve reliable generation systems in the event of unexpected delays in the installation of new capacity.

It has been shown in this study that the inclusion of additional surrounding, interconnected areas which have available generation reserves will further improve the reliability of the study areas. Thus, the results obtained were generally conservative since some surrounding areas were omitted in the studies.

The studies have also demonstrated the feasibility of applying the multi-area reliability computation methods as reported in References 1 and 2 to large interconnected regions. This method does not have any theoretical constraint on the number of areas and the configuration of the system that may be studied. However, more extensive systems with a larger number of separate areas may increase the computational effort severely. System equivalents may be used to reduce the computational effort.

Inter-area transfer limitations are observed at all times in this method and the computation of all area LOLP values is performed simultaneously. Area LOLP data for both complete sharing and no sharing of load losses are obtained. Both sets of values have been found to be useful in interpreting the results. The probability that each tie line (or "link") is limiting identifies locations in the interconnecting network that can be strengthened in order to facilitate the sharing of generation reserves and reduce the LOLP values. The multi-area reliability program results may be used to study the benefits, scheduling and selection of both generation and transmission capacity installations.

REFERENCES

(1) C.K. Pang and A.J. Wood, "Multi-Area Generation System Reliability Calculations," IEEE Trans., Vol. PAS-94, No. 2, March-April 1975, pp. 508-517.

(2) C.K. Pang and A.J. Wood, "Multi-Area Power System Reliability Calculations: Extensions and Examples," IEEE paper no. A75 535-5, presented at Summer Power Meeting, July 1975.

(3) L.R. Ford and D.R. Fulkerson, "Flows in Networks," Princetown University Press, 1962.

(4) R. Billinton, R.J. Ringlee and A.J. Wood, "Power System Reliability Calculations," The MIT Press, 1973.

(5) V.M. Cook, C.D. Galloway, M.J. Steinberg and A.J. Wood, "Determination of Reserve Requirements of Two Interconnected Systems." IEEE Transactions, Vol. PAS-82, pp. 18-33, 1963.

APPLICATION OF FOURIER METHODS
FOR THE EVALUATION OF CAPACITY OUTAGE PROBABILITIES

N. S. Rau, SMIEEE
National Energy Board

K. F. Schenk, SMIEEE
University of Ottawa

Ottawa, Ontario, Canada

Abstract - The LOLP index is used by North American utilities as a criterion to compare different generation expansion plans. This index is computed either by obtaining a capacity outage probability table or by convolving the load duration curve with the outage probabilities of each generator. This paper proves that a very accurate probability density function for the capacity outages can be obtained by using Fourier techniques. The proposed method is very efficient, speeds up calculation and can handle the derated outage states of generating units accurately. The proposed distribution for the capacity outages is shown to be very well suited for further analysis on load forecast uncertainty and other uncertainties in the evaluation of LOLP.

INTRODUCTION

Probabilistic methods have been used by utilities in North America to assess adequacy of generation and future expansion plans. The extensive amount of analysis of this problem and various philosophies used by different utilities is documented in Reference 1. The most common index used is that of the probability of loss of load (LOLP). While different utilities use different methods and judgement to calculate the LOLP index, the methods can be categorized under two broad philosophies. The first is that of obtaining a capacity outage table indicating the probability of losing a certain capacity. This is obtained by using the documented statistics or the anticipated forced outage rate (FOR) of generating units. The load is modelled as the daily peak loads [2] and the average probability that the reserve is ≤ 0 over a given month gives the LOLP. In another variation of this method [3], all loads over a given period are considered and the resulting LOLPs for reserves ≤ 0 for different loads are weighted in accordance with the probability of such a load occurring. In a second philosophy called the Frequency and Duration method (FAD), the generators are modelled in their up and down states and the mean residence time in these states is considered. Further, by considering the frequency of changing from one state to the other, the LOLP is obtained. This method is ideal for spinning reserve calculations. Again the data such as mean residence time and frequency of changing states is obtained by statistical documentation of unit performance and a judgement of such data for future units. The FAD method is admirably suited for spinning reserve calculations [8]. For long range system planning, however, one is concerned with the FORs of units and these can be obtained from FAD methods as $t \rightarrow \infty$. This paper is concerned with long range system planning and therefore addresses itself to the capacity outage probability distribution.

The electric utility industry is capital intensive. The use of larger units by the utilities has resulted in a reserve margin which is an increasing percent of the installed capacity for the same LOLP criteria. This, coupled with the increasing participation by the public in the utility planning process, has given rise to several questions about the reliability index used by the utility industry. Common questions asked are about the value of LOLP index to be chosen as a criterion for planning and about the fact that this index by itself does not indicate the probable duration of the expected outage. Although some of the logic used may be questionable, Telson [4] made some efforts to link the unserved energy to social costs. There is a general concensus among the system planners that, for long range capacity planning, further refinements and additions to the LOLP index are desirable.

With the above background, the motivation for this work was to establish a theoretical continuous probability distribution for the capacity outage probabilities. Such a distribution allows quicker calculations to be performed but would also offer a better insight to planners about the effect of different generation schemes on the LOLP index. These advantages are absent in the well-perfected programs based on recursive algorithms or otherwise, to calculate the LOLP. Further, a mathematically expressible distribution lends itself easily to further analysis such as the effects of load forecast uncertainties and of FORS, the effect of multi-area interconnections, cost of interruptions and the like. In a subsequent publication, we shall point out the advantages of such distributions in arriving at loss of energy predictions in a generating system.

In this paper we have fulfilled the above motivation. The first section contains the results of the Fourier techniques used to express the capacity outage table as a sum of two normal distibutions. The accuracy of the sum of two normal distributions has been shown to be unacceptable, except for rough calculations [7]. In the subsequent section, the Fourier technique has been extended to obtain a very accurate distribution in the form of Gram Charlier's expression. The accuracy of the resulting distribution has been shown to be excellent, in the same order as the conventional outage tables obtained by recursive relations. Two model power systems of 4190 MW and 23,000 MW have been used to confirm the accuracy of the proposed method. It is believed that this method extends the work of [9] to obtain a continuous and accurate distribution. Therefore, the proposed method offers a faster algorithm as an alternative to the conventional methods and is amenable to hand calculations.

PROBABILITY DISTRIBUTION VIA FOURIER TECHNIQUES

Preamble

For a preliminary analysis, the n generators in a system with capacity C_i for $i=1,\ldots,n$, can be modelled as a two state (UP and DOWN) process with an availability p and a forced outage rate q_i $(p_i + q_i = 1)$. These p_i and q_i are obtained from the conventional FORs. This model can be represented as two impulses: one of magnitude p_i at $x=0$ and another of magnitude q_i at $x=C_i$, as shown in Figure A-1 of Appendix I. It is

A 79 103-3   A paper recommended and approved by the IEEE Power System Engineering Committee of the IEEE Power Engineering Society for presentation at the IEEE PES Winter Meeting, New York, NY, February 4-9, 1979. Manuscript submitted September 1, 1977; made available for printing December 6, 1978.

Reprinted from *Proc. IEEE Winter Power Mtg.*, Paper A79, Jan./Feb. 1979.

clear that a convolution of such models of n machines will result in 2^n impulses. The magnitude of these impulses indicates the outage probability of a certain capacity x. This train of impulses represents the density function of outage probabilities. Note that the sum of the magnitudes of these impulses is equal to unity and that the process of convolution is equivalent to a multiplication of Fourier transforms in the Fourier domain. In the following paragraphs, this distribution of impulses will be referred to as "absolute" density function.

For large systems, it is a formidable task to obtain the conventional capacity outage table by the above convolution process. Instead, a very efficient recursive method[5] is used (it should be observed that recursive methods are valid if $p_i + q_i = 1$. This aspect will be discussed in a subsequent publication). In effect, these methods divide the capacity (x) axis into discrete step sizes, typically 50 MW. The probability of outage for a certain capacity entry in the outage table thus generated will contain a value proportional to the summation of all the impulses of the absolute density function in that step interval of capacity. Evidently, the accuracy of such an outage table is dependent on the step size used during the computation and the distribution of impulses in the absolute density function. Unless the step size in the recursive algorithm is chosen to be very small, the resulting density function may not have any resemblance to the absolute density function.

The accuracy of the probability density function (PDF) of capacity outage, however, is not important to the calculation of the LOLP. If R is the reserve margin, LOLP is given by $\Pr.\{R \le 0\}$ and this is evaluated from a cumulative distribution function (CDF) obtained by integrating (addition of impulses) the PDF from operating or available capacity to 0. In other words, for any operating capacity OC and peak load L_p, one has that $R = OC - L_p$ and therefore

$$\Pr.\{R \le 0\} = \Pr.\{(OC - L_p) \le 0\} \qquad (1)$$

which is identical to $\Pr.\{\text{capacity outage} \ge (OC-L_p)\} =$ LOLP. This is obtainable directly from the CDF. Therefore, in the compilation of LOLP from capacity outage tables, the purpose has been that of obtaining a table the yields LOLPs, calculated from the CDF, with reasonable accuracy.

In the following analysis, we have addressed ourselves to obtaining a continuous density function for the outage probabilities. It should be realized that such a distribution as well as the outage table discussed above have no physical existence (unlike the absolute distribution). The rationale for this work had been that the CDF obtained by integrating the continuous distribution function should give values for the probability (of a certain capacity outage or more) that are identical to conventionally computed values or to those from the absolute density function. In statistical analysis, this procedure is called "fitting a distribution" to a process.

GAUSSIAN APPROXIMATION

Appendix III shows the details of the theory in obtaining a Gaussian distribution. In this process, terms only up to the second order in the series expansion for the Fourier transform have been used. For machines with 2 operating states (UP and DOWN) the PDF for any capacity x is given by

$$P(x) = (1/\sqrt{2\pi}\sigma)\{\exp[-(x-\eta)^2/2\sigma^2] + \exp[-(x+\eta)^2/2\sigma^2]\} \quad (2)$$

where

$$\eta = \sum_{i=1}^{n} \eta_i \qquad (3)$$

$$\sigma = \sum_{i=1}^{n} \sigma_i \qquad (4)$$

and

$$\eta_i = C_i q_i \quad ; \quad i = 1, \ldots, n \qquad (5)$$

$$\sigma_i^2 = C_i^2 p_i q_i \quad ; \quad i = 1 \ldots, n \qquad (6)$$

The corresponding changes which can be easily considered for multiple and derated states have been shown in Appendix III. Equation (2) represents two densities as shown in Figure 1a. Therefore, the probability of a capacity outage of amount x MW or more may be obtained as follows

$$\Pr.\{\text{Capacity Outage} \ge x\} = \text{Area 1} + \text{Area 2}$$

Note that the above calculation can be performed easily and quickly by hand calculations and probability tables.

The density shown in (2) can be standardized by defining the standard variables

$$Z_1 = (x - \eta)/\sigma \qquad (7)$$

$$Z_2 = (x + \eta)/\sigma \qquad (8)$$

In terms of the standard variables, Area 1 and Area 2 of Figure 1a correspond to the areas Area 1 and Area 2 of Figure 1b.

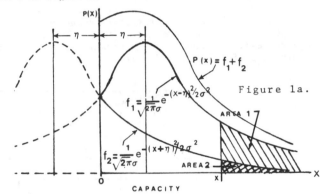

Figure 1a.

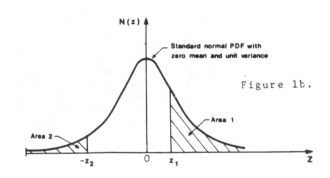

Figure 1b.

Figure 1. The Gaussian distributions

Example:

To help the reader in understanding the above procedure, we give the following simple example.

Consider a four machine system with capacities of 50, 100, 150 and 200 MW. Let the FORs of the first three machines be 0.1, 0.15 and 0.17 respectively. Let the fourth machine have an overall FOR of 0.2 with a derating state at 100 MW capacity. Let this FOR of 0.2 be divided to 0.1 at 100 MW and 0.1 at 200 MW. Then, for each machine, we have the following .

$C_1 q_1 = 50 \times 0.1 = 5$. Similarly, we have 15 for $C_2 q_2$ and 25.5 for $C_3 q_3$. For the forth machine, $\eta_i = 100 \times 0.1 + 200 \times 0.1 = 30$. Thus, the sum of the means is 75.5.

Use of (A-22) and (A-20) to calculate σ_i will give the following values.

$\sigma_1^2 = 225$, $\sigma_2^2 = 1275$, $\sigma_3^2 = 3174$ and $\sigma_4^2 = 4500$.

165

The total of these quantities represents the square of the standard deviation of the density function and is equal to $\sigma = 95.75$.

If one desires to calculate the probability of loss of 110 MW, say, using (7) and (8) one gets $Z_1 = 0.3682$ and $Z_2 = 1.98$. From probability tables, Areas 1 and 2 are 0.3567 and 0.0238 respectively. Hence, the Prob. of 110 MW or more being on outage is 0.3805.

Note that the total mean and standard deviation, once calculated, can be stored and considered as fundamental parameters of the system.

Accuracy:

We now examine the results obtained by this distribution with those obtained by using conventional capacity outage programs. Appendix II outlines the 23 machine generation model used to check the accuracy of this distribution and others. Figure 2 shows the CDF obtained by conventional methods and by the Gaussian distributions. Note that the accuracy is good at high risk levels and is unacceptable at low risk levels. Stated differently, the range of outage probabilities in which the system planners are interested is between 0.003 and 0.0003. In this range, for a given probability criterion, the desired reserve calculated from the distribution is quite in error compared to the conventional result. Therefore, it was found desirable to develop higher order methods to obtain accurate results for low risk levels.

We mention in passing that the Hermite polynomial correction suggested by Papoulis[6] gave somewhat better results. The result obtained is shown if Figure 2. However, it was realized that the Gram-Charlier expansion using four terms gave more accurate results and was amenable to hand calculations requiring the use of probability tables. This resulted in a simpler algorithm. Appendix IV shows the details of this method. For the standard variables of (7) and (8), the final result for the distribution is

$$P(z) = N(z) - \frac{G_1}{3!} N^{(3)}(z) + \frac{G_2}{4!} N^{(4)}(z) + \frac{10G_1^2}{6!} N^{(6)}(z) \quad (9)$$

where $N(z)$ is the Gaussian distribution and $N^{(k)}(z)$ is the kth derivative of the Gaussian.

To facilitate easy reading, we summarize this procedure clearly under "proposed algorithm". However, it is important to stress here that the areas under normal curves and their derivatives are tabulated in mathematical handbooks. A closer look at (9) will reveal that under some conditions, it degenerates into the Hermite polynomial suggested in [6].

The CDF obtained by using (9) is shown in Figure 2. Note the excellent accuracy at low risk levels. Therefore,

1. The probability of outage of a certain capacity or more can be accurately and quickly obtained by employing (9) from hand calculations using probability tables.

2. A Fourier transform exists in closed and simple

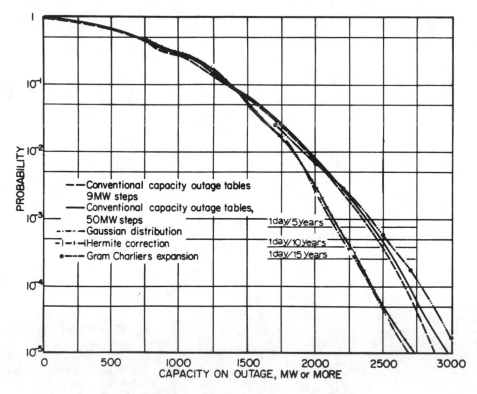

Figure 2. Comparison of results obtained

GRAM-CHARLIER EXPANSION

One obvious way of increasing the accuracy is to include higher order terms in the series expansion of Fourier transforms reported above. However, efforts in this direction proved futile due to problems in inverse transforming from the Fourier domain to the capacity domain. Juxtaposed with this, one could consider applying corrections in the capacity domain.

form for $P(z)$ in (9) making it admirably suited to further analysis such as convolution, etc.

PROPOSED ALGORITHM

Based on the above description of the Charlier's expansion method outlined in the Appendices, we summarize below the following procedure for calculating the probability of loss of a certain capacity of more in a generating system. This algorithm can be used for the calculation of LOLP by employing a calculator and probability tables. A digital computer can be employed, if desired. The latter has proven to be much faster than any of the existing algorithms for the computation of LOLP.

Algorithm

For a generating system with n machines of capacity C_i with forced outage rate q_i for the ith machine.

Step 1:
Calculate the following quantities for each machine:

a. $m_1(i) = C_i q_i$ (10)

b. $m_2(i) = C_i^2 q_i$ (11)

c. $m_3(i) = C_i^3 q_i$ (12)

d. $m_4(i) = C_i^4 q_i$ (13)

e. $V_i^2 = m_2(i) - m_1^2(i)$ (14)

f. $M_3(i) = m_3(i) - 3m_1(i)m_2(i) + 2m_1^3(i)$ (15)

g. $M_4(i) = m_4(i) - 4m_1(i)m_3(i) + 6m_1^2(i)m_2(i) - 3m_1^4(i)$
                                   (16)

For machines with derated states, the above modifies to:

$m_1(i) = C_{ai}q_{ai} + C_{bi}q_{bi} + \cdots$ (17)

$m_2(i) = C_{ai}^2 q_{ai} + C_{bi}^2 q_{bi} + \cdots$ (18)

$m_3(i) = C_{ai}^3 q_{ai} + C_{bi}^3 q_{bi} + \cdots$ (19)

$m_4(i) = C_{ai}^4 q_{ai} + C_{bi}^4 q_{bi} + \cdots$ (20)

where q_{ai} is the FOR corresponding to a capacity C_{ai}. Similarly for q_{bi} and C_{bi}.

Step 2:
Calculate the following for the system of n units:

h. $M = \sum\limits_{i=1}^{n} m_1(i)$ (21)

i. $V^2 = \sum\limits_{i=1}^{n} V_i^2$ (22)

j. $M_3 = \sum\limits_{i=1}^{n} M_3(i)$ (23)

k. $M_4 = \sum\limits_{i=1}^{n} [M_4(i) - 3V_i^4] + 3V^4$ (24)

l. $G_1 = M_3/V^3$ (25)

m. $G_2 = (M_4/V^4) - 3$ (26)

Note: The values obtained in Step 1 and 2 above can be stored and need not be repeated for each calculation of LOLP. Addition or deletions of units can be easily studied by adding or deleting the appropriate quantities.

Calculation of capacity outage probability

For a capacity x, calculate the standard variable

$Z_1 = (x - M)/V$ (27)

$Z_2 = (x + M)/V$ (28)

Case 1: $Z_2 \leq 2.0$

Calculate the areas Area 1 and Area 2 under the normal curve (Figure 1) by using probability tables. Then.
$$\text{Pr.(cap. outage} \geq x) = \text{Area 1} + \text{Area 2}$$
This case corresponds to risk levels of values 0.3 to 1.0 in probability.

Case 2: $2 \leq Z_2 \leq 5.0$
Calculate Area 1 and Area 2 as above. In addition, calculate:

$N(Z_1) = \dfrac{1}{\sqrt{2\pi}} e^{-Z_1^2/2}$ (29)

$N(Z_2) = \dfrac{1}{\sqrt{2\pi}} e^{-Z_2^2/2}$ (30)

$N^{(2)}(Z_1) = (Z_1^2 - 1)N(Z_1)$ (31)

$N^{(3)}(Z_1) = (-Z_1^3 + 3Z_1)N(Z_1)$ (32)

$N^{(5)}(Z_1) = (-Z_1^5 + 10Z_1^3 - 15Z_1)N(Z_1)$ (33)

$N^{(2)}(Z_2) = (Z_2^2 - 1)N(Z_2)$ (34)

$N^{(3)}(Z_2) = (-Z_2^3 + 3Z_2)N(Z_2)$ (35)

$N^{(5)}(Z_2) = (-Z_2^5 + 10Z_2^3 - 15Z_2)N(Z_2)$ (37)

$K_1 = \dfrac{G_1}{6}N^{(2)}(Z_1) - \dfrac{G_2}{24}N^{(3)}(Z_1) - \dfrac{G_1^2}{72}N^{(5)}(Z_1)$ (38)

$K_2 = \dfrac{G_1}{6}N^{(2)}(Z_2) - \dfrac{G_2}{24}N^{(3)}(Z_2) - \dfrac{G_1^2}{72}N^{(5)}(Z_2)$ (39)

Then
$$\text{Pr.(cap. outage} \geq x) = \text{Area 1} + \text{Area 2} + K_1 + K_2$$
This case corresponds to risk levels of values 0.01 to 0.3 in probability.

Case 3: $Z_2 > 5.0$
For this case, only Area 1 and K_1, as previously defined, are needed. Therefore
$$\text{Pr.(cap. outage} \geq x) = \text{Area 1} + K_1$$

Example:

To elucidate the above mentioned algorithm further, we shall apply the Gram Charlier's expansion to the four machine system considered earlier. Using the terminology indicated above, one gets
$m_1(1) = 50 \times 0.1 = 5.0$
Similarly $m_1(2) = 15$, $m_1(3) = 25.5$ and $m_1(4) = 30$.
The second moments $m_2(i)$ are given by
$m_2(1) = 50^2 \times 0.1 = 250$

$m_2(2) = 100^2 \times 0.15 = 1500$

$m_2(3) = 150^2 \times 0.17 = 3825$

$m_2(4) = 100^2 \times 0.1 + 200^2 \times 0.1 = 5000$

The corresponding third and fourth moments are given by
$m_3(1) = 50^3 \times 0.1 = 12500$
$m_3(2)$ and $m_3(3)$ are 0.15×10^6 and 0.573×10^6 respectively.
$m_3(4) = 100^3 \times 0.1 + 200^3 \times 0.1 = 0.9 \times 10^6$
The fourth moments
$m_4(1) = 50^4 \times 0.1 = 0.625 \times 10^6$
$m_4(2) = 0.15 \times 10^8$
$m_4(3) = 0.86 \times 10^8$
$m_4(4) = 100^4 \times 0.1 + 200^4 \times 0.1 = 1.7 \times 10^8$
The values for V_i are given by
$V_1^2 = m_2(1) - m_1^2(1) = 225$
$V_2^2 = m_2(2) - m_1^2(2) = 1275$
$V_3^2 = 3174.7$
$V_4^2 = 4100$

The values for M_3 and M_4 are given by
$M_3(1) = 12500 - 3 \times 5 \times 250 + 2 \times 5^3 = 9000$
$M_3(2) = 0.15 \times 10^6 - 3 \times 15 \times 1500 + 2 \times 15^3 = 89250$
$M_3(3) = 0.573 \times 10^6 - 3 \times 25.5 \times 3825 + 2 \times 25.5^3 = 314300$
$M_3(4) = 0.9 \times 10^6 - 3 \times 30 \times 5000 + 2 \times 30^3 = 504000$
$M_4(1) = .625 \times 10^6 - 4 \times 5 \times 12500 + 6 \times 5^2 \times 250 - 2 \times 5^4 = 4.112 \times 10^5$
$M_4(2) = 0.15 \times 10^8 - 4 \times 15 \times 0.15 \times 10^6 + 6 \times 15^2 \times 1500 - 2 \times 15^4 = 7.923 \times 10^6$
$M_4(3) = 0.86 \times 10^8 - 4 \times 25.5 \times 0.573 \times 10^6 + 6 \times 25.5^2 \times 3825 - 2 \times 25.5^4 = 4.1618 \times 10^7$
$M_4(4) = 1.7 \times 10^8 - 4 \times 30 \times 0.9 \times 10^6 + 6 \times 30^2 \times 5000 - 2 \times 30^4 = 8.738 \times 10^7$
Thus
$M = M_1(1) + M_1(2) + M_1(3) + M_1(4) = 75.5$
$$V^2 = \sum_{i=1}^{4} V^2 = 8.774 \times 10^3$$
$$M_3 = \sum_{i=1}^{4} M_3(i) = 9.1655 \times 10^5$$
$$M_4 = \sum_{i=1}^{4} [M_4(i) - 3V^4] + 3V^4 = 2.8262 \times 10^8$$

Therefore,
$G_1 = M_3/V^3 = 1.115$
$G_2 = (M_4/V^4) - 3 = 0.6706$
For the probability of loss of 110 MW or more, we have
$Z_1 = (110 - 75.5)/93.67 = 0.368$
$Z_2 = (110 + 75.5)/93.67 = 1.98$
since $2 \leq Z \leq 5.0$, case 2 holds and from tables
Area 1 = 0.356
Area 2 = 0.0238
and the values of K_1 and K_2 as in (38) and (39) turn out to be
$$K_1 = \frac{1.115}{6}(-0.322) - \frac{0.6706}{24}(0.3932) - \frac{1.115^2}{72}(-1.876)$$
$$= -0.0384$$
$$K_2 = \frac{1.115}{6}(0.164) - \frac{0.6706}{24}(-0.1025) - \frac{1.115^2}{72}(0.9826)$$
$$= 0.0163$$
Therefore,
Pr.$\{$ outage ≥ 110 MW $\} = 0.356 + 0.0238 - 0.0384 + 0.0163 = .358$

From the above, the reader might get a feeling that this procedure is rather complicated compared to the conventional capacity outage table method. To the contrary, this procedure is very much simpler than the conventional methods. This will be obvious by considering that the values for M, V and M_3, once calculated, can be stored as fundamental parameters. The changes in M, V and M_3 due to unit additions or deletions can be quickly and easily calculated. From this point on, the calculations of M_4, G_1, G_2 and the referencing of the tables to obtain the probability of loss of a certain capacity is a very simple procedure. Further, the calculation effort in the conventional method increases enormously as the number of machines are increased. Using the above algorithm, we have been able to compute the LOLP index in about 25% of the time taken by conventional methods employing recursive algorithms on the digital computer.

Accuracy and Results of Computation

It has been pointed out earlier that the Gram Charlier's expansion gives accurate results at low risk levels. In Figure 2, one can obtain the reserve requirement for an LOLP of 3.85×10^{-4} (expectation of 1 day in 10 years) from various distributions. For the 23 machine models studied, the difference in the reserve requirements between the conventional methods and the Gaussian distribution is of the order of 400 MW. This difference between the conventional method

and the Gram Charlier's expansion is only 100 MW. Several computations were made to ascertain the maximum error one can expect by using Charlier's method. For practical power systems of varying sizes and unit sizes, it was confirmed that the reserve requirement will not be more than 150 MW different from those obtained by conventional methods.

The Gram-Charlier expansion was used to calculate the probability of loss of load for the generation Model 2 shown in Appendix II. Table I below shows a sample of the results obtained.

Table I. Results obtained by conventional method and proposed method

Capacity on Outage MW	Probability of Capacity outage of X MW or Greater			
	Conventional Method 50 MW Steps	Conventional Method 150 MW Steps	Proposed Method	Proposed Method With Deratings
0	1.0	1.0	1.0	1.0
3900	3.4 (-2)	5.0 (-1)	3.17 (-2)	4.15 (-3)
4500	1.14 (-2)	1.6 (-2)	9.26 (-3)	6.23 (-4)
5100	2.8 (-3)	4.5 (-3)	2.32 (-3)	4.85 (-5)
5400	1.3 (-3)	2.2 (-3)	1.05 (-3)	1.04 (-5)
5700	5.9 (-4)	1.0 (-3)	4.38 (-4)	1.85 (-6)
6000	2.5 (-4)	4.9 (-4)	1.65 (-4)	2.76 (-7)

Notation: $(-1) = 10^{-1}$

Table II. Arbitrarily chosen derated levels for 48 generating unit system.

STATE	GENERATING UNIT STEADY-STATE PROBABILITIES							
	4 x 525 MW Units	4 x 547 MW Units	8 x 531 MW Units	4 x 547 MW Units	685 MW Unit	685 MW Unit	685 MW Unit	644 MW Unit
UP	0.940	0.920	0.900	0.880	0.850	0.840	0.820	0.790
5% der.	-----	-----	-----	-----	0.030	0.030	0.040	0.050
10% der.	0.015	0.025	0.025	0.040	0.010	0.010	0.010	0.010
20% der.	0.005	0.005	0.015	0.010	0.007	0.007	0.007	0.007
30% der.	-----	-----	0.010	0.005	0.005	0.005	0.005	0.005
40% der.	-----	-----	0.003	0.003	0.004	0.004	0.004	0.004
50% der.	-----	-----	0.002	0.002	0.004	0.004	0.004	0.004
DOWN	0.040	0.050	0.045	0.060	0.090	0.100	0.110	0.130

Table III. Reserve requirements corresponding to a LOLP = 3.85×10^{-4}.

Method	Reserve MW
Gram Charlier with no deratings	5820
Gram Charlier with derating	4700
Conventional Cap. outage table method, 150 MW steps	6050
Conventional Cap. outage table method, 50 MW steps	5860

Notice in Table I that the risk levels fall when units are modelled with derated capacities. The deratings of Table II were arbitrarily chosen. Note that any values based on operating records may be selected. The Appendices and the section "proposed algorithm" show the relative ease with which such multistate models can be incorporated into the proposed method. Note in table III that the reserve MW calculated by Gram Charlier's expansion is only 40 MW less than that obtained by conventional method.

CONCLUSIONS

A new method to obtain the capacity outage probability by using the Gram-Charlier expansion has been proposed. This method can handle derated operating states of generating units and has been shown to be accurate. The accuracy of this method has been demonstrated by comparing the results for two model generating systems with those obtained from a conventional method. The conventional method used for comparison utilised a recursive technique to obtain capacity outage tables. In addition, the method is simple, amenable to hand calculations and yields accurate results in a very short time. A distribution in a closed form for the generation outage probability should facilitate greatly the analysis of load forecast and FOR uncertainties, effect of interconnections etc. In a subsequent report, we show the necessity of the application of this method for the calculation of the probabilities associated with the loss of a given amount of energy.

ACKNOWLEDGEMENT

This work was carried out at the National Energy Board during the summer of 1977 when Dr. K.F. Schenk was working as a consultant to the Board. We wish to express our appreciation to·the Board for the facilities provided.

Also, our sincere thanks are due to R.W. Graham of the Board who performed many computer calculations and benefitted us by his many interesting discussions about this work.

APPENDIX I

CONVOLUTION BY FOURIER TECHNIQUES

Consider the Fourier transform pair

$$F(w) = \int_{-\infty}^{\infty} f(t)e^{-jwt}dt \tag{A-1}$$

$$f(t) = \frac{1}{2\pi}\int_{-\infty}^{\infty} F(w)e^{jwt}dw \tag{A-2}$$

The outage of any generator can be represented by impulses as in Figure A1.

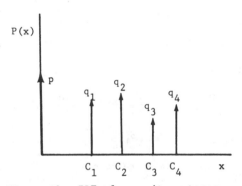

Figure A1. PDF of capacity outages.

The values of p, q_1, q_2 . . represent the outage probabilities of capacity outage equal to 0, C_1, C_2 MW respectively. If a generator has only two states, up and down, this can be represented by two impulses, p at x=0 and q at x equal to the full capacity of the unit. Evidently, all the machines in a system can be represented by such a diagram and the convolution (multiplication of their Fourier transforms) of all such representations give the density function of outage of capacity in the system. If all the machines have identical capacities, it can be shown[6] that their repeated convolution tends in the limit to a lattice with a Gaussian envelope. This limit theorem, unfortunately, cannot be used for a generating system. In power systems, since all the machines could have different capacities (although some subset of units can be identical), we had to address ourselves to develop a method which considers units with different capacities in the repeated convolution. To our knowledge, no documented methods exist to achieve this.

APPENDIX II

GENERATION MODELS

Two generation models were used in this study. The first is a hypothetical system with 23 units and the other was a model representing a large North American utility. The unit capacities followed by the corresponding FORS are indicated below:

Model 1: 23 Machines, Total Capacity 4190 MW Machine capacities in MW,
45, 45, 45, 50, 75, 75, 100, 100, 100, 100, 100, 125, 140, 140, 150, 150, 150, 200, 300, 500, 500, 500, 500

Forced Outage Rates:
0.1, 0.1, 0.2, 0.1, 0.1, 0.2, 0.1 0.1 0.1 0.1 0.1, 0.1, 0.1, 0.1, 0.05, 0.15, 0.1, 0.1, 0.2, 0.2, 0.2, 0.25, 0.25

Model 2: 48 Machines, Total Capacity 23070 MW
100% reliable generation = 4621 MW
Machine capacities in MW:
200, 400, 600, 4x100, 4x200, 4x64, 8x287, 4x525, 4x547, 8x531, 206, 4x514, 685, 685, 685, 644

Forced Outage Rates:
.0885, .065, .120, 4x0.080, 4x0.090, 4x.060, 8x0.13, 4x0.06, 4x0.08, 8x0.1, 0.2, 4x0.12, 0.15, 0.16, 0.18, 0.21

APPENDIX III

REPEATED CONVOLUTION

The symbols and the discussions below are taken from Papoulis[6]. The new approximations and variations will be pointed out during this development.

The Fourier transform of any density can be expressed as a product of an even and odd function as

$$P_i(w) = A_i(w)\exp[j\phi_i(w)] \tag{A-3}$$

where

$$A_i(w) = 1 + a_2w^2/2! + \ldots \tag{A-4}$$

$$\phi_i(w) = b_iw + \ldots \tag{A-5}$$

and thus

$$P_i(w) = A(w)\exp[j\phi(w)] = 1+jb_1w+(a_2-b_1^2)w^2/2!+\cdots \tag{A-6}$$

It can be shown the $P_i(w)$ can also be represented in terms of the moments as

$$P_i(w) = \sum_{n=0}^{\infty}(-j)^n m_n w^n/n! \tag{A-7}$$

where the moments are defined by

$$m_n = \int_{-\infty}^{\infty} x^n P(x)dx \tag{A-8}$$

Considering only two moments, define

$$\sigma^2 = m_2 - m_1^2 \tag{A-9}$$

and

$$\eta = m_1 \tag{A-10}$$

By equating the coefficients of the series in (A-6) and (A-7)

$$A_i(w) \simeq 1 - \sigma_i^2 w^2/2! \tag{A-11}$$

$$\phi_i(w) \simeq -\eta_i w \tag{A-12}$$

We now make an approximation for (A-11). For $|w| < w_0$ and w_0 sufficiently small, we have

$$A_i(w) \simeq \exp[-\sigma_i^2 w^2/2!] \tag{A-13}$$

Thus a repeated convolution of many machines will give

$$P(w) = A_1(w) \cdot A_2(w) \cdots \cdot \exp[-j\eta_1 w] \cdot \exp[-j\eta_2 w] \cdots \tag{A-14}$$

or

$$P(w) = \exp[-\sigma^2 w^2/2! - j\eta w] \qquad (A-15)$$

where

$$\sigma^2 = \sum_{i=1}^{n} \sigma_i^2 \qquad (A-16)$$

and

$$\eta = \sum_{i=1}^{n} \eta_i \qquad (A-17)$$

The density function shown in Figure A1 is causal and therefore a repeated convolution should yield a causal distribution. Thus

$$P(x) = (2/\pi)\int_0^\infty R(w) \cos wx \, dw \qquad (A-18)$$

in which $R(w)$ is the real part of $P(w)$ given by $R(w) = P(w) \cos \eta x$, therefore,

$$P(x) = (1/\sqrt{2\pi}\sigma)\{\exp[-(x-\eta)^2/2\sigma^2]+\exp[-(x+\eta)^2/2\sigma^2]\} \qquad (A-19)$$

By defining the standardized variables as in (27) and (28) in the main body of this paper, (A-19) represents the summation of two areas under the normal curve with a mean η and a standard deviation σ where the values of η_i and σ_i for a two-state outage representations of the machine are given by

$$\eta_i = C_i q_i \quad ; \quad i = 1,\ldots\ldots,n \qquad (A-20)$$

$$\sigma_i^2 = C_i^2 p_i q_i \quad ; \quad i = 1,\ldots\ldots,n \qquad (A-21)$$

and for a multi-state representation as in Figure A-1

$$\eta_i = C_1 q_1 + C_2 q_2 \cdots \qquad (A-22)$$

$$\sigma_i^2 = m_2(i) - m_1^2(i) \qquad (A-23)$$

Observe that (A-19) can be evaluated by using probability tables (see Figure 1) and its accuracy has been discussed in the text of this paper. This method and other more accurate methods shown below enables one to calculate the probability of outage of a certain capacity directly without having to compute all outage states in the form of a capacity outage table. Further, the values of η and σ, once computed for a system, can be stored and deemed as fundamental parameters for the set of generating units.

APPENDIX IV

GRAM CHARLIER'S EXPANSION

The Gram Charlier's expansion for a distribution $P(Z)$ is

$$P(Z) = N(Z) - \frac{G_1}{3!}N^{(3)}(Z) + \frac{G_2}{4!}N^{(4)}(Z) + \frac{10}{6!}G_1^2 N^{(6)}(Z) \qquad (A-24)$$

where the normal distribution $N(Z)$ and its derivatives are given by

$$N(Z) = \frac{1}{\sqrt{2\pi}}\exp[-Z^2/2] \qquad (A-25)$$

$$N^{(r)}(Z) = \frac{d^r(Z)}{dZ^r}N(Z) \quad ; \quad r = 1, 2, \ldots \qquad (A-26)$$

In terms of the standard variables

$$Z_1 = (x-\eta)/\sigma \qquad (A-27)$$

$$Z_2 = (x+\eta)/\sigma \qquad (A-28)$$

The probability of a capacity outage of x MW or greater is then

$$\text{Pr.}\{\text{cap. outage} \geq x\} = \int_{Z_1}^{\infty} P(Z)dZ + \int_{Z_2}^{\infty} P(Z)dZ \qquad (A-29)$$

and our goal is to make the result obtained by (A-29) the same as those obtained by conventional methods.

From (A-24) and (A-29) we have

$$\int_{Z_1}^{\infty} P(Z)dZ = \int_{Z_1}^{\infty} N(Z)dZ + \int_{Z_1}^{\infty} [-\frac{G_1}{3!}N^{(3)}(Z_1)+\frac{G_2}{4!}N^{(4)}(Z_1)+ \frac{10}{6!}G_1^2 N^{(6)}(Z_1)]dZ_1 \qquad (A-30)$$

Carrying out the required integration yields,

$$\int_{Z_1}^{\infty} P(Z)dZ = \text{Area } 1+\frac{G_1}{3!}N^{(2)}(Z_1)-\frac{G_2}{4!}N^{(3)}(Z_1)-\frac{10}{6!}G_1^2 N^{(5)}(Z_1) \qquad (A-31)$$

Similarly, since $N^{(r)}(\infty) = 0$, for $r = 0, 1, 2, \ldots$

$$\int_{Z_2}^{\infty} P(Z)dZ = \text{Area } 2+\frac{G_1}{3!}N^{(2)}(Z_2)-\frac{G_2}{4!}N^{(3)}(Z_2)+\frac{10}{6!}G_1^2 N^{(5)}(Z_2) \qquad (A-32)$$

Therefore,

$$\text{Pr.}\{\text{cap. outage}\geq x\} = \text{Area } 1 + \text{Area } 2 + K_1 + K_2 \qquad (A-33)$$

where K_1 and K_2 are defined in (A-31) and (A-32). Note that (A-33) can be evaluated by a pocket or desk calculator and by probability tables. The required derivatives of the Gaussian density defined in (32) to (37) are also found in handbooks.

The only unknown parameters that are to be established in (A-33) are G_1 and G_2.

Evaluation of G_1 and G_2

The Fourier transform of (A-25) is

$$N(w) = \exp[-w^2/2] \qquad (A-34)$$

Moreover, it is well known that

$$\frac{d^n N(Z)}{dZ^n} \longleftrightarrow (jw)^n N(w) \qquad (A-35)$$

Hence, the Fourier transform of (A-24) is given by

$$F(w) = N(w) - \frac{G_1}{3!}(jw)^3 N(w) + \frac{G_2}{4!}(jw)^4 N(w) + \frac{10}{6!}(jw)^6 N(w)$$

or

$$\qquad (A-36)$$

$$= \exp[-w^2/2] \cdot [1+j\frac{G_1}{6}w^3 + \frac{G_2}{24}w^4 - \frac{G_1^2}{72}w^6] \qquad (A-37)$$

Equation (A-37) is in reference to standardized variable Z. In terms of x, in the capacity domain, the transform must be phase shifted by $\exp(-j\eta w)$ where η is the equivalent phase and w must be replaced by σw. Thus the Fourier transform of the distribution resulting from a repeated convolution is represented in terms of x as

$$P(w) = e^{\frac{-\sigma w^2}{2}} e^{-j\eta w} [1+j\frac{G_1}{6}\sigma^3 w^3 + \frac{G_2}{24}\sigma^4 w^4 - \frac{G_1^2}{72}\sigma^6 w^6] \qquad (A-38)$$

Expanding (A-38) in a series and retaining powers of w up to w^4 one gets

$$P(w) = 1-j\eta w - \frac{w^2}{2}[\eta^2+\sigma^2] + \frac{jw^3}{3!}[\eta^3+\sigma^3+3\eta\sigma^2] + \frac{w^4}{4!}[3\sigma^4+G_2\sigma^4+ 6\eta^2\sigma^2+4G_1\eta\sigma^3+\eta^4] + \cdots \qquad (A-39)$$

Equating the coefficients of w in the above to those of w in the series expansion of $P(w)$ in terms of its moments m_i, $i=0, 1, \ldots$, given by

$$P(w) = m_o - jm_1 w - m_2 w^2/2! + jm_3 w^3/3! + m_4 w^4/4! \qquad (A-40)$$

one gets

$$m_o = 1 \qquad (A-41)$$

$$m_1 = \eta \qquad (A-42)$$

$$m_2 = \eta^2+\sigma^2 \qquad (A-43)$$

$$m_3 = G_1 \sigma^3 + \eta^3 + \eta\sigma^2 \qquad (A-44)$$

$$m_4 = 3\sigma^4 + G_2\sigma^4 + 6\eta^2\sigma^2 + 4G_1\eta\sigma^3 + \eta^4 \qquad (A-45)$$

In Appendix III, a Gaussian distribution was obtained by considering terms up to w^2. If one considers terms up to w^4, the Fourier transform of each machine outage (p_i and q_i in the two-state model) can be represented as [7]

$$A_i(w) = 1 - \sigma_i^2 w^2/2! + \alpha_i w^4/4! \qquad (A-46)$$

$$\phi_i(w) = -\eta_i w + \beta_i w^3/3! \qquad (A-47)$$

with the usual definition of the moments

$$m_n = \int_{-\infty}^{\infty} x^n p_i(x) dx \qquad (A-48)$$

one gets

$$\eta_i = m_1 \qquad (A-49)$$

$$\sigma_i^2 = m_2(i) - m_1^2(i) \qquad (A-50)$$

$$\alpha_i = m_4(i) - 4m_1(i)m_3(i) + 6m_1^2(i)m_2(i) - 3m_1^4(i) \qquad (A-51)$$

$$\beta_i = m_3(i) - 3m_1(i)m_2(i) + 2m_1^3(i) \qquad (A-52)$$

For the repeated convolution of n generating units one has

$$A(w) = A_1(w) \cdot A_2(w), \ldots A_n(w) \qquad (A-53)$$

$$\phi(w) = \phi_1(w) + \phi_2(w) + \ldots \phi_n(w) \qquad (A-54)$$

Substituting the relations for $A_i(w)$ and $\phi_i(w)$ in the above two equations, after some algebra [7], one obtains

$$A(w) = 1 - \sigma_{eq}^2 w^2/2! + \alpha_{eq} w^4/4! + \ldots \qquad (A-55)$$

$$\phi(w) = -\eta_{eq} w + \beta_{eq} w^3/3! \ldots \qquad (A-56)$$

in which

$$\sigma_{eq}^2 = \sum_{i=1}^{n} \sigma_i^2 \qquad (A-57)$$

$$\alpha_{eq}^2 = \sum_{i=1}^{n} (\alpha_i - 3\sigma_i^4) + 3\sigma_{eq}^4 \qquad (A-58)$$

$$\eta_{eq} = \sum_{i=1}^{n} \eta_i \qquad (A-59)$$

$$\beta_{eq} = \sum_{i=1}^{n} \beta_i \qquad (A-60)$$

From (A-53) to (A-56), the complete transform of the repeated convolution cn be obtained as

$$P(w) = 1 - j\eta_{eq}w - (\eta_{eq}^2 + \sigma_{eq}^2)w^2/2! + j(\beta_{eq} + \eta_{eq}^3 + 3\eta_{eq}$$

$$\sigma_{eq}^2)w^3/3! + (\alpha_{eq} + 4\eta_{eq}\beta_{eq} + 6\sigma_{eq}^2\eta_{eq}^2 + \eta_{eq}^4)w^4/4! \qquad (A-61)$$

Clearly, (A-61) must be identical to (A-39) if the Gram-Charlier's expansion must represent the repeated convolution. Therefore

$$\eta = \eta_{eq}, \quad \sigma^2 = \sigma_{eq}^2 \qquad (A-62a-b)$$

and

$$G_1 = \beta_{eq}/\sigma_{eq}^3 \qquad (A-63)$$

and finally

$$3\sigma^4 + G_2\sigma^4 + 4G_1\sigma^3\eta + 6\eta^2\sigma^2 + \eta^4 = \alpha_{eq} + 4\eta_{eq}\beta_{eq} + 6\eta_{eq}^2\sigma_{eq}^2 + \eta_{eq}^4 \qquad (A-64)$$

which gives, by referring to (A-62a-b) and (A-63)

$$G_2 = (\alpha_{eq}/\sigma_{eq}^4) - 3 \qquad (A-65)$$

In summary, we have proved that the expansion of (A-24) with the parameters G1 and G2 defined by (A-63) and (A-65), respectively, is equivalent to the PDF of the convolution of n-machine capacity outage states. Note that η, σ^2, etc., in the above equations have been expressed in terms of the generating unit capacities and based on a two-state probabilistic model. The corresponding changes for units with derated capacity states can easily be obtained by referring to (17) to (20).

We wish to point out that the calculations of the parameters of the Gram-Charlier expansion by equating its Fourier transform to that of a repeated convolution has not been, to our knowledge, attempted before.

REFERENCES

[1] R. Billinton, "Bibliography on the Application of Probability Methods in Power System Reliability." IEEE Transactions, Vol. PAS-91, pp. 649-660, 1972

[2] A.K. Ayoub, J.D. Guy and A.D. Patton, "Evaluation and Comparison of some Methods for Calculating Generating System Reliability." IEEE Transactions Vol. PAS-89, pp. 537-544, April 1970.

[3] R.R. Booth, "Power System Simulation Model based on Probability Analysis," IEEE Transactions, Vol. PAS-91, pp. 62-69, 1972.

[4] M.L. Telson, "The Economics of Alternative Levels of Reliability for Electric Power Generation Systems." Bell Journal of Economics, No. 16, pp. 2204-2212, Sept/Oct. 1972.

[5] R. Billinton, Power System Reliability Evaluation New York: Gordon and Breach, Science Publishers, 1970.

[6] A. Papoulis, The Fourier Integral and Its Applications, New York: McGraw-Hill, 1972.

[7] K.F. Schenk, "Analysis of Reliability Criteria for Generation Planning," Report submitted to the National Energy Board, Ottawa, Summer 1977.

[8] R. Billinton, R.J. Ringlee and A.T. Wood, "Power-System Reliability Calculations", The MIT Press, 1973.

[9] M. Bhavaraju, An Analysis of Generating Capacity Reserve Requirements, C-74-135-0, Abstract in IEEE Transactions, Vol. PAS-93, July 1974.

THE IEEE RELIABILITY TEST SYSTEM - EXTENSIONS TO AND EVALUATION OF THE GENERATING SYSTEM

R.N. Allan, Senior Member
UMIST
Manchester, England

R. Billinton, Fellow
University of Saskatchewan
Saskatoon, Canada

N.M.K. Abdel-Gawad
UMIST
Manchester, England

Abstract - This paper outlines some of the restrictions which currently exist in the generation data of the IEEE Reliability Test System (RTS). The paper extends the RTS by including more factors and system conditions which may be included in the reliability evaluation of generating systems. These extensions create a wider set of consistent data. The paper also includes generation reliability indices for the base and extended RTS. These indices have been evaluated without any approximations in the evaluation process and therefore provide a set of exact indices against which the results from alternative and approximate methods can be compared.

INTRODUCTION

Reliability is one of the most important criteria which must be taken into account during the design and planning phases of a power system. This need has resulted in the development of comprehensive reliability evaluation and modelling techniques [1,2]. One particularly frustrating aspect associated with the wide range of material published [3-5] on this subject is that, until 1979, there was no general agreement of either the system or the data that should be used to demonstrate or test proposed techniques. Consequently it was not easy, and often impossible, to compare and/or substantiate the results obtained from various proposed methods.

This problem was recognized by the IEEE Subcommittee on the Application of Probability Methods (APM) which, in 1979, published [6] the IEEE Reliability Test System (RTS). This is a reasonably comprehensive system containing generation data, transmission data and load data. It was intended to provide a consistent and generally acceptable set of data that could be used both in generation capacity and in composite system reliability evaluation. This would enable results obtained by different people using different methods to be compared.

This primary objective has been satisfied in most respects but, from experience, it has become evident that certain particular limitations exist. These are discussed in the next section. The purpose of the present paper is to address and reduce these limitations for the generating system and thus make the RTS more useful in assessing reliability models and evaluation techniques.

LIMITATIONS OF EXISTING RTS DATA

System Input Data

The original RTS was developed in order to provide system data that was perceived to be sufficient at

86 WM 038-4   A paper recommended and approved by the IEEE Power System Engineering Committee of the IEEE Power Engineering Society for presentation at the IEEE/PES 1986 Winter Meeting, New York, New York, February 2 - 7, 1986. Manuscript submitted August 6, 1984; made available for printing November 12, 1985.

that time. Experience has shown that certain additional data in both the generation system and the transmission system would be desirable. This has become evident because of developments that have taken place subsequently in both modelling and evaluation and also because of the type and scope of more recently published analyses. This particular paper is concerned with expanding the information relating to the generation system.

Although the system input data is already comprehensive, several important aspects are omitted. These include unit derated (partial output) states, load forecast uncertainty, unit scheduled maintenance and the effect of interconnections. It is desirable for these factors to be specified for the RTS in order that users of the RTS may all consider the same data and therefore evaluate results which can easily be compared. This additional data is specified in Appendix 1. There are many other aspects which could also be considered including start-up failures and outage postponability. Inclusion of these, however, requires additional modelling assumptions which are outside the scope of the present RTS and therefore, should be included only when the RTS is revised.

System Reliability Indices

Although the original RTS paper [6] included the input data, no information was included concerning the system reliability indices. Experience has shown that this was an important omission which should be rectified. The main reason is that most practical evaluation techniques include approximations and modelling assumptions regarding the generating capacity model, the load model and/or the evaluation algorithm.

Consider first the generating capacity model. This has 1872 states if no rounding is used and the $model_8$ is truncated at a cumulative probability of 1×10^{-8}. In practice, rounding and higher truncation values of cumulative probability are frequently used. Also, other approximate methods, such as the cumulant method [7] have been developed. All of these aspects introduce approximations.

Consider now the load model. The specified RTS load model has 364 levels if only the daily peak loads are used and 8736 levels if the hourly loads are used. In many practical applications, the load model is usually represented, not by the actual daily or hourly levels, but by a smoothed curve depicted by a restricted number of coordinate points. These practical models are known [2] as the daily peak load variation curve (DPLVC) in the case of daily peak loads and the load duration curve (LDC) in the case of hourly loads. These two curves (DPLVC and LDC) for the RTS are shown in Figure 1. This modelling aspect also leads to approximations.

It is evident from the above reasoning that a result obtained from a particular analysis, in which one or more of the above approximations are incorporated, will not be exact. The degree of error however is unknown unless the result can be compared against an exact value. Therefore it was decided that a series of results should be evaluated for the RTS in which no approximations in the evaluation process and no assumptions, additional to those already associated

Reprinted from *IEEE Trans. Power Syst.*, vol. PWRS-1, no. 4, pp. 1–7, Nov. 1986.

172

with the RTS, were made. These indices could then form base values against which results from alternative and approximate methods can be compared. All the results in the following sections, except those showing the effect of rounding, are therefore exact (for the given data) since no approximations have been made in either the capacity model or the load model. They are therefore reproducible within the precision limitations of a particular computer. The methods used to ensure the exactness are described in the relevant section that follows.

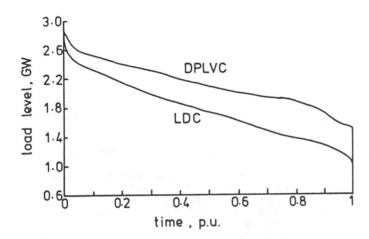

Figure 1. Load Models

LOLE ANALYSIS OF BASE CASE

The base case is considered to be the system as published in the original RTS [6]. In order to evaluate the exact loss of load expectation (LOLE) indices, the complete capacity model was developed with no rounding and truncated at a cumulative probability of 1×10^{-8}, i.e. it consisted of 1872 states. It was assumed that there were no energy or capacity limitations associated with the hydro units. The load model was represented by all 364 daily peak loads in order to evaluate the exact LOLE in day/yr and by all 8736 hourly peak loads in order to evaluate the exact LOLE in hr/yr. The LOLE indices were evaluated by deducing the risk for each of these load levels and summing over all load levels, i.e.:

$$LOLE = \sum_{i=1}^{n} P(C < L_i) \qquad (1)$$

where $P(C < L_i)$ = probability of loss of load on day i or during hour i. This value is given directly by the capacity model.

n = 364 or 8736 as appropriate.

Using this technique, the exact LOLE indices are:

LOLE = 1.36886 day/yr using daily peak loads
LOLE = 9.39418 hr/yr using hourly loads

These two values must be considered as two fundamental indices. They can be used to compare results obtained using the DPLVC and the LDC respectively. Any deviation from these values will be due to precision (or lack of it) in the computer system.

EFFECT OF ROUNDING

If the generation model and/or load model is rounded, the values of LOLE will differ from those given in the previous section. This effect is illustrated in this section in which both the generation model and the load model have been rounded separately and concurrently.

The generation model was rounded in steps between 20 and 100 MW using conventional techniques [2]. In practice, rounding is normally done during the development of the model. However the results are then dependent on the order of adding units and how frequently rounding is performed. In order to ensure a set of consistent and reproducible results, the table was rounded after the complete capacity model was evaluated. The results then become independent of the order of adding units to the model.

The load model was rounded by first dividing the load into n equally spaced load levels. The number of days that each of these load levels is exceeded was deduced. This process created n coordinate points for the DPLVC, which was used with the capacity model to give the value of LOLE in day/yr using conventional techniques [2]. If a capacity level existed between two sets of coordinate points, interpolation was used to find the number of corresponding days.

The results are shown in Table I for the cases of rounded generation/exact load, exact generation/rounded load and rounded generation/rounded load. It is seen that, as severity of rounding is increased, the values of LOLE tend to increase. It should be noted that the 364 point rounded load model is represented by 364 equally spaced load levels which is not the same as the exact 364 individual daily peaks.

TABLE I
Effect of Rounding

Capacity Model Rounding Interval (MW)	Load Model (no. of points)	LOLE d/yr.
20	exact	1.38587
40	exact	1.37978
60	exact	1.39806
80	exact	1.37687
100	exact	1.41622
exact	10	1.74649
exact	100	1.42843
exact	200	1.38993
exact	364	1.37256
20	100	1.43919
20	200	1.39869
20	364	1.38967
40	100	1.45041
40	200	1.41514
40	364	1.39415

EFFECT OF DERATED STATES

Derated or partial output states can have a significant effect on the LOLE, particularly units of large capacity. There are various ways in which such units can be included in the analysis; the only exact way being to represent the unit by all its states and to add the unit into the capacity model as a multi-state unit. An EFOR (equivalent forced outage rate) representation is not an equivalent and gives pessimistic values of LOLE [2]. It is not normally necessary however to include more than one or possibly two derated states [2] to obtain a reasonably exact value of LOLE.

For these reasons, the 400 MW and 350 MW units of the RTS have been given a 50% derated state. The number of service hours (SH), derated hours (DH) and forced outage hours (FOH) are shown in Appendix 1 and were chosen so that the EFOR [8] of the units are identical to the FOR specified in the original RTS [6]. Using the evaluation concepts of Reference 8:

400 MW unit   SH = 1100 hr, DH = 100 hr   FOH = 100 hr
              derated capacity = 200 MW   EH = $\frac{200}{400}$ DH
                                              = 50 hr

$$\therefore \text{EFOR} = \frac{DH+EH}{SH+DH+EH} = \frac{100+50}{1100+100+50}$$

$$= 0.12$$

350 MW unit   SH = 1150 hr   DH = 60 hr   FOH = 70 hr
              derated capacity = 175 MW   EH = $\frac{175}{350}$ DH
                                             = 30 hr

$$\text{and EFOR} = \frac{70+30}{1150+70+30} = 0.08$$

These values of state hours give the limiting state probabilities [1] shown in Table II.

TABLE II
Limiting State Probabilities

Unit	State Probability		
	Up	Derated	Down
400 MW	0.846154	0.076923	0.076923
350 MW	0.898438	0.046875	0.054687

The LOLE was evaluated using the exact generating capacity and load models i.e. no rounding, for three cases; when one 400 MW unit, when both 400 MW units and when both the 400 units and the 350 MW unit was represented by 3 states. The results are shown in Table III.

TABLE III
Effect of Derated States

Units Derated	LOLE d/yr.
1 x 400 MW	1.16124
2 x 400 MW	0.96986
2 x 400 + 1 x 350 MW	0.88258

These results show that the value of LOLE decreases significantly when derated states are modelled. This clearly demonstrates the inaccuracies that can be created if EFOR values are used, particularly for the larger units.

EFFECTS OF LOAD FORECAST UNCERTAINTY

Load forecast uncertainty was modelled using a normal distribution divided into seven discrete intervals [2]. The probabilities associated with each interval can be evaluated [1] as the area under the density function and these are shown in Appendix 1. It is suggested that a load forecast uncertainty having a standard deviation of 5% should be associated with the RTS. This is the value specified in Appendix 1. In the present analysis however, standard deviations from 2-15% have been considered. The results using the exact capacity and load models are shown in Table IV.

TABLE IV
Effect of Load Forecast Uncertainty

Uncertainty %	LOLE d/yr.
2	1.45110
5	1.91130
10	3.99763
15	9.50630

These results clearly show the very significant increase in LOLE as the degree of uncertainty is increased.

EFFECT OF SCHEDULED MAINTENANCE

There are two main aspects relating to scheduled maintenance. The first is to ascertain or deduce the schedule. The second is its effect on LOLE. The value of LOLE will increase when maintenance is considered because of the reduced and variable reserve at different times of the year.

The schedule selected is Plan 1 of Reference 9. This complies with the maintenance rate and duration of the original RTS [6], and was derived using a levelized risk criteria. The schedule is shown in Appendix 1.

The analysis proceeds by using the exact capacity model and the exact load model for each week of the year. The LOLE for each week is evaluated using Equation 1. The annual LOLE is deduced by summing all the weekly values. The details of this exercise is shown in Table V together with the overall annual LOLE. These results show that the risk is approximately doubled when this maintenance schedule is included.

EFFECT OF PEAK LOAD

One particular criticism levelled against the RTS is that the transmission system is too reliable compared with that of the generation system. This is because the generation is particularly unreliable: the LOLE is 1.36886 day/yr compared with a frequently quoted practical value of 0.1 day/yr. The reason for this high level of risk can be viewed as being due to a load level that is too great for the generating capacity or a generating capacity that is too small for the expected load.

The first of these reasons, i.e. the effect of the peak load on the LOLE, is considered in this section. Taking the RTS peak load of 2850 MW as 1 pu, a range of peak loads between 0.84 and 1.1 pu were studied. In each case, all 364 daily peak loads were multiplied by the same pu factor and the LOLE evaluated using Equation 1 and the exact capacity and load models. The results are plotted in Figure 2 and some are tabulated in Table VI. These results show that the peak load carrying capability, PLCC [2] for a risk level of 0.1 day/yr is 2483.5 MW which is 0.8714 pu of the specified [6] RTS peak load.

174

TABLE V
Effect of Scheduled Maintenance

Week Nos.	LOLE d/yr.
1,2,19,23-25,44-52	1.12026
3-5	0.11395
6,7	0.06801
8	0.07424
9	0.02122
10	0.04624
11	0.07223
12,13	0.04632
14	0.03701
15	0.04654
16,17	0.07203
18	0.04392
20	0.06214
21,22	0.07202
26	0.06483
27	0.02015
28	0.06718
29	0.03259
30	0.04878
31,32	0.08787
33	0.05896
34	0.02059
35	0.11809
36	0.02266
37	0.07039
38,39	0.05062
40	0.02819
41,42	0.03858
43	0.04098
Total LOLE	**2.66659 d/yr.**

TABLE VI
Effect of Peak Load

Multiplying Factor p.u.	Peak Load MW	LOLE d/yr.
1.10	3135	6.68051
1.06	3021	3.77860
1.04	2964	2.67126
1.00	2850	1.36886
0.96	2736	0.65219
0.92	2622	0.29734
0.88	2508	0.12174
0.84	2394	0.04756

EFFECT OF ADDITIONAL GENERATION

The second reason mentioned in the previous section for the unreliable generating system can be alleviated by adding generating units to the system. This was achieved by adding a number of gas turbines each rated at 25 MW and having a FOR of 0.12.

Using the exact capacity and load models together with Equation 1 gives the results plotted in Figure 3 some of which are shown in Table VII. These results show that 15 such gas turbines are required in order to achieve a PLCC of 2850 MW with a LOLE of about 0.1 day/yr.

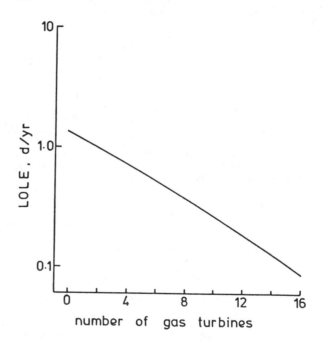

Figure 3. Effect of Added Generation

It is therefore suggested that the generating system of the RTS as originally specified [6] should be used as the base but that additional 25 MW gas turbines as specified in Appendix 1 should be included if a smaller risk index is required, e.g. in order to make transmission and generation more comparable or to achieve an LOLE that is nearer to frequently quoted practical values. In order to be consistent with the RTS and to conduct network analysis, it is necessary to specify the busbars to which additional generation must be attached. This requires a composite reliability study which is beyond the scope and objective of the present paper. This decision is, therefore, left to subsequent studies which will lead to revisions of the RTS.

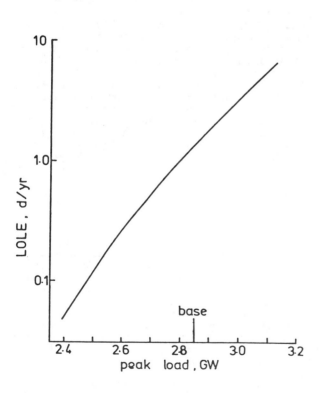

Figure 2. Effect of Peak Load

175

TABLE VII
Effect of Adding Gas Turbines

No. of Gas Turbines	LOLE d/yr.
1	1.18298
3	0.86372
5	0.62699
8	0.38297
10	0.27035
12	0.18709
15	0.10674
16	0.08850

TABLE VIII
Energy Supplied by Each Unit

Merit Order	Type	Size MW	Energy Supplied GWh
1	hydro	50	432.432
2	hydro	50	432.432
3	hydro	50	432.432
4	hydro	50	432.432
5	hydro	50	432.432
6	hydro	50	432.432
7	nuclear	400	3075.072
8	nuclear	400	3067.682
9	coal 1	350	2521.737
10	coal 2	155	963.742
11	coal 2	155	833.633
12	coal 2	155	677.731
13	coal 2	155	527.295
14	coal 3	76	217.557
15	coal 3	76	186.101
16	coal 3	76	153.884
17	coal 3	76	122.912
18	oil 1	197	196.003
19	oil 1	197	96.639
20	oil 1	197	40.645
21	oil 2	100	9.859
22	oil 2	100	5.661
23	oil 2	100	3.119
24	oil 3	12	0.268
25	oil 3	12	0.248
26	oil 3	12	0.229
27	oil 3	12	0.210
28	oil 3	12	0.194
29	GT	20	0.265
30	GT	20	0.234
31	GT	20	0.205
32	GT	20	0.181

ENERGY BASED INDICES

The most popular generation reliability index is the Loss of Load Expectation (LOLE) as derived in the previous sections. Energy based indices are now receiving more attention particularly for systems that have energy limitations (e.g. pumped storage) or for studying replacement of thermal energy by novel forms of generation (e.g. solar, wind). It is useful therefore to evaluate relevant energy indices for the RTS, these include Loss of Energy Expectation (LOEE) and Energy Index of Reliability (EIR). An additional advantage given by energy based evaluation methods is that the energy generated by each unit can be evaluated [2]. This enables production costs to be found.

The principle used to evaluate exact values of energy not supplied in these studies was as follows. Each hourly load level is numerically equal to the energy demanded during that hour. Consequently the total energy demanded by the system is numerically given by the summation of all 8736 load levels. The energy not supplied can be found using a similar principle. For each state of the capacity model C_k, the energy not supplied E_k is given numerically by summing all positive values of $(L_i - C_k)$ where L_i is the i-th load level and i=1 to 8736. The expected energy not supplied is then given by:

$$EENS = \sum_{k=1}^{n} E_k P_k \qquad (2)$$

where P_k = probability of capacity state C_k

and n = number of capacity states

This value of EENS can be evaluated after adding each unit into the system capacity model. Hence the expected energy produced by each unit is given [2] by the difference in EENS before and after adding the unit. The order of adding units is important and must follow the merit order. When all units have been added, the final value of EENS gives the system LOEE. Also the EIR is given by

$$EIR = 1 - LOEE/energy\ demanded \qquad (3)$$

Using the above principle, the expected energy produced by each unit is shown in Table VIII. The merit order was assumed to be that shown in the table. Also:

energy demanded = 15297.075 GWh

LOEE = 1.176 GWh

EIR = 0.999923

EFFECT OF INTERCONNECTION

The effect of interconnection is modelled using two identical RTS's joined with a single tie line. It is suggested that the tie line should have a rating of 300 MW, a length of 55 miles and be energized at 230 kV. The terminal busbars for the interconnections are required in order to be consistent with the RTS and to perform network studies. However, as in the case of the additional generators, this is beyond the present scope and objective and is, therefore, left to subsequent studies. Using the RTS data [6] means that the permanent outage rate is 0.477 f/yr. It is also suggested that it should have a repair rate of 364 repairs/yr (since the RTS has 364 days in its year, this repair rate is equivalent to an average repair time of one day or 24 hr). The unavailability of the tie line is therefore 0.00130873. These are the values specified in Appendix 1. In the present analysis tie line ratings from zero (no tie line) to 555 MW have been considered.

The evaluation technique used was the equivalent assisting unit approach [2]. This models the assisting system by a single multi-state unit which is moved through the tie line and added to the capacity model of the assisted system. The LOLE for the assisted system can then be found using Equation 1. No rounding was performed in the analysis and therefore exact models were used for the equivalent assisting unit and the capacity model of the assisted system. The results are, therefore, exact for the conditions stated below.

It was assumed that the assisting system would share its reserves but would not share the deficiencies of the assisted system. Consequently, the effective capacity of the equivalent assisting unit is equal to the reserve in the assisting system or the capacity of the tie line, whichever is smaller.

Two sensitivity analyses were performed. The first considered a single day only when the peak loads in each system were 2850 MW (the maximum value). The maximum reserve of the assisting system is therefore 555 MW.

The second analysis used the maximum peak load reserve approach [2]. This assumed that the load in the assisting system remained constant at 2850 MW whilst that in the assisted system was represented by the 364 daily peak loads. The equivalent assisting unit therefore remained a constant for a given tie line capacity.

The results for both analyses, assuming the tie line can and cannot fail, are shown in Table IX. It is seen that the values of LOLE decrease as the tie line capacity is increased to 555 MW above which no further change occurs. This is because no further benefit can be achieved from the assisting system. It is also seen that the effect of tie line reliability is very small; an effect often observed in practice.

TABLE IX
Effect of Interconnection

Tie Capacity MW	LOLE Tie Does Not Fail	LOLE Tie Can Fail
a) single peak load of 2850 MW (LOLE in day/day)		
0	0.08458	–
100	0.07007	0.07009
200	0.05928	0.05931
300	0.04483	0.04488
400	0.03578	0.03584
500	0.02477	0.02485
555	0.02104	0.02112
600	0.02104	0.02112
b) complete load model in assisted system (LOLE in day/yr)		
0	1.36886	–
100	1.09762	1.09798
200	0.85594	0.85661
300	0.64707	0.64802
400	0.51094	0.51206
500	0.34434	0.34571
550	0.27404	0.27547
600	0.27404	0.27547

CONCLUSIONS

This paper has extended the data and available information relating to the IEEE Reliability Test System. In so doing, it increases the value of the RTS in two important respects.

The first is that an increased range of data is now available. This will enable users of the RTS to employ a consistent set of data even with extended techniques and ensure that comparison of results is much easier.

The second is that generically exact indices for a wide range of conditions have been evaluated. These

will enable the results from alternative and approximate methods to be compared since the indices quoted should be reproducible within the precision limitations of the computers used.

REFERENCES

1. R. Billinton and R.N. Allan, "Reliability Evaluation of Engineering Systems: Concepts and Techniques". Pitman Books, 1983.

2. R. Billinton and R.N. Allan, "Reliability Evaluation of Power Systems", Pitman Books, 1984.

3. IEEE Committee Report, "Bibliography on the Application of Probability Methods in Power System Reliability Evaluation". IEEE Trans. on Power Apparatus and Systems, PAS-91, 1972, pp. 649-660.

4. IEEE Committee Report, "Bibliography on the Application of Probability Methods in Power System Reliability Evaluation, 1971-1977". IEEE Trans. on Power Apparatus and Systems, PAS-97, 1978, pp. 2235-2242.

5. R.N. Allan, R. Billinton and S.H. Lee, "Bibliography on the Application of Probability Methods in Power System Reliability Evaluation, 1977-1982". IEEE Trans. on Power Apparatus and Systems, PAS-103, 1984, pp. 275-282.

6. IEEE Committee Report, "IEEE Reliability Test System". IEEE Trans. on Power Apparatus and Systems, PAS-98, 1979, pp. 2047-2054.

7. N.S. Rau and K.F. Schenk, "Application of Fourier Methods for the Evaluation of Capacity Outage Probabilities". IEEE Winter Power Meeting, 1979, New York, paper A79 103-3.

8. IEEE Std 762, "Definitions for Use in Reporting Electric Generating Unit Reliability, Availability and Productivity".

9. R. Billinton and F.A. El-Sheikhi, "Preventive Maintenance Scheduling of Generating Units in Interconnected Systems". International RAM Conference, 1983, pp. 364-370.

APPENDIX 1

ADDITIONAL DATA TO BE USED WITH THE RTS

Generating System

TABLE X
Generating Unit Derated State Data

Unit Size MW	Derated Capacity MW	SH(1) hr	DH(2) hr	FOH(3) hr	EFOR (4)
350	175	1150	60	70	0.08
400	200	1100	100	100	0.12

Notes: (1) SH = service hours
(2) DH = derated state hours
(3) FOH = forced outage hours
(4) EFOR = equivalent forced outage rate

See Reference [8] for more detail of these terms.

TABLE XI
Maintenance Schedule

Weeks	Units on Maintenance			
1,2	none			
3-5	76			
6,7	155			
8	197	155		
9	197	155	20	12
10	400	197	20	12
11	400	197	155	
12,13	400	155	20	20
14	400	155		
15	400	197	76	
16,17	197	76	50	
18	197			
19	none			
20	100			
21,22	100	50		
23-25	none			
26	155	12		
27	155	100	50	12
28	155	100	50	
29	155	100		
30	76			
31,32	350	76	50	
33	350	20	12	
34	350	76	20	12
35	400	350	76	
36	400	155	76	
37	400	155		
38,39	400	155	50	12
40	400	197		
41,42	197	100	50	12
43	197	100		
44-52	none			

Additional Generating Units

These additional gas turbines can be used with the RTS in order to reduce the LOLE of the system to a level frequently considered acceptable.

TABLE XII
Additional Gas Turbines

Unit Size MW	Forced Outage Rate	MTTF hr	MTTR hr
25	0.12	550	75

All other data may be assumed to be identical to the existing gas turbines of the RTS.

Load Forecast Uncertainty

The load levels are assumed to be forecasted with an uncertainty represented by a normal distribution having a standard deviation of 5%. This is equivalent to a load difference of 142.5 MW at the peak load of 2850 MW. Using a load model having 7 discrete intervals, the discretised peak loads are shown in Table XIII.

TABLE XIII
Data For Load Forecast Uncertainty

Std. Deviations from Mean	Load Level MW	Probability
-3	2422.5	0.006
-2	2565.0	0.061
-1	2707.5	0.242
0	2850.0	0.382
+1	2992.5	0.242
+2	3135.0	0.061
+3	3277.5	0.006
		1.000

Tie Line

The information shown in Table XIV should be used to connect two identical RTS.

TABLE XIV
Tie Line Data

Voltage kV	Rating MW	Length Miles	Permanent Outage Rate f/yr	Outage Duration hr	Unavailability
230	300	55	0.477	24	0.00130873

APPENDIX 2

SUMMARY OF RTS RESULTS

The following results were evaluated using the exact capacity and load models together with Equations 1, 2 and 3 as appropriate. They can therefore be considered exact and can be used to compare the results of approximate methods.

The details of each "case" can be found in the appropriate section of the main text.

LOLE Indices

base (as per RTS [6])	1.36886 day/yr [9.39418 hr/yr]
with derated states	0.88258 day/yr
with 5% load forecast uncertainty	1.91130 day/yr
with maintenance	2.66659 day/yr
with 15 x 25 MW gas turbines	0.10674 day/yr
with interconnection	0.64802 day/yr

Energy Indices for Base System

LOEE	1.176 GWh
EIR	0.999923

178

A MONTE CARLO SIMULATION APPROACH TO THE RELIABILITY MODELING OF GENERATING SYSTEMS RECOGNIZING OPERATING CONSIDERATIONS

by

A. D. Patton
Texas A&M University

J. H. Blackstone
Univ. of Georgia

N. J. Balu
Electric Power
Research Institute

Abstract - The paper describes Monte Carlo simulation models which have been developed under EPRI projects RP1534-1,2 (1,2) for the reliability analysis of generating systems with explicit recognition of those unit and system operation considerations, rules, and constraints which influence system reliability indices. The results of sample systems using the new models are given and compared with results obtained using more conventional models which do not explicitly model the influences of operation considerations. These studies show that explicit modeling of operating considerations can have a very significant effect on computed system reliability indices.

INTRODUCTION

A number of generating unit and system operating considerations, constraints, and policies which influence system reliability performance have been identified through analyses of utility system data and through sensitivity studies (1,2). The most important general consideration in the accurate modeling of system reliability performance seems to be proper treatment of generating unit duty cycles. Proper modeling of unit duty cycles permits accurate consideration of unit exposures to failure due to running failures and to starting failures. This separate treatment of running and starting failures together with recognition of opportunities to repair failures during times of relative safety (outage postponability) is central to accurate system reliability performance modeling. Unit duty cycles are complex functions of system operating rules and policies as well as transmission network constraints. Some of the more important operating considerations are:

1. Operating reserve policy including "hard" and "soft" spinning reserve objectives;

2. Unit commitment policy including economic minimum unit running and shutdown times;

3. Postponability of unit unplanned outages and the management of postponable outages;

87 SM 492-2   A paper recommended and approved by the IEEE Power System Engineering Committee of the IEEE Power Engineering Society for presentation at the IEEE/PES 1987 Summer Meeting, San Francisco, California, July 12 - 17, 1987. Manuscript submitted January 28, 1987; made available for printing May 15, 1987.

4. Unit starting times, the probabilities of starting failures, and the durations of starting failure repairs;

5. Area interchanges of power on an economic or contractual basis including entitlements from jointly-owned generating units;

6. Rules governing emergency interchanges of power between areas;

7. Operation and management of energy-limited resources.

Operating considerations such as those mentioned above introduce complex temporal correlations between generating units, between generating units and load, and between generating units and the transmission network. Thus, traditional models which assume independence between generating units, transmission elements, and load (10, 11) cannot properly reflect many important operating considerations. EPRI projects RP1534-1,2 pursued two approaches to the explicit modeling of operating considerations and to the relaxation of idealizing independence assumptions: one analytical and one based on Monte Carlo simulation. The analytical models, described elsewhere (3,4), were successful in overcoming many of the modeling deficiencies contained in traditional analytical models. However, the most detailed and accurate modeling of complex operating considerations presently requires the use of Monte Carlo simulation. The Monte Carlo models which have been developed and their results are described in the sections which follow.

MONTE CARLO SIMULATION MODELS

Monte Carlo simulation models have been applied in a variety of generating system reliability applications for many years (5,6,7). Often the impetus for use of the Monte Carlo simulation approach was accurate modeling of limited-energy hydro resources whose presence introduces temporal correlation. In 1982 a Monte Carlo simulation model called GENESIS (2) was developed for the reliability evaluation of single area generating systems with explicit recognition of operating considerations other than energy limitations. This model is based on a modified version of an event-oriented simulation code called GASP and may be termed a sequential simulation model in that events are modeled sequentially through time. The GENESIS model has now been generalized and extended for studying the reliability performance of interconnected systems with recognition of those additional operating considerations which arise in interconnected system operation. (1)

Reprinted from *IEEE Trans. Power Syst.*, vol. 3, no. 3, pp. 1174–1180, Aug. 1988.

Generating Unit Model

Generating units are committed and decommitted to serve the hourly load cycle according to a specified unit commitment priority list while seeking to maintain a specified spinning reserve level. In general units are committed in economic priority order, but in emergencies quick-starting units may be committed out of economic order. The unit commitment process recognizes unit minimum reserve shutdown times as well as minimum running times. Further, unit starting times are recognized and modeled as functions of how long the unit has been shutdown when units must be started to replace capacity lost unexpectedly. Unit starting time is not modeled when a unit is started in response to a load increase under the assumption that load is forcastable thereby allowing the unit starting process to begin prior to actual need for the unit.

It is assumed that generators may reside in one of three states: fully available, unavailable, or partially available. The fully available and unavailable states are discrete, but the magnitude of the derating existing in the partially available state is treated as a random variable with a specified probability distribution. Two modes of failure are considered: (1) failures and deratings while running, and (2) starting failures. Running failures and deratings are modeled by drawing random sequences of state residence times from exponential distributions defined by specified state transition rates. Running failures and deratings reflect all unplanned outages, both forced outages and maintenance outages, and are coupled with explicit treatment of outage postponability. Starting failures are modeled by random draws in accordance with specified starting failure probabilities. Repairs following starting failures are modeled separately from running failure repairs by random draws from specified repair time distributions.

An important feature in the Monte Carlo simulation models is explicit modeling of unplanned outage postponability. Here, probability distributions of the time from the onset of trouble until the unplanned outage must be taken are specified. These postponability time distributions are sampled to obtain a random maximum outage postponement time each time an unplanned running failure is encountered. Model logic then seeks to schedule the impending unit outage at a time of relative safety within the maximum postponement period considering the load and capacity status of the system thereby simulating the actions of the system operator. This explicit treatment of outage postponability is in contrast to the ad-hoc approach of most analytical models wherein all forced outages are assumed to occur instantaneously and maintenance outages are ignored.

Generating unit planned outages are modeled by removing units from service in accordance with a specified planned outage schedule.

Interconnected System Operation

The multi-area simulation models recognize two basic modes of operation: (1) area commitment and dispatch of generation, and (2) pool commitment and dispatch of generation. These two modes are abbreviated here as area-mode and pool-mode. The area-mode model has been implemented in a program called GENAREA while the pool-mode model is called GENPOOL. In the area mode of operation, areas are assumed to commit and dispatch units as required to satisfy internal loads plus interchanges associated with firm contracts or jointly-owned units or emergency interchanges. In the pool commitment and dispatch mode of operation, all units are assumed to be committed from a pool-wide commitment priority list and dispatched to satisfy pool load and spinning reserve requirements. If transmission constraints are encountered, the model reverts to area commitment priority lists for commitment of needed units.

Area Interchanges: Area interchanges of power can have important influences on unit duty cycles and hence upon unit availabilities and system reliability indices. Further, area firm interchanges may load up the transmission network and thereby block or impede emergency transfers - thus influencing system reliability indices. Thus, area interchanges are an important operating consideration influencing reliability performance in interconnected systems. The models treat the following types of interchanges where the types of interchanges are listed in priority order. In the event of transmission limitations the higher priority interchanges are satisfied first.

Area mode of operation:

1. Firm contracts. Firm contract interchanges are specified by given MW transfers between sending and receiving areas for given time intervals. Firm contracts are treated as definite obligations to be fulfilled by any means possible.

2. Unit contracts or jointly-owned units. Unit contracts or percentage entitlements to the output of a jointly-owned unit in another area are modeled. These interchanges depend on the status of the particular unit and are not recovered by use of other resources should the unit in question be out of service.

3. Emergency interchanges. Emergency interchanges are scheduled depending on need and nature of emergency operation: reserve sharing, risk sharing, or loss sharing.

Pool mode of operation:

1. Economy interchanges. Units are generally committed from a pool priority list and are dispatched to supply pool loads regardless of area. The interchanges which result from this mode of operation carry definite capacity commitments.

2. Emergency interchanges. Emergency interchanges in the pool mode of operation are basically similar to economy interchanges. That is, units are started anywhere in the pool as required to satisfy needs in any area subject to the pool commitment priority list, transmission constraints, and other constraining factors.

Spinning Reserve Rules: For the purposes of the simulation model spinning reserve in an area is defined as: (capacity in service in area) - (area native load) - (net exports from area)

Note that an export from an area is regarded as a positive quantity and an import is regarded as a negative quantity. Note further that only those exports (or imports) which carry a capacity obligation are counted for the purposes of spinning reserve calculation. Similarly, the spinning reserve in the pool is defined as: (capacity in service in pool) - (pool load).

Spinning reserve objectives for each area and for the pool are specified both in terms of a "hard" objective and a "soft" objective. The "soft" spinning reserve objective specifies the desired spinning reserve level in normal operations while the "hard" spinning reserve objective defines the spinning reserve below which emergency operating procedures prevail.

Emergency Operation: Three possible modes of available capacity sharing during emergencies are modeled for both the area and pool modes of operation. These are: (1) reserve sharing, (2) risk sharing, and (3) loss sharing. In the reserve-sharing mode, areas with capacity surplus will assist capacity deficient areas to the extent possible without violating "hard" spinning reserve objectives in the assisting areas or "hard" transmission constraints. Similarly, in the risk-sharing mode, capacity surplus areas assist capacity deficient areas down to the point of zero margin in the assisting areas. In the loss-sharing mode all areas participate in the minimization of pool load loss by fully committing and transferring capacity up to the limits of "hard" transmission constraints. In the absence of transmission limitations, area load losses are in proportion to area loads.

The pool is assumed to be homogeneous in that all areas of the pool operate according to the same capacity sharing policy. However, it is realized that preferential arrangements may exist between some areas for the sharing of available capacity. Therefore, the simulation models permit a preferred receiving area for surplus capacity to be specified for each area. Then, in the event of capacity shortages, the preferred areas have first call on the resources of their preferred partners after any actual load-loss conditions are covered. The non-preferred areas with capacity shortages, but no actual loss of load, receive capacity assistance only after the margins in the preferred areas have been restored to their "hard" spinning reserve objectives.

Response to Loss of Unit or Line

This section discusses the general logic for responding to the unplanned outage of a generating unit or transmission line. The postponability of unplanned unit outages is modeled in detail with units taken out of service for repairs at times of relative safety within constraints imposed by the degree of outage postponability. Transmission line outages are assumed to have no postponability.

First consider an interconnected system operating in the area mode. In this mode of operation areas operate independently except for firm contract interchanges, unit contract interchanges, and emergency interchanges. An emergency is assumed to prevail if an area cannot meet its "hard" spinning reserve objective without emergency imports.

1. If following loss of unit or line the spinning reserve in an area lies between "hard" and "soft" objectives, start unit(s) in area in priority order as required to recover "soft" reserve objective in area. Note, however, that units

will not be started if the load is decreasing so that the "soft" reserve objective will be met within one hour.

2. If following loss of unit the spinning reserve in an area lies below "hard" objective, start quick-start units in area of unit loss in attempt to recover "hard" objective. If necessary, also start quick-start units in other areas to help recover "hard" objective with due regard for "hard" transmission constraints. Simultaneous with starting of quick-start units, initiate starting of economic units in commitment priority order in area of unit loss to recover "soft" spinning reserve objective. Shut down quick-start units (other areas first) as more economical units come on line.

3. If a jointly-owned generating unit fails, each participating area must respond to the capacity loss it will incur.

Next, consider an interconnected system operating in the pool mode. In this mode of operation areas share capacity resources on an economic basis under the control of the pool composite commitment priority list. An emergency prevails if area or pool "hard" spinning reserve objectives cannot be met without relaxing transmission security (soft) constraints.

1. If following loss of unit or line the spinning reserve in one or more areas and/or the pool lies between "hard" and "soft" objectives, start unit(s) from composite priority list to recover "soft" spinning reserve objectives while enforcing transmission security (soft) limits.

2. If following loss of unit or line spinning reserve in an area or the pool lies below "hard" objective, start quick-start units in area of deficiency in attempt to recover "hard" spinning reserve objective; also start quick-start units in other areas if necessary while observing "hard" transmission constraints. Simultaneously, initiate starting of unit(s) from composite priority list to recover "soft" spinning reserve objectives. Shut down quick-start units as more economical units come on line and as excess capacity occurs.

Transmission Network Model

The transmission network is modeled as a multi-state channel between each pair of areas. Each channel state is characterized by an admittance and by a basic capacity. The basic capacity of a transmission channel can be modified by various factors to reflect the fact that transmission transfer capabilities are functions of various aspects of system status.

Each transmission channel is assumed to have multiple states corresponding to various line and equipment statuses. Channel state changes are considered possible once each hour with transitions possible in general between any pair of channel states. Thus, each channel is viewed and modeled as a discrete state, discrete step, Markov process or chain. Transmission network flows are calculated using the prevailing channel admittances and the desired area interchanges.

Transmission channels are considered to have two types of capacities: capacities based on physical limitations or constraints; and capacities based on administrative factors, such as security considerations. Capacities based on physical limitations are viewed as "hard" constraints which cannot be violated under any circumstances while capacities based on security considerations are viewed as "soft" constraints which can be violated during periods of actual emergencies. Soft constraints can be violated for short periods during normal operation, but must generally be observed during periods not classified as emergencies. Hard (physical) constraints reflect: (1) line thermal limitations, and (2) load flow feasibility constraints. Soft or administrative constraints reflect security considerations such as transmission flow margins or limits adopted to avoid overload or loss of stability in the event of possible failures.

Loads

Loads are modeled by specifying hourly loads for each area for the study year. This approach allows the correlation between area loads to be completely and specifically considered in the simulation.

Reliability Indices

The simulation models compute the following reliability indices for each area and for the pool: expected number of hours of load loss per year, expected number of load loss events per year (frequency), expected duration of a load loss event, expected magnitude of a load loss event, expected unserved energy per year. Each of these statistics is given in total as well as broken down by cause: generation capacity shortage or transmission constraint.

The basis on which the various reliability indices are computed requires some explanation. The area indices reflect all constraints and are physically measurable indices in the areas. Two types of pool indices are computed. One set of indices, called the "pool" indices, ignores any transmission constraints and are obtained by comparing the sum of area generating capacities in service to the sum of area loads. Thus, "pool" indices amount to treating the pool as a single-area without transmission limitations. it follows, therefore, that "pool" indices are an indication of the reliability performance obtainable without any transmission limitations. The second set of pool indices are called "union" indices. These indices reflect the union of load loss events in the areas. That is, a load loss event occures in the pool union if a load loss event occurs in one or more areas.

The GENAREA and GENPOOL programs also produce histograms (probability distributions) of a number of measures of performance for each area and for the pool. These histograms provide valuable supplementary information to the usual expected value indices. The ability to produce probability distributions (histograms) as well as expected values of reliability performance measures is an important ability of Monte Carlo simulation models.

SAMPLE SYSTEM STUDIES AND RESULTS

A series of studies of three-area interconnected systems was made using the GENAREA and GENPOOL simulation models to investigate the effects of different operating considerations and policies on calculated reliability indices. These studies showed all operating considerations to be important in particular circumstances. Shown here are results reflecting detailed treatment of operating considerations in comparison with results obtained using more conventional assumptions. Also shown are sensitivity studies which provide insight into the relative importance of different modeling issues and assumptions.

Sample System

The sample interconnected system used in the studies reported here is composed of three identical areas with interconnections between each pair of areas. The installed generating capacity in each area totals 10,300 MW and is composed of the generating units shown in Table 1. Typical generating unit parameters were used. These parameters are shown in detail in Reference (1).

The peak load in each area was assumed to be 7923 MW which corresponds to a 30 percent installed capacity reserve in each area. Loads were modeled using identical hourly load cycles for each area. The hourly load cycle shape used in the studies is that of scenario A of EPRI report EM-285 (8) and is representative of a summer-peaking utility.

Transmission links between areas were modeled using a four-capacity-state model. State capacities were 100, 66.7, 33.3, and 0 percent of maximum transmission link capacity. Each transmission link was assumed to have a "hard" or physical maximum capacity of 600 MW and a "soft" or administrative capacity of 80 percent of the physical capacity. Transmission state transition probabilities used are typical of bulk power transmission networks. Transmission link admittances were assumed to be proportional to transmission state capacities.

Spinning reserve objectives were assumed to be identical for each of the three areas and to be constant over the study year. The area "soft" spinning reserve objective was 1000 MW and the "hard" spinning reserve objective was 800 MW.

Study Results

Naturally, a key question is the relationship between reliability indices computed with full modeling of operating considerations and those computed under more typical, or conventional assumptions. Here by "conventional" assumptions we mean the following. First, all units are assumed to operate continuously thereby ignoring all considerations of unit duty cycle and operating reserve policy. Second, the postponability of unplanned outages is treated in an approximate, ad hoc, manner. The usual approach in practice is to assume that forced outages (unplanned outages not postponable beyond the weekend) occur instantaneously and without possibility of postponement. Maintenance outages (unplanned outages postponable beyond the weekend) have been treated in practice in various fashions: (1) ignored altogether under the assumption that they can be taken at times of relative safety due to their degrees of postponability, (2) added in whole or in part to forced outages and assumed to occur instantaneously, or (3) added to planned outages and modeled deterministically by removing generating units from service during periods and seasons of relative safety according to a planned outage schedule. In the studies reported here, the "conventional" approach assumed that forced outages occur instantaneously and without any possible postponement and maintenance outages were ignored. Comparisons of area reliability indices computed using detailed models and models with conventional assumptions are shown in Table 2. Note that the results under "conventional" assumptions were also obtained using the GENAREA simulation model. This was done by appropriately altering the input data to GENAREA. Observe the substantial differences in the indices computed using the detailed and conventional models.

A series of sensitivity studies was also conducted to assess the importance of different modeling details. These studies were conducted assuming area-mode operation and a risk-sharing policy of capacity sharing in emergencies. The results of these studies are shown in Table 3. The cases are summarized as follows.

Case 1: Fully detailed modeling of all operating considerations.

Case 2: Full modeling of all operating considerations except that unit starting failures are assumed not to occur and unit starting times are assumed to be zero.

Case 3: This case is similar to Case 2 except that unit maintenance outages (unplanned outages postponable beyond the weekend) are assumed not to occur. Unit forced outages are modeled with full consideration of their postponability up to, but not beyond the weekend.

Case 4: This case extends the assumptions of Case 3 by assuming that unit forced outages occur instantaneously. Note that unit duty cycles are still being modeled.

Case 5: This case extends the assumptions of Case 4 by assuming that all units operate continuously. That is, this case contains the conventional modeling assumptions which have been described previously.

Case 6: This case amounts to Case 1 with the usual ad-hoc modeling of unplanned outages, namely maintenance outages are ignored and forced outages are assumed to occur instantaneously. Thus, this case corresponds, in large part, to conventional approaches using peaking-service models for generating units. (9)

Referring to Table 3, the following observations are made.

1. Comparison of Cases 1 and 2 shows that unit start-up failures and unit starting times have a moderate effect on reliability indices.

2. Comparison of Cases 2 and 3 shows that ignoring maintenance outages has a very large impact on reliability indices and is highly optimistic.

3. Comparison of Cases 3 and 4 shows that modeling unit forced outages as occurring instantaneously gives results which are not much different from results obtained with explicit postponability modeling of forced outages. Therefore, the postponability modeling of forced outages does not seem to be critical, at least for the postponability time distributions used for the synthetic system studies. Here most forced outages are only postponable up to six hours.

4. Case 6 results are basically those which would be obtained using conventional peaking-service unit models provided unit duty cycles were known a priori. A comparison of Case 6 and Case 1 shows the Case 6 reliability indices to be highly optimistic. This seems to be primarily due to the omission of all maintenance outages in Case 6.

5. Comparing Cases 4 and 5 shows the effects of unit duty cycles when unplanned outages are modeled using the usual ad hoc postponability assumptions. In this sample case ignoring the effect of unit duty cycles changes reliability indices by a factor of more than two.

6. Comparison of Case 5 results (conventional assumptions) with Case 1 results shows the conventional assumptions to result in optimistic reliability indices - in this case by a factor of more than two to one.

Based on the above observations it appears that accurate postponability modeling of unplanned outages with a high degree of postponability is the most critical modeling issue. Accurate modeling of postponability requires that unit duty cycles be properly modeled which implies explicit modeling of those operating considerations, constraints, and policies which influence unit duty cycles. Further, the effect of unit duty cycles on reliability indices is seen to be important regardless of the method of unplanned outage postponability modeling. Therefore, it seems that the key to accurate system reliability modeling is proper recognition and treatment of unit duty cycles. once unit duty cycles are modeled, the other issues such as postponability can be readily resolved.

ACKNOWLEDGEMENT

The research reported here was supported by the Electric Power Research Institute under projects RP 1534-1 and RP 1534-2 (1,2). The EPRI project manager was N. J. Balu.

REFERENCES

(1) "Reliability Models of Interconnected Systems That Incorporate Operating Considerations", Report EL-4603, Electric Power Research Institute, Palo Alto, CA., (August 1986).

(2) "Modeling of Unit Operating Considerations in Generating Capacity Reliability Evaluation", Report EL-2519, Electric Power Research Institute, Palo Alto, CA, (July 1982).

(3) "Operating Considerations in Generation Reliability Modeling - An Analytical Approach", C. Singh, A. D. Patton and M. Sahinoglu, *IEEE Transactions on Power Apparatus & Systems*, pp. 2656-2663, May 1981.

(4) "Operating Considerations in Reliability Modeling of Interconnected Systems - An Analytical Approach", C. Singh, et al, IEEE Summer Power Meeting, 1987.

(5) "Quantitative Evaluation of Power System Reliability in Planning Studies", P. L. Noferi, L. Paris, *IEEE Transactions on PA&S*, pp. 161-169, March/April 1972.

(6) "Optimization of the Basic Characteristics of Pumped-Storage Plants", L. Paris and L. Salvaderi, IEEE Paper C74-158-2.

(7) "Monte Carlo Methods for Power System Reliability Evaluations in Transmission and Generation Planning", P. L. Noferi, L. Paris, and L. Salvaderi, *Proceedings of 1975 Annual Reliability and Maintainability Symposium*, pp. 449-459.

(8) "Synthetic Electric Utility Systems for Evaluating Advanced Technologies", Report EM-285, Electric Power Research Institute, Palo Alto, CA., (February 1977).

(9) IEEE Task Group, "A Four State Model for Estimation of Outage Risk for Units in Peaking Service", *IEEE Transactions on Power Apparatus and Systems*, pp. 618-627, (March/April 1972).

(10) J. Endrenyi, *Reliability Modeling in Electric Power Systems*, J. Wiley and Sons, Chichester, 1978.

(11) R. Billinton and R. N. Allan, *Reliability Evaluation of Power Systems*, Pitman, Boston, 1984.

Table 1. Area Generation Mix

Unit Type in Commitment Priority Order	Number of Units	Unit Cap., MW	Total Cap., MW
Nuclear	2	800	1600
Coal, Fossil > 500 MW	1	800	800
Coal, Fossil > 500 MW	2	600	1200
Coal, Fossil 250-499 MW	1	400	400
Coal, Fossil 100-249 MW	1	200	200
Gas, Fossil > 500 MW	1	800	800
Gas, Fossil > 500 MW	2	600	1200
Gas, Fossil 250-499 MW	2	400	800
Gas, Fossil 100-249 MW	11	200	2200
Oil, Fossil 250-499 MW	1	400	400
Oil, Fossil 100-249 MW	1	200	200
Combustion Turbine	10	50	500
			10300

Table 2. Comparison of Area Indices with Detailed Modeling
and with Conventional Assumptions

	Area Indices				
	f No./Yr.	D Hrs.	HLOLE Hrs./Yr.	XLOL MW	EUE MWH/Yr.
Reserve Sharing					
Detailed Modeling	7.46	2.90	21.80	373	8113
Conventional Assumptions	2.86	2.85	8.19	361	2988
Risk Sharing					
Detailed Modeling	3.38	2.77	9.40	361	3390
Conventional Assumptions	1.43	2.64	3.80	320	1227
Loss Sharing					
Detailed Modeling	5.82	2.26	13.16	274	3624
Conventional Assumptions	2.87	2.17	6.13	280	1717

Table 3. Sensitivity of Model Assumptions

Study Case	Area Indices				
	f No./Yr.	D Hrs.	HLOLE Hrs./Yr.	XLOL MW	EUE MWH/Yr.
(1) Detailed Modeling	3.38	2.77	9.37	361	3390
(2) Case 1 without starting failures and delays	2.68	2.72	7.23	343	2463
(3) Case 2 without maint. outages	0.58	2.81	1.64	339	543
(4) Case 3 without postpona- nability of forced outages	0.68	2.66	1.80	333	599

(5) Case 4 except all units oper-
    ate continuously, conventional
    modeling assumptions
	1.43	2.64	3.80	320	1277

(6) Case 1 without maint. outages
    or postponability of forced
    outages
	0.78	2.80	2.21	393	879

CRITERIA USED BY CANADIAN UTILITIES IN THE PLANNING AND OPERATION OF GENERATING CAPACITY

Roy Billinton
Power Systems Research Group
University of Saskatchewan
Saskatoon, Canada

Abstract - This paper reviews the reliability criteria presently used by Canadian utilities in regard to planning and operating generating capacity. These results have been obtained from surveys conducted by the Power System Reliability Subsection of the Canadian Electrical Association. These surveys cover both operating and planning generating capacity criteria. In the case of adequacy assessment, a comparison of the different utility criteria is presented using the IEEE Reliability Test System.

Keywords - Adequacy evaluation, loss of load expectation, loss of energy expectation, operating reserve requirements.

INTRODUCTION

System reliability can be grouped into the two distinct aspects of system security and system adequacy [1]. System security involves the ability of the system to respond to disturbances arising within the system while system adequacy relates to the existence of sufficient facilities within the system to satisfy the customer load demand.

System reliability is usually predicted using one or more indices which quantify expected system reliability performance and implemented using criteria based on acceptable values of these indices. A complete reliability evaluation of a power system involves a comprehensive analysis of its three principal functional zones, namely generation, transmission and distribution [2]. These functional zones can be combined to give the Hierarchical Levels (HL) under which the various techniques used in reliability assessment are grouped. Reliability assessment at HLI [1] is concerned only with the generation facilities. The transmission and distribution facilities are asssumed to be fully reliable and capable of moving the generated electrical energy from the generating stations points to the customer load points. At HLI therefore, only the total system generation is examined to determine its ability to meet the total system load requirements. HLII assessment includes a composite appraisal of both the generation and transmission facilities and HLIII involves all three functional zones in an assessment of customer load point reliability. This paper is restricted to the criteria used by Canadian utilities in HLI adequacy and security assessment.

System adequacy at HLI depends on many factors such as the amount of installed capacity, unit size, unit availability, maintenance requirements, inter-connections, load forecasting errors and the shape of the load curve. In order to maintain the desired level of adequacy and to ensure against excessive shortages, additional generating capacity

above the peak demand, called reserve must be maintained. Security considerations also require that electric power systems also maintain operating or spinning reserve generating capacity margins against the fluctuations in the operating capacity and deviations in customer demands. In the case of both adequacy and security, the higher the reserve margin, the higher the system reliability but at a substantial economic cost. Both deterministic and probabilistic methods have been applied extensively to determine the required level of capacity reserve to be maintained by a system. The most common deterministic criterion relates the reserve margin to the size of the largest generating unit or to some percentage of the peak demand [2]. The selection of the actual value for these criteria has been largely based on past experience and judgement.

One of the responsibilities of the Power System Reliability Subsection of the Power System Planning and Operating Section is to review the techniques and criteria used by Canadian utilities in the assessment of generation and transmission system reliability. In this regard, a number of surveys have been carried out dealing with planning and operating generation capacity criteria, and reliability aspects of major transmission and terminal station planning.

ADEQUACY CRITERIA AT HLI

Surveys on generating capacity adequacy criteria were conducted in 1964, 1969, 1974, 1977 and 1979. The identities of the individual utilities providing the data were not revealed in the early reports and therefore only trends in Canadian utility application can be determined. The 1979 survey provided the first report in which the individual utilities were identified together with their criteria and methods of application. Table 1 presents a summary taken from the 1964, 1969, 1974 and 1977 surveys which shows the criteria in use at the time [1].

Table 1

Criteria Used in Reserve Capacity Planning

		Survey Date		
	1964	1969	1974	1977
1. Percent Margin	1	4	2	2
2. Loss of Largest Unit	4	1	1	1
3. Combination of 1 and 2	3	6	6	6
4. Probability Methods	1	5	4	4
5. Other Methods	2	1	-	-
	11	17	13	13

As can be seen from Table 1, only one utility indicated that it used a probabilistic approach in 1964. The technique used was a standard loss of load expectation method. In the surveys shown in Table 1, all the utilities which indicated using probability methods utilized a loss of load expectation approach.

It became obvious in the 1977 survey that although a number of utilities were using a basic loss of load expectation approach with similar numerical criteria, there were considerable differences in planned adequacy due to the assumptions and factors incorporated. The 1979 survey therefore attempted to

88 WM 150-5   A paper recommended and approved by the IEEE Power System Engineering Committee of the IEEE Power Engineering Society for presentation at the IEEE/PES 1988 Winter Meeting, New York, New York, January 31 - February 5, 1988. Manuscript submitted July 20, 1987; made available for printing December 29, 1987.

Reprinted from *IEEE Trans. Power Syst.*, vol. 3, no. 4. pp. 1488-1493, Nov. 1988.

further identify these assumptions and list them for each participating utility. Table 2 shows the basic criteria utilized by these utilities.

The total installed capacity of the utilities responding to the 1977 survey was approximately 60,000 MW, of which approximately 55,000 MW was located in the utilities using a probabilistic approach. In order to obtain at least a relative indication of the differences in methodology and assumptions used by each utility, a comparative adequacy evaluation study was conducted as a Subsection Working Group activity. This study was performed using the IEEE Reliability Test System (IEEE-RTS) which has a total installed capacity of 3405 MW [2]. Each utility was requested to compute two parameters; the calculated risk at a peak load of 2850 MW and the allowable peak load using the individual utility reserve criterion. The results supplied by the participating utilities showed a difference between the lowest and highest allowable

Table 2

Criteria Used In Reserve Capacity Planning

1979 Survey

Utility/System	Technique
Probabilistic Techniques	
Ontario Hydro	LOLE
Hydro Quebec	LOLE
BC Hydro and Power Authority	LOLE
Alberta Interconnected System	LOLE
Manitoba Hydro	LOLE
Nova Scotia Power Corp.	LOLE

Utility/System	Technique
Non-Probabilistic Techniques	
NBEPC	Capacity Reserve (20%)
SPC	Capacity Reserve (11%) _or_ Contingency Outage Reserve
NLH	Contingency plus 5% Reserve _or_ Capacity Reserve (15%)
NLPC	Capacity Reserve (15%)
MEC	Capacity Reserve (15%)

peak load of 462 MW. Since the 1979 survey, two utilities have changed from a deterministic to a probabilistic approach and there have been some changes in methodology and criteria.

Table 3 shows the basic criteria and the indices used by the participating utilities. This information was obtained from a survey completed in 1987. In order to simplify the remaining tables in this paper the following abbreviations are used in each case.

British Columbia Hydro and Power Authority	BCHPA
Alberta Interconnected System	AIS
Saskatchewan Power Corporation	SPC
Manitoba Hydro	MH
Ontario Hydro	OH
Hydro Quebec	HQ
New Brunswick Electric Power Commission	NBEPC
Nova Scotia Power Corporation	NSPC
Newfoundland and Labrador Hydro	NLH

It can be seen from Table 3, that the LOLE method is still the most popular technique. The two utilities (SPC and OH) using the EUE approach utilize different procedures to normalize the expected unsupplied energy. In the case of SPC, the

normalizing factor is the annual energy requirement, while in the case of OH, the annual peak load is used. Table 4 indicates the basis of selection for the criteria and the numerical indices together with a statement regarding the need for an independent assessment of energy supply adequacy.

A number of questions were asked regarding the generating unit models used in each evaluation procedure. The questions and responses are shown in Table 5.

Tables 6 and 7 show the questions and responses regarding load modelling considerations.

Table 3

Basic Criteria and Indices

System	Type of Criterion	Index
BCHPA	LOLE	1 day/10yrs.
AIS	LOLE	0.2 days/yr
SPC	EUE	200 Units per million (UPM)
MH	LOLE	0.003 days/yr (with connections) 0.1 days/yr (without interconnections)
OH	EUE	25 system minutes (S.M.)
HQ	LOLE	2.4 hours/year
NBEPC	CRM*	Largest unit or 20% of the system peak (whichever is larger)
NSPC	LOLE**	0.1 days/yr (under review)
NLH	LOLE	0.2 days/yr

LOLE - Loss of load expectation
EUE - Expected unsupplied energy
CRM - Capacity Reserve Margin
* With supplementary checks for LOLE
**With supplementary checks for CRM

Table 4

Basis of Selection and Need For an Independent Assessment of Energy Supply Adequacy

System Assesment	Basis of Selection	Independent Energy
BCHPA	Experience and Judgement	Yes
AIS	Experience and Judgement	Yes
SPC	Minimized expected customer cost	Not necessary
MH	Experience and Judgement	Yes
OH	In-house analysis	Yes
HQ	Experience and Judgement	Yes
NBEPC	Experience and Judgement	Yes
NSPC	Experience and Judgement	No
NLH	Experience and Judgement	Yes

Table 5

Capacity Modelling Considerations

System	Higher FOR Used for Immature Units	Multistate Unit Representation	Energy Limited Units Modelled	Planned Maintenance Included
BCHPA	Yes	Yes	No	Yes
AIS	Yes	No	No	Yes
SPC	Yes	Yes	Yes	Yes
MH	No	No	No	No
OH	Yes	No	No	Yes
HQ	Yes	Yes	Yes	Yes
NBEPC	Yes	No	Yes	Yes
NSPC	Yes	No	No	Yes
NLH	Yes	No	No	Yes

Table 6

Load Modelling Considerations

System	Historical Basis for Load Shape	Incorporation of Future Changes in Load Shape	Period for Which Risk is Evaluated
BCHPA	4 previous yrs, average	No	April - March
AIS	5 previous yrs, average	Yes	October - September
SPC	Yes	Yes	January - December
MH	10 previous yrs, average	No	January - December
OH	5 previous yrs, average	Yes	All year
HQ	11 previous yrs, average	Yes	All year
NBEPC	5 previous yrs, average	Yes	April - March
NSPC	1 typical yr.	No	January - December
NLH	7 previous yrs, average	No	April - March

Table 7

Load Modelling Considerations

System	Time Periods Used in the Calculation	Load Types Used in the Calculation	Interruptible Loads Considered
BCHPA	12 months	Daily peaks on all days	Not required
AIS	12 months	Daily peaks on all days	Yes for LOLE calculations
SPC	52 weeks	Hourly loads on all days	Yes
MH	12 Januarys	Daily working days peaks in January	Not required
OH	12 months	Hourly loads on all days	Yes
HQ	12 months	Half hourly at peaks / Hourly at other times	Yes
NBEPC	12 months	Hourly loads on all days	Yes
NSPC	52 weeks	Daily peaks on all days	Yes
NLH	13-4 week intervals	Daily peaks on all days	Not required

Table 8

Interconnection Considerations

System	Capacity Support from Interconnections Considered
BCHPA	Alberta Tie only
AIS	Yes
SPC	Firm contracts only
MH	Yes - asociated with risk criterion of 0.003 days/yr
OH	Yes
HQ	Yes
NBEPC	Yes (NPCC criterion)
NSPC	Yes
NLH	None available

Table 9

Interconnection Assistance Modelling

System	Not Modelled Explicitly	Equivalent Generating Units N	O	R	Multi-Area Probabilistic Method N	O	R
BCHPA							X
AIS			X			X	
SPC	X						
MH							X
OH			X			X	
NBEPC	-	-	-	-	-	-	-
NSPC	X						

N - Never
O - Occasionally
R - Regularly

Note: NBEPC use a deterministic approach. See Table 3.

The calculated reliability index in a LOLE or EUE calculation can be affected by interconnection considerations and therefore each utility was asked to indicate how these are incorporated into their procedure. Table 8 shows the responses provided from the basic questionnaire.

In order to obtain a better appreciation of how interconnection considerations are incorporated in the assessment, a further questionnaire was developed and forwarded to the participating utilities. Table 9 shows the response to the question "when evaluating the reserve value of interconnections, how are the interconnections modelled?"

The participating utilities were asked to describe their "standalone" and "interconnected" reliability criteria. Table 10 shows the responses to this question.

Table 10

Standalone and Interconnected Reliability Criteria

System	Interconnected Criterion	Standalone Criterion
BCHPA	1 day in 10 yrs.	None
AIS	0.2 days/yr	None
SPC	None	200 UPM
MH	0.003 days/yr	0.1 days/yr
OH	25 S.M.	None
NBEPC	Agreement between NBEPC	None
NSPC	NSPC regarding Assistance	None

COMPARATIVE STUDY USING THE IEEE-RTS

In order to obtain a relative indication of the differences in methodology, assumptions used and the indices adopted by the various utilities, each utility was asked to conduct an evaluation of the IEEE-RTS [1] using their own procedures, criteria and computer programs. The utilities were asked to perform the following studies:

1. What is the risk at a peak load of 2850 MW?
2. What is the maximum peak load that can be carried by the IEEE-RTS using your risk criterion?

The answers to these questions as provided by the participating utilities are shown in Table 11.

It can be seen from Table 12 that there is considerable variation in the calculated risk and the allowable peak loads determined by the various utilities. The calculated risks are different due to modelling variations in the individual methodologies. The maximum allowable peak load in each case provides a valuable comparison of the relative reliability of the various utilities in terms of the maximum peak load that they would carry in the IEEE-RTS before requiring additional capacity. Table 12 shows a rearrangement of the results in Table 11. The results for those utilities using a probabilistic approach are ranked in terms of decreasing maximum allowable peak load.

Table 11

IEEE-RTS Study Results

System	Risk at a Peak Load of 2850 MW	Maximum Allowable Peak Load (MW)
BCHPA	3.36 days/year	2385
AIS	4.00 days/year	2410
SPC	87 UPM	2980
MH	2.95 days/year	2389
	3.72 days/year*	2352*
OH	52 SM	2766
	177 SM*	2577*
HQ	3.7 days/year	2275
NBEPC	-	2760
NSPC	2.96 days/year	2365
NLH	3.49 days/year	2440

* Including load forecast uncertainty

Table 12

IEEE-RTS Study Results

System	Maximum Allowable Peak Load (MW)	
SPC	2980	
OH	2766	2577*
NLH	2440	
AIS	2410	
MH	2389	2352*
BCHPA	2385	
NSPC	2365	
HQ	2275	

*Including load forecast uncertainty.

The results shown in Table 11 can be generally divided into four zones as noted in Table 12. The criterion adopted by SPC would permit a much higher load to be carried by the IEEE-RTS generation system than that allowed by any other system. The Ontario Hydro value is approximately 200 MW lower than the SPC value without considering load forecast uncertainty and approximately 400 MW lower if load forecast uncertainty is included. The largest group of utilities, ie. Newfoundland and Labrador Hydro, Alberta Interconnected System, Manitoba Hydro, British Columbia Hydro and Power Authority and the Nova Scotia Power Corporation can be considered to be generally comparable in regard to the maximum allowable peak load. Table 12 shows that the Hydro Quebec criterion will permit the lowest allowable load to be carried for the test system. The results shown in Table 11 and 12 provided a valuable indication of the relative levels of generation adequacy associated with the respective utility methodologies and indices. It should be realized, however, that the respective methodologies and indices were developed by each specific utility with regard to their own system and not for the IEEE-RTS. As shown in Table 4, the methodologies and resulting adequacy indices were mostly determined using "experience and judgement" by the specific utility. This "experience and judgement" obviously pertains to the utility in question and not to a hypothetical system.

SECURITY CRITERIA AT HLI

The data on the methods used by Canadian utilities to assess generation system operating reserve requirements are based on responses to a questionnaire distributed in 1983 [3]. The results indicate that most utilities determine operating reserve requirements based on a "largest contingency" criterion. This method has generally been tailored to suit each system's particular needs. Probabilistic methods are not employed, and no trend is indicated toward their future use.

Classification of Operating Reserve

There is a wide range of approaches in regard to how the operating reserve is actually maintained in the operation of the responding utilities' systems. Generating unit operating reserve may be in the form of synchronized or non-synchronized unit reserve. Operating reserve may also include system reserve, which consists of underfrequency relaying, load shedding, and interruptible loads. Interconnections may also supply a portion of operating reserve requirements. Only two utilities rely heavily on system reserve while synchronized unit reserve is very common. Use of nonsynchronized unit reserve is also widespread, particularly among utilities with peaking hydro and gas-fired generation.

There is general agreement on utility philosophy with regard to load shedding as operating reserve. Aside from Newfoundland and Labrador Hydro, which use underfrequency relaying exclusively for operating reserve, no other utility relies on load shedding or underfrequency relaying.

In regard to how quickly operating reserve must be restored after its depletion, most utilities indicate that operating reserve should be restored as soon as possible but do not commit to accomplishing this in a specified period of time.

In regard to the use of neighbouring systems for assistance, seven of the ten utilities to which this question applies purchase operating reserve in some form. Three utilities share operating reserve over relatively short duration peaks.

Operating Reserve Requirements

There is considerable variation in the specific techniques implemented by the utilities in their determination of operating reserve. The widespread use of the "largest contingency" method is apparent and several of the utilities complement this reserve assessment technique with a megawatt margin of some form. None of the utilities polled assess operating reserve requirements in terms of a "constant risk" or probabilistic analysis.

Other constraints and system parameters which are accounted for in the techniques are detailed in Reference [3].

The information is summarized in Table 13 which shows that the majority of the utilities consider system parameters (i.e., ratio of instantaneous to hourly average load, load control capacity, and low forecast error) in the determination of operating reserve.

Transmission limitations are considered by seven utilities in determining operating reserve, while thermal unit ramp rates are considered by five utilities. Other system constraints considered include transmission losses, and environmental limitations. Generally, these factors affect the location, rather than the amount, of operating reserve.

Assessment of Current Practices

Each utility was asked to assess the suitability of its operating reserve criteria, and to discuss any changes being considered in determining operating reserve requirements. In general, utilities are satisfied with their current operating reserve practices. Each has apparently tailored its criteria to meet its particular needs. No utilities are planning any changes in their methods of calculating required operating reserve.

Summary

Probabilistic methods are not employed for operating reserve assessment, and no trend is indicated towards their future use. The "largest contingency" method, or some variation on this method, appears to be the most popular way of determining reserve requirements. A "fixed MW" margin is only used as a supplement to the "largest contingency" method. No changes in operating reserve assessment practices are foreseen by any of the utilities which replied to the survey. The current practices have apparently been tailored to each system, and meet the needs of these systems.

Unit ramp rates and transmission constraints are generally considered in determining operating reserve. Transmission constraints affect the amount or location of reserves. Synchronized unit reserve is the most common type of operating reserve. Only two utilities rely heavily on system reserve, i.e. voltage reduction and load shedding for first contingency outages.

All utilities specifying load pickup periods require over 50% of the contingency to be regained in the first ten minutes following a contingency. There is a direct relationship between the initial load pickup and the amount of synchronized unit reserve.

Interconnections play an important role in operating reserve assessment. Utilities in pools tend to use the same operating reserve policy as other pool members. Of the ten utilities able to purchase reserve, seven do so in some way. This could entail either a reserve purchase or an outright capacity purchase.

CONCLUSION

This paper has presented a summary of the generating capacity adequacy criteria utilized by Canadian utilities and systems. The historical review provided, illustrates the changes which have occurred over the past twenty years. It can be conclusively stated that virtually all Canadian utilities now use probabilistic techniques in planning future generating capacity requirements. The data provided and the benchmark study provided by each utility provides a valuable reference regarding the relative adequacy associated with the respective utility methodologies and indices.

ACKNOWLEDGEMENT

The cooperation provided by the member utilities of the Canadian Electrical Association is very gratefully acknowledged.

Table 13

RESPONSE TO THE QUESTION - WAS THE OPERATING RESERVE CRITERIA DEVELOPED ALLOWING FOR

	YES	NO	NOT APPLICABLE	NOT STATED
Ratio of instantaneously to hourly average load?	7 utilities	2 utilities	1 utility	0 utilities
Load Control Capacity?	8 utilities	0 utilities	1 utility	1 utility
Load forecast error?	8 utilities	1 utility	1 utility	0 utilities

REFERENCES

1. "Generating Capacity Adequacy Criteria Used By Canadian Utilities For Planning Purposes." R. Billinton, J.J. Kirby, M.L. Asgar-Deen. CEA Power System Planning and Operating Section. Engineering and Operating Division. Spring Meeting, Vancouver, March 1987.
2. "IEEE Reliability Test System", Reliability Test System Task Force of the Application of Probability Methods Subcommittee. IEEE Transactions PAS-98, No. 6 Nov./Dec. 1979, pp. 2047-2054.
3. "Generating System Operating Reserve Assessment Practices Used By Canadian Utilities," J.I. Nish, N.A. Millar. CEA Power System Planning and Operating Section. Engineering and Operating Division, Spring Meeting, Toronto, 1984.

Discussion

Mark G. Lauby (Electric Power Research Institute, Palo Alto, CA): The discussor would like to compliment the author for presenting an important treatment of generating capacity criteria. It provides a novel approach for the relative comparisons of a multitude of planning/operating criteria of generating capacity used by a number of utilities in Canada.

As mentioned in the paper, experience and judgment is many times the technique used by various utilities/regions to identify criteria. Quantifying how well the criteria is performing, as well as what changes may be needed, is very important in today's utility environment. However, it is not clear how transportable criteria developed for one utility/region may be to others. For example, utilities/regions currently attempt to operate their system at the N-1 guideline created at NERC. However, it is common knowledge that the length of a line, location, season, and environment all play a role in its certainty of being forced out from service. Deterministic criteria, therefore, are not truly transportable from one system to the next. The performance/results may not be consistent because they ignore many of the factors affecting the performance of the power system which are not consistent from one system to the next.

Are probabilistic criteria transportable? Namely, if a criteria is utilized based on the engineering judgment of a utility, it may be applicable only to that utility. Carrying the criteria to another system may produce conservative/optimistic results. Though it is true that all criteria were tested on the Reliability Test System (RTS), their comparison may not reflect the actual relative differences. The degree of transportability of the probabilistic criteria to the RTS may be not consistent for each utility system. The author's comments on this would be of great interest.

The work outlined in this paper is timely and of great interest to the industry. It provides another application to the RTS, which has been of great benefit to the industry for both benching new programs and now for evaluating the operating/planning criteria for generating capacity. Within this analysis, have weaknesses been identified with the RTS? Should they be addressed?

The discussor would like to thank the author for an interesting paper.

Manuscript received February 19, 1988.

Roy Billinton: The author would like to thank Mr. Lauby for his comments. As noted by Mr. Lauby, deterministic criteria are not transportable from one system to another as they do not repond to the actual factors that influence the reliability of the system. Deterministic criteria are not even consistent when applied at different locations within a single system. Probabilistic criteria are consistent in that their evaluation can include those factors that a utility deems to be important and probabilistic indices respond to the actual factors that influence the system reliability. It is also possible to examine the sensitivity of the calculated indices and the resulting criteria to these system factors. There is no obvious reason, however, why all utilities should utilize the same criterion, other than convenience. All utilities should, however, understand their own criteria and what they imply in terms of system reliability. As noted in Table II of the paper there are large differences in the peak load carrying capabilities due to the various utility criteria (including the calculation process) when applied to the RTS. Table 12 ranks the peak load carrying capabilities and clearly shows the relative design reliabilities of the participating utilities.

The objective should not be to design all systems with the same reliability level but to attempt to determine and therefore design the system with a reliability level that is suitable for that system. This involves the recognition of reliability cost and reliability worth. It should be realized that certain probabilistic criteria do not readily lend themselves to the evaluation of reliability worth. The most suitable means to incorporate the cost of service interruptions into the process is to use an energy based index. The two Canadian utilities which utilize an energy based index arrived at their design levels using economic considerations in addition to "experience and judgement". This is a sound approach which I believe will receive much more attention in the future. In conclusion, I would once again like to thank Mr. Lauby for his comments.

Manuscript received April 25, 1988.

Application of Probability Methods to the Determination of Spinning Reserve Requirements for the Pennsylvania–New Jersey–Maryland Interconnection

L. T. ANSTINE
SENIOR MEMBER IEEE

R. E. BURKE

J. E. CASEY

R. HOLGATE
MEMBER IEEE

R. S. JOHN

H. G. STEWART
MEMBER IEEE

Table I. Typical Values of the Standard Deviation of the Load Forecasting Error for the Various Seasons of the Year, Days of the Week, and Periods of the Day, in Per Cent of Estimate

	Summer*	Spring and Fall†	Winter‡
Morning Period:			
Monday	2.280	1.368	1.368
Tuesday through Friday	1.976	1.368	1.368
Saturday	2.736	1.976	1.976
Sunday	2.736	2.736	2.736
Afternoon Period:			
Monday through Friday	1.368	1.064	1.064
Evening Period:			
Monday through Friday	1.976	1.064	1.064
Saturday	2.736	1.368	1.368
Sunday	2.736	1.976	1.976

* Summer: Includes months of June, July, August, and September.
† Spring: Includes months of March, April, and May. Fall: Month of October.
‡ Winter: Includes months of November, December, January, and February.

Summary: A procedure is presented for the determination of spinning reserve requirements such that the reliability of service remains constant from hour to hour, day to day, and season to season. The effect of such factors as changes in load level, changes in short-time load forecasting error probabilities and changes in the size of units scheduled to operate are taken into account. The procedure described has been adopted by the Pennsylvania–New Jersey–Maryland Interconnection.

THE RAPID INCREASE in the size of electrical generating units and the construction of high-capacity interconnections has added greatly to the importance of the problem of scheduling of operating reserve generating capabilities. This paper presents a method of determining operating reserve requirements developed by a Task Force of Pennsylvania–New Jersey–Maryland (PJM) Inter-connection. The authors are the members of the Task Force.

The system studied has an installed capacity of approximately 16,000 mw (megawatts) and a peak hour integrated load of about 13,000 mwh (megawatt-hours). The interconnecting transmission between the member companies is of sufficient capacity so that spinning reserve capacity in any part of the system can generally be considered as fully available to any other part of the system.

Development of the Basic Concept

In order to evaluate exactly the probability of failure to carry the load during any period such as a day or a year it would be necessary to determine the probability of failure at each instant during the period and then to integrate these probabilities to find the total probability of failure.

In a study of installed capacity requirements the problem is to find the prob-

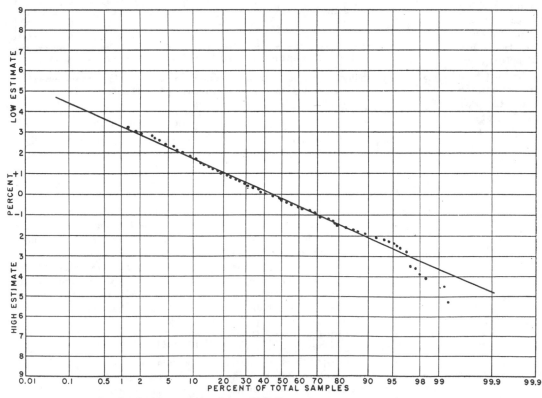

Fig. 1. Load forecasting error experienced during period of January 1 to December 31, 1957—weekday morning peaks

Reprinted with permission from *AIEE Trans. Power Apparatus Syst.*, vol. 82, pp. 726–735, Oct. 1963.

ability that the load will exceed the installed capacity less maintenance outages and forced outages. Since the installed capacity remains constant throughout any day the probability of such failure occurring wholly outside the peak load hour is quite remote and is generally taken to be negligible. Thus the day is taken as the smallest unit of time to be considered and the computations are based on the load for the daily peak hour. Any forced outage during the day is assumed to occur prior to the peak hour for the day and to continue beyond the peak hour. Since the average duration of a forced outage is about 5 days[1] this is a reasonable assumption. The probability of failure for the year is then the summation of the daily probabilities. If this summation should come out to be 0.1, the failure rate is said to be 1/10 day per year or 1 day in 10 years.

In the hour-to-hour operation of the electric utility system the objective is to operate at any time during the day only such capacity as is required to supply the load and to provide reasonable protection against load forecasting errors and generating equipment failure, except where it may be more economical to operate additional equipment. Thus the probability that the actual load will exceed the capacity scheduled to operate less forced outages may be equally as great at times other than the peak load hour for the day as it is for the peak load hour. On the PJM Interconnection the peak hour load is estimated and capacity is scheduled for each of the three heavy load periods on weekdays (morning, afternoon, and evening) and the two heavy load periods on Saturday and Sunday (morning and evening). During the other hours of the week there is generally considerable excess capacity operating and the probability of failure to carry the load at such times is considered to be negligible. Thus, as a practical expedient, the "heavy load period" has been taken as the smallest unit of time to be considered in deter-

Paper 63-210, recommended by the AIEE Power System Engineering Committee and approved by the AIEE Technical Operations Department for presentation at the IEEE Winter General Meeting, New York, N. Y., January 27–February 1, 1963. Manuscript submitted October 29, 1962; made available for printing November 29, 1962.

L. T. ANSTINE is with the Baltimore Gas and Electric Company, Baltimore, Md.; R. E. BURKE is with the Public Service Electric and Gas Company, Newark, N. J.; J. E. CASEY is with the Pennsylvania Power and Light Company, Hazleton, Pa.; R. HOLGATE is with the Philadelphia Electric Company, Philadelphia, Pa.; R. S. JOHN is with the General Public Utilities System, Reading, Pa.; and H. G. STEWART is with the Pennsylvania–New Jersey–Maryland Interconnection Office, Philadelphia, Pa.

The authors wish to acknowledge the contribution of H. D. Limmer of the Public Service Gas and Electric Company, to the development of the digital computer program described in the text.

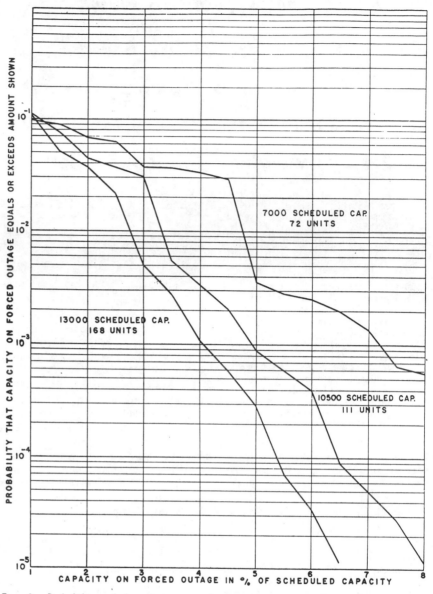

Fig. 2. Probability of forced outage of generating capacity for various values of scheduled capacity; installed capacity 15,489 mw

mining the probability of failure to carry the load. The ideal from a reliability standpoint would be to schedule spinning reserve for each heavy load period such that the probability of failure for all such periods would be the same (uniform risk). As for the installed capacity problem, the probability of failure for a year is the the sum of the probabilities for the base periods. Thus, if the spinning reserve is scheduled for each of the approximately 1,000 heavy load periods per year at a level of risk of 1 in 1,000 (0.001 probability of failure) the yearly loss of load probability is 1.0 or 1 period per year. However, if during half of the 1,000 heavy load periods extra capacity is operated for economy or other reasons and in such amount that the risk of failure during

these periods is reduced to a negligible amount then the yearly loss of load probability is 0.5 or 1 period in 2 years. Thus, it can be seen that probable exposure (number of periods operated) at the design level of risk plays a large part in establishing what is a satisfactory design level of risk. However, once a design level of risk has been established, probability methods provide a convenient means of varying the spinning reserve from period to period so as to maintain a uniform level of reliability.

Parameters Involved

The spinning reserve required to provide a given degree of reliability of service for a given period is a function of

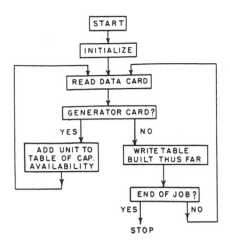

Fig. 3 (left). Simplified flow diagram of Part I of computer program; it develops tables of probability of availability of capacity

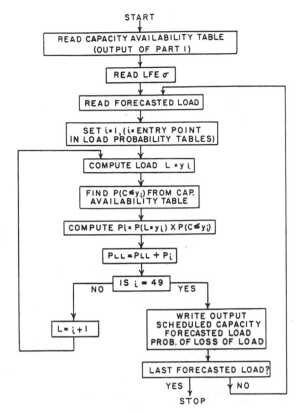

Fig. 4 (right). Simplified flow diagram of Part II of computer program; it combines load forecasting error and capacity forced outage probabilities

size of the individual units, the number of units and the reliability of the units scheduled to operate during the period. It is also a function of the load forecasting error probabilities for the period and the time required to start up marginal equipment.

Mathematical Theory

The probability of just carrying or failing to carry the load during any period can be expressed as

$$PLL = \sum_{Y=0}^{\infty} P(C \leq) \, P(L = Y)$$

where

$P(C \leq Y)$ is the probability that, of the capacity scheduled to operate, the amount available (not forced out) will be equal to or less than Y

$P(L = Y)$ is the probability that, for a given estimated load the actual load, will be an amount Y

Table II. Load Forecasting Error Probability Versus Magnitude of Error

Load Forecasting Error in σ's	Probability of Load Equal to Amount Shown
−6.00	2.1142160×10^{-9}
−5.75	7.1611820×10^{-9}
−5.50	2.9015942×10^{-8}
−5.25	1.1047753×10^{-7}
−5.00	3.9527343×10^{-7}
−4.75	1.3289497×10^{-6}
−4.50	4.1986320×10^{-6}
−4.25	1.2465114×10^{-5}
−4.00	3.4775612×10^{-5}
−3.75	9.1168370×10^{-5}
−3.50	2.2459773×10^{-4}
−3.25	5.1994685×10^{-4}
−3.00	1.1311122×10^{-3}
−2.75	2.3123108×10^{-3}
−2.50	4.4420267×10^{-3}
−2.25	8.0188310×10^{-3}
−2.00	1.3948602×10^{-2}
−1.75	2.1339370×10^{-2}
−1.50	3.2484443×10^{-2}
−1.25	4.5728795×10^{-2}
−1.00	6.0492436×10^{-2}
−0.75	7.5198576×10^{-2}
−0.50	8.7844705×10^{-2}
−0.25	9.6431542×10^{-2}
0.00	9.9476450×10^{-2}

In order to facilitate a comprehensive study of the effect of variations in the parameters involved the computations have been programmed for a digital computer. The program is discussed in a later section of this paper.

Forced Outage Probabilities

In studying the effect of forced outages on spinning reserve requirements, it is necessary to assume that there is ample installed capacity so that as units are forced out of service other units can be brought on to take their place. Thus, the longer the start-up time of the marginal equipment the greater will be the chance of a second outage occurring before the first has been replaced. The part of the duration of a single outage which extends beyond the start-up time of the marginal equipment can be ignored in determining the probability of multiple forced outages. Since the duration of most forced outage is greater than the time required to start up marginal equipment, the effective duration of all forced outages can be taken as equal to the start-up time. Thus, the probability that a particular unit will be effectively on forced outage (forced out and not yet replaced) is the rate of occurrence of forced outages (outages per hour of operation) multiplied by the start-up time in hours of the marginal equipment.[2] In selecting

a start-up time to be used for this purpose it should be borne in mind that reducing the start-up time does not change the total amount of equipment likely to fail in the course of a day's operation. Thus, the amount of marginal equipment having the selected start-up time or better should approximate the difference between the spinning reserve requirement for a 24-hour start-up time and the spinning reserve requirement for the selected start-up time.

The required forced outage rates can be developed from the basic data on forced outage[1] since the probability of occurrence of an individual outage is equal to the probability of its existence divided by its average duration.[2] These rates based on a 4-hour start-up time are as follows:

Pressures less than 1,000 psi (pounds per square inch), 0.001168
Pressures from 1,000 to 1,350 psi, 0.002032
Pressures more than 1,350 psi, 0.001544

The rates used in the study were adjusted slightly to conform more closely to the interconnection experience.

Experience has indicated a higher outage expectancy for the new equipment during the first years of operation. During recent years there has been operating on the Interconnection an average of 12 units with less than 2 years' service and an average of three units with less than 6 months' service. Because these units

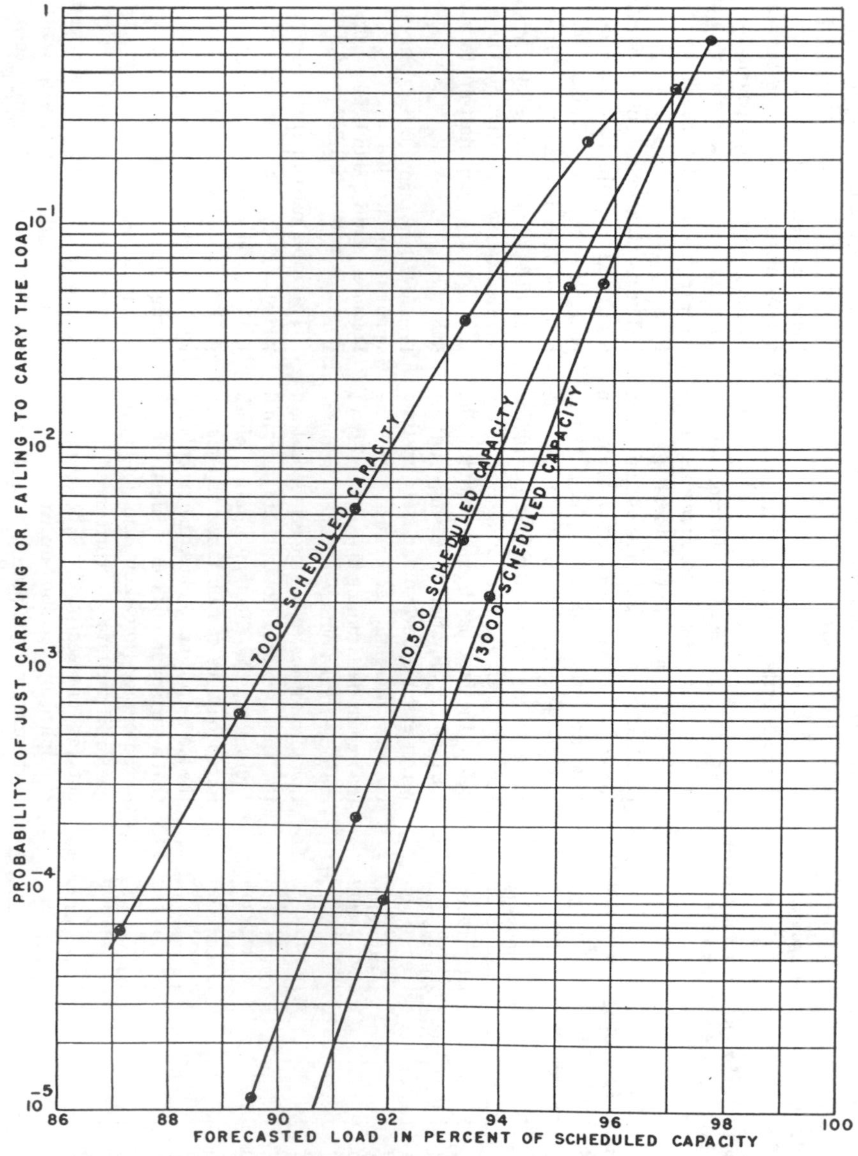

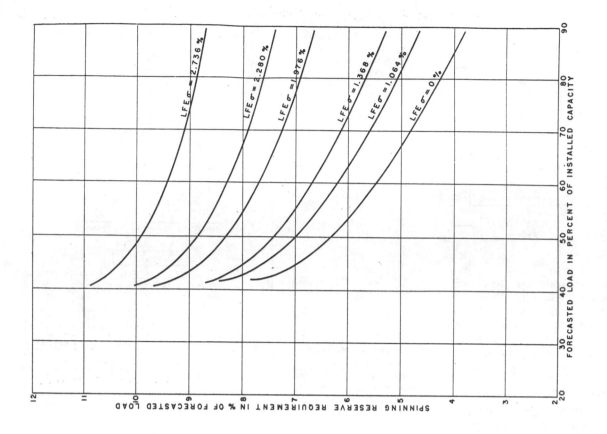

Fig. 6. Spinning reserve requirement versus forecasted load for various values of load forecasting error sigma; risk of failure 1 in 1,000; installed capacity 15,489 mw

Fig. 5. Probability of just carrying or failing to carry the load versus forecasted load for various values of scheduled capacity

represent a large part of the total capacity operated, particularly at light loads, it was thought desirable to give special consideration to the outage rates of such units. Based on a review of the performance for new units installed on the PJM interconnection since January 1956, the outage probability for units in service less than 6 months was taken to be 0.008 and for units in service 6 months to 2 years was taken to be 0.004.

Load Forecasting Error Probabilities

As mentioned previously, an estimate is made of the peak hour load for each heavy load period of the week. Since there are approximately 1,000 such periods in a year a sizable data file can be accumulated in a relatively short period of time. Such a record lends itself readily to probability analysis. For the purpose of such analysis the deviations of the estimates from the actual are first converted to per cent of estimate to eliminate the effect of load growth. An estimate less than actual is considered to produce a positive forecasting error.

It has been found that these errors when plotted on arithmetic probability paper very closely approximate a straight line, indicating that they follow the normal probability distribution. A sample plot of such data is shown in Fig. 1. Thus the probability characteristic of the

load forecasting errors can be fully defined by the algebraic mean M and the standard deviation σ of the errors. It has further been found that there is a significant difference in the standard deviation of the forecasting errors for the different seasons of the year, days of the week, and periods of the day. Typical values of the standard deviation of the load forecasting errors are shown in Table I.

Computer Program

The computations have been programmed for an IBM *7090* computer. The program is in two parts. Part I computes $P(C \leq Y)$ for the given set of units scheduled to operate for a range of values of Y from Y equal to the total capacity scheduled to operate down to a value of Y where $P(C \leq Y)$ becomes infinitesimal, while Part II computes $\sum_{Y=0}^{\infty} P(C \leq Y) P(L = Y)$.

The method used in Part I is sufficiently well described elsewhere[3] and will not be discussed in detail. In applying this method each unit is represented by a separate data card giving the capacity of the unit and the probability of its being forced out. The cards are arranged generally according to operating priority with the more efficient units first. Thus, they also fall generally according to size with

the larger units first. If, because of local area requirements, certain of the less efficient units are normally required to operate at the lighter loads this effect can be simulated by moving the appropriate cards toward the front of the deck. Maintenance outages can be simulated by removing cards from the deck. To limit the number of entries in the capacity table, and thereby conserve computer storage space, all unit capacity values are rounded to the nearest 5 mw.

The data cards are read one at a time and added into the capacity availability table. A control card can be inserted in the data deck at any point, corresponding to a given scheduled capacity, to instruct the computer to write out the capacity availability table built thus far and then continue to add units. Thus, in a single pass a series of tables can be built, the first including those units which would normally be scheduled to operate under light load conditions, the next including those units which would normally be scheduled at some higher load level, and the next for some still higher load level, etc. Thus, a set of tables is produced which properly reflect the effect of reduction in average size of unit operating with increase in load level. Typical results of Part I are shown in Fig. 2 for three values of scheduled capacity plotted, however, in terms of capacity on forced outage in per cent of scheduled capacity. A simplified

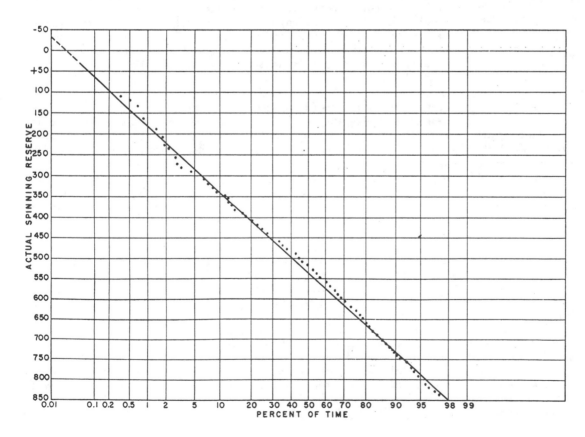

Fig. 7. Actual spinning reserve at time of daily weekday peak, for the period of November 4, 1956, to March 11, 1959

flow diagram of Part I is shown in Fig. 3.

Part II of the program combines the load forecasting error probability with the probability of availability of capacity determined by Part I to obtain the probability of just carrying or failing to carry the given forecasted load. Fig. 4 gives a simplified flow diagram for one capacity availability table and a series of related forecasted loads.

The normal probability distribution, as used in connection with the load forecasting error is represented as a step curve and is presented to the computer in tabular form as shown in Table II. The argument of the table is load forecasting error in intervals of 0.25 σ out to a total deviation of $\pm 6\ \sigma$. The function of the argument is the probability that the forecasting error will fall in a particular interval. The probability that the error will exceed $\pm 6\ \sigma$'s is extremely small and is lumped at 6 σ's. The following example will illustrate the use of the table. Assume that the forecasted load is 10,000 mwh and the standard deviation of the load forecasting errors is 0.5%. From the table the probability that the actual load will be high by 6 σ's (an error of -6 σ's) is 2.1142160×10^{-9}. The magnitude of the actual load for this deviation is

$$L = (1.0 + 6 \times 0.5 \div 100) \times 10,000$$
$$= 10,300 \text{ mwh}$$

Likewise, the probability of a 5.75 to 6 σ deviation is 7.1611820×10^{-9} and the corresponding magnitude of the actual load is

$$L = (1.0 + 5.75 \times 0.05 \div 100) \times 10,000$$
$$= 10,287 \text{ mwh}$$

Results

Fig. 5 shows the results of a computer run involving three capacity availability tables; one for light load, one for medium load, and one for heavy load. Five load values were used with each capacity table. Thus, a set of curves is obtained giving the variation in the risk of failure to carry the load versus forecasted load for a given load forecasting error sigma and three values of scheduled capacity. At least two additional sets of computer runs, each for a different load forecasting error sigma, are needed in the final development of the spinning reserve requirement.

To proceed with the determination of the spinning reserve requirement, a selection of the level of risk must be made. No numerical calculations can alone establish the proper level of risk for a particular system. The selection of a satisfactory level of reliability requires the exercise of

Table III. Spinning Reserve Requirements Based on Uniform Risk of Just Carrying or Failing to Carry the Load of 1 in 1,000

Forecasted Load		Load Forecasting Error, Per Cent				
From Mw	To Mw	1.064	1.368	1.976	2.280	2.736
0	7,000	575	593	639	670	732
7,001	7,500	576	599	655	690	761
7,501	8,000	577	606	673	712	791
8,001	8,500	586	620	692	736	823
8,501	9,000	596	634	715	762	857
9,001	9,500	609	649	738	789	891
9,501	10,000	619	665	760	817	926
10,001	10,500	630	681	785	846	962
10,501	11,000	642	696	808	874	999
11,001	11,500	652	710	832	901	1,035
11,501	12,000	662	724	857	931	1,070
12,001	12,500	669	736	878	958	1,106
12,501	13,000	677	748	900	986	1,144
13,001	13,500	684	761	924	1,014	1,179
13,501	14,000	689	773	945	1,041	1,214
14,001	14,500	694	784	968	1,069	1,253
14,501	15,000	699	794	988	1,097	1,291

informed judgment.[2] Comments concerning the selection of the level of reliability are included later. Once the selection of a level of risk has been made the set of curves shown in Fig. 6 can be drawn. Fig. 6 shows, for a selected level of risk, the spinning reserve requirement in per cent of forecasted load versus forecasted load in per cent of total installed capacity for various values of load forecasting error sigma. By expressing forecasted load in per cent of installed capacity the curves become more or less universal and can be used over a period of several years without the need of rerunning the computations, provided no new unit is installed which appreciably increases the size in per-cent of total installed capacity, of the largest unit installed on the system.

If adequate records have been kept of actual spinning reserves (that is, the excess of the actual capacity remaining in service at the times of the various daily peaks over the actual load at such times) this information can be of considerable value in the selection of a satisfactory level of risk. Fig. 7 is a plot, on arithmetic probability paper, of the actual spinning reserve on the PJM Interconnection for a period of years prior to the adoption of uniform risk. By extending this curve to the left (dotted portion) the probability of zero spinning reserve is found to be 0.025% which is equivalent to one in 4,000. Since there are approximately 1,000 heavy load periods per year this corresponds to a failure rate of one period in 4 years. Since this is a plot of actual experience it includes the effect of the extra capacity which was carried for economy reasons and for local area protection and is a true indication of the reliability of service provided. If this past

experience is considered to represent a satisfactory level of reliability then the problem is to find the uniform level of risk which if it had been in effect during this past period would have given approximately the same over-all result. A similar plotting of actual spinning reserves can be used to monitor the results obtained under uniform risk and can be used as a basis for changing the design level of risk from year to year, as may be indicated by changes in the amount of extra capacity which is operated for economy reasons as well as for local area requirements.

Table III shows the practical application of the results of the study to the scheduling of spinning reserve on the PJM Interconnection.

Conclusions

Considerable savings in operating costs can be realized by a group of interconnected systems by pooling spinning reserve requirements. However, full realization of the potential savings while still maintaining a desired level of reliability requires a systematic approach to the problem of evaluating the spinning reserve requirements of the pool.

The paper presents a method of evaluating the spinning reserve required to maintain a uniform level of risk in the day-to-day operation of an interconnected system having sufficient transmission capacity so that spinning reserve capacity in any part of the system can generally be considered as fully available in any other part of the system. The method takes into account such factors as changes in load level, changes in the variability of the load, and changes in the size of units scheduled to operate.

References

1. FORCED OUTAGE RATES OF HIGH-PRESSURE STEAM TURBINES AND BOILERS, AIEE Committee Report. *AIEE Transactions*, pt. III-B (*Power Apparatus and Systems*), vol. 73, Dec. 1954, pp. 1438–42. Also, *Combustion*, New York, N. Y., Oct. 1954, pp. 57–61.

2. APPLICATION OF PROBABILITY METHODS TO GENERATING CAPACITY PROBLEMS, AIEE Committee Report. *Ibid.*, pt. III, vol. 79, 1960 (Feb. 1961 section), pp. 1165–77.

3. DIGITAL COMPUTER AIDS ECONOMIC-PROBABILISTIC STUDY OF GENERATION SYSTEMS—I, M. K. Brennan, C. D. Galloway, L. K. Kirchmayer. *Ibid.*, vol. 77, Aug. 1958, pp. 564–71.

———————◆———————

Discussion

S. J. Litrides (Pennsylvania Power and Light Company, Philadelphia, Pa.): The authors are to be commended for developing a method of determining the daily spinning reserve requirements on the basis of a standard of service reliability in much the same manner that is used for the determination of the planned installed capacity requirements. Both the paper and the subject stimulate many questions of which only a few will be asked. Basically, the problem breaks down into two major areas: (1) how to treat the changing conditions that occur during operation to permit the calculation of probability of loss of load; and (2) the determination of the correct or desired standard of service reliability that is to be met.

In essence, the paper deals with the first area and only touches on the second but equally important problem. The paper also gives the impression of being highly condensed, making the reading difficult for those foreign to system operation and probability determinations. Greater discussion of the principles being applied, particularly as to the assumptions being made and their relative effect on calculations, would have enhanced the value of the paper. The following comments and questions are intended, therefore, to bring out this type of supplementary information.

1. It would help if the authors had distinguished which definition of probability they were using throughout the paper, i.e., probability in terms of absolute probability (a pure fraction or decimal) or in terms of days per year, etc. Some of the operations indicated would be correct for one but not the other. For example, the definition of probability of loss of load for any period as given in the paper would have to be divided by the number of subsets being considered if the period of absolute probability is being used. Also, the limits of summation used, $K=0$ to $Y=$ infinity, are only for mathematical convenience whereas in actuality, the limits cannot exceed the total installed capacity or total connected load without diversity, whichever is greater.

2. The meaning of the last sentence of the second paragraph under "Results" needs clarification. Although the use of one set of curves for several years is being recommended (Fig. 6), could more material be presented on the changes caused by different combinations of units and the effects caused by the addition of new units?

3. Would it not be better if the term "load forecasting deviations" were used rather than "load forecasting errors" for the rather obvious reason that they are not errors but expected probability deviations? This point also raises the question as to what deviates from what, the forecast from the load or the load from the forecast and next in turn, leads to the question as to how the forecasts are made. This is a crucial point, since the complete study is based on it. Would the authors care to outline the forecasting method being used and the corresponding load model?

4. The term "design level of risk" or similar terms using "risk" should be tactfully avoided. A manufacturer selling his product never talks of risk of failure, but, on the contrary, sells on the basis of reliability. Would it not be preferable to use "standard of service reliability" or "reliability criterion" in the power industry?

5. The mathematical model being used appears to differ somewhat from the written description in the text. The method of calculation used in the paper actually assumes that a peak load will *exist* for a certain duration during the day, called a period. During this period, the probabilities of the existence of forced outages will be in effect for the operating capacity. This is the simplest and most direct model for determining the probability of loss of load for the period, assuming that capacity is not replaced. The modification to the outage rates which the authors have made in an attempt to evaluate the probability that machine outages are replaced by *like* capacity prior to the peak load of only 1 hour duration. The actual determination is much more complicated than the one used by the authors if the possibility of a peak occurring early in the period is considered. It appears then, that the timing of the peak and its duration are involved, in addition to start-up times. Some consideration should also be accorded to the remaining hourly loads. Do the authors plan further development along these lines?

6. Although the authors mention the use of three daily periods, their hourly duration is not mentioned. This information would be of some value, particularly as to the hours they start and end, since the adjustment factors for machine outage rates are based on them. Here again, a question arises as to the proper ratio, assuming temporarily that the factor method is correct; should it be 4/8 or 4/24 if 4-hour start-up times are used and 8 hours is the duration of the period under consideration?

The authors have taken a step in the direction of the determination of the correct reliability level to be used by trying to correlate their results to past experience, using Fig. 7. This figure is troublesome for several reasons. First, it does not have sufficient points for the period indicated. Second, these points are not symmetrical about the 50% line as correct plotting would have them, but rather they indicate a bias. Third, it is not specified whether these points include the existence of a forced outage. Consequently, it would be interesting to know on what basis this figure was obtained. A more appropriate figure, it would seem, would be one in which load or forecast deviations and load growth were accounted for. Would the plotting of these points as per-cent spinning reserve capacity using peak load as a basis be more meaningful and useful?

Some such correlation with the past, if correctly done, will help determine the level desired, but the relative economics involved, which is the crux of the whole problem, need to be included. This can be done in a manner similar to that which was suggested for use in determining installed capacity requirements. (See C. W. Watchorn's discussion of reference 2 of the paper.) Once this is done, knowledgeable cost decisions can be made as to system spinning reserve operation versus reliability.

This last set of questions regards actual system operation using this method. If a forced outage occurs during a period, is the spinning reserve adjusted back to its original specified level for the remaining portion of the period or is the chance event accepted on the basis that the purpose of spinning reserve has been fulfilled? Also, what adjustments are made if the required amount of capacity having a 4-hour start-up time is not available for a specified period? And finally, how are hydroelectric units accounted for in the spinning reserve determination and what is their actual operation in practice?

C. W. Watchorn (Pennsylvania Power and Light Company, Philadelphia, Pa.): The word "error," which is used throughout the paper, could result in considerable misunderstanding, particularly to the uninitiated since neither of the two general meanings that could be understood to be intended is applicable.[1] The first of these is that the difference referred to as errors could be taken to mean the result of carelessness and inattention and the second is the mathematical sense which is the difference between the true and measured values of a quantity.

If the first of these meanings were intended, which I am sure it is not, the paper could be taken to be proposing a method of implementing the defeatist attitude of living with ineptness. The question could then be asked if it rather would not be a better approach to develop ways of minimizing the effect of or of eliminating the errors, since to do so, according to the results shown in Figs. 6 and Tables I and III, could result in reductions in operating costs of possibly several hundred thousand dollars per year.

On the other hand, it is incorrect to intend the second meaning, since the differences referred to in the paper as "errors" are actually the differences between forecasts and measured quantities, which are more like the differences between the calls and falls in throwing dice, and which are certainly not errors, but rather a natural phenomenon about which nothing can be done so long as the dice are honest and are of the same basic nature as the differences between the falls of dice and their mean. Thus, the fact that some differences are errors and appear to follow the same mathematical laws as these types of differences, does not make them all errors.

These comments are more than a matter of semantics, since often the loose use of terms can lead to improper applications of principles.

Fig. 1 raises several questions. First, it purports to be for the weekday morning peaks for a calender year, yet there are points shown plotted for only 59 or approximately 25% of the number of such days, of which 30 are shown on one side of 50% line and 28 on the other side with one on the 50% line itself. It would be of interest to learn both the reason for this different distribution about the 50% line, when there should be the same number on each side, and also what treatment was accorded the other 75% of the differences between the actual and forecast loads.

Second, Fig. 1 shows an unsymmetrical plotting position for the end points shown there; the maximum low estimate deviation is shown plotted at about 48.6 probability points from the 50% line and the maximum high estimate deviation is shown plotted at about 49.2 probability points. Correct plotting would require these end points to be the same value of probability points from the 50% line, with the best approximation to the correct plotting position being determined by the equation:[2]

$$P = \frac{m - 0.5}{n}$$

where P is the plotting position, between zero and unity, or the accumulated probability as indicated by the experienced data that the load level will be equal to or smaller than the corresponding deviation from trend; m is the integer number in the ascending order of magnitude of the value of the deviation under consideration; and n is the total number of values of deviations involved, 51 for our problem here.

Third, Fig. 1 shows a bias in the direction of high load estimates. It does not appear that this fact was taken into account in the determination of the spinning reserve requirements. If this is so, it could result in a loss of possibly in the order of $50,000 to $100,000 per year for the operation of extra capacity that is really not required.

Next, Fig. 1 shows that the larger magnitude high-load forecasts deviate from the mean by much larger amounts than the deviations from the mean of the larger magnitude low load forecasts. This characteristic results in a larger standard deviation than otherwise, and thus, as the result of large deviations of high load forecasts, gives rise to the anomaly of determining a larger spinning reserve requirement than actually needed for protection against the contingency of the occurrence of large deviations of low load forecasts. This could also result in further unjustified increased operating cost.

Further, the authors, by assuming the deviations of the actual loads from the forecasts to follow a normal frequency distribution curve, have thereby automatically assumed that such deviations are limitless. This, of course, is not the case and can result in an increase in the determined spinning reserve requirements over that actually required for the desired standard of service reliability at a further substantial increase in operating cost, particularly for the larger values of standard deviation. This could be recognized by truncating the normal frequency distribution curve, at possibly three times sigma, which should be ample, or possibly by using a beta frequency distribution for load forecasting deviations.

It would be amiss, however, not to compliment the authors on the preparation and presentation of a comprehensive paper on a subject that is long overdue, and that they saw fit to make the probability calculations on a generally rigorous basis as to method rather than use theoretical approximations over the range of small values of probabilities to which the approximations are generally not applicable.

REFERENCES

1. Reference 2 of the paper.

2. Use of Normal Probability Paper, H. Chernoff, G. L. Lieberman. *Journal*, the American Statistical Association, New York, N. Y., vol. 49, 1954, pp. 778–84.

G. R. Tebo (Hydro-Electric Power Commission of Ontario, Toronto, Ont., Canada): The authors have made a valuable extension of the methods used in assessing capacity reserve requirements to analyze the problem of spinning reserve requirements.

Fig. 7 shows a plot of the actual spinning reserve on the PJM interconnection over the years 1956 to 1959 before the adoption of the "constant risk" principle. Would the authors outline the policy for establishing spinning reserve requirements which was in effect during this time?

The system the authors deal with is essentially an all-thermal system. In a combined hydraulic and thermal generating system, the theoretical problem may be more complex because the start-up times and start-up costs associated with hydraulic units are much smaller than those associated with thermal units and these differences would have to be recognized in any solution.

On the other hand, optimum solution of the problem in a combined hydraulic thermal system may be less critical because the cost of spinning reserves can be very low whenever hydraulic capacity is available. Thus, some simplifications in the computation may be justified.

It would be of interest to know the effects of the following simplifications applied to the method presented in the paper:

1. Assume all small capacity units are 100% reliable.

2. Assume that the probability of more than one or two simultaneous outages is negligible.

3. Assume larger capacity steps than the 5-mw steps indicated in the paper, to reduce the number of calculations.

Have the authors used their complete program to investigate the validity of these or other simplifications, and would they offer any comments on the problem of spinning reserve requirements in a combined hydraulic-thermal system?

M. J. Steinberg (Consolidated Edison Company of New York, Inc., New York, N. Y.): This paper extends the application of probability techniques into a new area which has not received much attention. The paper is therefore welcome as a timely contribution when the interest in the interconnection of electric systems is high. We wish to commend the authors for making available to the industry, their analysis and end results for the PJM pool.

Although there may be some areas for differences of opinion with respect to application, there should be general approval of the philosophy presented by the authors.

The operation of spinning reserve still is, to a large extent, determined on a rule-of-thumb basis, under which the objective is the operation of minimum spinning reserve necessary to maintain continuity of service upon loss of the largest single source of energy supply. In practice, more than the objective minimum spinning reserve would be operated as dictated by economy and requirements for local area load protection. Now, rule-of-thumb procedure can be replaced by sound mathematical principles to resolve the problem of operating spinning reserve, as was done to resolve the problem of installed reserve capacity requirements.

The problem of how much spinning reserve should be operated becomes more complex and increasingly important with expansion of system interconnections into pools. Use of rule-of-thumb procedure becomes less desirable, and it is in this respect that we consider the paper to be of constructive value to the industry.

The authors have been most generous in supplying detailed step-by-step information which makes analysis of their procedure a simple matter for those who are familiar with the subject matter. There are, however, several items of interest not treated in the paper which we hope will be furnished by the authors in their closure. They are,

1. By use of Fig. 7 of the paper, the authors have established an acceptable on-peak level of risk defined as a failure rate of 1 day in 4 years.

(a). How does this objective service reliability compare with that used to determine the requirements of installed reserve capacity?

(b). If a different reliability standard is used for determination of installed reserve capacity, how do the authors reconcile the different reliability standards?

(c). Have the authors considered using the same standard of service reliability as the basis for operation of spinning reserve and for providing installed reserve capacity?

2. On the assumption of the existence of a centralized operating organization, what control is exercised in regard to operation of spinning reserves in excess of that shown in Table III of the paper? To what extent may a member company of the pool exercise independent judgment in the operation of capacity as spinning reserve on its own system?

3. The authors state that interconnections between member companies are adequate so that spinning reserve capacity in any part of the system can generally be considered as fully available to any other part of the system.

(a). Does every member company of the pool accept the possibility that it may be called upon to operate turbine-generator capacity totalling less than the load in the member company's system? This would in effect place reliance upon the interconnection(s) for local spinning reserve.

(b). Has any member company been subjected to such an operating condition since pool operation became effective?

(c). If so, the details would be appreciated.

4. Assuming that Table III and Fig. 6 of the paper indicate objective values, inclusion of experience on one or more typical days showing objective and actual operating spinning reserve capacities, together with reasons for the variations, would be very helpful.

R. J. Ringlee and **A. J. Wood** (General Electric Company, Schenectady, N. Y.): We extend our congratulations to the authors for presenting a rational and easily mechanized method for the determination of spinning reserve levels on the basis of service reliability requirements. This technique combined with available digital computers should permit the rapid establishment of spinning and ready reserve requirements.

It would be helpful if the authors would illustrate the computation of the forced outage existence rates which they have presented in their section on forced outage probabilities.

It appears to us that analysis of the spinning reserve model of the generation system can be expedited by using a Markov process to describe the probabilities of the possible states. When this is done, a set of simultaneous linear differential equations results. These equations describe the transient behavior of the state probabilities. The probabilities determined by the authors are the steady-state values of the state probabilities ordered in terms of increasing outage.

If an assumed State 1 is taken as the initial condition and the question is asked, what is the likelihood of the system being in another State 2, at some later time T, the answer will be found by the integration of the differential equations in the state probabilities from $t=0$ to $t=T$. As T becomes very large, the state probabilities reach steady-state values.

While direct solution of such a problem would be formidable, some useful information can be gained from a study of the characteristic roots of the system of differential equations. Under the assumptions made by the authors, it appears that the smallest root (slowest time constant) is approximately equal to the inverse of the average start-up time of a machine. This means, in effect, that the steady-state solutions, that is, the outage existence probabilities, may be used to evaluate the generating system reliability as long as the time interval between evaluations is about twice the slowest time constant.

For example, with a mean start-up time of 4 hours, the system state predictions may be made for a period of 8 hours in the future with the steady-state solution of the state probability functions. The steady-state values give the outage existence probabilities.

L. T. Anstine: The authors wish to thank the discussers for their many suggestions and challenging questions. Since many of the questions are closely related we shall try to answer them in a logical sequence, referring directly to a specific discusser or question only where such reference appears desirable.

The paper states that the required forced outage rates can be developed from the basic data on forced outage. As pointed out by Mr. Anstine in his discussion of the report by the AIEE Subcommittee on Application of Probability Methods,[1] forced outage rates based on hours of operation are required for a probability study of spinning reserve requirements. Since the published data give only days of operation it is necessary to convert to hours by assuming an average number of hours per unit day of operation. Assuming the average day of operation to be 15 hours, we obtain the outage occurrence rate for units operating at pressures less than 1,000 psi as follows:

Turbine[2]:

$$\frac{338 \text{ Occurrences}}{289{,}793 \text{ Turbine Days} \times 15 \text{ Hours}} =$$
0.0000777 failure per hour of operation

Boilers[2]:

$$\frac{1119.5 \text{ Occurrences}}{348{,}432 \text{ Boiler Days} \times 15 \text{ Hours}} = 0.0002139$$
failure per hour of operation

Combined:

$(1-.0000777)(1-.0002139)=0.999708$

$(1-0.999708)=0.000292$ occurrences per hour of operation, or $4\times0.000292=0.001168$ occurrences per 4 hours (the start-up time of marginal equipment as used in our study)

This brings us to Mr. Litrides' question about the duration of the various heavy load periods of a day. The weekday heavy load periods are as follows: morning, 10 a.m. to 12 noon; afternoon, 1 p.m. to 4 p.m.; and evening, 5 p.m. to 7 p.m. in winter, shifting with the seasons to 8 p.m. to 10 p.m. in summer. However, this duration is not a factor in our computations since we do not integrate the probabilities over the duration of the heavy load periods. We compute only the probability that the capacity remaining in service at the time of the peak will be less than actual peak load. Since we are assuming a start-up time of 4 hours for marginal equipment and, further, that as units are forced out other units are started up to take their place, the probability of outage of capacity at any time is the probability of failure of equipment during the previous 4-hour period. The capacity scheduled to operate during the previous 4-hour period is assumed to be constant and equal to the capacity scheduled for the peak of the period. Thus, we approximate a steady-state condition as suggested by Messrs. Ringlee and Wood.

The basis for the load forecasts is, of course, very important. We frequently hear it said that a particular system carries spinning reserve for the loss of the largest generating unit. If this is based on a middle-of-the-road forecast (that is, there is a 50% probability that the actual load will exceed the forecast) then 50% of the times that the largest unit fails during a heavy load period the system must either curtail load or lean on its neighbors. If the estimates are high-side estimates (say only a 10% or 20% probability that the actual load will exceed the estimate) then the system is actually providing spinning reserve greater than the largest unit. The PJM load forecasts are middle-of-the-road forecasts, as demonstrated by Fig. 1 of the paper. For the set of data plotted the mean deviation of

the estimates is approximately -0.25%. For another set of samples this difference might be slightly more or less, or even positive, but always by some insignificant amount. If it should be found from a review of the load forecasting experience that the load forecasters are providing some spinning reserve by consistently high forecasts, then the spinning reserve requirements as presented in the paper should be reduced by this amount, as suggested by Mr. Watchorn. The converse is also true.

The policy for establishing spinning reserve requirements which was in effect prior to the adoption of the "constant risk" principle as set forth in the paper was, in effect, a limited application of this same principle. However, since the inability to forecast peak loads exactly constituted the greatest single hazard the spinning reserve requirement was based on load forecasting experience alone. A set of curves similar to Fig. 1 of the paper was prepared to show the forecasting experience for the various seasons of the year, days of the week, and periods of the day. These curves were each extended to a 0.05% probability (1 in 2,000 chance) and the indicated percentages were applied to the PJM Interconnection peak load of the previous winter to obtain the mw of spinning reserve requirement for the various periods of the current year.

Mr. Tebo has questioned the validity of certain simplifications of the program as might be justified for use with a system which is primarily hydroelectric. In this connection we would like to point out that the total IBM 7090 computer time required in connection with Fig. 6 of the paper was only approximately 5 minutes. Thus we see no need to consider possible simplification of the program. The use of larger capacity steps than 5 mw might be justified if it were desired to fit the program into some smaller computer. However, it is doubtful that a detailed study such as we have presented would be justified for a system on which the marginal equipment is quick-starting hydro. On such a system little consideration need be given to load forecast accuracy since capacity can be brought on as the load develops. Furthermore, the probability that a second forced outage will occur before the first is replaced is negligible because of the short start-up time. Thus, the spinning reserve requirement should be little more than the capacity of the largest unit.

Cost is a very important factor in establishing a design level of risk and, obviously, if we had found the cost of maintaining a level of 1 in 1,000 to be prohibitive some lower level would have been proposed. Full economic consideration would probably lead to the adoption of a design level of risk which varies from period to period according to cost. However, such a procedure would be very difficult to administer, especially under an arrangement for sharing spinning reserve requirements with a neighboring system.

Mr. Steinberg has misinterpreted our established "acceptable on-peak level of risk" which he says is "defined as a failure rate of 1 day in 4 years." In our treatment of the subject we deal in heavy load periods rather than days and since there are three such periods per day one might argue that 1 period in 4 years is equivalent to 1 day in 12 years. Actually we are using a different yardstick to measure day-to-day operating reliability and know of no way to relate it exactly to the yardstick used for installed capacity purposes.

In the paper it was implied, but not specifically stated, that at times the capacity scheduled to operate for economy or other reasons more than covers the forecasted load plus spinning reserve requirement. Actually there is no obligation on the part of any member to operate less capacity than it deems necessary for protection of its own system. Whether or not a member is willing to operate capacity less than its own load is largely a matter of how the member's system is situated electrically with respect to the remainder of the interconnection. Several of the member companies regularly schedule capacity which is less than their own load. However, the amount of capacity which remains to be scheduled by the central dispatching office to meet the interconnection spinning reserve requirement generally does not exceed 300 mw.

Messrs. Watchorn and Litrides have each questioned the use of the term "load forecasting errors." This is a term which was adopted nearly 30 years ago by people closely associated with the operation of the PJM interconnection. The authors recognize that the technically correct term is "load forecasting deviations."

As for the plotting of Figs. 1 and 7, Fig. 1 is made up from a total of 250 data samples and Fig. 7 from more than 500 data samples. Consider the problem of trying to plot each of these samples as an individual point (25 points in the small interval from 40% to 50%) and you have the answer. All samples were included in arriving at the proper plotted positions but only enough of the points were plotted to define the curve.

Mr. Watchorn has pointed out that by assuming a normal frequency distribution we have assumed that the deviations are limitless. Theoretically speaking this is true, but from a practical standpoint deviations beyond four sigmas are of little consequence since the total probability beyond four sigmas, for the normal frequency distribution, is only $3.\times10^{-5}$ or 1 in 33,000. Mr. Watchorn suggests truncating the normal frequency distribution "at possibly three times sigma." This we feel is too low since we have actually experienced deviations in excess of three times sigma. We have attempted to develop a correlation between the level of the load forecasts relative to the seasonal load trend curve and the magnitude of the deviations with the thought that the actual loads should tend to fall below the forecasted load when the forecasted load is above the trend curve and above the forecasted load when the forecasted load is below the trend curve. Any such correlation would support Mr. Watchorn's argument for truncation. However, our results have thus far been inconclusive.

No detailed analysis has been made of the actual results under the principle of uniform risk; however, we feel that there has been a considerable improvement in the reliability of service with little or no change in the megawatt-hours of capacity operated. That is, the capacity has been operated where it does the most good.

We do not consider our study of spinning reserve requirements a closed subject. We plan to extend our study of load forecasting deviations to see if we can find where the technique employed by any one company or group gives consistently better results than that employed by the others. In this connection it must be recognized that the variability of the different types of load is a limiting factor on the accuracy of short-range forecasts. We also plan to extend our computer program so as to be able to take into account the effect of transmission limitations between areas.

REFERENCES

1. See reference 2 of the paper.
2. See reference 1 of the paper.

A PROBABILITY METHOD FOR BULK POWER SYSTEM SECURITY ASSESSMENT, I—BASIC CONCEPTS

by

A. D. Patton

Electric Power Institute

Texas A&M University, College Station, Texas

Abstract—The paper is the first of a series describing a probability method for the "on-line" assessment of the near-term future security of an operating bulk power system. The present paper presents the basic concepts and applications of the method; future papers will describe probability models for implementing the method.

INTRODUCTION

The operator of a bulk power system is continually faced with the problem of assessing the near-term future security of the system and then taking any control action which may be required to reasonably assure the continued satisfactory operation of the system. Note that the concern here is with preventive control rather than with control actions required after the onset of actual system trouble. Control action may involve starting additional generators, reallocating generator loads, modifying tie line flows, returning transmission lines to service, etc.

Today, system security is commonly assessed using some form of a contingency rule. That is, if the system is within a certain specified number of contingencies of an unsatisfactory operating condition some security control action is required. The system consequences of contingencies may be evaluated using on-line computer techniques or may be estimated by the operator based on off-line studies of the system and experience. The probabilities that contingencies causing unsatisfactory system operation will occur are commonly evaluated, if at all, only through operator judgement. Thus, security assessment using a contingency rule cannot be expected to give a consistent assessment of the likelihood or probability of unsatisfactory system operation in the future.

The method of security assessment presented here supplies, quantitatively, the probability ingredient missing from the contingency rule method to create a consistent index of future system security: the probability of unsatisfactory system operation at future times. The method presented not only permits a consistent assessment of the need for security control action, but also indicates those control actions which are likely to yield the greatest benefits in terms of reduction of probability of system trouble.

The basic idea of a probability index for security assessment was presented in an earlier paper[1]. This paper extends and refines that earlier work.

DEFINITIONS OF BREACHES OF SECURITY

The first step in any assessment of system security is the development of a set of events whose occurrence would be considered to be a breach of security. A breach of security is defined to be some intolerable or undesirable operating condition. For reasons that will be made clear in the next section it is desirable to separate breaches of security into two categories; steady state and transient.

Steady-state breaches of security are those which result because the system is in a particular state* at a particular time (this implies some load level) without regard for how long the system has been in that state or what sequence of events caused the system to be in that state. Steady-state breaches of security include:

1. steady-state voltage at load busses outside tolerance,
2. steady-state overload of transmission lines or other equipment,
3. insufficient generating capacity (real and reactive) including tie line contributions.

It is believed that the rating of transmission lines or other equipment used to determine overload should be the long-term thermal limit. The use of a lower rating such as 80 percent of long-term thermal limit is believed to be too conservative. The use of a conservative rating is appropriate for many purposes, but here the intent is to define situations which are actually problems requiring remedial control action and not potential problems. Similarly, the rating used should probably not be an emergency rating since operation at such a level would be undesirable for any extended period of time. Generating capacity used in connection with (3) should be the maximum available capacity of the generators which are running.

Transient breaches of security are those which result due to a *change* in system state at a particular time (load level). Transient breaches of security include:

1. instability following some shock to the system,
2. shedding of load due to excessive frequency drop following loss of a generator or tie line,
3. tie line tripping due to excessive transient flow following some outage.

While all of the above can be evaluated by making a stability study of the system, it appears that reasonable approximations for frequency drop and tie line loading following an outage can also be obtained using simplified calculations.

THE SECURITY FUNCTION

Reference [1] suggested that an appropriate figure-of-merit for system security is the probability of a breach of system security at times in the near-term future given a known operating condition at the time calculations are made. Such a probability function, called the "security function", could be compared with a maximum tolerable insecurity or "risk" level to determine if and when some control action is required to maintain the risk of system insecurity at acceptable levels. To illustrate the concept consider Fig. 1. Here, S(t), the probability of system insecurity at time t into the future, is plotted versus time. The amount of time into the future for which S(t) is computed is in general the time for which system operations are planned or the lead time required by system operators for orderly modification of the system operating configuration. Note that at time t = 0, S(t) is zero. This follows because at time zero, the present in real time, the condition of the system is

Paper 71 C 26-PWR-II-A, recommended and approved by the Power System Engineering Committee of the IEEE Power Society for presentation at the 1971 PICA Conference, Boston, Mass., May 24-26, 1971. Manuscript submitted January 6, 1971; made available for printing June 9, 1971.

*A system state is defined to be a particular operating configuration of the system consisting of certain generators running, certain generators unavailable, certain generators on stand-by, certain transmission lines in service and others out of service, etc.

Reprinted from *IEEE Trans. Power Apparatus Syst.*, vol. PAS-91, no. 1, pp. 54–61, Jan./Feb. 1972.

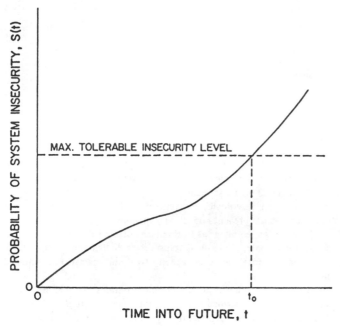

Fig. 1. A security function plot

$$S(t) = \sum_i P_i(t-\Delta t) [1 - \sum_{\substack{j \\ j \neq i}} R_{ij}(t-\Delta t, t)] Q_i(t)$$

$$+ \sum_j \sum_{\substack{i \\ i \neq j}} P_j(t-\Delta t) R_{ji}(t-\Delta t, t) V_{ji}(t) \qquad (2)$$

where the summations are over all possible system states and where:

$P_i(t-\Delta t)$ = probability system is in state i at time $t-\Delta t$,

$R_{ij}(t-\Delta t, t)$ = probability of transition from state i to state j in the time interval $\{ t-\Delta t, t \}$ given system in state i at time $t-\Delta t$,

$Q_i(t)$ = probability state i constitutes a steady-state breach of security at time t,

$V_{ij}(t)$ = probability that transition from state i to state j during the time interval $\{ t-\Delta t, t \}$ results in a transient breach of security. (Note that $V_{ij}(t)$ will be non-zero only for state transitions caused by failure or loss of some system component. The repair and restoration of some previously failed component also results in a transition from one system state to another but clearly such a transition could not cause a breach of security.)

known with certainty—the system is either secure or it is insecure. Since the goal of the control criterion is to avoid breaches of system security rather than to suggest corrective measures once a breach of security has actually occurred, the initial condition of interest is that the system is secure. Now, as time into the future increases S(t) increases in a manner dictated by the failure and repair characteristics of generators, transmission lines, and other bulk power system components and by the load at these future times. Suppose, as illustrated in Fig. 1, that the security function S(t) rises above the maximum tolerable insecurity level (MTIL) at time t_0. This indicates to the system operator that some remedial control action may be required. If the lead time required for implementation of the appropriate action, e.g., starting another generator, is greater than or equal to t_0, the control action should be initiated immediately. Otherwise, no immediate action need be taken. In practice, S(t) plots would be up-dated periodically, say once an hour, or any time system state (operating configuration) changed due to equipment failure or deliberate action. Thus, the system operator would be provided with a constantly revised picture of the likely future security of the system as affected by current system state and possible future contingencies.

A later section of the paper will illustrate the application of the security function to the security control of a small sample system. Also discussed in a later section is the matter of selection of the appropriate maximum tolerable insecurity level to be used in connection with the security function.

A number of different formulations of the security function are possible. Reference [1] suggested a security function of the form

$$S(t) = \sum_i P_i(t) Q_i(t) \qquad (1)$$

where the summation is over all possible system states i and where:

$P_i(t)$ = probability system is in state i at time t,

$Q_i(t)$ = probability state i constitutes a breach of security at time t. This form of the security function gives equal weight to all breaches of security regardless of their severity. The security function of equation (1) seems also to be limited to consideration of steady-state breaches of security only. That this is so comes from the fact that the state probability $P_i(t)$ yields no information as to how long the system has been in state i or by what sequence of events the system arrived at state i.

Another form of the security function which takes into account both steady-state and transient breaches of security is given as follows:

Note that equation (2) has two sets of terms. The first set of terms gives the probability that the system will suffer some steady-state breach of security at time t into the future. Consider the i th term of the first set of terms. This term gives the probability that the system suffers a steady-state breach of security due to being in state i at time t. Understanding why this term relates to steady-state breaches of security requires that one realize that the probability of the system being in state i at time t is composed of two probabilities: (1) the probability that the system is in state i at time $t-\Delta t$ and remains in state i during the time interval $\{ t-\Delta t, t \}$, and (2) the probability that the system is not in state i at time $t-\Delta t$, but enters state i during the time interval $\{ t-\Delta t, t \}$. The first of these probabilities is called the "steady-state" probability of state i at time t because the system has been in state i for at least a time Δt. The second probability is called the "transient" probability of state i at time t because the system has suffered the shock of transition from some other state to state i in the time interval $\{ t-\Delta t, t \}$. Returning to the i th term of the first set of terms of equation (2), the first factor of this term is the probability that the system is in state i at time $t-\Delta t$ and the second factor is the probability of no transition out of state i in the time interval $\{ t-\Delta t, t \}$ given the system in state i at time $t-\Delta t$. Hence, the product of these two factors is the probability that the system is in state i at time t and has been in that state for at least a time Δt. The second set of terms of equation (2) gives the probability that the system will suffer some transient breach of security at time t into the future. To explain the second set of terms consider the term associated with system transition from state j to state i in the time interval $\{ t-\Delta t, t \}$. The product of the first two factors of this term gives the probability of transition from state j to state i during the time interval $\{ t-\Delta t, t \}$.

Unfortunately, the answers for S(t) obtained using equation (2) depend in part on the size of the time increment Δt which is arbitrarily chosen. This is so because the transition probabilities $R_{ij}(t-\Delta t, t)$ depend on Δt. $R_{ij}(t-\Delta t, t)$ is zero for Δt equal to zero. It is believed that a Δt of one hour will give good results, but no detailed studies were made of the effect of the size of Δt. Equation (2) presumes that a system state cannot constitute a steady-state breach of security unless the system has been in that state for at least a time Δt. In some cases, however, it is likely that a system state could constitute a breach of security by persisting for a time less than Δt if Δt is equal to one hour. Hence, it would be desirable in some cases to eliminate the requirement that a state persist for Δt in order to constitute a steady-state breach of security In such cases the following formulation for the security function can be used.

$$S(t) = \sum_i [P_i(t)Q_i(t)$$

$$+ \sum_{\substack{j \\ j \neq i}} P_j(t-\Delta t)R_{ji}(t-\Delta t, t)V_{ji}(t)] \qquad (3)$$

where $V_{ji}(t) = 0$ if $Q_i(t) = 1$. The restriction on the $V_{ji}(t)$'s simply avoids the possibility of saying that the probability that a state constitutes a breach of security, regardless of how or when reached, is greater than unity.

The expressions of equations (1), (2), and (3) all share the property of being true probability expressions. In each case S(t) can only take on values between zero and unity. This point is raised to underscore the fact that the security function as formulated in (1), (2), and (3) has physical meaning and is not merely a numerical index. The fact that these formulations of the security function give a result with physical meaning is thought to be important. This physical meaning should be of help in evaluating the significance of the security function.

The three formulations for the security function presented so far give equal weight to all breaches of security regardless of the nature or magnitude of the security breach. It is certainly true, however, that some kinds of security breaches are more serious than others and that the degree by which some condition violates a security criterion is significant. Consequently, in some cases it may be desirable to incorporate a severity weighting factor into the security function expressions. This presents no problem in the formulation for the security function. For example, the security function of equation (1) modified by a severity weighting factor becomes

$$S(t) = \sum_i P_i(t)Q_i(t)w \qquad (4)$$

where: w = a weighting factor whose value depends on the nature and severity of a breach of security.

The security function expressions of equations (2) and (3) can be modified to include a severity weighting factor in an exactly similar manner. The problems lie in choosing values for the weighting function and in interpreting the modified security function. It appears that values for the weighting function w must be chosen entirely by judgement. The modified security function of equation (4) is no longer a probability function (although it does involve probabilities) and is now just a numerical index of security. Hence, S(t) no longer has physical meaning and can only be interpreted in the light of experience with the behavior of the function under various conditions. The present work has concentrated on the development and interpretation of security functions of the form of equations (1), (2), or (3), but the models and methods which have been developed could also be used to compute security indices such as equation (4).

The equations given for the various forms of security functions indicate summations which are, in theory, to be carried out over all possible system states. Recall that a system state is defined to be a particular (unique) combination of the states of the components which make up the bulk power system. Hence, in a bulk power system with a large number of component the number of system states would be very large. Clearly then summation over all system states would be a very time-consuming task. Fortunately, it is possible in practice to limit the system states which need be considered to those which can be formed from the system state which exists at the time the security function is computed by some limited number of component outages — usually two or three. The probabilities of system states requiring loss of more than this number of components within the time span of interest are usually entirely negligible. Discussion of methods for calculating system state probabilities and state transition probabilities will be deferred to Part II of the paper series.

The probabilities $Q_i(t)$ and $V_{ji}(t)$ in equations (1) through (3) require some additional discussion. These quantities function as true probabilities only when uncertainty exists as to whether or not a certain system state or state transition constitutes a breach of security at time t. Uncertainty may arise due to the fact that the load at the future time t is not known precisely at the time the security function is calculated or because the methods used to evaluate the effects on the system of the various contingencies are approximate rather than exact. In such cases the values for $Q_i(t)$ and $V_{ji}(t)$ depend on the load forecasting method used and the nature of the approximate system contingency evaluation methods. In the event that load forecasts can be considered as exact and without error and if exact system contingency evaluations are used then $Q_i(t)$ and $V_{ji}(t)$ can only take on values of either zero or unity. That is, either a system state or state transition constitutes a breach of one or more security criteria or it does not.

To be most useful a security assessment scheme should not only provide an indication of the need for security control action but also indicate what control action should be taken to reasonably assure continued satisfactory operation of the system. Proper interpretation of the security function provides this second function. Recall that the security function is composed of a sum of probabilities of system states and of system state transitions (possibly multiplied by a severity weighting function) which result in some breach of security. Hence, by identifying those terms which contribute the most, numerically, to the security function it is possible to identify the control action which would be most effective in improving system security. In practice the procedure might be as follows. When the security function indicates the need for security control action request the computer to provide a list of system states and/or state transitions which constitute breaches of security ordered by their probabilities. A review of this list by the operator should suggest a reasonable if not the optimum control action.

RELATIONSHIP OF MAXIMUM TOLERABLE INSECURITY LEVEL TO LONG-TERM AVERAGE SYSTEM RELIABILITY

Application of the security function as a security assessment criterion requires that a maximum tolerable insecurity level (MTIL) be established. Specification of the MTIL amounts to specifying the desired level of system security. The question then is, what value of MTIL will result in the desired level of system security? It seems reasonable to measure system security by some long-term average measure of security such as loss-of-load probability (LOLP). Long term average measures of security (or reliability) such as LOLP have been used for many years by system planners and are generally understood and accepted. Presuming that a value for long-term average security can be agreed upon, the problem then is to select a value of MTIL which if used in hourly operation would result in the desired long-term average security.

The problem of finding a relationship between MTIL and LOLP, the most commonly used long-term average measure of reliability, was investigated using a small sample system consisting of ten generators. Data for these generators are given in Table I. The effects of transmission lines and all other system components on system security were ignored for the purposes of the investigation. The only breach of security considered in the study was insufficient generating capacity operating to supply the load. The annual load cycle for the sample system has a shape typical of load cycles in the Southwestern United States. The load magnitude is such that LOLP based on *installed* capacity is less than one day in ten years — a typical criterion for adequate installed capacity.

A range of MTIL values from about 0.0004 to about 0.0009 were studied for the sample system. The procedure was to select a value of MTIL; then to determine the optimum generation schedule for each day of the year. The method of optimum generating scheduling used is

Table I
Sample System Data

Unit No.	MW	Forced Outage Rate Outages/Year	Average Outage Duration, Hours
1	60	1.2	5.8
2	80	1.2	5.8
3	100	1.2	5.8
4	120	2.5	15.0
5	150	2.5	15.0
6	280	2.6	13.7
7	320	2.6	13.7
8	445	2.6	13.7
9	520	4.0	13.7
10	550	4.0	40.0

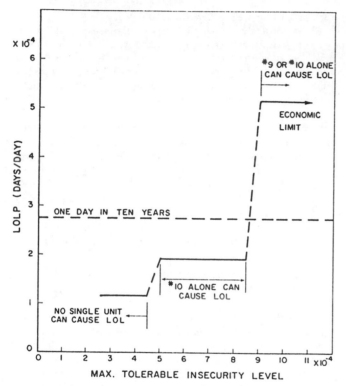

Fig. 2. Relationship of maximum tolerable insecurity level (MTIL) to long-term average performance as measured by LOLP for a sample system.

described in Reference [2]. The generation scheduling method used a security function expression like that of Equation (1). All generators were represented by two-state models assuming constant state transition rates. The start-up time of stand-by generators was assumed to be one hour. After the generation schedule corresponding to a certain MTIL was determined, the operation of the system was simulated using Monte Carlo methods to determine the long-term average probability of insufficient generating capacity *operating* to supply the load. This probability is also called LOLP. In conformity with the usual practice in planning studies the system was assumed to have insufficient capacity for an entire day if capacity was insufficient at any time during a day for the purposes of calculating the LOLP.

Figure (2) shows LOLP versus MTIL for the sample system. It should be noted that the plateaus shown in Figure (2) actually represent narrow bands within which there are slight variations in LOLP. The upper plateau corresponds to a generation schedule having minimum operating costs. The associated LOLP value of about 5.2×10^{-4} (about one day in five years) reflects the amount of spinning reserve inherent in the minimum operating cost schedule. Thus, the upper plateau represents an economic limit on LOLP in the sense that any reduction

in capacity would constitute a deviation from the minimum cost schedule and would raise operating costs while at the same time reducing system reliability. Clearly then there are no incentives to reduce capacity below that called for by the minimum operating cost schedule.

The loss of either of the two largest units can cause a loss of load when the system is operating according to the minimum operating cost schedule. For an MTIL range from about 0.0005 to about 0.00085 schedules result such that loss of the largest generator can result in a loss of load. For MTIL less than about 0.00045 only multiple contingencies can result in loss of load. It is of interest to note that the lower plateau corresponds to the rule of providing enough spinning reserve to cover the loss of any single unit. The LOLP for this plateau is about one-half that of the widely used LOLP criterion of one day in ten years. This illustrates the fact that simply providing enough capacity to cover for a specified contingency without regard for the probability of that contingency may not yield the desired degree of security.

The results of Figure (2) illustrate that it is possible to relate LOLP and MTIL for a specific system. The author has not, however, discovered a general relationship between LOLP and MTIL. It appears that each system would have to be studied on its own merits to discover the value of MTIL corresponding to the desired long-term average performance.

EXAMPLE OF SECURITY FUNCTION CALCULATION AND INTERPRETATION FOR A SAMPLE SYSTEM

A small sample system has been chosen to illustrate the application of the security function to security assessment and control. The sample system used is the familiar Ward and Hale system shown in Fig. (3). The system has an installed generating capacity of 400 MW in two steam plants. Generators number 1, 2, and 3 are rated 100 MW and 80 per cent power factor while generators number 4 and 5 are rated 50 MW and 80 per cent power factor. Each generator is assumed to have physical parameters typical of generating units of that size and to have a governor with 5 per cent droop. For probability calculations it is assumed that all generators behave in the same manner and can be modeled using two-state models. It is assumed that all forced outages are total forced outages. The forced outage rate of a generator is taken

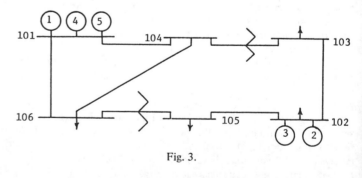

Fig. 3.

Table II
Transmission Line and Transformer Parameters

Bus	Bus	puR*	puX*	Length Miles	Rating Mva
101	104	.03	.12	30	160
101	106	.06	.18	50	160
102	103	.15	.45	40	100
102	105	.09	.27	20	100
104	106	.04	.15	40	160
103	104	0	.20	–	60
105	106	0	.20	–	60

*Per unit on 100 Mva base

206

to be 0.000666 outages per hour. Generator forced outage durations are assumed to be Weibull distributed, see Appendix, with scale parameter $\alpha = 23.0$ and shape parameter $\beta = 4.8$.

Stand-by generators are assumed to have a 5 per cent probability of failure to start and Weibull distributed start-up times with parameters $\alpha = 3.6$ and $\beta = 1.17$.

The physical parameters of transmission lines and transformers are given in Table II. Line charging is neglected. Transmission lines are modeled by two-state models. Transient forced outages of transmission lines are ignored. The persistent forced outage rate of a line is assumed to be directly proportional to line length. Line forced outage rate is 0.000114 outages per hour per 100 miles. Forced outage durations for all lines are assumed to be Weibull distributed with parameters $\alpha = 4.6$ and $\beta = 0.55$. It is assumed that transformers and busses cannot fail.

At the instant in time at which the security function is to be computed the projected load cycle is as shown in Fig. (4). It is assumed

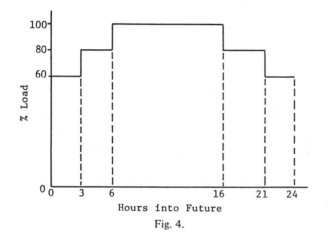

Fig. 4.

Table III
Bus Loads at 100% Load Level

Bus	MW	Mvar
101	0	0
102	60	20
103	30	10
104	0	0
105	30	10
106	70	25

that the load forecast is error-free. All bus loads conform to the same load pattern and load factor is constant regardless of load level. The loads for each bus at the 100% load level are given in Table III. The load-frequency characteristic of the load is taken to be 0.01 per unit change in load per Hz change in frequency.

Operating Rules and Security Breach Definitions for Sample System

The following operating rules are assumed to apply to the sample system.

1. If a normally-operating generator suffers a forced outage, stand-by generators having capacity equal to that of the generator lost are started if available.

2. Assuming that the generation schedule calls for generators number 1, 2, and 3 to be running and for generators 4 and 5 to be in a stand-by status, one stand-by generator is started upon the loss of either of the lines 102-103 or 102-105.

3. Transformer taps are adjusted following any outage for best system performance under the contingency condition.

Steady-state security breaches are defined to be: (1) insufficient operating generating capacity, active and reactive; (2) steady-state line or transformer loading greater than rating; (3) steady-state voltages at load buses outside an acceptable range of 0.9 to 1.10 per unit.

Transient security breaches are defined to be: (1) instability following some outage, and (2) frequency deviation in excess of 1.0 Hz due to loss of load and/or generation. The tolerable frequency deviation for the sample system must be large because of the small size of the system.

Computation of Security Function

Suppose that at some time the sample system is operating normally with generators 1, 2, and 3 in service, generators 4 and 5 in a stand-by status, and all transmission lines and transformers in service. The load forecast at this time is that shown in Figure (4). Generator voltages are set at 1.10 per unit. The dispatch is such that generator 1 carries 38% of the load while generators 2 and 3 each carry 31%. This dispatch is scheduled to be maintained for future load levels.

The system operator estimates that start-up of a stand-by generator, if required, would take four hours. Management has set the maximum tolerable insecurity level (MTIL) at 0.0007. It is desirable to compute the security function, then take any control action which may be indicated.

The first step in computing the security function involves studying the system to determine those system states which constitute breaches of security according to the definitions which have been adopted. This is done by performing the required flow studies, stability studies, etc. In the sample system only single and double contingencies were studied. The probabilities of higher order contingencies are quite small compared to lower order contingencies. Hence, higher contingencies can be neglected providing at least some lower order contingencies result in security breaches. A useful rule-of-thumb is to study the contingency level which first results in a breach of security plus the next higher order level of contingencies and neglect all higher order contingencies.

A summary of those cases found to result in a steady-state breach of security is given in Table IV. The columns headed Q100, Q80, and Q60 indicate whether a case results in a security breach at the 100%, 80%, or 60% load levels. A "1" indicates a breach of security. Since the

Table IV
Cases Resulting in Steady-State Breach of Security

Case	Description	Q100	Q80	Q60
1	Single contingency: generators			
	Unit #1, no stand-by up	1	0	0
	Double contingency: generators			
	Unit #1 and Unit #2 or Unit #3			
2	No Stand-by up	1	1	1
3	One stand-by up	1	1	0
	Units #2 and #3			
4	No stand-by up	1	1	1
5	One standy-by up	1	1	0
6	Two stand-bys up	1	0	0
	Double contingency: lines			
7	101-106 and 101-104	1	0	0
8	101-106 and 104-106	1	0	0
9	101-106 and 102-105	1	0	0
10	102-103 and 102-105 and no stand-up up	1	1	0
	Double contingency: generator/line			
11	Unit #1 and 102-103 and no stand-by up	1	1	1
12	Unit #1 and 102-105 and no stand-by up	1	1	1

the load forecast is assumed to be error-free, the quantities $Q_i(t)$ needed in the security function calculation are those given in the above table. The Q's for all system states other than those corresponding to the cases in the table are zero.

The security function formulation of Equation (3) is to be used for the sample system. This formulation requires that $V_{ji}(t) = 0$ if $Q_i(t) = 1$. Recall that $V_{ji}(t)$ is the probability that a transition from state j to state i during a time interval $[t-\Delta t, t]$ results in a transient breach of security while $Q_i(t)$ is the probability that state i constitutes a steady-state breach of security at time t. Studies of the sample system revealed that all state transitions resulting in transient breaches of security were to system states which constituted steady-state breaches of security. Hence, $V_{ji}(t)$ is zero for all i and j for the sample system. Under these conditions the security function for the sample system is given by

$$S(t) = \sum_{i=1}^{12} P_i(t) \; Q_i(t).$$

In this equation system state i corresponds to case i in Table IV. The system state probabilities were obtained using methods which will be discussed in Part II of the paper.

Figure (5) shows a plot of the security function. Note that the security function exceeds the MTIL at a time seven hours into the future. Now, because the lead time needed to start one of the 50 Mw stand-by generators (the only effective control action available to the operator of the sample system) is estimated to be four hours, no immediate action is required. Now suppose that real time has advanced three hours and the operating configuration is the same as before. The load forecast is that of Figure (4) with the time origin moved three hours to the right. The security function as calculated at this time is shows in Figure (6). Notice now that the security function exceeds the MTIL at a time four hours into the future. Since this time equals the estimated start-up time of a stand-by generator, some control action

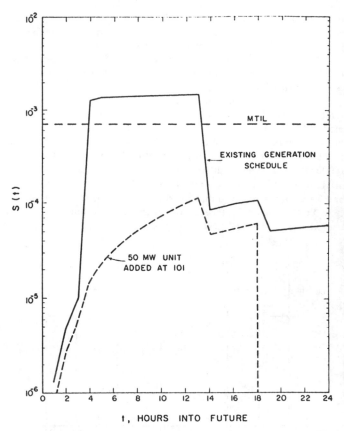

Fig. 6. Security function plot for system sample at a three hours later than in Fig. 5.

Table V

System State	$P_i(4) \; Q_i(4)$
1	1.23×10^{-3}
2	3.28×10^{-6}
3	2.67×10^{-6}
4	1.11×10^{-6}
5	2.67×10^{-6}
6	1.28×10^{-6}
7	6.42×10^{-8}
8	3.71×10^{-9}
9	1.01×10^{-8}
10	1.35×10^{-8}
11	6.77×10^{-9}
12	6.57×10^{-6}

may be required to reasonably insure the continued proper operation of the system. The proper control action can be found by examining the contributions to the security function attributable to each of the system states at the time the security function exceeds the MTIL. This data for the sample system is given in Table V. Examination of the table reveals that system state number 1 is by far the largest contributor to the value of the security function. (This of course was obvious from the start for the simple sample system). Hence, some control action which will cause state 1 not to result in a breach of security is indicated. One available control action is to place one of the 50 Mw generators at bus 101 in service. Evaluation of the effectiveness of this action requires that system load flow studies, stability studies, etc. be performed once again to identify those system states and state transitions which constitute security breaches. The resulting modified security function is plotted as a dashed line in Figure (6). The security of the system with the modified generation schedule is shown to be adequate.

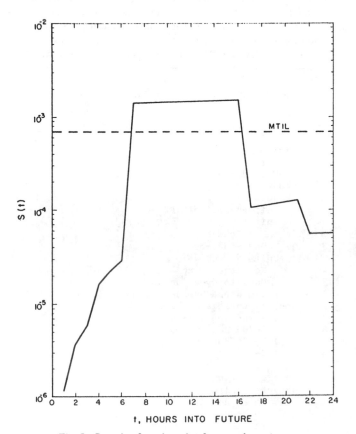

Fig. 5. Security function plot for sample system.

CONCLUSIONS

The paper has begun a description of a new method for assessing the near-term future security of an operating bulk power system. Future papers will describe in detail methods for computing the probabilities needed to implement the security assessment method.

The primary application of the method is as an aid to the bulk power system operator. The system operator at the time of a change in system configurations, or on a periodic basis, can obtain the calculations of probability of insecurity—the security function. Comparison of the output of the security function calculations with an established maximum tolerable risk of insecurity permits the operator to make consistent judgments as to the need for security control action. Interpretation of the security function also identifies those conditions contributing most to the probability of trouble and thereby guides the choice of control action employed.

ACKNOWLEDGEMENT

The work reported in this paper was supported by the Electric Research Council under Edison Electric Institute Project RP 90-6, Bulk Power System Security Assessment (Probability Approach) as part of a general research effort aimed at improving bulk power system security. The author wishes to acknowledge the contributions and cooperation of the project steering committee: R. A. Lamley, Consumers Power Co.; T. L. Hatcher, Texas Power and Light Co.; V. H. Tompkins and B. R. Clausen, Gulf States Utilities Co. The contributions of the author's co-worker A. K. Ayoub and other members of the project team are also gratefully acknowledged.

APPENDIX

The Weibull distribution is a particularly useful distribution in reliability or security work. The Weibull distribution, as used here, has two parameters: α, the scale parameter; and β, the shape parameter. The usefulness of the distribution comes about primarily due to the fact that through manipulation of the shape parameter β, the Weibull distribution can be made to resemble a wide range of distributions. For example, if $\beta = 1$, the Weibull distribution reduces to the exponential distribution; while if β is large the distribution approximates a normal distribution. The probability density function of the Weibull distribution is

$$f(t) = \frac{\beta}{\alpha^\beta} t^{\beta-1} e^{-(t/\alpha)^\beta}, \quad t \geq 0. \tag{5}$$

REFERENCES

[1] A.D. Patton, "Short-term reliability calculation", *IEEE Trans. on Power Apparatus and Systems,* vol. PAS - 89, pp. 509-513, April 1970.
[2] A. K. Ayoub and A. D. Patton, "Optimal thermal generating unit commitment", IEEE Paper No. 71-TP94-PWR.
[3] Final report, "Edison Electric Institute Project RP 90-6, Methods of Bulk Power System Security Assessment (Probability Approach)", prepared by Electric Power Institute, Texas A&M University, College Station, Texas.

Discussion

W. R. Christiaanse (Stagg Systems, Inc., New York, N.Y. 10017): I would like to compliment the author on his clear and pragmatic approach to a probabilistic evaluation of security on a real-time basis.

Manuscript received June 7, 1971.

I agree with the author's comments on the value of a probabilistic approach. Probabilistic techniques, which are used routinely in system planning, can provide a basis for rational and consistent operating decisions in the face of uncertainties about future load and outages. As also noted by the author, these methods provide a more refined measure of security than contingency rules. Increases in overall system reliability, due to a more balanced operation of facilities, are possible. Dollar savings are also possible due to refined control of short-term purchases of capacity, for example.

I would appreciate the author's comments in response to the following remarks.

Formulation of Security Function. Presumably the security function formulated by the author has two functions: (a) a measure of system security, and more importantly, (b) a guide to control action. With these functions in mind, some aspects of s(t) seem undesirable.

The security function s(t) appears to be affected by system size. This may not be a desirable characteristic. The installation of new lines and transformers would cause an increase in the function, i.e., the insecurity, even though the probability of unacceptable voltage variations, for example, for each bus remained constant. Also, size-dependence makes a comparison of the s(t)'s for a large and small system difficult. The problem arises because the function defined by the author includes the sums of the probabilities of (1) every possible steady-state line or transformer overload and (2) every steady-state variation in bus voltage outside the range of 0.9 to 1.10 per unit. Therefore, even though two systems may have an equal probabilities of overload on every line and equal probabilities of an unacceptable bus voltage at every bus, the larger system would have a higher value of s(t), simply because the larger system contains more lines, transformers and buses. It does not seem reasonable, therefore, to simply sum the probabilities of overloads and unacceptable bus voltage variations over an entire system. The effect of system growth on a size-dependent s(t) also makes it difficult to interpret the long-run consequences of a fixed MTIL, the maximum tolerable insecurity level.

Another criticism of the formulation of s(t) is that the sum itself does not provide a direct guide to the specific control action required. The value of the sum seems less relevant for determining control action than the contribution from each contingency illustrated in Table V.

An alternative approach would be to set limits on the probabilities of each of three (or more) steady-state "failure modes," e.g., shortage of generating capacity, overload of a line or transformer, or an unacceptable bus voltage. Control action would be initiated when the probability of any single line overload, for example, exceeds the limit for overloads. Admittedly, it would be necessary to establish a number of probability limits rather than a single value. These alternative limits, however, would be applicable on a bus-by-bus and line-by-line basis, as the familiar voltage and capability limits are applied. Control action would be more straightforward and neither control action nor the measurement of security would be affected by the number of devices being monitored. This approach may also eliminate the need to introduce compensating subjective weights, as in Equation (4), to reflect the relative severity of various failure modes.

Apparently the author is not in favor of multiple security limits, and feels that there is a need for a meaningful, single index of security.

There seems to be another difference in the author's approach and our own approach, since in our work we treat reliability as a constraint applicable to each type of contingency, and to operate and plan the system to minimize costs within these constraints. The author, on the other hand, allows a trade-off between contingencies by summing the probabilities of each contingency and constraining the sum. Perhaps the author would comment on these ideas.

Establishment of Security Standards. As noted by the author, the establishment of probabilistic security standards is a difficult problem. We agree with the author that standards should be understood in terms of long-term results. On the other hand, we have also found that the lack of a numerical frame of reference is the chief difficulty in establishing and applying probabilistic standards. Consequently our approach has been to use actual system operating experience to monitor probabilities as they actually exist on the current system. Plotting these values over time indicates the range of the numerical values of probability that describe the actual level of reliability that currently exists. Assuming that the current operating configuration and policies are providing an acceptable level of reliability most of the time, some reasonable numerical frame of reference will evolve from these data. Then, using these standards based on actual past experience, we can measure incremental improvements and achieve a refined control of reliability.

No doubt the author has considered this approach, that is, collecting actual hour-by-hour system data and using it to evolve standards, in addition to using theoretical relationships to long-term performance. Are there plans to present hour-by-hour data from an actual system?

Calculation of Probabilities. The author has described his plans for a companion paper which will describe the method of calculating the security function. I do have, however, some brief questions on the sample problem in this paper.

First, it seems that the plot of s(t) in Figure 6 seems to be very close to an horizontal translation of the plot in Figure 5. Also, no failure rates and repair rates are specified for units, lines, and transformers in the problem, although forced outage rates (steady-state probability of an outage) are specified. These observations seem to indicate that the author is using steady-state probabilities for some or all of unit, line and transformer outages. Is this true?

For short-term probability calculations from a known initial condition, it is very important to use time-dependent probabilities. For example, the probability of a unit (or line) being out of service at time t, assuming it is in service at t=0, is

$$\text{F.O.R. } (1 - \exp(-\lambda t - \mu t)), \qquad \text{D-1}$$

where F.O.R. is the steady-state probability of an outage and λ and μ are the repair and failure rates. As we would expect, the probability in D-1 approaches zero as t approaches zero. (For a lead time of t=6 hours, for example, the probability in D-1 would typically be 20 to 50% of the steady-state value.) If time-dependent values were used, we would expect s(t) in Figure 6 for a given hour of the day to be significantly lower than in Figure 5, in spite of the fact that we still have the same units and lines in service three hours later. The use of steady-state device probabilities instead of time-dependent probabilities would ignore the known initial condition of the devices in the system and would be less accurate because this information is unused.

In practice, therefore, the value of s(t) for specific future hours would change over time, regardless of whether or not the operating configuration changes.

In conclusion I would again like to compliment the author for suggesting a workable role for probablistic techniques in real-time securing monitoring.

B. F. Wollenberg (Leeds & Northrup Company, North Wales, Pa. 19454): Let me commend the author on presenting a very clear and interesting paper on this pertinent topic. I would like to question the adequacy of the "state" of the power system used for Security Assessment as used in this paper. The state should include both the configuration (as used by the author) and the numerical values of the loads, flows, voltages, etc. present with that configuration. If this type of definition of state is used it will fall in line with the present use of "state estimation" techniques to determine these quantities. However, this adds considerable complexity to the system in that:

a) The "state" of the system at "time present" cannot be known with certainty. In which case the security function is not nessarily zero initially, and

b) The number of state possibilities is now a continuous and not just a discrete set as would be the case with a set of configuration changes.

The implication here is that the probability of a failure is influenced by the loading of the system. A breaker may trip or a voltage may dip below limit purely as a function of a bus load — independently of the configuration of the system. Similarly, an overload or low voltage may exist at the present time without the operator's knowing it due to inaccurate measurements (or an incorrect state estimate).

I would also like to suggest that a possible use of the security function would simply be to incorporate it as a functional constraint into the scheduling programs such as unit commitment, Economy B, or maintenance scheduling which are being incorporated into present day dispatch computers. I think the author's idea of having the program set out a list of those terms which contribute most to security as the most effective controls would be better applied within a true optimum scheduling program where the cost of taking such control action would be properly weighted.

Finally, one other aspect of the security function should be noted, namely that the probabilities, $Q_i(t)$, involved are themselves random variables in the sense that they are not known for sure — but rather only within rough limits. This would suggest a *Bayesian* approach to the problem wherein the probabilities are expressed as distributions — with the resulting security function also expressed as a time dependent distribution rather than a number. I think the probabilities are indeed very poorly known and that a Bayesian approach is needed to correctly reflect this.

Manuscript received June 7, 1971.

The reviewer looks forward to the future papers growing out of this research. I believe such an approach as outlined here is quite important to correct monitoring and control in applications at on-line security assessment.

A. D. Patton: The author wishes to thank Mr. Christiaanse and Mr. Wollenberg for their discussions of the paper. Their comments contribute significantly to the value of the paper.

Mr. Christiaanse is concerned about the influence of system size (number of components) on the security function S (t). It is certainly true that S (t) is influenced by the number of components in a system. Moreover, the author regards this as a strength rather than a weakness of the proposed method. The fact that S (t) is influenced by system size is evidence that a simple contingency rule for security assessment applied without regard for system size does not result in a consistent security level as measured by probability of trouble on the system. An analogy can be drawn to generation planning where it is widely understood that the risk of insufficient capacity is influenced by the number of generators installed as well as the forced outage rate of each. Further, it does not necessarily follow that adding components to a system results in larger values of S (t). If the number of component failures necessary to cause a breach of security is increased by an increase in the number of components, it is virtually certain that lower values of S (t) will result. If on the other hand the number of failures necessary to cause a breach of security is not increased by an increase in the number of components, then clearly S (t) will be increased.

Mr. Christiaanse points out that S (t) itself provides no guide to the specific control action required. This fact was noted in the paper where it was pointed out that the probabilities associated with each of the various system states which constitute breaches of security provide a guide for choosing the appropriate control action.

Mr. Christiaanse suggests an alternative to the approach of using S (t) as a single, all inclusive, measure of security. He would break S (t) into several parts corresponding to different types of security breaches and assess each such part for adequacy. This might well be a good approach, but it does have the drawback of requiring several risk limits — all of which are difficult to obtain. The author does not agree that this alternative approach would, or should, be free of the influence of system size. The only correct way to compute system state probabilities requires simultaneous consideration of all system components.

The establishment of security standards or limits is a difficult and important part of any security assessment and control scheme. One way of determining such standards would be to examine past performance as suggested by Mr. Christiaanse. There are no present plans to make such a study.

It is of course necessary to use time-dependent probabilities for computing S (t) as Mr. Christiaanse notes. A detailed treatment of the time-dependent probability models which have been developed and which were used for the sample problem of the paper will be given in a future paper. The outage and repair parameters of all components of the sample system of the paper are given in the section dealing with the sample problem. Probably some confusion exists because of synonymous use of "failure rate" and "forced outage rate" in the paper.

Mr. Wollenberg suggests an alternative definition of system "state" in which a state would be defined by a vector of system conditions such as bus loads, line flows, and bus voltages as well as component statuses. The major incentive would seem to be to permit recognition of the possible influence of system conditions on component failure rates. A basic assumption underlying the method proposed in the paper, and virtually all other probabilistic reliability calculations made in power systems, is that component failure rates are independent of system conditions. This assumption is widely believed to be very good.

The author agrees with Mr. Wollenberg that the security function can be used as a constraint in scheduling problems. It was so used in a unit commitment program discussed in Reference [2] of the paper. Mr. Wollenberg's idea of choosing appropriate security control action based upon the cost of the control action as well as the probability of the various system states constituting security breaches is good and worthy of further investigation. Mr. Wollenberg mentions that the variables $Q_i(t)$ used in the security function are probably not known with much precision and should therefore be treated as probabilities with appropriate distributions. This is also pointed out in the paper. The distributions for $Q_i(t)$ are probably determined primarily by the distributions associated with forecasted loads.

Manuscript received June 7, 1971.

THE EFFECT OF RAPID START AND HOT RESERVE UNITS
IN SPINNING RESERVE STUDIES

Roy Billinton   Adarsh V. Jain
Power System Research Group
University of Saskatchewan
Saskatoon, Canada

INTRODUCTION

The application of probability methods to the determination of generating capacity spinning reserve requirements has been illustrated in several publications [1,2,3]. The basic difference between a static capacity reserve study and a spinning reserve study is in the time period considered. The basic statistics in the first case are the probabilities of finding the existing and proposed units in various states at some time in the relatively distant future. In the spinning reserve case, the probabilities of having load carrying capability within the next few hours are required given that the system resides in a particular state at time zero. In each case, the actual system load is a variable and can be described by a probability distribution.

The simple two state Markov model for a generating unit in which λ and μ designate the failure and repair rates can be used to find the probabilities of finding the unit available or unavailable at some time t. If the unit is available at time t = 0 and P(t) is the probability of finding the unit on outage at time t.

$$P(t) = \frac{\lambda}{\lambda + \mu} - \frac{\lambda}{\lambda + \mu} e^{-(\lambda+\mu)t} \qquad (1)$$

$$P(t) = 0 \text{ at } t = 0$$

$$P(t) = \frac{\lambda}{\lambda + \mu} \text{ at } t = \infty \quad \text{i.e. the unit forced outage rate.}$$

The time period t is usually quite short in a spinning reserve study. If the assumption is made that no repairs can be accomplished during such a short period

$$P(t) = 1 - e^{-\lambda t} \simeq \lambda t \qquad (2)$$

This approximation is normally quite good for short lead times. The time period t in equation (2) has been designated as the delay time for additional equipment, i.e. the period during which no additional capacity can be made available to carry the load. The probability P(t) is therefore the probability of a unit failing and not being replaced by additional equipment. This interpretation can also be applied to equation (1) in which repair is included. The basic risk criterion is normally the probability of having sufficient capacity at time t to just carry or to fail to carry the system load. The effect of unit deratings[4,5] and the load forecast uncertainties[2] can be added to the analysis and these applications are illustrated in the available literature.

The delay time associated with additional generation is quite variable and is dependent upon many factors. The most important, however, is the type of additional generation available. Gas turbine peaking units and hydraulic equipment can usually be made available in a relatively short time as compared to more conventional thermal or nuclear generating capacity. Pumped storage hydraulic units operating in the pumping mode have a different delay time associated with their load carrying capability. This is also true for conventional thermal equipment operating in a hot reserve mode. All these factors should be considered in a spinning reserve study and the reliability method should respond to the factors that influence the actual reliability of the system.

Paper 71 TP 506-PWR, recommended and approved by the Power System Engineering Committee of the IEEE Power Engineering Society for presentation at the IEEE Summer Meeting and International Symposium on High Power Testing, Portland, Ore., July 18-23, 1971. Manuscript submitted February 18, 1971; made available for printing April 26, 1971.

AREA RISK CURVES

Equation (1) gives the probability of finding the unit on outage at some time t given that the unit is available at time t = 0. The probability of finding the unit on outage as a function of time can be expressed in the form of an area as shown in Figure 1.

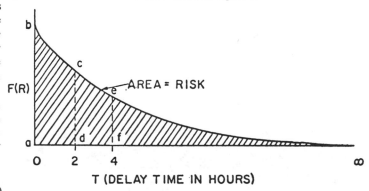

Fig. 1. A Single Unit Area Risk Curve

In a spinning reserve study it is often assumed that sufficient additional generation is always available and that it is only a matter of time before this capacity is placed in service. If this delay time is 2 hours then the risk is given by the area abcd in Figure 1, for 4 hours by the area abef. The entire area out to t = ∞ is equal to the generating unit forced outage rate.

In an actual system a number of units would be operating with additional generation available after a finite delay time. As previously noted, this additional generation can take many forms. The delay time associated with conventional thermal generation may vary from 2 to 24 hours depending upon when the equipment last operated. Hydro and gas turbine units require a very short lead time to start, synchronize and carry load. The loading rate in these cases in also quite different from that of large thermal equipment. The delay time of a conventional thermal unit can be reduced by maintaining the boiler in a banked state. Units in this condition are designated as hot reserve. The increase in cost must be balanced against the decrease in risk to the system which is a function of the state of readiness of the units expressed in the form of delay time.

Consider a theoretical system with eight generating units for which the load requirements dictate that five units be operated. Assume that of the remaining three units, two can be placed in service after 4 hours and the third requires 8 hours notice. The risk after 4 hours is of having less generation than load with the seven available units and after 8 hours, of having less generation than load with all the units. The area risk curve for this condition is shown in Figure 2.

The area under the curve at any particular point is dependent upon two factors, the available generation and the load level. The assumption has been made that the load is a constant value. If after the peak has been realized, the system load level decreases then the risk contribution decreases from that point on. The area risk curve may be truncated at that point. If the operating capacity is also reduced then there will be a change in the capacity composition and an increase in risk. At any time t = 0, the question becomes what is the risk associated

Reprinted from *IEEE Trans. Power Apparatus Syst.*, vol. PAS-91, no. 2, pp. 511–516, Mar./Apr. 1972.

with the known generation state and the expected load situation. If this risk is too high then the known generation state must be modified. Figures 3 and 4 show symbolic area risk curves for seven units operating out of eight and for all eight units respectively.

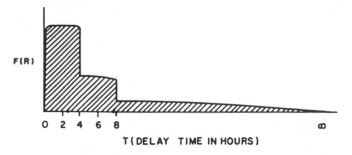

Fig. 2. Area Risk Curves for Five Units Operating Out of Eight

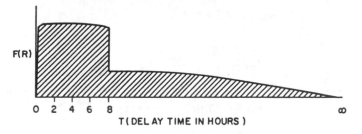

Fig. 3. Area Risk Curves for Seven Units Operating Out of Eight

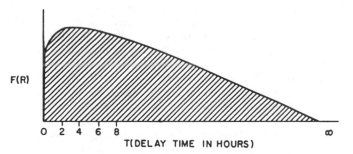

Fig. 4. Area Risk Curves for All the Units Operating

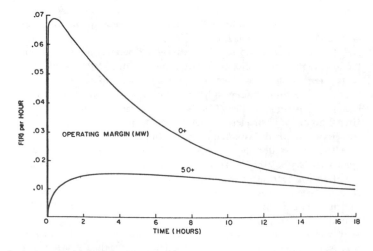

Fig. 5. Area Risk Curves for Different Operating Margins in a Hypothetical System.

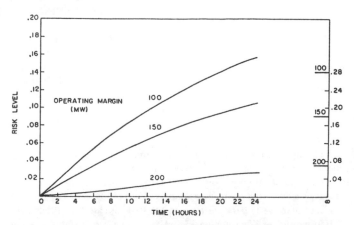

Fig. 6. Variation of Risk Levels for Different Operating Margins for Increasing Lead Times in a Hypothetical System.

The individual units may include one or more derated states and load forecast uncertainty can be included in the analysis. It can clearly be seen from the area risk curves that the risk is a function of the delay time associated with the additional equipment. The risk is often negligible after additional generation becomes available. The effect of lead time and operating margin on the area risk curves are shown in Figure 5 for a 2900 MW hypothetical system. The system composition is given in the Appendix. The effect on the risk level of increasing the lead time for operating margins of 100, 150 and 200 MW are shown in Figure 6.

RAPID START AND HOT RESERVE UNITS

The area risk curve concept can be used to illustrate the effect on the risk level of rapid start and hot reserve units. Consider a hypothetical system which contains some rapid start units which can be placed in service after a five minute delay. If the decision is made, the risk function F(R) will decrease after 5 minutes to a new value dependent upon the quantity of rapid start capacity. This condition is illustrated in Figure 7 where it has been assumed that additional conventional generation is available after 4 hours.

The periods t_q and t are the times required to bring the rapid start and the conventional equipment into service. If the system also contains some hot reserve units with a delay time t_h taken to be in the order of 1 hour, the area risk curve takes the form shown in Figure 8.

In the limiting cases, an infinite amount of rapid start capacity would reduce the system delay time to 5 minutes after which there would be a negligible risk. In the case of an infinite amount of hot reserve this change would occur at the assumed 1 hour mark.

The generation available to carry load and exposed to failure in the rapid start and hot reserve case shown in Figure 7 is not constant. The system lead time can be divided into three parts and treated individually.

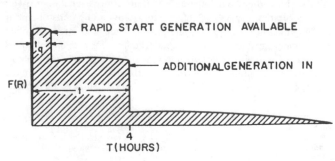

Fig. 7. Effect of Rapid Start Units on the Risk Level

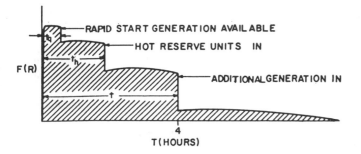

Fig. 8. Effect of Rapid Start and Hot Reserve Units on the Risk Level

Period 1 – Only the actual generation in service is available. This will extend over the period t_q. If no quick start unit is available this will extend up to t_h and up to t if there is no hot reserve.

Period 2 – This will extend from t_q to t_h when both rapid start and hot reserve units are available.

Period 3 – This will extend from t_h to t. The risk level will be quite small in this region if large quantities of rapid start and hot reserve exist, however, it may extend over a longer period.

It has been assumed that additional conventional generation becomes available after time t and that the risk function F(R) becomes quite small. If this contribution to the area risk is not small then an additional region must be added. In most cases, the required lead time is less than 12 hours and the duration of the peak load is sufficiently small that extensive time spans are not required.

The risk level contribution during the first time period can be computed using the standard procedure described in detail in the literature. In the second time period, the possibility arises that the rapid start units may fail to start and therefore become unavailable to the system. This condition can be included using a standard conditional approach.

Probability of System Failure =

(Probability of the system generation and the quick start units just carrying or failing to meet system load | quick start units becoming available.)
X(Probability of quick start units becoming available.)
+(Probability of the operating capacity just carrying or failing to carry the system load | quick start units not becoming available).
X(Probability of quick start units not becoming available.)

In the case of a unit on hot reserve, the probability of being unavailable is the probability of failure while banked plus the probability of failure to take up load. The hot reserve units are added to the capacity available in the third period.

The risk level for the entire lead time is:

$$R = \int_0^{t_q} F(R_1)dt + \int_{t_q}^{t_h} F(R_2)dt + \int_{t_h}^{t} F(R_3)dt$$

Where $\int_0^{t_q} F(R_1)dt$ = Risk level calculated at the operating capacity alone for the time interval 0 to t_q.

$\int_{t_q}^{t_h} F(R_2)dt$ = Risk level calculated at the operating capacity + quick starts units for the time interval t_q to t_h.

$\int_{t_h}^{t} F(R_3)dt$ = Risk level caluclated for the same load at available capacity equal to operating capacity, quick start units and hot reserve for the time interval t_h to t.

The time dependent probabilities for the time periods t_q, t_h and t can be easily obtained[5]. The applicable model is used for each unit

in operation, i.e. thermal, hydraulic, rapid start and hot reserve units, etc.

RAPID START AND HOT RESERVE UNIT MODELS

A rapid start unit is normally required at relatively short notice and at that time can therefore behave in two ways. It can fail to start or it can go into operation. When it fails to start, the unit is considered to have failed and is then checked for faulty components, repaired and either brought into service if required or placed back in the waiting for service mode. A Markov model of such a system is shown in Figure 9.

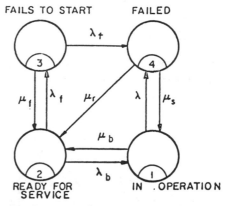

Fig. 9. State Space Diagram for a Rapid Start Unit

A model for peaking units has recently been introduced by Calsetta et al[6], proposing a four state model which takes into account the effect of reserve shutdown of a peaking unit. This model gives the conditional probability of a rapid start unit not being able to serve when called into service. The model considers that all repairs do not take place immediately. In the case of spinning reserve studies, the probabilities of a unit not being able to start when it is called upon and of its moving to the failed state from the operation state during the lead time, can be calculated from the model given in Figure 9. During the short lead time, the rapid start units if available are always considered for possible operation in the event of any outage or contingency. A large number of rapid start equipment is remote controlled and failures can take place due to the failure of the communication equipment. These repair rates will have very different average durations and distributions as compared to the repair rate of the rapid start units itself. The repair rate μ_f has been inserted to take this condition into account.

The departure or the transition rates from one state to the other can be obtained as follows:

N_S = Number of times the unit fails to start

N_F = Number of times the unit is forced out.

N_T = Number of successful transitions from the ready for service to the operation state.

N_B = Number of times the unit is shut down though it is not actually forced out.

N_R = Number of transitions from the failed to ready for service state.

N_K = Number of times the unit is repaired back to service.

N_A = Number of times the unit goes from the fails to start condition directly to the ready for service condition.

N_D = Number of times the unit goes from the fails to start to the failed state.

It can be seen that

$N_S + N_F = N_R + N_K + N_A$

T_1 = Total time spent in the operating state.

T_2 = Total time spent in the ready for service state.

T_3 = Total time spent in the failure to start state (it will be very small).

213

T_4 = Total time spent in the failed state.

Therefore the transition rates are

$$\lambda_f = \frac{N_S}{T_2} \qquad \lambda_b = \frac{N_T}{T_2}$$

$$\lambda_t = \frac{N_D}{T_3} \qquad \lambda = \frac{N_F}{T_1}$$

$$\mu_r = \frac{N_R}{T_4} \qquad \mu_s = \frac{N_K}{T_4}$$

$$\mu_b = \frac{N_B}{T_1} \qquad \mu_f = \frac{N_A}{T_3}$$

The state space diagram shown in Figure 9 gives the following stochastic differential probability matrix.

$$\begin{vmatrix} P_1'(t) \\ P_2'(t) \\ P_3'(t) \\ P_4'(t) \end{vmatrix} = \begin{matrix} 1 \\ 2 \\ 3 \\ 4 \end{matrix} \begin{vmatrix} -\lambda-\mu_b & \lambda_b & 0 & \mu_s \\ \mu_b & -\lambda_b-\lambda_f & \mu_f & \mu_r \\ 0 & \lambda_f & -\lambda_t-\mu_f & 0 \\ \lambda & 0 & \lambda_t & -\mu_s-\mu_r \end{vmatrix} \cdot \begin{vmatrix} P_1(t) \\ P_2(t) \\ P_3(t) \\ P_4(t) \end{vmatrix}$$

FAILS TO TAKE UP LOAD   **FAILED**

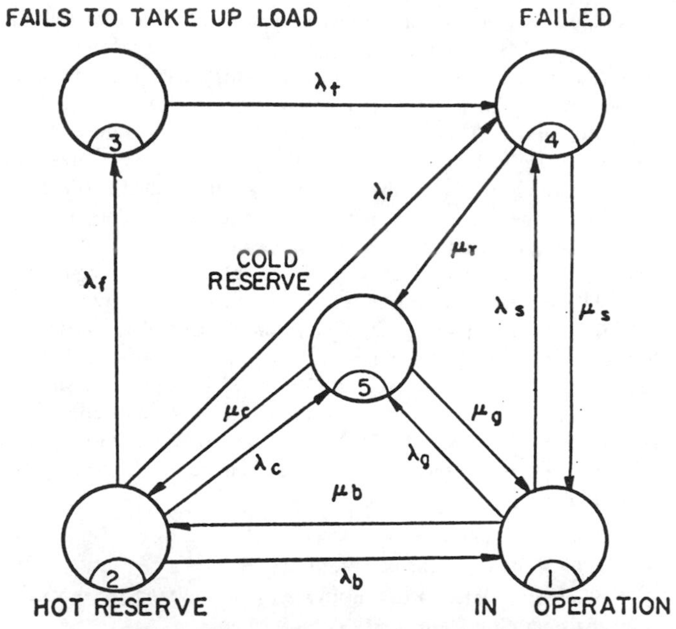

Fig. 10. State Space Diagram for a Unit on Hot Reserve

The time dependent probabilities can be easily calculated from this differential probability matrix[5].

The state space diagram for the units on hot reserve is basically the same as that shown previously for quick start units. The differences are that while the units are banked they can also fail and after a unit is repaired it will go back to the cold reserve state. Repairs on thermal units are normally made in the cold state and the unit cannot return to the hot reserve state immediately. A new state representing cold reserve must, therefore, be added. The state space diagram is given in Figure 10 where it has been assumed that the unit will not be derated. If deratings can occur, the state space diagram will be modified accordingly. In the case of a unit on hot reserve, the probability of residing in State 3 where the unit fails to take up load, will be extremely small. This state may be omitted entirely in many cases.

The additional transfer rates in this diagram over a rapid start unit are given as follows:

T_5 = Total time spent in the cold reserve state.

λ_r = Transition rate from the hot reserve state to the failed state.

μ_r = Transition rate from the failed state to the cold reserve state (repair rate).

μ_c = Transition rate from the cold reserve state to the hot reserve state.

λ_c = Transition rate from the hot reserve state to the cold reserve state.

μ_g = Transition rate from the cold reserve state to the operation state.

λ_g = Transition rate from the operation rate to the cold reserve state.

N_P = Total number of transitions from the failed state to the cold reserve state.

N_F = Number of times the unit is forced out.

N_p = Number of times the unit moves from the cold reserve state to the hot reserve state.

N_Q = Number of times the unit moves from the hot reserve state to the cold reserve state.

N_C = Number of transitions from the cold reserve state to the operation state.

N_L = Number of transitions from the operation state to the cold reserve state.

N_W = Total number of transitions from the hot reserve state to the failed state.

$$\lambda_s = \frac{N_F}{T_1} \qquad \mu_r = \frac{N_R}{T_4}$$

$$\mu_c = \frac{N_P}{T_5} \qquad \lambda_r = \frac{N_W}{T_2}$$

$$\lambda_c = \frac{N_Q}{T_2}$$

$$\mu_g = \frac{N_C}{T_5}$$

$$\lambda_g = \frac{N_L}{T_1}$$

And the stochastic differential probability matrix is:

$$\begin{vmatrix} P_1'(t) \\ P_2'(t) \\ P_3'(t) \\ P_4'(t) \\ P_5'(t) \end{vmatrix} = \begin{matrix} 1 \\ 2 \\ 3 \\ 4 \\ 5 \end{matrix} \begin{vmatrix} -\lambda_s-\mu_b-\lambda_g & \lambda_b & 0 & \mu_s & \mu_g \\ \mu_b & -\lambda_b-\lambda_f-\lambda_r-\lambda_c & 0 & 0 & \mu_c \\ 0 & \lambda_f & -\lambda_t & 0 & 0 \\ \lambda_s & \lambda_r & \lambda_t & -\mu_s-\mu_r & 0 \\ \lambda_g & \lambda_c & 0 & \mu_r & -\mu_c-\mu_g \end{vmatrix} \begin{vmatrix} P_1(t) \\ P_2(t) \\ P_3(t) \\ P_4(t) \\ P_5(t) \end{vmatrix}$$

Value of transfer rates λ_c and λ_g should be substituted as zero for spinning reserve studies.

The time dependent probabilities can be obtained from these simultaneous linear differential equations[5].

APPLICATION TO PRACTICAL PROBLEMS

The concepts developed in the previous sections were applied to the system given in the Appendix. The operating capacity in the system was assumed to be 2100 MW and the load 1900 MW. The effects of different amounts of rapid start generation and units on hot reserve on the risk level were considered. The times to put the rapid start units and the hot reserve units into service were taken as 10 minutes and 1 hour respectively. The capacities of the hot reserve units and the rapid start units together with the various transition rates used in this study are given in the Appendix. It can be seen that rapid start units have a greater effect on the risk levels than the units on hot reserve. Table I shows the risk levels in the hypothetical system as a function of the hot reserve capacity at different levels of rapid start capability.

APPLICATION TO THE S.P.C. SYSTEM

The concepts developed in this paper have been applied to the Saskatchewan Power Corporation (SPC) System. The on line operating capacity was taken to be 1128.9 MW. The system was assumed to have gas turbine and gas engine units at Estevan and Success with a total capacity of 25 MW. Three units at the A.L. Cole generating station with a total capacity of 46.3 MW were assumed to be on hot reserve. The risk levels were then computed including the effect of rapid start generation and the units on hot reserve. The start up time of the hot

reserve units was considered to be one hour. Two values of lead time, five minutes and ten minutes, were considered for the rapid start generation. The risk levels obtained in each case are slightly different and are shown in Table II. The risk levels for various reserve margins are shown in Figure 11 for rapid start generation lead times of five minutes and ten minutes.

Table I
Variation in Risk Level for Different Amounts of
Generation on Hot Reserve

Operating Capacity = 2100 MW
Load = 1900 MW
Risk without any rapid start generation = .00239083

(a) Rapid start generation available = 30 MW

Hot Reserve MW	Risk in Periods I	II	III	Total Risk
0	.00008399	.00004199	.00054262	.00066860
50	.00008399	.00004199	.00019673	.00032271
100	.00008399	.00004199	.00005107	.00017705
150	.00008399	.00004199	.00000714	.00013312
200	.00008399	.00004199	.00000150	.00012748
250	.00008399	.00004199	.00000043	.00012641
300	.00008399	.00004199	.00000005	.00012603

(b) Rapid start generation available = 60 MW

0	.00008399	.00001389	.00021113	.00030901
50	.00008399	.00001389	.00005543	.00015331
100	.00008399	.00001389	.00000844	.00010632
150	.00008399	.00001389	.00000168	.00009956
200	.00008399	.00001389	.00000043	.00009831
250	.00008399	.00001389	.00000004	.00009792
300	.00008399	.00001389	0	.00009788

(c) Rapid start generation available = 90 MW

0	.00008399	.00001316	.00019249	.00028964
50	.00008399	.00001316	.00003867	.00013582
100	.00008399	.00001316	.00000319	.00010034
150	.00008399	.00001316	.00000126	.00009841
200	.00008399	.00001316	.00000033	.00009748
250	.00008399	.00001316	0	.00009715
300	.00008399	.00001316	0	.00009715

(d) Rapid start generation available = 150 MW

0	.00008399	.00000245	.00003545	.00012189
50	.00008399	.00000245	.00000292	.00008936
100	.00008399	.00000245	.00000124	.00008768
150	.00008399	.00000245	.00000032	.00008676
200	.00008399	.00000245	0	.00008644
250	.00008399	.00000245	0	.00008644
300	.00008399	.00000245	0	.00008644

(e) Rapid start generation available = 300 MW

0	.00008399	0	.00000030	.00008429
50	.00008399	0	0	.00008399
100	.00008399	0	0	.00008399
150	.00008399	0	0	.00008399
200	.00008399	0	0	.00008399
250	.00008399	0	0	.00008399
300	.00008399	0	0	.00008399

Table II
Risk Levels in the S.P.C. System Including the Effects
of Rapid Start and Hot Reserve Generation

System on-line capacity   1128.9 MW
Available rapid start capacity   25.0 MW
Hot reserve capacity   46.3 MW

Risk Level

System Load in MW	Hot Reserve and Rapid Start Not Included	Hot Reserve and Rapid Start Included Rapid Start Lead Time - 5 minutes	Rapid Start Lead Time - 10 minutes
958.9	.00005447	.00000102	.00000109
968.9	.00009197	.00000118	.00000127
978.9	.00014055	.00000215	.00000230
988.9	.00200689	.00004587	.00008745
998.9	.00203303	.00004587	.00009293
1008.9	.00204807	.00004587	.00010684
1018.9	.00207299	.00051579	.00051588
1028.9	.00217013	.00052873	.00052888
1038.9	.00232486	.00055991	.00056026
1048.9	.00256661	.00060937	.00061003
1058.9	.00691403	.00218039	.00233463
1068.9	.01249314	.00224137	.00245489
1078.9	.01293508	.00228208	.00249584
1088.9	.01872309	.00342444	.00366383
1098.9	.03302748	.00507825	.00550574
1108.9	.05376441	.00566850	.00655241
1118.9	.07970780	.00774418	.00909073
1128.9	1.00000000		

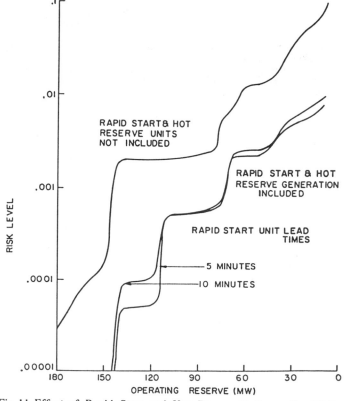

Fig. 11. Effect of Rapid Start and Hot Reserve Units on the S.P.C. Risk Levels

215

CONCLUSIONS

As noted previously, the system load cannot be considered to be a single value. The area risk curves are modified by both load and operating capacity changes. The role of rapid start, conventional hydro and pumped storage hydro units and conventional thermal generation can be included in a spinning reserve analysis. This approach should prove useful in practical utility studies where the lead time associated with additional or standby capacity cannot be considered as a single constant value. The economics associated with a given generation composition can be readily determined and the benefits associated with rapid start equipment evaluated using a predetermined risk index for the system. The effects of unit deratings and load forecast uncertainties can be included in the analysis using previously published techniques.

ACKNOWLEDGEMENT

The authors gratefully acknowledge the financial assistance provided by the Saskatchewan Power Corporation in this project.

REFERENCES

[1] Application of Probability Methods to the Determination of Spinning Reserve Requirements for the Pennsylvania — New Jersey — Maryland Interconnection, L. T. Anstine, R. E. Burke, J. E. Casey, R. Holgate, R. S. John, H. G. Stewart, IEEE Transactions, PAS-82, October 1963, pp. 720-735.

[2] Spinning Reserve Criteria in a Hydro Thermal System by the Application of Probability Mathematics, R. Billinton, M. P. Musick, Journal of the Engineering Institute of Canada, Vol. 48, pp. 40-45, October 1965.

[3] Determination of Spinning Spare Requirements Using Probability Mathematics, R. Billinton, M. P. Musick, IEE Conference Record on Economics of Reliability of Supply, London, England, 1967, p. 86.

[4] The Markov Process as a Means of Determining Generating Unit State Probabilities for Use in the Spinning Reserve Application, B. E. Biggerstaff, T. M. Jackson, IEEE Transactions, PAS-88, No. 4, April 1969, pp. 423-430.

[5] Unit Derating Levels in Spinning Reserve Studies, R. Billinton, A. V. Jain, IEEE Transactions, PAS-90, No. 4, July/August 1971, pp. 1677-1687.

[6] A Four State Model for Estimation of Outage Risk for Units in Peaking Service, Report of IEEE Task Group on Models for Peaking Service Units, Application of Probability Methods Subcommittee, IEEE Paper No. 71 TP 90-PWR, IEEE Winter Power Meeting, New York, N. Y.

APPENDIX

Hypothetical System Used to Determine the Effects of Rapid Start and Hot Reserve Units

The system was assumed to consist of 2900 MW of conventional thermal capacity consisting of the following units. The transfer rates given in the table refer to the state space diagram shown in Figure 12.

Number of Units	Size MW	Transfer Rates/Hour					
		a	b	c	d	e	f
15	100	0.0003	0.0010	0.0225	0.0350	0.0008	0.0004
8	150	0.0006	0.0050	0.0400	0.1000	0.0004	0.0004
1	200	0.0005	0.0002	0.0240	0.0430	0.0001	0.0001

The derated state shown in Figure 12 was assumed to be 80% of the full capacity rating.

At the 2100 MW capacity level it was assumed that 7 — 100 MW units, 8 — 150 MW units and 1 — 200 MW units were in operation.

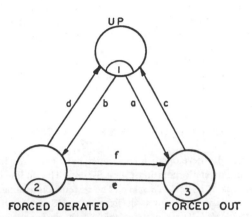

Fig. 12. Biggerstaff and Jackson's[4] state space model for a generating unit with one derated state

Outage Statistics Used for the Rapid Start and Hot Reserve Units

Rapid Start Unit Size = 30 MW
Refer to Figure 9
Transfer Rates/Hour

$\lambda_f = .0001$ $\lambda_b = .0040$

$\lambda_t = .0250$ $\lambda = .0006$

$\mu_r = .0250$ $\mu_s = .0150$

$\mu_b = .0050$ $\mu_f = 0$

Probability of Failure in State 2 for

$(P_{2-3} + P_{2-4})$

Lead Time (2 Hours) = .000203
Lead Time (4 Hours) = .000412
Lead Time (6 Hours) = .000627
Lead Time (8 Hours) = .000845
Lead Time (12 Hours)= .001290

Units on Hot Reserve - Capacity = 50 MW

Refer to Figure 10

Transfer Rates/Hour

$\lambda_f = .00002$ $\lambda_s = .00200$ $\lambda_c = 0$

$\lambda_t = .03000$ $\mu_r = .02500$ $\mu_c = .0025$

$\lambda_b = .02000$ $\lambda_r = .002100$ $\lambda_g = 0$

$\mu_s = .03500$ $\mu_b = .0240$ $\mu_g = .0030$

Probability of Failure in State 2

$(P_{2-3} + P_{2-4} + P_{2-5})$ for

Lead Time (2 Hours) = .0041595
Lead Time (4 Hours) = .0080157
Lead Time (6 Hours) = .0116029
Lead Time (8 Hours) = .0148479
Lead Time (12 Hours)= .0210208

INTERCONNECTED SYSTEM SPINNING RESERVE REQUIREMENTS

Roy Billinton   Adarsh V. Jain

Power System Research Group
University of Saskatchewan
Saskatoon, Canada

INTRODUCTION

There has been a considerable amount of work done in the area of static capacity reliability evaluation in interconnected systems. The first important paper in this subject[1] utilized a two dimensional probability array upon which the tie line constraints were imposed. The basic generating unit statistics in static and spinning reserve studies are fundamentally different and in the latter case the problem is a question of will the capacity immediately available to the system be capable of satisfying the load in the event of possible capacity and load changes. The possible assistance under these conditions from an interconnected system cannot therefore be neglected in determining the risk of just carrying or failing to carry the system load. In a static capacity interconnected system study, the tie line is normally rated at its nominal value. In a spinning study, the tie line rating or inter-system transfer capability will depend upon the immediate and known conditions between each system, i.e., the existing transmission configuration, the scheduled generation pattern, etc. etc. These could be quite variable even over the course of a day.

The two basic elements in any spinning reserve study are unforeseen generating capacity changes and the inherent variability in the system load. A recent paper[2] illustrated a general method of including the possible derated unit levels in any spinning reserve study and these concepts have been included in the studies reported in this paper. The effects of load forecast uncertainty in each of the two interconnected systems have also been studied.

The approach used to determine the benefits of interconnection has been designated as the "Capacity Assistance Method" and is quite fast and efficient in regard to computer storage[3,4]. Interconnected system risk levels have been calculated for different operating conditions using the Saskatchewan Power Corporation (S.P.C.) and the Manitoba Hydro (M.H.) systems as models. In each case only the risk levels in the S.P.C. System have been found.

TWO SYSTEM PROBABILITY ARRAY APPROACH

Cook et al in their paper[1] proposed the use of a two dimensional capacity outage array to find the probability of simultaneous outages in the two systems. If the outage in the first system is less than the operating reserve, it can assist the second system to the extent of the effective assistance or the tie line capability, which ever is less. In the spinning reserve sense, the operating reserve is the additional capacity synchronized to the bus and available to pick up load. The effective assistance in this case is the operating reserve minus the capacity on outage. If the outage in the second system is less than the effective assistance, the risk for that particular simultaneous event is zero. If, however, the effective assistance fails to satisfy the given capacity outage, the probability of occurrence of the simultaneous outages in the two systems will contribute to the risk level. If the operating reserve in the second system is greater than the capacity on outage, the event does not contribute to the risk level. The probabilities of all simultaneous conditions which lead to capacity deficiencies are computed and added. This sum is the risk level for the particular operating capacities and the corresponding loads in the two systems. This method becomes a little unwieldly for the calculation of system risk levels over a long period of time for varying operating capacities in the two systems.

CAPACITY ASSISTANCE PROBABILITY METHOD

The individual system risk levels can be obtained using the capacity assistance probability method. A probability table for the assisting system is first obtained. This assistance table is a two dimensional array representing the amount of assistance and the corresponding probability of assistance. This table can be converted into a capacity outage probability table with total capacity equal to the maximum assistance available and replacing each level of capacity available by a capacity equal to the maximum assistance minus the actual capacity assistance. The effective assistance as modified by the tie line capability is essentially equivalent to a single unit with many derated states. This equivalent single unit with a capacity equal to the maximum possible assistance can be added to the capacity outage probability table of the assisted system. The resulting capacity outage probability table can be used to determine the risk levels in the assisted system.

The basic algorithm for this approach is as follows:

If system A has

$$\text{Operating capacity} = CA \text{ MW}$$
$$\text{Load} = LA \text{ MW}$$
$$\text{Operating reserve } SA = CA - LA \text{ MW}$$
$$\text{Tie line capacity} = TL \text{ MW}$$

$P(C)$ is the cumulative outage probability and $T(C)$ is the individual outage probability of C MW.

Capacity Outage Probability Table

Capacity Out MW	Cumulative Probability	Individual Probability
$C_0 = 0$	$P(C_0)$	$T(C_0)$
C_1	$P(C_1)$	$T(C_1)$
C_2	$P(C_2)$	$T(C_2)$
-	-	-
-	-	-
-	-	-
C_n	$P(C_n)$	$T(C_n)$

Two distinct cases arise,

Case (a) $SA \leqslant TL$

Now equivalent assistance unit of System A

Capacity Out MW	Probability
C_0	$T(C_0)$
C_1	$T(C_1)$
C_2	$T(C_2)$
-	-
-	-
C_k	$P(C_k)$

Paper 71 TP 576-PWR, recommended and approved by the Power System Engineering Committee of the IEEE Power Engineering Society for presentation at the IEEE Summer Meeting and International Symposium on High Power Testing, Portland, Ore., July 18 - 23, 1971. Manuscript submitted January 4, 1971; made available for printing May 4, 1971.

Reprinted from *IEEE Trans. Power Apparatus Syst.*, vol. PAS-91, no. 2, pp. 517–525, Mar./Apr. 1972.

Where C_k is less than or equal to TL.

Case (b) SA > TL

Let C_L = SA = TL

The equivalent assistance unit of System A is

Capacity Out MW	Probability
$C_m - C_L = 0$	$1 - P(C_{m+1})$ where $C_m = C_L$
$C_{m+1} - C_L$	$T(C_{m+1})$
$C_{m+2} - C_L$	$T(C_{m+2})$
-	-
-	-
-	-
-	-
$C_{m+k} - C_L = TL$	$P_k(C_{m+k})$

Where m indicates the position in the capacity outage probability table of the assisting system at which the capacity outage is C_L MW. k is the number of steps in the equivalent assistance unit of system A.

This algorithm may be used to develop the equivalent assistance unit of the assisting system. A numerical example is shown in the Appendix.

FUNDAMENTALS OF INTERCONNECTED SYSTEM ASSISTANCE

There are two basic factors determining the extent of the interconnection assistance from one system to the other. They are the operating reserve in the assisting system and the tie line transfer capability.

The assumption has been made that any negative capacity contingency in the assisting system does not affect the assisted system. Positive capacity margins in either system do, however, increase reliability of the other system. When the tie line capacity increases beyond the operating reserve of the assisting system, no further gain in reliability is obtained by increasing the tie line capability. This limiting tie line capacity is called the infinite tie capability. If the tie line capability is reduced beyond the operating reserve, it limits the maximum assistance to the maximum transfer capability. This is visually seen from the capacity assistance probability table, which does not change after TL > SA. The operating reserve in the assisting system determines the extent of the assistance up to the tie line transfer capability. After which, if the operating reserve is further increased, the reliability of the assistance increases, i.e. the probabilities of higher assistance are increased while those of lesser assistance are decreased.

The maximum possible reliability, i.e. the minimum possible risk level in the assisted system is basically determined by the assisted system itself. If the assisted system has a tie line whose maximum capacity is T_M MW, then the lowest risk ever possible for that system with infinite interconnected system assistance at a load of LA MW will be $P(M_{MIN})$, where:

$P(M_{MIN})$ is the cumulative outage probability of M_{MIN} MW,

M_{MIN} = CA – LA + T_M

CA = Operating capacity of the assisted system

and

LA = Load on the assisted system

Inclusion of the Tie Line Forced Outage Rate

The discussion presented earlier has assumed a perfect tie line which is assumed to be always available. In actual practice the tie line may be forced out of service due to a variety of reasons. The tie line availability can be readily incorporated into the equivalent assistance unit of the system.

Let the tie line forced outage probability = p and the tie line availability q = 1 – p.

When the tie line is out, no assistance is possible and assistance to the other system is available only when the tie line is available.

The equivalent assistance unit can be modified by the tie line availability.

Case (a) SA ≤ TL

The Equivalent Assistance Unit of System A

Capacity Out MW	Probability
$C_0 = 0$	$T(C_0) \cdot q$
C_1	$T(C_1) \cdot q$
C_2	$T(C_2) \cdot q$
-	-
-	-
$C_k = SA$	$P(C_k) \cdot q + p$

Case (b) SA > TL

Let C_L = SA – TL

The Equivalent Assistance Unit of System A

Capacity Out MW	Probability
$C_m - C_L = 0$	$(1 - P(C_{m+1})) \cdot q$
$C_{m+1} - C_L$	$T(C_{m+1}) \cdot q$
$C_{m+2} - C_L$	$T(C_{m+2}) \cdot q$
-	-
-	-
$C_{m+k} - C_L = TL$	$P(C_{m+k}) \cdot q + p$

Where m indicates the position in the capacity outage probability table of the assisting system at which the capacity outage is C_L MW.

This modified table will include the effect of tie line outages.

APPLICATION TO THE PRACTICAL PROBLEMS

In order to illustrate the two system interconnection case, the Manitoba Hydro System was considered to assist the S.P.C. System.

The operating capacity level of the S.P.C. System was considered as 1130 MW, while the operating capacity level of Manitoba Hydro was considered to be 1361.4 MW. The most important factors affecting the interconnection benefits in the order of their importance are:

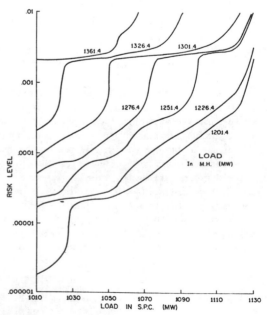

Fig. 1. Effect on the S.P.C. Risk Levels of Variation in the Operating Reserve in the M.H. System, Tie Line Transfer Capability 135 MW.

218

(1) Operating reserve in Manitoba Hydro.
(2) Tie line capability.
(3) Tie line forced outage rate.

In study 1, the tie line capability was fixed at 135 MW, while the load in the Manitoba Hydro System was changed from 1201.4 to 1326.4 MW in steps of 25 MW. The variation of the S.P.C. risk level versus peak load is given in Figure 1. It can be seen that the operating reserve in the Manitoba Hydro System significantly affects the risk in S.P.C. As the load increases in the Manitoba Hydro System, the risk levels become higher in the S.P.C. System.

In study 2, the operating reserve in the Manitoba Hydro System was fixed at 160 MW, while the transfer capability of the tie line was varied from 75 to 300 MW. When the tie line transfer capability was increased from 75 to 100 MW and 100 to 135 MW, significant decreases in the risk level are visible. As previously noted the tie line transfer

capability may vary over the course of a day as the conditions within the two systems vary. The gain in the risk level is relatively small when the transfer capability is further increased from 135 to 175 MW. As explained earlier, the infinite tie capacity in this case is 160 MW. No gain at all in the risk level was obtained after the transfer capability was increased from 175 MW to 300 MW. These curves are shown in Figure 2. The curves shown in Figure 3 are when the operating reserve in the Manitoba Hydro System was fixed at 135 MW, while the tie line transfer capability was varied from 75 to 100 MW and higher. It is again visible from this figure that significant improvement in the risk level is affected as the tie line capability is increased from 75 to 135 MW. No gain was made after increasing the tie line capability beyond that. The risk levels in the S.P.C. System obtained at the different operating capacities of the M. H. System for tie line transfer capabilities ranging from 75 MW to 160 MW are given in Table I.

Figure 4 shows the variation in the S.P.C. risk levels when the tie line forced outage rate was varied. As the tie line forced outage rate increases, the risk level in the S.P.C. System naturally increases. The tie line transfer capability was fixed at 135 MW for these studies. These results are given in Table II.

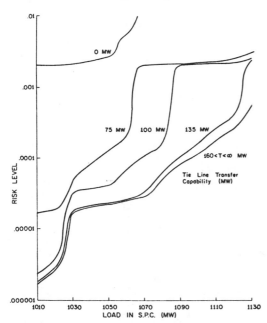

Fig. 2. Effect on the S.P.C. Risk Levels of Variation in Tie Line Transfer Capability, Operating Reserve in the M.H. System 160 MW.

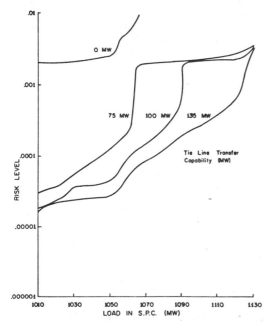

Fig. 3. Effect on the S.P.C. Risk Levels of Variation in Tie Line Transfer Capability, Operating Reserve in the M.H. System 135 MW.

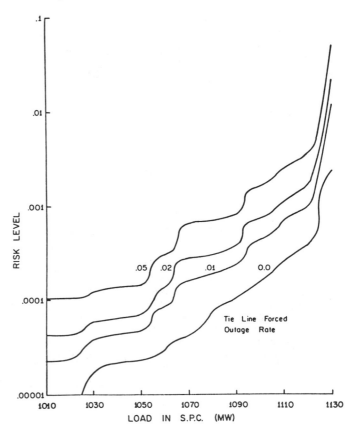

Fig. 4. Effect on the S.P.C. Risk Levels of Variation in Tie Line Forced Outage Rate, Tie Line Transfer Capability 135 MW, Operating Reserve in the M.H. System 160 MW.

INCLUSION OF LOAD FORECAST UNCERTAINTY IN THE INTERCONNECTED SYSTEM ASSISTANCE

If load forecast uncertainty is included, the effect is to alter the risk levels, i.e. cumulative outage probabilities. In most cases it will reduce the probability of lesser outages and increase the probability of higher outages.

The process utilized to incorporate load forecast uncertainty in interconnected systems is analogous to the calculation of risk levels in a single system when load forecast uncertainty is present. Risk levels can be calculated for the assisting system from low load values up to the

Table I — Risk in S.P.C. System for Variation in M.H. Operating Capacity

S.P.C. Operating Capacity = 1130 MW
M.H. Operating Capacity = 1361.4 MW
Tie Line Capacity = 75 MW

(a)   Risk in S.P.C. System for M.H. Load (MH)

S.P.C. Load MW	1201.4	1226.4	1251.4	1276.4	1301.4	1326.4
980	.00000061	.00000103	.00000181	.00000261	.00000487	.00002093
990	.00000139	.00000265	.00002141	.00002365	.00007127	.00010159
1000	.00000768	.00000960	.00002810	.00003116	.00007375	.00015060
1010	.00001670	.00003211	.00003752	.00005954	.00008584	.00020897
1020	.00001822	.00003676	.00003943	.00007371	.00013764	.00031855
1030	.00005243	.00005710	.00007958	.00009502	.00018855	.00201690
1040	.00008123	.00008305	.00011876	.00012910	.00026665	.00205159
1050	.00012351	.00012663	.00016232	.00019658	.00201769	.00206854
1060	.00019274	.00021601	.00026708	.00031342	.00212541	.00220599
1070	.00201514	.00202348	.00211932	.00213270	.00234323	.00262094
1080	.00204746	.00209783	.00215638	.00222848	.00237903	.00288814
1090	.00205858	.00214921	.00218624	.00233864	.00256352	.00338643
1100	.00212224	.00220705	.00241037	.00257320	.00325123	.00933398
1110	.00231163	.00242193	.00279532	.00300276	.00403673	.01402607
1120	.00248960	.00271374	.00318992	.00364692	.00760554	.01570892
1130	.00301213	.00371251	.01302113	.01425188	.04044907	.05572797

(b)   Tie Line Capacity = 100 MW

S.P.C. Load MW	1201.4	1226.4	1251.4	1276.4	1301.4	1326.4
980	.00000022	.00000066	.00000145	.00000245	.00000487	.00002093
990	.00000080	.00000208	.00002085	.00002338	.00007127	.00010159
1000	.00000145	.00000348	.00002201	.00002572	.00007375	.00015060
1010	.00000227	.00001793	.00002358	.00005164	.00008584	.00020897
1020	.00000353	.00002220	.00003522	.00007301	.00013764	.00031855
1030	.00003162	.00003665	.00005977	.00007662	.00018855	.00201690
1040	.00003612	.00003869	.00007502	.00010578	.00026655	.00205159
1050	.00004035	.00004469	.00008168	.00015782	.00201769	.00206854
1060	.00006999	.00009513	.00014770	.00025009	.00212541	.00220599
1070	.00010584	.00014714	.00027500	.00043792	.00234323	.00262094
1080	.00015615	.00022053	.00032757	.00220078	.00237903	.00288814
1090	.00201305	.00210428	.00214222	.00223179	.00256352	.00338643
1100	.00206404	.00214951	.00235359	.00254074	.00325123	.00933998
1110	.00214994	.00226301	.00263810	.00288611	.00403673	.01402607
1120	.00218303	.00241191	.00289267	.00350243	.00760554	.01570892
1130	.00250052	.00320861	.01252466	.01399317	.04044907	.05572797

(c)   Tie Line Capacity = 135 MW

S.P.C. Load MW	1201.4	1226.4	1251.4	1276.4	1301.4	1326.4
980	.00000020	.00000063	.00000145	.00000245	.00000487	.00002093
990	.00000068	.00000196	.00002076	.00002338	.00007127	.00010159
1000	.00000120	.00000320	.00002184	.00002572	.00007375	.00015060
1010	.00000174	.00001740	.00002333	.00005164	.00008584	.00020897
1020	.00000268	.00002136	.00002490	.00007301	.00013764	.00031855
1030	.00001741	.00002259	.00004643	.00007662	.00018855	.00201690
1040	.00002140	.00002427	.00007437	.00010578	.00026655	.00205159
1050	.00002272	.00002754	.00007860	.00015782	.00201769	.00206854
1060	.00003031	.00005613	.00012128	.00025009	.00212541	.00220599
1070	.00004220	.00008452	.00024450	.00043792	.00234323	.00262094
1080	.00006562	.00013221	.00030125	.00220078	.00237903	.00288814
1090	.00010194	.00020345	.00039769	.00233179	.00256352	.00338643
1100	.00015637	.00027546	.00233310	.00254074	.00325123	.00933398
1110	.00026814	.00042501	.00261753	.00288611	.00403673	.01402607
1120	.00038351	.00067580	.00288560	.00350423	.00760554	.01570892
1130	.00238797	.00309821	.01245772	.01399317	.04044907	.05572797

(d)   Tie Line Capacity ≥ 160 MW < ∞

S.P.C. Load MW	1201.4	1226.4	1251.4	1276.4	1301.4	1326.4
980	.00000020	.00000063	.00000145	.00000245	.00000487	.00002093
990	.00000068	.00000196	.00002076	.00002238	.00007127	.00010159
1000	.00000116	.00000320	.00002184	.00002572	.00007375	.00015060
1010	.00000170	.00001740	.00002333	.00005164	.00008584	.00020897
1020	.00000262	.00002136	.00002490	.00007301	.00013764	.00031855
1030	.00001721	.00002259	.00004643	.00007662	.00018855	.00201690
1040	.00002103	.00002427	.00007437	.00010578	.00026655	.00205159
1050	.00002216	.00002754	.00007860	.00015782	.00201769	.00206854
1060	.00002430	.00005613	.00012128	.00025009	.00212541	.00220599
1070	.00002840	.00008452	.00024450	.00043792	.00234323	.00262094
1080	.00005177	.00013221	.00030125	.00220078	.00237903	.00288814
1090	.00008189	.00020345	.00039769	.00233179	.00256352	.00338643
1100	.00011314	.00027546	.00233310	.00254074	.00325123	.00933398
1110	.00018881	.00042501	.00261753	.00288611	.00403673	.01402607
1120	.00026601	.00067580	.00288560	.00350423	.00760554	.01570892
1130	.00056177	.00309821	.01245772	.01399317	.04044907	.05572797

Table II

Risk in S.P.C. System for Variation in Tie Line Forced Outage Probability

```
                S.P.C. Operating Capacity  =  1130   MW
                 M.H. Operating Capacity   =  1361.4 MW
                       Tie Line Capacity   =  135    MW
(a)      Tie Line Forced Outage Probability =  0.01
```

Risk in S.P.C. System for M.H. Load (MW)

S.P.C. Load MW	1201.4	1226.4	1251.4	1276.4	1301.4	1326.4
980	.00000129	.00000172	.00000253	.00000352	.00000591	.00002182
990	.00002070	.00002197	.00004058	.00004317	.00009058	.00012060
1000	.00002143	.00002341	.00004187	.00004571	.00009326	.00016934
1010	.00002219	.00003770	.00004356	.00007159	.00010545	.00022735
1020	.00002320	.00004170	.00004519	.00009283	.00015681	.00033591
1030	.00003851	.00004364	.00006724	.00009713	.00020793	.00201801
1040	.00004399	.00004683	.00009643	.00012753	.00028679	.00205388
1050	.00004721	.00005198	.00010253	.00018095	.00202223	.00207257
1060	.00008510	.00011067	.00017516	.00030268	.00215925	.00223902
1070	.00016585	.00020775	.00036613	.00055761	.00244387	.00271881
1080	.00019282	.00025874	.00042609	.00230663	.00248309	.00298712
1090	.00024588	.00034637	.00053867	.00245343	.00268285	.00349752
1100	.00045154	.00056944	.00260651	.00281207	.00351545	.00953737
1110	.00072537	.00088067	.00305127	.00331717	.00445628	.01434573
1120	.00102310	.00131246	.00350016	.00411261	.00817290	.01619525
1130	.01236408	.01306722	.02233315	.02385323	.05004457	.06517065

(b)   Tie Line Forced Outage Probability = 0.02

S.P.C. Load MW	1201.4	1226.4	1251.4	1276.4	1301.4	1326.4
980	.00000238	.00000280	.00000360	.00000459	.00000696	.00002270
990	.00004071	.00004197	.00006040	.00006296	.00010989	.00013961
1000	.00004167	.00004363	.00006190	.00006570	.00011277	.00018808
1010	.00004264	.00005799	.00006380	.00009155	.00012506	.00024573
1020	.00004371	.00006203	.00006549	.00011265	.00017598	.00035327
1030	.00005961	.00006468	.00008805	.00011763	.00022732	.00201911
1040	.00006657	.00006939	.00011849	.00014927	.00030693	.00205617
1050	.00007169	.00007642	.00012646	.00020409	.00202677	.00207660
1060	.00013989	.00016520	.00022904	.00035528	.00219309	.00227206
1070	.00028950	.00033097	.00048775	.00067731	.00254451	.00281667
1080	.00032002	.00038528	.00055093	.00241248	.00258716	.00308609
1090	.00038982	.00048930	.00067965	.00257507	.00280217	.00360862
1100	.00074672	.00086342	.00287991	.00308340	.00377967	.00974077
1110	.00118261	.00133634	.00348501	.00374822	.00487582	.01466538
1120	.00166268	.00194912	.00411472	.00472099	.00874027	.01668159
1130	.02234020	.02303624	.03220857	.03371330	.05964008	.07461339

(c)   Tie Line Forced Outage Probability = 0.05

S.P.C. Load MW	1201.4	1226.4	1251.4	1276.4	1301.4	1326.4
980	.00000565	.00000606	.00000684	.00000779	.00001009	.00002535
990	.00010077	.00010199	.00011985	.00012233	.00016782	.00019663
1000	.00010237	.00010427	.00012198	.00012567	.00017129	.00024430
1010	.00010400	.00011888	.00012451	.00015141	.00018389	.00030087
1020	.00010527	.00012303	.00012639	.00017210	.00023349	.00040535
1030	.00012291	.00012782	.00015048	.00017915	.00028548	.00202242
1040	.00013434	.00013707	.00018647	.00021451	.00036733	.00206303
1050	.00014516	.00014974	.00019825	.00027350	.00204038	.00208869
1060	.00030426	.00032880	.00039069	.00051306	.00229461	.00237116
1070	.00066045	.00070066	.00085264	.00103639	.00284643	.00311026
1080	.00070162	.00076488	.00092546	.00273002	.00289936	.00338301
1090	.00082165	.00091808	.00110260	.00294000	.00316015	.00394190
1100	.00163223	.00174536	.00370012	.00389738	.00457234	.01035095
1110	.00255431	.00270333	.00478623	.00504138	.00613447	.01562434
1120	.00358144	.00385911	.00575546	.00654612	.01044236	.01814057
1130	.05226856	.05294329	.06183483	.06329346	.08842659	.10294151

221

Table III (a) — Risk Level in S.P.C. System for Load
Forecast Uncertainty in M.H.
(Assisting System)

Operating Capacity in S.P.C. = 1130 MW
Operating Capacity in M.H. = 1361.4 MW
Forecast Load in M.H. = 1201.4 MW
Tie Line Capacity = 135 MW

Load In S.P.C. MW	Risk Level in S.P.C. for % Load Forecast Uncertainty in M.H. Forecast Load			
	0	2	4	6
980	.00000020	.00000033	.00000110	.00000449
990	.00000068	.00000298	.00001241	.00003123
1000	.00000120	.00000349	.00001347	.00003830
1010	.00000174	.00000658	.00002474	.00014601
1020	.00000268	.00000899	.00002808	.00015705
1030	.00001741	.00001410	.00003687	.00016429
1040	.00002140	.00002055	.00005556	.00017423
1050	.00002272	.00002240	.00016588	.00019399
1060	.00003031	.00004390	.00019048	.00025954
1070	.00004220	.00006795	.00025057	.00039615
1080	.00006562	.00009199	.00028826	.00101205
1090	.00010194	.00014501	.00038534	.00146441
1100	.00015637	.00031863	.00093892	.00174086
1110	.00026814	.00041624	.00112858	.00214963
1120	.00038351	.00058203	.00160128	.00339974
1130	.00238797	.00343882	.00755208	.01657006

Table III (b) — Risk Level in S.P.C. System for Load
Forecast Uncertainty in S.P.C.

Operating Capacity in S.P.C. = 1130 MW
Operating Capacity in M.H. = 1361.4 MW
Forecast Load in M.H. = 1201.4 MW
Tie Line Capacity = 135 MW
% Load Forecast Uncertainty in M.H. = 4%

Load In S.P.C. MW	Risk Level in S.P.C. for % Load Forecast Uncertainty in S.P.C. Forecast Load			
	0	2	4	6
980	.00000110	.00000542	.00001979	.00012918
990	.00001241	.00001050	.00003125	.00018215
1000	.00001347	.00001521	.00003855	.00026197
1010	.00002474	.00003048	.00012379	.00108711
1020	.00002808	.00004065	.00016803	.00115776
1030	.00003687	.00007536	.00023104	.00159297
1040	.00005556	.00009762	.00028894	.00190054
1050	.00016588	.00016232	.00081083	.00255942
1060	.00019048	.00022086	.00104397	.00460255
1070	.00025057	.00032484	.00181505	.00973839
1080	.00028826	.00053213	.00209849	.01398740
1090	.00038534	.00108159	.00422594	.01626954
1100	.00093892	.00146625	.00527937	.01822224
1110	.00112858	.00358190	.01132695	.03317743
1120	.00160128	.00408579	.01248629	.03665758
1130	.00755208	.00802749	.02038815	.08236330

Table IV

Effect of Not Considering the Derated States on Risk Level

Operating Capacity in S.P.C. = 1130 MW
Operating Capacity in M.H. = 1361.4 MW
Tie Line Transfer Capability = 75 MW

(a)

Risk in S.P.C. System for M.H. Load (MW)

S.P.C. Load MW	1201.4	1226.4	1251.4	1276.4	1301.4	1326.4
980	.00000472	.00000563	.00000903	.00001138	.00002190	.00014140
990	.00000811	.00001152	.00006875	.00007589	.00029488	.00035594
1000	.00005095	.00005598	.00011192	.00012008	.00029920	.00048403
1010	.00013325	.00017879	.00019490	.00025889	.00033543	.00063392
1020	.00013696	.00019202	.00019943	.00030068	.00045459	.00085752
1030	.00021261	.00022660	.00029169	.00033691	.00058188	.00602120
1040	.00028077	.00028573	.00038837	.00041706	.00079508	.00612856
1050	.00037993	.00038522	.00048916	.00058647	.00603122	.00620688
1060	.00047028	.00053382	.00060123	.00073066	.00605879	.00623512
1070	.00601454	.00602787	.00622831	.00625118	.00678054	.00710784
1080	.00612370	.00618978	.00633965	.00643360	.00679924	.00756819
1090	.00619815	.00637479	.00642899	.00670502	.00697813	.00820494
1100	.00621830	.00641941	.00653170	.00690024	.00763978	.01530766
1110	.00651352	.00661021	.00701085	.00725910	.00847248	.02836746
1120	.00678608	.00693554	.00749419	.00788034	.00981286	.02930795
1130	.00734412	.00790747	.01738060	.01855839	.05432025	.06480342

(b)   Tie Line Transfer Capability = 100 MW

S.P.C. Load MW	1201.4	1226.4	1251.4	1276.4	1301.4	1326.4
980	.00000248	.00000342	.00000686	.00001010	.00002190	.00014140
990	.00000432	.00000779	.00006508	.00007410	.00029488	.00035594
1000	.00000651	.00001230	.00006847	.00008119	.00029920	.00048403
1010	.00000972	.00005740	.00007558	.00018355	.00033543	.00063392
1020	.00001485	.00007095	.00008137	.00029921	.00045459	.00085752
1030	.00017822	.00019281	.00025816	.00030669	.00058188	.00602120
1040	.00019017	.00019663	.00030042	.00036367	.00079508	.00612856
1050	.00020161	.00020962	.00031619	.00049641	.00603122	.00620688
1060	.00025121	.00031794	.00038865	.00064791	.00605879	.00623512
1070	.00033468	.00044614	.00074209	.00109282	.00678054	.00710784
1080	.00046712	.00057457	.00086946	.00633566	.00679924	.00756819
1090	.00602292	.00620248	.00625969	.00664070	.00697813	.00820494
1100	.00604943	.00625220	.00636834	.00689076	.00763978	.01530766
1110	.00636568	.00646493	.00686697	.00713222	.00847248	.02836746
1120	.00641411	.00656952	.00713320	.00766804	.00981286	.02930795
1130	.00671617	.00778897	.01677190	.01826347	.05432025	.06480342

(c)   Tie Line Transfer Capability = 135 MW

S.P.C. Load MW	1201.4	1266.4	1251.4	1276.4	1301.4	1326.4
980	.00000143	.00000240	.00000685	.00001010	.00002190	.00014140
990	.00000290	.00000641	.00006474	.00007410	.00029488	.00035594
1000	.00000466	.00001050	.00006746	.00008119	.00029920	.00048403
1010	.00000652	.00005426	.00007392	.00018355	.00033543	.00063392
1020	.00001002	.00006622	.00007975	.00029921	.00045459	.00085752
1030	.00005488	.00007059	.00014504	.00030669	.00058188	.00602120
1040	.00006719	.00007615	.00029947	.00036367	.00079508	.00612856
1050	.00007258	.00008438	.00030824	.00049641	.00603122	.00620688
1060	.00011712	.00018728	.00034412	.00064791	.00605879	.00623512
1070	.00020359	.00031704	.00067083	.00109282	.00678054	.00710784
1080	.00026909	.00038104	.00080428	.00633566	.00679924	.00756819
1090	.00034512	.00055263	.00103409	.00664070	.00697813	.00820494
1100	.00041449	.00071526	.00635433	.00689076	.00763978	.01530766
1110	.00068989	.00091725	.00671894	.00713222	.00847248	.02836746
1120	.00089294	.00124219	.00713021	.00766804	.00981286	.02930795
1130	.00648586	.00706535	.01671671	.01826347	.05432025	.06480342

operating capacity in convenient steps. This load versus risk level table is equivalent to a capacity availability probability table and an equivalent capacity outage probability table can be then obtained. This table can be used to obtain the assistance probability table, which can then be added to the assisted system.

The following equation is useful in the computation of an equivalent assistance unit.

$$\text{Operating Capacity} = CA \text{ MW}$$
$$\text{Load} = LA \text{ MW}$$

Risk level for the load

$$= \sum_{n=1}^{n=k} P_L (1 + b_n t) \cdot p_n$$

Where

$P_L (1 + b_n t) = $ the cumulative outage probability of $CA - LA (1 + b_n t)$ capacity.

k = number of steps in the stepped approximation of load forecast uncertainty.

p_n = probability of existence of the nth step of the load forecast uncertainty distribution curve.

t = load forecast uncertainty equal to one division of the stepped approximation.

b_n = difference between the nth step and the mean value.

To include the effect of load forecast uncertainty in the assisting and the assisted systems, 2%, 4% and 6% load forecast uncertainty

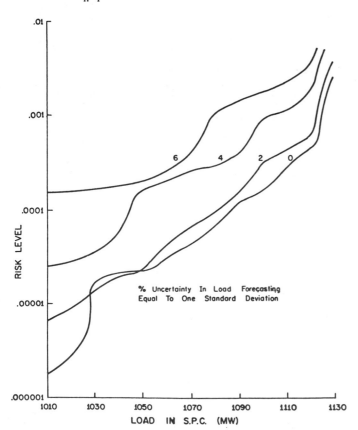

Fig. 5. Effect on the S.P.C. Risk Levels of Various Amount of Load Forecast Uncertainty in the M.H. System, Tie Line Transfer Capability 135 MW. Operating Reserve in the M.H. System 160 MW.

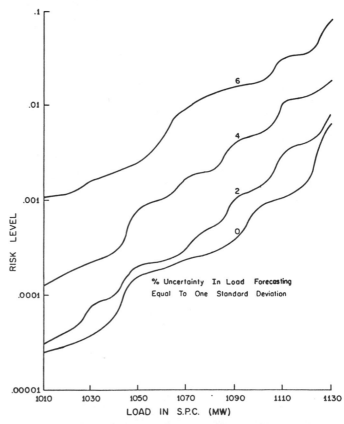

Fig. 6. Effect on the S.P.C. Risk Levels of Load Forecast Uncertainty in the S.P.C. System. Standard Deviation of Load Forecast Uncertainty in the M.H. System 4%, Tie Line Transfer Capability 135 MW, Operating Reserve in the M.H. System 160 MW.

equal to one standard deviation was applied to the Manitoba Hydro System. The operating reserve in the M.H. System was fixed at 160 MW, while the tie line transfer capability was maintained at 135 MW. Figure 5 shows the effect of load forecast uncertainty in the assisting system on the risk level. The results are given in Table III (a). It can be seen that load forecast uncertainty in the assisting system has a prominent effect on the risk level in the assisted system.

When load forecast uncertainties in both assisting and the assisted system are present, the effective risk levels are further increased. Figure 6 shows the effect of load forecast uncertainty in the S.P.C. System when 2%, 4% and 6% load forecast uncertainties equal to one standard deviation were applied to the S.P.C. System with the uncertainty in the M.H. System maintained constant at 4%. The results for 2%, 4% and 6% load forecast uncertainties in S.P.C. are given in Table III (b). The calculated risk levels can be considerably in error if load forecast uncertainty in either of the systems is neglected.

EFFECT OF NOT CONSIDERING EXACT EQUIVALENT DERATED MODELS IN INTERCONNECTED SYSTEM STUDIES

The quantitative effects of including forced derated states into the capacity model for a single system have been discussed in a recent paper[2]. It was observed that representation of some fraction of the partial outages as full outages gives pessimistic results. Derated states were considered for the S.P.C. thermal units in all the interconnected system studies in this paper. This section investigates the effect of not representing the derated states in interconnected system studies.

The M.H. System was considered to be assisting the S.P.C. System. All the risk levels were computed for M.H. operating reserves of 160 MW, 135 MW, 110 MW, 85 MW, 60 MW and 35 MW and tie line transfer capabilities of 75 MW, 100 MW and 135 MW. These results are given in Table IV. A comparison of Tables I and IV indicates that the effect of considering derated states as full outages is an increase in the risk

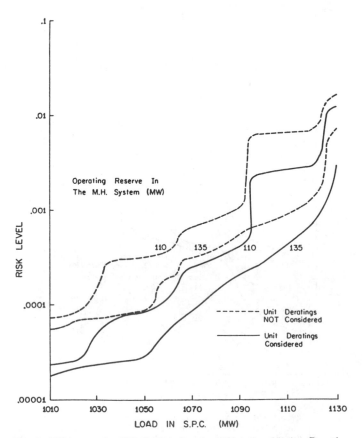

Fig. 7. Effect on the S.P.C. Risk Levels of Not Considering Deratings in the S.P.C. System, Tie Line Transfer Capability 135 MW.

levels. The assisted system characteristics still dominate the risk levels when interconnection benefits are available. The effect of not considering the equivalent derated states for the assisting system tends to diminish the reliability of the assisting system and therefore the assistance.

Figure 7 compares the risk levels at a tie line transfer capability of 135 MW and operating margins in the M.H. System of 110 MW and 135 MW. The solid curves are for equivalent derated state models and the dotted curves indicate the risk levels when deratings are not considered. It can be seen that the risk levels are substantially higher when deratings are not considered.

CONCLUSIONS

Methods have been presented for calculation of the spinning reserve interconnection benefits in two interconnected systems. Algorithms have been developed for rapid calculation of the risk levels in the interconnected systems in the case of load forecast uncertainty being present in either of the systems and inclusion of the tie line forced outage rate. The concept of single system being equivalent to one gigantic unit having a large number of derated states has been utilized to incorporate the practical limitations of the above noted factors, tie line constraints and the generating unit derated states. The methods presented in this paper can be used to obtain an economic appraisal of the benefits associated with ensuring various tie capacity levels. The nature of the agreement between the systems influences the interconnection benefits and certain agreements may favour only one particular system. With some modifications the effect of these agreements can be incorporated.

REFERENCES

[1] Cook, V. M., Galloway, C. D., Steinberg, M. J. and Wood, A. J., "Determination of Reserve Requirements of Two Interconnected Systems", IEEE Trans. Vol. PAS - 82, pp. 18 - 27, April 1963.

[2] Billinton, R. and Jain, A. V., "Unit Derating Levels in Spinning Reserve Studies", IEEE Trans. Vol. PAS-90, pp. 1677-1687, July/August 1971.

[3] Billinton, R., "Power System Reliability Evaluation", Gordon and Breach Science Publishers, New York, N.Y., 1970. (book)

[4] Billinton, R., Bhavaraju, M. P. and Thompson, P., "Power System Interconnection Benefits", Transactions of the Canadian Electrical Association, 1969.

ACKNOWLEDGEMENT

The authors gratefully acknowledge the financial assistance provided by the Saskatchewan Power Corporation in this project.

APPENDIX

Consider the capacity outage probability table of System A.

Operating capacity = 200 MW
Tie line capability = 50 MW
Peak load = 160 MW

Capacity Out MW	Cumulative Probability	Individual Probability
0.0	1.0000	0.9642
10.0	0.0358	0.0062
15.0	0.0296	0.0008
20.0	0.0288	0.0012
30.0	0.0276	0.0170
40.0	0.0106	0.0018
60.0	0.0088	0.0001
70.0	0.0087	0.0005
80.0	0.0082	0.0072
90.0	0.0010	0.0005
100.0	0.0005	0.0005

Maximum assistance = Min {200 - 160, 50}
= 40 MW

Assistance Probability Table

Assistance from System A MW	Probability
40.0 - 0.0 = 40.0	0.9642
40.0 - 10.0 = 30.0	0.0062
40.0 - 15.0 = 25.0	0.0008
40.0 - 20.0 = 20.0	0.0012
40.0 - 30.0 = 10.0	0.0170
40.0 - 40.0 = 0.0	0.0106

Equivalent Multiderated State Assistance Unit

Capacity Out MW	Probability
0.0	0.9642
10.0	0.0062
15.0	0.0008
20.0	0.0012
30.0	0.0170
40.0	0.0106

Consider the case when

Peak load = 140 MW

Tie line capability = 50 MW

Equivalent Multiderated State Assistance Unit

Capacity Out MW	Probability
0.0	0.9704
5.0	0.0008
10.0	0.0012
20.0	0.0170
30.0	0.0018
50.0	0.0088

225

Part 2
Composite Systems
Hierarchical Level II

QUANTITATIVE reliability evaluation using probability methods began with the evaluation of system adequacy at hierarchical level I. Considerable effort has been expended during the last two decades in developing techniques and criteria for composite generation and transmission system assessment. The term *composite* refers to the consideration of both generation and transmission system contingencies, including the modeling of the operating policies necessary in order to dispatch generating units, assessment of power flows on transmission system components, alleviation of network violations, and load shedding if required. This general area of evaluation is known as HLII assessment.

Two basic approaches have been applied in the development of the computing tools used to evaluate composite system adequacy. These are contingency enumeration (analytical) and Monte Carlo simulation.

The contingency enumeration technique represents the system by simplified mathematical models and evaluates the reliability indices using analytical solutions. References [1] and [2] illustrate an initial approach. These papers present the basic idea of replacing the traditional approaches by the introduction of reliability criteria based on the application of conditional probability. They also present hypothetical system studies in which the reliability indices of each system load point are obtained. The utilization of the indices in system planning is illustrated in [A1]. The proposed procedure used ac load flow solutions of the network to examine possible voltage and overload problems. Additional examples of alternative contingency enumeration approaches are presented in [A2]–[A5]. The application of contingency enumeration in HLII assessment appears to have been started in North America.

The Monte Carlo technique simulates the actual system behavior and the random processes involved. It obtains the required results by averaging the values obtained in repeated probabilistic sampling of the input data. Reference [3], published in the same period as [1] and [2], as well as [4] present the general philosophy and some applications of this technique. The application of Monte Carlo simulation in HLII assessment appears to have been initiated in Europe and particularly in Italy, although extensive work has also been done in France [A6] and more recently in Brazil [A7].

There has been considerable debate regarding the relative merits of the two techniques. Reference [5] attempts to address this problem by presenting the results for the IEEE Reliability Test System [6] using computer programs based on the two approaches. This comparison indicates the conceptual differences in modeling and problem perception and allows a better understanding of the merits and demerits of each approach. The IEEE Reliability Test System (RTS) [6] was established in

1979 as a reference configuration for testing and comparing alternative techniques for power system reliability evaluation. Reference [7] presents a recent application of the Monte Carlo technique to the RTS and illustrates how reliability evaluations can be used in the actual planning of composite systems in order to obtain a better—that is, more economical, system structure.

References [8]–[10] and [A8] examine the impact of terminal station failure events (failures of breakers, transformers, and bus sections) in HLII adequacy assessment. In particular, [10] illustrates the concept by proposing an extended version of the RTS which includes all the switching and terminal stations. The quantitative results show that station-originated events can create significant increases in the load point and system adequacy indices and particularly at load levels lower than system peak. The impact of dependency effects due to common cause and weather-related outages are illustrated in [A9] and [A10].

References [11]–[13] present a summary of the risk indices used for reliability assessment at HLII. Reference [11], published in 1978, discusses the two aspects of reliability, namely adequacy and security. It should be appreciated that probabilistic indices are still only derived in the adequacy domain. Security or dynamic indices expressed in probabilistic terms are still not available because the modeling and network reduction problems are not yet solved.

References [14]–[16] present an overview of the major concepts recently debated by experts in the field of HLII adequacy assessment. The first two papers review the goals, the time frame, and methods of reliability evaluation as well as the conflicting opinions sometimes occurring between American and European utility planners. Reference [16] describes the requirements for composite system reliability evaluation and concentrates on philosophy and purpose rather than on modeling details. There is still considerable developmental work being done in composite system reliability evaluation [A11].

Early work in measuring and reporting overall reliability at HLII is described in [A12]. Reference [13] presents the first stage of an activity sponsored by the IEEE Application of Probability Methods Subcommittee on identifying techniques and approaches for monitoring, measuring, predicting, and applying reliability indices in the planning and operation of power systems. The second stage of this activity, which will be published in the future, is aimed at identifying indices that can be calculated, the degree to which they are used in practice, commonality and/or lack of it between utilities, the perceived problems with their calculation or practical use, computor resources, and management understanding.

Progress in the area of HLII evaluation has been relatively

slow as many conceptual, modelling and computational difficulties have had to be resolved. Many utilities apply a deterministic approach which is based on an assessment of the system adequacy for certain predetermined "dangerous" situations selected according to experience. This can be considered as an extension into the planning phase of criteria usually adopted in operation where the system already exists and traditional load flows, in the presence of prespecified contingencies, are used. In the planning phase, the expansion or reinforcement of the system does not exist, and therefore contingency preselection presents many difficulties. The probabilistic approach recognizes the random nature of the phenomena involved and examines, at least in principle, all possible deterministic situations.

Most of the utilities around the world are in the transition phase from the utilization of the "old" deterministic procedures to the more modern probabilistic approaches. The paper selection attempts to reflect the historical development of the analytical techniques, indices, and criteria for HLII adequacy assessment and to illustrate the present level of utilization.

References

[1] R. Billinton, "Composite system reliability evaluation," *IEEE Trans. Power Apparatus Syst.*, vol. PAS-88, no. 4, pp. 276–280, Apr. 1969.

[2] R. Billinton and M. P. Bhavaraju, "Transmission planning using a reliability criterion: Part I—a reliability criterion," *IEEE Trans. Power Apparatus Syst.*, vol. PAS-89, no. 1, pp. 28–34, Jan. 1970.

[3] P. L. Noferi and L. Paris, "Quantitative evaluation of power system reliability in planning studies," *IEEE Trans. Power Apparatus Syst.*, vol. PAS-91, no. 2, pp. 611–618, Mar./Apr. 1972.

[4] P. L. Noferi, L. Paris, and L. Salvaderi, "Monte Carlo method for power system reliability evaluation in transmission and generation planning," *Proc. Annu. Reliab. Maintainab. Symp.*, Jan. 1975, pp. 449–459.

[5] L. Salvaderi and R. Billinton, "A comparison between two fundamentally different approaches to composite system reliability evaluation," *IEEE Trans. Power Apparatus Syst.*, vol. PAS-104, no. 12, pp. 3486–3493, Dec. 1985.

[6] IEEE Subcommittee on the Application of Probability Methods, "IEEE reliability test system," *IEEE Trans. Power Apparatus Syst.*, vol. PAS-98, no. 6, pp. 2047–2054, Nov./Dec. 1979.

[7] O. Bertoldi, L. Salvaderi, and S. Scalcino, "Monte Carlo approach in planning studies: An application to IEEE RTS," *IEEE Trans. Power Syst.*, vol. 3, no. 3, pp. 1146–1154, Aug. 1988.

[8] R. Billinton and T. K. P. Medicherla, "Station originated multiple outages in the reliability analysis of a composite generation and transmission system," *IEEE Trans. Power Apparatus Syst.*, vol. PAS-100, no. 8, pp. 3870–3878, Aug. 1981.

[9] R. N. Allan and A. N. Adraktas, "Terminal effects and protection system failures in composite system reliability evaluation," *IEEE Trans. Power Apparatus Syst.*, vol. PAS-101, no. 12, pp. 4557–4562, Dec. 1982.

[10] R. Billinton, P. K. Vohra, and S. Kumar, "Effect of station originated outages in a composite system adequacy evaluation of the IEEE reliability test system," *IEEE Trans. Power Apparatus Syst.*, vol. PAS-104, no. 10, pp. 2649–2656, Oct. 1985.

[11] IEEE Subcommittee on the Application of Probability Methods, "Reliability indices for use in bulk power supply adequacy evaluation," *IEEE Trans. Power Apparatus Syst.*, vol. PAS-97, no. 4, pp. 1097–1103, July/Aug. 1978.

[12] R. Billinton, T. K. P. Medicherla, M. S. Sachdev, "Adequacy indices for composite generation and transmission system reliability evaluation," IEEE Paper A79 024-1.

[13] IEEE Working Group on Measurement Indices: C. C. Fong, R. Billinton, R. O. Gunderson, P. M. O'Neil, J. Raksani, A. W. Schneider Jr., and B. Silverstein, "Bulk system reliability—measurement and indices," *IEEE Winter Power Meeting,* Paper no. 89 WM155-3 PWRS.

[14] J. Endrenyi, P. F. Albrecht, R. Billinton, G. E. Marks, N. D. Reppen, and L. Salvaderi, "Bulk power system reliability assessment—why and how? Part I: why?," *IEEE Trans. Power Apparatus Syst.*, vol. PAS-101, no. 9, pp. 3439–3445, Sept. 1982.

[15] J. Endrenyi, P. F. Albrecht, R. Billinton, G. E. Marks, N. D. Reppen, and L. Salvaderi, "Bulk power system reliability assessment—why and how? Part II: how?," *IEEE Trans. Power Apparatus Syst.*, vol. PAS-101, no. 9, pp. 3446–3456, Sept. 1982.

[16] M. P. Bhavaraju, P. F. Albrecht, R. Billinton, N. D. Reppen, and R. J. Ringlee, "Requirements for composite system reliability evaluation models," *IEEE Trans. Power Syst.*, vol. 3, no. 2, pp. 149–157, Feb. 1988.

Additional Papers

[A1] M. P. Bhavaraju and R. Billinton, "Transmission planning using a reliability criterion: Part II—transmission planning," *IEEE Trans. Power Apparatus Syst.*, vol. PAS-90, no. 1, pp. 70–78, Jan. 1971.

[A2] P. L. Dandeno, G. E. Jorgensen, W. R. Puntel, and R. J. Ringlee, "Program for composite bulk power electric system adequacy assessment," *Proc. IEEE Conf. Reliab. Power Supply Syst.*, vol. 148, Feb. 1977.

[A3] G. E. Marks, "A method of combining high speed contingency load flow analysis with stochastic probability methods to calculate a quantitative measure of overall power system reliability," IEEE paper A78 053-1.

[A4] T. A. Mikolinnas, W. R. Puntel, and R. J. Ringlee, "Application of adequacy assessment techniques for bulk power systems," *IEEE Trans. Power Apparatus Syst.*, vol. PAS-101, no. 5, pp. 1219–1228, May 1982.

[A5] K. A. Clements, B. P. Lam, D. J. Lawrence, and N. D. Reppen, "Computation of upper and lower bounds on reliability indices for bulk power systems," *IEEE Trans. Power Apparatus Syst.*, vol. PAS-103, no. 8, pp. 2318–2325, Aug. 1984.

[A6] J. C. Dodu and A. Merlin, "New probabilistic approach taking into account reliability and operation security in EHV power system planning at EDF," *IEEE Trans. Power Syst.*, vol. PWRS-1, no. 3, pp. 175–181, Aug. 1986.

[A7] S. H. F. Cunha, M. V. F. Pereira, L. M. V. G. Pinto and G. C. Oliveira, "Composite generation and transmission reliability evaluation in large hydroelectric systems," *IEEE Trans. Power Apparatus Syst.*, vol. PAS-104, no. 10, pp. 2657–2664, Oct. 1985.

[A8] R. N. Allan and J. R. Ochoa, "Modelling and assessment of station-originated outages for composite system reliability evaluation," *IEEE Trans. Power Syst.*, vol. 3, no. 1, pp. 158–165, Feb. 1988.

[A9] R. Billinton, T. K. P. Medicherla, and M. S. Sachdev, "Application of common-cause outage models in composite system reliability evaluation," *IEEE Trans. Power Apparatus System*, vol. PAS-100, no. 7, pp. 3648–3657, July 1981.

[A10] R. Billinton and L. Cheng, "Incorporation of weather effects in transmission system models for composite system adequacy evaluation," *Proc. IEE*, vol. 133 part C, no. 6, pp. 319–327, Sept. 1986.

[A11] S. Kumar and R. Billinton, "Adequacy equivalents in composite power system evaluation," *IEEE Trans. Power Syst.*, vol. 3, no. 3, pp. 1167–1173, Aug. 1988.

[A12] W. R. Winter, "Measuring and reporting overall reliability of bulk electricity systems," CIGRE, paper 32-15, 1980.

Composite System Reliability Evaluation

ROY BILLINTON, MEMBER, IEEE

Abstract—This paper illustrates the application of a conditional probability approach to the determination of a reliability index at any point in a composite system. A general design criterion is postulated in terms of quality of service rather than continuity. Using a Markov approach, it is shown that the effect of storm associated failures on the system failure probabilities is dependent upon the degree of redundancy in the configuration under study. The effects of shunt compensation, on-load tap changing, and variations in allowable voltage levels on the reliability of a simple configuration are illustrated. Using the techniques described in this paper, it is possible to arrive at a measure of steady-state adequacy for any point in a system and, particularly, at those points at which major transmission terminates and subtransmission begins.

Introduction

THERE IS a considerable amount of published material available on the subject of generating capacity reliability evaluation [1], and an increasing number of papers are being presented dealing with the area of transmission system reliability [2]. There is, however, very little material available on the subject of composite system reliability evaluation. The basic problem in this regard is to decide just what does constitute a failure at any point within a system. If all the components in a transmission system are fully redundant, then reliability is only a matter of continuity with success and failure easily defined. It has been suggested [3] that a possible approach is to write a Boolean description of the network configuration, then reduce it to a success or failure conclusion with respect to a particular node. This approach is readily adaptable to simple series-parallel configurations but becomes complicated and virtually unworkable when applied to a networked system with several supply points.

Continuity is not an acceptable single criterion, as complete component redundancy is not economically feasible in a modern power system. The definition of a breach of continuity can be extended to include a breach of quality (i.e., if, due to line outages, a low-voltage condition exists at a load point, this is not an actual breach of continuity though the voltage level may be considerably lower than a desired minimum). If bounds are placed upon desirable voltage levels at each point in the system, any departures from these ranges can be classed as a breach of service continuity. This would not include voltage transients caused by system disturbances unless the voltage remained for a defined period of time in the unacceptable region. Customers can be served at reduced voltage levels but this should be considered as a last resort rather than a design criterion.

A clear definition of quality of service at each point must be made and any departure from these requirements classed as a breach of continuity of that service. The reason often given for the construction of an additional infeed to an area is increased transmission reliability. It is obvious that the addition of a line to supply a station will increase the reliability of that station, as any duplication of facilities will do this to some extent. The construction of a line entirely on reliability grounds implies extremely high outage rates for the remaining lines supplying the station. A general design criterion could be as follows:

1) If, with all transmission and generation facilities in service, station voltage levels are outside the defined limits, then new facilities are required to meet the quality of service standards.

2) If, with the various possible combinations of system components out of service, the reliability index for the station is below an acceptable minimum, then additional facilities are required to meet the reliability standards. A station could still meet quality of service standards for several years of load growth but not meet present reliability standards if a minimum number of system components with high outage rates were installed.

It can be clearly seen that the determination of a reliability index based upon service quality standards involves considerably more effort than the determination of a success or failure probability based only on continuity. It will, however, provide a valuable tool in assessing the adequacy of proposed system alternatives from a planning, design, and operating viewpoint.

Conditional Probabilities of System Failure

In almost all probability applications in reliability evaluation, component failures within a fixed environment are assumed to be independent events. It is entirely possible that component failure can result in system failure in a conditional sense. This can occur in parallel facilities that are not completely redundant. If the load can be considered as a random variable and described by a probability distribution, then failure at any station due to component failure is conditional upon the load exceeding some value at which a satisfactory voltage level at the load point can be maintained.

If two events designated A and B are considered to be independent,

$$P(A \cap B) = P(A) \cdot P(B).$$

If the events are not independent and $P(B \mid A)$ denotes the conditional probability that B occurs given that A has occurred,

$$P(A \cap B) = P(A) \cdot P(B \mid A).$$

Also

$$P(A \cap B) = P(B) \cdot P(A \mid B).$$

If the occurrence of A is dependent upon a number of events B_j, which are mutually exclusive,

$$P(A) = \sum_{i=1}^{j} P(A \mid B_i) \cdot P(B_i).$$

Paper 68 TP 76–PWR, recommended and approved by the Transmission and Distribution Committee of the IEEE Power Group for presentation at the IEEE Winter Power Meeting, New York, N. Y., January 28–February 2, 1968. Manuscript submitted June 26, 1967; made available for printing November 22, 1967.

The author is with the Power System Research Group, University of Saskatchewan, Saskatoon, Sask., Canada.

Reprinted from *IEEE Trans. Power Apparatus Syst.*, vol. PAS-88, no. 4, pp. 276–281, Apr. 1969.

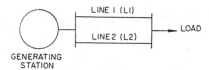

Fig. 1. System configuration.

If the occurrence of A is dependent upon only two mutually exclusive events for component B, success and failure, designated B_x and B_y, respectively, are

$$P(A) = P(A \mid B_x) \cdot P(B_x) + P(A \mid B_y) \cdot P(B_y).$$

With respect to reliability this can be expressed in a simpler form [4]:

$$P(\text{system failure}) = P(\text{system failure if } B \text{ is good}) \cdot P(B_x)$$
$$+ P(\text{system failure if } B \text{ is bad}) \cdot P(B_y). \quad (1)$$

The complementary situation is similar in form

$$P(\text{system success}) = P(\text{system success if } B \text{ is good}) \cdot P(B_x)$$
$$+ P(\text{system success if } B \text{ is bad}) \cdot P(B_y).$$

SIMPLE SYSTEM APPLICATION

Consider a simple system consisting of a generating station with two parallel transmission lines feeding a single load as shown in Fig. 1. Define

P_g probability of generation inadequacy as determined by the loss of load probability approach

P_c probability of transmission inadequacy, i.e., $P_c(1)$, $P_c(2)$ are the probabilities that the load will exceed the carrying capabilities of lines 1 and 2, respectively, and $P_c(1, 2)$ is the probability that the load will exceed the combined carrying capability of lines 1 and 2

R_{L1}, Q_{L1} probabilities of line availability and line outage, respectively, for line 1

R_{L2}, Q_{L2} probabilities of line availability and line outage, respectively, for line 2

Q_s probability of system failure.

Start with

$$Q_s = Q_s(L1_{\text{in}}) \cdot R_{L1} + Q_s(L1_{\text{out}}) \cdot Q_{L1}.$$

For $L1_{\text{in}}$ (i.e., given that line 1 is available),

$$Q_s = Q_s(L2_{\text{in}}) \cdot R_{L2} + Q_s(L2_{\text{out}}) \cdot Q_{L2}.$$

For $L1_{\text{in}}$ and $L2_{\text{in}}$

$$Q_s = P_g + P_c(1, 2) - P_g \cdot P_c(1, 2).$$

The probabilities of capacity deficiencies and transmission inadequacies are independent. For $L1_{\text{in}}$ and $L2_{\text{out}}$

$$Q_s = P_g + P_c(1) - P_g \cdot P_c(1).$$

Therefore, for $L1_{\text{in}}$

$$Q_s = R_{L2}(P_g + P_c(1, 2) - P_g \cdot P_c(1, 2))$$
$$+ Q_{L2}(P_g + P_c(1) - P_g \cdot P_c(1)).$$

For $L1_{\text{out}}$

$$Q_s = Q_s(L2_{\text{in}}) R_{L2} + Q_s(L2_{\text{out}}) Q_{L2}$$
$$= R_{L2}(P_g + P_c(2) - P_g \cdot P_c(2)) + Q_{L2}.$$

For the complete system

$$Q_s = R_{L1}\{R_{L2}[P_g + P_c(1, 2) - P_g \cdot P_c(1, 2)]$$
$$+ Q_{L2}[P_g + P_c(1) - P_g \cdot P_c(1)]\}$$
$$+ Q_{L1}\{R_{L2}[P_g + P_c(2) - P_g \cdot P_c(2)] + Q_{L2}\}.$$

If the two lines are identical, this reduces to

$$Q_s = R_L^2[P_g + P_c(1, 2) - P_g \cdot P_c(1, 2)]$$
$$+ 2R_L Q_L[P_g + P_c(1) - P_g \cdot P_c(1)] + Q_L^2 \quad (2)$$

where

$$R_{L1} = R_{L2} = R_L$$

and

$$Q_{L1} = Q_{L2} = Q_L.$$

The solution for this simple system could have been obtained directly from the binomial expansion of $(R_L + Q_L)^2$. Equation (2) expresses the reliability at the load in terms of the probabilities of adequacy of the generation and transmission facilities.

To illustrate the effect of various system parameters the following system has been analyzed. A hydraulic generating station contains six 40-MW generating units. It is connected to a load 150 miles away by two 230-kV transmission lines. The voltage at the receiving end of the line was assumed to be fixed at 230 kV and any deviation from this, other than transient, was considered to be a breach of quality. The sending-end voltage of the 230-kV line was allowed to vary within limits of 5 percent, i.e., 218.5–241.5 kV.

The load carrying ability of the transmission system can be shown graphically using the conventional circle diagram. A modified graph is shown in Fig. 2, obtained from a digital computer solution using the conventional long-line equations. This graph shows the carrying capacity of a single line with the assigned voltage restrictions. In addition to the 230-kV receiving-end voltage, 225 kV and 220 kV are also shown. For two identical lines the load carrying capability is double the single-line values. The system load was represented by a straight-line load duration curve with a 75 percent load factor. Using this load distribution, the loss of load probabilities at various peak loads for a plant containing six 40-MW hydraulic generating units and for a plant containing seven 40-MW units is shown in Table I.

In each case the unit forced outage rate was assumed to be 0.005. Neglecting the system transmission losses and the effect of transmission reliability, the figures shown in Table I represent the conventional reliability assessment for the system. The probability of system failure was given in (2) as

$$Q_s = R_L^2[P_g + P_c(1, 2) - P_g \cdot P_c(1, 2)]$$
$$+ 2R_L Q_L[P_g + P_c(1) - P_g \cdot P_c(1)] + Q_L^2$$

where

R_L^2 probability of both lines being available

$2R_L Q_L$ probability of one line being available

Q_L^2 probability of no lines being available.

Parallel outdoor facilities such as the two lines in Fig. 1 may be subjected to both normal and stormy weather conditions.

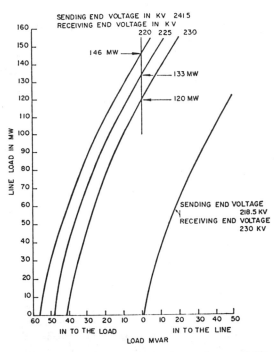

Fig. 2. Transmission system load carrying capability.

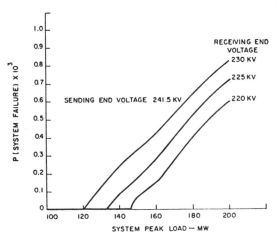

Fig. 3. Probability of system failure as a function of system peak load (for various receiving-end voltages).

TABLE I
GENERATING SYSTEM LOSS OF LOAD PROBABILITIES

Peak Load, MW	Loss of Load Probability
240-MW Capacity 6 40-MW Units	
160	1×10^{-6}
170	44.6×10^{-6}
180	83.0×10^{-6}
190	117.7×10^{-6}
200	148.8×10^{-6}
280-MW Capacity 7 40-MW Units	
200	2×10^{-6}
210	50×10^{-6}
220	95.4×10^{-6}
230	126.5×10^{-6}
240	173.7×10^{-6}

TABLE II
SYSTEM FAILURE PROBABILITIES FOR THE
240-MW CAPACITY SYSTEM

Percent Failures During Storms	Probability of System Failure × 1000 at Peak Load				
	120 MW	140 MW	160 MW	180 MW	200 MW
0	0.0002	0.2448	0.4289	0.6534	0.8332
10	0.0002	0.2448	0.4288	0.6532	0.8331
20	0.0003	0.2448	0.4287	0.6531	0.8329
30	0.0005	0.2448	0.4285	0.6528	0.8325
40	0.0008	0.2448	0.4283	0.6524	0.8314
50	0.0012	0.2447	0.4279	0.6517	0.8310
60	0.0016	0.2446	0.4275	0.6510	0.8301
70	0.0022	0.2445	0.4270	0.6502	0.8291
80	0.0027	0.2445	0.4263	0.6481	0.8276
90	0.0034	0.2444	0.4257	0.6480	0.8262
100	0.0041	0.2448	0.4258	0.6480	0.8260

These effects have been considered using a Markov approach [5] to find the line outage existence probabilities for varying degrees of storm associated failures, assuming that for each line

λ_{av} annual failure rate is 0.5 failure per year
N expected duration of normal weather periods is 200 hours
S expected duration of stormy weather periods is 1.5 hours
R expected repair time is 7.5 hours.

It is interesting to note that while the probability of having both lines out of service increases rapidly as the percentage of failures during storms increases, the probability of one line being available actually decreases slightly over this range. It must be remembered that the annual failure rate is constant and therefore while the stormy failure rate is increasing, the normal weather failure rate is decreasing. With regard to the probability of occupying a given state, the inclusion of the storm condition materially affects only the catastrophic condition of both lines being unavailable. The probability of system failure for various

percentages of storm associated failures are tabulated in Table II for 240 MW of capacity and peak loads of 120 MW to 200 MW in 20-MW increments. The effect of storm associated failures on the probability of system failure is virtually negligible in those cases in which the system is not fully redundant. As noted above, this is due to the fact that the probability of occupying the single-line state is virtually unaffected by the degree of storm associated failure. In the 120-MW peak load case, the probability of system failure is the probability of losing both lines and is governed almost entirely by the degree of storm associated failures.

The assumption was made that 50 percent of the transmission failures for this system occurred during stormy periods. The probability of system failure for 240 MW of capacity as a function of peak load is shown in Fig. 3. As previously noted, system transmission losses have been neglected in evaluating the generating capacity failure probability component. This graph illustrates the change in system failure probability due to the ability to relax the receiving-end conditions from 230 kV–225 kV and 220 kV. Reliability considerations then become part of the economic evaluation of on-load tap changing facilities at the receiving end. These results were obtained assuming a unity power factor load. Similar results at any other power factor could be obtained by finding the correct power carrying capabilities of a single line using Fig. 2.

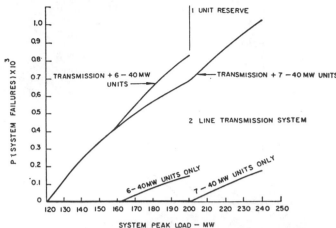

Fig. 4. Probability of system failure as a function of system peak load.

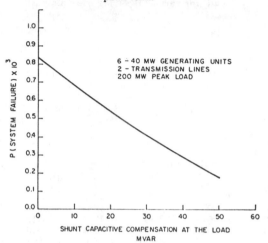

Fig. 5. Probability of system failure as a function of shunt compensation at load point.

If a generating capacity reserve margin of at least one unit is maintained, then neglecting transmission losses, the maximum peak load is 200 MW in the 240-MW capacity case. Fig. 4 shows the variation in the probability of system failure with peak load for the 240-MW and 280-MW installed capacity conditions. It can be clearly seen that for the failure values chosen, the probability of system failure is clearly dominated by the probability of transmission inadequacy.

The load carrying capability of a line with fixed receiving- and sending-end voltages can be increased by the addition of shunt capacitive compensation at the receiving end. Depending, of course, on the system, this can have a considerable effect on the probability of system failure. This condition is shown in Fig. 5 where a maximum of 50 Mvar of capacitance is available at the receiving end. The capacitor installation was assumed to be 100 percent reliable. The load was considered to be at unity power factor. The loading limitations were obtained from Fig. 2. As in the case of on-load tap changing transformers, reliability assessment can be an integral part of the economic evaluation of shunt compensation requirements.

The results obtained from a Markov solution for three parallel facilities were used to assess the reliability of three 230-kV lines. For a peak load of 240 MW there must be at least two lines out of service for there to be any possible transmission curtailment. In this case the probability of system failure is governed almost entirely by the probability of generating capacity inadequacy.

The system considered is an extremely simple one. It illustrates,

however, that reliability evaluation can be an integral part of system planning, design, and operation. The probability of system failure is the probability that the generation and transmission facilities will not satisfy the required load condition. No time period has been considered and as previously illustrated this could be any consistent period for which the load probability distribution is applicable. The generating capacity failure probability was calculated on an installed capacity basis using the loss of load approach. A similar study could have been performed to evaluate spinning requirements though it would be trivial in this simplified case.

TWO-PLANT TWO-LOAD SYSTEM

The same basic approach illustrated in the simple series system case can be applied to any system. Consider the system shown in Fig. 6. In a practical system many of the transmission curtailment probabilities would be unity or zero, thus eliminating many terms. The system shown in Fig. 6 is still an extremely simple one compared to a practical power system. The form of analysis is, however, logical and sequential and therefore amenable to digital computer application. The probabilities of transmission curtailment are extremely important as was illustrated in the simple series system analysis. These probabilities can be obtained by an automated load flow program on a digital computer or by load flow studies on an ac network analyzer using predetermined service quality levels for each bus in the system. Quality of service may not be a matter of voltage only, and tripping due to overload or to steady-state instability can be incorporated into the probabilities of transmission curtailment as the different configurations are analyzed.

In an installed capacity investigation the entire system capacity is used to evaluate the generation component of system failure probability. In a spinning capacity study, the capacity in operation is dependent upon the load level, and the failure probabilities are dependent upon the time required to put additional capacity into service. If the transmission system is assumed to be completely adequate and reliable, it is immaterial where the spinning reserve is located within the system. This is not true, however, if the transmission system is not fully reliable. A transmission adequacy analysis in this case may involve the utilization of several generation schedules for each load condition. In many cases the best schedule is reasonably obvious due to economic constraints, however, it may be offset by reliability benefits.

The probability of system failure for load 1 in Fig. 6 is

$$
\begin{aligned}
Q_s = &\ R_{L1}[R_{L4}(R_{L2}\{R_{L3}[P_g(1,2) + P_c(\text{all}) - P_g(1,2) \cdot P_c(\text{all})] \\
&+ Q_{L3}[P_g(1,2) + P_c(1) - P_g(1,2) \cdot P_c(1)]\} \\
&+ Q_{L2}\{R_{L3}[P_g(1,2) + P_c(2) - P_g(1,2) \cdot P_c(2)] \\
&+ Q_{L3}[P_g(1) + P_c(3) - P_g(1) \cdot P_c(3)]\}) \\
&+ Q_{L4}(R_{L3}\{R_{L2}[P_g(1,2) + P_c(4) - P_g(1,2) \cdot P_c(4)] \\
&+ Q_{L2}[P_g(1,2) + P_c(5) - P_g(1,2) \cdot P_c(5)]\} \\
&+ Q_{L3}[P_g(1) + P_c(6) - P_g(1) \cdot P_c(6)])] \\
&+ Q_{L1}[R_{L4}(R_{L2}\{R_{L3}[P_g(1,2) + P_c(7) - P_g(1,2) \cdot P_c(7)] \\
&+ Q_{L3}[P_g(2) + P_c(8) - P_g(2) \cdot P_c(8)]\} + Q_{L2}) + Q_{L4}].
\end{aligned}
$$

A similar expression can be obtained for load 2. The transmission curtailment configurations are shown in Fig. 7. Certain configurations do not require load flow analysis as the load is completely isolated. The conditional probability of load point failure

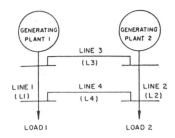

Fig. 6. Two-plant two-load system configuration.

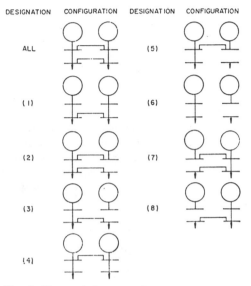

Fig. 7. Transmission curtailment configurations.

in these cases is therefore unity. An actual system with this configuration would not normally be symmetrical and therefore the probability of system failure at the two load points would not be identical. The minimum reliability acceptable to the system is a management decision.

Operating considerations can be easily included in the system adequacy analysis. In certain cases loads may be tripped off the system rather than allow other load points to experience undesirable voltage levels. These conditions can all be incorporated into the analysis of any particular configuration. It should be realized that the addition of shunt compensation or on-load tap changing transformers in general affects the reliability of the entire networked system rather than just the point of application. The effect, of course, may be negligible at points removed from the immediate locale. Maintenance considerations can be included in the analysis by placing the probability of component availability to zero and the probability of component unavailability to unity.

As previously noted in the analysis of the simple series system configuration, the effect of storm associated failures on the system failure probabilities depends almost entirely on the degree of redundancy. This effect will be further diminished in the evaluation of major transmission reliability over relatively long distances as storms become more local in nature.

CONCLUSIONS

Using the techniques described it is possible to arrive at a measure of steady-state adequacy for any point in a system, particularly those points at which major transmission terminates and subtransmission begins. For an actual system involving many components it may not be necessary to consider all the possible cases. The digital computer program can be instructed

to consider up to a fixed number of components out of service at any one time. The maximum number will depend upon the component probabilities. Further work is presently being done at the University of Saskatchewan, Saskatoon, Canada, in the application of these concepts to more complicated systems.

REFERENCES

[1] R. Billinton, "Bibliography on application of probability methods in the evaluation of generating capacity requirements," presented at the IEEE Winter Power Meeting, New York, N. Y., January 30–February 4, 1966.
[2] R. Billinton, "Transmission system reliability evaluation," *Trans. Canadian Elec. Assoc.*, vol. 6, 1967.
[3] C. F. DeSieno and L. L. Stine, "A probability method for determining the reliability of electric power systems," *IEEE Trans. Power Apparatus and Systems*, vol. 83, pp. 174–181, February 1964.
[4] I. Bazovski, *Reliability Theory and Practice*. Englewood Cliffs, N. J.: Prentice-Hall, 1961.
[5] R. Billinton and K. E. Bollinger, "Transmission system reliability evaluation using Markov processes," *IEEE Trans. Power Apparatus and Systems*, vol. PAS-87, pp. 538–547, February 1968.

Discussion

R. J. Ringlee and **A. J. Wood** (General Electric Company, Schenectady, N. Y.): This refreshingly brief, yet penetrating probe of a significant contemporary problem in power system analysis deserves careful attention. The author has offered both a criterion and an approach to the analysis of generation–bulk transmission reliability.

Manuscript received February 12, 1968.

It is believed that considerable insight for extending generation–bulk transmission reliability models can be gained from the author's results. With two well-chosen examples, the author has illustrated the reliability gain with shunt compensation, Fig. 5, and the negligible effect of the independent two-state environmental model, Table II.

Hoping that the author may wish to offer additional comments on ways to recognize maintenance considerations in the reliability model, we offer the following question and observation. Would it be desirable to include recognition of maintenance outage effects in 1) and 2) of the design criterion? The author's suggestion for including maintenance considerations appears to cover the scheduled outage events and incidents of random outages during scheduled outages. The random maintenance model appears to require an additional parameter, the probability that the component is in the maintenance outage state.

Our sincerest compliments accompany this discussion.

Roy Billinton: I would like to thank Dr. Ringlee and Dr. Wood for their comments. It was intended that maintenance outage effects would be included in the design criterion, as noted by the discussers, in terms of random behavior occurring during scheduled outages. This presumably would not affect 1) of the criterion, if maintenance can be scheduled in a relatively light load period. Part 2) of the criterion would then be determined by considering the various scheduled network configurations and the corresponding load models. The inclusion of a random maintenance model introduces additional complexity to the problem. It also requires the assumption that the model is truly random, which may not be entirely valid. In regard to the initial Markov model for two lines in parallel, the random maintenance model would presumably be added only to the normal weather side as it is unlikely that maintenance would be conducted if stormy weather is likely to exist. This is an interesting area of investigation though probably only of academic interest. I rather doubt that the utilization of a random maintenance parameter is valid. I would prefer to think of maintenance effects in terms of specific configuration and applicable load models in a similar manner to that conventionally used in a generating capacity study. We have developed a digital computer program to analyze a relatively complicated system using the approach outlined in this paper. It is expected that maintenance considerations will be included in these studies. In conclusion, I would again like to thank the discussers.

Manuscript received March 20, 1968.

Transmission Planning Using a Reliability Criterion, Part I: A Reliability Criterion

ROY BILLINTON, MEMBER, IEEE, AND MURTY P. BHAVARAJU, MEMBER, IEEE

Abstract—The utilization of a quantitative reliability criterion in long range transmission planning is proposed. The application of a conditional probability approach to the determination of load point reliability indices in practical systems is discussed and illustrated by a simple hypothetical system study. Failure at a bus is defined by three planning criteria and the reliability level obtained in terms of a probability and an expected frequency of bus failure. The method requires ac load flow analysis at several load levels under possible system component outages. The technique is quite general, and any known operating conditions can be included. Planning based on acceptable load bus reliability levels results in optimum utilization of the investment placed in transmission facilities.

Introduction

THE ADEQUACY of major transmission planning is normally tested by considering the effects on the system of selected critically loaded component outages [1], [2]. The actual outages considered are determined by the system planner and are influenced by previous system experience. A method of evaluating the reliability at any point in a composite system including both generation and transmission facilities was presented in a recent paper [3]. This publication proposed the use of a service quality standard as the reliability criterion rather than simple continuity between sources and load points. A general design criterion was postulated in which "if with the various possible combinations of system components out of service the reliability index for the station is below an acceptable minimum, then additional facilities are required to meet the reliability standards." The effect of failures during storms, acceptable bus voltage levels, shunt capacitive compensation, and transmission redundancy were illustrated for a simple system [3]. The basic technique for analyzing a more complicated network was also illustrated. A digital-computer program for logical transmission planning has been developed using the previously noted design criterion. This work breaks down into two separate phases, the evaluation of the reliability index in a practical system network, and the logical addition of subsequent facilities. This paper describes the digital-computer program developed to determine the reliability index at any point in a system and discusses the necessary assumptions and the results obtained for a practical configuration.

Conditional Probability Approach

In a power system network there are a number of possible outage combinations of lines, transformers, and generating units. Each outage condition has a probability of existence and a frequency of occurrence. Under each outage condition there is a maximum load at each bus that can be supplied without violating the service quality criterion. The probability that the load will exceed this maximum can be determined from the load probability distribution for the bus in question. This is a conditional probability as the maximum load is determined given that a certain outage condition has occurred. In [3] the system generating facilities were included by developing a capacity outage probability table for all the units within the system. The maximum load that can be supplied at a bus can be obtained for any given outage condition in the transmission system. The probability of the load at the bus exceeding this maximum value is then combined with the probability that the available system generating capacity is insufficient to meet the total system load. Generation and transmission outage conditions are considered as two independent events resulting in failure at a bus:

probability of failure at bus k

$$= \sum_j [P(B_j)(PG_j + PL_j - PG_j PL_j)] \quad (1)$$

where

B_j outage condition in the transmission network (lines and transformers)

$P(B_j)$ probability of existence of outage B_j

PG_j probability of the generating capacity outage exceeding the reserve capacity (a cumulative probability figure obtained from the capacity outage probability table)

PL_j probability of load at bus k exceeding the maximum load that can be supplied at that bus without failure.

The probability of existence of outage B_j is obtained by assuming that the individual component outages are independent. It has been shown previously [3] that the effect of storm-associated failures on the probability of system failure is virtually negligible in those cases in which there is a reasonable probability that the system load will exceed the remaining transmission capability. The effect of storms would be further diminished in major transmission reliability evaluation over relatively long distances as storms become more local in nature.

It is also possible to determine a reliability index at any bus in the system in terms of an average or expected frequency of failure. The frequency of occurrence of an outage condition is equal to the product of the probability of existence of the outage and the rate of departure from that condition [4]. If the generating unit outages and the load variation are considered in terms of probability only and not in terms of frequency of occurrence, then the expected frequency of failure at bus k is given by

$$\sum_j [F(B_j)(PG_j + PL_j - PG_j PL_j)] \quad (2)$$

Paper 69 TP 121-PWR, recommended and approved by the Power System Engineering Committee of the IEEE Power Group for presentation at the IEEE Winter Power Meeting, New York, N. Y., January 26–31, 1969. Manuscript submitted September 3, 1968; made available for printing November 19, 1968.

The authors are with the Power System Research Group, University of Saskatchewan, Saskatoon, Sask., Canada.

Reprinted from *IEEE Trans. Power Apparatus Syst.*, vol. PAS-89, no. 1, pp. 28–34, Jan. 1970.

where $F(B_j)$ is the frequency of occurrence of outage B_j. If the rates of departure associated with the individual generating unit states and the load model states for each bus in the system are included, the evaluation becomes extremely complicated. This was shown in a recent publication in which system capacity and system load models were combined [5].

In the approach described by (1), the generation schedule used in the load-flow analysis is not modified to include the outage of individual units. The assumption is made that adequate generation is available, and that any breach of quality is due to line or transformer outages or to the total available generation exceeding the system load. This assumption is not required if the generating units are considered individually together with the transmission lines and transformers to determine each outage condition B_j. The generation schedule is then modified for each generation outage condition. The number of individual outage conditions in this case may be much greater than those considered using (1). This approach is, however, more accurate as the bus voltage and line loadings are affected by the generation schedule.

The equations for this case are as follows:

probability of failure at bus k $\qquad = \sum_j [P(B_j)\, PL_j]$ (3)

expected frequency of failure at bus k $\qquad = \sum_j [F(B_j)\, PL_j].$ (4)

In (4) as in (2) the rates of departure associated with the load model states are not included in the evaluation.

COMPUTER PROGRAM

The determination of the maximum load that can be supplied without failure at each load bus requires several basic assumptions together with the necessary system data. The required system data are listed in the Appendix. It was assumed in this program that the load variation at each bus was represented by a normalized load duration curve, approximated by a single straight line. The individual bus load levels used in each load-flow analysis are those corresponding to a specified probability of exceedance.

A load bus is assumed to be failed under the following conditions:

1) the voltage at the bus is less than a specified minimum value (not meeting the quality standards at the bus);
2) a line or transformer supplying power to the bus is overloaded;
3) the generating capacity required to meet the total load exceeds the available capacity (this includes transmission losses).

A computer program has been developed to create the possible outage conditions and to perform a load-flow analysis for each outage condition and specified load level. Under each outage condition B_j, if a load bus fails at any of the increasing load levels, PL in (1) is taken as the average of the probability value of the load level at which the load bus failed and the previous lower level. PL is zero if the load bus does not fail even at the peak load level. PG in (1) is the cumulative probability of the capacity outage in the system exceeding the reserve.

If a load bus is isolated from the network due to some outage condition, PL is equal to 1 for that bus. If a generating bus is isolated, the capacity outage probability table for the system is modified by removing [6] the generating units at the isolated bus before using it for computing PG. If there is a load at the

isolated generating bus, the contribution to the bus risk is computed by combining the load and the generating capacity models at the bus as in a normal loss of load probability study. If the isolated generating bus is the swing bus, another bus is selected for this purpose.

If the load-flow solution does not converge under an outage condition within a specified maximum number of iterations, the system is divided into subsystems, with each generating bus supplying one or more neighboring loads. The risk contribution is then calculated for each bus using the capacity outage probability table at the generating bus, and a combined load duration curve is calculated for the loads being supplied.

If a line or transformer is overloaded at any load level it is assumed to be tripped by the protective equipment and therefore removed from the network. The probability steps to increase the load level are specified in the initial data. The maximum load level for any given outage condition can be determined more accurately by increasing the number of steps; however, the computation time also increases. It was observed that sufficiently accurate results can be obtained by considering up to a maximum of two simultaneous independent outages. Outages of higher order contribute negligible quantities to the total risk.

The risk evaluation is performed only if the base case load-flow solution with all the system components in service is satisfactory. The system load-flow solution is obtained using the conventional Gauss–Seidel method.

The system loads can be represented by either constant power or constant admittance models. The loads can also be represented by constant power models, and if the voltage falls below a specified minimum, the model is modified to a constant admittance representation. A constant admittance representation increases the diagonal dominance of the nodal admittance matrix and results in a significant saving in solution time.

SYSTEM STUDY

The system shown in Fig. 1 was studied to compare the two approaches given by (1) and (3) and also to investigate the effect of the number of load probability steps, the load representation, the bus voltage limits and the addition of selected lines. The data used for this study are shown in the Appendix. Table I shows the risk levels obtained in terms of the probability of failure and the expected frequency of failure for loads at buses 2–5. Case 1 of Table I can be considered as the base case. Equations (1) and (2) have been used and the loads represented as constant admittances. A maximum of two simultaneous independent outages were considered for the system consisting of lines 1–6. The load model represented by a straight line from the 100- to the 40-percent load points was approximated by 10 equal probability steps. The minimum acceptable system voltage was 0.97 pu, and the maximum generation voltage was 1.05 pu.

The results of case 2 were obtained using (3) and (4). The risk at bus 2 was found to be extremely low using this approach. This is partially due to the use of a maximum of two simultaneous independent outages. The risk at bus 2 in case 1 is basically a system generation contribution rather than an actual failure at bus 2. This effect is eliminated in case 2 in which the complete system capacity outage probability table is not used. Case 3 shows the increase in risk at buses 3 and 5 when the system load models are changed to a constant power representation. The effect of modifying the minimum acceptable voltage levels are shown in cases 4a–h and are plotted in Fig. 2. The

TABLE I

RISK LEVELS FOR LOADS IN THE SYSTEM SHOWN IN FIG. 1

Case	Changes from the normal case	Bus 2 Probability	Bus 2 Expected Frequency	Bus 3 Probability	Bus 3 Expected Frequency	Bus 4 Probability	Bus 4 Expected Frequency	Bus 5 Probability	Bus 5 Expected Frequency
1	Normal case	0.000 062	0.0013	0.000 937	0.8060	0.001 729	1.5147	0.002 075	1.8335
2	Generating unit outages considered individually (3) and (4)	0.000 000	0.0000	0.000 866	0.8114	0.001 654	1.5315	0.001 876	1.7335
3	Load representation as constant power	0.000 063	0.0025	0 001 500	1.3067	0.001 730	1.5156	0.002 638	2.3332
4a	Minimum acceptable voltage = 1.02 pu	0.000 062	0.0016	0.004 150	3.6707	0.006 236	5.5331	0.005 622	4.9929
4b	Minimum acceptable voltage = 1.01 pu	0.000 062	0.0015	0.002 910	2.5652	0.003 872	3.4272	0.004 047	3.5900
4c	Minimum acceptable voltage = 1.00 pu	0.000 062	0.0015	0.001 727	1.5111	0.002 747	2.4239	0.002 864	2.5364
4d	Minimum acceptable voltage = 0.99 pu	0.000 062	0.0013	0.001 727	1.5096	0.001 732	1.5195	0.002 864	2.5361
4e	Minimum acceptable voltage = 0.95 pu	0.000 062	0.0013	0.000 376	0.3082	0.000 942	0.8142	0.001 514	1.3364
4f	Minimum acceptable voltage = 0.93 pu	0.000 062	0.0013	0.000 092	0.0544	0.000 377	0.3112	0.001 231	1.0825
4g	Minimum acceptable voltage = 0.92 pu	0.000 062	0.0013	0.000 090	0.0505	0.000 097	0.0626	0.001 229	1.0787
4h	Minimum acceptable voltage = 0.91 pu	0.000 062	0.0013	0.000 090	0.0505	0.000 097	0.0587	0.001 228	1.0787
5	Line 6 removed	0.000 064	0.0043	0.005 108	4.5116	0.005 561	4.9148	0.006 242	5.5278
6	Line 7 added	0.000 061	0.0022	0.000 069	0.0155	0.000 069	0.0169	0.001 207	1.0552
7	Line 7 and 8 added	0.000 061	0.0024	0.000 067	0.0126	0.000 067	0.0121	0.000 068	0.0149
8	Generating capacity 100-percent reliable	0.0	0.0	0.000 875	0.8040	0.001 667	1.5128	0.002 013	1.8315
9	Same as case 8 all loads increased by 10 percent	0.0	0.0	0.001 439	1.3070	0.001 669	1.5164	0.002 576	2.3334
10	Same as case 9 line 8 added	0.0	0.0	0.000 877	0.8095	0.001 440	1.3118	0.000 881	0.8165
11	Single outages only considered	0.000 062	0.0013	0.000 903	0.7470	0.001 689	1.4432	0.002 025	1.7458
12	Load distribution approximated to 5 steps	0.000 062	0.0013	0.000 657	0.5586	0.002 233	1.9598	0.001 795	1.5856

Normal Case: Lines 1 to 6 in service; up to two simultaneous independent outages considered; load distribution approximated to 10 steps; loads represented as constant admittances; minimum acceptable voltage at all buses = 0.97 pu; maximum allowed voltage at generating buses = 1.05 pu; equations (1) and (2) are used.

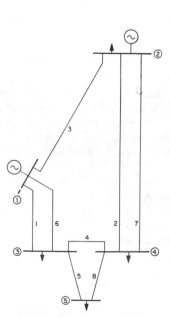

Fig. 1. System studied for composite system reliability evaluation.

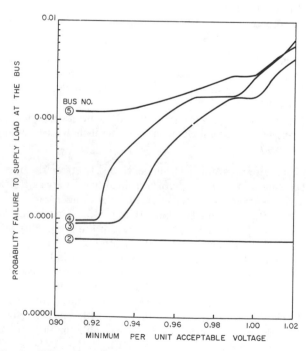

Fig. 2. Effect of minimum acceptable bus voltage on risk level.

TABLE II

GENERATION DATA

Bus	Number of Units	Capacity of Each Unit (MW)	Total Bus Capacity (MW)	Type of Units	Failure Rate Per Unit (failures per year)	Repair Rate Per Unit (repairs per year)	Probability of Outage
1*	4	20	80	thermal	1.1	73	0.015
2	7	5	130	hydro	0.5	100	0.005
	1	15		hydro	0.5	100	0.005
	4	20		hydro	0.5	100	0.005

* Swing bus. If bus 1 is isolated from the network due to an outage condition bus 2 is selected as swing bus.

TABLE III

LOAD DATA

Bus	Peak Load (MW)	Power Factor	Generation Allotted under Peak Load, (MW)	VAR Limits (Mvar)	Voltage Limits (pu) Maximum	Minimum
1	0		swing bus	−10 to +10	1.05	0.97
2	20	1.0	110	0 to 40	1.05	0.97
3	85	1.0			1.05	0.97
4	40	1.0			1.05	0.97
5	10	1.0			1.05	0.97

Load-probability steps: 1.00 0.80 0.60 0.40 0.20 0.0 ——(5 steps)
1.00 0.90 0.80 0.70 0.60 0.50⎫
0.40 0.30 0.20 0.10 0.0 ⎬——(10 steps)

TABLE IV

LINE DATA

Line	Length (miles)	Impedance (pu)	Susceptance $(b/2)$	Failure Rate	Probability of Failure
1,6	30	$0.0342 + j0.1800$	0.0106	1.5	0.001 713
2,7	100	$0.1140 + j0.6000$	0.0352	5.0	0.005 710
3	80	$0.0912 + j0.4800$	0.0282	4.0	0.004 568
4,5,8	20	$0.0228 + j0.1200$	0.0071	1.0	0.001 142

Tolerance for load-flow solution = 0.0001.
Maximum number of iterations for convergence = 200.
Maximum number of simultaneous independent outages of any combination considered = 2.

effect of removing a line and adding additional lines are shown in cases 5–7.

In case 8 the generating capacity was assumed to be completely reliable. The risk at bus 2 is therefore zero in this case. The risk levels at other buses in the system are reasonably close to those obtained in case 1. Case 9 shows the effect of increasing the system loads. If the increase in risk violates the system design criterion, then the system must be modified to reduce the risk to an acceptable value. This concept is the basis of the logical transmission planning program to be described [7]. The assumption of complete generation reliability may be necessary in certain cases to eliminate the effect of the generation risk on the total risk upon which logical transmission planning is based.

The results obtained by considering only single independent outages as shown in case 11 are very close to those obtained in case 1. This approach results in a significant saving in computer solution time over that required in case 1. The bus risk levels using this approach in a system designed for single contingency outages would be approximately equal to the values obtained by considering the generation adequacy only.

Case 12 shows the effect on the calculated values of using a five-step representation of the load characteristics. The results are significantly different than those obtained in case 1. A reduction in the number of steps may be required in a large system to reduce the computation time to an acceptable value.

Preliminary studies on a reduced major transmission network of the Saskatchewan Power Corporation have indicated that there is relatively little difference between the results obtained using single independent outages and two simultaneous independent outages. The system consisted of 230- and 138-kV components with 15 buses, 21 lines, and 4 transformers. The relative computer solution time, however, was in the order of 13 to 1 for two outages and for single outages.

CONCLUSION

A digital-computer program has been developed for the planning of transmission facilities using a probability approach to the evaluation of a reliability criterion. The paper describes the subprogram developed for composite system reliability evaluation of practical networked systems using the conditional probability approach [3]. The risk levels obtained at the various system load points in terms of probability of failure and the expected frequency of failure give a consistent measure of transmission adequacy.

The reliability evaluation performs ac load-flow solutions of the system at specified load levels under all possible outage conditions up to a maximum of two simultaneous independent outages. Line and transformer overloads due to these outage conditions are included in the analysis by removing the overloaded component from the network.

The reliability evaluation of single systems has shown that the risk levels calculated by the two different methods of considering generation inadequacy are not very different. The risk levels obtained by considering only single outages were found to be reasonably close to those obtained by considering two simultaneous independent outages. Considerable saving can be obtained in large systems using single outages only and also by reducing the load probability steps to an acceptable number. The effect on the bus risk levels of acceptable voltage limits, the load representation, and the removal or addition of lines are shown in the paper. The computation time required to obtain the reliability indices for each bus in a system is a function of the technique used. The variation in calculated risk with selected changes in the approach are shown in Table I. It is believed that the use of any one of these techniques will lead to more consistent transmission planning than that obtained by the use of rule-of-thumb reliability criteria.

APPENDIX

The data required for composite system reliability evaluation are given as follows:

1) all the data required for load-flow analysis,

2) var limits at generating buses and voltage limits at all buses,

3) maximum current carrying capability of lines and transformers,

4) the distribution of the system load at various buses and the normalized load duration curves,

5) the generation schedule at the increasing load levels being considered in the method (this is approximated to a straight-line variation),

6) failure rate and repair rate of the generating units, lines, and transformers in the system.

The data for evaluating the reliability of the system, as shown in Fig. 1, are given in Tables II–IV. In Tables II and III generation and load data are given. Load and generation variation is approximated to a straight line joining the peaks at zero probability value and 0.4 peak at unity probability value. The base MVA = 100 and the base kV = 110. In Table IV line data is given. Lines are assumed to be 795 ACSR 54/7. The current-carrying capability = 900 amperes = 1.71 pu. The failure rate = 0.05 failures per year per mile. The expected repair duration = 10 hours.

REFERENCES

[1] J. L. Whysong, R. Uran, H. E. Brown, C. W. King, and C. A. DeSalvo, "Computer program for automatic transmission planning," AIEE Trans (Power Apparatus and Systems), vol. 81, pp. 774–781, 1962 (February 1963 sec.).

[2] C. A. DeSalvo and H. L. Smith, "Automatic transmission planning with ac load flow and incremental transmission loss evaluation," IEEE Trans. Power Apparatus and Systems, vol. PAS-84, pp. 156–163, February 1965.

[3] R. Billinton, "Composite system reliability evaluation," IEEE Trans. Power Apparatus and Systems, vol. PAS-88, pp. 276–281, April 1969.

[4] J. D. Hall, R. J. Ringlee, and A. J. Wood, "Frequency and duration methods for power system reliability calculations, pt. I: generation system model," IEEE Trans. Power Apparatus and Systems, vol. PAS-87, pp. 1787–1796, September 1968.

[5] R. J. Ringlee and A. J. Wood, "Frequency and duration methods for power system reliability calculations, pt. II: demand model and capacity reserve model," IEEE Trans. Power Apparatus and Systems, vol. PAS-88, pp. 375–388, April 1969.

[6] R. Billinton and M. P. Bhavaraju, "Generating capacity reliability evaluation," Trans. Engrg. Inst. Canada, vol. 10, October 1967.

[7] M. P. Bhavaraju and R. Billinton, "Transmission planning using a reliability criterion, pt. II: transmission planning," IEEE Trans. Power Apparatus and Systems (submitted for publication).

Discussion

W. S. Ku and **V. Thomas Sulzberger** (Public Service Electric and Gas Company, Newark, N. J.): The authors should be complimented for their work in the field of power system reliability. Their paper presents an interesting proposal for combining generation and transmission outage conditions in evaluating the service quality at any point in a bulk-power system. This proposal is certainly worth further development and investigation. However, we feel some additional explanation and clarification of several of the assumptions and mathematical approaches are required in order to fully understand and evaluate the authors' reliability calculation method. These areas are as follows:

1) We presume that the calculation of the probability that the available system generating capacity is insufficient to meet the total system load is determined from a capacity outage probability table without regard to the transmission system limitations. If the probability of the generating capacity shortage PG_j is so determined without consideration of system transmission conditions, then it is doubtful whether relating the probability of the existence of a transmission outage PB_j with PG_j in evaluating the reliability at any point in a power system, as indicated in (1), is valid. The supply to a bulk-power system load bus is subject to failure for any one of the following conditions:

a) As a result of multiple-generator overlapping outages, the total available system generating capacity, including available capacity backup from neighboring systems, is less than the total system load. Automatic or manual selective load dumping or system-wide load reduction by means of voltage reduction then becomes necessary.

b) As a result of certain overlapping generator outages or overlapping generator and transmission outages, transmission system components for supplying certain load buses become overloaded. Even though there is enough generating capacity available in the system to supply the total system load, no transmission relief can be provided other than curtailing loads at certain buses.

c) Due to overlapping transmission and/or transformer outages, a certain amount of generation is "bottled," resulting in loss of load even though the generating capacity outage probability table would indicate an adequate capacity margin.

d) Due to overlapping transmission and/or transformer outages, transmission system components for supplying certain load buses become overloaded. Load curtailment is needed to relieve overloaded facilities. It does not appear that the authors' method adequately covers the previously-described conditions in evaluating the service quality of a bulk-system load bus.

2) The authors' do not mention whether generator and transmission system maintenance or scheduled outages have been taken into consideration in evaluating the service quality. In particular, what assumptions have been made pertaining to component scheduled outages which are usually performed only during certain times in the day, during certain load levels, and in certain seasons?

Manuscript received January 31, 1969.

Our own system reliability evaluations have indicated that the effect of overlapping scheduled and forced outages on system reliability is more significant than that of two overlapping forced outages because the frequency of occurrence of component scheduled outages is much higher than the frequency of forced outages.

3) The authors' results indicate that for their sample system single- and double-contingency outages have very nearly the same consequences on system reliability as only single-contingency considerations. We do not feel that this will be the experience in reliability evaluations of an actual large-system network. Single contingencies generally do not result in any direct risk of loss of load. In fact, most systems are planned to withstand certain double contingencies. For risk evaluation, it is therefore essential to consider the consequences of multiple overlapping outages since the consideration of single contingencies only is somewhat meaningless.

4) In evaluating the voltage quality criterion, the authors apparently consider the load-power factor of the bus being fixed for various transmission outages and voltage conditions. In a real system such is usually not the case. The net Mvar load supplied by the bus could vary for different system conditions depending on the reactive capacity status in the area supplied by the bus.

5) The reliability indices should include the duration of the expected failure and magnitude of the load affected in addition to the average or expected frequency of the failure. It is not clear whether these factors have been taken into account in the authors' example.

The authors' clarifications and comments on these areas will be appreciated.

V. Burtnyk, J. H. Cronin, and R. L. Schmid (Westinghouse Electric Corporation, East Pittsburgh, Pa.): The authors are to be complimented on their continued efforts toward the development of a quantitative measure of transmission system reliability. Such a measure would provide planners with an additional tool for evaluating alternate transmission plans on a more consistent basis similar to the technique now widely used in generating capacity planning studies.

The authors have applied their analysis to a simplified generation–transmission system and have illustrated the effect of various assumptions, criteria, and calculation methods. One conclusion they reach is that the generating system can be considered independently from the transmission system [as per (1)] with only a small error as compared to the more accurate but more involved method (3). Do the authors feel that the effect would be similar for all systems, particularly larger systems with more widely distributed generating stations and larger units?

It is indicated in the paper that computation time increases rapidly as the number of the simultaneous outages considered is increased. To keep within practical limits, the authors suggest that not more than two simultaneous outages need be considered and that "considerable savings can be obtained in large systems using single outages only . . . ." The effect on accuracy is demonstrated to be small for a six-line transmission network. We question whether such a general conclusion can be drawn from this simplified example. On a six-element system, neglecting three or even two simultaneous outages may be justified. However, on a larger system, the probability of simultaneous outages is much greater, and it would appear that the larger the system, the greater the number of simultaneous outages that must be considered.

In addition, the number of possible outage conditions in the load-flow analysis would increase dramatically, due to both system size and the necessity of considering a larger number of simultaneous outages. We would appreciate the authors' comments on the maximum system size that can be handled by this technique and on any possibilities they foresee in overcoming some of these size barriers.

We would also appreciate additional clarification concerning the method of handling the load and the necessity of incorporating a load duration curve in the analysis. It appears that the entire system generating capacity is assumed to be available during light load periods. If this is so, the probability of failure during light load periods should be low enough to be neglected. Planning studies are usually based on peak load conditions only as long as sufficient off-peak periods are available to perform the required planned maintenance without degrading system reliability. It would appear that the calculations would be simplified if peak load conditions only were considered.

It is recognized that systems can fail for reasons other than random independent line and generator failures, and by means other than low voltage, overload, and insufficient generating capacity included in the paper. Bus faults, multiple line failures due to single incidents, failure of protection equipment to operate properly, and system instability may also result in system failure. Do the authors feel that these additional factors contribute very little to system risk levels or do they plan to include them in some of their future work? Do they plan to include aid from interconnections?

We think the authors have made a worthwhile contribution to the field of power system reliability. Hopefully, programs such as this will serve as guidelines for gathering outage data on components and will make possible the establishment of acceptable levels of system performance.

R. J. Ringlee (General Electric Company, Schenectady, N. Y.): This work represents a substantial step toward the realization of a quantitative reliability criterion for transmission planning. The three criteria proposed by the authors are certainly necessary for a bulk-power supply reliability criteria. One would like to add stability as a fourth criterion when reasonable means are found to evaluate it for all the contingencies. This writer supports the contingency approach employed by the author and wishes to note that expressions for event probability (1) and for event frequency (2) are quite general. In their examples, the authors have illustrated the calculations for overlapping outages of circuits arising from statistically independent causes. The probability and frequency expressions (1) and (2) are equally applicable to the dependent events such as circuit forced outage during maintenance on other circuits in the system and multiple-circuit outages due to common causes arising, for example, in the substations.

Manuscript received February 12, 1969.

Roy Billinton and Murty P. Bhavaraju: The authors are grateful for the interesting and detailed discussions. We feel that the basic method proposed is capable of including any of the operating details usually known to the planning engineer and pertaining to his particular system. Conditions 1a), b), and d) and 4) of the discussion by Mr. Ku and Mr. Sulzberger illustrate such possible operating conditions. The object of the digital-computer program developed by the authors was only to illustrate the basic approach, and therefore all the possible operating details were not included.

As noted, (1) and (2) do not consider the effect of individual generating unit outages on voltage levels and therefore are not exact. Equations (3) and (4), however, include these outages and provide a more accurate approach. The error involved in the approach using (1) and (2) depends on the generating capacity reserve, the distribution of the capacity at various points in the system, and the individual generating unit sizes. The error should be examined for a system of any size before applying (1) and (2). If applicable, however, the utilization of these equations results in a significant saving in computer solution time. Condition 1c) of the discussion by Mr. Ku and Mr. Sulzberger is automatically considered by (3) and (4).

Maintenance can be considered by dividing the period of study into intervals and evaluating the reliability levels for the network existing during the interval. Risk levels in each interval should be examined separately, as combining the weighted values in the various intervals into an annual risk tends to overlook short intervals with high-risk levels.

The duration of an outage state can be weighted by the probability of bus failure in that state to determine an expected duration

Manuscript received February 7, 1969.

Manuscript received March 24, 1969.

similar to that given for expected frequency. This approach, however, does not consider chronological load variation.

In smaller systems which are not designed for adequacy under all possible single contingency outages, the risk levels obtained by single outages alone will be high at some load points, and any further contribution of two simultaneous independent outages to these risk levels will generally be negligible. The maximum number of independent outages which should be considered in any study depends on the individual system. It may be necessary to consider up to three independent outages or at least all those whose probabilities of occurrence exceed some selected value. The basic approach has no limitations in regard to the inclusion of multiple outage conditions. The computer solution time will of course increase as the number of multiple contingencies increases. As suggested by Mr. Butnyk, Mr. Cronin, and Mr. Schmid, if it is known that the risk contribution under light load conditions is negligible, then a reduction in the number of load probability steps in this light load region will reduce the solution time to some extent. The authors cannot see any theoretical limitations on the system size that can be handled by this approach. System experience may be used to simplify the network by the elimination of components, which results in negligible error in the final reliability indices.

Assistance from interconnections can be represented in the system network as a generating unit with an equivalent capacity and an outage rate or by representing the interconnected system in more detail. Bus faults, failure of protective equipment, and system instability may have considerable effect on bus voltage levels even though the condition is of a transient nature in many cases. The approach given, has been postulated as a measure of the steady-state adequacy of the various system facilities. If, however, sufficient data are available to determine the probability of various system disturbances, these factors can be included utilizing available system analysis techniques. The addition of stability as a fourth criterion as suggested by Dr. Ringlee is a valuable addition especially in short-range planning studies. He also emphasized a point that we feel is extremely important. The method is quite general and any known operation conditions in the system under study can be included in the analysis.

In conclusion, we would again like to thank the discussers for their valuable comments and constructive suggestions.

QUANTITATIVE EVALUATION OF POWER SYSTEM RELIABILITY IN PLANNING STUDIES

Pier L. Noferi — Luigi Paris
Ente Nazionale per l'Energia Elettrica (ENEL)
Italy

Abstract—In considering the problem of quantitative evaluation of the reliability of a power system for planning purposes, the authors show the advisability of making use of a couple of risk indices in terms of curtailed energy and abruptly disconnected power.

The authors then show that these indices can be evaluated by means of computers, using direct simulation methods (Montecarlo). Numerical examples confirm the applicability of the method and make it possible to come to certain conclusions; both as regards utilization of the results to obtain indications for the improvement of system reliability, and as regards methods to determine the capability limits of lines during the planning studies.

INDICES OF RISK

When considering the problem of the determination of risk in a power system that includes a transmission network, the first problem that arises is what risk indices to assume. The risk indices normally assumed in the case of power systems composed of generators directly connected to the load (i.e. systems where transmission is supposed to be infinitely safe and of sufficient capacity) are well known. They are: a) Probability of being unable to meet annual peak load [1]; b) Annual expected value of being unable to meet daily peak load [2]; c) Annual expected value of being unable to meet each hour's load; d) Expected value of energy unavailable in a year [3]; e) A linear combination of the indices mentioned above [4].

In every case these indices derive from the comparison at a given moment between the available power and the load, the available power being estimated on the basis of availability for service of the individual elements in the power system. We shall call these "the indices of risk due to unavailability".

Often, use is also made of the adoption of a double index of risk [5], taking into account not only the probability of being unable to meet the load, but also of the average probable time-interval between two consecutive events, which reduce the power available below that required. However, we consider that this information is not of sufficient importance to justify the more complicated calculations that it involves especially when one intends, as we do, to deal with more complex systems.

The mechanism through which, in practice, the corresponding reduction in load is achieved is, however, never taken into account with the indices of risk due to unavailability.

This reduction may occur in a planned way through the dispatcher who will vary the load to take account of the unavailability, or it can occur in an abrupt way at the moment in which a change in availability in the system occurs (outage of a generator, fault on a line, etc.). In this latter case, the load reduction may be significantly greater than that corresponding to the simple initial change in availability (Fig. 1a). The extremest form of this phenomenon is black-out (Fig. 1b).

It may be, therefore, that two systems having the same index of risk due to unavailability, mentioned above, can have different reliability due to the different policy followed in adjusting load to availability and to different dynamic responses of the systems.

For example, a system of generation that exploits availability to

Paper 71 TP 89-PWR, recommended and approved by the Power System Engineering Committee of the IEEE Power Engineering Society for presentation at the IEEE Winter Power Meeting, New York, N.Y., January 31-February 5, 1971. Manuscript submitted September 18, 1970; made available for printing November 20, 1970.

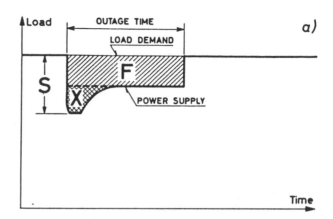

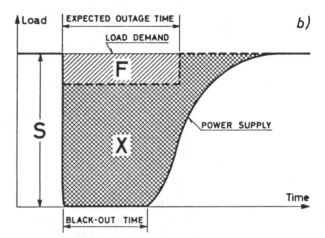

Fig. 1. Power supply trend during an outage of an element in the system

S = Abruptly disconnected power
F = Curtailed energy on the assumption of disregarding transient
F + X = Total curtailed energy

the maximum (thus reducing the spinning reserve or increasing the line overloads) in order to avoid scheduled load-reduction, will have high risks of load-shedding in the dynamic phase. Similarly, two generating systems made-up of units of very different sizes, but with total installed capacity such as to have the same risk of unavailability, will behave in different ways following the loss of a group. Such a loss will cause greater repercussions in the system with the larger units, and this system will consequently have bigger load-shedding in the dynamic phase.

To define the reliability of a power system, therefore, in a more realistic way, it is appropriate, in our opinion, to refer to two indices of risk expressed: firstly, by the expected value of the energy not supplied, over a certain period of time (we shall refer to the energy not supplied in one year), on the assumption that the adjustment of load to availability is carried out without transient (corresponding to the area marked "F" in Figs. 1); and secondly, by the expected value of the power disconnected without forewarning during the year, as a consequence of the transients caused by the unavailability of elements of the system (corresponding to the ordinate marked S in Figs. 1).

Reprinted from *IEEE Trans. Power Apparatus Syst.*, vol. PAS-91, no. 2, pp. 611–618, Mar./Apr. 1972.

Obviously, even these indices represent only indicatively the costs arising from breakdowns. For example, they do not take into account the effect of the extensions of breakdowns. For an equal amount of energy not supplied, a certain number of geographically and temporally separated breakdowns have a much less serious effect than one concentrated breakdown (of the black-out type), and this is because of the possibility, in the first case, of using the essential services that the surrounding areas (still supplied with electricity) can provide for the breakdown areas.

As a second risk index, we could alternatively have taken over the energy not supplied in the case of a transient phase, in excess of the energy needed to adapt load to availability (indicated by X in Fig. 1). But it seemed to us right that an abrupt load disconnection without forewarning should be penalized, even though limited to the necessary minimum [6]. On the other hand, evaluation of the aforementioned excess energy is very difficult, because it is closely related to the evolution, which is hard to simulate, of the load restoring after total or partial disconnection of the system.

It can be seen that the expected value of energy not available during the year (which characterizes the index of risk due to unavailability) is generally much less than the value of energy lost on an average during the year through load-reduction, and this above all because of the weight of the load-shedding caused by response of the system to a disturbance (cf. the case of a black-out). This explains why, to justify the high reliability of modern systems, it is necessary to penalise unavailable energy with economic figures very much greater than the weight derived from an equally effective load-reduction.

EVALUATION OF RISK INDICES

To evaluate the suggested risk indices, the authors decided to use a direct simulation method of the Montecarlo type.

The use of an analytical evaluation method would, really, have been possible for the first risk index; but as soon as the system becomes a little complex (in practice, more complex than a two point system), this method requires drastic simplifications which can be achieved by: a) Eliminating all the weak interactions (e.g., ignoring the effect of the lack of a line on the load of a node to which the line is not connected) [7]; b) Eliminating the dependence of power availability at the nodes of a network on the load situation (for example by using graph theory) [8]). The limits of applicability decrease considerably when the probabilities of breakdown of the elements are not independent of each other, as in the case of overhead lines.

Moreover, the use of the analytical evaluation method seems impracticable when it is desired to evaluate the second risk index. In this case, it would be necessary to evaluate the probability of all the possible situations of the system and of the occurring outage or faults, and analyze the transient in question. Considering the multiplicity of contingencies determining each case[1], and bearing in mind that the number of cases grows exponentially with the number of contingencies, it can easily be understood what a huge volume of work such a procedure would involve.

Under these circumstances, it would seem more practical to apply the Montecarlo method, which does not, as a rule, involve increasing the number of cases to be analyzed, however great may be the number of the random parameters affecting the formation and evolution of the transient. On the other hand, the risk indices we have assumed do not require consideration to be given to the temporal succession of events, so that the application of the Montecarlo method is greatly simplified.

The direct simulation method proposed by the authors is based on the following premises.

To evaluate the first index (curtailed energy F), it is assumed that the system works in a steady state, changing the situation and the

operating condition in a negligible space of time at the end of every hour; when the change in operating condition takes place, a notional dispatcher goes into actions, who shifts the generated power about in the most suitable way and disconnects some loads, in order to adapt the load to availability. A number of these hours, chosen at random, is examined and note is taken of disconnected loads. The average value for the energy added up in this way every 8.760 hours, as the number of hours tends towards infinity, provides the first risk index.

To evaluate the second risk index (abruptly disconnected power S), we take into consideration the situation changes that have occurred in the system during the same hours as above. It is assumed that these changes occur in the situation already established at the beginning of the hour, as a consequence of the situation changes that took place during the previous hour. All the transients that follow from this are analyzed, independently of each other, and note is taken of the loads disconnected during these transients. The average value for the power added up in this way for every 8.760 hours, as the number of hours tends towards infinity, provides the second risk index.

The random data for the reconstruction of the system's operating conditions in a given hour and for analyzing the transient occurring in the system during that hour, are generated by means of casual selection.

DESCRIPTION OF A PROGRAM FOR THE
EVALUATION OF RISK INDICES

An experimental computer program was set up by the authors in order to test the feasibility of the method suggested for the quantitative evaluation of the reliability of a complex power system. The program works as described below, in accordance with the block diagram in Fig. 2.

Generation of situations and situation changes

Reference Period — The week to which to relate the test is chosen by means of random generation of a number between 1 and 52. The selection of the week makes it possible to establish the season, the load, and the elements under planned maintenance.

Load — Information about the loads is given to the computer in the form of a deterministic weekly load duration curve, one for each week. The value of the load to be considered in each case is taken at random from that curve.

Maintenance — The experience acquired in estimating the reliability of bus-bar-type generation systems makes it possible to say that detailed simulation of the maintenance of individual units is of considerable importance to the estimation of reliability [9]. For this reason, maintenance of the various elements — in practice, the maintenance of the power stations — is fixed by input giving, week by week, the units that are out of service for planned maintenance. Such maintenance is optimized assuming the system to be bus-bar-type.

Outage of Elements and Faults — So far as the unavailability of elements is concerned, this is defined by two parameters: —a) The number of hours in the year in which the element is unavailable; —b) The number of times in the year that the element becomes unavailable, or a fault occurs. The first parameter serves to define the probable situation of the system in each hour, while the second serves to identify the variations in the situation of the system during the hour under examination.

For the generators and transformers, the outage rates are considered constant throughout the year, even though differentiated between unit and unit. Furthermore, the final hypothesis is made of considering the unavailabilities of these elements of the system as being independent of one another.

In the case of outage or fault of the lines, the situation is more complex, given that the probability of a line being or going out of service depends on atmospheric conditions that cause a close correlation between the outage probability of lines situated in areas subject contemporaneously to similar metereological conditions. This in adverse

---

[1])Various faults defined by type, position, cause; — behaviour of network protection and under-frequency relays; — behaviour of apparatus etc.; all these factors must be introduced in a probabilistic way.

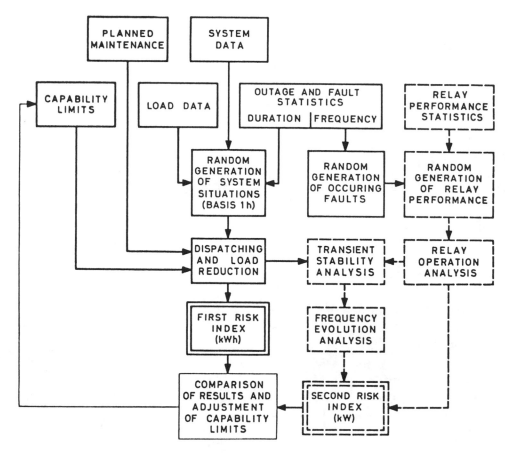

Fig. 2. Block diagram of the computer program for reliability evaluation

The dashed lines show blocks, the calculation of which is performed out of line of the main program.

metereological conditions (storms, salt storms, fog, ice or snow formation), causes a considerable increase in the probability of various line being out of service at the same time. The problem has been resolved by dividing the region concerned into areas in which metereological conditions are assumed to be uniform, and by dividing the total time considered into periods whose metereological conditions are assumed to be constant. Each period of each area has a corresponding outage or fault rate. In this way, a good simulation of reality is achieved as the contemporaneousness of outages is concerned. In the examples we give, we have divided the time into three periods: 90% of the total with a low outage rate (fine weather); 9% with an intermediate rate (bad weather); and 1% with a high rate (storm); it may be observed that the ratio between the outage rates of storm and fine weather has been taken of the order of 100.

For every type of weather, a number of trials, in proportion to the duration, are made for these conditions. The simplifying hypothesis is made that storm or bad weather in one area is accompanied by good weather in all the others.

A final improvement would be to consider the storm areas (or those with bad weather), moving at random in predetermined positions, so as not to limit the correlation within a given group of lines.

Then there is the correlation, caused by seasonal factors, between load conditions and a system situation on the one hand, and the line outage rate, on the other. The season in fact influences: the load in relation to the seasonal consumption diagram, the system situation in relation to the maintenance programmes, and the line outage rate as a function of atmospheric conditions.

The problem is solved by dividing the period of the year into seasons and giving each season a weight for outage on the various lines,

the season being defined by the week selected. In the examples that follow two seasons are considered.

Evaluation of the First Risk Index

The system situation is defined in accordance with what has been described in the preceding paragraph.

A comparison is then made of the load with the generation available, and a sufficient number of loads is disconnected to balance load with generation.

Then, with a d.c. load-flow calculation a check is made of the ability of the system to meet its loads. It has been preferred to accept the approximations of simplified load-flow calculation in order to have the possibility of considering a large number of cases in reasonable time for the purpose of achieving, at equal cost, a greater accuracy in the final result in estimating the expected value of curtailed energy. Checks using d.c. load-flow calculation to be made in the various cases are as follows:

a) Integral system: no checks

b) Integral network and one or more generators out of service; two cases can arise:

b_1) In all nodes the generating power available exceeds the generation demanded in that node to meet the load at that moment assuming the dispatching of loads to be proportional to that considered at peak in the integral system. In this case, no checks are made.

b_2) In all other cases, checks are made to see if any element in the system is overloaded. In this event, the program automatically acts to ensure the elimination of such an overload, first by shifting the generation (dispatching), and

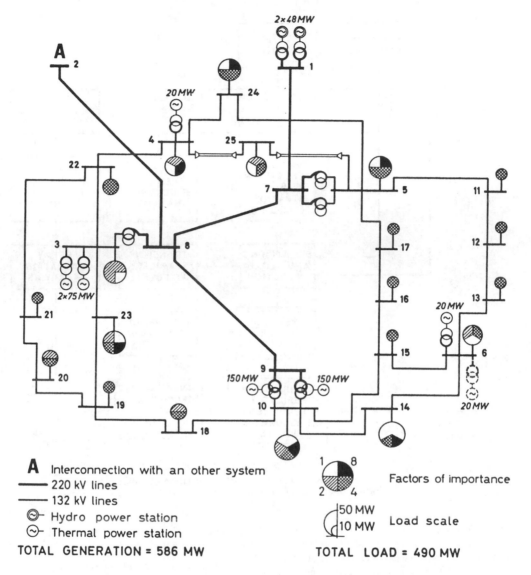

A Interconnection with an other system
— 220 kV lines
— 132 kV lines
⊙- Hydro power station
⊙- Thermal power station

TOTAL GENERATION = 586 MW

Factors of importance

Load scale

TOTAL LOAD = 490 MW

Fig. 3. Characteristics of System 1.

then by load-reduction.

c) Some element in the network is out of service (lines and interconnection transformers). The procedure as in b_2 is carried out.

As regards the amount of calculation to be done in each situation selected, it can be seen that this is greater, the greater is the potential risk:

— case a, in fact, requires no elaboration;

— case b_1 requires a simple comparison operation in each node;

— case b_2, in addition to the calculation above, involves load-flow calculations which require short computing time because it consists of a variation in power at the nodes in a unchanged overall network. If, in addition, an overload situation exists, the case requires dispatching and load-reduction calculations which are longer the more serious is the breakdown;

— case c requires the same operations as case b_2, with the difference that the load-flow requires more calculating time, because it is carried out on overall situations that vary from time to time.

The analysis, therefore, of the many situations that have no or little risk and that the Montecarlo system involves, does not therefore

substantially increase the weight of calculation to be made.

To carry out dispatching and load-reduction, use has been made of the "influence coefficients", i.e. those coefficients, easily obtained from the impedance matrix, which supply directly the information as to what shifts are to be made to have the maximum effect on the overloaded element. In carrying out dispatching, it is assumed, initially, that the load is uniformly distributed between the available groups. In the event of some elements in the system becoming overloaded, the overload is eliminated or reduced, without any consideration of the problem of maximum economy. In fact, the cost arising out of uneconomical dispatching in critical conditions, is negligible, compared with the costs arising from load-reduction. These do not influence the choice of loads to be reduced, and do not therefore alter the final result in which we are interested [1].

So far as load-reduction is concerned, the choice of loads to be reduced must take into account the importance of the various loads. There is always, in fact, a differentiation between loads in relation to the consequences of non-supply. The need is met by giving each load a different importance factor. The costs of reducing a certain load are meant to be proportional to this factor for that load. Usually, the least important loads are given a factor equal to 1. The out-put of the program supplies not only the total number of kWh, but also the total number of kWh weighted by their coefficients of importance. This

---

[1] It should be noted that in the event of more than one element being overloaded, we make use of the "weighted influence coefficients", which are the weighted averages of the "influence coefficients" relating to the overloaded elements, having taken the overload of the elements themselves as weight.

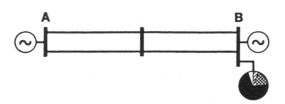

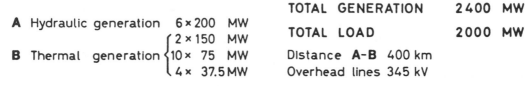

MAIN DATA:

A Hydraulic generation   6 × 200   MW

B Thermal   generation   { 2 × 150   MW
                                 10 × 75   MW
                                 4 × 37.5 MW

TOTAL   GENERATION   2400 MW

TOTAL   LOAD   2000 MW

Distance **A-B** 400 km
Overhead lines 345 kV

Fig. 4. Characteristics of System 2.

latter figure gives a more realistic index of risk than the former.

The total number of kWh lost in the year is not sufficient, however, to indicate what modification might most cheaply be made to the network, to improve its reliability. For this it is necessary to have some indications on *where* and *how* to make changes. To obtain indications on the part of the system most hardly hit, compared with another, by breakdowns, the curtailed energy is given sub-divided into nodes and to know the causes of the breakdown the curtailed energy in each node is classified by cause as follows: a) Shortage of generation; b) Shortage of local generation, due to the node being isolated; c) Overloading of lines and transformers; d) Interruption of supply.

Evaluation of the Second Risk Index

For each system situation considered in evaluating the first risk index, network faults and generator outages occurring in the hour to which the situation refers are selected in accordance with the criteria laid down above.

Analysis of the consequent transients enables the disconnected power to be calculated, which provides the second risk index. For simplicity, this analysis is done in two phases. In the first phase (transient stability analysis), maintenance of synchronism between the individual units of the system is checked. In the second phase (frequency evolution analysis), the trend of the frequency in the system as a function of time is determined, assuming that all the generating units are rigidly connected each other; the loads disconnected by the underfrequency relays and possible black-out are evaluated.

The system may become divided into sections, either by the event itself (a permanent breakdown on a single connecting line), or following the loss of synchronism among various groups of machines. In this case, the second phase of the study is carried out separately for every sub-system.

For the above analysis, the operation of the network protection relays and the underfrequency relays must be taken into account. These operations involve random events (measuring errors, undesired tripping, failure to trip) which are generated again, case by case, using the Montecarlo method.

As a result of the analysis, the following loads are added up in the risk index:

a) Loads isolated from generation in one of the parts into which the system has become subdivided.

b) Loads connected to the system or its parts when the frequency falls below a predetermined limit (black-out).

c) Loads disconnected from the system or from its parts by the underfrequency relays.

The loads under a) above arise directly from analysis of the event itself,

while loads b) and c) arise from the analysis of frequency evolution. The disconnected power is given sub-divided into nodes and classified by cause.

EXAMPLES OF APPLICATION OF THE METHOD

By way of an example, the method described was applied to quantitative evaluation of the reliability of two typical systems: one, characterized by a meshed network (System 1 – see Fig. 3), and the other by long-distance transmission (System 2 – see Fig. 4).

System 1

Only the first risk index was evaluated for this system; the main results, obtained by means of 100.000[1] trials are given in Table I. The total disconnected energy weighted with the importance factors works out at 171 MWh per annum. If we take the value of a kWh not supplied as $1, we get a total of $171,000 equal to about 1% of annual operational and investment costs; thus, the system can be regarded as one having a good reliability.

On examining the cause of breakdowns, it may be noted that the energy curtailed for shortage of generating power is considerably greater than that which is lost through overloading network elements. Examination of the distribution, node by node, of the loads disconnected through overloading of the network elements has shown that the disconnection has mainly taken place at the nodes on the right side of the network.

These results suggest that a considerable improvement in the system reliability can be obtained by adding generating power in one of the nodes on the right side of the network. This should achieve the dual purpose of reducing the energy disconnected, both for shortage of generating power and overloading of the network.

A new system derived from the preceding one, with the addition of a 20-MW turbogas unit in node 6, was analyzed; the results confirmed expectations (see Table I).

Comparison of the results also make it possible to see whether or not the operation was an economical one. If every kWh not supplied is valued at over $1.2, then the operation can be considered economical.

System 2

For this system, the method of evaluation proposed was applied in its entirety, both risk indices being evaluated. In this case, there is the problem mentioned earlier in the first paragraph: that of establishing what limits should be taken for line capability in determining load reduction and, therefore, the first risk index. In this case, the limits are identified with the stability limits. If the limits chosen are too high, the resulting operating conditions are often such that, in the event of a fault occurring on the lines, loads are disconnected in the transient phase. The choice of line capability limits therefore has the opposite

---

[1] The average computing time required was of about 33 rms per trial on a UNIVAC 1108 computer.

TABLE I

FIRST RISK INDEX IN SYSTEM 1

	Curtailed Energy (MWh/year)			Curtailed Energy Weighted with Importance Factors (MWh/year)		
	Due to Shortage of Generation	Due to Overloading of the Lines	Total	Due to Shortage of Generation	Due to Overloading of the Lines	Total
System 1	130	33	163	138	33	171
System 1 A	37	3	40	37	3	40

System 1 A is derived from the System 1 by adding a 20 MW turbogas unit in node 6.

TABLE II

RISK INDICES IN SYSTEM 2

Assumed Capability Limits of Lines (MW) Configuration			First Risk Index Curtailed Energy (MWh/year)		Second Risk Index Abruptly Disconnected Power (MW/year)	
a)	b)	c)	Not Weighted	Weighted	Not Weighted	Weighted
> 1200	450	380	2210	4920	148	765
> 1200	800	600	216	469	340	2195

Configuration : a) ⊢□─□⊣   b) ⊢□─⊣   ⊢─□⊣   c) ⊢─⊣

effect on the two risk indices, in the sense that increasing the limits involves reducing the first risk index and increasing the second.

The evaluation of these indices, obtained by means of 200.000 trials,[2] was made in two cases with different values for these limits, as shown in Table II. From the results of the evaluations, it can be seen how strongly the risk indices are influenced by the line overload limits chosen in dispatching; obviously evaluation of the reliability of the system based on the first risk index in this case might lead to serious errors.

It should be noted that application of the importance factors has only a slight effect on the first risk index, while it has considerable effect on the second. This is due to the fact that the second risk index is derived from a few load disconnections of considerable extent, which therefore also effect the most important loads.

Valuing a kWh[3] of the first index as $0.5 and a kW of the second index as $1.5 the total annual economic burden in both cases comes to about $3.5 .$10^6$, equal to approximately 2% of the total annual operating and equipment costs of the system.

[2] The average computing time required for the evaluation of the first risk index was about 9 ms per trial on a UNIVAC 1108 computer.
[3] The economic value of the kWh has been assumed equal to $0.5 instead of $1 as considered for System 1. This has been done because the economic burden of the disconnected power has been taken into account directly, while for System 1 has been considered implicitly by increasing the economic value of the energy.

CONCLUSIONS

1. The risk indices based on a comparison between load and power available, usually assumed in planning studies concerning generating systems, are unsuitable for defining, quantitatively, the reliability of a complex power system.

2. A better definition of the reliability of a power system for planning purposes may be obtained through two indices that will provide, firstly, the energy not supplied through unavailability and, secondly, the power abruptly disconnected as the result of faults or outages.

3. The two risk indices in terms of energy and power can be obtained by the Montecarlo method, by analysing a large number of network situations chosen at random.

4. The use of a direct simulation method makes it possible to consider a vast number of random contingencies and to take into account, perfectly simply, the interrelationship between the contingencies themselves (particularly the interrelationship between faults in the various lines caused by the contemporaneousness of adverse metereological conditions).

5. A programme of numerical calculation produced by the authors shows the method's possibilities of practical application.

6. The method suggested makes it possible to obtain risk indices sub-divided by locality and main cause, and so have some indication of

where and *how* action should be taken on the network design to make it more reliable.

7. One of the numerical examples shows how the capability limit of the lines that the dispatcher sets himself in reducing loads at critical moments may have a decisive bearing on risk indices. A variation in such a limit has opposite effects on the two indices; this limit should therefore be decided, so as to reduce to the minimum the total equivalent economic burden of the two risk indices.

ACKNOWLEDGEMENT

The authors would like to express their appreciation to Mr. O. Bertoldi and Mr. L. Salvaderi of ENEL for their help in conducting the study.

REFERENCES

[1] T. W. Berrie, "Further Experiences with Simulation Models in System Planning" Second Power System Computation Conference, Stockholm, 1966.

[2] C. J. Baldwin – C. H. Hoffman, "Results of a Two-Year Study of Long-Range Planning by Simulation", CIGRE 1962, Rep. No. 306.

[3] C. Concordia – L. Lalander – J. Cladé "Rapport d'activité du Comité d'études N. 13 – Conception et fonctionnement des réseaux" Annexe II, CIGRE 1968, Rep. 32-01.

[4] L. Paris – M. Valtorta, "Economic Aspects of Planning Large Generating Units in Interconnected Systems" – World Power Conference, Moscow 1968 – Rep. 101.

[5] H. Halperin – H. A. Adler, "Determination of Reserve Generating Capability" AIEE Transactions, Vol. 77, Aug. 1958.

[6] B. Mattsson, "Economy vs. Service Reliability in Sweden", IEEE Spectrum, Vol. 3, p. 90, May 1966.

[7] J. Bergougnoux – J. C. Arinal, "Valuation of a Power System Supply Reliability" PICA 1967.

[8] H. Baleriaux – E. Jamoulle – P. Doulliez – J. Van Kelecom, "Optimal Investment Policy for a Growing Electrical Network by a Sequential Decision Method", CIGRE 1970, Rep. 32-08.

[9] G. Manzoni – P. L. Noferi – M. Valtorta, "Planning Generating Units for Peak-Load in Power Systems-Relevant Parameters and their Relative Influence", CIGRE 1968, Rep. 32-06.

[10] C. F. DeSieno – L. L. Steine, "A Probability Method for Determining the Reliability of Electric Power Systems", IEEE Trans. Power Apparatus and Systems, vol. 83, pp. 174-181 – February, 1964.

[11] D. P. Gaver – F. E. Montmeat – A. D. Patton, "Power System Reliability – I. Measures of Reliability and Methods of Calculation", IEEE Trans. Power Apparatus and Systems, vol. 83, pp. 727-737, July, 1964.

[12] D. J. Cumming – F. E. Montmeat – A. D. Patton – J. Zemkowski, "Power System Reliability – II. Applications and a Computer Program," ibid., vol. 87, pp. 636-643.

[13] S. A. Mallard – V. C. Thomas, "A Method for Calculating Transmission System Reliability", ibid., vol. 84, pp. 824-834.

[14] R. Billington – K. E. Bollinger, "Transmission System Reliability Evaluation Using Markov Processes", ibid., vol. 87, pp. 538-547, February, 1968.

Discussion

A. J. Wood and **R. J. Ringlee** (Power Technologies, Inc., Schenectady, N. Y. 12301):
This paper presents an interesting addition to the growing body of data and approaches to the composite reliability evaluation of bulk power supplies. Utilizing simulation, the authors have attacked bulk power reliability supply problems both from the point of measure of matching load with static capacity, as well as an evaluation of transient phenomenon associated with stability. They have considered the possible switching and reclosing operations associated with under-frequency relaying, in addition to other forms of load matching and fault isolation. The authors are to be commended for undertaking this difficult task.

The paper cites a computer simulation of Figure 3 (system 1) which required 100,000 Monte Carlo trials with an average computing time of about .03 seconds per trial on the Univac 1108. The 1108 is a large, high speed computing system. It would appear that some 3000 seconds or slightly less than an hour was necessary to evaluate this

Manuscript received February 9, 1971.

relatively small system, presumably for one year's study. We presume the figures shown in Table I are the average or expected values. Have the authors made an estimate of the standard error inherent in these figures based upon the 100,000 trials?

It is difficult for us to accept Conclusion 1, that the classic approaches to planning studies are unsuitable for defining quantitatively the reliability of a complex power system. We would agree with the authors that static evaluations are necessary, but possibly not sufficient in a mathematical or analytical sense. However, it is difficult to reconcile the idea that the static analysis is unsuitable. We prefer to believe that the authors intended a term similar to the ". . . . necessary but not sufficient . . . ." term found widely in the literature of mathematical analysis.

It occurs to us that with the frequency and duration model of load and generation, the table of reserve margin states presently developed in these planning programs would yield an expected value of the megawatt hour's duration. The merit in the use of the megawatt hour loss index suggested here and by others is that it tends to reflect the degree of risk of the larger negative margin events occurring. Two systems with the same loss of load probability index may have markedly different megawatt hour curtailment indices due to the slope of the LOLP versus margin curve for the two systems.

We approve of the desire of the authors to make use of a more realistic, dynamic model of the power system. However, in our experience "the best Monte Carlo is the least Monte Carlo" since synthetic sampling techniques tend to be extremely expensive to use, and one wonders if they give sufficient added information.

M. P. Bhavaraju and **Virginia T. Sulzberger** (Public Service Electric and Gas Company, Newark, N.J.): We compliment the authors for a very interesting application of simulation methods for quantitative evaluation of power system reliability. Several new risk concepts applicable to reliability studies are presented. Of significance, the authors relate the calculated risk indices to economics in terms of evaluating the cost of disconnected electric system load or curtailed energy. This method provides management with a means of balancing economy and reliability.

We would like to suggest a possible ideal index of risk in addition to those indices considered by the authors. The ideal index would be a measure of customer reaction to interruptions. This index was investigated and reported to some extent in the "Powercasting" work done by Public Service and Westinghouse in the late fifties. It was developed as an empirical function of the customer's dependence on electric supply, the frequency, duration and magnitude of the interruptions, the cause of the interruptions, and the time of the day, week or year in which the interruptions occurred. Voltage reductions and voltage dips might also be treated as partial or total interruptions to some customers. Further work in this area would be of considerable value. The authors attempted the former approach by using weighting factors on loss of energy indices.

The authors have examined a breakdown of nodal risk indices by different causes to decide what changes should be made in the system in order to alleviate high risk. In previous work at the University of Saskatchewan on power system reliability methods for planning, a similar approach was used to obtain load point risk indices due to generating capacity and transmission system unavailability separately. These separate indices for generation and transmission were more meaningful for planning those facilities.

Since the authors indicated that both generation and transmission outage contingencies were investigated in their risk analysis, we would appreciate the author's comments on 1) the number of overlapping outage contingencies or the severity of the outages which were required to cause loss of system load; 2) if transmission line maintenance in addition to generation maintenance was considered in the analysis; and 3) what effect the transmission maintenance had on the risk indices compared to considering only forced outages of the transmission system. The authors also indicated that protection and underfrequency relay operations were considered in the analysis. We would appreciate if the authors would elaborate on the performance statistics used for the protection and underfrequency relays.

Previous reliability methods have ignored transient performance. Transient performance is a function of the operating reserve capacity, transmission system, and other factors. We would like to know the effect of ignoring the transients on the risk indices developed in the paper. Did the authors assume, in their analysis, all the installed capacity to be operating without any economy outages?

Manuscript received February 19, 1971.

Pier Luigi Noferi and **Luigi Paris:** We greatly appreciate the discussion of our paper by Wood, Ringlee, Bhavaraju, and Sulzberger, and would like to thank them for the many valuable comments and questions that help clarify some details of our study.

Ringlee and Wood: dealing with the computing time for 100,000 Montecarlo trials on System 1 — Fig. 3 (about 3000 seconds), we wish to point out that there are some possibilities of improving the computing means adopted: our program was in fact set up without particular care for the optimization of computing time. We are now carrying out useful experiments by combining simulation methods with direct evaluation methods, and this seems to remarkably reduce computing time. At present, it appears to us that there is no valid alternative to the Montecarlo method when the random variables affecting the results are so numerous, as in power systems especially dealing with transient analysis.

Results reported in our paper refer to data obtained by 10 groups, each of them consisting of 10,000 system situations; the standard deviations of the mean referring to the 4 typical power systems dealt with in our study are the following:

System	: 1	1A	2 with low cap lim.	2 with high cap. lim.
Standard deviation of the mean:	18%	34%	6%	23%

Higher standard deviation values are referring to the power system having lower risk indices.

As regards the adjective "unsuitable", we can agree with Mr. Wood and Mr. Ringlee on the advisability of adopting the expression "necessary but not sufficient" in a mathematical sense, but we think that, when dealing with reliability evaluation, "not sufficient" is somehow inadequate. This can be better understood when looking at Table II of the paper; if we examine only the first index we will obtain a very different reliability according to different operating conditions, whereas the actual reliability is the same as can be obtained by examining the two risk indices at the same time.

Bhavaraju and Sulzberger: Question No. 1—We presume you want

Manuscript received March 26, 1971.

to know the amount of load-shedding due to failure bunching of 1, 2, 3 or more lines. In our program this output is not available; we are therefore not in a position to answer your question. Nevertheless, this output could be easily obtained when required.

Questions Nos. 2 and 3—Transmission line maintenance was not taken into account so that we are unable to give information on its effect on the risk indices. Our program can easily take into account this maintenance.

Question No. 4—The protection relays are considered as two-state elements, operating or not, when a fault has to be cleared by them. The reliability assumed is 97% for the main protections and 95% for the reclosure. Among the improvements in our program is the introduction of the effect of uncorrected working of underfrequency relays, whose settings have been fixed at the moment at the following steps:

Hz	% of the total load shedded
49.5	7%
49	15%
48.5	30%
48	Black-out

Question No. 5: effect of ignoring transient—We presume you want to know whether we developed another index considering shedded kW by ignoring the transient; this would mean to treat kW due to occurring faults as those due to permanent ones. The information supplied by this new hypothetical index, along with our first one, would be similar to that obtained by the well-known frequency — duration indices. We have not, however, followed this approach since it does not seem complete to us.

In conducting the transient analysis we have assumed all the installed capacity to be operating without any economy outages; in fact, the program always considers the system in the best conditions from the reliability point of view, disregarding fuel expenses and losses. Thus the result is the maximum system reliability and not its economic optimum. Our approach is sufficient for planning purposes; we deem that the research of the optimum reliability is a question of refinement to be solved when in operation.

Montecarlo Methods for Power System Reliability Evaluations in Transmission and Generation Planning

P. L. Noferi - L. Paris - L. Salvaderi
Ente Nazionale per l'Energia Elettrica
Rome, Italy

ASQC Descriptors: 831

Key Words: Computer Programs, Failure Rates, Models, Operation, Probability, Reliability, Sampling, Simulation

Abstract

In planning studies of generating and transmission systems, simulation methods based on the Monte carlo technique allow the system reliability to be quantitatively evaluated even in the most complex cases. The numerical reliability indices can be expressed in monies and introduced into the overall economic optimization of the system.

The paper describes the fundamental criteria adopted in setting up computer programs by using this simulation technique The Montecarlo method is applied with two approachs: (a) the hours of the system lifetime are taken into consideration at random ("random approach"); (b) each hour in the year is examined in a chronological order ("sequential approach"). Either method can successfully be applied in the solution of different problems.

The "random approach" is used in planning studies of transmission systems and allows the evaluation of the reliability indices of a system as to both its capability of supplying the load and the quality of its voltage service.

The "sequential approach" is used in planning studies of mixed thermal-hydro generating systems. It affords the quantitative evaluation of the risk of not covering the demand because of lack of water. It is mainly adopted in the sizing of the reservoir and pumping capacity of pumped-storage stations and of the characteristics of other peak coverage means.

The paper points out the advantages resulting from the adoption of the simulation methods (easiness in dealing with complex problems, availability of a great deal of information) which render their computer times acceptable even though they exceed those usually required in other analytical approaches.

Foreword

In planning studies a considerable amount of the investments (20-30%) is spent for the constitution of the necessary reserve; this must allow the system to operate "correctly" also when the system components are subject to planned or forced outages.

Being the forced outage a random event ruled by probabilistic laws, the reserve requirement is at any moment a random variable. Therefore, the risk of not meeting the demand always exists.

The sizing of a power system reserve is obtained through a compromise between its capital cost and the risk of not being capable of performing the service requested of the system; probabilistic methods are therefore to be used in power system planning studies. This compromise between the reserve cost and the quality of the service offered requires a quantitative evaluation of the risk of not satisfying the demand and can be made on an economic basis by expressing the risk in monies according to adequate criteria.

This paper is concerned with the methods for quantitative evaluation of this risk, by means of risk indexes. Furthermore, it will be pointed out how similar evaluations are required not only by system planning but also by engineering of system components. Infact   the latter are frequently to give a performance defined only in probabilistic terms (for instance, for overhead lines the power flow can be defined only in a random way due to the dependence on the system state).

Probabilistic inputs of the problem are the distributions which define the availability of the system components, the load and, more generally, any random quantity affecting the system behaviour. Deterministic inputs are the system composition and the policies to be applied. The output of the study is again given in random quantities, defined by probabilistic distributions or at least by their average values. These quantities define the system behaviour or the performance requested of some components.

In order to solve the problem, two possible approaches exist: the analytical evaluation and the Monte carlo simulation.

The first is capable of supplying the planner with expected results, in a relatively short computing time. On the other hand, unfortunately, it is adoptable only when the links established between input data and desired output quantities by the system structure and operation policies are very simple. Consequently, it is generally applied in the evaluation of the risk of not meeting the demand in the case of an entirely thermal generating system and in a generating and transmission system only when the network is very simple (Refs. 1, 2, 3, 4).

For more complex systems, the required simplification for application of the analytical approach in modelling the system (structure and operation) would be so drastic that the resulting analysis would frequently lose significance.

Reprinted with permission from *Proc. 1975 Annu. Reliab. Maintainab. Symp.*, pp. 449–459, Jan. 1975.

For this case, we deemed it necessary to develop Montecarlo simulation methods, in addition to the analytical ones (Refs. 5, 6) notwithstanding that they supply only estimates of the investigated parameters. The advantages of this approach will be described in the following pharagraphs.

The computer programs run to date can be classified into two main groups:

(a) Programs intended for the detailed study of generating systems including hydro generating modulating plant and pumped-storage. These programs do not take into consideration the transmission system, assuming it of infinite reliability and transmitting capability. They are used for detailed examination when planning generating systems and when establishing the basic characteristics of peak duty plant.

(b) Programs intended for the study of thermal generating and transmitting systems. They are mainly used when planning transmission systems and in order to verify the correct allocation in the network of generating plant. They are also employed for the study of how and where to install the reactive power generating plant and for the determination of the basic characteristics of the main transmission-system components.

In the following paragraphs we shall examine the main features of our simulation approach and we shall present the main application of the computer programs developed according to this philosophy.

Risk Indexes

The quantitative evaluation of power system reliability is obtained through "risk indexes". These quantities generally express the average characteristics (value, frequency, duration) of system outages (Ref. 7).

Static indexes are used to express, in a probabilistic way, the comparison at given moments between the steady-state generated and transmitted capacity with the components available at that moment, and the load to be supplied. These static indexes do not take into account the transients of the system when forced outages occur.

Dynamic indexes are used in order to evaluate the dynamic performance of the system during the above mentioned transients. In fact, the load that could be shedded can be higher than the difference between system generating and transmitting capacity and load requested in each steady state. While the methods set up allow easy computation of the static indexes, studies are still carried on in order to obtain programs capable to computing the dynamic indexes on a statistic basis and with rather low computing times.

Of the well-known static indexes one of the most commonly used in ENEL practice is the expected yearly curtailed energy, given in kWh/yr. It corresponds to the yearly average not supplied energy due to deficiency of the generation and/or transmission system, and takes into account the active and reactive system constraints. As to the how of the power system deficiency: the deficiency in generation can be caused by lack of active power (risk of active power)

and/or lack of stored energy in the reservoir of hydro plants (risk of water) (Refs. 6, 9); the deficiency in transmission can be caused by lack of capacity of lines and/or transformers (risk due to overload), by cut-out nodes (risk due to isolated nodes) (Ref. 8), by the impossibility of meeting the reactive power demand of the loads and network (risk due to lack of reactive power), by the impossibility of satisfying the voltage constraints in the nodes and the reactive capability limits of generators (risk due to abnormal voltage). Furthermore, as regards the voltage service quality, a "voltage irregularity index" is used (Ref. 10) to penalize the voltage variations even if within the limits. All the above quoted indexes can be obtained for the system as a whole and for each network node where the deficiency occurs.

More detailed indexes are also used to obtain further information on the characteristics of system outages. They are: (a) frequency, the average rate at which the energy curtailments occur (outages/yr); (b) the expected amount, the average value of each energy curtailment (kWh/outage) (the product of the expected amount and frequency corresponds, of course, to the yearly expected curtailed energy); (c) the expected duration, the average residence time in an energy curtailment state (hours/outage).

For a better understanding of the phenomena, it may be necessary to know not only the average values of all the above-mentioned variables, to wit the risk indexes, but also their probability distributions.

As for dynamic indexes, the figure used is the "suddenly disconnected power" (indicated by S in Fig. 1) during transients following forced outages; the quantitative evaluation of this power is easier than that of the energy not supplied (X in the figure) during the transient. The total disconnected power S consists: (a) power required by load that gets isolated from the generation; (b) power disconnected by underfrequency relays during system frequency evolution; (c) power required by the loads of the whole system when the frequency drops below the critical value at which the system cannot operate.

Simulation Programs for Power System Reliability Evaluation

In order to evaluate the static reliability we apply the Montecarlo simulation method according to the

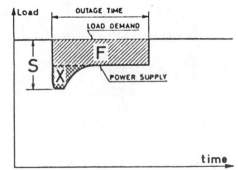

Fig. 1- Power supply trend during an outage
S=Abruptly disconnected power; F=Curtailed energy disregarding transient;
X + F=Total curtailed energy

following premises:

(a) The reference period of time during which we examine the system behaviour is divided into elementary intervals. During them we suppose that the state of the system is constant, pretending that all its changes are effected at the beginning of the interval, while naturally they take place during it.

(b) The power system state is characterized by the following elements: (i) load of each network node, (ii) availability of the system components (generators, lines, etc...), (iii) energy availability (hydraulic inflow, reservoir levels, etc...), (iv) operation assessments (generation committment, load shedding, network connections, etc...).

(c) The changes in the system state are caused by: (i) events that are generally random, and so unpredictable by the operations staff, which cause load and availability variations; (ii) dispatcher's actions in order to match the system state to the above mentioned changes. Such actions are carried out in order to obtain the best service quality (maximum security policies) or the best compromise between the service quality and the economy (security and economy policies).

(d) We consider a number of time intervals (sample) and in each one: (i) we simulate the random components of any element availability and of the load by means of casual generations starting from their probabilistic distributions, and the action of the Dispatching Center by means of the more or less complex fixed selected policies; both type of simulation define the system state: (ii) we evaluate any random variables of interest (energy curtailment, power flows in the lines, etc...) which will be utilized for the determination of their "parent distribution".

In order to evaluate the dynamic reliability, for each system situation previously considered, network faults and generation outages occurring in the hour to which the situation refers are considered separately. All these "contingencies" are recorded to be utilized in the analysis required for the dynamic risk index determination and in the quantitative evaluation of the dynamic services performed by the generators (Ref. 11).

The duration of the elementary time interval had to be chosen so that the hypothesis of a constant system state would be realistic enough according to our aims. In our studies an hourly basis was generally selected because the hour is the longest interval that allows us: (a) to consider the load constant without any important error in most of the cases examined; (b) to take into account outage durations of a few hours characterizing the overhead line and system components; (c) to deal with hydraulic generation systems employing reservoirs with few generating hours at full capacity, particularly the pumped storage plants for power service; (d) to assess the statistical characteristics of the service required of the thermal units for peak service (frequency of start-up, etc...); (e) to perform the studies related to the determination of the dynamic services of the generating plant, with quantitative evaluations done by system approach.

The total number of the hours considered (sample size) must be representative of the reference period, usually fixed in a "solar year", during which the system composition is assumed to remain constant and so extended that the random results of the effected simulation will be defined with the desired confidence. The total number of hours and the order of their examination can be obtained by:

(a) Random generation of any hour that constitutes the solar year: in the following we will refer to this type of Montecarlo approach as "random approach"[+]. We shall furthermore refer to any group of 8760 of these hourly operation simulations as "yearly sample".

(b) Examining all the hours of the solar year in their real chronological succession. In the following we will refer to this type of Montecarlo approach as "sequential approach". We shall furthermore refer to any group of consecutive 8760 of these hourly operation simulations as "simulated year".

The "random approach" is applicable when the system state in each hour does not depend on the occurrences in the preceding ones. This is the case, for instance, when studying with a maximum security policy the reliability of a transmitting and generating system with pure thermal generation, in which the fuel is always available. Unfortunately, this approach presents some limitation in the completeness of the result. When evaluating the system reliability it can infact produce only the expected yearly curtailed energy and not the expected characteristics (amount, frequency, duration) of the curtailments, which constitute other common risk indexes. Again, when evaluating the thermal stresses of overhead line conductors it allows only the determination the expected duration by which any temperature is overcome, but not the number and duration of each time interval during which the phenomenon occurs.

The "sequential approach" is really the most general. It must necessarily be applied when the system state in each hour strictly depends on its previous history. This is the case, for instance, of generating systems including hydro plants with reservoirs whose stored energy depends on the energy demand in the hours preceding the one under examination. It is also the case when the operation characteristics (number of start-ups, shut-downs etc...) of thermal units must be determined. This approach, as we shall discuss later on, involve larger computing time than the previous one so that in spite of its limitations the "random approach" is always to be adopted when the phenomenon considered allows its application.

Models utilized in the computing programs.

In our programs, models for the statistical description of system component behaviour and for the policies by which the system is operated can be, in general, different and more or less detailed according to the different purposes of the program set up.
~~~~~

(+) - The modifier "random" in this connection is of course, referred to the order of time interval examination and not to the Montecarlo method which, in any case, is based on random simulations.

Load Model. The active load of the network can be set by means of the hourly load diagram of each week in the "sequential method" application, or of the load duration curves of each week in the "random method" approach. To take into account the load randomness, each load level can be assigned a normal distribution by which the random component can be properly generated. The total load is subdivided into each network load. A further subdivision is made in each node, corresponding to the portions which it is subsequently possible to shed, according to their "factor of importance".

As for the reactive power requirements, each load is characterized by its $\cos \phi$; the voltage limits in each node are furthermore given.

Environment models. In each homogeneous area of the system, we consider three weather conditions, randomly alternating during the solar year: "fair", "bad", "stormy" weather. The duration of each condition is assumed to be exponentially distributed on the basis of an examination performed in a typical Italian area (Ref. 12). Other field distributions are used for ambient temperature, wind velocity, solar emission.

Component unavailability models. A two-state model (up-down) is considered for forced outages of generators, transformers and other main network components, by weighting the field data in order to take into account the occurence of derated states. Different rates of failure are considered to take into account the infant mortality or the aging of the components. The probability distributions of forced outages of generators, were found to be well represented by exponential laws.

The forced outage of overhead lines is described by taking into account its dependence upon weather conditions. A two-state model is considered inside any weather condition. The occurrence of permanent and transient faults can be well represented by exponential distributions; naturally, the rate of failure increases as soon as we pass from fair to stormy weather conditions (Ref. 12).

In the program adopting the "random approach", three samples are considered each sized to reflect the probability of occurrence of each weather condition. The state of any component is determined from its forced outage probability in each simulated hour by random generation.

In the program adopting the "sequential method" the system residence times in each weather condition are generated from their distributions; the transition from one condition to the other is assumed to be random. The state of any component is obtained by random generations of the "time to failure" and "time to repair" from the distributions corresponding to each weather condition, taking into account their evolution.

Generating Unit Models. Thermal units. They are characterized either by their rating when the program perform only reliability studies, or by the rating, the technical minimum, the hourly production cost (the hourly production cost model assumes constant marginal cost: $F = A_0 + \lambda W$; the start-up cost model is as-

sumed to be a linear function of the "out-of-parallel" residence time h (banked boiler) $C_{av} = q + mh$), when the program is used to perform reliability and economy studies.

Pumped storage units. These are pure and are characterized by: (a) generating capacity; (b) the pumping capacity (π) in p. u. of generating capacity; (c) the energy storable in the reservoir in p. u. of the generating capacity (h); (d) the pumping cycle efficiency.

Hydro systems. The most general hydro system model considered by the program is made up of three stations "in series" (Ref. 13). The first two are fed by reservoirs (lakes and pond respectively) and the third is of the run-of-river type. A fixed time delay is considered between pond and run-of-river stations. Three natural inflows are considered for the lake, pond and run-of-river stations, they can be defined both by historical series and by statistically generated series. Since the computer program can be applied to a relatively limited number of system models of this kind, in general there exists the problem of reducing a complex hydro system. This reduction is carried on by trying to assign each "equivalent station" of each system model the same stored energy as that of the stations represented by it and the same yearly production. The larger the number of system models, the greater the accuracy of the result to be expected.

Operation Policy Models. Generally speaking, the operation policies adopted in our programs must be sufficiently simple to obtain the result describing system behaviour in acceptable computing times, and flexible to allow changes in them, according particular aims purposed. The policies set up can be divided into two categories:- pure security policies and mixed security and economy policies.

The studies of transmission systems nearly always involve pure security policies, the main aspects of which are the following.

a. When the system reliability is determined only with reference to active power network requirement: (i) the load requested is matched to the total available generation by curtailing the load if necessary (risk index due to lack of active power); (ii) no load economic dispatching is made in the unit commitment, all the generators available are kept in service and loaded to the same per-unit value of their available power; (iii) the powers flows in the lines and transformers are obtained through d. c. load flows; (iv) when the load of a line is greater than a certain level W_G, called "alarm level", the loading of generating units is modified in order to minimize excessive line loading and, if possible to bring the load of such lines below this "alarm level"; (v) the line loading above the "alarm level" is then tolerated up to a limit called "absolute" W_A, at which some loads are curtailed, until the line loadings come within the limit itself (risk index due to overload of the components).

b. When the effects of voltage and reactive power on system reliability are to be assessed: (i) active power dispatching, shifting and load shedding are performed as in the previous case; (ii) voltage val-

ues in the nodes and reactive power network requirements, are obtained by linearized reactive load
flows; (iii) total network and load reactive requirements are fulfilled, if necessary, by load curtailment (risk index due lack of reactive power); (iv) the
Dispatcher changes the average voltage level, reactive power dispatch in order to avoid that the
voltage limits in some nodes are exceeded and to
minimize the voltage irregularity; if the voltage
limits are still exceeded the Dispatcher curtails
loads (risk index due to abnormal voltage limits);
(v) when the voltage value in all the nodes are within
the imposed limits the voltage irregularity index is
computed.

The studies of generating systems can involve
both pure security and mixed security-economy policies, the main aspects of which are the following.
a. When only the system reliability is to be determined,
the simplest model is assumed for the thermal
units and each of them is put in service at maximum
power when available according to an increasing
marginal cost order.
The maximum security policy adopted utilizes the
hydro generating means with a water conservation
policy, that is, the hydro modulating stations (pond,
lake and pumped storage stations) are put in servi
ce to face the load only when any thermal units, in
cluding gas turbines, have already been utilized or
to avoid spilling (run-of-river stations are, of
course, subjected to a fixed policy): they are there
fore utilized only for "power service".
Within each hour the priority order by which the
modulating means of hydro generation and pumped
storage are to be put in operation can be given as
an input datum. When dealing with a generating sys
tem having a relatively limited hydro capacity the
most suitable priority order seems to be "pond-
lake-pumped storage".
As for the refilling of pumped storage reservoirs,
according to the maximum security policy adopted,
it is carried out as soon as there is spare capacity

exceeding that necessary to the load. The program
allows both low and high marginal thermal energy
and also, optionally, the energy of the seasonal
reservoirs to be used for the refilling.
b. When not only the system reliability, but also the
operations costs and the main operating characteristics of peaking units are investigated, the more
detailed model for the thermal unit is to be taken
into account and the mixed security-economy policy
adopted. The solution of this problem implies, of
course, more computing time and is performed in
two steps.
Firstly we determine the system reliability employing the mixed-economy policy. The utilization of
the thermal peaking units having the highest marginal cost is reduced by adopting the "water conser
vation policy" only when strictly necessary to
avoid worsening the security.

The program singles out the "dangerous" weeks
(i.e. the weeks in which a system crisis would
occur if the thermal peaking units were not used be
fore the hydro modulating plants) and then applies
for them the appropriate security policy explained
above, both to generating units and pumped-storage
reservoir refilling. In the other "safe" weeks, the
thermal peaking units are utilized, if necessary,
only after the modulating hydro and the pumped
storage units. In this case only low marginal cost
energy is used for reservoir refilling. Figure 2
shows, as an example, the effect of the policy
described on the generation duration curve of the
same week as resulting in the simulation of a situa
tion causing system crisis, and of another in which
the system is "safe". It can be seen that the turbo
gas units are required either to operate a great
deal during the week (about 90 hrs) or not at all.
The total thermal capacity necessary hour by hour
to ensure system reliability when employing the
mixed security-economy policy is thus obtained.
Secondly, having the aforesaid total thermal capacity, the hour by hour economic load dispatching

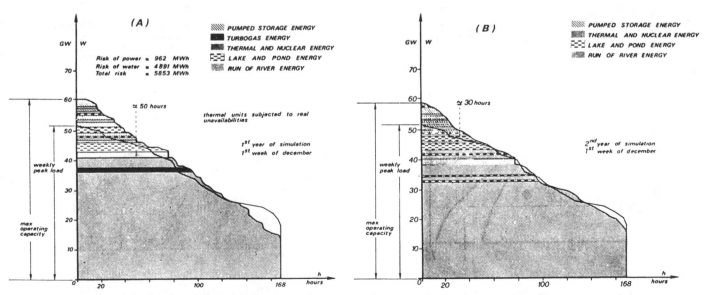

Fig. 2 - Effect of the mixed "security economy" policy on the weekly generation duration curves of two
simulated weeks having high (a) and low (b) unit availabilities.

is determined, taking into account also the technical minimum and the cost of start-up. The thermal unit operating characteristics and the operation detailed costs are obtained.

Typical Outputs

The main outputs of the simulation programs set up are:

a) <u>Estimators of risk indexes</u>. The average value of yearly curtailed energy is given both as a total figure and divided as follow.

In the case of <u>transmission system,</u> by: (i) the causes, which allow to be understood <u>how</u> to take the actions necessary to improve the network reliability; (ii) the network nodes, which allow to be known <u>where</u> these actions should be taken; (iii) the weather conditions <u>when</u> the energy curtailments occurred, which allow the evaluation of the influence of failure bunching.

In the case of <u>generating system</u> the figure is divided by causes, which allows us to understand whether the deficiency is due to <u>lack of power</u> or <u>lack of water.</u> Both the "sequential" and "random" approach can supply these indexes.

The frequency of load curtailment and the statistical distributions of their characteristics (amount, duration) can also be given. These latter risk indexes, as mentioned above, can be obtained only by the "sequential approach" (Fig. 3).

All the above-mentioned outputs are standard for the programs.

b) <u>Loading and temperature duration curves</u>. These outputs supply the durations of the power flowing and these of hourly and maximum yearly temperature in each system component.

In Fig. 4 are represented the typical loading

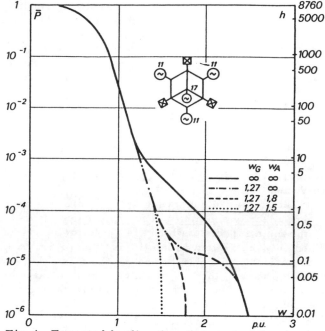

Fig. 4 Expected loading duration curves for the peripheral lines of the model.

curves of one line of a model of a meshed network as obtained by applying different operation policies. The first is related to the power "naturally" flowing, when no actions are taken by the Dispatching Center. The second corresponds to the generation shifting when the line loads overcome the set "alarm levels" (W_G). The third is obtained when shifting at W_G is concurrent will load shedding because the line load is still above the "absolute limits" W_A.

c) <u>Generation duration curve</u>. This output gives the duration of the utilization hours of each unit by a generating system. In Fig. 5 the generation duration curve when applying the mixed security-economy policy is shown for a system having characteristics similar to those of the future 1985 Italian system

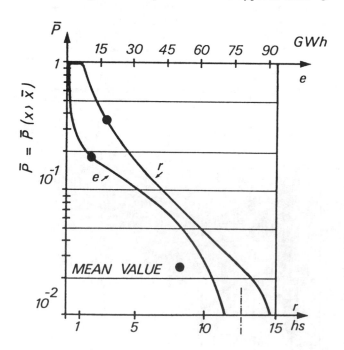

Fig. 3- Probability distributions of amounts (a) and durations (b) of load curtailments in a mixed hydro-thermal system.

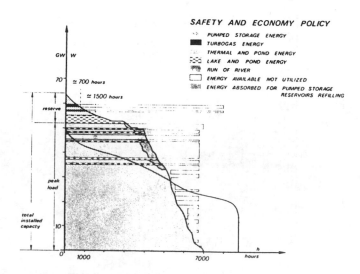

Fig. 5- Generation duration curve of a mixed hydro-thermal system as resulting from the application of a mixed "security-economy" policy.

(it is possible to notice, among other things, the energy supplied for pumping purposes).

d) <u>Detailed graphic descriptions.</u> This output, which can be obtained only when applying the "sequential method", thanks its visual synthesis is particularly useful when a progressive detail of information is desidered in order to better understand the nature of the phenomena considered and to check the correctness in setting the input data (load, operation policies, etc...). For instance, an output more detailed than the estimators of the risk indexes (which are synthetic) is given by the values of the curtailed energy in each of the "simulated years" (Fig. 6).

It is also possible to have, by means of a plotter

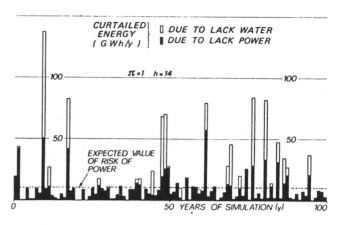

Fig. 6 - Subdivision of the curtailed energy of a generating system by causes, for each of the "simulated years".

connected to the computer, the hour-by hour variations of the main quantities of the system in each of the above-mentioned simulated years on various scales (Fig. 7).

Furthermore the same figure shows, on a larger scale, the details of a very critical week. The repeated crises in the week considered are due to prolonged outage of a considereble amount of nuclear capacity, which in turn caused the hydro stations to

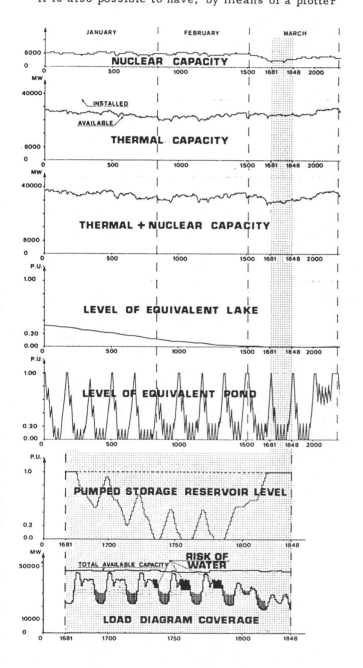

Fig. 7- Detailed reproduction of the evolution of the main quantities of a mixed hydro thermal system in one of the "simulated year"

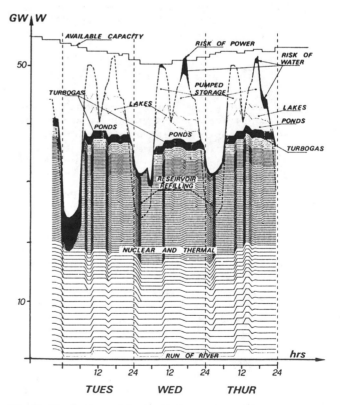

Fig. 8- Evolution of the load dispatching among the thermal units of a generation system

to operate most of the day with the consequent repeated emptying of the reservoirs.

If, in any one week, we are also interested in controlling the economic dispatching of the load among the units, a more detailed representation (Fig. 8) can be further obtained.

e) <u>List of contingencies for dynamic studies.</u> This output is standard for the programs studying transmission systems, and it is to be utilized in the analysis of dynamic system behaviour. For each hour in which the system state changes because of forced outages, it indicates: (i) the network state; (ii) the characteristics of the occurring faults (type, location); (iii) the occurring fault evolution taking into account the protection system behaviour. All this information can also be given in graphic form on a schematic network layout obtained by a plotter connected to the computer.

f) <u>Energy supplied and operation costs.</u> This output, obtained when examining generating systems, gives for each unit and for the overall system the energy supplied and the corresponding costs subdivided into production and start-up costs. Other types information can be obtained by dividing the total energy supplied by each unit between that supplied to the load and pumping or by dividing the hydro energy among the various hydro systems considered or by other similar procedures.

g) <u>Risk of breakdown on the overcrossed objects and loss of strenght of the conductors.</u> These outputs, which constitute the constraints when studying the possibility of encreasing the line loadings, are obtained starting by distributions of hourly and maximum yearly conductor temperatures, switching over voltages and by utilizing appropriate models for the loss of strenght as a function of heating periods (Ref. 19)

Computing Time and Confidence
in the Results

The amount of computing time constitutes the only remarkable limitation of the simulation methods. Infact the main concern of the Montecarlo work is to obtain relatively small standard error in the final results. Unfortunately, the standard error (σ_m) of the distribution of the mean values ($\bar{\eta}$) obtained by the Montecarlo method cannot substantially be reduced by increasing the sample size n, since it is inversely proportional to the square root of the size n of the sample ($\sigma_m = \sigma/\sqrt{n}$).

Rather, the best way to improve confidence in the results lies in the carefull choice of the manner in which the sample is obtained and/or analyzed, as we shall discuss later on (Ref. 15).

A first comment, at any rate, is necessary on the earlier mentioned absolute convenience of applying, whenever possible, the "random approach", even if the "sequential approach" is more extensive and flexible. Actually, the computation of an estimator of the expected system performance (for instance, the risk index) made by the sequential approach requires much longer computing times than those necessary with the random approach in order to achieve the same standard error. Naturally, in both cases the computing times

are strictly dependent on the intrisic characteristics of the system examined, particularly on its reliability. Fig. 9 a) shows the curves representing the percentage error of the progressive mean ($\bar{\eta}$) in respect to the expected value (ϑ), computed in this case by analytical approach, as a function of the sample size for a pure thermal generating system having rather high reliability and roughly the same characteristics on the Italian one. For a sample size n=100 the "random approach" implies an error ($\varepsilon = (\eta - \vartheta)/\vartheta$) of 3.4% while the "sequential approach" an error of 8.6%. The corresponding standard errors (σ_m/ϑ) are 4% and

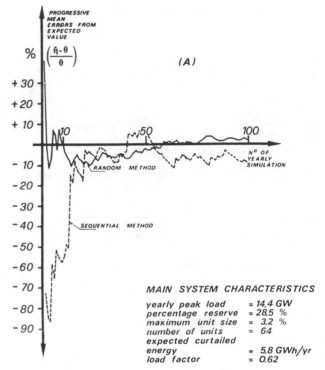

MAIN SYSTEM CHARACTERISTICS

yearly peak load	= 14,4 GW
percentage reserve	= 28,5 %
maximum unit size	= 3.2 %
number of units	= 64
expected curtailed energy	= 5.8 GWh/yr
load factor	= 0.62

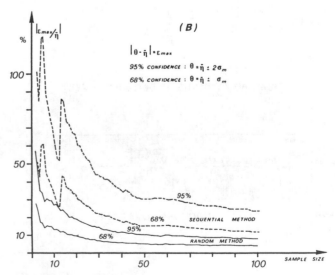

Fig. 9- Comparison of the progressive error (a) and X% confidence (b) of random and sequential approach, when increasing the sample size.

11.4%. This difference is due to the fact that an energy curtailment is determined by very large but very rare availability crises. The results offered by a series of "yearly samples" consisting of "8760 hourly operation simulations": therefore, in respect to the curtailment phenomenon they represent a lower statistical significance.

The ratio of the "efficiencies" of the two types of the Montecarlo method (given by the inverse ratio of the product of the variance and the computing time $K_e = e_r/e_s = n_s . \sigma^2 s/n_r . \sigma^2 r$) assumes, in the hypothesis of equal computing time for the same sample size, for the system considered the value of about $K_e = 7$.

Fig. 9b) shows, for the same case, the curves representing the "x% confidence" limits as obtained by assuming, according to the Central Limit Theorem, that the distributions of each progressive mean, relative to increasing sample sizes n, are normal.

If we had considered a system with lower reliability, for the same sample size the values of the errors for "random approach" and "sequential approach" would have decreased respectively to 10% and 14%. The corresponding standard errors should be respectively 2% and 7% and the ratio of the efficiencies should be $K_e = 12.5$

As earlier stated a lot of ways are used in order to improve the confidence in the results without increasing the computing time.

a) When comparing two slightly different situations S_1 and S_2 of a given system, instead of simulating each situation independently and then comparing the two sets of results, it is much more satisfactory to use the same random numbers for the simulations of the two situations. The precision of the estimated difference is greater since we avoid small differences in the results from being swamped by gross variations due to the finite size of the simulation. This method is referred to as "comparative simulation" . It is adopted when evaluating, for instance, the risk indexes of a mixed hydrothermal system for different sizes of pumped-storage reservoirs.

b) When the risk, or one of its components under study, is due to very rare outages of very important system elements (for instance the substation bus bar), their reliable evaluation would ask for a considerable high sample size n. In this case, in order to reduce the computing time, a method which is inspired by the so-called "stratified sampling" or "strata sampling" is used. We compute the risk of the system when the system elements under consideration are assumed to be safe (stratum 0) and the risk when the system elements under examination are out of service, one by one, weighting the results by the system element predetermined outage probability (stratum 1). The obtained values are finally added. In this case the proposed evaluation is sufficiently correct even without taking into account the further possible contingencies (stratum 2 and higher), being the phenomena considered very rare.

c) More interesting for the transmission system appears to be a method we are now developing, called "importance sampling". Its object is to concentrate the sample points in the time intervals that are of most importance instead of having them

spread. This can be obtained by the random generation of the "time to repair" and "time to failure" of network components. The hourly intervals with a safe network (stratum 0) are examined with the "random approach", while the few hourly intervals during which one or more network elements are out of service (strata higher than 0) are singled out by "sequential approach" and examined together.

As regards the structure of the programs set up we should like to point out that a conditioning factor of a set of integrated programs of the size used in our planning studies, is not only the cost of CPU time but also the interference that it may cause to an EXEC 8 Operating System. In order to solve this problem, according to the nature of the approach used, a "segment" (Ref.16) or a "cascade structure" (Ref. 14) has been used.

Finally, we want to stress that all the programs were set up with the principal aim to explore the possibility offered by the method but so far with no more than a normal care in their optimization from the computing time view point.

In any case, generally speaking, it is very difficult to state the amount of computing time for a program since it depends not only on the size of the system considered (number of network nodes, generators etc...) but also, and greatly, on the reliability of the system components (unavailability, capability limits of the lines, etc...), on operation policies and on the output required.

The Simulation Programs and their Applications to ENEL Power System

The main characteristics of the programs set up according to the philosophy and applying the models previously described are indicated in Tables I and II. For more details on them the reader is referred to the bibliography (Refs. 8, 9, 10, 12, 14).

Generally speaking we can state that, when the reasons for the application of the simulation methods exist, the "sequential approach" is to be applied when hydro generation is involved in the study. When the effect of the network is to be taken into account, it is possible to renounce the "sequential approach", resorting to the "random approach", only if the interactions of the transmission and generation systems are weak.

When the generation is pure thermal, the simulation approach must be "random" and its application is convenient only if transmission system planning is to be examined (otherwise, analytical evaluation methods are much more advantageous).

In order to complete the informations given in the Tables, we deem finally at least it interesting to list some Papers in which the main application of the programs to ENEL's trasmitting and generation systems are described:

a) for SICRET   program: Refs. 17, 18 ; 20
b) for SIREAT   program: Refs. 10, 19;
c) for SOVCON   program: Ref. 19;
d) for SICPOM   program: Ref. 9;
e) for SICIDR   program: Refs. 12, 14;
h) for CARTER   program: Ref. 14.

Conclusions

1)   The Montecarlo methods applied for the quantitative evaluation of power system reliability offer the following advantages:

a) They allow the examination of even complex systems without renouncing a system modelling very near the reality.

b) They constitute a versatile, tool allowing an easy modification of the number and the characteristics of input random quantities and allowing different output quantities to be related.

c) They offer to the planning engineer both the final results of the examination obtained in a synthetic way and the detailed description of the events that the results have caused. The latter output allows the planner engineer to get a sort of "operating experience" absolutely essential to facilitate the exchange of ideas between him and the engineer responsible for system operation.

2)   The application of Montecarlo methods, on the other hand, implies the expenditure of high computing time in order to obtain a sufficient confidence in the results. So far, this disadvantage has limited the application of these methods to the phases of planning refinements and to the first stage of engineering.

3)   The programs utilizing the Montecarlo methods have been set up until now with no more than a normal care in the computing time reduction. At any rate, the advantages pointed out in the conclusion 1), seem to us so fundamental that we think it useful to proceed to the computing time optimisation of the programs set up, since, so doing, we feel that these methods will repay the computing expeditures more and more.

References

1   Bayley E. , Hawking E. , Galloway C. , Wood A. "Generation Planning Programs for Interconnected Systems"-IEEE Trans. PAS Vol. 82, P761, 1963

2   Mallard S. , Thomas V. -"A Method for Calculating Transmission System Reliability"-IEEE Trans. Pas. Vol. 87, March 1968

3   Gaver D. , Montmeat F. , Patton A. -"Power System Reliability - Measures of Reliability and Methods of Calculation"-IEEE Trans. Pas. Vol. 83, July 1964

4   Billington R. -"Power System Reliability Evaluations"-Gordon, Breach Science Publishers, New York, 1970

5   Paris L. , Valtorta M. -"Economic Aspects of Planning Large Generating Units in Interconnected Systems"-World Power Conference, Moscow 1968

6   Manzoni G. , Noferi P. L. , Valtorta M. -"Planning Thermal and Hydraulic Power Systems Relevant Parameters and their Relative Influence"-CIGRE 1972, Rep. 32-16

7   Manzoni G. , Noferi P. L. , Paris L. , Valtorta M. "Power System Reliability Indexes"-NATO Advanced Study Institute of Generic Techniques in System Reliability Assessment, University of Liverpool 1973

8   Noferi P. L. , Paris L. -"Quantitative Evaluation of Power System Reliability in Planning Studies" IEEE Winter Meeting, New York 1971, 71 T. P. 89

9   Paris L. , Salvaderi L. -"Optimization of the Basis Characteristics of Pumped Storage Plants" -IEEE Winter Meeting, New York, C74 - 158.2

10   Noferi P. L. , Paris L. -"Effects of Voltage and Reactive Power Constraints on Power System Reliability" -IEEE Summer Meeting, Anaheim 1974 - T74 428-9

11   Paris L. -"Analisi delle funzioni degli impianti idroelettrici ad accumulo per pompaggio nel sistema di produzione dell'energia elettrica" Proceedings of the Seminario de Usinas Reversiveis -AECESP Sao Paulo 1973

12   Panichelli S. , Salvaderi L. , Scalcino S. -"Unavailability Statistical Models for Power System Reliability Studies" -NATO Advanced Study Institute of Generic Techniques in System Reliability Assessment, University of Liverpool 1973

13   Paris L. , Salvaderi L. -"Pumped Storage Plant Basic Characteristics: their Effects on Generating System Reliability" -American Power Conference, Chicago 1974

14   Panichelli S. , Salvaderi L. , Scalcino S. -"A Set of Programs Used for Detailed Operation Simulation in Power System Planning Studies" -UNIPEDE Informatique Conference -Madrid 1974

15   Hammesley M. , Hardscomb D. -"Montecarlo Methods" -Methnen Ltd, Norwich 1967

16   Noferi P. L. , Salvaderi L. -"A Set of Programs for the Calculation of the Reliability of a Transmission System" -UNIPEDE Informatique Conference - Lisboa 1971

17   Paris L. , Reggiani F. , Valtorta M. -"The Study of UHV System Reliability in Connection with Structure and Component Characteristics" CIGRE 1972, Rep. 31-14

18   De Falco E. , Reynaud C. , Salvaderi L. -"An Approach for Power System Station Reliability Evaluation" -AEI Annual Meeting, Turin 1972

19   Paris L. Reggiani F. , Valtorta M. -"Possibilities to Increase Transmission Line Loadings in Well Developed Electrical Networks"-CIGRE 1974 Rep. 31-13

20   Comellini E. , Noferi P. L. , Paris L. , Reggiani F. Salvaderi L. -"Evolution of the Primary Distribution System Related to Reliability and Short Circuit Currents" -CIRED, London 1972

TABLE I - MAIN FEATURES OF SIMULATION PROGRAMS USED IN TRANSMISSION PLANNING

Program code	Systems considered	Program aims	Type of simulation	Operation policies adopted	Main outputs
SICRET	-Transmission and thermal generation systems (only active power requirements)	-Transmission system planning -Substation layout selection	-Random on hourly basis	-Maximum security	-Risk index in terms of curtailed energy:how, where, when. -Line and transformer loading duration curves
SIREAT	-Transmission and thermal generation systems (both active and reactive power requirements)	-Determination of amount and allocation of reactive power generation -As above	-As above	-Maximum security -Minimum voltage irregularity	- As above - Network voltage irregularity: where
SOVCON	-Transmission and thermal generation system	-SICRET and SIREAT integration for the determination of transmission components mains characteristic	=====	=====	- Hourly and maximum yearly temperature duration curves for lines and transformers - Aging of lines and transformers - Risk of discarge beween line conductors and over crossed objects.

TABLE II - MAIN FEATURES OF SIMULATION PROGRAMS USED IN GENERATION PANNING

Program code	Systems considered	Program aims	Type of simulation	Operation policies adopted	Main output
SICPOM	-Busbar generating system:thermal basic and peaking units, pumped storage	-Generation system plannig -Determination of basic pumped-storage plant characteristics	-Sequential on an hourly basis	-Maximum security -Combined security and economy	-Risk indexes -Frequency, duration, amount of curtailments -Generation duration curve
SICIDR	-Busbar generating system:thermal basic and peaking units, complete hydro systems (lake ponds, run-of-river stations) and pumped storage	-Generation system planning -Determination of basic pumped storage plant characteristics -Basic studies for hydro plant modification and/or sizing	-As above	- As above	-As above
CARTER	-Busbar generating system:thermal basic and peaking units, complete hydro system (lake pond , run-of-river stations)and pumped storage	-Determination of basic chacacteristics of peaking thermal units -Evaluation of generating units operation costs	-As above	- As above	-Generation duration curve -Operation costs for each unit.

RELIABILITY INDICES FOR USE IN

BULK POWER SUPPLY ADEQUACY EVALUATION

A report prepared by the Working Group on Performance Records for Optimizing System Design, Power Systems Engineering Committee. M. B. Guertin, Chairman, Hydro-Quebec; P. F. Albrecht, General Elecric Co.; M. P. Bhavaraju, Public Service Electric and Gas Co.; R. Billinton, University of Saskatchewan; G. E. Jorgensen, Northeast Utilities Service Co.; A. N. Karas, Canadian National Energy Board; W. D. Masters, Cleveland Electric Illuminating Co.; A. D. Patton, Texas A and M University; N. D. Reppen, Power Technologies, Inc.; R. P. Spence, Alabama Power Co.

ABSTRACT

A number of reliability indices may be used to measure bulk power supply adequacy. This paper presents a classification and listing of a sampling of indices in use and under consideration in Europe and North America. The fundamental definition of each index is given. The paper is offered for review with the objective of promoting the use of reliability indices in bulk power system planning and design.

INTRODUCTION

Bulk power planning and operation procedures include reliability as one of the essential measures of system requirements. Criteria for bulk power systems assess reliability with two attributes: adequacy and security. Adequacy may be defined as a system's capability to meet system demand within major component ratings and in the presence of scheduled and unscheduled outages of generation and transmission components or facilities. Security may be defined as a system's capability to withstand disturbances arising from faults and unscheduled removal of bulk power supply equipment without further loss of facilities or cascading. This paper deals only with adequacy assessment.

To provide a quantitative evaluation of the reliability of an electrical system, it is necessary to have indices which can express system failure events on a frequency or a probability basis. System failure events must be defined for each specific purpose, which for example may include:

1. Load interruption
2. Capacity reserve deficiency
3. Energy deficiency
4. Reactive capacity deficiency and/or bus voltages outside design criteria
5. Limitations in transfer capability and/or circuit or component overload.

Bulk power supply systems are "distributed" in the sense that points of generation and of load span a region or territory. Evaluations may therefore, neces-

sitate system-wide comparisons of alternative configurations or, on the other hand, may require identification of "weak links" and potential trouble spots. Selection of appropriate indices may thus be constrained by satisfaction of the macro and micro aspects of bulk power supply evaluation requirements.

The indices offered in this document represent a sampling from indices used and under consideration in Europe and North America. These indices have been applied to system generation and transmission capability models and to composite models of system generation, transmission, and load to measure bulk power supply adequacy.

CLASSIFICATION OF RELIABILITY INDICES

Reliability indices for adequacy assessment may be classed generally under the categories Probability, Frequency, Duration, and Expectation.

Probability Indices measure risk (or assurance) under specified configurations, load, time and time-span conditions. For example, the risk of insufficient capacity reserve at annual peak load is a probability index.

Example: P {insufficient capacity reserve at annual peak} = 0.001

Frequency Indices measure the expected rate of recurrence of specific events per unit of time.

Example: Frequency of capacity deficiency = 0.2 events/year

Duration Indices are employed to indicate the expected time in residence in a specific state; i.e., load and configuration.

Example: Average duration of capacity deficiency = 0.25 days

Expectation Indices are formed from the average or expected value of a random variable such as reserve deficiency or interrupted load. Let C be the available capacity, and let $p_C(C)$ be the probability density function for the random occurrence of C. For a given load (L) the expected capacity deficiency would be given by

$$E\{Cap.\ Deficiency\} = \int_{C=0}^{C=L} (L-C)\ p_C(C)\ dc \qquad MW$$

These indices are by themselves mono-parametric indices; i.e., the indices employ a single statistical parameter. Bi-parametric indices[1] are expressed by two statistical parameters. As an example, a frequency and

F 77 590-3. A paper recommended and approved by the IEEE Power System Engineering Committee of the IEEE Power Engineering Society for presentation at the IEEE PES Summer Meeting, Mexico City, Mex., July 17-22, 1977. Manuscript submitted February 9, 1977; made available for printing April 6, 1977.

Reprinted from *IEEE Trans. Power Apparatus Syst.*, vol. PAS-97, no. 4, pp. 1097–1103, July/Aug. 1978.

TABLE 1

Failure Classification or Measure	Mono-Parametric Indices	Bi-Parametric Indices
a) Generation Reserve Deficiency	Probability of Deficit	Frequency and Duration of Reserve Deficit
b) Resource or Energy Deficiency	Expected Energy Curtailment	Expected Energy Curtailment per Occurrence and Frequency of Occurrence
c) Circuit or Transformer Overload	Probability of Overload Frequency of Overload	Expected Overload per Occurrence and Frequency of Occurrence
d) Bus Voltage Outside Design Criteria	Probability of Violation Frequency of Violation	Expected Violation per Occurrence and Frequency of Occurrence

duration index provide the following information concerning the occurrence of various capacity system states*.

Frequency: The average rate at which a specific rate is encountered.

Average Duration: The average residence time in a specific state.

Table 1 illustrates the format of several mono-parametric and bi-parametric indices employed for several categories.

Tri-parametric indices may be called for in certain applications such as overloads of circuits or transformers. (Expected overload per occurrence, frequency of occurrence and duration of overloads.)

GENERATING CAPACITY PLANNING INDICES

Generating capacity planning indices are used to assess the amount of generation capacity in terms of its sufficiency to supply the load. Since transmission limitation or operating problems are generally not considered the indices do not assess the ability of the system to provide uninterrupted supply of power at particular points in the system. Various models of generating capacity and loads are used in generating capacity assessment.

A. LOLP[2]

This index is defined as the long run average number of days in a period of time that load exceeds the available installed capacity. The index may be expressed in any time units for the period under consideration and, in general, can be considered as the expected number of days that the system experiences a generating capacity deficiency in the period. This index is commonly, but mistakenly, termed the "loss of load probability, (LOLP)". A year is generally used as the period of consideration. In this case, the LOLP index is the long-run number of days/year that the

hourly integrated daily peak load exceeds the available installed capacity.

Let $P_C(C)$ = P {Capacity $\leq$ C} and daily peak load for day j be L_j

then

$$LOLP = \sum_j P_c(C_j)$$

where

C_j is the greatest value of capacity which is insufficient to meet the load L_j.

Example: LOLP = 0.1 days/year.

$P_C(C)$ is influenced by the set of generator units available for service. $P_C(C)$ is stationary for an interval in which the units on maintenance is constant.

B. The Probability of Not Meeting the Annual Peak[3] Load

This index is defined as the probability of insufficient capacity available to meet the annual peak load. The index is evaluated by comparing the probabilistic distribution of the maximum annual load with that of the system capacity availability on the day of the year in which the peak occurs.

Let $P_C(C)$ = P {Capacity $\leq$ C} and $p_L(L)$ be the probability density function for annual peak load.

$$P \text{ \{not meeting annual peak load\}} = \int_{L_{min}}^{L_{max}} p_L(L) P_C(C_j) dL$$

where

L_{max} and L_{min} are the upper and lower extremes of the probability density function for annual peak load

C_j is the greatest value of capacity which is insufficient to meet the load

Example: Probability of not meeting the annual peak load = 0.001

*States possessing a common attribute, such as all states of capacity less than or equal to a given amount, may be combined or "merged" to form a "cumulative state". Definitions of frequency and average duration may then be extended to the events of occurrences of or residence in the cumulative states.

C. Expected Loss of Load (XLOL)[4]

This index is defined as the expected value of capacity deficiency given an event of capacity deficiency. The index is equal to the expected capacity deficiency divided by the probability of capacity deficiency. Let $p_c(C)$ be the probability density function for capacity valid for period(s) with a given set of generator units not on planned outage. Let $p_L(L)$ be the probability density function for load over the corresponding periods. Depending on the application $p_L(L)$ may represent annual peak load, daily peak load, hourly load, etc. Then for each time period within which $p_c(C)$ is stationary:

E {Capacity Deficiency} =

$$\int_{L_{min}}^{L_{max}} \int_{C=0}^{C=L} (L-C) p_L(L) p_c(C) dCdL$$

P {Capacity Deficiency} =

$$\int_{L_{min}}^{L_{max}} \int_{C=0}^{C=L} p_L(L) p_c(C) dCdL$$

XLOL =

E {Capacity Deficiency}/P {Capacity Deficiency}

Example:

E {Capacity Deficiency}   =   0.5 MW

P {Capacity Deficiency}   =   0.005

XLOL   =   100 MW

D. Expected Energy Curtailment/Year[1]

This index is defined as the long-run average energy curtailed per year due to deficiency of generation capacity. The index is evaluated by adding the expected energy deficiency for each capacity deficiency event considering the deficiency probability, magnitude, and duration. Let $p_{ci}(C)$ be the probability density function for capacity for time period i within which there is no change in generator units on planned outages. Let $p_{Li}(L)$ be the probability density function for load over this same period. $p_{Li}(L)$ may typically represent all hourly integrated loads. If there are N periods of equal length in a year, then:

E {Energy Curtailment/Year} =

$$\sum_{i=1}^{N} \int_{L=L_{imin}}^{L_{imax}} \int_{C=0}^{C=L} \frac{8760}{N} (L-C) p_{Li}(L) p_{ci}(C) dCdL$$

Example:

Expected Energy Curtailment/Year

= 10 MWh/year

E. Frequency and Duration of Capacity Deficiency[5-8]

This index is defined as the average rate at which a capacity deficiency is encountered and the average duration of capacity deficiency. The dimensions of this bi-parametric index are typically frequency/year and duration in days or hours. Two-state daily load models (peak and off-peak), multiple state, and continuous models have been used in computing this index.

Example:

Frequency of capacity deficiency

= 0.2 events/year

Average duration of capacity deficiency

= 0.25 day.

BULK POWER INTERRUPTION AND CURTAILMENT INDICES FOR COMPOSITE GENERATION AND TRANSMISSION ADEQUACY ASSESSMENT

Bulk power interruption and curtailment indices measure the magnitude of interruptions in terms of affected load. Interruption indices are defined in terms of the cumulative MW of load interrupted per year while the energy curtailment index is defined in terms of the cumulative MW minutes (or hours) of load interrupted per year. The following example illustrates key quantities in the computation of interruption and energy curtailment indices.

	No. of Bulk Power Supply Disturbances/Year	Annual Load Interrupted
Numerical Computation of Quantity	$\sum_{i=1}^{N} F_i$	$\sum_{i=1}^{N} F_i C_i$
Example:	19 year^{-1}	10,4000 MW/yr.

	Annual Peak Load	Annual Energy Curtailment
Numerical Computation of Quantity	C_{MX}	$\sum_{i=1}^{N} F_i C_i D_i$
Example:	346,425 MW	624,000 MW-min./yr.

where

F_i = Frequency of outage event i (year^{-1})

C_i = Load interrupted for outage event i (MW)

D_i = Duration of interruption (minutes)

C_{MX} = Annual peak load

A. Bulk Power Interruption Index

This index is defined as the ratio of annual load interrupted to annual peak load.

Example:

Annual Load Interrupted $\sum_{i=1}^{N} F_i C_i$ = 10,400 MW/yr.

Annual Peak Load $\quad C_{MX}$ = 346,425 MW

Bulk Power Interruption Index $\dfrac{1}{C_{MX}} \sum_{i=1}^{N} F_i C_i = \dfrac{10,400}{346,425}$

$$= 0.03 \text{ MW/MW-yr.}$$

This index may be interpreted as equivalent per unit interruption of annual peak load. Thus, one complete system outage during peak load conditions contributes 1.0 to this index. The index has its counterpart in the Load Interruption Index used to assess reliability at the distribution level[6].

B. Bulk Power Energy Curtailment Index

The bulk power energy curtailment index is defined as the ratio of the annual energy curtailment to the annual peak load.

Example:

Annual Energy Curtailment:

$$\sum_{i=1}^{N} F_i D_i C_i = 624,000 \text{ MW-min./yr.}$$

Annual Peak Load:

$$C_{MX} = 346,425 \text{ MW}$$

Bulk Power Energy Curtailment Index:

$$\dfrac{\sum_{i=1}^{N} F_i D_i C_i}{C_{MX}} = \dfrac{624,000}{346,425} = 1.8 \text{ MW-min./MW-yr.}$$

This index may be interpreted as equivalent duration (minutes) of loss of all load during peak load conditions. Thus, one complete system outage during peak conditions contributes with its duration (in minutes) to this index.

C. Bulk Power Supply Average Curtailment Per Disturbance

This index is defined as the average MW of load affected per disturbance. It is the ratio of the annual load interrupted to the number of annual bulk power supply disturbances.

Example:

Number of Bulk Power Supply Disturbances per year:

$$\sum_{i=1}^{N} F_i = 19 \text{ year}^{-1}$$

Annual Load Interrupted:

$$\sum_{i=1}^{N} F_i C_i = 10,400 \text{ MW/yr.}$$

Bulk Power Supply Average MW Curtailment per Disturbance:

$$\dfrac{\sum_{i=1}^{N} F_i C_i}{\sum_{i=1}^{N} F_i} = \dfrac{10,400 \text{ MW/yr.}}{19 \text{ disturbances/yr.}}$$

$$= 547 \text{ MW/disturbance}$$

Tests of generation and transmission system failure may necessitate system-wide and spot or local application. Thus, constraints upon selection of reliability indices for bulk power supply performance evaluation include the need for applicability to both system and spot or local application. It would appear that reserves assessment and/or load curtailment provide the most readily applied measures of system failure to use for formation of reliability indices. This is not to suggest that other measures are not of high applicability but rather that measures such as:

- Overload of line, transformer, or breaker

- Out of design limit bus voltage

- Generation capacity reserve deficiency

- Resource or energy deficiency

cannot be directly related to events involving load interruption. Bulk power reliability assessment approaches may use circuit overloads as a failure criterion rather than as an index in its own right. If for specific events, overloads cannot be eliminated by generation redispatch, remedial strategies are provided to curtail sufficient load to meet the criterion circuit ratings. The load curtailed by the strategy then carries the same dimension as the load affected by events leading to bus isolation or system separation. In this manner, overloads are translated into a contribution to the "Bulk Power Interruption Index" and/or the "Bulk Power Energy Curtailment Index."

Aspects of bulk power reliability evaluations are discussed in references 1, 10, 11, and 12.

ACKNOWLEDGEMENTS

Work on this document was initiated by the Working Group on Performance Records for Optimizing System Design in 1970. The effort was spearheaded by R. J. Ringlee of Power Technologies, Inc., in collaboration with R. Billinton, University of Saskatchewan, and A. D. Patton, Texas A and M University. The document is based in part on work performed by R. J. Ringlee under the sponsorship of the Ford Foundation[13] and the National Science Foundation.[14] The paper was finalized during 1975/76 by N. D. Reppen of Power Technologies, Inc., with major contributions from P. F. Albrecht of General Electric Co., M. P. Bhavaraju of Public Service and Gas Co., G. E. Jorgensen of Northeast Utilities Service Co., and L. Salvaderi of ENEL, Italy.

REFERENCES

1. "Risk Indices for Evaluating the Reliability of an Electrical System", by G. Manzoni, P. L. Noferi, L. Paris, M. Valtorta, 1973 NATO Conference, Liverpool, England.

2. "Application of Probability Methods of Generating Capacity Problems", AIEE Subcommittee on Application of Probability Methods, AIEE Transactions, PT. Ill. (Power Apparatus and Systems), Vol. 79, 1960, pp. 1794-1800.

3. "Generation Reserve Requirements - Sensitivity to Variations in System Parameters", E. M. Mabuce, R. L. Wilks, S. B. Boxerman. Paper PG 75-651-0, presented at the Joint Power Conference, Portland, OR, Sept. 29, 1975.

4. "Expected Value of Generation Deficit: A Supplemental Measure of Power System Reliability", J. T. Day, P. B. Shortley, J. W. Skooglund, IEEE Transactions, Vol. PAS-91, No. 5, September/October 1972, pp. 2213-2223.

5. "Frequency and Duration Methods for Power System Reliability Calculations - Part I - Generation System Model", J. D. Hall, R. J. Ringlee, A. J. Wood, IEEE Transactions, PAS-87, No. 9, September 1968, pp. 1787-1796.

6. "Frequency and Duration Methods for Power System Reliability Calculations - Part II - Demand Model and Capacity Reserve Model", R. J. Ringlee, A. J. Wood, IEEE Transactions, PAS-88, No. 4, April 1969, pp. 375-388.

7. "System Load Representation in Generating Capacity Reliability Studies; Part I - Model Formulation and Analysis", R. Billinton, C. Singh, IEEE Transactions, Vol. PAS-91, No. 5, September/October 1972, pp. 2125-2132.

8. "A Frequency and Duration Method for Generating System Reliability Evaluation", A. K. Ayoub, A. D. Patton, IEEE Transactions, PAS-95, November/December 1976, pp. 1929-1933.

9. "Definitions of Customer and Load Reliability Indices for Evaluating Electric Power System Performance", by Working Group on Performance Records for Optimizing System Design, Power System Engineering Committee, Paper A 75-588-4, presented at San Francisco, CA, July 20-25, 1975.

10. "Power System Reliability - I - Measure of Reliability and Methods Calculation", D. P. Gaver, F. E. Montmeat, A. D. Patton, IEEE Transactions (Power Apparatus and Systems), July 1964, pp. 727-737.

11. "Composite System Reliability Evaluation", R. Billinton, IEEE Transactions, PAS-88, April 1969, pp. 276-280.

12. "Transmission System Reliability Methods", M. P. Bhavaraju, R. Billinton, IEEE Transactions, Vol. PAS-91, No. 2, March/April 1972, pp. 628-637.

13. "An Assessment of Policies and Technical Factors Affecting the Degree of Interconnection of Electric Utilities in the U.S.", Final Report to the Energy Policy Project, sponsored by the Ford Foundation, by Power Technologies, Inc., December 1974.

14. "Adequacy Assessment of Interconnected Electric Power System", Report under NSF Grant AER74-19588, by Worcester Polytechnic Institute and Power Technologies, Inc., March 1976.

Discussion

Luigi Salvaderi (ENEL, Rome, Italy): The W.G. of Report represents an excellent review and presentation of the risk indices used in system planning and design.

To supplement the paper I think it useful to mention two indices used in Europe when mixed hydro-thermal systems are to be dealt with.

In systems including generating plant of limited energy resources (hydro plants or any other plant of the storage type), other risk indices are to be taken into account in addition to those related to the impossibility of facing the demand owing to the lack of installed capacity. These indices are usually taken to represent the "risk of energy" and are commonly evaluated and used in planning practice in Europe.

Two main indices are used:

$LOLP_E$ = loss of load probability due to lack of stored energy (days/yr)

E_E = yearly curtailed energy, due to lack of stored energy (MWh/yr)

They are to be added to the correspondent indices related to the lack of installed capacity, and the total indices to be considered are:

$$LOLP_{TOT} = LOLP + LOLP_E \qquad \text{(days/yr)}$$
$$E_{TOT} = E + E_E \qquad \text{(MWh/yr)}$$

Their evaluation in any period of the year depends, apart from the actual inflows, on the water (or other type of energy) available in the reservoir, that is on the "history" of the water in the previous periods. Furthermore, they depend on the operation policies by which any hydro plant is exploited, especially when the hydro plants are mutually interacting.

For these reasons the evaluation of these "energy risk indices" is obtained by direct probabilistic evaluations only in particular and simplified cases; in other cases it is necessary to resort to simulation of the Montecarlo type.

In a mixed hydro-thermal system the analytical direct calculation of these indices is possible in the case that: a) the behavior of the system can be considered on a weekly basis; b) the hydro subsystem is modeled with only one large (at least monthly) equivalent reservoir; c) the hydro subsystem is used for peaking purposes only after every thermal plant has been utilized, or to avoid waste of water ("defence of water" policy) [1].

In this case the possibility of meeting the load depends on the following five random variables:

C = thermal and hydro capacity available. Probability densities in each week within which there is no change in generating units on planned outages: $p_{AT}(C)$; $p_H(C)$

L = load. Weekly probability density: $p_L(L)$

E_{LH} = load energy to be supplied weekly by the hydro plant. Probability density: $p_H(E_{LH})$

W = volumes stored in the reservoirs at the beginning of each week. Probability density: $p_V(W)$

W = weekly inflows to the hydro plants. Probability density: $p_{IN}(W_a)$

$LOLP_E$ can be defined as the weekly probability that the energy stored in the reservoirs during the week, as resulting from the volume at the beginning of the week plus the inflows during the week, is insufficient as compared with the energy required by the hydro plant during the same week.

Its calculation can be obtained as follows. For a fixed week k, from the weekly load duration curve and the probability density of the available thermal capacity (having taken into account the thermal units on planned maintenance) it is easy to obtain the probability density of the energy required by the hydro plants in the week: $p_H(E_{LH})$. Furthermore, from the probability density of the reservoir level at the beginning of the week $p_V(W)$ and the probability density of the inflows during the week $p_{IN}(W)$, it is possible to evaluate the probability density of the water $p_S(W)$ stored during the week (and therefore energy, through the plant coefficient o).

For the week considered we will get:

$$LOLP_E(w_k) = \int_0^{E_{LH}} p_H(E_{LH}) dE_{LH} \, P_s(E_{LH}) \qquad \text{(days/year)}$$

Manuscript received July 19, 1977.

and the yearly correspondent risk will be:

$$LOLP_E = \sum_K^{52} LOLP_E(w_k) \qquad \text{(days/year)}$$

As regards the computation of E_E, in the same conditions and utilizing the same symbols as above, the expected value of the yearly curtailed energy due to lack of stored energy is:

$$E_E = \sum_K^{52} \int_0^{E_{LH}} \int_{\alpha W=0}^{\overline{E}_{LH}} p_H(E_{LH})dE_{LH}(E_{LH} - \alpha W)\, p_s(\alpha W)d\alpha W \qquad \text{(MWh/year)}$$

Fig. 1 illustrates the situation for a real system.

As said, when the hydro-system is complex—for instance, when they are constituted of lake stations, pond stations, run-of-river stations mutually interacting—or when pumped-storage plant having reservoirs of 4 to 14 hours, or any other type of storage plant, is considered the analytical approach is impossible because of the impossibility of taking into account the operation policies and the necessity of examining the system behavior on an hourly basis. In this case, the yearly curtailed energy can be evaluated only by resoirting to sequential Montecarlo simulation [2, 3]. All the hours of the year are examined and those when it's impossible to face the load, owing to both lack of available capacity and lack of stored energy, evidenced. The values of curtailed energy are added up for all these hours of the year. By repeated simulation of the same year, the average value of the yearly curtailed energies gives, with the accepted confidence in the results, an estimand of the expected value of the "risk of power" and "risk of energy."

REFERENCES

[1] Manzoni, G., Noferi, P. L., and Valtora, M, "Planning Thermal and Hydraulic Power system—Relevant Parameters and their Relative Influence," CIGRE, 1972, Rep. 32-16.
[2] Paris, L. and Salvaderi, L., "Optimization of the Basic Characteristics of pumped storage plants," IEEE Winter Meeting, New York, 1974, C 74 158-2.
[3] Panichelli, Salvaderi, L., and Scalcino, S., "A set of Programs Used for Detailed Operation Simulator in Power System Planning Studies," UNIPEDE Data Processing Conference, Madrid, 1974.

R. Hoseason (General Electricity Generating Board, London, England): There are two points I would like to make about the use of the various types of index. For a large interconnected system we can be certain that a significant percentage of plant will be unavailable at all times. The variation in availability will also be relatively small. In planning the generating capacity reserve margin the uncertainty in the estimated load is the most significant parameter.

Secondly, the duration of failures in hours and the energy curtailed are sensitive to the expected daily load shape. For the CEGB system a 1 percentage point change in the daily load factor distributed over the critical daytime load plateau results in a 25% change in the forecasts of hours and MWh of failure per century.

Because of the relative simplicity of the presentation and under-

Manuscript received August 11, 1977.

standing of the results, the probability of not meeting the annual peak is the preferred index at present for Great Britain when making the margin policy decision. This annual peak may occur at any time during the three winter months. The other indices are calculated as a back up to ensure that risks outside the peak half-hour are not increasing unduly.

G. E. Marks (Florida Power Corporation, St. Petersburg, FL): Of the indices presented in this paper, probably the two most widely used within the United States are the loss of load probability (LOLP) and the expected loss of load (XLOL). It should be pointed out that when the index is calculated in days per year that it equally weights insufficient capacity situations regardless of the number of hours during the day for which the capacity deficiency exists. Some companies have overcome this by calculating the index using an hourly load model and expressing the LOLP as the ratio of hours of insufficient capacity to total hours of exposure [1]. Unfortunately, the term "LOLP" has been applied to both approaches and the results are quite dissimilar. Documentation should always make it clear which method is being used. One approach would be to always quote the units in which the results are being expressed.

The expected loss of load (XLOL) is also commonly used. However, it can easily be misinterpreted by lay persons to mean the predicted magnitude of a capacity deficiency. To prevent this confusion, it might be better to quote the expected energy curtailment per year (expected loss of energy, XLOE). This index provides essentially the same measure of system performance without the above-mentioned disadvantage. The ratio of this value to the total system energy requirements has been defined as the loss of energy probability (LOEP) [1].

Numerous efforts are currently being made in the area of composite generation and transmission adequacy assessment. Hence, the presentation of bulk power interruption and curtailment indices is of major interest. The indices presented here appear to be valuable, particularly with respect to the comparison of composite systems and expansion plans. It is possible, however, to also calculate a set of indices based on the number of customers interrupted [2]. Since such data is already being collected by many companies within the United States, the latter may be of some advantage in the comparison of calculated to actual results to customer-oriented management executives and/or governmental agencies.

There has been some discussion recently concerning relative versus absolute indices. Relative indices being those by which systems or alternate plans may be compared, and absolute indices referring to those which form absolute standards of reliability or bulk power system criteria. If absolute standards for reliable electric service are to be formulated, they should center around the frequency and duration indices. These indices have the advantage of direct physical significance. They are directly measurable and can be easily comprehended by lay persons. Consequently, frequency and duration of load curtailment or interruption should be included under the topic of bulk power interruption and curtailment indices.

In summary, the indices proposed within this paper appear to be reasonable measures of power system performance. Adjustments will be made to incorporate the proposed indices in a computer program for power system reliability assessment now under development [2].

Manuscript received August 11, 1977.

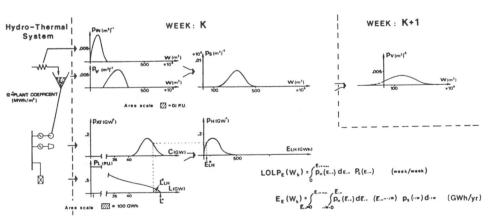

Fig. 1. Direct analytical evaluation of energy risk indices $LOLP_E$ and E_E in a mixed hydro-thermal system in which the hydro-plants have prevailing peaking duty.

REFERENCES

[1] "Power System Reliability Assessment, Phase I—Generation Effects," EEI—Reliability Assessment Task Force, presented to —System Planning Committee on February 15, 1977.
[2] Marks, G. E., Member, IEEE, Florida Power Corporation, St. Petersburg, Florida, "A Method of Combining High-Speed Contingency Load Flow Analysis With Stochastic Probability Methods to Calculate a Quantitative Measure of Overall Power System Reliability," May, 1977.

C. **Singh** (Ministry of Transportation & Communications, Ontario,* Canada): The Working Group on Performance Records for Optimizing System Design is to be complimented for this review paper. The discussor wishes to make a few comments.

Reference 1 provides an alternative interpretation of the "Expected Loss of Load (XLOL)" as

$$XLOL = \frac{\text{Expected Energy Curtailed/Yr.}}{\text{Expected Duration of Load Loss/Yr.}}$$

That is, XLOL is expected energy curtailed in a given period normalized by the expected duration of load loss in that period. An other measure called Energy Index of Reliability [2, 3] can be obtained by dividing the expected energy curtailed per year by the expected energy required per year and subtracting this ratio from unity.

The capacity deficiency state may be encountered either due to the loss of generation capacity for a given load level or by the increase in load for a given capacity level. Reference 4 raises some questions regarding the contribution to frequency from the multiple curtailment of load by a single capacity deficiency state, due to the daily variation of load between its peak and low values. The frequency and duration taken together, however, tend to compensate the picture.

The various indices would be used to provide an input into the process of planning, design or operation by comparing their values for alternative strategies and decisions or by relating to reference values for particular indices. Sensitivity studies should be performed to examine the effect of the choice of index on the decision process. Such sensitivity studies and analyses will provide guidance for the selection of appropriate indices under given circumstances. As an example, the effect of the choice of index (availability, frequency and energy index of reliability) on the decision regarding unit additions is studied in [2, 3, 5]. More studies of this kind, on typical systems, will provide a better insight into the behavior of various indices.

The choice of a target or reference value for an index is generally a difficult problem. Studies are needed both to examine the effect of the

Manuscript received August 12, 1977.

choice of target values on decision processes and also for providing guidelines in selecting appropriate values. Is this group contemplating any work in this direction?

REFERENCES

[1] Billinton, R., and Singh, C., Discussion to Reference 4 of the paper under discussion.
[2] Billinton, R., and Singh, C., "System Load Representation in Generating Capacity Reliability Studies, Part II—Applications and Extensions," IEEE Transactions, Vol. PAS-91, No. 5, Sept./Oct. 1972, pp. 2133-2143.
[3] Singh, C., "Reliability Modelling and Evaluation in Electric Power Systems," Ph.D. Thesis, August 1972, University of Saskatchewan.
[4] Singh, C., and Kankam, M. D., Discussion on "Load Modelling in Power System Reliability Evaluation," R. Billinton and J. Endrenyi, Paper A 76305-3, Summer Power Meeting 1976.
[5] Singh, C., "Generating Capacity Reliability Evaluation Using Frequency and Duration Methods," M.Sc. Thesis, August 1970, University of Saskatchewan.

*The opinions expressed in this discussion are author's own and may may not be attributed to M.T.C.

Working Group on Performance Records for Optimizing System Design: We thank Messrs. Salvaderi, Hoseason, Marks, and Singh for relating their experiences with the use of reliability indices. We feel this set of discussions is a most valuable contribution to the paper.

We concur wholeheartedly with Mr. Marks' comments on the importance of precise definition of reliability indices and we hope this paper will be helpful in this regard. Mr. Marks also suggests the use of indices based on the number of customers interrupted. Such indices were defined and classified as customer and load reliability indices in a previous PROSD paper (Reference 9). They are not as suitable for bulk power system adequacy assessments as those defined in the current paper.

Dr. Singh points out that further work is needed to examine the effect on the system expansion decision process of both the choice of index and of the choice of target value for indices used. The PROSD Working Group has not scheduled any work regarding the selection of criteria or target values of reliability indices, but we expect to do some work comparing various indices on a test system which is presently being prepared by the Application of Probability Methods Subcommittee.

Manuscript received October 31, 1977.

ADEQUACY INDICES FOR COMPOSITE GENERATION AND TRANSMISSION SYSTEM RELIABILITY EVALUATION

R. Billinton   T.K.P. Medicherla   M.S. Sachdev
Fellow, IEEE   Student Member, IEEE   Senior Member, IEEE

Power Systems Research Group
University of Saskatchewan
Saskatoon, Canada

Abstract - This paper defines a new set of bus and system indices for evaluating the adequacy of a composite generation and transmission system. These indices, together with those proposed by an IEEE Committee Working Group are illustrated by application to a practical system configuration. The new indices are based upon the load curtailment required to alleviate transmission line overloads. A decoupled line overload alleviation technique which is a computationally efficient and direct method of determining a new generation and load schedule is used. The proposed set of indices provide detailed information on system adequacy. Selection of the most appropriate index in a particular situation is dependent upon the system under study and the purpose for which the index is intended.

INTRODUCTION

A power system can be divided into the three basic segments of generation, transmission and distribution. This paper is concerned with the development of a range of adequacy indices for assessing the composite generation and transmission segments. The indices obtained for the individual bulk system load points can be used as basic infeed parameters for including the distribution segments to determine individual customer load point indices. A recent paper[1] prepared by the IEEE Working Group on Performance Records for Optimizing System Design defined a set of bulk power interruption and curtailment indices for composite generation and transmission adequacy assessment. These indices were defined in terms of global system adequacy and therefore are not basically suited to individual load point assessment. The adequacy of supply at a load bus can be expressed in terms of the probability of failure and the frequency of failure[2,3]. These indices have been calculated by defining unacceptable quality of service at a load point as the voltage at the bus being less than a specified minimum value and/or the available generating capacity not being adequate to supply the total system requirement.

One problem encountered in such studies is the overload of transmission lines and transformers. The simplest solution to this problem is to allow the operation of lines with an overloaded condition.[2] This assumption results in highly optimistic reliability indices especially when the line overload is heavy and persists for a long duration. An alternative solution[3] is to remove the overloaded element and continue to analyze the remaining system until no other line is overloaded or until the total system has failed. This approach results in pessimistic reliability indices which may suggest expensive and unnecessary investment in system improvement.

A practical solution is to alleviate the line overloads by:

(a) Generation rescheduling; and/or

(b) Curtailment of some of the interruptible loads.

Numerous approaches have been previously investigated[4-7] to determine a generation rescheduling pattern for alleviating line overloads. These techniques, which use optimization algorithms to minimize the operating cost while reducing the active power flow or current in critical lines, can require considerable computational time and may also prove to be unnecessary for general composite system reliability evaluation. The recent development of a computationally inexpensive and direct method[8-9] of determining a pattern of generation rescheduling and/or load curtailment to alleviate line overloads appears to offer a practical approach for reliability studies.

The opportunity to curtail the load at a bus in order to alleviate the line overloads necessitates modification of the bus failure definition to consider any part of the load curtailment at a bus under consideration, as a failure at that bus. This definition of bus failure reflects the effect of load curtailment in the probability and frequency of failure indices. These indices can also be extended to recognize the overall impact of load curtailment on customers in terms of a new set of indices which quantify the expected MW, MWh and hours of curtailment, expected number of curtailments, etc. These additional indices provide further information on desirable changes in the system in order to improve the reliability of power supply to a customer.

In this paper, a new set of bus indices based on the load curtailment necessary to alleviate line overloads is defined. An additional set of system indices are also defined to supplement the set of bulk power interruption and curtailment indices defined by the IEEE Working Group to provide a better appreciation of system adequacy. Bus and system indices are evaluated and presented in the paper for a 30-bus, 56-line model of the Saskatchewan Power Corporation system in order to illustrate the developed indices for a practical configuration. The decoupled line overload alleviation technique used in these studies is also briefly described.

NOTATION

J	is an outage condition in the network;
P_J	is the probability of existence of the outage J;
F_J	is the frequency of occurrence of the outage J;
P_{KJ}	is the probability of the load at bus K exceeding the maximum load that can be

A 79 024-1. A paper recommended and approved by the IEEE Power System Engineering Committee of the IEEE Power Engineering Society for presentation at the IEEE PES Winter Meeting, New York, NY, February 4-9, 1979. Manuscript submitted August 31, 1978; made available for printing November 2, 1978.

Reprinted from *Proc. IEEE Winter Power Mtg.*, Paper A79 024-1, July/Aug. 1979.

supplied at that bus during the outage J;

D_{KJ} is the duration in hours of the load curtailment arising due to the outage J; or the duration in hours of the load curtailment at an isolated bus K due to the outage J;

C is the number of load points in the system at which the reliability indices are calculated; .

L_S is the total system load;

L_{KJ} is the load curtailment at bus K to alleviate line overloads arising due to the contingency J; or the load not supplied at an isolated bus K due to the contingency J;

$J \varepsilon I$ includes all contingencies which result in an isolation of bus K;

$J \varepsilon \phi$ includes all contingencies resulting in line overloads which are alleviated by load curtailment at bus K;

$J \varepsilon V$ includes all contingencies which cause the voltage violation at a bus K;

ΔP_j is the active power mismatch at bus j; or the desired increment in scheduled generated power and/or load at bus j;

ΔQ_j is the reactive power mismatch at bus j; or the desired increment in the reactive load at bus j;

$\Delta \delta_j, \delta V_j$ are the increments in the phase angle and magnitude of the voltage at bus j;

ΔI_{ij} is the current overload of a line connecting buses i and j;

$\Delta P_{ij} + j \Delta Q_{ij}$ is the desired change in the line flow from bus i to j;

I_{rated} is the rated current of the line;

I_{actual} is the current flowing in the line before alleviating the line overload;

V_j is the magnitude of the voltage at bus j;

The superscript T denotes the transpose of a matrix

ADEQUACY INDICES

The probability and frequency of failure indices defined and used earlier[2,3] do not properly reflect system adequacy if load curtailment is used for alleviating line overloads. An additional set of bus indices are therefore defined in this section in order to quantify the effects of load curtailment. Overall system indices in addition to load point parameters are required to provide a realistic assessment of system adequacy.

The indices in this section have been calculated for a single fixed load level. In order to represent these on a base of one year, the indices have been modified by the assumption that this load level exists for a year. These indices are then referred to as annualized indices. The effect of a variable load level can be included in order to produce a more representative annual index but at the expense of considerable

computation time. This also depends on the degree to which the load variation is modelled. Annualized indices can be used to compare alternate system expansion plans and provide representative indices to assess the sensitivity of the system adequacy to configuration changes.

It is important to appreciate that the system and bus indices do not replace each other but actually complement each other to more completely represent system adequacy. The indices defined in References 1, 2 and 10 are also included in this section in order to make the list of indices more complete.

Indices at Bus K

Probability of failure $= \sum\limits_{J} P_J \cdot P_{KJ}$

Frequency of failure $= \sum\limits_{J} F_J \cdot P_{KJ}$

For a fixed load level considered for a specific period of time, P_{KJ} will be equal to zero if the total load at bus K can be supplied without line overloads. Otherwise P_{KJ} will be equal to unity.

Annualized number of voltage violations $= \sum\limits_{J \varepsilon V} F_J$

Annualized number of load curtailments $= \sum\limits_{J \varepsilon \phi, I} F_J$

Annualized load curtailed $= \sum\limits_{J \varepsilon \phi, I} L_{KJ} \cdot F_J$ MW

Annualized energy not supplied $= \sum\limits_{J \varepsilon \phi, I} L_{KJ} D_{KJ} F_J$ MWh

$= \sum\limits_{J \varepsilon \phi, I} L_{KJ} P_J \times 8760$ MWh

Annualized load curtailment duration or expected number of hours in a year (during which a fixed load level is assumed) that any curtailment would exist

$= \sum\limits_{J \varepsilon \phi, I} D_{KJ} F_J$ hrs

$= \sum\limits_{J \varepsilon \phi, I} P_J \times 8760$ hrs

Maximums

Maximum load curtailed

$= Max. |L_{K1}, L_{K2} \cdots L_{KJ} \cdots|$

Maximum energy curtailed

$= Max. |L_{K1} D_{K1}, L_{K2} D_{K2}, \ldots, L_{KJ} D_{KJ}, \ldots|$

Maximum duration of load curtailment

$= Max. |D_{K1}, D_{K2}, \ldots D_{KJ}, \ldots|$

The particulars of the contingencies which cause the above maximums are also desirable in order to appreciate their severity. A high probability of the contingency associated with any of the above maximums suggests a required improvement in the system whereas a low probability contingency may be ignored.

Averages

Load curtailed

$= (\sum\limits_{J \varepsilon \phi, I} L_{KJ} \cdot F_J) / (\sum\limits_{J \varepsilon \phi, I} F_J)$ MW/curtailment

Energy not supplied

$= (\sum\limits_{J \varepsilon \phi, I} L_{KJ} D_{KJ} F_J) / (\sum\limits_{J \varepsilon \phi, I} F_J)$ MWh/curtailment

Duration of curtailment

$$= \left(\sum_{J\epsilon\phi,I} D_{KJ} F_J \right) / \left(\sum_{J\epsilon\phi,I} F_J \right) \quad \text{hours/disturbance}$$

Indices due to the isolation of bus K

Annualized number of curtailments

$$= \sum_{J\epsilon I} F_J$$

Annualized load curtailed

$$= \sum_{J\epsilon I} L_{KJ} F_J \quad \text{MW}$$

Annualized energy not supplied

$$= \sum_{J\epsilon I} L_{KJ} D_{KJ} F_J \quad \text{MWh}$$

$$= \sum_{J\epsilon I} L_{KJ} P_J$$

Annualized duration of load curtailment

$$= \sum_{J\epsilon I} D_{KJ} F_J \quad \text{hours}$$

$$= \sum_{J\epsilon I} P_J$$

System Indices

The first set of three indices of this section were defined in Reference 1 by the IEEE Working Group on Performance Records for Optimizing System Design, Power System Engineering Committee.

Bulk power interruption index

$$= \sum_{K} \sum_{J\epsilon\phi,I} L_{KJ} F_J / L_S \quad \text{MW/MW-yr.}$$

Bulk power supply average MW curtailment/disturbance

$$= \left(\sum_{K} \sum_{J\epsilon\phi,I} L_{KJ} F_J \right) / \left(\sum_{J\epsilon\phi,I} F_J \right)$$

Bulk power energy curtailment index

$$= \sum_{K} \sum_{J\epsilon\phi,I} 60 L_{KJ} D_{KJ} F_J / L_S \quad \text{MW-min./yr.}$$

The bulk power energy curtailment index in MW-minutes/year can be a very large number when calculating annualized system indices. In this paper, this index is therefore expressed in MWh/yr. A system severity index has been defined in Reference 10 in terms of the sum of the severity values for all outage events. The severity associated with each outage event is defined as the total unsupplied energy because of that event, expressed in MW-Minutes, divided by the peak system load in MW. Severity is therefore expressed in "System Minutes". One system minute is equivalent to an interruption of the total system load for one minute at the time of system peak. The severity index for expressing the total unavailability of the system can thus be defined as:

Severity index

$$= \sum_{K} \sum_{J\epsilon\phi,I} 60 L_{KJ} D_{KJ} F_J / L_S \quad \text{System-Minutes}$$

It is interesting to note that the system severity index is equivalent to the bulk power energy curtailment index defined by the IEEE Working Group.

A useful modification of the bulk power energy curtailment index in MWh/yr. is possible by dividing this index by a factor of 8760. This modified index is the probable ratio of the load energy curtailed due to deficiencies in the composite generation and transmission system, to the total load energy required to serve the requirements of the system. This index is similar to the energy index of unreliability defined and used in generating system reliability evaluation. This index can thus be defined as:

Modified bulk power energy curtailment index or Energy index of unreliability including transmission

$$= \sum_{K} \sum_{J\epsilon\phi,I} L_{KJ} D_{KJ} F_J / (8760 \, L_S)$$

Averages

Average number of curtailments/load point

$$= \sum_{K} \sum_{J\epsilon\phi,I} F_J / C$$

Average load curtailed/load point

$$= \sum_{K} \sum_{J\epsilon\phi,I} L_{KJ} F_J / C$$

Average energy curtailed/load point

$$= \sum_{K} \sum_{J\epsilon\phi,I} L_{KJ} D_{KJ} F_J / C$$

Average duration of load curtailment/load point

$$= \sum_{K} \sum_{J\epsilon\phi,I} D_{KJ} / C$$

Average number of voltage violations/load point

$$= \sum_{K} \sum_{J\epsilon V} F_J / C$$

Maximums

Maximum system load curtailed under any contingency condition,

$$= \text{Max.} \left| \sum_{K} L_{K1}, \sum_{K} L_{K2}, \ldots, \sum_{K} L_J, \ldots \right|$$

Maximum system energy not supplied under any contingency condition,

$$= \text{Max.} \left| \sum_{K} L_{K1} D_{K1}, \sum_{K} L_{K2} D_{K2}, \ldots, \sum_{K} L_{KJ} D_{KJ}, \ldots \right|$$

The particulars of the contingencies, which cause the above maximum, are also desirable to appreciate their severity.

SYSTEM STUDIES

The adequacy indices previously defined were evaluated for a 30-bus model of the Saskatchewan Power Corporation system using a digital computer program[2] modified to incorporate the decoupled line overload alleviation technique. A brief description of the decoupled line overload alleviation technique is included in Appendices A and B. The single line diagram of the system model is given in Figure 1 and is composed of 56 transmission lines and transformers, 9 generator buses and 24 generating units. The generator and line outage data of this model is given in Reference 2. The criterion for defining unacceptable quality of service at a load point is the voltage at the bus being less than 0.95 per unit and/or the inability of the system to supply the total load connected to that bus after alleviating line overloads. The adequacy indices of this system were calculated by simulating all possible first and second order simultaneous independent outage combinations of transmission lines, transformers, and generating units. The annualized bus and system indices of the 30-bus model of the Saskatchewan Power Corporation system are listed in Tables I and II respectively.

DISCUSSION OF THE RESULTS

At the present time there is no consensus within the power industry on what constitutes a complete set of adequacy indices. This will depend almost entirely on the use to be made of the indices. It is also true to

TABLE I. Annualized bus indices for the 30-bus model of the SPC

Bus No.	Load MW	Load MVAr	Failure Probability x 1000	Failure Frequency	No. of Curtailments Total	Isolated	Load Curtailed MW Total	Isolated	Energy Curtailed (MWh) Total	Isolated	Duration of Curtailment Hrs. Total	Isolated	No. of Voltage Violations
1	2	3	4	5	6	7	8	9	10	11	12	13	14
10	0.7876	0.0676	2.1927	2.3804	2.38	0.0	99.95	0.0	828.22	0.0	19.20	0.0	0.0
11	0.6081	0.2194	0.0681	0.0846	0.07	0.0	2.14	0.0	15.80	0.0	0.52	0.0	0.01
12	0.6040	0.3370	0.8564	0.7627	0.29	0.0	13.02	0.08	171.43	0.43	3.48	0.01	0.49
13	0.6093	0.0986	0.0471	0.0694	0.07	0.0	5.72	0.09	22.05	0.48	0.41	0.01	0.01
14	1.9940	0.3852	15.7197	13.2317	13.22	0.0	861.69	0.0	9055.00	0.0	137.64	0.0	0.02
15	1.4010	0.5179	0.8463	0.8598	0.86	0.0	46.62	0.0	405.60	0.0	7.40	0.0	0.0
16	0.4303	0.1030	4.8020	4.0127	4.01	0.0	170.64	0.0	1796.78	0.0	42.06	0.0	0.0
17	1.2330	0.3923	0.0021	0.0032	0.0	0.0	0.04	0.0	0.25	0.0	0.02	0.0	0.0
18	0.8013	0.1866	0.2357	0.3186	0.31	0.0	7.18	0.25	45.86	1.45	2.02	0.02	0.01
19	0.0836	0.0212	0.2107	0.2620	0.26	0.0	2.19	0.0	15.43	0.0	1.85	0.0	0.0
22	0.3429	0.0435	0.2174	0.2546	0.25	0.0	6.35	0.0	47.77	0.0	1.90	0.0	0.0
24	0.1508	0.0577	0.0011	0.0017	0.0	0.0	0.0	0.0	0.0	0.0	0.0	0.0	0.0
25	0.1784	0.0030	0.0065	0.0079	0.01	0.0	0.13	0.02	0.97	0.14	0.05	0.01	0.0
26	1.9320	0.6899	4.0628	3.0045	3.00	0.0	71.78	0.0	918.36	0.0	35.59	0.0	0.0
27	0.5659	0.0785	4.7612	3.9584	3.96	0.0	87.28	0.0	909.02	0.0	41.71	0.0	0.0
28	0.4950	0.1350	7.4587	6.7043	6.70	0.0	243.01	0.0	2465.15	0.0	65.33	0.0	0.0
29	0.4645	0.1257	0.0312	0.0482	0.01	0.01	0.60	0.50	3.41	2.83	0.08	0.06	0.04
30	0.6676	0.1641	0.0504	0.0747	0.04	0.0	0.63	0.0	3.83	0.0	0.25	0.0	0.03

TABLE I. (Continued)

Bus No.	Max. Load Curtailed MW	Prob. x 1000	Max. Energy Curtailed MWh	Prob. x 1000	Max. Duration of Curtailment Hrs.	Prob. x 1000	Average Values/Curtailment Load Curtailed MW	Energy Curtailed MWh	Duration of Curtailment Hours
	15	16	17	18	19	20	21	22	23
10	78.76	0.0147	612.68	0.0147	9.05	1.1989	42.03	348.25	8.29
11	60.81	0.0213	473.43	0.0213	8.85	0.0002	29.60	218.93	7.40
12	60.40	0.0077	933.64	0.2988	17.26	0.2988	45.62	600.71	13.17
13	60.93	0.0077	623.78	0.0	10.24	0.0	54.02	320.48	5.93
14	199.40	0.0213	1763.64	0.0002	11.50	0.6713	65.18	684.91	10.51
15	140.10	0.0213	1236.34	0.0	13.85	0.0397	54.33	472.69	8.70
16	43.03	3.7546	491.23	3.7546	11.42	3.7546	42.53	447.79	10.53
17	123.30	0.0	1262.30	0.0	10.24	0.0	13.99	79.42	5.68
18	80.13	0.0064	524.48	0.0035	8.86	0.0002	23.01	146.92	6.39
19	8.36	0.0213	73.94	0.0002	8.85	0.0002	8.36	58.89	7.04
22	34.29	0.0213	268.64	0.0035	7.97	0.0508	25.06	188.45	7.52
24	0.0	0.0	0.0	0.0	0.0	0.0	0.0	0.0	0.0
25	17.84	0.0035	139.77	0.0035	7.83	0.0035	17.84	129.02	7.23
26	193.20	0.2988	3335.01	0.2988	17.26	0.2988	23.89	305.67	12.80
27	56.59	0.0118	371.42	0.0118	11.42	3.7546	22.05	229.65	10.42
28	49.50	3.7546	565.09	3.7546	11.42	3.7546	36.26	367.78	10.14
29	46.45	0.0070	263.06	0.0070	5.66	0.0070	40.80	231.03	5.66
30	33.04	0.0056	188.53	0.0056	7.82	0.0035	15.33	93.10	6.07

TABLE II. Annualized system indices

IEEE Indices	
Bulk power interruption index (MW/MW-yr)	1.14866
Bulk power energy curtailment index (MWh/yr)	11.86680
Bulk power supply average MW curtailment index (MW/disturbance)	69.15350
New Indices	
Energy index of unreliability including transmission	0.00135
Severity index (System-Minutes)	712.008
Average Indices	
Number of load curtailments/load point/year	1.96975
Number of voltage violations/load point/year	0.03453
Load curtailed/load point/year (MW)	89.83194
Energy curtailed/load point/year (MWh)	928.04980
Number of hours of load curtailment/load point/year	10.33096
Maximum Indices	
Load curtailed (MW)	533.51
Probability x 1000	0.02125
Contingency description	Lines 29 & 30 out
Energy curtailed (MWh)	4268.64
Probability x 1000	0.29877
Contingency description	Generators at buses 1 & 7 out

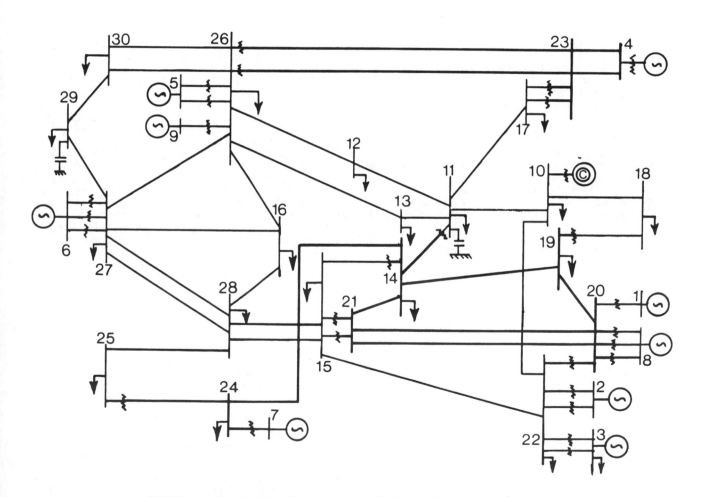

FIGURE 1. Single line diagram of the 30-bus model of the SPC system.

say that there has been little or no work done on defining what is an acceptable/unacceptable level of service adequacy in quantitative terms.

Attempts are being made to equate reliability worth with reliability cost by developing customer loss or damage functions. The ability to quantitatively evaluate service adequacy at a customer load point is an important requirement in this form of analysis.

Table I shows that buses 10, 14, 16, 26, 27 and 28 have annualized expected failure frequencies of 2.4, 13.2, 4.0, 3.0, 4.0 and 6.7 respectively. The loads at Yorkton, Regina and Condie, Hawarden, Saskatoon, Swift Current and Pasqua are on buses 10, 14, 16, 26, 27 and 28 respectively. These loads represent most of the domestic consumer loads of the principal cities and towns of the province of Saskatchewan.

An examination of the detailed study results, which are not included in this paper, indicated that about 70% of the failures at buses 16 and 27 are due to the outage of line 23 which is a 230 kV line connecting buses 24 and 14. The outage of this line resulted in an overload of lines 21 (connecting the buses 24 and 25) and 20 (connecting buses 25 and 28) which are alleviated by curtailing the loads at buses 16, 27 and 28. This overload arises because the generation connected to bus 24 is 280 MW, whereas the rating of line 20 is only 120 MVA. The remaining 30% of the failures at buses 16 and 27 are due to the independent, simultaneous double outage combinations of line 23 with other generating units and lines.

The single outages of lines 18, 19 and 23 contribute about 70% of the failures at bus 28. The rest of the failures are due to the outage combinations of these lines with the generating units and lines. Lines 18 and 19 form the 138 kV double circuit line connecting buses 15 and 28. The outage of one of these lines results in an overload of the other line which is alleviated by curtailing the load at bus 28. The effect of the outage of line 23 is already described.

Approximately half of the expected failures at bus 10 are contributed by the single outage of line 45.* The outage of this line overloads lines 46 (connecting buses 10 and 22) and 55 (connecting buses 14 and 19) which are alleviated by curtailing the loads at buses 10 and 14. The outage of line 45 in combination with the outage of generating units and lines contribute the other half of the failures at bus 10.

Single outages of lines 21, 29, 30, 45 and 47 contribute 70.4% of the annualized failures at bus 14. The outages of these lines in combination with the outages of other generating units and lines contribute the remaining portion of the expected failures. The outage of line 21 results in an overload of line 23 which is alleviated by curtailing the load at bus 14. Lines 29 and 30 (connecting buses 20 and 21) represent the 230 kV double circuit line connecting buses 20 and 21. Almost half of the total system installed generating capacity is connected to bus 20 and this double circuit is a vital link to transport energy to other load buses. Outage of any one line of this double circuit line results in an overload of the second line necessitating a 50% load curtailment at bus 14. Almost 50% of the failures at bus 14 are contributed by this double circuit line. The rest of the contribution is from lines 45 and 47 (connecting buses 19 and 20) which also form part of the transmission system transporting the power generated at bus 20 to the load buses. The effect of the outage of line 45 has been already mentioned while discussing the failures at bus 10. The outage of line 47 results in an overload of line 28 (connecting buses 14 and 21) which is alleviated by curtailing the load *(connecting buses 18 and 19).

at bus 14.

It is interesting to note that 94% of the failures occuring at bus 26 are contributed by generator outage combinations. These failures are due to the outage of generating units at buses 1 and 7 in combination with any of the three generating units connected to bus 8. The rest of the failures are due to line outages. Single generator and line outages do not cause any failures at this bus.

The annualized number of load curtailments at each bus due to load curtailment and bus isolation are tabulated in column 6 of Table I. This index and the expected failure frequency will be equal if voltage violations do not occur at a particular bus. The total number of load curtailments at bus 14 is a very high value of 13.22. The next higher value of 6.7 occurs at bus 28. The number of curtailments due to bus isolation are included in column 7. This index is a small number at all buses except at bus 29. The annualized load and energy curtailment indices are tabulated in columns 8 to 11. These indices provide an indication of the severity of the load curtailments. The annualized duration of curtailment in hours is available in columns 12 and 13. More than a day of load curtailment per year is occuring at buses 14, 16, 26, 27 and 28. Remedial measures are therefore necessary if this index is considered higher than the acceptable limit. Column 14 lists the number of voltage violations at each bus. The voltage problem appears to be severe at bus 12 compared with the rest of the buses in the system. The single outage of line 36 (connecting buses 12 and 26) as well as its outage combination with other lines creates the violation of the minimum voltage limit at bus 12.

The expected maximum load curtailed at each bus under any considered outage condition is given in column 15. The probability of the outage condition resulting in this load curtailment is available in column 16. This index at each bus is equal to the load connected to that bus except at buses 24 and 30. The highest probability of this maximum load curtailment is 0.0038 at buses 16 and 28. The maximum energy curtailed and its probability are given in columns 17 and 18. The highest energy curtailment in the system is 3335 MWh at bus 26 with a probability of occurence of 0.0003. A study of column 19, which lists the maximum duration of curtailment at each bus, indicates that the buses 12 and 26 suffer the highest number of hours of load curtailment of 17.26 hours due to a particular contingency. The probability of occurence of this outage condition is 0.0003 as can be seen from column 20.

The annualized system indices of the 30-bus model of the Saskatchewan Power Corporation system are given in Table II. In this table, the bulk power interruption index value of 1.15 indicates that the total system load curtailed per year is equivalent to 1.15 times the peak load of the system. Similarly, the bulk power energy curtailment index of 11.87 signifies that the total energy not supplied per year is equivalent to the total system shut down under peak load conditions for a period of 11.87 hours. The bulk power supply average MW curtailment index of 69.2 MW/disturbance is self explanatory.

The energy index of unreliability including transmission is 0.00135. This parameter indicates that the system is incapable of supplying 0.135% of the annual energy requirements of the total system either because of line overloads or due to bus isolations. The average system indices are self explanatory. A maximum system load of 533.5 MW will be curtailed due to the simultaneous independent outage of lines 29 and 30 with a probability of 2.1×10^{-5}. This index is 37.9% of the total system peak load of 1408 MW. The outage of

272

the 280 MW generating unit connected to bus 7 results in the maximum system energy curtailment of 4268.64 MWh.

The numerical values of the indices listed in Table I and the above discussion identify buses 10, 14, 16, 26, 27 and 28 as buses in the system which appear to have inadequate levels of reliability. Any improvements should therefore be directed towards raising the reliability of supply to these points. The adequacy indices therefore identify the weak points in the system and provide a system planner with a basis on which to consider improvements and modifications.

A detailed examination and discussion of the results of the 30-bus model of the Saskatchewan Power Corporation also indicate that double outages alone contribute all failures at more than 70% of the system buses. These studies clearly stress the importance of considering double outages and the inadequacy of single outages in evaluating the reliability of a composite generation and transmission system.

CONCLUSIONS

The basic function of an electric power system is to satisfy the customer energy requirements as economically as possible and at an adequate level of continuity and quality. The determination of what is an adequate level is basically a management decision. It is, however, definitely related to the customer expectation, the standard of living and the economic consequences associated with unreliability. Individual load point adequacy indices are necessary to identify weak points in the system and help establish optimum response to equipment investment. These indices when obtained using a valid load model can provide infeed values for use in overall customer adequacy evaluation and provide a quantitative indications of the relative contribution of the three basic segments of generation, transmission and distribution. Overall system indices provide an appreciation of global adequacy which may prove more appealing to management and useful in the comparison of one system with another. They may not, however, be as sensitive to the addition of a line or generating element as the individual bus indices in the proximity of the added element.

Line overloads arising in contingency studies of composite generation and transmission system reliability evaluation can be alleviated by generation rescheduling and/or load curtailment. The bus failure definition has therefore been extended in this paper to include any part of the load curtailment at a bus in order to alleviate line overloads, as a failure at that bus. A new set of bus and system indices based on the load curtailment are therefore defined. It is important to appreciate that the bus and system indices do not replace each other but complement each other. The set of calculated indices were annualized in the sense that the indices were calculated assuming the existence of a fixed load level for the period of a year. The definition of these indices can also include the load model of the system. There are no doubt, additional indices which can be created and used in system adequacy evaluation. It is believed however, that, this paper provides the most complete set available and a basis for further discussion and evaluation. In order to appreciate what these indices mean in a practical system configuration, the bus and system indices evaluated for a 30-bus, 56-line model of the Saskatchewan Power Corporation system are presented in this paper.

REFERENCES

1. Working Group on Performance Records for Optimizing System Design of the Power System Engineering Committee, IEEE Power Engineering Society, "Reliability indices for use in bulk power supply adequacy evaluation", IEEE Transactions on Power Apparatus and Systems, Vol. PAS-97, July/August 1978, pp. 1097 - 1103.

2. R. Billinton, T.K.P. Medicherla and M.S. Sachdev, "Composite generation and transmission system reliability evaluation", Paper No. A 78 237-0 presented at the IEEE PES Winter Meeting, New York, January 1978.

3. R. Billinton and M.P. Bhavaraju, "Transmission planning using a reliability criterion - Part I. A reliability criterion", IEEE Transactions on Power Apparatus and Systems, Vol. PAS-89, No. 1, January 1970, pp. 28 - 34.

4. J.C. Kaltenbach and L.P. Hajdu, "Optimal corrective rescheduling of power for system reliability", IEEE Transactions on Power Apparatus and Systems, Vol. PAS-90, No. 2, March/April 1971, pp. 843 - 851.

5. H. Daniels and M. Chen, "An optimization technique and security calculations for dispatching computers", IEEE Transactions on Power Apparatus and Systems, Vol. PAS-91, No. 3, May/June 1972, pp. 883 - 887.

6. A. Thanikachalam and J.R. Tudor, "Optimal rescheduling of power for system reliability", IEEE Transactions on Power Apparatus and Systems, Vol. PAS-90, September/October 1971, pp. 2186 - 2192.

7. K.R.C. Mamandur and G.J. Berg, "Economic shift in electric power generation with line flow constraints", Paper No. F 77 697-6, presented at the IEEE PES Summer Meeting, Mexico City, Mexico, July 1977.

8. T.K.P. Medicherla, R. Billinton and M.S. Sachdev, "Generation rescheduling and load shedding to alleviate line overloads - Analysis", Paper No. F 78 686-8 presented at the IEEE PES Summer Meeting, Los Angeles, July 1978.

9. T.K.P. Medicherla, R. Billinton and M.S. Sachdev, "Generation rescheduling and load shedding to alleviate line overloads - System studies", Paper No. F 78 685-0 presented at the IEEE PES Summer Meeting, Los Angeles, July 1978.

10. B.K. LeReverend, "Bulk electricity system performance indices used within Ontario Hydro", a paper presented at the Canadian Electrical Association Meeting, System Planning and Operating Section - Spring Meeting 1978.

11. B. Stott and O. Alsac, "Fast decoupled load flow", IEEE Transactions on Power Apparatus and Systems, Vol. PAS-93, May/June 1974, pp. 859 - 869.

APPENDIX A

Decoupled Line Overload Alleviation Technique

The overload of transmission lines and transformers can often be alleviated by generation rescheduling and/or load curtailment. A decoupled line overload alleviation technique[8,9] suitable for use with the fast decoupled load flow[8] approach is briefly described in this section. This technique uses a mathematical model described by Equations A1, A2 and A3, which are based on a set of linearized relationships between line currents and state variables, and between bus injected powers and state variables.

$$[\Delta P/V] = [B'] \quad [\Delta\delta] \qquad\qquad\text{A.1}$$

$$[\Delta Q/V] = [B''] \quad [\Delta V] \qquad\qquad\text{A.2}$$

$$\left[\frac{\Delta\delta}{\Delta V}\right] = [A]^{-1} \quad [\Delta I] \qquad\qquad\text{A.3}$$

Equations A.1 and A.2 are obtained[11] after applying the principle of decoupling to the linearized forms of nonlinear load flow equations used in the Newton-Raphson load flow technique. The initial load flow solution of the outage case under consideration is obtained through the successive solution of Equations A.1 and A.2. This set of equations can also be used to obtain the desired increments in generated powers and loads after the computation of initial load flow solution of the outage case. The use of Equation A.2 for calculating the changes in reactive loads differs from an earlier approach[8] in the sense that the reactive load curtailment at generator buses is not calculated as the reactive load requirements at generator buses can be satisfied by manipulating the generator terminal voltages. This feature saves some computer storage and solution time.

Linear relationships between line current overloads and increments in the phase angles and magnitudes of bus voltages are expressed in Equation A.3. Each row of matrix A consists of a maximum of four terms. These terms are the partial derivatives of line currents with respect to the magnitudes and phase angles of the terminal voltages of overloaded lines. The number of rows and columns of this matrix will be equal to the number of overloaded lines and the number of state variables respectively. Matrix A is highly sparse and rectangular, and its inverse does not exist. The pseudoinverse of this matrix is defined in Equation A.4 and can be used for the purposes of obtaining a solution of Equation A.3.

$$A^{+} = A^{T} [A \, A^{T}]^{-1} \qquad\qquad\text{A.4}$$

The decoupled line overload alleviation model is solved by obtaining the initial load flow solution of the outage case through the successive solution of Equations A.1 and A.2 until the mismatches ΔP and ΔQ are within a prespecified tolerance. A row and a term are added to matrix A and vector ΔI respectively for each line overload indicated by the initial load flow solution. The pseudoinverse of A is determined and post multiplied by the column vector ΔI for determining the desired changes in the state variables as shown in Equation A.3. These changes in the state variables are used to calculate the desired changes in generated powers and/or loads using Equations A.1 and A.2. A new load flow solution is then computed using the latest bus voltages, loads and generator outputs. This procedure is repeated until no lines are overloaded. Detailed description of this technique is given in References 8 and 9. If this technique fails to alleviate line overloads within a prespecified number of passes, an approximate method of load curtailment and generation rescheduling developed in Appendix B can be used.

APPENDIX B

Approximate Method of Line Overload Alleviation:

The approximate method of line overload alleviation is used if the generation rescheduling and load shedding technique fails to alleviate the line overloads within a prespecified number of passes. The mathematical equations used in the approximate method are developed in this Appendix, and are derived for maintaining constant power factor at any load bus.

Consider an overloaded line, connecting buses i and

j, in which the active and reactive powers are flowing from bus i to j. This line overload can be alleviated by reducing the power flow in the line by an amount $\Delta P_{ij} + \Delta Q_{ij}$. Assume that this desired change in power is possible by curtailing the load at bus j by an amount equal to $\Delta P_j + \Delta Q_j$, which reduces the current flowing in the line from I_{actual} to I_{rated}. Based on this assumption, the Equation B.1 is written:

$$V_i^2 \, I_{rated}^2 = P_{ij}^2 + Q_{ij}^2 + 2P_{ij} \cdot \Delta P_{ij} + 2Q_{ij} \cdot \Delta Q_{ij}$$
$$+ (\Delta P_{ij})^2 + (\Delta Q_{ij})^2 \qquad\text{B.1}$$

However, the following equation should be satisfied for maintaining constant power factor at bus j.

$$\frac{P_j}{Q_j} = \frac{P_j + \Delta P_j}{Q_j + \Delta Q_j} = \frac{\Delta P_j}{\Delta Q_j}$$

Assuming that $\Delta P_j/\Delta Q_j = \Delta P_{ij}/\Delta Q_{ij}$;

$$\Delta Q_{ij} = \Delta P_{ij} \cdot Q_j/P_j \qquad\qquad\text{B.2}$$

Also, $\qquad V_i^2 \, I_{actual}^2 = P_{ij}^2 + Q_{ij}^2 \qquad\qquad\text{B.3}$

Substituting Equations B.2 and B.3 in Equation B.1:

$$V_i^2 \, (I_{rated}^2 - I_{actual}^2) = 2P_{ij} \cdot \Delta P_{ij} + 2Q_{ij} \cdot (Q_j/P_j)$$
$$\cdot \Delta P_{ij} + (\Delta P_{ij})^2 + (Q_j/P_j)^2 (\Delta P_{ij})^2$$

or,

$$(\Delta P_{ij})^2 (1 + Q_j/P_j)^2 + 2\Delta P_{ij} (P_{ij} + Q_{ij} \cdot Q_j/P_j) - V_i^2$$
$$(I_{rated}^2 - I_{actual}^2) = 0 \qquad\qquad\text{B.4}$$

Equation B.4 is a quadratic equation which can be solved for ΔP_{ij} as:

$$\Delta P_{ij} = (P_j/S_j)^2 \, [(-P_{ij} + Q_j \cdot Q_{ij}/P_j) +$$
$$\sqrt{(P_{ij} + Q_j \cdot Q_{ij}/P_j)^2 + V_i^2(I_{rated}^2 - I_{actual}^2)(S_j/P_j)^2}]$$
$$\text{B.5}$$

The negative sign preceding the radical sign is not included in Equation B.5, since it can easily be proved that the negative sign is not valid.

In a few overloaded line cases, the quantity under the radical sign may become negative. The Equation B.6 is derived for use under such circumstances by neglecting $(\Delta P_{ij})^2$ and $(\Delta Q_{ij})^2$ of Equation B.1.

$$\Delta P_{ij} = |V_i|^2 \, (I_{rated}^2 - I_{actual}^2) \, /[2(P_{ij} + Q_{ij} \cdot Q_j/P_j)]$$
$$\text{B.6}$$

The desired change in reactive power flow ΔQ_{ij} can be calculated using Equation B.2 after determining ΔP_{ij}.

The line overload can be alleviated by reducing the active and reactive flows by the calculated amounts using Equations B.5 and B.2. This reduction is possible by curtailing the load at a bus of the overloaded line into which power is flowing (and/or at the buses into which the power is flowing from the bus). Any such load curtailment may result in excessive generation in the system. Under such circumstances, the generation at generating bus/buses, which are directly or indirectly connected to this line and are injecting power into this line, is reduced by an amount approximately equal to ΔP_{ij}.

This approximate method of line overload alleviation was found to be effective whenever the decoupled line overload alleviation technique failed to alleviate the line overloads. It is also possible to restrict the load curtailment at some special buses to the prespe-cified amounts. The load curtailment can also be attempted using a predetermined bus list of priority. However, the use of a single priority list for alleviating any line overload may result in excessive or unnecessary load curtailment.

IEEE RELIABILITY TEST SYSTEM

A report prepared by the Reliability Test System Task Force of the Application of Probability Methods Subcommittee*

ABSTRACT

This report describes a load model, generation system, and transmission network which can be used to test or compare methods for reliability analysis of power systems. The objective is to define a system sufficiently broad to provide a basis for reporting on analysis methods for combined generation/transmission (composite) reliability.

The load model gives hourly loads for one year on a per unit basis, expressed in chronological fashion so that daily, weekly, and seasonal patterns can be modeled. The generating system contains 32 units, ranging from 12 to 400 MW. Data is given on both reliability and operating costs of generating units. The transmission system contains 24 load/generation buses connected by 38 lines or autotransformers at two voltages, 138 and 230 kV. The transmission system includes cables, lines on a common right of way, and lines on a common tower. Transmission system data includes line length, impedance, ratings, and reliability data.

INTRODUCTION

There has been a continuing and increasing interest in methods for power system reliability evaluation. In order to provide a basis for comparison of results obtained from different methods, it is desirable to have a reference or "test" system which incorporates the basic data needed in reliability evaluation. The purpose of this report is to provide such a "reliability test system".

The report describes a load model, generation system, and transmission network. The objective is to define a system sufficiently broad to provide a basis for reporting on analysis methods for combined generation/transmission (composite) reliability methods. It is not practical to specify all the parameters needed for every application. The goal is to establish a core system which can be supplemented by individual authors with additional or modified parameters needed in a particular application. For example, the reliability test system as reported in this paper does not include data on the following:

- Substation configuration at load/generation buses

- Distribution system configuration
- Interconnections with other systems
- Protective relay configurations
- Future expansion, such as load growth, future unit sizes, types, and reliability.

The Electric Power Research Institute (EPRI) has recently reported data on synthetic electric utility systems [1]. These contain much larger systems than the one in this report. They are designed primarily for use in evaluation of alternate technologies.

A smaller test system was developed by the CIGRE Working group 01 of Study Committee No. 32 [2]. But that system was judged too small and incomplete to be applicable as a model in reliability analysis, especially when considering composite systems.

DESCRIPTION OF RELIABILITY TEST SYSTEM

Load Model

The annual peak load for the test system is 2850 MW.

Table 1 gives data on weekly peak loads in per cent of the annual peak load. The annual peak occurs in week 51. The data in Table 1 shows a typical pattern, with two seasonal peaks. The second peak is in week 23 (90%), with valleys at about 70% in between each peak. If week 1 is taken as January, Table 1 describes a winter peaking system. If week 1 is taken as a summer month, a summer peaking system can be described.

Table 1
Weekly Peak Load in Percent of Annual Peak

Week	Peak Load	Week	Peak Load
1	86.2	27	75.5
2	90.0	28	81.6
3	87.8	29	80.1
4	83.4	30	88.0
5	88.0	31	72.2
6	84.1	32	77.6
7	83.2	33	80.0
8	80.6	34	72.9
9	74.0	35	72.6
10	73.7	36	70.5
11	71.5	37	78.0
12	72.7	38	69.5
13	70.4	39	72.4
14	75.0	40	72.4
15	72.1	41	74.3
16	80.0	42	74.4
17	75.4	43	80.0
18	83.7	44	88.1
19	87.0	45	88.5
20	88.0	46	90.9
21	85.6	47	94.0
22	81.1	48	89.0
23	90.0	49	94.2
24	88.7	50	97.0
25	89.6	51	100.0
26	86.1	52	95.2

* Chairman, P.F. Albrecht, General Electric; M.P. Bhavaraju, Electric Power Research Institute; B.E. Biggerstaff, Federal Energy Regulatory Commission; R. Billinton, University of Saskatchewan; G. Elsoe Jorgensen, Northeast Utilities Service Company; N.D. Reppen, Power Technologies, Inc.; P.B. Shortley, New England Power Planning.

F 79 152-0   A paper recommended and approved by the IEEE Power System Engineering Committee of the IEEE Power Engineering Society for presentation at the IEEE PES Winter Meeting, New York, NY, February 4-9, 1979. Manuscript submitted September 14, 1978; made available for printing December 14, 1978.

Reprinted from *IEEE Trans. Power Apparatus Syst.*, vol. PAS-98, no. 6, pp. 2047–2054, Nov./Dec. 1979.

Table 2 gives a daily peak load cycle, in per cent of the weekly peak. The same weekly peak load cycle is assumed to apply for all seasons. The data in Tables 1 and 2, together with the annual peak load define a daily peak load model of 52x7 = 364 days, with Monday as the first day of the year.

Table 2
Daily Peak Load in Percent of Weekly Peak

Day	Peak Load
Monday	93
Tuesday	100
Wednesday	98
Thursday	96
Friday	94
Saturday	77
Sunday	75

Table 3 gives weekday and weekend hourly load models for each of three seasons. A suggested interval of weeks is given for each season. The first two columns reflect a winter season (evening peak), while the next two columns reflect a summer season (afternoon peak). The interval of weeks shown for each season in Table 3 represents application to a winter peaking system. If Table 1 is started with a summer month, then the intervals for application of each column of the hourly load model in Table 3 should be modified accordingly.

Table 3
Hourly Peak Load in Percent of Daily Peak

Hour	Winter Weeks 1-8 & 44-52 Wkdy	Wknd	Summer Weeks 18-30 Wkdy	Wknd	Spring/Fall Weeks 9-17 & 31-43 Wkdy	Wknd
12-1am	67	78	64	74	63	75
1-2	63	72	60	70	62	73
2-3	60	68	58	66	60	69
3-4	59	66	56	65	58	66
4-5	59	64	56	64	59	65
5-6	60	65	58	62	65	65
6-7	74	66	64	62	72	68
7-8	86	70	76	66	85	74
8-9	95	80	87	81	95	83
9-10	96	88	95	86	99	89
10-11	96	90	99	91	100	92
11-Noon	95	91	100	93	99	94
Noon-1pm	95	90	99	93	93	91
1-2	95	88	100	92	92	90
2-3	93	87	100	91	90	90
3-4	94	87	97	91	88	86
4-5	99	91	96	92	90	85
5-6	100	100	96	94	92	88
6-7	100	99	93	95	96	92
7-8	96	97	92	95	98	100
8-9	91	94	92	100	96	97
9-10	83	92	93	93	90	95
10-11	73	87	87	88	80	90
11-12	63	81	72	80	70	85

Wkdy = Weekday, Wknd = Weekend

Combination of Tables 1, 2, and 3 with the annual peak load defines an hourly load model of 364x24 = 8736 hours. The annual load factor for this model can be calculated as 61.4%.

Generating System

Table 4 gives a list of the generating unit ratings and reliability data. In addition to forced

Table 4
Generating Unit Reliability Data

Unit Size MW	Number of Units	Forced Outage Rate(3)	MTTF(1) hrs.	MTTR(2) hrs.	Scheduled Maintenance wks/year
12	5	0.02	2940	60	2
20	4	0.10	450	50	2
50	6	0.01	1980	20	2
76	4	0.02	1960	40	3
100	3	0.04	1200	50	3
155	4	0.04	960	40	4
197	3	0.05	950	50	4
350	1	0.08	1150	100	5
400	2	0.12	1100	150	6

NOTES:

(1) MTTF = mean time to failure

(2) MTTR = mean time to repair

(3) Forced outage rate = $\dfrac{MTTR}{MTTF + MTTR}$

outage rate, the parameters needed in frequency and duration calculations are given (MTTF and MTTR). Table 4 gives data on full outages only. Generating units can also experience partial outages, both forced and scheduled. Partial outages can have a significant effect on generation reliability. However, modeling of partial outages can be done in many ways; and no single approach has achieved widespread use over all others. Therefore the task force elected to leave partial outage data as a parameter to be specified for a particular application.

The generation mix is as shown below:

	MW	%
Steam:		
Fossil-oil	951	28
Fossil-coal	1274	37
Nuclear	800	24
Combustion Turbine	80	2
Hydro	300	9
Total	3405	100

Table 5 gives operating cost data for the generating units. For power production, data is given in terms of heat rate at selected output levels, since fuel costs are subject to considerable variation due to geographical location and other factors. The following fuel costs are suggested for general use (1979 base).

#6 oil	$2.30/MBtu
#2 oil	$3.00/MBtu
coal	$1.20/MBtu
nuclear	$0.60/MBtu

277

Table 5
Generating Unit Operating Cost Data

Size MW	Type	Fuel	Output %	Heat Rate Btu/kWh	O&M Cost Fixed $/kW/YR	Variable $/MWh
12	Fossil Steam	#6 oil	20	15600	10.0	0.90
			50	12900		
			80	11900		
			100	12000		
20	Combus. Turbine	#2 oil	80	15000	0.30	5.00
			100	14500		
50	Hydro		SEE TABLE 6			
76	Fossil Steam	Coal	20	15600	10.0	0.90
			50	12900		
			80	11900		
			100	12000		
100	Fossil Steam	#6 oil	25	13000	8.5	0.80
			55	10600		
			80	10100		
			100	10000		
155	Fossil Steam	Coal	35	11200	7.0	0.80
			60	10100		
			80	9800		
			100	9700		
197	Fossil Steam	#6 oil	35	10750	5.0	0.70
			60	9850		
			80	9840		
			100	9600		
350	Fossil Steam	Coal	40	10200	4.5	0.70
			65	9600		
			80	9500		
			100	9500		
400	Nuclear Steam	LWR	25	12550	5.0	0.30
			50	10825		
			80	10170		
			100	10000		

The operating and maintenance (O&M) costs are also intended to apply to 1979. For hydro units, data on capacity and energy limitations is given in Table 6.

Table 6
Hydro Capacity and Energy

Quarter	Capacity Available (1) %	Energy Distribution (2) %
1	100	35
2	100	35
3	90	10
4	90	20

NOTES:

(1) 100% capacity = 50 MW

(2) 100% energy = 200 GWh

Transmission System

The transmission network consists of 24 bus locations connected by 38 lines and transformers, as shown in Figure 1. The transmission lines are at two voltages, 138 kV and 230 kV. The 230 kV system is the top part of Figure 1, with 230/138 kV tie stations at buses 11, 12, and 24.

The locations of the generating units are shown in Table 7. It can be seen that 10 of the 24 buses are generating stations. Table 8 gives data on generating unit MVAr capability for use in load flow calculations.

Table 7
Generating Unit Locations

Bus	Unit 1 MW	Unit 2 MW	Unit 3 MW	Unit 4 MW	Unit 5 MW	Unit 6 MW
1	20	20	76	76		
2	20	20	76	76		
7	100	100	100			
13	197	197	197			
15	12	12	12	12	12	155
16	155					
18	400					
21	400					
22	50	50	50	50	50	50
23	155	155	350			

Table 8
Generating Unit MVAr Capability

Size MW	MVAr Minimum	Maximum
12	0	6
20	0	10
50	-10	16
76	-25	30
100	0	60
155	-50	80
197	0	80
350	-25	150
400	-50	200

The system has voltage corrective devices at bus 14 (synchronous condenser) and bus 6 (reactor). Table 9 gives the MVAr capability of these devices. These devices increase the ability of the test system to maintain rated voltage, particularly under some contingency conditions. The amount of such correction capability provided is a system design parameter, which depends partly on the criteria chosen for acceptable voltage limits.

Table 9
Voltage Correction Devices

Device	Bus	MVAr Capability
Synchronous condenser	14	50 Reactive
		200 Capacitive
Reactor	6	100 Reactive

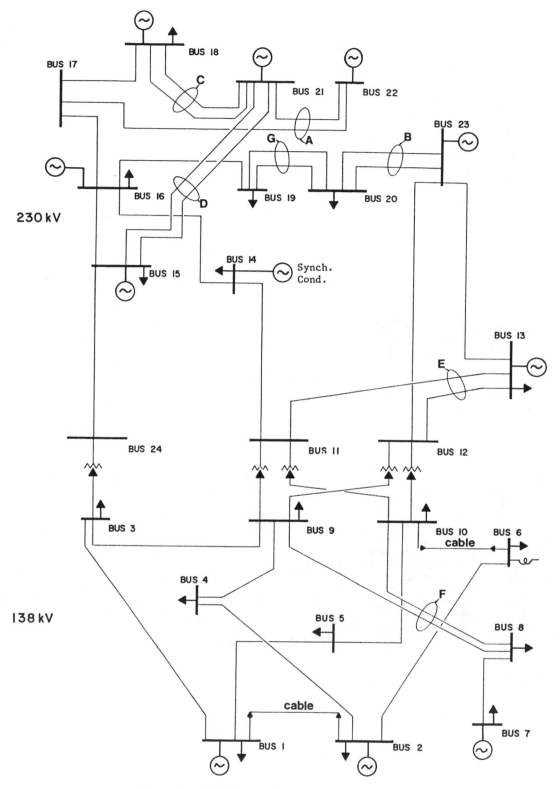

Figure 1 - IEEE Reliability Test System

Bus load data at time of system peak is shown in Table 10. No data on load uncertainty or load diversity between buses is provided. For times other than the annual system peak, the bus loads are assumed to have the same proportional relation to system load as at the peak load conditions. The per unit bus loads are given in the last column of Table 10. For MVAr requirements, a 98% power factor is assumed. This corresponds to an MVAr requirement of approximately 20% of the MW load at each bus. The 98% power factor is assumed to apply at all load levels. These restrictions on bus loads (no uncertainty, no diversity, constant power factor) are the assumptions usually made in reliability evaluations. It will be of interest to compare results obtained with these assumptions with those from less restrictive models.

Table 10
Bus Load Data

| | Load | | Bus Load |
Bus	MW	MVAr	% of System Load
1	108	22	3.8
2	97	20	3.4
3	180	37	6.3
4	74	15	2.6
5	71	14	2.5
6	136	28	4.8
7	125	25	4.4
8	171	35	6.0
9	175	36	6.1
10	195	40	6.8
13	265	54	9.3
14	194	39	6.8
15	317	64	11.1
16	100	20	3.5
18	333	68	11.7
19	181	37	6.4
20	128	26	4.5
TOTAL	2850	580	100.0

Transmission network connection data is defined by Figure 1. Although no attempt has been made to define actual geographical layout, the physical bus locations on Figure 1 are fairly consistent with the line lengths, which are shown in Table 11. Buses 9, 10, 11 and 12 are at a single physical location (step-down station); buses 3 and 24 are also at a single location. As noted on Figure 1, the connections from bus 1 to 2 and from bus 6 to 10 are 138 kV cables.

Transmission line forced outage data is given in Table 11. Permanent outages are those which require component repair in order to restore the component to service.[3] Therefore, for permanent outages both outage rate and outage duration are shown. Transient outages are those which are not permanent. These include both automatic and manual reclosing.[3] For transient forced outages, only the outage rate is given, since the outage duration is very short. In specific applications, transmission line forced outage rates (particularly for transient outages) are dependent on geographical location as well as other factors. The data in Table 11 is representative of experience in the United States and Canada.

The term "outage rate" has been applied in keeping with current industry practice. Unfortunately, the term has a different meaning for generating units than for transmission equipment. For generating units, forced outage rate refers to the probability of forced outage at a random point in time between scheduled outages. This is the meaning of forced outage

Table 11
Transmission Line Length and Forced Outage Data

From Bus	To Bus	Length miles	Permanent Outage Rate l/yr	Permanent Outage Duration Hours	Transient Outage Rate l/yr
1	2	3	.24	16	0.0
1	3	55	.51	10	2.9
1	5	22	.33	10	1.2
2	4	33	.39	10	1.7
2	6	50	.48	10	2.6
3	9	31	.38	10	1.6
3	24	0	.02	768	0.0
4	9	27	.36	10	1.4
5	10	23	.34	10	1.2
6	10	16	.33	35	0.0
7	8	16	.30	10	0.8
8	9	43	.44	10	2.3
8	10	43	.44	10	2.3
9	11	0	.02	768	0.0
9	12	0	.02	768	0.0
10	11	0	.02	768	0.0
10	12	0	.02	768	0.0
11	13	33	.40	11	0.8
11	14	29	.39	11	0.7
12	13	33	.40	11	0.8
12	23	67	.52	11	1.6
13	23	60	.49	11	1.5
14	16	27	.38	11	0.7
15	16	12	.33	11	0.3
15	21	34	.41	11	0.8
15	21	34	.41	11	0.8
15	24	36	.41	11	0.9
16	17	18	.35	11	0.4
16	19	16	.34	11	0.4
17	18	10	.32	11	0.2
17	22	73	.54	11	1.8
18	21	18	.35	11	0.4
18	21	18	.35	11	0.4
19	20	27.5	.38	11	0.7
19	20	27.5	.38	11	0.7
20	23	15	.34	11	0.4
20	23	15	.34	11	0.4
21	22	47	.45	11	1.2

rate in Table 4. For transmission equipment, the term "outage rate" is commonly used to describe the number of outages per unit of exposure time [3]. This is the meaning of outage rate in Table 11 and subsequent tables.

The permanent forced outage rates in Table 11 were calculated as follows:

138 kV lines: $\lambda_p = 0.52\ L + 0.22$

230 kV lines: $\lambda_p = 0.34\ L + 0.29$

138 kV cables: $\lambda_p = 0.62\ L + 0.226$

where L is the length of the line or cable in 100 miles. The constant in each equation accounts for faults on terminal equipment switched with the line (including bus sections, but excluding circuit breakers).

The permanent outage duration data in Table 11 is a combination of permanent outage duration data for lines (or cables) and terminal equipment. The separate outage durations used to obtain Table 11 were as follows:

Equipment	Permanent outage duration, hours	
	Line/Cable	Terminal
138 kV line	9	11
230 kV line	18	8
138 kV cable	96	9

The outage duration values in Table 11 were developed by use of the following equation:

$$R = (\lambda_1 R_1 + \lambda_2 R_2)/(\lambda_1 + \lambda_2)$$

where

λ_1, R_1 =   Line/cable outage rate and outage duration.

λ_2, R_2 =   Terminal outage rate and outage duration.

Rather than calculating a different repair time for each line, the average line length in the test system for each of the two voltages was used to calculate a single (average) value of λ_1. From this, the average outage duration for each voltage level was calculated. For the two cables, separate repair times were calculated by use of the actual cable length.

The transformer outage duration in Table 11 is 768 hours, which corresponds to 32 days. In a particular situation, transformer outage duration will be greatly influenced by whether or not a spare transformer is available.

The transient forced outage rates in Table 11 were calculated as follows:

138 kV lines: $\lambda_t = 5.28$ L

230 kV lines: $\lambda_t = 2.46$ L

It is assumed that transient outages occur only on transmission lines. Hence, no constant term for terminal outages is included, and the transient outage rate for transformers and cables is taken to be zero.

Outages on substation components which are not switched as a part of a line are not included in the outage data in Table 11. For bus sections, the following data is provided:

	138 kV	230 kV
Faults per bus section-year	0.027	0.021
Percent of faults permanent	42	43
Outage duration for permanent faults, hours	19	13

For circuit breakers, the following statistics are provided:

Physical failures/breaker year	0.0066
Breaker operational failure, per breaker year	0.0031
Outage Duration, hours	72

A physical failure is a mandatory unscheduled removal from service for repair or replacement. An operational failure is a failure to clear a fault within the breaker's normal protection zone.

As noted previously, this report does not give substation configurations for load and generation buses. However, for any assumed configuration, the foregoing data on bus sections and circuit breaker outages could be used to model substation reliability.

No data on scheduled outages of transmission equipment is given. This does not imply that scheduled outages are felt to have negligible effect on reliability. Like partial outages of generating units, scheduled outages of transmission lines can have a major impact on reliability. However, very little published data on scheduled outages is available. Therefore, the task force decided to leave this as another parameter to be specified for a particular application. Hopefully this will encourage publication of typical scheduled outage data by various organizations.

There are several lines which are assumed to be on a common right of way or common tower for at least a part of their length. These pairs of lines are indicated in Figure 1 by circles around the line pair, and an associated letter identification. Table 12 gives the actual length of common right of way or common tower. For example, lines from buses 22-21 and 22-17 are 47 and 73 miles long respectively. Table 12 shows that 45 miles of this distance is on a common right of way.

Table 12
Circuits on Common Right of Way
or Common Structure

Right-of Way Identification	From Bus	To Bus	Common ROW miles	Common Structure miles
A	22	21	45.0	
	22	17	45.0	
B	23	20		15.0
	23	20		15.0
C	21	18		18.0
	21	18		18.0
D	15	21	34.0	
	15	21	34.0	
E	13	11		33.0
	13	12		33.0
F	8	10		43.0
	8	9		43.0
G	20	19		27.5
	20	19		27.5

In addition to the exposure to outages shown in Table 11, the circuits on a common right of way or a common structure in Table 12 are exposed to "common mode" outages, in which a single event causes an outage of both lines. There is currently a great interest in data on the frequency of such common mode events. However, very little data of this type has been published. Therefore, as with scheduled outages, the task force elected not to publish arbitrary values of common mode outage rates, with the hope that users of the test system would publish data or assumptions used in particular studies.

Table 13
Impedance and Rating Data

From Bus	To Bus	Impedance P.U./100 MVA Base			Rating (MVA)			
		R	X	B	Normal	Short Term	Long Term	Equipment
1	2	.0026	.0139	.4611	175	200	193	138 kV cable
1	3	.0546	.2112	.0572	"	220	208	138 kV line
1	5	.0218	.0845	.0229	"	"	"	"
2	4	.0328	.1267	.0343	"	"	"	"
2	6	.0497	.1920	.0520	"	"	"	"
3	9	.0308	.1190	.0322	"	"	"	"
3	24	.0023	.0839		400	600	510	Transformer
4	9	.0268	.1037	.0281	175	220	208	138 kV line
5	10	.0228	.0883	.0239	"	"	"	"
6	10	.0139	.0605	2.459	"	200	193	138 kV cable
7	8	.0159	.0614	.0166	"	220	208	138 kV line
8	9	.0427	.1651	.0447	"	"	"	"
8	10	.0427	.1651	.0447	"	"	"	"
9	11	.0023	.0839		400	600	510	Transformer
9	12	.0023	.0839		400	"	"	"
10	11	.0023	.0839		400	"	"	"
10	12	.0023	.0839		400	"	"	"
11	13	.0061	.0476	.0999	500	625	600	230 kV line
11	14	.0054	.0418	.0879	"	"	"	"
12	13	.0061	.0476	.0999	"	"	"	"
12	23	.0124	.0966	.2030	"	"	"	"
13	23	.0111	.0865	.1818	"	"	"	"
14	16	.0050	.0389	.0818	"	"	"	"
15	16	.0022	.0173	.0364	"	"	"	"
15	21	.0063	.0490	.1030	"	"	"	"
15	21	.0063	.0490	.1030	"	"	"	"
15	24	.0067	.0519	.1091	"	"	"	"
16	17	.0033	.0259	.0545	"	"	"	"
16	19	.0030	.0231	.0485	"	"	"	"
17	18	.0018	.0144	.0303	"	"	"	"
17	22	.0135	.1053	.2212	"	"	"	"
18	21	.0033	.0259	.0545	"	"	"	"
18	21	.0033	.0259	.0545	"	"	"	"
19	20	.0051	.0396	.0833	"	"	"	"
19	20	.0051	.0396	.0833	"	"	"	"
20	23	.0028	.0216	.0455	"	"	"	"
20	23	.0028	.0216	.0455	"	"	"	"
21	22	.0087	.0678	.1424	"	"	"	"

Impedance and rating data for lines and transformers is given in Table 13. The "B" value in the impedance data is the total amount, not the value in one leg of the equivalent circuit. Three ratings are given; normal, short term, and long term. The normal rating indicates the daily peak loading capability of a circuit with due allowance for load cycles. The long-term rating means a circuit's capability to handle a 24 hour load cycle following a contingency. The short-term rating indicates the loading capability of a circuit following one or more system contingencies allowing for 15 minutes to provide corrective action. No attempt has been made to provide data on seasonal variation in line ratings. The data in Table 13 should be taken as the ratings at the time of annual system peak, which is week 51 (Table 1).

The data in the paper is sufficient to completely define a DC load flow for the test system. However, an AC load flow is not completely defined. Data on reactive impedances and loads are given, but complete specification of data for an AC load flow requires additional assumptions with regard to voltages at generator buses (regulated) and transformer tap information (tap ratio, fixed tap or LTC).

Reliability Test System Design Criteria

The predominant criteria in choice of the test system configuration was the desire to achieve a useful reference for testing and comparison of reliability evaluation methods. In light of this goal, the task force attempted to incorporate sufficient complexity and detail so that the test system would be representative of actual utility system applications. The test system was designed to have a lower reliability than is typically considered acceptable in utility planning. This was done to facilitate use of the test system in comparison of results from a wide variety of methods. In addition, the ability to evaluate alternatives for reliability can be considered. As experience is gained from study of the test system by various investigators, it may prove desirable to modify the system to be more useful as a means for evaluating and comparing reliability methods.

References

1. Synthetic Electric Utility Systems for Evaluating Advanced Technologies, EPRI Report EM-285, Final Report, Project TPS75-615, February 1977, Power Technologies Inc.

2. Working Group 01 of Study Committee 32 (System Planning and Operation), "Report on the Optimization of Power System Operation (CIGRE Exercise No. 2), "Electra, March, 1975 pp. 47-82.

3. IEEE Standard 346-1973 (Section 2), Terms for Reporting and Analyzing Outages of Electrical Transmission and Distribution Facilities and Interruptions to Customer Service.

Discussion

L.L. Garver (General Electric Company, Schenectady, NY): The test system will be a great help for illustrating power system measures and gaining new insights into their meaning.

One area where readers may wish to explore involves the loss-of-load probability quantity. An essential piece of information is the capacity outage table (1). This table is not easily calculated without the use of a digital computer program. Once the table is available, then maintenance scheduling ideas and new unit additions may be studied with a digital computer (2, 3). This publication will benefit from the inclusion of the capacity outage table.

REFERENCES

(1) R. Billinton, *Power System Reliability Evaluation,* New York: Gordon and Breach, 1970, pp. 97–102.
(2) L. L. Garver, "Adjusting Maintenance Schedules to Levelize Risk", *IEEE Transactions on Power Apparatus and Systems,* vol. PAS-91, pp. 2057–2063, September-October 1972.
(3) L. L. Garver, "Effective Load Carrying Capability of Generating Units", *IEEE Transactions on Power Apparatus and Systems,* vol. PAS-85, pp. 910–919, August 1966.

Manuscript received February 21, 1979.

Paul F. Albrecht on behalf of the **Test System Task Force:** We appreciate the comments by Dr. Garver, and we agree that a capacity outage table would be useful. A complete capacity outage table was prepared using the recursive equation given in reference (1) of Garver's discussion. The table was prepared without rounding unit capacities. Tables 1 and 2 give selected results from this "exact" (no roundoff) table. In the tables, x is the MW outage and P(x) is the probability of x or more MW on outage.

For the range 0-60 MW, Table 1 defines every change in the function P(x). For example, the minimum unit size in the test system is 12 MW. Hence, P(x) = 0.763604 for all positive values up to X = 12. Similarly, the table is constant from x = 12 to x = 20 since 20 MW is the second smallest unit size. The next change in P(x) is 24 MW (two 12 MW units out), and the next at 32 MW (20 + 12).

Beyond x = 60, Table 1 tabulates values of P(x) in increments of 20 MW. These values were extracted from the complete cumulative outage table. Therefore, the tabulated values of P(x) are exact. However, between successive values, P(x) is not constant (nor is the change linear with x).

Table 2 extends the range of Table 1 to 2450 MW, using an increment of 50 MW. The number in parenthesis in Table 2 is the negative exponent of 10 to be applied. For example, for
$$x = 1500, P(x) = 0.4043(10)^{-4}.$$
Tables 1 and 2 do not include any maintenance. All 32 units have been included in the capacity outage table. Further, the hydro units are included at full (100%) capacity (see Table 6 of paper). Therefore, Tables 1 and 2 are based on the full system capacity of 3405 MW.

Manuscript received April 16, 1979.

Table 1
Capacity Outage Table
0 - 1600 MW

x	P(x)	x	P(x)	x	P(x)
0	1.000000	420	0.186964	1020	0.003624
12	0.763604	440	0.151403	1040	0.003257
20	0.739482	460	0.137219	1060	0.002857
24	0.634418	480	0.126819	1080	0.002564
32	0.633433	500	0.122516	1100	0.002353
36	0.622712	520	0.108057	1120	0.002042
40	0.622692	540	0.101214	1140	0.001889
44	0.605182	560	0.084166	1160	0.001274
48	0.604744	580	0.075038	1180	0.000925
50	0.604744	600	0.062113	1200	0.000791
52	0.590417	620	0.054317	1220	0.000690
56	0.588630	640	0.050955	1240	0.000603
60	0.588621	660	0.047384	1260	0.000490
80	0.559930	680	0.044769	1280	0.000430
100	0.547601	700	0.042461	1300	0.000401
120	0.512059	720	0.040081	1320	0.000305
140	0.495694	740	0.038942	1340	0.000257
160	0.450812	760	0.030935	1360	0.000164
180	0.425072	780	0.026443	1380	0.000122
200	0.381328	800	0.024719	1400	0.000102
220	0.355990	820	0.018716	1420	0.000084
240	0.346093	840	0.015467	1440	0.000071
260	0.335747	860	0.013416	1460	0.000056
280	0.328185	880	0.012136	1480	0.000046
300	0.320654	900	0.011608	1500	0.000040
320	0.314581	920	0.009621	1520	0.000027
340	0.311752	940	0.008655	1540	0.000020
360	0.283619	960	0.006495	1560	0.000013
380	0.267902	980	0.005433	1580	0.000010
400	0.261873	1000	0.004341	1600	0.000008

Table 2
Capacity Outage Table
1500 - 2450 MW

x	P(x)	x	P(x)
1500	0.4044(4)	2000	0.7246(8)
1550	0.1490(4)	2050	0.2951(8)
1600	0.8064(5)	2100	0.8431(9)
1650	0.4076(5)	2150	0.3057(9)
1700	0.1583(5)	2200	0.9270(10)
1750	0.7216(6)	2250	0.2323(10)
1800	0.2912(6)	2300	0.7971(11)
1850	0.1529(6)	2350	0.1664(11)
1900	0.4692(7)	2400	0.4697(12)
1950	0.2151(7)	2450	0.1045(12)

STATION ORIGINATED MULTIPLE OUTAGES IN THE RELIABILITY ANALYSIS
OF A COMPOSITE GENERATION AND TRANSMISSION SYSTEM

R. Billinton, Fellow IEEE
Power System Research Group
University of Saskatchewan
Saskatoon, Sask.
Canada S7N 0W0

T.K.P. Medicherla,* Member IEEE
American Electric Power
Service Corporation
2 Broadway
New York, N.Y. 10004, USA

Abstract - Station originated outages resulting in line and/or generating unit outages can occur in a power system because of faults on breakers, stuck breaker conditions, bus faults, etc. They are normally of short duration but can result in multiple circuit and/or generating unit outages. The effect of such outages on reliability indices of a composite generation and transmission system can be significant.

This paper describes and illustrates the cause and effects of station originated outages using the configurations of two practical substations. Models suitable for including their effects in the reliability analysis of composite generation and transmission systems are also described. One of the proposed models is analysed to illustrate the effect of station originated outages on the down state probabilities of a two line configuration.

INTRODUCTION

The reliability evaluation of a composite generation and transmission system is concerned with the problem of determining the adequacy of the combined generation and transmission system in regard to providing a dependable and suitable supply at its terminal stations.[1,2] Such an evaluation in a complete sense involves the simulation and load flow analysis of each possible outage condition in the system in order to determine the ability of the system to supply individual bus loads without voltage violations, line and generator overloads etc., and to quantitatively express the deficiencies, if any, in terms of reliability indices.

A composite generation and transmission reliability evaluation technique [3-6], which includes a system representation of the form used in an ac load flow analysis was developed at the University of Saskatchewan. An important aspect of this technique is the calculation of individual load point reliability indices and overall system indices. Other techniques such as that reported in Reference 7 evaluate the overall system indices using the dc load flow analysis of each contingency while Reference 8 describes the basis of a digital computer program for evaluating both the system and load point reliability indices using an ac load flow.

The papers referenced earlier basically assume that the simultaneous or overlapping outages constituting a contingency situation are independent. Another set of outages which have been designated as common-cause or common mode outages can also occur in a power system. An IEEE Task Force [9] defined a common-cause

* Formerly with the University of Saskatchewan, Canada.

outage as an event having a single external cause with multiple failure effects where the effects are not consequences of each other. Investigations (10) show that a significant contribution to reliability indices can come from this set of outages. A recent paper (6) examined the effect of common-cause outages on the bus and system indices of a composite generation and transmission system.

Composite system reliability evaluation techniques at the present time use a representation of the system in which lines simply terminate at a bus without extensive representation of the bus switching and breaker configuration. The outage effects of terminal station components are reflected in reliability calculation by simply adding a terminal component to the failure and repair rates of lines and/or generators affected by the failure station component. This approach is accurate when only one element of the system is unavailable because of the failure of a terminal station component. However, when two or more elements of the system are unavailable, the approach assumes an unrealistic independence between the system element outages which are actually caused by the failure of a single terminal component. The correct approach is to regard the terminal component failure rate as the simultaneous failure rate of the relevant system elements.

Multiple failures described in the previous paragraph which result from station originated causes such as an active failure on a breaker, a stuck breaker condition etc. are referred to in this paper as station originated outages. The effect of such outages can be significant when included in the overall reliability analysis of a composite power system.

This paper examines the station originated outages and describes some models suitable for representing such failures in composite reliability analysis. These outages are examined using the configurations of two terminal stations from an actual system.

NOTATION

AF	active failure
M	on maintenance
λ	independent failure rate in failures/year
μ	independent repair rate in repairs/year
r	average repair duration for independent failures in hours
λ_c	common-cause failure rate in failures/year
μ_c	common-cause repair rate in repairs/year
r_c	average repair duration for common-cause failures in hours
λ_s	station-originated failure rate in failures/year
μ_s	station-originated repair rate in repairs/year
r_s	average repair duration for station originated outages in hours

81 WM 002-5   A paper recommended and approved by the IEEE Power System Engineering Committee of the IEEE Power Engineering Society for presentation at the IEEE PES Winter Meeting, Atlanta, Georgia, February 1-6, 1981. Manuscript submitted August 12, 1980; made available for printing October 22, 1980.

Reprinted from *IEEE Trans. Power Apparatus Syst.*, vol. PAS-100, no. 8, pp. 3870–3878, Aug. 1981.

284

CLASSIFICATION OF OUTAGES

A power system contains a number of generating units, transmission lines and transformers. These components are referred to in this paper as elements of the power system. Most of the failures of these elements can be grouped into the following four categories:

1. Independent outages,
2. Dependent outages,
3. Common-Causes or Common Mode outages, and
4. Station originated outages.

These four types of outages can be described as follows.

Independent Outages

Independent outages of two or more elements are referred to as overlapping or simultaneous independent outages. The outage of each element is caused by an independent event. The probability of such an outage is calculated as a product of the failure probabilities of each of the elements. The state-space diagram of Figure 1 shows all possible states for a two element configuration considering independent failures.

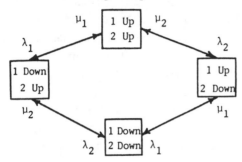

Figure 1. The basic simultaneous independent failure model.

Most of the presently available techniques for composite system reliability evaluation assume that the outages constituting a contingency situation are independent. The reliability evaluation of a composite generation and transmission system in a complete sense involves the investigation of all possible outage combinations of its elements. The computational cost associated with doing this is however prohibitive. A scheme for ordering the contingencies in order of their severity can be used for eliminating certain contingencies. The criterion of severity is, however, debatable. There is no simple answer applicable to all conditions and the cut-off point for the order of contingencies to be considered will be dependent upon the specific system characteristics, system size and the purpose for which the reliability indices are evaluated.

Dependent Outages

As the name implies, these outages are dependent on the occurrence of one or more other outages. An example is the removal from service of the second line of a double circuit line due to overload which resulted from an independent outage of the first line of the double circuit configuration. These outages are not normally included in the reliability evaluation of composite systems.

Common-Cause Outages

The probability of occurrence of an event consisting of two or more simultaneous independent outages is the product of the individual outage probabilities.

If the probabilities of individual outages are low, the product can become extremely small. The probability of a common-cause outage resulting in a similar event, can however, be many times larger. The effect of common-cause outages on reliability indices can be significant as compared with the effect of second and higher order outages. A common-cause outage is an event having an external cause with multiple failure effects where the effects are not consequences of each other.

The Task Force on Common Mode Outages of Bulk Power Supply Facilities from the IEEE Subcommittee on the Application of Probability Methods in the Power Engineering Society suggested a common mode outage model for two transmission lines on the same right-of-way or on the same transmission tower. This model is shown in Figure 2. It is similar to the model of Figure 1 except for the direct transition rate of λ_c from State 1 to State 4. This model assumes basically the same repair process for all failures including common-cause failures. Various other possible common-cause outage models are described and analyzed in Reference 6 which also examined the sensitivity of reliability indices with common-cause outages using studies of a practical system model.

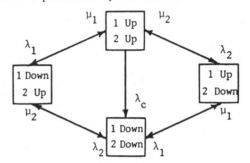

Figure 2. A common-cause outage model - The IEEE model.

Station Originated Outages

The outage of two or more transmission elements (not necessarily on the same right-of-way) and/or generating units can arise due to station originated causes. The station originated outages can occur due to a ground fault on the breaker, a stuck breaker condition, bus fault etc. or a combination of these outages. Such outages have been previously accounted for in the line and/or generator outage rates of the system by combining these outages with independent outage rates. Such a treatment cannot however recognize the situation in which more than one element of the system is simultaneously out because of a single event in the terminal station. It is, therefore, necessary to consider these outages as separate events. Station originated outages in regard to their effect on composite system reliability have not been extensively analyzed and can have an appreciable effect on load point reliability indices.

It is important to realize that common-cause transmission outages normally involve transmission lines on the same right-of-way, whereas, the station originated outages can involve system elements (which need not be on the same right-of-way) such as generating units and transmission lines. The effect of certain station-originated outages can, therefore, be more pronounced than common cause outages. The average duration of station originated outages will, however, be considerably less than common-cause outages.

Four fundamentally different types of outages which can occur in a power system have been described in this section. The station originated outages are considered in detail in the next section.

STATION ORIGINATED OUTAGES

The origin and effects of station originated outages is described and illustrated in this section using the configurations of two terminal stations from the Saskatchewan Power Corporation (SPC) system. Selected models for investigating the combined effects of independent, common-cause and station originated outages on a two element system are also presented. The effect of station originated outages is illustrated using hypothetical data.

The substation configurations of the Boundary Dam Station which has a 1½ breaker configuration and that of Squaw Rapids which is a ring bus configuration are analysed in this section in order to illustrate the station originated outage phenomenon. In these studies, active failures on breakers, and buses are considered independently and in combination with another stuck breaker condition or a breaker on maintenance. Higher order combinations can be considered, if necessary. The probability of such an outage state will, however, be very small.

Analysis of the Boundary Dam Station

The single line diagram of the Boundary Dam Station is given in Figure 3. Approximately 40% of the installed generating capacity of the Saskatchewan Power Corporation system is located at this station which can be divided into two major sections: the 230 kV and 138 kV sections. Two generators of 66 MW each are connected on the 138 kV side and four generators of 150 MW each are connected on the 230 KV side. Single breaker failures, single bus failures, and these failures in combination with another breaker in a stuck condition or under maintenance are considered in this analysis. Table 1 lists those conditions which result in an outage of two or more than two transmission lines, and/or generating units. This table shows that a large number of station originated events exist which cause an outage of two or more elements of the system. The probability associated with these events may be small but their total effect on bus and system reliability indices can be significant. A study of this table indicates that three of these events result in an outage of 8 or more elements which can have an adverse effect on the system. One of these events is an active failure on breaker 916 in combination with breaker 917 stuck. This event results in the outage of lines B3R, B1K, B2R, R7B and the outage of generators G3-6. Similar multiple outages occur due to an active failure on the breaker 917 in combination with the breaker 916 in stuck condition. The third event is an active failure (ground fault) on both buses A and B.

Table 1   Station originated outages - Boundary Dam Station

Case Description	Effect
AF on 914	B1K & B3R out
AF on 908	B2R & R7B out
AF on 905	G3 & G4 out
AF on 916 & 917 stuck	B3R, B1K, B2R, R7B G3, G4, G5, G6 out
AF on 916 & 915 stuck	G6 & B3R out
AF on 916 & 910 stuck	G6 & G5 out
AF on 916 & 909 stuck	G6 & R7B out
AF on 916 & 904 stuck	G6 & G3 out
AF on 917 & 916 stuck	B3R, B1K, B2R, R7B & G3,4,5,6 out
AF on 917 & 907 stuck	B2R & G6 out
AF on 917 & 906 stuck	G6 & G4 out

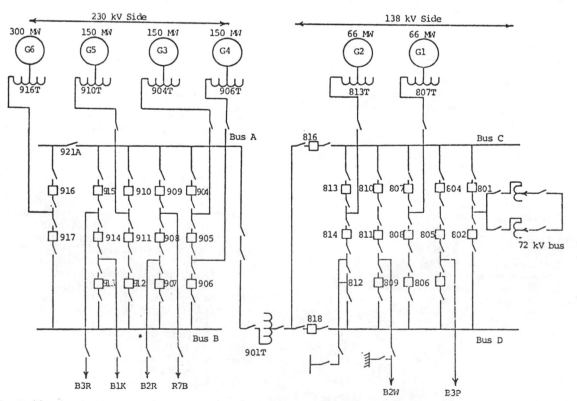

B3R & B2R = Double circuit line to Regina   R7B = Interconnector with Manitoba-Hydro system

Figure 3. Single line diagram of the Boundary Dam Station configuration.

Table 1 ... continued

Case Description	Effect
AF on 915 & 916 stuck	G6 & B3R out
AF on 915 & 914 stuck	B1K & B3R out
AF on 915 & 910 stuck	B3R & G5 out
AF on 915 & 909 stuck	B3R & R7B out
AF on 915 & 904 stuck	B3R & G3 out
AF on 914 & 915 stuck	B3R & B1K out
AF on 914 & 903 stuck	B1K & B3R out
AF on 913 & 914 stuck	B1K & B3R out
AF on 913 & 907 stuck	B1K & B2R out
AF on 913 & 906 stuck	B1K & G4 out
AF on 910 & 916 stuck	G5 & G6 out
AF on 910 & 915 stuck	G5 & B3R out
AF on 910 & 909 stuck	G5 & R7B out
AF on 910 & 904 stuck	G5 & G3 out
AF on 909 & 908 stuck	B2R & R7B out
AF on 909 & 916 stuck	R7B & G6 out
AF on 909 & 915 stuck	R7B & B3R out
AF on 909 & 910 stuck	R7B & G5 out
AF on 909 & 904 stuck	R7B & G3 out
AF on 908 & 909 stuck	B2R & R7B out
AF on 908 & 907 stuck	B2R & R7B out
AF on 904 & 905 stuck	G3 & G4 out
AF on 904 & 916 stuck	G3 & G6 out
AF on 904 & 915 stuck	G3 & B3R out
AF on 904 & 910 stuck	G3 & G5 out
AF on 904 & 909 stuck	G3 & R7B out
AF on 905 & 904 stuck	G3 & G4 out
AF on 905 & 906 stuck	G3 & G4 out
AF on 906 & 905 stuck	G3 & G4 out
AF on 906 & 917 stuck	G4 & G6 out
AF on 906 & 913 stuck	G4 & B1K out
AF on 906 & 907 stuck	G4 & R2R out
AF on buses A & B	G3, G4, G5, G6, B2R, B3R, B1K & R7B out
915 on M & AF on 914	B3R & B1K out
913 on M & AF on 914	B1K & B3R out
909 on M & AF on 908	R7B & B2R out

The durations of these outages will be equal to the switching time required for isolating the faulted breakers and to put the outaged components back into service. These durations will be dependent upon the operating practices of the individual stations and whether the switching is completed manually or automatically.

Analysis of the Squaw Rapids Station

The single line diagram of the Squaw Rapids Station is given in Figure 4. This is a generating station feeding power to the system through a radial configuration and is considered in this paper because of its ring bus structure.

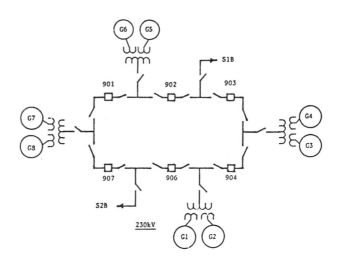

Figure 4. Single line diagram of the Squaw-Rapids substation configuration.

The Squaw Rapids Station is a generating station with six 33.5 MW generating units, and two 39.0 MW generating units. The generators are connected to a ring bus and the power from this Station is fed into the rest of the system through two 230 kV transmission lines which are also connected to this ring bus. Table 2 lists the station originated events which resulted in an outage of two or more than two transmission lines, and/or generating units. An examination of Table 2 shows that an active failure on any of the breakers causes an outage of more than two elements. The probability of such events could therefore be significant.

Table 2 Station originated outages - Squaw Rapids

Case Description	Effect
AF on 901	G5, G6, G7 and G8 out
AF on 902	G5, G6 and S1B out
AF on 903	G3, G4 and S1B out
AF on 904	G1, G2, G3 and G4 out
AF on 906	G1, G2 and S2B out
AF on 907	G7, G8 and S2B out
AF on 901 & 902 stuck	G5, G6, G7, G8 and S1B out
AF on 902 & 901 stuck	G5, G6, G7, G8 and S1B out
AF on 901 & 907 stuck	G5, G6, G7, G8 and S2B out
AF on 907 & 901 stuck	G5, G6, G7, G8 and S2B out
AF on 902 & 903 stuck	G3, G4, G5, G6 and S1B out
AF on 903 & 902 stuck	G3, G4, G5, G6 and S1B out
AF on 903 & 904 stuck	G1, G2, G3, G4 and S1B out
AF on 904 & 903 stuck	G1, G2, G3, G4 and S1B out
AF on 904 & 906 stuck	G1, G2, G3, G4 and S1B out
AF on 906 & 904 stuck	G1, G2, G3, G4 and S1B out
AF on 906 & 907 stuck	G1, G2, G7, G8 and S2B out
AF on 907 & 906 stuck	G1, G2, G7, G8 and S2B out
901 on M and AF on 902	G5, G6 and S1B out
902 on M and AF on 901	G5, G6 and S1B out

287

Table 2 ... continued

Case Description	Effect
901 on M & AF on 907	G7, G8 and S2B out
907 on M & AF on 901	G7, G8 and S2B out
902 on M & AF on 903	G3, G4 and S1B out
903 on M & AF on 902	G3, G4 and S1B out
903 on M & AF on 904	G1, G2, G3 and G4 out
904 on M & AF on 903	G1, G2, G3 and G4 out
904 on M & AF on 906	G1, G2 and S2B out
906 on M & AF on 904	G1, G2 and S2B out
906 on M & AF on 907	G7, G8 and S2B out
907 on M & AF on 906	G7, G8 and S2B out

Station originated outages are described and illustrated in this section with reference to two terminal station configurations in a practical system. The numerical evaluation of each of these events can be accomplished using the equations provided in Reference 12. This paper details the equations used in such an analysis and also the component and operating data needed to perform the calculations. These data and equations are not normally associated with composite system reliability evaluation but with isolated or individual terminal station or substation configurations. These stations are, however, the terminating points for the transmission lines and generating units which are represented in composite system evaluation. It is not possible to retain the stations in their entire complexity in a composite system appraisal and therefore it becomes necessary to combine the event data resulting from the failure modes and effect analysis detailed in Tables 1 and 2 into reduced models which can be used in a composite system reliability study. The inclusion of these outages in the reliability evaluation of a composite generation and transmission system requires the development of one or more suitable models. Selected possible models are presented and analysed in the next section of this paper. It is important to realize that the development of a single model suitable for all practical situations is not possible, and the models in this paper can be modified or new models can be created to suit the given data.

Models for Station Originated Outages

In the previous section, the origin and possible cases of station originated contingencies resulting in multiple outages have been described with reference to the Boundary Dam and Squaw Rapids stations. This section describes some possible models for studying the effects of station originated outages. These models can be used to account for outages affecting two elements and can be extended to more than two elements. Hypothetical data is used to illustrate the effect of station originated outages on the probability of outage of both elements in a two transmission line configuration.

The simplest possible model can be obtained by combining the station originated outages which result in both lines out with the common-cause outages. This model which is shown in Figure 5, is similar to that of Figure 2 except for the fact that the transition rate from the both lines up state to the both lines down state has been increased from λ_c to $\lambda_c + \lambda_s$ (where λ_s is the contribution of station originated outages). This model therefore assumes the same repair process for independent, common-cause, and station

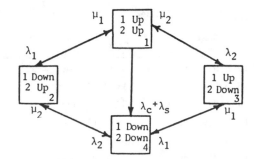

Figure 5. Model 1

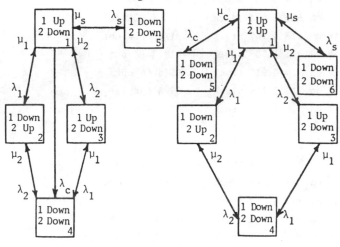

Figure 6. Model 2   Figure 7. Model 3

originated outages. A serious objection to this model is the fact that the repair duration for a station originated outage will be very short compared to the repair duration for common-cause outages, and independent outages.

Another possible model is shown in Figure 6. This model is a simple modification of Model 1. In this model, a separate state is created to account for the station originated outages. A more practical model is shown in Figure 7. In this model, the all lines down states due to independent outages, common-cause outages and station-originated outages are shown as three separate states. This model has been used in this paper for system studies. Other models can be created to suit the data and needs of a particular situation.

In Models 1 and 2, λ_c will be equal to zero, if the station originated outage involves a transmission line and a generator, or two generators which can be considered to be independent. Under such circumstances, state 5 of Model 3 does not exist.

Effect of Station-Originated Outages

Three possible models for including the effects of station originated outages in the reliability evaluation of a composite generation and transmission system are described in the previous section. The Model 3 shown in Figure 7 is considered in detail in this section to illustrate the effect of station originated outages.

Assuming that the lines 1 and 2 have identical failure and repair rates, the expression for the total probability for both lines down in Model 3 is as follows:

$$P(\text{Lines 1 and 2 down}) = P_4 + P_5 + P_6 = \frac{1}{D}[\,(\lambda^2 \mu_s \mu_c / \mu)$$
$$+ \mu_c \mu \mu_s + \lambda_s \mu \mu_c\,]$$

288

where $D = \mu_s(\lambda\mu_c + \mu\mu_c + \lambda_c\mu) + \lambda_s\mu_c\mu$

In order to illustrate the effect of station originated outages, two situations shown in Figure 8.a and 8.b are considered.

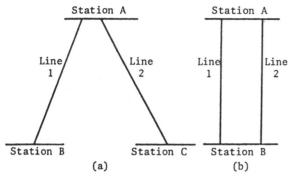

Figure 8. Two Line Configurations.

In both cases the line lengths were assumed to be 100 mile long each. In Figure 8.a both the lines originate at station 'A' but occupy different rights-of-way and terminate at stations B and C. These lines are therefore considered to be not vulnerable to common-cause outages. Lines 1 and 2 in Figure 8.b however, originate at station A and terminate at the same station B. They are on the same right-of-way or on the same transmission tower. Common-cause outages are therefore assumed to be possible on these lines.

The outage data provided in Reference 11 for five different line voltage ratings were used. These data are reproduced in Table 3 for ready reference. The probabilities for the all lines down state calculated for the configuration of Figure 8.a using Model 3 are listed in Table 4. The probabilities have been calculated for a 0 to 15% range of λ_s/λ ratio and a fixed $r = 1$ hour. This table indicates that the probability of the both lines down state increases by a factor of six for a change in λ_s/λ ratio of 2.5% and then the increase is relatively slow.

Table 3 Transmission line outage data.

Line Voltage kV	Failure Rate Outages/(100 miles.year)	Repair Duration hours
69	1.10	6
138	0.65	9
230	0.50	9
345	0.25	12
500	0.20	12

The probability for the all lines down state calculated for the configuration of Figure 8.b using Model 3 for a fixed λ_c/λ ratio of 0.1 and $r_c/r = 1.5$ are listed in Table 5. These probabilities are also calculated for a 0 to 15% range of λ_s/λ ratio and a fixed $r_s = 1$ hour. This table shows that the consideration of station originated outages when considered in addition to common-cause outages has negligible effect. In other words, the station originated outages may need not be considered for lines vulnerable to common-cause outages.

These studies show that station originated outages which involve two or more transmission lines and/or generating units should be considered in the reliability

Table 4 Variation of the two lines down state probability with the station-originated outage rate for the two line configuration of Figure 8.a.

λ_s/λ %	Probability of all lines down x 10^4				
	69 kV	138 kV	230 kV	345 kV	500 kV
0	0.0057	0.0045	0.0026	0.0012	0.0008
2.5	0.0370	0.0230	0.0169	0.0083	0.0065
5.0	0.0684	0.0415	0.0312	0.0154	0.0122
7.5	0.0997	0.0600	0.0454	0.0226	0.0179
10.0	0.1311	0.0786	0.0597	0.0297	0.0236
12.5	0.1624	0.0971	0.0739	0.0368	0.0293
15.0	0.1937	0.1156	0.0882	0.0440	0.0350

analysis of a composite generation and transmission system. They may not however need to be considered if the elements involved are themselves vulnerable to common-cause outages.

Table 5 Variation of the two lines down state probability with the station-originated outage rate for the two line configuration of Figure 8.b.

λ_s/λ %	Probability of all lines down x 10^4				
	69 kV	138 kV	230 kV	345 kV	500 kV
	1.1340	1.0047	0.7723	0.5145	0.4115
2.5	1.1653	1.0233	0.7866	0.5216	0.4172
5.0	1.1967	1.0418	0.8008	0.5288	0.4229
7.5	1.2280	0.1603	0.8151	0.5359	0.4286
10.0	1.2593	1.0788	0.8294	0.5430	0.4343
12.5	1.2907	1.0973	0.8436	0.5501	0.4400
15.0	1.3220	1.1159	0.8579	0.5573	0.4457

CONCLUSIONS

The presently available techniques for evaluating the reliability of composite generation and transmission systems consider independent overlapping outages, and common-cause outages. The outages originating in terminal and switching stations are accounted for by increasing the failure rate of transmission lines by some fixed amount. This is a valid addition for those station related failures which impact only on the transmission element concerned. This approach, however, does not recognize that multiple outages can occur due to a station originated cause and does not simulate such a situation.

This paper describes and illustrates the cause and effect of station originated outages using the configurations of two practical substations. Three possible models are examined for considering these outages in evaluating the reliability of composite generation and transmission systems. The sensitivity of the probability of the two element down state with station originated outage rate for one of these models is also examined. This study shows that the effect of station originated outages on elements already vulnerable to common-cause outages may be negligible. However, station originated outages which involve two or more transmission lines and/or generating units should be considered in the reliability analysis of a composite generation and transmission system.

It appears obvious that composite system reliability evaluation leading to individual load point and global system indices involves much more than the creation and examination of a large number of system outage conditions generated by independent removal of generation and transmission elements. The desire to examine high order independent events has dictated the need for approximate solution techniques for the network and for the selection of system and load point failure criteria. Recognition and classification of dependent events such as common mode failures and station induced outages may, however, obviate the need to examine high order independent outage events and also lead to a more realistic appraisal of practical systems. These events, however, require detailed analysis and data before they can be consigned as input to a system appraisal program. They cannot be included by simple addition to the independent failure rates of the transmission or generation elements but should be included as an additional level of element outage data.

REFERENCES

1. R. Billinton, Power System Reliability Evaluation, Gordon and Breach Science Publishers, New York, 1970.

2. R. Billinton, "Composite system reliability evaluation", IEEE Transactions on Power Apparatus and Systems, Vol. PAS-88, No. 4, April 1969, pp. 276-280.

3. R. Billinton and M.P. Bhavaraju, "Transmission planning using reliability criterion - Part I. A reliability criterion", IEEE Transactions on Power Apparatus and Systems, Vol. PAS-89, No. 1, January 1970, pp. 28-34.

4. R. Billinton, T.K.P. Medicherla and M.S. Sachdev, "Composite generation and transmission system reliability evaluation", Paper No. A 78 237-0 presented at the IEEE PES Winter Meeting, New York, January 1978.

5. R. Billinton, T.K.P. Medicherla and M.S. Sachdev, "Adequacy indices for composite generation and transmission system reliability evaluation", Paper No. A 79 024-1 presented at the IEEE PES Winter Meeting, New York, February 1979.

6. R. Billinton, T.K.P. Medicherla and M.S. Sachdev, "Application of common-cause outage models in composite system reliability evaluation", Paper No. A 79 461-5 presented at the IEEE PES Summer Meeting, Vancouver, B.C., July 1979.

7. P.L. Dandeno, G.E. Jorgensen, W.R. Puntel and R.J. Ringlee, "A program for composite bulk power electric system adequacy evaluation", A paper presented at the IEE Conference on Reliability of Power Supply Systems, February 1977, IEE Conference Publication No. 148.

8. G.E. Marks, "A method of combining high-speed contingency load flow analysis with stochastic probability methods to calculate a quantitative measure of overall power system reliability", Paper No. A 78 053-1 presented at the IEEE PES Winter Meeting, New York, January 1978.

9. Task Force on Common Mode Outages of Bulk Power Supply Facilities of the Application of Probability Methods Subcommittee of Power System Engineering Committee, IEEE Power Engineering Society, "Common mode forced outages of overhead transmission lines", IEEE Transactions on Power Apparatus and Systems, Vol. PAS-95, May/June 1976, pp. 859-863.

10. R. Billinton, T.K.P. Medicherla and M.S. Sachdev, "Common-cause outages in multiple circuit transmission lines", IEEE Transactions on Reliability, Vol. R-27, No. 2, June 1978, pp. 128-131.

11. E. Masud, "Automatic load flow contingency evaluation using a reliability index", Paper No. A 76 395-4 presented at the IEEE PES Summer Meeting, Portland, July 1976.

12. M.S. Grover and R. Billinton, "A Computerized Approach To Substation And Switching Station Reliability Evaluation", IEEE Transactions PAS-93, September/October, 1974, pp. 1488-1497.

Discussion

Robert J. Ringlee (Power Technologies, Inc., Schenectady, NY): Events involving faults and/or misoperation of transmission substation or switching station components often have significant impact on the security of bulk power systems. It is of interest to review the characteristics of bulk power system failures and to note the numbers which were affected significantly by station-initiated events. Three findings may be readily discerned concerning the characteristics of bulk power system failures. First, the majority of bulk power system failures stem from incidents arising in the transmission system; that is, most of the incidents may be identified with transmission line or station equipment failures, the minority are identified with generator incidents (Refs. 13-16). Of course, this finding must be tempered with the significant influence that generation dispatch, interchanges and transfers can have on the system disturbance preconditions. That is, unscheduled outages of generation or fuel supply curtailments or economic interchanges can result in heavy power transfers and set up a predisposition to bulk power system failure. Second, the majority of bulk power system failures were initiated by the incidents that were judged to be within the design criteria (Ref. 15). A survey of readily available summaries of major disturbances suggest that only one in three of these has been caused by what might be called an abnormal fault condition that is exceeding the bulk power system reliability criteria. About half of the faults classified as abnormal had been faults at bulk power system buses or breaker failures (Ref. 15). Here, though, the initiating incidents were within the design criteria, the effect of the precondition and the consequential system responses were clearly outside those design criteria. Thirdly, initiating events caused by equipment malfunction or human error may be more devastating to the system than those of natural origin because they often trip a number of lines or units simultaneously and in combinations for which no automatic protection has been provided (Ref. 14).

The significance of proper performance of the bulk power station in handling the more commonly arising faults and disturbances which the bulk power system is subjected to can be of little question. The authors are quite correct in emphasizing the importance of the effects of faults and abnormal operation in stations which can affect the status of more than one bulk power transmission element.

The discusser has found the procedures outlined in the paper of much value in the analysis of alternative designs of bulk power stations. Based upon these studies, the following comments seem in order. Reliability assessments of station alternatives should consider several points of view. Steady-state reliability assessment should be based on enumeration of the contingencies and classification of the effect of these contingencies in terms of both overloads and voltage problems. Overload problems may be expressed in terms of overload on specific equipments and in terms of power transfer curtailment necessary to avoid equipment tripping on overload. Voltage problems may be described in terms of power transfer curtailment required to avoid insecure operation and overloading voltage support equipment. Contingencies which cause both overload and failure due to lack of reactive support should be rated in terms of curtailment required to correct the most severe effect. Reliability assessments for steady-state post-fault conditions provide frequencies and durations of various degrees of power transfer curtailment. Dynamic assessment through disturbance simulation seeks contingencies which would cause cascading, instability, or voltage collapse with loss of generating facilities and loads. The evaluation of the frequency of occurrence of major system interruption events can be the most important aspect of the overall system reliability assessment. This evaluation requires disturbance simulation and interpretation of the results to establish whether the event caused isolaiton by instability or would lead to isolation by cascading (tripping of lines, transformers, generators, or condensers through overload), or controlled load curtailment due to operator intervention to relieve overloads, or due to voltage collapse. It should be clear, therefore, that station reliability evaluations must be made in parallel with system studies (Ref. 17).

Techniques, such as those described in the paper, may be readily automated to facilitate the analysis of large station configurations (Refs. 18-20). One of the categorizations that the discusser has found useful has been to distinguish line-initiated and station-initiated events from those events which involve overlapping outage of lines and station elements (Refs. 17, 21). For example, it is desirable to distinguish events involving faults on lines and associated loss of other station elements and other lines caused by protection or breaker misoperations from those events involving faults in the station and associated element losses.

REFERENCES

13. "Review of Major Power System Interruptions," National Electric Reliability Council, August, 1979, Research Park, Terhune Road, New Jersey.
14 "Long-Term Power System dynamics," Volume 1B, Electric Power Research Institute, Research Project 90-7, April, 1974.
15. D. R. Davidson, D. N. Ewart, L. K. Kirchmayer, "Long-Term Dynamic Response of Power Systems: An Analysis of Major Disturbances," *IEEE Transactions on Power Apparatus and Systems,* Vol. PAS-94, May/June 1975, pp. 819-825.
16. "Electricite de France, Power Failure of December 19, 1978," Report to CIGRE Study Committees 32/34, Paris, May 10, 1979.
17. R. J. Ringlee, C. A. MacArthur, G. E. Scott, "Reliability Criteria and Predictions for Transmission Stations on an 800 kV System," EPRI Report WS77-66, 1978.
18. C. R. Heising, R. J. Ringlee, and H. O. Simmons, "A Look at Substation Reliability," American Power Conference, April, 1969.
19. IEEE Tutorial Course, Probability Analysis of Power System Reliability, 71 M 30-PWR, pp. 51-57.
20. R. Billinton, R. J. Ringlee, A. J. Wood, "Power System Reliability Calculations," MIT Press, 1973, pp. 31-44.
21. B. P. Lam, N. D. Reppen, R. J. Ringlee, "Reliability Criteria and Predictions for the Transmission System of Itaipu," Fourth National Conference on Production and Transmission of Electric Energy, Rio de Janeiro, Brazil, September 18-24, 1977.

Manuscript received February 9, 1981.

G. L. Landgren, S. W. Anderson, and **A. W. Schneider, Jr.,** (Commonwealth Edison Company):
This paper is a significant contribution to the evolution of models for analysis of transmission system performance. It points out that "station originated outages" are an important component of the total rate of multiple line outages. Commonwealth Edison experience has been that over half of our multiple 345 kV transmission line outages are of this type. [1] The authors' Model 3 properly recognizes that the distribution of restoration times for "station originated outages" and common mode outages originating elsewhere on the line are different so the states must be distinguished.

Simple reasonableness checks which we applied to the formulas for state probabilities led to two apparent discrepancies, as described below. It would be helpful if the authors could review their derivation, furnish formulas for the probabilities of the remaining states, and generalize to the important case where λ_1 and λ_2, and μ_1 and μ_2, are not equal. This corresponds to a system of two lines which are not symmetrical but do have common exposures.

The resonableness checks to which we refer are:
1. Because of the symmetry of states 5 and 6 in Model 3, their probability formulas should be identical except for the exchange of subscripts "s" and "c." By setting λ and λ_s to zero, the model

simplifies to a two state model (states 1 and 5) and the probability of state 5 should be

$$\lambda_c/(\lambda_c + \mu_c)$$

(See paper reference (1), page 65).
Thus, we believe P_5 should be $\lambda_c\mu\mu_s/D$ rather than $\mu_c\mu\mu_s/D$.
2. By setting λ_c and λ to zero, the model simplifies to an independent four state model and P_4 should simplify to $\lambda^2/(\lambda + \mu)^2$ rather than to $\lambda^2/(\mu(\lambda + \mu))$. (*Ibid.,* p. 74).

As previously mentioned, our experience is that over half of our 345 kV multiple outages are station originated. This cannot be directly observed from the data for lockout outages in reference (1) for two reasons: First, (1) tabulates line outage rates, while the present model uses outage *event* rates which are half as great for common mode and station originated outage events causing outage of two lines. Second, a different hierarchy of aggregation was followed than is required for the present model. The following event rates may be developed for lines averaging 35.12 miles in length and having the common exposures described in reference (1):

$$\lambda = 0.600/\text{year}$$
$$\lambda_c = 0.091/\text{year (15\% of } \lambda)$$
$$\lambda_s = 0.109/\text{year (18\% of } \lambda)$$

This illustrates the importance of retaining sufficient detail in the data collection process to permit reaggregation as required by a new model.

Table 1 illustrates the impact of seemingly minor differences in bus layout upon reliability. For, if G6 were connected between breakers 911 and 912, two of the three double-contingency cases outaging four generating units would be eliminated but a single contingency, AF on breaker 911, outaging two units totaling 450 MW would be introduced. Which layout is preferable depends on whether loss of all four units will lead to system collapse and on the probability that a breaker will stick.

Table 2 omits some more severe and equally probable cases which can arise at the Squaw Rapids station, given an AF on one breaker and maintenance on another. If there is one breaker between the faulted and the maintained breakers (e.g., AF on 901 and 903 on M) three bus sections, rather than two, will be completely isolated. Furthermore, if any of the four breakers adjacent to a line is on maintenance and the one exactly opposite it (which is also adjacent to a line) suffers an AF, four bus sections will either be outaged or completely isolated from a transmission outlet. (E.g., 902 on M and AF on 906 outages S2B, G1, and G2, and leaves G5, G6, G7 and G8 without transmission outlet). The frequency of fault on a particular breaker during the maintenance outage of another would presumably be nearly the same regardless of which breaker is being maintained.

We note that the third voltage level in Table 3 is given as 230 kV in this paper, and 161 kV in paper reference (6) citing the same source, paper reference (11). The author of the latter has informed us that 161 kV is correct.

Analysis of networks of more than two lines, with full consideration of all relevant common physical exposures, will lead to consideration of increasingly complex models with much larger numbers of states. This paper is an important step in the development of such models. EPRI Project RP 1468 (2), which will be completed this year, also addresses some of the data collection, model development and model validation questions which must be considered in evolving models to predict the performance of larger systems.

REFERENCES

(1) G. L. Langren and S. W. Anderson, "Data Base for EHV Transmission Reliability Evaluation." 80 SM-721-1, IEEE PES Summer Meeting, Minneapolis, Minnesota, 1980.
(2) J. T. Day, *et. al., Component Outage Data Analysis,* EPRI Project RP 1468, in preparation.

Manuscript received March 9, 1981.

002-5d

R. Billinton and **T. K. P. Medicherla:** We thank the discussors for their interest and comments on the paper. We are particularly thankful to Messrs. Landgren, Anderson and Schneider for pointing out an error in the expression for the total probability for both lines down in Model 3. The correct expression is as follows:

P (Lines 1 and 2 down) = $P_4 + P_5 + P_6$

$$= \frac{1}{D} \{\lambda^2 \mu_c \mu_s + \mu^2 (\lambda_c \mu_s + \lambda_s \mu_c)\}$$

where $D = \mu_c \mu_s (\lambda + \mu)^2 + \mu_c \mu^2 \lambda_s + \lambda_c \mu_s \mu^2$

when $\lambda_s = 0$

$$P \text{ (Lines 1 and 2 down)} = \frac{\lambda^2 \mu_c + \lambda_c \mu^2}{\mu_c (\lambda + \mu)^2 + \lambda_c \mu^2}$$

when $\lambda_c = 0$

$$P \text{ (Lines 1 and 2 down)} = \frac{\lambda^2 \mu_s + \lambda_s \mu^2}{\mu_s (\lambda + \mu)^2 + \lambda_s \mu^2}$$

when $\lambda_s = \lambda_c = 0$

$$P \text{ (Lines 1 and 2 down)} = \left(\frac{\lambda}{\lambda + \mu}\right)^2$$

The expressions for rest of the states of the model are:

$P_1 = \mu^2 \mu_c \mu_s / D$

$P_2 = \lambda \mu \mu_c \mu_s / D$

$P_3 = P_2$

The errors in the equations of the paper do not affect the probabilities listed in Tables 4 and 5 because the basic equations using frequency balancing approach were formed using numerical values and then the matrix inversion approach was used for calculating the state probabilities.

For the generalized case where λ_1, λ_2 and μ_1, μ_2 are not equal, the expressions for state probabilities may not be simple to use and a better approach would be to solve the basic equations using numerical values.

It is reassuring to note that the common mode and station originated outages are 15% and 18% of the independent failure rate. This data supports our argument in the present and previous papers that common mode and station originated outages can have significant effect on reliability and have to be examined before attempting to include the effect of higher order independent outages.

We are thankful to the discussors for providing additional insight into the effect of station originated outages by pointing out the effects of minor changes in bus layouts at Boundary Dam and Squaw Rapids stations.

An oversight error occurred in Table 3 of this paper, as pointed out by the discussors. In the first column of this table, 230 should be replaced by 161.

Most of the presently available techniques for evaluating the reliability of composite generation and transmission systems attempt to assess the steady state adequacy of the system using contingency enumeration approach. We agree with Dr. Ringlee that the dynamic assessment of contingencies should also be included for a total assessment of the system.

Manuscript received April 27, 1981.

BULK POWER SYSTEM RELIABILITY ASSESSMENT — WHY AND HOW?
PART I: WHY?

A Panel Discussion sponsored by
the Application of Probability Methods Subcommittee of the PES
Panelists: J. Endrenyi (Chairman), P.F. Albrecht,
R. Billinton, G.E. Marks, N.D. Reppen, L. Salvaderi

Abstract — The first of the discussions presented in this paper reviews the special difficulties involved in bulk power system reliability evaluations. These are partly conceptual, partly concerning modeling, computational techniques, and data requirements. The second discussion analyzes the system failure modes which are of particular interest to system planners, including widespread cascading outages and generation bottling, and proposes a consistent set of bulk power system reliability definitions. The third discussion describes the principal uses of bulk power system reliability studies.

INTRODUCTION

A panel discussion on the uses and techniques of bulk power system reliability assessment was presented at the 1981 Winter Meeting of the PES, with an international assembly of panelists from utilities, consultants, and universities participating. Subsequently, it was decided to condense the material into a publication, and as a result, this paper and its companion came into being. In the following, each section represents the individual contribution of a panelist and is, therefore, published under his name.

While the entire material has been edited, some overlaps and repetitions were allowed to stand to ensure as much fidelity as possible to the original presentations. Some of the terms are given different definitions by different panelists - this also was allowed to stay, as a telling indication of the unsettled state in which the methodology of bulk system reliability evaluation finds itself at this early stage of its development.

BULK POWER SYSTEM RELIABILITY - AN OVERVIEW

J. Endrenyi
Ontario Hydro
Toronto, Canada

Introduction

Because of the many complexities involved in the reliability evaluation of even simple power networks, power system reliability studies have been traditionally performed "in parts", that is, the reliability performances of the major parts of the system have been evaluated separately. The sophistication and maturity of the methods applied in these studies show a wide variation; in the reliability evaluation of generating systems or station layouts, for example, they are high, in distribution system studies they are lower, while in reliability studies of transmission and bulk power systems (composites of generation and

transmission), methods are still in the developmental stage[1]. These disparities were well demonstrated in a presentation at the 1978 EPRI Workshop on Power System Reliability[2], and while efforts in developing transmission reliability approaches have speeded up since then, the above comparisons are still valid.

The difficulties in bulk system reliability methodology can be classified into the following categories:

o conceptual problems,

o modeling difficulties,

o computational difficulties,

o those related to data requirements.

In the following, brief comments will be made on each of these aspects.

Conceptual Problems

Before attacking the many technical problems involved, one ought to have a clear concept about what one wants to achieve by studying bulk power system reliability. The answer is far from obvious, and in trying to find it, one encounters a surprising number of questions with no clear answers. These include:

o defining the purposes and uses of the studies,

o selecting appropriate system failure events and reliability indices,

o selecting acceptable risk levels.

The purposes and uses of bulk power system reliability studies are widely debated. Most investigators agree, however, that such studies are conducted with the purpose of making assessments in the following areas:

o system adequacy - the capacity of the system for meeting the load demand within component ratings and voltage limits at any time,

o system security - the ability of the system to withstand the impact of sudden changes.

While both of these aspects are equally important, in many system planning studies only adequacy indices are sought, because of the greater complexities involved in security assessments.

The results of the above studies can be used in the following ways:

o to compare alternative designs,

o to assess system reliability against reliability standards (if available),

o to seek optimal balance between costs and benefits.

82 WM 147-7   A paper recommended and approved by the IEEE Power System Engineering Committee of the IEEE Power Engineering Society for presentation at the IEEE PES 1982 Winter Meeting, New York, New York, January 31-February 5, 1982. Manuscirpt submitted September 1, 1981; made available for printing November 16, 1981.

Reprinted from *IEEE Trans. Power Apparatus Syst.*, vol. PAS-101, no. 9, pp. 3439-3445, Sept. 1982.

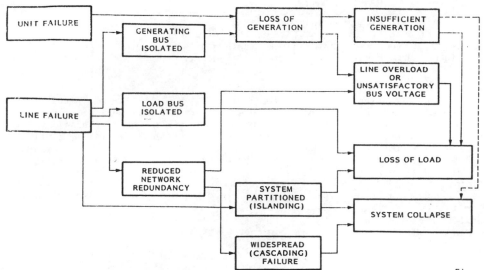

Figure 1. Possible consequences of component failures in bulk power systems.

The <u>system failure events</u> to be selected and the corresponding reliability <u>indices</u> depend, of course, on the purpose of the study. As to the system failure events and how they can arise as consequences of component outges, the sketch in Figure 1 offers a general orientation. According to this, <u>loss of load</u> and <u>system collapse</u> arising from widespread outages or islanding are the principal system failure events. Therefore, when the adequacy of the entire system is considered, the primary indices are the probability, frequency and mean duration of these events. In addition, similar indices may be calculated for the "secondary" failure events, including insufficient generation, line overloads and unsatisfactory bus voltages. A second group of indices may describe the severity of load losses in terms of the amount of energy not supplied, the amount of load cut, and so on. A third group of indices will measure the interruptions of supply at each load bus, and a fourth group will produce cost figures, such as the incremental fuel costs or transmission losses in contingencies.

Once the value of an index is computed, it ought to be compared with a standard value set for the index. This standard must reflect the <u>acceptable risk level</u> in terms of that index. At the present stage of development, however, such standards are rarely available. Most studies are, therefore, restricted to comparisons of alternatives and to cost-benefit evaluations. "Standard values" may be selected in the vicinity where the cost-benefit ratio is optimal, although this consideration may not be the only input in the selection.

Modeling

The goal in the modeling phase is to mathematically represent all relevant factors and effects without making the resultant model unmanageable. To what extent this is attainable is open to question. At any rate, a wide variety of factors needs to be included in the model, with the following list providing just a few of these:

o component failure modes,

o dependent failures (including common-mode failures),

o load characteristics (electrical),

o load demand model,

o energy constraints,

o system/operator response in contingencies,

o generation dispatch,

o weather effects on overhead lines,

o planned maintenance.

Some of these represent extremely tough tasks. Consider, for example, the modeling of system/operator response in contingencies. The events following an "incident" are shown in Figure 2. If the model accounts for only the steady-state condition in each contingency (a comparatively simple task), measures for system adequacy may be correctly obtained, but not security measures, and even some of the severity indices may be in error. If the model was capable of tracing the events back closer to the moment of "incident", more of the indices could be accurately determined, but the modeling itself (and its solution) would entail almost insurmountable difficulties.

Computational Difficulties

In principle, the solution of a power system reliability model is a simple task; the steps of the algorithms are known, and if a computer of infinite capacity and speed was at hand, there would be no computational difficulties. But such a device does

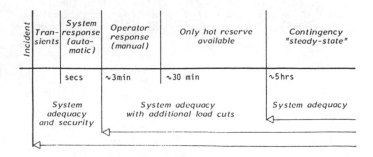

Figure 2. The aftermath of a component failure, and possible extents of coverage by models.

294

not exist, and with the computing power presently at the utilities' disposal, the reliability evaluation of a large system may take several hundred years if done without simplifications and approximations[3]. Since this is generally considered too long, every possible approximation, shortcut, and other trick in the book must be employed to obtain a program that is also practical.

Power system reliability programs consist of several algorithms representing the main steps of the program. These are the following:

o contingency selection,

o evaluation of the selected contingencies,

o corrective action through system or operator response,

o computation of indices.

Some of the ways that can make these algorithms more efficient are elaborated on in Part II of this paper. Quite obviously, many of the fast techniques involve approximations with the penalty of reduced accuracy. The challenge is to balance <u>speed</u> and <u>accuracy</u> so that both should still remain acceptable. Programs will have a chance of being accepted by users only if the reconciliation of these factors is successful.

Data Requirements

Even the best programs are only as good as the input data used in them. For the complex models describing the power system, the failures of its components, the load conditions, and so on, a multitude of data is required, some of which are readily available and some that are not. Much development is going on to identify the data needed, and to establish and implement suitable data collecting schemes.

The last discussion in Part II of this presentation elaborates on aspects of data requirements for bulk power system reliability studies.

Closing Comment

As already stated, there is still no satisfactory bulk power system reliability program available. However, several efforts are underway to develop one, and there is some hope that success may not be too far away. In the meantime, the discussions that follow provide some insight into the problems to be solved and identify some of the most promising avenues of solution.

BULK POWER SYSTEM RELIABILITY CONCEPTS

Paul F. Albrecht
General Electric Company
Schenectady, New York

Introduction

Two basic reasons can be given for the current interest in bulk power system reliability assessment.

One reason is the ever-increasing impact, or potential impact, of line outages or equipment failures. The 1965 Northeast Blackout brought this into sharp focus. Subsequent to 1965, major power interruptions have continued to occur periodically, even as recently as January, 1981. Murphy's Law guarantees that such interruptions will occur in the future. However, some aspects of major outages, such

as their frequency, duration, and magnitude, are a function of planning, design, and operating criteria.

A second reason for interest in reliability assessment is the increasing need to consider cost, environmental and political constraints. This concern has motivated research in more quantitative methods for analyzing bulk power system reliability, so that reliability considerations can be placed in an economic context.

The first factor - impact of interruptions - relates to the reliability of the entire system. Reliability assessment could lead to the conclusion that the bulk power system is in need of reliability improvement - or conversely, show that the existing system provides higher reliability than necessary. The second factor focuses on the "worth" of specific equipment additions, such as a new transmission line.

In summary then, current interest in quantitative methods stems from the fact that two questions are frequently asked today:

1. Are existing bulk power systems over- or under-designed in terms of reliability?

2. If a new line is constructed from point A to point B, how much will bulk power system reliability be improved?

How should the reliability of a bulk power system be measured? This question is not easy to answer. Before attempting to define bulk power system reliability, it will be reviewed how reliability has been defined for other sections of the electric power system.

Reliability Concepts for Distribution and Generating Systems

For the distribution system, a practical measure of system failure is individual customer service interruptions. Several different indices can be computed, such as frequency of customer interruption, duration of interruption per interruption, and fraction of time interrupted per customer. These measures of reliability are quite appropriate because a majority of equipment outages in the distribution systems do directly affect customer service. In other words, frequency and duration of customer interruption are, in fact, primary considerations involved in distribution system planning. This is a fundamental point - a reliability index is useful as a planning tool only if it actually reflects the goals the planner is trying to achieve.

Secondly, customer outages originating in the distribution system can be traced back to specific equipment outage events. Therefore, if historical records on distribution equipment outages are maintained, the reliability of a future distribution network can be calculated.

To summarize, the key points are that:

1. a distribution system <u>failure</u> is well defined - service interruption to customers;

2. the events that <u>cause</u> system failures are known - equipment outages.

Next, consider these two points with respect to the generation system. In the generation system the events which <u>cause</u> system failure are still well defined - generating system capacity is reduced only by outages of generating units. Typically, outages of single units do not cause service interruption. Concurrent outages of several units are necessary for that. The probability distribution of available capacity can be estimated by using simple probability methods. Several utilities have plotted actual vs. calculated capacity probability distributions and have found that actual and calculated results show good agreement.

What should be counted as a generation system failure? Historically, planners have again applied the "service interruption" concept, and thus, the generation system is considered "failed" when load demand exceeds available generation. With this definition, methods for calculation of generation system reliability in terms of outage characteristics of generating units have been developed. In fact, reliability evaluation of generation system capacity expansion plans has historically been the most important application of probability methods.

However, in recent years, planners have begun to question the adequacy of a single definition of failure for the generation system. It is realized - particularly as interconnections between areas, pools, and regions increase - that "capacity equals load" is not the point at which customer interruption due to generation shortage takes place. Additional margin exists beyond this point due to ability to import emergency power, reduce voltage, make voluntary appeals, etc. On the other hand, initiation of "emergency actions" occurs before the margin between capacity and load is reduced to zero. Thus, in planning generation systems, it is being recognized that actually several levels of capacity deficiency, and the associated system events (eg., voltage reduction), must be considered.

To summarize with respect to the same points used for distribution systems:

1. A generation system failure is less well defined than a distribution system failure. In particular, service interruption to customers is not the direct measure of generation system reliability. Rather, several specific events (eg., emergency power purchases, voltage reductions) must be considered.

2. The events that cause generation system failure are known - overlapping independent outages of generating units.

Reliability Concepts for Bulk Power Systems

With this background, consider the bulk power system. Again, the first reaction is to define failure in terms of load interruption or curtailment. The literature on transmission reliability reflects this tendency. The usual approach is to calculate frequency and duration of load interruption or curtailment. Various additional parameters may also be calculated, such as energy not served.

However, a review of reliability planning criteria used by various regions shows that customer load interruption, per se, is not the primary consideration in bulk power system planning. In practice, the most important concern of the bulk power system planner seems to be to avoid widespread, cascading, or uncontrolled interruptions. Occasional interruption of certain loads may be accepted, depending on criticality and location of the loads.

A second goal of bulk power system planners appears to be that equipment outages should not limit

ability to dispatch generation. Thus, in order to judge the "value" of adding a new transmission line, the planner does not ask: "How much will this line reduce the frequency of load interruption or curtailment?" Rather, the planner asks: "How much will this line reduce the risk of cascading outage and how much will this line reduce the probability of not being able to use available generation due to transmission outages?"

If this is the case, then it would appear that a method which calculates load interruption/curtailment indices does not qualify as a useful bulk transmission planning tool. As noted previously, the fundamental quality which an index must have to be useful is that it measures the same characteristic the planner is trying to achieve qualitatively without the index.

It does seem that there is agreement by planners on what constitutes system failure - cascading outages or "bottling up" of generation. However, while conceptually clear, the notion of "cascading" must be defined precisely in order to be applied in a quantitative evaluation. What are the essential features of a cascading outage? How many lines have to trip to be cascading? How much load must be interrupted, if any? Similarly, what constitutes bottling up of generation?

It is easy to classify the extreme cases. Therefore, historically, bulk power system planners have established a number of "test cases". One selects only the most severe outage conditions to test. Based on the results of load flow or stability runs, it is easy to classify the case as "success" or "failure". A network which did not experience cascading outages or bottle up generation in the test cases was judged acceptable - whether or not load interruption occurred.

This is a deterministic method to assess reliability. If quantitative rules are to be applied to calculate the probability of cascading outage, then a definite rule must be given which will permit every system state to be classified as success or failure.

Assuming that a satisfactory definition of failure can be made, the next question is what events cause failures. A cursory review of major outage reports reveals that most have two characteristics:

1. The initiating event was either a freak event not previously thought of, or a known event which it was thought the system could withstand.

2. The initiating event precipitated additional events which eventually led to widespread outage (cascading failure).

Major outages which do not have these characteristics are usually ones where severe environmental conditions have occurred to simultaneously outage many equipments.

Thus, there is not a clear relation between normal equipment outages and system failure for transmission systems as there is for generation and distribution systems.

To summarize the same two points as previously done for distribution and generation:

1. A transmission system failure is fairly well defined conceptually (cascading, widespread or uncontrolled), but there is lack of precise definition needed to classify every system state as success or failure.

2. The events that <u>cause</u> transmission system failure are very poorly defined. The sources of system failure are largely unknown until they occur.

Present Bulk System Failure Definitions

In the previous discussion, bulk system failure was defined as one involving three characteristics: being widespread, cascading and uncontrolled. The question arises whether or not all three must be present for an incident to qualify as a bulk power system failure.

Consider first how the National Electric Reliability Council (NERC) addresses these questions. NERC does not define bulk power system failure directly. Rather, NERC defines reliability, as follows[4]:

"<u>Reliability</u> involves the security of the interconnected transmission network and the avoidance of uncontrolled cascading tripouts which may result in widespread power outages.

It is apparent that the NERC definition does involve all three of the terms widespread, uncontrolled, and cascading. Therefore, the NERC definition would seem to require that all three characteristics be present in order to have a bulk power supply failure. However, the use of the word "may" indicates that uncontrolled cascading tripouts may be considered a failure even though widespread interruptions do not occur.

In addition to the definition of reliability, NERC has a second fundamental definition, called adequacy[4]:

"<u>Adequacy</u> refers to having sufficient generating capability to be able at all times to meet the aggregate electric peak levels of all customers and supply all their electric energy requirements."

It can be observed that whereas reliability, as defined by NERC involves the avoidance of "<u>widespread power outages</u>", adequacy involves the capability to meet the needs of "<u>all</u> customers".

Thus, for bulk power systems, there are two definitions of system failure. These can be summarized as follows:

<u>Adequacy</u> - sufficient capability to meet load

<u>Reliability</u> - avoidance of uncontrolled cascading tripouts

The choice of the terms "adequacy" and "reliability" by NERC is perhaps unfortunate. First, both definitions relate to bulk power system "reliability". Therefore, it is confusing to use the name reliability for one of the two terms. Secondly, the definition of adequacy used by NERC actually corresponds to the definition of reliability as used in most power system reliability evaluation models. Third, the NERC definition of reliability is close to the commonly accepted definition of security. In fact, the NERC definition of reliability uses the word security.

Proposed Definitions

It may be useful to revise the NERC definitions to remedy these disadvantages. Towards this end, some proposed definitions are offered as a starting point.

Some years ago, a definition for bulk power system reliability was proposed in the advisory Committee Report on Prevention of Power Failures. This report was published in June, 1967, as a consequence of the Northeast Blackout. Following is the definition in the advisory committee report[5]:

"Reliability is the degree of assurance of a bulk power supply in delivering electricity to major points of distribution."

This definition recognizes the probabilistic nature of reliability by the phrase "degree of assurance", but the definition does not tie down a specific measure of reliability. It is interesting to note that neither of the two NERC definitions - reliability or adequacy - contain the phrase "degree of assurance" or similar phrase implying uncertainty. Taken literally, each of the two NERC definitions defines an impossible criterion! However, the regional councils, in implementing the NERC criteria, have generally added a phrase like "most adverse credible conditions". This introduces the degree of assurance concept.

It appears desirable that the term reliability itself should be kept as broad as possible so as to provide an "umbrella" term which covers all aspects of interest. It seems that the advisory committee definition accomplishes this purpose for the bulk power system. It can be noted that the advisory committee definition does not qualify the extent of the interruption or the type of operation involved.

It is now proposed that the advisory committee definition be adopted as the definition of reliability for the bulk power system. If this definition of reliability were adopted, then the NERC definition of reliability could be called <u>security</u>. Adding the "degree of assurance" concept in the advisory committee definition, the NERC definition of reliability could be revised to read:

<u>Security</u> is the degree of assurance of a bulk power system in avoiding uncontrolled cascading tripouts which may result in widespread power interruptions.

Using the same concept, the NERC definition of adequacy could be revised to read:

<u>Adequacy</u> is the degree of assurance that a bulk power system has sufficient capability to meet the aggregate loads of all customers.

With these revisions, reliability remains tied directly to total service continuity, by the phrase "delivering electricity to major points of distribution". Under this definition, failure to supply any bulk power delivery point is a bulk power system failure.

Security and adequacy, under the revised definitions, deal with two broad classes of events which produce loss of service continuity. Security deals with loss of service arising from widespread interruptions involving many delivery points. Adequacy deals with loss of service arising from insufficient capability, involving one or a few delivery points.

A second change made in these revised definitions was to use "bulk power systems" in all three definitions. The NERC definitions use the phrase "interconnected transmission network" in the definition of reliability, while the word "generation" is used in the definition of adequacy. Thus, it could be argued that the NERC definition of reliability applies only to transmission, while the NERC definition of adequacy applies only to generation.

From a system planning viewpoint, this is precisely the case. That is, generation systems have historically been planned to satisfy "adequacy" criteria, like loss of load probability (LOLP); while interconnected transmission networks have been planned to meet "security" criteria. That would explain why generation planners associate the term reliability with adequacy, while transmission and bulk power supply planners associate reliability with security.

However, both adequacy and security are actually important to both types of planners. When generation systems are planned to meet adequacy criteria which have historically been deemed prudent, they have historically contained sufficient reserve to insure security. Conversely, when transmission networks are designed to meet the test conditions specified for avoiding complete system collapse, they have typically contained sufficient redundancy (reserve) so that the effect of local interruptions due to individual equipment outages have a very small impact on service reliability (compared to other sources of interruption). The bulk power system comprises both the generation system and the transmission network. Thus, the bulk power system must satisfy both criteria. This not only explains why NERC specifies both criteria; it also explains why each criterion focuses only on one aspect – generation or transmission.

The proposed definitions recognize the fact that both adequacy and security are "components" of reliability. Furthermore, one can replace the phrase "bulk power system" in the three definitions with "generation", and get three similar definitions for the generation system. The same can be done for distribution. Thus, a "unified structure" of definitions for reliability, security, and adequacy is proposed which can be applied to any section of a power system. The definition reliability is directly tied to service continuity, and the definitions of security and adequacy are derived by considering classes of service interruption.

ELECTRIC UTILITY APPLICATION OF RELIABILITY THEORY

G.E. Marks
Florida Power Corporation
St. Petersburg, Florida

Introduction

Electric utility planners, operators, and engineers are constantly weighing the cost of maintaining or improving system reliability. Whether it be a recommendation to construct a multi-million dollar power plant or to increase or decrease the frequency of tree trimming along a distribution feeder, reliability is a major consideration. In the majority of cases today, reliability considerations must be evaluated by qualitative comparisons of failure possibilities and their anticipated consequences. Although this has proven to be an acceptable approach, quantitative methods of reliability assessment are becoming available and gaining acceptance as a means of improving reliability evaluations. One area in which such applications have been made and new methods are being implemented is that of power system planning.

Electric utility planners are charged with the task of developing a power system facility plan based on projected needs and economic considerations. Such a facility plan includes a description of the generation, transmission, and/or distribution facility

required; its size, general location, and timing. It may also include a description of cost-effective modifications to existing facilities and projected equipment retirement dates. The determination that facility plan additions or changes are justified is usually based on economics, reliability, or a combination of the two.

Establishment of Reliability Criteria

Historically, reliability requirements have been assessed by comparing simulated system performance to established reliability criteria. For example, the reliability criteria might require that for an outage of any bulk power transmission line or transformer, no other transmission line or transformer should be overloaded. If a certain line outage caused other lines or transformers to overload, then facility additions to correct the problem would be considered justified. Additional evaluations might then be conducted to compare alternative solutions. These evaluations would normally include a comparison of the performance of the alternatives with respect to line flows and voltages under normal and contingency conditions, as well as a comparison of the cost of the alternatives. The final result would be a selection of what was considered to be the best alternative based on a judgemental comparison of the performance and cost factors. This approach has been used for decades in the continental U.S. and Canada and electric power system performance indicates it has not been an unreasonable approach.

Today with the severe financial restraints under which electric utilities must operate, as well as the increased public pressure to keep electric rates to a minimum, utility mangement and regulatory commissioners are looking for a more substantial method of determining what projects are justified based on reliability. It appears that they would prefer to quantify reliability in terms of the costs, savings, and customer impact. Questions which might be posed are:

1. Is the project justified, based on a comparison of project costs and customer savings?

2. What would be the impact on the customer in terms of customer interruption frequency and outage duration if the project were not built? (Adequacy)

3. What would be the impact on system integrity (chances of separation or system collapse) if the project were not built? (Security)

4. Taking into account the above items, can it be proven that the alternative selected is the optimum?

Ultimately, utility management and regulatory commissioners would like to see the final results of the entire process presented as an economic comparison of benefits and costs. If the reliability assessment process could be handled on this basis, the justification of projects before management and commissioners would be greatly expedited. Unfortunately, the methods and data available today limit the ability of the planner to supply this type of information.

In the following paragraphs, a discussion of how quantitative reliability assessment methods might be used is presented. The present state of the art, the desirability of reliability evaluation and the challenges which must be faced in the application of these methods are also reviewed.

298

Comparative Analysis

At the present state of the art, reliability assessment methods are best applied to comparative analysis. In comparative analysis, the relative reliability merits of one alternative is compared to that of another. In such computational comparisons, the absolute or true value of reliability is not so critical. This is an important feature, since reliability assessment is somewhat hampered today by the limited availability of historical data, and also due to certain simplifying assumptions which are often made to reduce the complexity of the calculations. The rationale in favor of comparative analysis is that although the data and method of calculation may not be sufficiently accurate for absolute reliability assessment, it may be sufficient for comparative analysis since the inaccuracies would "wash out" when one alternative is compared to another. Without a detailed analysis using a reasonably accurate data source and a theoretically precise methodology, there is no way to know whether this rationale is true in any given case. However, the premise appears reasonable and is generally accepted as true.

Comparative analysis is a very powerful tool and may continue to be the mainstay of utility reliability assessment practices for some time into the future. It can be used quite effectively for the comparison of similar plans and alternatives. It can also be used for priority ranking of projects which are competing for construction and operating funds. Likewise, when it becomes necessary to cut costs, comparative analysis can be used to select the items which are to be cancelled or delayed. Indices which might be used for comparative analysis are: loss of load probability, expected unserviced energy, IEEE bulk power reliability indices and EEI service reliability indices.

Absolute Reliability Assessment

A more sophisticated application of reliability assessment theory is needed for the calculation of absolute indices. Absolute indices could be compared to historical values or some agreed-upon standard. When the projected index falls short of the standard, corrective action would be deemed necessary or at least worthy of further evaluation. This approach has the advantage of reducing a very complicated task to simple terms. The problems, however, are buried deep within the calculations. First, if absolute values are to be calculated the data must be sufficiently accurate to produce meaningful results. Such data are very difficult to obtain. Secondly, the methodology used in the calculation would need to be exceptionally accurate since the results would be compared to historical values or standards of performance. Simplifying assumptions would not "wash out" as they would in comparative analysis.

Although these obstacles may be difficult to overcome, the benefits resulting from this type of calculation would be worth the effort. One approach might be to develop confidence limits or probability distributions around the index calculated. Another approach might be to adjust the criteria or standard such that it compensates for the uncertainty in the absolute value calculated. Examples of indices which might be used in an absolute evaluation are the anticipated frequency and duration of customer interruptions, the estimated number of customers affected, or the expected hours of service failure.

Benefit Cost Evaluation

In this type of evaluation, the savings resulting from reliability are compared to the cost of providing such reliability. This approach requires another source of data; namely, the worth of reliability to utility customers, or to look at it another way, the cost of outages to utility customers. Early work in this area has revealed that industrial customers can most readily determine the cost of outages. Although commercial customers can estimate direct costs such as food spoilage, etc., it is difficult to determine the impact of lost sales, if any. The cost of outages to residential customers, however, seems to be the most elusive since for the most part it is one of inconvenience, unless the outage is for an extended period. In addition, there are social costs involved with power system outages such as accidental injuries, increased crime and vandalism, and the cost of emergency procedures and programs. Attempts are being made to define such costs and preliminary estimates should become available as the work continues. With such information, the utility planner can weigh the cost of adding reliability to the system against the anticipated benefits.

When the benefit cost approach is utilized, recommendations are made based on the merits of the individual situation being evaluated and not based on generic norms or system criteria. Examples of indices which might be used in a benefit cost evaluation are the unserviced energy, unserviced customer hours, and frequency and duration of customer interruptions. These, then, would be multiplied by their respective cost components since the ultimate evaluation would be a comparison of the expense of the outage on one hand versus the expense of preventing the outage on the other.

Future Development

Although it has been emphasized that utility management and regulatory commissioners would in most cases prefer the benefit-cost approach for project justification, it should be pointed out that each of these methodologies will likely have their respective applications. Consequently, the reliability methods developed and data bases created should enable the production of indices which can be utilized in these types of evaluations.

REFERENCES

1. J. Endrenyi, Reliability Modeling in Electric Power Systems. J. Wiley and Sons, Chichester, 1978.

2. P.F. Albrecht, "Overview of Power System Reliability". EPRI Workshop Proceedings: Power System Reliability - Research Needs and Priorities, EPRI Report WS-77-60, October 1978.

3. C. Singh, Discussion of the paper "Computing Methods and Devices for the Reliability Evaluation of Large Power Systems" by K. Gallyas and J. Endrenyi. IEEE Transactions on Power Apparatus and Systems, Vol. 100, No. 3, p. 1258, March 1981.

4. 7th Annual Review of Overall Reliability and Adequacy of the North American Bulk Systems. National Electric Reliability Council (NERC), Princeton, NJ, July 1977.

5. Prevention of Power Failures. Vol. II, Advisory Committee Report, Reliability of Electric Bulk Power Supply: Report to the Federal Power Commission, June 1967.

BULK POWER SYSTEM RELIABILITY ASSESSMENT – WHY AND HOW?
PART II: HOW?

A Panel Discussion sponsored by
the Application of Probability Methods Subcommittee of the PES
Panelists: J. Endrenyi (Chairman), P.F. Albrecht,
R. Billinton, G.E. Marks, N.D. Reppen, L. Salvaderi

Abstract – The three discussions in this paper are all concerned with practical aspects of bulk power system reliability computations. The first describes promising methods for contingency selection, contingency evaluation and the determination of reliability indices. The second discusses some models and solution methods based on a Monte Carlo simulation of the system's behavior, an approach often favoured in Europe. The last discussion reviews the data required in bulk system reliability studies, and stresses the need for data on common-mode and station-originated line outages.

ANALYTICAL APPROACH TO SYSTEM RELIABILITY ASSESSMENT: CONTINGENCY ENUMERATION

N.D. Reppen
Power Technologies, Inc.
Schenectady, New York

Introduction

Probabilistic reliability procedures attempt to assess the risk of system failure on the basis of outage statistics of system components. The ambition of assessing risk in quantitative terms brings with it the requirement that the investigation cover all disturbances that can significantly contribute to the risk of system trouble – or at least all such disturbances within a specified subclass of the reliability problem.

The contingency enumeration approach includes the systematic selection and evaluation of disturbances, the classification of each disturbance according to failure criteria and the accumulation of reliability indices. Contingency enumeration techniques are structured so as to minimize the number of disturbances that need to be investigated in detail. This is achieved by testing, to the extent possible, only those disturbances which are sufficiently severe to cause trouble and sufficiently frequent to impact the risk indices to be computed.

The contingency enumeration approach is structured as shown in Figure 1. For a specific predisturbance condition a contingency (disturbance) is selected and tested to determine whether the contingency causes any immediate system problem such as a circuit overload or a bus voltage out of limits. If it does not, a new contingency is selected and tested.

The occurrence of a system problem may by itself be logged as a failure. However, in most cases, it will be possible to adjust generation or phase shifters to relieve overloads and to adjust generator

82 WM 148-5   A paper recommended and approved by the IEEE Power System Engineering Committee of the IEEE Power Engineering Society for presentation at the IEEE PES 1982 Winter Meeting, New York, New York, January 31–February 5, 1982. Manuscript submitted September 1, 1981; made available for printing November 16, 1981.

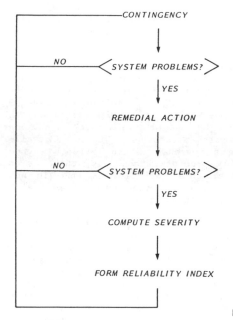

Figure 1. Contingency enumeration approach.

voltages or transformer taps to bring bus voltages back within range. It is thus of interest to determine whether it is possible to eliminate a system problem by such remedial actions. A failure is logged when remedial actions, short of curtailing consumer loads, are insufficient to eliminate the system problems. The severity of such system problems may be assessed by computing the amount and location of load curtailment necessary to eliminate the problem. In this way, it is possible to compute area or bus reliability indices that measure the frequency, duration and amount of expected load curtailment.

System Problems

Two categories of system problems are identified in Table 1. Circuit overloads and abnormal voltages can be corrected by remedial actions including load curtailment if necessary. Failures related to these system problems contribute to adequacy indices and contingency testing can be performed with steady-state analysis tools such as load flows. The second group of system problems consisting of voltage collapse,

TABLE 1

System Problems

o Circuit Overload	} Partial Curtailment
o Abnormal Voltage	
o Voltage Collapse	} Blackout Possibility
o Overload Cascading	
o Instability	

Reprinted from *IEEE Trans. Power Apparatus Syst.*, vol. PAS-101, no. 9, pp. 3446–3456, Sept. 1982.

overload cascading and instability may lead to shutdown of portions of the system. The outcome of such disturbances depends on system dynamics. Dynamic solution methods or appropriate steady-state tests that are capable of determining with reasonable approximation the severity of the dynamic problems are required. These types of disturbances contribute to security indices.

Practical applications of contingency enumeration techniques are presently limited to the computation of adequacy indices; that is, to system failures relating to thermal overloads and abnormal voltages.

Contingency Selection

Contingency selection involves the determination of possible and reasonably likely outage events and a means of preselecting the most severe of these events for testing.

Since a system as a minimum is designed to withstand the loss of a single major component, the contingency selection is generally concerned with the several classes of multiple outages listed in Table 2.

Common-mode and related forced outages (originated by a single cause) involve multiple components out in a local area and are almost always sufficiently severe to warrant testing. In contrast, independent overlapping outages or forced outages overlapping maintenance outages are severe only if the various components out have an adverse impact on the same system problem. The structure of a power system indicates that for each multiple outage of this latter category that causes trouble, there may be thousands of others that have no mutually reinforcing adverse effects. Therefore, while reasonably likely common-mode and related multiple outages may all have to be tested, there is a need for screening methods to identify the independent overlapping outages that need to be tested.

One attractive screening approach is the ranking of contingencies by severity. The testing of the ranked contingencies would start with the most severe contingency and proceed with progressively less severe contingencies until contingencies causing no system problems are encountered.

One technique capable of quickly ranking single-circuit outages according to overload severity employs a performance index[1]. The index, PI, is defined by the formula

$$PI = \sum_{\substack{i:all \\ circuits}} \left(\frac{P_i}{P_{\ell i}}\right)^2$$

where P_i is the power flow in circuit i, and $P_{\ell i}$ is the power flow rating of circuit i. The PI for all single-circuit outages in the system can be determined in a small fraction of the time it takes to compute the individual power flows for those outages. By ranking single outages for the weakened systems that result from each of the outages determined as severe based on the PI, rankings of pairs of outages are obtained. This procedure can be extended to multiple outages involving three or more components.

Contingency screening of independent overlapping outages is essential to reduce the large number of such outages to the several hundreds or several thousands of contingencies that can be tested in practical reliability index calculations. In addition to screening by severity, it is possible to establish a lowest level of probability or frequency of occur-

TABLE 2

Multiple Outage Events

1. Independent overlapping forced outages.

2. Forced outages overlapping scheduled outages.

3. Common-mode forced outages; e.g., double circuit outages.

4. Related forced outages; e.g., multiple circuit outages resulting from substation equipment or relay failures.

rence as a criterion for contingencies which will not significantly impact the computed reliability indices even if the contingencies are very severe. Contingency testing may also be limited to a specific number of independent overlapping outage events or contingency levels. The contingency-level criterion should be consistent with the probability or frequency cutoff criterion.

Contingency Testing - Network Solutions

In reliability index computations speed of network solutions is of essence, more so than solution precision. If the system problems to be detected are restricted to circuit overloads and network separation, dc load flow or distribution factor techniques are satisfactory. If bus voltages need be computed, approximate linearized solutions or decoupled load flow solutions are often applied. The decoupled load flow is often selected over the Newton load flow on account of speed for approximate solutions.

Contingency Testing - Remedial Action

Remedial actions may be taken automatically by the system or effected through operator actions. Automatic remedial action is presently restricted to special system situations of particular concern. Therefore, most remedial actions to be accounted for in reliability index computations involves the operator.

The extent to which operator actions should be considered is debatable and debated. Thus, there is little agreement on the extent to which system capability for generation shift, phase shifter adjustment and voltage and var adjustment should be included in the analysis. Reluctance to include the possible benefits of remedial operator actions is due in part to planning or system design philosophy and in part to the many, sometimes arbitrary, assumptions that must be made in defining the characteristics of the remedial actions.

The time factor represented by the time for which a system problem can be tolerated and the time it takes to diagnose the problem, decide on proper action and execute the action is a major complication. For example, the time factor gives significance to the sequence of component outages in a multiple outage contingency and to the time between successive component outages.

One practical assumption is to ignore the time factor and apply the maximum remedial action capability available in the system. The reliability indices resulting would be representative for the post-disturbance conditions but would not reflect the possibly more severe system conditions existing during the time remedial actions were being considered or in progress. It is clear that such indices would be optimistic as a measure of actual system reliability while indices based on the initial system problems would be pessimistic.

When allowable remedial actions have been defined, one of several mathematical optimization techniques can be applied to determine whether available remedial capability is sufficient to eliminate the system problems detected. Linear programming has been used to perform generation shifts and phase shifter adjustments to eliminate circuit overloads. Least squares techniques have been applied to adjustment of bus voltages.

Linear programming has also been used to minimize the amount of load curtailment necessary to eliminate circuit overlaods in cases where remedial actions short of load curtailment have been insufficient[2]. With this approach, necessary load curtailment is computed for specific load buses in a consistent manner providing the means for computing reliability indices by bus as well as for system areas.

Accumulation of Reliability Indices

In the contingency enumeration approach, event probabilities and frequencies are accumulated into system indices or bus indices. The indices formed by considering all tested contingencies that were logged as failures according to a specified failure criterion constitute a lower bound on the reliability indices (note that most indices are actually measures of unreliability). These lower bound indices would increase if additional, generally more severe and less likely contingencies were tested. The reliability index computed is, therefore, a function of the number of contingencies tested. Since this is somewhat unsettling, it is of interest to compute an upper bound of the indices as well, thus bracketing the indices that would be computed if all contingencies were tested and classified as failure or nonfailure. Such an upper bound can be obtained by considering all contingencies tested that did not fail as well as all contingencies not tested but presumed not to cause system failure based on the prescreening of contingences.

Figure 2 illustrates the convergence of the upper and lower bounds of the frequency of overloads on the 24-bus IEEE reliability test system[18] as a function of the number of contingencies tested*. The results are based on the peak load condition of 2850 MW. The frequencies stated in times per year are thus annualized frequencies; that is they are the frequencies that would result if the peak load condition was maintained for the entire year.

The frequency indices shown are the results of one single-level test, four 2-level tests and one 3-level test.

The single-level test included the most severe single-circuit outages and the common-mode (single event induced) two-circuit outages. Details of the latter are given in Table 3. This test required the testing of 13 contingencies.

The 2-level tests considering explicitly or implicitly all combinations of two independent overlapping outage events resulted in 66, 92, 101 and 119

---

*Figure 2 was prepared using an experimental computer program under development for the Electric Power Research Institute under Research Project 1530-1.

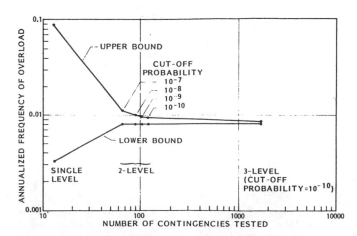

Figure 2. Upper and lower bounds of frequency of overload as a function of contingencies tested, for the 24-bus IEEE reliability test system.

contingencies tested using cut-off probabilities of 10^{-7}, 10^{-8}, 10^{-9}, and 10^{-10} respectively.

The 3-level test considering combinations of three independent overlapping outage events and a cut-off probability of 10^{-10} resulted in 1706 contingencies tested.

Circuit overloads were detected by means of a dc load flow solution. Contingencies causing separation of one or more buses from the system were also tested for overloads provided no load curtailment was required.

The upper bound in Figure 2 is obtained by excluding separations causing load curtailment from the analysis. The difference between the upper bound and the lower bound represents the frequency of con-

---

TABLE 3

Supplementary Data for IEEE Test System
as Used in the Computations

A. **Outage Data**

1. Single circuit outage frequencies and durations are the "permanent" outage statistics in Table 11 of Reference 18.

2. Transformer 3-24 is assumed tapped to ckt 15-24. Load at Bus 5 is assumed tapped to ckt 1-5-10.

3. Double-circuit common-mode outage statistics are assumed as follows:

Circuit Pair	Frequency [1/year]	Duration [hours]
22-21/22-17	0.0045	30
23-20/23-20	0.0045	15
21-18/21-18	0.0054	15
15-21/15-21	0.0034	30
13-11/13-12	0.0100	15
8-10/8-9	0.0130	15
20-19/20-19	0.0830	15

B. **Generation Dispatch**

Five 12 MW oil-fired units at bus 15 not available.

Four 20 MW combustion turbines at buses 1 and 2 in stand-by.

Other units dispatched economically.

302

tingencies for which the existence of overloads has not been established either explicitly or implicitly. That is, the difference includes the contribution to the failure frequency of independent overlapping contingencies that are of a higher level or have a lower probability of occurrence than the cut-off probability. For this system it may be considered that the frequency of overloads due to transmission outages as defined here can be determined with sufficient precision by analyzing about 100 contingencies. Also, it does not appear necessary to consider more than two levels of independent overlapping transmission outage events. The latter observation becomes even more evident as additional multiple dependent circuit outage events such as substation related events are considered.

When generation outages are considered in addition to transmission outages, the number of tested contingencies required for similar convergence of upper and lower bounds as shown in Figure 2 is at least an order of magnitude higher than when considering transmission by itself.

Upper and lower bounds may also be computed for various levels of load curtailment due to generation insufficiency, or as needed to eliminate overloads on circuits or to correct low voltage problems. In this case, the various system problems are expressed on a common basis, avoiding the complexities in interpretation of reliability indices caused by multiple, non-exclusive failure criteria.

The accumulation of probability and frequency indices is not a trivial undertaking. The main challenge is to perform the accumulation in such a way that the events considered are mutually exclusive, thus avoiding "double-counting". In addition, the treatment of maintenance outages and storm-related outages results in significant complexities.

MONTE CARLO METHODS IN TRANSMISSION PLANNING

L. Salvaderi
ENEL, Rome, Italy

Introduction

The subject of the panel dicussion includes two questions: "why" are reliability assessments necessary in power system studies, and "how" are they best done. This discussion attempts to give the "Italian answers" to these questions.

In response to the question "why?", reliability indices are computed in order to determine the cost of reliability which is one of the three components of the "total system cost" (the other two being the capital and operation costs). This total system cost is then, in ENEL's planning practice, the basis for power system optimization. The cost of reliability is obtained through a quantitative evaluation of system reliability using a suitable risk index. If the reliability study is confined to the evaluation of static system adequacy [3,4], the risk index chosen by ENEL is the "expected value of the yearly curtailed energy" (kWh/yr).

Such "yearly curtailed energy" is obtained: - i) for the system as a whole according to the causes: static deficiency of generation, transmission or transforming capacity of the system; - ii) for each network node where the deficiency occurs; - iii) for the weather conditions when the deficiency occurs.

The "yearly curtailed energy" is then converted into a cost value by an apropriate "risk unit cost"

($/kWh). A widely accepted value for this unit cost is about 1 $/kWh; at the usual level of reliability, it is near to the ratio of the GNP to the total electric energy demand, which is practically the same in many industrialized countries.

Regarding the question "how?", the answer depends on the purpose of the study. While for the long-term optimization of the generating system, reliability is evaluated by a direct analytical aproach [5], for the operational planning of both generation and transmission recourse is made at ENEL to the Monte Carlo method [5,6,7]. The latter can be applied in two ways: (a) each hour in the year is examined in chronological order (sequential approach); (b) the hours of the system lifetime are taken into consideration at random (random approach).

Both methods can be successfully applied in the solution of appropriate problems. The "sequential approach" is used in planning studies of mixed thermal-hydro generating systems. It provides a quantitative evaluation of the risk of not meeting the demand because of lack of water in the reservoirs. It is mainly applied for the sizing of reservoirs and the pumping capacity of pumped-storage stations or of any type of storage (battery systems, flywheels, etc.), and for determining the characteristics of other peak-duty plants. The "random approach" is used in planning studies of transmission systems and allows the evaluation of the reliability indices of a system, which describe the system's capability of supplying the load. Since in this discussion only transmission systems will be considered, only the last method will be dealt with.

The Monte Carlo method offers the following advantages:

a) It makes the examination of complex systems possible without forcing the system model to become unrealistic.

b) It constitutes a tool for allowing easy modification of the number and characteristics of input random quantities.

c) It offers the planning engineer a synthesis of the final results and a detailed description of the events that caused the results. The latter output allows the planner to get a sort of "operating experience" which can facilitate the exchange of ideas between him and the operating engineers.

The application of Monte Carlo methods implies the expenditure of considerable computing time in order to obtain sufficient confidence in the results. Nevertheless, it should be remembered that it is used for the reliability evaluation during the phase of transmission system planning refinements, and other tools based on techniques faster but less sophisticated (graph theory, etc.) are used for transmission system expansion. Also, the costs of even several hours of CPU for running such programs are well below the gains that a large utility can obtain from correct solutions of planning problems.

Main Features of the Transmission System Reliability Program

The Monte Carlo-based computing program (SICRET) set up at ENEL for transmission system reliability evaluations has been described in earlier papers [5,6,7]. The program has almost 10 years of "experience" and is currently employed in real network planning (in particular, of the large 420 kV ENEL network, with 150 buses, 250 lines and transformers, and 450 loads), both for network and network component optimization. In the following the main models describing the system states (components and environment) and the operating policies will be reviewed, and some planning applications described.

a) The planning of the transmission network is performed after the planning of the generation system. Therefore the load demand characteristics and the generating system composition are generally already fixed. The reference period of time for which the system behaviour is examined is usually a "solar year", during which the system composition does not change. It is subdivided into elementary intervals of one hour. It is assumed that the state of the system is constant during these elementary intervals, and all changes occur at the beginning of the interval.

Any group of 8760 hourly simulations of operation is referred to as "yearly sample". The total number of randomly generated hours considered (sample size) must be representative of the reference period and large enough so that the results will be obtained with the desired confidence.

b) The power system state is characterized by the following factors: (i) the load at each network node, (ii) the availabilities of the system components (generators, lines, etc), (iii) operator assessments (generation dispatch, load shedding, network connections, etc).

c) Changes in the system state are caused by: (i) events that are generally random, and therefore unpredictable, which cause variations in loads and component states; (ii) dispatcher's actions to match the system state to the above changes. Such actions are carried out by using one of the operating policies described below.

d) In each of the samples considered:

(i) Simulated are the states of all components and of the load by random selections from their probability distributions, and the action of the Dispatching Centre based on one of the more or less complex operating policies (see later). The two types of simulations define the system state.

(ii) Evaluated are the random variables of interest (power flows in the lines, energy curtailments, etc).

Models Used in the Computations

a) Load Model. The active load of the network can be determined from the hourly load diagram or from the load duration curves of each week, according to the target pursued. To take into account the randomness of load, each load level can be assigned a normal distribution. The total load is apportioned between the bus loads and further subdivision is made at each bus, indicating the percentage of load that can be interrupted.

b) Environment Model. In each homogeneous area of the system, three weather conditions are considered, randomly alternating during the solar year: "fair", "bad", and "stormy" weather. The duration of each condition is assumed to be exponentially distributed. Other distributions are used for the ambient temperature, wind velocity, and solar emission.

c) Component Failure Models. A two-state model (up–down) is considered for forced outages of generators, transformers, and other main network components. Different rates of failure are used to take into account the infant mortality or the aging of the components.

The probability distributions of forced outage durations of generators were found to be well represented by exponential laws. The forced outages of overhead lines are described by taking into account their dependence upon weather conditions: a two-state model is used "inside" any weather condition.

d) Generating Unit Technical Models. Units are characterized by their ratings, technical minimum loadings, and hourly production costs.

e) Operating Policy Models. Generally speaking, the operating policies adopted must be sufficiently simple to permit the results to be obtained within acceptable computing times, and they must be similar to those performed in real practice. They can be classified into two types:

- pure safety policy,

- mixed economy-safety policy.

Of these, the policy is selected that best suits the planner. The first is adopted when only the system reliability is to be evaluated; the second, when the system running costs are also required. They differ in the policies adopted for unit commitment and for generation dispatch in order to relieve component overload; the second policy is the more refined but requires nearly twice the computing time used by the first[8].

The main steps of the program are the following.

i) The system load is matched to the total available generation, by curtailing a portion of load if necessary (risk index due to lack of active power).

ii) If a mixed "economy-safety" policy is adopted, the generators available are loaded node by node, according to a priority list based on increasing specific running costs (mills/kWh) at full capacity, and taking into account local constraints (eg., limitation of primary energy availability in some sites). Depending on the balance between load demand and generation available, some units (those having higher specific running costs) will not be utilized, independently of their siting. If, on the other hand, only "pure safety" policy is followed, all the available units are charged in proportion to their available capacity.

iii) The power flows in the lines and transformers are obtained through dc load flow.

iv) When the load of a component is greater than a certain level W_G, called "alarm level", the loading of generating units is modified according to the policies described in the following, in order to reduce the overload, and if possible, to bring the load of such a component below this "alarm level".

v) Component overloads above the "alarm level" are then tolerated up to a limit called "absolute", W_A. If this level is exceeded, some loads are curtailed, until the component loadings come within this limit (risk index due to overload of the components). The hourly energy curtailments (kWh/hr), if any, are added up for all the 8760 hrs examined (yearly sample), to give a yearly value (kWh/yr); by repeating for many yearly samples and averaging, an estimate of the average yearly curtailed energy (kWh/yr) is obtained.

vi) When applying "mixed economy policy" in the process of reducing overloads, capacity shifting is carried out between nodes (in some increasing, in others decreasing the unit loading) in such a way that the "cost of the shifting" is minimum [8]. If, on the other hand, the "pure safety" policy is followed, the generation shifting is carried out only to produce maximal influence on overloaded components, with no attention to the effect on the system running cost[5,8].

Interconnection Simulation

Particular operating policies allow to handle the reliability evaluation of interconnected systems (in Europe always involving several countries), as necessary when programmed or emergency exchanges are under study. In this case, examination of the overall balance of load demand and available generation, which must be ensured before the execution of the load flow, is carried out according to the following criteria.

- Load shedding in one or more of the interconnected networks (should in one of these national networks the available generating capacity be lower than the demand) takes place only if the load-generation balance of the overall interconnected networks makes it necessary. The load-generation balance is possibly achieved within each national network, even considering possible capacity exchanges among different nations. Such exchanges are governed by contracts and supplied to the program as inputs. National deficits in an emergency situation are coped with by additional capacity exchanges not constrained by the contractual rules.

- Once the load-generation balance is reached in each national network (taking into account possible imports or curtailments), generation dispatch is carried out network by network (following the policies described previously), with the networks considered independent.

- Possible further dispatches of generation and load sheddings directed to overload elimination take place inside each national network, while the generating capacities imported (or exported) that made the load demand-generation balance possible at the outset are kept constant.

Adequacy and Security Evaluation

Many European systems, and in particular the Italian one, are densely meshed with relatively short lines: therefore normally no voltage or stability problems arise. This is the reason why the SICRET program has been particularly adapted to static reliability, or adequacy, evaluations. Nevertheless, it should be pointed out that a complete reliability evaluation cannot disregard the probabilistic analysis of dynamic reliability, or security, mainly in systems having point-to-point structures, or long lines carrying large loads.

The problem in such probabilistic evaluations is, once more, to limit excessive computing times and to find adequate simplified criteria to decide which cases are "dynamically" dangerous [5,9]. Such studies are on the agenda of the APM Subcommittee of the Power Engineering Society. It should be mentioned that also in this case, the Monte Carlo method can successfully generate all the events, originated by failures in the system or by the behaviour of the protection system, which are to be examined.

As a final result, a dynamic or security risk index can be computed, which is to be compared with the previously described static or adequacy risk index, to aid planning decisions.

Examples of Applications

To conclude this discussion, it may be of interest to review some of the most common planning applications of the static reliability (adequacy) evaluation procedure.

a) Determination of optimum system structure[10]. Quantitative evaluations of reliability through the yearly curtailed energy at each bus (option "where"), and according to the causes which have produced the load shedding (option "how"), allow to choose the network reinforcements in the most appropriate way and to compare the cost of the reinforcements with the gain arising from a reduction of yearly curtailed energy.

From another point of view, the gain in reliability obtained when increasing the density of the network can be compared with the concurrent increase of short circuit level and, therefore, with the cost of the necessary additional equipment (circuit breakers and so on). Variations of the system running cost are added to the value of reliability.

b) Choice of substation layout[11]. The performance of various possible substation layouts is expressed in terms of the energy curtailed, and the cost of the latter can be compared with the cost of the layouts.

c) Determination of the load carrying capacity of overhead lines[12]. Factors that limit the load carrying capacity of an overhead power line are the effects of conductor heating: loss of mechanical strength of the conductor, and risk of discharge due to a reduction of clearances. To determine the load capacities of lines, the following are required:

- yearly load duration curves,

- yearly duration curve of conductor temperatures,

- expected duration curve of the maximum annual temperature of the conductor.

The evaluations must be based on a probabilistic approach, since both the outages of the components (affecting the loading of the lines) and the meteorological conditions (affecting the temperature of the conductors) are random factors. Determination of the load duration curve is obtained from the optimization study of the overall transmission system, described above, by using the program based on the Monte Carlo method.

With such an approach it was possible to increase the capability limit of the 3×520 mm^2 aluminum conductors of the 380 kV lines in Italy from 1000 to 1300 MW, because it was found that in a well planned transmission system the temperature of 75°C, conventionally considered to be the maximum allowable temperature for the material of which the conductor is made, is exceeded for an average of only about 50 hr/yr, while the effect of loss of strength begins roughly at 95°C, a value for which the average yearly duration is even lower.

DATA REQUIREMENTS FOR COMPOSITE SYSTEM RELIABILITY EVALUATION

Roy Billinton
Power System Research Group
University of Saskatchewan
Saskatoon, Canada

Introduction

This presentation is concerned with the data requirements for composite (or bulk power) system reliability evaluation. Such a system contains both generating equipment and the necessary transmission facilities required to transport the generated energy to bulk or major system load points. While composite systems are invariably quite complex, for the purpose of discussing the data requirements of reliability evaluations it is perhaps advisable to consider a simplified system representation. Figure 3 shows a reduced system in which it is easier to see and therefore appreciate some of the basic requirements. The configuration shown is of the form used in load flow studies in which terminal station equipment is consolidated in the form of a bus and not represented in complete detail.

The data required to analyze a system such as that shown in Figure 3 can be broken down into two basic parts. In a simplistic sense these two requirements can be considered as deterministic system data and stochastic system data. Each set can be further subdivided into parts pertaining to components and to the system.

Deterministic Data

The component data includes known parameters such as line impedances and susceptances, current carrying capacities, generating unit parameters and other similar factors normally utilized in conventional load flow studies. This is not normally a difficult task as this data are used in a range of studies. The system data, however, are somewhat more difficult to appreciate and to include as they should take into account the response of the system under certain

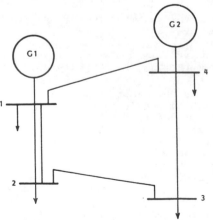

Figure 3. Reduced network configuration.

outage conditions. An example of this might be that if one of the lines between buses 1 and 2 in Figure 3 suffers an outage, the loading on the remaining line could be such that the line would be removed from service, or it would carry the overload, or some remedial action would be taken in the system in order to maintain overall system integrity. This form of system data is extremely important in a composite reliability study. The computer model must behave the same way as the actual system would behave or the results will not be appropriate. This is an important aspect of the data requirement problem for composite system reliability evaluation and one that has not been properly recognized up to this time.

Stochastic Data

Stochastic data describe random variables (quantities whose magnitude is subject to uncertainty). Typical are failure frequencies, and repair or maintenance durations. They can again be broken down into two parts with respect to component and system requirements. The component requirements pertain to the failure and repair parameters of the individual elements within the system. These data are generally available. There is also a need to consider and to include system events which involve two or more components. This type of data is system specific and has to be provided. This kind of data it appears will have to be inserted as a second and third level of data input in an overall composite system reliability analysis. System data might include multiple failures resulting from common transmission line configurations or station induced effects.

The different types of outages can be categorized as follows:

1. Independent outages.

2. Dependent outages.

3. Common cause or common mode outages.

4. Station originated outages.

Independent Outages

This form of outage is possibly the easiest to deal with and independent outages of two or more elements are referred to as overlapping or simultaneous independent outages[13]. The probability of such an outage is calculated as a product of the failure probabilities for each of the elements. The basic component model used in these applications is the simple two stage representation in which the component can be up or down.

The rate of departure from the up-state to the down-state is designated as the failure rate λ. The restoration process from the down-state to the up-state is however somewhat more complicated. It is normally designated by the repair rate μ. The restoration of a forced outage can take place in several distinct ways which can result in a considerable difference in the probability of finding the component in the down-state (usually designated as the unavailability). Some of these restoration processes are as follows:

(a) High speed automatically re-closed.

(b) Slow speed automatically re-closed.

(c) Without repair, by switching.

(d) With repair.

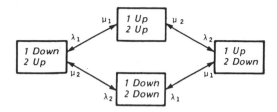

Figure 4. The basic simultaneous
independent failure model.

In addition to forced outages, the component may also be removed from service for a scheduled outage. The scheduled outage rate, however, should not be added directly to the failure rate as scheduled outages are not considered to be random events. The component is not normally considered to be removed from maintenance if the actual removal would result in customer interruption.

The state space diagram shown in Figure 4 shows all possible states for a two element configuration considering independent failures. Most of the presently available techniques for composite system reliability evaluation assume that the outages constituting a contingency situation are independent.

Dependent Outages

As the name implies, these outages are dependent on the occurrence of one or more other outages. An example is the removal from service of the second line of a double circuit line due to overload which resulted from an independent outage of the first line. These outages are not normally included in the reliability evaluation of composite systems and require acquisition of system data in addition to individual component data.

Common Cause Outages

As previously noted, the probability of occurrence of an event consisting of two or more simultaneous independent outages is the product of the individual outage probabilities. If the probabilities of the individual outages are low, the product can become extremely small. The probability of a common cause outage resulting in a similar event can however be many times larger. The effect of common cause outages on bulk system reliability indices can be far more significant than the effect of second and higher order independent outages. A common cause outage is an event having an external cause with multiple failure effects where the effects are not consequences of each other. The most obvious example of a common cause outage is the failure of a transmission tower supporting two or more transmission circuits. This event can be contrasted with the independent outages of two circuits which are on separate tower structures and possibly quite widely separated physically.

The Task Force on Common Mode Outages of Bulk Power Supply Facilities in the IEEE Subcommittee on the Application of Probability Methods suggested a common mode outage model for two transmission lines on the same right-of-way[14]. It is basically similar to the model shown in Figure 4, except for allowing direct transition, at a rate λ_C, from the no-failure state to the double-failure state. It is assumed that the same repair process applies for all failures including common cause failures. Various other possible common cause outage models have been described and analyzed in the available literature.

Station Originated Outages

The outage of two or more transmission elements not necessarily on the same right-of-way, or of two or more generating units, can arise due to station originated causes[15]. Station originated outages can occur due to a ground fault on a breaker, a stuck breaker condition, a bus fault, etc. or a combination of these. Such outages have been previously accounted for in the line and generator outage rates by combining these outages with independent outage rates. Such a treatment cannot, however, recognize a situation in which several elements of the system are simultaneously out because of a single event in the terminal station. It is, therefore, necessary to consider these outages as separate events. Station originated outages in regard to their effect on composite system reliability have not been extensively analyzed and can have an appreciable effect on load point reliability indices. The impact of these can be very clearly seen from the following example.

Consider a section of a ring bus where two generating units are tied to the bus between breakers A and B, and another two between breakers B and C. A failure to ground on breaker B would result in breakers A and C opening and isolating four generating units from the system. This type of event is not normally included in either generating capacity or composite system reliability studies. The duration of the outage in this case however might not be associated with the repair of breaker B but simply with switching action required to remove the breaker from the system and restore the four units to system service. It becomes important therefore to recognize that restoration in the case of terminal station faults may not involve repair directly but may be by switching action, and therefore a somewhat different model is required. Figure 5 shows a possible model which includes both common mode and station related events. In this, different restoration processes are assumed to take place after the three types of double failures.

It is important to appreciate the different impacts that common mode and station originated events can have on the system transmission components. Consider two line configurations. In the first, two lines start at a given station A and go to two different stations B and C. In addition to the likelihood of two overlapping independent failures, it may be possible for both these lines to be removed from service due to a single element failure in station A. Such a possibility would not be normally considered in the load flow type of representation shown in Figure 3. In the second line configuration, both lines go from station A to the same station B. In this case, the two lines may be removed from service either due to a terminal station failure or, if the lines are on

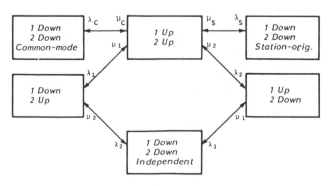

Figure 5. A more general model for common-mode, independent and station-associated events.

a common tower structure, by a common mode failure or by two overlapping independent events. All these factors must be included in a composite system reliability analysis and it appears that the most suitable way is to input them as separate levels of component and system data.

The stochastic data requirements for composite system reliability therefore include both individual component parameters and also higher levels of data which involve more than one component and are system specific. This form of input can be designated as system data requirements.

Collecting Data

There appears to be considerable activity at the present time within individual utilities in North America in regard to the establishment of data collection systems for transmission line outage performance. There is also some interest in the area of pooling data by collecting this information on a regional basis. By pooling, sample sizes are increased which results in enhanced confidence in the values obtained from the samples and used in the models.

An important part of the above activity is the EPRI research study entitled "Bulk Transmission System Component Outage Data Base"[16], which provides a survey of present practices and suggests possible models and data collection procedures. The Canadian Electrical Association have also recently implemented a reporting procedure by which all Canadian utilities are now collecting transmission outage data using a common format[17]. Similar systems exist in Great Britain, Europe, and other parts of the world.

Conclusions

It appears that there is a relatively large body of data related to individual component outages. Many companies either have established, or are in the process of setting up, comprehensive outage data collection procedures which should provide individual component outage data with an acceptable level of confidence. This is not the case in regard to common mode, dependent and station associated failures. This type of data tends to be system specific and therefore is not quite as susceptible to data pooling as the independent component outage data. However, increased awareness of the need for such data, arising as it does from the need to conduct overall system adequacy studies, should lead to better data collection.

REFERENCES

1. T.A. Mikolinnas, and B.F. Wollenberg, "An Advanced Contingency Selection Algorithm". IEEE Transactions on Power Apparatus and Systems, Vol. 100, No.2, pp. 608-617, February 1981.

2. G.C. Ejebe, W.R. Puntel, and B.F. Wollenberg, "A Load Curtailment Algorithm for the Evaluation of Power System Adequacy." Paper A77 505-1, IEEE-PES Summer Meeting, July 1977.

3. G. Manzoni, P.L. Noferi, L. Paris, M. Valtorta, "Power System Reliability Indexes". NATO Advanced Study Institute of Generic Techniques in System Reliability Assessment, University of Liverpool, 1973.

4. IEEE Working Group Report, "Reliability Indices for Use in Bulk Power Supply Adequacy Evaluation". IEEE Transactions on Power Apparatus and Systems, Vol. 97, No. 4, pp. 1097-1103, July/August 1978.

5. P.L. Noferi, L, Paris, "Quantitative Evaluation of Power System Reliability in Planning Studies". IEEE Transactions on Power Apparatus and Systems, Vol. 91, No. 2, pp. 611-618, March/April 1972.

6. P.L. Noferi, L. Salvaderi, "A Set of Programs for the Calculation of the Reliability of a Transmission System". UNIPEDE Informatique Conference, Lisbon, 1971.

7. P.L. Noferi, L. Paris, L. Salvaderi, "Montecarlo Methods for Power System Reliability Evaluations in Transmission or Generation Planning". Proceedings, 1975 Annual Reliability and Maintainability Symposium, Washington, 1975.

8. I. Facchinei, L. Salvaderi, E. Giampiere, S. Scalcino, "Network Reliability Evaluation: Peculiar Problems to be Faced when Planning Bulk Meshed Transmission Systems". In preparation: VII PSCC Conference, Lausanne, 1981.

9. R. Billinton, P. Kuruganty, "A Probabilistic Index for Transient Stability", IEEE Transactions on Power Apparatus and Systems, Vol. 99, No. 1, pp. 195-206, Jan./Feb. 1980.

10. L. Paris, F. Reggiani, M. Valtorta, "The Study of UHV System Reliability in Connection with Structure and Component Characteristics". CIGRE Report No. 31-14, 1972.

11. M. Giovannini, C. Reynaud, L. Salvaderi, "Selection of Substation Layout in Power Systems". IEE Conference Publication No. 148: Reliability of Power Supply Systems. London, February 1977.

12. L. Paris, F. Reggiani, M. Valtorta, "Possibilities to Increase Transmission Lines Loadings in Well Developed Electrical Network". CIGRE Report No. 31-13, 1974.

13. R. Billinton, Power System Reliability Evaluation. Gordon and Breach, New York, NY, 1970.

14. IEEE Committee Report, "Common Mode Forced Outages of Overhead Transmission Lines". IEEE Transactions on Power Apparatus and Systems, Vol. 76, No. 3, pp. 859-863, May/June 1976.

15. R. Billinton and T.K.P. Medicherla, "Station Originated Multiple Outages in the Reliability Analysis of a Composite Generation and Transmission System". Paper 81 WM 002-5, IEEE-PES Winter Meeting, February 1981.

16. "Bulk Transmission System Component Outage Data Base". EPRI Report EL-1797 (Research Project 1283-1 prepared by the General Electric Co.), April 1981.

17. "Instruction Manual for Reporting Component Forced Outages of Transmission Equipment", CEA, April 1978.

18. IEEE Task Force Report, "IEEE Reliability Test System". IEEE Transactions on Power Apparatus and Systems, Vol. 98, No. 6, pp. 2047-2054, Nov./Dec. 1979.

Combined Discussions[1,2]

Lester H. Fink, (Systems Engineering for Power, Inc., Oakton, VA): Mr. Albrecht's discussions of power system reliability are always enlightening and stimulating. His definitions of *failure* and of the *cause of failure* on the distribution and generation systems respectively do indeed suggest complementary definitions for the bulk power system which could turn out to be practicable and ultimately powerful. Accordingly, in the spirit of his discussion and to further that end, I would like to suggest some revisions to his formulation.

I

While agreeing thoroughly with Mr. Albrecht's suggested revisions of the NERC definitions, I have one reservation on terminology. The term "security" has a clearly defined meaning in an operating context [1] which does not accord with the problem of reliability in a planning context.

The planner, who is charged with designing a reliable system, has at his disposal components which can be chosen for that system, and the manner in which they can be mutually connected; he has no influence over the conditions to which the system will be subjected in service.

A definition of reliability accordant with the responsibilities of the system planner should deal with qualities that the planner can build into the system using parameters under his control. Such a quality might be termed "robustness" or "soundness," and will be inherent in the system itself, as designed and built [2], and not vary with random circumstances, beyond the planner's control, to which the system will ultimately be subjected. If "security" could be built into the system, there would be no need for "security assessment" in real time.

The operator has at his disposal a system which, due to scheduled or forced outages, is *not* as the planner has defined it, and which is subject to continually varying disturbance probabilities. In an operating context, security is an instantaneous, time-varying condition that is a function of the robustness of the system relative to imminent disturbances. Elsewhere [3], it has been suggested that this concept can be formalized in terms of the probability P that the system will enter the emergency state, given its current state, the forecast probability distribution of disturbance processes (e.g. weather), and available control actions. The security level, S, then would be the expected value of the system's remaining in the normal state, $S = E\{1-P\}$, and the system would be in the alert state when this time-varying quantity fell below an acceptable, given threshold value. Recognition of the accepted meaning of "security" in the operating arena [1] should lessen confusion between operating and planning problems and contribute to clearer definition of the latter.

II

Turning to Mr. Albrecht's discussions of *failure* and *cause of failure* on the bulk power system, it seems somewhat paradoxical to suggest that "failure is fairly well defined conceptually (cascading, widespread or uncontrolled)" while admitting that "there is lack of precise definition needed to classify every system state as success or failure." The difficulty is just that "cascading, widespread, uncontrolled" are qualitative terms which are thereby unsuitable for a practicable definition. Moreover, by focusing exclusively on events which are almost invariably triggered by "a freak event not previously thought of, or a known event which it was thought the system could withstand" and hence are unstudied by the system planner, it ignores the "bottling up" of generation which is of considerable concern to the system planner.

For the last half-dozen years, there has been increasing attention given to the analytical properties of the load flow problem, and in particular to the definition and calculation of "feasibility" or (pardon the term) "steady state security" regions [4-8]. In general terms, these "regions" contain all possible load flow solutions for a given network which do not violate given constraints on equipment loading: any combination of loads and generation falling outside the boundary of such a region represents an unacceptable condition.

Now the function of the transmission system is to provide a bridge between generating buses and load (distribution) buses, i.e. mathematically to map a generation pattern (vector) onto the load pattern (vector). If for some load vector such a permissable mapping does not exist, if there is some probability that not all of the load can be served, certainly that indicates some impairment of reliability, the transmission system has partially failed. (Cannot there be "deratings" of the transmission system as well as of generators?) The boundaries of the feasibility region depend, of course, on the lines in service, and for some (low probability) combinations of line outages, the volume of the region can shrink to zero, representing a total failure: *no* load can be carried. Would it not be useful, then, to measure transmission reliability in terms of the (prospective) joint probabilities of (a) feasibility region boundaries shrinking to given positions as a function of equipment failure rates, and (b) load vectors falling inside or outside those regions [9]? Once such a methodology was developed, the system planner would have a powerful tool for evaluating planned system reliability in terms of (a) equipment failure rates and (b) probability distributions of system loads without recourse to detailed load flows. If "widespread, cascading, uncontrolled" interruptions are also to be treated quantitatively, it would appear they should be redefined in terms of "loss of system integrity" which could be defined quantitatively (e.g., islanding, loss of one of more load buses, etc.) and/or treated statistically as rare events defined on classes of undesigned conditions, such as mis-set or malfunctioning breakers.

In summary then, from this perspective:

1. A transmission system *failure* is a (partial or total) inability to transfer power from generator buses to load buses on demand.
2. The events that *cause* transmission failure are equipment outages, scheduled and/or forced, which impair or destroy the ability of the transmission network to transfer power from generator buses to the load buses as demanded.

In terms of feasibility regions, such failure may be evidenced by (a) unusual load vectors which fall outside the feasibility region of the transmission system as designed (a very rare partial failure), (b) contraction of the feasibility region boundaries, due to scheduled or forced outages of equipment, to an extent that they exclude commonly occurring load vectors (a common partial failure), (c) total contraction of feasibility region boundaries, due to multiple scheduled and/or forced outages, such that the feasibility region is wholly wiped out (a total failure), or (d) loss of system integrity defined statistically on a set of classes of rare, sport events outside system design parameters (total failure, due to "cascading. . . etc."). Work now in progress suggests it may be possible to extend this approach to include robustness with regard to transient stability [10].

Such an approach to transmission failure definition, taken in conjunction with distribution and generating system failure definition may provide a consistent set of quantitative measures to serve as a basis for a treatment of overall power system reliability.

REFERENCES

[1] The Reliability of Large Power Systems, J. J. Archambault et al. (13 authors, 6 nations) PROC. CIGRE, 1980 Session, paper 32-16.
[2] FUNDAMENTALS OF RELIABILITY THEORY, A. M. Polovko, Academic Press, New York, 1968.
[3] Statistical Characterization of Power System Stability and Security, G. L. Blankenship, L. H. Fink, PROC. 2nd Lawrence Symp. on Systems and Decision Sciences, Berkeley, California, Oct. 1978, pp. 62-70.
[4] Steady-State Security Regions: Set-Theoretic Approach, E. Hnyilicza, S. T. Y. Lee, F. C. Schweppe, PROC. IEEE/PES PICA 9, 1975, pp. 347-55.
[5] Power-Voltage Limitations Imposed by the Network Structure of a Power System, F. D. Galiana, ibid. pp. 356-63.
[6] Identification of Power System Steady-State Security Regions Under Load Uncertainty, R. Fischl, G. C. Ejebe, J. A. DeMaio, IEEE/PES Summper Meeting, 1976 (A76 495).
[7] Feasibility Sets for Steady-State Loads in Electric Power Networks, P. Dersin, A. H. Levis, TRANS. IEEE, Vol. PAS-101, Jan. 1982, pp. 60-70.
[8] Limits on Power Injections for Security-Constrainted Power Flows, F. F. Wu, S. Kumagai, PROC. IEEE Symp. on Circuits and Systems, April 1981, pp. 755-58.
[9] Mathematical Theory of Power System Reliability Evaluation, F. F. Wu, et al., Memorandum UCB/ERL M79/67, November 1979, University of California, Berkeley.
[10] A Method of Deriving Dynamic Security Regions of Power Systems, R. J. Kaye, F. F. Wu, PROC. IEEE Conf. on Decision and Control, Dec. 1981, pp. 1331-39.

Manuscript received February 16, 1982.

---

[1] J. Endrenyi (Chairman), P. F. Albrecht, R. Billinton, G. E. Marks, N. D. Reppen, and L. Salvaderi, Bulk Power System Reliability Assessment—Why and How? Pt. 1: Why?, this issue pp. 3439-3445.

[2] J. Endrenyi (Chairman), P. F. Albrecht, R. Billinton, G. E. Marks, N. D. Reppen, and L. Salvaderi, Bulk Power Systems Reliability Assessment—Why and How? Pt 11: How?, this issue, pp. 3446-3456.

Norton Savage (U.S. Department of Energy, Washington, DC): The comments offered here are those of the writer, and do not necessarily reflect the views of the Department of Energy. The two papers presented by the Panel on Bulk Power Reliability Assessment form a clearly organized, concise and readable introduction to that topic. The authors have neatly described the problems one meets in attempting to evaluate power system reliability and have indicated several methods of attack. An especially useful feature of the presentation, in my view, is the attention given to the subject of data requirements. A number of papers have appeared in the last couple of decades, dealing with various aspects of generating capacity reliability and transmission system reliability. Of those that have come to my attention, few have stressed the need for a data base that could be used to implement the various mathematical methods.

The pioneering papers of Calabrese and the work of the AIEE Subcommittee on Application of Probability Methods, followed by the work of Billinton, Patton, Landgren and others, have provided a strong foundation in methodology. However, little attention has been paid to the need for an equally strong foundation of data--comprehensive, accurate, detailed and tied to system conditions. It would appear obvious that such a data base of information, both historic and current, and continuously updated, is necessary to the proper use of the mathematical procedures, if large-scale systematic studies are to be done. I strongly endorse the comments in both papers regarding the need for more and better basic data on the characteristics of equipment, components and systems.

Manuscript received February 16, 1982.

Harry W. Colborn (North American Electric Reliability Council, Princeton, NJ): The discussion of the use of reliability indices to assist in justification of adding system elements brings to mind the many discussions I have had regarding the postponement of a particular project because it was "only for reliability". To my mind, it is a rare addition that is *not* for reliability. If one installs two lines from a generating station, who is to say which is not needed for reliability? A line added anywhere in a mature network is always installed because the system cannot withstand the outage of some existing line. A second transformer is added for reliability, as well. However, as time goes on, the element that was added for reliability becomes loaded to the point of being required even with all elements in service. Before that point is reached, however, good system planning dictates the addition of yet another redundancy - again, added for reliability. Thus, an element added for reliability loses its ability to perform that function as time continues. The 'spilt milk' philosophy of economists says that once the element is in service you can forget why it was added, and how much it cost. I wonder if treating the decreasing function and value of reliability as an annuity with a present value would be more proper in choosing among alternatives?

Manuscript received February 22, 1982.

J. Endrenyi, P. F. Albrecht, R. Billinton, G. E. Marks, N. D. Reppen and L. Salvaderi: We wish to thank the discussors for submitting thought-provoking comments. All the discussions present ideas which supplement those contained in the papers.

Like the paper, Mr. Fink's discussion is divided into two parts. In the first part, Mr. Fink is concerned with the use of the term "security". His arguments underline the statement, made in the introduction of our paper, that "some of the terms are given different definitions. . . a telling indication of the unsettled state [of the] methodology. . .". In fact, security is differently defined by panelists Endrenyi and Albrecht, and now Mr. Fink proposes a third definition. Generally, security is contrasted with adequacy, and Endrenyi makes a distinction between the two according to the analytical method used to detect system failure, whereas Albrecht's definitions are based on differences in the system effects of contingencies. In the first case, the system failure states identified by steady-state analysis are considered those that affect system adequacy, and contingencies whose occurrences would cause instability (where the system cannot withstand the impact of sudden changes) are regarded to influence security; in the second approach, adequacy is affected by contingencies causing widespread failures.

Mr. Fink feels that a sharp distinction is needed between definitions used in planning and operating. In the planning context, only adequacy (he calls it "reliability") violations can occur and in the operating context only security violations. Perhaps the necessary distinction is more between models than between definitions. We agree that planning models should be responsive to parameters over which planners have control, and likewise for operating models. However, should not both organizations have the same definitions of reliability? If security is a goal of the operator, the planner must provide a system which can be operated in a secure mode. Thus, to rephrase one of Mr. Fink's comments: if "security" is not built into the system, one may as well forget about "security assessment" in real time.

The second part of Mr. Fink's discussion deals with the problem of defining and modelling system failure and failure causes. The definitions suggested are basically those upon which current reliability models are based. With respect to modelling, attention is called to several references dealing with formulation of feasibility regions. Again current methods are based on this concept. This is illustrated by the contingency enumeration approach discussed by Reppen in Part II of the paper. Those contingencies which do not cause "system problems" are the feasibility region. Some of the practical difficulties in applying this approach are discussed by Endrenyi in Part I (overview).

Mr. Colborn raises some interesting points concerning justification of new facilities based on reliability considerations. He suggests that in a mature (growing) system each new facility is initially added to meet reliability requirements. This leads to the suggestion that the "reliability value" contributed by a given facility is a decreasing function, as "loading" of the facility increases. Could one not argue the converse? Initially, the system almost meets reliability requirements without the facility. However, as time goes on the facility becomes more critical to system performance. This suggests that the value of reliability required initially is an increasing function of the time until the next system upgrade. Thus, the fact that many projects are deferred because they are "only for reliability" may result from failure to recognize the increasing reliability value with time.

Mr. Savage's emphasis on the importance of outage data collection is shared by many in the utility industry. However, we hope that limitations in available outage statistics will not discourage utilities from exploring the benefits of quantitative reliability analysis. Only through applications on utility systems can data requirements as well as study procedures be refined.

In conclusion, we thank the discussors again for their useful and stimulating comments.

Manuscript received April 21, 1982.

TERMINAL EFFECTS AND PROTECTION SYSTEM FAILURES

IN COMPOSITE SYSTEM RELIABILITY EVALUATION

R.N. Allan, Senior Member IEEE   A.N. Adraktas

UMIST
Manchester
England.

Abstract

This paper considers the effect of terminal station outages on the contingencies that should be studied in composite system reliability evaluation. It describes how these contingencies are affected by the operation and non-operation of the protection breakers in a terminal station. It also describes how the reliability indices of protection systems which contain commonly-shared components, redundancy and diversity can be evaluated using the event tree method and how these indicies are used to determine the associated indices of the contingency. The discussion illustrates that the effect of terminal station outages on the contingencies to be included in a composite system study can be severe and that it can be very important to model the protection system in a realistic manner.

INTRODUCTION

The main problem area in composite reliability evaluation is the enormous amount of computational effort required if an exhaustive evaluation is made of the complete system representation. Consequently, techniques have generally been concerned with reducing the number of states needed to be assessed and with realistic simplification of the system topology and representation. These aspects are an essential requisite of composite reliability evaluation but their effect must be examined in order to establish whether they can be justified.

In order to simplify the system topology, the circuits are usually represented as single elements and the terminal stations (substations and switching stations) as single busbars. The terminal stations are therefore not fully represented and the switching and breaker configurations are neglected. The detailed analysis then proceeds by considering combinations of line and generator outages as independent overlapping events. This approach is reasonable provided no events can occur which cause the simultaneous outage of more than one line or generator. Two types of events which can cause simultaneous outages however are common mode outages and terminal station outages. There is a fundamental difference between these two type of outages. A common mode outage has been defined[1] as "an event having a single external cause with multiple failure effects which are not consequences of each other". Therefore one primary requirement of a common mode outage is that the initiating event is "external" to the system being analysed and is not artificially made to appear as an external cause by drawing the system boundary around a restricted part of the system. An internal cause, on the other hand, is an event that occurs within the system being analysed and the system boundary should be constructed to include all possible events that may have a significant contribution to the system behaviour. A terminal station outage due to protection operation must therefore be defined as an internal event even though it may be caused by a component that is not fully represented in the topology being used to assess composite system reliability. A detailed discussion of these two types of outages can be found in Reference 2 which defined terminal station outages as station originated outages.

The effect and evaluation of common mode failures has been the subject of previous papers[3-5] and will not be discussed further in the present one. Terminal station outages on the other hand have received only little attention[2] although it is recognised that they can have a significant impact on the reliability evaluation of composite systems.

This paper is therefore concerned only with the effect of terminal station outages, the contribution made by protection operation and protection failures on the severity and probability of such station outages and the inclusion of these outages in composite system reliability evaluation.

TERMINAL STATION OUTAGES

Concepts

The types of failure events which can occur in a terminal station and which can cause one or more lines/generators to be outaged simultaneously are :
a) active failures - these are usually ground faults on breakers or busbars which cause the appropriate protection breakers to operate. Depending upon the protection scheme, the breakers which trip can simultaneously disconnect more than one line and/or generator.
b) stuck breaker conditions - if, following an active failure, one or more of the primary protection breakers fail to trip, other breakers will respond which may disrupt a larger section of the station and may cause a greater number of lines/generators to be disconnected.

These two concepts are not new and have been used widely[6,7] in the reliability evaluation of substations and distribution systems. It is only their application to composite systems that is much more recent. It is evident from the description of these failure events that the underlying concept is associated with the protection system within the terminal station. Correct modelling of this system, the ways in which it operates and can fail and the impact on the system is therefore fundamental in the reliability evaluation of the composite system in which the terminal station exists.

One assumption made consistently in previous papers[2,6,7] is that the probability of more than one simultaneously stuck breaker is negligible. Although this may be true in many instances, it is an assumption that cannot be justified without detailed examination. It will be a reasonable assumption if each

82 SM 428-1 A paper recommended and approved by the IEEE Power Systems Engineering Committee of the IEEE Power Engineering Society for presentation at the IEEE PES 1982 Summer Meeting, San Francisco, California, July 18-23, 1982. Manuscript submitted July 15, 1981; made available for printing May 25, 1982.

Reprinted from *IEEE Trans. Power Apparatus Syst.*, vol. PAS-101, no. 12, pp. 4557–4562, Dec. 1982.

311

breaker is actuated quite independently of all others. On the other hand, protection schemes sometimes share common components, e.g. the fault detection device, battery supplies, etc. If any of these common components fail, the entire protection scheme will fail and more than one breaker may be "stuck" simultaneously. These events are not common mode events, i.e. external events, because they are caused by failures of components that are inherent to the proper operation of the protection system, i.e. they are "internal" to the system being considered.

The remaining sections of this paper illustrate typical failure events that can occur in a terminal station, describe how the indices of these events can be assessed by performing a reliability analysis on the protection system of the terminal station and demonstrates how these events and their associated indices can be included in composite system reliability evaluation in order to assess more realistically the severity and indices of this type of reliability evaluation.

Outage Events

The single line diagram shown in Figure 1 is based on the system given in Reference 2. It contains the same features but has been modified in order to ensure

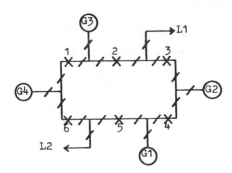

Figure 1 - Typical ring-bus terminal station

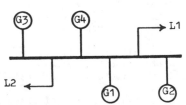

Figure 2 - Simplified composite system representation of terminal station

that the following discussion and evaluation is not necessarily indicative of the operating and failure modes of the real system on which it is based.

A detailed failure modes and effects analysis of the system is first required. A selected number of events for the system of Figure 1 is shown in Table 1 in which the symbol A represents an actively failed component, S represents a stuck breaker and M represents a maintenance outage which may be caused either by a scheduled maintenance or by the component requiring repair following a previous failure event. A more detailed list is given in Reference 2.

COMPOSITE SYSTEM RELIABILITY EVALUATION

Concepts

The general practise used in composite system rel-

Table 1   Selected Terminal station outages of Figure 1

failure mode	lines/generators outaged
1. 2A	G3, L1
2. 1A + 2S	G3, G4, L1
3. 2A + 1S	G3, G4, L1
4. 2A + 3S	G2, G3, L1
5. 2A + 1S + 3S	G2, G3, G4, L1
6. 1M + 2A	G3, L1
7. 3M + 1A	G3, G4, L1
8. 3M + 2A	G3, L1
9. 4M + 2A	G2, G3, L1
10. 5M + 2A	G1, G2, G3, L1
11. 6M + 2A	G3, G4, L1

iability evaluation is to deduce a set of contingencies based on a severity or liklehood of occurrence criterion. These contingencies include both single order, i.e one line or generator on outage, and higher orders, i.e. combinations of lines and/or generators on outage. The analysis would proceed by simulating the outage effect of these contingencies on the behaviour of the system using a load flow technique. If the contingency violated appropriate network or generation constraints, the outage would be classed as a failure event and relevant system and load point reliability indices evaluated.

The simplified method of representing the terminal station shown in Figure 1 in this type of analysis would be as a single busbar to which the generators and lines are connected. This is shown in Figure 2. This simplified representation, together with the rest of the system in which it exists, would be analysed using a composite system reliability evaluation technique and the system and load point indices would be evaluated assuming the component outages are independent.

The general method used to accommodate the effect of terminal station outages in this type of evaluation, when simplified station representations are used, is to modify the component indices of the lines and generators by including terminal component values. This approach is intended to compensate for the effect of terminal station outages by increasing the value of failure rate for those lines and generators which are subjected to a simulated outage during the composite reliability evaluation.

This is not a realistic approach when the indices of the overlapping events or combinations are evaluated assuming independent component outages because the method does not recognise the fact that a particular station outage may remove the selected combination of components by a single station event. The single terminal component factor is included in such cases as two or more independent events rather than one single event affecting more than one component (line and/or generator) of the simulated outage. This becomes a manipulative procedure therefore which does not truly reflect the behaviour and malfunction of the system. As an example of this problem, consider the system shown in Figure 1 and the information provided in Table 1. This Table shows that the active failure of breaker 2 removes G3 and L1 simultaneously. The indices of this event should therefore include not only those associated with the independent overlapping outage of G3 and L1 but also those associated with the event that initiates the simultaneous outage of G3 and L1, i.e. the indices associated with the active failure of breaker 2. Models for combining these events have been given in Reference 2.

Inclusion of Terminal Station Events

A detailed analysis that provides the information

shown in Table 1 can not be accommodated within a single composite system evaluation because of the large computational effort. Furthermore the information can not be provided by a generalised composite system evaluation technique because the consequences of terminal station events depend on the topology and operating characteristics of the station and are therefore system-specific. Instead, a separate assessment of the station can be performed before commencing the composite system reliability evaluation. This will produce a list of contingencies that should be studied, a ranking order in terms of severity or probability of occurrence and the relevant indices of the event. This information can then be considered as input data and included, as appropriate, with all the other contingencies that should be assessed in the composite reliability evaluation.

Table 1 shows that several failure modes lead to the same system effect, i.e. they cause the same set of lines and generators to be outaged. For example, failure modes 2,3,7, and 11 all lead to the simultaneous outage of G3,G4 and L1. It follows therefore that these groups and their indices can be combined together for the purposes of composite reliability evaluation.

The remaining problem is therefore to evaluate the indices of each of the contingencies to be assessed. These indices will depend upon whether the protection system operates as desired or whether one or more of the protection breakers fail to operate,i.e. are stuck. For instance, the outage of G3 and L1 (event 1 in Table 1) will occur with a rate equal to the failure rate of breaker 2 weighted by the probability of both breakers 1 and 3 operating successfully, the outage of G3, G4 and L1 (event 3 in Table 1) will occur with a rate equal to the failure rate of breaker 2 weighted by the probability of breaker 1 failing to operate and breaker 3 operating successfully. A similar deduction can be made for all the events of Table 1 and events which cause the same set of lines and generators to be outaged simultaneously can be combined. Further consideration of this numerical aspect is discussed in a later section of this paper.

It follows that the indices of each outage can be deduced from a knowledge of the reliability indices of each station component and a knowledge of the probabilities of the protection system operating and not operating. The last aspect is addressed in the next section of this paper.

MODELLING PROTECTION SYSTEMS

Concepts

The probability of a breaker responding or not responding to an active failure depends on the protection scheme, its construction and the quality of the components used. Conventionally, the probability of a breaker failing to operate, i.e. probability of being stuck P_S, has been derived by considering the protection system and its associated breaker as a single component and deriving P_S from a fault reporting scheme as:

$$P_S = \frac{\text{no. of failures to operate}}{\text{no. of requests to operate}} \qquad (1)$$

This method of evaluating P_S neglects the possibility of breakers sharing a protection system or sharing some common components. The protection system however is an integrated system and can be analysed independently. Unlike the rest of the power system, a protection system remains in a dormant state until it is called on to operate. Any failures other than inadvertent trips which occur during this dormant state do not manifest themselves until the operating request

is made. These failures have been defined[8] as unrevealed faults. In order to reduce the probability of an operating failure, the protection system is usually checked and proof-tested at regular inspection intervals.

There are a number of alternative techniques that can be used to assess the reliability of protection systems including fault trees, event trees and Markov modelling. A recent paper[11] considered the use of Markov models to these systems. The present paper however uses the event tree method. This method has been used[9,10] extensively in other applications and is an ideal technique because it can recognise the sequential logic of a system and can be easily extended to analyse the system at increasing depth.

There are many types of protection systems and therefore only a generalised form will be discussed; this being represented by the blocks shown in Figure 3. These blocks relate to most protection systems in which the fault detector (FD) includes appropriate CT's,VT's and comparators, the relay (R) includes operating and restraint coils, the trip signal (TS) contains the trip signal device and associated power supply and the breaker (B) is the actual fault breaking component. If more than one of the blocks or modules shown in Figure 3 share one or more common components, then the shared components must be represented by a separate module.

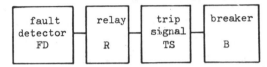

Figure 3 - Block diagram of a general protection scheme

Basic event tree

Consider a particular component that is protected by two breakers B1 and B2. Assume that both breakers are operated by the same fault detector (FD), relay (R) and trip signal device (TS). This is clearly an unrealistic operating procedure since considerably more redundancy, diversity and independence will generally be included in a practical system. This example however is intended to illustrate the modelling procedure, the basic evaluation method and the effect of common components. The event tree for this system, given an active failure on the component, is shown in Figure 4. This illustrates the sequence of events together with the outcomes of each event path, only one of which in this case leads to both breakers operating successfully and four of which lead simultaneously to both breakers not operating.

This event tree assumes that each component can reside in one of only two states; these being the operable state, i.e. it can respond to a system fault when one occurs, and the inoperable state, i.e. it is failed and cannot respond to a system fault. A further state exists in practise which represents an inadvertent operation and causes the breaker to trip. This state however is generally associated with a revealed fault[8] of the protection system or breaker, i.e. it manifests itself immediately. It has been neglected in the present example since it is not relevant to the conditions appertaining when a system fault occurs. If required it can be included however by creating three branches at each required node of the event tree instead of the two shown in Figure 4 and deducing the appropriate outcomes of each of the new event paths.

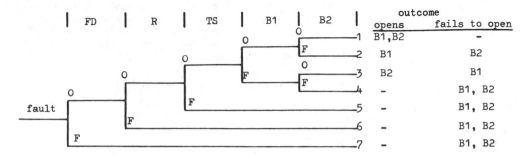

O = operates, F = fails to operate

Figure 4 - Basic event tree

Evaluating Event Probabilities

The event probabilities are the probabilities that each device in the protection system will and will not operate when required. These are time dependent probabilities. The time at which a device is required to operate however is a random variable and unknown in advance. The most useful single index is the average unavailability of the device between consecutive tests. This index has also been defined[8] as the mean fractional dead time.

If the time between consecutive tests is T_c, then the average unavailability U is :

$$U = \frac{1}{T_c} \int_o^{T_c} Q(t)\, dt \qquad (2)$$

where $Q(t)$ represents the time dependent failure probability of a component. In the special case when times-to-failure are exponentially distributed :

$$U = \frac{1}{T_c} \int_o^{T_c} (1 - e^{-\lambda t})\, dt$$

$$= 1 - \frac{1}{\lambda T_c} (1 - e^{-\lambda T_c}) \qquad (3)$$

$$\simeq \lambda T_c/2 \qquad \text{if } \lambda T_c \ll 1 \qquad (4)$$

A similar evaluation can be used if the times-to-failure follow any other type of distribution. If the distribution can not be integrated, a numerical integration method can be used to evaluate the average unavailability.

Consider the situation when the failure rates are 4×10^{-2} f/yr for the fault detector, 5×10^{-3} f/yr for the relay, 3×10^{-2} f/yr for the trip signal device and 8×10^{-3} f/yr for each breaker. These failure rates have been based on information given in Refs. 8 and 12. The average unavailabilities with an inspection interval of 6 months are given by Equation 4 as 0.010, 0.00125, 0.0075 and 0.0020 for the fault detector, relay, trip signal device and each breaker respectively; these being only very slightly different from those that would be given by Equation 3.

Evaluating Outcome Probabilities

The outcome probabilities are readily evaluated after the event-tree and event probabilities have been deduced. First the event paths leading to each outcome are identified. The probability of occurrence of each path is the product of the event probabilities in the path. The outcome probability is then the sum of the probabilities of each path leading to that outcome.

In the example shown in Figure 4:

prob (B1 and B2 operating)   = prob. of path 1
prob (B1 operating, B2 not operating) = prob. of path 2
prob (B1 not operating, B2 operating) = prob. of path 3
prob (B1 and B2 not operating) = Σ prob. of paths 4-7

Using the same data and inspection interval as in the previous section gives the path and outcome probabilities shown in Table 2.

Table 2-Path and outcome probabilities for Figure 4

path	outcome	probability of path	outcome
1	B1 operates, B2 operates	0.977424	0.977424
2	B1 operates, B2 fails	0.001959	0.001959
3	B1 fails, B2 operates	0.001959	0.001959
4)	B1 fails, B2 fails	0.000004)	
5)	"	0.007416)	0.018658
6)	"	0.001238)	
7)	"	0.010000)	

The results shown in Table 2 clearly depend upon the assumed operational mode, numerical reliability data and the sharing of components and devices between the two breakers. The results would be very different when different assumptions and operational characteristics are considered. The results indicate however the importance of correctly modelling the protection system and, in particular, any commonly-shared components.

If in the present example, only the failing to operate status of a breaker is considered, i.e. the status of the other breaker is ignored, then :
prob (B1 failing to operate) = Σ prob. paths 3-7
prob (B2 failing to operate) = Σ prob. paths 2, 4-7
which, in both cases, is equal to 0.020617. This would be the value recorded from a fault reporting scheme if the failure occurrences are recorded, or at least stored in a data bank, as independent events. Using this data and the assumption of independent overlapping failures, the probability of B1 and B2 not operating would be evaluated as $0.020617^2 = 4.25 \times 10^{-4}$.

Although it is assumed in the above discussion that the faulted component is protected by two breakers, the concepts can readily be extended to any number of breakers.

Separate and Redundant Operating Components

It was assumed in the event tree shown in Figure 4 that the same fault detector (FD), relay (R) and trip signal device (TS) operated both breakers (B1 and B2). Consequently, any failure in this system other than failure of the breaker itself, causes both breakers to malfunction. This possibility can be reduced by providing alternative channels to each breaker or including

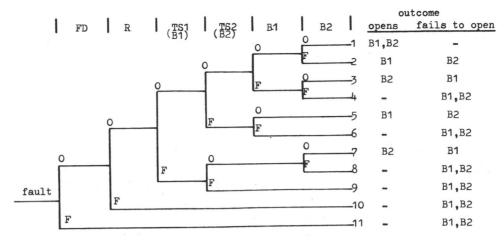

Figure 5 - Event tree with two trip signal devices

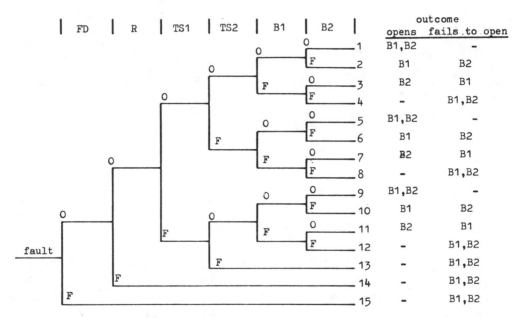

Figure 6 - Event tree with redundant trip signal devices

redundancy in the system. In order to illustrate this aspect, consider two trip signal devices and two case studies :
a) one trip signal device (TS1) actuates B1 and the other (TS2) actuates B2.
b) operation of either trip signal device actuates both B1 and B2.

The event trees for cases (a) and (b) are shown in Figures 5 and 6 respectively. If the previous values of failure rate and inspection interval are used, the modified probabilities of protection operation are shown in Table 3.

Table 3 - Outcome probabilities for Figures 5 and 6

| outcome | probability | |
	Figure 5	Figure 6
B1 operates, B2 operates	0.970094	0.984754
B1 operates, B2 fails	0.009289	0.001974
B1 fails, B2 operates	0.009289	0.001974
B1 fails, B2 fails	0.011328	0.011298

The results shown in Table 3 indicate that, as expected:
i) the probability of both breakers failing to operate is considerably reduced due to the diversity (case a) and redundancy (case b) in having two trip signal devices
ii) the probability of B1 failing irrespective of the status of B2 remains unchanged in case (a) - (0.009289 + 0.011328 = 0.020617) since the number of devices in the channel to B1 is unchanged but is reduced considerably in case (b) (0.013272) due to redundancy in the trip signal devices.

These values would be affected by an even greater amount if further redundancy and diversity is incorporated in the system, such as an increased number of relays. Additional features such as increased diversity, redundancy and majority vote (m-out-of-n) systems do not add any complexities to the fundamental techniques and the concept of event trees and subsequent evaluation methods remain identical.

315

INCLUSION IN COMPOSITE RELIABILITY EVALUATION

As discussed previously, the contingencies to be considered for inclusion in composite system reliability evaluation should include those caused by terminal station outages in addition to the conventional overlapping component outages. Typical station outages and associated contingencies for the system shown in Figure 1 are identified in Table 1. The indices associated with the above contingencies therefore consist of two distinct contributions. These are :
a) the direct contribution associated with the overlapping events causing the contingency
b) the indirect contribution associated with terminal station outages.

The second contribution is deduced from the component reliability indices of the initiating event (station component failure) and the probability of the protection system operating or not operating. As an example, consider the system in Figure 1, the protection probabilities shown in Table 2 and the following contingencies caused by the active failure of breaker 2. Let this breaker have an active failure rate of λ^a of 0.04 f/yr

a) contingency G3, L1 (event 1):
$$\lambda \ (G3, L1) = \lambda^a \ (\text{breaker } 2) \times \text{prob (breakers 1 and 3 both operate)}.$$
$$=0.04 \times 0.977424 = 3.91 \times 10^{-2} \ f/yr$$

b) contingency G3, G4, L1 (event 3) :
$$\lambda \ (G3, G4, L1) = \lambda^a \ (\text{breaker } 2) \times \text{prob (breaker 1 fails and breaker 3 operates)}$$
$$=0.04 \times 0.001959 = 7.84 \times 10^{-5} \ f/yr$$

c) contingency G2, G3, G4, L1 (event 5):
$$\lambda \ (G2, G3, G4, L1) = \lambda^a \ (\text{breaker } 2) \times \text{prob (breakers 1 and 3 both fail)}$$
$$=0.04 \times 0.018658 = 7.46 \times 10^{-4} \ f/yr$$

This concept can be applied to all the contingencies in Table 1. As explained previously, the indices of all station events leading to the same contingency can be grouped together to form a single set of indices. This can be achieved using the models of Reference 2. Each contingency is then simulated in a composite reliability assessment using standard techniques.

In the case of the above contingencies, the outage time will be associated with the switching time needed to isolate the actively failed component and reconnect the remaining system. Even when these switching times are relatively small compared with component repair times, the probability of a particular contingency caused by a terminal station outage can be more significant than the probability caused by independant overlapping outages. This however is not a general conclusion and depends on the topology of the system, the component reliability indices and the components involved in the contingency.

CONCLUSIONS

The reliability evaluation of composite generation/transmission systems has become an important concept and appropriate modelling techniques and evaluation methods are continuously being developed. The simplifications that are incorporated into these techniques must be carefully assessed in order to ensure that they do not incur unrealistic approximations. This paper has been concerned with one particular aspect of this problem, namely, the effect of terminal station outages on the list of contingencies that should be included in the composite reliability evaluation and, in particular, the effect of protection operation and

malfunction on the reliability indices of these contingencies.

It has been shown that the contingencies that should be studied can be made significantly more severe by protection system operation and failures in the terminal station; thus having a greater impact on the overall behaviour of the composite system.

The event tree method, which has been widely accepted in other areas of reliability evaluation but less so in power system reliability, is a very useful and convenient technique to assess the reliability of protection systems. This method can recognise the sequential logic of the system and permits a more realistic assessment of its reliability.

REFERENCES

1. IEEE PES Committee Report - Common mode forced outages of overhead transmission lines. IEEE Trans. on Power Apparatus and Systems, PAS-95, 1976, pp 859-863.
2. R. Billinton, T.K.P. Medicherla - Station originated multiple outages in the reliability analysis of a composite generation and transmission system. IEEE PES Winter Power Meeting, Atlanta, 1981, paper 81 WM002-5.
3. R. Billinton, T.K.P. Medicherla, M.S. Sachdev - Common-cause outages in multiple circuit transmission lines. IEEE Trans. on Reliability, R-27, 1978, pp 128-131.
4. R.N. Allan, E.N. Dialynas, I.R. Homer - Modelling common mode failures in the reliability evaluation of power system networks. IEEE PES Winter Power Meeting, New York, 1979, paper A79 040-7.
5. R. Billinton, T.K.P. Medicherla, M.S. Sachdev- Application of common-cause outage models in composite system reliability evaluation. IEEE PES Summer Power Meeting, Vancouver, 1979, paper A79 461-5.
6. M.S. Grover, R. Billinton - A computerized approach to substation and switching station reliability evaluation. IEEE Trans. on Power Apparatus and Systems, PAS-93, 1974, pp 1488-1497.
7. R.N. Allan, R. Billinton, M.F. de Oliveira - Reliability evaluation of electrical systems with switching actions. Proc.IEE, 123, 1976, pp 325-330.
8. A.E. Green, A.J. Bourne - Reliability Technology (John Wiley, 1972.).
9. R.N. Allan, I.L. Rondiris, D.M. Fryer, C. Tye - Computational development of event trees in nuclear reactor systems. Second National Reliability Conference, Birmingham, UK, March 1979, paper 3D/1.
10. R.N. Allan, I.L. Rondiris, D.M. Fryer - An efficient computational technique for evaluating the cut/tie sets and common-cause failures of complex systems. IEEE Trans. on Reliability, R-30, June 1981.pp 101-109.
11. C. Singh, A.D. Patton - Protection system reliability modelling, unreadiness probability and mean duration of undetected faults. IEEE Trans on Reliability, R-29, 1979, pp 339-340.
12. IEEE - Nuclear reliability data manual. IEEE Standard 500-1977.

EFFECT OF STATION ORIGINATED OUTAGES IN A COMPOSITE SYSTEM ADEQUACY EVALUATION OF THE IEEE RELIABILITY TEST SYSTEM

R. Billinton, Fellow, IEEE   P.K. Vohra   Sudhir Kumar
Power System Research Group
University of Saskatchewan
Saskatoon, Saskatchewan
Canada S7N 0W0

Abstract – The reliability evaluation of a bulk power system normally includes the independent outages of generating units, transformers and transmission lines. Station originated outages can, however, also have a significant effect on the composite system adequacy indices. Terminal station related failure events such as failures of breakers, transformers and bus sections are a major cause of multiple outages of major components (generators and/or transmission lines etc.). This paper illustrates the effects of terminal stations in composite system reliability evaluation by using the IEEE Reliability Test System (RTS). The paper also provides a valuable extension of the original RTS by including switching configurations at each bus.

INTRODUCTION

Recent attention to the role of terminal stations in system disturbances strongly supports the need to recognize the multiple outages of major system components due to terminal station failures, in composite system adequacy assessment. Landgren and Anderson reported [1] that over 40% of the multiple line outages in the Commonwealth Edison Company's 345 kV power system are caused by terminal related disturbances. The probabilities associated with station originated multiple line outages can be quite high compared to the corresponding event probabilities associated with independent outages. It is therefore not practical to consider higher level independent line outages and ignore the station originated multiple outages. The effect of stations is sufficiently dominant in most cases that their inclusion diminishes the need to consider high level independent line outages.

Reliability analysts have devoted relatively little attention to station originated outages in a composite power system due to the lack of data on station originated outages and because of the complexity involved in the analysis. Some work has been done on station reliability evaluation [2] but without considering the overall composite system. Initial work including terminal stations in composite system adequacy evaluation is described in References 3, 4, & 5. Present composite system reliability evaluation techniques normally include terminal stations by simply modifying the reliability parameters of lines and/or generators affected by the outages of station components. This approach is suggested in Reference 6 which describes the IEEE RTS. This is correct only if the failure of a station

component results in the outage of a single major element. When the unavailability of two or more elements in a system is considered, the approach assumes an unrealistic independence between these element outages. A more realistic approach has been developed by the authors to include terminal stations in composite system reliability evaluation and is used in the system studies presented in this paper. In this approach, the major component outages resulting from terminal station failures are treated as separate events. It is not feasible to solve all the contingencies simultaneously due to station initiated outages and independent outages. System event data can be obtained due to failures originating in the terminal stations. These event data include:

(i)   various sets of major components on outage due to station related failures.
(ii) probability of outage of each set of major elements.
(iii) frequency of outage of each set of major elements.

These event data are used as input to the basic composite system reliability evaluation program which gives the load point and system adequacy indices.

Assumptions

The following assumptions were utilized in the develpment of the station originated models used for the studies conducted in this paper.

(i)   The probability of a stuck breaker condition is sufficiently small that it can be assumed to be zero (as per information obtained from B.C. Hydro, Vancouver).
(ii) The effects of adverse weather are neglected.
(iii) The probability of overlapping outages of three or more station components is negligible and is therefore not considered.
(iv) A component is not taken out for preventive maintenance if it results in the outage of a major component.
(v)   Relays associated with all breakers are assumed to be non directional.

Station Originated Outage Events and Their Inclusion in Composite System Reliability Evaluation

The failure of a station component may result in multiple outages of generators and/or transmission lines. It may also isolate the load feeder completely. The procedure for evaluating the outages of major components due to station equipment failure is described in References 3,5. Several station related failure events could result in the same system effect i.e. they might cause the same set of major elements or load feeders to be removed from the system. In this case, the reliability indices (probabilities and frequencies of outage) associated with these failure events can be directly added for the purposes of composite system reliability evaluation. Outages of generating units, lines and

85 WM 014-6   A paper recommended and approved by the IEEE Power System Engineering Committee of the IEEE Power Engineering Society for presentation at the IEEE/PES 1985 Winter Meeting, New York, New York, February 3 - 8, 1985. Manuscript submitted August 17, 1984; made available for printing January 3, 1985.

Reprinted from *IEEE Trans. Power Apparatus Syst.*, vol. PAS-104, no. 10, pp. 2649-2656, Oct. 1985.

317

transformers are therefore considered to arise in this study from the following causes:

(i) outages due to their independent failures.
(ii) outages due to station related failure events.

It has been assumed that these two events are independent and mutually exclusive. The probability of an outage state is therefore the sum of the probabilities associated with each cause, i.e.

P(Components X and Y are out) = P(Components X and Y are out due to their own internal failures given that all other components are operating) + P(Components X and Y are out due to station related failure events given that all other system components are operating)

The frequency of the outage state is obtained by adding the frequencies of these two events.

The station originated models were developed for the following failure events originating in the stations:

(i)   forced outage (active and passive failure) of a breaker.
(ii) forced outage of a bus section.
(iii) forced outage of a transformer.
(iv) overlapping forced outages of breakers, transformers and bus sections.
(v)   forced outage of a component overlapping maintenance of another component.

A general list of the major component combinations removed from service due to station disturbances is as follows:

(1) one generating unit is out.
(2) two generating units are out.
(3) three generating units are out.
(4) four generating units are out.
(5) one line or transformer is out.
(6) two line(s) or transformer(s) are out.
(7) three line(s) or transformer(s) are out.
(8) one generating unit and one line or transformer is out.
(9) two generating units and one line or transformer is out.
(10) load feeder(s) are isolated.

The contribution of the above contingencies except for the outage of three lines (No. 7), the outage of two generating units and one line (No. 9), and the isolation of the load feeder (No. 10) is taken into account by adding the probabilities and the frequencies to the probability and the frequency of the respective contingencies resulting from the independent outages of these components. The remaining three outages are calculated separately.

The effect of station originated outages on the adequacy indices of a composite generation transmission system is illustrated in the next section. These studies have been conducted using the IEEE Reliability Test System.

SYSTEM STUDIES

A digital computer program has been developed to include the station originated outages in the adequacy evaluation of a composite generation and transmission system. The basic procedure is as discussed in the previous section.

The single line diagram of the IEEE RTS is shown in Figure 1. This is the original single line diagram presented in Reference 6. The original network has been extended in Figure 2 which shows the complete

network with station configurations at each bus. These station configurations have been adapted from existing configurations in actual systems. The IEEE RTS was developed to provide a practical reference system which could be used to compare alternate solution techniques. The complete network shown in Figure 2 is a basic extension of the original configuration and extends the usefulness of the RTS.

System studies were conducted at a load level of 2400 MW and also at a load level of 2850 MW in order to examine the effect of terminal stations at different load levels. A description of the load model for a peak load of 2850 MW is given in Reference 6. The load model for a peak load of 2400 MW was obtained by scaling down the basic load data given in Table 10 of Reference 6. The basic system data for the RTS is given in Reference 6. The additional data required to include the terminal stations is as follows.

Active failure rate of a breaker = 0.0066 failures/yr.

Passive failure rate of a breaker = 0.0005 failures/yr.

Maintenance rate of a breaker = 0.2 outages/yr.

Maintenance time of a breaker = 108 hours.

Switching time of a component = 1.0 hour.

The annualized system adequacy indices calculated using the digital computer program developed at the University of Saskatchewan are described in detail in Reference 7. Four basic annualized individual load point indices, i.e. the total number of load curtailments, the expected load curtailed in MW, the expected energy curtailed in MWhr and the duration of load curtailment in hours, are given in Tables I to IV respectively for the system load of 2400 MW. Independent outages of generating units are considered up to the 4th level. Outage contingencies up to the 2nd level are considered in the case of transmission lines/transformers. Under any outage event line overloads are alleviated by generation rescheduling and load shedding [8].

TABLE I

Expected Annualized Number of Load Curtailments at the Load Buses in the RTS System Load = 2400 MW

Load Bus No.	Without Stations	With Stations	% Increase
1	0.334	0.369	10.48
2	0.669	0.705	5.38
3	0.334	0.370	9.73
4	0.335	0.433	29.25
5	0.335	0.433	29.25
6	0.336	0.434	29.16
7	0.355	0.363	2.25
8	0.355	0.400	12.67
9	0.0	0.0367	
10	0.0	0.0369	
13	1.997	2.033	1.80
14	0.233	0.284	21.88
15	1.962	2.000	1.93
16	0.538	0.587	9.10
18	2.216	2.256	1.80
19	0.232	0.290	25.0
20	1.162	1.257	8.17

318

TABLE II

Expected Annualized Load Curtailed at the Load
Buses in the RTS
System Load = 2400 MW

Load Bus No.	Without Stations	With Stations	% Increase
1	1.62 MW	4.77 MW	194.4
2	3.22	6.23	93.47
3	2.70	8.06	198.52
4	1.24	7.12	474.2
5	1.10	7.00	536.37
6	2.34	13.19	463.67
7	2.64	3.46	31.06
8	4.16	10.65	156.00
9	0.0	5.33	
10	0.0	5.91	
13	41.86	52.23	24.77
14	2.64	11.09	320.07
15	50.31	63.16	25.54
16	3.43	7.65	123.03
18	67.15	81.18	20.89
19	2.34	10.97	368.80
20	19.17	29.98	56.39

TABLE III

Expected Annualized Energy not Supplied at the
Load Buses in the RTS
System Load = 2400 MW

Load Bus No.	Without Stations	With Stations	% Increase
1	21.63 MWhr	48.28 MWhr	123.2
2	43.06	73.80	71.38
3	36.04	100.27	178.22
4	16.36	97.22	494.25
5	14.49	99.71	588.12
6	30.52	195.22	539.64
7	33.40	34.71	3.92
8	53.27	134.12	151.77
9	0.0	85.25	
10	0.0	101.11	
13	585.73	768.50	31.20
14	30.67	184.51	501.59
15	708.77	949.00	33.89
16	41.64	125.04	200.28
18	928.77	1219.43	31.29
19	27.58	191.32	593.69
20	272.12	516.20	89.69

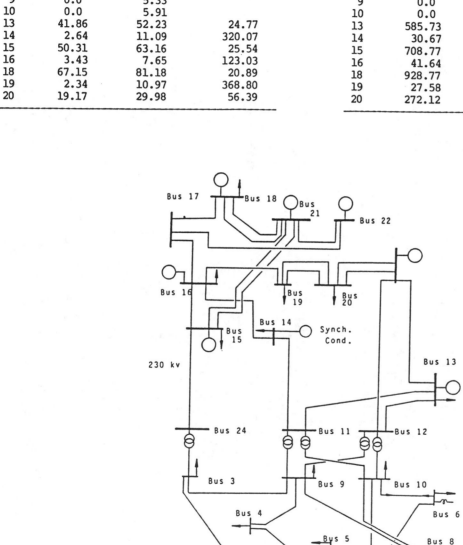

Figure 1. Single Line Diagram of the IEEE Reliability Test System (RTS). Reference [5]

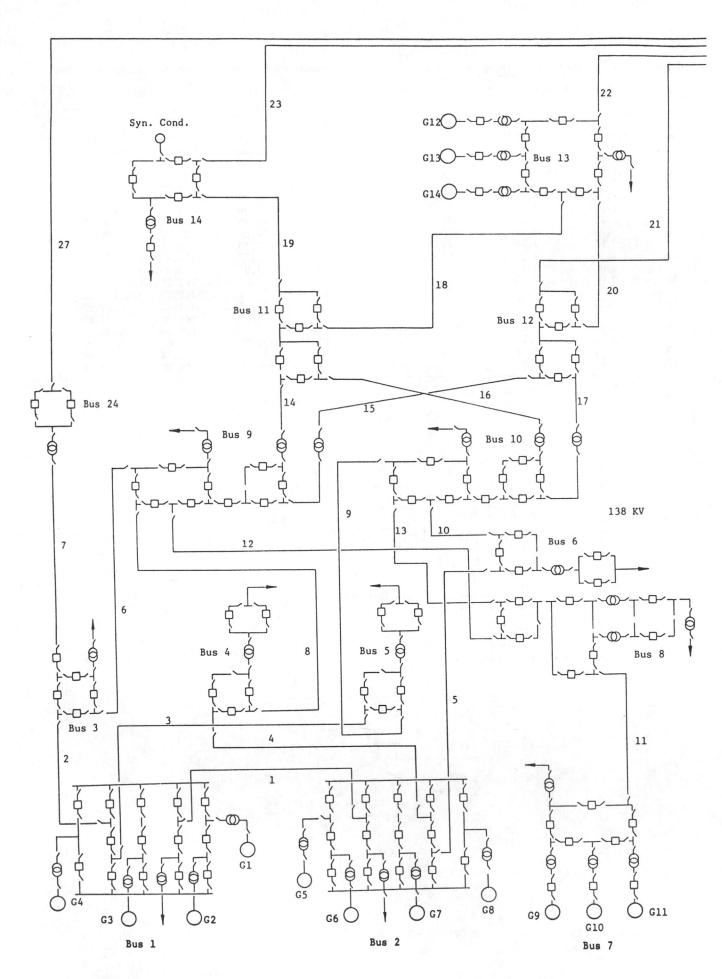

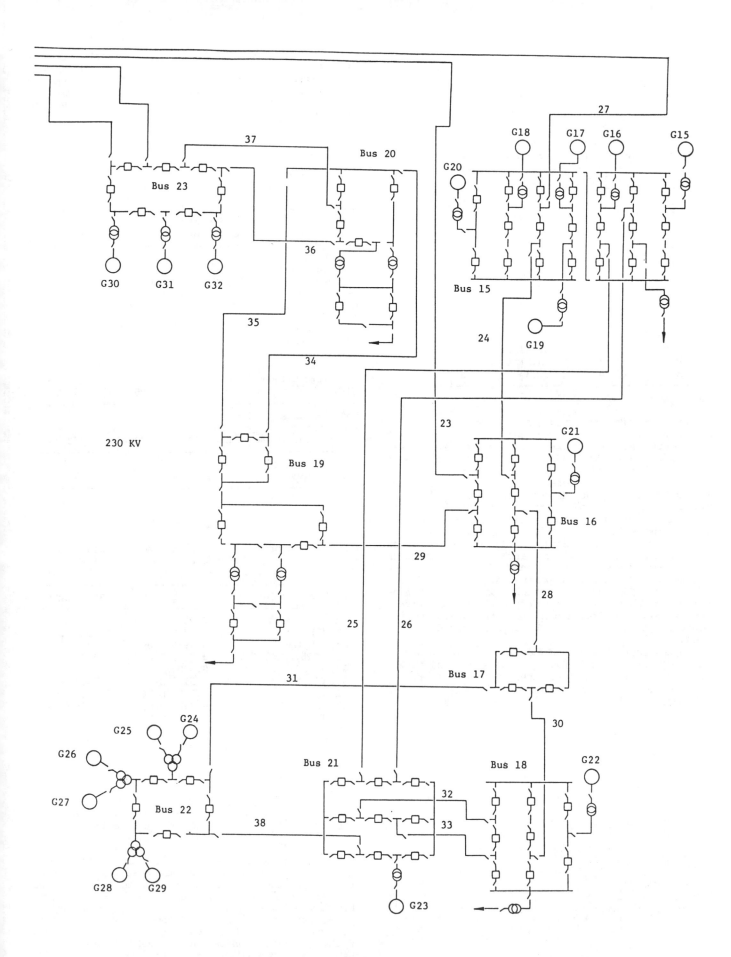

Figure 2. Extended Single Line Diagram of the IEEE Reliability Test System (RTS).

TABLE IV

Expected Annualized Duration of Load
Curtailment at the Load Buses in the RTS
System Load = 2400 MW

Load Bus No.	Without Stations	With Stations	% Increase
1	4.18 hr	4.48 hr	7.17
2	8.36	8.72	4.3
3	4.18	4.61	10.28
4	4.18	5.53	32.29
5	4.18	5.60	33.97
6	4.19	5.68	35.56
7	4.33	4.34	0.23
8	4.33	4.88	12.70
9	0.0	0.59	
10	0.0	0.63	
13	27.01	27.65	2.37
14	2.68	3.61	34.70
15	26.72	27.43	2.65
16	6.36	7.34	15.40
18	29.55	30.41	2.91
19	2.68	3.77	40.67
20	15.20	17.37	14.27

The indices reported in column 2 (without stations) in Tables I-IV are due to the independent outages of the generating units, transmission lines and transformers. The indices noted in column 3 (with stations) were obtained after including the effect of station originated outages together with the independent outages of the generating units, lines and transformers. The net increase in the indices is given in column 4 of each table. As seen from the tables, the effect of station originated outages on the bus indices is considerable. Buses 9 and 10 which do not encounter load curtailment problems with independent outages, experience load curtailment because of the failure of station components. It can be seen that the effect of station originated outages on the indices at each bus is not uniform. The effect depends upon the load curtailment philosophy for an outage event and the station configuration selected at a bus.

In the case of a capacity deficiency in the system under any outage condition, load is curtailed at the buses located in the problem area, i.e., in the region where an outage of a component(s) is reported. At each bus, load is classified into two categories:

(i)   firm load.
(ii) curtailable load.

A capacity deficiency situation is alleviated by interrupting the curtailable load proportionately at the buses in a problem area. If a system problem persists, after curtailing the total curtailable load, then the firm load is curtailed. In this study, curtailable load at each bus is assumed to be 20% of the total connected load at the bus.

Table V shows the system indices for the 2400 MW system load. The increment in system indices is quite significant and the contribution of station originated outages can be clearly seen. The most distinct contributing factor to this increase in the adequacy indices due to including station originated outages is the isolation of a load feeder at a bus. The load at a bus is isolated if there is a fault at the

associated bus-section and/or in the breaker of the load feeder. These single order outage events result in higher values of the probability and the frequency of load isolation at a bus. A sample of event data resulting from station related failures is given in Table VI.

TABLE V

Basic Annualized System Indices for the RTS
System Load = 2400 MW

	Without Stations	With Stations	% Increase
Bulk Power Interruption Index (MW/MW-Yr)	0.08579	0.13666	59.29
Bulk Power Supply Average MW Curtailment Index (MW/Disturbance)	92.81903	104.54426	12.63
Bulk Power Energy Curtailment Index (MWh/MW)	1.18503	2.05153	73.12
Modified Bulk Power Energy Curtailment Index	0.000135	0.000234	73.33
Severity Index (System Minutes)	71.102	123.092	73.12

TABLE VI

A Sample of Event Data Resulting From Station
Related Failures

Elements Out	Probability	Frequency
G1	0.2650 E-03	0.3787 E-01
G1, G2	0.1889 E-05	0.1507 E-02
G1, G2, G7	0.1916 E-09	0.1702 E-05
G24, G25, G26, G27	0.8140 E-06	0.1490 E-02
G2, L1	0.1280 E-06	0.1317 E-03
G1, G2, L1	0.3131 E-09	0.2807 E-05
L1	0.7666 E-04	0.6223 E-01
L1, L6	0.3200 E-07	0.5685 E-04
L1, L12, L13	0.1440 E-09	0.1387 E-05
LDB1	0.3380 E-04	0.3495 E-01
LDB1, LDB3	0.2411 E-09	0.2324 E-05

G - Generator
L - Line
LDB - Load on Bus
The numbers 1,2 etc. associated with G, L and LDB represent the number of a generator, a line or a bus.

The adequacy indices were also calculated for a system load of 2850 MW. In this case, the effect of station originated outages is overshadowed by the contribution due to independent outage contingencies. This can be seen from Table VII which gives various system indices for this load level. At this system load, the system indices are dominated by the outages of the generating units. At the 2850 MW load level, the outage of one large generating unit (300 MW or 400

MW) together with the outage of another medium unit results in a capacity deficiency in the system as the system static reserve at 2850 MW is only 555 MW. As seen from Tables V and VII, the severity index due to independent outage events has increased from 71 system-minutes to 3070 system-minutes, while with the station originated outages it increased from 123 system-minutes to 3094 system-minutes.

TABLE VII

Basic Annualized System Indices for the RTS
System Load = 2850 MW

Index	Without Stations	With Stations	% Increment
Bulk Power Interruption Index (MW/MW-Yr)	3.69510	3.71753	0.607
Bulk Power Supply Average MW Curtailment Index (MW/ Disturbance)	167.6755	168.41957	0.443
Bulk Power Energy Curtailment Index (Mwh/MW)	51.16847	51.57207	0.788
Modified Bulk Power Energy Curtailment Index	0.0058411	0.0058872	0.789
Severity Index (System Minutes)	3070.108	3094.324	0.788

This non-uniform increase in the system indices illustrates that the effect of independent outage events and station originated outage events is highly sensitive to the system load at which the indices are calculated. It is not possible to develop specific conclusions about the effects of these outage events for a particular system at specific load levels without conducting a detailed analysis. In addition to the system load and generation reserves, there are also a number of other factors such as system topology, component reliability parameters and components involved in the contingency which play an important role in the contribution of station originated outages in composite system reliability evaluation.

CONCLUSIONS

The inclusion of station originated outage events creates a significant increase in the load point and system adequacy indices, particularly at the lower system load levels. It is, therefore, necessary to examine these events prior to considering the inclusion of higher level independent outage events, particularly line outages. The total computation time does not increase significantly with the addition of station originated outages but can increase tremendously with the inclusion of higher level independent outages. The load at each bus and hence the system load does not remain at a constant value throughout the year. Generally, system load remains at its peak value for only a short time in a practical system. At lower load levels, the effect of station originated outage events on the system adequacy may be

comparable to that of independent outage events. It is, therefore, very important to include the effect of station originated events when calculating the annual indices for a practical system.

The single line diagram for the RTS shown in Figure 2 is a practical extension of the basic configuration shown in Figure 1 and can be used to compare alternate techniques for including station effects in composite system adequacy evaluation.

REFERENCES

1. G.L. Landgren, S.W. Anderson, "Data Base for EHV Transmission Reliability Evaluation", IEEE Trans. PAS-100, 1981, pp. 2046-2058.

2. Bibliography On The Application Of Probability Methods In Power System Reliability Evaluation. IEEE Trans. PAS-91. 1972, pp. 649-660, PAS-97, 1978, pp. 2235-2242; PAS-103, 1984, pp. 275-282.

3. R. Billinton, T.K.P. Medicherla, "Station Originated Multiple Outages in the Reliability Analysis of a Composite Generation and Transmission System", IEEE Trans. PAS-100, 1981, pp. 3869-3878.

4. R. Billinton, J. Tatla, "Composite Generation and Transmission System Adequacy Evaluation Including Protection System Failure Modes", IEEE Trans. PAS-102, 1983, pp. 1823-1831.

5. R.N. Allan, A.N. Adraktas, "Terminal Effects and Protection System Failures in Composite System Reliability Evaluation", IEEE Trans. PAS 101, 1982, pp. 4557-4562.

6. IEEE Committee Report "IEEE Reliability Test System", IEEE Trans. on Power Apparatus and Systems, PAS-98, 1979, pp. 2047-2054.

7. R. Billinton, R.N. Allan, "Reliability Evaluation of Power Systems", Plenum Press, New York and London, 1984.

8. T.K.P. Medicherla, R. Billinton, M.S. Sachdev, "Generation Rescheduling and Load Shedding to Alleviate Line Overloads - Analysis", IEEE Trans. on Power Apparatus and Systems, PAS-98, 1979, pp. 1876-1884.

Discussion

J. Endrenyi (Ontario Hydro, Toronto, Canada): The authors must be commended for a clear and thought-provoking exposition of the necessity for, and a method of, considering station-originated failures in composite system reliability evaluations. The presentation, however, gives rise to some questions. For example, while reviewing the station component outages proposed to be considered, and the effects of such outages in the composite system (both listed on the second page of the paper), it became apparent that really a choice between two possible approaches must be made. One could consider the incorporation of these contingencies into the composite system reliability evaluation process as, in fact, an extension of the composite system to include the station components as well. Alternatively, one could argue that only those station failures be considered which result in outages in the "original" composite system; ie, which cause generator or line outages. Entry (10) in the list of effects (load feeders isolated) would then not qualify for inclusion.

What are the pro's and con's of the two approaches? If the composite system concept is to include stations, really two design activities are combined into one, where reliability improvements can be effected by changes in the transmission network and/or in the station layouts and the protection system arrangements. Yet, these are tasks which are normally handled separately. On the other hand, if station-originated and transmission-system-originated contingencies are kept entirely separate, some significant system events may not be identified at all, such as line outages due to station component failures, which cause other lines to overload. One may make a case for completely omitting station-originated effects from composite system studies and, at the same time, extending station reliability assessments to include the transmission lines supplying the station in question. This would keep the investigations separate according to the origins of failures, but then again, a line outage originating in a station component failure would be difficult to represent if it is relieved by, say, generation redispatch.

It appears that a judiciously chosen part of the station-originated events must be included in composite system studies: those that result in transmission line or generating unit outages. Entry (10) is not an effect of this type and, therefore, it may be logical to remove it from the list. It should be noted that the effect of omitting (10) is not at all negligible; according to the paper itself, the increment in composite system indices is quite significant if station-originated outages are included, and the. "most distinct contributing factor to this increase . . . is the isolation of a load feeder at a bus." It goes without saying that in the arrangement described, composite system studies must be augmented by station reliability studies, as before.

Manuscript received February 26, 1985

M. G. Lauby (MAPP CC, Minneapolis, MN): This paper discusses the important task of evaluation of the effects of the station configuration on the evaluation of adequacy of the reliability test system. Its contribution pertains to both an understanding of the effect of the modeling of these outages and also, addresses the test system itself. Namely, is there enough detail within the current test system to perform adequacy assessment without modeling of the station originated outages. I have the following question.

Many times, in the past, the station originated outages have been ignored in typical adequacy assessment studies [1]. Though one can attempt to model as many of these multiple outages as possible, clearly there is no direct way to enumerate all of them. With the ideas of this paper in mind, one could enumerate all of the station originated outages for each contingency level and use ranking procedures as outlined in [2]. However, this additional analysis may not be required if the final decision would not be affected even if the station originated outages were modeled. Namely, have you performed analysis which compares the resultant bulk power indices between the analysis with station originated outages and those without? Adequacy assessment techniques many times are used to compare, in a relative manner, which network additions of a power system are more adequate. Could modeling of these outages and their effect on the bulk power indices suggest that one option may be more adequate than another, when ignoring them would reverse this decision?

REFERENCES

[1] T. A. Mikolinnas, W. R. Puntel, R. J. Ringlee, "Application of Adequacy Assessment Techniques for Bulk Power Systems," *IEEE Trans. Power App. and Syst.*, Vol. PAS-101, pp. 1219-1228, May 1982.

[2] S. Vermuri, R. E. Usher, "On-Line Automatic Contingency Selection Algorithms," *IEEE Trans. Power App. and Syst.*, Vol. PAS-102, pp. 346-352, February 1983.

Manuscript received March 1, 1985

R. Billinton, P. K. Vohra, and **S. Kumar:** We would like to thank the discussers for their observations and comments. The points raised confirm our own view that station related outages are an important aspect of composite system adequacy evaluation.

The question raised by Dr. Endrenyi regarding two possible approaches can be answered only by considering what is the intention behind the proposed composite system adequacy evaluation. overall system adequacy evaluation can be divided into three hierarchical levels as described in Ref. [1]. The first zone or Hierarchical Level I (HLI) simply involves the generation system and its ability to satisfy the total system load and energy requirements. This is the conventional capacity assessment procedure used by most utilities in North America. Hierarchical Level II is the composite generation - transmission problem addressed by this paper and involves the ability of the system to satisfy the major load points within the system. This is the point raised by Dr. Endrenyi. Composite system adequacy evaluation which does not consider station originated events basically uses a load flow representation at a bus and actually concerns itself not with load points indices but with bus related indices. In the hierarchical sense, these indices are incomplete and cannot be utilized as input values to an HLIII study which includes the distribution network to obtain customer related indices. Entry (10) in the list of effects is therefore very clearly part of this evaluation. Our studies lead us to believe that the only practical way to incorporate station originated events in a composite system adequacy assessment is to insert the multiple outages resulting from station related events as an additional level of data. These events when introduced may lead to an overload situation which is alleviated or responded to by load curtailment using the procedures and policies incorporated in the digital computer program. Entry (10) could be removed from the list if desired but should not be removed if the objective is to conduct a comprehensive assessment at the HLII level. The effect of omitting Entry (10) is not as clear cut as Dr. Endrenyi appears to suggest. The most "distinct" contributing factor is the isolation of the load feeder. This is the most observable factor as it results in total curtailment of the load point. It may have a very low frequency and probability relative to partial curtailment of the load due to line overload alleviation or load shedding due to inadequate capacity. In essence, Entry (10) is easy to see and to consider relative to other events which can impact on the same load point.

With respect to the points raised by Mr. Lauby, the intention of the paper was to conduct and illustrate the adequacy evaluation of the IEEE RTS with and without the inclusion of station originated outages. In order to do this we have augmented the original RTS by adding station configurations. The resulting indices are therefore predicted upon the station configurations selected for each load point. The effect of the configuration selected for each load point is also dependent upon many factors including the load level selected for the analysis. Additonal studies were also done using other station configurations and suggest that the analysis of a station performed in isolation rather than within the system context within which it resides may not be realistic.

REFERENCE

[1] R. Billinton and R. N. Allan, "Power System Reliability In Perspective," *IEE Electronics and Power*, p. 231-236, March 1984.

Manuscript received April 29, 1985

A COMPARISON BETWEEN TWO FUNDAMENTALLY DIFFERENT APPROACHES TO COMPOSITE SYSTEM RELIABILITY EVALUATION

Luigi Salvaderi, MIEEE
Planning Department
Ente Nazionale per l'Energia Elettrica
Enel. Italy

Roy Billinton, FIEEE
Power System Research Group
University of Saskatchewan
Canada

Abstract A wide range of techniques have been proposed for reliability evaluation in composite generation and transmission systems. Many of the proposed methods are basically variations on fundamental approaches to the problem. Over the years, however, it has become obvious that there are in general, basic conceptual differences between the techniques used in Europe, particularly Italy and France and those utilized in North America. This paper briefly illustrates some of the fundamental differences between these techniques for composite system reliability evaluation by application to the IEEE Reliability Test System.

INTRODUCTION

The IEEE Test System (RTS) was established in 1979 [1] as a reference configuration for the purpose of testing and comparing alternate techniques for power system reliability evaluation. The generation and load data and the transmission configuration are described in detail in Reference 1 and are therefore not duplicated in this paper. Only those data which are required to provide a better interpretation of the analysis conducted are introduced. The overall single line diagram for the RTS is shown in Figure 1. The two techniques presented in the paper are conceptually different. The first one utilizes a Monte Carlo approach and is used by ENEL in Italy in its planning practice. The second approach has been developed at the University of Saskatchewan and uses a contingency evaluation procedure. The two methods are representative of other evaluation methods found in Europe and North America respectively. The basic objective of this paper is to illustrate the comparative fundamental differences between the approaches and show the numerical results obtained using a representative test system.

THE IEEE RELIABILITY TEST SYSTEM (RTS)

The complete data for the RTS is given in Reference 1. There are however certain salient features which can be summarized.

Table 1
RTS System Generation Mix

Type	Capacity	Percentage
Nuclear	800 MW	24%
Coal	1274 MW	37%
Oil	951 MW	28%
Hydro	300 MW	9%
Comb-Turb	80 MW	2%
Total	3405 MW	100%

85 WM 071-6   A paper recommended and approved by the IEEE Power System Engineering Committee of the IEEE Power Engineering Society for presentation at the IEEE/PES 1985 Winter Meeting, New York, New York, February 3 - 8, 1985. Manuscript submitted January 27, 1984; made available for printing November 19, 1984.

The base duty capacity is 61% in nuclear and coal fired units. The 28% intermediate duty capacity is oil fired and the peaking capacity is a mixture of hydro and combustion-turbines for a total of 11%. In regard to the siting of the generation capacity, 8 out of the 10 stations are located relatively close to the load points and therefore no large power swings are expected.

There are only two voltage levels in the RTS. The system is geographically divided into two regions. The upper region is at 230 KV while the lower region is at 138 KV. The 230 KV network is used to feed the load in the north-west portion of the network and to transfer power to these loads from the generating stations located in the north-east portion of the system. In these cases, power transfers generally do not exceed 30 miles except for one case in which the 230 KV network is used to transfer roughly 700 MW from the north to the main step-down transformer station located in the middle of the system, approximately 70 miles south. The main transfer from the 230 KV network to the 138 KV network is concentrated at this transformer station. This is for both local load and to supply the lines carrying approximately 10% of the 138 KV load. The average distance in this case is somewhat less than 30 miles.

The system load is supplied at the two voltage levels of 230 KV and 138 KV. Approximately 53% is located on the 230 KV system with the remaining 47% on the 138 KV system. From a geographic point of view, a considerable (40%) portion of the total 53% supplied at the 230 KV level is located in the north-west portion of the system. This corresponds to an approximate area of 28 x 18 square miles. With respect to the 47% load supplied at the 138 KV level, 13% is supplied at one substation while the remaining 34% is nearly equally distributed at 7 other buses.

Method 1 - The ENEL Approach

The "cost of reliability" is one of three components used by ENEL in computing the total system cost. The other two components used in system optimization are the capital and operating costs. The cost of reliability is obtained by developing a risk index in the form of the expected yearly curtailed energy (kwhr/yr). This is then transformed into a cost value using an appropriate conversion factor. A widely accepted value for this unit cost is about 1\$/kwh. In regard to the composite system, the annual energy curtailed can be obtained

1. For the entire system - with respect to the causes ie. static deficiency of generation, transmission or transformation.
2. For each network node where a deficiency occurs.
3. For the weather conditions during which a deficiency occurs.

The basic technique used to accomplish this is the Monte Carlo method. The main advantage of this method is the feasibility of taking into account theoretically any random variable, any contingency and of adopting operation policies similar to the real ones. The disadvantage could be the computing time to be spent. It also allows a better "dialogue" within the source utility between planning and operation departments, since the factors/indices used are nearly

Reprinted from *IEEE Trans. Power Apparatus Syst.*, vol. PAS-104, no. 12, pp. 3486-3492, Dec. 1985.

325

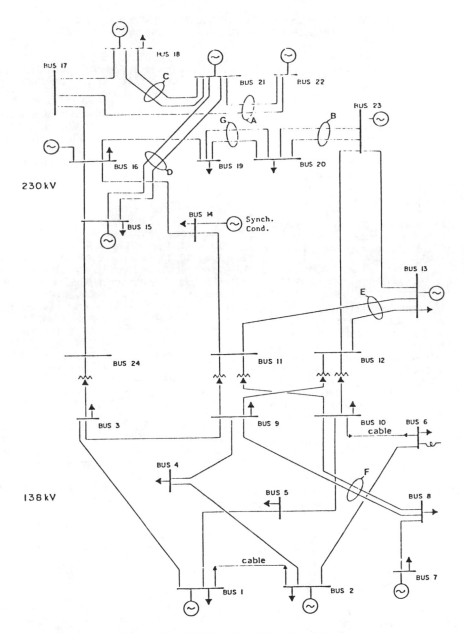

Figure 1. IEEE Reliability Test System

the same. A specified calendar year of system life is examined using repeated "yearly samples" each consisting of 8760 hours which are randomly selected. The computer program (SICRET) has been extensively described in the literature. It has been used for approximately 10 years [3-8] and is currently utilized in network planning of ENEL's system.

The models used for system components and dispatching actions are as follows:

a) <u>Load Model</u> – The real power load for the system and for each load point is determined for each hour. A further subdivision is made at each bus indicating the percentage of load that can be interrupted. No reactive load is considered in the analysis.

b) <u>Environment Model</u> – Three weather conditions can be considered in each homogeneous area of the system. These three conditions designated as fair, bad and stormy weather are considered to randomly alternate during the annual study period.

c) <u>Component Failure Model</u> – Forced outages of generators, transformers, and other network components are represented by two state (up-down) models. The weather conditions are incorporated in the transmission line models.

d) <u>Generating Unit Model</u> – Units are characterized by their ratings, technical minimum loadings and hourly production costs.

e) <u>Operating Policy Models</u> – The operating policies must be similar to those performed in practise but must be sufficiently simple to permit the results to be obtained within reasonable computing times. The following two types, selected by input, can be utilized

    (1) pure safety policy
    (2) mixed economy – safety policy.

The two types differ in the policies adopted for unit commitment and generation dispatch to relieve component overload. Type (1) is

326

utilized when only system reliability is required. Type (2) is utilized when the system operating costs are also required. Of course the corresponding computing times are different.

Program Procedure

The basic steps in the program are as follows:

1. The system load is matched to the total available generation, by curtailing if necessary a portion of the load. This provides an overall risk index due to "lack of real power".

2. If a mixed "economy-safety" policy is selected, the generating units are loaded in accordance with a priority list based on operating costs. Local constraints such as limited energy availability can be included. If the "pure-safety" policy is selected, the available units are loaded in proportion to their available capacity.

3. The system network solution is obtained using a dc load flow. Consequently, the SICRET program is particularly suited to large meshed systems with no voltage/stability problems, as is the case for many systems of Western Europe. Cautions are to be taken in other cases.

4. If the load on a component exceeds its capability, the generation dispatch is modified in an attempt to alleviate the overload. The overload relief is carried out according to the operation policy chosen (mixed economy-safety or pure safety) [7-8].

5. Component overloads above on "alarm level" are tolerated up to a preselected capability limit. If this level is exceeded, certain loads are curtailed until the component loads come within this limit (this gives rise to a risk index due to "component overload").

6. The resulting energy curtailments in kwh/yr are summed for all the 8760 hours examined in the yearly sample. An estimate of the "average annual energy curtailed" is obtained by repeating the analysis for many yearly samples of the same solar year.

7. When applying the mixed "economy-safety" policy and attempting to alleviate component overloads, capacity shifting is performed in such a way that the cost of shifting is a minimum. If the "pure-safety" policy is utilized, generation shifting (re-dispatch) is carried out only to produce maximum influence on the overload components with no attention to the system operating cost.

System Reliability

When the transmission network is assumed to be completely reliable and capable, the overall system characteristics are shown in Table 2.

Table 2
Generation System Adequacy Indices

Annual Peak	= 2,850 MW (at the load points)
Annual Energy	= 15,370 Gwh/yr
Installed Capacity	= 3,405 MW
Capacity Reserve	= 19.5%
Expected Curtailed Energy	= 4,000 MWH/yr

The analysis of overall system reliability has been performed utilizing a dc load flow. The static adequacy of the system is evaluated assuming that the system does not present any voltage problems. The cost figures presented below were obtained using the mixed "economy-safety" policy for generation dispatch and overload relief. Table 3 shows the annual production and running cost of each type of generating unit and for the system(+). The overall annual energy curtailed is also shown, subdivided in accordance with the causes which produced the curtailments.

(+) Fuel prices at Italian costs at 1.1.1983. Rate of change: 1US$ = 1400 lire.

Table 3
Annual Production Costs

Unit Type	Capacity MW	Fuel Cost mill/kwh	Energy Gwh/yr	Cost 10^6/yr
Nuclear	800	8	5,442	43.536
Coal	1274	22.7	8,242	187.093
Oil	951	40.0	1,646	65.840
Hydro	300	–	25	–
Combustion Turbine	80	95.7	14	1.340
Yearly Curtailed Energy				
Lack of generation			4.1	
Lack of capability			0.2	
Sub Total			4.3	
Total	3,405		15,373	297.809

Table 4 shows the yearly curtailed energy subdivided according to the causes and where the curtailments occurred.

Table 4
System Curtailments

Lack of Generation

Bus	Curtailment MWH/yr
18	2,782
16	560
15	270
13	210
7	200
1	65
2	55
8	–
Subtotal	4,142

Overload

Bus	Curtailment MWH/yr
5	66
4	50
19	–
20	–
8	100
7	–
Subtotal	216

Total Energy Curtailment = 4,358 MWH/yr.

The following comments can be made regarding the results and the RTS based upon the analysis conducted.

a) A great unbalance exists between the curtailments due to the lack of generating capacity and those due to overloads.

b) The critical buses are those with load and only one unit installed. In the case of forced outage, unbalance results between generation and load and the necessity of load shedding occurs. This is the case at buses 18 and 16. At bus 13, where three units of 197 MW each are installed, a reasonable generation margin (48%) exists even with one unit out of the three out of service. The likelyhood of load shedding in this case is much lower.

c) The transmission system in RTS appears to be somewhat oversized for the generation capacity involved. Curtailments only occur at buses 8, 4 and 5. If a critical generation situation occurs on the overall network which requires maximum output for the units at bus 7 (300 MW), this entails overload of the radial line between buses 7 and 8.

A logical extension of this study, in the manner normally conducted by ENEL would be to utilize the program to examine system modifications. The results shown in Tables 3 and 4 indicate that considerable load shedding occurs due to the lack of generation. The system generation composition has been tentatively modified by adding a combustion turbine of 20 MW at bus 8. In this case, the annual curtailed energy due to lack of generation drops to 3.7 MWH for a total of 3.9 MWH/yr: the main improvements are obtained in busses 18, 16, 8. Additional studies can be conducted using the program. The Monte Carlo approach is well suited to the determination of expected production costs and energy curtailments occuring in a reference system and under conditions of generation and transmission modifications. The incorporation of curtailed energy costs can be used to examine potential alternatives with the intention of developing an "optimum" configuration. The program is one in a family of programs used by ENEL in system planning. The conducted studies would be used in conjunction with those of load flow, stability, short circuit etc. etc. to arrive at investment decisions.

Method 2 - The U. of S. Approach

The primary objective in this approach is to evaluate adequacy indices for the total system and at every load point in the system [9]. Failure of service at any busbar in the system includes the violation of the minimum acceptable voltage at the busbar and/or failure of the system to supply the total load connected to that busbar after alleviating the line overloads, generator MVAr limit violations etc. Inadequacy of the available generation to meet system load requirements is also considered as a failure at each busbar. The evaluation of a composite system involves the analysis of all possible contingency states. This of course is not possible and therefore the number of contingencies considered must be limited. In addition to considering the independent events associated with the outage of generating units, lines and transformers either singly or in combination with others, common mode outages of multiple elements [10] or multiple outages due to station originated events [11] can be included in the list of contingencies studied.

The generating schedules at appropriate generating buses are modified for each generating unit outage case to compensate for the loss of generation. The violation of a specified service quality criterion are

determined for each outage condition by conducting an a.c. load flow analysis. The bulk of the computational time is utilized in the load flow analysis of each outage condition. The fast decoupled load flow technique is utilized in this approach. One problem encountered in composite system adequacy evaluation is the overload of transmission lines and transformers. The simplest solution to this problem is to allow these elements to operate in the overloaded condition. This results in optimistic reliability indices particularly when the line overload is heavy and persists for a long duration. An alternative solution is to remove the overloaded component and continue to analyze the remaining system until no other element is overloaded, or until the total system is failed. This approach results in pessimistic reliability indices which may suggest expensive and unnecessary investment in system improvement. The approach used in this regard [12,13] is to alleviate the line overloads by (a) generation rescheduling and/or (b) curtailment of some of the loads.

When load curtailment is required, the numerical indices calculated at each bus can vary widely when different load curtailment philosophies are used [14]. This makes it difficult to compare the results of one computer program with those of another.

The procedure used in COMREL (U of S program) to curtail load is a flexible, pre-specified approach. The load at each bus is classified into two types, firm and curtailable. In the case of a deficiency in generation capacity, the curtailable load is interrupted first followed if necessary by curtailment of firm load. The regional extent to which a particular capacity deficiency is applied can be varied using three topological classifications (designated as load curtailment pass 1, 2, or 3). The expected load curtailed at the individual bases in the network can therefore change quite considerably with the prespecified mix of firm/curtailable load and the regional constraints in load curtailment.

System Adequacy

The system adequacy indices obtained using this approach are in two forms. Individual bus indices are given in Tables 5 and the overall system indices are shown in Table 6.

Table 5
Annualized Bus Indices for the IEEE RTS
System load = 2850 MW

Bus No.	Failure Probability	Failure Frequency occ/yr	Load Curtailed MW	Energy Curtailed MW-hr
1	0.022446	16.59	171.5	2,086
2	0.040999	30.01	314.6	3,827
3	0.022640	16.73	369.2	4,560
4	0.022394	16.54	172.4	2,133
5	0.022446	16.54	145.2	1,794
6	0.022395	16.54	317.0	3,920
7	0.015922	11.98	160.4	1,905
8	0.015950	12.01	326.3	3,972
9	0.003171	1.98	32.4	425
10	0.003171	1.98	36.6	474
13	0.071273	45.83	1,769.2	23,662
14	0.009556	6.70	150.4	1,793
15	0.056509	35.38	1,961.7	28,069
16	0.026011	18.35	204.9	2,478
18	0.083433	51.51	3,377.9	50,912
19	0.011667	8.05	166.4	2,017
20	0.046213	29.97	853.4	11,792
				145,819

Table 6
Annualized System Indices

Index Description	Index Value
IEEE Indices	
Bulk Power Interruption Index (MW/MW-yr)	3.69
Bulk Power Energy Curtailment Index (MWH/yr)	51.16
Bulk Power Supply Average MW Curtailment Index (MW/Disturbance)	167.67
Energy Index of Unreliability	0.005841
Severity Index (System - minutes)	3070.1

The full range of indices described in Reference 9 can also be obtained.

Independent overlapping outages up to the fourth level for generating units and up to the second level in transmission elements were considered. The effects of higher order generating unit failures were included by using a cumulative probability and frequency for the highest level contingency considered. Common mode or common cause outages were not included in these studies. The indices have been evaluated considering a single step load model. A basic assumption with the single step load model is that the load remains constant for the period of study. In order to calculate annualized indices, the period of study is taken as one year. This load model provides a relative estimate of inadequacy at each load point and for the total system and provides representative indices to assess the sensitivity of the system adequacy to configuration changes. The adequacy indices are very sensitive to the load level. As an example, the Severity Index drops to 71.1 System-minutes at a load level of 2400 MW. A bus failure under any outage condition is defined as including the violation of the acceptable voltage limits at that bus and/or not meeting the load requirements at that bus, generator MVARS limit violations and non-convergent situations etc. The swing bus overload, if any, is alleviated by curtailing the load at various load buses. In these studies, curtailable load at each bus is assumed to be 20% of the total load of the bus and the number of load curtailment passes is specified as one. Line or transformer overload conditions are alleviated by generation rescheduling and/or load curtailment at the buses. Bus 6 experiences load curtailment because of the overloading of the connection between bus 6 and bus 10 when the line between bus 6 and bus 2 is out. A 400 MW generator outage at bus 18 results in unsatisfactory system performance because of generator MVAR limits. The effect of line outages except that of outage of line 5 on bus indices is negligible to moderate. Buses 6, 13, 15, 18 and 20 have high inadequacy indices compared to the indices of the other buses. Buses 3 and 6 also experience voltage violations. The maximum and minimum limits of bus voltages were assumed to be 1.05 per unit and 0.95 per unit respectively.

The outage of the largest generating unit (400 MW) alone does not cause load curtailment but the outage of one 400 MW generator either at bus 18 or bus 21 with other relatively large generating units causes load curtailment at the buses. Bus 18 has the lowest value of adequacy because of many outage combinations of the connected generator with any other relatively large generator in the system. Bus 13 and Bus 20 have low values of adequacy because of the outage of a generator at bus 23 with another larger generator outage in the system. The results can be summarized as follows:

Line overload is not a major problem except for the outages of a few lines in which case the connection between bus 6 and bus 10 is overloaded.

The contribution of line outages to reliability indices is negligibly small.

The major contribution to bus indices is due to generator deficiency in the system when large generating units are out.

COMPARISON

Both of the techniques described in this paper are attempts to evaluate composite or bulk system adequacy. These are, however, conceptual differences in modelling and problem perception between the two techniques. In summary some of the major differences in the models used, the risk indices obtained, the contingencies examined and the load alleviation techniques employed are shown in Table 7.

Table 7
A Comparison of the Major Factors

Factor	ENEL	U of S
Method	Simulation	Analytical
Risk Index	Yearly curtailed energy subdivided according to cause	Range of load point and system indices [9]
Situations Examined	For a fixed year, 8760 hourly samples resulting from the random combination of any load level and any availability of system components	Predetermined contingency level, up to four generating units and two transmission lines, at a specified load level
Load Model	For the year considered any load level (8760 load levels)	A specified load level
Generation Analysis	Planned maintenance of all units, predetermined for the year	Not included
Risk due to lack of generation	First component evaluted by curtailing load in accordance with a priority list	Curtailment by a firm/curtailable load policy and regional bounds
Initial Dispatch	Priority list - based on running costs	Pre-specified
Network analysis Load Flow	D.C. load flow, voltage limits not included	AC load flow, voltage, VAr limits included
Overload relief policy	D.C. load flow, overloads relieved using a "coefficient of influence" policy	alleviation at close proximity points [12]

The results shown in Tables 3-4 and Tables 5-6 cannot be compared directly primarily because of the difference in load models used in the two studies. This difference is quite obvious. In order to recognize other differences and to facilitate a direct comparison, several specific conditions were examined as follows:

Case 1

The load was held constant at 2850 MW for the entire year and no planned maintenance was considered.

ENEL Result
Expected energy curtailed = 125,215 MWh
U of S Result
Expected energy curtailed = 145,819 MWh

The expected energy curtailed is higher in the U of S calculation due to load curtailment resulting from line overload alleviation arising from A.C. load flow representation. This is illustrated specifically in Cases 2 and 3.

Case 2

The load was held constant at 2850 MW for the entire year. All network components were in service. One 400 MW unit at bus 18 and one 197 MW unit at bus 13 on forced outage.

ENEL Result
The overall available capacity is lower than the load to be supplied. Load shedding is necessary due to lack of generating capacity. No line overloads.
U of S Result
Load is curtailed due to inadequate generation capacity in accordance with the prespecified policy. No line overloads occur.

Case 3

The load was held constant at 2850 MW for the entire year. One 400 MW unit at bus 18 and one 197 MW unit at bus 13 on forced outage. Line between bus 2 and 6 removed from service.

ENEL Result
As in Case 2. No overload conditions determined by the D.C. load flow.
U of S Result
In addition to load shedding due to generation inadequacy, load curtailment occurs at bus 6 due to overloading of the line between bus 6 and bus 10.

The three cases considered illustrate some of the fundamental differences between the two methods. In Case 1, the expected energy curtailed is basically the same and attributable largely to generation deficiencies. It should be appreciated that the load bus components of the curtailed energy will be very dependent on the load curtailment philosophy employed. The additional component in the U of S result arises as shown in Cases 2 and 3 due to load alleviation to remove difficulties perceived by the A.C. load flow.

CONCLUSION

This paper has illustrated two conceptually different techniques for composite generation and transmission system adequacy evaluation. The first approach using Monte Carlo simulation is typical of the methods employed in Europe. The second approach uses a contingency enumeration approach and is representative of a number of techniques developed in North America.

The calculated results shown in this paper for the RTS using the two methods are not directly comparable but the differences are assignable. The major difference is due to the load models employed. The solution techniques, network representation and load curtailment policies employed all introduce further assignable differences in the calculated indices. In the case of the RTS, the load reduction philosophy utilized in the case of line overload and inadequate installed capacity is a major factor. The calculated individual load point adequacy indices will vary widely with the load curtailment policy employed. The calculated indices are relative values which respond to various factors that influence the adequacy of the system. They can be used to consider planning alternatives and assess the sensitivity of the system to configuration changes.

Acknowledgement

L. Salvaderi expresses his gratitude to Mrss. P.I. De Rossi and I. Facchinei of ENEL for the help offered in running the cases examined with the Monte Carlo method. R. Billinton expresses his appreciation to S. Kumar for his assistance in the analysis of the RTS using the COMREL program.

REFERENCES

1. Task Force of the IEEE Application of Probability Methods Subcommittee, "IEEE Reliability Test System" IEEE Transactions PAS, Vol. 1 98-6, Nov/Oct. 1979.

2. G. Manzoni, P.L Noferi, L. Paris, M. Valtorta, "Power System Reliability Indexes", Nato Advanced Study Institute of Generic Techniques in System Reliability Assessment, University of Liverpool, 1973.

3. P.L. Noferi, L. Paris, "Quantitative Evaluation of Power System Reliability In Planning Studies", IEEE Transactions PAS, Vol. 91, No. 2, March/April 1972.

4. IEEE Working Group Report, "Reliability Indices For Use In Bulk Power Supply Adequacy Evaluation", IEEE Transactions PAS, Vol. 97, No. 4, July/Aug. 1978.

5. P.L. Noferi, L. Salvaderi, "A Set Of Programs for the Calculation of the Reliability of a Transmission System", UNIPEDE Informatique Conference, Lisban, 1971.

6. P.L. Noferi, L Paris, L Salvaderi, "Monte Carlo Methods for Power System Evaluation in Transmission or Generation Planning", Proceedings 1975 Annual Reliability and Maintainability Symposium, Washington, 1975.

7. L. Salvaderi, I. Facchine, E. Gampiere, S. Scalcino, "Network Reliability Evaluation: Peculiar Problems to be Faced when Planning Bulk Meshed Transmission Systems", VII PSCC Conference, Lausanne, 1981.

8. J. Endrenyi, P. Albrecht, R. Billinton, G.E. Marks, N. Reppen, L. Salvaderi, "Bulk Power System Reliability Assessment - Why and How?" IEEE Transactions PAS-101, Sept. 1982, pp. 3439-3456.

9. R. Billinton, T.K.P. Medicherla, M.S. Sachdev, "Adequacy Indices For Composite Generation and Transmission System Reliability Evaluation", IEEE Winter Power Meeting, February 1979, Paper No. A 79 024-1.

10. R. Billinton, T.K.P. Medicherla, M.S. Sachdev, "Application of Common Cause Outage Models In Composite System Reliability Evaluation", IEEE Transactions PAS-100, July 1981, pp. 3648-3659.

11. R. Billinton, T.K.P. Medicherla, "Station Originated Multiple Outages In the Reliability Analysis of a Composite Generation and Transmission Systems", IEEE Transactions, PAS-100, August 1981, pp. 3869-3879.

12. T.K.P. Medicherla, R. Billinton, M.S. Sachdev, "Generation Rescheduling and Load Shedding to Alleviate Line Overloads - Analysis", IEE Transactions PAS-98, Nov./Dec., 1979, pp. 1876-1885.

13. T.K.P. Medicherla, R. Billinton, M.S. Sachdev, "Generation Rescheduling and Load Shedding to Alleviate Line Overloads - System Studies", IEEE Transactions PAS-100, January 1981, pp. 36-43.

14. R. Billinton, T.K.P. Medicherla, "Overall Approach to the Reliability Evaluation of Composite Generation and Transmission Systems", IEE Proc. Vol. 127, Pt. C, No. 2, March 1980.

Discussion

M. G. Lauby, (MAPP CC, Minneapolis, MN): This paper discusses the different approaches that are implemented on a single system and the final results. It is very important work as these differences which can influence the way a system may be designed should be recognized, as well as the test system should be investigated for its apparent strengths and weaknesses. I have the following questions.

Could the authors elaborate more on the type of contingency enumeration program that was employed in the University of Saskatchewan approach? Were contingency selection techniques employed? Could you elaborate further on the final indices that are calculated by these two methods and the deterministic meaning of these indices? Additional insights in the approaches of these two compared methods as well as the reliability test system would be greatly appreciated.

Manuscript received March 1, 1985.

L. Salvaderi and **R. Billinton:** We would like to thank Mr. Lauby for his discussion and would like to offer the following comments.

The COMREL program developed at the University of Saskatchewan does not use a contingency ranking approach. Independent simultaneous outages of generating units, transmission lines and transformers are considered up to certain fixed levels. Independent outages of generating units are considered up to the fourth level and transmission lines and transformers are considered up to the second level. A probability of occurrence cut-off can be used if desired to curtail the contingencies investigated.

A primary emphasis in the COMREL program is the ability to calculate individual load point indices and also overall system indices. These indices do not replace each other but are complementary and necessary in an overall appreciation of system adequacy. The calculation procedure for the range of indices produced by COMREL is described in detail in the text book *"Reliability Evaluation of Power System"* by Billinton and Allan.

The COMREL program therefore provides a detailed system specific analysis. Common mode outages and station originated major component (generating unit, transmission lines, transformer) outages can be included in the evaluation. The ENEL program provides a much more general analysis and is part of an overall approach to assess total system cost.

A recent study conducted for the Power System Reliability Subsection of the Canadian Electrical Association compares various existing computer programs developed for adequacy evaluation of composite systems. A paper [1] on this subject was presented at the 1985 Spring Meeting of the CEA Engineering and Operating Division and provides additional insight on the different approaches used in a number of available programs.

REFERENCE

[1] R. Billinton, and S. Kumar, "Adequacy Evaluation of a Composite Power System — A Comparative Evaluation of Existing Computer Programs," Power System Reliability Subsection CEA Spring Meeting, March 1985, Montreal. To be published in the 1985 CEA Transactions.

Manuscript received April 29, 1985.

REQUIREMENTS FOR COMPOSITE SYSTEM
RELIABILITY EVALUATION MODELS

M. P. Bhavaraju, Senior Member
Public Service Electric & Gas Company
P.O. Box 570, T14A
Newark, NJ 07101

P. F. Albrecht, Fellow
General Electric Company
One River Road
Schenectady, NY 12345

R. Billinton, Fellow
Dept. of Electrical Engg.
University of Saskatchewan
Saskatoon, Sask., S7J 2R6 Canada

N. D. Reppen, Senior Member
R. J. Ringlee, Fellow
Power Technologies, Inc.
P.O. Box 1058
Schenectady, NY 12301

Abstract - This paper describes the requirements of composite generation and transmission system reliability evaluation methods for use in power system planning. This work is based on inputs obtained from utility system planners. The paper concentrates on philosophy and purpose of composite system reliability evaluation rather than on the modeling details. This work is a part of an on-going Electric Power Research Institute sponsored project.

INTRODUCTION

Quantitative reliability evaluation using probability methods is widely accepted for evaluating installed generating capacity adequacy in planning power systems [1,2,3]. Quantitative methods provide indices of reliability which are sensitive to factors that influence reliability. These indices can be compared with accepted criteria for making decisions on adding system facilities. For example, a loss of load expectation of one day in ten years is a widely used criterion. Many computer programs are available to calculate loss of load expectation. These programs use a variety of input data assumptions and calculation algorithms.

At present, there is a significant interest in utilities in extending quantitative reliability evaluation models to include bulk transmission facilities. This can be done in several ways. For example, two area and multi-area generating capacity reliability programs include inter-area transfer capability in megawatts as inputs. These programs calculate area reliability considering the possible emergency capacity assistance between areas. Other programs include full transmission network representation and use dc or ac load flow analysis under outage states (contingencies). These network-based programs must include generating units at buses as basic input to define load flows. However, it appears that most of these programs consider only transmission outages as contingencies, and therefore, they are referred to as transmission reliability programs. Composite system reliability evaluation refers to assessments that consider both generation and transmission contingencies. Some of the existing network-based programs do have limited capability to model generating unit outages. However, there are several concerns in the industry regarding the limitations of the existing programs

87 WM 022-7   A paper recommended and approved
by the IEEE Power System Engineering Committee of
the IEEE Power Engineering Society for presentation
at the IEEE/PES 1987 Winter Meeting, New Orleans,
Louisiana, February 1 - 6, 1987. Manuscript submitted
August 29, 1986; made available for printing
November 13, 1986.

for composite system reliability evaluation. For example, generation contingencies are not considered accurately (sufficient "depth" and VAR representation) and efficiently (in terms of computation time). Thus, "bulk power system" reliability programs could be defined to include: (1) multi-area reliability programs (generating unit outages, no transmission outages, no load flows); (2) transmission system reliability programs (load flow analysis, transmission outages, no generating unit outages); (3) composite system reliability programs (load flow analysis, generation and transmission outages).

In the network-based programs (2 and 3 above), failure is defined in terms of line overloads and unacceptable bus voltage levels, and load curtailment needed to alleviate these conditions. Using the probability (or frequency) of contingencies that could cause these conditions, probability (or frequency) of different unacceptable conditions are computed as indices of reliability. A major difficulty in developing these programs has been to analyze a sufficient number of contingencies within a reasonable computation time in order to have "confidence" in the calculated reliability indices. This is particularly true when generation outages are included. The comprehensiveness of the load model also becomes important when period indices are to be calculated. In addition to the modeling and computational difficulties, much work is required in the industry to understand the philosophy of these approaches and the applications of the computed reliability indices before these methods can supplement - or replace conventional (deterministic) methods.

Previous efforts in considering bulk transmission in reliability evaluation include the computer program COMREL developed by University of Saskatchewan [4,5], PCAP by Power Technologies, Inc. [6], GATOR by Florida Power Corporation [7], and RECS by Georgia Power Company [8]. Recent efforts include EPRI Project RP1530-1 which resulted in the computer code SYREL, and project RP1530-2 which was initiated to develop a production grade computer code for studying large bulk transmission systems.

This paper summarizes the broad concerns of utilities and identifies general patterns in the responses obtained from utility interviews. In addition, this paper discusses the general scope and requirements of composite system reliability methods in system planning.

UTILITY SURVEY

Since the primary users of composite system reliability evaluation programs are expected to be electric utility system planners, it was considered important to seek inputs from utility planners in

Reprinted from *IEEE Trans. Power Syst.*, vol. 3, no. 1, pp. 149-157, Feb. 1988.

this work. The inputs from utilities were obtained using a Discussion Questionnaire. The questionnaire was prepared to list the issues of concern, and in most cases the questionnaire was given to the utilities at a meeting after a discussion of the issues. The questionnaire was also mailed for response in a small number of cases.

Existing Applications

The status of existing applications of bulk system reliability can be divided into four categories:

1. Some utilities are actively using probabilistic tools for bulk transmission system planning (e.g., Florida Power Corporation, Georgia Power Company, Ontario Hydro, Mid-continent Area Power Pool). These utilities have made a number of suggestions on the improvements needed in these tools.

2. Some utilities have reviewed the existing tools and concluded that they are not satisfactory. They have decided to use alternate approaches. For example, one utility is developing detailed probability distributions of transfer capability for use in making decisions on capacity transactions. One utility decided to use deterministic type studies but consider probabilities of multiple outages in applying the planning criteria. Another utility essentially decided not to use probability approaches. It feels that the deterministic approaches provide the minimum criteria for planning and the resulting systems have relatively high reliability. Probability approaches may suggest reduction of reliability levels, and the utility would be reluctant to relax the current planning criteria which were based on many years of experience.

3. Some utilities are actively looking for probabilistic tools and are following the industry developments. For example, these utilities have a strong interest in EPRI Project RP1530-2 "Bulk Transmission System Reliability Evaluation for Large Scale Systems."

4. Some utilities are not using any probabilistic tools other than for generating capacity reliability evaluation and they do not indicate strong interest in probability approaches for composite generation and transmission systems.

Those who are currently using probability methods for bulk transmission planning have identified the following applications:

- to evaluate transmission adequacy at distribution points
- to compare alternate facilities for area transmission planning; to justify transmission facilities
- to evaluate the effect of transfer capability on system reliability
- to evaluate benefits and costs of reliability improvement for use in decision making

Difficulties With Current Methods

The various reasons cited by respondents for not using probability methods for bulk transmission planning were as follows:

-- Difficulty in interpreting the expected load curtailment as calculated by existing models. Respondents felt that the load curtailment index as calculated by the models is not based on realistic operating practice. These respondents felt that the probability (or frequency) of operating problems (e.g., not being able to meet security dispatch) would be more meaningful.

-- System facilities are needed today to achieve economic operation or to meet new constraints such as wheeling for third parties. Hence, reliability is not the primary concern to justify the need for facilities.

-- Various difficulties with the current computer programs (e.g., not adequate for specific utility needs, too complex or cumbersome to use, not available for the available computer system, lack of awareness about what the programs can do, programs available in public domain do not have good support, and lack of man-power in utilities to try new methods).

-- Lack of transmission outage data.

Motivation for Composite System Reliability Evaluation

The questionnaire listed applications where composite system reliability evaluation could provide a more consistent analysis. Respondents indicated different applications and some indicated all the applications on the list.

These are listed in the order of significant interest:

- Need to consider internal transmission limitations in generating capacity reliability evaluation.

- Need to optimize relative investments in generation and transmission systems.

- Increased trends in resource sharing (economy transfers) and strategic planning (for coal strikes, etc.).

- Need for better representation of generation effects in transmission system reliability analysis.

- Increased transmission constraints on economic dispatch.

- Dispersed generation (battery storage, cogeneration) in the future systems. Also included in this category is the effect of load management on area transmission needs.

Adequacy vs. Security Assessment

A question was asked to find out if there is a clear role for adequacy assessment methods since security is generally a more significant concern in bulk system planning (the terms adequacy and security are discussed in Reference 9). Most responded that adequacy assessment should have a role in bulk system planning studies. One utility stated that probabilistic adequacy assessment is not a new concept since it is essentially a contingency load flow analysis which has been a common tool in bulk system planning.

333

Several respondents expressed a significant interest in developing adequacy assessment methods which are realistic. This implies that, ideally, assessment methods should recognize the sequence of events in system operation and time available for operator actions. Generator response times, outage durations which vary depending on the type of fault (not average duration), and applicable ratings should be recognized in evaluating the operator actions including load curtailments. The present adequacy methods cannot handle such details since they are based on analysis under contingency "existence" with no recognition of the sequence of outage occurrences. Because of this limitation, the credibility of the load curtailment indices calculated by the current adequacy methods was questioned by respondents.

Security was identified in the survey as an important issue. This was interpreted by respondents in two ways. One interpretation was the modeling of system dynamic performance under sudden disturbances. Utilities generally agreed that adequacy methods cannot include dynamic performance models. At best, the ratings used in the adequacy models could be based on limits to avoid dynamic performance problems (these limits should be determined from other studies). The second interpretation was modeling of security dispatch in the adequacy evaluation models. There was interest expressed in determining a reliability index as a probability of not being able to meet the security dispatch. One respondent suggested that facilities should be added if security dispatch cannot be maintained. If security dispatch is satisfied, then the addition of facilities should be based on economics (minimizing costs to customers).

This important concern regarding economics was expressed by several respondents but in different ways which can be summarized as follows:

Adequacy has been historically defined as the ability of the system to supply load considering component outages. However, adequacy should be assessed in a broader sense considering the ability to supply load "economically" under component outages. This has led to interest in using the adequacy assessment methods to determine the expected penalty due to deviating from economic dispatch or due to reducing economic transfers. In fact, some would prefer to use these penalties as indices of reliability rather than to use expected load curtailment because load curtailment is a rare event.

Reliability Indices

Respondents identified one or more of the following as good applications for composite system reliability indices:

- Justify addition of facilities
- Compare expansion alternatives
- Use to modify/defend deterministic criteria
- Use to communicate reliability issues to the management and the regulatory agencies
- Quantify reliability trends
- Evaluate benefit/cost ratio of reliability

Several suggested that the reliability index should be based on events which can be expected to be observed over a short period (e.g., one or two years). This implies that a reliability index based

on operating problems would be a good index. This also implies that reliability indices based on bulk system load curtailment events which occur infrequently may not be satisfactory. Some respondents, however, emphasized that load curtailment indices are essential even if the frequency associated with load curtailment is very low. These individuals feel that the load curtailment is the "bottom line" in adequacy evaluation and justification of facilities should be based on load curtailment indices.

Some respondents suggested that indices of very severe events and major disturbances would be useful even if their frequency of occurrence is extremely low.

Methods and Computational Tools

Several respondents indicated interest in a large system size (1500 bus), ac solution, recognition of security dispatch, representation of appropriate transfers and a comprehensive load model in a computer program for composite system reliability evaluation. There was equal interest in computing alert state, emergency and load curtailment indices. Significant interest was expressed for indices based on major disturbance events, in particular, cascading line outages and voltage collapse.

Use of "comprehensive" (to include all parameters) models could provide an index of reliability which is close to "observed" reliability, and such an index can be referred to as an "absolute" index. In contrast, known simplifications in the models can provide a "relative" index which "tracks" true reliability but with a consistent bias. A "relative" index may be sufficient for making certain decisions. To the question on preference of "absolute" vs. "relative" indices, the majority of responses indicated "relative" indices to be the most important. However, there was sufficient interest in developing "absolute" indices in the future.

The responses to a question on the acceptable level of effort varied over a wide range, as follows:

Acceptable level of effort for:
- Outage data collection and analysis   10-60 man-days/yr.
- Data preparation, analysis and setting up cases   30-130 man-days/yr.

Computational effort for one year simulation   10-30 CPU minutes (IBM 3083)

Respondents from large utilities stated that they would be willing to spend considerable effort (several man-years) and computer time if the results produced by any new method proves to be very useful. These respondents would consider probabilistic reliability evaluation similar to other typical system planning studies for which considerable effort is spent, once they are convinced of its usefulness.

A PROPOSED SCOPE FOR COMPOSITE SYSTEM RELIABILITY EVALUATION

Composite system is defined as the combined generation and transmission system. In the industry, it is more commonly referred to as bulk power electric system or simply bulk power system. It is the aggregate of electric generating plants, transmission lines, and appurtenant equipment.

Composite system reliability is a measure of the performance of the bulk power system. The term "reliability" refers to the degree to which that performance affects the continuity of power supply to customers. Specific definitions have been developed by the North American Electric Reliability Council (NERC) for "reliability" and the basic functional aspects "adequacy" and "security" for achieving reliability [9].

Several operating states can be defined in terms of the degree to which adequacy and security constraints are satisfied in a power system. The classification of operating states discussed in References [10,11] included five states--normal, alert, emergency, extreme emergency and restorative. It is typical in power system operations to take certain remedial actions while in alert or emergency states in order to avoid the extreme emergency (system separation or islanding) state. These actions include off-economic dispatch (also off-economic transfers) and controlled load curtailment (in contrast to uncontrolled load curtailment which could occur in the extreme emergency state). The original five and these additional states and also a state to identify local load curtailment due to distribution facility outages are shown in Figure 1.

Existing composite system reliability evaluation models typically include emergency and controlled load curtailment states. Frequency of the emergency (circuit overloads, low voltages) states is referred to as a "system problem" index, and controlled load curtailment to alleviate the "emergency" conditions as a "severity" index [12]. As utility industry respondents expressed strong interest in quantifying the risk of alert and off-economic dispatch states, it is proposed that these two states be included in the scope of future composite system reliability evaluation. The controlled load curtailment state should also include load curtailment required to transfer the system from the alert to the normal state in addition to load curtailment to alleviate "emergency" conditions. The authors feel that it would be impractical to include the extreme emergency state in the scope of future composite system reliability evaluation as the required analysis involves dynamic performance simulations for a large number of contingencies.

INDICES FOR COMPOSITE SYSTEM RELIABILITY EVALUATION

There are various aspects that should be considered in selecting indices for composite system reliability evaluation as follows:

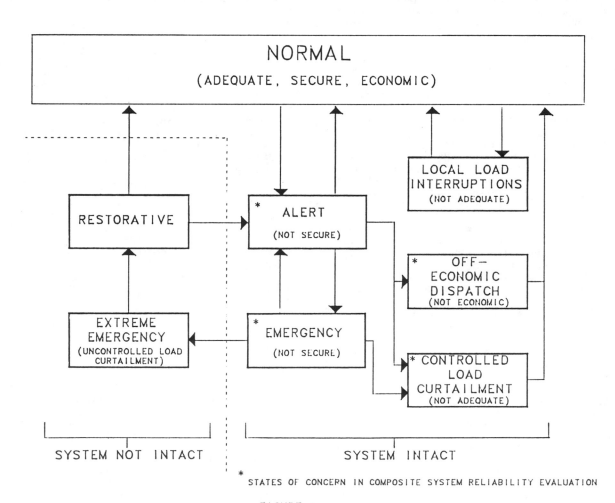

FIGURE 1
SYSTEM OPERATING STATES

System Failure Definition

The three conditions--alert, emergency and controlled load curtailment--were identified as being of equal interest to utilities in defining system failure. Alert and emergency (see Figure 1) can be referred to as "system problems" and controlled load curtailment as a measure of severity of system problems. The condition of off-economic dispatch is of significant interest to utilities, and it can also be defined as a system problem. Its severity can be expressed as a production cost penalty in a given period, however, the estimate may not be accurate unless good load models and production cost models are used.

Since it is relatively simple to identify several system problem conditions from the load flow analysis, it is proposed that the composite system reliability frame work include all the following conditions:

- Alert state (potential overload, low voltage, high voltage for specific contingencies)
- Actions without economic penalty (phase shifter adjustments, switch capacitors/reactors, line sectionalization, load transfer, interruptible load curtailment)
- Actions with economic penalty (generation redispatch, reduction in economy transfers)
- Actions affecting service to customers (voltage reduction, controlled load curtailment)
- Isolation of bus or buses due to contingencies

Alert state can be identified in terms of potential overloads, low and high voltages separately. Similarly, the various remedial actions can be identified separately to provide indices.

Reliability Measures

After defining the system failure and specifically identifying the conditions of interest for a study, there are several statistical measures that can be used to provide indices for these conditions. Not all measures are meaningful for every condition. These measures are listed below:

- frequency, occurrences/year
- average duration, hours/occurrence
- probability (for a given load level)
- severity, MW curtailed/occurrence
- unsupplied energy, MWh/year
- customer interruptions, number/occurrence, or number/year
- production cost penalty, dollars/year

The effort would be significantly greater to compute severity measures such as MW curtailed, unsupplied energy and production cost penalty as compared to computing the frequency measure alone.

Bus, Area and System Indices

The system problem indices discussed above are basically system indices. The alert state and the remedial actions needed to restore the system to the normal state (or to avoid the emergency state) can be expressed as system indices. However, some of these actions would affect only some load points in which case bus indices could be calculated. Specifically, controlled load curtailment would result in partial or complete loss of load at selected buses and indices can be calculated for these buses. Isolation of a bus or buses without generation would also result in complete loss of load at these buses and indices can be calculated for these buses. Thus, depending on the system failure condition, system and/or bus indices can be calculated.

An area index can also be defined to include a selected group of buses. The indices at these buses can be accumulated to calculate the area index. Voltage reduction as a remedial action may be applicable to an area or the entire system, and either an area or system index can be appropriately calculated.

Variation in weather is an important factor affecting the outage rates of transmission components. Weather models are typically included in the reliability studies involving transmission facilities. It is relatively straight forward to compute reliability indices (system, area or bus) separately for normal weather and adverse weather. This information is useful if customer perception of interruptions is different depending on the weather. Lower reliability may be tolerated by some customers during adverse weather. In such cases, the system planner may want to use different criteria for normal and adverse weather conditions rather than using the total index. Also, the criteria to define the alert state may be different during adverse weather. In this case, a separate index for the alert state may be calculated during adverse weather.

ROLE IN SYSTEM PLANNING

The studies required to determine facility additions in power system planning can have the following five broad objectives:

- Generation reserve objectives
- Transfer capability objectives (Interconnection requirements)
- Internal Transmission Reinforcements
- Generator siting
- Generation and Transmission Investment Trade-offs

Each of the categories has a subset of specific studies with different purposes. These various studies may require different models or different modeling details. The five basic studies, the alternates and the generic models required for these studies are illustrated in Table I.

The potential applications of a Composite System Reliability (CSR) Evaluation Model can be recognized in Table I. The basis for this table is that CSR models are needed only in studies where the complete interaction of generation and transmission systems are needed. It is suggested that in studies involving local area reliability needs or economic dispatch for a given generation system, Transmission System Reliability (TSR) models (which assume 100% generation availability) are sufficient. Also, in generating capacity reliability studies with interconnection representation Two or Multi-area Reliability (MGCR) models may be adequate.

Studies with objectives to minimize revenue requirements require Production Cost (MPC) and Optimum Generation Mix (OGM) models which are generally beyond the scope of reliability models. It can be argued that the study of transmission needs to improve economic dispatch should belong to economic

TABLE I
ROLE OF COMPOSITE SYSTEM RELIABILITY EVALUATION IN SYSTEM PLANNING STUDIES

STUDY OBJECTIVE	SPECIFIC STUDIES	GENERIC MODEL REQUIRED*
o Generation Reserve Objectives	-- Determine generating capacity reliability	GCR
	-- Determine economic generation reserve requirements	OGM
o Transfer Capability Objectives	-- Determine transfer capability requirements for reliability	MGCR
	-- Determine transfer capability requirements for economic transfers or diversity exchange	MPC
o Internal Transmission Reinforcements	-- Determine long range transmission needs	CSR
	-- Justify specific transmission line additions (investment decision)	CSR
	-- Compare transmission line addition alternatives	CSR
	-- Determine transmission needs to meet area reliability	TSR
	-- Determine transmission needs to improve economic dispatch	TSR
	-- Determine transmission needs (savings) with dispersed generation/storage devices	CSR/TSR
	-- Determine effects of load management on transmission system	TSR
o Generator Siting	-- Determine sites for large central generators	CSR
	-- Determine sites for dispersed generators/storage devices	CSR
o Generation and Transmission Investment	-- Determine trade-off between generation and interconnections	MGCR
	-- Determine trade-off between generation and internal transmission	CSR

*Legend for Models

GCR	- Generating Capacity Reliability	MPC	- Two or Multi-area Production Cost
OGM	- Optimum Generation Mix	CSR	- Composite System Reliability
MGCR	- Two or Multi-area Generating Capacity Reliability	TSR	- Transmission System Reliability

---

rather than reliability goals. Such a study requires a reasonably accurate estimate of the long range economic dispatch penalty and a trade-off between this penalty and installing transmission facilities. Typical TSR or CSR studies may not justify the use of good production cost models with a multiple year study period. However, a strong interest in such economic assessments was expressed in the utility survey. It appears that separate tools (which already exist in the industry) to evaluate production cost penalties should be used. These models include detailed load representation (e.g. chronological), generating unit heat rates models and fuel costs. It would be impractical, however, to include rigorous outage models which are typically included in TSR and CSR models in such production cost models.

MODELING NEEDS

The foregoing discussion leads logically to the following modeling needs for composite system reliability evaluations:

- Static assessment capability for generation and transmission.

- System problem and capability options with provision for remedial action, generation and phase shift or dc line flow control and load curtailment.
- Linearized network models, dc load flow solution, or in the ultimate, ac network models.
- Line limits defined as flow limits, but determined by consideration of both thermal and angle related (voltage and SIL) limits.
- Transformer limits defined by long term and contingency load carrying ability.
- Security constrained dispatch capability for system problem type evaluations.
- System load models capable of representing seasonal or discrete load level, bus loads determined by specification of load distribution factors.
- Load and resource models capable of meeting requirements for hydro and energy limited generation.
- Weather and station dependent effects for line and transformer outages.
- Generating unit capability and outage models with provision for multiple capacity state representation.
- Generating unit maintenance schedules.

It may be impractical to develop a model that includes all these features. More than one model may be needed depending on the applications. For example, if operating constraints, outage dependencies and hydro and energy limitations are important, a Monte Carlo simulation may have a partial or full role in the model. The detailed modeling requirements are beyond the scope of this paper.

OBSERVATIONS AND CONCLUSIONS

1. There has been considerable activity in the area of composite system reliability evaluation as seen in the available technical publications, however, only a small number of utilities are using the reported methods in their planning studies.

2. Utilities (except those who have developed their own programs to suit their specific needs) are not generally satisfied with the methods available in the industry for composite system reliability evaluation.

3. It appears that the available computer programs for composite system reliability evaluation are essentially being used for area transmission reliability and transfer capability evaluations because of unsatisfactory or inefficient (computationally) modeling of generation in these programs.

4. It appears there is a strong need to communicate with the industry on the potential use and applications of composite system reliability methods for practical problems while these methods are being developed under research projects or by consultants. This would assure a better match between the industry problems and the methods.

5. Combined generating unit and transmission line outage events could contribute significantly to the unreliability of the composite system, and this is the primary motivation for composite (combined) generation and transmission system reliability evaluation.

6. Composite system reliability evaluation should consider events leading to alert, emergency, controlled load curtailment and forced off-economic operation states and isolation of one or more buses.

7. Operating (security) constraints must be included in the composite system reliability evaluation scope; however, it is impractical to include system dynamic performance evaluation to identify the potential extreme emergency state.

ACKNOWLEDGEMENTS

This work has been funded by the Electric Power Research Institute (EPRI). The guidance and encouragement provided by N. J. Balu-EPRI, J. Endrenyi-Ontario Hydro, E. A. Schwerdt-Northeast Power Coordinating Council, R. E. Vandello-Iowa-Illinois Gas & Electric Co., and C. C. Young-Pacific Gas & Electric Co., are acknowledged. The authors are also grateful to the utilities for providing their valuable inputs to the survey.

REFERENCES

1. Bibliography on the Application of Probability Methods in Power System Reliability Evaluations, IEEE Committee Report, IEEE Trans. on Power Apparatus and Systems, PAS-91, 1972, pp. 649-660.

2. Bibliography on the Application of Probability Methods in Power System Reliability Evaluation, 1971-1977, IEEE Committee Report, IEEE Trans. on Power Apparatus and Systems, PAS-97, 1978, pp. 2235-2242.

3. Bibliography on the Application of Probability Methods in Power System Reliability Evaluation, 1977-1982, R. N. Allan, R. Billinton and S. H. Lee, IEEE Trans. on Power Apparatus and Systems, PAS-103, 1984, pp. 275-282.

4. Transmission Planning Using a Reliability Criterion - Part I - A Reliability Criterion, R. Billinton, M. P. Bhavaraju, IEEE Transactions PAS-89, No. 1, January 1970, pp. 28-34.

5. Composite Generation and Transmission System Reliability Evaluation, R. Billinton, T. K. P. Medicherla, IEEE Paper A78 237-0.

6. Program for Composite Bulk Power Electric System Adequacy Assessment, P. L. Dandeno, G. E. Jorgensen, W. R. Puntel, R. J. Ringlee, Transactions of the IEEE Conference on Reliability of Power Supply Systems, IEE Conference Publication No. 148, February 1977.

7. A Method of Combining High Speed Contingency Load Flow Analysis with Stochastic Probability Methods to Calculate a Quantitative Measure of Overall Power System Reliability, G. E. Marks, IEEE Paper A78 053-1.

8. Bulk Power Reliability Assessment with the RECS Program, A. P. Meliopoulos, A. G. Bakirtizis, Proceedings PICA, May 1985, pp. 38-46.

9. Reliability Concepts in Bulk Power Electric Systems, North American Electric Reliability Council, Reliability Criteria Subcommittee, February 1985, p. 8.

10. L. H. Fink and K. Carlsen, "Operating Under Stress and Strain," IEEE Spectrum, March 1978, pp. 48-53.

11. Reliability Indexes for Power Systems, Final Report, Project RP1353-1, EPRI Report EL-1773, March 1981, p. 4-10.

12. Transmission System Reliability Methods, EPRI Final Report EL-2526, July 1982.

Discussion

B. Porretta and **D. L. Kiguel** (Ontario Hydro, Toronto, Ontario, Canada): The authors are to be complimented for their well-written paper describing the requirements for composite system reliability evaluation.

Since the paper mentions Ontario Hydro among the utilities using probabilistic tools for bulk transmission system planning, we would like to note that the program used is PROCOSE (Probabilistic Composite System Evaluation) [1], [2]. This program was developed in Ontario Hydro and has been in use since 1980. The program, based on state enumeration, creates a large number of dc load flow states which are likely to exist during a given period of time. The program uses a novel algorithm to reduce the total load-flow state space to an equivalent smaller one. The generation-load pattern in each state represents the dispatch of the available generation which minimizes the total fuel costs and unsupplied load while obeying transmission constraints. From the state space of load flows the program obtains expected values and joint probability distributions of unsupplied load due to generation deficiencies and transmission limitations accounted separately, output of each generating unit, flow across monitored interfaces, additional fuel costs due to departure from economic dispatch as a result of transmission limits, MW, and dollar values of transmission losses, and the reactive power required at each bus to maintain a prespecified flat voltage profile. The current version of PROCOSE can model networks up to 2000 buses and 4000 lines. A larger version is under development.

References

[1] "Bulk Power System Reliability Evaluation. Part I: PROCOSE - A Computer Program for Probabilistic Composite System Evaluation," B. Porretta, D. L. Kiguel, 14th Inter-ram Conference for the Electric Power Industry. May 1987, Toronto, Canada.

[2] "Bulk Power System Reliability Evaluation. Part II: Application of PROCOSE for Reliability Assessment of the Ontario Hydro West System," G. Hamoud, E. G. Neudorf, 14th Inter-ram Conference for the Electric Power Industry. May 1987, Toronto, Canada.

Manuscript received February 23, 1987.

M. G. Lauby, D. D. Klempel, R. O. Gunderson, E. P. Weber, C. G. Dahl, M. A. Klopp, and **P. R. S. Kuruganty** (MAPP Transmission Reliability Task Force): The authors are to be commended for completing a survey of the industry on this timely topic. We have a number of questions.

The authors discussed some of the perceptions of the industry of composite system reliability evaluation models and tools. From your discussions with the industry, was there a standard accepted definition of adequacy and security? Is the line between these two definitions clear?

The discussers have great concern over the use of the definitions in reference [9] for adequacy and security, namely,

Adequacy is the ability of the bulk power electric system to supply the aggregate electric power energy requirements of consumers at all times, taking into account scheduled and unscheduled outages of the system components.

Security is the ability of the bulk power electric system to withstnd sudden disturbances such as electric short circuits or unanticipated loss of system components.

Specifically, the use of the phrase "at all times" in the adequacy definition completely redefines this word. Clearly, earlier attempts at adequacy assessment composite reliability programs have not addressed a time aspect of system operating conditions. In fact, it is questionable if contingency enumeration techniques can be applied in this way. LOLE calculations totally ignore applications of time pertaining to such factors as unit commitment, system condition, etc. Our question is, could the authors please give their views on the use and application of these new definitions in reliability assessment? Is the definition of adequacy, which implies a time component, correct?

Economics, as a measure of success and failure of the bulk power system, is of great interest. Perhaps the reason for this interest is the relatively small contribution of the bulk power system to total system inadequacy. Specifically, how often has bulk transmission or generation inadequacy caused system load curtailment in North America, compared to distribution outages? It may be for this reason that regional areas seek to identify other means to measure bulk power system inadequacy (i.e., economic effects of security dispatch requirements). Also, expected dollar values may be more palatable for management assessment of risk. Have

you reviewed the comments made about system inadequacy indices (load shed versus expected dollar savings) with this in mind?

Overall large bulk power system adequacy evaluation is difficult because large amounts of resources are made available to remove or eliminate system violations. Therefore, it may take many levels of contingency analysis, which is directly proportional to the additionally required CPU time, to identify failures which result in load curtailment. However, many times the solutions presented are not "realistic" or fall victim to the frailties of the tools that are implemented. For example, dc powerflow techniques are many times used to review MW inadequacy, as a dual to LOLE studies. Unfortunately, unrealistic solutions may be calculated by implementing too large an area to eliminate system violations. Perhaps a review of localization of composite system evaluation should be performed with the results of all areas reported separately as well as combined together for an indication of overall system adequacy. Clearly, this philosophy may overlook those situations that cause violations outside the local area of study, but these are a small subset of the overall set of system problems. However, all benefits of interconnection and pooling, much like a small subarea modeled in LOLE analysis, would not be realized. What are the thoughts of the authors on this localization of composite system reliability evaluation?

One of the difficulties mentioned with the current methods is the lack of transmission outage data. We agree. However, it is felt that the methods for collection are slowly being developed. For example, MAPP has been collecting transmission outage data since 1977. The MAPP Transmission Reliability Task Force has recently completed a joint effort with a MAIN task force to enhance our collection guide which in turn improves our capability to calculate certain indices. This enhancement mainly involved additional submittal for exposures to multiple related outage events, more outage cause codes, detailed information on the restoration process, etc. Most of these changes resulted from work that has been recently performed. However, after additional years of collection, more enhancements may again be required. One reason for this current dilemma is, though the models have been developed, the actual need for each when performing composite reliability evaluation has not been completely studied. This need is also different for each desired use of the data. Analysis is required to identify significant parameters within transmission performance that can appreciably affect the indices obtained from composite system reliability evaluation. What are the thoughts of the authors on this type of study?

One final question, from your interviews of the utility industry, what do you think the next step is? Are more programs needed? Or is more study required with the goal of identifying applications and implementations of the composite system reliability models?

The discussers would like to again congratulate the authors for an interesting paper. We look forward to their response to our concerns.

Manuscript received February 23, 1987.

M. P. Bhavaraju, P. F. Albrecht, R. Billinton, N. D. Reppen, and **R. J. Ringlee:** The authors appreciate the interest expressed by the discussers in the paper.

The PROCOSE program described by Messrs. Porretta and Kiguel appears to be an interesting approach to composite system reliability evaluation. It appears that only generation outages are given full consideration and transmission limitations are considered for one configuration only (rather than multiple outage states) in this method. Such a method could be useful for economic evaluations but its role in transmission reliability evaluations is uncertain. The authors will attempt to learn more about this approach in the future.

The MAPP Transmission Reliability Task Force raises a number of interesting questions.

The authors believe the phrase "at all times" is used only to imply that adequacy is a concern over a long period of time rather than, for example, at the annual peak time. Ignoring certain factors in the LOLE calculations as mentioned by the discussers would be acceptable if LOLE is interpreted as a relative index. It should be noted that the adequacy definition qualitatively describes the success condition while LOLE type of indices quantify adequacy failure conditions, and the authors do not believe there is an inconsistency between these two concepts.

The question on importance of reliability versus economics in bulk system planning cannot be resolved easily. In systems where the bulk system is generally strong and contributes little to loss of load, the concern is primarily economic operation of the system. In weak systems, load shed may be the primary reason to add facilities. The utility survey indicated both views.

339

It is important to examine the planning and operating criteria of a system and its neighboring systems before deciding how areas should be divided for a study. For example, in a pool operating as a single company, the analysis may ignore company boundaries and include the entire pool. In other cases, the study area may include only the company. The real problem and constraints cannot be changed because of modeling difficulties.

Regarding outage data, the authors agree that there are sufficient data to start using the existing composite system reliability programs. The difficulty is the lack of sufficient data for related outages which are rare but cause serious reliability problems if they occur. Special synthesis methods should be developed to estimate the probabilities of the rare events. The authors agree that new concepts or models sometimes need improvements in the data collection procedures and this problem will continue to exist.

The response to the last question is that the industry needs some effort to develop a new program (or modify an existing program) to consider generation outages accurately and efficiently and to model operating constraints. Parallel efforts are also needed in the industry to experiment with the existing tools applying them to realistic planning problems and to define specific model requirements for specific applications.

Manuscript received March 26, 1987.

MONTE CARLO APPROACH IN PLANNING STUDIES:
AN APPLICATION TO IEEE RTS

O. Bertoldi
R&D Department

L. Salvaderi, MIEEE
Planning Department

S. Scalcino
R&D Department

ENEL - Italy

Abstract - For probabilistic assessment of system adequacy and production cost when planning, ENEL has been using for many years a Monte Carlo-based program (SICRET). Two versions of the program were produced: - a "fast" one, to evaluate only system adequacy; - a "detailed" one, to conduct a refined evaluation of system operation costs. A report is provided here of the latest features of the two versions in terms of storage requirements and CPU time.

In the U.S.A., many system adequacy studies were performed on the IEEE RTS, mainly on the generation system. The application of ENEL methodology to the RTS shows that the investments related to the transmission system can be considerably reduced. A revised configuration of the RTS is thus proposed.

The paper provides RTS users with a series of indications on the behavior of the RTS transmission system and is intended to stimulate comparison with other alternatives that can be found by applying to the RTS other methodologies of composite system adequacy evaluation.

INTRODUCTION

When planning composite generation/ transmission systems, for quantitative evaluation of system adequacy and of fuel costs, ENEL uses the SICRET program, based on the Monte Carlo probabilistic simulation technique. The program, extensively described in various papers, /1/, /2/, /3/, has been developed for over 15 years and underwent continuous refinements over time, which were aimed at both reducing use of processing resources, /4/, and at modeling systems in such a way to increasingly match physical reality, /5/, /6/.

Recently, consideration has been given, not only to a closer connection with programs for the study of busbar generation systems (e.g. scheduled inputs of limited-energy hydro plants), but also to evaluation of additional production costs. In fact, as result of forced outages on the network, it is not always possible to apply the optimized dispatching schedule obtained on the corresponding busbar system.

The Monte Carlo method is mostly used in Europe whereas in other countries, especially in North America, composite system adequacy evaluation is performed by "State Enumeration" techniques. The two

approaches should not be looked at in a conflict sense but the planner should try to use the advantages and avoid the drawbacks of each approach.

These techniques were sometimes applied to the IEEE RTS, which was just developed as a "bench tool" for all the ones who, in the electric industries, deal with reliability. In particular, a recent paper /6/ tried to make an applicative comparison of the two methodologies (Monte-Carlo, State Enumeration). Differences and similarities, both in models used and in adequacy indices obtainable, were singled out and discussed. Such comparison was not particularly intended to the design aspects of RTS; with the present paper such further step is attempted.

Previous studies /6/, /7/, /8/, /9/ conducted on the IEEE RTS evidenced that, even if occasional use is made of the simplifying assumption of having constant voltage even under outage conditions, the risk pertaining to the RTS transmission network is practically much lower than the one due to the generating mix. It might thus be inferred that the transmission network is, to a certain extent, oversized. The purpose of this study was to explore a possible redesigning of the IEEE RTS to achieve the best compromise between reliability, optimized generating mix operation and amount of total investments in transmission.

The study was conducted by taking the maximum advantage of the existence of two SICRET versions: the "fast" one, for planning stages where the major consideration is system adequacy evaluation, and the "detailed" one, for project implementation, where an accurate operation cost evaluation is essential for decision making. The results of the study are reported in an attempt to stimulate discussion and receive possible feedbacks, with a mutual benefit for all the ones who deal with reliability evaluations, for power system analysis .

OUTLINE OF THE MONTE CARLO BASED SICRET PROGRAM

The Monte Carlo-based simulation technique allows the users to obtain a random sample of possible system states. Checking the operating conditions in each state enables to estimate the value of the selected risk index (Expected Energy Not Supplied = EENS), as well as other indices (utilization hours of generating units, distribution probability of line power flows, etc.) when necessary /11/, /12/.

As known, the main advantages of the sampling simulation techniques are high flexibility and detail in the simulation of complex system operation and/or configuration conditions. It is also possible to simulate the impact of weather conditions on the availability of overhead lines, etc. /12/, /13/.

Disadvantages may or may not be (according to different situations which the utilities must face as regards the cost of computing time), the rather long

87 SM 444-3   A paper recommended and approved by the IEEE Power System Engineering Committee of the IEEE Power Engineering Society for presentation at the IEEE/PES 1987 Summer Meeting, San Francisco, California, July 12 - 17, 1987. Manuscript submitted January 28, 1987; made available for printing April 28, 1987.

Reprinted from *IEEE Trans. Power Syst.*, vol. 3, no. 3, pp. 1146-1154, Aug. 1988.

CPU time which depends on the level of detail used in modeling the system and particularly its operation, and on the level of statistical convergence of the simulation process results; the number of samples generated is very wide and depends on the system reliability level.

During program running, two requirements may arise: on one hand, to minimize CPU time, by possibly sacrificing model completeness, when only adequacy is to be evaluated; on the other hand, not to give up a very accurate system examination, when this is essential.

The solution selected by ENEL consists in having two versions of the SICRET program, a Fast (F) one, generally to be used at the initial stage of the studies, and a Detailed (D) one, to be used generally at a later, more detailed stage.

This solution well suits planning studies, whose goal is to define a network configuration which achieves:

$$\text{min } Z = C + O + R \quad \begin{cases} \textbf{C} = \text{Capital cost} \\ \textbf{O} = \text{Operation cost} \\ \textbf{R} = \text{Risk cost} \end{cases}$$

Note: the latter two items are estimations of mean expected value of random variables

The minimization procedure can be split into 3 steps:

a) Select alternatives which fulfill a predetermined adequacy standard;
b) Test whether it is possible to improve the alternatives by comparing capital cost (C) and risk cost (R) of each
c) Test whether it is possible to improve furtherly the alternatives by taking into account also operation costs (O).

In steps a) and b), the quantitative risk index (Expected Energy Not Supplied) is computed on the assumption of operating the generating mix on the basis of a "pure safety" policy, by neglecting operation cost evaluation; dispatching is resorted to only in cases recognized as "dangerous" according to simplified rules; time is saved in the examination of cases and by the elimination of computation and storage of operation costs.

In step c), in addition to computation of the risk index, a detailed simulation of the generating mix operation policy is required, in order to minimize the risk cost (R) plus the operation cost (O).

The above procedure is justified by the fact that the unit cost of the kWh not supplied is usually higher than the difference between the maximum and minimum marginal costs by at least one order of magnitude. This means that if an alternative is selected on the basis of pure adequacy considerations, it does not generally conflict with further improvements selected by taking into account also the operation cost (O).

To attain the above goals, the SICRET program is available in two options, as previously pointed out.

Version F. Priority is given to processing speed, which is obtained without sacrificing accurate system modeling, but by organizing the entire computation procedure in accordance with the "pure safety" operation policy of the generating mix and of the possible exchanges with external systems.

Configurations are so identified for which load shedding should be a priori excluded and network configurations and load values are combined in an appropriate way, so as to minimize the number of checks to be carried out.

Additionally, redispatching and load shedding (in case of network overload) are simulated with a very fast heuristic procedure, based on node-link sensitivities; this allows to compute the minimum load shedding which is required and its location /1/, /2/,/4/.

Version D. Priority is given to a detailed examination of the system, with simulation of optimized and relatively complex operation policies. Particularly:

a) Dispatching is performed in order to meet, hour by hour, the full load at the minimum operation cost by complying with possible energy constraints; A d.c. load flow is accordingly calculated;
b) In case of network overloads, redispatching and load shedding are simulated by solving the L.P. problem:

$$\text{min } C = \sum_{i=1}^{NG} A_i \Delta P_i + K \sum_{i=1}^{NL} \Delta L_i$$

$$|F_j| \leq CAP_j;$$

$$F_j = \sum_{i=1}^{NBUS} S_{ij} (P_{io} - L_{io} + \Delta P_i + \Delta L_i)$$

where:

CAP_j	=	capability of line j
S_{ij}	=	sensitivity coefficient of node i-line j
P_{io}	=	production dispatching having min cost at bus i
NG, NL	=	generators and load nodes
L_{io}	=	value of load at node i in a given hour
$NBUS$	=	network busses
A_i	=	average production cost of generator i
ΔP_i	=	production variation at bus i
K	=	cost of kWh not supplied
ΔL_i	=	load shedding at node i
F_j	=	power flow on line j

This LP procedure represents the most recent refinement of the concept of mixed "safety-economy" policy already described in /5/,/6/,/9/.

In both versions, but prevailing for version D, various statistical data might be supplied (obviously with different levels of "accuracy"), such as:

1) probabilistic distribution of line power flows;
2) expected value of energy supplied by each generator, with related cost, both in relation to a ficticious dispatching, as would be the case in a busbar system, and taking into account transmission constraints.

Since the two versions are planned to be used in an integrated way, care was taken to obtain the same samples of random network configurations, in order to evaluate the effect of the phenomena, neglected in F and simulated in D, more accurately: this is required to prevent the scatter of the results intrinsic in the Monte Carlo method.

The computing resources used by the SICRET program are indicated below; they refer both to the IEEE RTS case and to systems examined in ENEL planning practice.

	IEEE RTS 24 Busses		BULK COMPOSIT SYSTEM 300 Busses 450 Links 300 Gen. Units	
	CPU TIME* (Minutes)	STORAGE (K Bytes)	CPU TIME* (Minutes)	STORAGE (K Bytes)
SICRET F	0.37	240	14	800
SICRET D	1.6	380	40	1 430

* IBM 3081 - 8760 Hourly Samples

APPLICATION TO THE IEEE RTS/EPRI

Purpose of the study

The purpose of this study is to explore the possibility of "redesigning" the IEEE RTS network, to curb investments without impairing system adequacy and optimized generating mix operation.

System features and study assumptions

The study is based on the EPRI version of the IEEE RTS, whose main features /10/ are:
- Yearly peak: 3,563 MW (3,650 MW with losses)
- Yearly energy demand: 19.6 TWh
- Installed capacity: 4,304 MW
- Reserve margin: 18%.

The network diagram is shown in Fig. 1 (dashed lines are the ones which will be deferred, as a result of the study). Respect to the original IEEE RTS version |14|:

- all the loads were increased by 25%, leaving the hourly diagram unaltered;
- 83% of the added generation capacity (899 MW) is concentrated in the plants that feed the 230 kV network, located on the righthand side (busses 22-23-13). This enhances the power tranfers from one region to the other previously reported in |6| and in particular (at peak load and with economic dispatching):
- The load of the 230-kV lines, along the East-West direction, rises to 60-80%;
- The power flow from the 230 to the 138-kV network rises of about 200 MW, with a higher load (40%) on the two long 230 kV lines, from Bus 23 to the south.

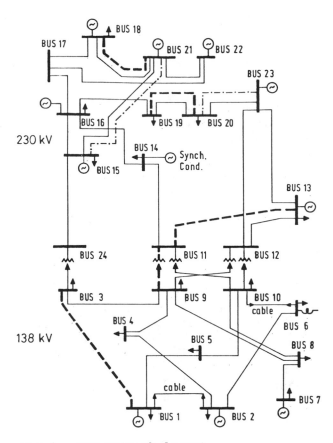

Fig.1 - RTS Network layout.

The EPRI version of the IEEE RTS was thus regarded as suitable for this study.

Transmission network

The network is the same as the original RTS (this study neglected the addition of a phase-shifting transformer). As regards to line transmission capability, the limits selected were <u>lower</u> than the ones specified for the original RTS system, taking into account that computations were made by considering only active power flows (d.c. load flow) and that a limit should be imposed on angular voltage phase differences between the extreme nodes of the lines, for static stability purposes (20 degrees).

The following values were thus assumed:

- 138 kV cables: 175 MW (instead of 193 MVA)
- 138 kV lines : 185 MW (instead of 208 MVA)

excluding
. line Bus 1 - Bus 3 (162 MW)
. line Bus 2 - Bus 6 (178 MW)
- 230 kV lines : 540 MW (instead of 600 MVA)
excluding
. line Bus 23 - Bus 12 (400 MW)
. line Bus 13 - Bus 23 (397 MW)
. line Bus 22 - Bus 17 (326 MW)

The above network is referred to as "Step 0 Basic network" configuration.

Generation system

The scheduled maintenance plan was selected in such a way as to make capacity reserve as uniform as possible during the 52 weeks of the year: in this way, high-load weeks (1-7 and 43-52) are excluded from maintenance. Energy limits specified for the hydro plant (Bus 22), were not considered.

Economic figures adopted (1984 constant money)

- Unit risk cost: 2 $/kWh
- Cost of plants (-life: 30 years - interest rate, inflation removed: 5%)
  . 230-kV line   13 k$/mile yr
  . 138-kV line   7 k$/mile yr
  . 230-kV bay   79 k$/mile yr
  . 138-kV bay   69 k$/yr
  . 138/220-kV transformer   30 k$/yr
- Running costs (Italian prices at exchange rate of
                 1 $ = 1,500 lire)
  . nuclear   28   mills/kWh
  . coal   38   mills/kWh
  . oil   54   mills/kWh
  . gas turbine 82,3 mills/kWh

Analysis of Basic Network (Step 0)

A first computation performed with SICRET/F for the "Step 0 = Basic Network", confirms that the RTS/EPRI has a lack of balance between the network adequacy (EENS due to overload = 8 MWh/yr,[*] i.e. 2.6 x 10^{-7} in p.u. of total energy demand) and the generation system adequacy (EENS = 1,060 MWh/yr vir. approx 5 x 10^{-5} in p.u.). EENS was computed on the assumption that, in each bus, the load can be decomposed into elementary loads, each of 1 MW.

A further accurate analysis carried out with SICRET/D also evidences that:

a) Many lines and transformers have a low utilization (flows never exceed 50% of the relevant capability) even in the "worst" network condition, that is when the machine dispatchings are based (as in the busbar system) simply on increasing order of specific cost and before that any possible overload relief should be attempted (see Table I);

b) The fuel cost after overload relief by application of LP technique (Table II/B = 541.1 M$/yr) is practically equal the one calculated for the busbar system (Table II/A = 541.03 M$/yr).

(*) No transmission line/transformer maintenance was taken into account.

LINE	CAPA-BILITY MW	FROM BUS	TO BUS	0 10%	10 20%	20 30%	30 40%	40 50%	50 60%	60 70%	70 80%	80 90%	90 100%	100 110%	110 120%	120 130%	130 140%	140 150%	150 160%	160 170%	AVERAGE POWER FLOW
1	193.00	BUS 1	BUS 2	4235	4141	1235	347	26	15	1	0	0	0	0	0	0	0	0	0	0	24.87
2	162.00	BUS 1	BUS 3	3126	3293	1743	1436	395	7	0	0	0	0	0	0	0	0	0	0	0	-28.68
3	185.00	BUS 1	BUS 5	380	1505	1999	2046	2326	1702	39	3	0	0	0	0	0	0	0	0	0	64.06
4	185.00	BUS 2	BUS 4	1137	2378	2566	2772	1145	1	1	0	0	0	0	0	0	0	0	0	0	47.02
5	178.00	BUS 2	BUS 6	570	2484	2605	2933	1397	5	1	2	.	0	0	0	0	0	0	0	0	48.01
6	185.00	BUS 3	BUS 9	278	479	641	2872	1683	4015	21	4	2	2	1	2	0	0	0	0	0	77.45
7	185.00	BUS 4	BUS 9	5512	3942	534	8	4	0	0	0	0	0	0	0	0	0	0	0	0	-18.14
8	185.00	BUS 5	BUS 10	6583	2812	559	39	7	0	0	0	0	0	0	0	0	0	0	0	0	15.81
9	193.00	BUS 6	BUS 10	115	696	3141	4730	1096	202	17	3	0	0	0	0	0	0	0	0	0	-61.78
10	185.00	BUS 7	BUS 8	346	304	349	2840	2411	2529	1116	81	14	7	3	0	0	0	0	0	0	-82.59
11	185.00	BUS 8	BUS 9	27	48	72	153	2651	2288	2484	1425	848	2	1	0	1	0	0	0	0	-110.51
12	185.00	BUS 8	BUS 10	30	53	88	259	3525	2038	2433	1226	346	0	0	1	0	0	0	0	1	-104.12
13	540.00	BUS 13	SLACK BUS	7870	1329	523	222	40	15	1	0	0	0	0	0	0	0	0	0	0	39.72
14	540.00	BUS 13	BUS 12	8397	1034	433	106	29	1	0	0	0	0	0	0	0	0	0	0	0	35.64
15	540.00	BUS 14	SLACK BUS	221	544	954	3191	4992	85	2	10	1	0	0	0	0	0	0	0	0	206.47
16	360.00	BUS 23	BUS 12	5	3	45	593	2366	1844	1736	1295	1330	767	9	5	0	1	1	0	0	225.56
17	397.00	BUS 13	BUS 23	32	50	123	825	2862	1749	1709	1410	960	274	6	0	0	0	0	0	0	-232.31
18	540.00	BUS 16	BUS 14	3	6	52	264	442	2426	2548	2762	1482	3	6	6	0	0	0	0	0	359.49
19	540.00	BUS 15	BUS 16	2427	2919	2860	1762	17	6	6	3	0	0	0	0	0	0	0	0	0	100.45
20	540.00	BUS 15	BUS 21	33	361	629	3085	975	4908	2	4	3	0	0	0	0	0	0	0	0	-237.01
21	540.00	BUS 15	BUS 21	31	361	629	3085	976	4907	3	4	4	0	0	0	0	0	0	0	0	-237.03
22	540.00	BUS 15	BUS 24	30	74	509	3021	2387	3955	19	3	2	0	0	0	0	0	0	0	0	241.11
23	540.00	BUS 16	BUS 17	60	184	199	332	1248	1557	890	1974	2321	1224	9	1	1	0	0	0	0	-366.15
24	540.00	BUS 16	BUS 19	1897	2320	2093	1344	1374	887	77	7	1	0	0	0	0	0	0	0	0	140.29
25	540.00	BUS 18	BUS 17	1265	1799	1121	2155	2687	958	11	4	0	0	0	0	0	0	0	0	0	167.89
26	326.00	BUS 22	BUS 17	11	0	1	72	694	990	8205	19	2	1	0	0	2	0	0	3	0	207.26
27	506.00	BUS 21	BUS 22	6	1	21	447	1770	7723	21	0	0	11	0	0	0	0	0	0	0	-267.98
28	540.00	BUS 18	BUS 21	5818	2266	1857	57	2	0	0	0	0	0	0	0	0	0	0	0	0	-62.46
29	540.00	BUS 18	BUS 21	5820	2262	1858	57	3	0	0	0	0	0	0	0	0	0	0	0	0	-62.46
30	540.00	BUS 19	BUS 20	5216	3753	853	167	11	0	0	0	0	0	0	0	0	0	0	0	0	-56.94
31	540.00	BUS 19	BUS 20	5218	3751	851	169	11	0	0	0	0	0	0	0	0	0	0	0	0	-56.95
32	540.00	BUS 23	BUS 20	4193	2702	1980	822	281	21	1	0	0	0	0	0	0	0	0	0	0	81.55
33	540.00	BUS 23	BUS 20	4191	2700	1982	822	280	23	2	0	0	0	0	0	0	0	0	0	0	81.61
34	510.00	BUS 9	SLACK BUS	149	7363	2432	29	25	2	0	0	0	0	0	0	0	0	0	0	0	-90.45
35	510.00	BUS 10	SLACK BUS	37	470	3900	5401	179	6	5	1	1	0	0	0	0	0	0	0	0	-154.23
36	510.00	BUS 9	BUS 12	1244	4392	3317	986	52	6	3	0	0	0	0	0	0	0	0	0	0	-98.68
37	510.00	BUS 10	BUS 12	14	303	4478	3266	1792	137	6	4	0	0	0	0	0	0	0	0	0	-162.80
38	510.00	BUS 3	BUS 24	30	48	486	2178	2189	4753	310	3	3	0	0	0	0	0	0	0	0	-241.11

Table I: Frequency of Power Flows in the "Step 0 Network" before overload relief.
(Total number of simulated hours: 10.000)

344

Naturally, the calculated results shown in this paper are not exactly comparable with the ones obtainable with other methods |6|. In fact, many differences might exist. The major difference may be due to the load models used. The solution techniques, network representation and load curtailment policies used introduce further assignable differences in the calculated indices. These considerations can be used in order to interpret the results better.

Initial Deferment of 7 Links (Step 1 Network)

The result described above provide useful indications to formulate a first attempt of network redesign.

Among the various low-loaded links, **seven** were initially removed (**five 230-kV lines, one 138-kV line, one transformer**) according to the criterion of not excessively weakening the network meshing.

The new network will be defined "Step 1 Network". The removed lines are the dashed ones in Fig. 1.

Fig. 2 shows the distributions of two links characterized by particularly low power flows: one transformer removed (busses 9-11) and one line (busses 20-23).

As previously noted, peak-load a.c. load-flow computations indicated that, under minimum-cost dispatching conditions, a correct operation of such "Step 1 Network" is possible, without violating the limits of:
- Line capability;
- Reactive power of the generators (including the Bus 14 synchronous condenser);
- Bus voltage (± 5% of nominal value).

For simplicity's sake losses were first supposed not to vary with respect to the original "Basic-Step 0 Network" configuration, subject to subsequent a.c. load-flow checks.

TABLE II: OPERATION COST FOR IEEE RTS/EPRI

T Y P E	TOTAL INSTALLED CAPACITY (MW)	FUEL COST (mills/kWh)	A) BUSBAR SYSTEM			B) COMPOSITE SYSTEM step 0 network			C) OPTIMIZED COMPOSITE SYSTEM step 4 network		
			ENERGY PRODUCED (°) (GWh/yr)	(%)	YEARLY OPERATION COST (K$/yr)	ENERGY PRODUCED (°) (GWh/yr)	(%)	YEARLY OPERATION COST (K$/yr)	ENERGY PRODUCED (°) (GWh/yr)	(%)	YEARLY OPERATION COST (K$/yr)
HYDRO	500	0.00	4 165	21	–	4 165	21	–	4 165	21	–
NUCLEAR	800	28.00	5 433	28	152 111	5 432	28	152 102	5 432	28	152 102
COAL	1 776	38.00	9 418	48	357 890	9 417	48	357 844	9 417	48	357 844
OIL	1 148	54.00	573	3	30 946	574	3	30 997	573	3	30 946
G.T.	80	83.30	1	–	82	2	–	164	3	–	246.
(°) Yearly Curtailed Energy = 1 060 MWh/yr			19 590	100	541 029	19 590	100	541 107	19 590	100	541 138.

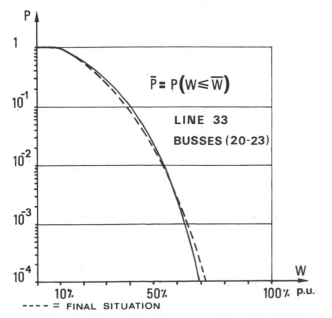

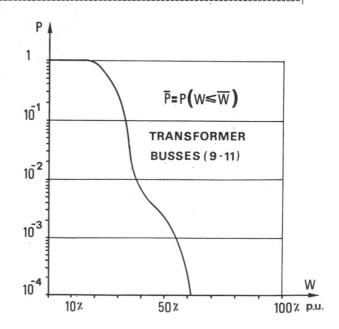

---- = FINAL SITUATION

Fig. 2 Examples of power flow distributions

Network Optimization (Steps 2-3-4)

A computation was made with SICRET/F to evaluate the reliability level of "Step 1 Network". It present a very limited EENS due to overload (245 MWh/yr), but its corresponding economic value (500 k$/yr) is such as to justify exploration of the possibility of **reintroducing one of the removed lines.** The new network will be called "Step 2 Network".

A study was so conducted on different configurations, by using SICRET/F and SICRET/D in an integrated way; the sequence of computations, results and decisions taken at the various "Step 2-3-4" is summarized in Table III.

To show the importance of the way of obtaining operation costs, results are reported on two types of computation /5/,/6/,/9/:

- "exact", by strictly following the policy of the best "safety-economy" compromise achieved through the LP technique
- "approximate", by following a "pure safety policy". EENS are detailed for the various Steps in Table IV. SICRET allows the derivation of the contributions due to the generation and transmission system ("**how**" option), subdivided node by node ("**where**" option) |1|, |3|. The most "critical" busses and the causes of their criticality are thus well singled out.

The final outcome of the procedure ("Step 4 Network") is that two lines should be added to the first attempt "Step 1 Network" (characterized by deferrement of 7 links). The two lines (see lightly dashed lines in Fig. 1) are:

WHERE	Lack of generat. capacity (*)	Basic network Step 0	Step 1 (-6lines) (-1trans)	Step 2 (-5lines) (-1trans)	Step 3 (°°) (-5lines) (-1trans)	Final situation Step 4 (-4lines) (-1trans)
bus 1	63.1	–	–	–	–	–
2	41.1	–	–	–	–	–
3	–	–	35.6	35.1	35.1	29.6
4	–	–	0.8	0.2	0.2	0.4
5	–	–	1.9	1.9	1.9	1.9
6	–	–	–	–	–	–
7	32.8	–	–	–	–	–
8	–	5.9	5.6	5.0	5.4	5.4
9	–	–	1.8	1.8	1.1	1.1
10	–	–	7.8	4.4	4.2	4.3
11	–	–	–	–	–	–
12	–	–	–	–	–	–
13	97.2	–	–	–	–	–
14	–	–	–	–	–	–
15	93.9	–	0.8	0.5	0.6	–
16	62.2	–	–	–	–	–
17	–	–	–	–	–	–
18	669.4	2.1	–	–	–	–
19	–	–	–	–	0.1	0.1
20	–	–	190.7	189.4	0.3	0.3
21	–	–	–	–	–	–
22	–	–	–	–	–	–
23	–	–	–	–	–	–
24	–	–	–	–	–	–
	1059.7	8.0	244.6	239.4	48.9	43.1

(*) Same value for all steps from 1 to 4

(°°) A different line in respect of the one of "step 2" is added

TABLE IV - EXPECTED ENERGY CURTAILMENTS (MWh/year) IN VARIOUS NETWORK CONFIGURATIONS (ENERGY DEMAND = 19 600

STEP	PROGRAM MODE	ACTION/QUALITATIVE RESULT	QUANTITATIVE RESULT	DECISION
0	Basic RTS/EPRI: Bus-Bar system running costs: 540.67 M$/yr. Expected Energy Not Supplied (EENS) – Due to line overload: 8 MWh/year; Due to lack of generation: 1 060 MWh/yr			
1	Use of "F" mode	6 lines + 1 transformer deferred. Adequacy still satisfactory but its cost high	EENS due to overload: 245 MWh/year	The amount of EENS could justify some reinforcement, to be chosen among previously removed links
2	Use of "F" mode	Line Bus 15 - Bus 21 added. Adequacy does not increase appreciably	EENS due to overload: 240 MWh/year	A different reinforcement will be tried
3	Use of "F" mode	Line Bus 20 - Bus 23 added. Adequacy increases	EENS due to overload: 49 MWh/year	Cost benefit analysis justifies line Bus 20 - Bus 23 and excludes any other reinforcement. System operation cost must be verified
	Use of "D" mode	Line Bus 20 - Bus 23 added. Operation cost is higher than that of bus-bar system	System operation costs: – "exact" 544 M$/yr – "approx." 655.8 M$/yr	Another line (Bus 21 - Bus 15) added
4	Use of "D" mode	By adding line Bus 20 - Bus 23 and line Bus 15 - Bus 21, operation cost decreases appreciably	System operation costs: – "exact" 541.14 M$/yr – "approx." 655.8 M$/yr	The system can be considered optimized with -4 lines and -1 transformer in respect of initial situation

TABLE III
SUMMARY OF THE CASES CONSIDERED

- Bus 20 - Bus 23 (15 miles), since the capital costs for its construction (350 k$/yr) are lower than the reduction in the risk cost (400 k$/yr) obtained by its installation (see Table III Step 1: EENS = 245 MWh; Step 3 EENS = 49 MWh/yr). No operation cost increase is clearly arising from the addition of one line:SICRET F version has been applied.
- Bus 15 - Bus 21 (34 miles), since it allows a fuel cost saving of 3.2 M$/yr, respect to "Step 3", as against a capital cost of 0.6 M$/yr. Practically no variation of system adequacy from Step 3 to Step 4 (Table IV) is found.

Table III displays that the decision to add the line is justified only if use is made of the "exact" computation of operation costs. In effect, the "approximate" computation, with dispatching based on the "pure safety" policy, practically gives the same value of the operation costs both with (step 4) and without (step 3) the line Bus 15 - Bus 21, while the "exact cost" shows a decrease. This correspond to a better utilization of the nuclear plant at bus 21. The network so obtained ("Step 4 Network") can be considered as optimized, because:

- the operation cost (Table II/C) practically matches the ones of the "Step 0 Basic Network" (Table II/B) which is much more meshed;
- the residual EENS due to overload is so low (43 MWh/yr) as to make its reduction (addition of some other lines) uneconomical;
- the extra costs due to the additional losses respect to the original EPRI/RTS system are in the range of 30 - 50% of the annual cost of deferred investments . (2.2 M$/yr) (*).

Tables V show power flows corresponding to those illustrated in Table I, referred to"Step 4 Network" (same minimum cost distpatching and absence of overload relief).

(*) - The difference of peak-load joule losses, evaluated with a.c. load flow, is about 5 MW. For their quantification, the annual utilization hours were supposed to be 3,500 and the annual costs (fixed + variable) ranging from 160 to 250 $/kW.yr.

| LINE | CAPA-BILITY MW | FROM BUS | TO BUS | POWER FLOW IN PERCENT | | | | | | | | | | | | | | | | | AVERAGE POWER FLOW |
				0 10%	10 20%	20 30%	30 40%	40 50%	50 60%	60 70%	70 80%	80 90%	90 100%	100 110%	110 120%	120 130%	130 140%	140 150%	150 160%	160 170%	
1	193.00	BUS 1	BUS 2	5956	3022	939	55	27	1	0	0	0	0	0	0	0	0	0	0	0	19.06
2	162.00	BUS 1	BUS 3	10000	0	0	0	0	0	0	0	0	0	0	0	0	0	0	0	0	0.0
3	185.00	BUS 1	BUS 5	1998	1625	1472	1676	1728	1099	379	22	1	0	0	0	0	0	0	0	0	54.76
4	185.00	BUS 2	BUS 4	2104	1776	2116	2124	1659	211	3	5	2	0	0	0	0	0	0	0	0	46.49
5	178.00	BUS 2	BUS 6	2209	2232	2031	2300	1190	32	1	2	3	0	0	0	0	0	0	0	0	41.44
6	185.00	BUS 3	BUS 9	363	504	546	1050	2096	1014	1880	2090	406	17	14	4	6	7	1	2	0	98.30
7	185.00	BUS 4	BUS 9	4518	3548	1783	141	8	2	0	0	0	0	0	0	0	0	0	0	0	- 22.89
8	185.00	BUS 5	BUS 10	4557	2816	1963	632	22	9	1	0	0	0	0	0	0	0	0	0	0	- 25.19
9	193.00	BUS 6	BUS 10	156	687	2221	2563	3918	422	27	6	0	0	0	0	0	0	0	0	0	- 69.31
10	185.00	BUS 7	BUS 8	346	304	349	2840	2411	2529	1116	81	14	7	3	0	0	0	0	0	0	- 82.58
11	185.00	BUS 8	BUS 9	27	51	84	225	3304	2072	2486	1274	471	2	2	1	1	0	0	0	0	-105.88
12	185.00	BUS 8	BUS 10	32	52	79	192	2813	2328	2495	1290	700	12	2	3	1	0	0	0	0	-108.73
13	540.00	BUS 13	SLACK BUS	10000	0	0	0	0	0	0	0	0	0	0	0	0	0	0	0	0	0.0
14	540.00	BUS 13	BUS 12	5564	2778	699	431	278	147	63	23	13	4	0	0	0	0	0	0	0	68.74
15	540.00	BUS 14	SLACK BUS	297	627	1518	2630	3943	966	9	10	0	0	0	0	0	0	0	0	0	202.75
16	360.00	BUS 23	BUS 12	4	6	137	1386	2233	1897	1337	1303	1212	471	7	5	1	0	1	0	0	236.64
17	397.00	BUS 13	BUS 23	71	93	150	815	2979	1866	1554	1394	900	174	4	0	0	0	0	0	0	-226.87
18	540.00	BUS 14	BUS 16	3	10	82	363	395	2788	1503	3624	1213	6	7	6	0	0	0	0	0	355.37
19	540.00	BUS 15	BUS 16	2127	2437	3092	2069	258	5	8	3	1	0	0	0	0	0	0	0	0	112.62
20	540.00	BUS 15	BUS 21	13	381	657	2351	1353	5233	5	2	5	0	0	0	0	0	0	0	0	-242.42
21	540.00	BUS 15	BUS 21	11	381	657	2351	1354	5233	5	2	6	0	0	0	0	0	0	0	0	-242.44
22	540.00	BUS 15	BUS 24	26	28	496	2732	3685	3002	22	7	2	0	0	0	0	0	0	0	0	239.69
23	540.00	BUS 16	BUS 17	114	186	288	681	1064	1222	1190	2170	2114	930	9	1	1	0	0	0	0	-355.26
24	540.00	BUS 16	BUS 19	1739	2310	2106	1471	1390	899	69	8	7	1	0	0	0	0	0	0	0	143.11
25	540.00	BUS 18	BUS 17	1850	1126	1565	2415	2513	521	8	2	0	0	0	0	0	0	0	0	0	159.86
26	326.00	BUS 22	BUS 17	11	0	4	69	646	1043	6588	1631	2	1	0	2	0	0	0	3	0	211.64
27	506.00	BUS 21	BUS 22	6	1	25	444	1781	7702	30	0	0	.11	0	0	0	0	0	0	0	-263.60
28	540.00	BUS 18	BUS 21	10000	0	0	0	0	0	0	0	0	0	0	0	0	0	0	0	0	0.0
29	540.00	BUS 18	BUS 21	1596	5563	913	114	1697	116	0	1	0	0	0	0	0	0	0	0	0	-106.90
30	540.00	BUS 19	BUS 20	10000	0	0	0	0	0	0	0	0	0	0	0	0	0	0	0	0	0.0
31	540.00	BUS 19	BUS 20	3064	2325	2130	1551	475	284	119	38	11	3	0	0	0	0	0	0	0	-111.32
32	540.00	BUS 23	BUS 20	4461	2619	1919	728	247	24	2	0	0	0	0	0	0	0	0	0	0	79.26
33	540.00	BUS 23	BUS 20	4459	2617	1920	729	247	25	3	0	0	0	0	0	0	0	0	0	0	79.32
34	510.00	BUS 9	SLACK BUS	10000	0	0	0	0	0	0	0	0	0	0	0	0	0	0	0	0	0.0
35	510.00	BUS 10	SLACK BUS	273	570	1041	2812	2507	2775	10	9	3	0	0	0	0	0	0	0	0	-202.76
36	510.00	BUS 9	BUS 12	40	1517	3121	2515	2058	637	83	12	8	7	2	0	0	0	0	0	0	-166.28
37	510.00	BUS 10	BUS 12	236	3307	2750	2271	880	356	155	32	9	1	3	0	0	0	0	0	0	-138.93
38	510.00	BUS 3	BUS 24	26	18	437	1998	3055	4284	161	17	4	0	0	0	0	0	0	0	0	-239.69

Table V: Frequency of Power Flows in the "Step 4 Network" before overload relief.
(Total number of simulated hours: 10.000)

A comparison of Tables I-V evidences that the percentage of usage of the lines, and thus of the network, rises sharply without too many capability limit violations. This confirms that the solution selected is more economical. Furthermore, Table II shows that the removal of four lines and of one transformer did not create such network constraints as to prevent a quasi-optimum dispatching (bus-bar system): some differences are observed only for peaking units. The operation cost of "Step 4 Network" (Table II/C) is very close to the one of the busbar system (Table II/A) obtained by dispatching generating units according to a priority order based on increasing specific cost. Such comparison confirms that the "Step 4 Network" can be considered as optimized in planning a "**new**" RTS.

When planning the IEEE RTS/EPRI, a decision could be taken to defer the construction of 55 miles of 138-kV lines and 78.5 miles of 220 kV lines, as well as to install a 230/138-kV transformer. With the unit costs shown in para. 3, this corresponds to an investment saving in the range of 22×10^6 US$ (about 70 miles of the standard ENEL 380-kV line, 3 bundle conductors, rated 1,500 MVA!)

CONCLUSIONS

When planning its generation and transmission system, ENEL has been following since many years the criterion to quantify system reliability through a risk cost, to be added to the operation cost and to the capital cost. To evaluate system adequacy, which on its turn gives rise to the "risk cost" in probabilistic terms, ENEL developed and continuously improved a Monte Carlo based computer program (SICRET).

In planning computations, it can be often supposed that the operation cost of a generation/transmission system does not change significantly when passing from a network configuration to another one. Some situations may nevertheless occur where, owing to particular constraints imposed on generation by network outages, such changes may have a crucial impact on the selection of network configuration. In this framework, ENEL worked out two versions of SICRET: a "fast" one (F), which does not include operation cost evaluation, thereby saving CPU time; and a "detailed" one (D), which affords "accurate" operation cost evaluations.

An example of application of the SICRET planning procedure to the IEEE RTS demonstrates that the RTS transmission network is redundant. The significant saving deriving from an appropiate deferrement of 4 lines and 1 transformer obtained by using the proposed method, was also quantified. A new configuration for the RTS/IEEE network is thus recommended.

The application example aims at increasing the usefulness of the RTS in two aspects. The first one is that an increased range of data on the RTS transmission system behaviour is now available (previous studies were mainly concerned with the generation system). The second one is that a "triggering" effect can be obtained for similar studies, that should be carried out by RTS users with computer programs based on other methodologies both for adequacy evaluation and for its quantification during planning.

Obviously, the numerical results obtained with different methodologies should be made as comparable as possible by unifying and comparing, in particular, load and network models, along with solution techniques and load shedding policies.

ACKNOWLEDGMENTS

The authors are indebted to Mr. Giampieri and Mr. Tagliabue for their helpful contribution to IEEE RTS studies.

REFERENCES

1   P.L. Noferi, L. Paris, "Quantitative evaluation of power system reliability in planning studies" IEEE Transaction PAS, Vol. 91; No. 2, March/April 1972.

2   P.L. Noferi, L. Salvaderi "A set of program for the calculation of the reliability of a transmission system", UNIPEDE Informatique Conference, Lisbon, 1971.

3   P.L. Noferi, L. Paris, L. Salvaderi, "Montecarlo methods for power system evaluation in transmission or generation planning", proceedings 1975 Annual Reliability and Maintainability Symposium, Washington, 1975.

4   O. Bertoldi "Network reliability evaluation: reduction in computer time and memory requirements 7th PSCC Lausanne 1981.

5   L. Salvaderi, I.Facchinei, E.Gamperi, S.Scalcino, "Network reliability evaluation: peculiar problems to be faced when planning bulk meshed transmission system", VII PSCC Conference Lausanne 1981.

6   L. Salvaderi, R. Billinton, "A Comparison between two fundamentally different approaches to composite system reliability evaluation "IEEE Transaction PAS, Vol. 104, December 1985.

7   R.Billinton, P.K. Vohra, Sudhir Kumer ""Effect of station originated outages in a Composite System Adequacy Evaluation of the IEEE Reliability Test System" IEEE W.M. N.Y. Febr. 198.

8   R. Billinton, S.Kumar " A Comparative Study of System versus Load Point Indices for Bulk Power System" IEEE W.M., N.Y. Febr. 1986

9   L. Salvaderi "ENEL Method for Reliability Evaluation EPRI, Workshop on Transmission System Reliability Methods. Washington, Dec. 1983, EL 3306, RP 1503.

10   EPRI: "Transmission System Reliability methods" Vol. 1, July 1982, Appendix Q.

11   IEEE Working Group Report "Reliability indices for use in bulk power supply adequacy evaluation", IEEE Transactions PAS, Vol. 97, No.4, July/Aug. 1978.

12   J.Endrenyi, P. Albrecht, R.Billinton, G.E. Marks, N. Reppen, L. Salvaderi, "Bulk Power System Reliability Assessment - Why and how?" IEEE transactions PAS-101, Sept. 1982, pp. 3439-3456.

13   S.Panichelli, L. Salvaderi, S. Scalcino "Unavailability Statistical Models for Power System Reliability Studies "Generic Techniques in System Reliability Assessment. Volum Pag. 295 NATO Advanced Study Institute, Liverpool 1983 Werdhoff Published.

14   T.F. of the Application of Probability Methods Subcommittee "IEEE Reliability Test System" IEEE Trans. P.A.S., Vol. PAS-98 pp.2047.

OSVALDO BERTOLDI, was born in Verona, Italy, on September 13th 1941 and received the Doctor's degree in Engineering from Politecnico of Milan in 1965.
After an initial activity in the field of polluted insulation, since 1969 he was involved as Research Engineer in the development of digital programs for the analysis of electrical power systems and in particular for the Reliability Assessment in network expansion planning and reliability evaluation.
In the framework of consultant services offered by ENEL and CESI he has been involved in execution and coordination of several planning and reliability studies of foreign countries transmission systems and he has held training courses on relevant methodologies.
He is member of CIGRE W.G. 38.03 "Power System Reliability Analysis" and he chairs the Task Force "Weather Modeling" of it.
He is author of about 10 technical papers published on national and international reviews.

LUIGI SALVADERI, was born in Ancona, Italy, June 1938. He received the Doctor's degree in Electrical Engineering from the University of Genova, Italy, in March 1962.
After his military service, he began working in June 1963 in the field of servosystems and automatic controls.
In november 1965 he joined ENEL (the Italian National Electricity Authority) and until 1968 his activity was concerned with the design and construction of 420-245 kV substations; he was also a members of the Italian staff involved in the project of Sardinia-Italy HVDC link.
Since 1969 he has been working by the Construction Dept. and by the Research and Development Dept. of ENEL in charge of the development of techiques for the evaluation of transmission and generation system reliability for power system design and planning.
Since 1978 he has been by ENEL Planning Dept. as responsible of the Generation Section. On behalf of ENEL, he has been acting as Senior Consultant Engineer in many contracts for foreign countries.
He partecipates in the activity of International Committees:
- CIGRE Working Group 37.01 "Planning Methods and Criteria"
- CIGRE Working Group 38.03 "Power System Reliability Analysis"
- IEEE "Application of Probability Methods" SubCommittee.
He is coauthor of more than 60 paper published on international reviews.

SERGIO SCALCINO, was born in Rome on december 4th 1945. He received the doctor degree in electrotechinical Engineering from the Static University in Rome in 1968.
He joined ENEL 1970, he began his activity in the National Engineering Center for Hydroelectric Station and Transmission System, where he was mainly involved in the development of the ENEL digital program for studies of generation and transmission planning and reliability.
Since 1975 he has been working in Electrical Research Center of Studies and Research Department of ENEL in the field of Power System Planning as responsable of Electric System Design Division.
From August 1976 to August 1977 he worked in Iran at Ministery of Energy as an advisor for a review of the Power System Planning of that country.
From 1980 to 1982 he was a member of a task force on batteries of CIGRE W.G.41.02; in 1982 he was the responsible of a study for CESP Company on reliability of various substation layout solutions.
Since 1983 is member of CIGRE W.G.38.03 (subject: reliability) and convener of T.F. 05.

BULK SYSTEM RELIABILITY - MEASUREMENT AND INDICES

A Report by IEEE Working Group on Measurement Indices*

C.C. Fong, Ontario Hydro (Chairman); R. Billinton, University of Saskatchewan; R.O. Gunderson, Nebraska Public Power District; P.M. O'Neill, Florida Power Corp; J. Raksany, Wisconsin Power and Light Co; A.W. Schneider, Jr. Commonwealth Edison Co; B. Silverstein, Bonneville Power Administration

Abstract:

This paper presents methods and indices for measuring actual bulk system reliability of individual utilities. Categories of unreliability events are identified for measurement, and locations in the bulk system where measurement can be made are suggested. Data requirements and indices calculation are discussed.

Measurement of bulk system reliability has a number of benefits for utilities. The proposed measurement system uses data available from utility operating reports so that the cost of implementation is expected to be modest.

INTRODUCTION

The quality of electricity supply consists of four aspects: reliability, frequency, voltage, and harmonics. Prediction techniques for bulk system reliability have been undergoing considerable development. This has resulted in models for evaluating generation system reliability, and composite generation and transmission system reliability [1 - 5, 7, 8]. Reliability Indices for the models have been proposed and discussed [1, 5, 13].

In order to effectively manage bulk system reliability for a utility, it is also necessary to develop methods and procedures for measuring the actual reliability of the bulk system. Measurement of actual reliability provides feedback to planners on the performance of the executed plans, and to operations personnel on the reliability effects of operating and maintenance practices.

For those developing or applying prediction techniques, measurement of bulk system reliability has the potential of being a validation tool for the predictive models. Information on actual reliability could also be of interest to some customers. Through measurement, it will be possible to determine the difference in reliability performance between the operating districts within a

---

* This group is under the direction of the Task Force on Bulk Power Indices (Chairman: R. Billinton, U of S) of the Working Group on Performance Records for Optimizing System Design (Chairman: M. Lauby, EPRI).

89 WM 155-3 PWRS A paper recommended and approved by the IEEE Power System Engineering Committee of the IEEE Power Engineering Society for presentation at the IEEE/PES 1989 Winter Meeting, New York, New York, January 29 - February 3, 1989. Manuscript submitted August 29, 1988; made available for printing November 17, 1988.

utility. The difference may be caused by climate, geography, load density, or system design. Comparison of performance between individual utilities, however, should be done with caution for data analysis efforts and practices as well as the years of collected data may vary.

Measurement of the actual reliability of bulk systems has in recent years been receiving more and more attention from utilities [11]. This has led to surveys of performance for specific types of unreliability phenomena [12].

The task assigned to the Working Group was to propose a method of measuring bulk system reliability for individual utilities, and to propose basic measurement indices. This paper describes the work performed by the Working Group, namely:

- identification of bulk system unreliability events,

- determination of the events to measure,

- determination of the points of measurement, measurement approach and required data, and

- recommendation of basic indices.

BULK SYSTEM RELIABILITY

A power system consists of three basic functional zones: generation, transmission and distribution. The functional zones can be combined for the purpose of reliability evaluation and measurement, as illustrated in the hierarchical levels shown in Figure 1 [3].

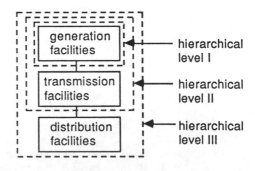

Figure 1 Hierarchical Levels

The bulk system is defined as hierarchical level II, i.e., consisting of integrated generation and transmission facilities. Some utilities may regard a portion of hierarchical level II as being of specific interest - the portion of their system consisting only of EHV transmission (230 kV and up) and the generation connected to it. The concept of a bulk system is therefore a somewhat relative one, and is dependent on the characteristics and

---

Reprinted from *IEEE Trans. Power Apparatus Syst.*, vol. 4, no. 3, pp. 829–835, Aug. 1989.

350

practices of individual utilities. The measurement approach to be presented, however, is not dependent on whether the interest is in the whole or a part of hierarchical level II.

Bulk system reliability performance is measured in terms of the amount of unreliability created by events in the bulk system. Unreliability denotes the inability to provide the required electricity to all ultimate customers supplied by the bulk system. It involves loss of load to the bulk system. The supply may be through a direct connection to the bulk system, or may be indirectly through the distribution system.

UNRELIABILITY EVENTS

A bulk system unreliability event is an event in the bulk system representing inability to supply electricity required by one or more customers. This inability may be a consequence of manual actions or of automatic operations of protective devices. In addition, operating procedures specify actions which should be taken to avoid violation of security limits which would jeopardize reliability.

Manual Actions to Avoid Unreliability

(a) Adjust power flow through phase shifters or DC lines.
(b) Re-dispatch generation.
(c) Start combustion turbine units.
(d) Reduce sales/increase purchases from interconnections.

Manual Actions to Confine
Unreliability by Reducing Load

(e) Cut interruptible loads.
(f) Reduce system-wide voltage.
(g) Appeal to public to reduce demand.
(h) Institute rotational load shedding.

Automatic Operations Signifying Unreliability

(i) Automatic, intentional load shedding by devices designed to prevent or minimize a System Disturbance [Item (k)].

(j) Interruption of supply due to bulk system equipment outages causing loss of continuity.

(k) System Disturbance which is a widespread load loss involving one or more of the following phenomena: system instability, cascading outages, formation of islands, or undesirable voltage or frequency.

While the ordering of the items generally reflects their significance to customers, variations in operating procedures and system design may cause some utilities to skip certain items or place them in a different hierarchy. Items (e) to (k) are classified as unreliability events.

The manual actions to avoid unreliability listed above are usually taken to avoid violation of operating security limits in anticipation of further outages which may cause unreliability. They are attempts to exit from alert states [1]. The manual actions to confine unreliability, on the other hand, are usually taken when a condition such as capacity or energy shortfall has already occurred, in order to minimize its consequences.

The proposed measurement system does not recommend collection of data on all unreliability events: Items (e), (f) and (g) are excluded. These are not included because their effect on the customer is less [Items (f) and (g)], or because the effect on the customer is voluntary [Items (e) and (g)], or because the unsupplied energy is difficult to measure accurately [Items (f) and (g)].

The cutting of supply from the bulk system by a customer's own protective devices, termed "load shaken off," is not included. Such devices may be responding to frequency or voltage fluctuations caused by events on the bulk system. However, this reflects the voltage and frequency aspects of supply quality, not reliability, and may be caused by overly sensitive settings of protective devices not under the control of the utility. In addition, the unsupplied energy is difficult for the utility to measure accurately.

Summarizing, the proposed measurement system concentrates on the following unreliability events:

• Interruption of supply due to loss of continuity [Item (j)]

• System Disturbance [Item (k)]

• Automatic and rotational load shedding [Items (i) and (h)].

MEASUREMENT SYSTEM

Owing to the differences in their impacts, the unreliability events selected in the previous section should be measured separately. A System Disturbance involves widespread load loss with important societal and economic impacts often absent in a localized interruption due to loss of continuity. The interruption of bulk system supply to a nuclear power station has important safety implications. Automatic and rotational load shedding involves tripping of distribution feeders, but the latter may allow enough time to notify customers.

In order to measure bulk system reliability, it is useful to define specific locations so that the reliability of supply at those locations can be conveniently observed. The locations are called measurement points. Although there are minor exceptions (to be discussed), measurement at these locations reflects bulk system reliability.

Measurement Points

Three types of measurement points have been identified. They are illustrated conceptually in Figure 2, and are defined as follows:

(a) Radial Delivery Point: A radial delivery point is:

• a low voltage bus of a transformer. The bus radially supplies a distribution system. Or

• a point in the bulk system which is an interface between utility-owned facilities and customer-owned facilities. The point radially supplies the customer.

(b) Meshed Delivery Point: A meshed delivery point is:

• a low voltage bus of a transformer. The bus supplies a meshed distribution (or sub-transmission) system. Or

• a point in the bulk system which is an interface between utility-owned facilities and customer-owned facilities. The point supplies the customer who also receives supply from other point(s).

(c) Significant Point: This is a bus or a point within the bulk system, where the utility deems it important to measure the reliability of supply. This type of measurement point recognizes that there may be special purpose measurement points which are different from radial or meshed delivery points.

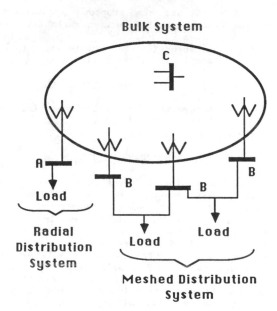

Bulk System

A: Radial Delivery Point
B: Meshed Delivery Point
C: Significant Point

Figure 2   Measurement Points

The distinction between radial and meshed delivery points is that interruption of radial delivery points results in load loss, whereas interruption of meshed delivery points does not always result in load loss. The need to distinguish between utility-owned and customer-owned facilities is based on the viewpoint that for the utility, the responsibility for reliability covers its own facilities only. An example of customer-owned facilities is an industrial customer directly connected to the bulk system.

Figure 3 illustrates the two types of delivery points. For example, Point A radially supplies a customer owned system directly connected to the bulk system. It is, therefore, a radial delivery point. Point B is a low voltage bus radially supplying a distribution system, and is also a radial delivery point. Point C is a meshed delivery point. It supplies a meshed distribution system which is also supplied from another meshed delivery point - Point D. Points E and F are meshed delivery points supplying a meshed system owned by the customer.

Although not common, it is possible that a low voltage bus supplies both a radial and a meshed distribution system. Such a bus may be classified for convenience as a radial delivery point.

An example of a significant point is the bulk system supply to a nuclear power station. The station may rely on bulk system supply in case of an accident on unit shutdown. An interruption of supply (whether the supply is needed by the station or not) is sometimes referred to as "loss of off-site power" [6]. The bus where measurement takes place may be on the low voltage side of the station service transformer associated with the unit. Another example is an important interconnection with a neighbouring utility or non-utility system. A third example could be the bulk system's supply to its own system control centre.

Measurement Approach

The measurement of bulk system reliability consists of separate measurement of the following.

(a) Delivery Point Interruption: This is the interruption of bulk system supply to a radial or meshed delivery point. The interruption of supply to the two types of delivery points should be separately identified as the impact on customer and the urgency of restoration can be different. Interruption of a meshed delivery point does not always involve load loss. If a radial delivery point is supplied from a meshed delivery point, then it is better to measure at the radial delivery point.

(b) Interruption of a Significant Point: This is the interruption of supply to a pre-selected bus or point within the bulk system. It does not necessarily result in load loss, but is deemed to reflect bulk system reliability.

(c) Automatic and Rotational Load Shedding: This is an unreliability event, parts or all of which may escape detection at the aforementioned measurement points. It involves the tripping of distribution feeders due to conditions in the bulk system. It can also involve tripping of delivery points (ie, Item (a)) in some cases.

(d) System Disturbance: This is a widespread load loss due to system instability, etc. In a System Disturbance, all of Items (a), (b) and (c) can happen. Owing to the important societal and economic impacts of a System Disturbance, this unreliability event is identified for separate measurement.

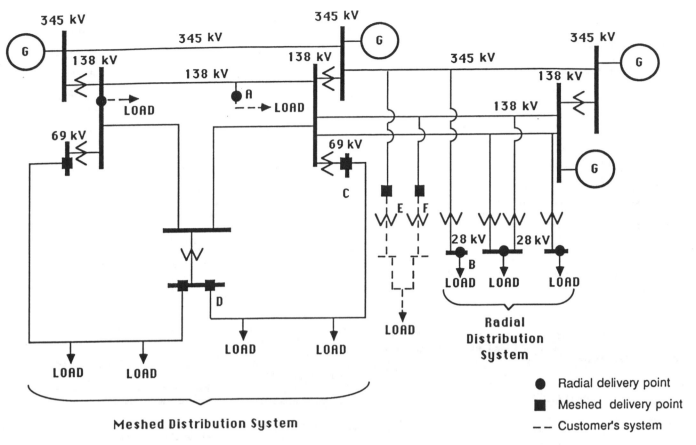

Figure 3   Delivery Points

Legend:
- ● Radial delivery point
- ■ Meshed delivery point
- – – Customer's system

The measurement approach focuses on customer load loss, through the monitoring of radial delivery point interruptions, System Disturbances, and automatic/rotational load shedding. It also accounts for events which may or may not involve load loss, namely, meshed delivery point interruptions, and interruptions of significant points. Although these two event types do not always cause load loss, they are regarded as desirable for inclusion in the measurement of bulk system reliability performance.

The use of the aforementioned measurement points simplifies the quantification of bulk system reliability, but it does not capture the tripping of individual distribution feeders due to bulk system problems. This shortcoming is overcome by the separate measurement of bulk system events which involves such trippings, i.e., automatic and rotational load shedding [Item (c)].

Data Requirements

Due consideration was given to keeping the cost of data collection to a reasonable level. The data described should be readily available from utility operating reports.

The data can be used to perform analysis on the cause of unreliability, the seasonal pattern of unreliability, etc. They are also required for calculating the measurement indices described in the next section. In all cases, unsupplied energy can be estimated as the product of duration and load loss.

(a)   Delivery Point Interruption:

The required data are date and time of occurrence, duration, load loss, number of customers affected, and cause.

If more than one meshed delivery point is interrupted and customer load loss results, then the load loss, unsupplied energy, and number of customers affected may be allocated amongst the meshed delivery points.

Two durations are required for a delivery point interruption: (i) duration of restoration of bulk system supply to the delivery point; (ii) duration of restoration of customer load which is the time taken to restore customer service either through the bulk system or through switching in the distribution system. Duration (i) reflects the restoration process of the bulk system whereas duration (ii) reflects customer impact.

Some utilities have a large number of delivery points, in which case it is worthwhile to set up a computer data base for delivery point data. Suggestions for data collection for a computer data base are provided in the Appendix.

(b)   Interruption of a Significant Point:

The required data are date and time of occurrence, duration, and cause.

The monitoring of this type of interruption depends on individual utility needs. The utility may wish to record more data than is indicated here.

(c) <u>Automatic and Rotational Load Shedding</u>:

The required data for an automatic load shedding are date and time of occurrence, load losses and respective durations, and cause.

For a rotational load shedding, the required data are date and time of start of event, load losses and respective durations, and cause.

Rotational load shedding is a complex unreliability event to monitor exactly. The unsupplied energy is accumulated from the effects of each round of load shedding.

(d) <u>System Disturbance</u>:

The data required are date and time of occurrence, load losses and respective durations, and cause.

Restoration is usually conducted in stages, and therefore the overall duration of a System Disturbance is not well defined and will not be used in indices calculation.

BASIC INDICES

Several fundamental indices are proposed. They can be expanded based on individual utility needs. All of these indices can be calculated for a defined recording period. The length of the period should be measured in years to facilitate comparison.

<u>Delivery Point Interruption Indices</u>:

(a) <u>Individual Delivery Point Indices</u>: For each delivery point (radial or meshed):

$$\text{Interruption Frequency} = \frac{\text{No. of Interruptions}}{\text{Period (Years)}}$$

$$\text{Annual Interruption Duration} = \frac{\text{Sum of Interruption Durations}}{\text{Period (Years)}}$$

$$\frac{\text{Average Duration}}{\text{Per Interruption}} = \frac{\text{Sum of Interruption Durations}}{\text{No. of Interruptions}}$$

$$\text{Annual Load Interrupted} = \frac{\text{Sum of MW Interrupted}}{\text{Period (Years)}}$$

$$\text{Annual Unsupplied Energy} = \frac{\text{Sum of Unsupplied Energy}}{\text{Period (Years)}}$$

$$\frac{\text{Delivery Point}}{\text{Interruption Severity}} = \sum_i \frac{\text{Total Unsupplied Energy (MW Minutes) in Year i}}{\text{Delivery Point Peak Load (MW) in Year i}}$$

(b) <u>System-Wide Delivery Point Indices</u>: For the bulk system and for each type of delivery point (radial or meshed):

$$\frac{\text{System Average}}{\text{Interruption Frequency}} = \frac{\text{No. of Interruptions}}{\text{Delivery Point-Years}}$$

$$\frac{\text{System Average}}{\text{Interruption Duration}} = \frac{\text{Sum of Interruption Durations}}{\text{Delivery Point-Years}}$$

$$\frac{\text{System Average}}{\text{Restoration Duration}} = \frac{\text{Sum of Interruption Durations}}{\text{Number of Interruptions}}$$

$$\text{System Average Load Interrupted} = \frac{\text{Sum of MW Interrupted}}{\text{Delivery Point-Years}}$$

$$\frac{\text{System Average}}{\text{Unsupplied Energy}} = \frac{\text{Sum of Unsupplied Energy}}{\text{Delivery Point-Years}}$$

$$\frac{\text{System Interruption}}{\text{Severity}} = \sum_i \frac{\text{Total Unsupplied Energy (MW Minutes) in Year i}}{\text{System Peak Load (MW) in Year i}}$$

The duration quantity used in the duration indices above may be either duration (i) or (ii) in the Data Requirements section.

If the number of customers interrupted is known, then the frequency and duration indices in the above can also be calculated in terms of customer-interruptions and customer-duration.

The System Average Interruption Frequency and Duration Indices are frequently referred to as SAIFI and SAIDI. These indices have been in use for distribution system reliability assessment [2].

The System Interruption Severity Index is expressed in System Minutes. One System Minute is equivalent to the interruption of the entire bulk system for one minute at the time of annual system peak. The index is synonymous with the Bulk Power Energy Curtailment Index in [13], except in this case it is applied to delivery point interruptions only.

When calculating system average indices for the entire bulk system, consideration must be given to the ultimate use of the indices. Such system averages give a general idea about overall performance. However, studies by the Mid-Continent Area Power Pool and Northeast Utilities have shown that care must be exercised when pooling data between different geographical areas [9]. System reliability for different areas may be quite different and pooling them may lead to a combined level of reliability that is not indicative of either area. There may therefore be a need for calculating geographical area based system average indices.

<u>Indices for Interruption of a Significant Point</u>

For each significant point:

$$\text{Interruption Frequency} = \frac{\text{No. of Interruptions}}{\text{Period (Years)}}$$

$$\text{Annual Interruption Duration} = \frac{\text{Sum of Interruption Durations}}{\text{Period (Years)}}$$

The monitoring of significant points is very specific to the needs of individual utilities. "System-wide" average indices are not suggested here but may be considered by the utility, for groups of significant points.

Automatic and Rotational Load Shedding Indices

For each type of load shedding:

$$\text{Frequency} = \frac{\text{No. of Events}}{\text{Period (Years)}}$$

$$\text{Load Shedding Severity} = \sum_i \frac{\text{Total Unsupplied Energy (MW Minutes) in Year } i}{\text{System Peak (MW) in Year } i}$$

These indices may also be calculated for a specific automatic scheme, to indicate its impact on bulk system reliability. The Load Shedding Severity Index is expressed in System Minutes.

System Disturbance Indices

For an individual System Disturbance:

$$\text{System Disturbance Severity} = \frac{\text{Unsupplied Energy (MW Minutes)}}{\text{Annual System Peak (MW)}}$$

This index is expressed in System Minutes. It is again fundamentally the same as the Bulk Power Energy Curtailment Index in [13], but is applied to an individual System Disturbance in this case.

The severities of individual System Disturbances can be grouped into ranges of severity (called Degrees of Severity). The purpose is to segregate major System Disturbances from minor ones. This approach also recognizes that societal and economic impacts are difficult to determine accurately. Therefore, it will suffice to classify System Disturbances only in terms of broad categories, i.e., Degrees of Severity.

The following judgmental scale can be used for measuring the Degree of Severity of individual events [5, 10 - 12].

	Severity
Degree 0	System Minutes < 1
Degree 1	1 ≤ System Minutes < 10
Degree 2	10 ≤ System Minutes < 100
Degree 3	100 ≤ System Minutes < 1000
Degree 4	1000 ≤ System Minutes

For overall bulk system assessment:

$$\text{Frequency} = \frac{\text{No. of System Disturbances}}{\text{Period (Years)}}$$

$$\text{Overall System Disturbance Severity} = \sum_i \frac{\text{Total Severity of System Disturbances in Year } i}{}$$

These two indices can also be calculated for System Disturbances of different Degrees of Severity, e.g., the frequency of Degree 1 System Disturbances.

CONCLUSION

Bulk system unreliability events are complex phenomena to monitor exactly. The Working Group has developed a measurement system with associated indices for measuring actual bulk system reliability. The emphasis of the measurement system has been on fundamental aspects, providing utilities with the approach, the data requirements, the indices, as well as the flexibility to expand on their own. The data requirements can largely be fulfilled by existing utility operating reports. The cost of implementing the measurement system is expected to be modest.

With the proposed measurement system, utilities are encouraged to consider the monitoring of additional events in the categories of manual actions to avoid unreliability and to confine unreliability via load reduction. Some of these events represent attempts to exit from alert states. These states have been suggested for inclusion in composite system reliability evaluation models[1]. Frequency and duration indices for the events are useful. The Working Group would like to hear from utilities which have developed measurement indices for these events.

ACKNOWLEDGMENT

We would like to thank the members of the Working Group on Performance Records for Optimizing System Design (PROSD), for their comments and suggestions. We would also like to thank M. Lauby, Chairman of PROSD, for his enthusiasm and support given to this work.

REFERENCES

1. Bhavaraju, M.P., R. Billinton, N.D. Reppen, R.J. Ringlee, and P.F. Albrecht, "Requirements for Composite System Reliability Evaluation Models," IEEE Transactions on Power Systems, Volume 3, No. 1, February 1988, pp. 149 - 157.

2. Billinton, R. and R.N. Allan, Reliability Evaluation of Power Systems, Pitman Books, 1984.

3. Billinton, R. and R.N. Allan, "Power-System Reliability in Perspective," Electronics and Power, March 1984, pp 231 - 236.

4. Billinton, R. and R.N. Allan, Reliability Assessment of Large Electric Power Systems, Kluwer Academic Publishers, 1988.

5. CIGRE Working Group 38.03, Power System Reliability Analysis - Application Guide, International Conference on Large High Voltage Electric Systems, 1988.

6. Electric Power Research Institute, Loss of Off-Site Power at Nuclear Power Plants: Data and Analysis, EPRI Report NP-2301, 1982.

7. Endrenyi, J., P.F. Albrecht, R. Billinton, G.E. Marks, N.D. Reppen, L. Salvaderi, "Bulk Power System Reliability Assessment - Why and How? Part I: Why?", IEEE Transactions on Power Apparatus and Systems, Vol. PAS - 101, No. 9, September 1982, pp 3439 - 3445.

8. Endrenyi, J., P.F. Albrecht, R. Billinton, G.E. Marks, N.D. Reppen, L. Salvaderi, "Bulk Power System Reliability Assessment - Why and How? Part II: How?," IEEE Transactions on Power Apparatus and Systems, Vol PAS - 101, No. 9, September 1982, pp 3446 - 3456.

9. Lauby, M.G., R.P. Ludorf, J.H. Doudna, D.D. Klempel, P.J. Lehman, C.G. Dahl, "Analysis of Pooling 345 kV Bulk Transmission Outage Data between the Mid-Continent Area Power Pool and Northeast Utilities," IEEE Transactions on Power Apparatus and Systems, Vol PAS - 104, No. 9, September 1985, pp 2427 - 2434.

10. LeReverend, B.K. and C.C. Fong, "The Measurement of Bulk Electricity System Reliability," Proceedings of the 10th Inter-Ram Conference, Montreal, 1983.

11. Winter, W.H., "Measuring and Reporting Overall Reliability of Bulk Electricity Systems," Paper No. 32-15, CIGRE 1980 Session - August 27 to September 4, Paris.

12. Winter, W.H. and B.K. LeReverend, "Disturbance Performance of Bulk Electricity Systems," Paper No. 37/38/39-02, CIGRE 1986 Session - August 27 to September 4, Paris.

13. Working Group on Performance Records for Optimizing System Design, "Reliability Indices for use in Bulk Power Supply Adequacy Evaluation," IEEE Transactions on Power Apparatus and Systems, Vol PAS-97, No. 4, July/August 1978, pp 1097-1103.

APPENDIX

Delivery Point Data Base

This Appendix describes the basic data requirements of a delivery point data base. Delivery point information is separated into two types: inventory data and interruption data.

Inventory data refers to information which describes the characteristics of individual delivery points. Interruption data refers to data collected when the delivery point suffers an interruption.

Both the inventory and interruption data listed in the following represent basic data with which a utility can analyze its delivery point performance. Utilities may expand this basic set to include other data of interest for a more comprehensive analysis.

(a) Inventory Data

• Delivery point identification.

• Delivery point type: Radial or meshed.

• Supply voltage: The high voltage supply to the delivery point (eg, 500 kV, 345 kV, 138 kV, etc).

• Supply type: Single-line or multi-line (more than one line) supplying the delivery point. Single-line supplied delivery points generally have worse reliability performance than multi-line supplied ones.

• Voltage of delivery point (kV).

• In-service date.

• Out-of-service date.

• Operating district/region.

• Capacity of transformer: MVA ratings.

• Annual peak load.

(b) Interruption Data

• Delivery point identification.

• Date and time of interruption.

• Durations of interruption: (i) the duration to restore bulk system supply to the delivery point, (ii) the duration that customer load is interrupted.

Delivery point interruptions restored by automatic reclosure facilities are referred to as momentary interruptions. The duration of these interruptions is very small (of the order of a few seconds). All other interruptions are referred to as sustained interruptions. Since the effect on customers is different from sustained interruptions, momentary interruptions can be analyzed separately. Momentary interruptions may be assigned zero duration to distinguish them from sustained interruptions.

• Unreliability type: Codes denoting whether the delivery point interruption was part of a System Disturbance, was attributable to EHV transmission, was attributable to a generation problem, or was part of an automatic load shedding operation, etc.

• Initial cause: The initial cause that resulted in the interruption. Examples are adverse weather, foreign interference (bird, animal, vehicle, etc), equipment failure (line, transformer, station problem, etc), emergency interruption (manual operation), incorrect relay operation, system configuration, unknown cause, etc.

• Extending cause: The cause (if different from the initial cause) that contributed to the duration of the interruption. For example, a delivery point may be interrupted due to lightning, but restoration is delayed because of a blacked-out supervisory board. The initial cause is lightning and the extending cause is station equipment failure.

• Weather condition: Weather condition at the time of interruption (eg, lightning, high winds, ice, snow, fog, freezing rain, and normal).

• Actual MW interrupted: From this quantity, the unsupplied energy can be approximated by the product MW x duration. If the interruption duration is very long, an assumed load curve for the delivery point may be used.

• Number of customers interrupted.

Part 3
Transmission and Distribution System Reliability

AN examination of the early literature on power system reliability evaluation clearly indicates the initial emphasis on generating capacity evaluation. There has also been considerable developmental work done in the transmission and distribution functional zones. The transmission functional zone is presently being absorbed within HLII because it is an integral part of composite system assessment. The techniques developed for the transmission and distribution functional zones will continue to be used and extended in the reliability assessment of distribution systems.

Quantitative assessment techniques in this area can be said to have been initiated in 1964 with the presentation of two papers [1], [A1] which, although quite different, presented important concepts. Reference [1] introduced the concept of failure bunching in parallel facilities due to storm-associated failures, together with some basic techniques which have proven to be useful in many areas of application. A major contribution was the introduction of procedures for calculating failure frequency (approximated as failure rate) and average outage duration in addition to the probability of failure. These indices give a practical basis for transmission, distribution, and customer reliability evaluation. A practical application of techniques available at the time was presented in [A2].

The application of Markov processes to transmission system evaluation was illustrated in [2] which considered the effect of storm-associated failures on simple parallel configurations and compared the calculated results with those of [1]. References [1] and [2] provided the basic framework for the material in [3] and [A3] which presented a consistent set of equations for series/parallel system reduction including adverse weather and permanent, temporary maintenance and overload outage modes. Reference [A3] also illustrated the concept of utilizing minimal cuts in complex configurations.

The incorporation of switching action in the evaluation of transmission circuits including protective elements was introduced in [4]. These ideas were formalized in [5], which presented a basic three-state model incorporating the switching after fault concept. This is the basic framework utilized in [6], which presented a procedure for evaluating substation and switching station reliability and a quantitative comparison of the reliability of fundamental station configurations. Reference [6] also introduced the concept of active and passive faults in systems containing protective elements. These concepts were extended under more general conditions in [7].

The utility of the techniques developed for transmission and distribution system evaluation including active and passive failures was illustrated in [8], which applied these concepts to the auxiliary systems of power stations. Reference [9] extended the basic concepts of distribution system reliability

evaluation by incorporating operational constraints such as the partial loss of continuity. Reference [A4] extended the concepts of minimal cut sets and the associated equations to include the effects of common mode failures in parallel and meshed systems. Many of the concepts presented in the references above and other publications are summarized and illustrated by simple numerical examples in [A5]. This includes the basic indices used by utilities to describe the past performance of distribution systems and the indices available for the prediction of future performance.

The question of data has always been of concern in quantitative reliability evaluation and although most utilities collect data, in some form, on transmission and distribution system components, there is relatively little generally published material. A basis assumption in most analysis has been that multiple component failures result from independent events.

The question of dependency is, however, a very important one and was addressed in an IEEE Committee Report in [10]. The approach utilized by one utility group in collecting transmission outage data is illustrated in [11] and provides an illustration of the data that can be collected. One of the main difficulties in developing a uniform data collection system is the definition of basic terms and indices for reporting and analyzing outages of transmission and distribution facilities. Early work in this regard is contained in [A6]. More recent work by an IEEE Working Group, which ultimately resulted in IEEE Standard 859-1987 [A7], is reported in [12].

Distribution system facilities contribute approximately 80 percent of the interruptions seen by the average customer. Typical North American statistics are illustrated in [A8]. The development of data collection systems and analytical techniques that can be used in optimizing planning, design, and operation of these facilities is therefore expected to continue and to provide important system benefits.

References

Reprinted Papers

[1] D. P. Gaver, F. E. Montmeat, and A. D. Patton, "Power system reliability: I—measures of reliability and methods of calculation," *IEEE Trans. Power Apparatus Syst.*, vol. 83, no. 7, pp. 727–737, July 1964.

[2] R. Billinton and K. E. Bollinger, "Transmission system reliability evaluation using Markov processes," *IEEE Trans. Power Apparatus Syst.*, vol. PAS-87, no. 2, pp. 538–547, Feb. 1968.

[3] R. Billinton and M. S. Grover, "Reliability assessment of transmission and distribution systems," *IEEE Trans. Power Apparatus Syst.*, vol. PAS-94, no. 3, pp. 724–732, May/June 1975.

[4] R. J. Ringlee and S. D. Goode, "On procedures for reliability evaluation

of transmission systems," *IEEE Trans. Power Apparatus Syst.*, vol. PAS-89, no. 4, pp. 527–537, Apr. 1970.

[5] J. Endrenyi, "Three state models in power system reliability evaluation," *IEEE Trans. Power Apparatus Syst.*, vol. PAS-90, no. 4, pp. 1909–1916, July/Aug. 1971.

[6] M. S. Grover and R. Billinton, "A computerized approach to substation and switching station reliability evaluation," *IEEE Trans. Power Apparatus Syst.*, vol. PAS-93, no. 5, pp. 1488–1497, Sept./Oct. 1974.

[7] R. N. Allan, R. Billinton, and M. F. de Oliveira, "Reliability evaluation of electrical systems with switching actions," *Proc. IEE,* vol. 123, pp. 325–330, Apr. 1976.

[8] R. N. Allan, M. F. de Oliveira, and R. Billinton, "Reliability evaluation of the auxiliary electrical systems of power stations," *IEEE Trans. Power Apparatus Syst.*, vol. PAS-96, no. 5, pp. 1441–1449, Sept./Oct. 1977.

[9] R. N. Allan, E. N. Dialynas, and I. R. Homer, "Modelling and evaluating the reliability of distribution systems," *IEEE Trans. Power Apparatus Syst.*, vol. PAS-09, no. 6, pp. 2181–2189, Nov./Dec. 1979.

[10] IEEE Committee Report, "Common mode forced outages of overhead transmission lines," *IEEE Trans. Power Apparatus Syst.*, vol. PAS-95, no. 3, pp. 859–864, May/June 1976.

[11] M. G. Lauby, K. T. Khu, R. W. Polesky, R. E. Vandello, J. H. Doudna, P. J. Lehman, and D. D. Klempel, "MAPP bulk transmission outage data collection and analysis," *IEEE Trans. Power Apparatus Syst.*, vol. PAS-103, no. 1, pp. 213–221, Jan. 1984.

[12] D. W. Forrest, P. F. Albrecht, R. N. Allan, M. P. Bhavaraju, R. Billinton, G. L. Landgren, M. F. McCoy, and N. D. Reppen, "Proposed terms for reporting and analyzing outages of electrical transmission and distribution facilities," *IEEE Trans. Power Apparatus Syst.*, vol. PAS-104, no. 2, pp. 337–348, Feb. 1985.

Additional References

[A1] Z. G. Todd, "A probability method for transmission and distribution outage calculations," *IEEE Trans. Power Apparatus Syst.*, vol. 33, no. 7, pp. 696–701, July 1964.

[A2] S. A. Mallard and V. C. Thomas, "A method for calculating transmission system reliability," *IEEE Trans. Power Apparatus Syst.*, vol. PAS-87, no. 3, pp. 824–834, March 1968.

[A3] R. Billinton and M. S. Grover, "Quantitative evaluation of permanent outages in distribution systems," *IEEE Trans. Power Apparatus Syst.*, vol. PAS-94, no. 3, pp. 733–741, May/June 1975.

[A4] R. N. Allan, E. N. Dialynas, and I. R. Homer, "Modelling common mode failures in the reliability evaluation of power system networks," IEEE Paper A79 040-7.

[A5] R. Billinton and R. N. Allan, "Reliability evaluation of power systems," New York: Plenum, 1984.

[A6] IEEE Standard 346-1973, Section 2. Standard Definitions in Power Operations Terminology Including Terms of Reporting and Analyzing Outages of Electrical Transmission and Distribution Facilities and Interruptions to Customer Service.

[A7] IEEE Standard 859-1987. Terms for Reporting and Analyzing Outage Occurences and Outage States of Electrical Transmission Facilities.

[A8] R. Billinton and J. E. Billinton, "Distribution system reliability indices," *IEEE Trans. Power Delivery,* vol. 4, no. 1, pp. 561–568, Jan. 1989.

Power System Reliability
I—Measures of Reliability and Methods of Calculation

D. P. Gaver   F. E. Montmeat, Senior Member IEEE   A. D. Patton, Member IEEE

An increasing amount of attention is being focused on the reliability or continuity of service afforded by transmission and distribution systems. This increased attention primarily stems from (1) the need to supply improved service as customers become more dependent on their electric service, and (2) the desire to use new system voltages and designs whose reliability is not well-known to supply the heavier loads of the future. In the past, the reliability of proposed systems has been estimated by extrapolating the experience obtained from existing systems and using rule-of-thumb methods. In the future, however, more precise methods of predicting and evaluating reliability will be required.

Recognizing the need for precise methods of predicting the reliability of transmission and distribution systems, Public Service Electric and Gas Company and Westinghouse Electric Corporation have developed analytical techniques employing probability theory which permit calculation of a number of important measures of reliability or service "goodness" in general power system networks. These techniques have built on and extend the work of other authors.[1-3] The techniques which are presented in this paper were developed as part of a general study of distribution system planning undertaken jointly by Public Service and Westinghouse. This general study, designed to assure economical future expansion of distribution systems, is introduced in a companion paper.[4]

The methods presented in the paper will permit the calculation of various measures of reliability in power systems from basic system component parameters and a characterization of environmental severity variations. This makes possible the comparison of alternative system designs to discover the lowest cost system yielding the desired reliability properties. The method will also permit a rational evaluation of alternative operating procedures, such as amount of tree trimming done and number of line crews available to perform repairs.

Measures of System Reliability

The determination of an adequate measure of reliability or service goodness for a transmission or distribution system is itself a difficult problem. Indeed, it seems certain that no single measure of system reliability is completely descriptive of the system's ability to supply satisfactory service. Satisfactory service as defined by the electric power customer has

been shown to be influenced by frequency and duration of outages as well as other more elusive factors.[5] Thus, at least two basic measures of reliability are significant. Any one customer is, of course, primarily interested in his own service goodness. That is, he is interested in system reliability as seen from his point of service. The utility may well be interested in other measures of reliability such as the average service goodness provided all customers on the system or the poorest service provided any customer. Such measures of reliability will have meaning when comparing alternative systems which serve many customers.

Listed here are some measures of reliability which are of interest and significance and which can be calculated by the methods of the paper:

$\bar{F}$ = average number of service interruptions per customer served per year
$\bar{R}$ = average customer restoration time
$\bar{H}$ = average total interruption time per customer served per year
F_{max} = maximum expected number of interruptions experienced by any one customer per year
R_{max} = maximum expected restoration time experienced by any one customer
P = probability that any customer will be out of service at any one time longer than a specified time

The first three quantities in this list express measures of average service reliability for a system serving large numbers of customers. Note that these quantities could also be expressed in terms of load units rather than customer units. The next two quantities indicate the poorest service reliability afforded any customer on the system. The last measure of reliability, P, is also a measure of the poorest service afforded any customer. The measure P may be of interest if a goal of system design is to assure, with some probability, that no customer will be out of service longer than a certain specified time.

The quantities $\bar{F}$, $\bar{R}$, $\bar{H}$, F_{max}, and R_{max}, can be estimated for existing systems if records are kept of the duration of each outage and the number of customers affected. The following are expressions for estimates of $\bar{F}$, $\bar{R}$, and $\bar{H}$. The values which should be assigned to F_{max} and R_{max} are obvious from their definitions.

$$\hat{F} = \frac{\text{No. of customer interruptions per year}}{\text{No. of customers served}} \tag{1}$$

$$\hat{R} = \frac{\text{No. of customer interruption hours during year}}{\text{No. of customer interruptions during year}} \tag{2}$$

$$\hat{H} = \frac{\text{No. of customer interruption hours during year}}{\text{No. of customers served}} \tag{3}$$

Note that the "hat" symbol ($\wedge$) denotes an estimate of the value of a parameter.

The average degree of customer satisfaction is a function of the quantities $\bar{F}$, $\bar{R}$, and $\bar{H}$. Ultimately, energy sales and profits are also a function of these quantities. Unfortunately, however, customer satisfaction and energy sales and profits cannot be evaluated for most utility systems given the various

Paper 64-90, recommended by the IEEE Power System Engineering Committee and approved by the IEEE Technical Operations Committee for presentation at the IEEE Winter Power Meeting, New York, N. Y., February 2-7, 1964. Manuscript submitted November 4, 1963; made available for printing December 2, 1963.

D. P. Gaver is with the Westinghouse Electric Corporation, Pittsburgh, Pa.; F. E. Montmeat is with Public Service Electric and Gas Company, Newark, N. J.; and A. D. Patton is with Westinghouse Electric Corporation, East Pittsburgh, Pa.

The authors wish to express their appreciation to R. M. Sigley of Lehigh University for his contributions toward the development of the methods described in the paper.

Reprinted from *IEEE Trans. Power Apparatus Syst.*, pp. 727–737, July 1964.

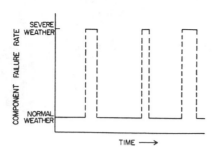

Fig. 1. A 2-state fluctuating environment

measures of reliability. The relationships involved are not well enough understood at the present time to permit the required analytical expressions to be written. In some industrial systems, however, relationships between reliability and product output have been formulated in such a way that a definite dollar value can be assigned to various degrees of reliability.[3] Since, in general, no definite economic value can be assigned to degree of reliability in utility systems, utilities are for the most part forced to rely on history and experience to establish acceptable levels of system reliability. Therefore, in studying utility systems, it is usually necessary to have estimates of the various measures of reliability for existing systems. These estimates of the various measures of reliability define levels of service goodness against which the calculated reliability of proposed systems can be compared measure for measure.

Calculation of Measures of Reliability

Power systems are composed of groups of elements or components that act in series or parallel or both with each other to carry power from generation sources to load buses. In a system link composed of several components in series, the failure of any component will result in an outage of the system link. The series link will then be out of service until the failed component can be replaced or repaired. In a system link composed of two or more components acting in parallel, however, a system link outage is experienced only when all parallel components are out of service or when load exceeds the capacity of components remaining in service. The parallel link will be out until a component capable of carrying the required load is restored to service. In the sections to follow, mathematical models for the calculation of various measures of reliability in such systems are developed. First, however, a unique feature of the power system reliability problem must be discussed.

Fluctuating Environment

A power system, at least that part of it that is exposed to the weather, is subjected to a fluctuating environment. During stormy periods, for example, environmental conditions may be so severe as to result in equipment and line failure rates much higher than those prevailing during nonstorm periods. Fig. 1 illustrates an environment which fluctuates between normal and severe states as a random process. The failure rate process illustrated in Fig. 1 is certainly simplified. One would expect the various occasions of "severe" conditions to result in failure rates with different values rather than the constant value assumed here. Not all storms have the same wind velocities, amount and kind of precipitation, amount of lightning, etc. Hence the value of failure rate prevailing during a severe period might more properly be a random draw from a population of values for "severe" failure rates.

If severe weather strikes a number of components making

up a system simultaneously, several component failures may occur during a short period of time. This bunching of failures due to a common event can have important effects on system reliability. To illustrate, consider a "system" made up of two parallel components equally capable of carrying full system load. This system will fail only if both components are down at the same time. Ignoring maintenance outages, both components will be down at the same time only if (a) both fail simultaneously, or (b) one component fails and is not repaired before the other component fails. It should be obvious that the chance of overlapping component outages and consequent system failure is greater when component failures are induced by severe environmental conditions to bunch than when component failures occur randomly and independently.

Previous methods of estimating power system reliability have presumed that constant average component failure rates apply at all times. That is, previous methods presume that components fail randomly and independently. In a later section, a numerical example will be presented to show that the independence assumption can yield an appreciable underestimate of the outage rate of parallel systems. The independence assumption in calculating the outage rate of series systems results in a slight overestimation of system outage rate. The error incurred due to an assumption of independence appears negligible, however, for reasonable numbers of components in series and for representative failure rates, since single component failures still greatly outnumber overlapping failures. Furthermore, a series system is at least partially de-energized as soon as a single component fails, thus reducing the chance of subsequent overlapping failures. Therefore, independence can usually be safely assumed in calculating the reliability of series systems.

The degree of sophistication used in representing the effect of fluctuating environment in parallel system reliability calculations is limited largely by the availability of required data. Field trouble reports usually state whether or not weather was a contributing or causative agent in the component failure. Hence failures can readily be separated into "normal-weather" or "random" failures and "storm-associated" failures. Trouble reports may also give a broad indication as to the type of failure-causing weather, such as lightning, wind, or ice, making possible classification of failures according to type of storm. Given these data, together with statistics on the number of components in service and the average time each year during which the various kinds of weather prevail, average component failure rates which prevail during normal weather and the different kinds of stormy weather may be estimated. It appears, however, that the additional accuracy in reliability calculations gained by identifying different types of storms rather than lumping all storms together does not warrant the considerable additional difficulty in data collection and cal-

Table I. Storm Definitions

Weather Bureau Basic Indicator	Minimum Wind Velocity, Knots	Minimum Temperature, F	Minimum Duration, Hours
Thunderstorm	..	..	..
Heavy thunderstorm	..	..	..
Tornado	..	..	..
Squall	..	..	..
Moderate rain	20	..	5
Heavy rain	20	..	..
Moderate freezing rain	..	..	10
Heavy freezing rain	..	..	4
Moderate snow	20	28	..
Heavy snow	20	28	..
Wind only	25	..	..

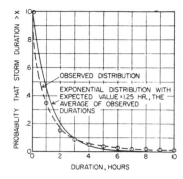

Fig. 2. Storm duration distribution for Newark, N. J.

tions of thunderstorm and "other" storm durations. Thus a possible refinement would be to consider separately different seasons or periods of the year; then combine the results of the separate calculations to put measures of reliability on an annual basis—their usual form. The methods of reliability calculation described in the paper are applicable to any period of time but require the approximate fulfillment of certain mathematical assumptions made in their derivation.

SYSTEM COMPONENT DATA FOR RELIABILITY CALCULATIONS

Two basic types of component data are required in making power system reliability calculations. These data are (1) component failure and maintenance outage rates and (2) distributions of component repair times. The detail with which component failure and repair time distributions are required depends somewhat on the portion of the system which is to be studied and the objective of the study, but in general the following data are required:

1. A normal weather (random) "permanent" forced outage (failure) rate should be estimated for all types of apparatus and lines which exhibit distinctive or characteristic failure rates. This failure rate, λ_i (where i labels the component type), is expressed in units of failures per year of normal weather per unit of apparatus or per mile of line. It may be estimated as follows:

$$\hat{\lambda}_i = \frac{C}{Y} \tag{4}$$

where

C = number of nonstorm-associated component failures during observation period
Y = summation of normal weather exposure times for each mile of line or piece of apparatus during the observation period

2. A stormy weather "permanent" forced outage rate should be obtained for components whose failure rate is affected by weather and which are used in a portion of the system, such as the subtransmission system, which is usually operated as a parallel system. The stormy weather failure rate, λ_i', has the units of failures per year of stormy weather per unit of apparatus or per mile of line. It may be estimated by the equation:

$$\hat{\lambda}_i' = \frac{C'}{Y'} \tag{5}$$

where

C' = number of storm-associated component failures during observation period
Y' = summation of stormy weather exposure times for each mile of line or piece of apparatus during observation period

culation methods required. Therefore, the method of this paper presumes the simple 2-state model of Fig. 1 in which the environment alternates between normal and severe conditions. The durations of normal and severe conditions are random variables drawn from distributions of durations obtained from historical weather data. Note that disaster storms such as major hurricanes and tornadoes should not be lumped with other less violent storms. Such storms are relatively rare, but give rise to component failure rates much higher than those experienced during "average" storms. Therefore, lumping disaster storms with average storms would greatly dilute the failure bunching effect of disaster storms. The effect of disaster storms on system reliability can be estimated by separate consideration of such high stress periods, again using the methods of the paper.

Historical weather records in the form of hourly and special observations, from which required weather statistics can be obtained, are available from the United States Weather Bureau for all major weather stations. Weather statistics should be obtained from data recorded at a station in the vicinity of the system being studied. For Public Service, data from the Newark and Trenton stations were analyzed. Weather data are scanned with a definition of weather conditions constituting a storm to determine (1) the total amount of time each year during which normal and storm conditions prevail, and (2) the distributions of normal and severe weather period durations.

Definitions of weather conditions constituting storms must be carefully correlated with conditions observed to cause an increased rate of component failure. If storm definitions are too stringent, too few storms will be recognized. On the other hand, if storm definitions allow too many mild, relatively undestructive storms to be recognized, the important failure-bunching effect of more severe storms will be diluted. Using the storm definitions given in Table I, Newark weather data were analyzed for the period 1955–1961. Storms were found to occupy an average of 0.65% of the time a year. Figs. 2 and 3 show, respectively, the distributions of durations of stormy and normal weather periods found for the 1955–1961 period. Note that these distributions do not differ greatly from the exponential. Thus it appears reasonable to utilize the exponential distribution to describe severe and normal weather period durations in making reliability calculations. For a discussion of the exponential distribution, see Feller.[6]

While stormy and normal weather period durations appear to be approximately exponentially distributed over an entire year, the storm rate and the kind of storms appearing are not the same for different seasons of the year. Fig. 4 shows the average frequency of occurrence of (1) thunderstorms and (2) all other storms for Newark. Fig. 5 shows the distribu-

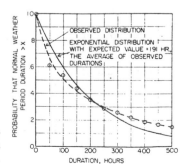

Fig. 3. Normal weather period duration distribution for Newark, N. J.

3. A temporary forced outage rate for various types of components is necessary if system temporary outages are to be calculated. In general, a temporary outage does not require repair or replacement of facilities but can be remedied by a reclosing operation or by replacing a fuse. Component temporary forced outage rate is estimated by dividing the number of component temporary outages during the observation period by the number of unit-calendar years of component exposure during the observation period. It is not considered necessary to separate temporary outages into normal and stormy weather components because temporary outages are usually quite short, making the chance of overlapping outages negligibly small.

4. Maintenance outage rates should be obtained for types of components which are used in portions of the system which operate as a parallel system. Maintenance outages of components in radial systems, if such occur, can be lumped with normal weather forced outages. Maintenance outage rate λ_i'' has the units of outages per calendar year per unit of apparatus or per mile of line. It is estimated by

$$\lambda_i'' = \frac{C''}{Y''} \tag{6}$$

where

C'' = number of component maintenance outages during observation period

Y'' = summation of observation periods for each mile of line or piece of apparatus (exposure to maintenance is assumed to be essentially the same each year)

5. Distributions of component repair times prevailing during normal and stormy weather and for maintenance are obtained from historical records. Repair time is here taken to mean the duration of a period during which a component is out of service being repaired or replaced following a forced outage, or the time a component is out of service for maintenance or other work. Repair time distributions seem to be exponential to a reasonable approximation. That is, probability (repair time$>t) = e^{-t/r}$. For example, see Fig. 6. The expected values of repair times during normal and stormy weather and for maintenance, denoted respectively by r_i, r_i', and r_i'', are estimated by the means (averages) of observed repair times. Because it appears that repair times during normal and stormy weather are very difficult to separate and are not greatly different, the calculation methods to follow presume that the same repair time distribution prevails during both normal and stormy weather. This combined repair time distribution is obtained from records of repair times following all forced outages.

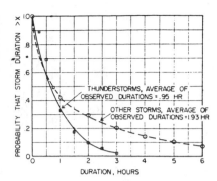

Fig. 5. Distributions of thunderstorm and "other" storm durations for Newark, N. J.

If it is desired to investigate the possible economies of parallel systems with less than 100% redundancy, that is, parallel components not individually capable of carrying system peak load, then certain additional data are required. These data consist of component capabilities, loadings under contingency conditions, and curves giving the probability of carrying load. A more complete discussion of overload outages in parallel systems is deferred to a later section of the paper.

ASSUMPTIONS IN RELIABILITY CALCULATIONS

Some of the assumptions which were made in deriving the reliability calculation methods of the paper have been mentioned in preceding sections. These assumptions together with other pertinent assumptions, are listed as follows for ease of reference:

1. Times to failure (periods between failures) and repair times are exponentially distributed during both normal and stormy weather. That is,

probability [time to failure (during normal weather)$>t] = e^{-\lambda t}$
probability [time to failure (during stormy weather)$>t] = e^{-\lambda' t}$

2. The durations of periods of normal and stormy weather are exponentially distributed.

3. Repair times are typically very short compared with times to failure and times between storms.

4. Storms are very short in duration compared with times to failure for components. Storms are also short compared with typical repair times.

5. Maintenance outages occur at random during normal weather periods except that components are not taken out for maintenance if (a) such action would cause the remaining components in a parallel system to become overloaded, or (b) maintenance could not be completed before a storm struck.

6. Maintenance down times are exponentially distributed.

7. In computing parallel system down times resulting from overload outages, it is assumed that once a line is overloaded, it remains overloaded and down until a failed parallel component is repaired.

SERIES SYSTEMS

In this and the following section, the necessary mathematical expressions for calculating the various measures of reliability in simple series systems and in simple parallel systems are presented. Because it is possible to regard most power system networks as a group of series and parallel subsystems, these expressions can be used to calculate measures of reliability in general systems.

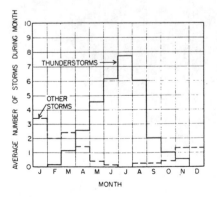

Fig. 4. Average number of storms per month in Newark, N. J.

Consider a system composed of n dissimilar components connected in series. A component may be a single piece of apparatus, several miles of line, or even an equivalent component composed of several other components connected in series or parallel or both. The following reliability parameters are required:

$\lambda_1, \lambda_2, \ldots, \lambda_n =$ normal weather component failure rate, failures/year of normal weather

$\lambda_1', \lambda_2', \ldots, \lambda_n' =$ stormy weather component failure rate, failures/year of stormy weather

$\lambda_1'', \lambda_2'', \ldots, \lambda_n'' =$ component maintenance outage rate, maintenance outages/calendar year

$r_1, r_2, \ldots, r_n =$ expected value of repair time for all forced outages, years

$r_1'', r_2'', \ldots, r_n'' =$ expected value of down time for maintenance outages, years

$N =$ expected value of normal weather period duration, years

$S =$ expected value of stormy weather period duration, years

The required values of component failure or outage rates just listed are obtained from normalized failure rate data by multiplying the normalized values by the appropriate number of units, such as miles of line.

The over-all forced outage rate (normal and stormy weather) for the ith component is

$$\lambda_{fi} = \frac{N}{N+S}\lambda_i + \frac{S}{N+S}\lambda_i' \quad \text{(forced outages/calendar year)} \quad (7)$$

This is an appropriate approximation when $\lambda_i N$ and $\lambda_i' S$ are very small compared with unity, as is usually the case in utility practice. The over-all forced outage rate of the series system is

$$\lambda_{fe} = \sum_{i=1}^{n} \lambda_{fi} \quad \text{(forced outages/calendar year)} \quad (8)$$

Again, this approximation is useful when $N\Sigma\lambda_i$ and $S\Sigma\lambda_i'$ are small compared with unity; see reference 7. Similarly, the maintenance outage rate of the series system is

$$\lambda_e'' = \sum_{i=1}^{n} \lambda_i'' \quad \text{(maintenance outages/calendar year)} \quad (9)$$

If the series system acts in parallel with other components, it is necessary to calculate normal and stormy weather failure rates for an element e which is the equivalent of the series system. This equivalent element is used in further calculations. For the equivalent element

$$\lambda_e = \sum_{i=1}^{n} \lambda_i \quad \text{(failures/year of normal weather)} \quad (10)$$

and

$$\lambda_e' = \sum_{i=1}^{n} \lambda_i' \quad \text{(failures/year of stormy weather)} \quad (11)$$

Expected values of down time for series system as a result of forced outages and maintenance outages are, respectively,

$$r_{fe} = \frac{\sum_{i=1}^{n} \lambda_{fi} r_i}{\lambda_{fe}} \quad \text{(years)} \quad (12)$$

and

$$r_e'' = \frac{\sum_{i=1}^{n} \lambda_i'' r_i''}{\lambda_e''} \quad \text{(years)} \quad (13)$$

Consideration of outages caused by overloading in a series system is not meaningful unless the series system operates in parallel with other components. If such is the case and if overload outages are to be evaluated, then the capacity and contingency loading of the equivalent element e, representing the series system, are required. The capacity of the equivalent element is obviously the minimum of the capacities of the components making up the series system. The contingency loading of the equivalent element depends on the configuration of the parallel system being studied.

If a load was attached to the last element of the series system and a source to the first element, then the significant measures of reliability for the load would be calculated as follows:

1. Annual outage rate:

$$\lambda_{SL} = \lambda_{fe} + \lambda_e'' \quad \text{(outages/calendar year)} \quad (14)$$

2. Expected value of outage duration (restoration time):

$$r_{SL} = \frac{\lambda_{fe} r_{fe} + \lambda_e'' r_e''}{\lambda_{SL}} \quad \text{(years)} \quad (15)$$

which may be converted to hours by multiplying by 8,760.

3. Average total outage time per year:

$$U_{SL} = \frac{r_{SL}}{r_{SL} + \frac{1}{\lambda_{SL}}} \cong \lambda_{SL} r_{SL} \quad \text{(years/year)} \quad (16)$$

which may be converted to hours per year by multiplying by 8,760.

4. Probability that a single outage will last longer than t hours

$$P(\text{outage} > t \text{ hours}) = \sum_{i=1}^{n} \frac{\lambda_{fi} e^{-t/(8760\,r_i)} + \lambda_i'' e^{-t/(8760\,r_i'')}}{\lambda_{SL}} \quad (17)$$

Parallel Systems

The equations to be given for parallel systems are limited to two components, or equivalent components, in parallel. If three or more components are in parallel, it appears that, for typical characteristics, they may be treated two at a time, i.e., two of the three parallel components are combined; then the equivalent combined with the third. This simplification implies a certain additional degree of independence between the three or more components not considered when only two components are in parallel. In other words, not quite as much failure bunching due to storms is recognized in the calculation when three or more components are in parallel.

Fig. 6. Repair time distribution for 26-kv open-wire sub-transmission circuits on the Public Service Electric and Gas Company system

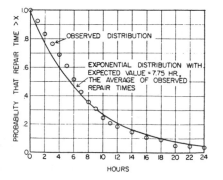

This is consistent with the fact that a system made up of three or more components in parallel is likely to be more dispersed geographically than a 2-component system with the result that all components are somewhat less likely to be affected by the same storm.

Component or equivalent component data required in making reliability calculations in parallel systems, excluding outages due to overloads, are the same as that listed previously for series systems.

The calculation of reliability in parallel systems may be approached in several ways. The most accurate method, at least theoretically, involves treatment of system experience as a Markov process. The mathematical background for this approach is discussed by Feller[6] and the approach used by De Sieno and Stine[2] in their paper on power system reliability. The Markov process approach generates a series of linear equations which must be solved simultaneously to find system properties such as long-run average availability and failure rate. Explicit consideration of a 2-state fluctuating environment for a parallel system composed of two nonidentical components requires the solution of eight simultaneous equations. Such a solution is, of course, easy on a high-speed digital computer, but reconsideration of the problem provides an intuitively appealing and simple approximation which can be quickly computed by hand. The approximate method can be expected to give results that are within a few per cent of those obtained by using the Markov method. It should be remarked that the Markov method itself is an approximation and does depend upon certain specific distributional assumptions.

The simplified approximate method yields the following expression for parallel system over-all outage rate due to normal and stormy weather forced outages and maintenance outages:

$$\lambda_{SL} = \frac{N}{N+S}\left[\lambda_1\lambda_2(r_1+r_2) + \frac{S}{N}(\lambda_1'\lambda_2 r_1 + \lambda_2'\lambda_1 r_2) + \frac{S}{N}(\lambda_1\lambda_2'r_1 + \lambda_2\lambda_1'r_2) + 2\frac{S^2}{N}\lambda_1'\lambda_2'\right] + \lambda_1''\lambda_2 r_1'' + \lambda_2''\lambda_1 r_2''$$

(outages/calendar year) (18)

A derivation of expression 18 is given in Appendix I. If the parallel system operates in parallel with other components, then normal weather, stormy weather, and maintenance outage rates are required for the equivalent component representing the parallel system in further calculations. These rates are expressed as follows:

$$\lambda_e = \lambda_1\lambda_2(r_1+r_2) + \frac{S}{N}(\lambda_1'\lambda_2 r_1 + \lambda_2'\lambda_1 r_2)$$

(failures/year of normal weather) (19)

$$\lambda_e' = \frac{N}{S}\left[\frac{S}{N}(\lambda_1\lambda_2'r_1 + \lambda_2\lambda_1'r_2) + 2\frac{S^2}{N}\lambda_1'\lambda_2'\right]$$

(failures/year of stormy weather) (20)

$$\lambda_e'' = \lambda_1''\lambda_2 r_1'' + \lambda_2''\lambda_1 r_2''$$ (maintenance outages/calendar year)

(21)

The failure rates λ_e, λ_e', and λ_e'' for the equivalent component representing the parallel system are derived from the various terms of the expression for λ_{SL}. The nature of the weather when the second component of the parallel system fails resulting in system failure determines whether a term contributes to equivalent component normal or stormy weather forced outage rate. The terms in the expression for λ_{SL} accounting for the effect of maintenance outages yield the maintenance outage rate of the equivalent component directly.

Given that the times to failure for components are exponentially distributed and that component repair times are independently and exponentially distributed, the expected value of parallel system down time as a result of system outages caused by overlapping component forced outages only is

$$r_{fe} = \frac{1}{\frac{1}{r_1}+\frac{1}{r_2}} = \frac{r_1 r_2}{r_1+r_2} \qquad \text{(years)} \qquad (22)$$

The expected value of system down time when a component forced outage overlaps a component maintenance outage is

$$r_e'' = \left(\frac{\lambda_1''\lambda_2 r_1''}{\lambda_1''\lambda_2 r_1'' + \lambda_2''\lambda_1 r_2''}\right)\left(\frac{r_2 r_1''}{r_2+r_1''}\right) + \left(\frac{\lambda_2''\lambda_1 r_2''}{\lambda_1''\lambda_2 r_1'' + \lambda_2''\lambda_1 r_2''}\right)\left(\frac{r_1 r_2''}{r_1+r_2''}\right) \qquad \text{(years)} \quad (23)$$

The reasoning used to derive r_e'' may be explained by considering the first term of the expression for r_e''. The first factor of this term is the fraction of system outages involving component maintenance outages in which component 2 failed while component 1 was out for maintenance. The second factor of the first term gives the expected system down time given that component 2 fails while component 1 is out for maintenance. The second term of the expression for r_e'' accounts in like manner for the situation where a forced outage of component 1 overlaps a maintenance outage of component 2.

The significant measures of reliability for a load fed by a parallel system are: annual outage rate, λ_{SL}; expected value of restoration time, r_{SL}; average total outage time per year, U_{SL}; and the probability that a single outage will last longer than t hours. The annual outage rate, λ_{SL}, is given by equation 18 of this section. The expected value of restoration time, r_{SL}, and the average total outage time per year, U_{SL}, are obtained using the same equations (15 and 16) as for series systems. The probability that a single outage will last longer than t hours is given by the following expression:

$$P(\text{outage} > t \text{ hours}) = \frac{1}{\lambda_{SL}}\left[(\lambda_{SL} - \lambda_e'')e^{-t/(8760 r_{fe})} + \lambda_1''\lambda_2 r_1'' e^{-t/\left(8760\frac{r_2 r_1''}{r_2+r_1''}\right)} + \lambda_2''\lambda_1 r_2'' e^{-t/\left(8760\frac{r_1 r_2''}{r_1+r_2''}\right)}\right] \quad (24)$$

To illustrate the difference in system failure rates calculated using the method of the paper and a method presuming independence, consider a "system" composed of two identical 10-mile sections of overhead circuit acting in parallel. Suppose that the average number of forced outages in both

Table II. Component Failure Rates

| % of Failures During Storms | Failures Per Calendar Year | | | |
	During Storms	During Normal Weather	λ_i'	λ_i
0	0	0.441	0	0.444
10	0.0441	0.3969	6.79	0.400
20	0.0882	0.3528	13.58	0.355
40	0.1764	0.2646	27.16	0.266
60	0.2646	0.1764	40.74	0.1775
80	0.3528	0.0882	54.32	0.0888
100	0.441	0	67.90	0

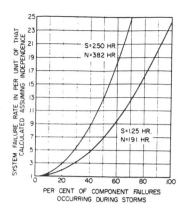

Fig. 7. A comparison of parallel system failure rates calculated (1) assuming independence, and (2) using the method of the paper

stormy and normal weather observed per year per mile for the type of circuit in question is 0.0441. Then

$$\lambda_{f1} = \lambda_{f2} = 10(0.0441) \doteq 0.441 \text{ failure/calendar year}$$

Let $r_1 = r_2 = 7.75$ hours $= 8.84 \times 10^{-4}$ years. Then, assuming independence and ignoring maintenance outages the system failure rate is

$$\lambda_{f1}\lambda_{f2}(r_1 + r_2) = 3.44 \times 10^{-4} \text{ outages/calendar year}$$

Let the expected values of stormy and normal weather periods, S and N, be, respectively, 1.25 hours and 191 hours. With these values of S and N storms occupy an average of 0.65% of the time each year. The values of component failure rates appropriate during stormy and normal weather periods when storms occupy 0.65% of time are given in Table II as a function of the percentage of total component failures that occur during storms. Parallel system failure rates may be calculated from the component failure rates given in Table II with equation 18. Maintenance outages are ignored. Expressing the resultant values of system failure rate in per unit of the value of system failure rate obtained assuming independence and plotting the per-unit values as a function of the percentage of total component outages occurring in storms yields Fig. 7. Fig. 7 indicates that the error in parallel system failure rate calculated assuming independence can be appreciable and is most serious when a sizable percentage of component failures occur during storms. Fig. 7 also shows the effect on parallel system failure rate of doubling storm durations while keeping the percentage of time that is stormy constant at 0.65%. The conclusion is that a smaller number of long storms is more likely to cause system outages than a larger number of short storms even though total storm time is the same in either case.

Outages Due to Overloads in Parallel Systems

If each component in a parallel system is capable of carrying the highest load to which it may be subjected in any contingency, the system will suffer an outage only if all parallel components are out of service at the same time. Such a system will never suffer an outage as a result of overloading of a sound component when a parallel component or components fail. A parallel system which can never fail because of component overload is termed fully or 100% redundant. That is, each parallel component of the system has a capacity which is greater than or equal to the highest load that component can ever have to carry. A fully redundant system while quite reliable is also relatively expensive because of high component capacities. A possible means

of reducing system cost is to reduce the capacities of parallel system components relative to the highest loads the components may be called on to carry under contingency conditions. Obviously, this procedure will introduce the possibility of system failure due to component overload and will, therefore, reduce system reliability to some extent. A component with less than 100% capacity may or may not be able to carry contingency load depending on time of day, day of week, season of year, duration of contingency, and component capacity relative to the peak value of load. The method which is to be described permits an estimate of the reduction in system reliability caused by overload outages to be calculated so that systems with less than 100% redundancy can be evaluated relative to 100% redundant systems.

The first step in the procedure for estimating the number of overload outages a parallel system is likely to experience is the determination of the probability that a component will be able to carry a given contingency load. Assuming that a contingency is equally likely to occur at any time and that component capacity is constant, the probability of carrying contingency load for a given time is found by randomly sampling the load cycle which would be appropriate if the component were called on to carry contingency load continuously. A given sample try is termed a success if load does not exceed component capacity during the assumed contingency period. The probability of carrying contingency load successfully is estimated by the ratio of the number of successes to the total number of samples. The total load cycle period sampled should, in general, be at least a year to insure that all types of load cycle variation are included: daily, weekly, and seasonal. If component capacity for design purposes is not constant throughout the year, it is necessary to sample the annual load cycle by seasons. A family of curves giving the probability of carrying contingency load as a function of contingency load duration and component capacity can be generated by sampling the load cycle with different values of contingency load duration and component capacity. Fig. 8 shows a typical family of probability curves. The Public Service system load cycle shape for the year 1961 was used in obtaining these curves. Component capacities were assumed constant over the year. Total system load cycle shape is thought to be representative of load cycle shapes in transmission and subtransmission systems serving a combination of industrial, commercial, and residential loads. It is worthwhile to note that the probability of carrying load for a given contingency load duration and component capacity increases as load cycle

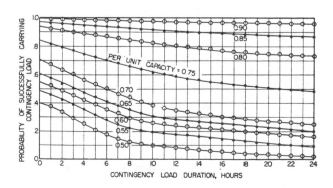

Fig. 8. Probability that contingency load can be carried successfully

variability increases and load factor decreases. This follows because load is below component capacity more of the time when the load cycle is found to be highly variable and the load factor low.

Consider a system composed of elements 1 and 2 in parallel. Presuming that component maintenance outages never precipitate system overload outages, system outage rates during normal and stormy weather due to overloads are approximated as follows:

$$\lambda_{oe} = \lambda_1 P_2 + \lambda_2 P_1 \quad \text{(overload outages/year of normal weather)} \tag{25}$$

and

$$\lambda_{oe}' = \lambda_1' P_2 + \lambda_2' P_1 \quad \text{(overload outages/year of stormy weather)} \tag{26}$$

where

P_i = probability that component i will not be able to carry contingency load

The system over-all outage rate due to overloads is λ_{ofe} where

$$\lambda_{ofe} = \frac{N}{N+S} \lambda_{oe} + \frac{S}{N+S} \lambda_{oe}' \quad \text{(overload outages/calendar year)} \tag{27}$$

This expression for system overload outage rates slightly overestimates these rates because some component forced outages overlap causing system forced outages and hence are not available to cause overload outages. Since the number of component forced outages that overlap is small compared with the total number of component forced outages, the error in the approximation should be negligible.

The probability P_i that a component will not be able to carry contingency load for a given load cycle shape is related to: (a) component capacity, (b) the value of contingency load at annual peak, and (c) the distribution of repair times for the component(s) which failed and thus precipitated the contingency. In a 2-element parallel system, peak contingency load is simply the sum of the normal peak loads on the two elements. In more complex systems, however, peak contingency load can only be found by a load flow analysis. Component capacity expressed in per unit of contingency load at annual peak, the distribution of failed component repair times (identical to contingency durations), and a set of curves such as those shown in Fig. 8 can be used to estimate P_i. If contingency load duration were a constant value rather than a distribution of values, the quantity $Q = (1 - P_i)$ could be read directly by entering the probability curves with component capacity and contingency load duration. If, however, contingency load duration is not constant but is exponentially distributed as appears reasonable, then P_1 and P_2 are given as follows:

$$P_1 = 1 - \int_0^\infty Q_1(X) dM_2(X) \tag{28}$$

and

$$P_2 = 1 - \int_0^\infty Q_2(X) dM_1(X) \tag{29}$$

where

$Q_i(X)$ = probability that component i can successfully carry contingency load for a time X

$M_i(X)$ = probability that repair of component i is completed in time X

$\quad = 1 - e^{-X/r_i}$ for exponentially distributed repair times

Since the function $Q_i(X)$ is difficult to work with, it is useful to approximate the integral in equation 28 or 29 as a summation over j. Thus, for example, P_1 becomes

$$P_1 = 1 - \sum_j Q_1(X_j) M_2(X_j) - M_2(X_j - 1) \tag{30}$$

A practical approach to the calculation of P_1 and P_2 would seem to be to split time into 1-hour intervals and to read values for $Q_i(X_j)$ off the probability curves that have been prepared. We can then put $X_j = j$ where $j = 0, 1, 2, \ldots$ Presuming repair times are exponentially distributed, P_1 is given as follows:

$$P_1 = 1 - \sum_{j=1}^\infty Q_1(j)[e^{-(j-1)/(8760 r_2)} - e^{-j/(8760 r_2)}] \tag{31}$$

Or perhaps better,

$$P_1 = 1 - \sum_{j=1}^\infty Q_1(j-1/2)[e^{-(j-1)/(8760 r_2)} - e^{-j/(8760 r_2)}] \tag{32}$$

Because there is reason to believe that $Q_1(X)$ should equal $Q_1(24)$ for all X greater than 24 hours,

$$P_1 = 1 - Q_1(24) e^{-24/(8760 r_2)} - (1 - e^{-1/(8760 r_2)}) \times \sum_{j=1}^{24} Q_1(j-1/2) e^{-(j-1)/(8760 r_2)} \tag{33}$$

and

$$P_2 = 1 - Q_2(24) e^{-24/(8760 r_1)} - (1 - e^{-1/(8760 r_1)}) \times \sum_{j=1}^{24} Q_2(j-1/2) e^{-(j-1)/(8760 r_1)} \tag{34}$$

Systems with three or more components in parallel are handled in much the same way as 2-component systems. For instance, in a 3-component system

$$\lambda_{oe} = \lambda_1 P_{2,3} + \lambda_2 P_{1,3} + \lambda_3 P_{1,2} + \lambda_{1,2} P_3 + \lambda_{1,3} P_2 + \lambda_{2,3} P_1$$
$$\text{(overload outages/year of normal weather)} \tag{35}$$

where

λ_i = normal weather failure rate of ith component

$\lambda_{i,j}$ = normal weather failure rate of equivalent component composed of components i and j in parallel (see equation 19 for method of calculation)

P_i = probability that ith component will not be able to carry load when other components are out

$P_{i,j}$ = probability that components i and j will not be able to carry load when other component is out

It appears that the terms in equation 35 involving forced outages of more than one component may be disregarded with an error of less than 5% in typical cases.

System failure rates due to all types of outages (overlapping forced outages, maintenance associated outages, and overload outages) can be obtained by adding the values found for each type of outage. Thus, defining $_T\lambda_{SL}$ as parallel system failure rate as a result of all types of outages,

$$_T\lambda_{SL} = \lambda_{SL} + \lambda_{ofe} \quad \text{(total outages/calendar year)} \tag{36}$$

The expected value of 2-component parallel system down time as a result of all types of outages including overloads is $_T r_{SL}$, where

$$_T r_{SL} = \frac{\lambda_{fe}}{_T\lambda_{SL}} r_{fe} + \frac{\lambda_e''}{_T\lambda_{SL}} r_e'' + \frac{\lambda_{f1} P_2}{_T\lambda_{SL}} r_1 + \frac{\lambda_{f2} P_1}{_T\lambda_{SL}} r_2 \quad \text{(years)} \tag{37}$$

The value of $_T r_{SL}$ for a parallel system with three or more components is not indicated here, but is found in a similar manner. The average total outage time per year for the

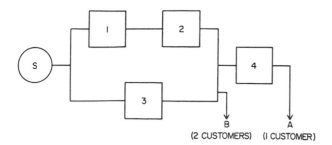

Fig. 9. A sample system

system as a result of all types of outages is $_T U_{SL}$. The quantity $_T U_{SL}$ can be obtained from equation 16 by substituting $_T r_{SL}$ for r_{SL} and $_T \lambda_{SL}$ for λ_{SL} in that equation.

Considering all types of system outages, the probability that a single system outage in a 2-component parallel system will last longer than t hours is

$$P(\text{outage} > t \text{ hours}) = \frac{1}{_T \lambda_{SL}} [\lambda_{fe} e^{-t/(8760 r_{fe})} +$$

$$\lambda_1'' \lambda_2 r_1'' e^{-t/\left(8760 \frac{r_2'' r_1''}{r_2 + r_1''}\right)} + \lambda_2'' \lambda_1 r_2'' e^{-t/\left(8760 \frac{r_1'' r_2''}{r_1 + r_2''}\right)} +$$

$$\lambda_{f1} P_2 e^{-t/(8760 r_1)} + \lambda_{f2} P_1 e^{-t/(8760 r_2)}] \quad (38)$$

The probability of a single outage longer than t for systems with more than two components is found in like manner.

EXAMPLE SOLUTION

The various measures of reliability for a simple series-parallel system will now be calculated to illustrate the method of applying the equations of the preceding sections to complex networks. The system to be studied is shown in Fig. 9. The source is considered to be 100% reliable. Two customers are served at B and one at A. The computations for the example system are contained in Appendix II.

Conclusions

The probability methods presented in the paper will permit important measures of reliability in general power system networks to be calculated from basic system component data. Thus, the reliability of alternative proposed systems can be compared to discover the system which yields the highest reliability or a desired level of reliability at lowest cost. Important features of the method of the paper include the ability to consider failure bunching caused by storms and outages as a result of component overloading in parallel systems.

Future papers will describe experience in application of the techniques which have been developed, data for reliability calculations, and a digital computer program to facilitate calculations.

Appendix I

When the conditions listed in the section of the paper titled "Assumptions in Reliability Calculations" are reasonable, it is possible to obtain an approximate expression for parallel system failure rate which will yield results very close to those obtained using the Markov approach. In the derivation of the simplified approximate expression for system failure rate to follow, discussion is limited to a 2-component system.

The derivation proceeds in the following stages:

1. INITIAL FAILURE IS DURING NORMAL WEATHER

Suppose that, at some initial instant, the system is entirely operative. Then if one component fails during normal weather, system

failure will follow if the second component fails during the repair time of the first component. Failure of the second component may occur during (a) normal or (b) stormy weather giving rise to the following two cases.

1.1. SECOND FAILURE IS DURING NORMAL WEATHER

Presuming that repair times are typically very short compared with times between storms such that, at most, one weather change is likely during a repair time, system failure rate per calendar year as a result of component failures during normal weather only is:

Failure rate = (long-run fraction of time that weather is normal) [(normal weather failure rate of component 1) (probability that a storm does not occur during repair of component 1) (probability that component 2 fails during repair of component 1) + (normal weather failure rate of component 2) (probability that a storm does not occur during repair of component 2) (probability that component 1 fails during repair of component 2)]

$$= \frac{N}{N+S} \left[\lambda_1 \left(1 - \frac{r_1}{N}\right)(\lambda_2 r_1) + \lambda_2 \left(1 - \frac{r_2}{N}\right)(\lambda_1 r_2) \right]$$

Since r_1 and r_2 are typically very small compared with N, it can be shown that

$$\text{Failure rate} = \frac{N}{N+S} \lambda_1 \lambda_2 (r_1 + r_2)$$

1.2. SECOND FAILURE IS DURING STORMY WEATHER

System failure rate per calendar year as a result of first component failure during normal weather and second component failure during stormy weather is

Failure rate = (long-run fraction of time that weather is normal) [(normal weather failure rate of 1) (probability a storm occurs during repair of 1) (probability 2 fails during that storm) + (term giving other possibility)]

$$= \frac{N}{N+S} \left[\lambda_1 \left(\frac{r_1}{N}\right)(\lambda_2' S) + \lambda_2 \left(\frac{r_2}{N}\right)(\lambda_1' S) \right]$$

$$= \frac{N}{N+S} \left(\frac{S}{N}\right) \quad (\lambda_1 \lambda_2' r_1 + \lambda_2 \lambda_1' r_2)$$

2. INITIAL FAILURE IS DURING STORMY WEATHER

If the initial component failure is during stormy weather, system failure will occur if the other component fails during (a) stormy or (b) normal weather before the failed component is restored to service. Thus, presuming that repair times are typically short compared with the time between storms, system failure may occur as a result of failure of one component during a storm and failure of the other component during normal weather following that storm.

2.1. SECOND FAILURE IS DURING STORMY WEATHER

Presuming that storm durations are typically short compared with repair times, system failure rate per calendar year as a result of component failures during the same storm is

Failure rate = (expected number of storms per year)[(probability 1 fails during a storm) (probability 2 fails during remainder of storm) + (term giving other possibility)]

$$= \frac{1}{N+S} [(S\lambda_1')(S\lambda_2') + (S\lambda_2')(S\lambda_1')]$$

$$= \frac{N}{N+S} \left(\frac{2S^2}{N} \lambda_1 \lambda_2\right)$$

Note that the probability of a second component failure during a storm is the same as the probability of the first failure because of the "memoryless" nature of the exponential distribution describing storm durations. If storm durations were a constant value rather than being exponentially distributed, the probability of a second component failure during a storm would be $\left(\frac{S}{2}\lambda_t\right)$ and the system failure rate would be half that given above.

2.2. SECOND FAILURE IS DURING NORMAL WEATHER

System failure rate per calendar year when the first component fails during stormy weather and the second during normal weather is

Failure rate = (expected number of storms per year) [(probability 1 fails during a storm) (probability 2 does not fail during that storm) (probability 2 fails during repair of 1)+(term giving other possibility)]

$$=\frac{1}{N+S}[(S\lambda_1')(1-S\lambda_2')(\lambda_2 r_1)+(S\lambda_2')(1-S\lambda_1')(\lambda_1 r_2)]$$

Usually $S\lambda_t'\ll 1$, hence

$$\text{Failure rate}=\frac{N}{N+S}\left(\frac{S}{N}\right)(\lambda_1'\lambda_2 r_1+\lambda_2'\lambda_1 r_2)$$

3. MAINTENANCE OUTAGES

It is presumed that a component will never be taken out for maintenance if this would precipitate a system outage. Likewise, a component is not taken out for maintenance if a storm is impending. A system failure can occur, however, if a component fails during normal weather while the other component is out for maintenance. System failure rate as a result of this situation is as follows:

Failure rate = (maintenance outage rate of 1) (probability 2 fails while 1 is out)+(term giving other possibility)

$$=\lambda_1''(\lambda_2 r_1'')+\lambda_2''(\lambda_1 r_2'')$$

Adding the system failure rates given under 1, 2, and 3 above gives system failure rate per calendar year due to all causes.

Appendix II

All components of the example system of Fig. 9 are overhead lines. The characteristics of the components are as follows:

(a). $\lambda_1=\lambda_2=\lambda_3=\lambda_4=0.4$ failure/year of normal weather
(b). $\lambda_1'=\lambda_2'=\lambda_3'=\lambda_4'=20$ failures/year of stormy weather
(c). $\lambda_1''=\lambda_2''=\lambda_3''=2$ maintenance outages/calendar year
(d). $r_1=r_2=r_3=r_4=8.76$ hours $=10^{-3}$ years
(e). $r_1''=r_2''=r_3''=10^{-3}$ years
(f). Component 4 is not taken out for maintenance; hence $\lambda_4''=0$ and r_4'' has no significance.
(g). The capacities of components 1, 2, and 3 are 90% of the peak contingency loads that they may be called on to carry.

The expected values of stormy and normal weather periods are

$S=1.25$ hours $=1.43\times 10^{-4}$ years

$N=191$ hours $=2.18\times 10^{-2}$ years

1. The first step in the reliability calculations for the example system is to compute an equivalent for the system link made up of components 1 and 2 in series. Call the equivalent component $e1$. From equations 9, 10, and 11 the failure rates of the equivalent are

$\lambda_{e1}''=4$ maintenance outages/calendar year

$\lambda_{e1}=0.8$ failure/year of normal weather

$\lambda_{e1}'=40$ failures/year of stormy weather

The over-all forced outage rates for single components is given by equation 7:

$\lambda_{f1}=\lambda_{f2}=0.527$ forced outage/calendar year

then

$\lambda_{fe1}=\lambda_{f1}+\lambda_{f2}=1.054$ forced outages/calendar year

---

The expected values of repair times following forced and maintenance outages for the equivalent component are given by equations 12 and 13.

$r_{fe1}=10^{-3}$ years

$r_{e1}''=10^{-3}$ years

2. An equivalent can now be found for the system link made up of components $e1$ and 3 in parallel. Call this equivalent component $e2$. From equations 19, 20, and 21, the failure rate of the equivalent, not involving overloads, are

$\lambda_{e2}=8.50\times 10^{-4}$ failures/year of normal weather

$\lambda_{e2}'=0.261$ failure/year of stormy weather

$\lambda_{e2}''=3.20\times 10^{-3}$ maintenance outages/calendar year

The over-all forced outage rate for the equivalent is given by equation 7:

$\lambda_{fe2}=2.55\times 10^{-3}$ forced outages/calendar year

The expected values of repair times followed forced and maintenance (associated) outages are given by equations 22 and 23.

$r_{fe2}=5\times 10^{-4}$ years

$r_{fc2}''=5\times 10^{-4}$ years

Since the capacities of parallel components $e1$ and 3 are not 100% of the contingency loads they may have to carry, the possibility of system link outages as a result of overloads exists. Because the system link does not operate in parallel with any other component, it is not necessary to separate overload outages occurring in stormy and normal weather. Therefore, equations, 25, 26, and 27 reduce to the following for this case:

$\lambda_{ofe2}=\lambda_{fe1}P_3+\lambda_{f3}P_{e1}$ (overload outage/calendar year)

The probabilities that components $e1$ and 3 cannot carry contingency load successfully are given by equations 33 and 34, making use of the curves of Fig. 8.

$P_{e1}=P_3=0.019$

then

$\lambda_{ofe2}=0.030$ overload outage/calendar year

The total number of outages per year of the parallel system link represented by equivalent component $e2$ is given by equation 36.

$_T\lambda_{SLe2}=3.575\times 10^{-2}$ outages/calendar year

The expected value of repair for component $e2$ considering all types of outages is given by equation 37.

$_Tr_{SLe2}=9.35\times 10^{-4}$ years

$\qquad =8.20$ hours

The customers served at point B in the system experience the same outage (interruption) rate and expected value of down time (restoration time) as equivalent component $e2$. The average total interruption time per year experienced by the customers at B as given by equation 16 is 0.293 hour. The probability that the customers at B will experience a single interruption longer than t, say 24 hours, is given by equation 38 as follows:

$P(B\text{ out}>24\text{ hours})=0.0553$

3. The customer at point A is served by components $e2$ and 4 in series. The interruption rate at A is given by equations 8, 9, and 14 and is 0.562 interruption per calendar year. The expected value of restoration time for the customer at A is given by equations 12, 13, and 15 and is 8.71 hours. The average total interruption time per year for the customer at A is given by equation 16 and is 4.9 hours. The probability that the customer at A will experience a single interruption longer than 24 hours as given by equation 17 is 0.0645.

4. The measures of reliability for the sample system then are as follows:

(a). Average number of interruptions per customer served per year

$$\bar{F} = \frac{2(3.575 \times 10^{-2}) + 1(0.562)}{3} = 0.211 \text{ interruption/year}$$

(b). Average customer restoration time

$$\bar{R} = \frac{2(8.20) + 1(8.71)}{3} = 8.37 \text{ hours}$$

(c). Average total interruption time per customer served per year

$$\bar{H} = \frac{2(0.293) + 1(4.9)}{3} = 1.83 \text{ hours}$$

(d). Maximum expected number of interruptions experienced by any one customer per year

$$F_{max} = 0.562 \text{ interruption/year}$$

(e). Maximum expected restoration time experienced by any one customer

$$R_{max} = 8.71 \text{ hours}$$

References

1. A Probability Method for Transmission and Distribution Outage Calculations, Z. G. Todd. *IEEE Transactions on Power Apparatus and Systems*, vol. 83, 1964.

2. A Probability Method for Determining the Reliability of Electric Power Systems, C. F. DeSieno, L. L. Stine. *Ibid.*, vol. 83, Feb. 1964, pp. 174–81.

3. Economic Evaluation of Industrial Power-System Reliability, W. H. Dickinson. *AIEE Transactions*, pt. II (*Applications and Industry*), vol. 76, 1957, pp. 264–72.

4. Automated Distribution System Planning, R. F. Lawrence, F. E. Montmeat, A. D. Patton, D. Wappler. *IEEE Transactions on Power Apparatus and Systems*, vol. 83, Apr. 1964, pp. 311–16.

5. The Application of Planning Criteria to the Determination of Generator Service Dates by Operational Gaming, C. A. DeSalvo, C. H. Hoffman, R. G. Hooke. *AIEE Transactions*, pt. III (*Power Apparatus and Systems*), vol. 78, 1959, pp. 1752–58.

6. An Introduction to Probability Theory and Its Applications, W. Feller. John Wiley & Sons, Inc., New York, N. Y., 1957, vol. I.

7. Random Hazard in Reliability Problems, D. P. Gaver. *Technometrics*, vol. 5, no. 2, May 1963.

Transmission System Reliability Evaluation Using Markov Processes

ROY BILLINTON, MEMBER, IEEE, AND KENNETH E. BOLLINGER, MEMBER, IEEE

Abstract—The basic concepts of stationary Markov processes and particularly their application to transmission system reliability evaluation are discussed and illustrated. Transmission components are assumed to operate within a 2-state fluctuating environment described by normal and stormy weather conditions. Markov processes are used to determine the system failure rate and the probabilities of failure for simple configurations and to illustrate the bunching effect of storm associated failures on parallel facilities. The variation in the failure probabilities is shown as a function of the expected repair time and the degree of storm associated failures. The results are compared with those obtained using a previously published approximate method.

Introduction

RELIABILITY theory is a subject that is receiving considerable attention at the present time[1] particularly in the area of space and military technology. A considerable amount of work has also been done in the field of power system reliability and some excellent papers have been published. Perhaps the most significant publications in the field of transmission system reliability within the last few years were two companion papers by a group of Westinghouse Electric Corporation and Public Service and Gas Company authors.[2], [3] These papers introduce the concept of a 2-state fluctuating environment to describe the failure rate of transmission system components and to illustrate the bunching effect of storm associated failures on parallel facilities. The techniques presented in the first of these two papers were stated to be approximations which would give results within a few percent of those obtained using more theoretical techniques, such as Markov processes. The application of Markov chains was briefly illustrated by DeSeino and Stine.[4] Within the bounds of the distributional assumptions made, the Markov approach is theoretically the most accurate for the prediction of system reliability.

This paper illustrates the basic concepts of Markov processes with respect to their application to power system reliability and in particular to the 2-state fluctuating environment condition for simple configurations. The results obtained by the Markov approach are compared to those obtained using the previously noted approximate method.[2]

Paper 31 TP 67–407, recommended and approved by the Power System Engineering Committee of the IEEE Power Group for presentation at the IEEE Summer Power Meeting, Portland, Ore., July 9–14, 1967. Manuscript submitted March 23, 1967; made available for printing May 2, 1967.

The authors are with the Power System Research Group, University of Saskatchewan, Saskatoon, Saskatchewan, Canada.

Markov Processes

The theoretical development of the Markov approach is discussed in considerable detail by Feller[5] particularly with regard to processes that are discrete in time and space. The reliability problem normally deals with a system that is continuous in time and discrete in space. The basic concepts are absolutely necessary if complicated systems are to be studied and, although documented elsewhere,[5], [6] these concepts have been included in the Appendix.

The failure rates of components located in relatively fixed environments can be considered to be constant during the useful life periods. For this type of equipment and assuming unit failure independence, complete time variant expressions for the probabilities of occupying the various states in multi-unit systems can be easily determined by correctly combining the individual unit expressions. The binomial expansion can be applied in identical unit systems as illustrated for the 2-unit case in the Appendix. In the case of a transmission line and other outdoor system components the failure rate is a function of its environment. The use of a 2-state fluctuating environment covering normal and stormy weather has been proposed[2] in which repair, failure, and weather duration distributions were assumed to be exponential. Using these distributional assumptions the Markov approach can be applied as follows to a single unit under a 2-state environment.

Single-Unit Case

The state space diagram for this case is shown in Fig. 1, where

λ, μ normal weather failure and repair rates
λ', μ' stormy weather failure and repair rates
m $1/S$, where S is the expected duration of a stormy weather period
n $1/N$, where N is the expected duration of a normal weather period.

Proceeding directly from the state space diagram in Fig. 1, the differential equations in matrix form are

$$\begin{bmatrix} P_0'(t) \\ P_1'(t) \\ P_2'(t) \\ P_3'(t) \end{bmatrix} = [P_0(t)\,P_1(t)\,P_2(t)\,P_3(t)] \begin{bmatrix} -(\lambda + n) & n & \lambda & 0 \\ m & -(m + \lambda') & 0 & \lambda' \\ \mu & 0 & -(\mu + n) & n \\ 0 & \mu' & m & -(\mu' + m) \end{bmatrix}.$$

The steady-state or limiting probabilities can be found from

$$\begin{aligned}
-(\lambda + n)P_0 &+ mP_1 &+ \mu P_2 & &= 0 \\
nP_0 &- (m + \lambda')P_1 & &+ \mu' P_3 &= 0 \\
\lambda P_0 & &- (\mu + n)P_2 &+ mP_3 &= 0 \\
&\lambda' P_1 &+ nP_2 &- (\mu' + m)P_3 &= 0
\end{aligned}$$

and

$$P_0 + P_1 + P_2 + P_3 = 1.0.$$

Reprinted from *IEEE Trans. Power Apparatus Syst.*, vol. PAS-87, no. 2, pp. 538–547, Feb. 1968.

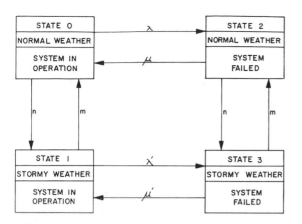

Fig. 1. State space diagram for the single-unit case with a 2-state fluctuating environment.

$$P(\text{system operating}) = P_0 + P_1$$

$$P(\text{system failed}) = P_2 + P_3.$$

Let $\mu = \mu'$, i.e., the repair rate is independent of the environment. Then

$$P(\text{system operating}) = \frac{\mu}{m + n}$$

$$\times \left[\frac{(m + n)^2 + m(\mu + \lambda') + n(\mu + \lambda)}{(\mu + \lambda)(\mu + \lambda') + m(\mu + \lambda) + n(\mu + \lambda')} \right].$$

$$P(\text{system failed}) = \frac{1}{m + n}$$

$$\times \left[\frac{n\lambda'(n + \mu) + m\lambda(m + \mu) + nm(\lambda' + \lambda) + \lambda'\lambda(m + n)}{(\mu + \lambda)(\mu + \lambda') + m(\mu + \lambda) + n(\mu + \lambda')} \right].$$

Consider the case of normal weather only; let $\lambda' = 0$, $m = 1$, and $n = 0$,

$$P(\text{system operating}) = \frac{\mu}{\lambda + \mu}$$

$$P(\text{system failed}) = \frac{\lambda}{\lambda + \mu}$$

as shown in the Appendix.

To obtain the mean time to failure (MTTF), it is not important which state the system failed in, only that it did fail. Both states 2 and 3 can be considered as absorbing states.

The stochastic transitional probability matrix obtained from Fig. 1 is

$$P =$$

$$
\begin{array}{c}
 \\
0 \\
1 \\
2 \\
3
\end{array}
\begin{array}{cccc}
0 & 1 & 2 & 3 \\
\end{array}
\left[
\begin{array}{cccc}
1 - (\lambda + n) & n & \lambda & 0 \\
m & 1 - (m + \lambda') & 0 & \lambda' \\
\mu & 0 & 1 - (\mu + n) & n \\
0 & \mu' & m & 1 - (\mu' + m)
\end{array}
\right].
$$

The truncated Q matrix is formed by deleting states 2 and 3.

$$
Q =
\begin{array}{c}
0 \\
1
\end{array}
\left[
\begin{array}{cc}
1 - (\lambda + n) & n \\
m & 1 - (m + \lambda')
\end{array}
\right].
$$

The fundamental matrix N is given by

$$N = [I - Q^{-1}] = \frac{1}{\lambda'\lambda + \lambda m + \lambda'n}\left[\begin{array}{cc} m + \lambda' & n \\ m & \lambda + n \end{array}\right].$$

Starting in state 0 the MTTF $(M_{0,23})$ is

$$M_{0,23} = \frac{m + \lambda' + n}{\lambda\lambda' + \lambda m + \lambda'n}.$$

Considering normal weather only, let $\lambda' = 0$, $m = 1$, and $n = 0$,

$$M_{0,23} = \frac{1}{\lambda}$$

as expected for a single unit.

Two- and Three-Unit Cases

$\lambda_1, \lambda_2, \lambda_3; \mu_1, \mu_2, \mu_3$	normal weather failure and repair rates for components 1, 2, 3
$\lambda_1', \lambda_2', \lambda_3'; \mu_1', \mu_2', \mu_3'$	stormy weather failure and repair rates for components 1, 2, 3
$S = 1/m$	expected duration of stormy weather
$N = 1/n$	expected duration of normal weather.

The state space diagram for the 2-unit case is shown in Fig. 2. Proceeding directly from this diagram, the differential equations in matrix form are

$$[P'(t)] = [P(t)]\left[\begin{array}{c:c} D_1 & n(I) \\ \hdashline m(I) & D_2 \end{array}\right]$$

where I is the identity matrix.

The complete matrices are shown in Table I. They apply to the case of two units in parallel or two in series. For two units in parallel with complete redundancy, the probability of the system occupying a failed state is given by summing the limiting state probabilities for states 7 and 3. The MTTF is obtained by truncating the stochastic transitional probability matrix by eliminating states 7 and 3. It is extremely laborious to obtain a general expression for the probability of occupying the failed states and for the MTTF. It is not too difficult, however, to obtain a numerical solution on a digital computer using specific values of failure, repair, stormy, and normal weather rates.

For two lines in series, the assumption is made that it is possible for a line to fail while the other line is being repaired. This is reasonable as many failures, particularly storm associated ones, are not contingent upon the line being energized. For this case the probability of the system occupying the failed state is obtained by summing the limiting state probabilities for states, 1, 2, 3, 5, 6, 7. The MTTF is obtained by truncating the stochastic transitional probability matrix by eliminating these states. This results in

$$[I - Q] =
\begin{array}{c}
 \\
0 \\
4
\end{array}
\begin{array}{cc}
0 & 4 \\
\end{array}
\left[
\begin{array}{cc}
\lambda_1 + \lambda_2 + n & -n \\
-m & \lambda_1' + \lambda_2' + m
\end{array}
\right]$$

$$N = [I - Q]^{-1} = \frac{\left[\begin{array}{cc} \lambda_1' + \lambda_2' + m & n \\ m & \lambda_1 + \lambda_2 + n \end{array}\right]}{\lambda_1\lambda_2' + \lambda_1\lambda_1' + \lambda_1 m + \lambda_2\lambda_2' + \lambda_2\lambda_1' + \lambda_2 m + \lambda_2'n + \lambda_1'n}.$$

TABLE I
PARTITIONED MATRICES FOR THE TWO-UNIT CASE

$$[D_1] = \begin{bmatrix} -\lambda_1 - \lambda_2 - n & \lambda_2 & \lambda_1 & 0 \\ \mu_2 & -\lambda_1 - \mu_2 - n & 0 & \lambda_1 \\ \mu_1 & 0 & -\mu_1 - \lambda_2 - n & \lambda_2 \\ 0 & \mu_1 & \mu_2 & -\mu_1 - \mu_2 - n \end{bmatrix}$$

$$[D_2] = \begin{bmatrix} -\lambda_1' - \lambda_2' - m & \lambda_2' & \lambda_1' & 0 \\ \mu_2' & -\lambda_1' - \mu_2' - m & 0 & \lambda_1' \\ \mu_1' & 0 & -\mu_1' - \lambda_2' - m & \lambda_2' \\ 0 & \mu_1' & \mu_2' & -\mu_1' - \mu_2' - m \end{bmatrix}$$

TABLE II
PARTITIONED MATRICES FOR THE THREE-UNIT CASE

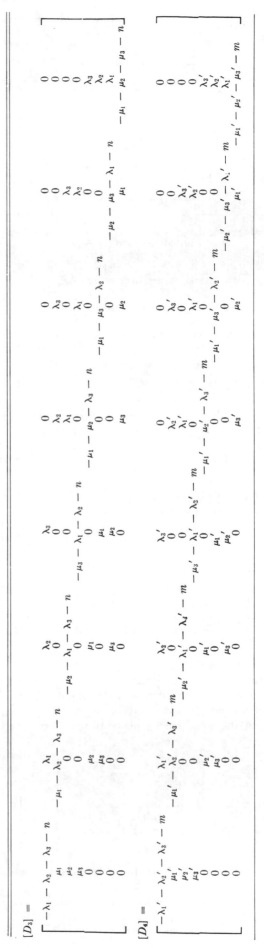

$[D_3] =$

$[D_4] =$

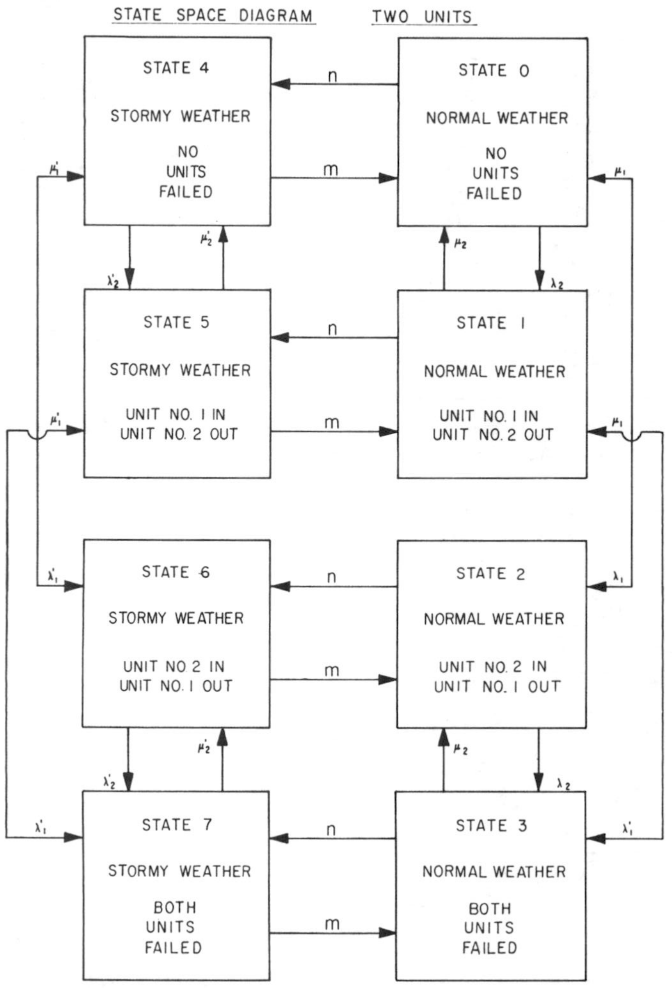

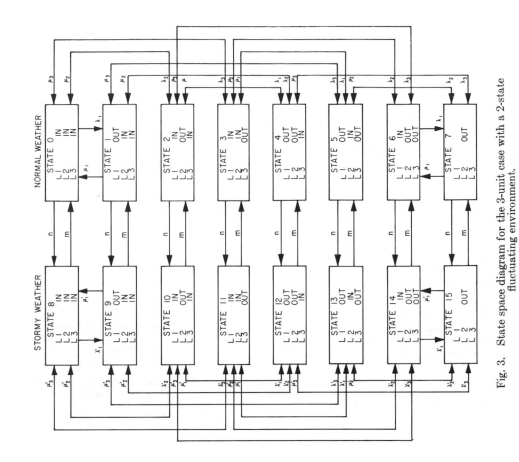

Fig. 3. State space diagram for the 3-unit case with a 2-state
fluctuating environment.

Fig. 2. State space diagram for the 2-unit case with a 2-state
fluctuating environment.

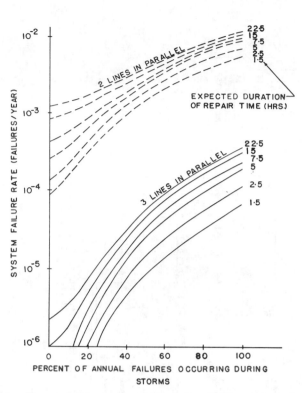

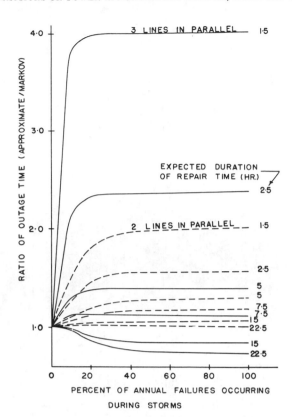

Fig. 4. Expected failure rate variation in 2 and 3 line systems due to variation in storm associated failures for system parameters of $N = 200$ hours, $S = 1.5$ hours, and λ_{Av} (component annual failure rate) = 0.5 failure per year.

Fig. 6. Comparison of the approximate and Markov methods for expected outage times over a range of repair times with system parameters of $N = 200$ hours, $S = 1.5$ hours, and λ_{Av} (component annual failure rate) = 0.5 failure per year.

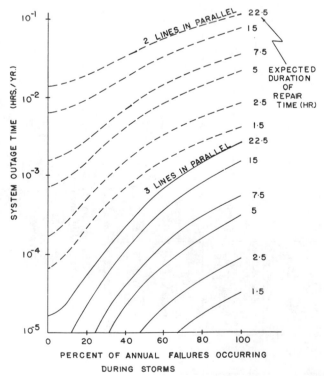

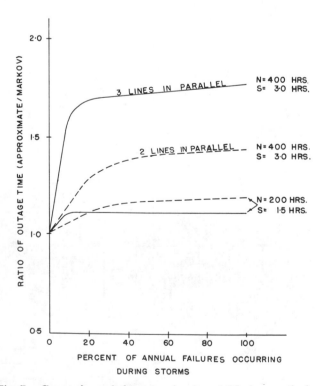

Fig. 5. Expected annual outage time variation in 2 and 3 line systems due to variation in storm associated failures for system parameters of $N = 200$ hours, $S = 1.5$ hours, and λ_{Av} (component annual failure rate) = 0.5 failure per year.

Fig. 7. Comparison of the approximate and Markov methods for expected outage time over a range of normal and stormy weather durations with system parameters of $R = 7.5$ hours and λ_{Av} (component annual failure rate) = 0.5 failure per year.

Starting in state 0 and letting $\lambda_1 = \lambda_2 = \lambda$ and $\lambda_1' = \lambda_2' = \lambda'$,

$$M_0 = \frac{2\lambda' + m + n}{2(2\lambda\lambda' + \lambda m + \lambda' n)}.$$

Considering normal weather only, $\lambda' = 0$, $m = 1$, and $n = 0$,

$$M_0 = \frac{1}{2\lambda}$$

as expected for two units in series.

The state space diagram for the 3-unit case is shown in Fig. 3. The differential equations in matrix form are similar to those shown for the 2-unit case. The complete matrices are shown in Table II.

$$[P(t)] = [P(t)] \left[\begin{array}{c|c} D_3 & n(I) \\ \hline m(I) & D_4 \end{array} \right].$$

System Studies

The MTTF for a single unit with a 2-state fluctuating environment was shown to be

$$M_{0,23} = \frac{\lambda' + m + n}{\lambda\lambda' + \lambda m + \lambda' n}.$$

Neglecting the short repair period compared to the MTTF, the expected failure rate

$$\lambda_{AV} = \frac{\lambda\lambda' + \lambda m + \lambda' n}{\lambda' + m + n}.$$

If

$$\lambda\lambda' \ll \lambda m + \lambda' n$$

$$\lambda' \ll m + n$$

then

$$\lambda_{AV} \simeq \frac{\lambda m}{m + n} + \frac{\lambda' n}{m + n}$$

$$\lambda_{AV} \simeq \frac{\lambda N}{S + N} + \frac{\lambda' S}{S + N}.$$

as given in Gaver et al.[2] The approximation is valid in most practical cases. The effect of storm associated failures on two and three line parallel configurations was examined for various storm failure percentage; the results are shown in Figs. 4 and 5. In each case the overall annual component failure rate was held constant at 0.5 failure per year and the expected duration of normal and stormy weather at 200 hours and 1.5 hours, respectively. The variation in expected failure rate and average annual outage duration due to variable expected repair duration at different percentages of storm associated failures is clearly shown.

It is interesting to compare the results obtained for expected failure rate and average annual outage duration by the Markov approach with those obtained by the approximate method.[2] The effect of changes in expected repair time is shown in Fig. 6 for the two and three parallel line configurations. The effect of varying the expected duration of normal and stormy weather periods is shown in Fig. 7. The effect of parallel component failure rate variation is shown in Fig. 8. The approximate method responds quite differently to changes in the system parameters and therefore no overall statement can be made

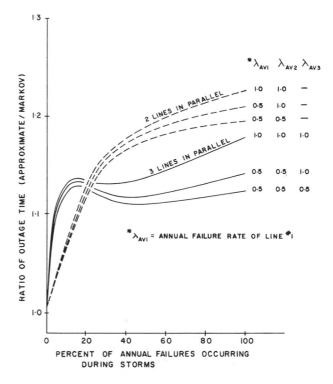

Fig. 8. Comparison of the approximate and Markov methods for expected outage times over a range of component annual failure rates with system parameters of $N = 200$ hours, $S = 1.5$ hours, and $R = 7.5$ hours.

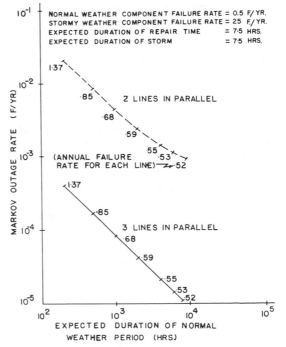

Fig. 9. Effect of variation in expected normal weather duration on system outage rate for two lines and three lines in parallel.

regarding the actual difference except in specific cases. This is clearly shown in Figs. 6 through 8.

As the transmission voltage increases, the expected duration of normal weather periods can also be expected to increase as a decreasing number of storms will produce storm associated failures. In some cases this may be only one or two storms per year. Fig. 9 illustrates this effect for two and three line parallel

Fig. 10. Series-parallel configuration.

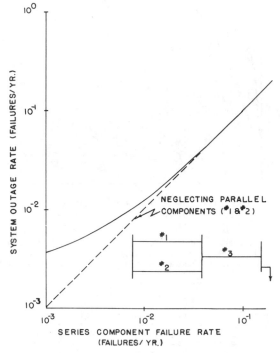

Fig. 11. System failure rate for series-parallel configuration showing the dominating effect of the series component on the overall outage rate of the system.

configurations. The variation in configuration failure rate is shown, together with the annual failure rate for each condition.

The state space diagram in Fig. 3 for the three line system can be applied to the case of two parallel lines in series with a third line. The system is shown in Fig. 10. For this configuration the failure rates for lines 1 and 2 were held at 0.5 failure per year. The expected duration of normal, stormy, and repair times for each component were held at 200 hours, 1.5 hours, and 7.5 hours, respectively. The failure rate for the series component was allowed to vary and the configuration failure rate calculated with the percentage of storm associated failures held at 50 percent for all components. The variation in configuration failure rate with the failure rate of the series component is shown in Fig. 11. It is interesting to note that the series component failure rate almost completely dominates the configuration value even for relatively small series component failure rates. It should be realized that this situation would be modified somewhat if the parallel system was not considered as being fully redundant.

CONCLUSIONS

The complete Markov approach can be applied without too much difficulty to relatively small systems. It becomes rather cumbersome when applied to larger systems. It can be applied to segments of a system, such as parallel facilities, that do suffer

the same environmental variations to obtain a theoretically exact solution. Certain system component failure rates are not greatly affected by the environment and therefore do not require the complete 2-state condition. These components simplify the analysis considerably. In a high-voltage system it is quite possible that a storm does not cover the entire system and therefore failure bunching for the entire system is unrealistic and only applies to local parallel facilities. In this case exact solutions by Markov methods can be applied to the parallel configurations and these segments linked by independent probabilities. The Markov approach does depend upon certain specific distributional assumptions though other distributions resulting in nonstationary processes could be considered. They would, however, become extremely complicated even for simple system configurations. The continuity of supply criterion has been used exclusively in this paper and with the added assumption of fully redundant parallel facilities. The state probabilities of the Markov approach can easily be applied to systems that are not fully redundant to obtain a quality of service criteria. Further work is being done in this regard by the authors.

APPENDIX
BASIC CONCEPTS OF MARKOV PROCESSES

The theory of Markov processes as applied to reliability problems deals generally with systems that are discrete in space and continuous in time. Systems that are discrete in space and in time are discussed in detail in Feller[5] and in Kemeny and Snell.[6]

Consider the case of a single repairable component.

$P_0(t)$ probability of the component being operable at time t

$P_1(t)$ probability of the component being failed at time t

λ failure rate

μ repair rate.

In this example and in all the cases in this paper it is assumed that failure and repair processes are characterized by exponential distributions. For the exponential case the conditional probability of failure during any unit time interval after time t, given that it has survived up to time t, is a constant.

Consider an incremental time interval dt, and assume that the probability of two or more events occurring during the increment of time dt is negligible.

$$P_0(t + dt) = P_0(t)(1 - \lambda dt) + P_1(t)\, \mu dt$$

$$P_1(t + dt) = P_1(t)(1 - \mu dt) + P_0(t)\, \lambda dt.$$

The state space diagram for this simple system is shown in Fig. 12.

$$\left. \frac{P_0(t + dt) - P_0(t)}{dt} \right|_{dt \to 0} = \frac{dP_0(t)}{dt} = P_0'(t).$$

In matrix form

$$\begin{bmatrix} P_0'(t) \\ P_1'(t) \end{bmatrix} = [P_0(t)\ P_1(t)] \begin{bmatrix} -\lambda & \lambda \\ \mu & -\mu \end{bmatrix}.$$

Solving for $P_0(t)$ and $P_1(t)$,

$$P_0(t) = \frac{\mu}{\lambda + \mu}[P_0(o) + P_1(o)] + \frac{e^{-(\lambda + \mu)t}}{\lambda + \mu}[\lambda P_0(o) - \mu P_1(o)]$$

$$P_1(t) = \frac{\lambda}{\lambda + \mu}[P_0(o) + P_1(o)] + \frac{e^{-(\lambda + \mu)t}}{\lambda + \mu}[\mu P_1(o) - \lambda P_0(o)]$$

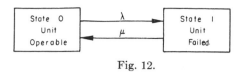

Fig. 12.

where $P_0(o)$, $P_1(o)$ are the initial conditions and $P_0(o) + P_1(o) = 1.0$.

Let $P_0(o) = 1.0$ and $P_1(o) = 0$,

$$P_0(t) = \frac{\mu}{\lambda + \mu} + \frac{\lambda e^{-(\lambda+\mu)t}}{\lambda + \mu}$$

$$P_1(t) - \frac{\lambda}{\lambda + \mu} + \frac{\mu e^{-(\lambda+\mu)t}}{\lambda + \mu}.$$

As $t \to \infty$

$$P_0(\infty) = \frac{\mu}{\lambda + \mu}$$

$$P_1(\infty) = \frac{\lambda}{\lambda + \mu}.$$

If

$$F = 1/\lambda = \text{mean time to failure}$$

$$R = 1/\mu = \text{mean repair time}$$

then

$$P_1(\infty) = R/(R + F)$$

is the conventional generating unit forced outage rate.

Designate

$$P_0(\infty) = P_0$$

$$P_1(\infty) = P_1.$$

The limiting state probabilities can also be obtained from the differential matrix since as $t \to \infty$

$$\frac{dP_0(t)}{dt} = 0.$$

The system equations are then

$$-\lambda P_0 + \mu P_1 = 0$$

$$\lambda P_0 - \mu P_1 = 0$$

$$P_0 + P_1 = 1.0.$$

The limiting state probabilities can also be obtained by considering the system as a Markov process that is discrete in time and in space. The stochastic matrix of transition probabilities that defines the process from t to $t + dt$ is designated as P and is as follows

$$P = \begin{bmatrix} 1 - \lambda & \lambda \\ \mu & 1 - \mu \end{bmatrix}.$$

The limiting state probabilities are constant. The process is ergodic as it is possible to go from one state to any other state in a finite number of steps.

It is necessary to find the probability vector $\alpha = P_0 P_1$ such

that $\alpha P = \alpha$. This is obtained from

$$P_0 + P_1 = 1.0$$

$$(1 - \lambda)P_0 + \mu P_1 = P_0$$

$$\lambda P_0 + (1 - \mu)P_1 = P_1$$

from which

$$P_0 = \frac{\mu}{\lambda + \mu}$$

$$P_1 = \frac{\lambda}{\lambda + \mu}.$$

In more complicated systems it becomes increasingly difficult to obtain a general time dependent expression. This is illustrated in the next case of a simple 2-unit system. The limiting state probabilities are much easier to obtain as the required simultaneous differential equations reduce to ordinary simultaneous equations. The transient component in all practical cases is negligible as the μ/λ ratio is usually quite large.

Consider the case of two identical units in parallel.

$P_0(t)$ Probability that both are in an operable state at time t

$P_1(t)$ Probability that only one is in an operable state at time t, and that the other is in a failed state at time t

$P_2(t)$ Probability that both are in a failed state at time t.

The system of differential equations is

$$\begin{bmatrix} P_0'(t) \\ P_1'(t) \\ P_2'(t) \end{bmatrix} = [P_0(t) P_1(t) P_2(t)] \begin{bmatrix} -2\lambda & 2\lambda & 0 \\ \mu & -(\lambda + \mu) & \lambda \\ 0 & 2\mu & -2\mu \end{bmatrix}.$$

Using $P_0(o) = 1.0$ and $P_1(o) = P_2(o) = 0$,

$$P_0(t) = \frac{\mu^2}{(\lambda + \mu)^2} + \frac{2\lambda\mu}{(\lambda + \mu)^2} e^{-(\lambda+\mu)t} + \frac{\lambda^2}{(\lambda + \mu)^2} e^{-2(\lambda+\mu)t}$$

$$P_1(t) = \frac{2\lambda\mu}{(\lambda + \mu)^2} + \frac{2\lambda(\lambda - \mu)}{(\lambda + \mu)^2} e^{-(\lambda+\mu)t} - \frac{2\lambda^2}{(\lambda + \mu)^2} e^{-2(\lambda+\mu)t}$$

$$P_2(t) = \frac{\lambda^2}{(\lambda + \mu)^2} - \frac{2\lambda^2}{(\lambda + \mu)^2} e^{-(\lambda+\mu)t} + \frac{\lambda^2}{(\lambda + \mu)^2} e^{-2(\lambda+\mu)t}.$$

It is interesting to note that in this system, component behavior are independent events and therefore the expressions for $P_0(t)$, $P_1(t)$, $P_2(t)$ could have been obtained from the results of the single-unit case using the binomial expansion. The limiting state probabilities can be obtained directly from the differential matrices by allowing $P'(t) \to 0$. If the availability A is defined as the probability that at least one unit is available

$$A(t = \infty) = P_0 + P_1$$

$$= \frac{\mu^2 + 2\lambda\mu}{(\lambda + \mu)^2}.$$

If the reliability of the system is defined as the probability that the system does not enter state 2, i.e., both units failed, then state 2 can be considered as an absorbing state. This is a state in which once entered it is not left until the process starts again.

$$\begin{bmatrix} P_0'(t) \\ P_1'(t) \\ P_2'(t) \end{bmatrix} = [P_0(t) P_1(t) P_2(t)] \begin{bmatrix} -2\lambda & 2\lambda & 0 \\ \mu & -(\lambda + \mu) & \lambda \\ 0 & 0 & 1 \end{bmatrix}.$$

377

The solution for this case is given by Epstein and Hosford[5] as

$$R(t) = \frac{S_1 e^{-S_2 t} - S_2 e^{-S_1 t}}{S_1 - S_2}$$

where

$$S_1 = \frac{1}{2}\left(3\lambda + \mu + \sqrt{\lambda^2 + 6\lambda\mu + \mu^2}\right)$$

$$S_2 = \frac{1}{2}\left(3\lambda + \mu - \sqrt{\lambda^2 + 6\lambda\mu + \mu^2}\right).$$

The mean time to failure (MTTF) for this system can be found by integrating the reliability function over the range 0 to ∞

$$\text{MTTF} = \int_0^\infty R(t)dt = \frac{S_1 + S_2}{S_1 S_2} = \frac{3\lambda + \mu}{2\lambda^2}.$$

As previously noted, it is virtually impossible to obtain a general time dependent expression for the reliability of more complicated systems. The MTTF can be obtained, however, using finite Markov chain concepts as outlined by Kemeny and Snell.[6]

The stochastic matrix of transition probabilities for the 2-unit case is

$$P = \begin{matrix} & \begin{matrix} 0 & \quad\;\; 1 & \quad\;\; 2 \end{matrix} \\ \begin{matrix} 0 \\ 1 \\ 2 \end{matrix} & \begin{bmatrix} 1 - 2\lambda & 2\lambda & 0 \\ \mu & 1 - (\lambda + \mu) & \lambda \\ 0 & 2\mu & 1 - 2\mu \end{bmatrix} \end{matrix}.$$

Designate state 2 as an absorbing state and create a new truncated matrix Q by eliminating the absorbed state.

$$Q = \begin{matrix} & \begin{matrix} 0 & \quad\quad\;\; 1 \end{matrix} \\ \begin{matrix} 0 \\ 1 \end{matrix} & \begin{bmatrix} 1 - 2\lambda & 2\lambda \\ \mu & 1 - (\lambda + \mu) \end{bmatrix} \end{matrix}.$$

Let I be the identity matrix,

$$[I - Q] = \begin{matrix} & \begin{matrix} 0 & \quad\;\; 1 \end{matrix} \\ \begin{matrix} 0 \\ 1 \end{matrix} & \begin{bmatrix} +2\lambda & -2\lambda \\ -\mu & \lambda + \mu \end{bmatrix} \end{matrix}$$

and let N be the fundamental matrix, where n_j is the time spent by the process in state s_j before being absorbed,

$$N = [I - Q]^{-1} = \frac{1}{2\lambda^2}\begin{bmatrix} \lambda + \mu & 2\lambda \\ \mu & 2\lambda \end{bmatrix}.$$

Starting in state 0 the MTTF (the time before entering the

absorbed state) of the process is $M_{0,2}$,

$$M_{0,2} = \frac{1}{2\lambda^2}[\lambda + \mu + 2\lambda] = \frac{3\lambda + \mu}{2\lambda^2}.$$

The matrix inversion approach is extremely useful when dealing with more complicated systems. The MTTF is usually very much longer than the time spent in the failed state and therefore the long term failure frequency can be expressed as 1/MTTF and designated as the system outage rate. It should be realized, however, that the MTTF or outage rate of a 2-unit system cannot be associated with a single exponential function to give an expression for the reliability of the system.

The case of two identical units in series can be examined using the same equations obtained for the two identical units in parallel system by redefining the failed states. It is assumed that it is possible for a unit to fail while the other unit is in a failed state.

Availability $A(t) = P_0(t)$

$P(\text{system failure}) = P_1(t) + P_2(t)$

$A(t = \infty) = P_0 = \mu^2/(\lambda + \mu)^2$

$MTTF = 1/2\lambda$

Failure rate $= 2\lambda$.

REFERENCES

[1] D. Kececioglu and J. McKinley, "Reliability engineering education activities in the United States and overseas," presented at the 5th Reliability and Maintainability Conf., New York, July 18–20, 1966.

[2] D. P. Gaver, F. E. Montmeat, and A. D. Patton, "Power system reliability: I—Measures of reliability and methods of calculation," *IEEE Trans. Power Apparatus and Systems*, vol. 83, pp. 727–737, July 1964.

[3] F. E. Montmeat, A. D. Patton, J. Zemkowski, and D. J. Cumming, "Power system reliability II—Applications and a computer program," *IEEE Trans. Power Apparatus and Systems*, vol. PAS-84, pp. 636–643, July 1965.

[4] C. F. DeSieno and L. L. Stine, "A probability method for determining the reliability of electric power systems," *IEEE Trans. Power Apparatus and Systems*, vol. 83, pp. 174–179, February 1964.

[5] W. Feller, *An Introduction to Probability Theory and Its Applications*, vol. 1. New York: Wiley, 1957.

[6] J. G. Kemeny and J. L. Snell, *Finite Markov Chains*. Princeton, N. J.: Van Nostrand, 1960.

[7] R. Billinton, "Bibliography on application of probability methods in the evaluation of generating capacity requirements," presented at IEEE Winter Power Meeting, New York, N. Y., January 30–February 4, 1966.

[8] Z. G. Todd, "A probability method for transmission and distribution outage calculations," *IEEE Trans. Power Apparatus and Systems*, vol. 83, pp. 695–702, July 1964.

On Procedures for Reliability Evaluations
of Transmission Systems

ROBERT J. RINGLEE, SENIOR MEMBER, IEEE, AND SHEILA D. GOODE

Abstract—A step-by-step procedure for reliability evaluations of transmission systems is described. Simple algebraic results for prediction of frequency and duration of transmission contingencies are obtained with the use of renewal process theory.

I. INTRODUCTION

A CONTINUING interest in quantitative reliability evaluations of transmission systems has prompted the preparation of this paper. With the substantial progress that has been made in the perfection of methods for simulating network power flows by fast methods on digital computing machines, the time for the general realization of quantitative transmission system reliability evaluations may be near. One limiting area appears to be the practical, rapid evaluation of stability limits. Techniques for predicting the frequency with which certain transmission contingencies will occur have been known for some time. The combination of rapid simulation methods with event prediction techniques may offer a practical method for carrying out transmission system reliability evaluations. An already substantial body of technique has been published on the subject of transmission system reliability, this paper offers a generalization of the concepts of the component run–fail–repair "life cycle" and describes a step-by-step procedure for performing reliability evaluations. Component reliability availability models have been investigated using renewal processes. The techniques described have proved quite feasible for hand as well as machine computations.

II. RESUME OF RECENT EFFORTS

Suggestions of numerical reliability criteria for transmission and distribution systems extend back at least three decades to Dean [1] who reported on interruption frequency and duration measures as an aid in decisions to modify circuit protection arrangements. Frequency and probability measures were suggested for generation and transmission systems nearly four decades ago by Lyman [2] and Smith [3].

Cone [4] suggested a criterion for generation and transmission reliability based upon the risk of overlapping transmission and generation outages. Todd [5] developed a probabilistic approach for transmission and distribution circuits. Treating circuit and component outages as independent events, he developed expressions for the probability of outage of one or more series-connected components and the probability of overlapping outage of

Paper 69 TP 654-PWR, recommended and approved by the Power System Engineering Committee of the IEEE Power Group for presentation at the IEEE Summer Power Meeting, Dallas, Tex., June 22–27, 1969. Manuscript submitted February 17, 1969; made available for printing April 9, 1969.

R. J. Ringlee was with the General Electric Company, Schenectady, N. Y. He is now with Power Technologies, Inc., Schenectady, N. Y. 12309.

S. D. Goode is with the General Electric Company, Schenectady, N. Y. 12305.

parallel paths. Todd's criterion was based upon each outage lasting 24 hours, or conversely, counting "one day" interruption for an outage event. Two outages overlapped if they occurred within the same 24-hour span. It is interesting to note that the component failure rate expressed in failures per day will be very closely estimated by Todd's component probability-of-outage. Parallel paths were assumed to be fully redundant such that the criterion of success could be measured by the existence of a path from the source to the load.

Modeling of repairable systems was greatly enhanced by the use of the Markov chain with continuous time parameter. Barlow and Hunter [6] gave a brief analysis of such an approach, and DeSieno and Stine [7] presented a state-description approach that employed Boolean algebra and Markov chains for the analysis of system reliability problems. The Boolean algebra approach has been applied by Egly *et al.* [8]. Effects of less than full redundancy in parallel circuits, effects of environmental states, and of scheduled outages were investigated in [9] and [10].

Formulas were developed for the prediction of the frequency and duration of outages in series and parallel paths. This work was extended by Mallard and Thomas [11] to multiple-path transmission circuits. Expressions were developed for the probability of path overload given that specified paths were out of service. Path overload contingency curves were developed for cases of single- and multiple-path forced outages. Patton, in a discussion of [11], questioned the method employed to estimate the risk of cascading outages, of overload outages precipitated by other overload outages. The essential point in question was the degree of correlation between loads at various points in the system and upon various components. It was Patton's observation that there was a definite correlation between loads which must be accounted for in probabilistic models of overload occurrence.

Methods of analysis and criteria discussed so far were directed to the prediction of interruptions at specific points in a network. The methods are readily adaptable to predictions of broader scope, but the preparation of numbers of individual "reliability diagrams" or diagrams of paths between sources and load points appears to be needlessly repetitive. Another approach, long used in reliability predictions, literally a state-enumeration technique, offers an alternative for developing predictions of system effect rather than load-point effect.

To the reliability engineer, the technique is termed the failure modes and effects analysis. The technique is well suited to simulation methods and is literally a mechanization of the contingency method long known to power system designers. Results employing this technique have been reported for distribution circuits [12], substation arrangements [13], and for transmission networks [14].

Employing a simulation method such as the load flow, the effects of each component outage upon the state of the system can be investigated in a macroscopic way. The voltage at each

Reprinted from *IEEE Trans. Power Apparatus Syst.*, vol. PAS-80, no. 4, pp. 527–536, April 1970.

TABLE I

SUMMARY OF FORMULAS FOR EVENT PROBABILITY, FREQUENCY, AND
DURATION FOR FOUR INDEPENDENT RENEWAL PROCESSES

Parameters for Each Renewal Process		
Up state availability $\quad A_i = m_i/(m_i + r_i)$		
Up state duration $\quad m_i$		
Outage state duration $\quad r_i$		
$m_i \gg r_i$ for transmission and distribution components		

Basic Independent Events, Four-Component Examples		
All Components Up	Component 1 Out	Components 1 and 2 Out
$P_0 = A_1 A_2 A_3 A_4$	$P_1 = (1 - A_1)A_2 A_3 A_4$	$P_{12} = (1 - A_1)(1 - A_2)A_3 A_4$
$1/M_0 = 1/m_1 + 1/m_2 + 1/m_3 + 1/m_4$	$1/M_1 = 1/r_1 + 1/m_2 + 1/m_3 + 1/m_4$	$1/M_{12} = 1/r_1 + 1/r_2 + 1/m_3 + 1/m_4$
$f_0 = P_0/M_0$	$f_1 = P_1/M_1$	$f_{12} = P_{12}/M_{12}$

Events Occurring During Exposure Period, Single-Contingency Example
Let E be per-unit exposure time and r_e be average duration of exposure period; assume all components are operating and only component 1 fails during exposure period leading to interruption $f_{1e} = Ef_1$

Duration of Interruption Event Depends Upon Means for Restoration
Restoration by automatic transfer: set interruption duration equal to switching time s (component 1 still in outage mode)
Restoration by repair of component 1: set interruption duration equal to M_1
Restoration by repair of component 1 or by completion of exposure period: set interruption duration equal to $M_1 r_e/(M_1 + r_e)$

Note: Only the effect of the single contingency has been accounted for; the overlapping double-contingent events must be investigated separately.

bus and the load on each component are accessible with such an approach.

As was shown in [14], the model that assumes completely correlated loads is easily represented with this method. With such a technique, it is possible to define system reliability criteria in terms such as the overload on any component and/or the violation of voltage limits at any bus, as well as the interruption of load at any point. Failure modes and effects analysis will be incorporated within a procedure for reliability and availability evaluations for substations and transmission circuits. A discussion of reliability models for repairable systems is offered first.

III. RELIABILITY AND AVAILABILITY MODELS FOR REPAIRABLE SYSTEMS

Before the discussion of reliability procedures is undertaken, it seems appropriate to discuss the models employed for repairable systems. The Markov chain with continuous parameter (time) has been used with considerable success to represent the discrete states: in service and outage for components. One of the minor criticisms of the model has been the concern frequently voiced that the Markov chain does not readily permit accurate representation of the observed distributions of times to repair. The Markov chain with constant transition rates leads to a model with exponentially distributed times to failure and times to repair. Although the criticism is valid, the results-in-the-mean for frequency and duration predictions developed by the Markov chain with constant transition rates are quite general. It can be shown that, if transition rates based upon the mean transfer times are used in a Markov chain model, the mean frequency and duration of resulting states predicted by the model are valid. While this statement is obvious for a single component, it may not be so obvious for the composite of a system of a number of independent components. Hence Appendix I shows the development of the frequency and duration of overlapping outages of components modeled after independent renewal processes.

A renewal process is literally a process that "renews itself" after each failure. The probability distributions of times to failure and repair can be of quite general shape. All that is required is that the distributions remain the same for each successive run–fail–repair cycle. The useful conclusion is that the frequency and duration results using the renewal process are identical to the results obtained with Markov chains. The results for systems composed of independent components, or of independent failure modes, are quite simple in form and it is believed potentially useful. These results are illustrated in Table I. It should be stressed that these results must be used for the description of independent failure events. One will note the use of an exposure factor in Table I; such a term has been employed by others [11]. The purpose of this term is to permit the partitioning of time into periods in which exposure to the particular event is zero (cannot occur) and into periods when the particular event can occur. The presence or absence of maintenance is a prime reason for inclusion of this factor. Representation of scheduled maintenance and of undetected failures may be done with an exposure term as shown in Appendix II.

IV. PROCEDURE FOR RELIABILITY AND AVAILABILITY EVALUATIONS FOR SUBSTATIONS AND TRANSMISSION CIRCUITS

The procedure will be outlined first and then illustrated with a small example. There are five essential steps to the procedure.

A. Physical System Description

Specify the component and circuit ratings, impedances, and connections within the boundary of study. Specify circuit and component outage modes and rates and repair statistics and maintenance requirements, i.e., per-unit time out for maintenance and average duration of maintenance outage. Specify supply and load terminals.

B. Performance Criteria

Specify the performance criteria for successful system operation. The criteria may include component overload as well as system frequency and bus voltage limits on the one hand and circuit continuity on the other.

C. Reliability Goal

Establish a level of satisfactory system performance. Negatively, the level is expressed in terms of events that lead to the system not meeting the performance criteria. Positively, the level may be set qualitatively in terms of the contingency level (number of overlapping outage events) the system can withstand and yet meet the performance criteria. The level may also be set numerically in terms of the reliability of system performance measured in terms of the interval of time between events leading to system failure. The level may also be set in terms of the per-unit time the system meets the performance criteria, that is, the "availability" of the system.

D. Failure Modes and Effects Analysis

Decide on the sequence of failure events to be investigated and the level of contingencies to be investigated. At this step, specify the conditions of system load and the state of component maintenance to be investigated with the failure events. Given the load and maintenance state, the effect of the failure event may be checked as follows.

1) Determine the effect of the failure event upon the protective system and determine the resulting breaker action.

2) Determine the effect of the breaker action upon the power system. Determine if overload or out-of-limit bus voltage or load interruption has occurred.

3) Determine whether the performance criteria have been violated.

4) If the performance criteria have not been met, determine what steps are required to bring the system back to success state:

    a) automatic transfer possible?
    b) manual transfer possible?
    c) component and circuit tests and repairs required?

5) Record the effect of the failure event by the terminals affected; that is, for each affected terminal, store the failure mode, the event probability, and the event duration. (Table I provides a summary of formulas for computing event probability, duration, and frequency.)

E. Accumulation of Failure Effects and Summary

Prepare a list of failure events leading to violation of the performance criteria. Order the list by event probability (or event frequency) to expose the events that by prediction are expected to cause the most trouble. Combine the system failure event probabilities and frequencies by the rules for combining like capacity states. Since each failure event investigated can be described by an exclusive state, the probability of occurrence of the system failure state is equal to the sum of the availabilities of the failure events. The "availability" of the failure state is equal to the product of the frequency of occurrence of the failure event and the associated duration of the failure state. For nonconsecutive failure states, the frequency of the sum of the failure states is equal to the sum of the frequencies of occurrence of the separate states [15]. A small, usually negligible, pessimistic error results if all failure states are assumed nonconsecutive.

The system availability and/or reliability prediction may now be compared with the goal. Should the prediction not measure up to the goal, one logical procedure to find improvement is to look for ways to mitigate the causes underlying the events that contributed most to failure.

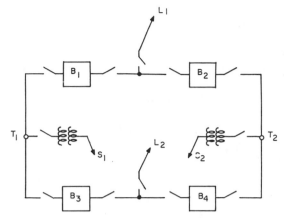

Fig. 1. Ring bus example.

TABLE II
COMPONENT OUTAGE DATA

	Failure Rate (years^{-1})	Outage Duration (hours)	Unavailability
Circuit breakers			
Circuit-breaker fault			
(backup required)	0.007	72	—
Maintenance	—	16	0.004
Probability of breaker			
found inoperative	—	—	0.0005
Transformer			
Forced outage	0.012	168	—
Maintenance	—	12	0.004
Bus section			
Forced outage	0.007	3.5	—
Maintenance (combined			
with line)	—	18	0.003
Line section			
Forced outage	0.05	23	—
Maintenance	—	15	0.005

V. SUBSTATION RELIABILITY PREDICTION

A. Physical System Description

Consider the ring bus configuration shown in Fig. 1. Assume that power is supplied over two widely separated lines $L1$ and $L2$ such that the supply and line capabilities are fully redundant. The lines will be assumed to perform as statistically independent components; environmental effects will be neglected. Lines $S1$ and $S2$ feed a multiple radial subtransmission network. Subtransmission breaker and switch equipment are outside the example. The rating of each transformer is two thirds the annual peak load of the station. The breaker ratings are assumed to be nonlimiting. Outage, repair, and maintenance data assumed for the example are shown in Table II. These data are patterned after the data reported in [11]. Disconnect switch failure events have been assigned to the bus, breaker, or transformer switched by the device.

B. Performance Criteria

Successful system operation is achieved 1) as long as either $L1$ or $L2$ can transmit power to either $S1$ or $S2$, and 2) the load on the substation transformers does not exceed the levels specified in [16]. A contingency curve has been prepared based

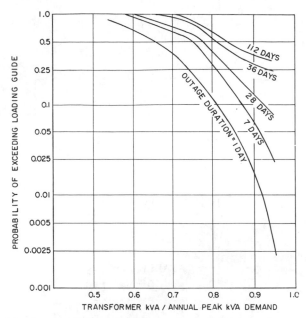

Fig. 2. Overload contingency curves for FOA transformer.

upon northeastern United States seasonal loads and temperatures. Contingency curves for an FOA transformer are shown in Fig. 2. The curves were determined by simulating the loading conditions (4380 bihourly loads) and daily mean temperatures that would be experienced by the transformer. Each day a check was made to see if the transformer of given rating could carry the demand and stay within the limits of the loading guide. The ratio of the number of days per year that the load exceeded the permissible values in [16] is defined as the probability of exceeding the loading guide for an outage duration of one day. Risks of exceeding the guide for longer outage durations were determined by simulation of longer outages and by counting the number of starting days upon which an outage of given duration would result in at least one violation of the loading guide. The ratio of starting days counted per year was plotted as the probability of exceeding the loading guide.

C. Reliability Goal

None was set for this example. The example will illustrate prediction of station interruptions.

D. Failure Modes and Effects Analysis

A failure modes and effects analysis was performed for single- and for double-contingent (overlapping) events. All events that resulted in loss of the substation are listed in Table III; the first event assumes that a breaker fault occurs, given that a stuck breaker condition already exists in the substation.

For example, suppose breaker $B1$ were to fail, given that breaker $B2$ is inoperative at the time of failure. A signal would be developed to trip breakers $B3$ and $B2$; $B2$ would not operate, hence $B4$ would be tripped by backup. Breakers at the sending end of $L1$ would also be tripped, of course. Service to $S1$ and $S2$ can be restored by disconnecting $B1$ and $B2$ and supplying the transformers from line $L2$. Assume one hour to analyze, test, and isolate $B1$ and $B2$. Note that column 2 in Table III lists restoration times for each failure event. With each breaker fault, there are two stuck breaker events that can cause the substation to be out. Accordingly, there are eight events of this type that can shut down the station, as is shown in column 1. The probability that at any instant the breaker is stuck was assumed to be 0.0005. The breaker failure rate was assumed to be 0.007 per year. The event of breaker $B1$ failure given breaker $B2$ stuck is expected to occur with frequency $0.0005 \times 0.007 = 0.0000035$ per year, as shown in column 5. Strictly speaking, this frequency is a measure of the events per exposure year; it is assumed that a breaker will not fail when it is out for maintenance. Also, the station cannot fail if it already has failed. Accordingly, the hours of maintenance per year and the hours of failure per year should be deducted from the period hours in the year.

Provision for the exposure has been incorporated in an exposure factor E, shown in column 2 of Table III. In this instance, the fraction of time out for maintenance and repairs is negligibly small compared to unity and has been neglected; E has been set to unity. The product of the combinations (col. 8), the exposure (col. 1), and the event frequency (0.0000035) gives the station outage frequency of 0.000028 events per year, or 0.028 events per thousand years, as shown in column 6. Column 7 of Table III lists the product of the event outage duration and the event frequency. Strictly speaking, this value should be divided by the factor $(1 + H/8760000)$ to give the expected hours out per thousand years.

Column 8 in Table III lists the ranking of contributions to unreliability (frequency). Thus the breaker fault given a stuck breaker is eighth in line of contributors.

The net effect of all the modes of failure tabulated on Table III is shown to be a station outage every 860 station years with an average duration of outage of 6.2 hours. The most likely events to lead to station interruption are line, breaker, and transformer outages during a maintenance outage of a line or transformer. Continuing the failure modes and effects analysis, another potential failure event is a transformer overload during forced outage of the other transformer. It was assumed that maintenance of transformers would be done during times of sufficiently reduced demand to permit the load to be carried on one unit without violation of [16]—assumed performance criteria.

An example of an approach to estimate the risk of transformer overload during forced outage of the other unit is illustrated in Table IV. A distribution of outage times $P_r(r)$: row 2 in the table, has been assumed. It gives the most likely outage duration to be one week and the average outage duration to be one week. The frequency of occurrence of outage events of given durations is shown in the third row. The probability of at least one overload for the given outage duration $P(0/r)$ is read from Fig. 2 at a ratio of transformer kVA/annual peak kVA demand equal to 0.667. The exposure to this type event is nearly unity on each transformer; hence an exposure factor of 2 has been assumed for the station.

The frequency of station events leading to transformer overload is predicted to be 15.5 per thousand station years, or reciprocally, one event per 65 station years.

E. Accumulation of Failure Effects and Summary

For the ring bus example, the two performance criteria resulted in widely disparate results. Station interruption, on the one hand, gave a predicted reliability of one interruption per 860 substation years. Component overload, on the other, gave a predicted reliability of one event per 65 substation years. It would not be meaningful to combine these measures into one composite, since the overload criterion does not necessarily imply an interruption.

TABLE III
Line, Transformer, Bus, Circuit-Breaker Persistent Forced Outages and Maintenance Events Leading to Station Interruption

Failure Event	Event Data					Station Effect		
	(1) Combinations	(2) Exposure	(3) Restoration Means	(4) Duration (hours)	(5) Frequency (1000 years^{-1})	(6) Frequency (1000 years^{-1})	(7) H (hours per 1000 years)	(8) Frequency Rank
Breaker fault (stuck breaker)	8	$\sim$1	S*	1	0.0035	0.028	0.028	8
Forced outage (maintenance on another component)								
Line and line bus maintenance								
Line forced outage (maintenance)	2	0.005	R†	9.1	50	0.5	4.52	1
Line bus forced outage (mainenance)	2	0.005	R	2.8	7	0.07	0.196	6
Breaker fault (maintenance)	4	0.005	S	1	7	0.14	0.14	2
Transformer and bus maintenance								
Transformer forced outage (maintenance)	2	0.004	R	11.2	12	0.096	1.07	5
Transformer Bus forced outage (maintenance)	2	0.004	R	2.7	7	0.056	0.152	7
Breaker fault (maintenance)	4	0.004	S	1	7	0.112	0.112	3
Breaker fault (breaker maintenance)	4	0.004	S	1	7	0.112	0.112	3
Overlapping independent forced outages								
(Transformer + bus)^2	2	$\sim$1	R	53.5	0.0044	0.0089	0.475	
(Line + bus)^2	2	$\sim$1	R	10.3	0.0076	0.0152	0.313	
Breaker^2	12	$\sim$1	S	1	0.00049	0.006	0.006	
Breaker·transformer	8	$\sim$1	S	1	0.0016	0.0129	0.0129	
Breaker·line	8	$\sim$1	S		0.00091	0.0073	0.0073	
Bus·breaker	16	$\sim$1	S		0.00002	0.0003	0.0003	
Station interruption:						1.16	7.14	
Average duration:						6.2 hours		

* Restoration by switching.
† Restoration by repair.

TABLE IV
Transformer Bank Outage Events Resulting in Overload on Remaining Transformer (Twin Transformer Bank Rating = 4/3 Annual Peak Demand)

Duration of outage (days)	r	1	7	28	86	112	Total
Outage duration (probability density function)	$P_r(r)$	0.402	0.540	0.0416	0.0082	0.0082	1.00
Outage event (occurrence frequency per thousand years)	$f_e(r) = 12P_r(r)$	4.8	6.5	0.5	0.1	0.1	12
Probability of at least one overload day during outage event	$P(O/r)$	0.47	0.75	0.82	0.97	1.0	—
Frequency of outage (events leading to overload per thousand years)	$f_0 = f_e(r)P(O/r)$	2.26	4.87	0.41	0.097	0.1	7.74
Exposure (2 units)	E	$\sim$2	$\sim$2	$\sim$2	$\sim$2	$\sim$2	$\sim$2
Frequency of station events leading to overload per thousand years	$f_s = Ef_e$	4.52	9.74	0.82	0.194	0.2	15.48

VI. Conclusions

Step-by-step procedures for reliability evaluations of complex systems, such as substations and transmission networks, are available. The logical and arithmetical operations for quantitative reliability evaluations are readily performed by hand for moderately complex systems and are amenable to programming for digital computers for the analysis of large, complex systems. Frequency and duration methods described here apply to quite general probability distributions of times to repair.

Appendix I
Frequency and Duration of Overlapping Events with Independent Renewal Processes
Renewal Processes

The renewal process [17] offers distinct advantages as a model for the run–failure–repair cycle of components and circuits. With such a model, rather general forms of probability distributions of times to failure t_f and times to repair t_r may be utilized [17, footnote, p. 182]. Such a process "renews" itself with each repair.

The mean time to failure m for each run length is by definition

$$m = E\{t_f\}$$

and the mean time to repair r is by definition

$$r = E\{t_r\}.$$

Each run–failure–repair cycle may be described by a mean cycle time

$$T_c = E\{t_f + t_r\} = m + r.$$

Parzen [17, theorem 3B, p. 182] has shown that in the interval of time t to $t + T$ the expected number of cycles of the run–failure–repair process $n(t + T) - n(t)$ tends to the limit

$$N(T) = \lim_{t \to \infty} [n(t + T) - n(t)] = T/(m + r).$$

The limit process removes the influence of initial state conditions. Likewise, Parzen has shown [17, example 3C, p. 185] that the availability A of the component run or upstate is given by $A = m/(m + r)$ within the same limit conditions. Note that the frequency of failure events is given by the reciprocal of the mean cycle time.

Overlapping Events: Heuristic Development

Consider two components with statistically independent renewal processes.

Component 1:

    mean time to failure $= m_1$
    mean time to repair $= r_1$
    availability of upstate $A_1 = m_1/(m_1 + r_1)$.

Component 2:

    mean time to failure $= m_2$
    mean time to repair $= r_2$
    availability of upstate $A_2 = m_2/(m_2 + r_2)$.

For the same limiting conditions described for the single component, consider an interval of time T units in duration. The expected down time on component 1 will be $(1 - A_1)T$, and for component 2 will be $(1 - A_2)T$. By Parzen [17, theorem 3B, p. 182] the expected number of events within interval T in which component 1 will fail during the time that component

Fig. 3.

2 is down for repair is $(1 - A_2)T/(m_1 + r_1)$; similarly, the number of times component 2 will fail when component 1 is down for repair is $(1 - A_1)T/(m_2 + r_2)$. The possibility of a simultaneous state change for both components is second-order small and rejected accordingly.

In the interval of time T, the total number of occurrences of overlapping outage events N_{12} is given by the sum

$$T(1 - A_1)/(m_2 + r_2) + T(1 - A_2)/(m_1 + r_1)$$

from which the frequency of overlapping events is determined as

$$f_{12} = N_{12}/T = (r_1 + r_2)/(m_1 + r_1)(m_2 + r_2).$$

Let T_o be the total expected time of overlapping outage during T; $T_o = T(1 - A_1)(1 - A_2)$. The average duration of overlapping outage may be determined as follows:

$$r_{12} = T_o/N_{12} = 1/(1/r_1 + 1/r_2).$$

The overlapping outage state and the state representing "not overlapping outage" form another renewal process with frequency and duration of overlapping outage f_{12}, r_{12}. The mean duration of the outage state r_{12} and the "availability" of the outage state $\bar{A}_{12}$ are readily shown to relate to the frequency f_{12} of the overlapping outage event as

$$f_{12} = \bar{A}_{12}/r_{12}$$

where $\bar{A}_{12}$ is determined from

$$\bar{A}_{12} = (1 - A_1)(1 - A_2) = \bar{A}_1\bar{A}_2$$

and

$$1/r_{12} = 1/r_1 + 1/r_2.$$

In like manner, it is readily shown that the mean duration of the "both components up" state is $1/m_{12} = 1/m_1 + 1/m_2$. The frequency of this event is equal to the event availability divided by the mean duration of the event.

Similarly, the event of component 1 down and component 2 up exists with mean duration of $1/(1/r_1 + 1/m_2)$.

The symmetry of these three events and the similarity to the results predicted for Markov processes is most encouraging. Generalization to more than two independent components can be obtained by recursive or by inductive methods. The results are illustrated in Table I for a system of four independent components. An exposure factor has been included in the table to convert time in a given configuration into period time.

Representation of Three-State Devices

Consider a component that has two independent modes of failure, and cannot "fail" or "repair" from one failure mode to the other. The theorem [17, theorem 3B, p. 182] and definition of availability [17, example 3C, p. 185] permit the development of the model of the three-state component.

As in Fig. 3, let state 1 be the up state, state 2 one down state, state 3 the second down state, and

r_2 = expected duration in state 2
r_3 = expected duration in state 3
m_{12} = expected duration of time between transfers from state 1 to state 2
m_{13} = expected duration of time between transfers from state 1 to state 3.

Then the availabilities of states 1–3 are, respectively,

$$A_1 = 1/(1 + r_2/m_{12} + r_3/m_{13})$$
$$A_2 = A_1 r_2/m_{12}$$
$$A_3 = A_1 r_3/m_{13}.$$

The mean duration of state 1 is $1/(1/m_{12} + 1/m_{13})$. The frequency of occurrence of state 1 is

$$f_1 = (1/m_{12} + 1/m_{13})/(1 + r_2/m_{12} + r_3/m_{13})$$

and the frequencies of occurrence of states 2 and 3 are

$$f_2 = (1/m_{12})/(1 + r_2/m_{12} + r_3/m_{13})$$

$$f_3 = (1/m_{13})/(1 + r_2/m_{12} + r_3/m_{13}).$$

Note that if state 3 is regarded as scheduled outage time and if the exposure is counted as all time not in state 3, $1 - A_3$, then

$$f_2 \equiv (1 - A_3)/(m_{12} + r_2).$$

Appendix II

Models for Exposure

Scheduled Outages

Procedures for handling randomly occurring repairs and maintenance have been presented for cases of independence [9] and dependence [11] upon system load level. Representation of scheduled outage periods is readily treated by the same methods. Let E_s be the fraction of time represented by the maintenance period:

$$E_s = \frac{\text{average maintenance period}}{\text{average interval between maintenance periods}}.$$

Then the incidence of an event occurring during the maintenance period must be weighted by the fraction of time in scheduled maintenance. The frequency of the event overlapping the maintenance period is the frequency of the independent event times E_s. Events occurring outside the maintenance period must be weighted by $1 - E_s$.

Modeling Undetected Failure Risks

Protective systems and standby systems may fail in ways that provide no warning of failure and no immediate effect upon the system. For example, a standby lube pump may be unable to start, but as long as no attempt is made to start it, it may never give an indication of unreadiness to serve. It is proposed to model standby system failures by the risk p that the system is not ready to serve at any randomly selected instant. Clearly, the magnitude of p would depend upon the types of periodic tests and maintenance performed on the standby system. The stuck-breaker condition seems to fit this model [13]. It is proposed that each breaker operation be weighted by the risk $1 - p$ that it is successful, and by p that it is not. In the event of a system fault, it is with probability p that

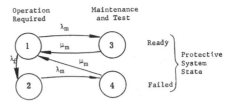

Fig. 4.   System requirement.

backup will be required to override a given device that has not functioned. Correspondingly, each success event will occur with probability $1 - p$. If f_f is the frequency of occurrence of the fault, the frequency of occurrence of backup operation required is pf_f.

A Markov chain has been proposed to model the undetected failure risk for protective systems. The simplest model (Fig. 4) contains four states. The long-term availability of the states for this model are

$$A_1 = 1/(1 + \lambda_f/\lambda_m + (\lambda_f + \lambda_m)/\mu_m)$$

$$A_2 = A_1 \lambda_f/\lambda_m$$

$$A_3 = A_1 \lambda_m/\mu_m$$

$$A_4 = A_1 \lambda_f/\mu_m$$

for

λ_m = 2 maintenance outages per year

μ_m = 2190 per year (4-hour outage)

λ_f = 0.001 failure per year.

Normal ready state is $A_1 \approx 1$; unprotected failure state $A_2 \approx (1)(0.001)/2 = 0.0005$. Hence, the estimate for p is 0.0005.

Acknowledgment

The authors wish to give credit to Mrs. R. B. Hall for the development of the computer logic to perform failure modes and effects analysis for substations.

References

[1] S. M. Dean, "Considerations involved in making system investments for improved service reliability," *EEI Bull.*, vol. 6, pp. 491–496, 1938.
[2] W. J. Lyman, "Fundamental consideration in preparing a master system plan," *Elec. World*, vol. 101, pp. 788–792, June 17, 1933.
[3] S. A. Smith, Jr., "Spare capacity fixed by probabilities of outage," *Elec. World*, vol. 103, pp. 222–225, February 10, 1934.
[4] H. P. Cone, "Probability and frequency of simultaneous forced outages of a transmission line and a steam generating unit coincident with critical levels of system loads," Paper DP 62-548, presented at the AIEE South Central District Meeting, Memphis, Tenn., April 4–6, 1962.
[5] Z. G. Todd, "A probability method for transmission and distribution outage calculations," *IEEE Trans. Power Apparatus and Systems*, vol. 83, pp. 695–701, July 1964.
[6] R. E. Barlow and L. C. Hunter, "Mathematical models for system reliability, pt. I," *Sylvania Technologist*, vol. 13, no. 1, pp. 16–31, January 1960.
[7] C. F. DeSieno and L. L. Stine, "A probability method for determining the reliability of electric power systems," *IEEE Trans. Power Apparatus and Systems*, vol. 83, pp. 174–181, February 1964.
[8] D. T. Egly, W. F. Esser, and S. G. Wasilew, "Computer programs for reliability," *Proc. 1967 IEEE PICA Conf.*, pp. 513–518.
[9] D. P. Gaver, F. E. Montmeat, and A. D. Patton, "Power system reliability, pt. I: measures of reliability and methods of calculation," *IEEE Trans. Power Apparatus and Systems*, vol. 83, pp. 727–737, July 1964.

[10] F. E. Montmeat, A. D. Patton, J. Zemkoski, and D. J. Cumming, "Power system reliability, pt. II: applications and a computer program," *IEEE Trans. Power Apparatus and Systems*, vol. PAS-84, pp. 636–643, July 1965.

[11] S. A. Mallard and V. C. Thomas, "A method for calculating transmission system reliability," *IEEE Trans. Power Apparatus and Systems*, vol. PAS-87, pp. 824–834, March 1968.

[12] M. W. Gangel and R. J. Ringlee, "Distribution system reliability performance," *IEEE Trans. Power Apparatus and Systems*, vol. PAS-87, pp. 1657–1665, July 1968.

[13] C. R. Heising, R. J. Ringlee, and H. O. Simmons, "A look at substation reliability," presented at the Minnesota Power Systems Conference, Minneapolis, Minn., October 15–17, 1968.

[14] M. P. Bhavaraju and R. Billinton, "Transmission planning using a reliability criterion, pt. I: a reliability criterion," *IEEE Trans. Power Apparatus and Systems*, vol. PAS-89, pp. 28–34, January 1970.

[15] J. D. Hall, R. J. Ringlee, and A. J. Wood, "Frequency and duration methods for power system reliability calculations, pt. I: generation system model," *IEEE Trans. Power Apparatus and Systems*, vol. PAS-87, pp. 1787–1796, September 1968.

[16] "Guide for loading oil-immersed distribution and power transformers," ANSI Appendix C57.92, January 1956.

[17] E. Parzen, *Stochastic Processes*. San Francisco: Holden Day, 1962, ch. 5.

Discussion

Mrs. V. Thomas Sulzberger (Public Service Electric and Gas Company, Newark, N. J.): This paper presents a step-by-step procedure, using reliability calculation methods, to effectively evaluate a power system substation design on a quantitative basis. The authors should be complimented for this contribution to the field of reliability as applied to power systems.

We are greatly interested in the paper and would appreciate some additional comments in three particular areas.

Performance Criteria: Having defined successful system performance criteria for a four-position ring bus as 1) either line $L1$ or $L2$ transmitting power to transformer $S1$ or $S2$ and 2) load on the substation transformers not exceeding certain specified transformer load levels, should not all possibilities of unsuccessful system operation be combined to effectively evaluate the overall substation design? Is not an overloaded transformer an unsuccessful system operation just as well as loss of continuity of service? How do the authors intend to use these two measures of system performance? Should one measure be weighted more heavily than the other?

Availability: In Table I, do the availability equations include the effect of forced as well as scheduled outages for the individual components? If so, the equation for P_{12} includes the possibility of overlapping scheduled outages, which is not usually the case in a real power system.

Computer Programming: The authors mention the development of computer logic to perform failure modes and effects analysis for substations. Was a computer program developed for this particular ring bus analysis, or was a general program written to evaluate several different bus designs? If so, what bus arrangements can be evaluated by this program? Is the program commercially available?

We would welcome the authors' explanations and clarifications on the above areas.

I would like to mention that we in Public Service made a similar type of quantitative reliability analysis of several 500-kV bus designs in 1967, and the results were made available to all companies of the PJM Interconnection. Our analysis compared the reliability and cost of three 500-kV bus arrangements for 500/230-kV stepdown stations. These bus arrangements included the Y bus (common terminal for a line and a transformer), ring bus, and breaker and one-half configuration. Each bus was assumed to be supplied by three 50-mile 500-kV circuits. The 500-kV bus, in turn, supplied a 230-kV bus via three 500/230-kV autotransformers.

The relative reliabilities of these three 500-kV bus arrangements considering forced, scheduled, and overload outages were as follows:

	Y	Breaker and One-Half	Ring
Frequency (per unit)	50	1	10
Duration (per unit)	8	1	1

These frequencies and durations are expressed in per unit of the breaker and one-half reliability parameters as base, or 1.0 per unit.

The outage frequency can be viewed as the number of times the supply to the 230-kV bus would be interrupted, or the number of times a portion of the 500-kV system would be inoperative or restricted in operation. The duration indicated the average period of time these conditions can be expected to exist.

As stated by the authors in their evaluation procedure, reliability goals and performance criteria for successful system operation must be defined so that a given substation design can be effectively evaluated in terms of given performance criteria. For different criteria, different reliability values are obtained.

Although the absolute reliability of each 500-kV bus arrangement may be somewhat uncertain because of the data assumptions and limited experience with 500-kV facilities, the relative reliabilities as indicated above do illustrate the improved overall reliability of a breaker and one-half bus arrangement.

J. Endrenyi (The Hydro-Electric Power Commission of Ontario, Toronto, Ont., Canada): The authors are to be congratulated for developing a methodical procedure for the evaluation of transmission system reliability, based on broad theoretical considerations yet practical enough to be useful in a wide variety of problems. Inevitably, the more general the method, the more difficulties are encountered when trying it on specific examples. Guidance on how to surmount a good many of these difficulties is given in the paper; other matters are still open for study.

One of the main problems encountered in reliability evaluations is caused by the sheer size of most systems to be investigated, that is, by the large number of components making up these systems. Even if this number is relatively low, as in the example offered by the authors, the selection of failure states for the system and their analysis (as given in Table III) requires experience, judgment, and skill. If the number of components in the system is large, the manual selection of failure states becomes impractical, and some kind of a mechanization of the process appears to be necessary. Some of the findings in the simple example in the paper may not apply in general; for example, all single-contingency states may not be success states, and possibly some higher contingency states may also acquire significance.

The amount of computations may become enormous not only because of the large number of system states to be scanned and the many failure states to be analyzed, but also because these analyses in many cases involve load flow studies—a generalization of the proceedings described in Table IV. The mathematical model describing the system may become further complicated if independence of failure events cannot be stipulated (as, for example, if weather effects are to be incorporated). It appears necessary that further research effort be directed toward finding suitable simplifications in the mathematical models describing larger systems and in the numerical techniques used to solve these models.

Another difficulty, this time in constructing the mathematical model, is encountered if, after the failure of certain components, system operation is restored through switching and the removal of the faulted devices. A Markov model describing a system of two such components is illustrated in Fig. 5. The solution in a given problem depends on which states represent system failure in that particular case. This, in turn, depends on the system configuration. Consider, for example, the ring bus arrangement shown in Fig. 1. If breakers B_1 and B_4 are the two components in question, the failure states are

Manuscript received July 7, 1969.

Manuscript received July 14, 1969.

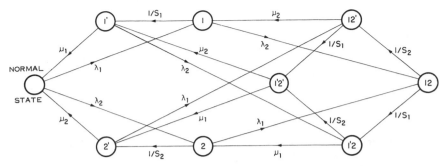

Fig. 5. Markov model of two-component system; faulted components disconnected. J—device J failed; J'—device J removed; S_J—mean switching time to remove device J.

12′, 1′2 and 12. For breakers B_1 and B_3, only 12 appears to represent system failure.

In some cases, shortcuts are possible in the computation of the frequency and mean duration of system failures. If, for example, $1/S_i \gg \mu_i$ for breakers B_1 and B_4, the system failure frequency can be calculated from a four-state Markov model derived from that in Fig. 5 by pooling all the failure states and 1′2′ together, and combining states 1′ and 1, and also 2′ and 2. The result is a simple model for two parallel components with the appropriate λ_i and μ_i values as transition intensities. The mean duration of system failure, however, cannot be derived from this model—it is one hour, by the very definition of the problem. It is basically this method that appears to have been employed by the authors.

Another four-state model with transition intensities λ_i and $1/S_i$ can be derived for breakers B_1 and B_3 if $\lambda_i \ll \mu_i$. Still other models can be constructed for other cases. The question is, how general are these techniques and how effective are the possible simplifications when it comes to the modeling of large systems?

L. O. Barthold (General Electric Company, Schenectady, N. Y.): This paper takes another step toward a very relevant and important engineering objective, i.e., the ability to evaluate in quantitative terms the influence of changes in system configuration and equipment reliability on the quality of service.

I would like to emphasize in this discussion some of the economic implications of this type of solution. To do so, it is worthwhile to cite a commonly used application of optimization principles. Fig. 6 shows the optimum commitment of generating units under varying system loads. In this case the optimum commitment occurs when the output of each unit is at a level where the incremental cost of energy is the same for each unit.

This principle is, of course, much more general. Fig. 7 shows that for any benefit attainable by alternative means, each of which can be evaluated in terms of incremental costs, the optimum deployment of investment is again that for which incremental costs of improved benefits are equal for each alternative.

Reliability is susceptible to the same interpretation. It is undeniably a benefit and is capable of improvement by increased investment in any one of several system areas.

This paper and others presented at the Summer Power Meeting, 1969, have dealt with reliability evaluation of different segments of the system—principally the generation segment on the one hand and the transmission segment on the other. Fig. 8 shows how appraisals might be made to determine how improved bulk power system reliability might be most economically improved.

Curves of the nature of those shown here might be expected to vary quite widely among systems, as would the actual operating points. I have based the generation curve on reasonable incremental costs of reserve and the generation operating point on the commonly accepted planning criterion that a deficiency once in ten years is acceptable. The transmission curve is harder to estimate. The operating point is based on transmission outage cause data published by the Federal Power Commission and the incremental cost curve

Manuscript received July 10, 1969.

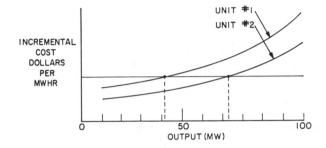

Fig. 6. Optimum unit commitment.

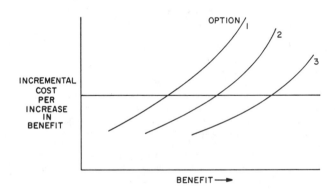

Fig. 7. Optimum deployment of investment among alternatives.

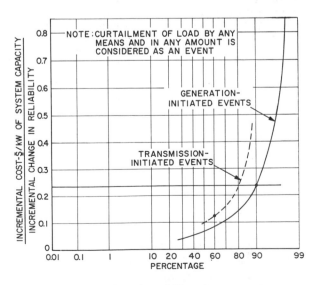

Fig. 8. System reliability, probability of no event in one year.

itself is based on estimates of the costs of altering system design practice in such a way as to reduce these causes.

Although these curves are quite approximate, they do give support to the increase in percent of total system investment being placed in transmission during the past decade and suggest perhaps a quantitative basis for applying the methods developed by the authors so as to achieve improved reliability at optimum cost.

Robert J. Ringlee and **Sheila D. Goode:** Permit us to express our gratitude to Mrs. Sulzberger and to Mr. Endreny and Mr. Barthold for their excellent discussions of our paper.

Mrs. Sulzberger raises three very interesting questions and provides a valuable summary of the results of her own investigations on 500-kV bus arrangements.

Responding to her first question concerning performance criteria, we chose to report the overload and interruption events separately for illustrative purposes; also, as noted in the paper, the criterion employed predicts the events that will result in loading conditions that exceed the transformer loading guide. These events may not necessarily result in interruption but may result in "loss of life" in accordance with [16]. A matter of degree is involved which should be considered in setting the criteria for unsuccessful operation.

Responding to her second question, P_{12} of Table I includes all independent overlapping events of outage of components 1 and 2, and, therefore, should not be used for such events as scheduled maintenance events. We have recommended that system maintenance events be viewed as a sequence of consecutive time periods; the system undergoes or is exposed to a specific maintenance state, S_i, for E_i per-unit time. Successive maintenance and nonmaintenance events appear as a nonoverlapping time sequence.

Responding to her third question, the computer program has been developed for analysis of arbitrary breaker, bus, and circuit arrangements. The program requires the electrical connection diagram as input. The program performs a failure modes and effects analysis for specified contingency levels and estimates the frequency and duration of interruption to specified buses and terminals. The present version of the program uses continuity of path as the basis of success. Yes, the program is commercially available.

We welcome the comments she has offered about the relative merits of the three 500-kV bus arrangements. In a parallel study of ring, breaker and one-half, and double-breaker double-bus six-terminal substations, our predictions assigned correspondingly superior performance to the breaker and one-half arrangement.

Mr. Endrenyi's comments and suggestions are most welcome. Regarding the nemesis of size, we entreat him not to lose heart but to continue his research in this matter! It is our belief that the problem of reliability evaluations for the bulk power supply can be treated as a hierarchy of problems such that the numbers of components treated within the boundary of a given problem are kept within feasible limits. Systematic investigations of failure modes and effects are well suited to programming for the digital computer. With exercise of prudence, the number of system states to be investigated can be kept within reason. One needs to keep in mind that the per-unit time that transmission components are inoperative or on forced outage is very small. Accordingly, the problem is most usually to be sure that the lowest order contingencies that can cause system failure are properly identified. The simple example was

Manuscript received August 14, 1969.

offered to illustrate method; the conclusions for this example should not be treated as general, but rather as specific to the configuration, criteria, and loads assumed.

One should attempt to keep the description of the problem as simple as possible, and to look for the underlying events that lead to failure. Little is served if the product states are formed from the states of independent components. Thus, the model in Fig. 5 is truly the product states of two independent three-state components. For example, component 1 has states N (normal), $1'$ (removed), and 1 (faulted):

$$P_N = 1/(1 + \lambda_1(S_1 + 1/\mu_1)) \approx 1.0$$

$$P_{1'} = (\lambda_1/\mu_1)P_N \approx \lambda_1/\mu_1$$

$$P_1 = \lambda_1 S_1 P_N \approx \lambda_1 S_1.$$

The product state of components 1 and 2 faulted, P_{12}, is, by his model

$$P_{12} = P_1 P_2 \approx \lambda_1 \lambda_2 S_1 S_2$$

$$\text{duration}_{12} = S_1 S_2/(S_1 + S_2).$$

This comment should in no way be regarded as criticism of the model proposed, rather the comment should be regarded as a plea to reduce each problem to the simplest description within the assumptions proposed. In this instance, Mr. Endrenyi chose, in accordance with the information given in Fig. 5, to assume that the states of component 1 and of component 2 are statistically independent.

The model posed by Mr. Endrenyi and the comments he has offered about the distinction of failure states involving breakers B_1–B_4 and B_1–B_3 are very pertinent. We claim his model to be superior to the conservative model we had used to hand calculate the risk of independent overlapping breaker fault events. Using his model, the "Breaker^2" row in part III of Table III would read as follows: column 6, frequency = 0.0017/1000 years; and column 7, H = 0.0017 hour/1000 years. Due to the insignificant contribution of this particular failure mode for the specific bus configuration, the sums of all failure modes for frequency and for duration of station interruption are changed negligibly. We might add that the computer program cited in the paper was used to perform the failure modes and effects analysis, and that a hand calculation was used to reconstruct the risk estimates. The program had properly identified and assessed the failure modes cited by Mr. Endrenyi.

Regarding the question of generality of method, we only wish to remark that the procedures for combining the effects of independent component states is general, of course; the care and specialization comes with the modeling of specific state sequences. The real problem is the proper identification of states associated with independent and with dependent events. We wish to add further that matrix methods are available to handle moderately large numbers of states [18], [19].

We wish to thank Mr. Barthold for his cogent remarks and insight on the steps beyond evaluation of transmission system reliability. We concur with his suggestion that a reliability criterion should serve as a constraint or a requirement to be used in the formulation of methods for seeking and evaluating investment deployment.

REFERENCES

[18] B. A. Kozlov, "Determination of reliability indices of systems with repair," *Engrg. Cybernetics*, no. 4, pp. 87–94, 1966.
[19] K. N. Stanton, "Reliability analysis for power system applications," *IEEE Trans. Power Apparatus and Systems*, vol. PAS-88, pp. 431–437, April 1969.

THREE-STATE MODELS IN POWER SYSTEM RELIABILITY EVALUATIONS

J. Endrenyi
The Hydro-Electric Power Commission of Ontario
Toronto, Ontario, Canada

Abstract — A reliability model is described which more closely simulates the states of power systems during faults than do the usual "two-state" models. This is achieved primarily by modelling the effects of component failures on the system through three-state cycles, which include a state following the fault and another after the failed device is isolated, through switching, for repair. The effects of component maintenance are also incorporated. A sample study is presented, wherein the average outage frequency and duration of a small 115-kV network are calculated. The findings include the observations that, in most cases, the effects of all but the lowest contingency system failures can be neglected, and that coincidences of failures and maintenance are often more frequent causes of system outages than coincidences of faults.

INTRODUCTION

Reliability techniques were originally developed for electronics and space applications. Their extension to other fields has occurred gradually and, apart from a very few pioneering efforts, only in the last five or six years did such techniques appear in power system studies. One reason for this delay was that the methods could not be adapted without modification; most power system applications really call for new reliability models that take the special characteristics of failure processes in power networks into account. The majority of publications in the field of power system reliability are, therefore, essentially proposals on improved reliability models designed to more accurately describe these processes.

The present paper is no exception. It undertakes to construct a probability model for systems of repairable components where component faults may initiate a whole sequence of events, such as the removal from service, and later the restoration, of unfaulted components and finally the repair and restoration of the faulted device — a chain of events common in power systems. The need for such a model has been recognized in previous publications[1, 2, 4] and the present discussion is an extension of some of the earlier results. The proposed model also incorporates the effects of component maintenance. The method of solving the model is based on the technique called "failure modes and effects analysis,"[1, 2, 3, 5] a method with the advantage of not being restricted to very small systems or simple system configurations. The procedure is illustrated by a network study, which involves the determination of the average failure frequency and duration in a three-station, four-line network.

The work, of course, is far from being complete. Reliability methods in power systems are still in an early stage of development, and no claims are made here or elsewhere of having all the answers to all the questions. Some of the assumptions in this paper that may appear to be oversimplifications will, no doubt, be replaced as studies progress to produce further refinements.

Paper 70 TP 693-PWR, recommended and approved by the Power System Engineering Committee of the IEEE Power Group for presentation at the IEEE Summer Power Meeting and EHV Conference, Los Angeles, Calif., July 12-17, 1970. Manuscript submitted February 27, 1970; made available for printing May 13, 1970.
This paper describes research carried out in partial fulfillment of the Ph.D. requirements at the University of Toronto. The work was partially supported by a grant from the National Research Council of Canada.

DESCRIPTION OF THE MODEL

Two-State Model

In most applications reported in the literature, that involve repairable components, reliability techniques are based on a two-state representation of components. In these models it is assumed that each component has an operating history made up of cycles of alternating periods spent in the "in-service" and "failed" states, as illustrated in Figure 1. The lines in the diagram indicate the possible transitions between states and each path is labelled with the mean frequency of that transition; these are the average failure frequency λ_{ri} (the reciprocal of the mean in-service time) for transitions into the failure state and the reciprocal of the mean repair duration T_{ri} for the transitions back to normal. In most calculations, these mean values sufficiently describe the "up and down" cycles and the findings are independent of the actual up-time and down-time distributions.

The state of the entire system at any given moment is determined by the prevailing states of its components. Thus the system state will change if, and only if, there is a change in the state of any of its components. For example, a system that consists of two independent two-state components, i and j, may assume any of the states illustrated in Figure 2.

In any given state, the system may be either successful or failed*; one of the main tasks in reliability analysis is the division of system states into success and failure states. It is by no means obvious to which category a given state belongs; this can be established only after applying suitable criteria for system failure, which themselves are not always easy to define. Diagrams like that in Figure 2, however, help in surveying the possibilities. In the given example, and depending on the circumstances, none of the states shown may represent system failure, or one of $i^{(r)}$ and $j^{(r)}$ may do so, or both (naturally in all three cases $i^{(r)}j^{(r)}$ becomes a failure state as well), or only $i^{(r)}j^{(r)}$ may constitute a (double-contingency) system failure state.

In power system applications the two-state model may provide a rather inaccurate representation, and in many cases a more sophisticated model is called for. Such a model is discussed in the following.

Three-State Model

Most power system components go through the two-state cycles described in the previous section. The failure of such components, however, will often put the system through a more complicated routine than that discussed above. When a high-voltage device fails, for example, first the system protection will isolate a number of "healthy" components along with the faulted one; as soon as possible after that, all but the minimum number of components that must be kept out of service for the isolation of the failed device will be restored to operation through appropriate switching. From the system's point of view, therefore, the fault of such a device is followed by a system state where several components are out of service and, after switching, this is followed by another state where possibly only the faulted device is out.

What actually occurs is that while the component is in the "failed" state, the system moves through two states, those "before switching" and "after switching." A model of this process can be constructed by

*Intermediate states are conceivable, but are not dealt with in this study.

Reprinted from *IEEE Trans. Power Apparatus Syst.,* vol. PAS-90, pp. 1909–1916, July/Aug. 1971.

considering such components to have three-state cycles consisting of an operating state, a state between the fault and switching (s state) and a "repair" state (r state) when the device is isolated for repair* (Figure 3). Obviously, the system effects of the s and r states are very different. It should be noted that the r states, lasting until repairs are completed, are usually much longer in duration than the s states and, also, that there are only very weak restrictions as to the time-distributions of any of the three states.

A system of two independent components i and j with three-state cycles will have a state transition diagram as shown in Figure 4. The diagram is a composition of two "single" diagrams of the type shown in Figure 3, and is constructed in such a way as to allow for complete cycles of j starting from any of the states of i and vice versa. Once again, the diagram will assist in recognizing the various possible groups of failure states; it is just the assistance in sorting out the failure possibilities that makes this diagram useful.

Considering, for example, the double contingencies, five possible sets of system failure states can be recognized from Figure 4; namely, $i^{(s)}j^{(s)}$ by itself, $i^{(r)}j^{(s)}$ and $i^{(s)}j^{(s)}$, $i^{(s)}j^{(r)}$ and $i^{(s)}j^{(s)}$, all three, and $i^{(r)}j^{(r)}$ plus the previous three. In addition to the appropriate double-contingency states, single-contingency states may join the sets of failure states in several combinations, such as $i^{(s)}$ alone, or both $i^{(s)}$ and $j^{(s)}$, or $i^{(r)}$ and $i^{(s)}$, and so on. Note, however, that if $i^{(r)}$ is a failure state then $i^{(s)}$, too, must be one, and similarly for $j^{(r)}$ and $j^{(s)}$.

Based on the observation that triple and higher contingency events occur with probabilities that are negligible in comparison with those of single and double contingency events, investigations of entire systems may be restricted to an analysis of single and double contingency states. Further, it is assumed that for components i and j the probability of occurrence of state $i^{(s)}j^{(s)}$ is negligible compared to that of the other states. The remaining possible sets of double-contingency failure states are illustrated in Figure 5.

It is quite possible, of course, for a system to contain a mixture of components with two-state and three-state effects. The state transition diagram of such a system is an appropriate combination of the diagrams in Figures 1 and 3. In a system made up of a "three-state" component i and a "two-state" component j, there are obviously fewer possible groups of failure states than if both were "three-state" types because the states containing $j^{(s)}$ are missing. With that in mind, all the previous findings can be extended to this case.

Maintenance

The maintenance of a component i is assumed to be a two-state (on-off) process independent of the failure cycles of the component. The process is characterized by the mean maintenance-frequency λ_{mi} and duration T_{mi}. As before, there are only weak restrictions on the up-time distribution; for the maintenance-times, however, it will be assumed that their standard deviations are negligible compared to T_{mi}.

While independence is assumed between the maintenance and failures of the same component, the maintenance of one component will be dependent on the repair (or maintenance) of another inasmuch as the former will not be started while the latter is in progress if the coincidence of the two events constitutes a system-failure. Once maintenance is in progress, however, it will be completed even if another component fails and a system-failure results from this coincidence.

A model that accounts for both the failure and maintenance of the components involved is a composite of two-state (for maintenance) and three-state (for failure) component representations. Although the model would admit, in principle, such physically meaningless states as $i^{(m)}i^{(s)}$ and $i^{(m)}i^{(r)}$, it would have them with a negligible frequency of

occurrence and, therefore, these states can simply be omitted from the calculations without the need for separate stipulations for their exclusion.

METHOD OF EVALUATION

As discussed before, the procedure of system reliability evaluation includes the screening of all single and double contingency states in order to find those which constitute system failure. In addition to finding the actual system failure states, each one of these must be classified according to the way system operation is restored after the fault. In some cases, the system can be restored through suitable switching (S-type failure), and in other cases, only after the repair of at least one of the faulted components (R-type failure). If, for example, the state $i^{(r)}j^{(r)}$ in Figure 5 is a system failure it is an R-type failure, whereas if only $i^{(r)}j^{(s)}$ constitutes system failure it is of the type S (switching will transfer the system into state $i^{(r)}j^{(r)}$ which is then not a failure state).

Since the probability of state $i^{(s)}j^{(s)}$ occurring is assumed to be negligible, the procedure of screening the events of two coinciding faults for system failure states basically reduces to an investigation of the interactions between $i^{(r)}$ and $j^{(s)}$, $i^{(s)}$ and $j^{(r)}$, and $i^{(r)}$ and $j^{(r)}$. It follows from Figure 5 that, for every pair of components (i, j), first the case $i^{(r)}j^{(r)}$ should be investigated and if it is not a system failure state, next $i^{(r)}j^{(s)}$ and $j^{(r)}i^{(s)}$ should be examined. On the other hand, if $i^{(r)}j^{(r)}$ is a failure state, states $i^{(r)}j^{(s)}$ and $j^{(r)}i^{(s)}$ shall not be taken into account (even though they, too, represent system failure in this case) because, being consecutive with $i^{(r)}j^{(r)}$, they must not show in the frequency count and, further, because their durations are negligible compared to that of $i^{(r)}j^{(r)}$.

As for the coincidences of maintenance and failures, obviously the states $i^{(m)}j^{(s)}$, $j^{(m)}i^{(s)}$, $i^{(m)}j^{(r)}$ and $j^{(m)}i^{(r)}$ must be screened for system failure. It should be observed, however, that whenever $i^{(r)}j^{(s)}$ is a system-failure state, $i^{(m)}j^{(s)}$ will be likewise (and vice versa), and similarly, if $i^{(r)}j^{(r)}$ constitutes system failure, so will $i^{(m)}j^{(r)}$ and $i^{(r)}j^{(m)}$. In studying the effects of maintenance, therefore, it is not necessary to re-examine all the double contingencies but it is sufficient to consider in the evaluation only those states whose inclusion is indicated by the above relations. The case $i^{(m)}j^{(m)}$ is omitted because, according to the assumptions, it can occur only if it is not a system-failure state.

For each failure state, its mean frequency f_F and duration T_F can be calculated by the equations derived in the Appendices and listed in Table IV. If there are several components with identical repair and maintenance data, their combinations producing the same type of fault can be pooled to form a single group. The mean frequency f_G of such a group of events can be calculated by computing f_F for only one combination in the group and multiplying it by the number of combinations (k) that make up the group.* The total outage duration per unit time for the group, T_{tG}, can be obtained by multiplying f_G and T_F. Finally, the system failure frequency f is the sum of all group failure frequencies; the system outage duration per unit time, T_t, is the sum of all the T_{tG} values; and the average system fault duration T equals T_t/f.

In the case of single-contingency failure events, it is assumed that no R-type failure can occur — from the practical point of view, such a fault would be most undesirable. Now, if state $i^{(s)}$ represents single-contingency system-failure, all higher-contingency events that include the S-type failure of component i (such as $j^{(r)}i^{(s)}$ or $j^{(m)}i^{(s)}$) can be omitted from the computations; the equations in Table IV to evaluate $i^{(s)}$ already account for these events. An explanation for this is given in Appendix I.

*The components described here should not be confused with three-state components, which can develop two kinds of faults and thus have two different failure states.

*For some relations between the k values in the various groups see notes (a) and (b) after Table II.

SUMMARY OF ASSUMPTIONS

- Components have either two-state or three-state system effects.
- It is assumed that $T_{si} \ll T_{ri} \ll 1/\lambda_{ri}$ and that $T_{sj} \ll T_{ri}$ for all i and j.
- The maintenance of component i has its own cycles described by λ_{mi} and T_{mi}. These cycles are independent of the component's own fault cycles. They are dependent, however, on the faults of other components in the sense that maintenance of i will not be started during a repair (or maintenance) of j if $j^{(r)}i^{(m)}$ is a system failure state. On the other hand, if maintenance of i is already under way when a fault of j occurs, the maintenance will be completed during the repair of j even if this results in a system failure. It is assumed that always $T_{mi} < T_{rj}$.
- It is assumed that all cycling processes have been going on for a long time and, therefore, the probability of change from any given state to any other is constant in time. Hence, only very weak restrictions need to be stipulated as to the distributions of all up-times and repair-times. The maintenance times, however, are assumed to have standard deviations negligible in comparison to their means.
- Only single and double contingencies are analyzed, higher contingencies are neglected. This implies that the total number of components is not too large so that

$$C = \sum_i \lambda_{ri} \ll 1/T_{ri} \, .$$

- It is assumed that no R-type single-contingency failure can occur; that is, in all single-contingency cases system operation can be restored through switching.
- For any two components i and j, the probability of state $i^{(s)}j^{(s)}$ occurring is neglected. Also excluded are such states as $i^{(m)}i^{(r)}$ and $i^{(m)}i^{(s)}$ which have no physical meaning.

A NETWORK STUDY

The sample system selected for reliability investigation is a network of 115-kV cable circuits, connecting three transformer stations in Toronto. The following discussion is arranged in such a way as to suggest a suitable sequence of steps in the evaluation of three-state reliability models.

Description of the System

A schematic diagram of the network under study is shown in Figure 6. For keeping the number of components in the system reasonably low it was decided to include only breakers, lines, transformers and buses in the study. Furthermore, the buses at the supply end (Station 1) were assumed to be perfectly reliable, thus the system "starts" with the five breakers B_1 to B_5 (all assumed to be of identical characteristics). In addition to these, there are four cable circuits (L_1 to L_4), the first three of which were considered identical in length and specifications ($L_1 = L_2 = L_3 = L$), four identical transformers (T_1 to T_4), five identical buses (P_1 to P_5) and the six breakers

on the low-voltage sides of the receiving stations (B_6 to B_{11}, all identical). There are 24 components altogether. It may have been more logical to limit the system to high-voltage components only; on the other hand, the inclusion of the low-voltage breakers leads to some interesting conclusions and, therefore, it was decided to leave them in.

The Effects of Component Failures

The effects investigated in this step include the action of system protection upon component failures, and the switching procedure through which a faulted component is isolated and removed for repair. In the present network study, only those component failures are taken into account that will activate the system protection to isolate the faulted device. Thus, ground faults in breakers are considered but not accidental trippings.*

The basic assumption concerning the action of system protection is that, when a fault occurs, the minimum number of breakers that are nearest to the fault and are necessary to clear the fault will operate. For example, upon a ground fault of B_2, breakers B_1, B_3 B_6 and B_9 are expected to operate; and a fault of L_4 would trip B_4, B_5, B_8 and B_{11}. Every element of the protective relay system is assumed to be perfectly reliable.

Faulted components are isolated for repair by opening the two (in some cases three) switches most adjacent to them. All other components that were de-energized upon the fault by system protection are then restored to service. It is also assumed that the normally open switch S_{12} is closed during the repair of L_3 or L_4 or P_4, and the normally open switch S_{34} is closed while L_1 or L_2 or P_5 is in repair.

All component failures result in three-state system cycles.

Component Data

Lacking sufficient information on failure and repair statistics pertaining to the components under study, the data used here were taken from the literature.[2, 6] This information is compiled in Table I. The switching times were assumed to be 1 hr in every case.

Criteria for System Failure

The system was considered failed in all cases where any one of the secondary buses G_1, G_2, D_1 or D_2 would remain without supply. Cable or equipment overloading was disregarded and a secondary bus was considered to be in service as long as the electric continuity between it and the buses in station 1 was not interrupted.

Single- and Double-Contingency System Failure Events

A manual count of such events was conducted and the results are compiled in Table II. There are six single-contingency failure events (all of them breaker failures on the low-voltage side) and dozens of groups

---

*Breakers and other components subject to several kinds of faults are dealt with in references 2 and 3.

TABLE I

Component Failure and Maintenance Data

Component	No.	λ_{ri} 1/yr	T_{ri} yr	T_{ri} hr	λ_{mi} 1/yr	T_{mi} yr	T_{mi} hr	T_{si} yr	T_{si} hr
B (HV)	5	0.012	0.008	70	3	0.0014	12	0.000114	1
B_L (LV)	6	0.0032	0.0031	27	2	0.0009	8	0.000114	1
L (3 mi)	3	0.005	0.0575	504	1.5	0.0016	14	0.000114	1
L_4 (1 mi)	1	0.0017	0.0575	504	0.5	0.0016	14	0.000114	1
T	4	0.012	0.0192	168	3	0.0016	12	0.000114	1
P	5	0.007	0.0004	$3\frac{1}{2}$	2	0.0016	14	0.000114	1

with hundreds of double-contingency failures, mostly of the S type. In listing the R-type failure events, the values of k are indicated separately for the $i^{(r)}j^{(r)}$ and the $i^{(m)}j^{(r)}$ cases under the headings of RR and MR, respectively; these data are related, of course, as indicated in the notes after Table II.

Computation of System Reliability Parameters

The organization of the steps involved and details about some of the more significant groups are shown in Table III. The entries in the table are samples only; in the full tabulation there is, of course, an entry for each line in Table II where $k \neq 0$. The equations used to calculate f_F and T_F are those listed in Table IV.

TABLE II

The Values of k in Each Group for Various Failure Types

Case	Total No. of Combinations	Failure Combinations (k)		
		S	MR	RR
B_L	6	6	–	–
Other Single Cont.	18	–	–	–
B B	20	2	–	–
$B_L B_L$	30	18*	12	6
L L	6	3	–	–
T T	12	2	4	2
P P	20	6	6	3
B B_L	30	30*	–	–
B_L B	30	16	–	–
B L	15	–	–	–
L B	15	6	–	–
B L_4	5	–	–	–
L_4 B	5	2	–	–
B P	25	–	–	–
P B	25	10	–	–
B T	20	–	–	–
T B	20	8	–	–
B_L L	18	8	–	–
L B_L	18	18*	–	–
$B_L L_4$	6	6	–	–
$L_4 B_L$	6	6*	–	–
B_L T	24	4	8	} 8
T B_L	24	16*	8	} 8
B_L P	30	8	8	} 8
P B_L	30	22*	8	} 8
L L_4	3	2	1	} 1
L_4 L	3	–	1	} 1
L T	12	9	–	–
T L	12	4	–	–
L P	15	10	2	} 2
P L	15	3	2	} 2
L_4 T	4	2	–	–
T L_4	4	2	–	–
L_4 P	5	1	1	} 1
P L_4	5	2	1	} 1
T P	20	4	4	} 4
P T	20	5	4	} 4
Total	576	239	70	35

Notes:
(a) The values of k in column S are the same for both the MS and RS cases.
(b) The values of k for $i^{(m)}j^{(r)}$, $j^{(m)}j^{(r)}$ and $i^{(r)}j^{(r)}$ must be identical. It follows that if i and j are of the same kind (as in the groups TT, PP, etc.), the k values under MR are twice those under RR.
(c) The groups marked by (*) are accounted for when the case B_L is evaluated, therefore they can be omitted from the calculations.

A summary of the results is as follows:
- the frequency of single-contingency system failures is 19.2 per 1000 yr, with an average duration of 19200/19200 = 1 hr (the latter result is, of course, obvious from the assumptions);
- the frequency of double-contingency failures due to overlapping faults is 0.16 per 1000 yr, with an average duration of 1495/160 = 9.36 hr;
- the frequency of double-contingency failures due to coinciding maintenance and fault is 4.085 per 1000 yr, with an average duration of 9862/4085 = 2.41 hr.

What Can the Results Be Used For?

The above results do not mean much unless they are used in forming some conclusions and judgments. Ways in which this can be done are suggested in the following:
- In applications where reliability standards are available, an assessment of the system can be made of whether or not it meets the standard requirements.
- Alternative solutions can be compared with respect to their reliability measures.
- The weakest components can be spotted and an insight can be gained as to the most feasible ways of improving system reliability.
- The system design can be perfected so as to optimize some cost function that includes the cost of system outages.

CONCLUSIONS

The three-state reliability model proposed in this paper is believed to provide a closer description of failure processes in power systems than the two-state models used in many previous applications. The procedures of reliability evaluation based on the three-state model and such techniques as the "failure modes and effects analysis" go very well hand-in-hand, as illustrated by the sample network study in the paper.

Although some of the underlying assumptions are highly hypothetical, the results of this sample study yield some interesting conclusions or at least some food for thought:
- Lower contingency failure events tend to outweigh those associated with higher contingencies. In the example, the single-contingency failures of breakers B_L make far more trouble than all the other component faults together. On the one hand, this observation appears to justify the decision not to investigate triple or higher contingencies. On the other hand, it underlines the importance of carefully defining the system limits and the criteria for system failure so that single-contingency failure events should only count if their inclusion is realistic.
- The coincidence of one component failing while another is being maintained appears to be a more frequent cause of system failures than the overlapping outages of two components. This suggests the idea that wider spaced maintenance intervals would perhaps have a beneficial effect on overall reliability even if this resulted in an increase in frequency of component failures.
- Apart from the above differences it appears that among the components of the given system the HV breakers and the transformers are prominent in causing system failures. This observation may provide some lead when ways of improving reliability are sought.

The need for further development is evident in a number of areas. It is, for example, most desirable to extend the presently available computer methods for failure modes analysis, based on very simple sets of system failure criteria, so as to include more sophisticated criteria, such as limits in overloading and in voltage or frequency variations. The mathematical model itself could be further refined to reduce the number of states to be analyzed in large systems, and to cover networks with less than perfect protection systems. Further research will, no doubt, make some of these refinements available.

TABLE III

System Failure Frequencies and Durations
($1\ Myr = 10^6$ years)

Case	Type	Maintenance Plus Failure					Coincident Failures				
		Event Parameters		k	Group Parameters		Event Parameters		k	Group Parameters	
		f_F	T_F		f_G	T_{tG}	f_F	T_F		f_G	T_{tG}
		1/Myr	hr		1/Myr	hr/Myr	1/Myr	hr		1/Myr	hr/Myr
B_L	S	–	–	–	–	–	3200	1.0	6	19200	19200
...		.....			.....		.....			.....	
T T {	S	49.3	1.0	2	99	99	2.74	1.0	2	5.48	5.48
	R	49.3	6.0	4	197	1182	5.47	83.3	2	10.95	911.0
B_L B	S	21.8	1.0	16	349	349	0.12	1.0	16	1.90	1.90
L B	S	28.8	1.0	6	173	173	3.48	1.0	6	20.88	20.88
P B	S	38.4	1.0	10	384	384	0.034	1.0	10	0.34	0.34
T B	S	49.3	1.0	8	394	394	2.74	1.0	8	21.89	21.89
L P {	S	16.8	1.0	10	168	168	2.03	1.0	10	20.30	20.30
	R	16.8	7.0	2	34	235	} 2.04	3.5	2	4.08	14.20
P L {	R	16.0	7.0	2	32	224					
	S	16.0	1.0	3	48	48	0.014	1.0	3	0.04	0.04
L T	S	28.8	1.0	9	259	259	3.48	1.0	9	31.32	31.32
T L	S	20.6	1.0	4	82	82	1.14	1.0	4	4.56	4.56
...		.....			.....		.....			.....	
Double Contingency Total:					4085	9862				160	1495

APPENDIX I

AVERAGE FREQUENCIES AND DURATIONS OF SYSTEM STATES

In this Appendix, equations for the frequencies and durations of single- and double-contingency system states are derived and tabulated. Only states defined through independent failure events are considered; thus state $i^{(m)}j^{(r)}$, which according to the main text does not qualify, is excluded – it will be discussed in Appendix II.

It is generally true for any state x that the probability $P^{(x)}$ of the system staying in it at any time, its mean duration $T^{(x)}$, and the mean time between its occurrences, $M^{(x)}$, are related by the equation

$$P^{(x)} = \frac{T^{(x)}}{M^{(x)}} = f^{(x)}T^{(x)}, \qquad (1)$$

where $f^{(x)}$ is the average frequency of occurrence for state x. In the following, $P^{(x)}$ and $T^{(x)}$ will be determined for the various cases and then $f^{(x)}$ calculated from the above equation.

First, the double-contingency states $i^{(y)}j^{(z)}$ will be considered, where y may be either m or r, and z either r or s. In equation (1), the superscripts (x) will now be changed to (yz). Let the system consist of N components and let K denote the set of these components. Also, let $P_{\delta k}$ be the probability of component k being in state δ at any given time; it is easy to show that

$$P_{sk} = \frac{\lambda_{rk}T_{sk}}{1 + \lambda_{rk}(T_{sk} + T_{rk})}, \quad P_{rk} = \frac{\lambda_{rk}T_{rk}}{1 + \lambda_{rk}(T_{sk} + T_{rk})}$$

and

$$P_{mk} = \frac{\lambda_{mk}T_{mk}}{1 + \lambda_{mk}T_{mk}}.$$

Finally, let $P_{nk} = 1 - P_{sk} - P_{rk}$ and $\overline{P}_{mk} = 1 - P_{mk}$. Since $P^{(yz)}$ is the probability of i and j being in the specified failure modes and, at the same time, all other components being in operating condition, it is given by

$$P^{(yz)} = P_{yi}P_{zj} \prod_{\substack{k \in K \\ k \neq i, j}} P_{nk}\overline{P}_{mk} = B\lambda_{yi}\lambda_{rj}T_{yi}T_{zj},$$

where

$$B = \prod_{k \in K} P_{nk}\overline{P}_{mk} = \prod_{k \in K} \frac{1}{1 + \lambda_{rk}(T_{sk} + T_{rk})}\frac{1}{1 + \lambda_{mk}T_{mk}}$$

If N is not too large, $B \approx 1$.

Since $T^{(x)}$ is the mean duration of state x, $1/T^{(x)}$ is the frequency of departure from that state (given that the system is in state x). Thus, $1/T^{(yz)}$ can be determined by the equation*

$$\frac{1}{T^{(yz)}} = \frac{1}{T_{yi}} + \frac{1}{T_{zj}} + \sum_{\substack{k \in K \\ k \neq i, j}} \lambda_{rk}$$

$$\approx \frac{1}{T_{yi}} + \frac{1}{T_{zj}} + C,$$

where

$$C = \sum_{k \in K} \lambda_{rk}$$

If N is not too large, $C \ll 1/T_{rk}$; that is, transitions to triple-contingency states occur with only negligible frequency. Finally,

$$f^{(yz)} = \frac{P^{(yz)}}{T^{(yz)}} = B\lambda_{ri}\lambda_{rj}(T_{yi} + T_{zj} + T_{yi}T_{zj}C).$$

*Assuming that the state $i^{(y)}j^{(x)}$ represents system failure; thus, no transitions are permitted from it to triple-contingency states that include maintenance.

TABLE IV

Equations for $f^{(x)}$ and $T^{(x)}$
($B \approx 1$ and $C \ll 1/T_{rk}$)

Case	$f^{(x)}$	$T^{(x)}$	Remarks
$i^{(s)}$	λ_{ri}	T_{si}	
$i^{(r)}j^{(s)}$	$\lambda_{ri}\lambda_{rj}T_{ri}$	$\dfrac{T_{ri}T_{sj}}{T_{ri}+T_{sj}} \approx T_{sj}$	
$\left.\begin{array}{l} i^{(r)}j^{(r)} \\ j^{(r)}i^{(r)} \end{array}\right\}$	$\lambda_{ri}\lambda_{rj}(T_{ri}+T_{rj})$	$\dfrac{T_{ri}T_{rj}}{T_{ri}+T_{rj}}$	
$i^{(m)}j^{(s)}$	$\lambda_{mi}\lambda_{rj}T_{mi}$	T_{sj}	
$i^{(m)}j^{(r)}$	$\lambda_{mi}\lambda_{rj}T_{mi}$	$\dfrac{1}{2}T_{mi}$	$i^{(m)}$ and $j^{(r)}$ are not independent events

Note: For system failure states, the notations
$f^{(x)} = f_F$ and $T^{(x)} = T_F$ are used.

The actual equations for the various cases are compiled in Table IV. In deriving the equations, it was assumed that $B \approx 1$, $C \ll 1/T_{rk}$, and that $T_{sk} \ll T_{rk}$ or T_{mk}.

The corresponding equations for single-contingency states where (z) denotes either (r) or (s) can now be easily derived. These are:

$$P^{(z)} = P_{zi} \prod_{\substack{k \in K \\ k \neq i}} P_{nk}\overline{P}_{mk} = B\lambda_{ri}T_{zi} , \qquad \frac{1}{T^{(z)}} \approx \frac{1}{T_{zi}} + C ,$$

and

$$f^{(z)} = B\lambda_{ri}(1 + T_{zi}C) . \qquad (2)$$

The following remarks will be restricted to single-contingency failures of type S. In this case, equation (2) assumes the form

$$f^{(s)} = B\lambda_{ri}(1 + T_{si}C) \approx B\lambda_{ri} . \qquad (3)$$

Now, if this equation is replaced by

$$f^{(s')} = \lambda_{ri} , \qquad (4)$$

all $j^{(r)}i^{(s)}$ and $j^{(m)}i^{(s)}$ cases can be omitted from the calculations since (4) yields their frequencies combined with that of $i^{(s)}$. This is because the frequency of all failures of component i is λ_{ri} including those which coincide with the faults of other components, whereas equation (3) relates to only those failures of i which occur while all other components are faultless. Therefore, as long as λ_{ri} is used for the frequency of $i^{(s)}$, such failure events as $j^{(r)}i^{(s)}$ and $j^{(m)}i^{(s)}$ do not have to be counted again. The rule would not be quite as simple if triple or higher contingencies were admitted, but under the given assumptions it holds.

APPENDIX II

EQUATIONS FOR THE STATE $i^{(m)}j^{(r)}$

The two events defining this state are not independent since, as explained before, $i^{(m)}$ will not be started when $j^{(r)}$ is in progress if $i^{(m)}j^{(r)}$ is a system failure state. In discussing this case, it should be recalled that (a) the standard deviations of maintenance-times are negligible in comparison with their means; and (b) $C \ll 1/T_{mi}$. It is also assumed that the probability of component j failing in any dt time interval during the maintenance of component i is constant. From the sketch in Figure 7 it is then apparent that, since all relative positions of T_{mi} and T_{rj} have the same probability of occurrence,

$$T^{(mr)} = \frac{1}{2} T_{mi} . \qquad (5)$$

Strictly speaking, another assumption implicit in the above reasoning is that the repair time t_{rj} of component j is never less than T_{mi}; that is, the probability density function of t_{rj} is zero for $t < T_{mi}$ (Figure 7, dotted line). The removal of this restriction, as well as the removal of condition (b) above, will result in a $T^{(mr)}$ value slightly lower than that given by (5); hence $1/2\ T_{mi}$ is an upper bound for $T^{(mr)}$. It is also a good approximation in most cases. The exact value of $T^{(mr)}$ will depend on the distribution of t_{rj}. As an example, the effects of an exponential distribution for t_{rj} (full-line p.d.f. in Figure 7) were investigated and it was found that under this assumption the ratio $T^{(mr)}/T_{mi}$ equals $1/2$ if $T_{mi} \ll T_{rj}$ and $1/2.718$ if $T_{mi} \approx T_{rj}$, whereas equation (5) yields $1/2$ in both cases. The approximation appears to be reasonably close and will be even closer for most other distributions.

In order to obtain the value of $f^{(mr)}$, all $j^{(r)}$ events that start during the maintenance of i must be counted, and the rest discounted. This, however, is just the measure provided by $f^{(ms)}$; therefore one can conclude that $f^{(mr)}$ is, with very good approximation, equal to $f^{(ms)}$. The same result is obtained if $f^{(mr)}$ is derived as the product of $f^{(r)}$ for component j, as given by (2), and a factor of exposure to the maintenance of i, which is $P_{mi} \approx \lambda_{mi}T_{mi}$.

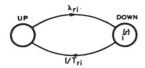

Fig. 1. Model of two-state component i.

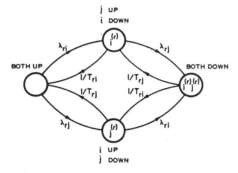

Fig. 2. States of independent two-state components i and j.

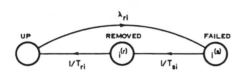

Fig. 3. Model of component i with three-state cycles.

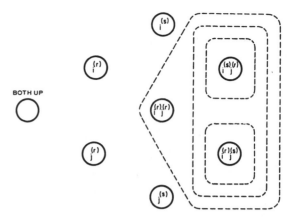

Fig. 5. Sets of double-contingency failure states in the model of Fig. 4 if state $i^{(s)}j^{(s)}$ is omitted.

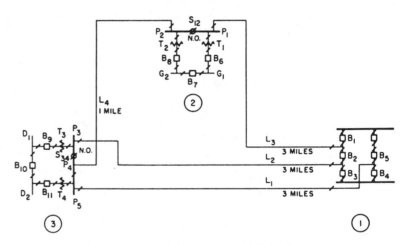

Fig. 6. A sample network of 115-kV circuits.

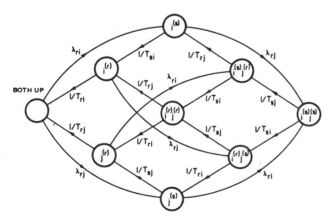

Fig. 4. States of independent components i and j, both with three-state cycles.

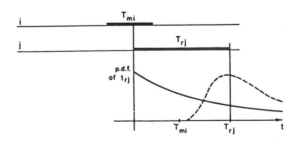

Fig. 7. Sketch for determining $T^{(mr)}$. p.d.f. = probability density function.

395

ACKNOWLEDGMENT

The author is indebted to Prof. J. A. Buzacott, Department of Industrial Engineering, University of Toronto, for valuable discussions and suggestions.

REFERENCES

1. M. W. Gangel and R. J. Ringlee, "Distribution System Reliability Performance," *IEEE Trans. Power Apparatus and Systems,* vol. 87, pp. 1657-1665, July 1968.

2. R. J. Ringlee and S. D. Goode, "On Procedures for Reliability Evaluations of Transmission Systems," *IEEE Trans. Power Apparatus and Systems,* vol. 89, pp. 527-537, April 1970.

3. C. R. Heising, R. J. Ringlee and H. O. Simmons, "A Look at Substation Reliability," presented at the Minnesota Power Systems Conference, 1968.

4. F. E. Montmeat, A. D. Patton, J. Zemkoski and D. J. Cumming, "Power System Reliability, II — Applications and a Computer Program," *IEEE Trans. Power Apparatus and Systems,* vol. 84, pp. 636-643, July 1965.

5. M. P. Bhavaraju and R. Billinton, "Transmission Planning Using a Reliability Criterion," *IEEE Trans. Power Apparatus and Systems,* vol. 89, pp. 28-34, January 1970.

6. S. A. Mallard and V. C. Thomas, "A Method for Calculating Transmission System Reliability," *IEEE Trans. Power Apparatus and Systems,* vol. 87, pp. 824-834, March 1968.

Discussion

R. J. Ringlee (Power Technologies, Inc., Schenectady, N.Y. 12301): Mr. Endrenyi has provided a most useful addition to the reliability and outage models for transmission and distribution systems. The three-state model he has proposed and has illustrated in this paper, provides the most accurate means offered thus far to represent the fault and switching operations which are performed on power circuits and apparatus.

Table 4 of the paper provides a convenient and very practical way of implementing the use of the three-state model for reliability calculations. This writer would like to emphasize the comment made at the end of Appendix II that the significant parameters for representing unscheduled outage overlapping conditions of maintenance and other apparatus is the exposure to maintenance P_{mi} and the duration of maintenance T_{mi}.

Stress should be added to the point that the most significant and critical conditions to be investigated for any failure modes and effects analysis are those which require the smallest number of unscheduled contingencies to lead to system failure.

Protective system dependability or reliability appears to play an equally important role with power apparatus reliability in bulk power supply systems. It is to be hoped that further publications will appear evaluating various protective system arrangements from a reliability standpoint.

It is to be hoped that this interesting paper will receive wide distribution among engineers in the power industry.

J. Endrenyi: The encouraging remarks by Dr. Ringlee are much appreciated. The three-state model appears to be advantageous over the more traditional two-state representation whenever a significant number of s states exist that represent system failure while the corresponding r states do not. As noted by Dr. Ringlee, the number of contingencies in these states should be the smallest that will produce system failure at all, although in certain cases states of the next "contingency level" may also be significant. All in all, the use of a three-state instead of a two-state model will likely increase the calculated system failure frequency and, since the additional outages are relatively short in duration, decrease the calculated mean system downtime.

Manuscript received April 28, 1971.

Manuscript received August 10, 1970.

A COMPUTERIZED APPROACH TO SUBSTATION AND SWITCHING STATION RELIABILITY EVALUATION

M.S. Grover
Student Member IEEE

R. Billinton
Senior Member IEEE

Power Systems Research Group
University of Saskatchewan
Saskatoon, Canada

Abstract—This paper presents a new approach for the evaluation of substation and switching station reliability performance in terms of outage frequencies and durations. All realistic component failure modes are included in the reliability predictions. The computer program described in the paper is fairly general. It performs failure modes and effects analysis and provides a concise and orderly description of the various combination of occurrences within the system that could result in an interruption. The application of the program is illustrated by considering two practical system examples. The concepts described are quite general and can be easily extended to the reliability evaluation of other power systems configurations.

INTRODUCTION

Considerable attention has been devoted by power utilities, in the past few years, to the evaluation of the effectiveness of their systems from a reliability viewpoint. The bulk of the work has been done in the field of generation capacity reliability evaluation. During the last decade, considerable attention has been focussed on the reliability studies of transmission and distribution schemes[1,2,3]. One of the main concerns has been the development of accurate and consistent models to represent true component and system behaviour. A two state weather model was developed to include environmental effects in the reliability predictions of overhead transmission and distribution systems[1,2,3]. In regard to the inclusion of circuit breakers and protective elements in transmission system analysis, a three state component model was described[4] which gives a more realistic representation to power system components than that given by previous two state component models. In this model, when a component fails, the system protection may isolate a number of unfaulted components. Following which, as soon as possible, through appropriate switching operations, all but the minimum number of components that must be kept out of service for the isolation of the failed component are restored to service. Thus a system component has three possible states namely, operating, before switching and after switching. The program as described in the paper[4] does not appear to be applicable to practical systems when

(i) normally open breakers or switches are present in the system.
(ii) all circuit breaker failures are not ground faults.
(iii) the protective system is not perfectly reliable.
(iv) the weather conditions have significant effect on the component failure and repair rates.
(v) component overload outages are to be evaluated.

This paper describes a computer program which provides a more realistic component and system representation. The effects of weather conditions, different component failure modes, normally open breakers or switches, unreliable protective systems etc. are included in the reliability predictions. The equations described in references 1 and 2 and in the paper are used to evaluate the outage frequency and duration indices due to the various modes of service interruption at the designated load point.

Paper T 74 133-5, recommended and approved by the IEEE Power System Engineering Committee of the IEEE Power Engineering Society for presentation at the IEEE PES Winter Meeting, New York, N.Y., January 27-February 1, 1974. Manuscript submitted September 4, 1973; made available for printing November 12, 1973.

Component Failure Modes

Power system components can have many types of failures. For the purpose of this paper, system components are classified into two categories according to their modes of failure. The first category includes components such as transmission lines, transformers, reactors, buses etc. These components can be in any of the following states: (i) Operating (ii) Faulted (iii) Out for repair or preventive maintenance.

In addition, if appropriate protection is not available, these components can have undetected open failures.

The second category includes components such as circuit breakers, reclosers, disconnect switches, carrier equipment etc. These components can be in any of the following states
(i) Operating (ii) Faulted (iii) Out for repair or preventive maintenance (iv) Stuck when called upon to operate or not closing when called upon to do so. (v) Undetected open failure.

It should be noted that the faulted state in the first category of components can be quite different from that in second category. In the latter case, a second level of system protection should operate to isolate the faulted component whereas in the former case the first level of system protection is required to operate.

A further classification of component failure modes can be obtained from this component classification. Component outages can be divided into active and passive failures. All component faults which result in the removal of certain other healthy components from service are classified as active failures. This class of failures includes component faults which cause operation of circuit breakers or disconnect switches. All component outages which do not remove any healthy components from service are classified as passive failures. These include undetected open failures, components out for repairs etc. It should be noted that all component active failures are included within the component passive failures.

Load Point Failure Modes

The simple system shown in Figure 1 has been utilized to illustrate various modes of load point failures. The various modes of interruption as considered in this paper are as follows:

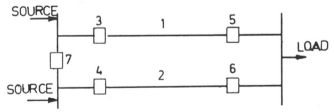

Fig. 1. System For Illustration of Failure Modes

(i) Active failures of breakers 5, 6 and 7.
(ii) Active failures of breaker 3 when breaker 5 or 7 is stuck.
(iii) Active failures of breaker 4 when breaker 6 or 7 is stuck.
(iv) Active failure of line 1 when breaker 5 is stuck.
(v) Active failure of line 2 when breaker 6 is stuck.
(vi) Passive failure of breakers 4 or 6 or line 2 overlapping the passive failure of breakers 3 or 5 or line 1.

Reprinted from *IEEE Trans. Power Apparatus Syst.*, vol. PAS-93, no. 5, pp. 1488–1497, Sept./Oct. 1974.

(vii) Passive failure of breakers 3 or 5 or line 1 overlapping the passive failure of breakers 4 or 6 or line 2.

It is assumed in the above that source and load buses are completely reliable. Circuit breakers are considered to be provided with isolators and the interruptions due to their active failures last only for the time required to operate the isolators. Continuity between source and load points is considered to be the criterion for successful system operation. The complexity of the problem can, therefore, be quickly realized from the large number of interruption modes associated with a simple system configuration such as that of Figure 1.

Reliability Analysis of Simple Substation Configurations

It is essential to perform reliability analysis on various possible alternative substation designs before the selection of any particular configuration is made. The simple substation configurations shown in Figure 2 have been evaluated manually in terms of their outage frequencies and durations considering continuity of supply to the load point as the requirement for successful operation. The various component parameters used in these studies are given in Table 1[5].

Table 1
Component Data For Systems in Figure 2

Component	Failure Rate, f/yr	Repair Time, Hrs
Line	0.050	23.0
Transformer	0.012	168.0
Circuit Breaker	0.007	72.0
Bus	0.007	3.5

For simplicity, it is assumed that all circuit breaker failures are active and that the probability of a stuck breaker condition is zero. Normally open breakers and switches are considered to be perfectly reliable and can be closed as and when required. No maintenance conditions are included in the reliability predictions. The following two criteria were considered for successful system operation.

(i) Continuity of supply to any one of the two load points.
(ii) Continuity of supply to both the load points.

The results obtained in terms of the outage rate and the total outage time per year for the above two modes of successful operation are shown in Figures 3 and 4. It is clear from these figures that the selection of a particular design from a reliability viewpoint is dependent upon how the successful system operation is defined. The first criterion of successful operation gives preference to the configuration shown in Figure 2f both from outage rate and total outage time point of view. The second criterion, from the outage rate graph, gives preference to systems 2c and 2d and from the total outage time graph, to system 2g.

This clearly illustrates the need to define very specifically the criterion of successful system operation and the important indices of reliability for the judicious section of a particular substation design.

The Computer Program

As noted earlier, the manual solution of reliability models of systems, involving components having different modes of failure, is extremely labourious and becomes quite unmanagable as the number of components increases. The inclusion of weather dependent failures and different repair and switching procedures requires a solution of many simultaneous equations. In this section a computer program is described which performs failure modes and effects analysis and computes systems reliability indices by the use of the simple equations described in references 1 and 2 and at the end of this section. The program is written in Fortan IV language for the IBM 370/155 computer. The algorithms are programmed to select the various possible combinations of occurrences within the system which can cause interruption to the designated load point. The only assumptions involved are that the equations described in references 1 and 2 are applicable.

A simplified flow chart for the computer program is shown in Figure 5. The basic steps involved in performing the failure analysis and the computation of the reliability indices are explained below.

The first step consists of reading the input data which requires information on the number of components in the system, the normally

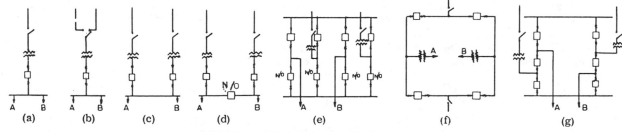

Fig. 2. Some Simple Substation Configurations

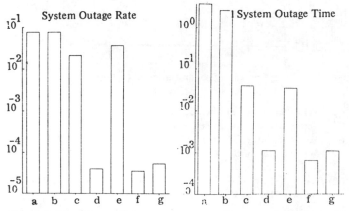

Fig. 3. System Outage Rate and Total Outage Time for Criterion(i)

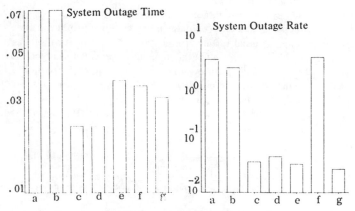

Fig. 4. System Outage Rate and Total Outage Time for Criterion(ii)

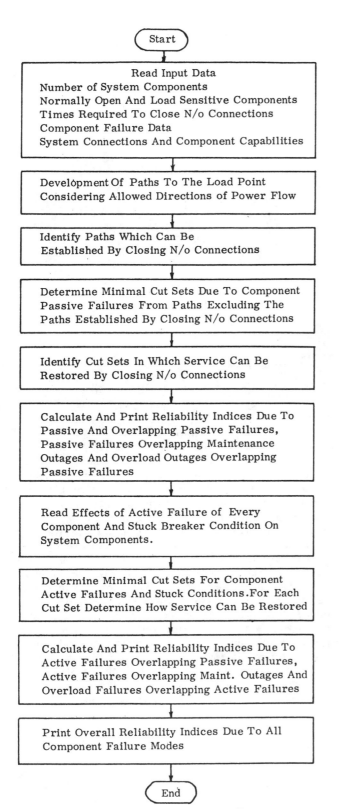

Fig. 5. Flow Chart of the Computer Program

open connections, times required to close normally open connections etc. The following information, which is also read in this step, is required on component reliability parameters.

(i) Component passive failure rate

This outage rate represents the total number of times in a year the component has to be removed from service for repairs due to any of the failure modes. This failure rate also includes component active failures.

(ii) Component average repair time

This time represents the average of times required to repair all kinds of component failure modes. The repair action may be warranted due to component fault, a breaker or switch stuck condition, undetected open connection, breaker false operation etc.

(iii) Component maintenance outage rate

This rate is the average number of times in a year that a component is taken out of service for preventive maintenance.

(iv) Component average maintenance time

This value represents the average of all times spent in performing preventive maintenance actions on the component.

(v) Component active outage rate

This value is that fraction of the component passive failure rate which corresponds to active failures of the component. The active outage rate is expressed in terms of the number of component active failures per year.

(vi) Component switching time

This is the time starting from the active failure of a component and lasting up to the time when the faulted component is removed from service and all other healthy components on the direct paths to the load point are restored to service. When other unfaulted components cannot be restored to service until the faulted component is repaired and put back into operation, the switching time value becomes the time required to repair the faulted component.

(vii) Stuck probability

This value represents the probability of a breaker or a switch being stuck when called upon to operate. In the case of normally open breakers or switches, this value is the probability of a breaker or switch not closing when called upon to do so. The stuck breaker probability is estimated from the ratio of the number of times the breaker fails to operate when called upon to do so to the total number of times the breaker is called upon to operate.

If weather conditions are to be included in the reliability calculations by using a two state weather model[1,2,3], the normal and adverse weather active and passive outage rates should be separately defined in the input. The average durations of normal and adverse weather periods are also specified in the input data.

There can be some components in the system which are not on any direct transmission path to the load point under consideration bu their active failures can interrupt some or all the direct paths. For such components, the passive outage rates values are assigned zero values in the program. The average outage time is also assigned a zero value. Maintenance outages of such components cannot interrupt the direct transmission paths and thus the maintenance rate and outage time are assigned zero values. The active outage rate and switching time parameters, used in the program, are the actual values associated with the component. If the component cannot be switched out, then switching time is assigned a value equal to component repair time.

In the next step, information about the configuration of the system under study is supplied. This is done by describing the components that immediately precede the component under consideration in the line of power flow.

The program execution starts with the establishment of paths between the source point and the designated load point. These paths are established taking into consideration the allowed directions of power flow. (In certain components the reverse flow of power may not be allowed). The paths which can be established by closing the normally open connections are also recognized.

The program execution then proceeds on to determining the minimal cut sets for the load point under consideration. This, in fact, consists of evaluating the various combinations of component passive failures within the system that can cause interruptions. The algorithm used in determining the minimal cut sets is based on boolean algebra and is very similar to that described in reference 6. It should be noted that these minimal cuts sets are determined from only those paths which are normally closed. A simple algorithm is then used to recognize

Table 2
Equations for the Evaluation of Outages Due to Component Active Failures

Comps. in the Cut Set	Component Actively Failed ,	Comp. Stuck	Event	Contribution to System Failure Rate	Average Outage Duration	
					Switching	Nonswitching
i	i	—	(a)	λ_{ig}	S_i	S_i
i,j	i	—	(a)	$\lambda_{ig}\lambda_j r_j + \lambda_j\lambda_{ig}S_i$	S_i	$\dfrac{S_i r_j}{S_i+r_j}$
			(b)	$\lambda_{ig}\lambda''_j r''_j$	S_i	$\dfrac{S_i r''_j}{S_i+r''_j}$
i	i	j	(a)	$\lambda_{ig}Pr(j)$	S_i	S_i
i,j	i	j	(a)	$\lambda_{ig}Pr(j)$	S_i	S_i
i,k	i	j	(a)	$\lambda_{ig}\lambda_k r_k Pr(j)$	S_i	$\dfrac{S_i r_k}{S_i+r_k}$
			(b)	$\lambda_{ig}\lambda''_k r''_k Pr(j)$	S_i	$\dfrac{S_i r''_k}{S_i+r''_k}$

NOTES: In the above table,

Equations are described for cut sets upto second order and similar equations can be written for higher order cut sets.

Equations similar to those described in references 1 and 2 can be written to include failures in normal and adverse weather conditions.

λ_{ig} and S_i are respectively the active failure rate and the switching time of component i.

λ_j and r_j are respectively the passive failure rate and the average repair time of component j.

λ''_k and r''_k are respectively the maintenance outage rate and the average maintenance time of component k.

$Pr(j)$ is the probability of a breaker or switch j being stuck at any time.

Event (a) includes component active failures and component active failure overlapping component passive failures.

Event (b) includes component active failures overlapping component maintenance outage periods.

those cut sets in which service can be restored by closing normally open paths.

When all the minimal cut sets have been determined, the computations are made for the contribution of each of the cut set to the following system reliability indices.

(i) The outage rate, the average duration and the total outage time due to passive failures and overlapping passive failures of components present in the cut set under consideration.

(ii) The outage rate, the average duration and the total outage time due to component passive failures overlapping the maintenance outage periods of components present in the cut set under consideration.

The above indices are computed by using the appropriate equations described in references 1 and 2. These equations are stored in the computer as subroutines. If for a particular cut set, the service can be restored by closing a normally open path, the average outage duration for that cut set equals the time required to close the open path. When these computations have been made for all the cut sets, the total contributions to the load point reliability indices due to passive failures and passive failures overlapping maintenance outages are evaluated separately.

In the next step, additional input information is required. This consists of describing the effect of an active failure of each component on other system components. The unfaulted components which are isolated as a result of a fault on the component under consideration are identified. The information is also provided on unfaulted components which are isolated because of a combined component fault and stuck breaker condition present in the system. Corresponding to each component fault, minimal cut sets for the designated load point are then determined by interrupting all those paths which contain unfaulted components but are on outage because of fault on the component under consideration. If any of these cut sets has already been evaluated, it is not considered any further. The contribution to outage frequency and duration indices due to various cut sets associated with the component fault is evaluated by equations described in Table 2. When all the components have been considered, the contribution to the reliability indices due to combined component fault and stuck breaker conditions is evaluated on the same basis as is done for component fault conditions. At the end of this step, the total contribution to the load point reliability indices due to component active failures and breaker or switch stuck conditions is determined for

Table 3

Component Data For System in Figure 6

Components	Passive Outage Rate, f/yr.	Average Outage Duration, Hrs.	Maint. Rate o/yr.	Maint. Time Hrs.	Active Outage Rate, f/yr.	Switching Time, Hrs.	Stuck Prob.
Lines (1, 2)	0.09	7.33	1.0	8.0	0.09	1.0	—
H.V. Breakers (3, 4, 5)	0.23	11.13	0.25	24.0	0.03	2.0	0.005
Disconnect Switches (6, 7)	0.22	2.09	0.25	4.0	0.02	3.0	—
Transformers (8, 9)	0.10	1000.0	0.50	48.0	0.10	1.0	—
L.V. Breakers (10, 11)	0.02	3.0	0.25	12.0	0.01	1.0	0.06
L.V. Bus (12)	0.024	2.0	0.00	0.0	0.024	2.0	—

Table 4
Detailed Reliability Analysis of Substation Configuration of Figure 6
(Contingencies only up to second order are listed)

FAILURE EVENTS Overlapping passive failures and maint. outages	Contribution due to non-maintenance events			Contribution due to maintenance events		
	Outage Rate F/Yr.	Av. Duration Hrs.	Total Outage Time Hrs./Yr.	Outage Rate F/Yr.	Av. duration Hrs.	Total Outage Time Hrs./yr
L.V. Bus, 12	0.240E-01	2.00	0.480E-01	0.	0.	0.
Lines 1 & 2	0.135E-04	3.66	0.496E-04	0.164E-03	3.83	0.628E-03
Line 1 & Bkr. 4	0.436E-04	4.42	0.193E-03	0.271E-03	4.87	0.132E-02
Line 2 & Bkr. 3	0.436E-04	4.42	0.193E-03	0.271E-03	4.87	0.132E-02
Bkr. 3 & Bkr. 4	0.134E-03	5.56	0.748E-03	0.315E-03	7.60	0.239E-02
Disconnects 6 & 7	0.231E-04	1.04	0.241E-04	0.502E-04	1.37	0.689E-04
Disconnect 6 & Trans. 9	0.252E-02	2.08	0.525E-02	0.614E-03	2.04	0.125E-02
Disconnect 6 & Bkr. 11	0.255E-05	1.23	0.315E-05	0.776E-04	1.78	0.138E-03
Disconnect 7 & Trans. 8	0.252E-02	2.08	0.525E-02	0.614E-03	2.04	0.125E-02
Disconnect 7 & Bkr. 10	0.255E-05	1.23	0.315E-05	0.776E-04	1.78	0.138E-03
Transformers 8 & 9	0.228E-02	500.00	0.114E+01	0.548E-03	45.80	0.251E-01
Trans. 8 and Bkr. 11	0.229E-03	2.99	0.685E-03	0.890E-04	6.30	0.561E-03
Trans. 9 and Bkr. 10	0.229E-03	2.99	0.685E-03	0.890E-04	6.30	0.561E-03
Breaker 10 and Bkr. 11	0.274E-06	1.50	0.411E-06	0.137E-04	2.40	0.328E-04
Total Contributions	0.320E-01	37.53	0.120E+01	0.319E-02	10.87	0.348E-01
Component Active failures **overlapping passive & maint.** **failures**						
Bkr. 3 A.F., Disconnect 7 out	0.383E-05	2.00	0.767E-05	0.342E-05	2.00	0.685E-05
Bkr. 3 A.F., Trans. 9 out	0.343E-03	2.00	0.686E-03	0.822E-04	2.00	0.164E-03
Bkr. 3 A.F., Bkr. 11 out	0.274E-06	2.00	0.548E-06	0.103E-04	2.00	0.205E-04
Bkr. 4 A.F., Disconnect 6 out	0.383E-05	2.00	0.767E-05	0.342E-05	2.00	0.685E-05
Bkr. 4 A.F., Trans. 8 out	0.343E-03	2.00	0.686E-03	0.822E-04	2.00	0.164E-03
Bkr. 4 A.F., Bkr. 10 put	0.274E-06	2.00	0.548E-06	0.103E-04	2.00	0.205E-04
Bkr. 5 A.F.	0.300E-01	2.00	0.600E-01	0.	0.	0.
Disconnect 6 A.F., Line 2 out	0.171E-05	2.13	0.364E-05	0.183E-04	2.18	0.398E-04
Disconnect 6 A.F., Bkr. 4 out	0.689E-05	2.36	0.163E-04	0.137E-04	2.67	0.365E-04
Disconnect 7 A.F., Line 1 out	0.171E-05	2.13	0.364E-05	0.183E-04	2.18	0.398E-04
Disconnect 7 A.F., Bkr. 3 out	0.689E-05	2.36	0.163E-04	0.137E-04	2.67	0.365E-04
Trans. 8 A.F., Line 2 out	0.856E-05	1.00	0.856E-05	0.913E-04	1.00	0.913E-04
Trans. 8 A.F., Bkr. 4 out	0.344E-04	1.00	0.344E-04	0.685E-04	1.00	0.685E-04
Trans. 9 A.F., Line 1 out	0.856E-05	1.00	0.856E-05	0.913E-04	1.00	0.913E-04
Trans. 9 A.F., Bkr. 3 out	0.344E-04	1.00	0.344E-04	0.685E-04	1.00	0.685E-04
Bkr. 10 A.F.	0.100E-01	1.00	0.100E-01	0.	0.	0.
Bkr. 11 A.F.	0.100E-01	1.00	0.100E-01	0.	0.	0.
Active failures and Stuck bkrs.						
Bkr. 3 A.F., Bkr. 5 stuck	0.150E-03	2.00	0.300E-03	0.	0.	0.
Bkr. 3 A.F., Bkr. 10 stuck	0.180E-02	2.00	0.360E-02	0.	0.	0.
Bkr. 4 A.F., Bkr. 5 stuck	0.150E-03	2.00	0.300E-03	0.	0.	0.
Bkr. 4 A.F., Bkr. 11 stuck	0.180E-02	2.00	0.360E-02	0.	0.	0.
Disconnect 6 A.F., Bkr.11 stuck	0.120E-02	1.50	0.180E-02	0.	0.	0.
Disconnect 6 A.F., Bkr. 5 stuck	0.100E-03	2.36	0.236E-03	0.	0.	0.
Disconnect 7 A.F., Bkr.11 stuck	0.120E-02	1.50	0.180E-02	0.	0.	0.
Disconnect 7 A.F., Bkr. 5 stuck	0.100E-03	2.36	0.236E-03	0.	0.	0.
Trans. 8 A.F., Bkr. 10 stuck	0.600E-02	1.00	0.600E-02	0.	0.	0.
Trans. 8 A.F., Bkr. 5 stuck	0.500E-03	1.00	0.500E-02	0.	0.	0.
Trans. 9 A.F., Bkr. 11 stuck	0.600E-02	1.00	0.600E-02	0.	0.	0.
Trans. 9 A.F., Bkr. 5 stuck	0.500E-03	1.00	0.500E-03	0.	0.	0.
Total contributions	0.703E-01	1.51	0.106E+00	0.575E-03	1.49	0.856E-03

Overall reliability indices

Failure rate = 0.106112 f/yr., Av. outage duration = 12.67 hrs., Total outage time = 1.344 hrs./yr.

the following two conditions.

(i) Component active failures and component active failures overlapping component passive failures.

(ii) Component active failures overlapping component maintenance outage periods.

When all the above computations are completed, the overall reliability indices are determined by combining the outage contributions of all active, passive and maintenance failure modes. These indices are (i) Total interruption rate (ii) Average outage time and (iii) The total average outage time.

It should be noted that the limit to the number of components which can be handled by the program is determined only by the size of the computer available. The dimensions of the arrays in the program can be increased, if the computer size can accomodate a larger number of components.

System Studies

The computer program described in the previous section has been used in a number of practical system studies. Two examples are given here to illustrate the basic concepts. In the first example, the distribution substation shown in Figure 6 is considered. Continuity of supply is considered as the criterion of successful operation. The component data used in this study is given in Table 3. The output of the program is shown in Table 4. This table lists contingencies only up to second order for simplicity. The program, however computes higher order contingencies also and their contribution is added to the overall results.

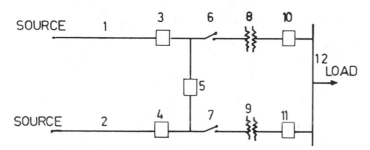

Fig. 6. A Distribution Substation Configuration

The overall reliability indices can be affected by changing the system configuration. The following variations for the system shown in Figure 6 were considered.

(a) Disconnects on the h.v. side of the transformer are replaced by h.v. breakers.

In this case failure parameters for h.v. breakers were assumed to be the same as those of the disconnects. The following events, will not cause interruption if this change is made.

Breaker 3 suffers an active failure when breaker 10 is stuck.
Breaker 4 suffers an active failure when breaker 11 is stuck.
Transformer 8 suffers an active failure when breaker 5 is stuck.
Transformer 9 suffers an active failure when breaker 5 is stuck.
Transformer 8 suffers an active failure when line 2 is being repaired or maintained.
Transformer 9 suffers an active failure when line 1 is being repaired or maintained.

The system outage rate decreases from 0.106112 f/yr to 0.101110 f/yr. with this change. The total outage time decreases from 1.344 hrs./yr. to 1.336 hrs./yr. It is, therefore, evident that this change in the configuration brings about an insignificant change in the system reliability indices.

(b) Breaker 5 is normally open

If it is assumed that breaker 5 in the normally open mode cannot have any active failure, some of the events listed in Table 4 are

eliminated while a few new failure modes are added. The system overall reliability indices for this change are

λ_{SL} = 0.08396 f/yr., r_{SL} = 15.50 hrs., $\lambda_{SL} \cdot r_{SL}$ = 1.3016 hrs./yr. where λ_{SL}, r_{SL} and $\lambda_{SL} \cdot r_{SL}$ are respectively the system outage rate, the average outage duration and the total outage time.

It is, therefore, evident that this change produces a significant decrease in the system outage rate whereas the total outage time is not affected very much. If in the normally closed configuration, breaker 5 is assumed to be completely reliable, the system overall reliability indices are

λ_{SL} = 0.076112 f/yr., r_{SL} = 16.88 hrs., $\lambda_{SL} r_{SL}$ = 1.284 hrs./yr. This shows that the reliability of breaker 5 has a significant effect on the system outage rate. This is because all active failures of breaker 5 will cause interruption at the load point.

(c) Breakers on the h.v. side of transformers and disconnects on l.v. side

The overall reliability indices resulting from this change are

λ_{SL} = 0.3733 f/yr., r_{SL} = 4.645 hrs., $\lambda_{SL} r_{SL}$ = 1.734 hrs./yr.

It is clear from these results that there is a considerable increase in system outage rate and total outage time. The average outage duration decreases because of increase in the number of relatively short duration outages.

(d) L.V. breakers are replaced by disconnects and a sectionalizing breaker is connected in the L.V. bus

The overall reliability indices associated with this configuration are as follows.

λ_{SL} = 0.2361 f/yr., r_{SL} = 6.50 hrs., $\lambda_{SL} r_{SL}$ = 1.53 hrs./yr.

There is again, a considerable increase in system outage rate and total outage time indices.

The second example utilizes the Regina switching station of the Saskatchewan Power Corporation. The single line diagram for this station is shown in Figure 7. The criterion of successful operation is

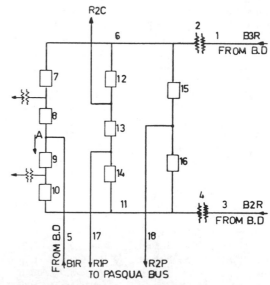

Fig. 7. Regina Switching Station of the Saskatchewan Power Corporation

considered to be the continuity of supply at the far end of lines R1P and R2P (i.e. the Pasqua bus). All breaker failures are assumed to be active and the stuck breaker probability is considered to be zero. The following component parameters were used in the reliability study.

Lines 1, 3 and 5

λ = 0.028 f/yr., r = 5.0 hrs., λ' = 1.0 o/yr., r'' = 10.0 hrs.

Buses 6 and 11

λ = 0.007 f/yr., r = 3.5 hrs., λ'' = 1.0 o/yr., r'' = 15.0 hrs.

Transformers 2 and 4

λ = 0.012 f/yr., r = 168.0 hrs., λ'' = 2.0 o/yr., r'' = 13.0 hrs.

Table 5
Reliability Indices Due to Component Overloads For System in Figure 6

Failure Event	Outage Rate, f/yr	Av. Duration, Hrs.	Total Outage Time,Hrs./yr.
Disct. 6 P.F, Trans. 9 overloaded	0.107 E + 00	1.49	0.160 E + 00
Disct. 7 P.F, Trans. 8 overloaded	0.107 E + 00	1.49	0.160 E + 00
Trans. 8 P.F, Trans. 9 overloaded	0.100 E + 00	5.17	0.517 E + 00
Trans. 9 P.F, Trans. 8 overloaded	0.100 E + 00	5.17	0.517 E + 00
Bkr. 11 P.F, Trans. 8 overloaded	0.113 E - 00	1.90	0.215 E - 01
Bkr. 10 P.F, Trans. 9 overloaded	0.113 E - 01	1.90	0.215 E - 01
Bkr. 3 A.F, Trans. 9 overloaded	0.128 E - 01	1.44	0.185 E - 01
Bkr. 4 A.F, Trans. 8 overloaded	0.128 E - 01	1.44	0.185 E - 01
Total Contributions	0.438 E + 00	3.25	0.142 E + 00

Lines 17 and 18

$\lambda = 0.05$ f/yr., r = 4.0 hrs., $\lambda'' = 0.5$ o/yr, $r'' = 4.5$ hrs.

Circuit breakers

$\lambda = 0.007$ f/yr., r = 70.0 hrs., $\lambda'' = 2.0$ o/yr, $r'' = 11.5$ hrs.
Switching time = 1.5 hrs.

Assuming the allowed direction of power flow at point A as shown in Figure 7, the system reliability indices are

$\lambda_{SL} = 0.000268$ f/yr., $r_{SL} = 3.17$ hrs., $\lambda_{SL}r_{SL} = 0.000849$ hrs./yr.

When line B2R is considered to be normally open, with switching time equal to one hour, the following results are obtained

$\lambda_{SL} = 0.000621$ f/yr, $r_{SL} = 1.936$ hrs., $\lambda_{SL}r_{SL} = 0.001202$ hrs./yr.

If the breaker active failure rate is 0.002 f/yr and 0.007 f/yr is its passive failure rate, then with B2R normally open,

$\lambda_{SL} = 0.000530$ f/yr., $r_{SL} = 2.00$ hrs., $\lambda_{SL}\cdot r_{SL} = 0.00107$ hrs./yr.
If power flow at point A is in the direction opposite to that shown in Figure 7, then

$\lambda_{SL} = 0.000578$ f/yr., $r_{SL} = 1.80$ hrs., $\lambda_{SL}\cdot r_{SL} = 0.001045$ hrs./yr.
If power flows at point A are allowed in both directions, then

$\lambda_{SL} = 0.000314$ f/yr., $r_{SL} = 1.24$ hrs., $\lambda_{SL}\cdot r_{SL} = 0.00039$ hrs./yr.
If power flows at point A can be in any direction, then

$\lambda_{SL} = 0.000306$, $r_{SL} = 1.26$ hrs., $\lambda_{SL}\cdot r_{SL} = 0.000386$ hrs./yr.

The above two examples clearly illustrate the usefulness of the program in performing the reliability studies. The relative economic and reliability benefits associated with various configuration changes can be evaluated in quantitative terms. This provides useful input to the judicious selection of a particular design keeping in view the economic and reliability constraints.

Application of the program for evaluation of overload outages

The program described in this paper has been extended to consider cases where discontinuity of connection between the source and the load points is not the only failure mode. Under certain system conditions, interruptions can occur at the designated load point due to component overload outages. The program proceeds to consider this condition in the same sequence as described earlier. An additional input is provided to identify the components such as transformers, lines etc. which are liable to suffer overloads. The capabilities of these components are also provided. As the program determines the cut sets for the designated load point, a simple algorithm scans those cut sets which contain components liable to suffer overload outages. These cut sets are then evaluated for their contributions to outage indices due to component overloads using the equations described in reference 2.

The distribution substation shown in Figure 6 is used as an example to illustrate this outage aspect. The capability of transformers and lines were assumed to be 10Mw and 20Mw each respectively. A Saskatchewan Power Corporation substation load[2] with a peak of 13.2 Mw was assumed at the load point. Table 5 lists the various double contingency events and their contribution to the outage indices due to

component overload outages. (The program, however, also evaluates higher order contingencies). It is assumed that component maintenance outages will not cause any overload outages. This situation, if applicable, can also be evaluated using the equations of reference 2.

It should be noted that the procedure for evaluation of overload outages as described above is valid when there is only one load point in the system. When more than one load point is present in the system, load flow studies are required to evaluate the limits at which the overload commences. The application of the program to these systems will probably be illustrated in future publications.

CONCLUSIONS

A computer program is described for the evaluation of reliability performance of substations and switching stations in terms of their outage frequencies and durations. The program is quite general and can be used for reliability evaluations of other parts of a power system. A new concept of component active and passive failures novel to the method of calculation, is introduced. This permits the inclusion of all the realistic component failure modes in the reliability predictions. The program, in addition, is capable of handling weather dependent failures, unreliability of protective schemes, normally open switches and breakers and component overload outages. The output of the program provides a concise and sequential description of various system contingencies that cause interruptions. The computational efficiency of the program arises from the evaluation of only the failure related events. The algorithms for performing failure modes and effect analysis are quite fast. The simple equations[1,2] used for calculations of reliability indices involve basic mathematical operations and provide additional computational efficiency to the program.

The program has been applied successfully to many practical system configurations and two substation and switching station examples are presented in this paper. The effects of varying the system configuration and successful mode of system operation on reliability indices are illustrated. This form of analysis provides a quantitative basis for the judicious selection of a reliable and economic design.

ACKNOWLEDGMENTS

The authors would like to thank the Saskatchewan Power Corporation for their assistance and financial support for these studies.

REFERENCES

[1] Grover, M.S. and Billinton R., "Quantitative Evaluation of Permanent Outages in Distribution Systems". A paper presented at the IEEE summer power meeting, 1972, Vancouver, Canada.
[2] Grover, M.S. and Billinton R., "Reliability Assessment of Transmission and Distribution Schemes". A paper presented at the IEEE summer power meeting, 1972, Vancouver, Canada.
[3] Gaver, D.P., Montmeat, F.E. and Patton, A.D., "Power System Reliability I-Measures of Reliability and Methods of Calculation". IEEE Trans. PAS, July 1964, pp. 727-737.

[4] Endrenyi, J., Maenhaut, P.C. and Payne, L.E., "Reliability Evaluation of Transmission Systems With Switching After Faults - Approximations and A Computer Program". A paper presented at the IEEE Winter Power meeting, 1972, New York, N.Y.

[5] Ringlee, R.J. and Goode, S.D., "On Procedures for Reliability Evaluations of Transmission Systems". IEEE Trans. PAS, April 1970, pp. 527-537.

[6] Nelson, A.C., Batts, J.R. and Beadles, R.L., "A Computer Program for Approximating System Reliability". IEEE Trans. on Reliability, May 1970, pp. 61-65.

[7] Grover, M.S. "Distribution System Reliability Assessment" M.Sc. Thesis, College of Graduate Studies, University of Saskatchewan, Saskatoon, Canada, Oct. 1972.

APPENDIX

Simplifications in equations and reduction in computer time

The computational time of the program can be reduced by combining certain component failure modes. Component maintenance outages have the same system effect as do the component passive failures. As such, these failure indices can be combined to obtain equivalent passive outage rates and durations. This is shown below for a cut set of two components 1 and 2.

Equivalent passive outage rate of component $1 = \lambda_1 + \lambda''_1 = \lambda_{e1}$

Equivalent average outage duration of component 1 =

$$\frac{\lambda_1 r_1 + \lambda''_1 r''_1}{\lambda_1 + \lambda''_1} = r_{e1}$$

Similar expressions can be written for component 2.

Then, the contribution to system failure rate due to this two component cut set is

$$\lambda_s = \lambda_{e1}\lambda_2 r_{e1} + \lambda_{e2}\lambda_1 r_{e2} \qquad (a)$$

This equation takes into consideration that a maintenance outage cannot occur if there is some outage already existing in the system. The average duration associated with the failure rate λ_s is given by:

$$r_s = \lambda_{e1}\lambda_2 \frac{r_{e1}^2 r_2}{\lambda_s(r_{e1}+r_2)} + \lambda_{e2}\lambda_1 \frac{r_1 r_{e2}^2}{\lambda_s(r_{e2}+r_1)} \qquad (b)$$

Equations (a) and (b) yield the following results

$$\lambda_s = \lambda_1\lambda_2 r_1 + \lambda''_1\lambda_2 r''_1 + \lambda_2\lambda_1 r_2 + \lambda''_2\lambda_1 r''_2$$

and

$$r_s = (\lambda_1 r_1 + \lambda''_1 r''_1)\frac{\lambda_2}{\lambda_s}\left[\frac{\lambda_1 r_1 r_2 + \lambda''_1 r''_1 r_2}{\lambda_1 r_1 + \lambda''_1 r''_1 + \lambda_1 r_2 + \lambda''_1 r_2}\right] + \qquad (c)$$

$$(\lambda_2 r_2 + \lambda''_2 r''_2)\frac{\lambda_1}{\lambda_s}\left[\frac{\lambda_2 r_2 r_1 + \lambda''_2 r''_2 r_1}{\lambda_2 r_1 + \lambda''_2 r''_1 + \lambda_2 r_2 + \lambda''_2 r''_2}\right] \qquad (d)$$

Expressions c and d are similar to those obtained by using the combined equations of maintenance and passive failures of references 1 and 2. Similar considerations are applicable for cut sets involving larger number of components. When weather conditions are to be included by using a two state fluctuating environment model, the component maintenance- parameters are combined with component normal weather associated indices. This method of calculation also reduces the computational time for events involving component active failures overlapping component passive and maintenance outages.

It should, however, be pointed out that by forming equivalent components the identity of maintenance and passive failures is lost. The individual contributions of these failure modes to system reliability indices cannot be evaluated.

Discussion

J. Endrenyi and **L. E. Payne** (Ontario Hydro, Toronto, Canada): It was with great interest that we have studied the description of this latest computer program for the reliability evaluation of transmission systems and transformer stations. We congratulate the authors for developing a program that can account for so many of the peculiar characteristics of these systems. It is particularly gratifying for us to see that the procedure of switching after faults, and the distinction between the states before and after switching, have been recognized, as we have been advocating this for some time. These concepts form the basis of our own transmission system model and program described in Reference 4 of the paper.

Since the authors list the limitations of the model and program in Reference 4, we would like to note that the various circuit breaker failure modes are now incorporated in our model, as described in Reference 1 below. An extension of the program, already discussed in the original reference, makes it capable of accounting for certain overload outages.

In connection with the program described in the present paper, we would like to put forward the following remarks and questions:

1. There appears to be a discrepancy in the definitions given for the active and passive failures. If during passive failures healthy components are not removed from service and during active failures they are, how can the latter be "included within the component passive failures"? Perhaps a different terminology would help — one could speak about a total failure rate which is the sum of the passive and active failure rates, and when in the course of computations the passive failures are tested, this total failure rate would be used. The passive failure rate would be defined as the difference between the total (formerly passive) and the active rates.

2. When the effects of active failures are considered (seventh box in Figure 5), are these effects read in as an input to the program or is the program capable of determining them? It would appear from the description that a manual input of the protection zones around the system components is required, which would considerably increase the volume of input information to be prepared.

3. The policy adopted for closing the normally open breakers appears to be that they will be closed whenever the system can, after having failed, be restored thereby to a working condition. However, there are other policies that could be applied; for example, a normally open breaker could be closed upon the failure of a given component as a preventive measure, even if there is no ensuing system failure. How could such a policy be modelled?

4. Is the "stuck-open" failure mode recognized for normally open breakers?

5. When considering protection system malfunctions, how are relay, communications channel, etc. failures evaluated that affect several breakers simultaneously?

6. Although the number of components handled by the program is limited only by the size of the computer available, we would be interested in knowing how core usage varies with the number of components handled since this is a very real constraint on many computer systems. Also an indication of execution times would provide additional insight into the efficiency of the program.

REFERENCE

[1] J. Endrenyi, "Reliability Models for Circuit Breakers with Multiple Failure Modes". IEEE Paper C74 138-4, PES Winter Meeting, New York, N.Y., January 1974.

Manuscript received February 7, 1974.

M. B. Guertin and **Y. Lamarre** (Hydro-Quebec, Montreal, Canada): The authors have presented an interesting and novel approach to substation reliability evaluation. The main advantage of this approach is to reduce the number of elements to be represented by requiring only the isolating time of major apparatus as input to the program instead of having to represent all the disconnect switches. This is a model of the substation and, as in any model, there is a trade-off between exactness, ease of use, computing time and completeness of results. This trade-off is always open to judgement and it is in this spirit that the following points are raised; we would appreciate comments from the authors on the following:

Manuscript received February 11, 1974.

— The concept of active and passive failure rates, while advantageous in some respects, does not permit the full representation of all operating conditions in substations. A certain number of assumptions must be made, in particular when dealing with less "standard" substations. For example, no difference is made between disconnects and load break switches, which might lead to undetected temporary outages.

— It is not possible from the output to immediately distinguish between temporary and permanent outages; most of the failure modes must be analysed and their effect ascertained before a classification can be made.

— The method requires prior analysis of the effects of single-contingency failures in the active mode in order to determine the elements which are affected in this case since this information is required as input to the program. In a large substation, this work would become quite appreciable.

— The load carrying capability of transformers depends of course on the actual overload on the transformer, but also depends on the load preceding the overload and on the ambient temperature. Taking into account only the actual overload to evaluate the reliability of the station might lead to completely erroneous conclusions.

— The authors give no indication of the computing time required to solve the example given in the paper. Would the computing time increase greatly for a substation of, say, one hundred elements?

R. J. Becker (Union Electric Co., St. Louis, Mo.): The authors have provided a most interesting paper on a timely subject.

It has been our observation that although circuit breaker failure rates are reported on an outages per year basis, these outages have a pronounced seasonal variation, at least in a four season climate such as ours. The effect of this seasonal variation is to produce "clustering" of the outages during particular climatic conditions.

It would be most interesting to know if the outage statistics used by the authors also showed a seasonal variation, and if so, how this was considered in the programs described.

Manuscript received February 13, 1974.

P. F. Albrecht (General Electric Company, Schenectady, N.Y.): In the first part of the paper two basic categories of components are distinguished on the basis of their differing failure modes and resulting system effects. We have used a similar classification in our work.

After starting with a definition of component states based on physically meaningful failure modes and effects, the authors immediately define a second categorization of failures—active and passive— and the remainder of the paper is expressed completely in terms of these failure modes. I find this classification much less satisfying than the one which preserves identification of individual component categories and physical failure modes. Most of the physical failure modes initially defined translate directly into active or passive failures, but the distinction between switching and nonswitching components is lost. It is not clear whether the stuck breaker failure mode is considered active or passive by the authors. In fact, it seems to be really carried along as actually a third type of failure. In terms of the authors' definitions, a stuck breaker would seem to be passive until it is manifested by calling the breaker to open, at which time it becomes active. We would be interested in the authors' comments regarding why the physical failure modes were not preserved in the computer model and tabulated results. What advantages are gained by use of the active and passive convention?

The procedure described provides for normally open switches or breakers. The computer program flow chart does not indicate that normally closed switches can be specified as input data. However, the distribution substation in Figure 6 shows normally closed disconnect switches on the high side of the transformer. Apparently, the computer program does recognize such devices.

When switching is allowed, both momentary and sustained interruptions can occur. In our work, we have attempted to preserve the distinction between momentary and sustained interruption in the computer output. Have the authors considered such classification of results in their work? We have found that the results of the failure mode and effects analysis are frequently useful in themselves, independently of

Manuscript received February 19, 1974.

the calculated reliability indices. For this reason, we have structured our programs into two separate programs. The first program only performs the failure mode and effects analysis, giving a list of component outage combinations which cause load interruption. Since no reliability indices are calculated, there is no need for component outage rates as input data.

M. S. Grover and **R. Billinton:** We would like to thank all the discussors for their valuable comments and for their interest in our work.

It should be emphasized at the outset that our objective in this project was to develop a computer program which could realistically represent component and system performance and at the same time be computationally efficient. The program will be improved upon in the near future to consider many other features of substation and switching station configurations.

Dr. Endrenyi and Mr. Payne have noted the capabilities of their modified program. We agree that our terminology regarding active and passive failures may not appear to be very clear. The effect of an active failure after the failed component is switched out of service becomes passive, and therefore, the active and passive failures were combined to determine the overall passive outage rate. Perhaps, a better name for this outage rate, as the discussors suggest, would have been the total failure rate.

In our program, the breakers which are protecting different components of the system have to be identified. This, as also noted by Messrs. Guertin and Lamarre, will increase the volume of data required. It should, however, be pointed out that this form of data does not require any extensive effort in its formulation. To eliminate this effort completely, we are now looking into a procedure which will computerize the determination of different protection zones.

Dr. Endrenyi's and Mr. Payne's next inquiry is concerned with the policy adopted in closing the normally open connections. The program, in its present form, considers the closing of normally open breakers only when an interruption has occurred at the load bus. The policy of closing a normally open breaker upon the failure of a given component even if there is no load point failure, can be easily incorporated in the program. This, however, will require the determination of new cut sets corresponding to the changed configuration. The increase in computational time due to this modification can be considerable.

The stuck open failure mode for normally open breakers is considered by the present program. The probability of breaker not closing when called upon to do so is specified in the input.

We have not considered the possibility of the malfunctions of the protection system affecting several breakers simultaneously. This failure mode, could possibly be included by specifying the breakers which are simultaneously affected and the associated probability of such an event.

The variation in core usage with the number of components on the IBM 370/158 computer is as follows:

NO. OF COMPONENTS	CORE USAGE
25	30K
50	70K
80	110K

It should be noted that the program was not written with the intention of minimizing the core usage and therefore, some of the arrays in our program may have been overdimensioned. The core usage can be reduced by decreasing the size of the arrays for the paths, cut sets etc., according to the requirements of the system configuration and also by using the one array storage for different purposes. Dr. Endrenyi and Mr. Payne, and Messrs. Guertin and Lamarre have asked about the computational time of the program. The execution times for the substation and switching station examples given in the paper were 1.50 and 1.75 seconds respectively. It is to be noted that the computational time does not only depend upon the number of system components but also on the system complexity.

In reply to Messrs. Guertin and Lamarre we would like to point out that the concept of active and passive failures is quite general as it models the different types of component failures according to their system effects. The program can distinguish between disconnects and load break switches.

Mr. Albrecht, Messrs. Guertin and Lamarre have asked if we considered distinguishing between permanent outages and the so called temporary or momentary outages (in which service is restored by switching action). We did not attempt to use this classification in our program. This was done intentionally because some of the permanent

Manuscript received May 6, 1974.

outages (in which service is restored by repairs) have average durations comparable to the manual switching times required for restoration of service in the case of momentary outages. Our program, however, can indicate for each failure event whether the service is restored by switching action or by component repair. This information is obtained while performing failure modes and effects analysis.

The comment by Messrs. Guertin and Lamarre regarding the transformer capability is obviously valid. We agree that the transformer load carrying capability depends not only on the actual overload but also on the ambient temperature and the load preceeding the overload. It is quite difficult to incorporate all these factors in the analysis as this will require a knowledge of the temperature distributions and the evaluation of hot spot temperature. To simplify matters, we have chosen to determine the frequency and duration of exceeding a preassigned carrying capability. These indices then provide an indication of "how often and for how long" on the average the transformer is subjected to a condition where its carrying capability is exceeded.

Mr. Becker has noted some very interesting points regarding breaker failure rates. We are not aware of any consistent statistics in which breaker failure rates are segregated according to the seasons. In our program, we have, therefore, not considered the variation in breaker failure rates according to different seasons of the year. If, however, wide variations are observed, as has been noted by the discussor, an appropriate approach would be to divide the year into different seasons and to use fail rate values applicable to a particular season.

Mr. Albrecht has provided some valuable comments in describing some of the peculiar features of his program. As noted earlier, our intent in using active and passive failure modes was to classify many types of component failures according to their system effects. We have treated component passive failures and maintenance outages (which have the same system effect as do the passive failures) separately to keep their identity. Similarly a stuck breaker condition, which is an active failure when need for operation arises, has been treated separately in the program. We have, in fact, not lost the identity of the important component failure categories. The advantage is using active and passive failure modes is the reduction in computational time. If component passive failures or overlapping passive failures can cause interruption at the load point then the active failures or active failures overlapping the passive failures for the component under consideration are not evaluated. Another reduction in computer time is obtained when performing sensitivity studies on component active failure rates while keeping the overall failure rates constant. The segment of the program which evaluates component active failures only can be used, in this case, many times without evaluating the passive failures.

Our program, in fact, performs reliability analysis in the following four phases.

(i) Perform failure modes and effects analysis (FMEA) for component passive failures.
(ii) Calculate the reliability indices due to component passive failures.
(iii) Perform the failure modes and effects analysis for component active failures.
(iv) Calculate the reliability indices due to component active failures and active failures overlapping the passive failures.

The steps (i) and (iii), and (ii) and (iv), perhaps, correspond to the two separate programs used by Mr. Albrecht. We fully agree with Mr. Albrecht that the results of a failure modes and effects analysis [steps (i) and (iii)] are very useful in themselves. This analysis is perhaps the most important part of a reliability study.

In conclusion we would again like to thank all the discussors for their interest and for their contributions.

RELIABILITY ASSESSMENT OF TRANSMISSION AND DISTRIBUTION SCHEMES

R. Billinton   M.S. Grover
Senior Member IEEE   Student Member IEEE
Power System Research Group
University of Saskatchewan
Saskatoon, Saskatchewan

ABSTRACT

This paper presents a further step in the development of accurate and consistent techniques for the reliability evaluation of transmission and distribution systems. A theoretically accurate Markov approach[6] is limited in application because of computer storage requirements and the rounding errors which occur in the solution of relatively large systems. This paper extends the modified approximate technique presented in reference (1) to permit evaluation of the contribution to the system reliability indices of different modes of component outage. The equations developed give results which compare very closely those predicted by the Markov approach. A three state weather model is developed to illustrate the effect of severe adverse weather periods on the system reliability performance. The load point failure modes considered are those which arise due to component permanent, temporary, maintenance and overload outages.

Introduction

A transmission or distribution system can be composed of either overhead or underground facilities or both. Overhead schemes operate in a fluctuating weather environment. Components tend to fail more frequently under severe weather conditions than they do in a normal weather situation. Distribution schemes which are concentrated in small areas are therefore greatly affected by adverse weather conditions. The initial development which included weather conditions in reliability predictions was done by Gaver et al.[2] The proposed two state weather model assumed that the weather fluctuated between normal and adverse periods. The results predicted by this method were compared with those obtained by a theoretical accurate Markov approach[6] and it was illustrated that the two techniques do not give consistent results. The modified equations presented in reference(1) to evaluate the permanent outage condition give results which compare well with those obtained by a Markov approach.

Component maintenance is another event which can cause load point interruption. In this paper, this form of maintenance action is considered to be preventive maintenance rather than the corrective maintenance which follows a permanent outage. Equations have been suggested in the literature[2,3,4,5] for the evaluation of outages in which maintenance outage period of a component is overlapped by permanent outages of other components. This paper investigates the existing equations by comparing the results with those obtained by a Markov approach.

In distribution schemes, temporary interruptions cause considerable irritation to the customers. In some industrial processes where continuity of supply is essential, considerable financial loss can occur in the form of spoiled material, lost production, etc. due to temporary supply interruptions. This aspect of component outages have not been developed in any of the previous publications. This paper presents some simple equations to evaluate temporary load point interruptions.

In those cases in which the parallel facilities are not redundant, component overload outages can result in high unreliability values at the system load points. A great deal of work has been done in the representation of system load models[2,4,5,7,8]. This paper extends the two state load model[4] to include the weather conditions.

Failure Modes of a Component

The failure modes of a component as considered in this paper are shown in Figure 1. The definitions of the various failure modes are given in reference (9). The permanent outage of a component requires that it be taken out of service for a period of time during which it is repaired. The actual outage time may be the replacement time for a spare component. Many of the outages on distribution schemes are of a temporary nature. If a component fault is cleared by a reclosing operation of a circuit breaker or by an automatic switching operation, a temporary outage is said to have occurred. The duration associated with such component outages is generally of the order of a few minutes. Components are also taken out of service for periodic inspection and maintenance. Such an action is desirable to forestall the occurrence of a failure or malfunction by preventive methods. Since the component is not available to perform its function during the maintenance period, it is another mode of failure. The final mode of component failure as illustrated in Figure 1 is the

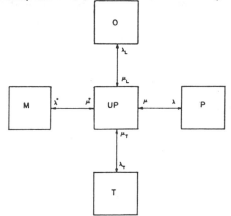

P ---- PERMANENT OUTAGE
T ----- TEMPORARY OUTAGE
M —---MAINTENANCE OUTAGE
O ---- OVERLOAD OUTAGE

Figure 1: Distinct Failure Modes of a Component.

Paper C 73 477-7 recommended and approved by the IEEE Power System Engineering Committee of the IEEE Power Engineering Society for presentation at the IEEE PES Summer Meeting and EHV/UHV Conference, Vancouver, B.C. Canada, July 15-20, 1973. This paper was upgraded to Transactions status, T 73 447-7, for presentation by title for written discussion at the 1975 Winter Meeting. Manuscript submitted February 20, 1973; made available for printing November 11, 1974.

Reprinted from *IEEE Trans. Power Apparatus Syst.*, vol. PAS-94, no. 3, pp. 724–732, May/June 1975.

overload outage of a component. Under certain outage and system conditions, components may be called upon to carry loads which exceed their capability. This can result in overload outage of the components. In an actual system depending upon the amount of overload and the system philosophy, the components may be called upon to carry the overload or be removed from service to prevent loss of life or permanent damage. In some cases the system load may be reduced to relieve the overload.

It should be appreciated that an overload outage of a component is different from the other failure modes described above. The overload of a component may or may not result in an actual outage. The other three modes of failure illustrated in Figure 1 do represent actual failure of the component.

Failure Modes of a Load Point

Component outages may or may not result in load point interruption depending upon the configuration of the system. The various basic events which lead to load point failure and are evaluated in this paper are listed below using the system of Figure 2.

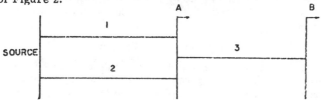

Figure 2: System for Illustration of Load Point Failure Modes

(1) The permanent outage of component 1 overlaps the permanent outage of component 2 or vice versa, causing interruption of supply at load points A and B.

(2) The permanent or temporary outage of component 3 causes an interruption of supply to the load point B.

(3) The maintenance outage period of component 1 (or component 2) is overlapped by a permanent or a temporary outage of component 2 (or component 1) resulting in the interruption of supply at load points A and B.

(4) The permanent outage of component 1 (or component 2) is overlapped by a temporary outage of component 2 (or component 1) causing an interruption of supply to the load points A and B.

(5) Component 1 is on permanent outage and during its repair period, the load at the load point A and B rises to a value which causes overload on component 2. Under these circumstances, component 2 may be tripped out of service or the load at any or both of the load points reduced to relieve the overload. In the former case, both load points will be interrupted while in the latter case, a fraction of of the total system load will be interrupted. A similar event can occur if components 1 and 2 are interchanged in the above description.

The probability of an event in which two independent component temporary outages overlap is very small and therefore such an event is not included. The probability of component overload occurring during the small duration associated with a temporary outage can also be neglected. One basic assumption made here is that the system is capable of satisfying its function with all components in service.

Evaluation of Load Point Failure Modes
A. Evaluation of interruptions due to overlapping permanent outages:-

The appropriate assumptions and equations for evaluating this mode of load point interruptions have been presented in reference (1). The equations included the weather conditions by considering the weather to fluctuate between a normal weather period and an adverse weather period. It should be recognized, however, that very severe weather periods, or the so-called "disaster" adverse weather periods, such as severe hurricanes and tornados, have a significant effect on the reliability of the systems although their occurrence is very rare.[11] These periods cannot be lumped with other less violent periods of adverse weather. In order to represent the effect of disaster adverse weather periods, a three state weather model is postulated. According to this model, the component failure rate fluctuates between three levels corresponding to normal, adverse and disaster adverse weather periods. It is assumed that overlapping outages cannot occur in two different types of adverse weather periods. This, in fact, implies that there is enough time between adverse weather and disaster adverse weather periods to perform repairs. It is also assumed that repairs are possible during adverse weather periods and no repair can be performed during disaster adverse weather periods.

Let N = The average duration of a period between adverse weather periods.

   P = The average duration of a period between disaster adverse weather periods.

   S = The average duration of an adverse weather period.

   T = The average duration of a disaster adverse weather period.

   λ_i = The failure rate of component i during normal weather.

   λ_i' = The failure rate of component i during adverse weather.

   λ_i'' = The failure rate of component i during a disaster adverse weather period.

   r_i = The average repair time of component i.

There are seven distinct modes of failure for a system of two parallel components and equations can be formulated for each of the failure modes in the same manner as described in reference (1). The resulting equation for the system failure rate is[11].

$$\lambda_{SL} = \frac{N}{N+S}[\lambda_1(\lambda_2 r_1) + \lambda_2(\lambda_1 r_2) + \lambda_1(\frac{r_1}{N})(\lambda_2'\frac{Sr_1}{S+r_1}) +$$

$$\lambda_2(\frac{r_2}{N})(\lambda_1'\frac{Sr_2}{S+r_2})] + \frac{S}{N+S}[\lambda_1'\lambda_2\frac{Nr_1}{N+r_1}) +$$

$$\lambda_2'(\lambda_1\frac{Nr_2}{N+r_2}) + \lambda_1'\lambda_2'S(\frac{r_1}{S+r_1} + \frac{r_2}{S+r_2})] +$$

$$\frac{T}{P+T}[\lambda_1''(1-\lambda_2''T)(\lambda_2 r_1) + \lambda_2''(1-\lambda_1''T)(\lambda_1 r_2) +$$

$$2\lambda_1''\lambda_2''T] + \frac{P}{P+T}[\lambda_1(\frac{r_1}{P})(\lambda_2''T) +$$

$$\lambda_2(\frac{r_2}{P})(\lambda_1''T)] \qquad (1)$$

The average duration associated with λ_{SL} is the weighted duration associated with each of the failure modes and is given by:-

$$r_{SL} = \frac{\lambda_n}{\lambda_{SL}} \left(\frac{r_1 r_2}{r_1 + r_2}\right) + \frac{\lambda_d}{\lambda_{SL}} \left(\frac{r_1 r_2}{r_1 + r_2} + T\right) \qquad (2)$$

where

λ_n = The fraction of system failures which occur in normal and adverse weather,

λ_d = The fraction of system failures which occur in disaster adverse weather.

It should be appreciated that equations 1 and 2 are not only applicable to a system of two parallel components but any two permanent overlapping outages. Similar equations can be written for three or more overlapping outages.

The Markov technique can also be applied to the three state weather model. The reliability indices obtained from the above equations have been compared with those predicted by the Markov technique and good agreement in the results observed.

Using the above equations, the effects on the outage rate of a two component parallel system of varying component disaster adverse weather failure rate and disaster adverse weather period are shown in Figure 3. The system failure rate and the average outage duration increase with increases in the disaster adverse weather component failure rate and the average duration of a disaster adverse weather period.

B. Evaluation of interruptions due to component permanent outages overlapping component maintenance outages:-

The following assumptions were made to evaluate the interruptions due to component maintenance outages[2].

(i) No maintenance is carried out if there is some outage already existing in a related portion of the system.

(ii) Maintenance occurs in normal weather and no maintenance is carried out if adverse weather is expected to occur before it could be completed.

(iii) No maintenance is performed if the removal of the component results in the overload of another component in the system.

(iv) Maintenance up times and down times are exponentially distributed. This assumption is not necessary for long term reliability predictions.

The contribution to the failure rate of a two component parallel system due to component permanent outages overlapping component maintenance outage periods is given by[2]:

$$\lambda_{mL} = \lambda_1'' \lambda_2'' r_1'' + \lambda_2'' \lambda_1'' r_2'' \qquad (3)$$

where

λ_i'' = The maintenance outage rate of component i in occurrences/yr.

r_i'' = The expected time required for maintenance of component i.

The expected down time for a system with two parallel components is given by:

$$r_{mL}'' = \frac{\lambda_1'' \lambda_2'' r_1''}{\lambda_{mL}''} \left(\frac{r_1'' r_2}{r_1'' + r_2}\right) + \frac{\lambda_2'' \lambda_1'' r_2''}{\lambda_{mL}''} \left(\frac{r_2'' r_1}{r_2'' + r_1}\right) \qquad (4)$$

The Markov approach can also be applied to evaluate the contribution to the system reliability indices due to component maintenance. It involves the extension of the state space diagram on the normal weather side to include the maintenance states. The Markov approach is, however, difficult to apply in a large system due of computer storage requirements and rounding error limitations.

The results given by equations 3 and 4 are compared with those obtained by the Markov approach in Table 1, considering a two state weather model.[11]

Table 1

Reliability Indices Due to Maintenance for a Two Component Parallel System

$\lambda_{av_1} = \lambda_{av_2}$ = 0.5 f/year	$r_1 = r_2$ = 7.5 hours
$\lambda_1'' = \lambda_2''$ = 3.0 o/year	$r_1'' = r_2''$ = 5.84 hours
N = 200.0 hours	S = 1.5 hours

Percentage Component Failures During Adverse Weather	Approximate Equations		Markov Technique	
	Failure Rate	Average Duration	Failure Rate	Average Duration
	f/year	Hours	f/year	Hours
0	0.00201	3.280	0.00199	3.280
10	0.00181	3.280	0.00179	3.280
20	0.00161	3.280	0.00159	3.280
30	0.00143	3.280	0.00139	3.280
40	0.00121	3.280	0.00119	3.280
50	0.00100	3.280	0.00099	3.280
60	0.00080	3.280	0.00080	3.280
70	0.00060	3.280	0.00060	3.280
80	0.00040	3.280	0.00040	3.280
90	0.00020	3.280	0.00020	3.280
100	0.00000	—	0.00000	—

This table illustrates that equations 3 and 4 do give results which are very close to those obtained by the Markov approach

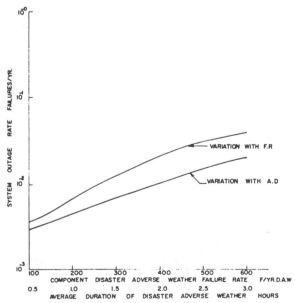

Figure 3: Effects of Average Duration of Disaster Adverse Weather and Assoicated Component Failure Rate on the Two Component Parallel System Outage Rate.

An equivalent component formation approach has been suggested[3] for consideration of more than two overlapping outages in which one component is out for maintenance and two or more components are forced out during the maintenance period. In this technique, the contribution to the failure rate of a three component parallel system due to component permanent outages overlapping component maintenance periods is:

$$\lambda''_{ML} = \lambda''_1 \lambda''_{2\text{-}3} r''_1 + \lambda''_2 \lambda''_{3\text{-}1} r''_2 + \lambda''_3 \lambda''_{1\text{-}2} r''_3 \quad (5)$$

where

$\lambda''_{i\text{-}j}$ = The overall failure rate of components i and j.

The system average duration is given by[3]:

$$r''_{ML} = \left(\frac{\lambda''_1 \lambda''_{2\text{-}3} r''_1}{\lambda''_{ML}}\right)\left(\frac{r_{2\text{-}3} r''_1}{r_{2\text{-}3}+r''_1}\right) + \left(\frac{\lambda''_2 \lambda''_{1\text{-}3} r''_2}{\lambda''_{ML}}\right)\left(\frac{r_{1\text{-}3} r''_2}{r_{1\text{-}3}+r''_2}\right) +$$

$$\left(\frac{\lambda''_3 \lambda''_{1\text{-}2} r''_3}{\lambda''_{ML}}\right)\left(\frac{r_{1\text{-}2} r''_3}{r_{1\text{-}2}+r''_3}\right) \quad (6)$$

where $r_{1\text{-}j}$ = The expected overlapping outage duration of components i and j.

The reliability indices obtained from equations 5 and 6 were compared with those predicted by the Markov approach considering a two state weather model[11]. The results obtained from the two techniques are given in Table 2. It is evident from this table that the system outage rate predictions as given by equations 5 and 6 are quite different from those obtained by the Markov approach.

The following section presents some simple modified equations which provide a more accurate representation of the maintenance outage event.[11]

The outage rate resulting from an event in which the maintenance period of a component is overlapped by two permanent outages is given by:

$$\lambda''_{ML} = \lambda''_1 (\lambda''_2 r''_1) \left(\lambda_3 \frac{r''_1 r_2}{r''_1+r_2}\right) + \lambda''_1 (\lambda_3 r''_1)\left(\lambda_2 \frac{r''_1 r_3}{r''_1+r_3}\right) \quad (7)$$

where λ''_1 = The maintenance outage rate of component 1.

$\lambda''_2 r''_1$ = The probability that component 2 suffers a permanent outage during the maintenance period of component 1.

$\lambda_3 \dfrac{r''_1 r_2}{r''_1+r_2}$ = The probability that component 3 suffers a permanent outage during the maintenance period of component 1 and the repair of component 2.

The second term in equation 7 follows the same argument but with components 2 and 3 interchanged. The expected outage duration associated with the above outage rate is:

$$r''_{ML} = \frac{r''_1 r_2 r_3}{r''_1 r_2+r''_1 r_3+r_2 r_3} \quad (8)$$

The contribution to the system failure rate due to maintenance of a component and two permanent outages overlapping it for a system with three components in parallel is therefore given by:

$$\lambda''_{ML} = \lambda''_1 \lambda''_2 r''_1 \lambda_3 \frac{r''_1 r_2}{r''_1+r_2} + \lambda''_1 \lambda''_3 r''_1 \lambda_2 \frac{r''_1 r_3}{r''_1+r_3}$$

$$+ \lambda''_2 \lambda''_1 r''_2 \lambda''_3 \frac{r''_2 r_1}{r''_2+r_1} + \lambda''_2 \lambda''_3 r''_2 \lambda''_1 \frac{r''_2 r_3}{r''_2+r_3}$$

$$+ \lambda''_3 \lambda''_1 r''_3 \lambda''_2 \frac{r''_3 r_1}{r''_3+r_1} + \lambda''_3 \lambda''_2 r''_3 \lambda''_1 \frac{r''_3 r_2}{r''_3+r_2} \quad (9)$$

The expected outage duration of λ''_{ML} is equal to the weighted outage duration of each of the terms in the above expression. If λ_i represents the outage rate due to one of the above terms, and r_i its associated expected outage duration, then the expected outage duration of λ''_{ML} is:

$$r''_{ML} = \frac{\sum_{i=1}^{6} \lambda_i r_i}{\lambda''_{ML}} \quad (10)$$

Using equations 9 and 10, the reliability indices for a three component parallel system are shown in Table 2. A comparison of the results with those obtained by a Markov approach clearly shows the higher accuracy of equations 9 and 10.

Using equations 3, 4, 9 and 10, the contribution to the system failure rate due to component maintenance for simple configurations of two and three parallel components are plotted against the percentage of component failures during adverse weather periods in Figure 4. This figure also illustrates the effect of varying the component maintenance parameters. As the percentage of component failures during adverse weather periods increases, the component maintenance contribution to the system failure rate decreases. This is, perhaps, obvious because as the component failures during adverse weather periods increase, the risk of maintenance outages, which occur in normal weather, overlapping the component permanent outages decreases.

The error in the equivalent component formation can be appreciated by considering equations 5 and 7. From equation 5, the contribution to the system failure rate due to an event in which component 1 is on maintenance and components 2 and 3 suffer overlapping permanent outages during the maintenance period is:

$$\lambda''_{ML} = \lambda''_1 \lambda''_{2\text{-}3} r''_1 = \lambda''_1 \lambda_2 \lambda_3 (r_2+r_3) r''_1$$

Let $r_1 = r_2 = r_3 = r''_1 = r''_2 = r''_3 = r$

$$\lambda''_{ML} = 2 \lambda''_1 \lambda_2 \lambda_3 r^2 \quad (11)$$

From equation 7,

$$\lambda''_{ML} = \lambda''_1 \lambda_2 \lambda_3 \frac{r^2}{2} + \lambda''_1 \lambda_2 \lambda_3 \frac{r^2}{2} = \lambda''_1 \lambda_2 \lambda_3 r^2 \quad (12)$$

It is worth noting that in this case the component maintenance and repair times are equal and equation 11 gives an outage rate estimate which is exactly twice that given by the correct equation 12. This error arises because equivalent component formation does not represent the true performance in systems where dependent component behaviour is present.

Equations 3, 4, 9 and 10 were derived under the assumption that weather cannot become adverse during the component maintenance period. If this can happen, then for a two component parallel system the following outage rate equation holds if maintenance can be carried on in adverse weather.

$$\lambda''_{ML} = [\lambda''_1 \lambda''_2 r''_1 + \lambda''_1 (\frac{r''_1}{N}) \lambda''_2 (\frac{Sr''_1}{S+r''_1}) + \lambda''_2 \lambda''_1 r''_2 + \lambda''_2 (\frac{r''_2}{N}) \lambda''_1 (\frac{Sr''_2}{S+r''_2})]$$

Table 2

Maintenance Contribution to Reliability Indices For a Three Component Parallel System

$\lambda_{av_1} = \lambda_{av_2} = \lambda_{av_3} = 0.5$ f/year $\qquad$ $r_1 = r_2 = r_3 = 7.5$ hours

$\lambda_1'' = \lambda_2'' = \lambda_3'' = 3.0$ o/yr. $\qquad$ $r_1'' = r_2'' = r_3'' = 5.84$ hours

$N = 200.0$ hours $\qquad$ $S = 1.5$ hours

Percentage Component Failures During Adverse Weather	Equations 5 & 6		Markov		Equations 9 & 10	
	Failure Rate f/10^6 yrs	Average Duration Hours	Failure Rate f/10^6yrs	Average Duration Hours	Failure Rate f/10^6yrs.	Average Duration Hours
0	2.600	2.280	1.114	2.280	1.141	2.283
10	2.340	2.280	0.924	2.280	0.926	2.283
20	2.086	2.280	0.730	2.280	0.738	2.283
30	1.820	2.280	0.559	2.280	0.569	2.283
40	1.570	2.280	0.411	2.280	0.422	2.283
50	1.300	2.280	0.285	2.280	0.290	2.283
60	1.040	2.280	0.183	2.280	0.187	2.283
70	0.720	2.280	0.103	2.280	0.105	2.283
80	0.520	2.280	0.046	2.280	0.048	2.283
90	0.200	2.280	0.011	2.280	0.011	2.283
100	0.000	—	0.000	——	0.000	——

In the case, in which maintenance cannot be carried on in adverse weather,,

$$\lambda_{ML}'' = [\lambda_1'' \lambda_2'' r_1'' + \lambda_1'' (\frac{r_1''}{N})(\lambda_2'' S)' + \lambda_2'' \lambda_1'' r_2'' + \lambda_2'' (\frac{r_2''}{N})(\lambda_1'' S)']$$

The average outage duration of λ_{ML}'' is equal to the weighted duration of each of the above terms.

For a three component parallel system, there are four possible modes of failures as listed in Table 3. Equations for each of the failure modes are formulated to consider the occurrence and nonoccurrence of repair and maintenance in adverse weather. The contribution to the system failure rate

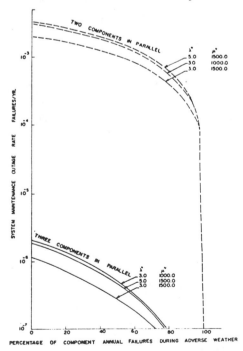

Figure 4: Contribution to System Failure Rate Due to Component Maintenance.

due to component maintenance is given by the summation of the appropriate terms in Table 3. The results given by these equations were compared with those predicted by Markov approach and good agreement in the results was observed.

The technique of failure modes and effects analysis can be applied to deal with complex system configurations. Consider the system shown in Figure 5. The various failure modes at load points 1 and 2 are listed below.

Load Point 1
 Line sections 1-2 and 3 out.
 Line sections 1-2 and 4 out.
 Line sections 1-2-3 and 4 out.

Load Point 2
 Line sections 3 and 4 out.
 Line sections 1, 2 and 4 out.
 Line sections 1, 2, 3 and 4 out.

The failure mode in which all the components are out makes very little contribution to the system failure rate because it is fourth order contingency and can be ignored.

The maintenance outage contribution to system failure rate was calculated by using the above failure modes and the equations described in this section. The results obtained are compared with those predicted by a 64 state Markov model in Table 4.[11] The close proximity in the results obtained by these two methods clearly illustrates the usefulness of the simple equations and the failure modes and effects analysis technique.

C. Evaluation of interruptions due to component temporary outages overlapping component maintenance outages.

Temporary outages in distribution facilities tend to occur more frequently than permanent outages. The incidence of temporary outages also tends to increase with preventive or corrective maintenance activity on neighbouring facilities. The mathematical modelling of this mode of load point interruption proceeds on the same basis as that used for permanent outages.

Table 3

Maintenance Contribution Associated With Different Modes of Failure of a Three Component Parallel System

No.	Failure Modes 1st	2nd	3rd	Contribution to the System Maintenance Outage Rate — Repair & Maint. During Adverse Weather	No Repair & Maint. During Adverse Weather
1	N	N	N	$\{\lambda_1''(\lambda_2 r_1'')(\lambda_3 R_1) + \lambda_1''(\lambda_3 r_1'')(\lambda_2 R_2)$ + Similar terms for components 2 & 3}	$\{\lambda_1''(\lambda_2 r_1'')(\lambda_3 R_1) + \lambda_1''(\lambda_3 r_1'')(\lambda_2 R_2)$ + Similar terms for components 2 & 3}
2	N	N	A	$\{\lambda_1''(\lambda_2 r_1'')(\frac{R_1}{N})(\lambda_3 R_6) + \lambda_1''(\lambda_3 r_1'')(\frac{R_2}{N})(\lambda_2 R_7)$ + Similar terms for components 2 & 3}	$\{\lambda_1''(\lambda_2 r_1'')(\frac{R_1}{N})(\lambda_3'S) + \lambda_1''(\lambda_3 r_1'')(\frac{R_2}{N})(\lambda_2'S)$ + Similar terms for components 2 & 3}
3	N	A	N	$\{\lambda_1''(\frac{r_1''}{N})(\lambda_2'R_3)(\lambda_3 R_4) + \lambda_1''(\frac{r_1''}{N})(\lambda_3'R_3)(\lambda_2 R_5)$ + Similar terms for components 2 & 3}	$\{\lambda_1''(\frac{r_1''}{N})(\lambda_2'S)(\lambda_3 R_1) + \lambda_1''(\frac{r_1''}{N})(\lambda_3'S)(\lambda_2 R_2)$ + Similar terms for components 2 & 3}
4	N	A	A	$\{\lambda_1''(\frac{r_1''}{N})(\lambda_2'R_3)(\lambda_3'R_6) + \lambda_1''(\frac{r_1''}{N})(\lambda_3'R_3)(\lambda_3'R_7)$ + Similar terms for components 2 & 3}	$\{\lambda_1''(\frac{r_1''}{N})(\lambda_2'S)(\lambda_3'S) + \lambda_2''(\frac{r_1''}{N})(\lambda_3'S)(\lambda_2'S)$ + Similar terms for components 2 & 3}

$$R_1 = \frac{r_1'' r_2}{r_1'' + r_2} \ ; \quad R_2 = \frac{r_1'' r_3}{r_1'' + r_3} \ ; \quad R_3 = \frac{S r_1''}{S + r_1''} \ ; \quad R_4 = \frac{N r_1'' r_2}{N r_1'' + N r_2 + r_1'' r_2} \ ; \quad R_5 = \frac{N r_2 r_3}{N r_2 + N r_3 + r_2 r_3} \ ; \quad R_6 = \frac{S r_1'' r_2}{S r_1'' + S r_2 + r_1'' r_2}$$

$$R_7 = \frac{S r_1'' r_3}{S r_1'' + S r_3 + r_1'' r_3}$$

In failure mode column:

N represents component failure in normal weather,   A represents component failure in adverse weather

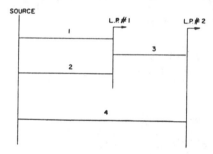

Figure 5: A simple Series-Parallel System.

Let λ_{iT} = the temporary outage rate of component i.

The temporary outage rate resulting from a temporary outage of component 1 overlapping the maintenance outage period of component 2 is given by[11]:

$$\lambda_{tL} = \lambda_2''(\lambda_{1T} r_2'')$$

where $\lambda_{1T} r_2''$ = The probability that component 1 suffers a temporary outage during the maintenance period of component 2.

For a system of two parallel components, therefore:

$$\lambda_{TL} = \lambda_1''(\lambda_{2T} r_1'') + \lambda_2''(\lambda_{1T} r_2'').$$

Equations similar to equation 9 can be written in this case for a three component parallel system. If component temporary outages are separated into normal and adverse weather failures, the λ_{iT} values in the above equations represent the component normal weather temporary outage rates.

D. Evaluation of interruptions due to component temporary outages overlapping component permanent outages.

A temporary outage can occur in one facility while the adjacent parallel component is being repaired due to a permanent outage. Depending upon the system configuration, such

---

Table 4

Reliability Indices Due to Maintenance For the System Configuration of Figure 5

$\lambda_{av_1} = \lambda_{av_2} = 0.5$ f/yr $\quad \lambda_{av_3} = 0.1$ f/yr $\quad \lambda_{av_4} = 0.6$ f/yr

$\lambda_1'' = \lambda_2'' = 2.0$ o/yr $\quad \lambda_3'' = 1.0$ o/yr $\quad \lambda_4'' = 3.0$ o/yr

$r_1 = r_2 = 30.0$ Hrs. $\quad r_3 = 2.5$ Hrs. $\quad r_4 = 12.5$ Hrs.

$r_1'' = r_2'' = 20.0$ Hrs. $\quad r_3'' = 5.0$ Hrs. $\quad r_4'' = 10.0$ Hrs.

N = 400.0 Hrs. $\qquad$ S = 3.0 Hrs.

Percentage of Component Failures During Adverse Weather		Approximate Equations Failure Rate f/10⁶yr	Average Duration Hours	Markov Technique Failure Rate f/10⁶yr	Average Duration Hours
0	Load pt. 1	8.634	5.71	8.230	5.87
	Load pt. 2	696.32	2.81	681.30	2.82
40	Load pt. 1	3.108	5.71	2.965	5.87
	Load pt. 2	416.29	2.81	406.93	2.81
80	Load pt. 1	0.345	5.71	0.330	5.87
	Load pt. 2	138.26	2.80	135.04	2.80
100	Load pt. 1	0.000	—	0.000	—
	Load pt. 2	0.000	—	0.000	—

an event can lead to service interruption to certain load points in the system. The probability of occurrence of a permanent outage during the period when the another component is on a temporary outage is considered to be negligible. For a series system of 'n' component, the temporary outage rate of the system is given by:

$$\lambda_{TS} = \sum_{i=1}^{n} \lambda_{iT}$$

If component temporary outages are not separated into normal and adverse weather failures, the outage rate of a system of two components in parallel, assuming repair occurs in adverse weather is given by[11]:

$$\lambda_{TL} = \frac{N}{N+S}[\lambda_{1}\lambda_{2T}r_{1} + \lambda_{2}\lambda_{1T}r_{2}] + \frac{S}{N+S}[\lambda'_{1}\lambda_{2T}r_{1} + \lambda'_{2}\lambda_{1T}r_{2}]$$

$$= (\lambda_{1}\frac{N}{N+S} + \lambda'_{1}\frac{S}{N+S})\lambda_{2T}r_{1} + (\lambda_{2}\frac{N}{N+S} + \lambda'_{2}\frac{S}{N+S})\lambda_{1T}r_{2}$$

$$= \lambda_{av_{1}}\lambda_{2T}r_{1} + \lambda_{av_{2}}\lambda_{1T}r_{2} \qquad (13)$$

For a system of three components in parallel, the system failure rate due to component temporary outages overlapping component permanent outages is:

$$\lambda_{TL} = \lambda_{1-2}\lambda_{3T}r_{1-2} + \lambda_{1-3}\lambda_{2T}r_{1-3} + \lambda_{2-3}\lambda_{1T}r_{2-3} \qquad (14)$$

It sometimes becomes necessary to separate the temporary outages into normal and adverse weather failures. (In many cases all component temporary outages occur during adverse weather periods). To incorporate this aspect, equations of the form described in reference (1) can be written. The failure rate due to component temporary outages overlapping component permanent outages, for a two component parallel system, when no repair is performed during the adverse weather period is,

$$\lambda_{TL} = \frac{N}{N+S}[\lambda_{1}\lambda_{2T}r_{1} + \lambda_{2}\lambda_{1T}r_{2} + \frac{S}{N}(\lambda'_{1}\lambda_{2T}r_{1} + \lambda'_{2}\lambda_{1T}r_{2})$$

$$+ \frac{S}{N}(\lambda_{1}\lambda'_{2T}r_{1} + \lambda_{2}\lambda'_{1T}r_{2}) + \frac{S}{N}(\lambda'_{1}\lambda'_{2T}S + \lambda'_{2}\lambda'_{1T}S)] \qquad (15)$$

where λ_{iT} = The normal weather outage rate of component i.

λ'_{iT} = The adverse weather outage rate of component i.

When repairs are performed during the adverse weather period, equation 15 is modified to the following form[11],

$$\lambda_{TL} = \frac{N}{N+S}[\lambda_{1}\lambda_{2T}r_{1} + \lambda_{2}\lambda_{1T}r_{2} + \frac{S}{N}(\lambda'_{1}\lambda_{2T}\frac{r_{1}^{2}}{S+r_{1}} + \lambda'_{2}\lambda_{1T}\frac{r_{2}^{2}}{S+r_{2}})]$$

$$+ \frac{S}{N+S}[\lambda'_{1}\lambda_{2T}r_{1} + \lambda'_{2}\lambda_{1T}r_{2} + \lambda'_{1}\lambda'_{2T}\frac{Sr_{1}}{S+r_{1}} + \lambda'_{2}\lambda'_{1T}\frac{Sr_{2}}{S+r_{2}}] \qquad (16)$$

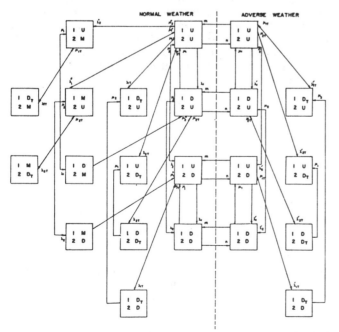

Figure 6: 22 State Model for a Two Component Parallel System.

Similar equations can be written for larger numbers of components[11]. The results obtained by the equations given in this section are compared with those obtained by constructing the 22 state model, shown in Figure 6 for a two component parallel system, in Table 5. This table clearly illustrates that the approximate equations described in this paper give results which are fairly close to those obtained by a complete Markov approach. In this table, the percentage of component failures during adverse weather applies to both permanent and temporary outages.

Table 5

Temporary Outage Rate for a Two Component Parallel System

$\lambda_{av_{1}} = \lambda_{av_{2}} = 0.5$ f/yr.　　　$r_{1} = r_{2} = 7.5$ hours

$\lambda''_{1} = \lambda''_{2} = 3.0$ o/year　　　$\mu''_{1} = \mu''_{2} = 1500$ r/years

$S = 1.5$ hours, $\lambda_{1T} = \lambda_{2T} = 3.0$ f/yr.　　N = 200.0 Hours

Percentage of Component Failure During Adverse Weather	Approximate Equations		Markov Technique	
	A f/yr	B f/yr	A f/yr	B f/yr
0	0.00258	0.01210	0.00259	0.01200
20	0.00524	0.00967	0.00530	0.00955
40	0.01321	0.00725	0.01340	0.00716
60	0.02649	0.00483	0.02676	0.00477
80	0.04509	0.00241	0.04528	0.00238
100	0.06875	0.00000	0.06885	0.00000

A = Component temporary outages overlapping component permanent outages.

B = Component temporary outages overlapping component maintenance outages.

E. Evaluation of interruptions due to overload outages of system components.

There is a finite conditional probability of loss of load at the load point in those systems in which the parallel components are not redundant. This probability is a function of the system load level, the capability of the remaining component of the system configuration and the duration of the contingency. As previously noted if the load during system outages exceeds the emergency capability of a component, several possible events can happen. The component may be called upon to carry the overload or it may be tripped out of service, or the system load reduced to relieve the overload. Each of these conditions will give different reliability indices.

The basic step in the estimation of overload outages is the determination of a probability value that a component will not be able to carry a given contingency load. A technique of contingency curves has been suggested in reference (2), in which curves representing the probability of carrying the contingency load under various outage conditions are precalculated for different contingency durations. This technique is quite accurate but cannot be easily applied to complicated networks. Another technique used in generating capacity reliability evaluation is to represent the system load model as a stationary Markov process. It is assumed that the system daily load consists of peak load period where it stays for a certain period of time and the remainder of the day it remains at a constant low load level[8].

This paper extends the two state load model[4] to include the weather conditions. This model attaches a frequency and duration interpretation to the system load. The system load is considered to exist in two states, one corresponding to a value greater than the load carrying capability of the component and the second corresponding to a value less than the load carrying capability. The two parameters which are estimated from the chronological hourly load variation curve are the rate of occurrence and the average duration of a load level greater than

413

the carrying capability of the components. Equations can be derived for the average rate and the average interruption duration using these load parameters.

Let λ_L = The rate of occurrence of a load level $\ell(t) > L$
and μ_L = The reciprocal of the average duration of $\ell(t) > L$.

where L is the capability of the remaining components and $\ell(t)$ in the system load at hour t.

The frequency of component 1 becoming overloaded when component 2 is forced out is given by[4]:

$$\lambda_{OL} = \{ Pr [\text{Component 1 is not forced out}] \{\lambda_L (1-Pr(\ell(t) > L)) $$
$$Pr(\text{Component 2 is forced out}) + \lambda_2 Pr(\ell(t) > L) (1-Pr $$
$$(\text{Component 2 is forced out}))]\}_2 \quad (17)$$

The expected outage duration associated with λ_{OL} is given by:

$$r_{OL} = \frac{r_2 r_L}{r_2 + r_L} \quad (18)$$

Using equations 17 and 18, the average outage rate and the associated expected outage duration due to component overload outages at various load points can be evaluated. When more than one permanent outage occurs which results in overload of the remaining components of the system, equation 17 can be modified as follows[11]:

$$\lambda_{OL} = Pr [\text{The remaining components are not forced out}] $$
$$x\{\lambda_L(1-Pr(\ell(t) > L))Pr \text{ (The components under consi-} $$
deration are forced out)] + [The overlapping outage rate of the components under consideration] x Pr($\ell(t)$ > L) (1-Pr(The components under consideration are forced out))$\}$

Since Pr [The remaining components are not forced out] $\simeq 1$ and Pr [The components under consideration are forced out] $\simeq \lambda_{SL} r_{SL}$

Carrying out this simplification

$$\lambda_{OL} = \lambda_L (1-Pr(\ell(t)>L)) (\lambda_{SL}r_{SL}) + \lambda_{SL} Pr(\ell(t)>L) (1-\lambda_{SL}r_{SL}) $$
$$(19)$$

where λ_{SL} is the overall annual outage rate of the components under consideration

r_{SL} is the overlapping outage duration of the components under consideration

Using equations 17, 18 and 19 results were obtained for the system outage rates of two and three component parallel systems[11] and are shown in Figure 7 for various percentages of component failures during the adverse weather period when no repair is performed in this period. The load parameters were obtained from hourly load values at a Saskatchewan Power Corporation substation. It can be seen from these curves that there is no variation in the system outage rate and expected outage duration with the component failures during the adverse weather period for a two component parallel system. This is due to the absence of the bunching effect of adverse weather associated component failures in this case. If both the components of a two component parallel system are out, no overloading is possible. For the three component parallel system of this example, however, two components are required to have overlapping outages for the occurrence of the overload of the remaining component. The probability of occurrence of overlapping outages of two components increases as the percentage of component failures during adverse weather periods increases.

Some questions may arise about the validity of the above results under the condition when all the component failures occur during the adverse weather period, as under these circumstances the system load cycle analysed would be confined to adverse weather periods and the small periods following the adverse weather during which repairs are performed. Since

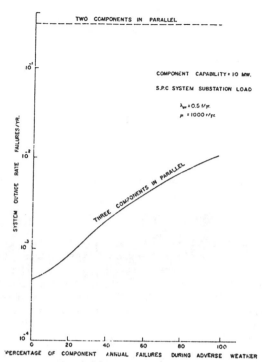

Figure 7: System Outage Rate Due to Component Overload Outages.

adverse weather periods are assumed to occur randomly, the annual load cycle of the system is therefore utilized.

In the previous examples, a single load point case has been considered. In the case of system configurations involving many load points, the evaluation of load point interruptions due to component overload outages becomes quite involved. A load flow analysis is required to determine the load values at which the component overload outage commence. This may require a prohibitive amount of work. One simplifying assumption which is quite justifiable in systems in which all the load points serve similar types of loads is that the individual loads are directly proportional to the system load. In simpler terms it implies that the normalized system load model is applicable at all the load points in the system. With this assumption, load flow analysis is necessary at different load levels (expressed in percentage of the peak load at each load point) to establish the load level at which the overload on the system components commences.

The system configuration of Figure 5 is used as an example to illustrate this. This system has two load points. The following carrying capabilities of the line sections are assumed.

Line section 1 - 20 MW	Line section 2 - 10 MW
Line section 3 - 15 MW	Line section 4 - 20 MW

Two of the many possible overload outage situations are

(i) If a line becomes overloaded, the load near to the vicinity of this line is curtailed.

(ii) If a line becomes overloaded, the load farthest away from the vicinity of this line is curtailed (if possible).

Table 6 lists the various double contingency outage events which result in the overload of the remaining system components under the situations described above.

The load levels at which overload outages commence are obtained by a load flow program. The results obtained by equations 17, 18 and 19 for the system shown in Fig. 5 are given

414

Table 6
Contingencies Considered for the System of Figure 5

No	Contingency Line Sections Out	Loss of Service			
		Load Pt. 1		Load Pt. 2	
		(i)	(ii)	(i)	(ii)
1	1-2	—	$\ell(t)>77\%$	$\ell(t)>77\%$	—
2	1-3	$\ell(t)>77\%$	$\ell(t)>77\%$	—	—
3	1-4	$\ell(t)>39\%$	$\ell(t)>77\%$	$\ell(t)>77\%$	$\ell(t)>39\%$
4	2-3	—	—	—	—
5	2-4	$\ell(t)>77\%$	—	—	$\ell(t)>77\%$
6	3-4	—	—	—	—

in Table 7 for both methods of relieving the overload. This table clearly illustrate that the reliability indices at different load points will be different for different contingency modes. An approach such as this can be used to logically decide the required capabilities of the system components in order to recognize the importance of various load points in a system.

Table 7
Reliability Indices Due to Component Overload
Outages for the System in Figure 5

PC	Policy	Load Point 1		Load Point 2	
		Failure Rate $f/10^3$yr	Average Duration Hours	Failure Rate $f/10^3$yr	Average Duration Hours
0	(i)	0.876	4.61	0.336	4.25
	(ii)	0.353	4.14	0.859	4.66
40	(i)	3.300	4.58	1.350	4.21
	(ii)	1.450	4.05	3.210	4.66
80	(i)	8.360	4.57	4.440	4.20
	(ii)	4.780	4.03	10.330	4.66

PC = Percentage of component failures during adverse weather.

It must be appreciated, however, that an event having a probability of unity associated with an overload condition is different from that in which there is complete discontinuity of the supply to the load points. In the former case, the load at the load point or points may be cut off or reduced to relieve the overload while in the latter case such an action is not possible. A complete disconnection of the load point implies a complete interruption until the components which caused the discontinuity are restored to service whereas an overload outage may imply a partial outage at the load point until the load falls below the capability of the remaining components, or the components capable of carrying the system load are restored to service. The reliability indices which result from the complete interruption of service to the load point cannot be simply combined with those obtained from the overload outages of the components. It is assumed, while calculating the reliability indices in Table 7 that the component maintenance outages will not lead to interruptions due to overload outages of the components. If necessary, these events can be included in the calculations.

CONCLUSIONS

The primary intent of this paper is to present some simple equations to evaluate the reliability indices of transmission and distribution schemes in terms of the outage frequency and duration at various system load points. As illustrated in reference (1), these simple equations can be used in conjunction with a standard failure modes and effects approach to analyze a relatively complex configuration in a sequential manner. The component failure modes recognized in this paper are permanent, temporary, maintenance and overload outages. Temporary and permanent outages have been recognized as two entirely separate conditions. Temporary interruptions of supply,

even though they may be of small duration, can have detrimental effects on customer-power company relationships. The separation of various modes of component failures can pinpoint the components and their parameters which make a significant contribution to system failure. The three state weather model developed recognizes the effects of very severe weather conditions on system reliability. This paper also compares the results obtained by the equations already existing in the literature with those obtained by a theoretically accurate Markov technique and it is illustrated that the results predicted by the two methods are quite different. The modified equations given in this paper give results which are very close to those obtained by the Markov approach. The frequency and duration indices that result from component overload outages depend on the operational philosophy of the system and are different from those resulting from complete interruption of supply. The simple equations presented provide a fast and inexpensive method of computing system reliability performance and are useful in determining the most effective means by which any selected level of security of supply can be realized in a practical system.

ACKNOWLEDGEMENT

The authors would like to thank the Saskatchewan Power Corporation for their assistance and the financial support for these studies.

REFERENCES

1. Grover, M.S., and Billinton, R., Quantitative evaluation of permanent outages in distribution systems. A paper presented at IEEE Summer Power Meeting, July, 1973.
2. Gaver, D.P., Montmeat, F.E., and Patton, A.D. Power system Reliability. I-Measures of Reliability and Methods of Calculation. IEEE Trans., PAS, July 1964, pp. 727-737.
3. Mallard, S.A., and Thomas, V.C. A Method for Calculating Transmission System Reliability. IEEE Trans., PAS, March, 1968, pp. 824-833.
4. Christiaanse, W.R. Reliability Calculations Including the Effects of Overload and Maintenance. IEEE Trans., PAS, July/Aug. 1971, pp. 1664-1676.
5. Ringlee, R.J., and Goode, S.D. On Procedures for Reliability Evaluation of Transmission Systems. IEEE Trans., PAS, April 1970, pp. 527-537.
6. Billinton, R., and Bollinger, K.E. Transmission System Reliability Evaluation using Markov Processes. IEEE Trans., PAS, March 1968, pp. 824-833.
7. Bhavaraju, M.P., and Billinton, R. Transmission System Reliability Methods. IEEE Trans, PAS., Mar/Apr. 1972, pp. 628-636.
8. Ringlee, R.J. and Wood, A.J. Frequency and Duration Methods for Power System Reliability Calculations. Part II Demand Model and Capacity Reserve Model. IEEE Trans. Aug. 1969, pp. 1216-1223.
9. IEEE Committee report on Proposed Definition of Terms for Reporting and Analyzing Outages of Electrical Transmission and Distribution Facilities and Interruptions. IEEE Trans., PAS, May, 1968, 1318-1323.
10. Billinton, R. Composite System Reliability Evaluation. IEEE Trans. PAS, April 1969, pp. 276-280.
11. Grover, M.S. Distribution System Reliability Assessment. M.Sc. Thesis, College of Graduate Studies, University of Saskatchewan, Saskatoon, Canada, Oct., 1972.

Reliability evaluation of electrical systems with switching actions

R.N. Allan, M.Sc. Tech., Ph.D., C.Eng., M.I.E.E., R. Billinton, M.Sc., Ph.D., Sen.Mem. I.E.E.E., P.Eng., and M.F. De Oliveira, M.Sc.

Indexing terms: Power systems, Reliability, Switching networks

Abstract

The paper presents a new approach for comparing the reliability of different designs for electrical systems with switching actions; this switching taking place following a fault in the system. An efficient program to evaluate the failure modes of each load point and its associated reliability indices has been developed. Also a concept of system minimal cutsets is introduced to assess the overall system-reliability indices. This is a powerful concept for systems that have many load points and also those that have a single overall objective. It permits comparison of the reliability between systems that are different in design, although meeting all the operating requirements. The program incorporates this technique and was used to obtain the results presented in the paper.

1 Introduction

It is well recognised that to evaluate the quality of a system or its ability to perform a required function, it is necessary to quantify the reliability of that system. Considerable literature has been published, particularly in North America, which describes many techniques for such quantification.

Recent literature[1,2] describes computer programs that assess the reliability of electrical systems and permit the effect of switching actions to be included. Grover and Billinton[2] describe a technique that incorporates all realistic component failure modes in the reliability predictions and state the limitations of the computerised approach presented by Endrenyi et al.[1] However, the technique[2] was not designed for, and therefore cannot easily handle, systems with more than one load point. Also, the input data required increase dramatically as the number of nonunidirectional system components increases.

This paper describes computational techniques to overcome these limitations and introduces a completely new concept of overall system-reliability indices. These techniques permit easy and computationally efficient reliability analyses of systems that have many load points, and also those that have a single overall objective. The overall system indices permit the overall incremental reliability cost to be evaluated. The usefulness of these system indices is shown by the numerical examples considered in this paper. A program based on these techniques does not limit the number of load points and is very efficient computationally.

2 Reliability concepts

The reliability techniques are based on the concept of expected failure rate and average-outage-duration method.[3,4] For each load point, the expected failure rate, average outage duration and average annual outage time are evaluated. These indices are easily understood and have a much more meaningful engineering significance than the mathematical concept of probability of failure.

To indicate the critical areas of failure, it is necessary to deduce the individual contributions of each failure event to the reliability indices of each load point. These contributions are the expected failure rate, the average outage duration and the average annual outage time due to

(a) overlapping forced outages

Paper 7648 P, first received 28th October 1975 and in final form 22nd January 1976

Dr. Allan and Mr. De Oliveira are with the Department of Electrical Engineering & Electronics, University of Manchester Institute of Science & Technology, Sackville Street, Manchester M60 1QD, England, and Prof. Billinton is with the Power Systems Research Group, University of Saskatchewan, Saskatoon, Sask., Canada

(b) forced outages overlapping a maintenance outage.

The individual contributions and the overall load-point indices can be evaluated, using equations based on those published previously,[2-4] after the failure modes for the load point under consideration have been deduced. The minimal cutset theory is a powerful computational tool for evaluating the load-point failure modes because the failure modes can be related to the minimal cutsets. The application of cutset theory to reliability evaluation has been described previously.[5-7]

It is not easy to compare the reliability of different system designs from a knowledge of the load-point indices only, particularly when many load points exist and when their indices change in different ways by different amounts. Furthermore, the indices at different load points may not be independent owing to common-mode failures and therefore cannot be realistically summated. To overcome these difficulties, a concept of system reliability in contrast to load-point reliability has been developed. To apply this concept rigorously, it is necessary to define a system failure. The present algorithm, as described in Appendix 10, considers that the system fails when at least one of the load points fails. If only one kind of failure mode exists, this concept becomes a simple extension of the minimal cutset theory applied to a multiple-output system. However, in an electrical system, a number of different failure modes exist, as described in Section 4, and the technique becomes much more complex.

3 Reliability input data

To evaluate the reliability indices described in Section 2 for each failure event, for each load point and for the overall system, the following component data are required:

(a) total failure rate, i.e. the average total number of component failures per year that require the removal of the component from service for repair due to any of its failure modes

(b) active failure rate, i.e. the average number of component failures per year that cause breakers to open and subsequent tripping of other healthy components. The healthy components can be re-energised after isolation of the actively failed component. It should be noted that an actively failed component may cause a temporary outage of several healthy components

(c) average repair time, i.e. the average time taken to repair all kinds of component-failure modes

(d) switching time, i.e. the average time between the occurrence of an active failure and the instant when the failed component is isolated and all possible healthy components are restored to service. It should be noted that the actively failed component itself can only be restored to service following a repair or replacement action

(e) maintenance outage rate, i.e. the average number of occasions per year that a component is taken out of service for preventative or scheduled maintenance. It is assumed that maintenance is not commenced if an outage already exists in a related part of the system.

(f) average maintenance time, i.e. the average duration of all preventative maintenance outages.

(g) stuck probability, i.e. the probability of a breaker or a switch failing to open or close when called upon to operate. It should be noted that, when a breaker fails to open following an active failure, a considerably greater number of healthy components may suffer a temporary outage due to backup protection operating.

4 Failure modes of an electrical system

As discussed in Section 2, minimal cutset theory is a powerful tool for deducing the failure modes of a system. Briefly, a cutset is defined as a set of components, the removal of which from the system causes failure of the load point. A minimal cutset however, is a cutset in which the load point fails if all the components of the cut fail and the load point does not fail if any one of the components does not fail.

In an electrical system with switching actions, there are four kinds of minimal cutset, if the probability of two overlapping active failures and two stuck breakers are considered to be negligible:

(i) a cutset of which all components are out of service, either for repair or for maintenance. Service can be restored only by replacing at least one component

(ii) a cutset of which all components are out, either for repair or maintenance, but service can be restored by closing a normally open path

(iii) a cutset of which one component is actively failed and the other components are out, either for repair or maintenance. Service is restored by isolating the actively failed component and re-energising the rest of the system

(iv) a cutset similar to type (iii) but including a stuck breaker.

To illustrate these failure modes, consider the system shown in Fig. 1.

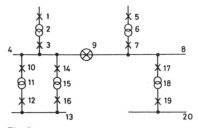

Fig. 1
System used to illustrate failure modes

All the load-point-failure modes for this system are shown in Table 1. These load-point-failure modes can be deduced using the following procedure:

(a) deduce all minimal paths from the sources to the load point including normally open paths

(b) separate the normally closed paths and the normally open paths

(c) deduce the minimal cutsets associated with the normally closed minimal paths. Identify which of these failure events can be eliminated by closing a normally open path

(d) simulate an active failure in every component unless it constitutes a 1st-order-failure event. Simultaneously, consider operation of, first, the primary protection breakers, and, secondly, the backup protection breakers. In both simulations identify the normally closed paths that are broken. If all these paths are broken, the active failure, or active failure plus stuck-breaker event, is a load-point-failure mode. If none of the normally closed paths is broken, the event is ignored. If some paths are broken, the minimal cutsets associated with the remaining paths are evaluated. These, together with the primary event, form a load-point-failure mode. Each time such a failure event is detected it is checked for independence; if it is already included in previous events, it is ignored

The reliability indices of these load-point-failure modes can be evaluated using equations based on those published previously.[2-4] It is evident from Table 1 that some failure modes are common to more than one load point. The load-point indices are therefore dependent. To evaluate the system-reliability indices, the independent system-failure modes must be deduced from the independent failure modes of all the load points. The algorithm developed to achieve this objective is described in Appendix 10. Briefly, it involves comparing two load-point-failure modes at a time and deciding whether they are independent events. Using this algorithm the system-failure modes

Table 1
FAILURE MODES FOR SYSTEM IN FIG. 1

Load point	Failure modes
4	1*, 2*, 3*, 4, <u>10</u>, <u>11</u>(10), <u>12</u>(10), <u>14</u>, <u>15</u>(14), <u>16</u>(14)
8	5*, 6*, 7*, 8, <u>17</u>, <u>18</u>(17), <u>19</u>(17)
13	1*, 2*, 3*, 4, 13, 10 + 14, 10 + 15, 10 + 16, 11 + 14, 11 + 15, 11 + 16, 12 + 14, 12 + 15, 12 + 16, <u>10</u>, <u>11</u>(10), <u>11</u>(12), <u>12</u>, <u>14</u>, <u>15</u>(14), <u>15</u>(16), <u>16</u>
20	5*, 6*, 7*, 8, 17, 18, 19, 20
system	1*, 2*, 3*, 4, <u>10</u>, <u>11</u>(10), <u>14</u>, <u>15</u>(14), 5*, 6*, 7*, 8, 13, 10 + 14, 10 + 15, 10 + 16, 11 + 14, 11 + 15, 11 + 16, 12 + 14, 12 + 15, 12 + 16, <u>12</u>, <u>16</u>, 17, 18, 19, 20

Underlined numbers are actively failed components
Numbers in parentheses are stuck breakers
Asterisked numbers are failed components but service can be restored by closing a normally open path (breaker 9)

can be determined for the system shown in Fig. 1. These are also shown in Table 1. It is clearly evident that there is a total of 47 cuts associated with all the load points, but only 28 of these are independent events for the system.

5 Computational aspects

A computer program, that incorporates all the previous concepts, has been developed. The following aspects are evaluated:

(i) the failure events for every system node from a description of the network topology and of the protection zones. The most important and novel features of the present computerised approach are that the system can be either radial or highly meshed, can have unidirectional and/or bidirectional components, can have multi-ended components that are not 100% reliable (busbars) and can be a multiple-input/multiple-output system.

(ii) the expected failure rate, average outage duration and average annual outage time of each load-point-failure event and for each load point

(iii) the overall system independent failure events, assuming that the system fails when at least one load point fails. It should be noted that other criteria of system failure are at present being considered and investigated.

(iv) the overall system-reliability indices.
The computer storage requirement is given approximately by

$$storage \simeq k + 4n_c + 29n_e + 3n_c n_n$$

where

$k = $ 12k words

$n_c = $ maximum number of minimal cutsets for a system load point

$n_e = $ number of system components

$n_n = $ number of system load points

For a system having 100 components, 250 failure modes per output node and 10 load points, the required storage is approximately 23k words. The computing time is a function of the system complexity, the total number of failure modes detected and the number of output nodes. The time taken to analyse the systems shown in Figs. 1 and 2 were 0·185 s and 6·481 s, respectively.*

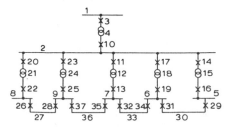

Fig. 2
Typical distribution network

* Using a CDC 7600 computer.

417

Table 2
COMPONENT RELIABILITY DATA

Component	Annual failure rate		repair time	Switching time	Stuck probability	Maintenance	
	total	active				annual rate	time
1, 2, 5, 6, 7, 8, 9.	0·024	0·024	h 3	h 2	–	–	h –
3, 10	0·23	0·10	15	2	0·005	–	–
4	0·10	0·10	24	1	–	–	–
11, 13, 14, 16, 17, 19, 20, 22, 23, 25, 26, 28, 29, 31, 32, 34, 35, 37,	0·23	0·10	20	1	0·005	0·4	12
27, 30, 33, 36.	0·12	0·12	10	1	–	0·1	2
12, 15, 18, 21, 24.	0·10	0·10	500	1	–	0·5	48

Table 3
SUMMARY OF FAILURE EVENTS

Busbar number	No. of paths	Failure events of order 1, 2 and 3												Total
		Overlapping outages				Active failures				Active plus stuck breaker				
		1	2	3	total	1	2	3	total	1	2	3	total	
2	1	5	–	–	5	5	–	–	5	10	–	–	10	20
5, 8	5	6	12	36	54	7	6	18	31	12	9	27	48	133
6, 7, 9	5	6	–	84	90	8	–	42	50	14	–	63	77	217
System	–	10	18	135	163	18	–	–	18	18	3	33	54	235

6   Analysis of typical system

6.1   Busbar analysis

To illustrate the application of the concepts described in earlier Sections and the type and depth of the results that can be obtained, the system shown in Fig. 2 was analysed. This system may reasonably be considered as basically a radial distribution network with interconnections between the low-voltage busbars. The component reliability data used in this analysis are shown in Table 2.

Using the reliability data shown in Table 2, all the failure events up to third order, i.e. up to three overlapping outages plus, if applicable, a stuck breaker, were deduced for each busbar and the reliability indices of each event and the overall indices of each busbar were calculated. As discussed previously, it was assumed that a component is not taken out for maintenance if it causes a busbar failure or if an outage already exists in a related part of the system. Maintenance details are therefore not given for some components in Table 2, as these were not required.

From this analysis, 20 failure events up·to third order were deduced for busbar 2, 133 failure events for busbar 5 and busbar 8, and 217 failure events for busbar 6, busbar 7 and busbar 9; i.e. a total of 937 events. Clearly, it is not possible to specify all these in detail, but a summary of the type and order of cut is shown in Table 3.

To illustrate the types of failure event and reliability index obtained from this part of the analysis, consider the 1st- and 2nd-order-failure events associated with busbar 5 (details of these events are shown in Table 4). The overall indices given for this busbar are those for the 1st- and 2nd-order events only and also those including the 3rd-order events. These overall indices will also apply to busbar 8.

A similar set of results to those shown in Table 4 can be obtained for each busbar in the system. The overall reliability indices for each busbar are shown in Table 5.

From the results shown in Table 4, it is evident, at least in this example, that the lower-order events dominate higher-order events. This is made clear by the overall indices, which indicate that the 3rd-order events contribute very little to the overall values. Several other interesting features can be observed from the results in Table 4.

The failure rate due to 1st-order active failures (0·7 per year) is greater than the failure rate due to 1st-order total failures (0·632 per

year). This stresses the importance of studying the impact of active failures on healthy components in the system. On the other hand, the average duration of these active failures (1 h) is significantly less than that of the total failures (15·057 h), because the former is associated with switching actions and the latter with repair or replacement actions. It is therefore evident that the two types of failure events should be categorised separately otherwise an average outage duration not applicable to either would be obtained.

It can be seen from Table 4 that the failure rate due to an active failure plus a stuck breaker (0·61 x 10^{-2} per year is greater than that for two overlapping forced outages (0·48 x 10^{-2} per year).

Considering the results shown in Table 5, it is evident that the total outage rates of busbars 5–9 are greater than that of busbar 2 owing to the increased number of possible failure events. This could reasonably be expected. The average outage duration of busbars 5–9 are, however, less than that of busbar 2. This is because an increased number of events have a relatively small duration; the weighted average is then less than that of busbar 2. A more surprising observation from Table 5 is that the annual outage time due to overlapping forced outages of busbars 6, 7 and 9 is greater than that of busbars 5 and 8, although there are more feeders (increased reliability?) to these three busbars. This is due to the increased number of components and therefore the increased number of active failures to these busbars. Although the average duration decreases because of a larger number of small-duration events, this decrease is outweighed by the increased number of failures. It cannot be concluded that an increased number of feeders automatically increases the reliability; it also depends on the number of associated failure events.

These aspects, together with many other observations that can be made from the Tables, are clearly related to the assumed data. However, it is evidently necessary to include all appropriate and realistic failure modes in the analysis and to deduce the breakdown of the constituent events to ensure a reasonable appraisal of the reliability behaviour of the system. At the present time, component-reliability data still lack the precision that many feel is necessary. However, the development of efficient computational techniques and realistic modelling of the system and its mode of operation as presented in this paper are vital to facilitate the determination of the failure modes of a complex system. These, by themselves, can be considered fundamentally important, so that a designer can detect

Table 4

RELIABILITY INDICES ASSOCIATED WITH BUSBAR 5

Failure event	Forced outages			Maintenance overlapping forced outages		
	yr^{-1}	h	h/yr	yr^{-1}	h	h/yr
1	0·24000E(−01)	3·0000	0·72000E(−01)	−	−	−
2	0·24000E(−01)	3·0000	0·72000E(−01)	−	−	−
3	0·23000E(+00)	15·0000	0·34500E(+01)	−	−	−
4	0·10000E(+00)	24·0000	0·24000E(+01)	−	−	−
5	0·24000E(−01)	3·0000	0·72000E(−01)	−	−	−
10	0·23000E(+00)	15·0000	0·34500E(+01)	−	−	−
Subtotal	0·63200E(+00)	15·0570	0·95160E(+01)	−	−	−
6 , 14	0·14493E(−04)	2·6087	0·37808E(−04)	0·13151E(−04)	2·4000	0·31562E(−04)
6 , 15	0·13781E(−03)	2·9821	0·41096E(−03)	0·65753E(−04)	2·8235	0·18566E(−03)
6 , 16	0·14493E(−04)	2·6087	0·37808E(−04)	0·13151E(−04)	2·4000	0·31562E(−04)
14, 29	0·24155E(−03)	10·0000	0·24155E(−02)	0·25205E(−03)	7·5000	0·18904E(−02)
14, 30	0·94521E(−04)	6·6667	0·63014E(−03)	0·71005E(−04)	5·1856	0·36820E(−03)
14, 31	0·24155E(−03)	10·0000	0·24155E(−02)	0·25205E(−03)	7·5000	0·18904E(−02)
15, 29	0·13653E(−02)	19·2310	0·26256E(−01)	0·68493E(−03)	1·3926	0·95382E(−02)
15, 30	0·69863E(−03)	9·8039	0·68493E(−02)	0·33105E(−03)	8·2325	0·27254E(−02)
15, 31	0·13653E(−02)	19·2310	0·26256E(−01)	0·68493E(−03)	13·9260	0·95382E(−02)
16, 29	0·24155E(−03)	10·0000	0·24155E(−02)	0·25205E(−03)	7·5000	0·18904E(−02)
16, 30	0·94521E(−04)	6·6667	0·63014E(−03)	0·71005E(−04)	5·1856	0·36820E(−03)
16, 31	0·24155E(−03)	10·0000	0·24155E(−02)	0·25205E(−03)	7·5000	0·18904E(−02)
Subtotal	0·47513E(−02)	14·8950	0·70770E(−01)	0·29432E(−02)	10·3120	0·30349E(−01)
11	0·10000E(+00)	1·0000	0·10000E(+00)	−	−	−
14	0·10000E(+00)	1·0000	0·10000E(+00)	−	−	−
16	0·10000E(+00)	1·0000	0·10000E(+00)	−	−	−
17	0·10000E(+00)	1·0000	0·10000E(+00)	−	−	−
20	0·10000E(+00)	1·0000	0·10000E(+00)	−	−	−
23	0·10000E(+00)	1·0000	0·10000E(+00)	−	−	−
29	0·10000E(+00)	1·0000	0·10000E(+00)	−	−	−
Subtotal	0·70000E(+00)	1·0000	0·70000E(+00)	−	−	−
34, 14	0·55137E(−04)	0·95238	0·52511E(−04)	0·54795E(−04)	0·92308	0·50580E(−04)
34, 15	0·57192E(−03)	0·99800	0·57078E(−03)	0·27397E(−03)	0·97959	0·26838E(−03)
34, 16	0·55137E(−04)	0·95238	0·52511E(−04)	0·54795E(−04)	0·92308	0·50580E(−04)
19, 14	0·55137E(−04)	0·95238	0·52511E(−04)	0·54795E(−04)	0·92308	0·50580E(−04)
19, 15	0·57192E(−03)	0·99800	0·57078E(−03)	0·27397E(−03)	0·97959	0·26838E(−03)
19, 16	0·55137E(−04)	0·95238	0·52511E(−04)	0·54795E(−04)	0·92308	0·50580E(−04)
Subtotal	0·13644E(−02)	0·99063	0·13516E(−02)	0·76712E(−03)	0·96345	0·73908E(−03)
12, (11)	0·50000E(−03)	1·0000	0·50000E(−03)	−	−	−
13, (11)	0·50000E(−03)	1·0000	0·50000E(−03)	−	−	−
15, (14)	0·50000E(−03)	1·0000	0·50000E(−03)	−	−	−
15, (16)	0·50000E(−03)	1·0000	0·50000E(−03)	−	−	−
18, (17)	0·50000E(−03)	1·0000	0·50000E(−03)	−	−	−
19, (17)	0·50000E(−03)	1·0000	0·50000E(−03)	−	−	−
21, (20)	0·50000E(−03)	1·0000	0·50000E(−03)	−	−	−
22, (20)	0·50000E(−03)	1·0000	0·50000E(−03)	−	−	−
24, (23)	0·50000E(−03)	1·0000	0·50000E(−03)	−	−	−
25, (23)	0·50000E(−03)	1·0000	0·50000E(−03)	−	−	−
30, (29)	0·60000E(−03)	1·0000	0·60000E(−03)	−	−	−
31, (29)	0·50000E(−03)	1·0000	0·50000E(−03)	−	−	−
Subtotal	0·61000E(−02)	1·0000	0·61000E(−02)	−	−	−
18, 14(19)	0·27568E(−06)	0·95238	0·26256E(−06)	0·27397E(−06)	0·92308	0·25290E(−06)
18, 15(19)	0·28596E(−05)	0·99800	0·28539E(−05)	0·13699E(−05)	0·97959	0·13419E(−05)
18, 16(19)	0·27568E(−06)	0·95238	0·26256E(−06)	0·27397E(−06)	0·92308	0·25290E(−06)
32, 14(34)	0·27568E(−06)	0·95238	0·26256E(−06)	0·27397E(−06)	0·92308	0·25290E(−06)
32, 15(34)	0·28596E(−05)	0·99800	0·28539E(−05)	0·13699E(−05)	0·97959	0·13419E(−05)
32, 16(34)	0·27568E(−06)	0·95238	0·26256E(−06)	0·27397E(−06)	0·92308	0·25290E(−06)
33, 14(34)	0·33082E(−06)	0·95238	0·31507E(−06)	0·32877E(−06)	0·92308	0·30348E(−06)
33, 15(34)	0·34315E(−05)	0·99800	0·34247E(−05)	0·16438E(−05)	0·97959	0·16103E(−05)
33, 16(34)	0·33082E(−06)	0·95238	0·31507E(−06)	0·32877E(−06)	0·92308	0·30348E(−06)
Subtotal	0·10915E(−04)	0·99064	0·10813E(−04)	0·61370E(−05)	0·96344	0·59127E(−05)
Total for 1st or 2nd order	1·3442	7·6581	10·294	0·37165E(−02)	8·3666	0·31094E(−01)
Total for 1st, 2nd and 3rd order	1·3443	7·6582	10·295	0·37268E(−02)	8·3632	0·31168E(−01)

419

critical areas and components in his systems. In addition, sensitivity analyses permit the range of reliability indices to be evaluated, and many such studies can be performed if the computational effort is minimal.

6.2 System analysis

Using the algorithm in Appendix 10, it can be shown that, of the 937 total 1st-, 2nd- and 3rd-order-failure events associated with the 6 busbars, only 235 events are independent system events. A summary of these events is also shown in Table 3. From this Table it can be seen that, although there are a significant number of 2nd- and 3rd-order active failure events associated with the busbars, there are no such events associated with the system. This is because these events are already included in other failure events. For example, the failure event (breaker 19 actively failed with breaker 14 out) associated with busbar 5 is not an independent event because the active failure of component 19 as a 1st-order event causes failure of busbar 6.

Details of the various 1st-order system-failure events are shown in Table 6.

Combining the indices for the events shown in Table 6 and those for the appropriate 2nd- and 3rd-order events gives the following system-reliability indices:

expected failure rate = 2·5528 per year
average outage duration = 4·6286 h
average annual outage time = 11·816 h/yr

Once again, it is seen that the failure rate is greater than that of any one busbar owing to the increased number of failures, but the average outage duration has decreased owing to an increased number of short-duration outages.

Table 5
OVERALL INDICES FOR EACH BUSBAR

Busbar	Type of failure	Annual outage rate	outage duration	annual outage time
			hr	hr
2	a	1·1130	8·9389	9·9490
	b	0·0	0·0	0·0
	total	1·1130	8·9389	9·9490
5, 8	a	1·3443	7·6582	10·2950
	b	0·0037	8·3632	0·0312
	total	1·3480	7·6601	10·3262
6, 7, 9	a	1·4393	7·1730	10·3240
	b	$1·5921 \times 10^{-5}$	6·6809	$1·0636 \times 10^{-4}$
	total	1·4393	7·1730	10·3241

a = overlapping forced outages
b = maintenance outage overlapping forced outages

Table 6
1ST-ORDER SYSTEM-FAILURE EVENTS

Forced outages	Active failures		Active failures plus stuck breaker	
1	11	25	12(11)	24(25)
2	13	26	12(13)	27(26)
3	14	28	15(14)	27(28)
4	16	29	15(16)	30(29)
5	17	31	18(17)	30(31)
6	19	32	18(19)	33(32)
7	20	34	21(20)	33(34)
8	22	35	21(22)	36(35)
9	23	37	24(23)	36(37)
10				

7 Conclusions

The inclusion of switching actions in reliability analyses of electrical systems is a vital and important feature. It necessarily imposes a 3-state model of the system; the state before a fault, the state after a fault but before switching, and the state after switching. The second state has a small average annual outage time because its residence time is dependent on switching times, although its rate of occurrence can be very large. If this state can be neglected, the reliability analysis is much simpler but can only be done with the following assumptions:

(a) the component switching times are negligible compared with the component average outage time and
(b) the rate of occurrence of a down state is not required.

Generally, these assumptions are not valid, particularly because the rate of occurrence of a down state is an important reliability index. The inclusion of this state, as described in this paper, is therefore considered necessary.

The predicted reliability indices of an electrical system in its design stage is an important feature for the objective comparison of different alternative designs. The concept of overall-system-reliability indices facilitates these comparisons, particularly when the system has many load points.

8 Acknowledgments

The authors are indebted to Prof. C. B. Cooper of the Department of Electrical Engineering & Electronics, UMIST., for providing the facilities necessary to conduct these studies, and to the Central Electricity Generating Board for their financial assistance. M. F. De Oliveira is also indebted to INVOTAN of Lisbon for providing him with financial support.

9 References

1 ENDRENYI, J., MAENHAUT, P.C., and PAYNE, L.C.: 'Reliability evaluation of transmission systems with switching after faults', *IEEE Trans.*, 1973, **PAS-92**, pp. 1863–1875
2 GROVER, M.S., and BILLINTON, R.: 'A computerised approach to substation and switching station reliability evaluation', *ibid.*, 1974, **PAS-93**, pp. 1488–1497
3 GROVER, M.S., and BILLINTON, R.: 'Quantitative evaluation of permanent outages in distribution systems', *ibid.*, 1975, **PAS-94**, pp. 733–741
4 BILLINTON, R., and GROVER, M.S.: 'Reliability assessment of transmission and distribution schemes', *ibid.*, 1975, **PAS-94**, pp. 724–732
5 JENSON, P.A., and BELLMORE, M.: 'An algorithm to determine the reliability of a complex system', *ibid.*, 1969, **R-18**, pp. 169-174
6 NELSON, A.C., BATTS, J.R., and BEADLES, R.L.: 'A computer program for approximating system reliability', *ibid.*, 1970, **R-19**, pp. 61–65
7 BILLINTON, R., and GROVER, M.S.: 'Reliability evaluation in distribution and transmission systems', *Proc IEE*, 1975, **122**, pp. 517–523

10 Appendixes

Algorithm to calculate the reliability indices of the system.

Assumption: It is assumed that the loss of any busbar causes the loss of the system.

Identification of a minimal cutset: Each cut is defined by a set of five values having the following meaning:

x	N	a	b	c

x defines the kind of cut
N defines the busbar to which the minimal cut is attached
a, b, c are the elements of the cut, where b and/or c may be zero.
The value of x may be
—1 for a cut in which every component is out, and service is restored by a repair action.
—2 for a cut in which every component is out, but the cut may be eliminated by closing a normally open path.
—3 for a cut in which component a is actively failed and every other component is out; this cut may be eliminated by switching component a out of the system.
k for a cut as defined by —3, but where breaker k is stuck.

10.1 Aim of the algorithm

Given two busbar minimal cuts, it is necessary to decide

whether or not one is included within the other, i.e. whether both are system minimal cuts. The complexity of the procedure is due to the need to compare cuts of different kinds.

10.2 Algorithm

Consider two minimal cuts, say cuts i and j

| x_i | N_i | a_i | b_i | c_i | define i |

and

| x_j | N_j | a_j | b_j | c_j | define j |

Let I and J be a set of three numbers defined as follows:

$$I = (a_i, b_j, c_j) \quad \text{and} \quad J = (a_j, b_j, c_j)$$

n_i = number of nonzero elements of I

n_j = number of nonzero elements of J

n_m = minimum of n_i and n_j

Define the set $A = I \cap J$ and consider that n_a is the number of non-zero elements of A, which is necessarily less than or equal to n_m.

The algorithm can then be described by the following steps:

(i) If $N_i = N_j$, the cuts are considered independent.
(ii) If $n_a < n_m$ the cuts are considered independent.

So, if $N_i \neq N_j$ and $n_a = n_m$, the analysis proceeds in the following sequence:

(iii) $x_i = x_j$
    (a) $x_i = x_j = -1$ or -2, $n_i = n_j$

reject either of the cuts; the program retains the cut that minimises the execution time in future comparisons of different pairs of cuts
(b) $x_i = x_j = -1$ or -2, $n_i \neq n_j$
    reject the cut with most elements
(c) $x_i = x_j \neq -1$ or -2
    (1) $a_i = a_j$, $n_i = n_j$
        as for case a
    (2) $a_i = a_j$, $n_i \neq n_j$
        as for case b
    (3) $a_i \neq a_j$
Reject the cut with minimum total outage time per year.
(iv) $x_i \neq x_j$
    (a) $n_i \neq n_j$
The cuts are considered independent.
    (b) $n_i = n_j$
The decision taken according to Table 7, below.
The cut indicated in the Table is eliminated.

Table 7

	-1	-2	-3	k	← values of x_i
-1	▨	i	i	i	
-2	j	▨	i	i	
-3	j	j	▨	i	
k	j	j	j	a	

↑
values of x_j
(a) both i and j are independent and are retained.

COMMON MODE FORCED OUTAGES OF OVERHEAD TRANSMISSION LINES

Task Force on Common Mode Outages of Bulk Power Supply Facilities of the Application of Probability Methods
Subcommittee of the Power System Engineering Committee, IEEE Power Engineering Society

Task Force Members:
H. G. Saddock, Chairman
Rochester Gas and Electric Corp.

M. P. Bhavaraju
Public Service Electric & Gas Co.

R. Billinton
University of Saskatchewan

C. F. DeSieno
American Electric Power Service Corp.

J. Endrenyi
Ontario Hydro

G. E. Jorgensen
Northeast Utilities Service Co.

A. D. Patton
Texas A&M University

D. L. Piede
Rochester Gas and Electric Corp.

R. J. Ringlee
Power Technologies, Inc.

J. A. Stratton
Rochester Institute of Technology

ABSTRACT

The studies undertaken by the Task Force addressed the definition and collection of data for common mode outages of multiple transmission lines which are on the same right-of-way for a measurable part of their length. A line-originated common mode outage is defined as an overlapping outage of two or more circuits sharing a right-of-way, from a common event, and due to their proximity on the right-of-way. A preliminary investigation of the available data shows that common mode outage events could be significant in transmission reliability calculations and that most of the data collection procedures are inadequate to provide data needed to evaluate these events. The additional information required to calculate indices of common mode outages is suggested in this paper. A method of calculating these indices is also illustrated.

INTRODUCTION

The prevailing social-economic pressures on the electric utility industry have increased the desire to utilize new or existing rights-of-way more heavily than has been commonly done in the past. This has resulted in an increased construction of two or more transmission circuits on a common right-of-way. The physical closeness of transmission lines on such a right-of-way can result in multiple dependent outages whose nature is different from the more usual independent overlapping outages, and therefore require a somewhat different analytical approach. Multiple dependent failure representations in power system studies have been limited mainly because of the lack of relevant failure statistics. Probabilistic bulk power reliability programs which presently are in various stages of development will require more precise measures of the reliability (forced outage rates, forced outage duration) of specific multiples of lines on a common right-of-way. The creation of this task force was for the purpose of providing sufficient understanding and to propose statistical analysis of the common mode forced outage problem.

Definition

A common mode forced outage is an event having a single external cause with multiple failure effects, where the effects are not consequences of each other. (Unlike, for example, the forced outage of one transmission line due to a lightning stroke followed by an overlapping forced outage of an adjacent line due to protective relaying failure.) The recognition of common mode forced outages is important because considering specific outage causes they could occur nearly as frequently as single contingencies do, rather than at much lower frequencies implied by models of multiple statistically independent forced outages. (e.g., aircraft interference with two lines on the same right-of-way). The frequency of a given common mode forced outage depends largely on the frequency of the singular outage causing event. The state-space diagram, Figure # 1, illustrates the possible failure and repair mode for the two components, 1 and 2.

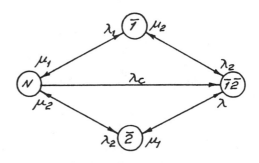

Figure # 1

A Two-Component System's State-Space Diagram
Indicating Independent and Common Mode
Failures and Repair

Where:

λ_i — ith component independent failure rate.

λ_c — component common mode failure rate.

μ_i — ith component repair rate.

$\overline{1}$ — component 1 unavailable.

$\overline{2}$ — component 2 unavailable.

$\overline{12}$ — components 1 and 2 unavailable.

N — components 1 and 2 available.

Paper F 76 116-4, recommended and approved by the IEEE Power System Engineering Committee of the IEEE Power Engineering Society for presentation at the IEEE PES Winter Meeting and Tesla Symposium, New York, N. Y., January 25-30, 1976. Manuscript submitted September 29, 1975; made available for printing November 4, 1975.

Reprinted from *IEEE Trans. Power Apparatus Syst.*, vol. PAS-95, no. 3, pp. 859–863, May/June 1976.

The common mode failure rates are expressed "per mile of parallel run." To obtain these rates, it is necessary in principle to establish equivalent mileages for the various types of rights-of-way with multiple circuits, because the rates depend on the number of circuits on the right-of-way and their arrangement. They may also depend on the multiplicity of failures considered (double, triple, etc.). To simplify matters in this first attempt to model common mode failures, right-of-way miles of multiple circuits will be used to measure exposure.

Survey Conducted

A survey of several utilities, drawn from a list of members of the Power System Engineering Committee, was conducted during the summer and fall of 1974, with 5 years of operating data requested. Of 69 requests to organizations, some 40 replied, yielding 18 useful pieces of information. The Task Force is deeply indebted to many concerned organizations.

The data did strongly suggest that the probability of two lines being forced out on an overlapping outage when they are on the same right-of-way is much larger than the probability of two lines being forced out on an overlapping basis when they are on separate independent rights-of-way. However, the amount of data is not significant enough to warrant release of figures at this time.

Scope of Task

Based on the currently available outage data, (λ_1 and λ_2 in Figure # 1) indices for outages of single lines can be calculated.[2] Data of this variety can be used to measure the probability of one line being out, and also the probability of more than one line being out simultaneously, if there are no common mode effects, by the product rule of probabilities. However, these simultaneous outage probabilities are incorrect if there are common mode effects between the lines. Thus, to compute the common mode forced outage probabilities, the failure rates λ_c are needed. Similarly, the modeling of separation effects may not be easily defined due to common mode effects.

Several assumptions were necessary to limit the scope of the problem to workable proportions. They include the following:

Those needed for the future as well as for this study:

1. Only transmission lines on a common right-of-way were analyzed. Previous studies have dealt with only the failure rates of individual transmission lines.

2. Both lines on common towers and lines on individual towers were analyzed. The information was kept separately for purposes of statistical analysis.

3. Simultaneous outage of two or more lines on the same right-of-way was assumed to be of common mode cause unless otherwise specified. Only the overlapping forced outage duration was used.

Assumptions for conducted survey only, which proved to be too limiting:

4. Lines with more than 50% of their right-of-way common were considered. A more realistic figure would be all parallel lines, for most are diverging for portions of the distance. Accurate information must be kept to discern common mode outages.

5. The number of lines on a right-of-way was not considered because of a lack of data. That is, the forced outage rates were calculated as the frequency per mile of right-of-way miles. (See definition section).

6. No attempt was made to define arbitrary boundaries for discerning which lines were common and which were not due to lack of data. A rule of common sense by the reporter was assumed. Care should be exercised by operating utilities in the future.

SUGGESTED FUTURE REPORTING FORMAT

The survey forms, which were used in the 1974 study have been revised, with much needed information requested, and with further definitions of the data, in Figures # 2 and # 3 of this report. Figure # 2 should be used for circuits on individual tower lines, while Figure # 3 should be used with multiple circuits on the same towers. A large amount of information is requested for appropriate statistical studies, but should not result in large amounts of additional record keeping for operating utilities.

Clarifications of specific entries are as follows:

- The line length should be expressed as miles of common right-of-way rather than miles of line.

- Specific records should indicate both the number of lines on a right-of-way and the number of lines out.

- The common mode outage duration is defined as the overlapping time period only, and should be noted in the appropriate unit.

- Permanent and Transient outages should all be recorded, allowing for future sorting.

  — Permanent forced outages for this report were taken as forced outages lasting at least one hour.

  — Transient forced outages for this report were subdivided into a) those that are reclosed automatically, and b) those forced outages not reclosed automatically but which last less than one hour.

- Only outages of line origination will be analyzed, as others are not dependent upon line proximity.

- The yearly summary of outage data and line lengths are of critical importance to this and any future study. Accurate information is of utmost importance.

- All common mode outages should be recorded, based on commonality of right-of-way. Future studies will need as much information as possible.

Outage causes could be classified as follows:

Natural Events.

1. Fire in Right-of-Way. (forest, brush, tall grasses; Agricultural: cane).

2. Foundation or Anchor Failure. (flood, landslide, ground subsidence).

3. Severe Environmental Conditions. (hurricane, tornado, icing).

Interference.

4. Interference with other circuits. HV crossing of lower voltage circuits.

5. Aircraft interference.

6. Rail and road vehicle interference.

Additional Data Suggested

The area isokeraunic level should be added to the information kept on record to isolate future storm plagued areas.

Use of Transient and Permanent Forced Outage Data

Transient outage data should be kept separate from permanent outage data in the study. Since the different durations have different system impacts, and since, often, the location of the outage cause cannot be defined it is recommended to classify transient outages into short duration or restoration by switching, and momentary duration or automatic reclosure.

ILLUSTRATION OF CALCULATING COMMON MODE OUTAGE INDICES

The method of calculating common mode outage indices is illustrated using hypothetical data of Figure # 4. The indices are calculated in Figure # 5. The data should not be assumed to be representative of any system.

The indices calculated in Figure # 5 are the failure rate in outages per 100 right-of-way miles per year, the average duration of an outage, the probability of more than one line being out simultaneously per 100 right-of-way miles for a common mode failure on a 230 KV system. Similar methods can be used to calculate indices for the following categories by sorting the data as required:

> Transient and permanent faults,
> Voltage classification,
> Tower type, and
> Outage cause.

CONCLUSION

The results of this preliminary study strongly suggest that the occurrence of line-originated common mode outages have not been sufficiently quantified in transmission reliability assessments. Data collected in this survey did support the expectation that the probability of a common mode outage is greater for transmission lines sharing a right-of-way than for those lines on separate rights-of-way. Although quantitative results are not given, a method for calculating probabilistic indices was shown.

The outage data collection survey undertaken provided insight into the variety of outage-event recording procedures in use by electric utilities. Indeed, the lack of any quantitative probabilistic failure indices are directly related to the magnitude and complexity of this data collection process. In many cases, common mode outages were not recorded as such, only as outages of circuits x and y. If future studies are to determine probabilistic failure indices, electric utilities must improve on existing line outage reporting. This improved documentation must recognize common mode outages as defined in this paper. The attached sample data reporting formats provide examples of acceptable methods. The Task Force urges that electric utilities immediately review their line outage reporting procedures, and revise them to include pertinent common mode outage data.

REFERENCES

(1) R. Billinton, "Bibliography on Application of Probability Methods in Power System Reliability Evaluation," IEEE Transactions on Power Apparatus and Systems, pp. 649-660, March/April 1972.

(2) S. A. Mallard, V. C. Thomas, "A Method for Calculating Transmission System Reliability," Ibid., March 1968, pp. 824-833.

(3) IEEE Std. 346-1973 "IEEE Standard Definition in Power Operation Terminology Including Terms for Reporting and Analyzing Outages of Electrical Transmission and Distribution Facilities and Interruptions to Customer Service".

SUGGESTED REPORT FORM   PAGE # 1

SEVERAL INDIVIDUAL TRANSMISSION LINES ON THE SAME RIGHT-OF-WAY

Year	Circuit Identification				Number Lines Out	Number Lines on R/W	Overlapping Outage Duration	Outage Location	Outage Cause
	Name or Number	Voltage	Structure Type	Line Length (R/W)					
1974 (to submission date) 1973 Etc.			See Note A	Common R/W Miles	2, 3, etc. See Note E	2, 3, etc. See Note E	See Note B	See Note C	See Note D

YEARLY SUMMARY SEVERAL INDIVIDUAL TRANSMISSION LINES ON THE SAME RIGHT-OF-WAY, R/W MILEAGE IN OPERATION

Year	Voltage	One Line per R/W				Two Lines per R/W				Three Lines per R/W
		Wood Pole	Steel Tower	Steel Pole	Total	Wood Pole	Steel Tower	Steel Pole	Total	Etc.
1974 (to submission date)	115 KV 230 KV Etc.									
1973	115 KV 230 KV Etc.									
Etc.										

ALL TRANSMISSION LINES

Year	Permanent Outages Duration ≥ 1 Hr.			Transient Outages					
	115 KV	230 KV	Etc.	"O" < Duration < 1 Hr.			Duration = "O"		
				115 KV	230 KV	Etc.	115 KV	230 KV	Etc.
1974 1973	Number of Outages			Number of Outages					

Notes: Give as much information as possible.
A. Such as Wood pole, Steel Tower, etc.
B. Full time of Outage in seconds, minutes, hours, or days. (Estimate if necessary).
C. Location such as Line, Breaker, Bus Section, etc.
D. Cause such as: Lightning, overheating, floods, hurricanes, foreign objects, aircraft, bullets, unexplained, other (label).
E. Where a portion of the lines are on common towers, they should be so noted.

Figure # 2

424

SUGGESTED REPORT FORM PAGE # 2

UPDATE OF MULTIPLE CIRCUIT (ON SAME STRUCTURE) OUTAGE INFORMATION

Year	Circuit Identification				Number Lines Out	Number Lines on R/W	Overlapping Outage Duration	Outage Location	Outage Cause
	Name or Number	Voltage	Structure Type	Line Length					
1974 (to submission date)			See Note A		2, 3 etc.	2, 3 etc.	See Note B	See Note C	See Note D
1973									
Etc.									

YEARLY SUMMARY MULTIPLE CIRCUIT LINES, MILEAGE IN OPERATION

Year	Voltage	Double Circuit					Triple Circuit					Etc.
		Wood Pole	Steel Tower	Steel Pole	Other	Total	Wood Pole	Steel Tower	Steel Pole	Other	Total	
1974 (to submission date)	115 KV 230 KV Etc.											
1973	115 KV 230 KV Etc.											
Etc.												

Notes: See Figure # 2

Figure # 3

Hypothetical Data Used in Calculating Common Mode Outage Indices in Figure # 5
(Permanent Outages Only)

SUGGESTED REPORT FORM PAGE # 1

SEVERAL INDIVIDUAL TRANSMISSION LINES ON THE SAME RIGHT-OF-WAY

Year	Circuit Identification				Number Lines Out	Number Lines on R/W	Overlapping Outage Duration	Outage Location	Outage Cause
	Name or Number	Voltage	Structure Type	Line Length (R/W)					
1974 (to submission date)									
1973	# 1	230 KV	SSC	100 mi.	2	2	5 Hr.	Line	Flood
1972									
1971	#17	230 KV	SSC	70 mi.	2	2	3 Hr.	Line	Lightning
1970									

YEARLY SUMMARY SEVERAL INDIVIDUAL TRANSMISSION LINES ON THE SAME RIGHT-OF-WAY, R/W MILES IN OPERATION

Year	Voltage	One Line per R/W				Two Lines per R/W				Three Lines per R/W	
		Wood Pole	Steel Tower	Steel Pole	Total	Wood Pole	Steel Tower	Steel Pole	Total	Wood Pole	Steel Tower
1974 (to submission date)	230 KV	1500					900				900
1973	230 KV	1500					900				900
1972	230 KV	1500					900				900
1971	230 KV	1000					650				400
1970	230 KV	1000					650				400

Figure # 4

ALL TRANSMISSION LINES

Year	Permanent 230 KV	Transient 230 KV
1974	12	
1973	12	
1972	12	
1971	12	
1970	12	
	Total 60	

Figure # 4 (continued)

Basis: Hypothetical data for 230 KV System shown in Figure # 4 for 1970-74 period.

Data

1974: Total circuit miles = 1500 X 1 + 900 X 2 + 900 X 3 = 6,000

Circuit miles of common right-of-way = 900 X 2 + 900 X 3 = 4,500

Common right-of-way miles = 900 + 900 = 1,800

Total common right-of-way mile Years = (900 + 900 + 900 + 650 + 650)
+ (900 + 900 + 900 + 400 + 400)
= 7,500

Number of all permanent (including C.M.) outages = 60 [2]

Number of common mode permanent outages = 2

Calculations

Estimated permanent outages on common right-of-way [1] = $\frac{4500}{6000}$ X 60 = 45

Common mode permanent outages as a percent of total permanent outages on common right-of-way = $\frac{2}{45}$ X 100 = 4.5%

Common mode permanent failure rate, λ_c = $\frac{2}{7500}$ X 100

= 0.027 outages / 100 ROW miles/year

Average duration, $\frac{1}{\mu_c}$ = $\frac{5+3}{2}$ = 4 hours = $\frac{4}{8760}$ year

Probability of common mode permanent outage

, = $\frac{\lambda_c}{\lambda_c + \mu_c}$ $\frac{0.027 \text{ X } 4}{8760}$ = 0.000012/100 ROW miles

Notes: (1). Actual data can be used if available instead of estimating as shown.

(2). Similar calculations can be performed if data on temporary and transient outages are available.

Figure # 5 Calculation of Common Mode Outage Indices

RELIABILITY EVALUATION OF THE AUXILIARY ELECTRICAL SYSTEMS OF POWER STATIONS

R.N. Allan
UMIST
Manchester
England

R. Billinton
University of Saskatchewan
Saskatoon
Canada

M.F. DeOliveira
UMIST
Manchester
England

ABSTRACT

This paper presents a novel approach to the reliability assessment of different designs of power station auxiliary electrical systems. This assessment evaluates the impact of the auxiliary system on the availability of the unit in terms of derated states of the unit. This technique permits the reliability cost to be calculated easily for different system designs. All realistic failure modes were taken into account. An efficient computer program using these techniques has been developed and was used to obtain the results presented in this paper.

INTRODUCTION

There has been considerable interest in recent years to quantify the effect of generating capacity on the reliability and long term availability of power systems. The availability of the generating unit is a fundamental element in such an analysis. However it is well known that the generating unit availability has decreased as the unit rating has been increased; large units having unavailabilities as high as 20%. Therefore a realistic quantitative appraisal of the reliability of the unit and the components of its subsystems is necessary in the design phase from which their impact on the output availability can be evaluated. One component contributing to this unavailability is the electrical auxiliary system and its design. Although this contribution is only one factor of the overall unavailability it is nevertheless one of the most difficult to assess due to the complexity of the system and the consequences of losing parts of it. The present studies have therefore been centred on assessing the impact of such auxiliary systems on the availability of the unit.

Basically these systems do not differ greatly from a large substation. Reference 1 presents a computerised approach for assessing the reliability of substations and switching stations. The basic concepts contained in this reference were utilized in the present study with certain improvements and modifications being made.

One of the assumptions in reference 1 is that the average outage time of a minimal cut set which is eliminated by closing a normally open path is the time taken to close the normally open path. This is equivalent to assuming that the system does not fail while the repair is being carried out and service is returned to normal by restoring at least one component of the minimal cut set. This assumption was eliminated in the present work by using a conditional probability approach.

The program described in reference 1 was not intended to analyse easily a system having more than one load point. The auxiliary electrical system of a power station has many load points all of which may affect the unit output by different amounts. The reliability indices of each load point of the auxiliary system does not allow the impact of each load point or combination of load points on the unit availability to be evaluated very easily. The overall reliability of the system is therefore very difficult to assess. The problem of analysing multi-load points was overcome by introducing a concept of system minimal cut sets and associating with each of these cuts either a partial capacity state or a total outage state of the unit. These overall system minimal cut sets are obtained from the minimal cut sets of the individual load points. For each of the feasible derated states of the unit due to the unreliability of the auxiliary system, the expected encounter rate and average duration are evaluated using the equations described by Billinton and Grover[1,2]. From these values an energy and capacity replacement cost as described by Billinton and Ringlee[3] can be evaluated and assigned to any design changes in the auxiliary system.

LOAD POINT FAILURE MODES

The analysis is commenced by calculating the reliability indices of each busbar in the electrical auxiliary system. The component failure modes that are considered are identical to those described by Grover and Billinton[1]. To deduce the busbar failure modes it was assumed that the probability of having simultaneously more than three faulty components is negligible and therefore only events up to third order were considered. Also two failure events at a busbar were considered independent if the set of components defining one failure event did not include the set of components defining the other failure event. This assumption is valid because the failure probability of any electrical component is numerically very small. It was also assumed that, after a busbar failure, no other healthy component in the paths from any source to the busbar can subsequently fail during the time the paths are de-energised. This is acceptable because the components will generally have a much smaller failure rate when they are out of service.

All the above assumptions permit the busbar failure modes to be identified with the minimal cut sets associated with the minimal paths from all the sources of supply to the busbar under consideration. An efficient algorithm was developed to deduce these minimal paths and cut sets from a simple description of the topology of the network. This method of deduction facilitated the evaluation of the minimal cut sets for every system busbar using the same set of input data.

The system configuration as seen from each load point or busbar was considered to exist at any instant of time in one of the three following states:

Paper A 76 018-2, recommended and approved by the Power Generation Committee of the IEEE Power Engineering Society for presentation at the IEEE PES Winter Meeting and Tesla Symposium, New York, NY January 25-30, 1976. This paper was upgraded to transactions status, F77 008-6, for presentation by title for written discussion at the IEEE PES Winter Meeting, New York, NY January 30-February 4, 1977. Manuscript submitted August 11, 1975; made available for printing September 30, 1976.

Reprinted from *IEEE Trans. Power Apparatus Syst.*, vol. PAS-96, no. 5, pp. 1441–1449, Sept./Oct. 1977.

427

1 - system without any failure, i.e normal operation

2 - system with a fault but before switching

3 - system with a fault but after switching

The state space diagram shown in Fig. 1 illustrates this three state model

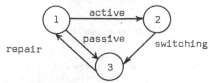

Fig. 1 - State space diagram of 3-state model

Each time states 2 or 3 are encountered by a component with either one, two or three overlapping outages or failures, the event is checked to establish whether it leads to a busbar failure. If it does the components defining this event are considered a busbar cut set and then checked to establish whether they form a minimal cut set.

The minimal cut sets were grouped into four types; each type being associated with distinctly different ways of restoring the supply to the busbar under consideration. These types are:

a) a cut set for which a busbar failure occurs when all the components of the cut are out of service, i.e. being maintained or repaired. Service is returned to normal after restoring to service at least one of the components.

b) a cut set for which a busbar failure occurs when all the components of the cut are out of service but service can be restored by closing a normally open path.

c) a cut set for which a busbar failure occurs when one component is actively failed and the others are out of service. Service is restored by isolating the actively failed component and closing the breakers that were opened due to the active fault.

d) a cut set for which a busbar failure occurs when one component is actively failed, the others are out of service and one breaker of the protection zone of the actively failed component is stuck. Service is restored by isolating the actively failed component and closing the breakers that were opened during the fault.

All four failure events were considered in reference 1. However the reliability indices associated with the second type of these cut sets were calculated assuming that the normally open path which was closed to restore the supply did not fail during the repair times of the components of the cut set. This assumption was eliminated using the following technique.

Consider:

event A; a failure event that can be eliminated by closing any of the normally open paths k, j, l .... m.

event B; an event defined as, "at least one of the normally open paths k, j, l .... m, does not fail before restoring to service one component of failure event A".

Using the conditional probability approach, the following equation can be written:

$$P(A) = P(A|B) \cdot P(B) + P(A|\overline{B}) \cdot P(\overline{B}) \qquad (1)$$

$P(A|B)$ is the probability of event A given that at least one of the normally open paths is always available and $P(A|\overline{B})$ is the probability of event A given that none of the normally open paths are available to restore service. For each of the events $A|B$ and $A|\overline{B}$ the contribution to the average failure rate, the average outage duration and average annual outage time due to forced outages (λ, r, U) and forced outages overlapping a maintenance outage $(\lambda", r", U")$ are evaluated. These indices are:

for $A|B \rightarrow \lambda_1$ and $\lambda_1"$, r_1 and $r_1"$, U_1 and $U_1"$

for $A|\overline{B} \rightarrow \lambda_2$ and $\lambda_2"$, r_2 and $r_2"$, U_2 and $U_2"$

noting that $\lambda_1 = \lambda_2$ and $\lambda_1" = \lambda_2"$

To calculate the same indices for the event represented by $\overline{B}$, the minimal cut sets for the paths k, j, l .... m are evaluated. Any of these that are also load point minimal cut sets are eliminated. The minimal cut sets of event $\overline{B}$ are evaluated up to second order if event A is of first order, up to first order if event A is of second order and are not evaluated if event A is of third order. From these minimal cut sets the reliability indices of the "element" equivalent to the normally open paths are evaluated; these being defined as λ_e and $\lambda_e"$, r_e and $r_e"$, U_e and $U_e"$.

The normally open breakers, which are first order cut sets of the minimal paths k, j, l .... m are also considered in order to include the stuck breaker probability P_s of this equivalent element.

Noting that the average annual outage time of event A, when expressed in years/year, represents $P(A)$ the following equations are obtained from equation 1:

$$U_A = U_1 \left[1 - (U_e + P_s) \right] + U_2 (U_e + P_s) \qquad (2)$$

$$U_A" = U_1"(1 - U_e") + U_2" \cdot U_e" \qquad (3)$$

The following equations can also be obtained:

$$\lambda_A = \lambda_2 (1 + \lambda_e r_2) \qquad (4)$$

$$\lambda_A" = \lambda_2"(1 + \lambda_e" r_2") \qquad (5)$$

$$r_A = \frac{U_A}{\lambda_A} \qquad (6)$$

$$r_A" = \frac{U_A"}{\lambda_A"} \qquad (7)$$

Having calculated the reliability indices of all the independent failure events of each busbar, the overall busbar indices are evaluated by simply adding the expected failure rates and the annual outage times of each failure event.

REPRESENTING BRANCHES OFF THE BUSBARS

To calculate the reliability indices of each busbar, the branches off the busbars cannot be ignored since they could have a significant impact on the busbar indices, particularly on its failure rate. This is due to the effect of active failures and active failures overlapping a stuck breaker condition of the components in these branches being reflected back to the breakers on the input side of the busbars. In a typical power station the number of branches off a busbar may exceed 60 or so with each branch containing at least three components, e.g. a breaker, a cable and a motor. Therefore in a typical station the number of additional components that would need to be introduced to adequately represent these branches could exceed several hundred. The computer time and storage requirements of a computer program is a function of the number of components. Therefore to include all these additional components would seriously degrade computational efficiency.

This difficulty was overcome by representing the branches off each busbar by an "equivalent branch". The active failure rate of each component in this equivalent branch is obtained by summing the active failure rates of the components existing in the real branches. The switching time and stuck breaker probability are the same as any one of the real components. Using this technique the reliability indices of each busbar are calculated precisely and the increase in computer time and storage requirements is a minimum.

REPRESENTING ALTERNATIVE SOURCES OF SUPPLY

The electrical auxiliary system of each unit is normally fed from its own generator and/or from the grid system. However alternative sources of supply such as diesel generators and gas turbines may be available either for black starts or for use when the normal feeding points are not available. In the present system model these alternative sources of supply were represented as normal system branches connected to the system through a normally open breaker; the switching time of this normally open breaker including the start-up time of the alternative source of supply.

For each busbar any such source is included as one of the normally open paths. For each failure event that does not include a component actively failed, the program checks whether it is possible to eliminate it by closing a normally open path. When the failure event under consideration involves the loss of the normal sources of supply the normally open paths that are considered include that of the alternative source. Because these alternative sources often have very high unreliabilities it is generally not acceptable to assume that they cannot fail when required, i.e. during the outage time of the minimal cut set. Therefore the conditional probability techniques described previously were incorporated for these sources of supply and in fact was the main reason for their development.

LOSS OF POWER DUE TO UNRELIABILITY OF SYSTEM

It is not possible to compare objectively different designs of electrical auxiliary systems from only a knowledge of the reliability indices of each busbar. This is because the requirement is a knowledge of the impact of the auxiliary electrical system on the availability of the generating unit and not simply the indices of each busbar. Objective comparison becomes difficult because the system busbars have different importance as far as the output of the unit is concerned and many failure events are common to more than one busbar. However this difficulty can be alleviated if all the independent failure events of the system busbars are considered as a set of events that include the unit independent failure events. Therefore this set of events are the unit cut sets from which the unit minimal cut sets must be obtained. The complexity of this procedure is due to the following reasons:

(a)   cut sets of different types have to be compared in order to decide whether or not one is included in the other

(b)   even when cut sets associated with busbars that cause derated states of the unit are included in others, all have to be taken into account because the simultaneous loss of two of these busbars may cause a derated state that is different from that associated with the loss of each busbar separately.

To calculate the indices of each feasible derated state of the unit due to the unreliability of the electrical auxiliary system, the loss of each busbar is considered to cause one of the following states:

(a)   total loss of the unit

(b)   no influence on the output of the unit

(c)   a loss of x% of the output of the unit where x may have any value and be different for each busbar and combination of busbars.

It was assumed that if an event causes the loss of three busbars, each of which creates individually a derated state, this event will cause total loss of the unit. Generally however the simultaneous loss of two such busbars is assumed to create a derated state that may or may not be equal to the derated states that are created by the individual loss of these busbars.

The algorithm developed for analysing this problem is described in the Appendix assuming that the independent failure events of each busbar are already known. In the associated computer program these failure modes are stored in a compact form which prevents excessive increase in computer storage. The data required to calculate the reduction of power due to the unreliability of the electrical auxiliary system are as follows:

(a)   consequences of losing each busbar

(b)   consequences of losing all combinations of two busbars both of which separately lead to a derated state.

Use of the algorithm described in the Appendix gives the average rate of occurrence or encounter, the average duration and the average annual residence time of each feasible derated state. From these indices it is possible to assign a reliability cost to every design and an incremental cost of each design change. This permits the "quality" of different design alternatives to be assessed objectively.

ANALYSIS OF TYPICAL AUXILIARY SYSTEM

The techniques and concepts described in the previous sections were applied to the typical auxiliary system shown in Fig. 2.

This system consists of a set of unit busbars shown on the left hand side of the Figure and a set of station busbars shown on the right hand side. At each voltage level the busbars are interconnected with normally open breakers. From each busbar a number of motors are connected. A standby generator (56) is connected to the high voltage unit busbar (4) through a normally open breaker (57).

The component reliability data used in this analysis is shown in Table 1 and the consequences of busbar outages in Table 2.

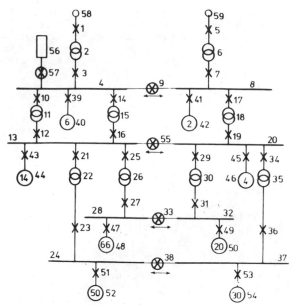

Fig. 2 - Typical auxiliary system

double arrows indicate bidirectional branches, all others assumed unidirectional. Encircled numbers indicate number of motors. ⊗ represent normally open breakers, all others assumed normally closed. Component 56 is a standby generator, normal supply surces are 58 and 59.

Using the reliability data shown in Table 1, all the failure events for each busbar were deduced and the reliability indices of each event and the overall indices of each busbar were calculated. It was assumed that a component is not taken out for maintenance if it causes a busbar failure or if an outage already exists in the system. Therefore maintenance details are not given for some components in Table 1 as these are not required.

In general the modes of failure were assumed to be forced outages, overlapping forced outages or forced outages overlapping a maintenance outage. All the techniques discussed in the preceding Sections were included in the analysis, that is, service could be restored (if possible) by closing normally open paths which themselves could subsequently fail, service could be restored using the standby plant when normal sources of supply failed and service could be restored following an active failure by isolating the failed component and reclosing the appropriate breakers.

From this analysis, the total number of failure events found for all busbars was **183** of which some would cause failure of more than one busbar. A summary of the number of paths and number of failure events associated with each busbar is shown in Table 3.

To illustrate the type of results obtained from this part of the analysis, consider the **32** failure events associated with busbar 13. It is evident that there are two normally closed paths from source 58 and seven normally open paths from sources 56, 58 and 59 through the normally open breakers 9, 55 and 57. Also

Table 1 - Reliability data

component	failure rate f/yr		repair time hr	switching time hr	stuck probability	maintenance	
	total	active				rate o/yr	time hr
1, 5	0.020	0.015	20	1.0	0.001	-	-
2, 6, 11, 15, 18, 22, 26, 30, 35	0.020	0.015	800	1.0	-	0.5	40
3, 7, 10, 14, 17, 39, 41	0.012	0.008	20	1.0	0.001	0.25	20
4, 8, 13, 20, 24, 28, 32, 37	0.005	0.005	10	1.0	-	-	-
9, 33, 38, 55	0.020	-	20	1.0	0.013	0.25	20
12,16,19,21,23,25,27,31,34,36,43,45,47,49,51,53	0.010	0.007	20	1.0	0.001	0.25	20
29	0.010	0.007	10	1.0	0.001	-	-
40, 42	0.050	0.040	-	1.0	-	-	-
44, 46	0.025	0.02	78	1.0	-	-	-
56	12.0		78	1.0	-	1.6	315
57	0.020	-	20	1.0	0.113	0.25	20
48, 50, 52, 54	0.035	0.03					

Table 2 - Consequences of busbar outages

number of failed busbars	busbar number	reduction of output power, %
1	4, 13, 24, 28	100
1	8, 37	30
1	20	20
1	32	0
2	8 and 37	50
2	20 and 37	45
2	8 and 20	40
3	any combination	100

if active failure modes are ignored, there are 14 minimal cut sets only one of which, failure of busbar 13, cannot be eliminated using a normally open path. These cuts are shown in Table 4.

Consider now the failure of breaker 1 and the subsequent restoration of supply to busbar 13 (a similar analysis exists for failures of transformer 2 and breaker 3). From Table 1 the failure rate of this event is 0.02 f/yr and the outage time if a normally open path was not available (event $A|\bar{B}$) would be 20hr. Therefore:

$$\lambda_2 = 0.02 \text{f/yr}, \quad r_2 = 20\text{hr}, \quad U_2 = 0.4\text{hr/yr}$$

Table 3 - Summary of failure events

| busbar number | number of paths | | failure events | | | | |
| | normally closed | normally open | overlapping forced outages | | active failures | active failures & stuck breaker | total |
			total	closed by n/o path			
4	1	2	4	3	3	5	12
8	1	2	4	3	2	3	9
13	2	7	14	13	8	10	32
20	1	8	8	7	4	6	18
24	2	16	18	17	8	9	35
28	2	16	18	17	8	9	35
32	1	17	12	11	4	5	21
37	1	17	12	11	4	5	21
							183

Table 4 - Minimal cuts associated with busbar 13

first order cuts	second order cuts	
1	10 and 14	11 and 16
2	10 and 15	12 and 14
3	10 and 16	12 and 15
4	11 and 14	12 and 16
13	11 and 15	

If however at least one normally open path is available (event A|B) then the failure rate remains unchanged but the outage time becomes the switching time for restoring the supply. Therefore:

$$\lambda_1 = 0.02 f/yr, \quad r_1 = 1hr, \quad U_1 = 0.02 hr/yr$$

The minimal cut sets of the normally open paths to busbar 13 that are not already included in the failure modes of busbar 13 are overlapping outages of 5 and 56, 5 and 57, 6 and 56, 6 and 57, 7 and 56, 7 and 57, 8 and 56, 8 and 57. Combining the reliability indices for these eight overlapping events gives the reliability indices for the equivalent element discussed previously. Therefore:

$$\lambda_e = 0.2899 \times 10^{-1} f/yr, \quad r_e = 0.7024 \times 10^{-2} yr,$$
$$U_e = 0.2036 \times 10^{-3} yr/yr$$

$$\lambda_e'' = 0.3312 \times 10^{-2} o/yr, \quad r_e'' = 0.1027 \times 10^{-1} yr,$$
$$U_e'' = 0.3401 \times 10^{-4} yr/yr$$

$$P_s' = 0.0$$

Then:

$$\lambda_A = 0.02 \ (1 + \frac{0.2899 \times 10^{-1} \times 20}{8760}) = 0.20001 \times 10^{-1} f/yr$$

$$U_A = 0.02(1 - 0.2036 \times 10^{-3}) + 0.4 \times 0.2036 \times 10^{-3}$$

$$= 0.20077 \times 10^{-1} hr/yr$$

$$r_A = \frac{U_A}{\lambda_A} = 1.0038 \ hr$$

Applying the same concepts and techniques to the other minimal cuts associated with busbar 13 and including the active failure modes with and without stuck breakers gives the failure modes and effects analysis shown in Table 5.

Sets of indices similar to those shown in Table 5 can be computed for all busbars. The overall indices for each busbar cannot be used directly to determine the output reduction of the unit because, as discussed previously, failure events may be common to more than one busbar. Therefore those that are unit minimal cuts must be detected from the busbar minimal cuts. Of the 183 busbar minimal cuts only 65 are unit minimal cuts, i.e. cuts that cause the output power to be reduced. These unit minimal cuts are shown in Table 6.

Having deduced the unit minimal cuts shown in Table 6, the percentage output reduction that is caused by these events must be evaluated. If the event causes failure of only one busbar then that event creates an output reduction equivalent to the effect of that busbar. If the event however causes failure of more than one busbar then the output reduction is dependent on the combined effect of those busbars. Using this technique, the busbar consequences shown in Table 2 and accumulating the appropriate indices gives the reliability indices shown in Table 7 for each feasible state. The details of the six failure events that contribute to the 30% derated state are shown in Table 8.

From Table 7 it is evident that some derated states defined in Table 2, i.e. 20%, 40%, and 50%, do not appear as feasible derated states. The reason for this is that the failure events causing the loss of the appropriate busbar(s) also cause the loss of other busbar(s). Consequently the derated state is greater than the above values and the relevant events are absorbed in either the 30%, 45%, or 100% derated states.

The indices shown in Table 7 enable different system designs to be compared easily and enable other risk indices to be evaluated. One such index is the expected energy lost each year due to failure of the

Table 5 - Reliability indices associated with busbar 13

failure event	λ f/yr	r hr	U hr/yr	λ'' o/yr	r'' hr	U'' hr/yr
1 out	0.20001×10^{-1}	1.0038	0.20077×10^{-1}	-	-	-
2 out	0.20053×10^{-1}	1.1597	0.23255×10^{-1}	-	-	-
3 out	0.12001×10^{-1}	1.0038	0.12046×10^{-1}	-	-	-
4 out	0.50007×10^{-2}	1.1513	0.57575×10^{-2}	-	-	-
13 out	0.50000×10^{-2}	10.00	0.50000×10^{-1}	-	-	-
10 + 14 out	0.65759×10^{-6}	1.0874	0.71507×10^{-6}	0.13703×10^{-4}	0.95727	0.13117×10^{-4}
10 + 15 out	0.22469×10^{-4}	1.2739	0.28624×10^{-4}	0.38830×10^{-4}	0.97645	0.37915×10^{-4}
10 + 16 out	0.54799×10^{-6}	1.0874	0.59589×10^{-6}	0.12561×10^{-4}	0.95727	0.12024×10^{-4}
11 + 14 out	0.22469×10^{-4}	1.8739	0.28624×10^{-4}	0.38830×10^{-4}	0.97645	0.37915×10^{-4}
11 + 15 out	0.73300×10^{-4}	6.9346	0.50830×10^{-3}	0.91423×10^{-4}	0.99571	0.91032×10^{-4}
11 + 16 out	0.18724×10^{-4}	1.2739	0.23853×10^{-4}	0.34262×10^{-4}	0.97567	0.33428×10^{-4}
12 + 14 out	0.54799×10^{-6}	1.0874	0.59589×10^{-6}	0.12561×10^{-4}	0.95727	0.12024×10^{-4}
12 + 15 out	0.18724×10^{-4}	1.2739	0.23853×10^{-4}	0.34262×10^{-4}	0.97567	0.33428×10^{-4}
12 + 16 out	0.45666×10^{-6}	1.0874	0.49657×10^{-6}	0.11416×10^{-4}	0.95727	0.10931×10^{-4}
10 active	0.800×10^{-2}	1.0	0.800×10^{-2}	-	-	-
12 active	0.700×10^{-2}	1.0	0.700×10^{-2}	-	-	-
14 active	0.800×10^{-2}	1.0	0.800×10^{-2}	-	-	-
16 active	0.700×10^{-2}	1.0	0.700×10^{-2}	-	-	-
21 active	0.700×10^{-2}	1.0	0.700×10^{-2}	-	-	-
25 active	0.700×10^{-2}	1.0	0.700×10^{-2}	-	-	-
39 active	0.480×10^{-1}	1.0	0.480×10^{-1}	-	-	-
43 active	0.980×10^{-1}	1.0	0.980×10^{-1}	-	-	-
11 active + 10 stuck	0.150×10^{-4}	1.0	0.150×10^{-4}	-	-	-
11 active + 12 stuck	0.150×10^{-4}	1.0	0.150×10^{-4}	-	-	-
15 active + 14 stuck	0.150×10^{-4}	1.0	0.150×10^{-4}	-	-	-
15 active + 16 stuck	0.150×10^{-4}	1.0	0.150×10^{-4}	-	-	-
22 active + 21 stuck	0.150×10^{-4}	1.0	0.150×10^{-4}	-	-	-
23 active + 21 stuck	0.700×10^{-5}	1.0	0.700×10^{-5}	-	-	-
26 active + 25 stuck	0.150×10^{-4}	1.0	0.150×10^{-4}	-	-	-
27 active + 25 stuck	0.700×10^{-5}	1.0	0.700×10^{-5}	-	-	-
40 active + 39 stuck	0.240×10^{-3}	1.0	0.240×10^{-3}	-	-	-
44 active + 43 stuck	0.280×10^{-3}	1.0	0.280×10^{-3}	-	-	-
total	0.25283	1.1959	0.30237	0.28783×10^{-3}	0.97904	0.28181×10^{-3}

Table 6 - Failure events causing power reduction of unit

passive outages restored by repair	passive outages restored by n/o paths	active failures		active failures plus stuck breaker	
4	1, 2, 3, 5, 6, 7, 17	10	41	11 + 10,	15 + 16
8	18, 19, 21, 22, 23, 25	14	43	15 + 14,	18 + 17
13	26, 27, 34, 35, 36,	16	45	19 + 17,	30 + 29
20	10 + 14, 10 + 15,	17	47	31 + 29,	35 + 34
24	10 + 16, 11 + 14,	29	51	36 + 34,	40 + 39
28	11 + 15, 11 + 16,	34	53	42 + 41,	44 + 43
37	12 + 14, 12 + 15,	39		46 + 45,	48 + 47
	12 + 16	12		52 + 51,	54 + 53
				11 + 12,	
7	27	14		17	

Table 7 - Encounter rate and duration of unit states

derated state %	expected encounter rate/yr	average duration hr	annual residence time, hr/yr	number of events
30	0.24901	2.161	0.53812	6
45	0.81159×10^{-1}	4.1927	0.34027	12
100	1.2260	1.7640	2.1627	45

Table 8 - Events causing 30% derated states

failure event	expected failure rate, f/yr	average duration hr	annual outage time, hr/yr
37 out	0.50×10^{-2}	10	0.50×10^{-1}
34 out, n/o path closed	0.30016×10^{-2}	1.9445	0.58366×10^{-2}
35 out, n/o path closed	0.20113×10^{-1}	12.855	0.25855
36 out, n/o path closed	0.99946×10^{-2}	1.2836	0.12830×10^{-1}
53 active	0.210	1.0	0.210
54 active plus 53 stuck	0.90×10^{-3}	1.0	0.90×10^{-3}
total	0.24901	2.161	0.53812

auxiliary system. As an example consider the system shown in Fig. 2 being the auxiliary supply of a 500 MW unit. The expected energy lost is then shown in Table 9.

From Table 9 it is evident that the single value

Table 9 - Expected energy lost with a 500 MW unit

derated state, %	capacity out, MW	annual residence time, hr/yr	MWhr lost/yr
30	150	0.53812	80.718
45	225	0.34027	76.561
100	500	2.1627	1081.350
			1238.629

of 1238 MWhr lost per year represents one method of defining the "quality" of the auxiliary system and is suitable for comparing systems that may be vastly different in concept and design.

CONCLUSIONS

To associate a cost to the reliability of a proposed design is one of the main objectives of long term reliability or availability studies. This paper has proposed techniques that assist these objectives to be fulfilled with regard to the specific area of electrical auxiliary systems of power stations. The concept of system minimal cut sets derived from nodal minimal cut sets which leads to the evaluation of the reliability indices of the unit derated states is very powerful and can be applied to many different problems.

The computer program that has been developed to incorporate these techniques is very efficient and requires only 23.5 kwords for a system having 100 elements and up to 1000 failure events. Furthermore the computing time to analyse the system considered in this paper was 2.9 sec on a CDC7600. A description of this program and its application to further practical studies such as power stations with more than one unit will be the subject of future publications.

ACKNOWLEDGEMENTS

The authors would like to thank Professor C.B. Cooper of the Department of Electrical Engineering and Electronics, UMIST for providing the facilities necessary to conduct these studies, the Central Electricity Generating Board for their financial assistance and particularly Mr. J.A. Chambers, Mr. A. Kozlowski and Mr. A. Mackrell of the CEGB Generation Construction and Development Division for their keen interest and invaluable assistance. Mr. De Oliveira is also indebted to INVOTAN, Lisbon for providing him with financial support.

APPENDIX

Assumptions

The loss of any busbar causes one of the following consequences:

(a) 100% reduction of output power
(b) x% (0 < x < 100) reduction of output power
(c) no reduction of output power

The loss of more than two busbars causes 100% reduction of output power.

The loss of any combination of two busbars (from those defined by b) causes a known loss of output power.

Identifying a Minimal Cut Set

Each cut set is defined by a set of six values as follows:

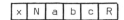

where

x defines the type of cut

N defines the busbar to which the cut is attached

a,b,c are the components of the cut and b and/or c may be zero

R is the reduction of power output of the unit due to the cut

The value of x may be:

-1: a cut for which every component is out but service is restored by a repair action

-2: a cut for which every component is out but service may be restored by closing a normally open breaker

-3: a cut for which component a is actively failed and all other components are out. Service is restored by isolating component a.

k: a cut as defined by -3 but one in which breaker k is stuck

Algorithm

Consider the two minimal cuts:

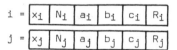

Let I and J be a set of three numbers defined as:

$$I = \{a_i, b_i, c_i\}, \quad J = \{a_j, b_j, c_j\}$$

n_i = number of non-zero elements of I

n_j = number of non-zero elements of J

n_m = minimum of n_i and n_j

Define the set $A = I \cap J$ and let n_a be the number of non-zero elements of A, then $n_a \leqslant n_m$.

Let $R(N_i)$ and $R(N_j)$ be the output power lost when busbars N_i and N_j are lost respectively.

Let $\ell=1$ when $0 < R(N_i) < 100$ and $0 < R(N_j) < 100$ and in this case let $R(N_i,N_j)$ represent the output power lost when both busbars N_i and N_j are lost. Otherwise let $\ell = 0$.

The algorithm then proceeds as follows:

(i) If $R(N_i) = 0$ neglect i and if $R_j = 0$ set $R_j = R(N_j)$

if $R(N_j) = 0$ neglect j and if $R_i = 0$ set $(R_i = R(N_i)$

(ii) if $N_i = N_j$ the cuts are independent. The values of R_i and R_j are as in (i)

(iii) if $n_a < n_m$ the cuts are independent as in (ii).

If $N_i \neq N_j$ and $n_a = n_m$ the algorithm proceeds as follows:

(iv) $x_i = x_j = x$

(a) $x = -1$ or -2 <u>and</u> $n_i = n_j$

Reject the cut associated with the busbar causing minimum power loss, i.e. if $R(N_i) < R(N_j)$ reject cut i.

If $\ell = 0$ set $R_j = -100$ if i is rejected and set $R_i = -100$ if j is rejected.

If $\ell = 1$ and j has been rejected then, if R_i or $R_j < 0$ set $R_i = -100$ but if R_i <u>and</u> $R_j \geqslant 0$ set $R_i = -R(N_i, N_j)$.

Similarly if i has been rejected the value of R_j is suitably set.

(b) $x = -1$ or -2 <u>and</u> $n_i \neq n_j$

Reject the cut associated with the busbar causing the minimum loss of energy per year (product of output power lost and average annual outage time).

If $\ell = 0$ or $\ell = 1$ use appropriate conditional statement in (a) above.

(c) $x \neq -1$ or -2

If $a_i = a_j$ <u>and</u> $n_i = n_j$: as in case (a)

If $a_i = a_j$ <u>and</u> $n_i \neq n_j$: as in case (b)

If $a_i \neq a_j$ then the cuts are considered independent

The values of R_i and R_j are set as in (i) above.

(v) $x_i \neq x_j$

(a)   if $n_i \neq n_j$ then the cuts are considered independent and the values of R_i and R_j are set as in (i) above.

(b)   if $n_i = n_j$, a decision is taken according to the following table

		-1	-2	-3	$k_i \rightarrow x_i$
	-1	-	i	i	i
	-2	j	-	i	i
	-3	j	j	-	i
x_j	k_j	j	j	j	(X)

If $\ell = 0$, the cut indicated in the table is eliminated and the reduction of power of the remaining cut is set to 100% but if reduction of power of the remaining cut is x%, proceed as for $\ell = 1$.

If $\ell = 1$ (in the following assume that cut i is indicated; a similar reasoning exists if it is cut j), then

R_i is set to -100 if previously either R_i or R_j is negative

R_i is set to $-R(N_i, N_j)$ if previously R_i and R_j are not negative

R_j is set to -100 if its previous value is negative

R_j is set to $R(N_j)$ if its previous value is not negative

The average annual outage time of cut j is decreased by the average annual outage time of cut i.

(X) - in this case the cuts are considered

independent and the values of R_i and R_j are set as in (i) above.

(vi) Scan all the cuts not rejected in the above logic and make $R_i = |R_i|$. Combine all the cuts that lead to the same reduction of output power of the unit.

REFERENCES

(1)   M.S. Grover and R. Billinton, "A computerised approach to substation reliability evaluation", IEEE Trans. Power Apparatus and Systems, Vol. PAS-93, 1974, pp. 1488-1497.

(2)   R. Billinton and M.S. Grover, "A sequential method for reliability analysis of distribution and transmission systems", Proceedings 1975 Annual Reliability and Maintainability Symposium, pp. 460-469, January 1975.

(3)   R. Billinton and R.J. Ringlee, "Models and techniques for evaluating power plant auxiliary equipment reliability", 1971 Power Generation Conference St. Louis, IEEE paper 71 CP702-PWR.

Discussion

M. S. Grover (Ontario Hydro, Toronto, Canada): We congratulate the authors for presenting a very interesting and useful paper. The operation of a generating unit is vitally dependent on the availability of its associated auxiliaries. This paper has presented some algorithms for quantitatively evaluating the effect of failures in auxiliary electrical system on the reliability of a generating unit. We would like to note that the performance of auxiliary electrical system (and of other systems) depends on the reliability and maintainability (R&M) characteristics of its constituents and the configuration in which they are connected. The authors have not illustrated these aspects. Perhaps the authors may like to comment on this.

In regard to the other material presented in the paper, we offer the following comments and questions.

1. The cost of unreliability is mentioned in the paper but no cost estimates have been associated with the calculated reliability indices. The authors may like to explain why this was not done. How is the cost of unreliability expressed? Is it expressed in terms of MWhr not supplied or 1% increase in Forced Outage Rate (FOR) of a particular unit? We, in Ontario Hydro, use the latter method and consider cost of additional reserves required and energy replacement costs.

2. The results given for the MWhr/yr not supplied by a generating unit (due to the failures in its auxiliary electrical system) are based on the assumption that the auxiliary electrical system is required to operate throughout the year. This may not be true when the generating unit is on a planned outage, on a reserve shutdown or on a forced outage due to failures in other auxiliary systems. The actual values of MWhr/yr not supplied due to the failures in auxiliary electrical system will be, therefore, less than those given in the paper.

3. The authors have considered the case of a single unit station. For a multi-unit station, the MWhrs not supplied cannot be calculated by multiplying the single unit values by the number generating units in the station. Some of the buses can be supplied from the station common bus or the buses of other units. For this reason, the whole auxiliary electrical system of the station will have to be considered for a reliability analysis. This may require a very large computer storage with the procedures described in the paper.

4. It is to be noted that in those cases where many single contingency events exist, a very little effort is required to evaluate the deratings or outages of a generating unit caused by the failures of its auxiliary system. This is evident from tables in the paper. We have used a very simple manual approach in our work on evaluating the effect of reliability performance of 4.16 kV in-station breakers on the reliability of a 4x500MW fossil station[1]. It was concluded that considerable savings in capitalized costs can be obtained if breaker failure rates were decreased by an order of magnitude.

We fully endorse the viewpoint of the authors that there is much to be gained by cost-benefit analysis of auxiliary equipment in stations. The capitalized cost of 1% increase in FOR of a 500MW nuclear unit at present is of the order of $10M. This cost figure clearly shows the magnitude of savings to be obtained by improvement in reliability of various systems and subsystems associated with a unit. It also provides a designer with a dollar figure which can be used in evaluating the worth of improvement of various equipment and systems.

In conclusion, we again compliment the authors for a useful contribution.

REFERENCE

[1] Grover, M. S. and M. T. G. Gillespie, "A Cost Oriented Reliability and Maintainability Assessment of Circuit Breakers in the Design of Auxiliary Electrical Power Systems of a Generating Station", Proceedings of the IEEE/ASQC Third Reliability Engineering Conference for the Electric Power Industry, September 1976.

Manuscript received February 24, 1977.

R. N. Allan, R. Billinton, and M. F. De Oliveira: We would first like to thank Mr. Grover for his kind remarks and for the very pertinent comments he has made.

Mr. Grover is quite correct in noting that the performance of the auxiliary electrical system is dependent on the R & M characteristics of the components and their configuration. Although we have considered only one set of characteristics and one configuration, the importance of these techniques is that they permit a wide range of sensitivity studies[1] to be performed by varying the R & M characteristics and thus the component indices used and by varying the configuration of the components. The ability to assess objectively the impact of both aspects by evaluating one overall index is considered of great benefit when comparing different systems and characteristics, all of which may produce very different sets of busbar indices and derated states.

We deliberately did not extend the overall index (MWhr not supplied) into one of cost since, at present, there is no universally accepted method for costing the unreliability, although the one used by Ontario Hydro seems to be an important and practical measure. We considered it more appropriate in the discussion of the techniques however to limit the analysis to evaluating the overall unreliability index which can be subsequently translated into one of cost depending on the policy and philosophy of individual utilities.

Mr. Grover is again correct in noting that the MWhr not supplied per calendar year may be less than the values quoted in the paper due to the reasons he quotes. However the results should be interpreted as the MWhr not supplied per year of operation and therefore they should be weighted by the proportion of a year that the unit is required to operate in order to evaluate the MWhr not supplied/calendar year.

Although the techniques considered in the paper relate to a single unit, the authors have extended these to consider multi-unit stations. These additional techniques permit stations having up to 5 units or 4 units plus a common station supply to be evaluated and include the assessment of busbars fed from common station supplies or from busbars of other units. Although computational effort clearly will increase both the storage requirements and computational time remain within very practical limits. The CEGB in the UK are currently using a program based on these techniques for evaluating multi-unit stations and it is anticipated that this will be the subject of a future paper.

We were very pleased to note that Ontario Hydro have made similar, albeit simpler, analyses and have found that considerable savings can be obtained. This is a view which we endorse and justifies the development of realistic reliability evaluation techniques that permit utilities to perform cost-benefit analyses.

REFERENCES

[1] Allan, R. N., Billinton, R. and De Oliveira, M. F., "Evaluating the reliability of auxiliary electrical systems of generating stations". Canadian Electrical Association, Thermal Power Section, Spring Meeting, Toronto 1976.

Manuscript received April 5, 1977.

MODELLING AND EVALUATING THE RELIABILITY
OF DISTRIBUTION SYSTEMS..

R.N. Allan, Member, IEEE
UMIST
Manchester
England

E.N. Dialynas

I.R. Homer
Electricity Council
London
England

ABSTRACT

Comparatively, distribution systems have received least consideration in the reliability evaluation of power systems although paradoxically they have the greatest average effect on loss of supply to the consumers. This paper discusses the inherent features of typical distribution systems and describes and evaluates models that more truly reflect the operational behaviour of such systems and their effect on the consumers. The analysis of a typical system is included which illustrates the additional and more meaningful results that can be achieved from the use of these improved system models.

INTRODUCTION

Reliability evaluation of power systems has been the subject of many recent papers[1] and the modelling and evaluation techniques have improved considerably. A continuing need to develop the techniques still remains however because complete system behaviour and operation are often not truly reflected in the techniques. This is usually due to an incomplete understanding of the complexities of system random behaviour or the need to incorporate various levels of approximations in order to make the evaluation process possible.

In the UK, the power system is divided into generation, primary transmission (400 kV), subtransmission (275 kV) and distribution (132 kV down to 11 kV). It is known that the distribution system accounts for about 99% of the consumers' unavailability although in the past, the emphasis has generally been directed towards the generation and transmission sections of the system. This focus of attention still continues with considerable emphasis being devoted to solving the bulk (composite generation/transmission) system. Although it is known that this part of the system contributes very little to the average load-point indices, the emphasis has been created due to the possible total collapse of the system that can be caused by certain system states in the generation and transmission area. The ability to assess these conditions have added considerable impetus to reliability evaluation.

Total collapse of the system due to events in the distribution system are most unlikely and the main purpose of evaluating the reliability of such systems is to deduce the consumers' load-point indices and the cost of reinforcing the load-points. Neglecting the transmission and generation in this analysis will create only very small imprecision in the results. However relatively little has been done in this area with the exception of a few papers [2-6] in which three basic

indices, expected failure rate (λ), average outage time (r) and expected annual outage time (U) have been evaluated using a criterion of total loss of continuity. It is now known that these indices are inadequate for a full understanding of the system performance and the criterion is adequate only for simple radial systems. To realistically assess all parts of the distribution system (132 kV-11 kV) it is necessary to extend the indices to include average load disconnected (L) and the average energy not supplied (E). This problem area has been discussed conceptionally[7], the application to distribution systems has been documented[8] and several computational techniques have been described[9].

The purpose of this paper is to discuss the modelling and evaluation techniques that are necessary to translate the conceptual ideas into practical analysis. These modelling and evaluation techniques are derived from a consideration of the relevant state-space diagrams and a Markov-approach solution to give a set of approximate equations that can be used with the established failure-mode-analysis or minimal-cut-set-analysis conventionally used for the reliability evaluation of power system networks. It is quite possible to apply these equations to simple systems using hand calculations although for more complex networks the use of the computational techniques discussed[9] previously would be required.

PRACTICAL DISTRIBUTION SYSTEMS

The hypothetical 33/11kV system shown in Fig.1 is based on a real UK distribution system. It illustrates the complexity of likely features in a system and serves as a useful example to demonstrate the modelling and evaluation techniques.

Both the 33kV network (superior voltage and the 11kV network (inferior voltage) are shown, the main features being:

a) each 11kV feeder has a number of load-points distributed along its length

b) many of the 11kV feeders are connected through normally-open points to other feeders fed from the same or other sources

c) loads may be transferred from one substation to another by disconnecting feeders at their normal source end and closing one or more normally-open points. This is done only in the event of a failure causing loss of supply at the source substation and it is not normal practice to parallel sources

d) a failure such as in (c) may cause total loss of continuity in which case all loads are disconnected until either repair is completed or load is transferred or it may cause partial loss of continuity because of unsatisfactory system conditions. In the latter case, only sufficient loads necessary to regain satisfactory system conditions are disconnected and these remain disconnected until system conditions recover, repair is completed. or load is transferred. The authors only consider overloads in this assessment since maintenance of statutory voltage limits is not as important during emergency conditions although this and other system conditions can easily be incorporated

F 79 276-7   A paper recommended and approved by the IEEE Power System Engineering Committee of the IEEE Power Engineering Society for presentation at the IEEE PES Winter Meeting, New York, NY, February 4-9, 1979. Manuscript submitted August 2, 1978; made available for printing November 14, 1978.

e) several alternative switching actions may exist to transfer the load of any given feeder.

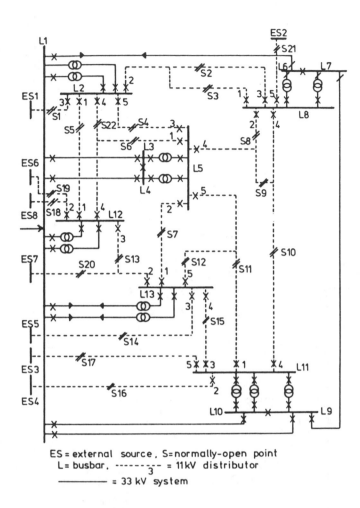

ES = external source, S = normally-open point
L = busbar, --------3-------- = 11kV distributor
———————— = 33 kV system

Fig.1 Hypothetical 33/11kV system

From consideration of these features, it is evident that total loss of continuity is an adequate criterion only for single feeder radial networks capable of supplying the maximum demand. Hence this criterion has limited application. The presently available techniques for evaluating distribution systems are limited to a total loss of continuity criterion and therefore the models and evaluation techniques presented in the subsequent sections enable a more realistic assessment of systems to be made.

MODELLING TECHNIQUES

The analysis of a distribution system that includes the features described previously can involve four types of assessment:

a) total loss of continuity without transferable load
b) partial loss of continuity without transferable load
c) total loss of continuity with transferable load
d) partial loss of continuity with transferable load

The modelling techniques required for each of these types are described in the following sections.

Total Loss of Continuity Without Transferable Loads

The models used in this assessment for evaluating

λ, r and U are identical to those published previously[2-6] and are best evaluated using minimal-cut-set theory. In this assessment the established effects of passive, active and temporary failures, stuck breakers, maintenance, repair, replacement, isolation, switching, reclosing and normal and adverse weather can be taken into account without any changes to the models.

The value of L and E are easily evaluated since L is the average load connected to the load-point and E is the product of L and U.

Partial Loss of Continuity Without Transferable Loads

In this assessment it is necessary to simulate branch and busbar outages and to determine whether any of the remaining system components become overloaded. This normally requires the application of a load flow routine and the authors currently use an ac fast-decoupled load flow program. If any of the remaining system components do become overloaded it is necessary to evaluate the maximum busbar loadings that can be allowed on the system to prevent this overloading; the excess load being shed by the disconnection of an appropriate number of feeders. If this excess load is shed at a load-point of interest, then the outage condition being simulated causes partial loss of continuity at that load-point.

There are a number of ways to accomodate this assessment and the methods chosen by the authors have been published previously[9]. The outage conditions that are simulated are all first order and a restricted number of second order branch and busbar outages. The second order branch and busbar outages are limited to all second order combinations of each third order minimal cut set obtained for the load-point of interest. If any of these outages cause an overload in the system, load is disconnected at the receiving end of the overloaded branch. If this proves insufficient, load is then disconnected at the receiving end of a branch whose sending end is the receiving end of the overloaded branch.

The reliability indices of each load-point of interest due to this type of event can be evaluated from the relevant state-space diagram shown in **Fig.2**, in which it is assumed that the system can withstand all load demands without the outage condition, λ_e is the rate of occurrence of the outage condition, μ_e is the reciprocal of the average duration of the outage condition (r_e), L_s is the maximum load that can be supplied to the load-point of interest during the outage condition, P is the probability of the load being greater than L_s, λ_H is the transition rate from load levels greater than L_s to load levels equal to or less than L_s and λ_L is the transition rate between load levels equal to or less than L_s to load levels greater than L_s.

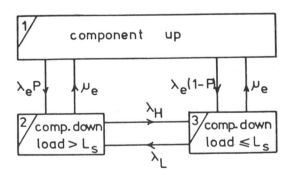

Fig.2 State Space diagram including load transitions

The down-state or failure-state of the load-point

will be state 2 in which the outage condition and a load level greater than L_S has occurred. The steady-state probability of being in this state can be evaluated by constructing the stochastic transitional probability matrix and solving the steady-state probability equations. This gives :

$$P \text{ (down)} = \frac{\lambda_e P (\mu_e + \lambda_L) + \lambda_e (1-P) \lambda_L}{(\lambda_e + \mu_e)(\mu_e + \lambda_L + \lambda_H)} \qquad 1.$$

Letting $r_H = \dfrac{1}{\lambda_H}$ = average duration of load level $> L_S$

$\qquad r_L = \dfrac{1}{\lambda_L}$ = average duration of load level $\leqslant L_S$

$\qquad\qquad = r_H (1-P)/P$

and assuming $(\lambda_e + \mu_e) \approx \mu_e$

and $\quad \mu_e (\mu_e + \lambda_L + \lambda_H) \approx (\mu_e + \lambda_L)(\mu_e + \lambda_H)$

reduces equation 1 to

$$P \text{ (down)} = \left[\lambda_e P + \lambda_e (1-P) \lambda_L \cdot \frac{r_e \cdot r_L}{r_e + r_L} \right] \frac{r_e \cdot r_H}{r_e + r_H} \quad 2.$$

The average duration of this down state r is given by:

$$\frac{1}{r} = \mu_e + \lambda_H$$

$$= \frac{1}{r_e} + \frac{1}{r_H}$$

$$\therefore \quad r = \frac{r_e \cdot r_H}{r_e + r_H} \qquad 3.$$

Equation 2 applies to the operating situation in which the excess load is connected and disconnected each time a load transition between states 2 and 3 occur, i.e. both λ_H and λ_L are non-zero values. An alternative operating strategy and one frequently used in practise to prevent many manual load transfer operations, is that once disconnected, the load remains disconnected until repair is completed. In this case $\lambda_H = 0$ and equation 2 reduces to :

$$P \text{ (down)} = \left[\lambda_e P + \lambda_e (1-P) \lambda_L \frac{r_e \cdot r_L}{r_e + r_L} \right] r_e \quad 4.$$

and from equation 3, the average duration of this down state is

$$r = r_e \qquad 5.$$

Since the probability of occurrence of a failure event is approximately equal to the product of its rate of occurrence and its average duration, the rate of occurrence λ of the down state in both of the above cases is :

$$\lambda = \lambda_e P + \lambda_e (1-P) \lambda_L \frac{r_e \cdot r_L}{r_e + r_L} \qquad 6.$$

Equation 6 is similar but not identical to that

published[2,6] previously for use with the minimal cut set approach. It reduces to that published previously if, and only if, $r_L \gg r_e$ whence :

$$\lambda = \lambda_e P + \lambda_e (1-P) \lambda_L r_e \qquad 7.$$

The method used to calculate the values of λ_e and r_e in equations 2–7 will depend upon the order of the outage event. If it is a first order outage event then the indices λ_e and r_e are those of the component itself. If it is a second or higher order event then the indices λ_e and r_e are equivalent indices obtained from the conventional equations[2-6] for second or higher order overlapping failure events which may include the effect of maintenance. The effect of normal and adverse weather may also be included for any order of outage event.

In the case of partial loss of continuity, the average load disconnected is not the average load connected to the load-point but is the average load in excess of the maximum level L_S that can be sustained. This can be obtained by evaluating the area under the load duration curve for load levels greater than L_S to give the energy that cannot be supplied given the outage condition and dividing by the time for which load L_S is exceeded. Essentially this is an application of conditional probability and the details are shown in Fig.3.

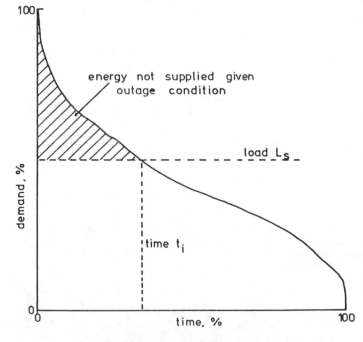

Fig.3 Load duration curve for evaluating L and E

Consequently, the complete set of expressions that give the five reliability indices (λ,r,U,L,E) for each partial loss of continuity event of the load-point of interest are :

$$\lambda = \lambda_e \lambda_L (1-P) \frac{r_e \cdot r_L}{r_e + r_L} + \lambda_e P \qquad 8a$$

$r = r_e$ if excess load remains disconnected   8b
   until repair is completed

$r = \dfrac{r_e \cdot r_H}{r_e + r_H}$ if load is disconnected and   8c
   reconnected each time a load
   transition occurs

$$U = \lambda r \qquad\qquad 8d$$

$$L = \left[\int_0^{t_1} L(t)dt - L_s t_1 \right] / t_1 \qquad 8e$$

$$E = LU \qquad\qquad 8f$$

where $L(t)$ is the load duration curve and t_1 is the time for which load level $> L_s$.

If a number of such events exist which is likely in most systems, the overall partial loss of continuity indices for the load-point can be evaluated using the principle of series events such that :

$$\lambda_p = \Sigma\lambda \qquad\qquad 9a$$

$$U_p = \Sigma U \qquad\qquad 9b$$

$$r_p = U_p / \lambda_p \qquad\qquad 9c$$

$$E_p = \Sigma E \qquad\qquad 9d$$

$$L_p = E_p / U_p \qquad\qquad 9e$$

Loss of Continuity with Transferable Loads

If a system has transferable load facilities at the inferior voltage level, it is necessary to evaluate the impact of these on the load-point reliability indices in order for a complete assessment to be made. In the event of a failure event in the primary voltage system due to either total or partial loss of continuity, it may be possible to recover some or all of the disconnected load by transferring the disconnected feeders to other sources. Therefore, although the previous values of λ, r and U may still be correct, the values of L and E could be in considerable error if these facilities are ignored.

Without transferable facilities, all the inferior voltage loads can be lumped together and considered as a single point load connected to the load-point of interest. In the case of transferable facilities, this cannot be done and instead the individual inferior voltage feeders must be identified so that the sequence of transfer actions and the load and energy recovered following each transfer action can be modelled.

The state-space diagram for transferable load switching associated with either total or partial loss of continuity events is shown in Fig.4. State 1 represents the system in its normal operating state when

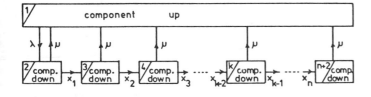

Fig.4 State space diagram including transfer switching states

it is assumed that all load demands can be supplied. State 2 represents the system after the initiating failure in the primary voltage system has occurred and all inferior voltage feeders have been disconnected (total loss of continuity) or sufficient inferior voltage feeders have been disconnected to relieve network overloads (partial loss of continuity). State 3 represents the system after the first transfer operation

has occurred and state k after the (k-2) transfer operation has occurred.

In the case of total loss of continuity, Fig.4 is a simple extension of the well-established state-space diagram for such an event in which λ is the rate of occurrence and μ is the reciprocal of the average duration of the event. Both can be evaluated from the standard equations for single or overlapping component failures [2-6]. In the case of partial loss of continuity, Fig.4 is an extension of the state-space diagram shown in Fig.2, in which the transition into the DOWN state (State 2) through State 3 and the direct transition into the DOWN state from the UP state have been combined. This is a realistic combination when transferable load facilities are available since, in this case, the excess load remains disconnected from the normal source of supply until repair of the initiating event is completed, i.e. the transition rate λ_H between state 2 and state 3 in Fig.2 is zero. The value of λ is therefore given by equation 8a and the value of μ by the reciprocal of r, i.e. by equation 8b.

In both the total and partial loss of continuity cases, x_1 represents the switching rate from state 2 to state 3 of Fig.4, i.e. it is the reciprocal of the switching time taken for the first transfer operation. In general x_k is the switching rate from state (k + 1) to (k + 2). Each of these switching rates, or in practise the switching times, can be deduced knowing the importance or priority of each inferior voltage feeder and therefore the order in which each will be transferred. There are likely to be some feeders that cannot be transferred and these will remain disconnected until repair is completed and will not be represented by one of the states in Fig.4. Since all the states in Fig.4 except state 1 are associated with the failure state of the initiating event, they represent the DOWN state of the system. Consequently the transition rate into the DOWN state (failure rate of the load-point) and the average duration of the DOWN state remain the same as that if transferable load facilities were not available. The essence of the problem when transferable load facilities exist therefore, is to establish the probability of residing in each of the DOWN sub-states, the load that can be transferred and therefore the energy that cannot be supplied given residence in the sub-states and then, using a conditional probability approach, to evaluate the average load disconnected and energy not supplied during the repair time of the initiating event.

The number of states required to represent the system behaviour with transferable load facilities is dependent upon the system and the number of possible transfer operations. To produce a set of equations suitable either for computer implementation or for hand calculations that eliminates the need to solve the problem by the more difficult and time-consuming Markov analysis, it is necessary to consider the general case of n transfer actions giving (n + 2) states.

It was found, after considerable manipulation of the stochastic transitional probability matrix and the resulting steady-state probability equations, that a very convenient recursive technique could be used to evaluate the state probabilities of a system containing any number of transfer states. The number of such states to be simulated in practise would be equal to the number of such states or the number necessary to recover all the disconnected load if this is smaller. From the state-space diagram shown in Fig.4, its stochastic transitional probability matrix and associated steady-state equations, the following steady-state probabilities can be derived:

$$P_1 = \mu / (\lambda + \mu) \qquad\qquad 10a$$

$$P_2 = \lambda\mu / (\lambda + \mu)(\mu + x_1) \qquad\qquad 10b$$

439

$$P_k = \lambda\mu \prod_{i=1}^{k-2} x_i/(\lambda + \mu) \prod_{i=1}^{k-1} (\mu + x_i) \qquad 10c$$

$$\text{for } k = 3 \ldots (n + 1)$$

$$P_{n+2} = \lambda \prod_{i=1}^{n} x_i/(\lambda + \mu) \prod_{i=1}^{n} (\mu + x_i) \qquad 10d$$

$$P(\text{down}) = \sum_{i=2}^{n+2} P_i \qquad 10e$$

$$= \lambda/(\lambda + \mu)$$

It is seen that equation 10c can be written as a recursive formula such that :

$$P_k = P_{k-1} \, x_{k-2}/(\mu + x_{k-1}) \qquad 10f$$

$$\text{for } k = 3 \ldots (n + 1)$$

It follows therefore that all state probabilities are easily and quickly evaluated with the possible exception of state (n + 2). However, if desired, this can be evaluated from :

$$P_{n+2} = 1 - \sum_{i=1}^{n+1} P_i \qquad 10g$$

Having evaluated each state probability P_i using equations 10, the average load that cannot be supplied in each of the down states L_i, the total energy that is not supplied during the period of study (t), which will frequently be one year, is given by :

$$E = \sum_{i=2}^{n+2} L_i \, P_i \, t \qquad 11a$$

and the average load disconnected is :

$$L = E/U \qquad 11b$$

If a number of these events exist for a given load-point the overall average indices can be found using equations 9.

COMPUTATIONAL TECHNIQUES

There are a number of ways in which the preceding modelling techniques can be applied to distribution systems; the method chosen must clearly reflect the operating characteristics of the distribution system being evaluated and the logic used by system operators given that a failure event has occurred in the system. The computational techniques employed by the authors for implementing the models to suit the operational behaviour of UK distribution systems has been described[9] recently. These techniques are briefly summarised in the flowchart shown in Fig.5.

Loss of load caused by an active failure event during its switching time, i.e. whilst in the switching state, is not considered as a transferable load situation since it is assumed that priority will be to isolate the actively failed component and reinstate the remaining healthy components before transfer switching is commenced. The effect of the failure however is simulated whilst in the repair state during the assessment of partial loss of continuity events.

The amount of load that can be recovered during each of the transfer switching states shown in Fig.4 may be less than the maximum demand on the inferior voltage feeder being transferred due to limitations caused by the capacity of the feeder itself, of the feeders to which it is being transferred or of the new source to which it will be connected at the time switching is being simulated.

It is therefore necessary to define a loading profile for each inferior voltage feeder if loads are distributed along the feeder that represents the distribution of loads and a capacity profile of each feeder that represents the thermal rating of the feeder as a function of its length. When evaluating the load that can be recovered by the transfer operation, it is necessary to compare the appropriate loading and capacity profiles and reducing the average load that can be recovered by the transfer if there is a probability that the loading level exceeds the capacity at any point along the length of the feeder. The loads that can not be recovered due to these restrictions are those nearest to the load-point of interest and would be achieved in practise by the creation of a temporary open-point. The profiles used by the authors are discretely distributed profiles for the capacity and either discretely or uniformly distributed profiles for the loading whichever is most appropriate.

In addition it is also necessary to ensure that the transfer action is not limited by the source to which the feeder becomes connected. When this new source is a load-point within the group being analysed, i.e. is normally supplied by the bulk supply point of that group, no load can be transferred to this source if the outage event being simulated causes it to suffer either total or partial loss of continuity. When this new source is external to the group, the only limitation is its capacity to supply a required additional amount of load.

ANALYSIS OF TYPICAL SYSTEM

To illustrate the additional and more meaningful results that can be achieved from the use of the previous models, the reliability indices of load-point L5 of the system shown in Fig.1 are shown in Table 1. The data used for this system is shown in the Appendix and four different case studies were examined:

case A - total loss of continuity without transferable loads

case B - total and partial loss of continuity without transferable loads

case C - total loss of continuity with transferable loads

case D - total and partial loss of continuity with transferable loads

In this analysis, the effect of active failures were neglected since, for the reasons discussed previously, these have an identical effect on each category of failure.

A considerable number of comments and conclusions can be drawn from the results shown in Table 1 but only the most pertinent will be discussed.

If the system shown in Fig.1 had had no transferable load facilities (cases A and B), considerable errors would have occurred in all of the indices if partial loss of continuity events had been ignored. The latter criterion considers many failure events which are not identified by the total loss of continuity criterion. Therefore the failure rate, the unavailability and the energy not supplied at the load point of interest increases. The average load not supplied, however, decreases because these additional events cause loads less than the average loading level of the busbar to be disconnected. Depending on the amount by which the failure rate and the unavailability increase, the average outage time may increase or decrease.

The above comments also apply if the system had

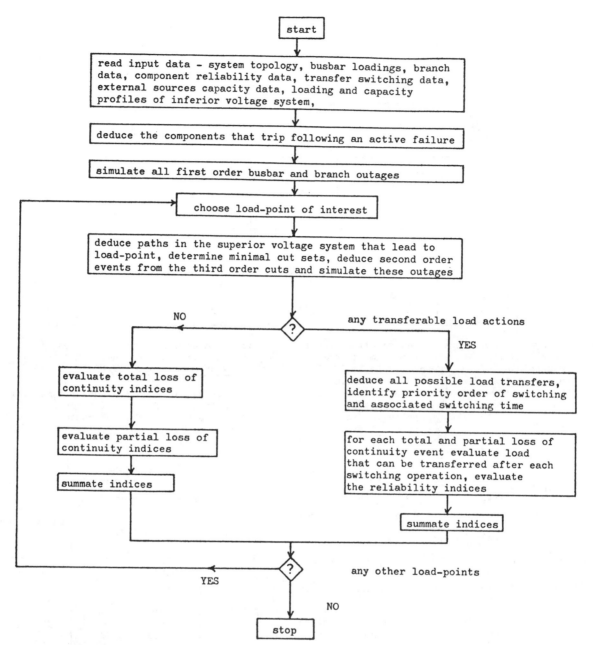

Fig.5 <u>Flowchart showing computational techniques</u>

had transferable load facilities (cases C and D). In both of the above cases for the system shown in Fig.1 and for the data used, all the load point indices of busbar 5 are dominated by the partial loss of continuity criterion and considerable increases in λ, r, U and E are found. Therefore the effect of including the partial loss of continuity criterion is substantiated and its contribution to the total indices can be very significant.

If the system shown in Fig.1 had had transferable load facilities (cases C,D) the values of λ, r and U would be identical to those if no such facilities existed (cases A,B respectively). Therefore, although such facilities do not influence these indices due to failure events occurring in the superior voltage system, they do affect the load disconnected and the energy not supplied. Since the objective of a distribution system is to satisfy load demand and supply energy, it is considered important to assess these indices if :

a) loads can be recovered due to transfer actions in the inferior voltage system

b) various reinforcement schemes are being considered to improve the reliability of a load-point since one possible cost-effective scheme is the use of transfer facilities.

<u>CONCLUSIONS</u>

The primary purpose of this paper was to describe two additional models for evaluating the reliability indices of distribution networks and to discuss the associated evaluation techniques. The main merits of these models are :-

a) their simplicity which makes them suitable not only for computer implementation but also for hand calculations

b) their efficiency in recognising real operational features of a system.

The paper also analysed a hypothetical but typical

Table 1   Reliability Indices for Load-Point 5

case	λ f/yr	r h	U h/yr	L MW	E MWh/yr
(a) total loss of continuity					
A,B	0.00033	65.0	0.021	4.063	0.0875
C,D	0.00033	65.0	0.021	0.280	0.0060
(b) partial loss of continuity					
B	0.0028	322	0.900	0.602	0.5415
D	0.0028	322	0.900	0.003	0.0026
(c) total results					
A	0.00033	65.0	0.021	4.063	0.0875
B	0.00313	294	0.921	0.682	0.6290
C	0.00033	65.0	0.021	0.280	0.0060
D	0.00313	294	0.921	0.009	0.0086

distribution system and compared the results obtained for four case studies. The execution time to analyse all six load-points of the system (L2,L5,L8,L11,L12, L13) for the longest study (case D) was only 6.65 sec. on a CDC 7600 computer. This comparison demonstrated the increased knowledge which can be gained about the reliability of a system load-point using the models described in the paper. Furthermore, they indicate that, if the techniques are not used, the reliability assessment may lead to inappropriate management decisions.

These improved techniques also enable a system planner to include a number of new reinforcement options in his reliability assessment which is not possible using previously published models and techniques.

REFERENCES

1. IEEE Sub-Committee Report, "Bibliography on the application of probability methods in power system reliability evaluation", IEEE Winter Power Meeting 1978, Paper F78 073-9

2. Billinton, R., Grover,M.S, "Reliability evaluation in transmission and distribution systems", Proc IEE, 122,1975,pp.517-523

3. Billinton,R.,Grover,M.S, "Reliability assessment of transmission and distribution schemes", IEEE Trans PAS-94,1975,pp.724-733

4. Billinton,R., Grover,M.S. "Quantitative evaluation of permanent outages in distribution systems", ibid,pp.733-742

5. Allan,R.N., Billinton,R., De Oliveira,M.F., "Reliability evaluation of electrical systems with switching actions",Proc.IEE, 123,1976,pp.325-330

6. Allan,R.N., De Oliveira,M.F., "Reliability modelling and evaluation of transmission and distribution systems", Proc IEE,124,1977,pp.535-541

7. Allan,R.N., Homer,I.R., Dialynas,E.N, "Reliability indices and reliability worth in distribution systems", EPRI Workshop on Power System Reliability Research Needs and Priorities, Asilomar,California March 1978. EPRI Pub. WS-77-60,pp.6.21-6.28.

8. Cartwright, W., Coxson, B.A.,"Reliability engineering and cost-benefit techniques for use in power system planning and design", Electricity Council Research Memorandum ECR/M966, Oct 1976

9. Allan,R.N.,Dialynas,E.N.,Homer,I.R, "Partial loss of continuity and transfer capacity in the reliability evaluation of power system networks", Power System Computational Conference, PSCC 6, 1978, Seeheim, West Germany.

APPENDIX

Table 2   Branch Data

branch (bus-bus)	R (p.u.)	X (p.u.)	rating (MVA)	length of line(cable) (km)
1-2,1-2	.04699	.19797	7	
2-5,4-5	.05403	.22304	12	
9-11	.01938	.05917	12	
1-12,1-12	.17093	.34802	12	
6-7	0	0	30	
9-10	0	0	18	
3-4	0	0	9	
1-3,1-4	.05695	.17388	18	1.96
6-8,7-8	.01335	.42110	18	
10-11,10-11	.08205	.19207	18	
1-6	.22092	.19988	36	6.04 (1.0)
7-9	.05403	.22304	18	4.52
1-9,1-10	.06701	.17103	30	4.64
1-13,1-13	.19178	.37491	12	1.96 (1.0)

Table 3   Capacity Data of External Sources

source number	capacity(MVA)
ES1	2.0
ES2	1.0
ES3	1.5
ES4	1.0
ES5	1.5
ES6	1.9
ES7	3.6
ES8	2.5

Table 4 Component Reliability Data

component	failure rate (f/yr) *	repair time (h)	maintenance rate (outs/yr)	maintenance time (h)
transformer	0.020	343	0.25	20
line **	0.025	212	0.25	8
cable **	0.032	33.5	0.25	8
hv breaker	0.005	75	0.25	8
lv breaker	0.002	18	0.25	8
isolator	0.001	75	0.25	8

** f/yr.km,   * total failure rate = active failure rate

Table 5 Busbar Loading Data

busbar no.		2	3,4,6	5	7,9,10	8	11	12	13
peak load (MVA)		10.41	3.12	12.5	5.2	5.2	6.25	12.5	12.5
max	1	2.3	–	1.9	–	1.1	1.1	2.4	2.1
demand	2	2.6	–	2.1	–	0.9	0.8	3.6	3.6
on	3	2.5	–	1.6	–	0.9	1.8	2.2	1.7
feeder	4	1.7	–	3.7	–	1.3	0.9	3.8	1.9
number	5	1.3	–	2.7	–	0.8	1.4	–	2.7

Table 6 Transfer Switching Data

normally open switch	busbar *	time (h)	busbar *	time (h)
S1	ES1	5.2	–	–
S2	L2	3.1	L8	4.2
S3	L2	4.0	L8	3.7
S4	L2	0.7	L5	0.8
S5	L2	1.0	L12	0.5
S6	L5	0.5	L12	1.5
S7	L5	1.5	L13	0.5
S8	L5	2.1	L8	1.8
S9	L5	2.7	L8	2.9
S10	L8	1.8	L11	3.5
S11	L5	3.1	L11	4.8
S12	L5	2.4	L13	3.1
S13	L12	2.2	L13	5.1
S14	ES5	3.3	–	–
S15	L13	4.1	L11	2.6
S16	ES4	0.7	–	–
S17	ES3	1.2	–	–
S18	ES8	4.3	–	–
S19	ES6	3.4	–	–
S20	ES7	2.1	–	–
S21	ES2	1.9	–	–
S22	L2	1.7	L12	1.9

* busbar to which load is transferred

Discussion

A. J. Wood and **R. J. Dinning** (South of Scotland Electricity Board, Glasgow, Scotland): The paper by Messrs. Allan, Dialynas and Homer has been examined from the point of view of future use of the model as an aid to planning of reinforcements to the distribution network. At present reliability calculations are carried out manually on single lines only using average failure rate and repair statistics. No attempt is made to evaluate reliability when considering the possibility of alternative routes of supply. Reliability thus calculated is used as a method of comparing alternative reinforcement schemes. No attempt is made to relate absolute costs of loss of load to the cost of providing reinforcement.

The Markov approach to the mathematical solution of the model through the use of failure mode analysis or minimal cut set analysis is based on sound and established principles and is deemed to be an improvement on existing methods. It is felt, however, that the underlying assumptions of these approaches require that the model be used with caution. In particular the following assumptions need some consideration.

1. *Faults are Independent and Failure Rates Remain Constant*
Clearly in a network with a high percentage of overhead transmission lines a very high percentage of faults will occur over relatively short periods when there is particularly adverse weather. It is also likely that certain parts of the network, not necessarily adjacent, will tend to be affected in a similar manner for a particular type of adverse weather, e.g. high winds from one direction, while other parts of the network will remain unaffected. Even "two weather" models may not adequately represent this apparent "interdependence" of faults. Possibly a conditional probabilistic approach might represent this more satisfactorily, i.e. an attempt could be made to estimate the probability that one line will have failed, given that another has just failed.

On overhead networks a single bolt of lightning may cause several faults to occur and could easily "bridge" an open point. This type of fault, again common in SSEB is arguably a situation where faults are inter-dependent.

The probability that fault rates remain constant is also somewhat doubtful. From experience it is known that certain areas may become "trouble spots" for limited periods of time due to vandalism or constructional activities but yet these causes can constitute a large percentage of faults in some areas.

Risk of failure will thus be dependent in some complex fashion on weather and temporary events which may not be best represented as a constant fault rate. It may fairly be argued that the model can be given data for each component in the network individually and this varied as appropriate, but the task of estimating these reliabilities and altering them to consider different possibilities could be massive. This might limit the usefulness of the model as an interactive planning tool.

2. *Failure Statistics may be Estimated from Historic Data*
There may be considerable difficulty in using historic data for failure statistics because of changing maintenance constructional and operational policy. There is always a danger that historic data will be used simply because no other data is available. The use of national averaged data for a localised area is also questionable. It may be difficult to obtain enough satisfactory data to place confidence in the results.

3. *Fault Occurrences and Loss of Load Probability may be Represented as Excepted Values*
There is likely to be a considerable spread in the distribution of fault occurrences for different types and construction of plant. The expected value approach tends to lead to the corresponding spread in reliability being ignored. It is, however, probably not realistic for a computational viewpoint to calculate it.

4. *Repair Time of Fault will be Predictable and Constant*
Clearly the time to repair a fault will be dependent on the resources available. As suggested in 1 above experience would suggest that at times of severe weather there will be a particular pressure on resources. There will thus be considerable inter-dependence on the 'time to repair' of faults. No allowance is made for the difficulties in location of a fault etc., which will mean that distribution of 'time to repair' of a fault will tend to have a large standard deviation. (See 3 above.)

Having considered some of the difficulties in the use of the model, it is wise to examine the use of the reliability statistics thus produced. Reliability is a complex issue and tolerable reliability levels will vary from consumer to consumer, e.g. the loss of supply in the early morning to domestic consumers is a much less serious event than to a hospital. Clearly the model cannot be expected to take this into consideration. There is a danger in considering loss of energy in terms of revenue and attempting to relate this directly to a proposed cost of reinforcement. In SSEB some attempts have been made to place different values on loss of energy to different consumer classes in an attempt to make some comparison possible. It is probable that the effects on the consumer of a loss of supply will require more careful consideration than the loss of revenue.

In conclusion, it seems that the model proposed is a considerable step forward in assessing reliability in distribution systems, but that there may be difficulties in its use as a day to day planning tool from the point of view of providing good input data. It is felt that because of this, and because of assumptions in the model it could not be used to compare different systems in any absolute fashion, but should be a very valuable guide to the system designer on which solutions to consider in more detail. Reliability is in our view a qualitative matter but that quantitative tools are very useful in its assessment.

Manuscript received March 19, 1979.

R. N. Allan, E. N. Dialynas, and **I. R. Homer:** We would first like to thank Messrs. Wood and Dinning for their very detailed comments and interest in the techniques we have proposed.

Most of their comments concern component reliability data and, in particular, the accuracy of such data and the interdependence between component failures and repair. Considerable work is in progress at the present time in various parts of the world to improve data collection schemes and to isolate and identify some of the problems posed by the discussors. The processes of data collection and of modelling and evaluation techniques must go hand-in-hand however in order to establish what data needs to be collected, in what form, how it should be processed and how it should be stored in data banks. It is all too easy to collect data and subsequently find that it is inappropriate to the mathematical models that can be derived and formulated. As this hand-in-hand process continues, both data collection and modelling will become more refined and many of the problems posed by the discussors will be solved.

The results we have included in the paper demonstrate clearly the impact that partial loss of continuity and transferable loads have on load point reliability indices and, therefore, indicates the need to include these effects in the analysis rather than using simple network evaluation techniques. We agree with the discussors, however, that no present-day evaluation method should be considered as a means of calculating absolute reliability. On the other hand, we are convinced that the techniques are a necessity in order to establish relative merits between one system of reinforcement and another and are, therefore, a sound and valuable guide in the objective decision-making process.

Manuscript received April 27, 1979.

MAPP BULK TRANSMISSION OUTAGE DATA
COLLECTION AND ANALYSIS

M. G. Lauby, Member K. T. Khu, Member
Mid-Continent Area Power Pool
Minneapolis, Minnesota

R. W. Polesky, Member R. E. Vandello, Member
Iowa-Illinois Gas and Electric Company
Davenport, Iowa

J. H. Doudna, Member
Nebraska Public Power District
Columbus, Nebraska

P. J. Lehman, Member
Northern States Power Company
Minneapolis, Minnesota

D. D. Klempel, Member
Basin Electric Power Cooperative
Bismarck, North Dakota

ABSTRACT - A bulk transmission line outage collection technique is described. An analysis of MAPP transmission system performance based on five years of historical outage data to yield pertinent statistics is presented.

INTRODUCTION

The benefits of bulk transmission line outage data collection and analysis stem from the potential for application in areas such as System Performance Evaluation, Operation and Maintenance Planning, System Planning et cetera. The Transmission Reliability Task Force (TRTF) of the Mid-Continent Area Power Pool (MAPP) has been collecting Bulk Transmission System outage data since 1977. Annual bulk transmission performance reports have been published within MAPP since 1980 [1]. There has been very little industry standardization of outage data analysis methods for bulk transmission lines in the past, and therefore, the annual MAPP reports have been subject to constant change. The Electric Power Research Institute (EPRI) recently provided a forum for standardizing the collection and analysis of bulk transmission outage data [2,3]. The MAPP TRTF recognizes the need for standardized collection and/or analysis methods and has adopted the guidelines suggested in [2] where applicable. The MAPP collection technique is a modified version of MAIN's (Mid-America Interpool Network) technique [4]. This paper presents MAPP bulk transmission line performance statistics for both Planned and Forced outage categories.

Categorization schemes have been suggested for bulk transmission outages in the past [4, 5] and data has been presented [5, 10, 11]; however, they did not have the benefit of the recent work by EPRI [2] or industry comments [6]. Philosophical differences become evident when comparing the various categorization schemes of any groups. For this reason, a glossary is given in Appendix A which includes certain important terms used in this paper. Review of the terms is strongly suggested for the definitions may differ from those suggested in [9].

METHODOLOGY OF BULK TRANSMISSION OUTAGE DATA COLLECTION AND ANALYSIS

The collected data consists of two types: component population and outage data.

83 SM 388-6   A paper recommended and approved by the IEEE Power System Engineering Committee of the IEEE Power Engineering Society for presentation at the IEEE/PES 1983 Summer Meeting, Los Angeles, California, July 17-22, 1983. Manuscript submitted January 28, 1983; made available for printing May 2, 1983.

o Physical characteristics of the bulk transmission line population were collected (e.g. line section and segment information, number of terminals, voltage, common ROW or common tower and their in-service date and, if any, their retirement date).

o Outage data pertaining to these lines were collected (Procedure shown in Figure 1).

The bulk transmission population data was updated as the bulk transmission system changed. Data is submitted on a monthly basis to the MAPP Coordination Center.

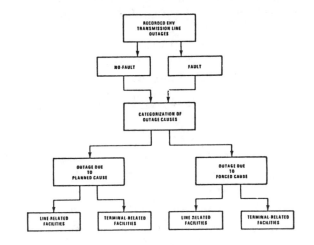

Figure 1
Categorization of Submitted Outage Data

Each record of outage data includes fault code, line identification, cause of outage code, time-in and time-out, and a remark statement. From the fault code and cause of outage code, the outage is categorized as either a Forced or Planned outage.

The categorization scheme the TRTF has adopted is presented in Figure 2.

Extensive analysis has been done under the Forced outage category. The reason for this is the recognition that Forced outages can be the most taxing on the bulk transmission system because of their randomness, unpredictability, and bunching. Multiple outages (Related outages) such as Associated and Common Mode outages have been incorporated into this scheme.

Reprinted from *IEEE Trans. Power Apparatus Syst.*, vol. PAS-103, no. 1, pp. 213–221, Jan. 1984.

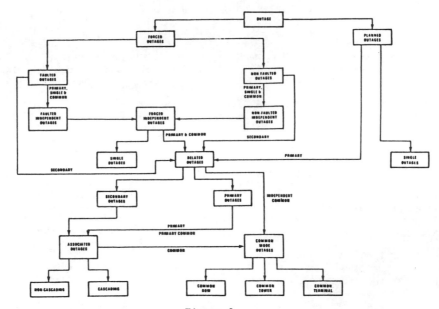

Figure 2
Categories for MAPP Bulk Transmission Outage Data Analysis

MAPP BULK TRANSMISSION SYSTEM

Table I below indicates the total number of transmission line circuit-miles, by voltage, for which outage data was reported from January 1, 1977 through December 31, 1981.

Table I
MAPP Bulk Transmission Lines
Circuit-Miles Reported

At the End of	230 kV	345 kV	500 kV	+250 kV DC	+400 kV DC
1976	6,568	2,552			
1977	6,590	2,928		465	
1978	6,807	3,235		465	426
1979	6,701	3,510	139	465	426
1980	6,769	3,909	442	465	426
1981	6,584	4,435	442	465	426

Because of conversion of some 230 kV lines to 345 kV lines, the number of 230 kV circuit-miles can be observed to decrease for some years. Specifically, total circuits-miles have dropped from previous years for 230 kV lines for two years(1979 and 1981).

ANALYSIS OF SYSTEM PERFORMANCE

Transmission outage data collected from January, 1977 through December, 1981 were studied. Some general trends can be observed from the performance indices developed from this cumulative data.

A. MTTF and MTTR Data Analysis

In Table II, the values of Mean-Time-To-Failure (MTTF) and Mean-Time-To-Repair (MTTR) are tabulated for three timespans (1977-1980, 1981, 1977-1981) and two voltage levels (230 kV and 345 kV). The timespans were selected in order that indices for the most recent year's data (1981) can be compared with indices for cumulative data for previous years. The 1981 data was then included into the cumulative indices for all the years of data collection (5 years). Bulk transmission outage data is multiply censored for it consists of failure times intermixed

with running times of the bulk transmission. System-wide indices were calculated using multiply censored data techniques. These techniques keep the data population relatively constant between years and thereby a comparison between various sets of data from different timespans can be performed [12, 13].

Table II shows that in 1981 the 230 kV line performance was better than the cumulative performance of earlier years. Conversely, because of numerous extended outages in the Spring of 1981, 345 kV lines performed less reliably than the cumulative performance of earlier years.

Since the number and frequency of Forced Secondary outages are a function of the number of Forced Independent outages, it is not surprising to find that the MTTF indices in Table II also suggest that 230 kV lines in 1981 were not subjected to Forced Secondary outages as often as they cumulatively had in previous years. Another more important reason the MTTF of the 230 kV lines increased was the installation of a 500 kV line which tied the Canadian system to the United States in a stronger fashion. This increased the availability of the 230 kV ties to Canada which were previously highly prone to Forced Secondary outages (usually 'Out of Step' outages).

The availabilities of bulk transmission lines of MAPP are displayed in Table III for the three previously described timespans.

B. Fault Data Analysis

Table IV shows fault types broken down into percentages of total line-related Faulted outages for the timespans of 1977-1980, 1981, and 1977-1981. The majority of faults in 1981 for 230 and 345 kV lines were Line-To-Ground; however, 230 kV lines experienced a large number of Line-To-Line faults probably due to conductor galloping caused by ice accumulation.

For both voltage levels there was a large number of Unknown fault types. One of the reasons for this problem was that a number of bulk transmission line terminals were not monitored by equipment which can differentiate fault types.

446

Table II

MTTF and MTTR Values for the Bulk Transmission System
(In Hours per Line Section)

VOLTAGE	TIMESPAN	OUTAGE CLASSIFICATION				
		FORCED			PLANNED	COMBINED
		INDEPENDENT	SECONDARY	BOTH		
		MTTF				
230 kV						
	1977-1980	6,135	25,000	4,923	2,056	1,437
	1981	6,800	125,560	6,451	2,955	2,016
	1977-1981	6,266	29,990	5,180	2,196	1,530
345 kV						
	1977-1980	4,586	50,042	4,199	2,466	1,535
	1981	2,831	43,313	2,655	2,644	1,304
	1977-1981	3,955	48,132	3,653	2,519	1,472
		MTTR				
230 kV						
	1977-1980	18	4	15	47	37
	1981	7	1	7	42	31
	1977-1981	16	4	14	46	36
345 kV						
	1977-1980	20	2	19	60	45
	1981	32	1	30	52	41
	1977-1981	24	1	23	58	44

Table III

Availability of MAPP Bulk Transmission Lines

VOLTAGE	TIMESPAN	AVAILABILITY FOR ALL OUTAGE CATEGORIES
230 kV		
	1977-1980	97.46%
	1981	98.48%
	1977-1981	97.68%
345 kV		
	1977-1980	97.18%
	1981	96.97%
	1977-1981	97.13%

Table IV

Percentages of Total Line-Related Faults
for MAPP Bulk Transmission Lines

VOLTAGE	TIMESPAN	FAULT TYPE				
		L-G	L-L	L-L-G	L-L-L	UNKNOWN
230 kV						
	1977-1980	69.5	17.9	1.9	1.9	8.9
	1981	37.8	34.2	0.9	0.0	27.0
	1977-1981	64.5	20.5	1.7	1.6	11.8
345 kV						
	1977-1980	67.7	3.2	0.4	4.3	24.4
	1981	70.1	5.7	0.0	2.9	21.3
	1977-1981	68.7	4.2	0.2	3.8	23.2

C. Related Outage Data Analysis

Related outages concern the loss of more than one line simultaneously and can be broken down into two types:

1) Those lines which are outaged together resulting from a single cause in which the outage of either line is not a direct consequence of the other outage (Common Mode).

2) Those lines which are outaged together which are sequential in nature where one fails from some failure or deficiency as a result of an independent outage (Associated).

In Table V, a summary of Common Mode outages in MAPP is displayed for the timespan 1977-1981. Many Common Terminal outages occured because of metering and relaying equipment failures, breaker failures, or design deficiencies. In addition, the majority of these failures in MAPP involve 345 kV facilities (69%).

Table V

Summary of Common Mode Outages for the Timespan 1977-1981

COMMON TOWER

Number of Outages	8
Average Number/100 Miles-Year	.20
Total Duration (Hours)	383.62
Average Duration (Hours)/Outage	47.95
Average Duration (Hours)/100 Miles-Year	9.48

COMMON ROW

Number of Outages	3
Average Number/100 Miles/Year	.14
Total Duration (Hours)	170.52
Average Duration (Hours)/Outage	56.84
Average Duration (Hours)/100 Miles-Year	7.79

COMMON TERMINAL

Number of Outages	25
Average Number/Year	5
Total Duration (Hours)	74.58
Average Duration (Hours)/Outage	2.98
Total Number of Affected Facilities	59
Average Number of Facilities/Outages	2.36

The second type of Related outage is the Associated outage. Table VI displays data for Associated outages. Associated outages can be caused by a latent failure on the system disclosed when a device is called to operate, line pairs sharing a common tie breaker, protection misoperation, or design deficiencies. Sometimes these outages can be initiated by a Planned outage.

Many of the Associated outages were initiated by transformer outages and therefore the transformer outage time was included to recognize the duration of the corresponding Associated outage. Since no meaningful procedure for calculation of statistics relating to transformers has been developed at MAPP, transformer outage durations were not included in the analysis of Secondary outages. Over 18.1% of all line sections on Forced outage were involved in Associated outages and of all the line sections involved in Associated outages, 71.0% were 230 kV line sections.

Cascading outages usually involve many facilities of multiple voltages. For this reason, they were displayed separately.

Table VI

Summary of Associated Outages for the Timespan 1977-81

	SUSTAINED	MOMENTARY	TOTAL
CASCADING			
Number of Outages	1	2	3
Average # Outages/Year	.20	.40	.60
Total Duration (Minutes)	7	0	7
Average Duration(Minutes)/Outage	7	0	2.33
Total Number of Facilities	5	22	27
Average # of Facilities/Outage	5	11	9
NON-CASCADING			
Number of Outages	111	61	172
Average # Outages/Year	22.20	12.20	34.40
Total Duration (Hours)	114.11	0.00	114.11
Average Duration(Hours)/Outage	1.03	0.00	0.66
Total Number of Facilities	244	141	385
Average # of Facilities/Outage	2.20	2.31	2.24

Results of general interest and which useful for bulk transmission reliability programs are compactly presented in Appendix B.

The coincident outage plots (Appendix C: Exhibits 1-3) indicate that 93.6% of the study time there was at least one bulk transmission line section out on a Planned outage. This suggests that the system was rarely fully available because of maintenance or construction. Over 41.4% of the study time at least one bulk transmission line section was on Forced outage. When all types of outages were taken into account, at least 50% of the study time five (5) or more bulk transmission line sections were not available.

D. Seasonal Performance Analysis

Data was broken down into two seasons, summer (April-October) and winter (November-March), and treated on a seasonal basis. This additional information is presented in order to further analyze the performance of bulk transmission lines in MAPP. The bulk transmission line performance during different seasons of the year is an important characteristic to recognize in order to appreciate the effect that seasonal differences in environment have on them [8, 13]. Only one timespan (1977-1981) was analyzed. Table VII contains the values of MTTF and MTTR for the two seasons.

Table VII

MTTF and MTTR for All of the Bulk Transmission Lines by Season for the Years 1977-1981 (In Hours per Line Section)

		OUTAGE CLASSIFICATION				
		FORCED			PLANNED	COMBINED
VOLTAGE	SEASON	INDEPENDENT	SECONDARY	BOTH		
		MTTF				
230 kV						
	SUMMER	5,605	25,881	4,617	1,996	1,381
	WINTER	7,337	38,400	6,159	2,466	1,749
345 kV						
	SUMMER	3,162	41,201	2,935	2,354	1,288
	WINTER	5,995	63,085	5,474	2,665	1,774
		MTTR				
230 kV						
	SUMMER	17	5	14	43	35
	WINTER	13	0	11	50	39
345 kV						
	SUMMER	28	1	26	50	39
	WINTER	13	2	12	70	51

As expected, the statistical performance of lines were poorer in the summer than winter. Responsible for this performance, could be the higher frequency of violent weather in summer (e.g. thunderstorms, tornados, etc.). Because of the high frequency of Forced Independent outages in the summer, there was a higher frequency of Related outages which is illustrated by a lower MTTF of summer Forced Secondary outages. Forced outages were more frequent in the summer than in the winter (e.g. 53.7% of the time at least one line was on Forced outage in the summer as compared to 24% of the time in winter).

Table VII also illustrates that the outage duration on bulk transmission lines experiencing a Planned outage was longer in the winter than in the summer. Planned outages extensively overlap (e.g. 98.1% of the time at least two lines were on Planned outage in the winter as compared to 90.4% in the summer). These above two points demonstrate that even though Forced outages were more frequent in the summer, the system seldom was fully available in the winter.

Another way to compare the seasonal differences is to observe the number of line sections outaged in the two seasons (see Table VIII). Since there are seven months in the summer season (April to October) and five months in winter (November to March), a normalized rate per seasonal month was calculated in order to eliminate the bias that the uneven seasonal exposure would cause. The rate of Forced outages in summer was 1.8 times higher than the rate of Forced outages in winter.

Table VIII

Summary of the Number of Sections on Forced Outage by Season Along with Rate per Seasonal Month Including Momentaries (1977-1981)

	TOTAL	RATE
SUMMER		
Number of Sections	1,675	47.9
WINTER		
Number of Sections	676	27.0

CONCLUSION

The indices presented are intended to aid in evaluation of the MAPP bulk transmission system performance. They provide comparison of the system performance during the year 1981 to the cumulative performance of the previous years. The analysis shows:

- Twenty-two percent (22%) of all line section Forced outages in MAPP for all study years (1977-1981) were Related outages. About seventy percent (70%) of the Associated outages involved 230 kV line sections. Conversely, seventy percent (70%) of the Common Mode outages involved 345 kV line sections.

- The system performance in MAPP during the summer season (April-October) was less reliable than during the winter season (November-March).

- The outage duration on bulk transmission lines in MAPP experiencing a Planned outage is longer in the winter than in the summer.

ACKNOWLEDGEMENT

The authors wish to express their thanks and appreciation to the MAPP Reliability Subcommittee for their support and suggestions. Also, a very special thanks goes to those nameless and hardworking people who collect each companies bulk transmission outage data and work closely with the authors to assure complete data. Sincere thanks goes to D. W. Bowen, Manager of the Engineering Section of the Mid-Continent Area Power Pool Coordination Center for the support and consultations. Appreciation goes to S. K. Lauby for making clear and concise flow charts. A final but very important thanks goes to K. S. Matthees who has diligently worked with the authors to insure quality data and workmanship.

REFERENCES

[1] 'Mid-continent Area Power Pool EHV Transmission System Outage Report', Since 1980 Annual, MAPP Transmission Reliability Task Force, Vol. I, II, III.

[2] 'Bulk Transmission System Component Outage Data Base', Electric Power Research Institute, EPRI EL-1797, Project 1283, April, 1981.

[3] 'Component Outage Data Analysis Methods', Electric Power Research Institute, Volume 1 & 2, EPRI EL-1980, Project 1468-1, August, 1981.

[4] G. L. Landgren, 'Transmission Line Performance of the Mid-America Interpool Network System', 1978 Reliability Engineering Conference for the Electric Power Industry Proceedings.

[5] G. L. Langren, S. W. Anderson, 'Data Base for EHV Transmission Reliability Evaluation', IEEE Trans. on Power Apparatus and Systems, Vol. PAS-100, No. 4, pp. 2046-2058, April, 1981.

[6] R. J. Ringlee, Discussion of 'Data Base for EHV Transmission Reliability Evaluation', IEEE Trans. on Power Apparatus and Systems, Vol. PAS-100, No. 4, pp. 2046-2058., April, 1981.

[7] R. Billinton, R. J. Ringlee, A. J. Wood, Power-System Reliability Calculations, MIT Press, Copyright 1973

[8] R. Billinton, Power System Reliability Evaluation, Gordon and Breach Science Publishers, Copyright 1970, Revised 1980

[9] IEEE Standard 346-1973, IEEE Standard Definitions in Power Operations Terminology Including Terms for Reporting and Analyzing Outages of Electrical Transmission and Distribution Facilities and Interruptions to Customer Service, November 2, 1973.

[10] AIEE Committee Report, 'Report of joint AIEE-EEI Subject Committee on Line Outages,' AIEE Transactions (Power Apparatus and Systems), vol. 71, pp 43-61, January 1952.

[11] IEEE-EEI Committee Report, 'Extra High Voltage Line Outages', IEEE Transactions on Power Apparatus and Systems, PAS-96, May 1967, pp 547-562

[12] W. Nelson, Applied Life Data Analysis, John Wiley and Sons, Copyright 1982

[13] R.J. Ringlee, 'Notes on Transmission Reliability Assessment', Power Technologies Incorporated, October, 1981

APPENDIX A
Glossary

AVAILABILITY: Is the fraction of time that a facility is in service during a given study period. Simply,

$$\text{Availability} = \frac{\text{Total Running Time}}{\text{Total Exposure Time}} \times 100$$

CASCADING: The uncontrolled successive loss of system elements.

COINCIDENT OUTAGE: An outage which occurs during a period in which one or more like facilities in the system are on outage.

COMMON R.O.W.: A Right-of-Way shared by two or more circuits on separate structures.

COMMON TERMINAL EQUIPMENT: Equipment which share the same terminal.

COMMON TOWER: A tower shared by two or more circuits.

EXPECTED VALUE: The average value for any random variable over an indefinite number of samplings. It is also commonly known as the mean or weighted average.

EXPOSURE TIME : The sum of the total in-service durations and the total outage durations within the study period.

FAILURE: The inability of a facility to perform its required function or functions.

FAULT: Any type of short-circuit or open-circuit which prohibits a facility from performing its required function or functions.

FAULTED OUTAGE: An outage that results from emergency conditions caused by a fault on a facility and is categorized as a Forced Outage.

FAULTED INDEPENDENT OUTAGE: Faulted outages which occur as a matter of chance and which could occur exclusive of any other disturbance or outage on the system even though they may be part of a multiple sequence of outages. When such outages occur as part of a multiple sequence of outages, they are usually the first outage in the sequence. Lightning striking a line and causing a fault which outages the line is an example of a faulted independent outage.

NON-FAULTED OUTAGE: An outage which occurs because of certain system conditions or the occurance of another outage or event. It is applied to a facility not being faulted and is categorized as a Forced Outage.

NON-FAULTED INDEPENDENT OUTAGE: Non-faulted outages which occur as a matter of chance and which could occur exclusive of any other disturbance or outage on the system even though they may be part of a multiple sequence of outages. When such outages are part of a multiple sequence of outages, they are usually the first outage in the sequence. False trips or workmen in a substation inadvertently tripping a line are examples of a non-faulted independent outage.

FORCED INDEPENDENT OUTAGE: An outage which results from a single cause but which is not a consequence of an outage of another facility.

FORCED OUTAGE: An outage that results from emergency conditions on a facility requiring that it be taken out of service immediately, either automatically or as soon as switching operations can be performed.

LINE-RELATED FACILITIES: Facilities which make up a transmission line, such as conductors, shield wires, insulators, cross-arms, structures, dampers etc...

LINE-RELATED OUTAGE: An outage as a result of the failure of a line-related facility.

MILE-YEAR: The product of the line length in miles and the line's exposure time in years.

MILE-SEASON: The product of the line length in miles and the line's exposure time in seasons (summer or winter).

MOMENTARY OUTAGE: Any outage whose duration is negligible.

MTTF: The Mean-Time-To-Failure is the duration within the cycle time during which the facility is performing its required function or functions.

$$MTTF = \frac{Total\ Running\ Time}{Number\ of\ Failures}$$

MTTR: The Mean-Time-To-Repair is the duration within the cycle time during which the facility is prohibited from performing its required function or functions.

$$MTTR = \frac{Total\ Down\ Time}{Number\ of\ Failures}$$

OPERATING ERROR: A mis-operation of a facility that causes an outage.

OUTAGE: The state of a facility in which it is prohibited from performing its intended function.

PLANNED OUTAGE: A scheduled outage that results when a facility is deliberately taken out of service at a selected time usually for purposes of construction or preventive maintenance.

RELATED OUTAGES: A multiple outage of transmission units or power transfer equipments which are directly related to each other either electrically or physically.

COMMON MODE OUTAGE: An outage which results from a single cause in which the outage of either line is not a direct consequence of the other line outage events. These can be broken down into three types: Common Right-of-Way (ROW), Common Tower, and Common Terminal (e.g. P.T. Failure).

ASSOCIATED OUTAGES: Primary and secondary outages which are sequential in nature where the primary outage is the results of an independent cause or event and where the secondary outage results from some failure or deficiency as a result of the primary outage.

PRIMARY OUTAGE: An outage in which the initiating incident would have caused the transmission unit or power transfer equipment outage in the absence of outages of other transmission units or power transfer equipment involved in the outage event.

SECONDARY OUTAGE: An outage in which the initiating incident resulted in outage of the transmission unit or power transfer equipment only as a result of the outage occurrence of another transmission unit or power transfer equipment involved in the same outage event.

SECTION-YEAR: The product of the number of line sections and their exposure time in years.

SECTION-SEASON: The product of the number of line sections and their exposure time in seasons (summer or winter).

STATISTICS: Statistical numbers pertaining to the state (or states) of a facility which happens by chance and upon which an estimate of the availability (or unavailability) of the facility can be made with a certain level of confidence and within a given set of engineering, environmental and climatic conditions to which the facility is exposed.

SUSTAINED OUTAGE: Any outage whose duration is not negligible.

TERMINAL-RELATED FACILITIES: Facilities located at substation or at a line's termination which support the operation of the line and provides a tie function between that line and other lines terminating at the same substation (e.g. disconnect switches, breakers, bus work, current transformers, potential transformers, relays, etc...)

TERMINAL-RELATED OUTAGE: An outage as a result of the failure of a terminal-related facility.

TERMINAL-YEAR: The product of the number of terminals of a line and its exposure time in years.

TERMINAL-SEASON: The product of the number of terminals of a line and its exposure time in seasons (summer or winter).

General Statistics for Outages of MAPP Bulk Transmission Line Sections

From January 1977 Through December 1981

	FORCED						PLANNED		COMBINED	
	INDEPENDENT		SECONDARY		BOTH					
	230 kV	345 kV	230 kV	345 kV	230 kV	345 kV	230 kV	345 kV	230 kV	345 kV
1. Number of Terminal-Related Outages (Excluding Momentaries)	201	137	43	30	244	167	1,741	589	1,985	756
2. Number of Line-Related Outages (Excluding Momentaries)	593	422	129	16	722	438	694	380	1,416	818
3. Number of Outages Due to Unknown Cause	75	72	10	6	85	78	NA	NA	85	78
4. Number of Momentary Outages	371	163	36	21	407	184	14	8	421	192
5. Average Number of Terminal-Related Outages per Terminal-Year	0.2	0.2	0.0	0.1	0.2	0.3	1.6	1.1	1.8	1.4
6. Average Number of Line-Related Outages per 100 Mile-Year	1.8	2.5	0.4	0.1	2.2	2.5	2.1	2.2	4.3	4.7
7. Average Terminal-Related Outage Duration in Hours per Terminal-Year	0.6	2.4	0.1	0.1	0.7	2.5	73.7	65.0	74.4	67.5
8. Average Line-Related Outage Duration in Hours per 100 Mile-Year	38.4	80.6	1.7	0.0	40.0	80.7	95.8	119.1	135.8	199.8
9. Average Terminal-Related Outage Duration in Hours per Occurrence	3.5	9.7	2.3	1.5	3.3	8.3	46.3	53.9	41.0	49.7
10. Average Line-Related Outage Duration in Hours per Occurrence	21.6	32.8	4.3	0.4	18.5	31.6	46.0	53.9	31.9	41.9

NA - Not Applicable

APPENDIX C

Coincident Outage Plots

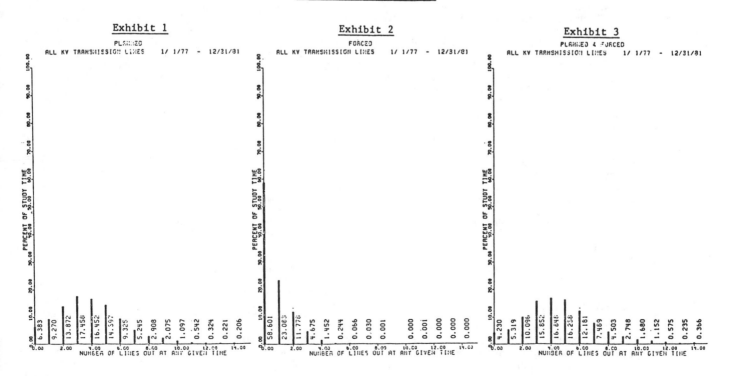

Exhibit 1	Exhibit 2	Exhibit 3
PLANNED	FORCED	PLANNED & FORCED
ALL KV TRANSMISSION LINES 1/ 1/77 - 12/31/81	ALL KV TRANSMISSION LINES 1/ 1/77 - 12/31/81	ALL KV TRANSMISSION LINES 1/ 1/77 - 12/31/81

Discussion

C. C. Fong and **J. N. Harris**, (Ontario Hydro, Toronto, Canada): The authors are to be congratulated for the analysis of outage data and calculation of statistics reported in this paper. The importance of outage data and statistics to power system reliability analysis cannot be emphasized more.

The discussers would like to ask the following questions:

1. In Appendix B, the Terminal-Related and Line-Related outage duration per occurrence are identical for the 345 kV planned outages (53.9 hrs). They are nearly identical for 230 kV line sections (46.3 and 46.0 hrs). Does it imply that for planned outages, the line section and its associated terminals are generally outaged as a group? This fact could be of interest to those who wish to use the general statistics for system reliability studies.

2. In describing the results of Table VI, it is stated that sometimes, the associated outages can be initiated by a planned outage. Since data for this phenomenon are seldom available, could the authors indicate approximately what percentage of the associated outages is of this type?

3. It is sometimes useful to obtain the duration of the switching performed to isolate a faulted component which needs repair. For example, when a line section is damaged by a fault, the terminals are also tripped if they are in the same protection zone. The line section could then be isolated and the terminals retured to service. Is the MAPP data system amenable to the identification and recording of this switching duration? If so, are there any duration statistics presently available?

Manuscript received August 5, 1983.

P. B. Anders, **P. L. Dandeno**, and **E. G. Neudorf** (Ontario Hydro, Toronto, Canada): The presentation of this paper is timely, since the industry at large should be made aware of the necessity of collecting transmission outage data. Examples of successful reporting of outages and useful management and compilation of such data are, we believe, most welcome. Ontario Hydro has had a system in place for over a dozen years, and it has proven most helpful in our planning, design, and operating activities.

Would the authors reply to the following questions, if they are in a position to repond:

1. Does the MAPP collect outage data for tranformers, busses, and breakers, as well as for lines?
   Perhaps the answer to the above question is related to their general classification of "line-related facilities" and "terminal-related facilities."

2. We believe an explanation of the term "Facilities" would be helpful to readers interested in this subject (see Table VI, for example, and Line Sections (see Table VIII). What significance does the term "number of facilities" have?

3. In Table IV, do the authors feel that any transmission outages involving no faults could have possibly been included in the "unknown" category?

4. Are partial outages counted? For example, if only one terminal of a three-terminal transmission unit were to become unavailable, or open, how is this line debited with an outage?

5. How are durations for two circuit outages, associated outages, and cascading outages defined?

6. How is inventory information compiled for common terminals at a station?

7. What is the reasoning behind dividing duration by 100 miles-year in Table V and the number of line sections in Table II?

We note that the general classification of faults by percentages agrees reasonably well with our 230 kV experience. Ontario Hydro has virtually no 345 kV transmission (except for international border crossings) and our outage experience at 500 kV is still too limited to provide a reliable data base. We expect that our percentage of three-phase faults, at the higher voltage, will be lower than that encountered at 230 kV.

In summary, we feel that the authors have provided a useful and thought-provoking review of the general area of transmission outage data collection and analysis.

Manuscript received August 8, 1983.

Norton Savage (U.S. Department of Energy, Washington, DC): This paper describes a useful data collection effort that provides information pertinent to system reliability studies. Further, it can point the way to improvement in transmission planning and construction by highlighting concentration of outages and directing attention to those areas that require improvement. "Areas" is used in the geographical sense and also in the sense of type of contruction and facilities. The systematic collection and analysis of data on a large scale, by competent people, is the only way in which the electric power industry can improve its performance. This was recognized when the Edison Electric Institute began collecting and publishing data on generating unit outages, over twenty years ago.

The transfer of the generating unit data collection to the North American Electric Reliability Council (NERC) several years ago should result in a wealth of practical, useful information being made readily available to students of electric power systems. "Students," of course, includes not only those formally enrolled in college engineering courses, but those whose professional concerns are related to the study, analysis and planning of electric power supply. It is to be hoped that the transmission data efforts of MAPP and MAIN are continued and expanded, and spread to the other seven Regional Electric Reliability Council areas. Without experience as a guide and a means of verification, theoretical analyses lack a firm foundation.

I note that the definition of "Availability" in the Glossary differs from the definition used in regard to generating units. Generating units are considered available when they can be operated, whether or not they are actually run. The definition of availability in this paper says that a line is available only if it is in use. Are there not times when a line could be put in service, but is not for some reason unrelated to the line or its associated facilities (circuit-breakers, tranformers, etc.)? The definition of "Non-Faulted Outage," it seems to me, penalizes a line by assigning it a Forced Outage when the outage is due to some event not related to the line itself. This also at variance with the use of the term Forced Outage in relation to a generating unit. It is possible to make the terminology uniform in its application to generating units and transmission facilities?

I would suggest that the percentage data of Tables III and IV would be more useful if the information were to be given in the form of distributions as well as totals and averages. That is, what were the month-by-month outages for the time periods indicated? This is not to say that the data as given are not useful, but it is helpful in some instances to know the range of a set of values and the grouping with in the range. Further work and publication in this area would be welcome.

These comments are those of the writer and do not necessarily reflect the views of the Department of Energy.

Manuscript received August 12, 1983.

G. L. Landgren and **A. W. Schneider, Jr.** (Commonwealth Company): The MAPP Transmission Reliability Task Force is to be complimented for their effort in providing a useful description of their procedure in compiling and analyzing bulk transmission performance data. Although we understand that the outage information is compiled on a unit basis, (complete transmission line as an operating entity), it is interesting to note the performance is also summarized in Appendix B on a component basis (100-mile line sections and terminations).

The performance of the MAPP transmission network appears to be similar to that of the MAIN network which includes the Commonwealth Edison system. It would be helpful in making comparisons on a unit basis to know the physical characteristics of the average line in MAPP, i.e., length, number of terminals and common-mode exposure.

One fundamental question to be asked is "What level of pooling statistical data is justified—by company, council or nationally?" Has the Task Force analzyed the individual Company performance to show the degree of variation?

Would the authors please comment on why the "faulted" and "nonfaulted" categories are distinguished? It appears from Fig. 2 that the categories are combined, making the fault/nonfault separation unnecessary.

MAPP has made consistant use of terminology as suggested by Refs. 2 and 9, and is also consistent with the PROSD* effort to update the IEEE Standard contained in the latter reference. One exception is the term Mean-Time-*To*-Failure which Military Standard 721C limits to nonrepairable systems; that reference applies the term Mean-time-*Between*-Failures (MTBF) to repairable systems. We suggest that MTBF be used for reporting transmission component performance which always involves failure and Mean-Time-Between-Outages (MTBO) for unit (line) performance which includes outages not caused by failure. This subject is being studied by the PROSD Working Group.

*A working group of the Application of Probability Methods Subcommittee, IEEE Power Systems Engineering Committee.

Manuscript received August 22, 1983.

M. G. Lauby, K. T. Khu, R. W. Polesky, R. E. Vandello, J. H. Doudna, P. J. Lehman, and D. D. Klempel: The authors wish to thank the discussers for their interest and their valuable comments. Their observations give new perspectives to the subject of bulk transmission outage data collection and analysis.

To answer Messrs. C. C. Fong and J. N. Harris:

1) The discussers point out an apparent similarity between Line-Related and Terminal-Related average durations. The authors wish to thank them for mentioning this for we discovered a typographical error. The Planned outage average duration for Terminal-Related outages is actually *61.4* not 53.9 as shown in the Appendix B. All other durations are correct. It is the experience of the authors that these durations are not necessarily the same. As mentioned in the glossary, the Terminal-Related outages are line outages due to the failure of terminal facilities (e.g., breaker relaying, etc.) while Line-Related outages are line outages due to the failure of line facilities.

2) Additional information on Associated events caused by Planned outages is requested. The MAPP bulk transmission system has experienced 9 such outages which is 5.1 percent of all the Associated events between the years 1977-1981.

3) Identification of "switching" duration can be done with the collection technique employed at MAPP. This collection technique is designed to collect information on each individual section. Therefore, the "switching" times can be gleaned from the collected data. However, this has not been done directly and the outage durations of each section are summed along with all other durations.

Messrs. G. L. Landgren and A. W. Schneider Jr. request that average line characteristics be displayed. The average line length of 230 kV transmission is 62.1 miles with 2.2 terminals. The 345 kV characteristics are 60.8 miles with 2.06 terminals. When calculating systems indices, it should be recognized that the statistics accrued are averages of all the outage data that are being analyzed. Bulk transmission system outage statistics must therefore be used carefully. Certainly, comparisons can be made when the basic procedures are similar. The MAPP and MAIN techniques are quite similar and it would be interesting to compare these statistics.

Messrs. G. L. Landgren and A. W. Schneider Jr. ask an important question, that is, "What level of pooling statistical data is justified—by company, council, or nationally?." Recent work performed by the authors indicates that there is a need to identify the dominant parameters which are generic to geographic regions that affect the performance of the bulk transmission system (e.g., weather, local hazards, construction, operating voltage, etc.). Company data can and have been studied at request from the member systems of MAPP along with recent weather analysis by isokeraunic level and mean annual snowfall level for weather-associated line outages.

The "Faulted" and "Non-Faulted" categories are subdivided in order to allow for another analysis point. Both are considered Forced outages and can cause Associated events. Messrs. G. L. Landgren and A. W. Schneider Jr. point out that the Military Standard 721C uses MTBF for repairable systems. Though this is true, the MTTF has become a commonly used term which is the reciprocal of the failure rate [7, 8]. However, if the IEEE PROSD group does standardize these two terms, the authors would definitely use this terminology.

Answers to the questions of Messrs. P. B. Anders, P. L. Dandeno, and E. G. Neudorf are as enumerated.

1) MAPP has been collecting transformer outage data for over three years but has not yet begun analysis. We are currently contemplating expanding the menu of outage causes. Also, the authors are currently studying the collection of breaker arrangements and busses. At this time, if a bus or breaker is responsible for a line outage, that outage is classified as Terminal-Related. However, the authors recognize that breaker and bus outages which are not responsible for transmission line outages are then ignored.

2) Because of the confusion between facilities and sections in Table VI and VIII the terminology was changed from Associated outages to Associated *Events* this year (1982). "Events" has the definition: An occurrence of one or more outages which can be attributed to an initiating cause.

The facilities were meant to indicate the number of line units which where outaged during the Associated events. The "Number of Sections" is simply as stated.

3) Because there is a "No fault" code offered in the menu for fault identification, it is doubtful that respondents would submit outages which had no fault along with those with unidentified fault types.

4) As mentioned earlier, data collection is based on line sections, and therefore one line section can have an outage of extended length while others are performing the intended function.

5) For both the Common Mode and Associated events the durations were defined to be the shortest outage duration of all the outaged lines.

6) Detailed information of each bus type and breaker arrangment is collected and a determination is made based on this information.

7) In Table II, the desired index of MTTF for all lines of a given voltage class was calculated as the ratio of the sum of the "up" time of each section and the sum of the number of outages of all the sections. The MTTR was calculated similarly with the numerator being the sum of the "down" time of all the sections. This index was used to compare each year and season. In Table V, the average per 100 mile-year exposure was desired for it is useful for reliability programs and analysis of outage rates of Common Mode events.

The EEI-IEEE paper of 1967 [11] also demonstrated a larger percentage of L-L-L and L-L-L-G faults for 345-360 kV than for 220-240 kV. The distribution was as follows:

Voltage (kV)	L-G	L-L	L-L-G	L-L-L	Unknown
220-240	60.9%	6.8%	3.8%	2.5%	26.0%
345-360	81.5%	8.1%	3.4%	4.0%	3.0%

This distribution which groups L-L-L-G and L-L-L together compares favorably to the distribution we've encountered. The reasoning for this is not immediately apparent and we cannot fully respond to the discussers' comments on this reference.

The authors appreciate the comments by Mr. Savage. Planned outages for voltage control, line loading control, etc., are the only situations that ocur with any frequency where a line could be put in service but is not. That is, there is no gain in reliability and perhaps a decrease if the line is put into service. The collection system does account for this and we do categorize these.

Mr. Savage further notes that the "Non-Faulted Outage" may penalize the line as a Forced outage. Non-Faulted outages usually occur because of some equipment failure or operational conditions. They are not Planned and can be automatically or manually initiated. If terminal facilities cause the outage, the outage is assigned to the terminal (Terminal-Related) and likewise, if line facilities cause the outage, the outage is assigned to the line (Line-Related).

Mr. Savage asks if the terminology could be uniform with that of generating units. The major problem with this desire is the differences between the operation of transmission and generating units. Also, weather and environmental factors cause numerous differences in design and operation. The IEEE PROSD group is working on standardization of terminology for bulk transmission collection and analysis. An example of the possible confusion is the usage of Forced Outage Rate (FOR). That is, will the FOR be used for the probability of failure as with generating units or as the frequency of outage as with transmission? It is the authors' opinion that the Forced outage rate of bulk transmission be considered the frequency of outage. However, as stated above, the authors will rely on and accept those standardized terms that are suggested by the IEEE PROSD group.

The month-by-month outage distribution requested by Mr. Savage was best demonstrated in the summer-winter subdivision which illustrates the major differences between summer and winter performance. We will study the suggested breakdown at a future date.

In conclusion, the authors would like to again thank the discussers for their excellent and well written discussions. Only in this way can common terminology and analysis be performed which enables the much needed comparison of data from different geographic regions.

Manuscript received September 14, 1983.

PROPOSED TERMS FOR REPORTING AND ANALYZING OUTAGES OF

ELECTRICAL TRANSMISSION AND DISTRIBUTION FACILITIES

IEEE T&D Outage Definitions Task Force

Application of Probability Methods Subcommittee

Power System Engineering Committee

Members of Task Force are: D. W. Forrest (Chairman), P. F. Albrecht, R. N. Allan

M. P. Bhavaraju, R. Billinton, G. L. Landgren, M. F. McCoy, N. D. Reppen

ABSTRACT

This paper proposes definitions of terms and indices for reporting and analyzing outages of transmission and distribution facilities. The definitions cover facilities from the generator terminals to the customer service entrance but they do not cover customer interruptions. Depending on comments received from the industry, these definitions could be proposed as an IEEE Standard.

Outage definitions and indices are given for two general types of facilities - units and components. Units are groups of components functionally related, while components are specific physical equipment.

Outage definitions are given for describing the outage history of a particular facility. Additional definitions are given for describing groups of individual outages which are related in some way.

INTRODUCTION

Data on outages of transmission and distribution facilities have been collected for many years. Initially, reporting was generally limited to reporting the frequency of outage occurrence on transmission lines and distribution feeders. Total outage frequency was classified into several general cause categories, but there were no formal definitions for the events and statistics reported beyond those contained in the data forms and instructions. Early efforts did not generally collect data on outage duration (except to classify outages as temporary or permanent according to type of restoration used).

84 WM 042-8   A paper recommended and approved by the IEEE Power System Engineering Committee of the IEEE Power Engineering Society for presentation at the IEEE/PES 1984 Winter Meeting, Dallas, Texas, January 29 - February 3, 1984. Manuscript submitted September 12, 1983; made available for printing November 11, 1983.

In the 1960's, methods for calculating reliability of transmission and distribution "systems" in terms of the reliability of the individual "components" were first proposed [1,2]. This led to the need for more formal definition of terms to foster uniformity and standardization of language among engineers engaged in reporting, analyzing, and predicting outages of transmission and distribution facilities and interruptions to customers. This resulted in development of IEEE Std. 346-1973 [3].

More recent developments have shown the need to include definitions for a broader scope of outage events. For example, two general categories of facilities must be recognized to support presently available models. In one type of model, a transmission or distribution system is considered as a group of functional units which represent the capability to transfer power between specified points. The term "unit" is proposed to describe this type of facility. In the second type of model, a system is viewed as being composed of a number of distinct facilities which partition the system physically into mutually exclusive and exhaustive sets of facilities. The term "component" is used to describe this type of facility. These definitions are included in the section entitled Equipment Classifications.

Another recent development is the increasing recognition of the importance of "related outages" in which outages on two or more facilities are related in some way due to the physical construction of equipment (e.g., common structure), the electrical interconnection of equipment (e.g., common bus), or exposure to a common environment (e.g. storms). Achieving clear and unambiguous definitions for related outages is important for reporting outages at the transmission unit level. At the component level, an important consideration is the recognition of several modes of failure, particularly for switching, protection, and automatic reclosing equipment. These concepts are included in the section entitled Component and Unit States and Events.

No attempt is made here to propose procedures for the collection of outage data. What is attempted is to specify certain common terms and indices to

Reprinted from *IEEE Trans. Power Apparatus Syst.*, vol. PAS-104, no. 2, pp. 337-348, Feb. 1985.

provide a basis of information exchange. The task force has attempted to keep the list of terms and indices as brief as possible. Brevity introduces the risk of differing interpretations. Therefore detailed reporting instructions should be carefully reviewed before indices reported by different agencies are compared.

The indices defined here are intended to serve as estimates of parameters for models which are used in system reliability evaluation. The indices may also be used directly to guide system operation or in assessment of equipment performance. No attempt is made to recommend acceptable equipment reliability levels.

The definitions in this paper are presented in the following five categories, Equipment Classifications, Equipment Outage States and Events, Equipment State Times, Exposure Parameters and Basic Outage Indices.

I. EQUIPMENT CLASSIFICATIONS

1.0 Component

"A device which performs a major operating function and which is regarded as an entity for purposes of recording and analyzing data on outages."

Some examples of components are line sections, transformers, AC/DC converters, series capacitors or reactors, shunt capacitors or reactors, circuit breakers and bus sections. Sometimes it is necessary to record data on the line segments which constitute a specific line section to allow proper calculation of failure rates and exposure data. For example, if a line section is composed of an overhead line segment and an underground line segment, failure and exposure data for each line segment should be recorded separately.

1.1 Line Section

"A portion of an overhead line or a cable bounded by two stations, a station and a tap point or two tap points."

1.2 Line Segment

"A portion of a line section that has a particular type of construction or is exposed to a particular type of failure, and therefore may be regarded as a separate entity for the purpose of reporting and analyzing failure and exposure data."

2.0 Unit

"A group of components which are functionally related and are regarded as an entity for purposes of recording and analyzing data on outages."

The components included in a unit are often protected by a common primary protective relay scheme. For example, a unit might consist of an overhead line and the terminal equipment switched with the line by circuit breakers during normal fault clearing operation. The concept and definition of unit was discussed by the task force at length. The EPRI project [4] proposed a definition for a transmission unit as a facility which transfers power between two or more points in a transmission system and is functional rather than a physical group of components. The task force concluded that it would be easier to identify a unit as a group of components with physical boundaries in collecting outage data. It is recognized that certain components (e.g. circuit breakers) could be part of more than one unit. Different types of units include transmission unit (overhead or cable), transformer unit, bus unit, and special units that consist of any equipment such as shunt capacitors protected by separate breakers.

II. COMPONENT AND UNIT STATES AND EVENTS

Except as noted, the terms in this section apply to either component or unit performance reporting.

States

Component or unit state is a particular condition or status of a component or unit which is important for outage reporting purposes.

1.0 In-Service State

"The component or unit is energized and fully connected to the system."

2.0 Outage State

"The component or unit is not in the In-Service State; that is, it is partially or fully isolated from the system."

The definition of outage state was discussed at length by the task force. Two extreme viewpoints were: 1) outage state is in the general sense of out-of-service, for any reason, whether it can perform its function or not. 2) outage state is due to some type of failure so that the facility cannot function. This second viewpoint is too narrow and it cannot properly accommodate related outages. A third viewpoint was to define unit outage state in the sense of out-of-service, and component outage state to imply only failures. This third viewpoint appears logical but it could cause some confusion because each outage state needs the explanation of the basis - either unit or component. The task force decided to use the first approach for both units as well as components so that outage state implies simply out-of-service. Additional definitions such as primary and secondary, will distinguish between outages due to failure of some type so that the facility cannot function, and outages when the facility is healthy but out-of-service due to other reasons. A separate definition of failure is provided later in the paper.

Examples:

A two-terminal transformer disconnected on one or both sides is in the outage state.

A three-terminal overhead transmission
line disconnected from one station is, as
a unit, in the outage state. However, the
two overhead line sections still carrying
power are, as components, in the in-service
state. (see 2.1 and 2.2)

A circuit breaker which is not energized
on either side is in the outage state.
Whether the circuit breaker is open or
closed does not affect its in-service/outage
state status.

A component which is disconnected from the
system for operating reasons, but is
available to perform its intended function
when called upon, is also in the outage
state.

2.1   Complete Outage State

"The component or unit is completely
de-energized or is connected at only one
terminal so that it is not serving any of
its functions within the power system."

2.2   Partial Outage State

"The component or unit is energized but is
not connected to all of its terminals so
that it is not serving some of its functions
within the power system."

A unit composed of a three-terminal line
would be in the partial outage state if it
is disconnected from one station but two
line sections are carrying power.

Outage Occurrences

3.0   Outage Occurrence

"A change in the state of one component or
unit from the In-Service State to the
outage state."

The term "outage" is used in the industry
to mean outage occurrence or outage state.
Proper interpretation depends on the
context.

4.0   Outage Event

"A single outage occurrence or multiple
outage occurrences which are grouped
together because they are related."

4.1   Single Outage Event

"An outage event involving only one
component or one unit."

4.2   Multiple Outage Event

"An outage event involving a) two or more
components or b) two or more units."

4.2.1   Independent Multiple Outage Event

"A multiple outage event where the outage
occurrences have distinct and separate
initiating incidents and no outage occurrence is the consequence of any other."

4.2.2   Related Multiple Outage Event

"A multiple outage event in which one
outage is the consequence of another
outage and/or in which multiple outages
are initiated by a single incident."

Each outage occurrence in a related
multiple outage event is classified as a
primary outage or a secondary outage
depending on the relationship between that
outage occurrence and the initiating
incident.

4.2.2.1   Primary Outage

"An outage occurrence, within a related
multiple outage event, which occurs as a
direct consequence of the initiating
incident and is not dependent on any other
outage occurrence."

A primary outage of a component or a unit
may be caused by:

    -Fault on equipment within the unit or
     component
    -Repair of a component within a unit

4.2.2.2   Secondary Outage

"An outage occurrence which is the result
of another outage occurrence."

Secondary outages of components or units
may be caused by:

    -Fault on a bus or circuit breaker
     resulting in outages of other compo-
     nents or units.

    -Repair of other components or units
     requiring additional outages to
     provide physical clearance.

    -Failure of a circuit breaker to clear
     a fault.

Primary outages have been sometimes
referred to as independent outages, and
secondary outages as dependent or related
outages.

4.2.2.3   Common-mode Outage Event

"A related multiple outage event consisting
of two or more primary outage occurrences
initiated by a single incident where the
outage occurrences are not consequences of
each other."

Primary outage occurrences in a common-mode
outage event are referred to as common-mode
outage occurrences or simply common-mode
outages.

Examples of common-mode outages are:

    -A single lightning stroke causing
     tripouts of two circuits on a common
     tower.

    -An external object causing the outage
     of two circuits on the same right-of-
     way.

-A tornado causing the outage of multiple circuits even if they are not on a common tower or a common right-of-way.

Outage Initiation

4.3   Automatic Outage

"An outage which results from automatic operation of switching devices."

4.4   Manual Outage

"An outage which results from intentional or inadvertent operator-controlled opening of switching devices."

Outage Type

An Outage is classified according to the urgency and the need for the outage and depending on how the equipment is restored to service.

5.0   Forced Outage

"An automatic outage or a manual outage that cannot be deferred."

5.1   Transient Forced Outage

"A forced outage where the unit or component is undamaged and is restored to service automatically."

5.2   Temporary Forced Outage

"A forced outage where the unit or component is undamaged and is restored to service by manual switching operations without repair but may be with on-site inspection."

5.3   Permanent Forced Outage

"A forced outage where the unit or component is damaged and cannot be restored to service until repair or replacement has been completed."

5.4   System-Related Outage

"A forced outage which results from a system disturbance and is not caused by an event directly associated with the component or unit being reported."

Typically system-related outages are not considered in developing outage indices for the use in system reliability evaluation.

5.5   Planned Outage

"A manual outage that could have been deferred without increasing risk to life, risk to property or equipment damage."

A manual outage is classified as planned if it were possible to defer the outage when such deferment is desirable. Otherwise, the outage is a forced outage. Deferring an outage may be desirable, for example, to prevent overload of facilities or an interruption of service to customers.

5.5.1   Scheduled Outage

"A planned outage for purposes of testing, maintenance, construction or repair."

5.5.2   Operations-Related Outage

"A planned outage in which the unit or component is removed from service to improve system operating conditions."

Typically operations-related outages are not considered in developing outage parameters for the use in system reliability evaluation.

Failures

6.0   Failure

"The termination of the ability of a component to perform its required function."

Components may have functions of two types: those which are continuously required and those which are required in response to a condition. Continuously required functions include carrying current, providing electrical isolation, and abstaining from tripping in the absence of a signal. Examples of inabilities to perform continuously required functions are:

-Component short circuit
-Component open circuit
-Switching equipment opening without proper command
-Switching equipment closing without proper command

Response functions include responding to fault conditions (protective systems), to command (circuit breakers), and to manual operation (disconnect switches). Inabilities to perform a response function do not cause an immediate interruption of power flow. They are disclosed by subsequent inspection or by failure to respond to conditions as intended. This type of failure has been refered to as dormant failure, latent failure and unrevealed failure. Examples are:

-Switching equipment failure to open on command
-Switching equipment failure to close on command
-Protection system tripped incorrectly (over-reach during fault)

Outage Duration

7.0   "The period from the initiation of an outage until the affected component or unit or its replacement once again is in the In-Service State."

Outage durations may be defined for specific types of outages, for example, permanent forced outage duration, temporary forced outage duration, and scheduled outage duration.

Outage duration is normally equal to the sum of switching time, repair time, travel and material procurement time. These items account for manpower, equipment or material. Outage duration may also increase for other reasons, for example, due to court injunctions or because the component or unit on outage is not needed. This excess of outage duration may be excluded in a typical analysis of outage duration.

III. EQUIPMENT STATE TIMES

1.0   Service Time

"The accumulated time a component or unit is in the In-Service State during the reporting period."

2.0   Outage Time

"The accumulated time a component or unit is in the Outage State during the reporting period."

Outage Times can be identified for each type of outage, for example:

Forced Outage Time
Scheduled Outage Time
Common-Mode Outage Time
Partial Outage Time

3.0   Period Time

"The duration of the reporting period."

IV.   EXPOSURE PARAMETERS

1.0   Exposure Time

"The time during which a component or unit is exposed to failure."

Time is the major measure of exposure for most failure modes of lines and transformers and the circuit breaker failure mode "opened without command."

Exposure time may include only Service Time, or it may also include Outage Time depending on the type of component or unit and mode of failure.

2.0   Exposure Operations

"The specific operations during which a component is exposed to certain types of failure."

For example, the number of commands to open the breaker is the exposure parameter in the case of circuit breaker failure to open.

3.0   Exposure Length

"The length of a line section, common structure or common right-of-way which influences the failure rate."

4.0   Weather

Exposure may be subdivided according to the type of weather to which a component or unit is exposed.

4.1   Adverse Weather

"Designates adverse weather conditions which cause an abnormally high failure rate for exposed components during the periods such conditions persist, but do not qualify as major storm disasters. Adverse weather conditions can be defined for a particular system by selecting the proper values and combinations of conditions: thunderstorms, tornados, wind velocities, precipitation, temperature, etc."

4.2   Major Storm Disaster

"Designates adverse weather which exceeds design limits of plant and which satisfied all of the following:

1.   Extensive mechanical damage to plant.

2.   More than a specified percentage of customers out of service.

3.   Service restoration times longer than a specified time."

The percentage of customers out of service and restoration times typically used by the industry are 10 percent and 24 hours. Percentage of customers out of service may be related to a company operating area rather than to an entire company. Examples of major storm disasters are hurricanes and major ice storms.

4.3   Normal Weather

"Includes all weather not designated as adverse or major storm disaster."

V   BASIC OUTAGE INDICES

Basic Rate Indices

1.0   Outage Rate

"The number of outages per unit of Service Time."

$$= \frac{\text{Number of Outages}}{\text{Service Time}}$$

The unit of time is usually one year.

1.1   Scheduled Outage Rate

"The number of scheduled outages per unit of Service Time."

$$= \frac{\text{Number of Scheduled Outages}}{\text{Service Time}}$$

In some studies scheduled outage rate may be defined as number of outages per unit of Period Time.

1.2   Normal Weather Outage Rate

"The number of outages per unit of Service Time in normal weather."

$$= \frac{\text{Number of Outages During Normal Weather}}{\text{Normal Weather Service Time}}$$

458

1.3 Adverse Weather Outage Occurrence Rate

"The number of outage occurrences per unit of Service Time in adverse weather."

$$= \frac{\text{Number of Outages During Adverse Weather}}{\text{Adverse Weather Service Time}}$$

Outage rate can be expressed as related to a specific component or unit, or it can be expressed as occurrences per unit length in the case of a line, common structure or common right-of-way.

2.0 Failure Rate

"The number of failures per unit time exposed to failure."

$$= \frac{\text{Number of Failures}}{\text{Exposure Time}}$$

Rates for different failure modes can be calculated. The exposure time for each failure mode may be different.

2.1 Transmission Line Section Failure Rate

"The number of transmission line section failures per unit length per unit time."

$$= \frac{\text{Number of Failures}}{(\text{Length of Line Section})(\text{Exposure Time})}$$

2.2 Protective System False Operation Rate

$$= \frac{\text{Number of False Operations}}{\text{Exposure Time}}$$

Basic Duration Indices

3.0 Mean Time to Outage

"The mean time between outages of a specified type."

$$= \frac{\text{Service Time}}{\text{Number of Outages of the Specified Type}}$$

4.0 Mean Outage Duration

"The mean duration of outages of a specified type."

$$= \frac{\text{Outage Time for the Specified Type}}{\text{Number of Outages of the Specified Type}}$$

Basic Probability Indices

5.0 Availability

$$= \frac{\text{Service Time}}{\text{Period Time}}$$

6.0 Unavailability

$$= \frac{\text{Outage Time}}{\text{Period Time}}$$

6.1 Forced Unavailability

$$= \frac{\text{Forced Outage Time}}{\text{Period Time}}$$

6.2 Scheduled Unavailability

$$= \frac{\text{Scheduled Outage Time}}{\text{Period Time}}$$

7.0 Failure Probability

"The probability of a component failing to respond to a command"

$$= \frac{\text{Number of Failures to Respond}}{\text{Exposure Operations}}$$

7.1 Probability of Failure to Open on Command

$$= \frac{\text{Number of Failures to Open}}{\text{Number of Commands to Open}}$$

This index can be calculated either under fault or no fault conditions.

7.2 Probability of Failure to Close On Command

$$= \frac{\text{Number of Failures to Close}}{\text{Number of Commands to Close}}$$

7.3 Probability of Failure to Operate on Command

$$= \frac{\text{Number of Failures to Operate}}{\text{Number of Commands to Operate}}$$

ACKNOWLEDGEMENTS

The task force acknowledges the comments received from the several members of the Application of Probability Methods Subcommittee, in particular, A. D. Patton.

REFERENCES

1. IEEE Committee Report, "Bibliography on the application of probability methods in power system reliability evaluation: 1971-77", IEEE Trans. (PAS), Nov./Dec. 1978.

2. IEEE Committee Report "Bibliography on the application of probability methods in power system reliability evaluation: 1978-82," IEEE Paper No. 83 WM 053-6, PES Winter Meeting, 1983.

3. IEEE Standard 346-1973, Section 2, "Standard definitions in power operations terminology including terms for reporting and analyzing outages of electrical transmission and distribution facilities and interruptions to customer service."

4. EPRI Final Report, "Bulk transmission system component outage data base, "EL-1797, Project 1283-1, April 1981.

Discussion

Richard P. Ludorf (Northeast Utilities, Hartford, CT): For many years, utilities, power pools, and reliability councils have been collecting data on transmission outages to identify the lines with poor performance, to develop a data base to facilitate projections of future performance, and to be in a position to respond to questions on performance raised in regulatory proceedings. Because of the many uses for such data, more utilities are seeing the need for establishing outage data collection methods. With this increased activity in the area of outage data collection, the need for coordinated data collection procedures and the definitions is apparent. I would like to commend the taks force on their worthy efforts to establish consistent definitions for use by reliability engineers in collecting outage data. I view this paper as the first step in establishing IEEE Standards and believe that the comments expressed by those in industry who are actively collecting data should be incorporated into the definitions.

Regarding Section I, Equipment Classifications, the definitions for both component and unit are excellent descriptions of the two types of facilities on which outages are presently being reported. However, the use of the word "line" in Subsections 1.1 and 1.2 is confusing, and therefore should be replaced with the word "circuit" because it would be more appropriate as related to outage data collection. A transmission "line" may consist of one or more circuits supported by the same transmission structures. For outage reporting purposes, it is generally the transmission circuit which is of interest.

Under Section II, Component and Unit States and Events, I believe that the definition for "In-Service State" should be modified to state that the component or unit is energized and fully connected to the system and *performing its intended function*. In addition, under Subsection 4.2.2.2, Secondary Outages should be subdivided into two separate categories: Direct secondary outages and Indirect secondary outages. Direct secondary outages are outage occurrences that are the *consequence* of an outage of another component or unit due to system design or system configuration. Indirect secondary outages are outage occurrences that result from the *failure* of other components or units requiring additional outages. Indirect secondary outages are the result of failed components or units such as relay overtrips or breaker failures.

In reference to the term common mode outage event in Subsection 4.2.2.3, there are a number of items which should be evaluated when deciding whether the multiple outage of two or more facilities are common mode. The first of these is that the components or units involved in the common mode event must have been outaged as a result of a single major cause. Next, common mode outages will usually be the result of the physical proximity of two or more facilities such as two transmission circuits on a common tower; the electrical connection between two or more facilities such as a common bus; or the environmental conditions in which two or more facilities exist such as the transmission circuits along a seacoast. Furthermore, the outages of each of the components or units in a common mode outage event should be overlapping although they may not necessarily be simultaneous. All of the above items have a bearing on whether a multiple outage can be considered to be common mode or not, and these items should be incorporated into an industry definition of the common mode outage event. There is a need to subdivide common mode outage events into two separate categories: Simulteneous Common Mode Outages and Overlapping Common mode Outages. Simultaneous common mode outages are outages of those components or units that occur within one second of each other. Overlapping Common Mode Outages are those that occur at a time interval of more than one second, thus allowing time for some possible operator intervention.

Also, the words "single incident" when referring to common mode outage events should be clarified. In the case of storm-related outages, it is unclear whether the words "single incident" refer to a single lighting stroke that causes two circuits to trip out or whether they refer to the entire storm in which there could be numerous overlapping outages in various parts of the transmission system. The related nature of storm-initiated, overlapping outages should be captured in the outage statistics regardless of the elapsed time between initiation of the outages. This interpretation should also apply to any other situations in which there exist overlapping outages due to such major causes as icing, salt contamination, etc.

Manuscript received February 6, 1984.

Harry W. Colborn (North American Electric Reliability Council, Princeton, NJ): The North American Electric Reliability Council (NERC) contains utilities from all segments of the industry, and from almost all of the States and Provinces of the USA and Canada. NERC has both a regional and a technical committee structure. The regional structure consists of nine Reliability Councils encompassing the forth-eight contiguous States and the Provinces having common boundaries with States. The main technical committees are the Operating Committee and the Engineering Committee. The Engineering Committee has decided to follow the efforts of IEEE in standardizing transmission outage definitions and, to that end, has influenced that this discussion be assembled from comments made by utilities to their regional representatives on the Engineering Committee. It is uncertain whether the attempt to secure comments has resulted in a valid industry view. Utilities in the MAIN and MAPP Councils responded, and those Councils will also be presenting their own discussions. Part of the Southeast contribution is from a Council Committee made up of persons involved in reliability matters. However, there has been no input from NPCC, MAAC, ERCOT, or SPP. Comments were received from only seven utilities in the eleven western states. Certain of these Councils may be represented by other discussers. This discussion will be a digest of the comments received and will assume that those comments represent the industry view.

There is a support for the concept of standardizing terms and definitions regarding outages, and most look upon this paper as a commendable beginning, a first step, a foundation on which to build. Several saw benefit in the establishing of a common language for the comparison of data and the results of analyses. At the same time, one utility questioned the benefit of comparing data on a wide basis.

There is a wide disparity in views of the detail implied by the paper. One comment indicated gratification that the third level of reporting (that is, the parts of equipment) had been left out while another utility complained that the paper fell short in identifying the outage detail the commenter felt was required. Some comments indicated a concern that, even if a collection procedure is not being specified in the paper, the detail of collection would be affected by the structure of the definitions.

The question of the required level of detail, and the hierarchical nature of the problem will be handled by the discussant from MAIN, and so I'll leave that point to him.

Several writers felt that more clarity should be achieved with examples and explanation, and some recommended a tutorial type appendix.

One writer stated that compatibility is lacking between the definitions of indices in the paper and similar definitions in long use in the generation reliability discipline; and that such compatibility is necessary for reaching the goal of bulk power system indexes.

There were several references made to word usage in the paper without definition or without consistency. Some of these are 'Common-mode Outage Time', 'station' versus 'terminal', 'line', 'failure mode', 'plant,' and 'restoration time'. When the word 'related' is used, in what characteristic is the outage related to another outage? Also, the word 'unit' is defined in Section I, but apparently used differently in Section V. One writer recommended a list of general or major outage causes for exemplifying. I have included more specific comments on definitions later in the discussion.

One writer suggested that a distinction be made in all outages as to whether there was proper or improper functioning of equipment or personnel. He pointed out the increasing importance of this as systems, particularly control and protective systems, become more complex. He noted that 30% of the major disturbances of the last four years had involved some inappropriate action by equipment or people.

One provocative statement was made which raised a question in my mind as to whether the authors are working toward a U.S. Standard, a U.S.-Canadian Standard or an International Standard. If it is to be more than a U.S. Standard, have the authors adequately included terms and definitions from other reporting systems? NERC is particularly interested in seeing that conflicts do not arise in definitions and meanings between the U.S. and Canadian systems.

The indices defined in Section V do not include any aggregate system-wide index, such as outages due to termination problems. The discussant from MAPP will cover this in more detail.

This concludes the summary of general comments received. The remainder of the discussion is of specific, and sometimes editorial, comment.

The Task Force may feel that some of these comments reflect misunderstanding by the readers. I hope they accept those rather as being indicative of requiring a clearer treatment.

Section I 1.1	The bounding of a line section could be better defined by relating to switches or other isolating devices rather than by stations.
Section II	The heading for Section II does not agree with the categories listed immediately before Section I.

460

2.0	An open breaker is not necessarily isolated from the system. Is it considered out of service?
2.0	In the second example, add, "Refer to definitions of Complete and Partial Outage States."
2.2	Before the second paragraph, ad the word, 'Example'. Certain three-terminal lines are operated normally in a two-terminal mode. How can one avoid having this considered as being a partial outage?
3.0	The difference between 'Outage Occurrence' and 'Outage Event'(4.0) is unclear.
4.0	The use of the word 'related' in describing occurrences is inconsistent with its use in 4.2.2, where it is associated with events.
4.2	This definition seems to eliminate outage events which include both components and units. A fault on a line terminating in a three-breaker ring having one breaker (a component) out for maintenance results in secondary outages of two additional units.
4.2.1	Insert 'of' after 'consequence'. Possibility re-word to ". . . the outage occurrence being examined is not the consequence of any other." An example would be appropriate here.
4.2.2	'Are' should follow 'outages'.
4.2.2.	In cascading, an outage can be both primary and secondary.
4.2.2.1	The distinction between primary and secondary outages (4.2.2.2) caused by repair procedures is unclear.
4.2.2.1	Because of the wide use of the terms 'primary' and 'secondary' (in voltage-level components and in relay systems, for examples), the words 'direct' and 'related' might be substituted. This should tie into 5.4 and 5.5.2.
4.2.2.2	For the example in the comment on 4.2, the secondary outages are the result of both initial outage events; but those are unrelated. That is, a secondary outage can be dependent on the existence of more than one independent primary outage. Possibily defining two types of dependency (single and multiple outage dependence) would help.
4.2.2.3	The tornado example is poor because it is very difficult to identify individual tornadoes in a storm system. A similar difficulty in identifying adverse weather at points remote from weather station or control centers was noted. This applies also to Section V.
4.3	Does a relay technician error fit in this category or in 4.4?
4.3	There are outages resulting from failed components where neither of the terms 'automatic' or 'manual' applies. A possible example is that of a breaker falling open.
5.1	The terms 'transient forced outage' and 'temporary forced outage' (5.2) are difficult because they both have the same abbreviation. Also, the distinction between them is in the *method* of restoration, not the *time.*One suggestion, 'automatically' and 'manually restored forced outage' is also difficult.
5.1	Reclosing, and particularly multiple reclosing, will recall more explanation and possible examples for clarity.
5.2	The distinction between 'temporary' and 'permanent' (5.3) should be one based on time, not the requirement of repair. Remote installations can result in long outage times even though no repair is required.
5.2	The proper action of a syncho-check relay may delay automatic reclosing for quite some time. It should be clarified as to whether 'manual' includes remote, supervisory controlled switching.
5.4	Insert 'with the' following 'associated'. Insert 'system-related' after 'typically'.
5.4	Examples of system disturbances would be helpful.
5.4	Following 'associated', add 'with the'. After 'reported', add 'as outaged'. In second paragraph, change 'These' to 'system-related' for specificity.
5.5	Outages may be extended for reasons of convenience or economics. Possibility the term 'discretionary' or 'extended' could apply to those situations. Similar extensions are handled in generator performance data collection.
5.5.1	Outages in preparation for special tests such as staged fault tests or ammeter tests should be allowed for.
5.5.1	The 'maintenance, construction or repair' referred to should be specifically described as being of the system component and not (e.g.) a highway.
5.5.2	The second paragraph applies to all planned outages and should have 'planned' inserted after 'typically'. Possible the

	paragraph should be in 5.5.
5.5.2	In second paragraph, change 'these' to 'operations-related'.
7.0	'Outage Duration' seems more appropriate for Section III.

Section III

| 2.0 | One writer recommended breakdown definitions for the subintervals of restoration time. |
| 3.0 | Add the word equation: 'Period Time equals Service Time plus Outage Time'. |

Section IV

1.0	It is difficult to envision an exposure time less than the period time.
2.0	Add 'specified' after 'certain'.
3.0	One writer suggested consideration of lines routed through different weather regions in 3.0 and 4.0. In 4.0, he specifically suggests stating "Exposure time and length may be . . .".
4.1	Quantification appears necessary.

Section V

One writer suggested consideration of defining all outage rates in terms of period time instead of service time, which would be very close to the period time. Sufficient accuracy should result.

This Section should make a clear distinction between forced outage rates and others. Once the 'forced' rates are separated, they can be broken down to conform to the outage types in Section II.

1.0	The inference here is that Service Time should be expressed as fractions of a year. It could be stated that "Outage Rates are generally expressed as occurrences per year."
2.2	The Protective System False Operation Rate was noted as not involving the number of protective and auxiliary relays involved.
2.2	What does the protective system include? Does it include communication devices, sensors, transducers, etc.?
5.0	Substitution of 'Basic Performance Indices' for 'Basic Probability Indices' would reinforce the idea that they are of past history and predictive.
7.0	In equation, replace 'report' with 'respond'.
7.0	Through 7.3 it is unclear how much of the control and protective system is included.

Manuscript received February 15, 1984.

J. G. Anders, P. L. Dandeno, E. G. Neudorf and **P. Semotiuk** (Ontario Hydro): We at Ontario Hydro fully support the purpose of this paper — to put before the electric utility industry technical terms and definitions relating to Transmission and Distribution Outage collections procedures.

As the Task Force notes in the Abstract of this paper, it is believed that if a sufficient number of favorable comments are received on such terms and definitions, they could well become a standard. Perhaps as a second important step (the first positive step being the presentation and discussion of *this* document), a set of recommended terms and definitions could be published, also in the Transactions, after the Task Force has collectively responded to all the discussions. Alternatively, the Task Force might also, to avoid confusion, *delay* publication of this document for a few months in PA&S, and resubmit it, with constructive criticisms included, as *RECOMMENDED TERMS* for publication after presentation again at the 1985 Winter Power Engineering Society meeting. Then the final step, after trial use for a one or two-year period (for example), it could then be sent to the Standards Board. We leave the final disposition and format of it up to the appropriate committee and subcommittee.

As some may know, Ontario Hydro has been in the business of collecting and analyzing transmission outage data for almost 15 years, through the offices of its Protection and Control Department. This Department is responsible for maintaining and operating the Protective relay schemes in our Bulk Electricity System. While the terms suggested here are by and large generic enough to cover the distribution system (in our case 50 kV and below), most of our thinking and discussions will relate to what we call the bulk electricity transmission system—in our case virtually all the 115 kV, 230 kV and 500 kV transmission lines of Ontario Hydro.

After 15 years of reporting on the outage experience, we are now in the process of revising our own Transmission Outage Data System (TODS) and wish to ensure that our thinking and terminology agree in general with that of IEEE.

Our only criticism of the paper of any major concern is that a typical station and transmission one-line diagram, showing the protection zones which might be used to define transmission units, should be given. This would be helpful in the paper in discussing some of the concepts which we propose.

We agree in general with the use of the term *unit* (transmission unit or distribution unit) and, as a subset of a unit, the idea of a component. We also feel that a term such as subcomponent might be used to describe individual items of electrical equipment which comprise the component.

In the Introduction, the authors introduce the term "unit", followed by the term "component". We believe that in Section I (equipment classification), the term unit should also be introduced first followed by the term component. Finally, a new additional term called subcomponent could then follow. Examples of each classification are given below after certain suggested modifications. We believe that our rationale for the modifications, and choices of words are logical.

The following specific changes or additions are proposed:

(a) For *Unit:* We agree with the wording as given, except that the word "often" could be removed from before the word "protected" in the explanation of the definition. Examples of a unit could also include shunt reactor and circuit breaker.

(b) For *Component:* This wording should be changed to read "A device which performs a major operating function and which is regarded as part of a unit for purposes of recording and analyzing data on outages.

Examples of a component are: Line Section, Line Terminal, Line Bus, Bus Section, Transformer, Transformer Terminal.

(c) For *Subcomponent* (New): Our proposed definition for this is "A part or portion of a component which is only identified if it is the cause of the outage or failure".

Examples of Subcomponents are: (1) As they would apply to a Line Section: Tower(s), Phase Conductor (s), Shield Wire(s), Insulators.

(2) As they would apply to a Line Terminal: Disconnect Switch(es), Potential Transformer(s), Current Transformer(s).

(3) As they would apply to a Transformer: Transformer Bushing(s) and Transformer Winding(s).

The definition of "in service state" could be further clarified by the addition of the phrase: "in other words it is fully functional."

Another basic philosophical difference exists in the Working Group's concept of outage state; we also observe that two extreme viewpoints were noted in the commentary which follows the formal definition.

In the final paragraph of this commentary in Section 2.0, under "Examples", it is stated that "A component which is disconnected from the system for operating reasons, but is available to perform its intended function when called upon, is also in the outage state." Ontario Hydro believes this paragraph should be deleted from any permanent document.

As an explanatory note immediately after the formal definition of "outage state" (Definition 2.0), a paragraph should be added to the effect that:

"This is not to imply that an outage state would be reported where a unit or component is taken out of service for operational reasons."

In Definition 2.1, we would like to see the following revised wording of *COMPLETE OUTAGE STATE:*

EXAMPLES OF TRANSMISSION UNITS

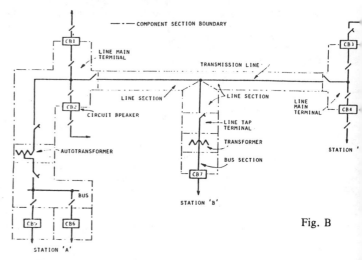

Fig. A

EXAMPLES OF TRANSMISSION COMPONENTS

Fig. B

"The component or unit is completely deenergized."

This would mean the elimination of the remainder of the words in 2.1 as presently written.

Then, if this wording is believed acceptable by the Working Group, the wording of 2.2 regarding *PARTIAL OUTAGE STATE* could be adjusted or revised to read as follows:

"The component or unit is energized but is not connected to all of its terminals, *or one of its terminals has a breaker or disconnect switch open,* so that it is not serving some of its functions within the power system." (The underlined words are the suggested addition which Ontario Hydro proposes).

The attached two Figures are typical of many protection schemes and are offered to the Task Force as additions to their next document, which as noted above should, for the permanent Transactions, be "Recommended Definitions." The first Fig. (A) indicates how "Unit definitions apply to a complete transmission line, with its associated protections and terminals. The second Fig. (B) shows different components which make up various transmission units.

Manuscript received February 23, 1984.

Harry Bonser and **Dennis Goodman** (Carolina Power and Light Co., Raleigh, NC): a) Page 3, Paragraph 2.2: The example stated does not discuss a three-terminal unit normally operated as a two-terminal unit. It should be qualified that reporting should be with respect to the normal mode of operation. Otherwise, because of unique situations, some three-terminal units whose normal mode of operation is to only have two units closed, would always be considered in a partial outage state.

b) Page 3, Paragraph 4.2.2: "Are" is omitted, and should be inserted to read "and/or in which multiple outages *are* initiated by a single incident."

c) Page 3, Paragraph 4.2.2.1 & 4.2.2.2: Suggest changing "Primary" and "Secondary" to "Initial" and "Related". Primary and Secondary are common terms often used to refer to Distribution components.

d) Page 3, Paragraph 4.2.2.3: Suggest changing "Common Mode" to "Same Cause", which would increase personnel understanding of the definition.

e) Page 4, Paragraph 4.3: Unswitched outages need to be included. These are outages which are initiated due to a failed component where neither of the terms automatic or manual apply.

f) Page 4, Paragraph 5.1: Does not specify whether the short interruptions associated with reclosing would apply. Also there may be single or repeated outages, such as a breaker reclosing two times. We suggest examples and/or explanation of how to count and record such outages.

g) Page 4, Paragraph 5.4: Suggest adding "with", which will then read "an event directly associated *with* component or unit being reported."

h) Page 4, Paragraph 5.5: Suggest specifying a minimum time for the deferment. In some cases deferment could be possible, but the permissible time to deferment might be so short that the outage would still occur when adverse consequences could not be prevented.

i) General: These indices are time-frame or equipment/outage-type oriented. CP&L currently uses customer-oriented indices for distribution outages. The Company indices are based on customer-minutes and, therefore, do not reflect the equipment responsible for the outage. However, some of the proposed standards could be used with CP&L indices to enhance our recording of the system's reliability.

Manuscript received February 24, 1984.

MAIN Transmission Outage Task Force: R. R. Matulis (Central Illinois LIght Co.), **J. A. Raksany** (Wisconsin Power & Light Company), **A. W. Schneider, Jr., R. A. Kossuth** (MAIN Coordination Center) and **G. L. Landgren,** Chairman (Commonwealth Edison Company): The Task Force on T&D Outage Definitions is to be praised for its effects to update terms for reporting outages. We recognize how formidable this task is. The MAIN Task Force has been keenly aware of the difficulties in comparing outage data reported by groups using different definitions. In our opinion, the proposed terms will be a decided improvement over the present IEEE Standard 346. It represents a concerted effort to include those fundamental terms required for a broad understanding of transmission outage data.

Although the proposed terms are judged to be complete and satisfactory for broad understandind of the subject, more details are required for unambiguous reporting of outage experience data and indices. Such detail was intentionally omitted from the proposed Standard 859; however, we feel that it is needed for effective communication. If the additional terms required for unambiguous reporting are not standardized by the IEEE or another organization representing the entire industry, regional differences will inexorably grow to the point that effective exchange of data by the industry is precluded. As a minimum, potential ambiguities should be identified in the Standard by examples so that the effects of significant differences in reporting procedures can be recognized and evaluated by users of published performance data.

The following are some of the terms having major potential for ambiguity:

1. Terminal versus termination. Is the terminal based on geographic or electrical factors? What components are in a termination?
2. Total versus partial outage events and durations for multi terminal units. Are the outage duration data based on restoration or repair? Are switching steps (outage and restoration) included in the outage rates?
3. Transmimssion unit. What physical components are included? Is a transformer tapped to an overhead line without high side breaker a separate unit or part of a combination line/transformer unit?
4. Outage. Are "open ended unit" events counted as an outage (e.g., one termination of a unit open, but the unit otherwise energized)?

Many of the potential ambiguities could be highlighted by increased use of examples in the proposed IEEE Standard. We suggest a tutorial section as an appendix containing diagrams and examples illustrating the concepts and definitions.

Following are specific suggestions for changes in the definitions which would eliminate some of the ambiguities mentioned above:

a. For an analysis of a multi-terminal line, the term "Restoration Time" should be defined and included in Section III. This definition will distinguish the outage time for a complete outage state caused by an outage event from the outage time (restoration time) to restore a "Unit" to in-service state from a partial outage state.
b. Use of the word "related" in the definition II-4.0 is confusing with the definition II-4.2.2 of Related Multiple Outage Event. A question also arises (for the definition II-4.0) as to related in what sense? Are they occurring in the same (a narrow) time frame or related in some other sense? An example of multiple outage occurrences in an outage event would be very helpful.
c. Primary outage (II-4.2.1) and secondary outage (II-4.2.2.2) have been defined as sub-categories of the Related Multiple Outage Event (II-4.2.2). Is it the intent of the Task Force to imply that a single outage or Independent Multiple Outage Event should no longer be referred to as a Primary Outage?
d. Definitions and methods of calculating "Normal Weather Service Time" and "Adverse Weather Service Time" require some clarification.

Manuscript received February 24, 1984.

J. H. Doudna, P. J. Lehman, D. D. Klempel, C. G. Dahl, and **M. G. Lauby** (Mid-Continent Area Power Pool Transmission Reliability Task Force): The development of terms which adequately, completely, and clearly define the basic aspects associated with the collection and analysis of electrical facility outage data is a complex task. The discussers recognize the large amount of effort and time which the authors have expended in the preparation of the terms presented in this paper and feel that their work is an important contribution in the area of system reliability evaluation.

The discussers agree that there is a need to establish standard terms and definitions for use in the system reliability area. We support the effort and intent of the authors to propose the terms presented in their paper as an IEEE Standard. However, prior to such proposal, we would like the authors to consider and incorporate the following comments:

Under Section I, Item 1.0, the phrase "performs a major operating function" is somewhat subjective. We propose the following wording for this definition:

"A device or piece of equipment which could be expected to significantly affect or influence the reliability of the system and which can be regarded as an entity for the purposes of recording and analyzing outage data."

Some of the examples given for components, particularly line sections, power transformers and ac/dc converters, can be treated and viewed as units depending on the particular collection scheme. There is not a black and white boundary which distinguishes a "Component" from a "unit" and this should be noted when preparing the standard. The examples cited effectively define a collection technique, which was not the expressed purpose of the paper.

Under Section I, Item 1.1, the use of "station" is not consistent with the use of "terminal" in other portions of the paper when delineating the bounds of a transmission facility. We propose the following definition for a Line Section:

"A portion of overhead line or a portion of cable bounded by two terminals. Such terminals may be at a substation or a tap point."

Also, Under Section I, the term "terminal" is not defined. The discussers feel that the inclusion of a definition for "terminal" might be useful and eliminate any chance of confusion.

Under Section I, Item 1.2, we feel that a line segment should be delineated only by its physical construction characteristics not by its exposure characteristics. If the authors desire to keep the exposure concept in delineating a line segment, we propose the use of two terms, "Physical Line Segment" and "Exposure Line Segment."

Under Section II, since these terms apply to units and components, we suggest the use of "unit/component" in the definition rather than "unit or component." "Unit/Component" should also be inserted as part of the defined term (i.e., Unit/Component In-Service State, Unit/Component Outage State, etc.). We feel this change more clearly indicates that these terms apply either to Components or Unit's separately, but not to Units and Components collectively (i.e., mixed mode accounting of outages). This will hopefully help potential users such that, once components and units are defined, it will be clear for any number of parties to understand what is meant (e.g., Unit Related Outage Event as opposed to Component Related Outage Event).

Under Section II, Item 1.0, for In-Service State, the phrase "fully connected" could be expressed differently. We suggest the following wording:

"The condition where the component/unit is energized and is performing its intended function."

Under Section II, Item 2.0, we suggest the following wording:

"The condition where the component/unit is not performing its intended function."

Under Section II, Item 2.1, the phrase "or is connected at only one terminal" should not be in the definition as this describes a Partial Outage State not a Complete Outage State. Thus the definition should read:

"The condition where the component/unit is not performing its intended function and is completely de-energized."

For Section II, Item 2.2, we suggest wording similar to 2.1. The use of the word "terminals" in this definition may not be appropriate as some components, such as relays, may not normally be considered as having "terminals" as would a line or transformer.

Under Section II, Item 4.0, we prefer a little different wording although the concept is that same. We suggest:

"An occurence of one or more outages which can be attributed to a single initiating cause."

Under Section II, Item 4.1, we suggest the use of the word "Independent" instead of the word "Single," that is, "Independent Outage Event."

Under Section II, if the proposed wording is used for 4.0, it may not be necessary to have the definitions in 4.2 and 4.2.1.

Under Section II, Item 4.2.2, the use of the word "Multiple" in the phrase heading is redundant as Related Outage always involved two or more unit/component outages, and thus are multiple outages.

Under Section II, after Item 4.2.2, we feel a definition for "Associated Outage Event" should be added. Related Outages can be subdivided into Associated or Common Mode, thus this addition would make it clear that Related Outages which are not Common Mode Outages must be Associated Outages. This allows one to draw a clearer logic tree of how Related Outages are analyzed.

Under Section II, Item 6.0, we feel definitions should be added to explicitly describe both types of failures. That is add definitions for:

Latent (or passive) Failure and Known (or active) Failure.

Under Section III, Item 2.0, the word "event" should be added to each classification to be consistent with previous usage in the paper. (i.e., Forced Outage *Event* Time).

Also under this same item, we feel the term "Related Outage Event Time" should be included. Such an Outage Time can be identified and calculated. At MAPP, for example, a Related Outage Event duration is defined as the outage duration which is the smallest. This criteria is used because it was assumed that at the point the first outage has completed is also the point at which the Related Outage Event ceases to be "related."

Also under this same item, we feel there should be some statement made as to how each of these outage event times are calculated. That is, when do they begin and when do they end?

Under Section IV, Items 4.1 an 4.2, are somewhat subjective in nature, and as such, may not be appropriate for inclusion in a standard.

For Item 4.1, we do not agree that a tornado is a good example of "Adverse Weather" as it could be and often is a "Major Storm Disaster."

For Item 4.2, the use of the word "facility" is preferred over the use of the word "plant." We further disagree with the three criteria which must be satisfied in order for an outage to be classified as a "Major Storm Disaster." The outage of a transmission line may be lengthy due to extensive storm damage, but such an outage may not and often does not cause any interruption in service to customers.

Under Section V, it is not apparent that calculations can be applied to "pooled" data of more than one facility. The general case would be to use Σ to indicate the summation of quantities. This is then applicable to one or more facilities. For example, for a pooled data set, the Mean Outage Duration (MOD) is calculated by

$$\text{MOD} = \frac{\Sigma \ \text{Outage times}}{\Sigma \ \# \ \text{of Outages}}$$

This can be generalized for each desired data set.

Manuscript received February 24, 1984.

Proposed Terms for Reporting and Analyzing Outages of Electrical Transmission and Distribution Facilities

Prepared by the T&D Outage Definitions Task Force, Application of Probability Methods Subcommittee, D. W. Forrest, Chairman: The objective of our paper was to solicit, from industry, comments on our proposed definitions. This objective was achieved and the task force wishes to thank all the discussers for their comments. We will consider each comment in the development of the IEEE Standard. Because of the conflicting nature of the comments, some compromises will have to be made. We expect that a Standard will be prepared for ballot within a year.

We considered the suggestion by J. G. Anders et al., that we publish a second paper with recommended definitions. However, the existing Standard 346-1973 has been partially replaced by a new Standard 858. This leaves the definitions in Standard 346-1973 section 2 in a state of limbo. Therefore, we believe a new Standard for definitions must be published as soon as possible.

Many of the questions which we received will be answered by describing the task force's philosophy on important issues. In other cases, we will respond to specific comments to explain the intent of our definitions. Errors which were pointed out by the discussers have been corrected.

The task force wishes to emphasize that the proposed Standard will not be a Standard for a data collection system. Rather it will define terms to facilitate both dialogue within the industry and data collection. The design of a data collection system must be determined by the user after consideration of what their ultimate goal is, what data are required to achieve that goal, and the economics of collecting those data.

The task force believes that the IEEE Standard it will propose should define terms and methods, which are in general use. Unfortunately there are many variations in the terms and methods in industry. It will be necessary for us to make compromises to achieve a standard terms. Also since the Standard is for terms and methods in general use, it will not contain all the definitions which some groups might desire. This Standard cannot be all emcompassing and should not explore new ground. As the state of the art progresses, this Standard will be revised just as Standard 346-1973 is being revised.

The task force deleted several terms from this paper for the purpose of brevity. Several discussers suggested that these terms be added. They will be reconsidered. However, we generally feel that the level of detail of the definitions is about right for a Standard since comments were received requesting both more and less detail. The proposed Standard will not define terms involving customer interruptions or protective relaying. These topics will be addressed by the distribution Reliability Task Force of the Transmission and Distribution Committee and the Protective Relay System Availability Working Group of the Relay Committee respectively.

Several discussers requested diagrams to help define terms such as unit, component, and subcomponent. The proposed Standard will contain some diagrams although they may not be the same as those proposed.

We received several comments on the definition of partial outage. We recognize that more examples are needed to explain this definition. Our assumption was that a partial outage can be represented in a load flow program by removing one branch. A three-terminal line energized from only one terminal could be represented in a load flow program by removing two or three branches and thus is considered a complete outage. One could argue that a three-terminal line energized at one terminal is exposed to faults and thus this state must be recorded. We do not believe that this is necessary. However, one could define the term "energized complete outage" to cover this case.

We find that there are two concepts of the term unit. The first concept is that a unit is a group of components or more simply a "super" component. The second concept is that a unit is a capability to transfer power. In this second concept, a unit is dynamic, that is, its components change with time. The proposed definition leans heavily towards the first concept. Development of the second concept requires further efforts in industry.

In their comment on section 2.0, we believe J. G. Anders et al., are confusing the outage state with the failure state. Our definitions allow for a component or unit to be in the outage state for reasons other than failure. A unit taken out of service for operational reasons is in the outage state (5.5.2); however, it is not failed. The type of outages to be considered in projecting outage rates would depend on the application. For example, operational outages would not be included in forced outage rate projections.

We have the following specific responses to H. Bonser and D. Goodman:

    A. We agree that the condition described would be considered an operations-related outage. Any component or unit in the operations-related outage state can be restored to the in-service state when system conditions require.

    F. Short outages where the restoration is by automatic reclosing are transient forced outages. If the first reclosure is unsuccessful and the second reclosure is successful, the component or unit is in the outage state until the second reclosure. However, it is desirable to have a data collection system that is able to record unsuccessful reclosures, especially when the unsuccessful reclosure resulted in a second fault on the system.

We have reviewed the comments from H. W. Colborn of NERC and will consider them in the preparation of the new Standard. We have specific responses on the following comments:

    II-2.0 An open breaker that is not isolated from the system and that has not failed is considered to be in-service.

    II-4.3 If a relay technician caused a relay system to operate improperly, we would call the resulting outage an automatic outage.

    II-5.2 The need for repair is potentially significant information which cannot be derived from the outage and restoration times. Since durations of outages are usually recorded, it may be desirable to record subclassifications such as switching or need to repair.

    Remote, supervisory-controlled switching is to be classified as manual.

    V We feel that the proposed definitions of indices in this paper are simply a formalization of indices presently in use. It is beyond the scope of this effort to develop bulk power system indices.

We have five specific responses to the comments by J. H. Doudra et al.

    1. We disagree with the suggested changes to the definition of component. We propose that components are devices which perform major operating functions because these devices are

what we model in system studies. Many pieces of equipment, for example, the subcomponents proposed by J. G. Anders et al., can significantly influence reliability.

2. We disagree with the comments that listing examples of components defines a data collection technique. We are simply identifying and naming the hierachy which usually is considered when collecting outage data: unit, component, subcomponent etc. We are not advocating collecting data at any level of detail. One may wish to collect data at the component and subcomponent level or one may wish to collect data at the unit level.

3. In Section I.1.2, we believe line segments must be delineated by physical construction characteristics *and* by exposure. For example, one may wish to know the outage rate for a 345 kV line, horizontal configuration on wood H-frames which is exposed to salt spray. When collecting data, one must review each line to determine where significant changes in physical construction or exposure occur. The places where these changes occur are the beginning and ending points of line segments.

4. We agree that the definitions of weather in section IV are subjective; however, they are the same as in Standard 346-1973. Outage rates vary greatly between normal weather and adverse weather. We think that these terms must be included in this standard to emphasize that point. Also, we use these terms in section V.

5. We agree that the formulas in section V can be used with "pooled data."

Manuscript received September 28, 1984.

THE general planning problem consists, traditionally, of a comparison between various alternatives for system development made on the basis of "system cost." Two methods can, at least theoretically, be followed in this approach to determine the system cost. The first method compares the cost of the investments for constructing the facilities and the expenses necessary for running them (fuel + operation and maintenance), under the assumption that each alternative should have practically the same reliability based on whatever risk index or procedure is used to obtain the index. The second procedure incorporates reliability in the costing process by comparing the overall (investment + running + reliability) cost. The costs in both methods are those incurred by the utility in each year of the time span considered, and equated using present worth analysis.

The first method implies that the reliability level is selected, a priori, without detailed qualifications of the advisability of the choice. In reality an increase in reliability (lower risk index) might be advisable even if this results in a slight increase in the cost. A slight reduction in reliability (higher risk index) might, on the other hand, be acceptable if it allows significant savings.

The second method does not involve a predetermined and fixed reliability level, instead it makes the comparison by attempting to minimize the overall cost. The reliability level can therefore be considered as the optimal level. This approach requires the assignment of a unit value or "failure cost" to the risk index selected for reliability assessment. Theoretically, the failure cost corresponds to the change in socioeconomic costs corresponding to a change of one unit in the risk index.

Two classical approaches exist for relating the socioeconomic costs to the risk index. These are the implicit cost and the explicit cost methods. With respect to the implicit cost, it may be argued that the value of the risk indices adopted by utilities in response to public needs as shaped by economic and/or regulatory forces, should reflect the optimum trade-off between the cost of achieving the value and the benefits derived by society. In the constrained problem of power system overall cost optimization, at the optimum for each active constraint (which may represent either a physical or a performance limit), a shadow price gives the effect on the total (optimal) cost for a unit change in the constraint. Assuming that the risk index is an active constraint, the shadow price for a given system plan in any year of its development is the implicit value of the failure cost (c_r).

For example, consider that the risk index selected is the expected energy not supplied (EENS) and that for a given development plan, the addition of a peaking gas turbine unit is necessary in order to reduce the corresponding EENS constraint. At the optimum, the variation of the overall system cost due to the additional capital and operating costs associated with the gas turbine addition corresponds to the decrease in the "risk cost" ($c_r \times$ EENS). The unit cost c_r, which is used to express the index EENS in money, must therefore be implicitly determined.

The explicit cost approach uses subjective and objective measures of customer "losses" arising from curtailments of supply. This method has been applied extensively to the risk index EENS. The unit cost of losses due to energy not supplied is a composite parameter formed from the losses of various classes of customers affected by an interruption. Investigations of the customer perception of the cost of the interruption have shown that figures vary widely from one customer to another but also depend on considerations that are purely subjective. Cost of losses due to interruptions are not always readily deduced from the direct economic consequences of interruptions. There are also indirect effects, which are often difficult to observe, analyze and predict.

Utilities are continually expanding and refining their techniques and analyses in order to apply the previously noted concepts in their planning practices. Reference [A1] provides a comprehensive view of the approach used by Electricite de France, which has considerable experience in the field of reliability evaluation using the implicit cost approach.

The results of explicit cost evaluation are available from six major surveys, conducted from 1969 to 1984 in various countries [A2]–[A4] as follows:

Sweden (1969), by the Committee on Supply Interruption Cost;

Canada (1978), by Ontario Hydro;

Finland (1979), by the Finnish Power Producers Coordinating Council (STYV);

Denmark (1981), by the Research Association of the Danish Electricity Supply Undertakings (DEFU);

Sweden (1982), by the Swedish Association of Electricity Supply Undertakings;

Canada (1984), by the Power System Research Group, University of Saskatchewan

The sectors examined in the surveys are indicated in the following table:

Sector	Sweden 1969	Canada 1978	Finland 1979	Denmark 1981	Sweden 1982	Canada 1984/87
Industry	*	*	*	*	*	*
Domestic	*		*	*	*	*
Tertiary	*		*		*	*
Agriculture	*		*	*	*	*

The table above shows that the tertiary sector was surveyed in only four of the six studies. This sector was subdivided as shown in the next table.

TERTIARY SECTOR

	Sweden 1969	Finland 1979	Sweden 1982	Canada 1984
Commercial	*	*	*	*
Offices	*	*		
Services		*	*	*
Public administration		*		
Transport	*	*	*	
Communications		*		
Harbors	*			

Only macro-interruptions of energy supply from seconds to many hours were investigated in these studies. Other possible malfunctions, such as micro-interruptions, temporary (milliseconds) voltage fluctuations, and waveshape variations, affecting computers were not considered. Only direct costs were identified.

The papers included in this book present some of the work done in Canada on the evaluation of explicit failure costs. Reference [1] is a bibliography of papers and publications on this subject prepared in 1983.

Reference [2] presents the results of surveys made in 1976 by Ontario Hydro on customers with demands in excess of 5 MW, in order to determine the effect of various proposed levels of reliability. Most of the customers were engaged in mining or manufacturing, but other activities ranging from public transit to universities were included. The results include the customers' estimates of interruptions of nine durations, from less than one minute to one week. The estimates were reduced to a cost in $/kW of load interrupted; individual annual peak and noncoincident annual peak demands for groups were used to normalize the costs. Substantial variation in such costs, within and among industry groups, was found. The effect of advance warning, frequency of occurrence, season and time of occurence on the interruption cost was also investigated. The tolerance of accepting voltage reductions of about 5 percent without being forced to curtail production was found. Reference [2] highlights many aspects that similar questionnaires could cover, and the related difficulties, and presents a considerable amount of quantitative data. The utilization of the results in system planning and operation was, at that time, still a subject of considerable reflection. An indication of the early concern was provided in [A5]. Reference [3] presents the results and methodology of a more recent survey performed to estimate the cost associated with interruptions in residential service. Additional work in the commercial and small industry sectors is presented in [A6]. These concepts are extended in [A7] and [A8] to the development of customer damage functions.

Reference [4], published almost ten years after [2], illustrates the application of the customer cost data and shows how these results can be utilized in the evaluation of an interrupted energy assessment rate (IEAR)—that is, the "cost of failure" in ($/kWh curtailed) of a particular utility service area. The IEAR for a given service area is obtained using the sector customer damage functions [A7]–[A9] for that area.

The assessment of reliability worth is a concept in continuous evolution. In the recent years of volatile energy costs and uncertain load growth, work on many proposed generating plants has been delayed or curtailed for various reasons ranging from public opposition to unwillingness to plan further capital-intensive-capacity additions. Privatization can also add uncertainties to the overall framework. As a result, there is the possibility that new planning approaches will be implemented which offer more explicit cost-reliability choices to consumers. Such programs, beginning with demand side management options, may involve the direct control by the consumer and the unbundling of electric services into discrete levels of reliability, with corresponding rates for each. Investigations leading to a better understanding and identification of the value that each customer sector places on reliability are already the subject of electric utility research programs.

REFERENCES

Reprinted Papers

[1] R. Billinton, G. Wacker, and E. Wojczynski, "Comprehensive bibliography on electric service interruption costs," *IEEE Trans. Power Apparatus Syst.,* vol. PAS-102, no. 6, pp. 1831–1837, June 1983.

[2] E. M. Mackay and L. H. Berk, "Costs of power interruptions to industry-survey results," *CIGRE,* Paper 32-07, 1978.

[3] G. Wacker, E. Wojczynski, and R. Billinton, "Interruption cost methodology and results—A Canadian residential survey," *IEEE Trans. Power Apparatus Syst.,* vol. PAS-102, no. 10, pp. 3385–3392, Oct. 1983.

[4] R. Billinton, J. Oteng-Adjei, and R. Ghajar, "Comparison of two alternate methods to establish an interrupted energy assessment rate," *IEEE Trans. Power Syst.,* vol. PWRS-2, no. 3, Aug. 1987, pp. 751–757.

Additional Papers

[A1] "Cost of interruption in electric power supply," Document transmitted by the Government of France Economic Commission for Europe. EP/GE/R.61/Add.1, 27.2878.

[A2] Committee on Supply Interruption Costs, "Cost of interruption in electricity supply," The Electricity Council, London, 1969.

[A3] The Finnish Power Producers Co-ordinating Council, "Value of not-distributed energy," Helsinki 1979.

[A4] Committee on Supply Interruption Costs, "An evaluation of the interruption in the electricity supply," The Swedish Association of Electricity Supply Undertakings, Stockholm, 1982.

[A5] R. B. Shipley, A. D. Patton, and J. S. Denison, "Power reliability cost vs. worth," *IEEE Trans. Power Apparatus Syst.,* vol. PAS-91, no. 5, pp. 2204–2212, Sept./Oct. 1972.

[A6] E. Wojczynski, R. Billinton, and G. Wacker, "Interruption cost methodology and results–A Canadian commercial and small industry survey," *IEEE Trans. Power Apparatus Syst.,* vol. PAS-103, no. 2, pp. 437–443, Feb. 1984.

[A7] R. Billinton, G. Wacker, and R. K. Subramaniam, "Factors affecting the development of a commercial customer damage function," *IEEE Trans. Power Syst.,* vol. PWRS-1, no. 4, pp. 28–33, Nov. 1986.

[A8] R. Billinton, G. Wacker, and R. K. Subramaniam, "Factors affecting the development of a residential customer damage function," *IEEE Trans. Power Syst.,* vol. PWRS-2, no. 1, pp. 204–209, Feb. 1987.

[A9] R. K. Subramaniam, R. Billinton, and G. Wacker, "Factors affecting the development of an industrial customer damage function," *IEEE Trans. Power Apparatus Syst.,* vol. PAS-104, no. 11, pp. 3209–3215, Nov. 1987.

COSTS OF POWER INTERRUPTIONS TO INDUSTRY
SURVEY RESULTS

by

E.M. MACKAY, L.H. BERK

Ontario Hydro

(Canada)

SUMMARY

Ontario Hydro has surveyed customers with demands in excess of 5 MW to determine the effects of various levels of reliability as perceived by them. Most of these were engaged in mining or manufacturing but other activities ranging from public transit to universities were included.

The report presents results from 172 establishments with peak demands totalling 3 900 MW in 1975. The results include the customers' estimates of the cost of interruptions of nine durations, and of other factors relating to reliability. The latter include effects of system voltage reductions, interruptible loads, use of stand-by generation and load shedding.

Costs, Customer evaluation, Industry, Interruption, Reliability.

REPORT

1. INTRODUCTION

The Ontario Hydro power system includes some 19 000 MW of generating capacity and an extensive bulk power transmission network, with interconnections to neighbouring power systems in Canada and the U.S.A. The system has been designed to criteria evolved over the years which provide a high and generally satisfactory level of reliability of electricity supply to users. In turn, it is believed that users have progressively adapted their production operations and their way of life to depend on this level of reliability.

In recent years, however, the level of reliability has come under public examination in Ontario, as well as elsewhere, because of concern about rising costs for electrical energy, and the impact of additional electrical supply facilities on the environment.

Some groups favour maintenance of the present level. Others have suggested that this is too high and that a lower level, which would require fewer additional facilities and in particular a lower level of generation reserve, should be accepted. These argue that major reductions in capital expenditures might thereby be possible, and that the resulting lower costs for supply of electricity would offset the costs and inconveniences imposed on users by more frequent interruptions. They also argue that the probability of unplanned interruptions and the cost to the

Reprinted with permission from *CIGRE,* Paper 32-07, Aug. 30–Sept. 7, 1978.

economy of those which do occur could be reduced ; electric utilities and industry could and should adapt themselves to regular use of methods of reducing load which are now used infrequently and only in severe emergencies. This would be a major departure from present practice.

Some work has already been done elsewhere [1, 2, 3] to determine the cost of interruptions to users. However, there is still too little information of this type available to assess the effects of such a change on customers individually or in groups.

To provide a basis for studies of this matter Ontario Hydro has planned surveys of various classes of customers to determine :

1.1. — Their estimates of the cost of interruptions in their power supply for various durations and conditions.

1.2. — How their production would be affected if system voltage were lowered to reduce demand.

1.3. — The amount of their load which might be interrupted without major disruption to production.

1.4. — The amount of stand-by generation they now have and its present purposes, and how these might change if interruptions increase.

1.5. — Their preferences if regular load-shedding should be necessary.

This report is based on results of a survey of all establishments supplied by Ontario Hydro having maximum demands in excess of 5 MW, carried out in 1976 [4]. Most members were engaged in such industries as mining or manufacturing, but the group included other activities ranging from public transit to universities.

2. SURVEY METHOD AND RESPONSE

The survey used a detailed 14-page questionnaire designed and tested by Ontario Hydro's market research staff. Customer service representatives discussed the questionnaire with the management of each establishment. It was then left for them to complete. Respondents were promised that individual replies would be held in confidence ; this contributed to a high rate of response although it limited sub-division of some industry groups.

Replies were received from 172 establishments, about 86 percent of the group membership, although some responses were incomplete, particularly on questions about costs. In 1975 the peak demands those replying totalled 3 900 MW and their energy consumption totalled 17 500 GWH, about 25 percent of the total primary energy sold by Ontario Hydro. Responses were classified into 12 industry groups including mines, iron and steel mills and various types of manufacturing.

3. COST OF INTERRUPTIONS – INDUSTRY ESTIMATES

There were 115 useable responses on this question from establishments with non-coincident peak demands totalling 2 658 MW. The total cost estimated by these customers amounted to $1.8 million for an interruption of less than one minute and $7.0 million for an interruption of one hour.

3.1. Scope of Questions – Users were asked to estimate the costs to them of interruptions without warning in their electrical supply, on the basis of full production conditions. Nine durations, from less than one minute to one week, were specified. Costs were to include the value of lost production, out-of-pocket expenses for idle labour, spoiled material, cleanup, etc., and repair of damaged equipment. Although the value of lost production may overstate the net cost of an interruption in those cases where it can be made up using idle capacity, it is a commonly available figure ; to have requested a net figure would have involved serious difficulties in definition and the probability of lower response. Users were also asked to indicate the effect which advance warning, frequency of occurrence, time of day and season would have on these costs.

3.2. $/kW of Load Interrupted – The respondents varied greatly in size, as well as other characteristics, and their estimates varied accordingly. To place them on a comparable basis, all estimates were reduced to a cost in $/kW of load interrupted. The cost estimates for an individual respondent for each duration of interruption were divided by that respondent's annual peak demand. Similarly, the sum of the estimates for each industry group were divided by the sum of the non-coincident peak demands of the members of the group.

The results have been plotted using logarithmic scales for both cost and duration because of the wide range of values. This permits more detailed observation of the cost variations for the interruptions of shorter durations.

3.3. Cost Estimates – General Results – Figure 1 shows the results for all 115 respondents as a group and for each of the 12 industry groups.

For all 115 respondents, the estimates were $0.68/kW for an interruption of less than one minute, and $2.69/kW for an interruption of one hour. Estimates for various industry groups varied widely about these values, from almost zero to $1.76/kW for a momentary interruption and from $0.20/kW to $8.95/kW for a one-hour interruption.

3.4. Cost Estimates – Industry Groups – Results for five industry groups are shown in Figures 2-6. The maximum (a) and minimum (c) individual cost values found in the group for each duration of interruption are also shown on these Figures ; these values form a cost band (d) about the value for the group as a whole (b), as illustrated on Figure 2.

The cost band is relatively narrow for Rubber Products Manufacturers (Fig. 2), Newsprint Mills (Fig. 3) and Iron Foundries (Fig. 4). In these industry groups the individual members employ generally similar processes although the products may be different, and costs of interruptions are broadly similar for all members. In contrast, the Iron and Steel Mills (Fig. 5) and Petroleum and Chemicals (Fig. 6) groups include a wide range of different processes and this is reflected in the broad cost bands for these groups. In the latter group for example, the cost for a momentary interruption ranged from 0 to $15/kW and for a one-hour interruption from about $.04 to $25./kW.

3.5. Advance Warning To Reduce Interruption Costs – About one-third of the respondents considered that adequate warning would reduce the cost of interruptions substantially but slightly more than one-half thought it would have no effect. Opinion was divided within most industry groups. Iron and Steel Mills was the only group in which a majority of respondents considered that advance warning would reduce costs of interruption.

3.6. Effect of Frequency of Occurrence on Interruption Cost – About one-quarter of respondents considered that more frequent interruptions would increase the cost per interruption over that of a single interruption of the same total duration. However, more than one-half indicated that this would cause no increase. Again, opinion was divided within industry groups.

3.7. Effects of Season and Time of Occurrence on Cost – About half of the respondents reported that the cost of interruptions would not vary with the season of occurrence. The majority of the others reported costs would be lowest in July and August, and highest from November to March, some because that is their period of greatest production, others because of damage from cold. Abour two-thirds reported costs would be highest for interruptions during the day shift and the normal working week ; the others reported no difference.

4. EFFECT OF VOLTAGE REDUCTION

More than 80 percent of the respondents considered that they could tolerate voltage reductions of 5 percent or less without being forced to curtail production. Acceptance diminished as the magnitude of the prospective reduction was increased further : only about 20 percent considered that production would not be curtailed by a reduction of 10 percent.

This is in accordance with experience from system tests and in emergencies which has shown that voltage reductions of 5 percent or less do not cause problems for most customers.

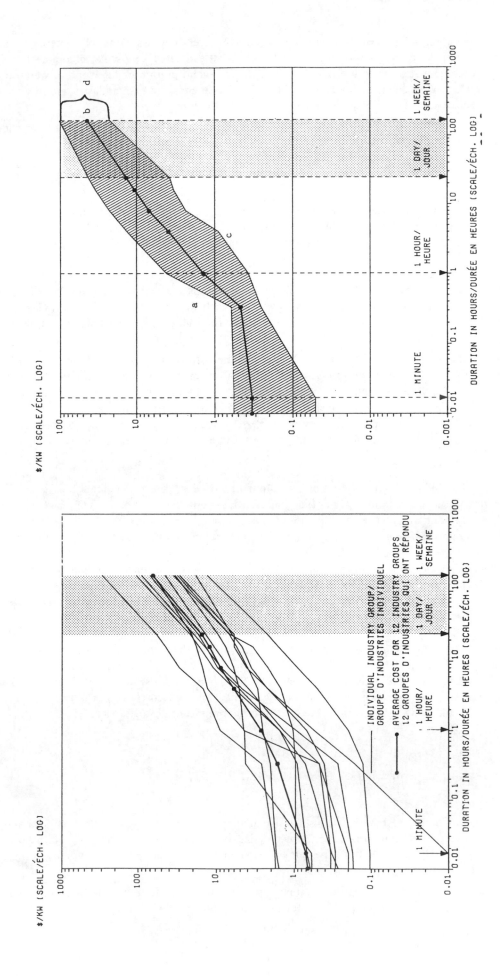

Figure 2 – Rubber products manufacturers.

Figure 1 – $/KW interrupted for specified durations.

472

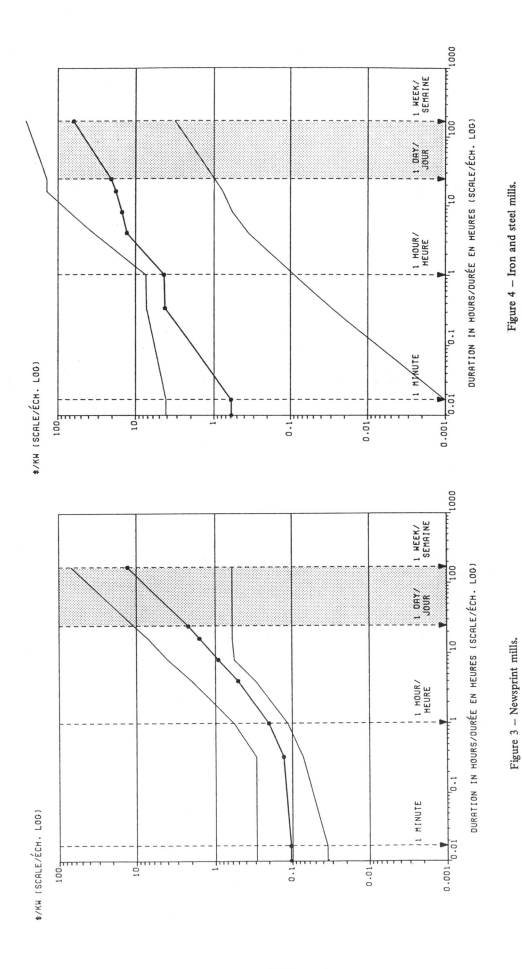

Figure 4 – Iron and steel mills.

Figure 3 – Newsprint mills.

473

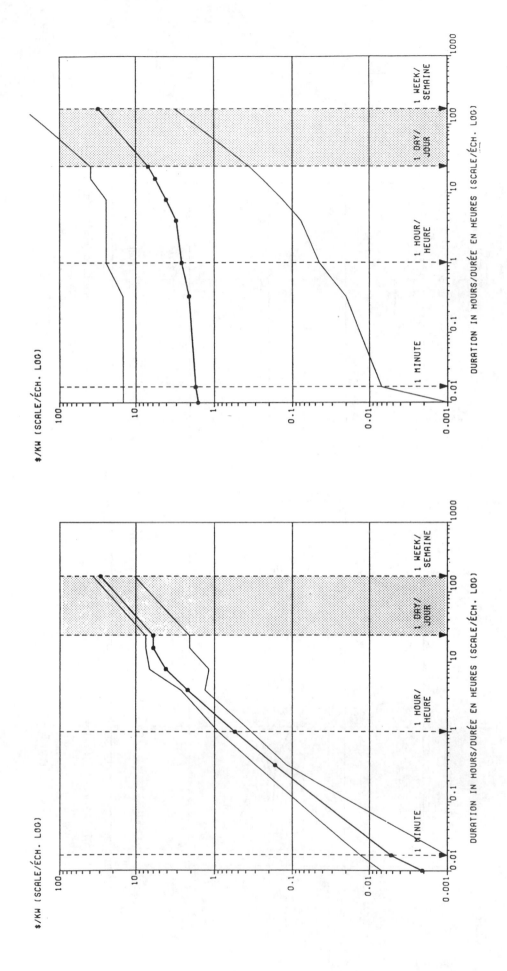

Figure 6 – Petroleum and miscellaneous chemicals.

Figure 5 – Iron foundries.

474

5. INTERRUPTIBLE POWER

Ontario Hydro now has contracts with customers for some 550 MW of interruptible power. This is sold at a lower price than firm power. In return, the customer must agree to reduce demand to the firm power contract amount on request within specified limits for duration of reduction and frequency of occurrence. The minimum acceptable amount is 5 MW ; smaller amounts have not been considered to warrant the problems involved in giving notice of reduction, billing etc.

The survey addressed only the physical possibility of interrupting part of the users' loads and did not attempt to determine under what terms interruptible power might be commercially acceptable to them.

A total of 64 respondents now without contracts for interruptible power indicated they had loads which could be interrupted. These totalled 388 MW, about 2 1/2 percent of the peak demand on the Ontario Hydro system. The total included 110 MW divided among 43 respondents in amounts of less than 5 MW.

6. STAND-BY CAPACITY

The effect of an interruption to the utility supply could conceivably be offset by use of customer-owned generation, defined as Stand-by capacity and intended for use only during interruptions to replace the normal electrical supply or part of it. How much was available ? The survey found that :

6.1. – Only 60 percent of these large users had any stand-by generating capacity ; 40 percent had none at all.

6.2. – The total capacity of those with stand-by generation amounted only to 4.2 percent of the total of their peak demands.

6.3. – Some installations consisted of several small generating units at different locations.

6.4. – The two largest installations had capacities of 15 MW and 10 MW. Most installations were much smaller.

6.5. – Almost 40 percent of respondents also reported stand-by mechanical equipment with a rated capacity totalling 28 MW. Including this total, stand-by capacity was about 3.3 percent of the total peak demands of all respondents.

6.6. – Most stand-by equipment was assigned to maintain essential services, e.g. emergency lighting and fire protection, to protect vulnerable equipment, e.g. by maintaining the supply of cooling water for furnaces or to keep process material from settling or solidifying. Because of its capacity and the arrangement of most installations it appears unlikely that this equipment could be used to reduce system demand or to maintain production.

6.7. – Some respondents indicated that they would add stand-by equipment to reduce hazards and damage if the frequency of interruptions increased significantly. However, many indicated that it would be uneconomic for them to provide sufficient stand-by capacity to maintain production.

7. ORGANIZED LOAD SHEDDING

This question referred to a series of planned load interruptions affecting all customers in rotation which might be necessary to ration power. It was found that :

7.1. – About three-quarters of the respondents would prefer fewer interruptions of longer duration to more frequent interruptions of the same total duration, but about one-fifth would prefer the opposite. A few said they could accept neither.

7.2. — Most preferred a partial reduction in supply of longer duration to a total interruption of shorter duration but about one-fifth preferred the latter.

7.3. — Estimates of the amount of advance warning required for planned interruptions varied widely, with an average requirement of 3 hours before a total interruption and of 1.6 hours before a partial interruption.

8 — CONCLUSIONS

For these customers with demand in excess of 5 MW :

8.1. — Interruptions in electricity supply are generally reported to impose large costs on industrial users ; there are substantial variations in the cost of interruptions within and between industry groups.

8.2. — The costs obtained from this survey are not precise, but can be used to determine the range of cost of interruptions for a group of industries.

8.3. — Such factors as advance warning, frequency of occurrence and time of year have little effect on the cost of interruptions for most industrial respondents.

8.4. — Industries can accept a reduction of 5 percent in system voltage in emergencies without curtailing production. Greater reductions could curtail production.

8.5. — There is no large quantity of readily available interruptible load not already served, although some industries have load which can be interrupted in emergencies. About one-third of this is in amounts of less than 5 MW.

8.6. — Existing stand-by installations generally could not be used to reduce system peak demands or to maintain production during interruptions. Most of these installations are neither large enough nor designed for that purpose.

8.7. — Many respondents believe it would be uneconomic to provide sufficient stand-by generation to maintain production during interruptions.

8.8. — If a regular load shedding program should be required, different industries would prefer different combinations of magnitude and duration for the same total reduction.

8.9. — Most of the cost estimates are for conditions which the respondents have seldom or never experienced ; if they have, restoration rather than cost was probably their primary concern. The validity of the cost results therefore remains a question (and one which has been raised most often by the interested groups to whom this information has been distributed). Accordingly, it would be desirable to compare these results with actual costs for these or similar establishments in any area which has experienced extensive interruptions. It would also be useful to compare them with the results of any similar surveys which may be undertaken by other utilities.

8.10. — Application of the results to planning and operating the power system is a complex problem and some time may be required to develop suitable methods.

8.11. — The engineering and operating functions should determine the information to be sought but market research specialists are best suited to plan and direct the process of obtaining it. A pilot survey to test the survey method is very important ; this led to many improvements in the survey under discussion. Where possible, a visit by a well-briefed representative to explain the survey to the management of each establishment can help to improve the rate of response and ensure the quality of information.

REFERENCES

[1] **Jonson G. et al.** – Cost of Interruptions In Electricity Supply, Committee on Supply Interruption Costs, Sweden, Sept. 1969, (Electricity Council Overseas Activities Translation Service, OA Trans 450 pp. 1-50).

[2] **Lundberg, Lennart.** – Report of the Group of Experts on Quality of Service from the Consumer's Point of View (Tariff Studies Committee, International Union of Producers And Distributors of Electrical Energy, 1972, Publication 60/D.1, pp. 1-27).

[3] **Dickinson W.H. et al.** – Report on Reliability Survey of Industrial Plants, Part II : Cost of Power Outages. (IEEE Transactions on Industry Applications, March/April 1974, pp. 236-241).

[4] **Ontario Hydro, Power Market Analysis Department.** – Ontario Hydro Survey on Power System Reliability : Viewpoint of Large Users. (Report No. PMA 76-5, April 1977).

Comprehensive Bibliography On
Electrical Service Interruption Costs

R. Billinton, G. Wacker, E. Wojczynski
Power System Research Group
University of Saskatchewan
Saskatoon, Saskatchewan
Canada S7N 0W0

ABSTRACT

This paper contains a comprehensive list of publications relating to the theory and results of research into the worth of reliability. The emphasis is on research and surveys concerned with determining the impacts and estimating the costs to customers of electrical service interruptions. The application of cost estimates in system design and other related topics are included. The first four categories in the bibliography are germane to the topic and the lists are intended to be exhaustive. The remaining categories are peripheral to the bibliography topic but are relevant in terms of power system planning or otherwise; hence their lists are more or less representative of existing knowledge rather than comprehensive.

INTRODUCTION

Increased attention is being given to numerical evaluation of the worth of reliability in regard to electric power supply. The material contained in this bibliography illustrates the state of the art in the collection of information on electrical service interruption costs and provides an awareness of the utilization of these data in the evaluation of power system reliability worth.

Short descriptions are provided for approximately one third of the publications listed in the bibliography. These references can be considered as the major publications in the respective categories and could therefore be studied first before proceeding if necessary with a more detailed examination.

1. IMPACTS OF INTERRUPTIONS

This section contains references investigating and describing the specific impacts to customers and society from electrical service interruptions and the more general impacts resulting from reduction in the reliability of supply. Part 1(a) references emphasize a qualitative description of the impacts to customers during interruptions and the possible response of customers and society to further interruptions. Part 1(b) references investigate in a qualitative and quantitative manner the impacts of blackouts in New York, while Part 1(c) investigates the impacts of blackouts in general.

83 WM 066-8   A paper recommended and approved by the IEEE Power System Engineering Committee of the IEEE Power Engineering Society for presentation at the IEEE/PES 1983 Winter Meeting, New York, New York, January 30-February 4, 1983. Manuscript submitted August 24, 1982; made available for printing November 10, 1982.

1(a) General Impacts of Interruptions and Reduced Levels of Reliability

1. D. McWilliams, "Users Needs: Lighting, Start-up Power, Transportation, Mechanical Utilities, Heating, Refrigeration and Production", IEEE Transactions on Industry Applications, vol. IA-10, no. 2, March/April 1974, pp. 199-204.

2. Technical Advisory Committee on the Impact of Inadequate Electric Power Supply, "The Adequacy of Future Electric Power Supply: Problems and Policies", Federal Power Commission, March 1976.

This report discusses the evaluation of electric power supply adequacy; methods of dealing with shortages and the effect of the shortages as they pertain to government and utility decision making in the U.S.A. The favoured short term method during shortages is rationing amongst customers on either an equal share or an economic basis. The effects of shortages are discussed mainly in a qualitative manner.

3. D. Myers et al., "Impacts from a Decrease in Electric Power Service Reliability", Bonneville Power Administration, June 1976.

This 49 page report attempts to qualitatively describe the impacts to customers of a reduction in their electrical service reliability. Methodologies using global economic indices (e.g. GNP/KWHR), as a means to estimate customer interruption cost are critically evaluated while the wages paid/KWHR approach is applied to the Pacific Northwest Region. The possible adaptive responses of customers to decreased reliability levels are discussed.

4. W. Turner, "Problems with Lifts during Power Cuts", Electrical Review, vol. 201, no. 20, November 18, 1977, pp. 57-58.

5. R. Bird, "Power Cuts ... the Cost to Industry", Electrical Engineer (Australia), December 1977, pp. 23-24.

6. J. Hooper, "Standby-Generating Systems, Power Cuts - Can Industry Afford Them?" Electrical Review, vol. 202, no. 15, April 1978, pp. 23-25.

7. T. Key, "Diagnosing Power Quality-Related Computer Problems", IEEE Transactions on Industry Applications, vol. IA-15, no. 4, July/August 1979, pp. 381-393.

8. K. Neilson, "Throwing a New Light on Security", Electrical Review, vol. 205, no. 11, September 21, 1979, pp. 47-50.

1(b) Impacts of Large Scale Interruptions (Blackouts) in New York

There are several studies not listed which investigate the New York area blackouts but their emphasis is on the engineering aspects of the causes of the event rather than on the impacts.

Reprinted from *IEEE Trans. Power Apparatus Syst.*, vol. PAS-102, no. 6, pp. 1831–1837, June 1983.

9. Federal Power Commission, "Northeast Power Failure, Nov. 9 and 10, 1965", U.S. Government Printing Office, Washington, D.C., December 1965.

10. G. D. Friedlander, "The Northeast Power Failure a Blanket of Darkness", IEEE Spectrum, February 1966, pp. 54-73.

11. Department of City Planning, City of New York, "Blackout Commercial Damage Survey", July 1977.

12. Joint Committee on Defense Production, Congress of the United States, "Emergency Preparedness in the Electric Power Industry and the Implications of the New York Blackout for Emergency Planning (Hearings)", U.S. Government Printing Office, Washington, D.C., August 1977.

13. A. Kaufman and B. Daly, "The Cost of an Urban Blackout, The Consolidated Edison Blackout, July 13-14, 1977", The Library of Congress Congressional Research Service, Washington, D.C., June 1978.

This study investigates the losses associated with the 1977 blackout which are estimated to consist of at least 172.7 M$ of economic costs and 136.8 M$ of social costs. Economic costs are estimated from the variation in regional business activity indices while social costs are estimated using figures from government agencies.

14. J. Corwin and W. Miles, "Impact Assessment of the 1977 New York City Blackout", U.S. Department of Energy, Washington, D.C., July 1978.

This study collects and investigates data on the blackout and defines a framework in order to estimate the value of reliability with knowledge of blackout impacts. The main distinction drawn between the impacts is between direct impacts (56 M$) and indirect impacts (290 M$). social and organizational impacts are qualitatively analyzed.

15. P. D. Blair, "Planning for Power Failures and the New York City Experience", Canadian Conference on Communication and Power, 1978.

16. W. Miles, J. Corwin and P. Blair, "Cost of Power Outages - The 1977 New York City Blackout", 1979 Reliability Conference for the Electric Power Industry, April 1979, pp. 193-197.

1(c) Impacts of Large Scale Interruptions (Blackouts) - General

17. Law Enforcement Assistance Administration Emergency Energy Committee, "Preliminary Report on Rolling Blackouts", U.S. Department of Justice, March 1, 1974.

18. D. Myers, "The Economic Effects to a Metropolitan Area of a Power Outage Resulting from an Earthquake", Earthquake Engineering Systems Inc., San Francisco, February 1978.

This report develops a methodology and an analytical model for determining the costs to consumers of an electric power outage. Regional economic data are used as the basis. Residential willingness to pay as a measure of cost is estimated by means of the residential price elasticity (yielding $.50/KWHR).

2. THEORY OF RELIABILITY WORTH

This section contains references dealing with the determination and use of reliability worth. Some of the more abstract economics papers dealing with this topic are included in Section 5. Papers reporting and discussing the results of surveys and not the underlying theory are included in Section 3.

19. S.M. Dean, "Considerations Involved in Making System Investments for Improved Service Reliability", Edison Electric Institute Bulletin, November 1938, pp. 491-498.

20. S. Lalander and U. Sandstrom, "Costs for Disturbances and Their Influence on the Design of Power Systems", CIGRE Proceedings, 1952, paper 329.

21. W.H. Dickinson, "Economic Evaluation of Industrial Power-System Reliability", AIEE Transactions, November 1957, pp. 264-271.

22. K. Goldsmith, "The Price of Reliability in Electricity Supply" IEE Conference on the Economics of the Security of Supply, 1967, pp. 50-52.

23. F.S. Brown, "Economics of Reliability of Bulk Power Supply in the United States", IEE Conference on Economics of the Security of Supply, Pt. 1, pp. 150-154.

24. H. Lindner, "Economic Evaluation of the Reliability of the Electricity Supply of Industrial Installations", Electricity Council Translation O.A. Trans. 482, from Energietechnik, vol. 18, August 1968.

25. C. Starr, "Social Benefit versus Technological Risk", Science, September 19, 1969, pp. 1232-1238.

26. R.B. Shipley, A.D. Patton and J.S. Denison, "Power Reliability Cost vs Worth", IEEE PAS, 1972, pp. 2204-2212.

This paper proposes a method to roughly estimate the worth of reliability by estimating the loss of GNP due to unsupplied electrical energy. A preliminary evaluation of the economics of the USA power system is made using the resultant estimate of $.60/KWHR (1967). The paper concludes that the USA system was probably being planned with too high a target reliability standard.

27. W.G. Watson, "Rapporteur's Report on Session 1. - System and Component Reliability", Cired 73 Pt. II, Record of Discussion, 1973.

28. M.L. Telson, "The Economics of Reliability for Electric Generation Systems", Massachusetts Institute of Technology Energy Laboratory Report MIT-EL 73-106, May 1973.

This 266 page report investigates reliability evaluation methods and indirectly discusses customer impacts from interruptions, attempts to develop a methodology to quantify these impacts as well as include the costs in an economic study of the system reliability. Wages paid during interruptions divided by the non-residential energy consumption yields an estimate of $1.13/KWHR for the New York area. The present system reliability was found to be roughly 100 times more stringent than the optimum.

29. A. Kaufman, "Reliability Criteria — A Cost Benefit Analysis", Office of Economic Research, New York State Public Service Commission, August 1975.

This report discusses a cost-benefit analysis of the New York Power Pool generation reserve level. The interruption cost was estimated using value added techniques to be $.77/KWHR. The results indicate that a 1 day/year LOLE target reliability level is optimum.

30. M. L. Telson, "The Economics of Alternative Levels of Reliability for Electric Power Generation Systems", The Bell Journal of Economics, Autumn 1975, pp. 679-694.

This paper discusses customer impacts from interruptions and the quantification of the economic costs incurred. GNP/KWHR is proposed to be a conservative upper bound while the reasonable upper bound of wages/KWHR yields a figure for the USA of $.57. The optimum reliability standard was determined to be much less stringent than the target level presently used.

31. W.B. Shew, "Costs of Inadequate Capacity in the Electric Utility Industry", Energy Systems and Policy, vol. 2, no. 1, 1977, pp. 85-110.

This paper studies the cost to the USA economy of insufficient capacity in the 1980s by means of an optimization between planned rationing and reduction of reserve. The costs of rationing are estimated using average price elasticities while interruption costs are estimated by means of GNP/KWHR. The total losses are found to be large but relatively insensitive to variation in the interruption cost estimate.

32. T.W. Berrie, "What Quality of Electricity Supply Can We Afford", Electrical Review, vol. 202, no. 1, January 1978, and "How To Work Out What Quality of Electricity Supply We Can Afford", Electrical Review, vol. 202, no. 3, January 20, 1978.

33. H. Khatib, "Economics of Reliability in Electrical Power Systems", Technicopy Limited, England, 1978.

This 157 page book investigates power system reliability evaluation techniques and the impacts of interruptions on customers. Previous studies attempting to determine the worth of reliability are critically evaluated and the results of surveys by the author of residential and industrial customers reported. Methodologies are presented to determine time of day variation of the cost of interruptions and to aggregate the interruptions caused by generation and distribution systems.

34. M.P. Bhavaraju and R. Billinton, "Cost of Power Interruptions - A User's Viewpoint", EPRI Workshop Proceedings, WS-77-60, March 1978.

This paper critically overviews cost-benefits evaluation of electric power supply reliability. Published research and data on customer interruption costs is discussed. Present data are concluded to be applicable only to local small random interruptions rather than larger wide scale blackouts.

35. R.L. Sullivan, "Worth of Reliability", EPRI Workshop Proceedings, WS-77-60, March 1978, pp. 6-40 to 6-50.

36. D.R. Myers, "Two Examples and Some Research Needs of Reliability/Worth Tradeoffs", EPRI Workshop Proceedings, WS-77-60, March 1978, pp. 6-29 to 6-39.

37. M.E. Samsa, K.A. Hub, G.C. Krohm, "Electrical Service Reliability: The Customer Perspective", U.S. Department of Energy, September 1978.

This 110 page report critically reviews studies concerned with the quantification of customer valuation of power system reliability. A customer classification scheme is proposed together with a discussion of the internal/external interruption costs of customers in the classification. Options for adjusting the reliability of the customers are discussed and categorized.

38. M. Munasinghe, "The Costs Incurred by Residential Electricity Consumers Due to Power Failures", World Bank Research Study, December 1978.

This paper presents a theoretical methodology for measuring residential interruption cost and the results of a survey of urban residential customers in Cascavel, Brazil. The main interruption cost is hypothesized to be the loss of evening leisure time which can be evaluated at the household income earning rate.

39. J. Nordin, "A Subscription Method of Determining Optimal Reserve Generating Capacity for an Electric Utility", The Engineering Economist, vol. 24, no. 4, 1979, pp. 249-257.

This paper proposes a methodology to optimize generation reliability by establishing a reserve capacity subscription system for customers. The principle involved is that system reliability would be determined and paid for by the individual customer subscriptions who could then be supplied at the level of reliability they opted for. Subscribing customers would have priority in being supplied before non-subscribers.

40. G. Wacker, E. Wojczynski, and R. Billinton, "Cost/Benefit Considerations in Providing an Adequate Electric Energy Supply", Third Symposium on Large Engineering Systems, July 1980.

41. K. Powell, "Valuing Reliability for Utility Planning, 1980". Reliability Conference for the Electric Power Industry, 1980, pp. 338-342.

42. Systems Control Incorporated, "Analysis of 12 Electric Power System Outages/Disturbances Impacting the Florida Peninsula", U.S. DOE report number DOE/RG 106359-1, December 1980.

43. R. Billinton, G. Wacker, and E. Wojczynski, "Quantitative Assessment of Power Supply Reliability - Reliability Cost/Reliability Worth Considerations", Alternative Energy Sources and Technology Symposium: Modelling Policy and Economics of Energy and Power Systems, May 1981. (Sponsored by The International Association of Science and Technology for Development - IASTED.)

44. Jack Faucett Associates, "Power Energy and Capacity Shortages - The 1976-77 Natural Gas Shortage", EPRI EA-1215, Volume 1, November 1979.

45. Jack Faucett Associates, "Power Shortage Costs and Efforts to Minimize: An Example", EPRI EA 1241, December 1979.

46. Jack Faucett Associates, "Analytical Framework for Evaluating Energy and Capacity Shortages", EPRI EA-1215, Volume 2, April 1980.

47. R. Billinton, G. Wacker and E. Wojczynski, "Power System Reliability Indices and Their Application in the Assessment of Reliability Worth", IEEE Conference Mexicon 1981, Mexico, CEA Transactions, Vol. 21, 1982.

48. A.P. Sanghvi, "Customer Outage Costs In Investment Planning Models for Optimizing Generation System Expansion and Reliability", CEA Transactions, Volume 21, 1982.

49. A.K. Lee and J.K. Snelson, "A Reliability Criterion For Generation Planning", CEA Transactions, Volume 21, 1982.

3. SURVEYS OF CUSTOMER COSTS

This section contains references which report and discuss surveys of customers attempting to quantify the worth of reliability. Section 3(a) deals with surveys other than those by Ontario Hydro which are included in section 3(b).

3(a) General

50. Swedish Committee on Supply Interruption Costs, "Costs of Interruptions in Electricity Supply", The Electricity Council, O.A. Translation 450, December 1969.

This Swedish survey was the first major investigation of consumer costs of interruptions. Direct questioning was preferred but only industrial customers were so surveyed. Residential and other cost estimates were obtained by discussion with representative organizations and worked examples. Residential costs were as significant as industrial costs. The cost was investigated as a function of interruption duration.

51. L. Lundberg, "Report of the Group of Experts on Quality of Service from the Consumer's Point of View", International Union of Producers and Distributors of Electrical Energy, Report 60/D.1, 1972.

This report is a collection and discussion of the experience and survey results of European utilities concerning the consumers' costs of interruptions. Only Swedish, French, and British utilities had performed studies and surveys. The Swedish survey is discussed above (ref. 50).

52. A. Jackson and B. Salvage, "Costs of Electricity-Supply Interruptions to Industrial Consumers-Summary", Proc. IEE, vol. 121, no. 12, December 1974, pp. 1575-1576.

53. A. Jackson and B. Salvage, "Costs of Electricity-Supply Interruptions to Industrial Consumers", IEE Library, 1974.

54. IEEE Committee, "Report on Reliability Survey of Industrial Plants, Part I: Reliability of Electrical Equipment", IEEE TPAS, March/April 1974, pp. 213-235.

    IEEE Committee, "Report on Reliability Survey of Industrial Plants, Part II: Cost of Power Outages, Plant Restart Time, Critical Service Loss Duration Time, and Type of Loads Lost Versus Time of Power Outages", IEEE PAS, March/April 1974, pp. 236-241.

    IEEE Committee, "Report on Reliability Survey of Industrial Plants, Part V: Plant Climate, Atmosphere, and Operating Schedule, the Average Age of Electrical Equipment, Percent Production Lost, and the Method of Restoring Electrical Service after a Failure", IEEE PAS, July/August 1974, pp. 463-466.

These papers report on a detailed survey of the reliability of electrical supply to industrial plants in the USA and Canada. Thirty companies with a total of 68 plants in nine main categories responded. The objective was to obtain information on the characteristics of failures of electrical supply equipment in the plants and the impact and cost of interruptions to help aid in the design of industrial power distribution systems. The forms used and the detailed response data are included.

55. P. Gannon, "Costs of Electrical Interruptions in Commercial Buildings", IEEE I & CPS Conference, 1975, pp. 123-129.

This IEEE sponsored survey of the cost to customers of interruptions in commercial buildings and offices received responses for 55 buildings from 48 companies. The survey form used was a simplified version of the 1972 IEEE industrial survey (ref. 54) and is included in the paper. The cost per KWHR unsupplied energy, per sq. ft./hr., and per employee/hr. was surveyed as a function of duration (15 min., 1 hour, > 1 hour) and computer usage.

3(b) Ontario Hydro

56. L.H. Berk, "Report on the Study of the London Power Interruption", OH Report No. PMA-76-2, March 1976.

A number of reports in this section each concern a segment of the Ontario Hydro customer market and have much in common in the way of objectives, approach and report features.

The surveys (refs. 57, 59, 61, 62, 63 and 64) were carried out in response to a suggestion of the Ontario Energy Board as expressed in their report on the hearings into the Ontario Hydro system expansion program. These surveys are part of a series of studies to investigate, amongst other topics, "the evaluation of alternate levels of reliability of power and energy supply from the viewpoint of the particular classes of customers and of the Province of Ontario as a whole." The purpose and scope of the surveys are stated to be:

1) To obtain customer estimates of costs and other effects of electric supply interruptions, voltage variations, frequency variations and rationing.
2) To gather data on customer groups for use in planning and operating the Ontario Hydro system.
3) To obtain information for use in seeking the cooperation of customer groups to reduce the adverse effects of operating problems which might occur in the future.

The customers in each sector (with the exception of residential customers) are classified into categories appropriate with their functions and their sensitivity to interruptions. Cost estimates for various durations of interruptions are reported on a $/KW of peak demand basis. Preferences for frequent but short as opposed to less frequent but longer interruptions are reported. Results concerning standby generation, variation of cost with time of season or day, non-monetary impacts (especially hazards), effects of voltage variation and advance warning and other factors are also reported. Each survey concerns the customer sector as the title of the reports suggest.

57. "Ontario Hydro Survey on Power System Reliability: Viewpoint of Large Users", OH Report No. PMA 76-5, April 1977.

58. L.V. Skof, "Customer Interruption Costs Vary Widely", Electrical World, July 15, 1977, pp. 64-65.

59. Market Facts of Canada Ltd., "Research Report: Residential Monitoring Survey of Energy Conservation: Extract on Supply Reliability", Ontario Hydro Customer Viewpoint Committee on Supply Reliability, August 1977.

60. L.H. e-k, "Up-Date on Ontario Hydro Surveys on Supply Reliability: Viewpoint of Selected Customer Groups", CEA Meeting, March 1978.

61. "Ontario Hydro Survey on Power System Reliability: Viewpoint of Small Industrial Users (Under 5000 KW)", OH Report No. R & U 78-3, April 1978.

62. "Ontario Hydro's Surveys of Power System Reliability: Viewpoint of Government and Institutional Users (Pilot Survey in the City of Guelph)", OH Report No. R & U 78-1, May 1978.

63. "Ontario Hydro Survey on Power System Reliability: Viewpoint of Farm Operators", OH Report No. R & U 78-5, December 1978.

64. "Survey of Power System Reliability: Viewpoint of Customers in Retail Trade and Service", OH Report No. R & U 79-7, July 1979.

65. "Energy Use in the Food and Beverage Industry in Ontario", OH Report No. R & U 79-4, July 1979.

66. "Ontario Hydro's Survey of Power System Reliability: Viewpoint of Customers in Office Buildings", Ontario Hydro Report No. R & U 80-5, March 1980.

67. "Ontario Hydro Survey of Power System Reliability: Viewpoint of government and Institutional Users", Ontario Hydro Report No. R & U 80-6, March 1980.

68. "Ontario Hydro Survey on Power System Reliability: Summary of Customer Viewpoints:, Ontario Hydro Report No. R & MR 80-12, Dec. 1980.

4. OPTIMAL DESIGN AND PLANNING USING WORTH OF RELIABILITY

This section contains references mainly concerned with the application of cost of interruption data in system planning and design. Engineering and economic considerations are delved into. Some of the references attempt to determine the optimum level of reliability using the cost estimates.

Section 4(a) considers the application to generation systems, 4(b) to transmission and distribution systems, and 4(c) to power systems in general.

4(a) Optimal Design and Planning using Worth of Reliability - Generation Systems

69. D. Farrar et al., "A Model for the Determination of Optimal Generating System Expansion Patterns", M.I.T. Energy Lab Report MIT EL 73 009, February 1973.

70. M.G. Webb, "The Determination of Reserve Generating Capacity Criteria in Electricity Supply Systems", Applied Economics, 1977, pp. 19-31.

71. D. Dees et al., "The Effect of Load Growth Uncertainty on Generation System Expansion Planning", Proceedings of the American Power Conference, vol. 40, 1978, pp. 1223-1232.

72. M.H. Bensky et al., "The Cost Benefits of Alternative Generation Reserve Levels", Proceedings of the American Power Conference, vol. 40, 1978, pp. 1323-1332.

73. L. Guth and H. Zellner, "Costs and Benefits of Systems Reliability", Proceedings of the Sixth Annual Illinois Energy Conference, September 1978, pp. 121-137.

74. E.G. Cazalet, "Costs and Benefits of Over/Under in Electric Power System Planning", Proceedings of the Sixth Annual Illinois Energy Conference, September 1978, pp. 106-120.

75. Decision Focus Inc., "Costs and Benefits of Over/Under Capacity in Electric Power System Planning", EPRI EA-927, October 1978.

76. Ontario Hydro, "The SEPR Study; System Expansion Program Reassessment Study", Six Interim Reports and the Final Report, February 1979.

This series of 7 reports presents a summary of an extensive study carried out by Ontario Hydro to re-evaluate possible generation system expansion plans. The effect of changes in load growth on Ontario Hydro and its customers, and of nuclear-coal capacity mix, generating unit size and amount of reserve generation capacity are investigated. The cost of interruptions to customers as determined from surveys (see Section 3(b)) is used to estimate the impact of interruptions.

4(b) Optimal Design and Planning Using Worth of Reliability - Transmission and Distribution Systems

77. H.J. Sheppard, "The Economics of Reliability of Supply - Distribution (Great Britain)", The Economics of Reliability of Supply, IEE Publication 34, 1967 pt. 1, pp. 248-266.

78. W. Dickinson, "Economic Analysis of Reliability in Industrial Power Plant Power Systems", IEEE I & CPS Conference, May 1971, pp. 27-31.

79. P. Gannon, "Fundamentals of Reliability Techniques as Applied to Industrial Power Systems - Part 5 Assumed Costs", IEE I & CPS Conference, May 1971, p. 26.

80. C.R. Heising, "Reliability of Electric Power Transmission and Distribution Equipment", 1974 ASQC Technical Conference Transactions - Boston, pp. 314-319.

81. P. Gannon, "Costs of Interruptions; Economic Evaluation of Reliability", IEEE 1976 I & CPS (Industrial and Commercial Power Systems) Conference, pp. 105-110.

82. E. Dahl and J. Huse, "The Level of Continuity of Electricity Supply and Consequent Financial Implications", CIRED 77, 1977, pp. 138-141.

83. L. Lunden and L. Hammarson, "The Level of Continuity in Urban Distribution Systems and the Consequent Financial Implications", CIRED 77, 1977, pp. 142-145.

84. H. Persoz, et al., "Taking into Account Service Continuity and Quality in Distribution Network Planning", CIRED 77, 1977, pp. 123-126.

85. R. Billinton and D.O. Koval, "Evaluation of Reliability Worth in Distribution Systems", PSCC, vol. 1, 1978, pp. 218-225.

86. R.N. Allan et al., "Reliability Indices and Reliability Worth in Distribution Systems", EPRI Workshop Proceedings, WS-77-60, March 1978, pp. 6-20 to 6-28.

87. M. Munasinghe and W. Scott, "Long Range Distribution System Planning Based on Optimum Economic Reliability Levels", IEEE Summer Meeting, 1978, Paper A 78 576-1.

88. D.O. Koval and R. Billinton, "Statistical and Analytical Evaluation of the Duration and Cost of Consumers Interruptions", IEEE Winter Power Meeting, Paper No. A79 057-1, February 1979.

89. R.S. Tsai, "Reliability Worth Guides Distribution System Design", IEEE Transactions on Industry Applications, vol. IA-15, no. 4, July/August 1979, pp. 368-375.

4(c) Optimal Design and Planning Using Worth of Reliability - Power Systems in General

90. B. Mattsson and J. Nuder, "Simplified Use of Failure Statistics for Optimizing System and Equipment Design", CIGRE 1972, Paper 31-06.

91. W.G. Watson and K.R. Jarret, "Reliability Investment Strategies for HV and EHV Networks Based on an Incremental Approach", 1977 CIRED, pp. 134-137.

92. L. Markel, N. Ross, and N. Badertscher, "Analysis of Electric Power System Reliability", California Energy Resources Conservation and Development Commission", October 1976.

93. M. Munasinghe, "A New Approach to Power System Planning", IEEE 1979 PES Winter Meeting, Paper F 79 155-3.

5. ECONOMIC THEORY CONCERNING WORTH OF RELIABILITY AND POWER SYSTEM PLANNING

This section contains references in the economics field pertaining to Power System Planning and Worth of Reliability. There are many others but the ones most relevant were included. Many economics publications dealing with peak pricing discuss reliability but it is only of late that economics research considers reliability as a variable to be optimized rather than as an outside constraint.

94. M.L. Visscher, "Welfare-Maximizing Price and Output with Stochastic Demand: Comment", American Economic Review, vol. 63, 1973, pp. 224-229.

95. E.J. Mishan, "Cost-Benefit Analysis", Allen and Unwin, London, Second Edition, 1975.

96. E.K. Smith, "Reliability in the Demand for and Supply of Electricity", Ph.D. Dissertation, Texas A & M University, December 1975.

97. J. Stabler and L. St. Louis, "A Cost - Benefit Model for the Evaluation of Public Power Development Proposals", Western Regional Science Association Meeting, San Diego, February 1976.

98. L. Higgins, "Load Forecast Error as an Element in Planning Optimal Capacity in an Electrical Supply System", Public Utilities Forecasting Conference at Bowness-on-Windermere, March 1977.

99. B. Fischoff, "The Art of Cost-Benefit Analysis", The U.S. Department of Commerce, National Technical Information Service, February 1977.

100. D.T. Nguyen, "Public Utility Pricing with Stochastic Demands: A Note", Applied Economics, vol. 10, 1978, pp. 43-47.

101. M.A. Crew and P.R. Kleindorfer, "Reliability and Public Utility Pricing", The American Economic Review, vol. 68, no. 1, 1978, pp. 31-40.

102. M. Munasinghe and M. Gellerson, "Optimum Economic Power Supply Reliability", World Bank Staff Working Paper no. 311, January 1979.

103. M. Munasinghe and M. Gellerson, "Economic Criteria for Optimizing Power System Reliability Levels", The Bell Journal of Economics, vol. 10, no. 1, spring 1979, pp. 353-365.

104. J. Tschirhart and F. Jen, "Behaviour of a Monopoly Offering Interruptible Service", The Bell Journal of Economics, vol. 10, no. 1, Spring 1979, pp. 244-258.

6. COST OF RELIABILITY AND COST EFFECTIVENESS ANALYSIS

This section contains papers mainly concerned with determining some estimate of the relative cost and merit of design schemes. No estimate of the cost of interruptions is used.

This is only a partial bibliography, only a few of the pertinent papers on this topic are included to give some indication of the work in this area.

105. G.F.L. Dixon and H. Hammersley, "Reliability and Its Cost on Distribution Systems", IEE Conference on the Economics of Reliability of Supply, October 1967, pp. 81-84.

106. W. Cartwright and B.A. Coxson, "Reliability Engineering and Cost-Benefit Techniques for use in Power System Planning and Design", Electricity Council Research Memorandum ECR/M966, October 1976.

107. N.E. Chang, "Evaluate Distribution System Design by Cost Reliability Indices", IEEE PAS, vol. 96, no. 5, September/October 1977, pp. 1480-1490.

108. S. Cluts, M.K. Ravindra, and E. Filstein, "Transmission Reliability - What's it Worth?", 1979 Reliability Conference for the Electric Power Industry, pp. 23-29.

7. MISCELLANEOUS PAPERS RELATED TO THE EVALUATION OF WORTH OF RELIABILITY

109. W.D. Rowe, "An Anatomy of Risk", John Wiley & Sons Ltd., New York, 1977.

This book discusses the determination of acceptable levels of risk for technological systems and programs.

110. G. Schofield and W. Marsh, "The Effect of Reliability on Utility Financial Results", Public Utilities Fortnightly, 1970, September 24, pp. 15-20.

111. H. Legler, "Modelling the Impact of Electricity Supply on Economic Growth and Employment", Progress in Cybernetics and systems Research, vol. IV, pp. 278-292.

112. A. Levis et al., "Large Scale System Effectiveness Analysis", U.S. Department of Energy, September 1978.

This report discusses the initial results of a theoretical study into the analysis of large scale effectiveness.

113. H.A. Cavanaugh, "How Much Will a KW Shortfall Cost?", Electrical World, January 15, 1979, pp. 43-46.

CONCLUSIONS

In compiling a bibliography of related publications it is not possible to include all the relevant material. The authors would like to apologize for any errors or omissions and hope that additional relevant publications will be advanced as discussion to this publication.

ACKNOWLEDGEMENT

The authors would like to thank the Canadian Electrical Association for providing the financial support for this work.

INTERRUPTION COST METHODOLOGY AND RESULTS - A CANADIAN RESIDENTIAL SURVEY

G. Wacker   E. Wojczynski   R. Billinton
Power System Research Group
University of Saskatchewan
Saskatoon, Saskatchewan

ABSTRACT

This paper presents the results of an investigation of the direct, short term impacts and costs incurred by residential electrical consumers resulting from local random supply interruptions. The study used a mail survey to determine the residential user's assessment of the cost of electric service interruptions. The survey obtained user's cost valuation using three approaches. Two of these approaches asked respondents to indicate rate changes that would be commensurate with specified changes in reliability, while the third approach was an indirect worth evaluation based on the preparatory actions a consumer predicted he would take, given a specified (un)reliability. User's valuations are reported for a number of frequency and duration failure scenarios. Variations of user cost estimates are discussed as functions of several user characteristics. Comparisons with the results of other researchers are presented.

INTRODUCTION

Electrical utilities have always been conscious of the necessity to optimize their expenditures and use of resources while supplying customers with electric power. Reliability is an important measure of the adequacy of this supply. Until recently, an acceptable level of reliability has often been rather arbitrarily assigned on the basis of judgment and experience. Significant increases in energy costs, construction costs, and interest rates, recognition of conservation and environmental concerns, and the impacts of government and public groups have resulted in the need for a more rational and consistent approach to determining acceptable reliability levels. A major aspect of this recent approach is the assessment of the worth of power system reliability so as to be able to compare it with the costs of obtaining this reliability. The most obvious means of assessing the worth of alternative levels of reliability is to evaluate the user and societal costs attributable to interruptions in the electrical supply.

The impact on a customer due to a cessation in the electrical supply depends on characteristics of both the customer and the interruption. Customer characteristics include demographic aspects (e.g., dwelling type, number of people in the household, etc.), and what functions the electricity performs (e.g., space heating, lighting, motor drives, etc.). Interruption characteristics include the frequency and duration of the interruption, and the time of day, week, or year when it occurs. Additional factors such as outside temperature or the occurrence of the interruption during special events also affect the impact.

Obtaining cost of interruption data is a complex and rather subjective task. A comprehensive literature search reveals a range of methods that have been used to determine such data. These methods can be broadly grouped (1) into three categories, namely: customer surveys (2); analytical methods such as price elasticities (3) and gross economic indices (4); and blackout case studies (5). Of the methods available, use of customer surveys is most applicable to the residential sector. This is because a significant attribute of the impact of interruptions to residential users is its intangible nature: the "value" of the "loss" due to inconvenience or disruption of housekeeping or leisure activities involves the perception of the individual(s) affected.

The residential survey results reported herein are based on 4740 useable responses resulting from a mailing sample of 13,359 questionnaires. The survey involved eight Canadian utility service areas located in the maritimes, the prairie provinces, and interior British Columbia. The survey was conducted as separate surveys in the various service areas over an 18 month period beginning in June, 1980.

RESIDENTIAL SURVEY METHODOLOGY

The specific methodology and questionnaire used in the survey underwent an extensive developmental process. This involved an iterative approach consisting of the identification of factors to be included, design and development of the questionnaire, and small scale testing of the questionnaire using interviews with sample users. Consultants skilled in questionnaire techniques were used throughout. The final pretest was a mail survey of 100 representative households.

Factors Investigated in the Survey

A comprehensive list of factors hypothesized to affect the cost of interruptions was prepared. While it would have been desirable to investigate all factors, the length of a questionnaire is limited by the degree of effort that respondents are willing to engage in. This limitation was found to be especially important due to the difficulty respondents encountered in the central set of questions: those asking for a monetary evaluation of the effects of interruptions. The following factors were selected for inclusion in the final questionnaire: interruption characteristics - duration and frequency of occurence, season, day of week and time of day; user characteristics - number and age of household members, sex and education of respondent, type of dwelling, (e.g. apartment or bungalow), location of dwelling, (e.g. urban, rural), electric heat usage, attitude towards utilities, energy consumption, and monthly electrical bill, income earning business at home, and user experience with interruptions.

83 WM 067-6   A paper recommended and approved by the IEEE Power System Engineering Committee of the IEEE Power Engineering Society for presentation at the IEEE/PES 1983 Winter Meeting, New York, New York, January 30-February 4, 1983. Manuscript submitted August 27, 1982; made available for printing November 11, 1982.

Reprinted from *IEEE Trans. Power Apparatus Syst.*, vol. PAS-102, no. 10, pp. 3385–3392, Oct. 1983.

Development of Questionnaire Design

The first major consideration in the development of the questionnaire approach was whether the questions are to concern hypothetical situations or ones that have been experienced by the respondent. Limitations of the hypothetical situation approach are that a person's predicted response may differ from his actual response and that the respondent may not give extreme scenarios serious consideration. Limitations of investigating actual interruptions are that large survey samples would be necessary since actual cost per interruption varies considerably, and that people remember poorly the effects of interruptions since actual failure frequency is low. A common limitation to both approaches is that long or frequent interruptions have not been experienced by the respondent and the nature and extent of the effects are unknown. It was concluded that the actual experience approach is not feasible and that the questionnaire design could assist respondents in making reasonable predictions as to the degree of the effects and what their reactions might be.

The second major questionnaire design consideration involves the choice of valuation methodology; three were considered and are discussed. An obvious approach is a rate change approach, wherein respondents are asked what changes in rates are acceptable for specified changes in reliability. The most serious drawback with this approach is the antagonism many people have towards price increases, especially those related to essential services such as electrical supply. Another drawback is that some people think that utilities should provide a reliable supply regardless of rates or other factors. Further aggravating these problems is the tendency for respondents to view with suspicion the survey as a means for utilities to determine user susceptibility to rate increases. Any responses to questions concerning rates may be affected by these factors to some degree.

The direct worth evaluation approach asks users to place a monetary value on the effects of certain interruption scenarios or to assess the worth to them of not having to experience an interruption. If this approach were feasible it would be ideal since it is the user's valuation of the worth of reliability that is desired. This estimate would provide the valuation without any intermediary transformation. This approach is normally used in surveys of non-residential users. Direct worth evaluation was attempted during the development phase and the pretest survey but it suffered from the difficulty residential respondents encountered when attempting to give a meaningful answer to direct valuation questions. This difficulty stems from the lack of respondent experience with markets in which intangible benefits, such as electric service reliability, are exchanged as a commodity.

An indirect worth evaluation approach poses questions which do not require a direct valuation; rather, responses to indirect questions are used to derive a value. By devising a suitable approach it is possible to decrease the problems associated with a lack of experience in rating the worth of reliability. This is achieved by asking questions which the respondents can relate to in the context of their previous experience. A limitation is that there exists a possibility that the derived value is not an estimate of the worth but some other entity associated with the indirect approach. Possible question forms that were considered for inclusion in the questionnaire are:

(i) Cost of hypothetical insurance policies to compensate for possible interruption effects, and the appropriate compensation payable to them in the event of an interruption claim.
(ii) Respondents' opinions as to the appropriate interruption cost figures utilities should use in planning.
(iii) Respondents' predictions of what preparatory actions they might take in the event of interruptions.

The preparatory action question was found to be the most promising. A list of actions that a householder might conceivably take in preparation against power interruptions was developed. A reasonable cost figure for purchase and usage of each action was assigned and included in the list. During analysis, the costs of the chosen action(s) are summed to provide an estimate of the value respondents place on not having to experience the effects of the interruption. This valuation is then taken to be an indirect estimate of the worth of reliability, making use of the economic concept of substitution wherein the user's willingness-to-pay for equivalent alternatives is a measure of worth. During the pretest interviews it was found that respondents considered the selection of actions adequate and the quoted costs reasonable.

The decision as to which of the interruption cost valuation approaches to employ was difficult to make. After final pretesting, the direct worth approach was deleted from the questionnaire and both the rate change and indirect worth approaches were retained.

Description of the Questionnaire

The first several questions are relatively easy to answer, inoffensive to the respondent and encourage the respondent to continue. Attitudes concerning electric service, a rating of the undesirability of various effects of interruptions, and respondents' estimates of their interruption history remind them of experiences they may have had with interruptions and stimulates them to consider what their experience might be during a power failure. A rating of the undesirability of various interruption scenarios follows, which provides a relative measure of the effects of interruptions as a function of interruption characteristics. A secondary purpose is to encourage the respondent to consider a range of scenarios in preparation for the interruption cost questions. A relative scale of 0 to 6 is used to indicate various degrees of undesirability, from 0 meaning "no undesirable effect" to 6 being "extremely undesirable". The indirect worth assessment using the preparatory action question follows. It precedes all rate change questions to provide an estimate free of rate related biases. The rate change questions begin by asking respondents to estimate their monthly electrical bill so as to encourage a reasonable response. The rate change questions seek both the permissible rate increase to maintain a reliable supply and the expected rate decrease to revert to a less reliable one. The last questions concern demographic and personal characteristics of the respondents in order to correlate variation of responses with user characteristics.

Survey Procedures

The survey procedures included sample selection, initial mailing, reminder letter, second mailing (in some cases), energy consumption information retrieval, and data analysis. Participating utilities selected the survey sample from their billing list using a simple random sampling process (every nth user) and performed the initial mailing. This mailing included the questionnaire, a release statement requesting permission to obtain the respondent's consumption information from the utility, and a letter on University letterhead which explained the survey objectives, promised confidentiality of responses from the utility, and offered complimentary energy booklets as appreciation for completing the questionnaire. A stamped, self-addressed envelope was used to return the responses to the University, and the utility was not involved in the reminder letter or second mailing (if used).

Additional Developmental Work

The following additional tests and validating investigations were performed using portions of the Saskatoon survey of 1000 households:

(i) In-person interviews with those who had completed and returned the questionnaire. The interviews focussed on the respondents' opinions concerning power interruptions and on the suitability of the questionnaire, but it also included questions which were directly comparable with the mail survey questions. The questionnaire was found to be understandable and applicable, interview responses were generally consistent with mail responses, and most importantly, the cost estimates obtained from the mail responses and from the interviews did not have a statistically significant difference.

(ii) A non-response bias investigation. Since the surveys had an average response rate just higher than one third, it is possible that those who did not respond hold different opinions or have different characteristics than those who did. Non-response bias was investigated by making a concerted effort using in-person contact to obtain responses from individuals who had not responded to earlier mailed contacts. These responses were then compared with those from the original respondents using statistical techniques. The results indicate that some statistically significant differences in attitude existed between the two groups, but that cost estimate differences were not statistically meaningful. It was concluded that non-response bias is not a significant problem for this survey.

RESIDENTIAL SURVEY RESULTS

The amount of residential cost data generated from the survey results is prohibitive in terms of providing a comprehensive report in this paper. Therefore, most of the results obtained from the survey are presented in a general qualitative way, but the quantitative cost estimates derived from the preparatory action question and rate change questions are presented more fully. All values quoted are for the total survey except as otherwise indicated. Significant general trends in the cost estimates as functions of user or interruption characteristics are reported. Variations within and among the various service regions are discussed. Correlations and variations among parameters were determined using standard statistical techniques. Both univariate and multivariate analyses were performed using the SPSS Breakdown and Analyses of Variance program and Discriminant program respectively (6). The F-test significance level was determined for all tests of difference and only correlations which are statistically meaningful are reported.

Effective survey response rates (excluding inapplicable selections from the sample) for the 10 separate surveys constituting the residential survey varied from a low of 30% to a high of 49%, with a response rate of 37% for the entire survey. This is considered to be quite acceptable, given the length and complexity of the questionnaire.

Opinions Regarding Electric Service and Interruptions

In response to the attitudinal questions, an overwhelming majority of the respondents indicated that the number of failures is very low or low, that the price of electricity compared to other commodities is moderate to high, and that the service provided by electric power companies is good to very good. Concerning the frequency of failures, the mean value of the estimates of the number of failures in the last two years is 5.49. This appears to be quite reasonable in comparison with a System Average Interruption Frequency (SAIFI) of 2.79 interruptions for customer per year for all utilities who reported to CEA for 1980 (7). Respondents indicated that only thirty percent of these interruptions caused problems or were disruptive. The estimated number of four hour failures was much lower, with most respondents answering that they had not experienced such a long failure in the last two years. Concerning their preference if it became necessary to ration electric power for four hours each week during the day-time, 75% of the respondents prefer the more frequent but short interruptions while 25% prefer the less frequent but long interruptions. The respondents to the Ontario Hydro residential survey (8) answered a virtually identical question with the same tendency.

When asked to rate how undesirable an effect would result for them and their household if a power failure 1 to 4 hours long occurred, discomfort such as might be due to change in home temperature was rated as being the most undesirable effect. This was followed closely by kitchen appliances being not usable and less closely by loss of lighting. Of considerable less importance are fear of accidents in the home, fear of crime, clothes care or cleaning appliances not useable, and hobby and leisure equipment not useable listed in slight decreasing order of importance.

Variation of Undesirability with Interruption Characteristics

Respondents are asked to rate how undesirable various interruption scenarios would be for their household. By varying only one parameter within each set of scenarios, a relative indication is obtained of the undesirability as a function of that parameter. The results are illustrated in Figure 1. Not shown is that weekday failures are rated slightly more undesirable than weekend failures. Use of a scenario common to every set enables general trends to be established between sets so as to begin to establish an overall residential customer cost function relative to interruption characteristics. However, the ratings for one set of scenarios cannot be compared directly with another set because they involve different dimensions of interruption characteristics and because the scale rating is a nonlinear process.

Cost Estimates from Preparatory Actions

Respondents are directed to suppose that they have been told by their power company that power failures will occur after 4:00 p.m. on winter weekdays but that the exact days or times are not known. They are then

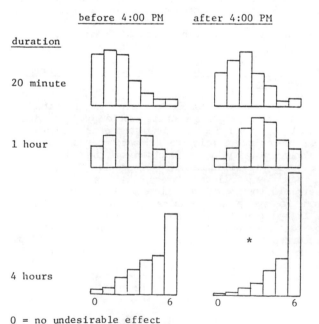

Variation with Frequency
Scenario: 4 hour failures after 4:00 PM on winter weekdays

yearly

monthly *

weekly

daily

Variation with Season
Scenario: 4 hour monthly failures after 4:00 PM weekdays

winter *

spring

summer

fall

Variation with Duration & Time of Day
Scenario: monthly failures on winter weekdays

before 4:00 PM   after 4:00 PM

duration

20 minute

1 hour

4 hours

0 = no undesirable effect
6 = extremely undesirable
* = similar scenario
ordinate magnitude = frequency of response
                    in each category

Figure 1. Residential Interruption Cost Variations

asked to predict which actions their household might take in preparation for the failures. Respondents are asked to choose one or more of the following six actions: make no preparation; purchase and use a candle at $0.25 per hour, an emergency lantern at $0.50 per hour, an emergency stove at $1.50 per hour; purchase or rent and use a small generator at $5.00 per hour, or a larger generator at $20.00 per hour. The costs quoted in the list of actions are used to compute an estimate of the cost of preparations that respondents indicate that they are willing to undertake to reduce or eliminate the adverse effects of the interruptions. The mean values of the costs per interruption are computed with the following results: 20 minute, 1 hour and 4 hour monthly failures result in cost estimates of $0.22, $1.18 and $11.87 per interruption, while the 4 hour weekly failure yields a cost estimate of $18.36 per interruption.

Cost Estimates from Rate Change Questions

The respondents are asked to estimate their monthly electricity bill to remind them of its approximate amount. They are then asked whether they prefer an increase in rates with a corresponding decrease in the number of failures, a decrease in rates with an increase in the number of failures, or for the situation to remain as present. 86% of respondents chose the status quo with the remaining respondents split between the other two options.

The first rate change question instructs respondents to suppose that the normal electric system has become subject to more frequent power failures. They are then asked whether they would choose an alternative assured electric power supply without any failures at various premiums above the cost of the normal system. A significant number of respondents were not willing to pay even two dollars per month more for a reliable supply. However, over 20% of the respondents were willing to pay at least ten dollars per month more to prevent monthly 4 hour failures and this rises to over 40% to prevent weekly 4 hour failures. The average value of the maximum amount per month that respondents are willing to pay for an assured system is $5.29; $8.05, and $8.06 per month if 4 hour monthly, 4 hour weekly, and one hour daily failures occurred on the normal system, respectively.

An approach symmetrical to the one of willingness to pay for an increase in reliability is that of willingness to accept a decrease in reliability for a decrease in rates. Respondents were asked to indicate the minimum percentage decrease in their rates for them to choose reduced reliability. The average value of the minimum rate decrease for respondents to choose the option of a cut in rates along with an increase in the number of failures to a 4 hour failure occurring once a month is $11.00 per month.

Demographic Characteristics of Respondents

The respondents were asked for information concerning themselves and their households to investigate whether, as a group, they were representative of the population. This also enables the cost estimates and other results to be analyzed for their relationship with the demographic characteristics. The representativeness of the sample was checked by comparing demographic variables with Statistics Canada information (see for example, ref. (9)). The general overall result is that all groups of residents are present in the data set but that

488

people living in apartments, people living alone, people between twenty one to thirty, women, and people with pre-high school or some high school are under-represented. These distortions are judged not to be so serious as to jeopardize the validity of the survey results.

Discussion

The cost estimates obtained from each approach measure different aspects of the worth of reliability and are not strictly comparable. The action cost measures the reliability worth by making use of the economic concept of substitution. The worth of one good (reliability) is equivalent to the amount the user is willing to pay for some other goods (preparatory actions) which provide the same benefits as the primary good. Of course the benefits from the actions are not exactly equivalent to those of the reliable electrical supply. However the measure is meaningful because the actions are reasonable preparations that individuals might take given the interruption scenarios. The rate increase estimates measure the willingness-to-pay for a reliable supply. However, the measure is influenced not only by the users' needs but by their attitudes and suspicions of the utilities. The rate decrease estimate is a complementary measure of the willingness-to-pay but is influenced by the users desire for no change from the present situation and by the tendency to include a compensation factor in the decrease. Each of these estimates is a measure of the worth of reliability but none of them can be said to be a complete and reliable measure. Obtaining a single complete measure of reliability is difficult if not impossible in the residential sector because the majority of the benefits are intangible.

The three approaches shared the common scenario of four hour monthly failures. The action cost and rate decrease estimates for that scenario are similar but the rate increase estimate is about half the other two. Respondents indicated that on a per month basis they were willing to pay significantly more to avoid weekly failures as compared to monthly ones. However, when translated to a cost per interruption basis, the rate increase estimate became less for weekly failures than monthly ones. This indicates the presence of a nonlinear saturation process when respondents are considering the rate increases. The respondents' indicated willingness-to-pay appears to vary non-linearly with interruption frequency. Unfortunately there is no conclusive evidence or explanation as to why this non-linear variation occurs. The same process occurs with 1 hour daily failures except that an additional factor may be that respondents perceive four 1 hour failures in a month as causing less than four times the inconvenience and undesirability of a single 4 hour failure in that month.

In contrast to the rate increase estimates, the action cost estimates continue to climb as the frequency increases to weekly failures. It is somewhat difficult to interpret this anomaly. Obviously at some point the saturation effect occurs but possibly it occurs more readily with the rate increase estimates because of rate antagonism. On the other hand, the action cost estimates are inherently on a per interruption basis because of the structure of the approach. Perhaps the anomaly is an artifact created by this structure. In either case, this saturation effect indicates the uncertainty inherent in obtaining an estimate with a scenario of very frequent failures and applying it in a situation of less frequent failures. Evidence appears to indicate,

however, that monthly interruption scenarios do not sufer appreciably from this saturation effect.

As suggested earlier, a single complete measure of the worth of reliability of supply to residential users is difficult if not impossible to obtain. The question that begs to be answered is: "Which set of estimates should an analyst use as the measure of reliability worth?" The answer depends on the specific purpose and theoretical structure of the analysis. In general, it is the authors' opinion that the most suitable estimates of reliability worth are the preparatory action cost estimates. The action cost estimates are as theoretically sound as the rate increase or decrease estimates and do not suffer from any taint of rate related antagonism.

The cost per interruption estimates quoted above are probably the most meaningful for the residential user group but the cost/KWHR and cost/KW are determined as well for comparison with other studies and for planning purposes. These are presented in Table 1. The $/MWHR estimates were obtained by normalizing with respondent annual energy consumption, (not by the energy unsupplied during interruptions). As well, the estimates which were on a per month basis have been converted to a per interruption basis to make the estimates consistent. The second part of the table lists the costs divided by the estimated peak demand which was determined from the user's one year consumption history by assuming a load factor of 23%.

Table 1 - Residential Interruption Cost Estimates

Cost Per Interruption Divided by Annual Energy Consumption ($/MWHR)

Interruptions	Action Cost	Rate Increase	Rate Decrease
20 min. monthly	$ .028	–	–
1 hour monthly	$ .156	–	–
4 hour monthly	$1.566	$ .727	$1.286
4 hour weekly	$2.437	$ .276	–
1 hour daily	–	$ .036	–

Cost Per Interruption Divided by Annual Peak Demand ($/KW)

Interruptions	Action Cost	Rate Increase	Rate Decrease
20 min. monthly	$ .06	–	–
1 hour monthly	$ .31	–	–
4 hour monthly	$3.16	$1.46	$2.59
4 hour weekly	$4.91	$ .56	–
1 hour daily	–	$ .07	–

Variation of Interruption Cost Estimates with Respondent Characteristics

The interruption cost estimates were analyzed for variation of the mean with answers to the attitudinal, interruption frequency, and demographic questions and the respondent consumption variables. Although often appreciable differences in means are involved, in no case was the amount of variance of cost estimates due to respondent characteristics a large part of the total variance of the cost estimates. However, the following trends were observed.

489

The preparatory action costs and rate increase estimates are positively correlated with the respondents estimates of their monthly bills. The estimates are somewhat positively correlated with consumption. Both of these correlations are weak. There is some tendency for apartment dwellers to have lower cost estimates than others. Part of this may be due to apartment dwellers thinking that generator sets are not feasible preparatory actions for them to take. Households with only one or two residents had a statistically significant lower mean than those with more than two residents. Respondents with electric heat had a higher mean interruption cost than those without. Respondents who reported home business had a higher mean cost than those who did not. Older respondents tended to give lower cost estimates especially for the rate decrease and increase estimates. This was not just due to lower average monthly bills because the percentage rate decrease was significantly lower as well. There was little significant difference with the sex of the respondent except that female respondents tended to be less willing to pay a higher rate increase. The education of the respondent only slightly affected the means of the action cost estimates but the mean values for the rate increase and decrease estimate are much higher for the more educated respondents. The mean rate decrease estimate is twice as high for respondents with a university education as for those with a pre-high school education.

Some variations may not be independent of each other so a number of correlations between respondent characteristics were investigated in pairs. Although some correlation effects were evident, generally it was found that the variation in cost occurred with each variable independent of the other variable.

Comparison of Results Between Service Areas and Within Service Areas

Analysis performed comparing results between participating utility service areas resulted in a number of statistically significant correlations. the mean values of the questions asking for estimates of the number of failures is much higher in service areas which are known to have a relatively low service reliability than in service areas with a high service reliability. Information concerning "known" service reliability of the utilities was obtained from the CEA 1980 Service Continuity Report (7). Similarly, the mean responses to the question on the opinion of the electricity price and on the estimate of the monthly bill is related to the relative rates and consumptions in the service areas. For example, utilities which have relatively inexpensive power rates also have low mean responses to the question concerning the opinion on the price of electricity and the monthly bill estimate question. Another important trend is that interruption cost means tend to be highest for some regions and lowest for others. The rate decrease cost estimate is somewhat of an exception in this respect because of its dependency on the consumption and monthly electric bill. However, the percentage rate decrease does not have this dependency and did have consistent correlations with the other cost estimates. This consistency supports the use of these estimates as measures of reliability worth.

The study also investigated whether a relationship exists between the level of electric service reliability that users experience and the valuation they place on that reliability. Respondents from some regions with high actual and perceived reliability tended to have the highest mean interruption cost estimates while respondents from regions with the lowest reliability tended to have the lowest mean cost estimates. Although other factors may be operative as well, this comparison provides some empirical evidence that the valuation residential users place on reliability is affected by the level of reliability they have experienced. This result is not definitive since there were two significant exceptions of service areas which had low perceived reliability but high cost estimates.

The analyses indicate that there are significant differences in the mean interruption cost estimates normalized with respect to annual energy consumption ($/KWHR) between utility service areas. Correlations of cost estimates expressed as a percentage of monthly bills were also investigated. The result is that cost estimates are not strongly correlated with either consumption or monthly bills, and the resulting ratios are strongly dependent on the regional differences in consumptions and bills rather than on interruption costs.

Results obtained from identifiable regions within individual service areas were also subjected to analysis; in particular urban/rural correlations were investigated. Trends which are utility specific are evident, but definitive overall trends were not apparent.

Limitations of Results

The potential limitations of the survey results are as follows:

i) The results concern respondent predictions of their actions in response to interruptions and the expected costs and effects arising from various interruption scenarios; their actual actions and cost may differ from their predictions.
ii) Respondents may not have experienced some of the scenarios.
iii) The estimates of interruption cost pertain to specific interruption scenarios. While indications of the variation in cost with factors such as time of occurrence were obtained, dollar cost estimates could not be obtained for every possible scenario.
iv) Interruption cost likely varies with the level of reliability. The given estimates assume a reliability level not too different from that presently experienced.

Comparison of Results with Previous Studies

Figure 2 depicts residential interruption frequency cost estimates obtained by this survey (U. of S.), by Ontario Hydro's survey(s), and by the non-survey approaches of a British study (10), the Swedish study (11), Meyers (3), and Markel et al (12). The Great Britain study used a value for lost leisure time equal to the wage rate. The monetary value was determined using average residential loads and occupancy. The Swedish Study determined domestic loss of supply costs on the basis of estimates, discussions, and worked examples. The derivations presumed to include actual losses, household activities, and leisure time. Myer's residential loss evaluation is based on the customers' willingness to pay. Markel, in a U.S.A. study, derived residential consumer losses on the basis of the depreciation of all household electrical appliances for the duration of supply interruptions. The comparison indicates that the Ontario study has yielded estimates much below those of the other studies.

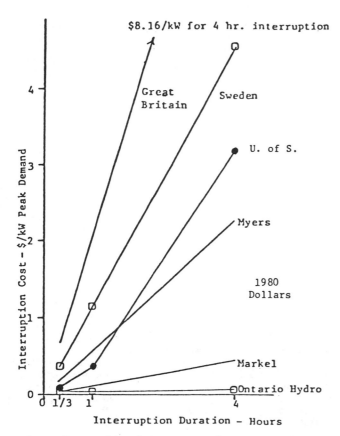

$8.16/kW for 4 hr. interruption

Figure 2. Residential Interruption Cost Estimates

The estimates obtained in this research ought to be most directly comparable with those obtained by Ontario Hydro but in fact are much higher. The 0.06, 0.31, and 3.16 $/KW values obtained in this study compare with 0.04, 0.04, and 0.07 $/KW from Ontario Hydro's results for 20 minute, 1 hour, and 4 hour monthly interruptions, respectively. The differences can not just be due to the U. of S. estimates being based on preparatory action costs while the Ontario Hydro estimates are based on willingness-to-pay rate increases. Specifically, the rate increase question used in this study yielded an average cost for a 4 hour monthly interruption of $1.46 per interruption ($/KW) which is much higher than the Ontario Hydro estimate using the rate increase approach. The authors are of the opinion that Ontario Hydro's residential estimates, which were based on daily interruption scenarios, may have been considerably lower than this study's estimates based on monthly interruptions, because of the nonlinear variation or saturation effect discussed earlier.

CONCLUSIONS

This paper has presented a summary of research conducted to determine estimates of the perceived costs or losses to residential consumers attributable to electric service interruptions. Using an indirect evaluation approach, cost estimates of $0.06, $0.31, and $3.16 per KW (annual peak demand) for 20 minute, 1 hour, and 4 hour monthly failures, respectively, have been determined. The major contribution of this work is the compilation of residential cost of interruption information as functions of both user and interruption characteristics. Another significant contribution is the improvement of interruption costing methodology.

ACKNOWLEDGEMENTS

The authors wish to thank the Canadian Electrical Association for providing the financial support for this work. The contribution of the CEA member utilities that participated in the survey is gratefully acknowledged.

LIST OF REFERENCES

(1) G. Wacker, E. Wojczynski and R. Billinton, "Cost/Benefit Considerations in Providing an Adequate Electric Energy Supply", Third Symposium on Large Engineering Systems, July 1980.

(2) "Ontario Hydro Surveys on Power System Reliability: Summary of Customer Viewpoints", Ontario Hydro Rates & Marketing Division, December 1980.

(3) D. Myers, Stanford Research Institute, "The Economic Effects to a Metropolitan Area of a Power Outage Resulting from an Earthquake", February 1978.

(4) R.B. Shipley, A.D. Patton, and J.S. Denison, "Power Reliability Cost vs Worth", IEEE TPAS, 1972, pp. 2204-2212.

(5) J. Corwin and W. Miles, Systems Control Inc., "Impact Assessment of the 1977 New York City Blackout", U.S. Department of Energy, Washington, D.C., July 1978.

(6) Nie, Hull, Jenkins, Steinbrenner and Brent, "Statistical Package for the Social Sciences (SPSS)", Second Edition, McGraw-Hill Book Company, New York, 1975.

(7) Vienneau, E.R., "1980 Annual Service Continuity Report on Distribution System Performance in Canadian Electric Utilities", Distribution System Reliability Engineering Committee, CEA, September 1981.

(8) Market Facts of Canada Ltd., "Research Report: Residential Monitoring Survey of Energy Conservation: Extract on Supply Reliability", Ontario Hydro Customer Viewpoint Committee on Supply Reliability, August 1977.

(9) "1976 Census of Canada: Dwellings and Households; Occupied Private Dwellings by Structural Type and Tenure", Statistics Canada 93-802, Ottawa.

(10) Lundberg, L. (1972): "Report of the Group of Experts on Quality of Service from the Customer's Point of View", UNIPEDE, Report 60/D.1.

(11) Swedish Committee on Supply Interruption Costs, "Costs of Interruptions in Electricity Supply", Electricity Council O.A. Translation 450, London, England, Dec. 1969.

(12) L. Markel, N. Ross and N. Badertscher, Systems Control Inc., "Analysis of Electric Power System Reliability", California Energy Resources Conservation and Development Commission, October, 1976.

Discussion

Norton Savage (U.S. Department of Energy, Washington, DC): These comments express only the views of the writer, and do not necessarily reflect the views of the Department of Energy.

The paper presents an informative and intelligent discussion of a survey that was carefully planned to ascertain the value, placed on reliable electricity supply, by residential customers in several areas of Canada. The full results of the survey would appear to merti close attention when (if ?) they are published in detail. As the authors note, the confines of an IEEE paper do not permit the full presentation of the details, data and conclusions to be drawn from a large-scale study such as the one described. Economists as well as electric utility planners would probably benefit from access to "hard" data, obtained in this study, in a field that has been little explored in quantitative fashion. Wassily Leontieff, in a letter to SCIENCE,[1] recently, made the point that economists have come to lean heavily on theory instead of fact; this paper provides some "fact".

The authors are careful not to make any claims regarding the universal applicability of the results they have obtained, and rightly so. Like the five other studies whose results are plotted in Figure 2 of the paper, this study contributes but one element, fixed in time and space, to the description of an entity that, in my opinion, has time-variant and space-variant characteristics. As the paper states, among the factors that determine the worth of power to the user are the season, the day and the hour of possible interruptions. I would add to this catalogue the year in which the survey is made and its relative position in the economic cycle. I would also add the region and country as major factors in the nature of the responses.

It seems to me (and of course this thesis would require resort to more studies of the nature described here) that in times of economic difficulty a residential user would place a lower value on continuous electric service than he would in prosperous times. For instance, in cities like Flint, Michigan and Detroit, Michigan, where unemployment is currently high, residential users would be less likely to consider expensive options for power supply than would have been the case in those cities during the late 1960's, in my opinion. Again, even in times of relative prosperity, a user's view of reliability would probably be conditioned by his use of electricity. One accustomed to the carefeee use of many electric appliances would probably place a higher value on uninterrupted service than one whose use is limited to lighting. The study, of course, is not to be faulted because it did not extend over half-a-dozen countries and several decades. But my thought is that studies of the worth of reliable electric power cannot be considered as one-time affairs; like a census of the population they should be made periodically, and should be considered truly valid only for the area in which made.

There is one element of the study whose practicality concerns me. Is it reasonable to suppose that a residential user would consider installing generating unit, small or large, to guard against failure of the central supply system? The mechanics of such an installation, the investment cost, the questions of fuel supply and storage, the safety aspects, the insurance problems, are among the many factors that would appear to make such installations impractical in most cases, for residential users.

REFERENCE

[1] SCIENCE, Vol 217, pp. 105-107, 9 July 1982.

Manuscript received February 8, 1983.

G. Wacker and **R. Billinton:** The authors are in general agreement with the comments offered above, particularly the suggestions (i) that other factors such as the economic climate at the time of the study may be significant and (ii) that the cost of interruption has both time- and location-variant characteristics. With reference to publication of detailed results, the authors were remiss in not specifically referencing their final report[1] to the Canadian Electrical Association, (the study had been conducted under contract with CEA). On the matter of the practically of actually using stand-by generating units on a large scale for residential consumers, the authors recognized (from the outset) the concern raised. Also, in the analysis of the residential responses it was noted[2] "There is some tendency for apartment dwellers to have lower cost estimates than others. Part of this though may be due to apartment dwellers thinking that generator sets are not feasible preparatory actions for them to take." However, on the basis of both pre-test respondent interviewers and comments on completed questionnaires, this particular question did not appear to raise concerns about credibility as often as did the more frequent or longer duration failure scenarios or indeed the justification for doing the study itself. Part of its acceptance may be because in some utility service areas, the availability of such generating units is advertised, often with apparent support from the utility.

The authors are appreciative of the discussors perceptive and inciteful comments.

REFERENCES

[1.] Canadian Electrical Association R&D Project 907 U 131 "Customer Damage Resulting from Electric Service Interruptions: Volume One -Report and Volume Two - Bibliography", Prepared by R. Billinton, G. Wacker and E. Wojczynski.
[2.] Ibid, p. 47, Vol. 1.

Manuscript received May 26, 1983.

COMPARISON OF TWO ALTERNATE METHODS TO ESTABLISH AN INTERRUPTED ENERGY ASSESSMENT RATE

R. Billinton, FIEEE   J. Oteng-Adjei   R. Ghajar

Power Systems Research Group
University of Saskatchewan
Saskatoon, Saskatchewan
Canada   S7N 0W0

Abstract - Quantitative reliability evaluation is an important aspect of power system planning and operation. The indices produced in these applications are utilized in a wide range of management decisions throughout a utility. One issue which is often debated is the cost associated with a particular level of reliability. In order to make this argument complete, it is necessary to also thoroughly examine the worth or benefit associated with a particular level of reliability. Estimates of the impact of interruptions in service can be obtained by assessing the losses incurred by different customer types and classes. These data can then be used to generate a composite customer damage function for a given utility service area. In order to create a practical tool for assessing reliability worth, interruption costs must be related to the calculated indices used in system planning and operation. This paper uses two different approaches and a customer damage function to evaluate a factor designated as the Interrupted Energy Assessment Rate (IEAR) which can be used in conjunction with the calculated expected energy not supplied in the assessment of reliability worth.

INTRODUCTION

The primary function of a modern electric power system is to supply the customers requirements at a reasonable level of reliability and quality. The specification of what constitutes a reasonable level of reliability is a difficult problem of balancing the need for continuity of power supply and the cost involved. In recent years, there has been considerable interest in attempts to relate the incremental worth or benefit of calculated electric service reliability levels to the incremental costs associated with providing those levels of reliability.

The general term "reliability", when applied to a complex power system involves two basic aspects: system adequacy and system security [1]. Adequacy relates to the existence of sufficient facilities within the system to satisfy the consumer load demand. Security, on the other hand, relates to the ability of the system to respond to disturbances arising within the system. Security is therefore associated with the response of the system to whatever perturbations it is subjected.

The basic techniques for adequacy assessment can be categorized in terms of their application to segments of a complete power system. Most utilities are divided into zones for purposes of organization,

87 WM 023-5   A paper recommended and approved by the IEEE Power System Engineering Committee of the IEEE Power Engineering Society for presentation at the IEEE/PES 1987 Winter Meeting, New Orleans, Louisiana, February 1 - 6, 1987. Manuscript submitted August 18, 1986; made available for printing November 21, 1986.

planning, operation and analysis. The three basic functional zones are generation, transmission and distribution. Different combinations of these functional zones define appropriate hierarchical levels in adequacy assessment. Hierarchical Level I (HLI) is concerned only with the generation facilities. The transmission and distribution facilities are assumed to be fully reliable and capable of transferring the energy from any generation point to a load point. Hierarchical Level II (HLII) includes both generation and transmission facilities and Hierarchical Level III (HLIII) includes all three functional zones in an assessment of consumer load point adequacy. This paper is primarily concerned with adequacy assessment at HLI.

System adequacy studies are an integral part of overall reliability assessment. The evaluation of reliability cost through the identification and analysis of criteria and methods used to predict and quantify reliability levels has progressed significantly during the past decade. By comparison, the assessment of reliability worth is in its infancy with most approaches used providing only an indirect or a boundary evaluation. This is because the assessment of the societal worth of electric service reliability is an extremely complex task. Some of the various approaches which have been used are reported in the literature [2]. One method which has been used to successfully establish reliability worth estimates is to survey electrical consumers, sector by sector in order to determine the costs or losses resulting from electric service interruptions [3].

Power system reliability assessment has been a topic of much research and is in common usage by most utilities. The most popular indices used in generating capacity adequacy evaluation are the Loss of Load Expectation (LOLE) and the Loss of Energy Expectation (LOEE) indices. The LOLE index measures the expected time in which the generation available will be insufficient to meet the demand. The LOLE index as normally calculated does not measure the severity of deficiencies. It is therefore difficult to relate this index directly to consumer interruption cost. The LOEE index specifies the expected energy that will not be supplied by the generation system due to those occasions when the load demanded exceeds the available generating capacity. This index includes a measure of the severity of deficiencies rather than only the amount of time that a deficiency exists. Some utilities are now commencing to utilize the LOEE in generating capacity adequacy assessments. This index can also be used together with a customer cost function to obtain a factor which relates the customer losses to the worth of electric service reliability. This factor is referred to in this paper as the Interrupted Energy Assessment Rate (IEAR).

Most of the data used in this paper were obtained from a study conducted by the Power Systems Research Group at the University of Saskatchewan during 1980-85 [3]. The remaining data were extracted from material published by Ontario Hydro. These data were used to create a composite customer damage function for the IEEE Reliability Test System (RTS) [4] which is then

Reprinted from *IEEE Trans. Power Syst.*, vol. PWRS-2, no. 3, pp. 751–757, Aug. 1987.

converted into an IEAR. The two techniques presented in the paper are conceptually different. The first one utilizes the Frequency and Duration method. The second one uses the Monte Carlo simulation technique. The basic objective of this paper is to illustrate the application of both methods in estimating the IEAR of the IEEE-RTS and to compare the results obtained.

CUSTOMER DAMAGE FUNCTION

There is an increasing utilization of cost/benefit considerations in the assessment of system reliability. In the past, the determination of what is a reasonable level of reliability was based largely on experience and judgement. Recent significant increases in energy costs and impacts of government and environmental groups have resulted in the need for more detailed justification of new system facilities and operating reliability levels. A major aspect of this justification is the assessment of the benefit or worth of reliability.

Actual or perceived costs of interruptions can be utilized to determine the benefit or worth of reliability and a variety of approaches have been used to investigate the cost of power interruptions [3]. One approach which is considered to yield acceptable results in an assessment of service interruptions is the customer survey method [5-7]. Studies of this type provide data which can be used to create a customer damage function. The generation of a composite customer damage function for a service area is an attempt to define the total customer costs for that area as a function of the interruption duration. The procedures involved in developing a composite customer damage function are discussed in detail in Reference 8. A brief summary of the procedure is as follows.

The interruption costs ($/KW) are a function of the various customer categories and other variables such as standby equipment, etc., etc. These costs are weighted in proportion to their respective energy utilization within the area. Weighting by the annual peak demand is used for short duration interruptions and weighting by the energy consumption is used for interruptions longer than one-half hour [3]. These weighted costs are then summed to provide the total cost for the area for each specified duration. The variation of this total cost with duration is considered to be the composite customer damage function for the service area.

Despite the uncertainties affecting the development of a composite customer damage function, it is the most suitable tool available for determining monetary estimates of reliability worth. The composite customer damage function can be tailored to reflect the individual nature of the system, the region within it and in the limit, any particular customer.

ESTIMATION OF THE INTERRUPTED ENERGY ASSESSMENT RATE

Method 1 - Frequency and Duration Approach

By far the broadest application of a customer damage function is its use to relate the composite customer losses to the socioeconomic worth of electric service reliability for an entire utility service area. Reliability worth can be evaluated in terms of expected customer interruption cost. This cost estimate can be obtained by multiplying the expected energy not supplied to customers due to power interruptions by a suitable factor. In this paper, this factor has been designated as the Interrupted

Energy Assessment Rate (IEAR) expressed in $/KWhr. The expected energy not supplied is a basic generating system adequacy assessment index which can be calculated in a number of ways. The severity and duration of discrete load loss events at HLI can be readily assessed using the frequency and duration approach [9]. This method in conjunction with the appropriate customer damage function can be used to estimate the IEAR.

Mathematical Formulation

The basic models required in this approach are as follows:

a) Generation Model Units are characterized by their capacity, forced outage rates, failure rates and repair rates.

b) Load Model The exact-state type load model was used in this study. This model represents the actual system load cycle by approximating it by a sequence of discrete load levels. The daily peak loads were assumed to exist for 12 hours giving an exposure factor of 0.5.

c) Cost Model This is represented either by the sector costs of interruption with their distribution of energy and peak demand for the service area or by the composite customer damage function.

The exact-state load model can be combined with an exact-state capacity model to yield the frequency and duration associated with each load loss event. The Expected Energy Not Supplied (EENS) for each load loss event is given by:

$$EENS = m_i f_i d_i \quad (KWhr/day) \quad (1)$$

where:

m_i = margin state in KW of load loss event i
f_i = frequency in occ/day of load loss event i
d_i = duration in hours of load loss event i.

The probability, p_i of a load loss event i is given by the product of the frequency, f_i (occ/day), and duration, d_i (days) of this load loss event. The total expected energy not supplied within the considered period for all the load loss events is as follows:

$$Total \ EENS = \sum_{i=1}^{N} m_i p_i = \sum_{i=1}^{N} m_i f_i d_i \quad (2)$$

where N is the total number of load loss events.

In deriving the cost associated with this unserved energy, attention should be drawn to the units of the composite customer damage function [10]. A major assumption in the estimation of the cost component is that the magnitude of each load loss event is assumed to be related to the annual peak demand of a particular customer. The total expected cost for all the load loss events of the system is given by the following equation:

$$Total \ expected \ cost = \sum_{i=1}^{N} m_i f_i c_i(d_i) \quad (\$/day) \quad (3)$$

where $c_i(d_i)$ is the interruption cost in $/KW for a duration d_i in hours of a load loss event i.

The frequency and duration method is used to evaluate the probability, frequency and duration of each load loss state. For each interruption duration, the corresponding cost is obtained from the costs of interruption data or the composite customer damage function.

$$\text{Estimated IEAR} = \frac{\sum\limits_{i=1}^{N} m_i f_i c_i (d_i)}{\sum\limits_{i=1}^{N} m_i f_i d_i} \quad (\$/KWhr) \qquad (4)$$

Numerical Example

The generation data of the RTS is given in Table A-1. The daily peak loads as obtained from the RTS [4] were arranged in descending order and then grouped in class intervals. The mean of each class was taken as the load level and the class frequency as the number of occurrenes of that load level. This load model is shown in Table 1. The period of study is 364 days and the peak demand is 2714 MW.

TABLE 1
Exact-state Load Data

Peak Load Level (MW)	No. of Occurrences (days)
2714.00	12.0
2458.00	82.0
2191.00	107.0
1928.50	116.0
1608.30	47.0
1485.40	364.0

The various assumed sector costs of interruption for the RTS system are given in Table 2 and their distribution of energy and peak demand are given in Table 3. The composite customer damage function is given in Table 4. The interruption duration was extended to 16 hours to cover all expected interruption durations. This extension was assumed to be of the same slope as that between 4 hours and 8 hours. Figure 1 shows the complete composite customer damage function.

TABLE 2
Interruption Cost Data Deemed Applicable to Example Service Area (1985 $/KW)

| User Sector | Interruption Duration | | | | |
	1 min	20 min	1 hr	4 hr	8 hr
Residential	0.004	0.09	0.56	5.17	15.47
Large Users	1.47	3.52	5.68	12.91	18.97
Small Users	0.72	3.38	5.27	19.66	29.37
Gov't & Inst.	0.04	0.36	1.45	6.35	25.23
Commercial	0.86	5.29	16.54	58.66	148.00
Office Buildings	4.65	9.61	20.50	67.00	116.12
Farm	0.03	2.80	14.00	116.80	326.80

TABLE 3
Distribution of Energy and Peak Demand of Example Service Area

User Sector	Energy (%)	Peak Demand (%)
Residential	31.0	34.0
Large Users	31.0	30.0
Small Users	19.0	14.0
Gov't & Inst.	5.5	6.0
Commercial	9.0	10.0
Office Buildings	2.0	2.0
Farm	2.5	4.0

TABLE 4
Composite Customer Damage Function

Interruption Duration	Interruption Cost (1985 $/KW)
1 min	0.73
20 min	2.42
1 hr	5.27
4 hr	19.22
8 hr	41.45

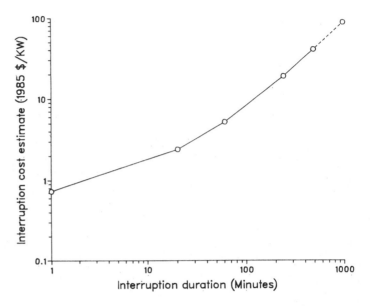

Figure 1. Composite Customer Damage Function for Example Service Area.

Results - Case A:

The composite customer damage function shown in Table 4 was used as the cost model. The generation and load models for this example are given in Tables A-1 and 1 respectively. The frequency and duration method [9] is used to generate the variables m, p, f and d of Equation (4) for each load loss state. Using the duration d, the cost associated with the load loss event is obtained from the composite customer damage function. Equation (2) is used to evaluate the total expected unserved energy. The total expected interruption cost is also evaluated by Equation (3). Finally, Equation (4) is used to estimate the value of IEAR. In this example, the estimated IEAR is 5.02 $/KWhr for an assumed exposure factor of 0.5.

The exact-state load model [9] is a function of the peak load and exposure factor. Studies were conducted to evaluate the impact of these variables on the estimated IEAR. First, the peak load was varied for a fixed exposure factor of 0.5. The results are shown in Figure 2. It can be seen that, a 125 percent change in peak load leads to only a 5 percent change in the estimated IEAR. It can be concluded that the estimated IEAR for a particular system (generation, load and composite customer damage function) does not vary significantly with the peak load.

The result of varying the exposure factor at a fixed peak load is shown in Figure 3. A 350% change in exposure factor produces only about 6% change in the estimated IEAR. The estimated IEAR can therefore be considered to be relatively constant with variations in exposure factor at a fixed peak load.

495

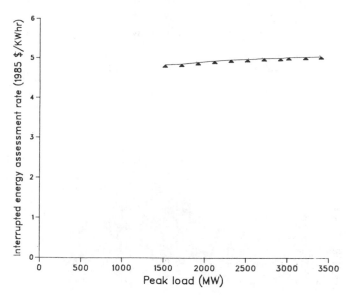

Figure 2. Variation of IEAR With Peak Load.

The value of 5.02 $/KWhr is therefore the best estimate of the overall system IEAR for the IEEE-RTS using the composite customer damage function given in Table 4.

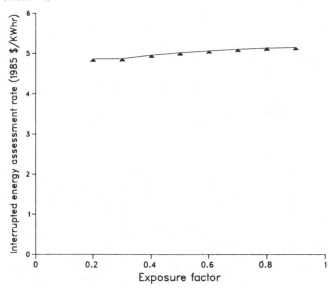

Figure 3. Variation of IEAR With Exposure Factor.

Results - Case B:

The cost model utilized in the previous example was the composite customer damage function for the entire system. The analysis can also be conducted using each customer sector damage function. In order to illustrate this consider the residential sector. If the service area is made up of only the residential sector, then the residential interruption cost data given in Table 2 becomes the composite customer damage function of the service area. A residential sector IEAR can be calculated by applying the same procedure described in Case A. A similar approach is employed to evaluate the IEAR of the other sectors within the service area. The IEAR of each sector is weighted by its percentage of energy consumption to give that sector's expected IEAR. The sum of the expected IEAR of all the sectors gives the composite IEAR of the service area. The results are given in Table 5. The estimated value of the IEAR for the entire service area is 4.98 $/KWhr.

TABLE 5
Sector IEAR for an Exposure Factor of 0.5

User Sector	Energy(%)	Sector IEAR ($/KWhr)	Sector Expected IEAR ($/KWhr)
Residential	31.0	1.64	0.51
Large Users	31.0	2.71	0.84
Small Users	19.0	4.16	0.79
Gov't & Inst.	5.5	2.39	0.13
Commercial	9.0	16.79	1.51
Office Buildings	2.0	15.43	0.31
Farm	2.5	35.54	0.89
			4.98

The IEAR values obtained from the two cases under study are slightly different due to interpolation of the respective customer damage functions. One advantage associated with the approach of Case B is that the IEAR for the various sectors within the service area of study are also available. These values can be used to assess the customer interruption costs for any particular sector and can be used to analyze the consequences associated with different load shedding policies.

Method 2 - Monte Carlo Simulation Approach

Monte Carlo simulation provides a method of analysis which permits relaxation of many of the traditional assumptions incorporated in the analytical techniques used to calculate reliability indices [11,12]. It also provides a benchmark for comparison of modelling assumptions associated with analytical techniques. The main advantage of Monte Carlo simulation is that it offers the opportunity to include, theoretically at least, any random variable and to include operation policies similar to the real ones. The main disadvantage is the excessive computer time required over analytical methods.

One limitation of the analytical techniques normally used for HLI adequacy assessment is that only the average or expected value of the reliability indices are determined and considered. Interruption costs often vary non-linearly with duration. Utilization of the average interruption duration to calculate interruption cost/reliability worth can result in large errors [15]. Knowledge of the probability distribution associated with this index can be used to more accurately estimate the system interruption costs and to more usefully apply the interruption cost estimates.

Model Description

In the simulation model, the power system is modelled by specifying a set of "events" where an "event" is a random or deterministic occurrence that changes the "state" of the system. The following events are recognized in the simulation model presented in this paper:

1. Change in load
2. Change in reserve requirements
3. Failure of a generating unit
4. Completion of repair of a generating unit
5. Derating of a generating unit
6. Completion of a derating repair.

Each of these events produces a change in the state of the system. There are a number of ways in which the system state can be defined. The central measure used in this simulation model is the "available margin". This is the difference between the available capacity (installed capacity less failed units and capacity loss due to derating) and load. Planned outages are not considered in this model.

The simulation model examines the system life during a specified calendar year using repeated "yearly samples" each consisting of 8736 hours which are selected in their chronological succession (sequential approach). The advantage of using this approach is the output that it provides in the form of probability distributions. These distributions are particularly useful when the interruption costs have to be estimated from the distributions of interruption durations.

The models used for the system generating units, system load and system operating policies are as follows:

a) Generating Unit Models   In the studies illustrated in this paper it is assumed that base-loaded generating units can reside in one of three states: (1) fully available; (2) partially available; and (3) unavailable. Peaking units are represented by a four-state model basically similar to that described in Reference [13]. In the model used, the unit duty cycle is not fixed and is estimated from the operating history of the unit for the entire simulation period.

b) State Residence Times   In the simulation model, the up time is assumed to be exponentially distributed, and the down time Weibull distributed [14].

c) Reserve Requirements   The spinning reserve available at each hour is used to identify the state of the system. This variable is computed each time the load changes by taking the difference between the available capacity and the load. This state variable is used to make decisions regarding placing units on line and removing them from service according to the operating reserve policy.

d) Unit Commitment   Units are committed in a fixed order in response to load increases, sudden unit outages or deratings. A priority list is created by the order in which the units are organized in the input data set. This ordering is assumed to be an economic ordering, such that units will be committed in this specified order and decommitted in the reverse priority order.

e) Load Model   The simulation model assumes that the load changes discretely every hour and is constant throughout the hour. Thus, the system load is represented in the simulation model by specifying an hourly load cycle for the specific period to be simulated.

f) Cost Model   This model is represented by the composite customer damage function of the service area in the form of the cost per KW peak demand for 1 minute, 20 minute, 1 hour, 4 hour, 8 hour and 16 hour interruptions. An interruption of zero minutes duration was assumed to have a zero cost. An interruption greater than 16 hours was assumed to result in the same cost as a 16 hour interruption. The costs resulting from interruptions of intermediate duration are computed using linear interpolation.

This section describes how the simulation model responds to sudden outages, deratings, repairs and load changes. The simulation process starts by placing all base load units on line and the peaking units in an idle file. At the beginning of every hour, the reserve requirements are computed to ensure that the proper amount of capacity is on-line. If insufficient capacity is on-line, additional peaking units are brought up from the idle file according to the economic order specified by the input listing of generating units. If, on the other hand, excess capacity exists, peaking units are decommitted in the reverse priority order.

The IEAR is estimated from the entire distribution of interruption durations using Equation (5).

$$ IEAR = \frac{\sum\limits_{i=1}^{n} c_i(r_i) \times l_i}{\sum\limits_{i=1}^{n} e_i} \qquad (\$/KWhr) \qquad (5) $$

where:

n = total number of interruptions
$c_i(r_i)$ = interruption cost in \$/KW for duration r_i
r_i = duration in hrs of interruption i
l_i = load loss in KW of interruption i
e_i = energy loss in KWhr of interruption i

Numerical Example

The simulation model was used to calculate the IEAR of the RTS. The generation data of this system is given in Table A-1. The units were committed according to the commitment priority order specified in this table. The hourly load cycle data are given in Reference 4. This load model specifies the value of system load for every hour of the year (8736 points). The yearly peak load was considered to be 2850 MW. The composite customer damage function is given in Table 4. The results of the simulation study give an estimated IEAR of 5.27 \$/KWhr. This value is quite close to the 5.02 \$/KWhr calculated using the frequency and duration method.

Sensitivity studies were conducted using the generation data given in Tables A-1 and A-2 and the yearly load cycle given in Reference 4 to examine the effect of modelling assumptions and operating policies on the estimated IEAR. The various cases studied in this paper are summarized as follows:

Case 1   This is the base case. All units are assumed to be running continuously except when in a state of total outage.
Case 2   This is the same as Case 1 except that the 350 MW and the 400 MW units are assumed to have one derated state.
Case 3   This is the same as Case 1 except that the 20 MW units are assumed to be peaking units. The repair times for these units are assumed to be exponentially distributed.
Case 4   This is the same as Case 3 except that the repair times of the 20 MW units are assumed to be Weibull distributed.
Case 5   This is the same as Case 3 but with different operating reserve requirements.

Table 6 presents the results for the five cases studied. It can be seen from this table that the IEAR is very stable and does not vary significantly with modelling assumptions and system operating considerations. The IEAR for the IEEE-RTS is 5.27 \$/KWhr. It could be suggested that this is a more

497

accurate figure than the 5.02 $/KWhr value obtained using the frequency and duration approach as fewer assumptions were made regarding the capacity and load models and the load interruption distribution data utilized. The difference between the two values is, however, relatively small. Practical estimates can therefore be made using the relatively simple frequency and duration approach.

TABLE 6
IEAR Calculated for Each Simulation Run

Case	IEAR ($/KWhr)
1	5.27
2	5.24
3	5.27
4	5.27
5	5.27

UTILIZATION OF IEAR IN GENERATING CAPACITY PLANNING

Conventional adequacy assessment at Hierarchical Level I is normally done using a Loss of Load Expectation (LOLE) criterion. There is, however, an increasing awareness that an energy based index such as Expected Energy Not Supplied (EENS) is a more responsive and perhaps a more valid index. This paper extends the utilization of an energy based index by presenting a basic methodology for adding a monetary component to the conventional EENS. This factor, the IEAR, can be used in capacity adequacy assessment to produce an estimate of reliability worth. The comparison of reliability worth with reliability cost is an important component in the assessment and determination of an appropriate planning index.

Customer interruption costs have been used to analyze the impact of over/under capacity planning [16]. The inherent uncertain nature of long term capacity planning creates a situation in which there is the potential for mismatch in supply and demand. An important component in the assessment of the implications associated with an inadequate supply is the customer costs associated with that condition. The IEAR as presented in this paper can be tailored to any particular system and restricted to any particular class or classes of customers. It responds to the factors, such as the customers in the system, that should be present in any determination of what is an acceptable level of reliability.

CONCLUSION

This paper has illustrated two conceptually different techniques for the evaluation of the Interrupted Energy Asessment Rate of a utility service area. The first approach using the frequency and duration method gives an IEAR value of 5.02 $/KWhr. The second method using Monte Carlo simulation gives a value of 5.27 $/KWhr. These values are quite comparable and the difference is basically due to the load models employed. The actual magnitude of the IEAR will depend on the particular system under study. This paper shows that a practical estimate can be obtained using the basic frequency and duration approach. The calculated results shown in this paper for the RTS indicate that the IEAR does not vary significantly with peak load, exposure factor and operating considerations of the system. This suggests that a single IEAR value can be obtained and utilized in a wide range of system studies. When combined with a basic LOEE index, the IEAR provides a fundamental tool for assessing adequacy worth in HLI adequacy evaluation.

REFERENCES

1) Billinton, R. and Allan, R.N., "Power-System Reliability in Perspective", IEE Electronics & Power, pp. 231-236, March 1984.

2) Wacker, G., Wojczynski, E. and Billinton, R., "Cost/Benefit Considerations in Providing an Adequate Electric Energy Supply", Third International Symposium on Large Engineering Systems, St. John's Newfoundland, pp. 3-8, July 10-11, 1980.

3) Billinton, R., Wacker, G. and Wojczynski, E., "Customer Damage Resulting from Electric Service Interruptions", Canadian Electrical Association, R&D, Project 907 U 131 Report, 1982.

4) IEEE Task Force, "IEEE Reliability Test System", IEEE Transaction, PAS-98, pp. 2047-2054, December 1979.

5) Wacker, G., Wojczynski, E. and Billinton, R., "Interruption Cost Methodology and Results - A Canadian Residential Survey", IEEE Winter Power Meeting, Paper No. 83 WM 067-6, January 1983.

6) Wojczynski, E., Billinton, R. and Wacker, G., "Interruption Cost Methodology and Results - A Canadian Commercial and Small Industrial Survey", IEEE Summer Powr Meeting, Los Angeles, Calif., July 1983.

7) IEEE committee, "Report on Reliability Survey of Industrial Plants, Part II, Cost of Power Outages, Plant Restart Time and Type of Loads Versus Time of Power Outages", IEEE Transactions, PAS, pp. 236-241, March/April 1974.

8) Wacker, G., Billinton, R. and Subramaniam, R.K., "Using Cost of Electric Service Interruptions Survey in Determination of a Composite Customer Damage Function", Proceedings of the International Association of Science and Technology for Development, IASTED, Energy Symposia, San. Francisco, Calif., June 4-6 1984.

9) Billinton, R., and Allan, Ronald N., "Reliability Evaluation of Power Systems", Plenum Press, New York and London 1984.

10) Wacker, G., Subramaniam, R.K. and Billinton, R., "Factors Affecting the Development of a Composite Customer Damage Function", Proceedings RAM Conference, Baltimore, Maryland, 1985.

11) Noferi, P.L., Paris, L. and Salvaderi, L., "Monte Carlo Methods for Power System Evaluation in Transmission or Generation Planning", Proceedings 1975 Annual Reliability and Maintainability Symposium, Washington, 1975.

12) "Modeling of Unit Operating Considerations in Generating Capacity Reliability Evaluation. Volume 1: Mathematical Models, Computing Methods, and Results", Report EL-2519, Electric Power Research Institute, Palo Afto, Ca., July 1982.

13) IEEE Task Group, "A Four-State Model for Estimation of Outage Risk for Units in Peaking Service", IEEE Transactions, PAS-91, pp. 618-627, April 1972.

14) Billinton, R., Chowdhury, A.A. and Hart, G., "Analysis of Generating Unit Outage Data Using the CEA Equipment Reliability Information System", Proceedings RAM Conference, Baltimore, Maryland, 1985.

15) Wojczynski, E. and Billinton, R., "Effects of Distribution System Reliability Index Distributions Upon Interruption Cost/Reliability Worth Estimates", IEEE Summer Power Meeting, Paper No. 85 SM 408-0, July 1985.

16) "Cost and Benefits of Over/Under Capacity in Electric Power System Planning", Decision Focus Incorporated (DFI), prepared for EPRI, Report EA-927, October 1978.

APPENDIX A

THE IEEE RELIABILITY TEST SYSTEM

The generating system of the RTS is comprised of 32 units of 9 distinct sizes with a total capacity of 3405 MW. This generation model is shown in Table A-1. The commitment priority list indicates the order in which the units were considered to be loaded in the system.

TABLE A-1
RTS Generating Unit Reliability Data

Commitment Priority	Number of Units	Unit Size(MW)	m (hrs)	r (hrs)	FOR
1	6	50	1980	20	0.01
2	2	400	1100	150	0.12
3	1	350	1150	100	0.08
4	3	197	950	50	0.05
5	4	155	960	40	0.04
6	3	100	1200	50	0.04
7	4	76	1960	40	0.02
8	5	12	2940	60	0.02
9	4	20	450	50	0.10

Table A-1 provides data only on full generating unit outages. Generating units can also experience partial outages which can have a significant effect on generation reliability, particularly units of large capacity. For this reason, the RTS has been supplemented by giving the 400 MW and 350 MW units a 50% derated state. The additional data for the three-state models are given in Table A-2. In this table, state 1 is the full capacity up state, state 2 is the derated capacity state and state 3 is the down state.

TABLE A-2
Generating Unit Derated State Data

Unit Size	Derated Cap.(MW)	λ_{12} (f/yr)	λ_{13} (f/yr)	μ_{21} (r/yr)	μ_{21} (r/yr)
350	175	1/1150	1/1150	1/60	1/70
400	200	1/1100	1/1100	1/100	1/100

Appendix
IEEE Bibliographies

THE Subcommittee on the Application of Probability Methods of the IEEE Power Engineering Society has actively supported the publication of periodic bibliographies on the application of probability methods in power system reliability evaluation. Four bibliographies have been published. The first one appeared in 1972 and the most recent one in 1988. These bibliographies cite nearly 700 published papers in the various areas of power system reliability assessment. Most of the papers are IEEE or IEE publications with some selected conference material. All of the publications are in English and therefore the bibliographies do not include some of the excellent papers available in other languages.

The four bibliographies are reprinted in this Appendix in order to bring together this important source of reference material.

REFERENCES

[1] R. Billinton, "Bibliography on the application of probability methods in power system reliability evaluation," *IEEE Trans. Power Apparatus Syst.,* vol. PAS-91, no. 2, pp. 649–660, Mar./Apr. 1972.
[2] IEEE Subcommittee on the Application of Probability Methods, "Bibliography on the application of probability methods in power system reliability evaluation 1971–1977," *IEEE Trans. Power Apparatus Syst.,* vol. PAS-97, no. 6, pp. 2235–2242, Nov./Dec. 1978.
[3] R. N. Allan, R. Billinton, and S. H. Lee, "Bibliography on the application of probability methods in power system reliability evaluation 1977–1982," *IEEE Trans. Power Apparatus Syst.,* vol. PAS-103, no. 2, pp. 275–282, Feb. 1984.
[4] R. N. Allan, R. Billinton, S. M. Shahidehpour, and C. Singh, "Bibliography on the application of probability methods in power system reliability evaluation 1972–1987," *IEEE Trans. Power Syst.,* vol. 3, no. 4, pp. 1555–1564, Nov. 1988.

BIBLIOGRAPHY ON THE APPLICATION OF PROBABILITY METHODS
IN POWER SYSTEM RELIABILITY EVALUATION

Roy Billinton, Member IEEE
Power System Research Group, University of Saskatchewan
Saskatoon, Saskatchewan

INTRODUCTION

The following bibliography of published material dealing with the application of probability techniques in the evaluation of power system reliability does not contain all the material available on this subject. It does, however, include most of the publications readily available in a power utility or technical reference library together with some of the earlier papers in the field. The many excellent publications clearly indicate the increasing usage and interest in the application of probability methods in the evaluation of power system reliability.

The assistance of the IEEE Subcommittee on the Application of Probability Methods in the compilation of this bibliography is gratefully acknowledged and in particular the suggestions and comments provided by V. T. Sulzberger, B. E. Biggerstaff, P. F. Albrecht and A. D. Patton.

STATIC GENERATING CAPACITY RELIABILITY EVALUATION

A considerable number of papers have been published in this area and the historical development which had lead to the techniques in use at the present time is extremely interesting. Although it is rather difficult to determine just when the first published material appeared it is safe to say that it was at least thirty five years ago. Interest in the application of probability methods to generating capacity requirements became evident about 1933 but it was not until 1947 that the first major group of papers were published. These papers, by Calabrese[11], Lyman[12], Seelye[13], Loane and Watchorn[14] proposed some of the basic concepts upon which the methods in use at the present time are based. Shortly after this in 1948, the first AIEE Subcommittee on the Application of Probability Methods was organized. This Subcommittee chaired

Paper 71 TP 93-PWR, recommended and approved by the Power System Engineering Committee of the IEEE Power Engineering Society for presentation at the IEEE Winter Power Meeting, New York, N.Y., January 31–February 5, 1971. Manuscript submitted August 18, 1970; made available for printing November 17, 1970.

by Calabrese recognized the need for consistent statistical data and produced a report in 1949 containing some comprehensive definitions for equipment outage classifications. (Reference 1 in the Equipment Outage Data section). Later reports on this subject by the Subcommittee appeared in 1954 and 1957.

The 1947 group of papers produced the methods which with some modifications are now generally known as the "Loss of Load Approach" and the "Frequency and Duration of Outage Approach". These two methods are described in considerable detail in a 1960 AIEE Committee Report[65]. The "Loss of Load Approach" is still sometimes referred to as the "Calabrese Method". The effect of interconnections and the determination and allocation of capacity benefits resulting from interconnections were discussed by Watchorn[19] and Calabrese[23] in 1950 and 1953 respectively. Until 1954 most probability studies had been done either by hand or using conventional desk calculators. The benefits associated with using digital computers to reduce the tedious arithmetic required in these investigations were noted by Watchorn[27] in 1954 and illustrated by Kirchmayer and his associates[28] in the evaluation of economic unit additions in system expansion studies.

Several excellent papers appeared each year until in 1958 a second large group of papers was published. This group of papers modified and extended the methods proposed by the 1947 group and also introduced a more sophisticated approach to the problem using "Game Theory" or "Simulation" techniques. Additional material in this area appeared in 1961 and 1962 but since that time interest in this approach appears to have declined.

The initial approach to the calculation of outage frequency and duration indices in generating capacity reliability evaluation has been recently modified by the introduction of a recursive approach. This technique is described in detail in a series of five recent publications[94, 97, 98, 99, 103]. The basic technique illustrated in the first paper of the series[94] is quite general and can be applied to virtually all areas of reliability evaluation. The basic concepts in the reliability

Reprinted from *IEEE Trans. Power Apparatus Syst.*, vol. 91, no. 2, pp. 649–660, March/Apr. 1972.

evaluation of interconnected systems were presented in a 1963 publication[75] and later illustrated in 1967[87, 89]. These concepts have also been extended to consider the interconnected benefits associated with more than two systems[96, 102].

As can be seen from the bibliography, there is a tremendous amount of published material available in the area of generating capacity reliability evaluation. It is difficult to single out any particular paper or group of papers for special reference. An arbitrary selection for anyone interested in commencing a general study of the area is as follows. The 1960 IEEE Subcommittee report[65] provides a basic starting point. It can be supplemented by References 91, 94, and 75 respectively in regard to the "Loss of Load Approach", the "Frequency and Duration Method" and interconnected system reliability evaluation.

BIBLIOGRAPHY – STATIC GENERATING CAPACITY RELIABILITY EVALUATION

1. Fundamental Consideration in Preparing Master System Plan, W. J. Lyman, Electrical World, New York, N.Y., Vol. 101, No. 24, June 17, 1933, pp. 788-92.

2. ˉare Capacity Fixed by Probabilities of Outage, S. A. Smith, Jr., Electrical World, New York, N.Y., Vol. 103, Feb. 10, 1934, pp. 222-25.

3. Service Reliability Measured by Probabilities of Outage, S. A. Smith, Jr., Electrical World, Vol. 103, March 10, 1934, pp. 371-74.

4. Probability Theory and Spare Equipment, S. A. Smith, Jr., Bulletin, Edison Electric Institute (New York, N.Y.) March 1934.

5. The Use of Theory of Probability to Determine Spare Capacity, P. E. Benner, General Electric Review, Schenectady, N.Y., Vol.37, No. 7, July 1934, pp. 345-48.

6. Considerations Involved in Making System Investments for Improved Service Reliability, S. M. Dean, Bulletin, Edison Electric Institute, New York, N.Y., Vol. 6, 1938, pp. 491-98.

7. Synthetic Determination of Frequency and Duration of Multiple Forced Outages of Electric Generating Units, E. S. Loane, C. W. Watchorn, Pennsylvania Electric Association, Harrisburg, Pa., November 1945.

8. Determination of Degree of Reliability of a Hypothetical Steam Electric Generating System, E. S. Loane, C. W. Watchorn, Pennsylvania Electric Association, Harrisburg, Pa., November 1945.

9. A New Approach to Probability Problems in Electrical Engineering, H. A. Adler, K. W. Miller, AIEE Transactions, Vol. 65, October 1946, pp. 630-32.

10. Reserve Requirements for Interconnected Systems, H. W. Phillips, Electrical World, May 11, 1946, p. 68; also Bulletin, Edison Electric Institute, Vol. 14, April 1946, pp. 117-20; also Mechanical World, Manchester, England, Vol. 120, September 27, 1946, pp. 351-53.

11. Generating Reserve Capability Determined by the Probability Method, G. Calabrese, AIEE Transactions, Vol. 66, 1947, pp. 1439-50.

12. Calculating Probability of Generating Capacity Outages, W. J. Lyman, AIEE Transactions, Vol. 66, 1947, pp. 1471-77.

13. Outage Expectancy as a Basis for Generator Reserve, H. P. Seelye, AIEE Transactions, Vol. 66, 1947, pp. 1483-88.

14. Probability Methods Applied to Generating Capacity Problems of a Combined Hydro and Steam System, E. S. Loane, C. W. Watchorn, AIEE Transactions, Vol. 66, 1947, pp. 1645-57.

15. Determination of Required Reserve Generating Capacity, W. J. Lyman, Electrical World, New York, N.Y., Vol. 127, No. 19, May 10, 1947, pp. 92-95.

16. Firm or Dependable Hydro Capacity, E. S. Loane, C. W. Watchorn, Pennsylvania Electric Association, Harrisburg, Pa., October 1947.

17. Probability Theory Helps Determine System Reserve Generating Capacity, G. Calabrese, Power, New York, N.Y., Vol. 92, July 1948, pp. 423-26.

18. A Convenient Method for Determining Generator Reserve, H. P. Seelye, AIEE Transactions, Vol. 68, Pt. II, 1949, pp. 1317-20.

19. The Determination and Allocation of the Capacity Benefits Resulting from Interconnecting Two or More Generating Systems, C. W. Watchorn, AIEE Transactions, Vol. 69, Pt. II, 1950, pp. 1180-86.

20. Determination of Reserve Capacity by the Probability Method, G. Calabrese, AIEE Transactions, Vol. 69, Pt. II, 1950, pp. 1681-89.

21. The Relation of Thermal Plan Design to Reserve Capacity Requirements, M. J. Steinberg, Electrical Engineering, Vol. 69, January 1950, pp. 64-67; also Electrical World, Vol. 133, January 2, 1950, pp. 58-60; also Power Generation, Vol. 54, February 1950, pp. 76-9.

22. Determination of Generator Stand-by Reserve Requirements, H. T. Strandrud, AIEE Transactions, Vol. 70, Pt. I, 1951, pp. 179-88.

23. Determination of Reserve Capacity by the Probability Method – Effect of Interconnections, G. Calabrese, AIEE Transactions, Vol. 70, Pt. I, 1951, pp. 1018-20; also Electrical Engineering, Vol. 72, 1953, pp. 100 for correction.

24. Elements of System Capacity Requirements, C. W. Watchorn, AIEE Transactions, Vol 70, Pt. II, 1951, pp. 1163-85.

25. Tables of Binomial Probability Distribution to Six Decimal Places, AIEE Committee Report, AIEE Transactions, Vol. 71, August 1952, pp. 597-620.

26. A Classification of Outages of Facilities in Electric Power Systems, A. Rich, Combustion, Vol. 24, March 1953, pp. 43-47.

27. Power and Energy Production, C. W. Watchorn, AIEE Transactions, August 1954, Vol. 73, Pt. III-B, pp. 901-908.

28. An Investigation of the Economic Size of Steam–Electric Generating Units, L. K. Kirchmayer, A. G. Mellor, J. F. O'Mara, J. R. Stevenson, AIEE Transactions, Pt. III (Power Apparatus and Systems), Vol. 74, August 1955, pp. 600-14; also Combustion, February 1955, pp. 57-64.

29. Evaluation of Unit Capacity Additions, M. J. Steinberg, V. M. Cook, AIEE Transactions (Power Apparatus and Systems),

Vol. 75, Pt. III, April 1956, pp. 169-79.

30. Probability Calculations for System Generating Reserves, C. Kist, G. J. Thomas, Electrical Engineering, Vol. 75, September 1956, pp. 806-11.

31. The Effect of Interconnections on Economic Generation Expansion Patterns, L. K. Kirchmayer, A. G. Mellor, H. O. Simmons, Jr., AIEE Transactions (Power Apparatus and Systems), Vol. 76, June 1957, pp. 203-14.

32. Operations Research Determination of Generation Reserves, E. L. Arnoff, J. C. Chambers, AIEE Transactions (Power Apparatus and Systems), Vol. 76, Pt. III, June 1957, pp. 316-28.

33. A Simplified Basis for Applying Probability Methods to the Determination of Installed Generating Capacity Requirements, C. W. Watchorn, AIEE Transactions (Power Apparatus and Systems), Vol. 76, October 1957, pp. 829-832.

34. The Use of a Digital Computer in a Generator Reserve Requirement Study, H. E. Brown, AIEE Transactions (Power Apparatus and Systems), Vol. 77, Pt. I, March 1958, pp. 82-85.

35. Game-Theory Techniques — A New Economic Tool, D. N. Reps, J. A. Rose, Electric Light and Power, May 15, 1958, pp. 56-62.

36. Economic Choice of Generator Unit Size, L. K. Kirchmayer, A. G. Mellor, American Society of Mechanical Engineers Transactions, Vol. 80, July 1958, pp. 1015-1023.

37. Probability in Planning, K. L. Hicks, Electric Light and Power, Chicago, Ill., July 15, 1958, pp. 43-46, 59.

38. Probability Calculations for System Generation Reserves, C. Kist, G. J. Thomas, AIEE Transactions (Power Apparatus and Systems), Vol. 77, August 1958, pp. 515-20.

39. Use of Probability Methods in the Economic Justification of Interconnecting Facilities Between Power Systems in South Texas, A. P. Jones, A. C. Mierow, AIEE Transactions (Power Apparatus and Systems), Vol. 77, August 1958, pp. 520-30.

40. Determination of Reserve — Generating Capability, H. Halperin, H. A. Adler, AIEE Transactions (Power Apparatus and Systems), Vol. 77, August 1958, pp. 530-44.

41. Determination of Reserve and Interconnection Requirements, H. D. Limmer, AIEE Transactions (Power Apparatus and Systems), Vol. 77, August 1958, pp. 544-50.

42. Details of Outage Probability Calculations, A. L. Miller, AIEE Transactions (Power Apparatus and Systems), Vol. 77, August 1958, pp. 551-57.

43. Generator Unit Size Study for the Dayton Power and Light Company, W. J. Pitcher, L. K. Kirchmayer, A. G. Mellor, H. O. Simmons, Jr., AIEE Transactions (Power Apparatus and Systems), Vol. 77, August 1958, pp. 558-63.

44. Digital Computer Aids Economic — Probabilistic Study of Generator Systems — I, M. K. Brennan, C. D. Galloway, L. K. Kirchmayer, AIEE Transactions (Power Apparatus and Systems), Vol. 77, August 1958, pp. 564-71.

45. Digital Computer Aids Economic — Probabilistic Study of Generation Systems — II, C. D. Galloway, L. K. Kirchmayer, AIEE Transactions (Power Apparatus and Systems), Vol. 77, August 1958, pp. 571-77.

46. Treatment of Hydro Capability Duration Curves in Probability Calculations, K. L. Hicks, AIEE Transactions (Power Apparatus and Systems), August 1958, Vol. 77, pp. 577-80.

47. Economic Selection of Generating Capacity Additions, T. W. Schroeder, G. P. Wilson, AIEE Transactions (Power Apparatus and Systems), Vol. 77, December 1958, pp. 1133-45.

48. Digital Program for the Economic Selection of Generating Capacity Additions, A. Cohen, L. E. Jenson, AIEE Transactions (Power Apparatus and Systems), Vol. 77, December 1958, pp. 1628-33.

49. New Analytical Tools Permit Integrated System Planning, L. K. Kirchmayer, A. G. Mellor, Electrical World, June 1959, pp. 52-54, 59, 85.

50. Mathematical Models for Use in the Simulation of Power Generation Outages; I — Fundamental Considerations, C. J. Baldwin, D. P. Gaver, C. H. Hoffman, AIEE Transactions, Pt. III (Power Apparatus and Systems), Vol. 78, 1959, pp. 1251-58.

51. Mathematical Models for Use in the Simulation of Power Generation Outages: II — Power System Forced Outage Distributions, C. J. Baldwin, J. E. Billings, D. P. Gaver, C. H. Hoffman, AIEE Transactions (Power Apparatus and Systems), Vol. 78, 1959, pp. 1258-1272.

52. The Interconnected Power Systems of Nova Scotia and New Brunswick, G. D. Mader, The Engineering Journal, October 1959, pp. 68-72.

53. Program for Planning, J. K. Dillard, H. K. Sels, Electric Light and Power, February 15, 1959, pp. 50-52.

54. Program for Planning, Part II — Game-Theory Techniques — A New Economic Tool, D. N. Reps, J. A. Rose, Electric Light and Power, May 15, 1959.

55. Program for Planning, Part III — Evaluating Chance in Planning, C. J. Baldwin, C. H. Hoffman, Electric Light and Power, August 15, 1959.

56. An Introduction to the Study of System Planning by Operational Gaming Models, J. K. Dillard, H. K. Sels, AIEE Transactions (Power Apparatus and Systems), Vol. 78, Part III, 1959, pp. 1747-52.

57. Mathematical Models for Use in the Simulation of Power Generation Outages: III — Models for a large Interconnection, C. J. Baldwin, D. P. Gaver, C. H. Hoffman, J. A. Rose, AIEE Transactions (Power Apparatus and Systems), Vol. 78, February 1960, pp. 1645-50.

58. Strategy for Expansions of Utility Generation, D. N. Reps, J. A. Rose, AIEE Transactions (Power Apparatus and Systems), Vol. 78 (February 1960 section), pp. 1710-20.

59. The Application of Planning Criteria to the Determination of Generator Service Dates by Operational Gaming, C. A. DeSalvo, C. H. Hoffman, R. G. Hooke, AIEE Transactions (Power Apparatus and Systems), Vol. 78, February 1960, pp. 1752-59.

60. Program for Planning (Part V) — Model Shows Generating Capacity Available, C. J. Baldwin, J. E. Billings, Electric Light and Power, February 15, 1960.

61. Program for Planning (Part VI) — Simulated Reserve Margins Show Generation Capacity Needs, C. J. Baldwin, C. H. Hoffman, Electric Light and Power, May 1, 1960.

62. Load and Capacity Models for Generation Planning by Simulation, C. J. Baldwin, C. A. DeSalvo, C. H. Hoffman, E. C. Plant, AIEE Transactions (Power Apparatus and Systems), Vol. 79, Pt. III, 1960, pp. 359-65.

63. The Use of Simulated Reserve Margins to Determine Generator Installation Dates, C. J. Baldwin, D. P. Hoffman, J. A. Rose, AIEE Transactions (Power Apparatus and Systems), Vol. 79, Pt. III, 1960, pp. 365-73.

64. An Approach to Peak Load Economics, C. D. Galloway, L. K. Kirchmayer, W. D. Marsh, A. G. Mellor, AIEE Transactions (Power Apparatus and Systems), Vol. 79, August 1960, pp. 527-35.

65. Application of Probability Methods to Generating Capacity Problems, AIEE Subcommittee on Application of Probability Methods, AIEE Transactions, Pt. III (Power Apparatus and Systems), Vol. 79, 1960 (February 1961 section), pp. 1165-82.

66. Theory of Economic Selection of Generating Units, K. L. Hicks, AIEE Transactions (Power Apparatus and Systems), Vol. 79, 1960, pp. 1794-1800.

67. Computers Aid System Generation Planning, L. K. Kirchmayer, C. D. Galloway, Paper presented at the American Power Conference, March 29, 1960, Chicago, Ill.

68. The Effect of Unit Size, Reliability and System Service Quality in Planning Generation Expansion, C. J. Baldwin, C. A. DeSalvo, H. D. Limmer, AIEE Transactions (Power Apparatus and Systems), Vol. 79, Pt. III, 1960, pp. 1042-50.

69. Evaluation of Tie Capacity Between Pools or Systems, H. E. Lokay, M. S. Schultz, Electric Light and Power, November 15, 1961.

70. Results of Long Range Planning by Simulation, C. J. Baldwin, Proceedings of the American Power Conference, Vol. 23, 1961, pp. 831-43.

71. Results of a Two-Year Study of Long Range System Planning by Simulation, C. J. Baldwin, C. H. Hoffman, Publication No. 306, CIGRE, Paris, France, 1962.

72. Determination of an Optimized Generation Expansion Pattern, R. J. Fitzpatrick, J. W. Gallagher, AIEE Transactions (Power Apparatus and Systems), Vol. 80, 1961 (February 1962 section), pp. 1052-59.

73. Generation Planning Programs for Interconnected Systems: Part I, Expansion Programs, E. S. Bailey, Jr., C. D. Galloway, E. S. Hawkins, A. J. Wood, AIEE Special Supplement, 1963, pp. 761-74.

74. Generation Probability Programs for Interconnected Systems: Part II, Production Costing Programs, E. S. Bailey, Jr., C. D. Galloway, E. S. Hawkins, A. J. Wood, AIEE Special Supplement, 1963, pp. 775-88.

75. Determination of Reserve Requirements of Two Interconnected Systems, V. M. Cook, C. D. Galloway, M. J. Steinberg, A. J. Wood, AIEE Transactions, Pt. III (Power Apparatus and Systems), Vol. 82, April 1963, pp. 18-33.

76. Load Growth Characteristics as Related to Generating Capacity Additions, C. W. Watchorn, AIEE Transactions, Pt. III (Power Apparatus and Systems), Vol. 82, April 1963, pp. 110-16.

77. Application of Probability Techniques in the Evaluation of Generating Capacity Requirements, R. Billinton, Transactions of the Canadian Electrical Association, Vol. 3, Part III, 1964.

78. Computer Design of Single-Area Generation Expansions, C. D. Galloway, L. L. Garver, IEEE Transactions (Power Apparatus and Systems), April 1964, Vol. 83, No. 4, pp. 305-10.

79. Economical Tie-Line Capacity for an Interconnected System, C. Szendy, IEEE Transactions (Power Apparatus and Systems), July 1964, Vol. 83, pp. 721-26.

80. A Review of Some Basic Characteristics of Probability Methods as Related to Power System Problems, C. W. Watchorn, IEEE Transactions (Power Apparatus and Systems), July 1964, Vol. 83, No. 7, pp. 737-43.

81. An Approach to the Analysis of Generating Capacity Reserve Requirements, G. S. Vassell, N. Tibberts, IEEE Transactions (Power Apparatus and Systems), January 1965, Vol. PAS-84, No. 1, pp. 64-79.

82. Effect of Generating Unit Transformers in System Reliability Evaluation, R. Billinton, Transactions of the Canadian Electrical Association, 1966.

83. Distributed Representation of the Forced Outage Probability Value in Generating Capacity Reliability Studies, R. Billinton, IEEE Conference Paper No. 31 CP 66-61.

84. The Effects of Availability upon Installed Reserve Requirements, A. K. Falk, IEEE Transactions, PAS 85, No. 11, November 1966, pp. 1135-1144.

85. Reliability and Optimum Design of the Interconnections of a Power System, H. Frank, S. L. Hakimi, IEEE Transactions, PAS-85, No. 12, December 1966, pp. 1191-96.

86. Effective Load Carrying Capability of Generating Units, L. L. Garver, IEEE Transactions, PAS-85, No. 8, August 1966, pp. 910-919.

87. Loss of Load Probability Approach to the Evaluation of Generating Capacity Reliability of Two Interconnected Systems, R. Billinton, M. P. Bhavaraju, Transactions of Canadian Electrical Association, 1967.

88. A Measure of Uncertainty in Probability Calculations for Generation Planning, C. Wasserman, W. G. Thompson, H. G. Houser, IEEE Paper No. 31 PP 67-35.

89. Reserve Planning for Interconnected Systems, L. L. Garver, IEEE Paper No. 31 PP 67-107.

90. Principles and Methods of Calculation of Probability of Loss of Generating Capacity, H. A. Adler, IEEE Transactions, PAS-86, No. 12, December 1967, pp. 1467-1479.

91. Generating Capacity Reliability Evaluation, R. Billinton, M. P. Bhavaraju, Transactions of the EIC, Vol. 10, No. C-5, October 1967.

92. The CAPCO Group Probability Technique for Timing Capacity Additions and Allocation of Capacity Responsibility, L. Firestone, W. D. Masters, A. H. Monteith, IEEE Transactions, PAS-87, No. 8, August 1968, pp. 1174-82.

93. Reliability of Generation Supply, A. D. Patton, D. W. Holditch, IEEE Transactions, PAS-87, No. 9, September 1968, pp. 1797-1803.

94. Frequency and Duration Methods for Power System Reliability Calculations – Part I – Generation System Model, J. D. Hall, R. J. Ringlee, A. J. Wood, IEEE Transactions, PAS-87, No. 9, September 1968, pp. 1787-96.

95. A Probability Analysis of the Use of Nuclear Capacity in a Hydro-Electric System, H. G. Houser, IEEE Paper No. 69 CP 128-PWR.

96. Power System Interconnection Benefits, R. Billinton, M. P. Bhavaraju, P. Thompson, Transactions of the Canadian Electrical Association, 1969.

97. Frequency and Duration Methods for Power System Reliability Calculations – Part II – Demand Model and Capacity Reserve Model, R. J. Ringlee, A. J. Wood, IEEE Transactions, PAS-88, No. 4, April 1969, pp. 375-88.

98. Frequency and Duration Methods for Power System Reliability Calculations – Part III – Generation System Planning, C. D. Galloway, L. L. Garver, R. J. Ringlee, A. J. Wood, IEEE Transactions, PAS-88, No. 8, August 1969, pp. 1216-23.

99. Frequency and Duration Methods for Power System Reliability Calculations – Part IV – Models for Multiple Boiler-Turbines and for Partial Outage States, V. M. Cook, R. J. Ringlee, A. J. Wood, IEEE Transactions, PAS-88, No. 8, August 1969, pp. 1224-1232.

100. Reserve Planning Using Outage Probabilities and Load Uncertainties, L. L. Garver, IEEE Transactions, PAS-89, No. 4, April 1970, pp. 514-521.

101. Evaluation and Comparison of Some Methods for Calculating Generation System Reliability, A. K. Ayoub, J. D. Guy, A. D. Patton, IEEE Transactions, PAS-89, No. 4, April 1970, pp. 521-527.

102. Probability of Loss of Load for Three Areas, H. T. Spears, K. L. Hicks, S. T. Lee, IEEE Transactions, PAS-89, No. 4, April 1970, pp. 521-527.

103. Frequency and Duration Methods for Power System Reliability Calculations – Part V – Models for Delays in Unit Installations and Two Interconnected Systems, R. J. Ringlee, A. J. Wood, IEEE Transactions, PAS-90, No. 1, January/February 1971, pp. 79-88.

AUTHOR INDEX – STATIC GENERATING CAPACITY RELIABILITY EVALUATION

Adler, H. A. 9, 40, 90
AIEE Committee Reports 25, 65
Arnoff, E. L. 32
Ayoub, A. K. 101

Bailey, Jr., E. S. 73, 74
Baldwin, C. J. 50, 51, 55, 57, 60, 61, 62, 63, 68, 70, 71
Benner, P. E. 5
Billings, J. E. 51, 60
Billinton, R. 77, 82, 83, 87, 91, 96
Bhavaraju, M. P. 87, 91, 96
Brennan, M. K. 44
Brown, H. E. 34

Calabrese, G. 11, 17, 20, 23
Chambers, J. C. 32
Cohen, A. 48
Cook, V. M. 29, 74, 99

Dean, S. M. 6
Dillard, J. K. 53, 56
DeSalvo, C. A. 59, 62, 68

Falk, A. K. 84
Firestone, L. 92
Fitzpatrick, R. J. 72
Frank, H. 85

Gallagher, J. W. 72
Galloway, C. D. 44, 45, 64, 67, 73, 74, 75, 78, 98
Garver, L. L. 78, 86, 89, 98, 100
Gaver, D. P. 50, 51, 57, 63
Guy, J. D. 101

Hall, J. D. 94
Halperin, H. 40
Hakimi, S. L. 85
Hawkins, E. S. 73, 74
Hicks, K. L. 37, 46, 66, 102
Hoffman, C. H. 50, 51, 55, 57, 59, 61, 62, 63, 71
Holditch, D. W. 93
Hooke, R. G. 59
Houser, H. G. 88, 95

Jenson, L. E. 48
Jones, A. P. 39

Kirchmayer, L. K. 28, 31, 36, 43, 44, 45, 49, 64, 67
Kist, C. 30, 38

Lee, S. T. 102
Limmer, H. D. 41, 68
Loane, E. S. 7, 8, 14, 16
Lokay, H. E. 69
Lyman, W. J. 1, 12, 15

Mader, G. D. 52
Marsh, W. D. 64
Masters, W. D. 92
Mellor, A. G. 28, 31, 36, 43, 49, 64
Mierow, A. C. 39
Miller, A. L. 42
Miller, K. W. 9
Monteith, A. H. 92

O'Mara, J. F. 28

Patton, A. D. 93, 101
Phillips, H. W. 10
Pitcher, W. J. 43
Plant, E. C. 62

Reps, D. N. 35, 54, 58
Rich, A. 26
Ringlee, R. J. 94, 97, 98, 99, 103
Rose, J. A. 35, 54, 57, 58, 63

Schroeder, T. W. 47
Schultz, M. S. 69
Seelye, H. P. 13, 18
Sels, H. K. 53, 56
Simmons, Jr., H. O. 31, 43
Smith, Jr., S. A. 2, 3, 4
Spears, H. T. 102
Steinberg, M. J. 21, 29, 75
Stevenson, J. R. 28
Strandrud, H. T. 22
Szendy, C. 79

Thomas, G. J. 30, 38
Thompson, P. 96
Thompson, W. G. 88
Tibberts, N. 81

Vassell, G. S. 81

Wasserman, C. 88
Watchorn, C. W. 7, 8, 14, 16, 19, 24, 27, 33, 76, 80
Wilson, G. P. 47
Wood, A. J. 73, 74, 75, 94, 97, 98, 99, 103

SPINNING GENERATING CAPACITY RELIABILITY EVALUATION

The publications listed in the area of static generating capacity reliability evaluation contain many comments on the subject of quantitative spinning reserve requirements. The first paper devoted entirely to this subject was published in 1963[1]. This publication presented the basic technique which has since been extended in subsequent papers. The difference between the basic statistic used in a spinning study and that used in a static capacity study is illustrated in Reference 83 of the static capacity bibliography. The concept of derated capacity levels was introduced in 1969 using a continuous Markov approach[5]. This has now been simplified using a discrete Markov approach to consider any number of derated levels[7].

BIBLIOGRAPHY – SPINNING GENERATING CAPACITY RELIABILITY EVALUATION

1. Application of Probability Methods to the Determination of Spinning Reserve Requirements for the Pennsylvania – New Jersey – Maryland Interconnection, L. T. Anstine, R. E. Burke, J. E. Casey, R. Holgate, R. S. John, H. G. Stewart, IEEE Transactions, PAS-82, October 1963, pp. 720-735.

2. Spinning Reserve Criteria in a Hydro Thermal System by the Application of Probability Mathematics, R. Billinton, M. P. Musick, Journal of the Engineering Institute of Canada, Vol. 48, pp. 40-45, October 1965.

3. Determination of Spinning Spare Requirements Using Probability Mathematics, R. Billinton, M. P. Musick, IEE Conference Record on Economics of Reliability of Supply, London, England, 1967, p. 86.

4. Outage Rate Confidence Levels in Static and Spinning Reserve Evaluation, R. Billinton, M. P. Bhavaraju, IEEE 1968 Summer Power Meeting, Chicago, June 1968, Paper 68 CP 608-PWR.

5. The Markov Process as a Means of Determining Generating Unit State Probabilities for Use in the Spinning Reserve Application, B. E. Biggerstaff, T. M. Jackson, IEEE Transactions, PAS-88, No. 4, April 1969, pp. 423-430.

6. Short Term Reliability Calculation, A. D. Patton, IEEE Transactions, PAS-89, No. 4, April 1970, pp. 509-514.

7. Quantitative Evaluation of System Spinning Reserve, R. Billinton, A. V. Jain, Transactions, Canadian Electrical Association, 1970.

8. Reliability Models and Measures for Power System Operation, R. J. Ringlee, A. J. Wood, American Power Conference, April 1970.

9. Security Constrained Unit Commitment, J. D. Guy, IEEE Transactions, PAS-90, No. 3, May/June 1971, pp. 1385-1390.

AUTHOR INDEX – SPINNING GENERATING CAPACITY RELIABILITY EVALUATION

Anstine, L. T. 1

Biggerstaff, B. E. 5
Billinton, R. 2, 3, 4, 7
Bhavaraju, M. P. 4
Burke, R. E. 1

Casey, J. E. 1

Guy, J. D. 9

Holgate, R. 1

Jackson, T. M. 5
Jain, A. V. 7
John, R. S. 1

Musick, M. P. 2, 3

Patton, A. D. 6

Ringlee, R. J. 8

Stewart, H. G. 1

Wood, A. J. 8

TRANSMISSION SYSTEM RELIABILITY EVALUATION

Several of the earlier papers noted in the Bibliography on Static Generating Capacity Reliability Evaluation included suggestions for numerical reliability criteria for transmission and distribution systems. This is particularly true of References 1, 2 and 6. After this initial period, the emphasis was clearly placed upon the generating capacity area and very few papers were published on transmission reliability for some time. The need for consistent methods has, however, been recognized within the past few years and some excellent papers have been published. The initial impetus to the present activity can be said to have been provided by two 1964 publications[4,5]. The two methods proposed are vastly different in concept and in the extent to which probability theory is applied. The second method introduced the concept of failure bunching in parallel facilities due to storm associated failures together with some basic techniques which can prove useful in many areas of application. These concepts have since been extended to the prediction of frequency and duration of outages in multiple path transmission circuits[10]. The application of Markov processes was introduced in 1964 (Reference 1 – General Section) and extended in 1968[9] to the transmission system problem. The effects of storm associated failures on simple parallel configurations were examined and the results compared with those obtained using the method of Reference 5.

A state-enumeration approach to more complicated configurations appears to have evolved from the literature. The technique is basically an enumeration of the failure modes and their effects. This general approach has been applied to distribution circuits[13,28] substation arrangements[19] and composite generation and transmission systems[18, 25, 26]. A general discussion of the failure modes and effects concept is given in Reference 29.

BIBLIOGRAPHY – TRANSMISSION SYSTEM RELIABILITY EVALUATION

1. The Probability of Coincident Primary Feeder Outages in Secondary Networks, D. N. Reps, AIEE Transactions (Power Apparatus and Systems), December 1954, pp. 1467-1478.

2. The Use of Probability in the Design and Operation of Secondary Network Systems, N. M. Neagle, D. R. Nelson, AIEE Transactions (Power Apparatus and Systems), August 1955, pp. 567-575.

3. Service Reliability on Rural Distribution Systems, L. B. Crann, AIEE Transactions (Power Apparatus and Systems), October 1958, pp. 761-764.

4. A Probability Method for Transmission and Distribution Outage Calculations, Z. G. Todd, IEEE Transactions (Power Apparatus and Systems), July 1964, pp. 696-701.

5. Power System Reliability – I – Measure of Reliability and Methods of Calculation, D. P. Gaver, F. E. Montmeat, A. D. Patton, IEEE Transactions (Power Apparatus and Systems), July 1964, pp. 727-737.

6. Power System Reliability – II – Applications and a Computer Program, F. E. Montmeat, A. D. Patton, J. Zemkoski, D. J. Cumming, IEEE Transactions (Power Apparatus and Systems), July 1965, pp. 636-43.

7. Steps in Measuring Service Continuity – Techniques for Comparing Feeder Outage Rates, E. S. Gardner, Jr., R. J. Ringlee, presented at EEI Transmission and Distribution Committee meeting, January 1968.

8. Transmission System Reliability Evaluation, R. Billinton, Transactions of the Canadian Electrical Association, 1967.

9. Transmission System Reliability Evaluation Using Markov Processes, R. Billinton, K. E. Bollinger, IEEE Transactions, PAS-87, February 1968, pp. 538-47.

10. A Method for Calculating Transmission System Reliability, S. A. Mallard, V. C. Thomas, IEEE Transactions on Power Apparatus and Systems, Vol. PAS-87, March 1968, pp. 824-833.

11. Reliability Calculations Help Evaluate Substation Supply, V. C. Thomas, F. E. Montmeat, Transmission and Distribution, June 1968, pp. 72-78.

12. Service Reliability as an Approach to Selecting an Optimum Underground Distribution System Design, Part I – Analysis of Service Continuity, R. D'Agostino, R. L. Adamson, R. G. Mann, M. A. Lozano, D. L. Hopkins, IEEE Paper No. 68 CP 77-PWR.

13. Distribution System Reliability Performance, M. W. Gangel,

R. J. Ringlee, IEEE Transactions PAS Vol. 87, July 1968, pp. 1657-1664.

14. A Method for Evaluating Redundant Capacity in Substations, A. D. Patton, Proceedings of Southwestern IEEE Conference, 1968.

15. Direct Current Transmission System Reliability Evaluation, R. Billinton, M. S. Sachdev, Transactions, Canadian Electrical Association, 1968.

16. Performance of URD Loops can be Predicted, J. W. Simpson, Electrical World, December 16, 1968, pp. 35-37.

17. Reliability of Overhead Distribution Systems, K. N. Stanton, E. F. Moore, IEEE Paper No. 68, CP 614-PWR.

18. Composite System Reliability Evaluation, R. Billinton, IEEE Transactions, PAS-88, April 1969, pp. 276-80.

19. A Look at Substation Reliability, C. R. Heising, R. J. Ringlee, H. O. Simmons, Jr., American Power Conference, April 1969.

20. A Probability Simulation Method for Determining the Reliability of Switching Stations, J. Sverak, IEEE Transactions, Vol. PAS-88, November 1969, pp. 1657-64.

21. Block Diagram Approach to Power System Reliability, M. Ramamoorty, Bal Gopal, IEEE Transactions, Vol. PAS-89, May-June 1970, pp. 802-811.

22. Transmission Planning Using A Quantitative Reliability Criterion, M. P. Bhavaraju, R. Billinton, Proceedings of the Sixth PICA Conference, May 1969.

23. Substation Reliability Estimation is a Useful Design Tool, A. D. Patton, Electric Light and Power, October 1969.

24. Underground Distribution System Design for Reliability, R. L. Capra, M. W. Gangel, S. V. Lyon, IEEE Transactions, Vol. PAS-88, June 1969, pp. 834-42.

25. Transmission Planning Using a Reliability Criterion – Part I – A Reliability Criterion, R. Billinton, M. P. Bhavaraju, IEEE Transactions PAS-89, No. 1, January 1970, pp. 28-34.

26. Transmission Planning Using a Reliability Criterion – Part II – Transmission Planning, M. P. Bhavaraju, R. Billinton, IEEE Transactions PAS-90, No. 1, January/February 1971, pp. 70-78.

27. Reliability Techniques Applied to Transmission Planning, R. Billinton, E. Zaleski, Transactions, Canadian Electrical Association, 1970.

28. A New Technique for Reliability Calculations, W. R. Christiaanse, IEEE Transactions PAS-89, No. 8, November/December 1970, pp. 1836-1847.

29. On Procedures for Reliability Evaluation of Transmission Systems, R. J. Ringlee, S. D. Goode, IEEE Transactions, PAS-89, No. 4, April 1970, pp. 527-37.

AUTHOR INDEX – TRANSMISSION SYSTEM RELIABILITY EVALUATION

Adamson, R. L. 12

Billinton, R. 8, 9, 15, 18, 22, 25, 26, 27
Bhavaraju, M. P. 22, 25, 26
Bollinger, K. E. 9

Capra, R. L. 24
Christiaanse, W. R. 28
Crann, L. B. 3
Cumming, D. J. 6

D'Agostino, R. 12

Gangel, M. W. 13, 24
Gardner, Jr., E. S. 7
Gaver, D. P. 5
Goode, S. D. 22
Gopal, B. 21

Heising, C. R. 19
Hopkins, D. L. 12

Lozano, M. A. 12
Lyon, S. V. 24

Mallard, S. A. 10
Mann, R. G. 12
Montmeat, F. E. 5, 6, 11
Moore, E. F. 17

Neagle, N. M. 2
Nelson, D. R. 2

Patton, A. D. 5, 6, 14, 23

Ramamoorty, M. 21
Reps, D. N. 1
Ringlee, R. J. 7, 13, 19, 29

Sachdev, M. S. 15
Simmons, Jr., H. O. 19
Simpson, J. W. 16
Stanton, K. N. 17
Sverak, J. 20

Thomas, V. C. 10, 11
Todd, Z. G. 4

Zaleski, E. 27
Zemkoski, J. 6

EQUIPMENT OUTAGE DATA

As previously noted in the section on static generating capacity reliability evaluation, the first AIEE Subcommittee on the Application of Probability Methods was organized in 1948. This committee chaired by Calabrese recognized the need for consistent statistical data and produced an important report in 1949 containing some comprehensive definitions of equipment outage classifications and some statistical data on equipment outages[1]. Later reports on this subject by this group appeared in 1954[3] and 1957[4]. These three reports were generally restricted to thermal unit equipment information with the exception of a short section on hydraulic equipment in the 1949 report[1]. The results of a statistical study of five years of data on 387 hydro-electric generating units was published in 1960[5]. This study used punched cards for the initial collection and sequential processing of the data. Shortly after this in 1961, the AIEE Subcommittee produced a manual[7] outlining reporting procedures and methods of analyzing forced outage data using digital equipment.

Detailed statistics on transmission equipment were published in 1961[8] and 1962[9] and on distribution facilities in 1966[16]. A useful paper on the application of statistical techniques in the determination and analysis of data for reliability studies was published in 1968[20] followed by an important IEEE Committee report entitled "Proposed Definition of Terms for Reporting and Analyzing Outages of Electrical Transmission and Distribution Facilities and Interruptions"[21].

Many organizations have task forces working in this important area of definition and compilation of outage statistics and it is expected that the results of their efforts will be published in the near future.

BIBLIOGRAPHY – EQUIPMENT OUTAGE DATA

1. Outage Rates of Steam Turbines and Boilers and of Hydro Units, AIEE Committee Report, AIEE Transactions, Vol. 68, Pt. I, 1949, pp. 450-57.

2. The Analysis of Some Failure Data, D. J. Davis, Journal American Statistical Association, Washington, D. C., Vol. 47, 1952, pp. 113-50.

3. Forced Outage Rates of High-Pressure Steam Turbines and Boilers, AIEE Committee Report, AIEE Transactions, Pt. III-B (Power Apparatus and Systems), Vol. 73, December 1954, pp. 1438-42; also Combustion, New York, N.Y., October 1954, pp. 57-61.

4. Forced Outage Rates of High-Pressure Steam Turbines and Boilers, AIEE Joint Sub-Committee Report, AIEE Transactions (Power Apparatus and Systems), Vol. 76, June 1957, pp. 338-43; also Combustion, Vol. 28, February 1957, pp. 43-46.

5. Forced Generation Outage Investigation for the Northwest Power Pool, H. U. Brown, III, L. A. Dean, A. R. Caprez, AIEE Transactions (Power Apparatus and Systems), Vol. 79, 1960, pp. 689-98.

6. Use and Availability of Outage Records in the Planning of System Generating Plant Additions, H. A. Adler, E. S. Loane, AIEE Paper No. CP 61-1103, 1961.

7. Manual for Reporting the Performance of Generating Equipment, AIEE Subcommittee on Application of Probability Methods, 1960, Revised January 3, 1961.

8. Recording of System Outage and Interruption Data, Electrical Systems and Equipment Committee, System Planning Subcommittee, Edison Electric Institute, June 1, 1961.

9. A General Program for Processing Distribution Data, G. P. Rhoten, Electrical World, New York, N.Y., Vol. 156, December 18, 1961, pp. 45-48.

10. Improvement of Electric Service Performance, Part I, Service and Facility Performance Records, J. J. Malone, R. K. Moore, N. N. Smeloff, AIEE Paper No. CP-61-790.

11. Report on Reliability of Electric Equipment in Industrial Plants, W. H. Dickenson, AIEE Transactions, Pt. III, July 1962, pp. 132-151.

12. Report of Joint AIEE-EEI Subject Committee on Line Outages, Joint AIEE-EEI Report, AIEE Transactions, Pt. III, Vol. 71, 1962, pp. 43-55.

13. Computer Analysis of Outage Data, J. P. Lynskey, R. J. Sloane, IEEE Transactions (Power Apparatus and Systems), April 1964, Vol. 83, No. 4, pp. 386-96.

14. Cable Operation 1963, EEI Publication No. 65-36.

15. Interruption Reporting, D. T. Egly, W. Chance, IEEE Paper No. 33-PP 66-457.

16. Operational Statistics in the Management of Large Distribution Systems, R. A. W. Connor, R. A. Parkins, Proceedings of the Institution of Electrical Engineers, November 1966.

17. Extra High Voltage Line Outages, IEEE-EEI Committee Report, IEEE Transactions on Power Apparatus and Systems, Vol. 86, No. 5, May 1967, pp. 547-562.

18. Techniques for Pooling Outage Data, R. J. Ringlee, Pennsylvania Electric Association (PEA), System Planning Committee Fall Meeting, Sharon, Pa., October 1967.

19. Report on Equipment Availability for the Seven Year Period 1960-66, EEI Publication No. 67-23, August, 1967.

20. Determination and Analysis of Data for Reliability Studies, A. D. Patton, IEEE, PAS-87, No. I, January 1968, pp. 84-100.

21. IEEE Committee Report on Proposed Definitions of Terms for Reporting and Analyzing Outages of Electrical Transmission and Distribution Facilities and Interruptions, IEEE Transactions PAS-87, May 1968, pp. 1318-1323.

22. Comments on Collection and Utilization of Outage Data, R. J. Ringlee, Canadian Electrical Association Spring Meeting, Toronto, 1970.

BIBLIOGRAPHY – EQUIPMENT OUTAGE DATA

Adler, H. A. 6
AIEE Committee Reports 1, 3, 4, 7, 12

Brown, III, H. U. 5

Caprez, A. R. 5
Chance, W. 15
Connor, R. A. W. 16

Davis, D. J. 2
Dean, L. A. 5
Dickenson, W. H. 11

EEI Committee Reports 8, 12, 14, 17, 19
Egly, D. T. 15

IEEE Committee Reports 17, 21

Loane, E. S. 6
Lynskey, J. P. 13

Malone, J. J. 10
Moore, R. K. 10

Parkins, R. A. 16
Patton, A. D. 20

Rhoten, G. P. 9

Ringlee, R. J. 18, 22

Sloane, R. J. 13
Smeloff, N. N. 10

GENERAL CONSIDERATIONS

There are a considerable number of publications which pertain to the general area of power system reliability evaluation but do not fit directly into the classifications previously considered. Some of the most recent and directly applicable ones are included in the following section.

BIBLIOGRAPHY – GENERAL CONSIDERATIONS

1. Economic Evaluation of Industrial Power System Reliability, W. H. Dickinson, AIEE Transactions, Pt. II (Applications and Industry), Vol. 76, 1957, pp. 264-72.

2. Evaluation of Alternative Power Distribution Systems for Refinery Process Units, W. H. Dickinson, AIEE Transactions, Pt. III, Vol. 79, pp. 110-124, April 1960.

3. A Probability Method for Determining the Reliability of Electric Power Systems, C. F. DeSieno, L. L. Stine, IEEE Transactions on Power Apparatus and Systems, Vol. 83, pp. 174-179, February 1964.

4. Maximal Electric Service Reliability at Minimal Cost, H. Greber, Conference Paper 31 CP 66-10, IEEE Winter Power Meeting, New York, 1966.

5. Computer Programs for Reliability, D. T. Egly, W. F. Esser, S. G. Wasilew, PICA Proceedings, pp. 513-518, May 1967.

6. Analysis of Distribution Customer Interruptions, R. E. Martin, PEA System Planning Committee Meeting, Sharon, Pa., October 1967.

7. Planning for System Reliability, W. E. Slemmer, American Power Conference, Chicago, Ill., April 1968.

8. Considerations in Planning for Reliable Electric Service, C. Concordia, American Power Conference, Chicago, Ill., April 1968.

9. Reliability Analysis for Power System Applications, K. N. Stanton, IEEE Transactions on Power Apparatus and Systems, Vol. 88, pp. 431-437, April 1969.

10. Quantitative Power System Reliability Evaluation, R. Billinton, Proceedings 1969 Power System Computational Conference, Rome, Italy.

11. How We Look Upon Power System Reliability, Lev S. Lindorf, Andrei M. Necrasov, IEEE Transactions on Industry and General Applications, Vol. IGA-5, No. 2, March/April 1969, pp. 112-120.

12. Power System Reliability, R. F. Lawrence, American Power Conference, April 23, 1969.

13. Prediction of Reliability and Availability of HVDC Valve and HVDC Terminal, C. R. Heising, R. J. Ringle, IEEE Transactions PAS-89, No. 4, April 1970, pp. 619-624.

14. Quantitative Reliability Assessment of Spare Single-Phase Transformers in Three Phase Installations, R. Billinton, V. Prasad, Transactions Canadian Electrical Association, 1970.

15. Quantitative Reliability Analysis of HVDC Transmission Systems – Part I – Spare Valve Assessment in Mercury Arc Bridge Configurations, R. Billinton, V. Prasad, IEEE Transactions PAS-90, No. 3, May/June 1971, pp. 1034-1046.

16. Quantitative Reliability Analysis of HVD Transmission Systems – Part II – Composite System Analysis, R. Billinton, V. Prasad, IEEE Transactions PAS-90, No. 3, May/June 1971, pp. 1047-1054.

17. Improved Reliability From Statistical Redundancy of Three Phase Operation of High Voltage Circuit Breakers, W. R. Wilson, C. E. Harders, IEEE Transactions PAS-90, No. 2, March/April 1971, pp. 670-681.

18. An Examination of Probability Methods in the Calculation of Power System Performance, R. C. Chan, S. B. Bram, IEEE Paper No. 70 CP 512-PWR.

19. Three State Models in Power System Reliability Evaluations, J. Endrenyi, IEEE Transactions PAS-90, No. 4, July/August 1971, pp. 1909-1916.

AUTHOR INDEX – GENERAL CONSIDERATIONS

Billinton, R. 10, 14, 15, 16
Bram, S. B. 18

Chan, R. C. 18
Concordia, C. 8

DeSieno, C. F. 3
Dickenson, W. H. 1, 2

Egly, D. T. 5
Endrenyi, J. 19
Esser, W. F. 5

Greber, H. 4

Harders, C. F. 17
Heising, C. R. 13

Lawrence, R. F. 12
Lindorf, L. S. 11

Martin, R. E. 6

Necrasov, A. M. 11

Prasad, V. 14, 15, 16

Ringlee, R. J. 13

Slemmer, W. E. 7
Stanton, K. N. 9
Stine, L. L. 3

Wasilew, S. G. 5
Wilson, W. R. 17

THEORY

It is impossible to include in this bibliography all the publications dealing with basic reliability theory. This would have to include the IEEE Transactions on Reliability, and the Proceedings of the Annual Symposia on Reliability sponsored by the IEEE, IES, SNT and ASQC. As of 1968, the Proceedings of this annual conference has a new title, "Annals of Assurance Sciences". There are also many other publications dealing with basic theory.

The following is a rather arbitrary list of text books which can prove useful in an appreciation of general reliability considerations. Recommended reading from this list are Reference 35 for general theory and Reference 36 for direct application power systems.

BIBLIOGRAPHY – THEORY

1. "Statistical Processes and Reliability Engineering", D. N. Chorafas, Van Nostrand Co., Inc., New York, 1960, 438 pp.

2. "Reliability Management and Mathematics", D. K. Lloyd and M. Lipow, Prentice-Hall, Englewood Cliffs, N. J., 1962, 528 pp.

3. "Understanding Reliability", C. E. Leake, United Testing Lab, Pasadena Lithographers, Inc., Pasadena, California, 1960, 151 pp.

4. "Reliability Principles and Practices", S. R. Calabro, McGraw-Hill, 1962, 371 pp.

5. "Semiconductor Reliability", J. Shwop, H. Sullivan, Engineering Publishers, Elizabeth, New Jersey, 1961.

6. "Reliability Theory and Practice", Igor Bazovsky, Prentice-Hall, Englewood Cliffs, N. J., 1961, 292 pp.

7. "Statistical Theory of Reliability", Marvin Zelen (ed), The University of Wisconsin Press, 1962, 166 pp.

8. "Engineering Applications of Reliability", Charles Lipson, Jal Kerawalla and Larry Mitchell, University of Michigan, 1963, 640 pp.

9. "Reliability Factors for Ground Equipment", Keith Henney (ed), McGraw-Hill, 1956.

10. "Reliability and Product Assurance", Richard R. Landers, Prentice-Hall, Englewood Cliffs, N. J., 1963, 516 pp.

11. "Reliability Control in Aerospace Equipment Development", SAE Technical Progress Series, Vol. 4, Society of Automotive Engineers, Inc., 485 Lexington Ave., New York 17, N. Y., 1963,

12. "System Reliability Engineering", Gerald H. Sandler, Prentice-Hall, Englewood Cliffs, N.J., 1963, 221 pp.

13. "Reliability Engineering", Technical Staff of ARINC Research Corporation, Prentice-Hall, Englewood Cliffs, N.J., 1964, 544 pp.

14. "Principles of Reliability", Eric Pierushka, Prentice-Hall, Englewood Cliffs, N.J., 1963, 365 pp.

15. "Reliability Engineering for Electronic Systems", edited by Richard H. Myers, Kam L. Wong, and Harold M. Gordy, John Wiley and Sons, Inc., New York, N.Y., 1964, 360 pp.

16. "Engineering Reliability and Long Life Design", Robert P. Haviland, D. Van Nostrand Co., Inc., Princeton, N.J., 1964, 205 pp.

17. "Structural Design of Missiles and Space Craft", Lewis H. Abraham, McGraw-Hill, New York, 1962.

18. "Reliability Training Text", F. M. Gryna, N. J. Ryerson and S. Zwerling (Editors), Institute of Radio Engineers, 1 East 79 St., New York 21, N.Y., 2nd Edition, 1960.

19. "Reliability of Military Electronic Equipment", AGREE Report, Office of the Assistant Secretary of Defense (Research and Engineering) U.S. Government Printing Office, Washington, D.C., 1957.

20. "Rome Air Development Center Reliability Notebook Supplement 1", Section 8 (AB 161894-1), U.S. Dept. of Commerce, Office of Technical Services, Washington 25, D.C., September 1, 1960.

21. "Quality Control and Reliability Handbook H 108", B. Epstein, Supt. of Documents, U.S. Govt. Printing Office, Washington 25, D.C., 1960. Sampling Procedures and Tables for Life and Reliability Testing.

22. "Quality Control and Reliability Technical Report No. TR-3", Good, H. P. and Kao, J. H. K., Office of Asst. Secretary of Defense (Installations and Logistics), Washington 25, D.C., Sampling Procedures and Tables for Life and Reliability Testing Based on the Weibull Distribution, 1961.

23. "Redundancy Techniques for Computing Systems", Wilcox, R. H., and Mann, W. C., Editors, Sparton Press, Washington, D.C., 1962.

24. "Probabilistic Models in Reliability Theory", Barlow, W. R., Hunter, L., and Proschan, F., Wiley and Sons, New York, N.Y., 1962.

25. "Mathematical Theory of Reliability", Barlow, Richard E., Proschan, Frank, John Wiley and Sons, 605 Third Ave., New York, 10016, 1965, 256 pp.

26. "Mathematical Methods in Reliability Engineering", Norman H. Roberts, McGraw-Hill, New York, N.Y., 1964, 300 pp.

27. "Electronics Reliability — Calculations and Design", G. W. A. Dummer and N. B. Griffin, Pergamon Press, Inc., 44-01, 21st Street, Long Island City, N.Y., 11101, 1966, 192 pp.

28. "Application of Reliability Techniques", Donald P. Kenney, Argyle Publishing Corp., 605 Third Avenue, New York, N.Y., 10016, 1966, 384 pp. (A self instructional programmed training course book).

29. "Manned Spacecraft: Engineering Design and Operation", Paul E. Purser, Maxime A. Paget and Norman F. Smith, Fairchild Publications, Inc., Book Division, 7 East 12th Street, New York, N.Y., 10003, 1964, 550 pp. (Chapter 38 — Reliability through Attention to Detail).

30. "Manual of Reliability", E. J. Tangerman (Editors), Product Engineering, Reader Service Department, 330 West 42nd Street, New York 36, N.Y., 1960, 32 pp.

31. "Maintainability: A Major Element of System Effectiveness", Allan S. Goldman and T. B. Stattery, John Wiley and Sons, New York, N.Y., 1964, 282 pp.

32. "Renewal Theory", D. R. Cox, John Wiley and Sons, New York, N.Y., 1962, 142 pp.

33. "Maintainability Design", F. L. Ankenbrandt, Engineering Publishers, Division of A. C. Book Company, Inc., Elizabeth, N.J., 1963, 247 pp.

34. "Reliability Theory and Practice", Frank M. Gryna, Jr. (Editor), Sixth Annual Workshop, June 22-24, 1965, Sponsored by Electrical Engineering Division of ASEE at the Illinois Institute of Technology, Chicago, Ill. (Available from ASEE Headquarters).

35. "Probabalistic Reliability: An Engineering Approach", M. L. Shooman, McGraw-Hill Book Co., New York, N.Y., 1968, 524 pp.

36. "Power System Reliability Evaluation", R. Billinton, Gordon and Breach Science Publishers, Inc., New York, N.Y., 1970, 299 pp.

CONCLUSION

It can clearly be seen from this bibliography that there is a considerable quantity of published literature on the subject of power system reliability evaluation. Many of the concepts presented in these publications have become standard engineering techniques. Since World War II, basic reliability theory has developed at a tremendous rate. This growth has been due largely to space and missile programs and the need for consistent quantitative evaluation of system reliability and effectiveness. It is expected that the utilization of these concepts in the power area will continue to increase in the future. It is interesting to note that the bulk of the publications in the first five sections of this bibliography have been generated by the power industry itself and only within the past few years has there been an appreciable university participation. It is to be hoped that this important aspect of power system engineering education will be further developed in the immediate future.

In compiling a bibliography of this type it is not possible to made it absolutely "complete". The author would like to apologize for any omissions and hope that they will be brought to his attention for inclusion at some future date.

Discussion

L. Paris (Ente Nazionale per l'Energia Elettrica, Rome, Italy): The author is to be complimented for the very interesting bibliography on the subject.

Note must however be taken of the fact that this bibliography does not contain European contributions on the application of probability methods in power system reliability evaluation.

It is therefore suggested to add the following references:

Bibliography — Static Generating Capacity Reliability Evaluation

1. Determinazione con elaboratore elettronico della potenza di riserva di una rete elettrica utilizzando un metodo probabilistico, P. Basso, R. Manni, G. Massini, Associazione Elettrotecnica Italiana Transactions, Alghero, 1966.

2. Further Experiences with Simulation Models in System Planning, T. W. Berrie, Second Power System Computation Conference, Stockholm, 1966.

3. Computer Simulation of Optimal Development and Operation of a Thermal Generating System for Planning Purposes, P. Basso, G. Manzoni, P. L. Noferi, L. Paris, M. Valtorta, Simpozionul de Analiza si Sinteza Retelelor Electrice, Bucuresti, 1967.

4. Planning the Installation of Large Units in Interconnected Systems: Relevant Parameters and their Relative Influence, L. Paris, M. Valtorta, Paper 32-12, CIGRE, 1968.

5. Planning Generating Units for Peak-Load in Power Systems: Relevant Parameters and their Relative Influence, G. Manzoni, P. L. Noferi, M. Valtorta, Paper 32-06, CIGRE, 1968.

6. Economic Aspects of Planning Large Generating Units in Interconnected Systems, L. Paris, M. Valtorta, Paper 101, World Power Conference, Moscow, 1968.

Manuscript received February 25, 1971.

Bibliography – Transmission System Reliability Evaluation

1. Valuation of a Power System Supply Reliability, J. Bergoughoux, J. C. Arinal, PICA, 1967.

2. Modèles probabilistes dan les études de réseaux et de production, J. Pouget, Paper 17, Helors-Hifors-Athenes, 1968.

3. Estimation de la defaillance d'un réseau de transport maillé: programme Peru; J. Auge, J. Bergoughoux, J. C. Dodu, J. Pouget, PSCC Proceedings, Rome, 1969.

4. Optimal Investment Policy for a Growing Electrical Network by a Sequential Decision Method, H. Baleriaux, E. Jamoulle, P. Doulliez, J. Van Kelecom, Paper 32-08, CIGRE, 1970.

Roy Billinton: I would like to thank Mr. Paris for his contribution to the bibliography and also Mr. N. H. Woodley who suggested the addition of the following paper.

"Substation Expansion, Reliability and Transformer Loading Policy Analysis", W. R. Ossman, N. H. Woodley, R. C. Craft, R. H. Miller, IEEE Transactions PAS August 1969, pp. 1195-1205.

In the interests of making the bibliography as complete as possible the following papers which were presented at the 1971 IEEE Winter Meeting are included.

Static Generating Capacity Reliability Evaluation

1. Generating Capacity Reliability Evaluation in Interconnected Systems Using a Frequency and Duration Approach – Part I – Mathematical Analysis, R. Billinton and C. Singh, IEEE Transactions PAS-90, No. 4, July/August 1971, pp. 1646-1654.

2. Generating Capacity Reliability Evaluation in Interconnected Systems Using a Frequency and Duration Approach – Part II – System Applications, R. Billinton and C. Singh, IEEE Transactions PAS-90, No. 4, July/August 1971, pp. 1654-1669.

Manuscript received March 26, 1971.

3. Factors Affecting Reliability Calculations for Generating Planning, M. P. Bhavaraju, J. K. Newton, IEEE Paper No. 71-CP-118-PWR.

4. A Four-State Model for Estimation of Outage Risk for Units in Peaking Service, IEEE Task Group on Model for Peaking Units of the Application of Probability Methods Subcommittee, IEEE Paper No. 71-TP-90-PWR.

5. Analysis of Generating-Capacity Reserve Requirements for Interconnected Power Systems, G. S. Vassel, N. Tibberts, IEEE Paper No. 71-TP-92-PWR.

Spinning Generating Capacity Reliability Evaluation

1. Unit Derating Levels in Spinning Reserve Studies, R. Billinton and A. V. Jain, IEEE Transactions PAS-90, No. 4, July/August 1971, pp. 1677-1687.

Transmission System Reliability Evaluation

1. Reliability Calculations Including the Effects of Overloads and Maintenance, W. R. Christiaanse, IEEE Transactions PAS-90, No. 4, July/August 1971, pp. 1664-1676.

2. Transmission System Reliability Methods, M. P. Bhavaraju, R. Billinton, IEEE Paper No. 71-TP-91-PWR.

General Considerations

1. Quantitative Evaluation of Power System Reliability in Planning Studies, P. L. Noferi, L. Paris, IEEE Paper No. 71-TP-89-PWR.

Theory

1. Probability Analysis of Power System Reliability, IEEE Special Publication, 71-M-30-PWR.

In conclusion, I would like to thank all those who assisted in the preparation of this Bibliography.

BIBLIOGRAPHY ON THE APPLICATION OF PROBABILITY METHODS IN POWER SYSTEM RELIABILITY EVALUATION

1971 - 1977

IEEE Subcommittee On The Application Of Probability Methods
Power System Engineering Committee

Abstract - This paper presents a bibliography of selected papers on the subject of power system reliability evaluation.

It includes material which has been made available following the publication of the initial "Bibliography On The Application Of Probability Methods In Power System Reliability Evaluation." (IEEE Transactions PAS-91, No. 2, Mar./Apr. 1972, pp. 649 - 660.) Several publications omitted from the initial bibliography are also included in this extension.

INTRODUCTION

There has been considerable activity in the development and application of probabilistic techniques for power system reliability evaluation since the initial "Bibliography" was published in 1972. The Subcommittee on the Application of Probability Methods has maintained an updated Bibliography since that time and in view of the interest expressed in this material it was decided to publish an extension of the earlier work.

The contributions and comments of all the members of the APM Subcommittee are gratefully acknowledged. The final version was assembled by R. Billinton and D. Taylor with assistance from A.D. Patton.

STATIC GENERATING CAPACITY RELIABILITY EVALUATION

1. "Practical Application of the Frequency and Duration Method of Generating Capacity Reliability Evaluation," C. Singh, R. Billinton, Proceedings Seventh Power Industry Computer Application Conference, Boston, Mass., May 1971, pp. 67 - 75, (IEEE Publication No. 71 C26-PWR).

2. "Planning Thermal And Hydraulic Power Systems - Relevant Parameters and Their Relative Influence," G. Manzoni, P.L. Noferi, M. Valtorta, CIGRE 1972, Paper 32-16.

3. "Analysis of Generating Capacity Reserve Requirements for Interconnected Power Systems, "G.S. Vassel, N. Tibberts, IEEE Transactions, Vol. PAS-91, No. 2, Mar./Apr. 1972, pp. 638 - 649.

4. "Pumped Storage Plant Basic Characteristics: Their Effect on Generating System Reliability," L. Salvaderi, L. Paris, American Power Conference Proceedings, Vol. 36, 1974, pp. 403 - 418.

5. "Probability Studies of Generation Capacity, Part I, " N. Kazak, B. Blagoveshensky, A. Rizk, IEEE Transactions on Reliability, Vol. R-23, No. 3, June 1974, pp. 123 - 125.

6. "Effect of Partial Outage Representation in Generation System Planning Studies," R. Billinton, A.V. Jain, C. MacGowan, IEEE Transactions, Vol. PAS-93, Sept./Oct. 1974, pp. 1252 - 1259.

7. "Peak Week Excess Load Probability," N. Savage, L.A. Wofsy, IEEE Winter Power Meeting, New York, January 1975, Paper C 75 019-5.

8. "A Reliability Model for Interconnected Electric Power Systems," J.R. Weiss, R.X. French, IEEE Transactions on Reliability, Vol. R-24, No. 2, June 1975, pp. 129 - 132.

9. "Reliability Evaluation for Application in Generation Planning," G. Irisarri, A.H. El-Abiad, IEEE Summer Power Meeting, San Francisco, July 1975, Paper A 75 429-1.

10. "Multi-Area Power System Reliability Calculations: Extensions and Examples," C.K. Pang, A.J. Wood, IEEE Summer Power Meeting, San Francisco, July 1975, Paper A 75 535-5.

11. "A Reliability Approach to Interconnection Planning," G. L. Landgren, S.W. Anderson, R. White, IEEE Transactions, Vol. PAS-94, No. 5, Sept./Oct. 1975, pp. 1644-1651.

12. "Computation of Loss-of-Load Probability, " G. Gambirasio, IEEE Transactions on Reliability, Volume R-25, No. 1, April 1976, pp. 54 - 55.

13. "Reliability of Generation Indices. A New Generation Reliability Index." II. Khatib, IEEE Paper No. A 76 321-0.

14. "An Efficient Method for Calculating Exact State Capacities and Probabilities," J.R. Weiss, IEEE Paper No. A 76 417-6.

15. "Frequency and Duration Method for Generating System Reliability Evaluation," A.K. Ayoub, A.D. Patton, IEEE Transactions, Vol. PAS-95, No. 6, Nov./Dec. 1976, pp. 1929 - 1933.

16. "The Combined Load Duration Curve and Its Derivation," J. Vardi, J. Zahari, B. Avi-Itzhak, IEEE Summer Power Meeting, Portland, July 1976, Paper F 76 426-7.

17. "Power Generation Reliability: Analytic and Optimization Approach," S.M. Yousif, H. Sklar, IEEE Winter Power Meeting, New York, Jan./Feb. 1977, Paper No. A 77 076-3.

18. "Effects of Uncertainties in Forced Outage Rates and Load Forecast on the Loss-of-Load Probability (LOLP), "L. Wang, IEEE Winter Power Meeting, New York, Jan./Feb. 1977, Paper No. F 77 045-8.

19. "Simplified Model of a Power Generation System For Use in Reliability Analysis," A.H. Maissis, IEEE Winter Power Meeting, New York, Jan./Feb. 1977, Paper No. A 77 081-3.

F78 073-9. A paper recommended and approved by the IEEE Power System Engineering Committee of the IEEE Power Engineering Society for presentation at the IEEE PES Winter Meeting, New York, NY, January 29-February 3, 1978. Manuscript submitted August 26, 1977; made available for printing November 2, 1977.

Reprinted from *IEEE Trans. Power Apparatus Syst.*, vol. PAS-97, no. 6, pp. 2235-2242, Nov./Dec. 1978.

20. "Graphical Method of Generating Capacity Determination," J.W. Guminski, M.M. Kumanowski, <u>Transactions of the IEE Conference on Reliability of Power Supply Systems</u>, IEE Conference Publication No. 148, February 1977.

21. "Generation Modelling in Power System Reliability Evaluation," R.N. Allan, F.N. Takieddine, <u>Transactions of the IEE Conference on Reliability of Power Supply Systems</u>, IEE Conference Publication No. 148, February 1977.

22. "Generating System Planning, Performance and Reliability Assessment," H. Khatib, <u>Transactions of the IEE Conference on Reliability of Power Supply Systems</u>, IEE Conference Publication No. 148, February 1977.

23. "Parameter Uncertainty in Generation Reliability Studies," A.D. Patton, N.H. Tram, <u>Transactions of the IEE Conference on Reliability of Power Supply Systems</u>, IEE Conference Publication No. 148, February 1977.

24. "Reliability and Sensitivity Analysis in Power System," J. Vardi, J. Zahavi, B. Avi-Itzhak, <u>IEEE Transactions</u>, Vol. PAS-96, No. 3, May/June 1977, pp. 478 - 484.

25. "Sensitivity of Generation Reliability Indices to Generator Parameter Variations," A.D. Patton, N.H. Tram, IEEE Summer Power Meeting, Mexico City, July 1977, Paper F 77 602-6.

26. "Confidence Intervals on the Frequency and Duration of Cumulative Margin Events for a Single Area System," A.D. Patton, N.H. Tram, IEEE Summer Power Meeting, Mexico City, July 1977, Paper A 77 612-5.

AUTHOR INDEX
STATIC GENERATING CAPACITY RELIABILITY EVALUATION

Allan, R.N. 21
Anderson, S.W. 11
Avi-Itzhak, B. 16, 24
Ayoub, A.K. 15

Billinton, R. 1, 6
Blagoveshensky, B. 5

El-Abiad, A.H. 9

French, R.X. 8

Gambirasio, G. 12
Guminski, J.W. 20

Irisarri, G. 9

Jain, A.V. 6

Kazak, N. 5
Khatib, H. 13, 22
Kumanowski, M.M. 20

Landgren, G.L. 11

MacGowan, C. 6
Maissis, A.H. 19
Manzoni, G. 2

Noferi, L. 2

Pang, C.K. 10
Paris, L. 4
Patton, A.D. 15, 23, 25, 26

Rizk, A. 5

Salvaderi, L. 4
Savage, N. 7

Singh, C. 1
Sklar, H. 17

Takieddine, F.N. 21
Tibberts, N. 3
Tram, N.H. 23, 25, 26

Valtorta, M. 2
Vardi, J. 16, 24
Vassel, G.S. 3

Wang, L. 18
Weiss, J.R. 8, 14
White, R. 11
Wofsy, L.A. 7
Wood, A.J. 10

Yousif, S.M. 17

Zahavi, J. 16, 24

OPERATING GENERATING CAPACITY RELIABILITY EVALUATION

1. "Unit Derating Levels in Spinning Reserve Studies," R. Billinton, A.V. Jain, <u>IEEE Transactions</u>, Vol. PAS-90, No. 4, July/Aug. 1971, pp. 1677 - 1687.

2. "Use of Outage Statistics to Develop Parameters for Use in Spinning Reserve Risk Assessments," L.T. Anstine, A. Street, R.J. Ringlee, S.F. Goode, IEEE Paper No. 71 CP 701-PWR.

3. "A Probability Method for Bulk Power System Security Assessment - I: Basic Concepts," A.D. Patton, <u>IEEE Transactions</u>, Vol. PAS-91, No. 1, Jan./Feb. 1972, pp. 54 - 61.

4. "The Effect of Rapid Start and Hot Reserve Units in Spinning Reserve Studies," R. Billinton, A.V. Jain, <u>IEEE Transactions</u>, Vol. PAS-91, No. 2, Mar./Apr. 1972, pp. 511 - 516.

5. "Interconnected System Spinning Reserve Requirements," R. Billinton, A.V. Jain, <u>IEEE Transactions</u>, Vol. PAS-91, No. 2, Mar./Apr. 1972, pp. 517 - 526.

6. "A Probability Method for Bulk Power System Security Assessment, Part II - Development of Probability Methods for Normally Operating Components," A.D. Patton, <u>IEEE Transactions</u>, Vol. PAS-91, No. 6, Nov./Dec. 1972, pp. 2480 - 2485.

7. "A Probability Method for Bulk Power System Security Assessment, Part III - Models for Stand-By Generators and Field Data Collection and Analysis," A.D. Patton, <u>IEEE Transactions</u>, Vol. PAS-91, No. 6, Nov./Dec. 1972, pp. 2486 - 2493.

8. "Reliability Methods for Generating Unit Peaking Equipment," R. Billinton, <u>CEA Transactions</u>, Vol. 11, Pt. 3, 1972.

9. "The Method of Availability Assessment of Power System Generating Capacity in Short Term Planning," A. Przyluski, Z. Reszczynska, <u>Proceedings Fourth Power System Computation Conference</u>, Grenoble, France, 1972, Vol. 2, Paper 2.2/1.

10. "Unit Commitment Reliability in a Hydro-Thermal System," A.V. Jain, R. Billinton, IEEE Paper No. C 73 096-6.

11. "Spinning Reserve Allocation in a Complex Power System," A.V. Jain, R. Billinton, IEEE Paper No. C 73 097-3.

12. "Power System Spinning Reserve Determination in a Multi System Configuration," R. Billinton, A.V. Jain, IEEE Transactions, Vol. PAS-92, No. 2, Mar./Apr. 1973, pp. 433 - 441.

13. "Reliable Loading of Generating Unit for System Operation," A.V. Jain, R. Billinton, Proceedings of the IEEE Power Industry Computer Applications Conference, June 1973. pp. 221 - 229, (IEEE Publication No. 73 CHO 740-1 PWR).

14. "Probabilistic Evaluation of the Operation Reserve in a Power System," J.A. Bubenko, M. Andersson, Proceedings of the IEEE Power Industry Computer Applications Conference, June 1973, pp. 240 - 249, (IEEE Publication No. 73 CHO 740-1 PWR).

15. "A Frequency and Duration Approach to Short Term Reliability Evaluation," C. Singh, R. Billinton, IEEE Transactions, Vol. PAS-92, No. 6, Nov./Dec. 1973, pp. 2073-2083.

16. "Reliability Cost in System Operation," A.V. Jain, R. Billinton, IEEE Paper No. C 74 314-1.

17. "Assessment of the Security of Operating Electric Power System Using Probability Methods," A.D. Patton, IEEE Proceedings, Vol. 62, No. 7, July 1974, pp. 892 - 901.

18. "Operating Reserve and Generation Risk Analysis for the PJM Interconnection," L.G. Leffler, R.J. Chambliss, G.A. Cucchi, N.D. Reppen, R.J. Ringlee, IEEE Transactions, Vol. PAS-94, No. 2, Mar./Apr. 1975, pp. 396 - 407.

19. "Generation Reliability Modelling for Static and Spinning Planning," S.B. Dhar, IEEE Paper No. A 76 433-3.

20. "Operating Capacity Reliability Assessment," R. Billinton, P.G. Harrington, A.V. Jain, Transactions of the IEE Conference on Reliability of Power Supply Systems, IEE Conference Publications No. 148, February 1977.

21. "Security Control in Network System Operation," T.O. Bernsten, A.T. Holen, Transactions of the IEE Conference on Reliability of Power Supply Systems, IEE Conference Publication No. 148, February 1977.

AUTHOR INDEX
OPERATING GENERATING CAPACITY
RELIABILITY EVALUATION

Andersson, M. 14
Anstine, L.T. 2

Bernsten, T.O. 21
Billinton, R. 1, 4, 5, 8, 10, 11, 12, 13, 15, 16, 20
Bubenko, J.A. 14

Chambliss, R.J. 18
Cucchi, G.A. 18

Dhar, S.B. 19

Goode, S.D. 2

Harrington, P.G. 20
Holen, A.T. 21

Jain, A.V. 1, 4, 5, 10, 11, 12, 13, 16, 20

Leffler, L.G. 18

Patton, A.D. 3, 6, 7, 17
Przyluski, A. 9

Reppen, N.D. 18
Reszczynska, Z. 9

Ringlee, R.J. 2, 18
Singh, C. 15
Street, A. 2

TRANSMISSION AND DISTRIBUTION SYSTEM
RELIABILITY EVALUATION

1. "Substation Bus-Switching Arrangements - Their Essential Requirements and Reliability," R.W. Davenport, E.M. Magidson, Yu A. Yakub, Elektra, p. 37, October 1969.

2. "Three State Models in Power System Reliability Evaluations," J. Endrenyi, IEEE Transactions, Vol. PAS-90, July/Aug. 1971, pp. 1909 - 1916.

3. "Transmission System Reliability Methods," M.P. Bhavaraju, R. Billinton, IEEE Transactions, Vol. PAS-91, No. 2, Mar./Apr. 1972, pp. 628 - 637.

4. "Reliability Evaluation in Large Transmission Systems," R. Billinton, C. Singh, IEEE Paper No. C 72 475-2.

5. "Reliability Estimation in the Design of Underground Distribution Systems for Commercial Load," L. Schwabel, W.E. Shula, S.R. Gilligan, Jr., R.D. Wood, IEEE Paper No. C 72 477-8.

6. "Economic and Reliability Assessments for Extra High Voltage Long Distance Transmission Lines," A.N. Karas, R. Billinton, CEA Transactions, Vol. 12, Pt. 3, 1973, Paper No. 73-SP-144.

7. "Reliability Evaluation of Transmission Systems With Switching After Faults-Approximations and a Computer Program," J. Endrenyi, P.C. Maenhaut, L.E. Payne, IEEE Transactions, Vol. PAS-92, No. 6, Nov./Dec. 1973, pp. 1863 - 1875.

8. "Reliability Models for Circuit Breakers With Multiple Failure Modes," J. Endrenyi, IEEE Paper No. C 74 138-4.

9. "A Computerized Approach to Substation and Switching Station Reliability Evaluation," M.S. Grover, R. Billinton, IEEE Transactions, Vol. PAS-93, 1974, pp. 1488 - 1497.

10. "Substation and Switching Station Reliability Evaluation," M.S. Grover, R. Billinton, CEA Transactions, Vol. 13, Pt. 3, 1974, Paper No. 74-SP-153.

11. "Reliability Evaluation of Substation Bus Arrangements," P.F. Albrecht, CEA Transactions, Vol. 13, Pt. 3, 1974, Paper No. 74-SP-152.

12. "A Computer Program for the Reliability Evaluation of Power Transmission Systems," J. Endrenyi, L. Wang, L.E. Payne, CEA Transactions, Vol. 13, Pt. 3, 1974, Paper No. 74-SP-151.

13. "Transmission System Reliability Studies," R. Billinton, M.S. Grover, CEA Transactions, Vol. 13, Pt. 3, 1974, Paper No. 74-SP-156.

14. "Reliability Evaluation of Power Transmission Networks - Models and Methods," J. Endrenyi, R. Billinton, CIGRE 1974, Paper No. 32-06.

15. "Reliability Evaluation In Distribution and Transmission Systems," R. Billinton, M.S. Grover, Proceedings IEE, Vol. 122, No. 5, 1975, pp. 517 - 523.

16. "Quantitative Evaluation of Maintenance Policies In Distribution Systems," M.S. Grover, R. Billinton, IEEE Paper No. C 75 112-8.

17. "A Sequential Method for Reliability Analysis of Distribution and Transmission Systems," M.S. Grover, R. Billinton, Proceedings 1975 Annual Reliability and Maintainability Symposium, Washington, D.C. Jan. 1975, Paper No. 1295, pp. 460-469, (IEEE Catalog No. 75-CHO-918-3 RQC).

18. "Reliability Assessment of Transmission and Distribution Schemes," R. Billinton, M.S. Grover, IEEE Transactions, Vol. PAS-94, No. 3, May/June 1975, pp. 724 - 733.

19. "Quantitative Evaluation of Permanent Outages in Distribution Systems," R. Billinton, M.S. Grover, IEEE Transactions, Vol. PAS-94, No. 3, May/June 1975, pp. 733-742.

20. "Quantitative Evaluation of Overload Outage Conditions in Substations and Subtransmission Systems," M.S. Grover, R. Billinton, IEEE Paper No. A 76 166-9.

21. "Distribution System Reliability Engineering Guide," Canadian Electrical Association Publication, March 1976.

22. "Unavailability Analysis of an Underwater Cable System," R. Billinton, S.Y. Lee, IEEE Transactions, Vol. PAS-96, No. 1, Jan./Feb. 1977, pp. 27 - 32.

23. "Evaluation of Distribution Circuit Reliability," D.O. Koval, R. Billinton, IEEE Winter Power Meeting, New York, Jan./Feb. 1977, Paper No. F 77 067-2.

24. "Reliability Analysis in the Design of Transmission and Distribution Systems," R.N. Allan, M.D. De Oliveira, Transactions of the IEE Conference on Reliability of Power Supply Systems, IEE Conference Publication No. 148, February 1977.

25. "Quantifying the Impact of Manual Emergency Transfer Switching Upon the Reliability of Distribution Systems," W. Cartwright, B.A. Coxson, Transactions of the IEE Conference on Reliability of Power Supply Systems, IEE Conference Publication No. 148, February 1977.

26. "Selection of Substation Layout in Power Systems," M. Giovannini, C. Reynaud, L. Salvaderi, Transactions of the IEE Conference on Reliability of Power Supply Systems, IEE Conference Publication No. 148, February 1977.

27. "Reliability Evaluation in the Planning of Distribution Systems," J.G. Lonsdale, G.B. Hitchen, Transactions of the IEE Conference on Reliability of Power Supply Systems, IEE Conference Publication No. 148, February 1977.

28. "Reliability and its Cost on Distribution Systems," G.F. L. Dixon, H. Hammersley, Transactions of the IEE Conference on Reliability of Power Supply Systems, IEE Conference Publication No. 148, February 1977.

29. "Urban Electricity Supply Systems," W.A. McNab, Transactions of the IEE Conference on Reliability of Power Supply Systems, IEE Conference Publication No. 148, February 1977.

30. "Evaluation of Elements of Distribution Circuit Outage Durations," D.O. Koval, R. Billinton, IEEE Summer Power Meeting, Mexico City, July 1977, Paper A 77 685-1.

AUTHOR INDEX
TRANSMISSION AND DISTRIBUTION SYSTEM RELIABILITY EVALUATION

Albrecht, P.F. 11
Allan, R.N. 24

Bhavaraju, M.P. 3
Billinton, R. 3,4,6,9,10,13,14,15,16, 17,18,19,20,22,23,30
Cartwright, W. 25
CEA 21
Coxson, B.A. 25
Davenport, R.W. 1
De Oliveira, M.D. 24
Dixon, G.F.L. 28
Endrenyi, J. 2,7,8,12,14
Gilligan, S.R., Jr. 5
Giovannini, M. 26
Grover, M.S. 9,13,15,16,17,18,19,20
Hammersley, H. 28
Hitchen, G.B. 27
Karas, A.N. 6
Koval, D.O. 23,30
Lee, S.Y. 22
Lonsdale, J.C. 27
Maenhaut, P.C. 7
Magidson, E.M. 1
McNab, W.A. 29
Payne, L.E. 7,12
Reynaud, C. 26
Salvaderi, L. 26
Schwabel, L. 5
Shula, W.E. 5
Singh, C. 4
Wang, L. 12
Wood, R.D. 5
Yakub, Y.A. 1

EQUIPMENT OUTAGE DATA

1. "The British Electricity Board's National Fault and Interruption Reporting Scheme - Objectives, Development and Operating Experience," D.V. Ford, IEEE Transactions, Vol. PAS-91, No. 5, Sept./Oct. 1972, pp. 2179 - 2188.

2. "Suggested Definitions Associated with the Status of Generating Station Equipment and Useful in the Application of Probability Methods for System Planning and Operation," IEEE Special Publication by IEEE Working Group in Performance Records for Optimizing System Design, IEEE Paper No. C 72 599-9.

3. "Reliability of Circuit-Breakers, Outage Data Evaluation," N.S. Selseth, CIGRE 1972, Paper 13-04.

4. "Circuit Breaker Reliability with Particular Reference to the Mechanical Behavior and Resistance to Climatic Effect, H.R. Wuthrich, CIGRE 1972, Paper 13-08.

5. "Interim Report on Study of the Reliability of High Voltage Circuit-Breakers," R. Michaka, CIGRE 1972, Paper 13-07.

6. "Reliability of Large Power Generating Units," P.O. Geszti, T. Frey, Z. Reguly, L. Raca, CIGRE 1972, Paper 32-01.

7. "Procedure for the Treatment and Analysis of the Data Necessary for Simulating System Operation," C. Dechamps, E. Jamoulle, H. Baleriaux, CIGRE 1972, Paper 32-09.

8. "Data Collection System for Analyzing Transmission and Distribution Performance, "S.S. Begian, IEEE Paper No. C 72 441-4.

9. "Generating Station Availability Data Requirements, " R. Billinton, J. Krasnodebski, IEEE 1975 Power Generation Conference, Portland, Sept./Oct. 1975, Paper PG 75 625-4, (IEEE Publication No. 75 CH 1042-1PWR).

10. "List of Transmission and Distribution Components for Use in Outage Reporting and Reliability Calculations, " Working Group on Performance Records for Optimizing System Design, IEEE Transactions, Vol. PAS-95, No. 4, July/Aug. 1976, pp. 1210 - 1215.

11. "Common Mode Forced Outages of Overhead Transmission Lines, "IEEE Committee Report, IEEE Transactions, Vol. PAS-95, No. 3, May/June 1976, pp. 859-864.

12. "Experience with Generation Outage Data Collection Procedures on a Large Scale Systems, " G.A. Cucchi, D.H. Pratzon, F.P. Witmer, IEEE Transactions, Vol. PAS-95, No. 2, Mar./Apr. 1976, pp. 542 - 549.

13. Report on Equipment Availability for the Ten-Year Period 1966 - 1975, EEI Publication No. 76-85, December 1976.

14. "An Appraisal of CEGB Transmission Network Fault Data, " J.W. Whines, Transactions of the IEE Conference on Reliability of Power Supply Systems, IEE Conference Publication No. 148, February 1977.

15. "Plant Operating Regime Analysis for Planning Purposes by Reliability Simulation, " S. Shields, W.G. Rattray, Transactions of the IEE Conference on Reliability of Power Supply Systems, IEE Conference Publication No. 148, February 1977.

16. "General Approach to Plant Availability Evaluation, " A. Harpur, E. O'Dwyer, Transactions of the IEE Conference on Reliability of Power Supply Systems, IEE Conference Publication No. 148, February 1977.

17. "Statistical Determination of the Availability of Important Components in the Electrical Power Supply, " K.W. Edwin, J. Nachtkamp, B. Siemes, Transactions of the IEE Conference on Reliability of Power Supply Systems, IEE Conference Publication No. 148, February 1977.

AUTHOR INDEX
EQUIPMENT OUTAGE DATA

Baleriaux, H. 7
Begian, S.S. 8
Billinton, R. 9
Cucchi, G.A. 12
Dechamps, C. 7
Edwin, K.W. 17
EEI 13
Ford, D.V. 1
Frey, T. 6
Geszti, P.D. 6
Harpur, A. 16
IEEE 2, 10, 11
Jamoulle, E. 7
Krasnodebski, J. 9

Michaka, R. 5
Nachtkamp, J. 17
O'Dwyer, E. 16
Pratzon, D.H. 12
Raca, L. 6
Rattray, W.G. 15
Reguly, Z. 6
Selseth, N.S. 3
Shields, S. 15
Siemes, B. 17
Whines, J.W. 14
Witmer, F.P. 12
Wuthrich, H.R. 4

INDICES AND CRITERIA

1. "Reliability Criterion for the Design of Power Installations and of Their Component Elements, " V.I. Nitu, Ch. Hortopan, H. Albert, M. Bogan, F. Lazariou, CIGRE 1972, Paper 32-17.

2. "Security Criteria of Power Supply in the Finnish Power System, " L. Nevanlinna, H. Viheriavaara, E. Voipio, CIGRE 1972, Paper 32-07.

3. "Distribution Protection and Restoration Systems: Design Verification by Reliability Indices, " IEEE Committee Report, IEEE Transactions, Vol. PAS-93, Mar./Apr. 1974, pp. 564 - 570.

4. "Definitions of Customer and Load Reliability Indices for Evaluating Electric Power System Performance, " IEEE Committee Report, IEEE Paper No. A 75 588-4.

5. "Reliability of Electrical Utility Supplies to Industrial Plants, " Power System Reliability Subcommittee, Record of 1975 Industrial and Commercial Power Systems Technical Conference, May 1975, pp. 123 - 129, (IEEE Publication No. 75-CHO 947-1-1A).

6. "Impact of Transmission Design Criteria on System Reliability, " R.J. Ringlee, N.D. Reppen, CEA Transactions, Vol. 15, Pt. 2, 1976, Paper No. 76-SP-149.

7. "Standards of Security in Network Planning, " M.G. Dwek, R.C. Bending, Transactions of the IEE Conference on Reliability of Power Supply Systems, IEE Conference Publication No. 148, February 1977.

8. "Reliability Indices for Use in Bulk Power Supply Adequacy Evaluation, "IEEE Committee Report, IEEE Summer Power Meeting, Mexico City, July 1977, Paper F 77 590-3.

9. "Reliability Criteria Used By Canadian Utilities In Generating Capacity Planning and Operation, " R. Billinton, IEEE Summer Power Meeting, Mexico City, July 1977, Paper F 77 601-8.

AUTHOR INDEX
INDICES AND CRITERIA

Albert, H. 1
Bending, R.C. 7
Billinton, R. 9
Bogan, M. 1
Dwek, M.G. 7
Hortopan, Ch. 1

IAS 5

IEEE 3, 4, 8

Lazariou, F. 1

Nevanlinna, L. 2
Nitu, V.I. 1

Reppen, N.D. 6
Ringlee, R.J. 6

Viheriavaara, H. 2
Voipio, E. 2

GENERAL CONSIDERATIONS

1. "Models and Techniques for Evaluating Power Plant Auxiliary Equipment Reliability," R. Billinton, R.J. Ringlee, 1971 Power Generation Conference, St. Louis, IEEE Paper No. 71 CP 702-PWR.

2. "Reliability Calculations Including the Effect of Overloads and Maintenance," W.R. Christiaanse, IEEE Transactions, Vol. PAS-90, July 1971, pp. 1664 - 1672.

3. Probability Analysis of Power System Reliability, IEEE Tutorial Text No. 71 M 30 PWR.

4. "Calculating the Reliability of Electrical Systems by a Statistical Modelling Method," N.V. Novella, B.I. Zotov, Electrical Technology (USSR), Vol. 4, 1971.

5. "Designing Reliability into the Electrical Power Supply for Power Plant Control Systems," R.L. Emerson, Instrumentation in the Power Industry, Vol. 14, 1971, pp. 72 - 78.

6. "Power System Reliability Indicators," V.I. Nitu, H. Albert, S. Ionescu, H. Cimpeanu, Proceedings Fourth Power System Computation Conference, Grenoble, France, 1972, Vol. 1, Paper 1.2/4.

7. "Reliability of Two Interconnected Systems," A. Holen, A. Fagerberg, S. Kroken, Proceedings Fourth Power Systems Computation Conference, Grenoble, France, 1972, Vol. 2, Paper 2.2/4.

8. "The Study of UHV System Reliability in Connection With Its Structure and Component Characteristics," L. Paris, F. Reggiani, M. Valtorta, CIGRE 1972, Paper 31-14.

9. "Risk of Failure in Power Supply and Planning of E.H.V. Systems," J. Auge, J. Barbey, J. Bergougnoux, J. de Calbiac, J. Delpech, J. Pouget, CIGRE 1972, Paper 32-21.

10. "Optimization of Power System Reliability Under Cost Constraint," O.P. Malik, A.S.A. Farag, IEEE Paper No. C 72 152-2.

11. "Evaluation of Service Reliability," G.P. Thoten, IEEE Paper No. C 72 153-0.

12. "Quantitative Evaluation of Power System Reliability in Planning Studies," P.L. Noferi, L. Paris, IEEE Transactions, Vol. PAS-91, No. 2, Mar./Apr. 1972, pp. 611-618.

13. "Power Reliability Cost vs. Worth," R.B. Shipley, A.D. Patton, J.S. Denison, IEEE Transactions, Vol. PAS-91, No. 5, Sept./Oct. 1972, pp. 2204 - 2212.

14. "Reliability Modelling in Systems With Non-Exponential Down Time Distributions," C. Singh, R. Billinton, IEEE Transactions, Vol. PAS-92, No. 2, Mar./Apr. 1973, pp. 790 - 800.

15. "Practical Applications of Reliability and Maintainability to Generating Station Design," R. Billinton, J. Krasnodebski, IEEE Transactions, PAS-92, No. 6, Nov./Dec. 1973, pp. 1814 - 1824.

16. Power System Reliability Calculations, R. Billinton, R.J. Ringlee, A.J. Wood, (Book), M.I.T. Press, Massachusetts, 1973.

17. "Reliability and Maintainability in the Design Phase of a Generation Station," J. Krasnodebski, R. Billinton, CEA Transactions, Vol. 12, Pt. 3, 1973, No. 73-SP-152.

18. "Reliability Modeling Using the Device of Stages," C. Singh, R. Billinton, S.Y. Lee, 1973 Power Industry Computer Applications Conference Proceedings, June 1973, pp. 22 - 30, (IEEE Pub. No. 73 CHO 740-1).

19. "Improvement in Quantitative Evaluation of Power System Reliability," O. Bertoldi, P.L. Noferi, F. Reggiani, IEEE Winter Power Meeting, New York, Jan./Feb. 1974, Paper No. C 74 137-6.

20. "Report on Reliability Survey of Industrial Plants, Part I: Reliability of Electrical Equipment," Power System Reliability Subcommittee, IEEE Transactions on Industry Applications, Vol. IA-10, No. 2, Mar./Apr. 1974, pp. 213 - 235.

21. "Report on Reliability Survey of Industrial Plants, Part II: Cost of Power Outages, Plant Restart Time, Critical Service Loss Duration Time, and Type of Loads Lost Versus Time of Power Outages," Power System Reliability Subcommittee, IEEE Transactions on Industry Applications, Vol. IA-10, No. 2, Mar./Apr. 1974, pp. 236 - 241.

22. "Monte Carlo Method for Power System Reliability Evaluation in Transmission and Generation Planning," P.L. Noferi, L. Paris, L. Salvaderi, Proceedings 1975 Annual Reliability and Maintainability Symposium, Washington, D.C., January 1975, Paper 1294, pp. 449 - 459, (IEEE Catalog No. 75-CHO-918-3 RQC).

23. "Maintenance Reserve Evaluation for Large Systems," L. G. Leffler, R.J. Chambliss, G.A. Cucchi, K.A. Clements, R.J. Ringlee, G.W. Woodzell, Proceedings 1975 Annual Reliability and Maintainability Symposium, Washington, D.C., January 1975, Paper 1293, pp. 444 - 448, (IEEE Publication No. 75-CHO-918-3 RQC).

24. "The Reliability Assessment of Emergency Electrical Supplies," A.E. Green, Proceedings 1975 Annual Reliability and Maintainability Symposium, Washington, D.C., January 1975, Paper 1296, pp. 470 - 475, (IEEE Publication No. 75-CHO-918-3 RQC).

25. "Effect of Voltage and Reactive Power Constraints on Power System Reliability," P.L. Noferi, L. Paris, IEEE Transactions, Vol. PAS-94, Mar./Apr. 1975, pp. 482 - 490.

26. "Application of Reliability Analysis Techniques in Increasing Power Plant Availability," P.A. Kales, W.N. Bley, American Power Conference Proceedings, Vol. 37, 1975, pp. 471 - 480.

27. "Cost of Electrical Interruptions in Commercial Buildings," Power System Reliability Subcommittee, Record of 1975 Industrial and Commercial Power Systems Technical Conference, May 1975, pp. 123 - 129.

28. "Optimization of the Storage Peak Duty Means in a Power System," L. Paris, L. Salvaderi, S. Scalcino, Joint IEEE 1975 Power Generation Conference, Portland, Sept./Oct.

1975, Paper PG 75-634-6, (IEEE Publication No. 75 CH 1042-1 PWR).

29. "Energy Storage Peak-Load Plants in the Future Energy Production Systems," L. Paris, L. Salvaderi, M. Valtorta, CIGRE 1976, Paper 31-10.

30. "Reliability Evaluation of Hierarchial Power Control Systems and Design of Component Redundancies," S. Narita, M.S.A.A. Hammam, IEEE Transactions, PAS-95, No. 3, May/June 1976, pp. 918 - 926.

31. "A Stochastic Approach to Reliability Design of Redundant Energy System," E. Balagurusamy, K.B. Misra, IEEE Paper No. A 76 322-8.

32. "Elements of Composite System Reliability Evaluation," R. Billinton, CEA Transactions, Vol. 15, Pt. 2, 1976, Paper No. 76-SP-148.

33. "Reliability in Generation and Transmission Planning," M.B. Guertin, Y. Lamarre, CEA Transactions, Vol. 15, Pt. 2, 1976, Paper No. 76-SP-150.

34. "Load Modelling in Power System Reliability Evaluation," R. Billinton, J. Endrenyi, IEEE Publication No. 76 CH 1135-3 PWR, New York, July 1976, Paper No. A 76 305-3.

35. "Reliability Planning and Analysis for Compressed Air Supply Systems for High-Capacity Airblast Circuit-Breakers," H. Frey, K. Holzinger, P. Lajda, IEEE Winter Power Meeting, New York, Jan./Feb. 1977, Paper No. A 77 084-7.

36. "Reliability Evaluation of the Auxiliary Electrical Systems of Power Stations," R.N. Allan, M.F. De Oliveira, R. Billinton, IEEE Winter Power Meeting, New York, Jan./Feb. 1977, Paper F 77 008-6.

37. "Application of Reliability Concepts in the Selection of Transformers for Large Generating Units and Stations," R. Billinton, D.O. Koval, D.R. Croteau, R.S. Weaver, V. Prasad, IEEE Winter Power Meeting, New York, Jan./Feb. 1977, Paper No. A 77 085-4.

38. "Program for Composite Bulk Power Electric System Adequacy Assessment," P.L. Dandeno, G.E. Jorgensen, W.R. Puntel, R.J. Ringlee, Transactions of the IEE Conference on Reliability of Power Supply Systems, IEE Conference Publication No. 148, February 1977.

39. "Reliability Effects of the Electrical Auxiliary Systems in Power Station," R.N. Allan, M.F. De Oliveira, J.A. Chambers, R. Billinton, Transactions of the IEE Conference on Reliability of Power Supply Systems, IEE Conference Publication No. 148, February 1977.

40. "Optimizing the Expenditure on Reliability of Supply in System Planning for the Benefit of the Consumer," H. Bocker, W. Kaufmann, Transactions of the IEE Conference on Reliability of Power Supply Systems, IEE Conference Publication No. 148, February 1977.

41. "Considerations Concerning The Choice of Optimal Risk in Medium Term Generation Scheduling," B. De Martino, G. Fusco, E. Mariani, L. Vergelli, Transactions of the IEE Conference on Reliability of Power Supply Systems, IEE Conference Publication No. 148, February 1977.

42. "Effect of Plant Reliability in Operational Planning Problems," R.N. Allan, C.H. Grigg, D.A. Newey, R.F. Simmons, Transactions of the IEE Conference on Reliability of Power Supply Systems, IEE Conference Publication, No. 148, February 1977.

43. "Costs Evaluation of Power Supply Reliability," Z. Reguly, Transactions of the IEE Conference on Reliability of Power Supply Systems, IEE Conference Publication No. 148, February 1977.

44. "Reliability Survey of Industrial Users," L.V. Skof, L.H. Berk, E.M. MacKay, CEA Transactions, Vol. 16, 1977.

45. "Calculation of Reliability Worth," R. Billinton, D.O. Koval, M.S. Grover, CEA Transactions, Vol. 16, 1977.

46. "Perception and Public Concern About Reliability," R.G. Rosehart, CEA Transactions, Vol. 16, 1977.

47. "Reliability Performance Evaluation of Power Circuit Breakers," J.S. Rai, V.H. Quintana, M.S. Grover, IEEE Summer Power Meeting, Mexico City, July 1977, Paper No. A 77 611-7.

AUTHOR INDEX
GENERAL CONSIDERATIONS

Albert, H. 6
Allan, R.N. 36, 39, 42
Auge, J. 9

Balagurusamy, E. 31
Barbey, J. 9
Bergougnoux, J. 9
Berk, L.H. 44
Bertoldi, O. 19
Billinton, R. 1, 14, 15, 16, 17, 18, 32, 34, 36, 37, 39, 45
Bley, W.N. 26
Bocker, H. 40

Chambers, J.A. 39
Chambliss, R.J. 23
Christiaanse, W.R. 2
Clements, K.A. 23
Croteau, D.R. 37
Cucchi, G.A. 23

Dandeno, P.L. 38
de Calbiac, J. 9
Delpech, J. 9
De Martino, B. 41
Denison, J.S. 13
De Oliveira, M.F. 36, 39

Emerson, R.L. 5
Endrenyi, J. 34

Faberberg, A. 7
Farag, A.S.A. 10
Frey, H. 35
Fusco, G. 41

Green, A.E. 24
Grigg, C.H. 42
Grover, M.S. 45, 47
Guertin, M.B. 33

Hammam, M.S.A.A. 30
Holen, A. 7
Holzinger, K. 35

IEEE 3, 20, 21, 27
Ionescu, S. 6

Jorgensen, G.E. 38

Kales, P.A. 26
Kaufmann, W. 40
Koval, D.O. 37, 45
Krasnodebski, J. 15, 17

Kroken, S. 7

Lajda, P. 35
Lamarre, Y. 33
Leffler, L.G. 23

MacKay, E.M. 44
Malik, O.P. 10
Mariani, E. 41
Misra, K.B. 31

Narita, S. 30
Newey, D.A. 42
Nitu, V.I. 6
Noferi, P.L. 12, 19, 22, 25
Novella, N.V. 4

Paris, L. 8, 12, 22, 25, 28, 29
Patton, A.D. 13
Pouget, J. 9
Prasad, V. 37
Puntel, W.R. 38

Quintana, V.H. 47

Rai, J.S. 47
Reggiani, F. 8, 19
Reguly, Z. 43
Ringlee, R.J. 1, 16, 23, 38
Rosehart, R.G. 46

Salvaderi, L. 22, 28, 29
Scalcino, S. 28
Simmons, R.F. 42
Singh, C. 14, 18
Skof, L.V. 44

Thoten, G.P. 11

Valtorta, M. 8, 29
Vergelli, L. 41

Weaver, R.S. 37
Wood, A.J. 16
Woodzell, G.W. 23

Zotov, B.I. 4

CONCLUSIONS

In compiling any "Bibliography" of related publications, it is not possible to make it absolutely complete. The APM Subcommittee would like to apologize for any errors or omissions and hope that additional suitable references will be advanced as discussion to this publication.

Discussion

R. N. Allan (UMIST, Manchester, England): Two of the greatest problems facing any engineer in any discipline is knowing what techniques are available to him for the analysis of his particular problem and what investigations similar to his own have been reported by other workers. These problems are particularly pertinent in the field of power system reliability evaluation because of its rapid devlopment in recent years and a greater tendency for papers to be much more specialised. This latter aspect is clearly indicated by the greatest number of papers appearing under the heading "General Considerations" which would suggest that the next issue of the bibliography should be divided into a larger number of section headings.

There are a growing number of abstracting services available nowadays, the efficiency of which rests with the abstracting editors in their choice of key words. However, even then, these are usually very time consuming to perform manually and the computer-based systems are often not readily available to the person requiring the information.

Consequently this bibliography is a first-class addition to the filing system of any power system engineer concerned with or interested in reliability evaluation and the IEEE Subcommittee should feel proud for producing such an exhaustive list of papers. One very pleasing point from a person working inthe U.K. is that the Subcommittee have included the papers presented at the I.E.E. Conference on Reliability of Power Supply Systems held in London in February 1977; a problem we often find difficult to answer is how widely read are the I.E.E. publications, particularly the Proceedings IEE, in North America.

I would like to raise two relatively small points, both of which tend to be personalised because of the nature of the paper. The first is that readers of the bibliography may also like to add two additional references [1, 2] to "Transmission and Distribution System Reliability Evaluation" and one reference [3] to "General Considerations". Secondly, the reference numbers allotted to me in the author index to "General Considerations" should be 36, 39, 42 and not 3, 6, 34, 42.

Once again may I compliment the Subcommittee on producing a valuable document in this topic area.

REFERENCES
1. "Reliability Evaluation of Electrical Systems with Switching Actions", R. N. Allan, R. Billinton, M. F. de Oliveira, *Proc. IEE*, 123, 1976, pp. 325–330.
2. "Reliability Modelling and Evaluation of Transmission and Distribution Systems", R. N. Allan, M.F. de Oliveira, *Proc. IEE*, 124, 1977, pp. 535–541.
3. "An Efficient Algorithm for Deducing the Minimal Cuts and Reliability Indices of a General Network Configuration", R. N. Allan, R. Billinton, M. F. de Oliveira, *IEEE Trans. on Reliability*, R-25, 1976, pp. 226–233.

Manuscript received February 14, 1978.

IEEE Subcommittee on the Application of Probability Methods: The subcommittee would like to thank Dr. Allan for both his comments and his contributions to the bibliography. In the interest of making the bibliography as complete as possible, the following papers which were presented at the 1978 IEEE Winter Power Meeting are included.

Static Generating Capacity Reliability Evaluation
1. "More Information from LOLP Calculations: Expected Value of Unsatisfied Energy Demand", E. Zadrzynski, Paper A 78 040-8.
2. "A Method for Estimating Equivalent Forced Outage Rates of Multistate Peaking Units", M. P. Bhavaraju, J. A. Hynds, G. A. Nunan, Paper F 78 008-5.
3. "Load Models for Bulk Power Supply Adequacy Assessment", T. A. Mikolinnas, K. A. Clements, Paper A 78 232-1.
4. "Reliability Evaluation in Energy Limited Generating Capacity Studies", R. Billinton, P. G. Harrington, Paper F 78 004-4.

Operating Generating Capacity Reliability Evaluation
1. "Outage Postponability Effects in Operating Capacity Reliability Studies", R. Billinton, M. Alam, Paper A 78 064-8.

Transmission and Distribution System Reliability Evaluation
1. "The Effect of Scheduled Outages in Transmission System Reliability Evaluation", L. Wang, Paper F 78 271-9.

General Considerations
1. "A Probabilistic Index for Transient Stability", R. Billinton, P.R.S. Kuruganty, Paper A 78 231-3.
2. "Linear Programming Network Flow Methods Applied to Bulk Power Supply Adequacy Assessment", K. A. Clements, G. C. Ejebe, B. F. Wollenberg, Paper 78 062-2.
3. "A Method of Combining High Speed Contingency Load Flow Analysis with Stochastic Probability Methods to Calculate a Quantitative Measure of Overall Power System Reliability", G.E. Marks, Paper A 78 053-1.
4. "Composite Generation and Transmission System Reliability Evaluation", R. Billinton, T. K. P. Medicherla, Paper A 78 237-0.

In conclusion, the Subcommittee would like to thank all of those who assisted in the compilation and preparation of this bibliography.

Manuscript received May 30, 1978.

BIBLIOGRAPHY ON THE APPLICATION OF PROBABILITY METHODS IN POWER SYSTEM RELIABILITY EVALUATION

1977 - 1982

R.N. Allan   R. Billinton   S.H. Lee
Sen. Member IEEE   Fellow IEEE   Member IEEE

Task Force Members of the IEEE Subcommittee on the Application of Probability Methods, Power System Engineering Committee.

Abstract

This paper presents a bibliography of selected papers on the subject of power system reliability evaluation.

It includes material which has become available since the publication of the two previous papers, "Bibliography of the Application of Probability Methods in Power System Reliability Evaluation", IEEE Transactions, PAS-91, 1972, pp. 649-660 and PAS-97, 1978, pp. 2235-2242. Several publications omitted from the second of these bibliographies are included in this extension.

INTRODUCTION

There has been considerable activity in the development and application of probabilistic techniques for power system reliability evaluation since the two previous bibliographies were published in 1972 and 1978. The Subcommittee on the Application of Probability Methods has maintained an updated Bibliography and in view of the interest expressed in this material, it was decided to publish a further extension.

The authors gratefully acknowledge the contribution and valuable comments of all members of the APM Subcommittee.

STATIC GENERATING CAPACITY

RELIABILITY EVALUATION

Papers

1. "Effects of Uncertainties in Forced Outage Rates and Load Forecast on the Loss-of-Load Probability (LOLP)". L. Wang, IEEE Transactions, PAS-96, 1977, pp. 1920-1927.

2. "Generator Maintenance Scheduling using Simplified Frequency and Duration Criteria". R.N. Allan, F.N. Takieddine, Proceedings IEE, 127, 1977, pp. 873-880.

3. "Reliability Evaluation in Energy Limited Generating Capacity Studies". R. Billinton, P.G. Harrington, IEEE Transactions, PAS-97, 1978, pp. 2076-2085.

83 WM 053-6   A paper recommended and approved by the IEEE Power System Engineering Committee of the IEEE Power Engineering Society for presentation at the IEEE/PES 1983 Winter Meeting, New York, New York, January 30-February 4, 1983. Manuscript submitted August 10, 1982; made available for printing November 10, 1982.

4. "More Information from LOLP Calculations: Expected Value of Unsatisfied Energy Demand". E. Zadrzynski, IEEE Paper A78 040-8.

5. "Reliability Criteria used by Canadian Utilities in Generating Capacity Planning and Operation", R. Billinton, IEEE Transactions PAS-97, 1978, pp. 1097-1103.

6. "Network Limitations on Generating Systems Reliability Evaluation Techniques". R.N. Allan, F.N. Takieddine, IEEE Paper A78 070-5.

7. "Theory and Applications of the Power System Probabilistic Simulation Method". D.W. Hilson, R.L. Sullivan, A.J. Wilson, IEEE Paper A78 530-8.

8. "Multi-Area Generation Reliability Studies". J.A. Bruggerman, C.K. Pang, R.L. Watt, A.J. Wood, IEEE Paper A78 546-4.

9. "Effects of the Generating Unit Design and Performance on the Expected Duration of Capacity Shortage and Energy Loss". N.S. Rau, K.F. Schenk, IEEE Paper A78 586-0.

10. "Maintenance Scheduling of Generating Facilities". H. Khatib, IEEE Transactions, PAS-98, 1979, pp. 1604-1608.

11. "Impact of Bulk Nuclear Capacity on Generating System Reliability". S. Panichelli, L. Salvaderi, M. Valtorta, IEEE Paper A79 030-8.

12. "Probability Models for Wind-Electric Conversion Systems and their Application to Reliability Studies". R.G. Deshmukh, R. Ramakumar, IEEE Paper A79 036-5.

13. "Considerations in Including Uncertainty in LOLE Calculations for Practical Systems". R. Billinton, G. Hamoud, IEEE Paper A79 075-3.

14. "Application of Fourier Methods for the Evaluation of Capacity Outage Probabilities". N.S. Rau, K.F. Schenk, IEEE Paper A79 103-3.

15. "Reliability and Sensitivity Studies of Interconnected Power Systems". C.K. Pang, S.B. Dhar, IEEE Paper A79 104-1.

16. "Probabilistic Simulation of Multi-Block Generating Units Subject to Partial Outages". D. Fieler, J. Zahavi, IEEE Paper A79 490-4.

17. "The Cumulant Method of Calculating LOLP". N.S. Rau, J.P. Stremel, IEEE Paper A79 506-7.

18. "Evaluation of Unconventional Sources in Mixed Hydro Thermal Systems". C. D'Amico, L. Salvaderi, M. Valtorta, L. Fracassi, G. Manzoni, S. Scalcino, Economic Commission for Europe - Symp. on the Prospects of Hydroelectric Schemes under the New Energy Situation and on Related Problems, Athens, 1979, EP/SEM/6/R6.

Reprinted from *IEEE Trans. Power Apparatus Syst.*, vol. PAS-103, no. 2, pp. 275-282, Feb. 1984.

523

19. "Pumped Storage Plant Evaluation in Planning Studies". S. Panichelli, L. Salvaderi, M. Valtorta, Economic Commission for Europe - Symp. on the Prospects of Hydroelectric Schemes under the New Energy Situation and on Related Problems, Athens, 1979, EP/SEM/6/R8.

20. "Pathology of Generation Planning Evaluation Indices". G.B. Hazen, T.L. Morin, A.H. El-Abiad, IEEE Paper A80 017-4.

21. "Multi-Area Reliability and Pool Operating Policies, New Concepts and Results". H. Duran, IEEE Paper A80 068-7.

22. "Interconnected Power Systems Planning Criteria Evaluation". S.B. Dhar, C.K. Pang, IEEE Paper A80 081-0.

23. "Generation System Planning under Load Forecast Uncertainty". J.P. Stremel, IEEE Transactions, PAS-100, 1981, pp. 384-393.

24. "Maintenance Scheduling under Uncertainty". J. Stremel, R. Jenkins, IEEE Transactions, PAS-100, 1981, pp. 460-465.

25. "Sensitivity Study of the Cumulant Method of Calculating Generation System Reliability". J.P. Stremel, IEEE Transactions, PAS-100, 1981, pp. 771-778.

26. "An Approximate and Practical Approach to Including Uncertainty Concepts in Generating Capacity Reliability Evaluation". G. Hamoud, R. Billinton, IEEE Transactions, PAS-100, 1981, pp. 1259-1265.

27. "Maintenance Scheduling for Generation System Planning". J. Stremel, IEEE Transactions, PAS-100, 1981, pp. 1410-1419.

28. "Uncertainty Aspects in LOLE Calcuations for Two Interconnected Systems". R. Billinton, G. Hamoud, IEEE Transactions, PAS-100, 1981, pp. 2320-2328.

29. "Operating Considerations in Generation Reliability Modelling - An Analytical Approach". A.D. Patton, C. Singh, M. Sahinoglu, IEEE Transactions, PAS-100, 1981, pp. 2656-2671.

30. "Application of Reliability Techniques to Mature, Near-term and Future Coal Utilization Technologies". S.N. Maruvada, IEEE Transactions, PAS-100, 1981, pp. 2813-2818.

31. "Effect of Partial Outage Representation on the Statistical Properties of Loss-of-Load Probability Estimates". M. Mazumdar, IEEE Transactions, PAS-100, 1981, pp. 4473-4478.

32. "Generation Planning Model with Reliability Constraints". G.J. Anders, IEEE Paper 81 SM 498-5.

33. "Digital Computer Algorithms for the Calculation of Generating Capacity Reliability Indices". R. Billinton, C.L. Wee, G. Hamoud, IEEE Transactions, PAS-101, 1982, pp. 203-211.

34. "Accuracy of the Edgeworth Approximation for LOLP Calculations in Small Power Systems". D.J. Levy, E.P. Khan, IEEE Transactions, PAS-101, 1982, pp. 986-996.

35. "A Frequency and Duration Approach for Interconnected System Reliability Evaluation". R. Billinton, C.L. Wee, IEEE Transactions, PAS-101, 1982, pp. 1030-1039.

36. "Discrete Convolution in Power System Reliability". R.N. Allan, A.M. Leite da Silva, A.A. Abu-Nasser, R.C. Burchett, IEEE Transactions, R-30, 1981, pp. 452-456.

37. "Interconnected System Reliability Evaluation - Applications", P.R.S. Kuruganty, P.R. Thompson, R. Billinton and C.L. Wee. CEA Transactions. Vol. 20, Part 3, Paper No. 81 SP-145. 1981.

38. "Interconnected System Reliability Evaluation - Concepts and Philosophy", R. Billinton, C.L. Wee, P.R.S. Kuruganty, P.R. Thompson. CEA Transactions. Vol. 20, Part 3, Paper No. 81-SP-144. 1981.

39. "Uncertainty Considerations in Frequency and Duration Analysis for Radial and Two Interconnected Systems". G. Hamoud, R. Billinton, IEEE Paper 82 WM 025-5.

40. "Reliability Models: The Influence of Model Specification in Generation Expansion Planning". J.P. Stremel, IEEE paper 82 WM 036-2.

41. "Reliability of Interconnected Power Systems with Correlated Demands". N.S. Rau, C. Necsulescu, K.F. Schenk, R.B. Misra, IEEE paper 82 WM 155-0.

42. "Reliability Evaluation in Hydrothermal Generating Systems". S.H.F. Cunha, F.B.M. Gomes, G.C. Oliveira, M.V.F. Pereira, IEEE Paper 82 SM 430-7.

Author Index

Abu - Nasser, A.A. 36
Allan, R.N. 2, 6, 36
Anders, G.J. 32
Billinton, R. ... 3, 5, 13, 26, 28, 33, 35, 37, 38, 39
Bruggerman, J.A. 8
Burchett, R.C. 36
Cunha, S.H.F. 42
D'Amico, C. 18
Deshmukh, R.G. 12
Dhar, S.B. 15, 22
Duran, H. 21
El-Abiad, A.H. 20
Fieler, D. 16
Fracassi, L. 18
Gomes, F.B.M. 42
Hamoud, G. 13, 26, 28, 33, 39
Harrington, P.G. 3
Hazen, G.B. 20
Hilson, D.W. 7
Jenkins, R. 24
Khan, E.P. 34
Khatib, H. 10
Kuruganty, P.R.S. 37, 38
Leite da Silva, A.M. 36
Levy, D.J. 34
Manzoni, G. 18
Maruvada, S.N. 30
Mazumdar, M. 31
Misra, R.B. 41
Morin, T.L. 20
Necsulescu, C. 41
Oliveria, G.C. 42
Panichelli, S. 11, 19
Pang, C.K. 8, 15, 22
Patton, A.D. 29

Pereira, M.V.F. 42
Ramakumar, R. 12
Rau, N.S. 9, 14, 17, 41
Sahinoglu, M. 29
Salvaderi, L. 11, 18, 19
Scalcino, S. 18
Schenk, K.F. 9, 14, 41
Singh, C. 29
Stremel, J.P. 17, 23, 24, 25, 27, 40
Sullivan, R.L. 7
Takieddine, F.N. 2, 6
Thompson, P.R. 37, 38
Valtorta, M. 11, 18, 19
Wang, L. 1
Watt, R.L. 8
Wee, C.L. 33, 35, 37, 38
Wilson, A.J. 7
Wood, A.J. 8
Zadrzynski, E. 4
Zahavi, J. 16

Author Index

Alam, M. 4
Allan, R.N. 8
Billinton, R. 4
Dillon, T.S. 3
Edwin, K.W. 3, 7, 9
Guven, C. 2
Hogg, G.L. 6
Insinga, F. 5
Invernizzi, A. 5
Kochs, H.D. 3, 7, 9
Manzoni, G. 5
Nunes, R.A.F. 8
Panichelli, S. 1, 5
Paris, L. 1
Patton, A.D. 6
Salvaderi, L. 1, 5
Spearman, M.L. 6
Taud, R.J. 3

OPERATING GENERATING CAPACITY

RELIABILITY EVALUATION

Papers

1. "Quantitative Evaluation of Operating Reserve Duty of Peak-Duty Storage Plants in Power System Planning". S. Panichelli, L. Paris, L. Salvaderi, Transactions of the IEE Conference on Reliability of Power Supply Systems, IEE Conference Publication No. 148, 1977, pp. 145-149.

2. "Model for Scheduling Electric Power Generation with Optimal Reliability". C. Guven, Operational Research Quarterly, 28, 1977, pp. 555-568.

3. "Integer Programming Approach to the Problem of Optimal Unit Commitment with Probabilistic Reserve Determination". K.W. Edwin, T.S. Dillon, H.D. Kochs, R.J. Taud, IEEE Transactions, PAS-97, 1978, pp. 2154-2166.

4. "Outage Postponability Effects in Operating Capacity Reliability Studies". R. Billinton, M. Alam, IEEE Paper A78 064-8.

5. "Integration of Direct Probabilistic Methods and Monte Carlo Approach in Generation Planning". F. Insinga, A. Invernizzi, G. Manzoni, S. Panichelli, L. Salvaderi, Proceedings 6th Power System Computation Conference, Darmstadt, W. Germany, 1978, pp. 48-58.

6. "Operating Considerations in Reliability Evaluation". A.D. Patton, G.L. Hogg, M.L. Spearman, Proceedings 6th Power System Computation Conference, Darmstadt, W. Germany, 1978, pp. 212-217.

7. "Determination of the Operational Fast Reserve for Bulk Power Generation Systems". K.W. Edwin, H.D. Kochs, Proceedings 6th Power System Computation Conference, Darmstadt, W. Germany, 1978, pp. 234-241.

8. "Modelling of Standby Generating Units in Short Term Reliability Evaluation". R.N. Allan, R.A.F. Nunes, IEEE Paper A79 006-8.

9. "Operational Power Reserve Determination and Application Experience". K.W. Edwin, H.D. Kochs, Electra, 76, 1981, pp. 29-42.

COMPOSITE GENERATION/TRANSMISSION

RELIABILITY EVALUATION

Papers

1. "Reliability Indices for Use in Bulk Power Supply Adequacy Evaluation". Committee Report, IEEE Transactions, PAS-99, 1978, pp. 1097-1103.

2. "A Method of Combining High Speed Contingency Load Flow Analysis with Stochastic Probability Methods to Calculate a Quantitative Measure of Overall Power System Reliability". G.E. Marks, IEEE Paper A78 053-1.

3. "Linear Programming Network Flow Methods Applied to Bulk Power Supply Adequacy Assessment". K.A. Clements, G.C. Ejebe, B.F. Wollenberg, IEEE Paper A78 062-2.

4. "Load Models for Bulk Power Supply Adequacy Assessment". T.A. Mikolinnas, K.A. Clements, IEEE Paper A78 232-1.

5. "Composite Generation and Transmission System Reliability Evaluation". R. Billinton, T.K.P. Medicherla, IEEE Paper A78 237-0.

6. "Load Supplying Capability of Generation - Transmission Systems". L.L. Garver, P.R. Van Horne, K.A. Wirgau, IEEE Transactions, PAS-98, 1979, pp. 957-962.

7. "Adequacy Indices for Composite Generation and Transmission System Reliability Evaluation". R. Billinton, T.K.P. Medicherla, M.S. Sachdev, IEEE Paper A79 024-1.

8. "An Approach for Bulk Power System Reliability Evaluation". I. Atiyyah, A. El-Abiad, IEEE Paper A79 108-2.

9. "Concepts for Composite System Reliability Applied to the Two-Area Planning Problem". B. Porretta, IEEE Paper A79 109-0.

10. "Overall Approach to the Reliability Evaluation of Composite Generation and Transmission Systems". R. Billinton, T.K.P. Medicherla, Proc. IEE, Part C, 127, 1980, pp. 72-81.

11. "Measuring and Reporting Overall Reliability of Bulk Electricity Systems". W.H. Winter, CIGRE, 1980, paper 32-15.

12. "Application of Common-Cause Outage Models in Composite System Reliability Evaluation". R. Billinton, T.K.P. Medicherla, M.S. Sachdev, IEEE Transactions, PAS-100, 1981, pp. 3648-3657.

13. "A Reliability Analysis Technique for Bulk System Planning". S.J. Ranade, R.L. Sullivan, IEEE Transactions, PAS-100, 1981, pp. 3658-3665.

14. "Station Originated Multiple Outages in the Reliability Analysis of a Composite Generation and Transmission System". R. Billinton, T.K.P. Medicherla, IEEE Transactions, PAS-100, 1981, pp. 3870-3878.

15. "TRAP: An Innovative Approach to Analysing the Reliability of Transmission Plans". P.R. Van Horne, C.N. Schoenberger, IEEE Transactions, PAS-101, 1982, pp. 11-16.

16. "Application of Adequacy Assessment Techniques for Bulk Power Systems". T.A. Mikolinnas, W.R. Puntel, R.J. Ringlee, IEEE Transactions, PAS-101, 1982, pp. 1219-1228.

17. "Bulk Power System Reliability Assessment - Why and How? Part I: Why?" J. Endrenyi, P.F. Albrecht, R. Billinton, G.E. Marks, N.D. Reppen, L. Salvaderi. IEEE Paper 82 WM 147-7.

18. "Bulk Power System Reliability Assessment - Why and How? Part II: How?" J. Endrenyi, P.F. Albrecht, R. Billinton, G.E. Marks, N.D. Reppen, L. Salvaderi. IEEE Paper 82 WM 148-5.

19. "Bulk Power Area Reliability Evaluation Considering Probabilistic Transfer Capability". PJM Transmission Reliability Task Force, IEEE Paper 82 WM 149-3.

20. "Terminal Effects and Protection System Failures in Composite System Reliability Evaluation". R.N. Allan, A.N. Adraktas, IEEE Paper 82 SM 428-1.

Author Index

Albrecht, P.F. 17, 18
Allan, R.N. 20
Atiyyah, I. 8
Adraktas, A.N. 20
Billinton, R. 5, 7, 10, 12, 14, 17, 18
Clements, K.A. 3, 4
Ejebe, G.C. 3
El-Abiad, A. 8
Endrenyi, J. 17, 18
Garver, L.L. 6
IEEE Committee 1
Marks, G.E. 2, 17, 18
Medicherla, T.K.P. 5, 7, 10, 12, 14
Mikolinnas, T.A. 4, 16
PJM Task Force 19
Porretta, B. 9
Puntel, W.R. 16
Ranade, S.J. 13
Reppen, N.D. 17, 18
Ringlee, R.J. 16
Sachdev, M.S. 7, 12
Salvaderi, L. 17, 18
Schoenberger, C.N. 15
Sullivan, R.L. 13
Van Horne, P.R. 6, 15
Winter, W.H. 11
Wirgau, K.A. 6
Wollenberg, B.F. 3

TRANSMISSION AND DISTRIBUTION SYSTEM
RELIABILITY EVALUATION

Papers

1. "Reliability Evaluation of Electrical Systems with Switching Actions". R.N. Allan, R. Billinton, M.F. de Oliveira, Proceedings IEE, 123, 1976, pp. 325-330.

2. "An Efficient Algorithm for Deducing the Minimal Cuts and Reliability Indices of a General Network Configuration". R.N. Allan, R. Billinton, M.F. de Oliveira, IEEE Transactions, R-25, 1976, pp. 226-233.

3. "Reliability Modelling and Evaluation of Transmission and Distribution Systems". R.N. Allan, M.F. de Oliveira, Proc. IEE, 124, 1977, pp. 535-541.

4. "Evaluate Distribution System Design by Cost Reliability Indices". N.E. Chang, IEEE Transactions, PAS-96, 1977, pp. 1480-1490.

5. "Evaluation of Elements of Distribution Circuit Outage Durations". R. Billinton, D.O. Koval, IEEE Paper A77 681-1.

6. "Application of Graph Theory in Estimating the Reliability of Electric Power Systems". D.L. Chan, Power Engineering, 15, 1977, pp. 21-29.

7. "Bibliography on Distribution System Reliability". IEEE Committee Report, IEEE Transactions, PAS-97, 1978, pp. 545-548.

8. "The Effect of Scheduled Outages in Transmission System Reliability Evaluation". L. Wang, IEEE Transactions, PAS-97, 1978, pp. 2346-2353.

9. "Common-Cause Outages in Multiple Circuit Transmission Lines". R. Billinton, T.K.P. Medicherla, M.S. Sachdev, IEEE Transactions, R-27, 1978, pp. 128-131.

10. "System Reliability Evaluation Considering Post-Fault Switching and Incomplete Redundancy". A.D. Patton, IEEE Transactions, IA-14, 1978, pp. 306-309.

11. "Generalised Computational Method for Reliability Analysis of Electric Power Installations". D.P. Papadopoulas, B.C. Papadias, Proc. IEE, 125, 1978, pp. 37-40.

12. "Evaluating the Effects of Isolation - Restoration Procedures on Distribution Circuit Reliability Indices". R. Billinton, D.O. Koval, IEEE Paper A78 512-6.

13. "Quantitative Method for Transmission Reliability Calculations". J.T. Day, T.P. Hayes Jr., R.P. Webb, IEEE Paper A78 566-2.

14. "Evaluation of Reliability Worth in Distribution Systems". R. Billinton, D.O. Koval, Proc. 6th Power Systems Computation Conference, Darmstadt, W. Germany, 1978, pp. 218-225.

15. "Partial Loss of Continuity and Transfer Capacity in the Reliability Evaluation of Power System Networks". R.N. Allan, E.N. Dialynas, I.R. Homer, Proc. 6th Power System Computational Conference, Darmstadt, W. Germany, 1978, pp. 250-257.

16. "Modelling and Evaluating the Reliability of Distribution Systems". R.N. Allan, E.N. Dialynas, I.R. Homer, IEEE Transactions, PAS-98, 1979, pp. 2181-2189.

17. "Modelling Common Mode Failures in the Reliability Evaluation of Power System Networks", R.N. Allan, E.N. Dialynas, I.R. Homer, IEEE Paper A79 040-7.

18. "Statistical and Analytical Evaluation of the Duration and Cost of Consumers Interruptions". D.O. Koval, R. Billinton, IEEE Paper A79 057-1.

19. "Integrated Distribution System Reliability Evaluation Part I - Current Practices", J.E.D. Northcote-Green, T.D. Vismor, C.L. Brooks, R. Billinton. CEA Transactions, Volume 19, 1980.

20. "Integrated Distribution System Reliability Evaluation Part II - The Models", T.D. Vismore, C.L. Brooks, S.J. Kostyal, R. Billinton. CEA Transactions, Volume 19, 1980.

21. "Reliability in the Design of Substations; Influence of the Location Within the Grid and the Kind of Connections to the Grid". K. Reichert, H.G. Schutte, P. Kulik, H. Rohsler, CIGRE, 1980, paper 23-01.

22. "Influence of Common Mode and Common Cause Failures on the Reliability of Switching Stations". F. Grutter, CIGRE, 1980, paper 23-03.

23. "Reliability Analysis of EHV Substations". J. Nahman, N. Mijuskovic, CIGRE, 1980, paper 23-05.

24. "Recent Progress in Substation Reliability Evaluation, the Method of the Degree of Damage". P. Van Migroet, P. Bulens, J.C. Declercq, CIGRE, 1980, paper 23-08.

25. "Markov Cut-Set Approach for the Reliability Evaluation of Transmission and Distribution Systems". C. Singh, IEEE Transactions, PAS-100, 1981, pp. 2719-2725.

26. "Transmission Line Reliability Models Including Common Mode and Adverse Weather Effects". R. Billinton, Y. Kumar, IEEE Transactions. PAS-100, 1981, pp. 3899-3910.

27. "Non-Markovian Models for Common Mode Failures in Transmission Systems". C. Singh, M.R. Ebrahimian, IEEE Transactions, PAS-101, 1982, pp. 1545-1550.

28. "Practical Calculations of Distribution System Reliability Indices and their Probability Distributions", R. Billinton, E. Wojczynski, M. Godfrey. CEA Transactions, Vol. 20, Part I, Paper No. 81-D-41. 1981.

29. "Reliability Evaluation of Distribution Systems using Graphic-Based Interactive Computational Methods". R.N. Allan, B. Barazesh, S. Sumar, IEEE Transactions, PAS-101, 1982, pp. 212-218.

30. "Customer Damage Resulting from Electrical Service Interruptions". R. Billinton, G. Wacker, E. Wojcynski, CEA Transactions, 1982.

Author Index

Allan, R.N. 1, 2, 3, 15, 16, 17, 29
Barazesh, B. 29
Billinton, R. 1, 2, 5, 9, 12, 14, 18, 19
.... 20, 26, 28, 30
Brooks, C.L. 19, 20
Bulens, P. 24
Chan, D.L. 6
Chang, N.E. 4
Day, J.T. 13
Declercq, J.C. 24
De Oliveria, M.F. 1, 2, 3
Dialynas, E.N. 15, 16, 17
Ebrahimian, M.R. 27
Godfrey, M. 28
Grutter, F. 22
Hayes, T.P. 13
Homer, I.R. 15, 16, 17
IEEE Committee 7
Kostyal, S.J. 20
Koval, D.O. 5, 12, 14, 18
Kulik, P. 21
Kumar, Y. 26
Medicherla, T.K.P. 9
Mijuskovic, N. 23
Nahman, J. 23
Northecote - Green, J.E.D. 19
Papadias, B.C. 11
Papadopoulos, D.P. 11
Patton, A.D. 10
Reichert, K. 21
Rohsler, H. 21
Sachdev, M.S. 9
Schutte, H.G. 21
Singh, C. 25, 27
Sumar, S. 29
Vismor, T.D. 19, 20
Van Migroet, P. 24
Wacker, G. 30
Wang, L. 8
Webb, R.P. 13
Wojccynski, E. 28, 30

EQUIPMENT OUTAGE DATA

Papers

1. "Operating Reliability of Vacuum Interrupters as Applied in Power Circuit Breakers". G.W.Alexander, D.R. Kurtz, C.H. Titus. Conference Record of the Annual Pulp and Paper Industry Technology Conference, 1977, IEEE Publ. No. 77 CH 1212-OIA, pp. 107-110.

2. "Reliability Performance Evaluation of Power Circuit Breakers". M.S. Grover, V.H. Quintana, J.S. Rai, IEEE Paper A77 611-7.

3. "Power Plant Computer Reliability Survey". J.O. Crowe, W. Leahy, T.C. Lee, C.L. Longenecker, W.P. Hamilton, H. Mehta, R.B. Squires, D.R. Swann, J.P. Wooley, A.B. Long, C. Moore, IEEE Transactions, PAS-97, 1978, pp. 1115-1123.

4. "A Method for Estimating Equivalent Forced Outage Rates of Multistate Peaking Units". M.P. Bhavaraju, J.A. Hynds, G.A. Nunan, IEEE Transactions, PAS-97, 1978, pp. 2067-2075.

5. "Power Plant Equipment Reliability". D.P. Hullinger, 1978 Joint Power Generating Conference, ASME Paper No. 78-JPGC-PWR-8.

6. "Estimation of Generating Unit Outage Parameters from Recorded Outage Data". L. Wang, IEEE Paper A79 485-4.

7. "Circuit Breaker Standards and the Statistical Approach". T.H. Dodds, J.J. Dodds, IEEE Transactions, PAS-98, 1979, pp. 60-65.

8. "Recording Failure Histories in Power Generating Stations" PES Availability Working Group, IEEE Transactions, PAS-100, 1981, pp. 1266-1277.

9. "Co-ordinating Equipment Reliability Activities". H.N. Scherer, A.J. McElroy, IEEE Transactions, PAS-100, 1981, pp. 1278-1287.

10. "Early Failure Occurrence Patterns in Power Generating Station Equipment". A.J. McElroy, J.N. Di Placido, R.E. Warner, J.L. Journey, IEEE Transactions, PAS-100, 1981, pp. 1288-1297.

11. "Data Base for EHV Transmission Reliability Evaluation". G.L. Landgren, S.W. Anderson, IEEE Transactions, PAS-100, 1981, pp. 2046-2058.

12. "Root Cause Analysis of Fossil Power Plant Equipment Failure: the EPRI Program". A.F. Armor, J.B. Parkes, D.E. Leaver, IEEE Transactions, PAS-100, 1981, pp. 2752-2758.

13. "Suggested Procedures for Generator Unit Status Reporting". W.J. Fernihough, IEEE Transactions, PAS-100, 1981, pp. 2775-2783.

14. "Evaluating Equipment Reliability Using Techniques Based on Kappa Square Statistics". S.R. Calabro, B.G. Harowitz, IEEE Transactions, PAS-100, 1981, pp. 4074-4082.

Author Index

Alexander, G.W. 1
Anderson, S.W. 11
Armor, A.F. 12
Bhavaraju, M.P. 4
Calabro, S.R. 14
Crowe, J.O. 3
Di Placido, J.N. 10
Dodds, J.J. 7
Dodds, T.H. 7
Fernihough, W.J. 13
Grover, M.S. 2
Hamilton, W.P. 3
Harowitz, B.G. 14
Hullinger, D.P. 5
Hynds, J.A. 4
IEEE Working Group 8
Journey, J.L. 10
Kurtz, D.R. 1
Landgren, G.L. 11
Leahy, W. 3
Leaver, D.E. 12
Lee, T.C. 3
Long, A.B. 3
Longenecker, C.L. 3
McElroy, A.J. 9, 10
Mehta, H. 3
Moore, C. 3
Nunan, G.A. 4
Parkes, J.B. 12
Quintana, V.H. 2
Rai, J.S. 2
Scherer, H.N. 9
Squires, R.B. 3
Swann, D.R. 3
Titus, C.H. 1
Wang, L. 6
Warner, R.E. 10
Wooley, J.P. 3

GENERAL CONSIDERATIONS

Papers

1. "Effect of Plant Reliability in Operational Planning Problems". R.N. Allan, C.H. Grigg, D.A. Newey, R.F. Simmons, IEE Int. Conf. on Reliability of Power Supply Systems, 1977, IEE Conf. Publ. 148, pp. 158-161.

2. "Reliability of Power Supplies". R. Billinton, Journal IEE - Electronics and Power, 24, 1978, pp. 307-310.

3. "Economic-Reliability Analysis of Simple Alternates for the Design of Auxiliary Electric Systems". G.R. Engmann, IEEE paper A78 540-7.

4. "Annotated Bibliography of Power System Reliability Literature - 1972-1977". S. Vemuri, IEEE Paper A78 548-0.

5. "Evaluating System Reliability". P.F. Albrecht, IEEE Spectrum, 15, 1978, pp. 43-47.

6. "Costs of Power Interruptions to Industry - Survey Results". E.M. Mackay, L.H. Berk, CIGRE, 1978, Paper 32-07.

7. "IEEE Reliability Test System". APM Subcommittee, IEEE Transactions, PAS-98, 1979, pp. 2047-2054.

8. "Cost-Reliability Evaluation of Commercial and Industrial Underground Distribution System Design". N.E. Chang, IEEE paper A79 058-9.

9. "Reliability Determination of Non-Markovian Power Systems Part 1: A Basic Analytical Procedure". K.W. Edwin, H.D. Kochs, IEEE Paper A79 502-6.

10. "Reliability Determination of Non-Markovian Power Systems Part 2: A Basic Simulation Technique". R. Danda, K.W. Edwin, H.D. Kochs, IEEE paper A79 503-4.

11. "Protection System Reliability Modelling: Unreadiness Probability and Mean Duration of Undetected Faults". C. Singh, A.D. Patton, IEEE Transactions, R-29, 1980, pp. 339-340.

12. "Evaluating the Reliability of Electrical Auxiliary Systems in Multi-Unit Generating Stations", R.N. Allan, M.F. de Oliviera, A. Kozlowski, G.T. Williams, Proc. IEE, Part C, 127, 1980, pp. 65-71.

13. "Application of Availability Analysis Techniques to Improve Power Plant Productivity". N.E. Chang, IEEE Paper A80 112-3.

14. "A Dual Interval Programming Approach to Power System Reliability Evaluation". G.J. Anders, E.A. Gunn, IEEE Transactions, PAS-100, 1981, pp. 1665-1673.

15. "Strategic Planning for Power System Reliability and Vulnerability: An Optimization Model for Resource Planning under Uncertainty". A.R. Sanghvi, I.H. Shavel, R.M. Spann, IEEE Transactions PAS-101, 1982, pp. 1420-1429.

16. "Computing Methods and Devices for the Reliability Evaluation of Large Power Systems". K. Gallyas, J. Endrenyi, IEEE Transactions, PAS-100, 1981, pp. 1250-1258.

17. "Consumer Interruption Costing for Reliability Cost/Benefit Evaluation". W.P. Poore, S.R. Greene, M.A. Kuliasha, IEEE Paper 82 SM 394-5.

18. "Evaluation of Importance and Related Reliability Measures for Electric Power Systems". G.J. Anders, IEEE Paper 82 SM 429-9.

19. "Power System Reliability Indices and their Application in the Assessment of Reliability Worth". R. Billinton, G. Wacker, E. Wojczynski, Bai Tongshuo, CEA Transactions, Vol. 22, 1982.

Author Index

Albrecht, P.F. 5
Allan, R.N. 1, 12
Anders, G.J. 14, 18
Berk, L.H. 6
Billinton, R. 2, 19
Chang, N.E. 8, 13
Danda, R. 10
De Oliveira, ...M.F. 12
Edwin, K.W. 9, 10
Endrenyi, J. 16
Engmann, G.R. 3
Gallyas, K. 16
Greene, S.R. 17
Grigg, C.H. 1
Gunn, E.A. 14
IEEE Subcommittee 7
Kochs, H.D. 9, 10
Kozlowski, A. 12
Kuliasha, M.A. 17
Mackay, E.M. 6
Newey, D.A. 1
Patton, A.D. 11
Poore, W.P. 17
Sanghvi, A.R. 15
Shavel, I.H. 15
Simmons, R.F. 1
Singh, C. 11
Spann, R.M. 15
Tongshuo, Bai. 19
Vemuri, S. 4
Wacker, G. 19
Williams, G.T. 12
Wojczynski, E. 19

CONCLUSIONS

In compiling any bibliography of related publications, it is not possible to make it absolutely complete. The APM Subcommittee would like to apologize for any errors or omissions and hope that additional relevant references will be advanced as discussion to this publication.

Discussion

M. Th. Schilling, K. H. Lee, J. C. G. Praca (ELETROBRAS, DEST/GPD, Rio de Janeiro, Brazil) and **J. F. de Queiroz** (Federal University, UFRJ/COPPE, Rio de Janeiro, Brazil): Messrs. Allan, Billinton and Lee and the Task Force Members of the IEEE Subcommittee on the Application of Probability Methods are to be commended for preparing this bibliography on the application of probability methods in power system reliability evaluation.

As commented by the authors, in compiling any bibliography it is not possible to make it absolutely complete. While the list presented is quite exhaustive, we find that some relevant references which relate to this subject, have not been included.

We would like to suggest the inclusion of the following titles:

REFERENCE
J. Endrenyi, *Reliability Modeling in Electric Power Systems,* Chichester, John Wiley, UK, 1978.

R. N. Allan, and R. Billinton, *Power System Reliability Evaluation,* University of Manchester, UK, UMIST, Sept. 1981

IEEE Power System Engineering Committee, *Tutorial Course: Power System Reliability Evaluation,* 82 EHO 195-8-PWR, San Francisco, CA, July, 1982

W. L. Rutz, P. B. Mirchandani, F. E. Wicks, and M. Becker, *A Reliability Treatment Applied to an Analysis of Optimal Generation Plant Size, IEEE Trans. on PAS,* 1982.

T. A. M. Sharaf, and G. J. Berg, *Reliability Optimization for Transmission Expansion Planning, IEEE Trans. on PAS,* Vol. PAS-101, No. 7, pp. 2243-2248, July, 1982 (ERRATA: Vol. PAS-101, No. 9, pp. 3592, Sept. 1982).

M. Munasinghe, and M. Gellerson, *Economic Criteria for Optimizing Power System Reliability Levels" The Bell Journal of Economics,* vol. 10, no. 1, Spring, pp. 353-365, 1979.

Manuscript received October 11, 1983.

Sastry Kuruganty and **Paul Thompson** (Manitoba Hydro, Winnipeg, Canada): The members of the task force must be complimented for producing this bibliography of selected papers on the subject of power system reliability evaluation. This bibliography is an excellent source of information for engineers working in the area of power system reliability. In view of this fact and in the interest of making the bibliography as complete as possible, we suggest the inclusion of the following references. We once again compliment the authors for producing a valuable document.

Static Generating Capacity Reliability Evaluation

Papers
1. R. F. Ramirez, "Adequacy of Supply for the National Interconnected Power System of Nicaragua—Selection of a Risk Level for Generation Planning," IEEE Paper A77 511-9.

2. C. R. Chowaniec and P. F. Pitman, "A Reliability Assessment Technique for Generating Systems with Photovoltaic Power Plants," IEEE Paper A77 655-4.

Equipment Outage Data
1. J. K. Shultis and N. D. Eckoff, "Selection of Beta Prior Distribution Parameters from Component Failure Data," *IEEE Transactions,* PAS-98, 1979, pp. 400-407.

General Considerations
1. K. D. Le, "Conventional Probability Methods and Monte Carlo Simulation Techniques—A Comparison of Results," IEEE Paper A78 238-8.

Manuscript received February 23, 1983.

R. N. Allan, R. Billinton, and **S. H. Lee:** The authors would like to thank the discussors for their comments and for their contribution to the bibliography. In the interest of making the bibliography as complete as possible, the following papers which were presented at the 1983 IEEE Winter Power Meeting and the 1983 IEEE Summer Power Meeting are included.

Static Generating Capacity Reliability Evaluation
1. "A Concise Frequency and Duration Approach to Generating System Reliability Studies." X. Wang, C. Pottle, 1983 IEEE WM 063-5.
2. "A Reliability Assessment of the Western Canada Grid." P. R. S. Kuruganty, P. R. Thompson, R. Billinton. 1983 IEEE WM 048-6.
3. "A Novel Approach to Frequency and Duration Analysis and Uncertainty Considerations for Radial and Two Interconnected Systems." G. J. Anders. 1983 IEEE WM 049-4.
4. "Reliability Modelling of Thermal Units for Peaking and Cycling Operations." L. Wang, N. Ramani, T. C. Davies. 1983 IEEE WM 050-2.
5. "An Investigation of Two Methods for the Probabilistic Energy Production Simulation." N. S. Rau, C. Necsulescu. 1983 IEEE WM 051-0.
6. "Use of the Cumulant Method in Computing the Frequency and Duration Indices of Single and Two Interconnected Systems." G. Hamoud, E. Neudorf. 1983 IEEE WM 052-8.
7. "A New Method for the Evaluation of Expected Energy Generation and Loss of Load Probability." K. F. Schenk, R. B. Misra, S. Vassos. 1983 IEEE SM 368-8.
8. "Generating Unit Maintenance Scheduling for Single and Two Interconnected Systems." R. Billinton, F. A. El-Sheikhi. 1983 IEEE SM 369-6.

9. "Modelling Intermittent Generation (IG) in a Monte-Carlo Regional System Analysis Model." Z. A. Yamayee. 1983 IEEE SM 370-4.
10. "A Comparison of Algorithms for Computing Power Generating System Reliability Indices." M. Mazumdar, D. P. Gaver. 1983 IEEE SM 371-2.

Composite Generation/Transmission Reliability Evaluation
11. "Composite Generation and Transmission System Adequacy Evaluation Including Protection System Failure Modes." R. Billinton, J. Tatla. 1983 IEEE WM 062-7.
12. "Power System Reliability Evaluation Using Stochastic Load Flows." A. P. Meliopoulos, A. G. Bakirtzis, R. Kovacs. 1983 IEEE SM 387-8.

Equipment Outage Data
13. "MAPP Bulk Transmission Outage Data Collection and Analysis." M. F. Lauby, K. T. Khu, R. W. Polesky, R. E. Vandello, J. H. Doudna, P. J. Lehman, D. D. Klempel. 1983 IEEE SM 388-6.

General Considerations
14. "A Probabilistic Approach to Power System Stability Analysis." P.M. Anderson, A. Bose. 1983 IEEE WM 064-3.

15. "Probabilistic Steady-State and Dynamic Security Assessment." F. F. Wu, Y. Tsai, Y. Yu. 1983 IEEE WM 065-0.
16. "Comprehensive Bibliography on Electrical Service Interruption Costs." R. Billinton, G. Wacker, E. Wojczynski. 1983 IEEE WM 066-8.
17. "Interruption Cost Methodology and Results-A Canadian Residential Survey." G. Wacker, E. Wojczynski, R. Billinton. 1983 IEEE WM 067-6.
18. "Interruption Cost Methodology and Results-A Canadian Commercial and Small Industry Survey." E. Wojczynski, R. Billinton, G. Wacker. 1983 IEEE SM 390-2.
19. "Reliability Modelling of Large Wind Farms and Associated Electric Utility Interface Systems." X. Wang, H. Dai, R. J. Thomas. 1983 IEEE SM 389-4.

Manuscript received September 22, 1983.

Note: Dr. J. Derald Morgan, Vice Chairman of the Power Systems Engineering Committee, reviewed the discussions with the Application of Probability Methods Subcommittee. Because of the length of the submitted discussions (44 pages), he has elected to reduce the discussion by selecting only the additional references printed here from the many provided by the discussors. He appreciates the efforts of the discussors in bringing to the committee's attention the large number of papers. Considering the scope of the original paper and the space available, he has selected only those papers in English that are related to the scope and intentions of the bibliography for publication with the discussion.

BIBLIOGRAPHY ON THE APPLICATION OF PROBABILITY
METHODS IN POWER SYSTEM RELIABILITY EVALUATION

1982-1987

R.N.Allan   R.Billinton   S.M.Shahidehpour   C.Singh
Senior Member   Fellow   Senior Member   Senior Member

Task Force Members of the IEEE Subcommittee on the Application of Probability Methods

Abstract

This paper presents a bibliography of selected papers on the subject of power system reliability evaluation.

It includes material which has become available since the publication of the three previous papers, "Bibliography on the Application of Probability Methods in Power System Reliability Evaluation", IEEE Trans, PAS-91, 1972, pp.649-660; PAS-97, 1978, pp.2235-2242 and PAS-103, 1984, pp.275-282.

INTRODUCTION

Considerable activity has continued in the development and application of probabilistic techniques for power system reliability evaluation since the three previous bibliographies were published in 1972, 1978 and 1984. The Subcommittee on the Application of Probability Methods has continued to maintain an updated bibliography and, in view of the interest expressed in this material, has decided to publish a further extension.

The authors gratefully acknowledge the contribution and valuable comments of all members of the APM Subcommittee.

STATIC GENERATING CAPACITY RELIABILITY EVALUATION

Papers

1. "Availability Modelling of Stochastic Power Sources". A.K.David. Proc IEE, 129 part C, 1982, pp.239-248.

2. "Cumulant Method Equivalent Load Curve Calculation Performance for Small Generation Systems". R.A.Smith, R.D.Shultz, T.M.Sweet. IEEE Trans on Power Apparatus and Systems, PAS-102, 1983, pp.1302-1307.

3. "Probability Distribution Functions for Generation Reliability Indices - Analytical Approach". M.Sahinoglu, M.T.Longnecker, L.J.Ringer, C.Singh, A.K.Ayoub. IEEE Trans on Power Apparatus and Systems, PAS-102, 1983, pp.1486-1493.

4. "Source Reliability in a Combined Wind-Solar-Hydro System". A.Traca-de-Almeida, A.Martins, H.Jesus, J.Climaco. IEEE Trans on Power Apparatus and

Systems, PAS-102, 1983, pp.1515-1520.

5. "Reliability Modelling of Thermal Units for Peaking and Cycling Operations". L.Wang, N.Ramani, T.C.Davies. IEEE Trans on Power Apparatus and Systems, PAS-102, 1983, pp.2004-2011.

6. "A Concise Frequency and Duration Approach to Generating System Reliability Studies". X.Wang, C.Pottle. IEEE Trans on Power Apparatus and Systems, PAS-102, 1983, pp.2521-2530.

7. "An Investigation of Two Methods for the Probabilistic Energy Production Simulation". N.S.Rau, C.Necsulescu. IEEE Trans on Power Apparatus and Systems, PAS-102, 1983, pp.2543-2551.

8. "A Reliability Treatment Applied to an Analysis of Optimal Generation Plant Size". W.L.Rutz, P.B.Mirchandani, F.E.Wicks, M.Becker. IEEE Trans on Power Apparatus and Systems, PAS-102, 1983, pp.2569-2577.

9. "Thermal Unit Availability Modeling in a Regional Simulation Model". Z.A.Yamayee, W.Robinett, J.Port. IEEE Trans on Power Apparatus and Systems, PAS-102, 1983, pp.3885-3891.

10. "Forced Frequency-Balancing Technique for Discrete Capacity States". C.Singh. IEEE Trans on Reliability, R-32, 1983, pp.350-353.

11. "Modelling and Reliability Evaluation of Generating Systems". R.N.Allan, I.G.Musa. Int Conf on Power Supply Systems, IEE Conf Publ 225, 1983, pp.22-26.·

12. "An Energy Based Approach to Generating Capacity Reliability Evaluation". R.Billinton, L.C.H.Cheung. Int Conf on Power Supply Systems, IEE Conf Publ 225, 1983, pp.27-31.

13. "Reliability Impacts of Wind-Generated Electricity on Californian Utilities". L.W.House. Int Conf on Power Supply Systems, IEE Conf Publ 225, 1983, pp.32-36.

14. "Effect of Size and Location of Conventional and Intermittent Generation on System Reliability". Z.A.Yamayee, F.S.Ma. Electrical Power and Energy Systems, 5, 1983, pp.94-100.

15. "Parameter Uncertainty in Generating Capacity Reliability Evaluation". G.Hamoud, R.Billinton. Electrical Power and Energy Systems, 5, 1983, pp.149-158.

16. "Preventive Maintenance Scheduling in Power Generation Systems Using a Quantitative Risk Criterion". R.Billinton, F.A.El-Sheikhi. Can Elec Eng J, 8, 1983.

17. "A Comparison of Algorithms for Computing Power Generating System Reliability Indices".

88 WM 172-9   A paper recommended and approved by the IEEE Power System Engineering Committee of the IEEE Power Engineering Society for presentation at the IEEE/PES 1988 Winter Meeting, New York, New York, January 31 - February 5, 1988. Manuscript submitted August 19, 1987; made available for printing November 17, 1987.

M.Mazumdar, D.P.Gaver. IEEE Trans on Power Apparatus and Systems, PAS-103, 1984, pp.91-98.

18. "Modeling Intermittent Generation (IG) in a Monte-Carlo Regional System Analysis Model". Z.A.Yamayee. IEEE Trans on Power Apparatus and Systems, PAS-103, 1984, pp.174-182.

19. "A New Method for the Evaluation of Expected Energy Generation and Loss of Load Probability". K.F.Schenk, R.B.Misra, S.Vassos, W.Wen. IEEE Trans on Power Apparatus and Systems, PAS-103, 1984, pp.294-302.

20. "Reliability Calculations for Dependent Plant Failures". G.R.Fegan. IEEE Trans on Power Apparatus and Systems, PAS-103, 1984, pp.1614-1620.

21. "A Technique for Approximating the Capacity Outage Table Based on the Modeling of Unit Outage Size". B.McNutt. IEEE Trans on Power Apparatus and Systems, PAS-103, 1984, pp.2891-2898.

22. "Evaluation of Load Management Effects Using the OPCON Generation Reliability Model". A.D.Patton, C.Singh. IEEE Trans on Power Apparatus and Systems, PAS-103, 1984, pp.3230-3238.

23. "Impact of Total Energy Systems in Generation System Planning". G.Manzoni, L.Salvaderi, M.Valtorta. Proc 8th Power Systems Computation Conf (PSCC), Helsinki, 1984, Butterworths (London), pp.20-25.

24. "The Computation of Equivalent Load Duration Curves using Mixtures of Normal Distributions". G.Gross, N.V.Garapic, B.McNutt. Proc 8th Power Systems Computation Conf (PSCC), Helsinki, 1984, Butterworths (London), pp.43-47.

25. "Operational Parameters in Generation Reliability Evaluation". R.N.Allan, D.Gomes. Proc 8th Power Systems Computation Conf (PSCC), Helsinki, 1984, Butterworths (London), pp.174-178.

26. "Markov Method for Generating Capacity Reliability Evaluation Including Operational Considerations". C.Singh, S.Asgarpoor, A.D.Patton. Electrical Power and Energy Systems, 6, 1984, pp.161-168.

27. "A Recursive Approach to the Cumulant Method of Calculating Reliability and Production Cost". H.Duran. IEEE Trans on Power Apparatus and Systems, PAS-104, 1985, pp.82-90.

28. "Reliability Modeling of Generation Systems Including Unconventional Energy Sources". C.Singh, A.Lago-Gonzales. IEEE Trans on Power Apparatus and Systems, PAS-104, 1985, pp.1049-1056.

29. "On the Application of Esscher's Approximation to Computation of Generating System Reliability and Production Costing Indexes". M.Mazumdar, Y.P.Wang. IEEE Trans on Power Apparatus and Systems, PAS-104, 1985, pp.3029-3036.

30. "Reliability and Economic Assessment of Generating Systems Containing Wind Energy Sources". R.N.Allan, P.Corredor-Avella. Proc IEE, 132 part C, 1985, pp.8-13.

31. "Incorporation of Large Stochastic Sources in the Power System". A.K.David. Proc IEE, 132 part C, 1985, pp.161-171.

32. "Maintenance Considerations in Generating Capacity Reliability Assessment - Methodology and Application". F.A.El-Sheikhi, R.Billinton. J of Maintenance Int, 5, 1985, pp.63-76.

33. "Equivalent Load Method for Calculating Frequency and Duration Indices in Generation Capacity Reliability Evaluation". Q.Chen, C.Singh. IEEE Trans on Power Systems, PWRS-1, No.1, 1986, pp.101-107.

34. "The IEEE Reliability Test System - Extensions to and Evaluation of the Generating System". R.N.Allan, R.Billinton, N.M.K.Abdel-Gawad. IEEE Trans on Power Systems, PWRS-1, No.4, 1986, pp.1-7.

35. "On Improving the Convergence and Accuracy of the Cumulant Method of Calculating Reliability and Production Cost". H.Duran. IEEE Trans on Power Systems, PWRS-1, No.4, 1986, pp.121-126.

36. "A New Fourier Method for Evaluating Generation System Reliability Indices". P.Mohan, R.Balasubramanian, K.S.Prakasa Rao. IEEE Trans on Power Systems, PWRS-1, No.3, 1986, pp.88-94.

37. "The Inclusion of Load Correlation in the Frequency and Duration Approach for Generating Capacity Reliability Evaluation". C.L.Wee, R.Billinton. Proc IEE, 133 part C, 1986, pp.6-15.

38. "An Error Correction Algorithm for Stochastic Production Costing". L.A.Sanabria, T.S.Dillon. IEEE Summer Power Meeting, 1986, paper 86 SM 354-5.

39. "A Procedure and Digital Computer Program for Derated State Modelling of Generating Units Using the CEA Data Base". R.Billinton, C.L.Wee, K.Debnoth. CEA Trans, 25, 1986.

40. "An Analytical Technique for the Reliability Modeling of Generation Systems Including Energy Limited Units". C.Singh, Q. Chen. IEEE Trans on Power Systems, PS-2, No.1, 1987, pp.123-128.

41. "Comparison of Two Alternate Methods to Establish an Interrupted Energy Assessment Rate". R.Billinton, J.Oteng-Adjei, R.Ghajar. IEEE Winter Power Meeting, New Orleans, 1987, paper 87 WM 023-5.

42. "The Mixture of Normals Approximation Technique for Equivalent Load Duration Curves". G.Gross, N.V.Garapic, B.McNutt. IEEE Winter Power Meeting, 1987, paper 87 WM 041-7.

43. "A Monte Carlo Simulation Approach to the Reliability Modeling of Generation Systems Recognizing Operation Considerations". A.D.Patton, J.H.Blackstone, N.J.Balu. IEEE Summer Power Meeting, 1987, paper 87 SM 492-2.

Author Index

Abdel-Gawad NMK	34
Allan RN	11, 25, 30, 34
Asgarpoor S	26
Ayoub AK	3
Balasubramanian R	36
Balu NJ	43
Becker M	8
Billinton R	12, 15, 16, 32, 34, 37, 39, 41
Blackstone JH	43
Chen Q	33, 40
Cheung LCH	12
Climaco J	4
Corredor-Avella P	30
David AK	1, 31

Davies TC		5
Debnoth K		39
Dillon TS		38
Duran H		27, 35
El-Sheikhi FA		16, 32
Fegan GR		20
Garapic NV		24, 42
Gaver DP		17
Ghajar R		41
Gomes D		25
Gross G		24, 42
Hamoud G		15
House LW		13
Jesus H		4
Lagos-Gonzales A		28
Longnecker MT		3
Ma FS		14
Manzoni G		23
Martins A		4
Mazumdar M		17, 29
McNutt B		21, 24, 42
Mirchandani PB		8
Mohan P		36
Misra RB		19
Musa IG		11
Necsulescu C		7
Oteng-Adjei J		41
Patton AD		22, 26, 43
Port J		9
Pottle C		6
Prakasa Rao KS		36
Ramani N		5
Rau NS		7
Ringer LJ		3
Robinett W		9
Rutz WL		8
Sahinoglu M		3
Salvaderi L		23
Sanabria LA		38
Schenk KF		19
Shultz RD		2
Singh C	3, 10, 22, 26, 28, 33, 40	
Smith RA		2
Sweet TM		2
Traca-de-Almeida A		4
Valtorta M		23
Vassos S		19
Wang L		5
Wang X		6
Wang YP		29
Wee CL		37, 39
Wen W		19
Wicks FE		8
Yamayee ZA		9, 14, 18

MULTI-AREA RELIABILITY EVALUATION

Papers

1. "Expected Energy Production Cost of Two Interconnected Systems with Correlated Demands". Q.Ahsan, K.F.Schenk, R.B.Misra. IEEE Trans on Power Apparatus and Systems, PAS-102, 1983, pp.2155-2164.

2. "A Novel Approach to Frequency and Duration Analysis and Uncertainty Considerations for Radial and Two Interconnected Systems". G.J.Anders. IEEE Trans on Power Apparatus and Systems, PAS-102, 1983, pp.2165-2172.

3. "Use of the Cumulant Method in Computing the Frequency and Duration Indices of Single and Two Interconnected Systems". G.Hamoud, E.Neudorf. IEEE Trans on Power Apparatus and Systems, PAS-102, 1983, pp.2391-2399.

4. "A Reliability Assessment of the Western Canada Grid". P.R.S.Kuruganty, P.R.Thompson, R.Billinton. IEEE Trans on Power Apparatus and Systems, PAS-102, 1983, pp.2826-2833.

5. "Probabilistic Flows for Reliability Evaluation of Multiarea Power System Interconnections". D.P.Clancy, G.Gross, F.F.Wu. Electrical Power and Energy Systems, 5 1983, pp.101-114.

6. "Single and Two Area Generation System Maintenance Scheduling Using the Moment Cumulant Method". R.Billinton, F.A.El-Sheikhi. CEA Trans, 22, 1983.

7. "Generating Unit Maintenance Scheduling for Single and Two Interconnected Systems". F.A.El-Sheikhi, R.Billinton. IEEE Trans on Power Apparatus and Systems, PAS-103, 1984, pp.1038-1044.

8. "The Segmentation Method Applied to the Evaluation of Loss of Load Probability of Two Interconnected Systems". K.F.Schenk, Q.Ahsan, S.Vassos. IEEE Trans on Power Apparatus and Systems, PAS-103, 1984, pp.1537-1541.

9. "A Frequency and Duration Evaluation for Looped Interconnected System Reliability". R.Billinton, C.L.Wee. Proc 8th Power Systems Computation Conf (PSCC), Helsinki, 1984, Butterworths (London), pp.147-154.

10. "Multi-Area Maintenance Scheduling for Generation System Planning Using the Gram-Charlier Series". R.Billinton, F.A.El-Shiekhi. Proc 8th Power Systems Computation Conf (PSCC), Helsinki, 1984, Butterworths (London), pp.290-297.

11. "DC Interconnection Reliability Using Frequency and Duration Evaluation". R.Billinton, C.L.Wee. Proc IEE Inter Conf on AC and DC Power Trans, London, Sept 1985.

12. "A Direct Method for Multi-Area Reliability Evaluation". G.C.Oliveira, S.H.F.Cunha, M.V.F.Pereira. IEEE Summer Power Meeting, 1986, paper 86 SM 351-1.

13. "Multi-Area Reliability - A New Approach". F.N.Lee. IEEE Summer Power Meeting, 1986, paper 86 SM 352-9.

14. "Reliability Evaluation of a Three-Area Power System". S.M.Shahidehpour, Electric Power System Research, 10, 1986, pp.227-233.

15. "Operating Considerations in Reliability Modeling of Interconnected Systems - An Analytical Approach". C.Singh, A.D.Patton, A.Lago-Gonzales. IEEE Summer Power Meeting, 1987, paper 87 SM 504-4.

16. "A New Multi-Area Production Costing Method". F.N.Lee. IEEE Summer Power Meeting, 1987, paper 87 SM 482-3.

Author Index

Ahsan Q	1, 18
Anders GJ	2
Billinton R	4, 6, 7, 9, 10, 11
Clancy DP	5
Cunha SHF	12
El-Sheikhi FA	6, 7, 10
Gross G	5
Hamoud G	3
Kuruganty PRS	4
Lago-Gonzales A	15

Lee FN	13, 16
Misra RB	1
Neudorf E	3
Oliveira GC	12
Patton AD	15
Pereira MVF	12
Schenk KF	1, 8
Shahidehpour SM	14
Singh C	15
Thompson PR	4
Vassos S	8
Wee CL	9, 11
Wu FF	5

COMPOSITE GENERATION-TRANSMISSION RELIABILITY EVALUATION

Papers

1. "Composite Generation and Transmission System Adequacy Evaluation Including Protection System Failure Modes". R.Billinton, J.Tatla. IEEE Trans on Power Apparatus and Systems, PAS-102, 1983, pp.1923-1930.

2. "Direct Method for Evaluation of Bulk Power System Reliability - Part 1 - Theoretic Foundation". K.Moslehi, F.F.Wu. Electrical Power and Energy Systems, 5, 1983, pp.3-7.

3. "Direct Method for Evaluation of Bulk Power System Reliability - Part 2 - Solution Algorithm". K.Moslehi, F.F.Wu. Electrical Power and Energy Systems, 5, 1983, pp.8-16.

4. "Power System Reliability Evaluation Using Stochastic Load Flows". A.P.Meliopoulos, A.G.Bakirtizis, R.Kovacs. IEEE Trans on Power Apparatus and Systems, PAS-103, 1984, pp.1084-1091.

5. "Computation of Upper and Lower Bounds on Reliability Indices for Bulk Power Systems". K.A.Clements, B.P.Lam, D.J.Lawrence, N.D.Reppen. IEEE Trans on Power Apparatus and Systems, PAS-103, 1984, pp.2318-2325.

6. "Unorthodox Reliability Evaluation". M.Th.Schilling, K.H.Lee, J.F.de Queiroz, J.C.G.Praca. Proc 8th Power Systems Computation Conf (PSCC), Helsinki, 1984, Butterworths (London), pp.184-190.

7. "Effect of Station Originated Outages in a Composite System Adequacy Evaluation of the IEEE Reliability Test System". R.Billinton, P.K.Vohra, S.Kumar. IEEE Trans on Power Apparatus and Systems, PAS-104, 1985, pp.2649-2656.

8. "Composite Generation and Transmission Reliability Evaluation in Large Hydroelectric Systems". S.H.F.Cunha, M.V.F.Pereira, L.M.V.G.Pinto, G.C.Oliveira. IEEE Trans on Power Apparatus and Systems, PAS-104, 1985, pp.2657-2664.

9. "A Comparison Between Two Fundmentally Different Approaches to Composite System Reliability Evaluation". L.Salvaderi, R.Billinton. IEEE Trans on Power Apparatus and Systems, PAS-104, 1985, pp.3486-3493.

10. "Modeling Simultaneous Outages for Bulk Power System Reliability Analysis". J.M.Nahman. IEEE Trans on Reliability, R-34, 1985, pp.554-559.

11. "Adequacy Evaluation of a Composite Power System - A Comparative Study of Existing Computer Programs". R.Billinton, S.Kumar. CEA Trans, 24, 1985.

12. "A Comparative Study of System Versus Load Point Indices for Bulk Power Systems". R.Billinton, S.Kumar. IEEE Trans on Power Systems, PWRS-1, No.3, 1986, pp.148-156.

13. "New Probabilistic Approach Taking into Account Reliability and Operation Security in EHV Power System Planning at EDF". J.C.Dodu, A.Merlin. IEEE Trans on Power Systems, PWRS-1, 1986, No.3, pp.175-181.

14. "Bulk Power System Reliability Assessment Experience with the RECS Program". A.P.Meliopoulos, A.G.Bakirtizis, R.R.Kovacs, R.J.Beck. IEEE Trans on Power Systems, PWRS-1, No.3, 1986, pp.235-242.

15. "Low Bus Voltage and Ill-Conditioned Network Situations in a Composite System Adequacy Evaluation". S.Kumar, R.Billinton. IEEE Summer Power Meeting, 1986, paper 86 SM 310-7.

16. "Pertinent Factors in the Adequacy Assessment of a Composite Generation and Transmission System". R.Billinton, S.Kumar. CEA Trans, 25, 1986.

17. "Incorporation of Weather Effects in Transmission System Models for Composite System Adequacy Evaluation". R.Billinton, L.Cheng. Proc IEE, 133 part C, 1986, pp.319-327.

18. "Bulk Power System Reliability Concepts and Applications". J.Endrenyi, M.P.Bhavaraju, K.A.Clements, K.J.Dhir, M.F.McCoy, K.Medicherla, N.D.Reppen, L.Salvaderi, S.M.Shahidehpour, C.Singh, J.A.Stratton (IEEE-PES Task Force Members). IEEE Winter Power Meeting, New Orleans, 1987, paper 87 WM 051-6.

19. "Requirements for Composite System Reliability Evaluation Models". M.P.Bhavaraju, P.F.Albrecht, R.Billinton, N.D.Reppen, R.J.Ringlee. IEEE Winter Power Meeting, New Orleans, 1987, paper 87 WM 022-7.

20. "Modelling and Assessment of Station Originated Outages for Composite Systems Reliability Evaluation". R.N.Allan, J.R.Ochoa. IEEE Winter Power Meeting, New Orleans, 1987, paper 87 WM 016-9.

21. "Station Initiated Outages in Composite System Adequacy Evaluation". R.Billinton, P.K.Vohra. Proc IEE, 134 part C, 1987, pp.10-16.

22. "Effect of Higher Level Independent Generator Outages in Composite System Adequacy Evaluation". R.Billinton, S.Kumar. Proc IEE, 134 part C, 1987, pp.17-30.

23. "Reliability Equivalents in Composite System Reliability Evaluation". R.Billinton, H.J.Koglin, E.Roos. Proc IEE, 134 part C, 1987, pp.224-233.

24. "Adequacy Equivalents in Composite Power System Evaluation". S.Kumar, R.Billinton. IEEE Summer Power Meeting, 1987, 87 SM 480-7.

25. "Monte Carlo Approach in Planning Studies: An Application to IEEE RTS". O.Bertoldi, L.Salvaderi, S.Scalcino. IEEE Summer Power Meeting, 1987, 87 SM 444-3.

Author Index

Albrecht PF	19
Allan RN	20
Bakirtizis AG	4, 14
Beck RJ	14
Bertoldi O	25
Bhavaraju MP	18, 19
Billinton R	1, 7, 9, 11, 12, 15, 16, 17, 19, 21, 22
	23, 24
Cheng L	17
Clements KA	5, 18
Cunha SHF	8
Dhir KJ	18
Dodu JC	13
Endrenyi J	18
Koglin HJ	23
Kovacs R	4, 14
Kumar S	7, 11, 12, 15, 16, 22, 24
Lam BP	5
Lawrence DJ	5
Lee KH	6
McCoy MF	18
Medicherla K	18
Meliopoulos AP	4, 14
Merlin A	13
Moslehi K	2, 3
Nahman JM	10
Ochoa JR	20
Oliveira GC	8
Pereira MVF	8
Pinto LMVG	8
Praca JCG	6
Queiroz de JF	6
Reppen ND	5, 18, 19
Ringlee RJ	19
Roos E	23
Salvaderi L	9, 18, 25
Scalcino S	25
Schilling MTh	6
Shahidehpour SM	18
Singh C	18
Stratton JA	18
Tatla J	1
Vohra PK	7, 21
Wu FF	2, 3

OPERATING RESERVE RELIABILITY EVALUATION

Papers

1. "Economic Allocation of Reserve in a Thermal Power System with Pumped Storage and Gas Turbine Plant". R.M.Dunnett. Int Conf on Power Supply Systems. IEE Conf Publ 225, 1983, pp.37-41.

2. "Power Reserve Computation for System Operation and Extension Planning". K.W.Edwin, R.Danda, W.Schaper. Int Conf on Power Supply Systems. IEE Conf Publ 225, 1983, pp.53-58.

3. "Reliability Assessment in Operational Planning for Large Hydro-Thermal Generation Systems". L.Wang, K.Gallyas, D.T.Tsai. IEEE Trans on Power Apparatus and Systems, PAS-104, 1985, pp.3382-3387.

Author Index

Danda R	2
Dunnett RM	1
Edwin KW	2
Gallyas K	3
Schaper W	2
Tsai DT	3
Wang L	3

TRANSMISSION AND DISTRIBUTION SYSTEM RELIABILITY EVALUATION

Papers

1. "Non-Markovian Models for Common Mode Failures in Transmission Systems". C.Singh, R.Ebrahimian. IEEE Trans on Power Apparatus and Systems, PAS-101, 1982, pp.1545-1550.

2. "Reliability Considerations for Distribution Automation Equipment". J.B.Bunch, H.I.Stalder, J.T.Tengdin. IEEE Trans on Power Apparatus and Systems, PAS-102, 1983, pp.2656-2664.

3. "Reliability Calculations for Electrical Transmission Systems on the Basis of Mean Failure Indices". S.K.Banerjee, B.R.Reddi. IEEE Trans on Reliability, R-32, 1983, pp.346-349.

4. "General Computation Method for Performing Availability Analysis of Power System Configurations". D.P.Papadopoulos. Proc IEE, 130 part C, 1983, pp.285-294.

5. "Application of Reliability Calculation Methods to Planning of High-Voltage Distribution Networks". H-J.Koglin, E.Roos, W.H.Wellssow. Int Conf on Power Supply Systems. IEE Conf Publ 225, 1983, pp.64-68.

6. "Reliability Investigations for 110kV Subtransmission Networks". K.W.Edwin, R.Dib, U.Niehage. Int Conf on Power Supply Systems. IEE Conf Publ 225, 1983, pp.73-78.

7. "Evaluation of Local Reliability Indices in a Composite System". J.Borges Gouveia, F.Maciel Barbosa, V.Miranda, A.Almeida Do Vale. Int Conf on Power Supply Systems. IEE Conf Publ 225, 1983, pp.84-88.

8. "Influence of Limited Transmission Capacity on the Reliability of Power Systems". K.W.Edwin, W.G.Lehr. Int Conf on Power Supply Systems. IEE Conf Publ 225, 1983, pp.93-98.

9. "Application of Graphic/Interactive Computation in Distribution System Planning". R.N.Allan, B.Barazesh, S.Sumar. Int Conf on Power Supply Systems. IEE Conf Publ 225, 1983, pp.99-103.

10. "The Degradation of the Reliability of Transmission and Distribution Systems During Construction Outages". A.J.Walker. Int Conf on Power Supply Systems. IEE Conf Publ 225, 1983, pp.112-118.

11. "Computation of Frequency of Severe Power System Faults". G.J.Anders, P.L.Dandeno, E.E.Neudorf. IEEE Trans on Power Apparatus and Systems, PAS-103, 1984, pp.2369-2374.

12. "Computation of Frequency of Right-of-Way Losses Due to Tornadoes". G.J.Anders, P.L.Dandeno, E.E.Neudorf. IEEE Trans on Power Apparatus and Systems, PAS-103, 1984, pp.2375-2381.

13. "Improving Transmission Design by Using Reliability Techniques". E.Ghannoum. IEEE Trans on Power Apparatus and Systems, PAS-103, 1984, pp.3144-3153.

14. "Transmission System Reliability Assessment by Contingency Evaluation". B.P.Lam, D.J.Lawrence, N.D.Reppen, N.J.Balu. Proc 8th Power Systems Computation Conf (PSCC), Helsinki, 1984,

Butterworths (London), pp.179-183.

15. "Probabilistic Transfer Capability – An Enhancement to Transmission System Performance Appraisal Methods". M.Chau, M.Heyeck. IEEE Trans on Power Apparatus and Systems, PAS-104, 1985, pp.2159-2167.

16. "A Practical Approach to Reliability Design". H.B.White. IEEE Trans on Power Apparatus and Systems, PAS-104, 1985, pp.2739-2747.

17. "Distributional Variation of Distribution System Reliability Indices". R.Billinton, E.Wojczynski. IEEE Trans on Power Apparatus and Systems, PAS-104, 1985, pp.3152-3160.

18. "Reliability Modeling of Multiple Overhead Transmission Lines". J.M.Nahman, N.Mijuskovic. IEEE Trans on Reliability, R-34, 1985, pp.281-285.

19. "Distribution System Reliability Evaluation and Data Collection in the Canadian Utility Industry". R.Billinton, S.J.Kostyal. CEA Trans. 24, 1985.

20. "Quantitative Reliability Evaluation of Terminal Stations". R.Billinton, P.K.Vohra. CEA Trans, 24, 1985.

21. "Status of Distribution System Reliability Evaluation in Canada". S.Kostyal, R.Billinton. IEEE Trans on Power Delivery, PWRD-1, No.2, 1986, pp.114-120.

22. "Local Generating Facilities in the Reliability Evaluation of Power Distribution Systems". E.N.Dialynas, R.N.Allan. IEEE Trans on Power Systems, PWRS-1, No.4, 1986, pp.62-67.

23. "An Analytical Approach to Evaluate Probability Distributions Associated with the Reliability Indices of Electric Distribution Systems". R.Billinton, R.Goel. IEEE Trans on Power Delivery, PWRD-1, No.3, 1986, pp.245-251.

24. "Reliability Calculation Under Planned Maintenance". D.Elmakis, P.Levy. IEEE Trans on Power Systems, PS-2, No.1, 1987, pp.1-7.

25. "Reliability Evaluation of Systems with After-Fault Switching". J.A.Buzacott, G.J.Anders. IEEE Summer Power Meeting, 1986, paper 86 SM 308-1.

26. "The Effect of Terminal Complexity and Redundancy on the Frequency and Duration of Forced Outages". M.G.Lauby, D.D.Klempel, C.J.Dahl, R.O.Gunderson, E.P.Weber, P.J.Lehman. IEEE Summer Power Meeting, 1986, paper 86 SM 309-9.

27. "Reliability Analysis of Transmission Lines with Common Mode Failures when Repair Times are Arbitrarily Distributed". C.Dichirico, C.Singh. IEEE Summer Power Meeting, 1987, paper 87 SM 458-3.

28. "Availability Analysis in a Cable Transmission System with Repair Restrictions". R.Billinton, P.Alukal. IEEE Summer Power Meeting, 1987, paper 87 SM 448-4.

Author Index

Allan RN	9, 22
Alukal P	28
Almeida do Vale A	7
Anders GJ	11, 12, 25
Balu NJ	14
Banerjee SK	3
Barazesh B	9
Billinton R	17, 19, 20, 21, 23, 28
Borges Gouveia J	7
Bunch JB	2
Buzacott JA	25
Chau M	15
Dahl CJ	26
Dandeno PL	11, 12
Dialynas EN	22
Dichirico C	27
Dib R	6
Ebrahimian P	1
Edwin KW	6, 8
Elmakis D	24
Ghannoum E	13
Goel R	23
Gunderson RO	26
Heyeck M	15
Klempel DD	26
Koglin HJ	5
Kostyal S	19, 21
Lam BP	14
Lauby MG	26
Lawrence DJ	14
Lehman PJ	26
Lehr WG	8
Levy P	24
Maciel Barbosa F	7
Mijuskovic N	18
Miranda V	7
Nahman JM	18
Neudorf EE	11, 12
Niehage U	6
Papadopoulos DP	4
Reddi BR	3
Reppen ND	14
Roos E	5
Singh C	1, 27
Stalder HI	2
Sumar S	9
Tengdin JT	2
Vohra PK	20
Walker AJ	10
Weber EP	26
Wellssow WH	5
White HB	16
Wojczynski E	17

EQUIPMENT OUTAGE DATA

Papers

1. "The Canadian Electrical Association's Equipment Reliability Information System". M.A.Baribeau, C.T.Driscoll, D.T.McGillis. Int Conf on Power Supply Systems. IEE Conf Publ 225, 1983, pp.150-154.

2. "MAPP Bulk Transmission Outage Data Collection and Analysis". M.G.Lauby, K.T.Khu, R.W.Polesky, R.E.Vandello, J.H.Doudna, P.J.Lehman, D.D.Klempel. IEEE Trans on Power Apparatus and Systems, PAS-103, 1984, pp.213-221.

3. "Effects of Pooling Weather Associated MAPP Bulk Transmission Outage Data on Calculated Forced Outage Rates". M.G.Lauby, J.H.Doudna, R.W.Polesky, P.J.Lehman, D.D.Klempel. IEEE Trans on Power Apparatus and Systems, PAS-103, 1984, pp.2345-2351.

4. "Breakdown by Cause of the Reliability Indices of Generating Units". M.A.Baribeau, D.T.McGillis.

Proc 8th Power Systems Computation Conf (PSCC), Helsinki, 1984, Butterworths (London), pp.278-283.

5. "Generating Unit Models Using the Canadian Electrical Association Data Base". R.Billinton, A.A.Chowdhury. CEA Trans, 23, 1984.

6. "Proposed Terms for Reporting and Analyzing Outages of Electrical Transmission and Distribution Facilities". D.W.Forrest, P.F.Albrecht, R.N.Allan, M.P.Bhavaraju, R.Billinton, G.L.Landgren, M.F.McCoy, N.D.Reppen. IEEE Trans on Power Apparatus and Systems, PAS-104, 1985, pp.337-348.

7. "Availability of Solar and Wind Generating Units". L.D.Jaffe. IEEE Trans on Power Apparatus and Systems, PAS-104, 1985, pp.1012-1017.

8. "Analysis and Prediction of Transmission Unit Performance Using Advanced Regression Models". A.W.Schneider, G.L.Landgren, M.P.Bhavaraju. IEEE Trans on Power Apparatus and Systems, PAS-104, 1985, pp.1084-1094.

9. "Analysis of Pooling 345kV Bulk Transmission Outage Data Between the Mid-Continent Area Power Pool and Northeast Utilities". M.G.Lauby, R.P.Ludorf, J.H.Doudna, D.D.Klempel, P.J.Lehman, C.G.Dahl. IEEE Trans on Power Apparatus and Systems, PAS-104, 1985, pp.2427-2434.

10. "A Model for Peaking Units Using the Canadian Electrical Association Data Base". R.Billinton, A.A.Chowdhury. IEEE Trans on Power Apparatus and Systems, PAS-104, 1985, pp.2972-2979.

11. "Transmission Outage Performance Prediction: Unit or Component Approach?". G.L.Landgren, A.W.Schneider, M.P.Bhavaraju, N.J.Balu. IEEE Trans on Power Systems, PWRS-1, No.2, 1986, pp.54-61.

12. "Unit Models for Capacity Reliability Studies Using the Canadian Electrical Association Data Base". R.Billinton, A.A.Chowdhury. Can Elec Eng J, 11, 1986.

13. "The Effect of Terminal Complexity and Redundancy on the Frequency and Duration of Forced Outages". M.G.Lauby, D.D.Klempel, C.J.Dahl, R.O.Gunderson, E.P.Weber, P.J.Lehman. IEEE Summer Power Meeting, 1986, paper 86 SM 309-9.

Author Index

Albrecht PF	6
Allan RN	6
Balu NJ	11
Baribeau MA	1, 4
Bhavaraju MP	6, 8, 11
Billinton R	5, 6, 10, 12
Chowdhury AA	5, 10, 12
Dahl CG	9, 13
Doudna JH	2, 3, 9
Driscoll CT	1
Forrest DW	6
Gunderson RO	13
Jaffe LD	7
Khu KT	2
Klempel DD	2, 3, 9, 13
Landgren GL	6, 8, 11
Lauby MG	2, 3, 9, 13
Lehman PJ	2, 3, 9, 13
Ludorf RP	9
McCoy MF	6
McGillis DT	1, 4
Polesky RW	2, 3

Reppen ND	6
Schneider AW	8, 11
Vandello RE	2
Weber EP	13

RELIABILITY COST/WORTH ANALYSIS

Papers

1. "Cost-Effectiveness Analysis of Substation Arrangements". W.J.Lannes. IEEE Trans on Reliability, R-31, 1982, pp.404-409.

2. "Consumer Interruption Costing for Reliability Cost/Benefit Evaluation". W.P.Poore, S.R.Greene, M.A.Kuliasha. IEEE Trans on Power Apparatus and Systems, PAS-102, 1983, pp.1361-1364.

3. "Comprehensive Bibliography on Electrical Service Interruption Costs". R.Billinton, G.Wacker, E.Wojczynski. IEEE Trans on Power Apparatus and Systems, PAS-102, 1983, pp.1831-1837.

4. "Interruption Cost Methodology and Results - A Canadian Residential Survey". G.Wacker, E.Wojczynski, R.Billinton. IEEE Trans on Power Apparatus and Systems, PAS-102, 1983, pp.3385-3392.

5. "Customer Perception of Interruption Costs". R.Billinton, G.Wacker, E.Wojczynski. Int Conf on Power Supply Systems, IEE Conf Publ 225, 1983, pp.12-16.

6. "Interruption Cost Methodology and Results - A Canadian Commercial and Small Industry Survey". E.Wojczynski, R.Billinton, G.Wacker. IEEE Trans on Power Apparatus and Systems, PAS-103, 1984, pp.437-443.

7. "Equipment and Reliability Costs for Residential A/C Load Control: One-Way VS. Two-Way Systems". H.E.Wegner. IEEE Trans on Power Apparatus and Systems, PAS-103, 1984, pp.1656-1661.

8. "Effects of Distribution System Reliability Index Distribution upon Interruption Cost/Reliability Worth Estimates". E.Wojczynski, R.Billinton. IEEE Trans on Power Apparatus and Systems, PAS-104, 1985, pp.3229-3235.

9. "Measurement of Supply Interruptions Related to Customer Impact". B.D.Caldwell, G.Wacker, R.Billinton, R.K.Subramaniam. CEA Trans, 24, 1985.

10. "Cost of Electric Service Interruptions to Industrial and Commercial Customers". R.Billinton, G.Wacker. Proc IEEE Industrial Applications Soc Conf, Toronto, Sept 1985.

11. "Factors Affecting the Development of a Commercial Customer Damage Function". R.Billinton, G.Wacker, R.K.Subramaniam. IEEE Trans on Power Systems, PWRS-1, No.4, 1986, pp.28-33.

12. "Economics and Reliability of Supply of developing Power Systems". A.O.Ekwue. Proc IEE, 133 part C, 1986, pp.373-377.

13. "Factors Affecting the Development of a Residential Customer Damage Function". R.Billinton, G.Wacker, R.K.Subramaniam. IEEE Summer Power Meeting, 1986, paper 86 SM 311-5.

14. "A Reliability Cost-Benefit Analysis for HVDC Transmission Expansion Planning". P.R.S.Kuruganty, D.A.Woodford. IEEE Winter Power Meeting, 1987,

paper 87 WM 151-4.

Author Index

Billinton R	3, 4, 5, 6, 8, 9, 10, 11, 13
Caldwell BD	9
Ekwue AO	12
Greene SR	2
Kuliasha MA	2
Kuruganty PRS	14
Lannes WJ	1
Poore WP	2
Subramaniam RK	9, 11, 13
Wacker G	3, 4, 5, 6, 9, 10, 11, 13
Wegner HE	7
Wojczynski E	3, 4, 5, 6, 8
Woodford DA	14

GENERAL CONSIDERATIONS

Papers

1. "Development of a New Procedure for Reliability Modeling of Wind Turbine Generators". P.Giorsetto, K.F.Ursurogi. IEEE Trans on Power Apparatus and Systems, PAS-102, 1983, pp.134-144.

2. "The Loss of Power Supply Probability as a Technique for Designing Stand-Alone Solar Electrical (Photovoltaic) Systems". E.Ofry, A.Braunstein. IEEE Trans on Power Apparatus and Systems, PAS-102, 1983, pp.1171-1175.

3. "Application of Availability Analysis Techniques to Improve Power Plant Productivity". N.E.Chang. IEEE Trans on Power Apparatus and Systems, PAS-102, 1983, pp.1521-1529.

4. "The Availability Improvement Methodology (AIM) for Evaluation Power Plant Productivity". C.C.Richard, M.E.Nelson, H.Singh. IEEE Trans on Power Apparatus and Systems, PAS-102, 1983, pp.1543-1551.

5. "Derated Capacity States of a Large Thermal Generating Unit". S.K.Banerjee, B.R.Reddi. IEEE Trans on Reliability, R-32, 1983, pp.7-8.

6. "Common Mode Failure Analysis in the Reliability Evaluation of Electrical Auxiliary Systems". R.N.Allan, N.M.Avouris, A.Kozlowski, G.T.Williams. Int Conf on Power Supply Systems. IEE Conf Publ 225, 1983, pp.132-136.

7. "Hierarchical Indices in System Adequacy Assessment". R.Billinton, J.S.Tatla. CEA Trans, 22, 1983.

8. "Reliability Equivalents - Power System Applications". R.Billinton, K.Hossain. Reliability Eng, 5, 1983.

9. "Reliability Modeling of Large Wind Farms and Associated Electric Utility Interface Systems". X.Wang, H.Dai, R.J.Thomas. IEEE Trans on Power Apparatus and Systems, PAS-103, 1984, pp.569-575.

10. "Reliability Modeling of Protective Systems". P.M.Anderson. IEEE Trans on Power Apparatus and Systems, PAS-103, 1984, pp.2207-2214.

11. "Power System Reliability in Perspective". R.Billinton, R.N.Allan. IEE Journal on Electronics and Power, 30, 1984, pp.231-236.

12. "Reliability Evaluation of Protection Devices in Electrical Power Systems". H.T.Yip, G.C.Weller, R.N.Allan. Reliability Eng, 9, 1984, pp.191-219.

13. "Efficient Calculation of Frequency and Duration Models for Large Composite Electric Generating Plants". W.R.Puntel, A.J.Wood, K.A.Clements, C.A.Dohner. IEEE Trans on Power Apparatus and Systems, PAS-104, 1985, pp.1617-1623.

14. "Power Plant Availability Prediction Using Standard Bayesian Analysis of NSERC/GADS Component Cause Code Failure Data". E.M.Gulachenski, W.K.Tsai. IEEE Trans on Power Apparatus and Systems, PAS-104, 1985, pp.1593-1601.

15. "Detection of Ageing in the Reliability Analysis of Thermal Generators". M.Th.Schilling, J.C.G.Praca, J.F.deQueiroz, C.Singh, H.Ascher. IEEE Winter Power Meeting, 1987, paper 87 WM 088-8.

16. "An Efficient Technique for Reliability Analysis of Power Systems Including Time Dependent Sources". C.Singh, Y.Kim. IEEE Summer Power Meeting, 1987, paper 87 SM 503-6.

Author Index

Allan RN	6, 11, 12
Anderson PM	10
Ascher H	15
Avouris NM	6
Banerjee SK	5
Billinton R	7, 8, 11
Braunstein A	2
Chang NE	3
Clements KA	13
Dai H	9
Dohner CA	13
Giorsetto P	1
Gulachenski EM	14
Hossain K	8
Kim Y	16
Kozlowski A	6
Nelson ME	4
Ofrey E	2
Praca JCG	15
Puntel WR	13
Queiroz JFde	15
Reddi BR	5
Richard CC	4
Schilling MTh	15
Singh C	15, 16
Singh H	4
Tatla JS	7
Thomas RJ	9
Tsai WK	14
Ursurogi KF	1
Wang X	9
Weller GC	12
Williams GT	6
Wood AJ	13
Yip HT	12

CONCLUSIONS

In compiling any bibliography of related publications, it is not possible to make it absolutely complete. The APM Subcommittee would like to apologise for any errors or omissions and hope that additional references will be advanced as discussion to this publication.

Discussion

Mark G. Lauby (Electric Power Research Institute, Palo Alto, CA): As a member of the Applied Probability Methods Subcommittee, I wish to thank the authors for their exhaustive bibliography which acts as a resource to those who desire to find material pertaining to probability applications. With this in mind, I wish to add two references resulting from my work which probabilistically measured the power system transfer capability based both on thermal and economic limits. I believe that they should be categorized under the *Transmission and Distribution System Evaluation* section:

References

[A1] "The Procedure Used in the Probabilistic Transfer Capability Analysis of the MAPP Region Bulk Transmission System", M. G. Lauby, J. H. Doudna, R. W. Polesky, P. J. Lehman, and D. D. Klempel, IEEE Transactions on Power Apparatus and Systems, November 1987, Vol. PAS-104, pp. 3013–3019.

[A2] "The Procedure Used to Assess the Potential Benefits of Additional Transmission Capacity in the Mid-Continent Area Power Pool", M. G. Lauby, D. A. Jepsen, H. V. Nguyen, C. J. Spargo, A. D. Burbach, and R. D. Lewis, III, IEEE Transactions on Power Systems, Vol. PWRS-1, pp. 113–120.

Manuscript received February 19, 1988.

M. T. Schilling (Ontario Hydro, Toronto, ON, Canada): Once again the authors are to be praised for this useful paper. It is our opinion that this already traditional work should not be discontinued.

We would appreciate to have the author's comments on Topic (A):

(A) We understand that this bibliography as well as the previous ones were compiled in accord to a set of established criteria in order to limit its length. Nevertheless, we would suggest that the next one includes all EPRI reports and CIGRE papers. Since they are in english, are worldwide available, are relevant, and are published on a regular basis by well known and established sources, we see no clear reason for this omission.

(B) Regarding our contribution to the previous bibliography (see note on page 282 of the last bibliography published on PAS-103, February 1984), and since it has not been completely accepted, we inform that part of it including 300 items may be found in the following reference:

"Power System Reliability: Available Bibliography in Brazil (1969–1985)", Feldman Neto, B.; Soares Ramos, D.; Nery, E. M. T.; de Queiroz, J. F.; Praca, J. C. G.; Miranda, L. F. S. A.; Morozowski, M.; Schilling, M. T. (Editor); Nunes, R. A. F.; Fontoura, R. N.; Cunha, S. H. F., *Revista Brasileira de Engenharia*, Caderno de Engenharia Eletrica, Volume 2, No. 2, December 1985, pp 23–51.

Manuscript received February 29, 1988.

R.N.ALLAN, R.BILLINTON, S.M.SHAHIDEHPOUR, C.SINGH: We would like to thank M.G.Lauby and M.T.Schilling for their respective discussions and welcome their contributions to this Bibliography.

In response to Mr Schilling's specific comment concerning criteria, the merits of including other publications, particularly EPRI reports, was discussed at considerable length at more than one meeting of the Application Of Probability Methods (APM) Subcommittee. It was decided not to include them for several reasons and our Task Force proceeded accordingly. It was recognised that, not only EPRI reports, but also reports of a similar nature from many other national and international organisations are of interest to the profession and therefore an activity which compiles all such reports, as opposed to papers, together with relevant books may be worth starting. We are sure that the encouraging comments made by Mr Schilling will be taken into account. Furthermore, if and when the APM Subcommittee considers it timely to produce another bibliography, the criteria will be reconsidered and could be amended.

Manuscript received April 20, 1988.

Author Index

A

Abdel-Gawad, N. M. K., 172
Adler, H. A., 30
Adraktas, A. N., 311
AIEE Subcommittee on Application of Probability Methods of the System Engineering Committee, 53
Albrecht, P. F., 293, 300, 332, 454
Allan, R. N., 1, 172, 311, 416, 427, 436, 454, 523, 531
Anstine, L. T., 193
Ayoub, A. K., 140

B

Baldwin, C. J., 45
Balu, N. J., 179
Berk, L. H., 469
Bertoldi, O., 341
Bhavaraju, M. P., 234, 332, 422, 454
Billinton, R., 1, 121, 145, 172, 186, 212, 217, 229, 234, 267, 284, 293, 300, 317, 325, 332, 350, 370, 397, 407, 416, 422, 427, 454, 478, 485, 493, 503, 523, 531
Blackstone, J. H., 179
Bollinger, K. E., 370
Bruggeman, J. A., 155
Burke, R. E., 193

C

Calabrese, G., 11
Casey, J. E., 193
Cook, V. M., 71

D

De Oliveira, M. F., 416, 427
DeSieno, C. F., 422
Dialynas, E. N., 436
Doudna, J. H., 445

E

Endrenyi, J., 293, 300, 389, 422

F

Fong, C. C., 350
Forrest, D. W., 454

G

Galloway, C. D., 71
Garver, L. L., 87
Gaver, D. P., 45, 359
Ghajar, R., 493
Goode, S. D., 379
Grover, M. S., 397, 407
Gunderson, R. O., 350

H

Hall, J. D., 97
Halperin, H., 30
Harrington, P. G., 145

Hoffman, C. H., 45
Holgate, R., 193
Homer, I. R., 436

I

IEEE Reliability Test System Task Force of the Application of Methods Subcommittee, 276
IEEE Subcommittee on the Application of Probability Methods, Power System Engineering Committee, 515, 523, 531
IEEE T&D Outage Definitions Task Force of the Application of Probability Methods Subcommittee, 454
IEEE Task Force on Common Mode Outages of Bulk Power Supply Facilities of the Application of Probability Methods, Subcommittee of the Power System Engineering Committee, 422
IEEE Task Group on Models for Peaking Service Units, Application of Probability Methods Subcommittee, 130
IEEE Working Group on Measurement Indices, 350
Ihavaraju, M. P., 332

J

Jain, A. V., 212, 217
John, R. S., 193
Jorgensen, G. E., 422

K

Khu, K. T., 445
Klempel, D. D., 445
Kumar, S., 317

L

Landgren, G. L., 454
Lauby, M. G., 445
Lee, S. H., 523
Lehman, P. J., 445

M

Mackay, E. M., 469
Marks, G. E., 293, 300
McCoy, M. F., 454
Medicherla, T. K. P., 267, 284
Montmeat, F. E., 359

N

Noferi, P. L., 241, 249

O

O'Neil, P. M., 350
Oteng-Adjei, J., 493

P

Pang, C. K., 155
Paris, L., 241, 249
Patton, A. D., 140, 179, 203, 359, 422
Polesky, R. W., 445

R

Raksany, J., 350
Rau, N. S., 164
Reppen, N. D., 293, 300, 332, 454
Ringlee, R. J., 97, 107, 332, 379, 422

S

Sachdev, M. S., 267
Saddock, H. G., 422
Salvaderi, L., 249, 293, 300, 325, 341
Scalcino, S., 341
Schenk, K. F., 164
Schneider Jr., A. W., 350
Shahidehpour, S. M., 531
Silverstein, B., 350
Singh, C., 121, 531

Steinberg, M. J., 71
Stewart, H. G., 193
Stratton, J. A., 422

V

Vandello, R. E., 445
Vohra, P. K., 317

W

Wacker, G., 478, 485
Watchorn, C. W., 23
Watt, R. L., 155
Wojczynski, E., 478, 485
Wood, A. J., 71, 97, 107
Working Group on Performance Records for Optimizing System Design, Power Systems Engineering Committee, 260

Subject Index

A

Absolute reliability assessment, electric utilities, 299
Additional unit, estimating effects of, 93–95
Adequacy evaluation:
  bulk power supply, 260–66
  at hierarchical level I, 3–4
  at hierarchical level II, 4–5
  at hierarchical level III, 5
  transmission system reliability, 305
  utility survey, 333–34
Adequacy indices, composite generation/transmission system reliability evaluation, 267–75
Alternative supply sources, representing, 429
Analytical approach, to bulk power system reliability, 300–303
Annual quantities, from interval data, 112
Area risk curves, spinning reserve studies, 212–13
Assessment rate, interrupted energy, 493–99
Automatic/rotational load shedding, 352
  indices, 355
Auxiliary electrical systems of power stations:
  analysis of, 429–33
  reliability evaluation, 427–35

B

Bulk power interruption/curtailment indices, 262–63
Bulk power supply adequacy evaluation, reliability indices for, 260–66
Bulk power system reliability, 293–310
  analytical approach, 300–303
    contingency selection, 301
    contingency testing, 301–2
    reliability indices, accumulation of, 302–3
    system problems, 300–301
  basic indices, 354–55
  computational difficulties, 294–95
  conceptual problems, 293–94
  data requirements, 295
  distribution/generating system concepts, 295–96
  electric utility application of reliability theory, 298–99

failure definitions:
    present, 297
    proposed, 297–98
  hierarchical levels, 350–51
  measurement/indices, 350–55
    data requirements, 353–54
    measurement approach, 352–53
    measurement system, 351–52
  modeling, 294
  unreliability events, 351
Bulk power system security, 203–10
  breaches of security, definitions of, 203, 207
  maximum tolerable insecurity level, relationship to long-term average system reliability, 205–6
  security function, 203–9
    computation of, 207–9
    operating rules, 207
Busbar analysis, 418–20
Busbars, representing branches off, 428–29

C

Canadian utilities:
  current practices, assessment of, 190
  operating reserve:
    classification of, 189–90
    requirements, 190
  planning/operation criteria of generating capacity, 186–91
    adequacy criteria at HLI, 186–89
    comparative study using IEEE-RTS, 189
    security criteria at HLI, 189–90
Capacity, 80
Capacity assistance probability method, 211–18
Capacity-outage probability:
  calculation of, 60–61, 63
  Fourier evaluation methods, 164–71
Capacity outages, calculation of frequency of, 63–64
Common-cause outages, 285, 307
Common mode forced outages, of overhead transmission lines, 422–26
Component, definition of, 455
Component failure modes, 397
Composite system reliability evaluation, 229–33
  bus, area, and system indices, 336
  conditional probabilities of system failure, 229–30

data requirements for, 306–8
    collecting data, 308
    common-cause outages, 307
    dependent outages, 307
    deterministic data, 306
    independent outages, 306–7
    station-originated outages, 307–8
    stochastic data, 306
  ENEL approach, 325–28
  indices for, 335–36
  modeling needs, 337–38
  modeling requirements, 332–40
  proposed scope for, 334–35
  reliability measures, 336
  simple system application, 230–31
  system failure, definition, 336
  system planning role, 336–37
  terminal effects/protection system failures in, 311–16
  two-plant two-load system, 232–33
  University of Saskatchewan (U of S) approach, 328–29
Constant risk principle, 32–33
Cumulative-event cycle time, 99–100

D

Delivery point interruption, 352–54
Dependent outages, 285, 307
Digital computer programs:
  interconnected system reserve requirements, 75–80
    expansion program logic, 76
    illustrative results, 76–80
Distribution systems:
  computational techniques, 440
  modeling, 436–44
    techniques, 437–40
    practical, 436–37
  reliability evaluation, 436–44
  typical system, analysis of, 440–41
Duration indices, 250, 260
Dynamic indices, 250

E

Effective load-carrying capability, 87–90
  determining, 87–88
  estimating, 89
    alternate procedure, 89–90
    change in reliability, 90
    examples of procedures, 89
    procedure, 89
  illustration, 88
  risk level, effect of, 88

Electrical systems with switching actions, reliability evaluation, 416–21
Electric utilities:
 reliability criteria, establishment of, 298
 reliability theory application, 298–99
  absolute reliability assessment, 299
  benefit/cost evaluation, 299
  future development, 299
Emergency/overload ratings, of equipment, 58
ENEL approach:
 to composite system reliability evaluation, 325–28
 models, 326–27
 program procedure, 327
 system reliability, 327–28
Energy limited generating capacity studies:
 reliability evaluation in, 145–54
 system example, 146–47
 units with limited storage, 148–50
 units with no storage, 147
 units partially base loaded/partially used for peak-shaving, 150–52
Exact capacity outage state:
 effective departure rate, 141
 frequency/duration, 141
 probability of, 140–41
Exact-state capacity model:
 frequency/duration methods, 140–41
 reliability measures, 142–43
Expectation indices, 250, 260–61
 expected loss of load (XLOL), 262
 expected energy curtailment/year, 262
 frequency/duration of capacity deficiency, 262

F

Failure modes, 397, 407–8
 electrical system, 417
 load point, 397–98, 408–15
Forced outages:
 average rate of, 12–11
 expectation of, 30–32
 frequency/duration of, 37
 independence of, 14–15
 See also Power generation outages
Fourier evaluation methods:
 capacity-outage probability, 164–71
  Gaussian approximation, 165–66
  Gram-Charlier expansion, 166–71
  probability distribution, 164–65
  proposed algorithm, 166–68

convolution by, 169
Frequency/duration approach:
 to generating system reliability evaluation, 140–44
  exact-state capacity model, 140–41
  interconnected systems, 121–29
  load model, 141–42
  numerical example, 142–43
  reliability measures, 142
 to interrupted energy assessment rate, 494–96
Frequency indices, 250, 260

G

Gaussian approximation, 165–66
Generating capacity planning indices, 261–62
 LOLP, 261
 probability of not meeting annual peak load, 261
Generating reserve capacity:
 determined by probability method, 11–22, 53–70
 hierarchical level I, 7–8
Generation expansion planning, 91–92
Generation system model, 97–106
Gram-Charlier expansion, 166–71

H

Hierarchical level I, 1–225
 adequacy evaluation at, 3–4
 generating capacity, 7–8
  operating reserve reliability assessment, 8
  static capacity reliability assessment, 7–8
Hierarchical level II, 227–354
 adequacy evaluation at, 4–5
Hierarchical level III, adequacy evaluation at, 5

I

Identical capacity states, 100
IEEE reliability test system (RTS), 172–78, 276–83
 additional generation, effect of, 175–76
 derated states, effect of, 173–74
 design criteria, 282
 energy based indices, 176
 EPRI version, 343–48, 343–48
  basic network, analysis of, 344–45
  deferment of seven links, 345
  generation system, 334
  network optimization, 346–48
  system features, 343

transmission network, 343–44
 existing RTS data, limitations of, 172–73
 generating system, 277–78
 interconnection, effect of, 176–77
 load forecast uncertainty, effects of, 174
 load model, 276–77
 LOLE analysis of base case, 173
  peak load, effect of, 174–75
  rounding, effect of, 173
  scheduled maintenance, effect of, 174
  system input data, 172
  system reliability indices, 172–73
  transmission system, 278–80
Independent outages, 285, 306–7
Independent renewal processes, frequency/duration of overlapping events with, 384–85
Interconnected generating systems:
 capacity benefits
  allocation of, 25–27
  calculations, 28–29
  determination of, 23–25
  tables, 26–27
 division of gains from, 39
 finite interconnection, 73–74
 gain from, 39, 60
 infinite interconnection, 73
 reliability evaluation, 121–29
 reserve requirements, determining, 71–86
 spinning reserve requirements, 211–25
 two systems, 72–73
Interconnected system assistance:
 fundamentals of, 218–24
 inclusion of load forecast uncertainty in, 219–24
Interconnected system studies, effect of not considering exact equivalent derated models in, 224
Interrupted energy assessment rate:
 customer damage function, 494
 establishing, 493–99
 estimation of, 494–98
Interruptions:
 costs of, 469–77
  bibliography on, 478–84
  industry estimates, 470–71
  interruptible power, 475
  methodology/results, 485–84
  organized load shedding, 475–76
  stand-by capacity, 475
  voltage reduction, effect of, 471–74
 due to component temporary outages overlapping component maintenance outages, 411–12
 due to component temporary

outages overlapping
component permanent
outages, 412–13
due to overload outages of system
components, 413–15

K

Kilowatt-hour output, probability of
loss of, 19

L

Line overload alleviation, approximate
method of, 274–75
Line section, definition of, 455
Line segment, definition of, 455
Load-carrying capability, 87–96
applications, 90–92
assumed unit size/outage rate,
effects of, 90–91
generation expansion planning,
91–92
effective capability, 87–90
determining, 87–88
illustration, 88
risk level, effect of, 88
See also Effective load-carrying
capability
Load characteristics, 59
Load-duration curve, 47–48
Load forecasts, uncertainties of,
59–60
Load point failure modes, 397–98,
408–15, 427–28
evaluation of, 408–11
Local reserve requirements, effect of
interconnecting groups of
units on, 19–21
Loss of energy probability, 62
Loss of load duration probability, 11
Loss of load probability, 11, 60–62,
87
calculation of, 21–22, 61–62,
74–75
capacity models, 75
load models, 74–75
for single system, 72
probability of capacity outages,
calculation of, 60–61

M

Machines in parallel, 99
Maintenance, added to load, 80
MAPP bulk transmission outage:
data collection and analysis, 445–53
methodology, 445–46
glossary, 449–50
MAPP bulk transmission system,
446
fault data analysis, 446

MITF and MTTR data analysis,
446
related outage data analysis, 447–48
seasonal performance analysis, 448
system performance, 446–48
Markov processes:
basic concepts, 376–78
single-unit case, 370–76
and transmission system reliability
evaluation, 370–78
Markov processes, frequency/duration
methods, of power-system
reliability, 98–100
Meshed delivery point, 352
Modeling:
bulk power system reliability, 294
composite system reliability
evaluation, 332–40
distribution systems, 436–44
techniques, 437–40
terminal station outage protection
systems, 313–15
Monte Carlo-based SICRET program,
341–43
Monte Carlo simulation, 179–85,
249–59
applications to ENEL Power
System, 257
component unavailability, 252
computing time/confidence in
results, 256–57
environment models/methods, 252
generating unit model, 180
generating unit models/methods,
252
interconnected system operation,
180–81
area mode of operation, 180
pool mode of operation, 180
and interrupted energy assessment
rate, 496–98
load model, 252
loads, 182
loss of unit/line, response to, 181
operation policy models/methods,
252–54
and planning studies, 341–49
reliability indices, 182–83
risk indexes, 250–51
sample systems, 182
study results, 182–83
transmission network model,
181–82
transmission planning, 303–6
typical outputs, 254–56
Multi-area generation reliability
studies, 155–63
definitions, 155–56
area failure/success, 156
load loss sharing, 156
system failure/success, 155
delay of units, 161
different modes of interconnection,

results from, 160–61
inclusion in additional area, 161–62
inter-area transfer tie requirements,
162
LOLP computation, 156
models/methods, 155–58
results, 158–63
sub-system LOLP in
interconnection, reduction
of, 162–63
system:
description, 156–58
equivalents, 159–60
models/methods, 156
states, effects of total number of,
158–59

N

Network, IEEE reliability test system
(RTS), EPRI version,
344–45, 346–48
Network study, power-system
reliability, 391–92
Numerical example, of
frequency/duration approach
to generating system
reliability evaluation,
142–43

O

Operating reserve reliability
assessment, 8
Operational gaming, 45–46
Outage duration/frequency, 47, 62–64
capacity-outage probability,
calculation of, 63
as function of peak annual load, 64
outage rate for machine exposure,
correction of, 62–63
Outage probability:
of combinations of units connected
in parallel, 14
effect of connections on, 16–19
in excess of reserve, 15–16
of a group of units, 12–14
Outage risk:
peaking service units, 130–39
applications, 133–35
eight-state Markov model, 131
four-state model development,
132–33
modified two-state model,
131–32
two-state model (classical
model), 131
Outages:
classification of, 285–86
common-cause outages, 285
dependent outages, 285
independent outages, 285

Outages (*continued*)
station-originated outages,
285–86
interval between, 64
Overhead transmission lines:
common mode forced outages of,
422–26
indices, 424
Overlapping events, with independent
renewal processes,
frequency/duration of,
384–85

P

Parallel systems:
reliability measures/calculation,
363–65
outages due to overloads in,
365–67
Peak annual load, outage
duration/frequency as
function of, 64
Pennsylvania-New Jersey-Maryland
Interconnection, spinning
reserve requirements,
193–202
Power generation outages, 45–46
mathematical models used in
simulation of, 45–52
new model, 48–51
past models/methods, 46–48
operational gaming, 45–46
system model, 46
Power interruptions, *see* Interruptions
Power station auxiliary electrical
systems:
analysis of, 429–33
reliability evaluation, 427–55
Power-system reliability, 1–6
adequacy evaluation:
at hierarchical level I, 3–4
at hierarchical level II, 4–5
at hierarchical level III, 5
definition of, 2–3
evaluation, need for, 1–2
frequency/duration methods,
97–106
demand model/capacity reserve
model, 107–20
generation system model, 97–106
numerical results, 100–103
by recursive techniques, 98–100
functional zones, 3
hierarchical levels, 3
comments on assessment, 3
load-carrying capability of
generating units, 87–106
reliability-cost/reliability-worth, 56
reliability criteria, 4
reliability data, 6
reliability measures/calculation,
54–55, 359–69

fluctuating environment, 360–61
parallel systems, 363–65
series systems, 362–63
system component data for,
361–62
simulation programs, 250–54
three-state evaluation models,
389–96
evaluation method, 390
model description, 389–90
network study, 391–92
summary of assumptions, 391
Probability distribution, 164–65
Probability indices, 250, 260

Q

Quantitative evaluation of system
reliability, 241–48
risk indices, 241–42
application of method, 245–47
evaluation of, 242
evaluation program, 242–45

R

Radial delivery point, 351
Reliability cost worth, 467–99
Reliability criterion, 234–40, 416
computer program, 235
conditional probability approach,
234–35
system study, 235–37
Reliability evaluation:
of auxiliary electrical systems of
power stations, 427–35
composite system, 229–33
of electrical systems with switching
actions, 416–21
in energy limited generating
capacity studies, 145–54
frequency/duration approach,
140–44
Monte Carlo simulation models,
179–85
Reliability indices, bulk power supply
adequacy evaluation,
classification of, 260–61
Reliability input data, 416–17
Reliability measures, exact state
capacity model, 142–43
Repairable transmission systems,
reliability/availability models
for, 380
Repeated convolution, Fourier
techniques and, 169–70
Reserve capability studies:
application of methods, 34–37
determination of reserve, 34–35
interconnections on reserve,
effect of, 35–36
size of units on reserve, effect
of, 36

methods used in, 30–34
constant risk principle, 32–33
emergency help, 33–34
expectation of forced outages,
30–32
monthly peaking factor, 33
Reserve-generating capability,
determination of, 30–44
Reserve margin states, 109–11
cumulative margin states, 110–11
margin-availability tables, 109–10
rationalizations/justifications, 111
Reserve requirements:
interconnected generating systems,
71–86
digital computer programs,
75–78
loss-of-load probability
calculations, 74–75
mathematical theory, 72–74
and unit size, 80
Residential survey:
interruption cost
methodology/results, 484–85
discussion, 489–91
factors investigated, 485
questionnaire description, 486
questionnaire design, 486
results, 487–89
survey procedures, 486–87
Risk, measure of, 80–81
Risk indices:
quantitative evaluation of system
reliability, 241–42
types of, 250–51

S

Seasonal variations, in capacity, 58
Security breaches, bulk power system,
203, 207
Series systems, reliability
measures/calculation,
362–63
Shedding:
automatic/rotational load shedding,
352
indices, 355
of interruptible loads, reductions in
loads from, 58–59
Significant point, definition of, 352
Significant point interruption, 352–55
Simple substation configurations,
reliability analysis of, 398
Simulation programs:
transmission planning/generation
planning, 249–59
main features of, 259
Simultaneous outages, derivation of
equation for duration of,
37–38
Single- /double-contingency system states,
states, average

frequencies/durations of, 393–94
Spinning reserve, 60
Spinning reserve requirements:
    interconnected systems, 211–25
        application to practical problems, 218–19
        capacity assistance probability method, 211–18
        interconnected system assistance, 218
        two-system probability array approach, 211
    Pennsylvania-New Jersey-Maryland Interconnection, 193–202
        computer program, 197–98
        conclusions, 198
        forced outage probabilities, 195–97
        load forecasting error probabilities, 197
        mathematical theory, 195
        parameters, 194–95
        results, 198
Spinning reserve studies:
    area risk curves, 212–13
    rapid start/hot reserve units, 213–14
    application to practical problems, 215–16
    application to S.P.C. system, 216
    effect of, 212–17
    models, 214–15
Static capacity reliability assessment, 7–8
Static indexes, 250
Station originated multiple outages, 284–92, 307–8, 317–24
    Boundary Dam Station, analysis of, 286–87
    effect of, 288–89
    events, inclusion in composite system reliability evaluation, 317–18
    models for, 288
    Squaw Rapids Station, analysis of, 287–88
    system studies, 318–23
    See also Terminal station outages
Stochastic processes, frequency/duration methods of power-system reliability, 98–100

Substations:
    reliability evaluation, 380–81
    computerized approach to, 397–406
    reliability prediction, 381–83
Switching stations, reliability evaluation, computerized approach to, 397–406
System disturbance, 352
    indices, 355

T

Terminal station outages, 311–16
    composite system reliability evaluation, 312–13
    concepts, 311–12
    modelling protection systems, 313–15
        basic event tree, 313–14
        evaluating event probabilities, 314
        evaluating outcome probabilities, 314
        separate/redundant operating components, 314–15
    outage events, 312
Transitions, frequency/duration methods of power-system reliability, 98–100
Transmission/distribution facilities:
    basic outage indices, 458–59
    reporting/analyzing outages, 454–65
        component and unit states/events, 455–58
        equipment classifications, 445
        equipment state times, 458
        exposure parameters, 458
Transmission/distribution schemes, reliability assessment, 407–15
Transmission line, effects of, 112–13
Transmission planning:
    Monte Carlo methods in, 303–6
    using a reliability criterion, 234–40
Transmission system reliability evaluation:
    adequacy and security evaluation, 305

application examples, 305–6
features of, 303–4
interconnection simulation, 305
and Markov processes, 370–78
models used in computations, 304–5
procedures, 379–88
Two-system probability array approach, 211

U

Unit, definition of, 455
Unit addition:
    estimating effects of, 93–95
        annual risk as function of reserve, 93
        generalized expressions, derivation of, 94–95
        risk function after addition, 93
        straight-line approximation, 93–94
Unit capacities/loads, variations in, 58–59
University of Saskatchewan (U of S) approach:
    to composite system reliability evaluation, 328–29
    compared to ENEL approach, 329–30
    index description, 329
    system adequacy, 328–29
Unreliability of system, loss of power due to, 429
Utility survey, 332–34
    adequacy vs. security assessment, 333–34
    difficulties with current methods, 333
    existing applications, 333
    methods/computational tools, 334
    motivation for composite system reliability evaluation, 333
    reliability indices, 334

V

Voltage reduction:
    costs of, 471–74
    reductions in loads from, 58–59

Editors' Biographies

Roy Billinton was born in Leeds, England and came to Canada in 1952. He obtained his B.Sc. and M.Sc. degrees from the University of Manitoba and his Ph.D. and D.Sc. degrees from the University of Saskatchewan.

He worked for Manitoba Hydro in the System Planning and Production Divisions and joined the University of Saskatchewan in 1964. Formerly Head of the Department of Electrical Engineering, he is presently the C.J. Mackenzie Professor of Engineering and Associate Dean of Graduate Studies, Research and Extension of the College of Engineering.

Dr. Billinton has authored or co-authored more than 300 papers on power system analysis, stability, economic system operation, and reliability. He has also authored or co-authored six books on reliability. He is a Fellow of the IEEE, a Fellow of the Engineering Institute of Canada, a Fellow of the Royal Society of Canada, a Fellow of the Safety and Reliability Society (UK), and is a professional engineer in the Province of Saskatchewan. He is an active member of the IEEE Subcommittee on the Application of Probability Methods, CIGRE WG 38.03 (Power System Reliability Analysis), and the Canadian Electrical Association of which he is chairman of the Consultative Committee on Outage Statistics and the Power System Reliability Subsection.

Ron Allan was born in Tidworth, England. He obtained the Dip. Tech. (Eng.) from Portsmouth Polytechnic and subsequently an M.Sc., a Ph.D. and a D.Sc. from the University of Manchester, England.

Dr. Allan joined UMIST (The University of Manchester Institute of Science and Technology) in 1966 and is presently Reader in the Electrical Energy and Power Systems Research Group with particular responsibilities for developing and promoting the group's research activities on a national and international scale. His personal research interests are the modeling, evaluation, and application of probability techniques to power system problems, particularly those concerned with reliability and probabilistic load flow.

Dr. Allan has authored or has co-authored more than 100 technical papers on these subjects and co-authored four textbooks, three on reliability and one on sparsity techniques. He is a Fellow of the IEE (UK), a Fellow of the IEEE (USA), a Fellow of the Safety and Reliability Society (UK), and a Chartered Engineer in the UK. He is a chairman and member of several committees associated with international organizations including the European Safety and Reliability Association (ESRA), the IEEE Subcommittee on the Application of Probability Methods, and CIGRE WG 38.03 (Power System Reliability Analysis).

Luigi Salvaderi was born in Ancona, Italy. He received his degree from the University of Genoa in 1962.

After his military service, he began his work in the automatic control field. Dr. Salvaderi joined ENEL, the Italian Electricity Board, in 1965. He worked in the Design and Construction Section and was involved in the projects of the 120–245 kV substations and of the SACOI DC link. He was then in the Research and Development Department in charge of the development of techniques to be used for generation and transmission system reliability evaluation. Since 1977, he has been in the Planning Department, responsible for the Generation Division.

On behalf of the ENEL, he has been acting as a Senior Consultant Engineer in various countries and has given lectures and tutorial courses in Argentina, Sweden, Venezuela, and the USA. He has authored and co-authored approximately 70 papers on power system reliability evaluation in planning. He has twice received (1972 and 1978) the award of the Italian Electrotechnical Association as co-authored of the best yearly Italian paper. He is an active member of CIGRE WG 37.01 (Planning Methods and Criteria), CIGRE WG 38.03 (Power System Reliability Analysis), UNIPEDE-WEC Working Group on International Availability Statistics, UNIPEDE WG SYSTREG (Frequency and Voltage Regulations), and the IEEE Subcommittee on the Application of Probability Methods.